T0313393

INVERTEBRATE ZOOLOGY

INVERTEBRATE ZOOLOGY

A TREE OF LIFE APPROACH

Edited by

Bernd Schierwater

Rob DeSalle

CRC Press is an imprint of the
Taylor & Francis Group, an **informa** business
A GARLAND SCIENCE BOOK

First edition published 2022
by CRC Press
6000 Broken Sound Parkway NW, Suite 300, Boca Raton, FL 33487-2742

and by CRC Press
2 Park Square, Milton Park, Abingdon, Oxon, OX14 4RN

Library of Congress Cataloging-in-Publication Data

Names: Schierwater, B. (Bernd), 1963– author. | DeSalle, Rob, author.
Title: Invertebrate zoology: a tree of life approach/Bernd Schierwater, Rob DeSalle. Description: Boca Raton: Taylor & Francis, 2022. | Includes
bibliographical references and index. | Summary: "This book presents a
comprehensive overview of invertebrate organismal zoology (IZ), defining
invertebrates as any multicellular animal without a backbone .
While there are many IZ textbooks on the
market, few combine the classical anatomical information with the newer
genome level DNA sequence data and evo-devo data for teaching purposes.
This book documents the latest research in these areas and combines it
with classical information, giving students a unique opportunity to
learn the biology of these organisms"– Provided by publisher.
Identifiers: LCCN 2020047984 | ISBN 9781482235814 (hardback) | ISBN
9780367685676 (paperback) | ISBN 9780429159053 (ebook)
Subjects: LCSH: Invertebrates. | Invertebrates–Anatomy. |
Invertebrates–Morphology. | Invertebrates–Classification.
Classification: LCC QL363. S35 2021 | DDC 592–dc23
LC record available at https://lccn.loc.gov/2020047984

ISBN: [978-1-4822-3581-4] (hbk)
ISBN: [978-0-367-68567-6] (pbk)
ISBN: [978-0-429-15905-3] (ebk)

Typeset in ITC Leawood Std
by KnowledgeWorks Global Ltd.

eResources can be accessed under Support Materials at:
https://www.routledge.com/9780367685676

We dedicate this book to our children, Nita and Bo, and to the many students who have worked with us on invertebrates as study organisms.

CONTENTS

PREFACE

When teaching our Invertebrate Zoology courses we start by asking our students the following question: Invertebrates are metazoan animals without backbones, but what are "metazoan animals?"

We have asked this question to several thousand biology students over the last 20 years in both, the United States and Germany, and received exactly one correct answer: Metazoan animals show *"intrasomatic differentiation,"* i.e., harbor *more than one* type of somatic cells (i.e. non-germ line cells).

All other attempts from comparative morphology or physiology to distinguish protists (single-cell organisms) from metazoans (multicellular animals) would fail. The most common wrong answers list cell number or the presence of a soma as the characters critical to the definition. However, several protists, like the Phytomonadea, are colonial. They are made of multiple cells (up to some 16,000) and often even show somatic cell differentiation. The flagellated trophic cells of the protist Volvox is another good example. But no protist harbors more than one somatic cell type.

There is one other correct means of characterizing metazoans: by using different molecular genetic characters in phylogenetic analysis which unambiguously separate metazoans from protists. But no student has answered our query in this way yet.

Consider this: The soma of metazoans quantitatively dominates and determines an animal's appearance and the intrasomatic differentiation is so rich that it provides a dazzling array of characters for taxonomic classification and systematics. For most metazoans the soma is mortal and has a specific life-expectation; i.e., soma dies at the latest stages of development with the endogenetically pre-programmed natural exitus. The left-over dead soma is called a "corpse." As an exception some invertebrate groups are capable of vegetative reproduction and, thus, have a soma that is "potentially immortal." Comparing soma differentiation patterns between animal groups is called "comparative zoology." This is a primary, centuries-approved concept and we are also following this approach in our textbook for classification and phylogenetic interpretation.

New forms of data are commonplace within all scientific discplines. Therefore, we have also added molecular datasets to the book, which allow phylogenetic tree building even in the absence of any morphological information. Students and researchers are obliged to broaden and sharpen their views for integrating morphological vs. molecular datasets or by using different algorithms for tree building and phylogenetic analysis. The views of how to do and how to teach Invertebrate Zoology can be different from author to author and instructor to instructor, and we have often found them strikingly different in Europe and the United States. We, the editors of this textbook, have enjoyed and benefited from exciting late-night debates discussing the pros and cons of slow old-school (sos) and ultra-fast data-rich (ufd) approaches.

Our team benefits from the best of both research approaches. Bernd was trained by one of the last genuine traditional Invertebrate Zoologists, Carl Hauenschild, in Germany. Rob comes at Invertebrate zoology from an equally valid direction. He was trained as a *Drosophila* geneticist by Alan Templeton and Dan Hartl in the United States. Our discussions were sometimes intense, but we survived and hope that our synthetic approach for this textbook adds to our overarching attempt to address both, anatomical intricacies of invertebrate animals and also genomic approaches to modern Invertebrate Zoology.

Rob has spent the last 30 years as a curator in the Division of Invertebrate Zoology at the American Museum of Natural History – the home of Libbie Hyman. So if a little of the Invertebrate Zoologist does not rub off on you there, it simply could not anywhere else. Early on in his career he was infatuated with a relatively small group of flies called the Hawaiian Drosophila. Their amazing variability and beautiful habitat enticed him to study their relationships using the "cutting-edge" tools of the 1980s. Since, he has branched out into many areas of evolutionary biology, but lately another even tinier invertebrate has infatuated him – the placozoon *Trichoplax*. Our shared interest in this weird beast is the tie that brought us together to organize this textbook. Bernd has spent his last 30 years as a researcher, lecturer and/or professor in different departments at Yale, Frankfurt, Freiberg, and Duke University and for 20 years has directed an Institute for Invertebrate Ecology & Evolution at TiHo University Hannover. He has worked on different groups, including mammals, insects, annelids, and cnidarians and has now focused his work on the Placozoa.

Our mutual interest in Placozoa has given us the opportunity to think about combining research approaches with the goal of finding concordance and identifying outstanding questions. We hope our modern synthesis attempt to organize and introduce all living invertebrate phyla in an up-to-date way allows the student to enjoy Invertebrate Zoology with the same humbleness and fascination for evolution we have been experiencing.

We are highly grateful to leading experts, who have written the chapters for the different groups, permitting students to quickly introduce themselves to the current knowledge on a group. All chapters use the same layout so the reader can quickly look up and compare details between groups. Once the reader is familiar with a certain group (or groups) he can use the molecular datasets provided here [www.routledge.com/9780367685676] under "Support Material" and exercise and examine relationship hypotheses. We believe that such training using phylogenetic trees and mapping characters onto these trees is extremely fruitful and a motivation for the student to understand open questions and controversies and to formulate cogent opinions and arguments for controversial subjects in Invertebrate Zoology.

The "Invertebrate Tree of Life" has attracted exceptional attention over the last three decades, for some good and some bad reasons. The good reasons include the steadily increasing number of molecular characters, that can help generate phylogenetic hypotheses worthy of further testing. The not-so-good reasons include the availability of both large numbers of characters and divergent tree building algorithms, resulting in some sensational and oftentimes nonsensical phylogenetic trees with high support values. Just for full disclosure we admit we are probably responsible for some of these ephemeral hypotheses. Unfortunately, publishing a solid molecular tree confirming traditional and established hypotheses is sometimes more of a hindrance than publishing an unexpected tree yielding unanticipated evolutionary scenarios. Thus it comes as no surprise that molecular systematics (evolutionary tree building based on molecular characters) has created the most short-lived (so-called) hypotheses natural sciences have ever seen. The conflict between the stability of classic data and the impermanence of modernity is obvious.

We here want to help the student to become critical and self-consistent and to be able to do their own analyses. Thus, we have added ancillary materials for the text that can be downloaded from the book product page under "Support Material" https://www.routledge.com/9780367685676. For students, the ancillary material consists of:

- Lists of the web links in each chapter, which can simply be clicked to access the web information from each chapter as desired. There is no need for the reader of this textbook to type a URL into their web browser to access websites cited in the text.
- Full color reproductions of all of the figures in each chapter.
- Several morphological and molecular data matrices relevant to the group discussed in a chapter. There are over 100 data matrices that are made available to the student, grouped by chapter. These data matrices are easily executed in RAXML, PAUP, TNT or any number of phylogenetic analysis programs. Each set of data matrices has a "README" file that should be consulted before working with the matrices. The student should explore these data matrices by first using the parameters of the original authors (references to the original authors are given in the README files), and then experiment by changing parameters or optimality criteria.

We hope that these ancillary materials and data-matrices will give the student an added perspective on the groups that the matrices address.

For the instructor, we make available lecture slides for each chapter (in some cases, because there are several subgroups within a major group, we include separate slideshows for those groups). Slides are available on the Instructor Download Hub (https://routledgetextbooks.com/textbooks/instructor_downloads/) where an instructor can register in order to obtain the materials. The slideshows were made by following the outline of each chapter, reproducing figures and tables from each chapter. We invite instructors to use these slides, alter them in any way they see fit and even to distribute to students if desired.

Please take advantage and enjoy!

Bernd Schierwater
TiHo University Hannover
Hannover, Germany

Rob DeSalle
American Museum of Natural History
New York City, New York

ACKNOWLEDGEMENTS

We first thank our families for their patience and indulgence during the production of this book. They were the first to suffer when we had a deadline and the last to know exactly what we were doing. We gladly acknowledge several other people for their untiring support and work during this project. We first acknowledge our editor Chuck Crumly, who generously approved our project and who has been a guardian of sorts for the book since the first outline. We acknowledge Jordan Wearing our project developer at CRC/Garland Science at Taylor Francis for his expert work at molding the project for us. We thank Elizabeth King at *KnowledgeWorks* our project editor for her hard work in producing the project. Her diligence, patience and original ideas were amazing and the appeal of the book and its execution would not have been possible without her hard work. We thank Kristin Fenske at Hannover University for literally herding cats - Bernd and Rob – during the production of the book. Without Kristin's organization, hard work and intellectual input we would simply not have a cogent product, in fact, we would have no product. The cheerfulness and enthusiasm from both Kristin and Elizabeth were inspirational and kept us going on all cylinders during the project. Finally, we thank all of our contributing authors. They all gladly and enthusiastically contributed to this project and we are immensely grateful to them.

EDITORS

Bernd Schierwater is Director of the Institute of Animal Ecology and Evolution, University of Veterinary Medicine Hannover, Foundation. Bernd does research in developmental biology, ecology, and evolutionary biology. His current projects are 'the urmetazoan puzzle', 'placozoan systematics and genomics', and 'next-generation biomonitoring of global change'.

Rob DeSalle is Curator at the Sackler Institute of Comparative Genomics in the Division of Invertebrate Zoology at the American Museum of Natural History. Rob works in molecular systematics, microbial evolution, and genomics. His current research concerns the development of bioinformatic tools to handle large-scale genomics problems using phylogenetic systematic approaches.

CONTRIBUTORS

Maja Adamska
Australian National University
Research School of Biology (RSB)
Canberra, Australia

Teresa Adell
Departament de Genètica, Microbiologia i Estadística
Institut de Biomedicina de la Universitat de Barcelona (IBUB)
Universitat de Barcelona
Barcelona, Spain

Massimo Avian
Department of Life Sciences
University of Trieste
Trieste, Italy

Maria Balsamo
Department of Biomolecular Sciences
University of Urbino Carlo Bo
Urbino, Italy

Neil W. Blackstone
Department of Biological Sciences
Northern Illinois University
DeKalb, Illinois

Mark Blaxter
Tree of Life Department
Wellcome Sanger Institute
Hinxton, United Kingdom

Mercer R. Brugler
Department of Natural Sciences
University of South Carolina Beaufort
Beaufort, South Carolina

Paulyn Cartwright
Department of Ecology and Evolutionary Biology
The University of Kansas
Lawrence, Kansas

Danielle M. DeLeo
American Museum of National History
New York City, New York

Shahan Derkarabetian
Department of Organismic and Evolutionary Biology
Harvard University
Cambridge, Massachusetts

Rob DeSalle
American Museum of Natural History
New York City, New York

Bahram Sayyaf Dezfuli
Department of Life Sciences and Biotechnologies
University of Ferrara
Ferrara, Italy

Kazuyoshi Endo
Department of Earth and Planetary Science
The University of Tokyo
Tokyo, Japan

Wiebke Feindt
Institute of Animal Ecology and Evolution
University of Veterinary Medicine Hannover, Foundation
Hannover, Germany

David H. A. Fitch
Department of Biology
New York University
New York City, New York

Diego Fontaneto
National Research Council of Italy
Water Research Institute, CNR-IRSA
Verbania Pallanza, Italy

Jonathan Foox
American Museum of National History
New York City, New York

Luisa Giari
Department of Life Sciences and Biotechnologies
University of Ferrara
Ferrara, Italy

José D. Gilgado
Department of Environmental Sciences
University of Basel
Basel, Switzerland

Paolo Grilli
Department of Biomolecular Sciences
University of Urbino Carlo Bo
Urbino, Italy

Vladimir Gross
Department of Zoology
University of Kassel
Kassel, Germany

Heike Hadrys
Institute of Animal Ecology and Evolution
University of Veterinary Medicine Hannover, Foundation
Hannover, Germany

Kenneth M. Halanych
Department of Biological Sciences
Auburn University
Auburn, Alabama

Steffen Harzsch
University of Greifswald
Zoological Institute and Museum
Cytology and Evolutionary Biology
Greifswald, Germany

Gerhard Haszprunar
Department Biology II, Systematic Zoology
Ludwig-Maximilians-University Munich
Munich, Germany

Masato Hirose
Kitasato University
School of Marine Biosciences
Kanagawa, Japan

Linda Z. Holland
Marine Biology Research Division
Scripps Institution of Oceanography (SIO)
University of California, San Diego
La Jolla, California

Tohru Iseto
Japan Agency for Marine-Earth Science and Technology (JAMSTEC)
Yokohama Institute
Yokohama, Japan

Hiroshi Kajihara
Faculty of Science
Hokkaido University
Sapporo, Japan

Kai Kamm
Institute of Animal Ecology and Evolution
University of Veterinary Medicine Hannover, Foundation
Hannover, Germany

Karin Kiontke
Department of Biology
New York University
New York City, New York

Sebastian Kvist
Department of Natural History
Royal Ontario Museum
Ontario, Canada

Department of Ecology and Evolutionary Biology
University of Toronto
Toronto, Canada

Michelle M. Leger
Institute of Evolutionary Biology
Barcelona, Spain

Sanna Majaneva
Department of Biology
Norwegian University of Science and Technology
Trondheim, Norway

Georg Mayer
Department of Zoology
University of Kassel
Kassel, Germany

Carsten H. G. Müller
University of Greifswald
Zoological Institute and Museum
General and Systematic Zoology
Greifswald, Germany

Hiroaki Nakano
College of Biological Sciences
University of Tsukuba
Tsukuba, Japan

Johannes S. Neumann
Richard Gilder Graduate School
American Museum of Natural History
New York, New York

Alejandro Oceguera-Figueroa
Department of Zoology
Institute of Biology
National Autonomous University of Mexico
Mexico City, Mexico

Otto M. P. Oliveira
Federal University of ABC
Center of Natural and Human Sciences (CCNH)
Santo André, Brazil

Ivo de Sena Oliveira
Department of Zoology
University of Kassel
Kassel, Germany

Hans-Jürgen Osigus
Institute of Animal Ecology and Evolution
University of Veterinary Medicine Hannover, Foundation
Hannover, Germany

Yvan Perez
Mediterranean Institute of Marine and Terrestrial Biodiversity and Ecology
Aix-Marseille University
Marseille, France

Michael Plewka
plingfactory
Hattingen, Germany

Andrea M. Quattrini
American Museum of National History
New York City, New York

Nicolas Rabet
BOREA Laboratory Biology of Aquatic Organisms and Ecosystems
Sorbonne University
Paris, France

Andreja Ramšak
National Institute of Biology
Marine Biology Station Piran
Piran, Slovenia

Marta Riutort
Departament de Genètica
Microbiologia i Estadística and Institut de Recerca de la Biodiversitat (IRBio)
Universitat de Barcelona
Barcelona, Spain

Iñaki Ruiz-Trillo
Institute of Evolutionary Biology
Barcelona, Spain

Scott Santagata
Department of Biology
Long Island University
New York City, New York

Bernd Schierwater
Institute of Animal Ecology and Evolution
University of Veterinary Medicine Hannover, Foundation
Hannover, Germany

Andreas Schmidt-Rhaesa
Leibniz Institute for the Analysis of Biodiversity Change
Center for Natural History, Invertebrate Collection
Hamburg, Germany

Martin V. Sørensen
University of Copenhagen Microbiologia i Estadística and Institut de Recerca de la Biodiversitat (IRBio) Denmark
Copenhagen, Denmark

Sabine Stöhr
Department of Zoology
Swedish Museum of Natural History
Stockholm, Sweden

Michael Tessler
Department of Biology
St. Francis College
New York City, New York

Kathrin Wysocki
Institute of Animal Ecology and Evolution
University of Veterinary Medicine Hannover, Foundation
Hannover, Germany

INTRODUCTION: SETTING THE STAGE FOR A RATIONAL TREATMENT OF INVERTEBRATE ANIMALS

Bernd Schierwater and Rob DeSalle

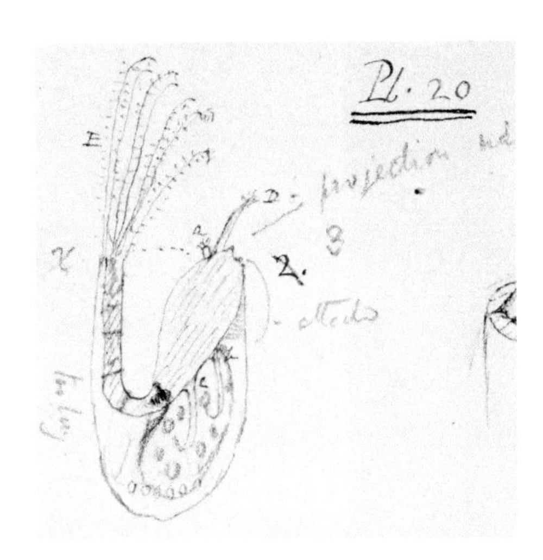

CONTENTS

SCOPE

This textbook on Systematic Zoology of Invertebrates is a systematic description of the main part of the animal kingdom, the metazoan animals excluding the sub-phylum Vertebrata, as illustrated in **Figure 1**. In this figure we have representatives of selected taxa in the form of a genealogical tree. The genealogical tree is a summary of the branching patterns from common ancestors of taxa both living and extinct. The traditional definition of a metazoan animal has often been forgotten, although it is clear, simple and completely unambiguous; related to the division of labor metazoans show "intrasomatic differentiation", i.e. harbor more than one type of somatic cells. The division of the Metazoa into Vertebrates and Invertebrates, as being indispensable in the zoological common speech, is solely based on a pragmatic footing. Only Vertebrates are a natural taxonomic unit, whereas the broad term Invertebrates is nothing but a collective name for all non-vertebrate metazoans; that is, the term Invertebrates does not imply any further relationship of all the different groups in it. Tunicates and lancet fish, to give an example, which are both Invertebrates, are less related to other Invertebrates than to Vertebrates, with which they are collated in the phylum Chordata. While we know about 1.2 million described Invertebrate species, estimates guess the real number of today's living species at anywhere between 1.5 million or even 10–20 million.

The basic principle of the zoological system on which this textbook is based is the well-established hierarchy of, in a certain manner, correlated super- or subordinated interleaved taxa, of which only the most important categories are represented in this scheme (namely using the concrete example of the systematic classification of the fruit fly *Drosophila melanogaster* into the recent animal kingdom as a whole).

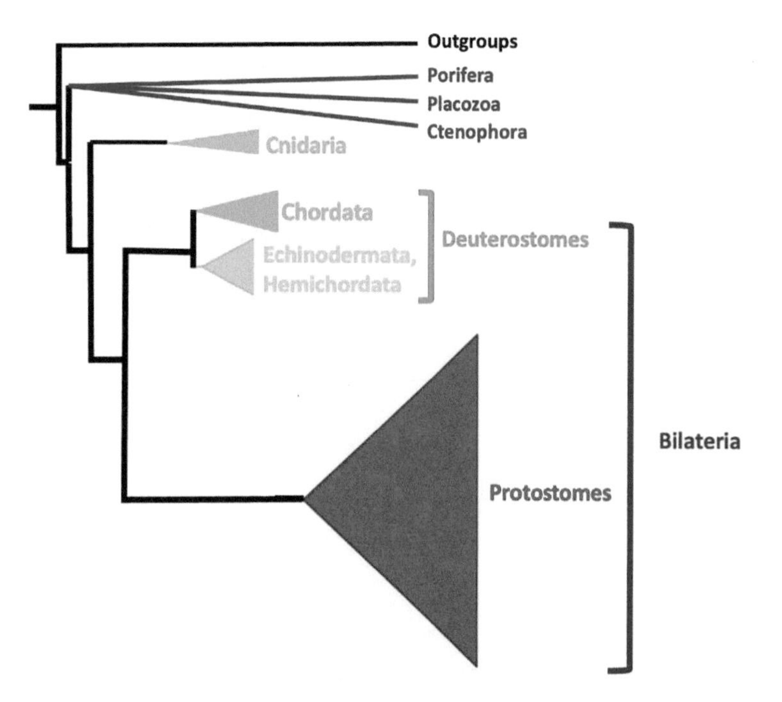

Figure 1 Genealogical tree showing the distribution of some of the major super-taxa and phyla relevant to Invertebrate Zoology. The lower metazoan taxa Porifera, Ctenophora and Placozoa (dark blue) are shown as an unresolved group at the base of the tree. While we show the relationships as unresolved, later we will comment at length on the current state of the art involving the relationships of these taxa. Cnidaria (light blue) branch next along with the bilaterians (red and green). The protostomes (red) are the most phylum-rich of these higher taxa. Deuterstomes are shown as comprised of two major groups – Echinodermata + Hemichordata and Chordata. Note that it is impossible to group the invertebrate groups in the figure (all but the dark green group of chordates) into a single monophyletic group without Chordata interfering.

LINNEAN TAXONOMIC HIERARCHIES

Linnaeus instituted a binomial nomenclature for species naming. This naming convention has a generic designation and a species-level designation where both names are italicized. Higher taxonomic names are applied to larger and larger groups of organisms in a hierarchical fashion. Different groups of organisms have different underlying hierarchies and there is some controversy as to how to arrange the very highest of categories because of shifting phylogenetic hypotheses at this level. For instance, Lynn Margulis (1974) introduced a five-kingdom system – prokaryotes, plants, animals, fungi and protists. Invertebrate zoology would then take on the following simple and traditional hierarchical structure.

Kingdom
Phylum
Class
Order
Family
Genus
Species

At the kingdom level recent phylogenetic work suggests that things are a bit more complex. Two major problems arise with five kingdoms. First, prokaryotes would be made of Archaea and Bacteria, but these two groups of microbes are not each other's closest relatives. In fact, eukaryotes appear to be the closest relative to archaeans. Second, protists or single-celled eukaryotes do not come from a single common ancestor. Some protists are more closely related to plants (algae) and others are more closely related to animals (choanoflagellates to name one). While we can define protists as single-celled eukaryotes, the name means little in a biological context. Also, this commonly used definition is misleading or even wrong (see Chapter 1). Perhaps a more biological way to organize the very highest part of the taxonomic scheme might be to use the name Domain to refer to each of Bacteria, Archaea and Eukaryota. Domains might then be followed by kingdoms (with protists being special cases of mixed kingdoms), and then the rest of the hierarchy which seems to be pretty stable:

Domain
Kingdom
Phylum
Class
Order
Family
Genus
Species

Many Protists consist of more than one cell, often several dozens of cells. The only valid definition, which unfortunately has never entered the English literature, is "the lack of intrasomatic differentiation". In "multicellular" protists all somatic cells are identical and only reproductive cells (germ line) differ. Some taxonomists have complicated the hierarchy even further by creating taxonomic entities in between the major levels described above. For animals the names for superfamily, family, subfamily, tribe and subtribe have specific endings to indicate these levels:

Domain
Kingdom
Phylum
Subphylum
Superclass
Class

Subclass
Infraclass
Cohort
Superorder
Order
Suborder
Infraorder
Superfamily (-oidea)
Family (-idea)
Subfamily (-inae)
Tribe (-ini)
Subtribe (ina)
Genus
Subgenus
Species
Subspecies

So for that taxon very familiar to all of us – *Drosophila melanogaster* – its taxonomic integration at the levels shown above would be

Domain	Eukaryota
Kingdom	Animalia
Subkingdom	Eumetazoa
Phylum	Arthropoda
Subphylum	Tracheata
Superclass	Hexapoda
Class	Insecta
Subclass	Pterygota
Infraclass	Neoptera
Cohort	Endopterygota
Superorder	Oligoneoptera
Order	Diptera
Suborder	Brachycera
Infraorder	Cyclorrhapha
Infraorder-Section	Schizophora
Infraorder-Subsection	Acalypterata
Superfamily (-oidea)	Ephidroidea
Family (-idae)	Drosophilidae
Subfamily (-inae)	Drosophilinae
Tribe (-ini)	Drosophilini
Subtribe (-ina)	Drosophilina
Infratribe (-iti)	Drosophiliti
Genus	*Drosophila*
Subgenus	*Drosophila*
Species	*melanogaster*
Subspecies	none

Some taxonomists furthermore divide the subgenus into species groups (e.g. *melanogaster* group), species subgroups (e.g. *melanogaster* subgroup) and species complexes (e.g. *melanogaster* complex). Having molecular data at hand providing *cum grano salis* infinite levels of relationships, one can make any scheme as complex and complicated as one wishes, and make them look very "precise". We may never forget, however, "whenever we put boundaries into the continuum of life forms, such boundaries will necessarily be artificial!" (Hauenschild, 1993).

Some large taxa with a great variety of morphological variation may require additional systematic in-between-taxonomic ranks in order to embody the presumptive degrees of kinship. Overall there is a widely unified zoological system which has been taken over for most recent textbooks in a basically compliant way. Diverging opinions about where a certain taxon should be classified within the system or how a taxon itself should best be subdivided often derive from the fact that our knowledge is still very incomplete or that the determination of homology allows different interpretations.

Certainly comparative zoology has still not reached the ultimate goal of showing a thoroughgoing natural system, despite substantial efforts for more than a century. Modern molecular approaches have added substantial progress in many areas but have also created some of the most short-lived and most conflicting hypotheses ever. In order to evade the difficulties which might occur by having various taxonomic determinations in different textbooks, you will find a firm structure for our purposes here. With each chapter written by an expert for the group, we leave the decisions completely up to the author of that chapter.

GETTING STARTED

In a first large-scale division, the animal kingdom can initially be divided into two sub-kingdoms, namely the unicellular animals or so-called Protozoa, and the multicellular animals or Metazoa. Only about 40,000 protist species have been described so far, although there is certainly a much larger number of them. Some 1.2 million metazoan species are currently known (i.e. animals with more than one somatic cell type), and more species are described every day. The more complex Metazoa must have emerged from some unicellular ancestor, some primeval animals or Protozoa. Recent Protozoa can of course not be considered for such an ancestral role, because the historical transition from the unicellular to the multicellular stage took place more than a billion years ago and because today's Protozoa are only the late descendants of those archaic unicellular organisms from which the Metazoa once evolved.

The second subkingdom of the animals, the Metazoa, comprises the vast majority of the modern animal world, namely at least 1.2 million described species. About 80% of them are Arthropods and another 10% are Molluscs; the other 22 phyla thus all together comprise only a little more than another 10% of all species. In a number of taxa, not all existing species have been described and named yet.

The multicellular body of the Metazoa consists of a number of cells (in extreme case trillions of cells), the majority being somatic cells, which always differentiate into several or even many structurally and functionally different types. Due to the phenomenon of intra-somatic differentiation, physiologically connected with a division of labor, all Metazoa differ fundamentally from Protozoa, even from those in which there is already a germ line separated from the soma. The soma of the metazoan individual, which dominates quantitatively and essentially determines its appearance, has in its entirety only a species-specific limited lifespan and ends its existence at the latest, if it is not violently destroyed beforehand, through endogenously programmed natural age. The dead soma is called the "corpse", so the formation of corpses is an inevitable consequence of the soma germ line differentiation. It should be noted that corpse formation is also seen in some protists, and thus is not an invention of the Metazoa. In Metazoa only the germ cells, which are used for sexual reproduction, are potentially immortal.

COMPLICATING FACTORS: SIMPLICITY AND COMPLEXITY

The evolution from a protist to the first metazoan is not documented in the fossil record and not mirrored in recent transitional stages. Thus we depend on inferences from the ontogeny of recent metazoans. The simplest blueprint of all Metazoa can be found in the phylum Placozoa, but its true meaning for understanding the evolution of the Metazoa has only recently been revealed. The animals resemble a hypothetically urmetazoon called Placula. The modern Placula hypothesis is explained in the chapter Placozoa. Other urmetazoan hypotheses have been rich on names and scenarios and often cause confusion. But leaving out some more aberrant scenarios and looking at the traditional urmetazoan hypotheses developed by Bütschli (1884) and Ernst Haeckel (1874), the interested student can easily extract that most metazoan bauplans can be derived either from a hypothetical Placula, Blastaea or Gastraea stage. There is continuous transition between stages and the Gastraea is a later stage, deriving either from a Placula or Blastaea.

Almost all scenarios start with a spherical flagellate (protist) colony, which automatically turns into a metazoan Blastaea if the somatic cells differentiate into ecto- and entoderm cells (intrasomatic differentiation). This stage is represented in Otto Bütschli's Placula and Blastaea hypothesis in the first drawings [**Figure 2 (1a-d)**]. The name Placula later leads to the phylum name Placozoa

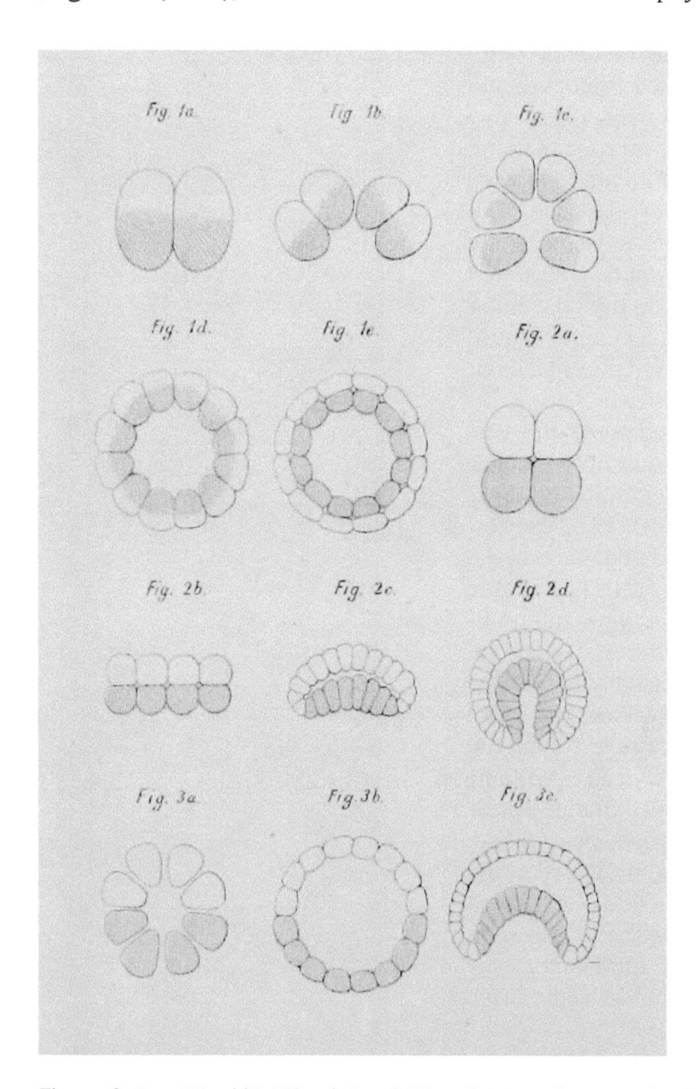

Figure 2 Otto Bütschli's "Placula" and "Blastula" Hypotheses on the Origin of Metazoan Animals. The figure summarizes the attempts to derive different Blastula and Gastraea forms from the same ancestor. The presumably most ancestral stages are shown in Figures. 1a (1st cleavage into 2 daughter cells) and 2a. (8-cell stadium with separated entoderm (red) and ectoderm (white) elements). The situation in 2a serves as the starting point for both, the Placula and the Blastula hypothesis. The "Placula" hypothesis is shown in Figure 2. In 2b the ecto- and entoderm cells have propagated and form an obvious Placula, which in 2c and 2d bends the entoderm inwards to form a Gastraea, which mirrors the typical metazoan organization. As an alternative to the hypothetical Placula a hypothetical "Blastula" is formed in 3a-c. The Blastula starts with the cells in 2a forming a cavity distancing the ecto- and entoderm cells before invagination of the entoderm forms a Gastraea similar to the one in 2d. (Drawing from Otto Bütschli (1884).)

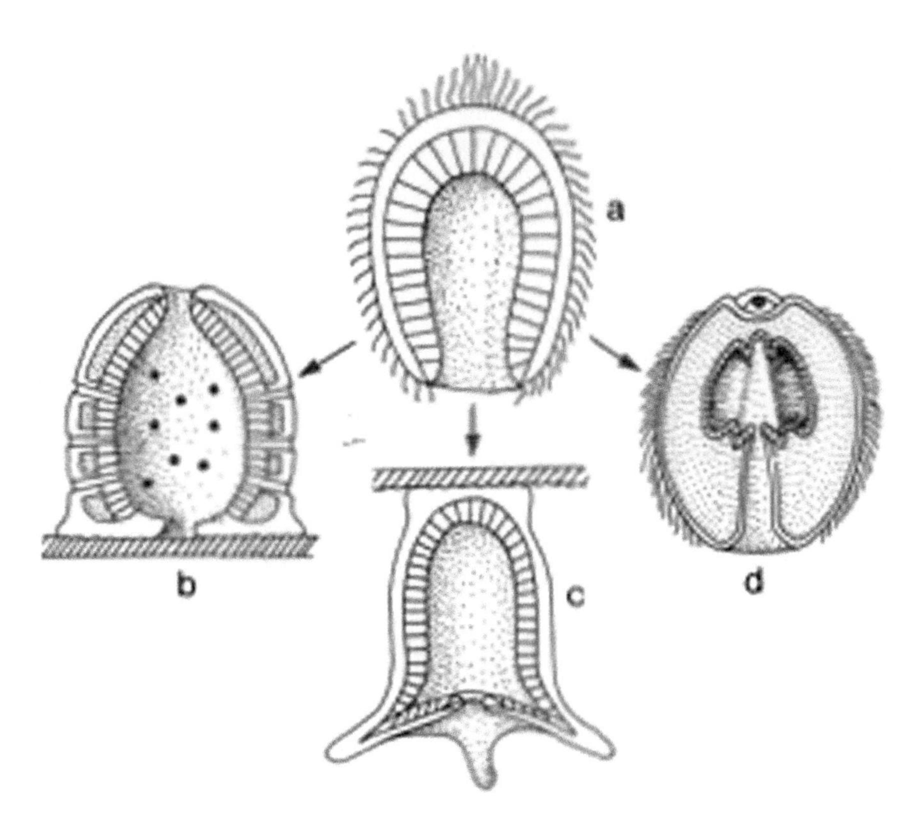

Figure 3 Based on Ernst Haeckel's Gastraea theory different metazoan bauplans can be derived from the hypothetical Gastraea, which possible origin has been shown in Figure 2. If the Gastraea (a) attaches with its oral pole it forms a sponge bauplan (b), if it attaches with its opposite (aboral) pole it forms a cnidarian bauplan (c), and if it does not attach the more complex ctenophore bauplan can be derived. (From Storch and Welsch 1994.)

and the term Blastaea is meant in analogy to the blastula in the ontogeny of a typical metazoan. Placozoans can be most simply derived from the hypothetical Placula, and all other basal metazoan phyla most simply from a hypothetical Gastraea (**Figure 3**).

There are also a number of less prominent hypotheses for the origin of the Metazoa. For example, as an alternative to the Gastraea hypothesis, some zoologists have discussed the possibility that an ancestral protozoon might have developed directly to the simplest bilaterian via compact multi-nucleated plasmodia, missing a Blastaea, Coelenterate or Porifera stage completely. The proponents of this view considered certain turbellarians to be the most primitive recent Bilateria and called the idea the Acoela hypothesis. This and other ideas are outlived and not supported by current knowledge.

A NOTE ON CHAPTER ICONS

We introduce each chapter in the text with an iconic drawing or figure. For the Prologue we include a drawing of "Mr. Arthrobalanus" a barnacle observed and drawn by Charles Darwin on board the Beagle. For Chapter 1, we include a page from the Arabic text Kitāb al-Hayawān by Al-Jahiz. This text is an Arabic translation of Aristotle's zoological works. To start out Chapter 2, we include Charles Darwin's famous "I think" diagram of hypothetical genealogical relationships drawn and explained in his notebooks. Chapter 3 discusses modern techniques, and here we include the original drawing of James Watson and Francis Crick's DNA double helix (from Watson, James D., and Francis HC Crick. "Genetical implications of the structure of deoxyribonucleic acid." Nature 171, no. 4361

(1953): 964-967). We discuss the difficult phylogenetic question of the sister to all other metazoans in Chapter 4, and in this context we include Ernst Haekel's exquisite drawing of Hexactinellae sponges (from Ernst Haeckel's Kunstformen der Natur, 1904). For Chapter 5 we return to Haekel and use his phylogenetic tree of life drawn in 1866. For Chapters 6 to 38 we use a phylogenetic tree and the position of the group discussed in the chapter marked by color. We use this general topology of invertebrates throughout the text and it is described in this prologue. This topology is a modified summary of the many recent molecular studies on the invertebrate animals. Finally, for the Epilogue we include the classical drawing of Placozoaire, Trichoplax adhaerens from the Encyclopædia Britannica, 1911 also known as "EB1911_Platyelmia Trichoplax_adhaerens" drawn by F. E Schultze in 1891.

REFERENCES

Bütschli, O. 1884. *Bemerkungen zur Gastraea-Theorie*. Volume 9, *Morphologische Jahrblatt*.

Haeckel, E. 1874. Die Gastraea-Theorie, die phylogenetische Classification des Thierreichs und die Homologie der Keimblatter. *Jenaische Z f Naturwiss* 8:1–55.

Hauenschild, C. 1993. *Fortpflanzung und Sexualität der Tiere. Volume MEYERS FORUM 11. Mannheim; Leipzig;Wien*;Zürich: B.I.-Taschenbuchverlag.

Linnaeus, C. 1758. Systema Naturae per regna tria naturae, secundum classes, ordines, genera, species, cum characteribus, differentiis, synonymis, locis. *Editio decima, reformata [10th revised edition], volume 1: 824. Laurentius Salvius: Holmiae.*

Margulis, Lynn. 1974. Five-Kingdom Classification and the Origin and Evolution of Cells. In *Evolutionary Biology: Volume 7*, edited by T. Dobzhansky, M. K. Hecht, and W. C. Steere. Boston, MA: Springer US.

Storch, V., and U. Welsch. 1994. *Kurzes Lehrbuch der Zoologie*. Volume 7. Auflage. Stuttgart, Jena, New York: Gustav Fischer Verlag.

CHAPTER 1

INVERTEBRATES AND INFORMATION

Rob DeSalle and Bernd Schierwater

CONTENTS

FINDING INFORMATION ABOUT INVERTEBRATE TAXA

We will organize our discussion of invertebrate taxa by focusing mostly on groups at the level of phylum. We have already discussed this taxonomic level in invertebrate zoology in then Prologue, and here we will simply use phyla (or specific subdivisions within phyla) as our organizing units. We first detail where information on the wonderful diversity and breadth of invertebrate animals can be found.

Several invertebrate zoology textbooks exist. In addition to the many textbooks available, there is a broad literature and a strong web presence addressing invertebrate zoology. In **Table 1.1** we list links to *Current Biology* synopses of most of the phyla we discuss in this text. These short articles are a good place to get an initial familiarity with these phyla. They are written by experts who focus on the phyla in the list and some of their authors are included in this text. Several websites exist that provide information on organisms, and in each chapter of this text specific websites for the individual groups are provided. One of the more comprehensive websites that provides information on

TABLE 1.1 ***Current Biology*** **short synopses of some of the major groups of invertebrates**

Placozoa[1]	https://www.cell.com/current-biology/fulltext/S0960-9822(17)31524-5
Porifer[2]	https://www.cell.com/current-biology/fulltext/S0960-9822(05)00143-0
Cnidaria[3]	https://www.cell.com/current-biology/fulltext/S09609822(13)00359-X
Cnidaria (Hydrozoa)[4]	https://www.cell.com/current-biology/fulltext/S0960-9822(10)01172-3
Ctenophora[5]	https://www.cell.com/current-biology/fulltext/S0960-9822(08)01291-8
Ctenophora[5A]	https://www.sciencedirect.com/science/article/pii/S0960982216309800
Mollusca[6]	https://www.cell.com/current-biology/fulltext/S0960-9822(12)00592-1
Platyhelminthes[7]	https://www.cell.com/current-biology/comments/S0960-9822(17)30152-5
Bryozoa[8]	https://www.cell.com/current-biology/fulltext/S0960-9822(14)00533-8
Chaetognatha[9]	https://www.cell.com/current-biology/fulltext/S0960-9822(06)01835-5
Gnathifera[10]	https://www.cell.com/current-biology/fulltext/S0960-9822(18)31541-0
Crustacea[11]	https://www.cell.com/current-biology/fulltext/S0960-9822(08)00658-1
Insecta[12]	https://www.cell.com/current-biology/fulltext/S0960-9822(15)00927-6
Chelicerata[13]	https://www.cell.com/current-biology/fulltext/S0960-9822(18)30672-9
Arthropods[14]	https://www.cell.com/current-biology/fulltext/S0960-9822(19)30486-5
Onychophora[15]	https://www.cell.com/current-biology/pdf/S0960-9822(11)00208-9.pdf
Tardigrada[16]	https://www.cell.com/current-biology/fulltext/S0960-9822(16)30007-0
Nematoda[17]	https://www.cell.com/current-biology/fulltext/S0960-9822(13)00985-8
Echinodermata[18]	https://www.cell.com/current-biology/fulltext/S0960-9822(05)01401-6
Xenoturbellida[19]	https://www.cell.com/current-biology/comments/S0960-9822(16)30191-9
Enteropneusta[20]	https://www.cell.com/current-biology/fulltext/S0960-9822(02)00491-8
Tunicata[21]	https://www.cell.com/current-biology/fulltext/S0960-9822(15)01521-3
Chordates[22]	https://www.cell.com/current-biology/fulltext/S0960-9822(05)01326-6
Choanoflagellata[23]	https://www.cell.com/current-biology/fulltext/S0960-9822(05)00142-9

The list also includes the URL of the work for quick access. References for the synopses are given in the superscript. 1. Schierwater and DeSalle, 2018; 2. Leys et al., 2005; 3. Katsuki and Greenspan, 2013 4. Glauber et al., 2010; 5. Pang and Martindale, 2008; 5A. Giribet, 2016; 6. Haszprunar and Wanninger, 2012; 7. Collins, 2017; 8. Leung, 2014; 9. Marlétaz et al., 2006; 10. Marlétaz et al., 2019; 11. VanHook and Patel, 2008; 12. Engel, 2015; 13. Sharma, 2018; 14. Giribet and Edgecombe, 2019; 15. Blaxter and Sunnucks, 2011; 16. Maderspacher, 2016; 17. Kiontke and Fitch, 2013; 18. Amemiya et al., 2005; 19. Telford and Copley, 2016; 20. Nübler-Jung and Arendt, 1996; 21. Holland, L., 2016; 22. Holland, N., 2005; 23. King, 2005.

all organisms is the Encyclopedia of Life (EoL; https://eol.org/). In **Table 1.2** we list links for each phylum addressed in this book (and several subcategories of phyla) to their EoL entries. We also list the number of species, genera and families in each higher taxon in the table.

Modern Invertebrate Systematics has been highly influenced by molecular data and the modern classification systems available for higher invertebrate

TABLE 1.2 Encyclopedia of Life (EoL) entries for the phyla (and subcategories within certain phyla) discussed in this text

Phylum	Web access to EOL	Sp	Gen	Fam
Choanoflagellata	https://eol.org/pages/42372141	85	46	4
Placozoa	https://eol.org/pages/8773	2	2	1*
Porifera	https://eol.org/pages/3142	9,049	747	143
Cnidaria (Scyphozoa)	https://eol.org/pages/46554113	196	62	21
Cnidarria (Cubozoa)	https://eol.org/pages/6681	47	18	8
Cnidaria (Hydrozoa)	https://eol.org/pages/1795	3758	557	133
Cnidaria (Anthozoa)	https://eol.org/pages/1746	7,496	1,112	172
Ctenophora	https://eol.org/pages/69	210	63	32
Mollusca	https://eol.org/pages/2195	85,844	11,213	1,161
Polyplacophora	https://eol.org/pages/2680	1,089	84	22
Annelida	https://eol.org/pages/36	20,481	2,382	171
Brachiopoda	https://eol.org/pages/1498	11,082	2,486	365
Bryozoa	https://eol.org/pages/2060	5,609	910	231
Phoronida	https://eol.org/pages/8867	19	3	1
Ectoprocta	https://eol.org/pages/49925179	not listed		
Platyhelminthes	https://eol.org/pages/2884	18,089	3,107	355
Nemertea	https://eol.org/pages/2855	1,351	315	43
Rotifera	https://eol.org/pages/6851	2,011	132	34
Cycliophora	https://eol.org/pages/1922	2	1	1
Gnathostomulida	https://eol.org/pages/8747	101	27	12
Micrognathozoa	https://eol.org/pages/5006388	1	1	1
Chaetognatha	https://eol.org/pages/1740	132	26	9
Gnathifera	https://eol.org/pages/46974508	not listed		
Acanthocephala	https://eol.org/pages/2	1,329	151	26
Entoprocta	https://eol.org/pages/2167	171	15	4
Gastrotricha	https://eol.org/pages/8728	851	68	18
Arthro (Crustacea)				
Branchiopoda	https://eol.org/pages/265	1,359	137	29
Cephalocarida	https://eol.org/pages/278	12	5	1
Malacostraca	https://eol.org/pages/1157	39,479	6,634	683
Copepoda	https://eol.org/pages/2625033	14,724	2,153	238
Ostrocoda	https://eol.org/pages/1456	6,588	942	91
Remipedia	https://eol.org/pages/1495	29	12	8
Arthro (Insecta)	https://eol.org/pages/344	942,651	84,973	1,686
Arthro (Chelicerata)	https://eol.org/pages/2579982	75,034	7,868	903
Arthro (Myriapoda)	https://eol.org/pages/2631567	18,804	3,280	186
Onychophora	https://eol.org/pages/6927	205	57	3

(Continued)

TABLE 1.2 (Continued)

Phylum	Web access to EOL	Sp	Gen	Fam
Nematoda	https://eol.org/pages/2715	3,452	954	170
Kinorhyncha	https://eol.org/pages/1526	188	21	9
Nematomorpha	https://eol.org/pages/1539	361	19	3
Priapulida	https://eol.org/pages/1533	22	7	5
Loricifera	https://eol.org/pages/1537	27	8	2
Tardigrada	https://eol.org/pages/3204	1,018	105	20
Echinodermata	https://eol.org/pages/1926	10,832	1,976	280
Xenoturbellida	https://eol.org/pages/3028428	6	1	1
Hemichordata	https://eol.org/pages/8854	139	25	7
Pterobranchia	https://eol.org/pages/47039079	see Hemichordata		
Enteropneusta	https://eol.org/pages/8859	see Hemichordata		
Tunicates	https://eol.org/pages/46582349	3,070	209	35
Cephalochordata	https://eol.org/pages/1585	30	3	1
Vertebrata	https://eol.org/pages/2774383	72,394	13,091	2,070

The number of species (Sp), of genera (Gen) and of families (Fam) are listed. The number of species, genera and families are taken from EoL.

* There are many more species in this phylum with a complex higher taxonomy, but these are yet to be described.

relationships have been affected with these kinds of studies in mind. We discuss this dynamic in more detail in Chapter 4. Unlike taxonomy (see below), systematic work in invertebrate zoology does not have a ruling body. There are many journals dedicated to invertebrate systematics, though (**Table 1.3**).

THE SCOPE OF INVERTEBRATE ZOOLOGY

This text will expand on this topic by going phylum to phylum. The critical aspects of invertebrates that we have focused on in each chapter concern their morphology, development, reproduction, ecology, distribution, genomics, behavior and where in the tree of life each phylum is placed. Hence, each chapter must and will address these aspects of the individual phyla as listed in **Table 1.4**. Since different aspects leave different degrees of room for personal interpretations and judgments, each chapter will end with a section called "Conclusion".

HISTORICAL INVERTEBRATES

Humans have probably been infatuated with describing invertebrate animals since the beginning of language. The Bible, perhaps one of the oldest written references, gives multiple nods to invertebrates such as moths, flies, bees, grasshoppers, hornets, locusts, spiders, ants, lice, cankerworms, earthworms, palmerworms, bald locusts, beetles, fleas, gnats, leeches and "all that have not fins and scales in the seas, and in the rivers" (these being things like shrimps). Most of the references to invertebrates were to what could be eaten or should be avoided. Records of invertebrates date back to 1200 BC in China (Sterckx, 2005) and to Sanskrit records around 500 BC in India (Kaur and Singh, 2018). Aristotle treated invertebrates at length a century later in his History of Animals (*Historia Animālium*) and Parts of Animals (*De Partibus Animālium*). He mentions over 500 vertebrate animals and nearly as many invertebrate animals in these

TABLE 1.3 List of journals that focus on taxonomy and systematics of invertebrates

Invertebrate Systematics	https://www.publish.csiro.au/is	ALL
Zootaxa	https://www.mapress.com/j/zt/	ALL
Journal of Zoological Systematics And Evolutionary Research	https://onlinelibrary.wiley.com/journal/14390469	ALL
Journal of African Invertebrates	https://africaninvertebrates.pensoft.net/	ALL
Marine and Freshwater Research	http://www.publish.csiro.au/mf	ALL
European Journal of Taxonomy	https://europeanjournaloftaxonomy.eu/index.php/ejt	ALL
Systematics and Biodiversity	https://www.tandfonline.com/toc/tsab20/current	ALL
Evolutionary Systematics	https://evolsyst.pensoft.net/	ALL
Zookeys	https://zookeys.pensoft.net/	ALL
Zoologische Mededelingen	https://www.zoologischemededelingen.nl/	ALL
Zoologica Scripta	https://onlinelibrary.wiley.com/journal/14636409	ALL
Sarsia	https://www.tandfonline.com/loi/ssar20	ALL
Ophelia	https://www.tandfonline.com/loi/smar19	ALL
Steenstrupia	https://zoologi.snm.ku.dk/english/about_the_zoological_museum/publications/steenstrupia/	ALL
Bulletin of Marine Science	http://bullmarsci.org/	ALL
Journal of Experimental Biology	https://jeb.biologists.org/	ALL
Organisms Diversity & Evolution	https://link.springer.com/journal/13127	ALL
The Biological Bulletin	https://www.journals.uchicago.edu/toc/bbl/current	ALL
Insect Systematics and Diversity	https://academic.oup.com/isd	Arth
Insect Systematics and Evolution	https://brill.com/view/journals/ise/ise-overview.xml	Arth
Systematic Entomology	https://onlinelibrary.wiley.com/journal/13653113	Arth
Journal of Entomology and Zoology Studies	http://www.entomoljournal.com/	Arth
International Journal of Entomology	http://www.entomologyjournals.com/	Arth
American Journal of Entomology	http://www.ajentomology.org/	Arth
Journal of Crustacean Biology	https://academic.oup.com/jcb/	Arth
Arthropoda Selecta	https://kmkjournals.com/journals/AS	Arth
Molluscan Research	https://www.mapress.com/mr/	Moll
Journal of Molluscan Studies	https://academic.oup.com/mollus/pages/About	Moll
Malacologia	https://instituteofmalacology.org/	Moll
Journal of Nematology	https://journals.flvc.org/jon	Nem
Journal of Nematode Morphology and Systematics	https://dialnet.unirioja.es/servlet/revista?codigo=1662	Nem
Journal of Arachnology	http://www.americanarachnology.org/JOA.html	Arac
Acaralogia	http://www1.montpellier.inra.fr/CBGP/acarologia/	Arac
Journal of Arachnologica	https://www.jstage.jst.go.jp/browse/asjaa/	Arac
Journal of Parasitology	https://www.journalofparasitology.org/	Para
Parasitology	https://www.cambridge.org/core/journals/parasitology	Para
Marine Biology	https://link.springer.com/journal/227	Para
Proceedings of the Biological Society of Washington	https://pbsw.org/	Annl

ALL indicates that the journal has published on many of the major invertebrate groups. Arth indicates that the journal focuses on arthropods. Moll means that the journal focuses on molluscs. Nem indicates a focus on nematodes. Arac indicates a focus on arachnids. Para indicates the journal focuses on parasites of which many of the taxa treated in this book are. Annl indicates a focus on annelids.

TABLE 1.4 Topics addressed for each phylum (or subdivision within a phylum)

I. History and Systematics
II. Morphology and Anatomy
III. Life Cycle, Development and Reproduction
IV. Distribution and Ecology
V. Physiology and Behavior
Vi. Genetics and Genomics
VII. Position in the Tree of Life (ToL)
VIII. Database and Collections
IX. Conclusion

volumes. His work was translated into Arabic and used by scholars from that region. Aristotle's writings were also used extensively by humans interested in nature during the Dark Ages, Middle Ages and Renaissance.

Scala Naturae: Wrong! but on the way to a rational classification of animals

Aristotle used several categories of animals based on the observational rationale he developed in his writing. Much of what Aristotle did in his writings is not terribly different from what we do today as invertebrate zoologists and systematists. We make observations, record them and use them in some way to make an inference that can advance our understanding of the organisms we have focused on. Aristotle used several categories of observations – some of them are still valid today (although the states of some of the observations are questionable). His method led him to the idea of *scala naturae* or the great chain of being, which was a method of organizing organisms on the planet all the way up to about three centuries ago.

Aristotle used a scoring system for his observations that relied on the following kinds of observations; "Not Said-Of and Not Present-In, Not Said-Of and Present-In, Said-Of and Not Present-In and Said-Of and Present-In" (Studtmann, 2008), which is very similar to the present-day presence/absence scoring system for morphological characters that we use in modern morphological systematics. He used several categories to make his observations. First, he considered the "Substance" of the organism, which includes movement, perception and destructibility. Next he used "Quantity", citing such aspects of organisms that have continuous qualities like surface, body, time and place, as well as discrete quantities like number of appendages or whether or not an organism has a trait like speech or blood. "Quality" was the next category of observations he used to understand organisms on our planet. These include attributes like habits, climate lived in and shape.

To this end, Aristotle collected significant data on vertebrates, invertebrates, plants and minerals that are still instructive for how to think about the natural world. Simply using three categories, "Substance, Quantity and Quality", Aristotle consistently described organisms and in some ways set the stage for more modern ways of examining animals. For instance, for the cephalopod shrimp, he observed that shrimps did not have blood, well at least no blood like we vertebrates have. It was a simple thing for him to count legs, and he deemed that shrimps have eight or more legs. As far as a soul is concerned, he reckoned that the shrimp had a vegetative and sensitive soul, and for its habits it lived in a wet and cold environment.

We can take this information and code it a matrix format (something Aristotle did not do; **Table 1.5**).

TABLE 1.5 Scoring presence of blood, number of legs, kind of soul and four qualities of squid

			Quantities				Soul			Qualities			
Group	Example	Blood	2 legs	4 legs	6 legs	>8 legs	R	S	V	hot	wet	cold	dry
Cephalopods	Squid	N	n	n	n	y	N	Y	Y	n	y	y	n

Y and y = yes; N and n = no.

TABLE 1.6 Data from Aristotle

			QUANTITIES				SOUL			QUALITIES			
Group	Example	Blood	2 legs	4 legs	6 legs	>8 legs	R	S	V	hot	wet	cold	dry
Man	Man	Y	y	n	n	n	Y	Y	Y	y	y	n	n
Live-bearing tetrapods	Cat	Y	n	y	n	n	N	Y	Y	y	y	n	n
Cetaceans	Dolphin	Y	n	n	n	n	N	Y	Y	y	y	n	n
Birds	Bee-eater	Y	y	n	n	n	N	Y	Y	y	y	n	n
Egg-laying tetrapods	Chameleon	Y	n	y	n	n	N	Y	Y	n	y	y	n
Snakes	Water snake	Y	n	n	n	n	N	Y	Y	n	y	y	n
Egg-laying fishes	Sea bass	Y	n	n	n	n	N	Y	Y	n	y	y	n
Placental selachians	Shark	Y	n	n	n	n	n	N	Y	Y	n	y	y
Crustaceans	Shrimp	N	n	n	n	y	N	Y	Y	n	y	y	n
Cephalopods	Squid	N	n	n	n	y	N	Y	Y	n	y	y	n
Hard-shelled animals	Cockle	N	n	n	n	n	N	Y	Y	n	n	y	y
Larva-bearing insects	Ant	N	n	n	y	n	N	Y	Y	n	n	y	y
Generates spontaneously	Sponge	N	n	n	n	n	N	Y	Y	n	n	y	y
Plants	Fig	N	n	n	n	n	N	N	Y	n	n	y	y
Minerals	Iron	N	n	n	n	n	N	N	N	n	n	y	y

Y and y = yes; N and n = no; R = rational, S = Sensitive, V = Vegetative. Each Y can be translated into a 1 and each N into a 0 for a full phylogenetic matrix for the groups in the table (Table 1.7).

It is possible, as Lovejoy (1936) and Mayr (1985) have discussed, to code data from Aristotle, as we show in **Table 1.6**. We can recode the data in a more modern format that can be used for phylogenetic analysis as in **Table 1.7**. While Aristotle did no formal analysis of the data, simple observation and summary led him to his ideas about *Scala Naturae*. This view of nature argues for arranging organisms in the order of their perceived complexity. We will return to this matrix in Chapter 2.

Invertebrates continued to be the subject of much study through the Renaissance, but their study was constrained by the *Scala Naturae* approach

TABLE 1.7 Data from Aristotle recoded as presence (1) and absence (0)

Man_Man	110001111100
Live_bearing_tetrapods_Cat	101000111100
Cetaceans_Dolphin	100000111100
Birds_Bee_eater	110000111100
Egg_laying_tetrapods_Chameleon	101000110110
Snakes_Water_snake	100000110110
Egg_laying_fishes_Sea_bass	100000110110
Placental_selachians_Shark	100000110111
Crustaceans_Shrimp	000010110110
Cephalopods_Squid	000010110110
Hard_shelled_animals_Cockle	000000110011
Larva_bearing_insects_Ant	000100110011
Spontaneous_generating_Sponge	000000110011
Plants_Fig	000000010011
Minerals_Iron	000000000011

Note: We will return to this example in Chapter 5.

begun by Aristotle and limited by the thinking of the time. With the invention of the microscope, a whole new world of structure concerning invertebrates was established. In the mid-1600s Jan Swammerdam observed and reported on minute structures of insects and other invertebrates and a whole new way of looking at these organisms was established. Just prior to Carolus Linnaeus' (Carl von Linne's) work in the mid-1700s, Rene-Antoine Ferchault de Reamur published his six-volume tome entitled "Memoirs Serving as a History of Insects" (*Mémoires pour servir à l'histoire des insects*). Aristotle's approach wasn't significantly modified until the work of Linneaus. Linnaeus developed the binomial classification system that is still useful today. Wonderful treatments of invertebrates followed Linnaeus' groundbreaking work. Charles Darwin even got in on the act with his taxonomic work on barnacles in the later 1800s. His work was quite sophisticated, and he re-emphasized the utility of diagnostic characters as articulated by Linnaeus for barnacle taxonomy. Other systematists of the time advanced the taxonomic work in invertebrates and as the student will see in many of the chapters in this text to come, the tradition of taxonomy and systematics is quite deep for almost all of the groups addressed here.

MODERN INVERTEBRATE TAXONOMY

One of the major developments in evolutionary biology coming from the capacity to analyze the genes and genomes of organisms concerns taxonomy. Taxonomy has traditionally been accomplished with morphological techniques using a diagnostic concept (**Sidebar 1.1**). Since DNA sequences can now be generated for taxa at any phylogenetic level, these genetic data have also been impacting how species are described and named.

Taxonomists do descriptive work that defines species boundaries. There are many ways that taxonomists can do their work, but they are all bound by the nomenclatural rules of their groups and by conventions of their various organismal areas. The taxonomic rule set for invertebrates is directed by the International Commission on Zoological Nomenclature (ICZN; https://www.iczn.org/). The ICZN itself describes its work in the following quote:

> The ICZN acts as adviser and arbiter for the zoological community by generating and disseminating information on the correct use of the scientific names of animals. The ICZN is responsible for producing the International Code of Zoological Nomenclature – a set of rules for the naming of animals and the resolution of nomenclatural problems.

In general, taxonomists look for diagnostic differences between the taxa they study, and diagnostic differences are usually the currency that a taxonomist uses in his descriptions to justify a new name for a new species. Taxonomic work is not just "stamp collecting". Rather it has a high intellectual content and is heavily slanted toward hypothetical deductivism (Sidebar 1.1). Biologists who study speciation as well as taxonomists who study the same phenomenon: species delimitation. Taxonomists use a very strict set of criteria to discover species boundaries. Their approach is a hard-wired rule-based diagnostic concept. Biologists who study speciation use appropriate species concepts to come to the most precise explanation for the processes involved in speciation. Taxonomists write species descriptions that are laden with the diagnostics for the species they describe. One way to organize taxonomic descriptions is in identification keys. Biologists who study species boundaries are more interested in the process of speciation, but many of these boundaries are also used in more process-based ways of delimiting species for taxonomy.

Species delimitation attempts to understand the speciation process from all kinds of perspectives. Species studies can be carried out in a wide variety of ways, including molecular and morphological approaches. Biologists interested in speciation and species boundaries employ many and varied concepts of what

Sidebar 1.1. The structure of taxonomy

A hypothesis of species existence is posited by a taxonomist on the basis of some a priori notion of what a species is. Usually this hypothesis is based on some geographical notion of species separation. Most species have separated on the basis of allopatric isolation and hence geography is important in how species hypotheses are developed. Doyen and Slobodchikoff authored a seminal paper on genetic (allozyme) that has relevance to evolutionary, speciation and taxonomic studies. In this paper, hypotheses of species existence were made on the basis of geographical evidence. Such hypotheses were tested by use of information about reproduction, ecology, physiology and genetics. If any of the information from these disparate sources provided diagnostics to corroborate the geographic hypothesis, then this was considered to constitute a species.

The process of determining a new species can be viewed as a circle with several contact points on the circle. To enter the circle means that a hypothesis of species existence has been offered up (again, such a hypothesis is usually based on a geographic perspective). The exit points on the border of the circle correspond to testing the hypothesis of species existence with different kinds of data, for example morphology, physiology, ecology or genetics (Figure 1.1). If any one of these sources of data corroborates the existence of a species by being diagnostic, then the circle is broken and a new species can be delimited or described by use of the diagnostic. There are several scenarios that this system allows, the most important of which is the potential to reject a geographic (or other) hypothesis.

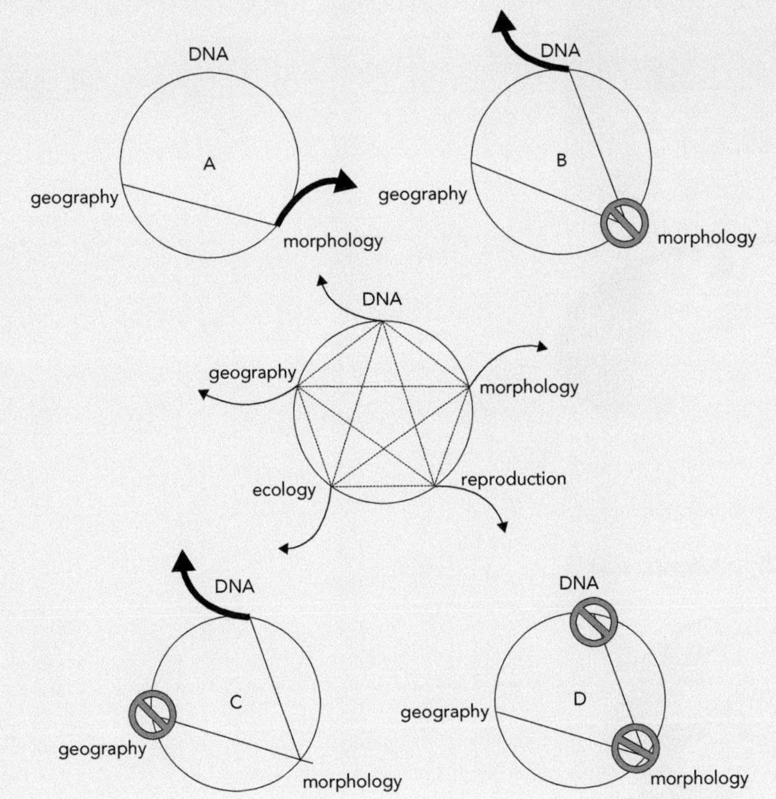

Figure 1.1 Modern taxonomic circles. General structure of modern taxonomic data (middle) and four different scenarios showing various ways to break out of the taxonomic circle, or not (bottom right). Scenario A (top right) represents the classical morphological approach to taxonomy, where a hypothesis of species existence is made and tested by anatomy (see arrow). The morphological data cannot reject the hypothesis, and so the analysis "breaks out" of the circle and species exist. Scenario B represents a situation where morphological data allow the rejection of the hypothesis of species existence, based on geography and the inability of DNA data to reject the hypothesis. Hence the "breakout" occurs when DNA sequence data are added to the analysis. This scenario is most commonly associated with cryptic species. Scenario C represents a situation where there is no geographical hypothesis. Instead the hypothesis is based on morphology and tested with DNA sequences (see arrow), which cannot reject the hypothesis of species existence and thus allows a "breakout." This scenario would represent a case of sympatric species. Scenario D shows that when all methods of testing result in rejection of the null hypothesis, there is no "breakout" and hence a single species exists. Circles are based on the work of Doyen and Slobodchikoff (1974). (Adapted from R. DeSalle, M. G. Egan, M. Siddall, *Philos. Trans. R. Soc. B* 360: 1905–1916, 2005.)

constitutes a species. The most common definition of species for a biologist is called the biological species concept, as formulated by Ernst Mayr in the 1940s: "A biological species is a group of individuals that can *breed* together (*panmixia*). However, they cannot breed with other groups. In other words, the group is reproductively isolated from other groups." There are dozens of species concepts that have been developed over the past half-century. The biological species concept since it is based on reproductive isolation has a sound objective basis to it. The problem is that it is not entirely operational. Other species concepts can be objective too, but the real problem is operationality of these concepts. How they are implemented in practice is the real problem in using these other concepts.

Sidebar 1.2.

Species description of *Zygothrica desallei* by O'Grady et al. (2002). Figures not included. Figure 1.1, is a posteroventral view of tibia and tarsi of *Z. desallei* (110X). Figure 1.2 is a wing of *Z. desallei* (25X). Figure 1.3 is a map of Ecuador showing the type locality of *Z. desallei.*

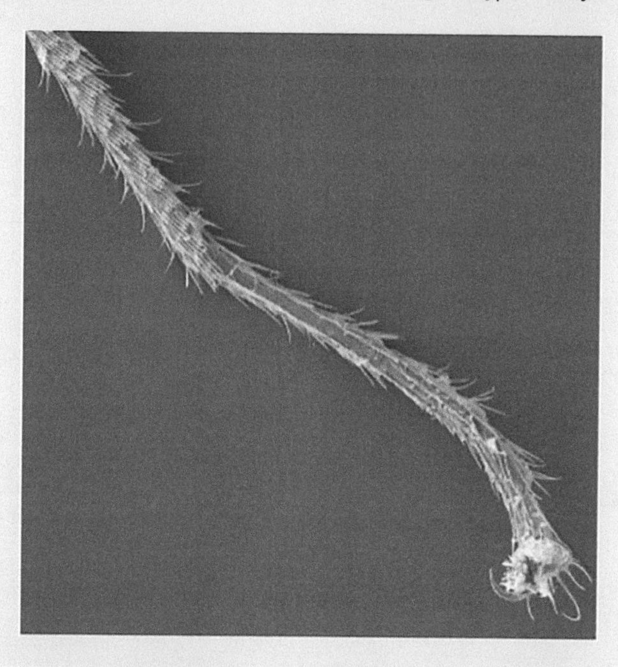

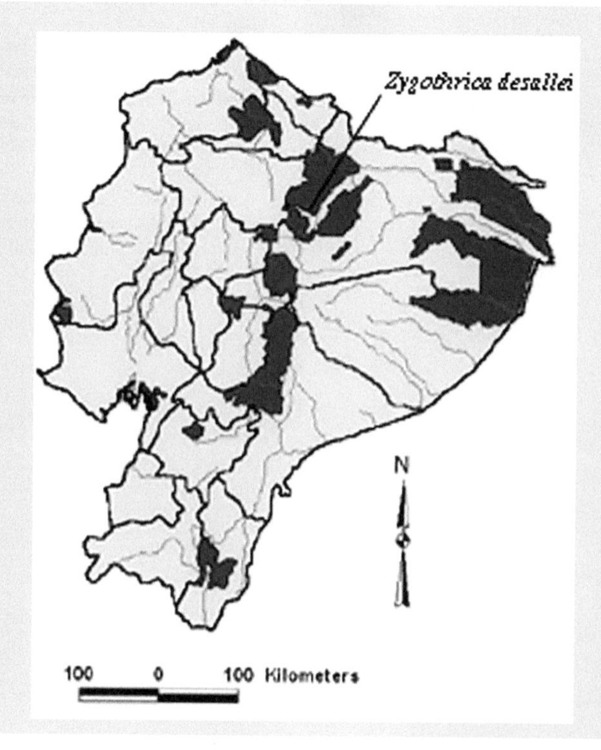

HOW TO WRITE AN INVERTEBRATE SPECIES DESCRIPTION

As we mentioned above, all invertebrate species descriptions are under the aegis of the ICZN. Species descriptions have very well-defined requirements (Winston, 1999). Once the new species has been delimited by the researcher, it is critical that it be described in a publication. The new species is not valid until the publication is accepted by a scientific journal. At the risk of pretentiousness, we discuss the outline of a simple species-description paper by O'Grady et al. (2002) that is a description of a member of the family Drosophilidae.

This standard insect description starts out with a very brief abstract that includes the new species name and short statements about the geographical location of the species and the diagnostics for the species. A short introduction to the description follows, which discusses the genus that the new species belongs to. This introduction should include any interesting or new aspects of the new species with regard to the genus it is placed into and its anatomy. This introductory section can also include a summary of the current taxonomy of the genus because taxonomic names can oftentimes be redundant or confusing. Hence a detailed exploration of the synonymy of the organisms involved is discussed at the outset of the description. The typical zoological description might also include a Materials and Methods section where the author describes the methodology used to accomplish the study. This is followed by the formal species description, which usually includes a header which is the formal new name followed by the name of the taxonomist who described the species.

In the O'Grady et al. description the authors describe the insect specimen's head, thorax, abdomen and wings in detail. They also describe the genitalia, which are the anatomical parts of the insect body that are most informative for diagnosis of the new species. Measurements of the specimen are then given, followed by details on the type material of the new species. The type specimen is the reference specimen that all future anatomical research will be referred to. The type specimen does not need to be collected by the authors of the description, and if it was not, then the original collector and collection need to be cited here. The location (i.e. which museum the specimen is archived in) and all archival numbers of the specimens used in the description are also given. This last task ensures that other researchers can access the specimens used to make the inferences in the description.

The distribution and ecology of the species follows in a short paragraph. Most modern descriptions use geospatial data to describe distribution and collection sites. The penultimate section of the paper describes the diagnostics of the new species. In the case of the O'Grady et al. (2002) description, wing morphology is the most conspicuous and diagnostic character for the species. The discussion of diagnosis is often followed by a fluid description of the new species which is a vivid, oftentimes colorful essay on the organism. This is followed by a discussion of how the organism is related to other close species including different aspects of its ecology. This discussion is also the author's opportunity to discuss overall evolutionary implications of the work reported in the paper. The last section of the paper discusses the etymology of the name of the new species; it is an explanation of why the authors chose the name they did. Any pertinent references are also included at the end of the paper.

The species description now follows.

Zygothrica desallei: A New Species of Drosophilidae (Diptera) from Ecuador

ABSTRACT: An unusual new species of *Zygothrica*, *Z. desallei*, is described from Ecuador. This species possesses three supernumerary crossveins extending from vein R2 ± 3 to the costa. Such a phenotype, while observed in other drosophilid genera such as *Jeannelopsis* and *Scaptomyza* (Tantalia), has not been previously observed in *Zygothrica*.

KEY WORDS: Zygothrica, Ecuador, systematics

Zygothrica Is predominantly a Neotropical genus, although some African and Indo-Pacific forms have been discovered recently (Grimaldi 1990). Although the first species in *Zygothrica* was described in 1830 by Wiedmann (1830), this name was not used as a generic designation until Loew (1873). Since then, many species have been added to this group. Wheeler (1981) listed 65 valid species names. Most recently, Grimaldi (1987, 1990) has added 62 new taxa and suggests that there may be as many as 60 more species remaining to be described in this genus. Here, we describe a single new species, *Z. desallei*, from the Ecuadorian Amazon.

This species is particularly interesting because it possesses between two and three extra crossveins extending from R2 ± 3 to the costa. This condition, while not unknown in the Drosophilidae, is only found in a few species and has never been reported from the genus *Zygothirca*.

Zygothrica desallei O'Grady, sp. nov.

Head. (from pinned material) Male. Flattened and elongate when seen in profile, longer than high; slightly hypercephalic, moderately wider second antennal segments tannish brown on dorsum, darker brown on venter. Third antennal segment long, 3 longer than second, brown in color. Arista with six dorsal and two ventral branches, not including the terminal fork. Frons tannish brown than thorax. Eyes bare, with few interfacetal setulae, with dark green tinges in pinned material. First and along 1/3 of anterior margin. Orbital plate tannish brown in color. Ocellar triangle, vertex and upper 2/3 of frons black, subshining. Ocelli pale yellowish white. Ocellar triangle extends anterior to 1/2 between proclinate and anterior reclinate orbitals. Orbital setae inserted equidistant from one another; proclinate approximately equal in length to posterior reclinate; anterior reclinate 2/3 length of posterior reclinate. Carina prominent; off-white in color. Gena wide, 1/6 width of eye at widest point; dark brown in color. Palps tannish brown; with single medioventrally directed apical seta and –4 strong setae on ventral surface. Clypeus and remainder of mouthparts dark brown to black.

Thorax. Shining dark brown to black on dorsum, pale off-white on venter. Acrostichal setulae in 6 irregular rows. Anterior katepisternal seta thin, 1/3 length of posterior. Humeral callus dark brown to black, with two strong subequal setae. Acrostichal setulae in 8 rows. Scutellum shining black; posterior scutellar setae cruciate Pleurae and halteres pale, yellow white. First pair of legs pale yellow brown; with short, curvate, indistinct cilia on tibia and tarsus (Figure 1.1'). Second and third set of legs darker yellow brown.

Abdomen. Shining dark brown to black on dorsal surface, off-white on venter.

Wings. Apex of wing pigmented on apical 1/5, extending just past intersection of costa and M1 (Figure 1.2). Two to three pigmented, supernumerary crossveins extending from costa to R2–3. Each specimen collected is asymmetrical with respect to this character, with two extra crossveins on one wing and three on the other. Crossvein dm-cu pigmented, sometimes with a short spur which entends into cell R4–5 (Figure 1.2). Crossvein r-m pigmented, and distinct from pigmentation covering basal_1/4 of wing, almost to level of r-m (Figure 1.2).

Measurements. A number of measurements traditionally used in Drosophilidae taxonomy have been made (after Grimaldi 1987). Definitions are as follows: costal index (CI) _ length of costa from subcostal break to R2–3/length of costa from R2-3 to R4-5, fourth vein index (4V) length of M1 from crossvein dm-cu to apex/length of M1 from crossvein r-m to crossvein dm-cu, 5_ index _ length of CuA1 from crossvein dm-cu to apex/length of crossvein dm-cu, 4C index _ length of costa from R2–3 to R4-5/length ofM1from crossvein r-m to crossvein dm-cu, and M index _ length of CuA1 from crossvein dm-cu to apex/length of M1 from crossvein r-m to crossvein dm-cu. N 3 males. TL (thorax length) 1.6 mm (1.5–1.8); WL (wing length)_3.3 mm (3.3–3.4); TL/WL _ 0.5; HW (head width) _ 1.3 mm (1.2–1.4); HW/TL _ 0.8; CI _ 3.3 (2.8–3.8); 4V _ 1.4 (1.0–1.6); 5X _ 0.6 (0.5–0.7); 4C _ 0.8 (0.7–0.8); M _ 0.3 (0.2–0.3).

Type Material. ECUADOR, HOLOTYPE, Estación Biológica Jatun Sacha (01_ 33_ S, 77_ 33_ W), 5 August 1997, O'Grady&Vela. TL_1.5mm; WL_3.3 mm; TL/WL 0.5; HW_1.2mm; HW/TL_0.8; CI_3.8; 4V 1.6; 5X _ 0.7; 4C _ 0.7; M _ 0.3. Two paratypes, both males, are also designated. All material is in the collection of the Museum of the Pontificia Universidad Católica del Ecuador (PUCE).

Distribution and Ecology. This species is known only from Ecuador (Figure 1.3), where it has been collected on fungus. This species has yet to be reared from any substrate.

Diagnosis. *Zygothirca desallei* is easily differentiated by the presence of between two and three supernumerary cross veins extending from R2–3 to the costa (Figure 1.2).

Etymology. This species is named in honor of Rob DeSalle, whose many systematic publications have helped determine the phylogenetic placement of the genus *Zygothrica* relative to the remainder of the family Drosophilidae.

ANATOMICAL TECHNIQUES FOR EXAMINING INVERTEBRATES

Understanding how organisms are related to each other would not be possible without some guiding principles. One of these guiding principles is homology, a term used to indicate that a feature (genetic or morphological) derives from the same evolutionary pathway (e.g. a gene or a forelimb) in the organisms being compared. Homology then becomes a term of particular evolutionary importance and can be confirmed via hypothesis testing. Analogy or cases where traits of different (parallel) evolutionary histories converge on one another may also be examined by hypothesis testing. Such traits are interesting in that they show how multiple organisms can arrive at the same evolutionary solution. However, because analogous characters are evolutionarily convergent, they are not useful in phylogenetics.

A number of authors have discussed homology with respect to phylogenetic analysis. De Pinna (1991) points out that assessing homology is a two-step process. The first step is basically the process of "getting in the ballpark" by assessing similarity of an organism's characteristics. If attributes appear similar enough by some criterion, then primary homology has been established. Primary homology is nothing more than a hypothesis that can be tested with further phylogenetic analysis. If the hypothesized attributes are shared and derived upon phylogenetic analysis, then they are said to be secondarily homologous. If they do not pass the second test, then the attributes are analogies. Homology is an absolute concept. Walter Fitch, a famous phylogeneticist, once said: "Homology is like pregnancy. Someone is either pregnant or not. A person cannot be 70% pregnant." In this context, two attributes can be homologous, but they cannot be 70% homologous.

As we have pointed out, most of invertebrate relationships prior to 1990 were based on morphological traits. Studying morphology is not easy. It is often time consuming and it takes a great deal of expertise to be useful and productive. It takes a keen eye and patience to scan the anatomy of an organism all the time keeping in mind characters that are relevant across wide ranges of taxa. The concept of homology is therefore always on the morphologist's mind. If one is interested in a particular group of invertebrates, then a great deal of expertise and familiarity with the anatomy of the members of that group is important. This is where not only a good handle on anatomy is important, but a working knowledge of the literature on the group is also required.

Some invertebrates are large enough that visualizing characters can be done with the naked eye. However, the grand majority of invertebrates are small, and microscopy becomes the major way that such organisms are examined. More recently invertebrate systematics and taxonomy have used X-ray, light microscopy and MRI (Qian et al., 2019; Jahn et al., 2018; Ziegler et al., 2011; Cameron and Whitfield, 2017; Wipfler, et al., 2016), electron microscopy (Scanning EM and Transmission EM), Computed Tomography (CT) scanning (Faulwetter et al., 2013; Du Plessis et al., 2017; Smith et al., 2016; Marcondes

Machado et al., 2019; Nguyen et al., 2017A; Chaplin et al., 2019; Krieger and Spitzner, 2020; Landschoff et al., 2018; Nguyen et al., 2017B; Keklikoglou, et al., 2019; Gusmão et al., 2018) and confocal microscopy (Corgosinho et al., 2018; Myles et al., 2019; Reier et al., 2019; Leon and Weirauch, 2016; Galli et al., 2006; Michels, 2007; Böhm et al., 2011; Culverhouse et al., 2006; Gonzalez et al., 2013; Lee, et al., 2009; Boistel et al., 2011; Wipfler et al., 2016; Klaus et al., 2003) all of which provide novel insights into the anatomical attributes of invertebrates. We dedicate an entire chapter (Chapter 4) to the collection of molecular data for invertebrate systematics and taxonomy.

DNA BARCODING, DNA TAXONOMY AND INTEGRATIVE TAXONOMY

The past decade has seen the development of many approaches to delimit species boundaries in a taxonomic context. This section examines some of the developments in this area. Since 2003, Paul Hebert at the University of Guelph has championed an approach called DNA barcoding that is intended to create a DNA sequence-based organismal identification system (Hebert et al., 2003). A single gene has been proposed as a marker for identifying organisms using their DNA. Cytochrome oxidase I (***COI*, *cox 1***), a mitochondrial gene, has been the target of the DNA barcoding field. This gene is conserved through all animals including invertebrates and is variable enough to discriminate between species, even very closely related ones. The DNA barcoding effort has grown into an international consortium called the Consortium for the Barcode of Life (CBoL; http://www.barcodeoflife.org/). It has a centralized website and exists to create DNA barcodes for all species of plants, animals and fungi. The BoLD database (http://www.boldsystems.org/) has also been developed for the storage and archiving of all DNA barcode data.

Several new ideas have arisen in the past two decades in the area of taxonomy. One of them is called DNA taxonomy which wants to use DNA sequences alone to do taxonomy. DNA taxonomists argue that because the process of naming species is so slow via classical morphological approaches, novel ways to name species are needed. For instance, for nematodes, only a handful of the million or so nematode species on the planet have been named even after a century of work on the taxonomy of nematodes. By scanning DNA sequences, DNA taxonomy would set up recognizable boundaries in the analysis of genetic boundaries as a means to delimit species. DNA Taxonomy suggests that the hypothesis testing of classical taxonomy needs to be replaced by an algorithm based on molecular yardsticks. In addition, the most expedient method for use of DNA sequences for DNA taxonomy is to use genetic distances and establish a specific distance where species may be delimited. DNA taxonomy is something very different from classical taxonomy, and it is also very different from speciation or species delimitation studies. Integrative taxonomy is the process of naming species with a broad range of tools, including DNA sequences and morphology. Integrative taxonomists use anatomical, behavioral, ecological, and molecular information to do taxonomy (Sidebar 1.1). Any character that is inherited becomes available to the integrative taxonomist as long as it can diagnose a new taxonomic entity, and hence integrative taxonomy usually employs the same species concepts or criteria that classical taxonomists use. While integrative taxonomy could use DNA sequences, an integrative taxonomist more than likely uses DNA sequences to assist in setting up hypotheses of species existence, which are then tested by use of some other information.

As pointed out above, COI data in the DNA barcoding system are analyzed with distance methods, usually neighbor joining (NJ; Chapter 4), to construct a tree which is visually inspected for clustering of individuals in the analysis. Inferences about species existence are made on the basis of the topology of the

NJ-generated tree; for instance, sometimes a distance of 2% between clusters of organisms is used as a general marker of species existence. The 2% cutoff established inductively by examining a large number of organismal groups for divergence between species and within species is not a universal cutoff and using it as such has many caveats.

Other methods may be used for analyzing DNA barcode raw data. There are two major ways one can analyze DNA barcoding data – character-based or distance-based. Within these two general ways of analyzing data for species delimitation, one can also either generate a tree or use some non-tree analysis. Classical taxonomists use a non-tree, character-based approach in their use of anatomy and other kinds of characters for species descriptions. As mentioned above, DNA barcoding takes the distance-based tree-building approach of constructing a NJ tree. Some DNA barcoding studies have used multivariate statistical analysis (a non-tree distance-based approach), and some systematists will prefer the character-based tree-building approaches that use ML or MP. Other species-delimitation approaches use Bayesian inference to delimit species boundaries (Chapters 2 and 3).

Davis and Nixon (1992) point out that there are two caveats that need to be discussed with respect to any species delimitation approach, whether it be distance or character or tree based. The first concerns underdiagnosis, a situation when the marker employed to test hypotheses evolves too slowly. If we used the slowly evolving 18S ribosomal RNA to test species status between populations of humans and chimpanzees, we would discover no differences in sequence between chimpanzees and humans for this marker. We would be led to sink the two populations together and to deny the existence of the two species. This would be a very silly and terribly erroneous conclusion. We would need only to use one of the many anatomical characters that are fixed and different between humans and chimpanzees to retest the hypothesis. The second phenomenon is called overdiagnosis, which occurs when too few individuals in a population or too few populations are included in the test of the hypothesis.

One of the really interesting aspects of using DNA barcoding approaches or DNA-based species-delimitation approaches is that they allow for very specific marking or flagging of entities for further tests of species status (Goldstein and DeSalle, 2011). For example, a set of animal specimens from a known species can be sequenced for COI or any other genetic marker used for species delimitation. The sequences are used in a phylogenetic analysis with other specimens and the accepted taxonomy of the group is overlaid on the phylogenetic result. If there is a good correspondence of the tree topology with the accepted taxonomy, then the DNA analysis can be used to establish identifiers for the species in the analysis. On the other hand, if a previously described single species appears not to be a simple unit under the species-delimitation approach used, then any anomalous entities can be "flagged" for future taxonomic work. What this means is that the specimen or specimens that are flagged do not have a formal name, but rather become specimens for taxonomists to examine and to describe.

CONCLUSION

We have discussed in detail in this chapter how invertebrate zoology and systematics has developed. We look in more detail in Chapter 2 at how invertebrate zoologists examine the organisms they are interested in with the organizing principles of systematics we have introduced here. In Chapter 3 we describe the new DNA sequence technology that has developed over the last decade to allow for broader phylogenomic analysis of genomes of invertebrates. Finally, in Chapter 4 we use several examples from the literature to describe how invertebrate systematics uses molecular and morphological characters for construction of phylogenies and how fluid the field really is.

REFERENCES

Amemiya, Chris T., Tsutomu Miyake and Jonathan P. Rast. "Echinoderms." Current Biology 15, no. 23 (2005): R944–R946.

Blaxter, Mark, and Paul Sunnucks. "Velvet worms." Current Biology 21, no. 7 (2011): R238.

Böhm, Alexander, Daniela Bartel, Nikolaus Urban Szucsich and Günther Pass. "Confocal imaging of the exo- and endoskeleton of Protura after non-destructive DNA extraction." Soil Organisms 83, no. 3 (2011): 335–345.

Boistel, Renaud, Jim Swoger, Uroš Kržič, Vincent Fernandez, Brigitte Gillet and Emmanuel G. Reynaud. "The future of three-dimensional microscopic imaging in marine biology." Marine Ecology 32, no. 4 (2011): 438–452.

Cameron, Sydney A. and James B. Whitfield. "Insect Systematics as a Central Discipline of Entomology." Insect Systematics and Diversity 1, no. 1 (2017): 1–2.

Chaplin, Kirilee, Joanna Sumner, Christy A. Hipsley and Jane Melville. "An Integrative Approach Using Phylogenomics and High-Resolution X-Ray Computed Tomography for Species Delimitation in Cryptic Taxa." Systematic Biology (2019).

Collins III, James J. "Platyhelminthes." Current Biology 27, no. 7 (2017): R252–R256.

Corgosinho, Paulo H. C., Terue C. Kihara, Nikolaos V. Schizas, Alexandra Ostmann, Pedro Martínez Arbizu and Viatcheslav N. Ivanenko. "Traditional and confocal descriptions of a new genus and two new species of deep water Cerviniinae Sars, 1903 from the Southern Atlantic and the Norwegian Sea: with a discussion on the use of digital media in taxonomy (Copepoda, Harpacticoida, Aegisthidae)." ZooKeys 766 (2018): 1.

Culverhouse, Phil F., Robert Williams, Mark Benfield, Per R. Flood, Anne F. Sell, Maria Grazia Mazzocchi, Isabella Buttino and Mike Sieracki. "Automatic image analysis of plankton: future perspectives." Marine Ecology Progress Series 312 (2006): 297–309.

Davis, Jerrold I. and Kevin C. Nixon. "Populations, genetic variation, and the delimitation of phylogenetic species." Systematic Biology 41, no. 4 (1992): 421–435.

De Pinna, Mario C. C. "Concepts and tests of homology in the cladistic paradigm." Cladistics 7, no. 4 (1991): 367–394.

DeSalle, Rob, Mary G. Egan and Mark Siddall. "The unholy trinity: taxonomy, species delimitation and DNA barcoding." Philosophical Transactions of the Royal Society B: Biological sciences 360, no. 1462 (2005): 1905–1916.

Doyen, John T., and C. N. Slobodchikoff. "An operational approach to species classification." Systematic Biology 23, no. 2 (1974): 239–247.

Du Plessis, Anton, Chris Broeckhoven, Anina Guelpa and Stephan Gerhard Le Roux. "Laboratory x-ray micro-computed tomography: a user guideline for biological samples." GigaScience 6, no. 6 (2017): gix027.

Engel, Michael S. "Insect evolution." Current Biology 25, no. 19 (2015): R868–R872.

Faulwetter, Sarah, Aikaterini Vasileiadou, Michail Kouratoras, Thanos Dailianis and Christos Arvanitidis. "Micro-computed tomography: Introducing new dimensions to taxonomy." ZooKeys 263 (2013): 1.

Galli, Paolo, Giovanni Strona, Anna Maria Villa, Francesca Benzoni, Stefani Fabrizio, Silvia Maria Doglia, and Delane C. Kritsky. "Three-dimensional imaging of monogenoidean sclerites by laser scanning confocal fluorescence microscopy." Journal of Parasitology 92, no. 2 (2006): 395–399.

Giribet, Gonzalo. "Zoology: At last an exit for Ctenophores." Current Biology 26, no. 20 (2016): R918–R920.

Giribet, Gonzalo and Gregory D. Edgecombe. "The phylogeny and evolutionary history of arthropods." Current Biology 29, no. 12 (2019): R592–R602.

Glauber, Kristine M., Catherine E. Dana and Robert E. Steele. "Hydra." Current Biology 20, no. 22 (2010): R964–R965.

Goldstein, Paul Z. and Rob DeSalle. "Integrating DNA barcode data and taxonomic practice: determination, discovery, and description." Bioessays 33, no. 2 (2011): 135–147.

Gonzalez, Victor H., Terry Griswold and Michael S. Engel. "Obtaining a better taxonomic understanding of native bees: where do we start?." Systematic Entomology 38, no. 4 (2013): 645–653.

Grimaldi, D. A. "Phylogenetics and taxonomy of Zygothrica (Diptera: Drosophilidae)." Bulletin of the American Museum of Natural History 186 (1987): 103–268.

Grimaldi, D. A. "Revision of Zygothrica (Diptera: Drosophilidae), Part II. The first African species, two new Indo-Pacific groups, and the *bilineata* and *samoaensis* species groups." American Museum of Natural History Novitates 2964 (1990): 1–31.

Gusmão, Luciana C., Alejandro Grajales and Estefania Rodríguez. "Sea anemones through X-rays: visualization of two species of Diadumene (Cnidaria, Actiniaria) using micro-CT." American Museum Novitates 2018, no. 3907 (2018): 1–47.

Hart, A. G., Bowtell R. W. and Ratnieks F. L. W. "Magnetic resonance imaging in entomology: a critical review." Journal of Insect Science 3, no. 1 (2003): 1–9.

Haszprunar, Gerhard, and Andreas Wanninger. "Molluscs." Current Biology 22, no. 13 (2012): R510–R514.

Hebert, Paul DN, Sujeevan Ratnasingham and Jeremy R. De Waard. "Barcoding animal life: cytochrome c oxidase subunit 1 divergences among closely related species." Proceedings of the Royal Society of London. Series B: Biological Sciences 270, no. suppl_1 (2003): S96–S99.

Holland, Linda Z. "Tunicates." Current Biology 26, no. 4 (2016): R146–R152.

Holland, Nicholas D. "Chordates." Current Biology 15, no. 22 (2005): R911–R914.

Jahn, Henry, Ivo De Sena Oliveira, Vladimir Gross, Christine Martin, Alexander Hipp, Georg Mayer and Joerg U. Hammel. "Evaluation of contrasting techniques for X-ray imaging of velvet worms (Onychophora)." Journal of Microscopy 270, no. 3 (2018): 343–358.

Katsuki, Takeo and Ralph J. Greenspan. "Jellyfish nervous systems." Current Biology 23, no. 14 (2013): R592–R594.

Kaur, Sagan Deep and Lakhvir Singh. "Indian Arthropods in Early Sanskrit Literature: A Taxonomical Analysis." Indian Journal of History of Science 53 (2018): 59–64.

Keklikoglou, Kleoniki, Sarah Faulwetter, Eva Chatzinikolaou, Patricia Wils, Jonathan Brecko, Jiří Kvaček, Brian Metscher and Christos Arvanitidis. "Micro-computed tomography for natural history specimens: a handbook of best practice protocols." European Journal of Taxonomy 522 (2019).

King, Nicole. "Choanoflagellates." Current Biology 15, no. 4 (2005): R113–R114.

Kiontke, Karin, and David H. A. Fitch. "Nematodes." Current Biology 23, no. 19 (2013): R862–R864.

Klaus, A. V., V. L. Kulasekera and V. Schawaroch. "Three-dimensional visualization of insect morphology using confocal laser scanning microscopy." Journal of Microscopy 212, no. 2 (2003): 107–121.

Krieger, Jakob and Franziska Spitzner. "X-Ray Microscopy of the Larval Crustacean Brain." In Brain Development, pp. 253–270. New York, NY: Humana, 2020.

Landschoff, Jannes, Anton Du Plessis, and Charles L. Griffiths. "A micro X-ray computed tomography dataset of South African hermit crabs (Crustacea: Decapoda: Anomura: Paguroidea) containing scans of two rare specimens and three recently described species." Gigascience 7, no. 4 (2018): giy022.

Lee, Sangmi, Richard L. Brown and William Monroe. "Use of confocal laser scanning microscopy in systematics of insects with a comparison of fluorescence from different stains." Systematic Entomology 34, no. 1 (2009): 10–14.

Leon, S. and C. Weirauch. "Small bugs, big changes: taxonomic revision of Orthorhagus McAtee & Malloch." Neotropical Entomology 45, no. 5 (2016): 559–572.

Leung, Tommy L. F. "Evolution: how a barnacle came to parasitise a shark." Current Biology 24, no. 12 (2014): R564–R566.

Leys, Sally P., Daniel S. Rohksar and Bernard M. Degnan. "Sponges." Current Biology 15, no. 4 (2005): R114–R115.

Loew, H. "Monographs of the Diptera of North America. Part III." Smithsonian Miscellaneous Collections 11 (1873): 1–351.

Lovejoy, Arthur O. The Great Chain of Being: A Study of the History of an Idea. William James Lectures, 1933. Cambridge, MA: Harvard University Press, 1936.

Maderspacher, Florian. "Zoology: the walking heads." Current Biology 26, no. 5 (2016): R194–R197.

Marcondes Machado, Fabrizio, Flávio Dias Passos and Gonzalo Giribet. "The use of micro-computed tomography as a minimally invasive tool for anatomical study of bivalves (Mollusca: Bivalvia)." Zoological Journal of the Linnean Society 186, no. 1 (2019): 46–75.

Margulis, Lynn. "Five-kingdom classification and the origin and evolution of cells." In Evolutionary Biology, pp. 45–78. Boston, MA: Springer (1974).

Marlétaz, Ferdinand, Elise Martin, Yvan Perez, Daniel Papillon, Xavier Caubit, Christopher J. Lowe, Bob Freeman et al. "Chaetognath phylogenomics: A protostome with deuterostome-like development." Current Biology 16, no. 15 (2006): R577–R578.

Marlétaz, Ferdinand, Katja TCA Peijnenburg, Taichiro Goto, Noriyuki Satoh, and Daniel S. Rokhsar. "A new spiralian phylogeny places the enigmatic arrow worms among gnathiferans." Current Biology 29, no. 2 (2019): 312–318.

Mayr, Ernst. "How biology differs from the physical sciences." Evolution at a Crossroads: The New Biology and the New Philosophy of Science (1985): 43–63.

Mayr, Ernst. "The biological species concept." In Species Concepts and Phylogenetic Theory: a Debate, pp. 17–29. New York, NY: Columbia University Press (2000).

Michels, Jan. "Confocal laser scanning microscopy: using cuticular autofluorescence for high resolution morphological imaging in small crustaceans." Journal of Microscopy 227, no. 1 (2007): 1–7.

Myles, Thomas A., Sabrina D. Eder, Matthew G. Barr, Adam Fahy, Joel Martens and Paul C. Dastoor. "Taxonomy through the lens of neutral helium microscopy." Scientific Reports 9, no. 1 (2019): 1–10.

Nguyen, Chuong V., David R. Lovell, Matt Adcock and John La Salle. "Capturing natural-colour 3D models of insects for species discovery." arXiv preprint arXiv:1709.02039 (2017).

Nguyen, Chuong, Matt Adcock, Stuart Anderson, David Lovell, Nicole Fisher and John La Salle. "Towards high-throughput 3D insect capture for species discovery and diagnostics." In 2017 IEEE 13th International Conference on e-Science (e-Science), 2017: 559–560. IEEE,

Nübler-Jung, Katharina and Detlev Arendt. "Enteropneusts and chordate evolution." Current Biology 6, no. 4 (1996): 352–353.

O'Grady, Patrick M., Doris Vela and Violeta Rafael. "Zygothrica desallei: A new species of Drosophilidae (Diptera) from Ecuador." Annals of the Entomological Society of America 95, no. 3 (2002): 314–315.

Pang, Kevin and Mark Q. Martindale. "Ctenophores." Current Biology 18, no. 24 (2008): R1119–R1120.

Qian, Jia, Shipei Dang, Zhaojun Wang, Xing Zhou, Dan Dan, Baoli Yao, Yijie Tong et al. "Large-scale 3D imaging of insects with natural color." Optics Express 27, no. 4 (2019): 4845–4857.

Reier, Susanne, Helmut Sattmann, Thomas Schwaha, Josef Harl, Robert Konecny and Elisabeth Haring. "An integrative taxonomic approach to reveal the status of the genus Pomphorhynchus Monticelli, 1905 (Acanthocephala: Pomphorhynchidae) in Austria." International Journal for Parasitology: Parasites and Wildlife 8 (2019): 145–155.

Schierwater, Bernd and Rob DeSalle. "Placozoa." Current Biology 28, no. 3 (2018): R97–R98.

Sharma, Prashant P. "Chelicerates." Current Biology 28, no. 14 (2018): R774–R778.

Smith, Dylan B., Galina Bernhardt, Nigel E. Raine, Richard L. Abel, Dan Sykes, Farah Ahmed, Inti Pedroso and Richard J. Gill. "Exploring miniature insect brains using micro-CT scanning techniques." Scientific Reports 6 (2016): 21768.

Sterckx, Roel. "Animal classification in ancient China." East Asian Science, Technology, and Medicine 23 (2005): 26–53.

Studtmann, Paul. "The foundations of Aristotle's categorial scheme." (2008).

Telford, Maximilian J. and Richard R. Copley. "Zoology: war of the worms." Current Biology 26, no. 8 (2016): R335–R337.

VanHook, Annalisa M. and Nipam H. Patel. "Crustaceans." Current Biology 18, no. 13 (2008): R547–R550.

Wheeler, M. R. "The Drosophilidae: a taxonomic overview." In M. Ashburner, H. L. Carson, and J. N. Thompson, Jr. (eds.), The Genetics and Biology of Drosophila, vol. 3a, pp. 1–97. Academic, New York, 1981.

Wiedemann, C. R. W. Achias Diptorum Genus. Kiliae Holsatorum, 1830.

Winston, Judith E. Describing Species: Practical Taxonomic Procedure for Biologists. Columbia University Press, 1999.

Wipfler, Benjamin, Hans Pohl, Margarita I. Yavorskaya and Rolf G. Beutel. "A review of methods for analysing insect structures—the role of morphology in the age of phylogenomics." Current Opinion in Insect Science 18 (2016): 60–68.

Ziegler, Alexander, Martin Kunth, Susanne Mueller, Christian Bock, Rolf Pohmann, Leif Schröder, Cornelius Faber and Gonzalo Giribet. "Application of magnetic resonance imaging in zoology." Zoomorphology 130, no. 4 (2011): 227–254.

THE BASICS OF PHYLOGENETICS

CHAPTER 2

Rob DeSalle, Heike Hadrys, and Bernd Schierwater

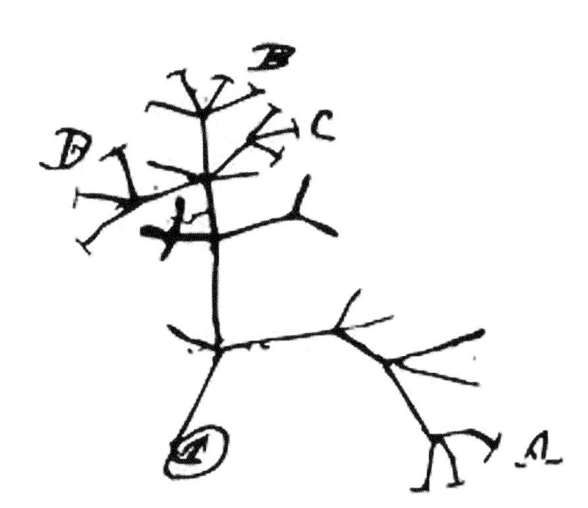

CONTENTS

Phylogenetics

Phylogenetic trees have been a part of evolutionary study since Lamarck produced the first tree of relatedness for vertebrates in 1809. Other famous trees are Darwin's "I think" phylogenetic tree which appeared in his notebooks in the 1830s and the only figure in his *On the Origin of Species* which was a phylogenetic tree. The first phylogenetic trees established several principles inherent in systematics (**Sidebar 2.1**). Phylogenetic trees are branching diagrams that represent the relationships of organisms. They are typically drawn with bifurcating branches from a common ancestor, but as Darwin's "I think" tree demonstrates, they don't necessarily have to be bifurcating to be informative. The history of tree drawing and generation is fascinating, but beyond the scope of this chapter. Students interested in exploring this topic should

Sidebar 2.1. Trees

In this sidebar we use Darwin's famous "I think" diagram to establish some of the basic vocabulary and principles of phylogenetic trees (Figure 2.1). While most modern phylogenetic trees are drawn as bifurcating some of the nodes in Darwin's Origin tree are trifurcating. Oddly enough there is only one bifurcating branch (marked with solid arrow) in this famous tree; the rest are trifurcating or greater. This does not prevent us from using the diagram to establish definitions and principles, though.

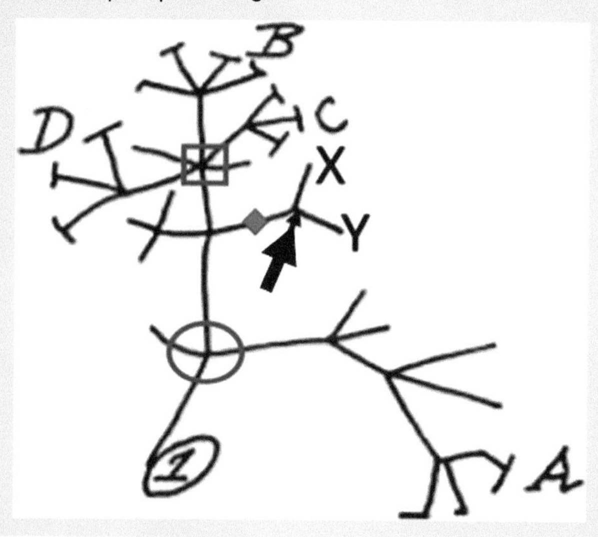

Figure 2.1 Darwin's "I think" tree with labels for text.

refer to David Morrison's website called the Genealogical World of Phylogenetic Trees (http://phylonetworks.blogspot.com/).

COMMON VOCABULARY TERMS

Bifurcation: The arrow points to the only bifurcation Darwin drew for this famous tree. The arrow indicates where two lineages come from an ancestor.

Node: The arrow also points to what is called a node. In bifurcating trees, the node has one branch leading into it and two emanating from it.

Branch: There are many branches in this tree. Darwin only labeled five of the terminals (see below) and none of the branches. The diamond labels the branch leading to the bifurcation. We have labeled two terminals as X and Y.

Common ancestor: The arrow points to the common ancestor of "X" and "Y".

Taxon/Terminal: A taxon (pl. taxa) is an observed or hypothetical entity, placed at the end of a branch. A terminal is the end of a branch where no more branching occurs. For instance, "A" is a terminal at the bottom of the tree.

Lineage: A lineage refers to a branch in the tree. The open box indicates a node where "B", "C" and "D" are members of the lineage leading upward.

Monophyletic: This term refers to a group of organisms with a single common ancestor to the exclusion of all others. It is a relative term so that when something is referred to as monophyletic it needs to be in reference to something else. "B", "C" and "D" are members of a monophyletic group relative to "A" because they have a single common ancestor (the node in the open box) to the exclusion of "A".

Polyphyletic: A set of taxa that do not have a single common ancestor to the exclusion of others by definition is polyphyletic. "A" and "B" are polyphyletic in reference to "C" and "D" because "A" and "B" do not share a single common ancestor to the exclusion of "C" and "D".

Paraphyletic: A group of organisms or taxa that are descended from a common ancestor, but that does not include all taxa in the descendent group. In the "I think" tree, the group descending from the node in the large circle would be paraphyletic if taxon "X" is not included in it when discussing it.

Sister group: A pair of taxa that share a common ancestor to the exclusion of all other taxa. "X" and "Y" are sister groups to each other because they share a common ancestor (indicated by the arrow). They are also said to be "sister to each other".

Newick topology: A concise way to express trees without drawing them. Parentheses are used to circumscribe taxa that are connected by a branch. If taxon A and taxon B are most closely related to each other then they are placed inside parentheses and separated by a comma [(A,B)]. If C is the sister taxon to A and B then C is connected to the expression by a comma and another set of parentheses [((A,B),C)]. In a Newick expression, the number of rightward parentheses [)] should equal the number of leftward parentheses [(].

Modern systematics began with two decades of intense discussion (1950 to 1970), expansion and exploration. Prior to the methodologies of modern systematics – the business of detailing the relatedness of organisms on the planet – researchers constructed phylogenetic trees, based on their collective knowledge about a group. This method was based on expert knowledge of the group being studied and trees were drawn with data in mind but without the data formally analyzed or optimized by the expert. This approach was called **evolutionary systematics** which in essence was tree drawing by ocular inference. Whatever the expert saw as important was incorporated into the process of drawing the tree.

In the 1950s, Robert Sokal and Peter Sneath introduced a novel way of thinking about tree building that they called **numerical taxonomy**. They were concerned about trees constructed by the subjective method of expert opinion, and proposed that the aggregate of data collected for specimens should be analyzed with an algorithm to remove the subjectivity. In doing so they used a methodology called **phenetics** that incorporates overall similarity as a yardstick to measure relatedness. The flip side of similarity is distance and so this approach is oftentimes also called **distance analysis** (**Sidebar 2.2**). Since the initial introduction of Sneath and Sokal's text, *Numerical Taxonomy*, several algorithms and

Sidebar 2.2. Distance analysis

The first step in any distance analysis is to obtain a matrix of distances or similarity for each pairwise comparison of the taxa in the study. These can be raw counts, or natural distances like immunological distances. For DNA sequence data a model of how the sequences evolve can be incorporated into the distance measure. For instance, the most commonly distance transformation model is the Kimura 2 parameter model which corrects for the unequal occurrence of kinds of changes via the mutation process. In DNA sequences there are four bases G, A, T and C. Due to the dynamics of the DNA double helix a G is always opposite a C and an A is always opposite a T in the DNA strands that make up the double helix. Because G and A (called purines) are physically larger than C and T (called pyrimidines), it is easier to substitute say a C for a T or an A for a G in the mutation process. Hence G can replace an A and C can replace a T (events called transitions) with higher probability than a G can replace a T, than a G can replace a C, than an A can replace a T and than an A can replace a C (all called transversions; see Figure 2.2). An equation can be developed to accommodate the higher frequency of transitions than transversions and this is then used to transform pairwise sequence comparisons to distances. For amino acids researchers can use the genetic code to estimate the frequency of changes between codons. For instance, it would be easier to change from CCC (proline) to CCA (also proline) than it would be to AAA (lysine) because the former requires only one base pair change and the latter requires three. In addition, researchers have used amino acid sequences from databases to calculate the actual transition frequencies of each pairwise amino acid change.

Once the distance matrix is generated the construction of a dendrogram (a diagram summarizing the distances) is constructed from the matrix. The technique of constructing this dendrogram uses a simple recursive process to compute the tree. The two taxa in the matrix with the smallest distance are linked together and the matrix is recalculated to accommodate that the two taxa are linked. The next-smallest distance in the matrix is targeted and the two taxa with this distance are linked together and so on, until all taxa have been attached to the tree. There are many algorithms for generating dendrograms, among them Unweighted Pair Grouping with arithMetic Averaging (UPGMA) and Neighbor Joining, the latter of which is the most popular method and used in almost all computer packages for evolution and systematics.

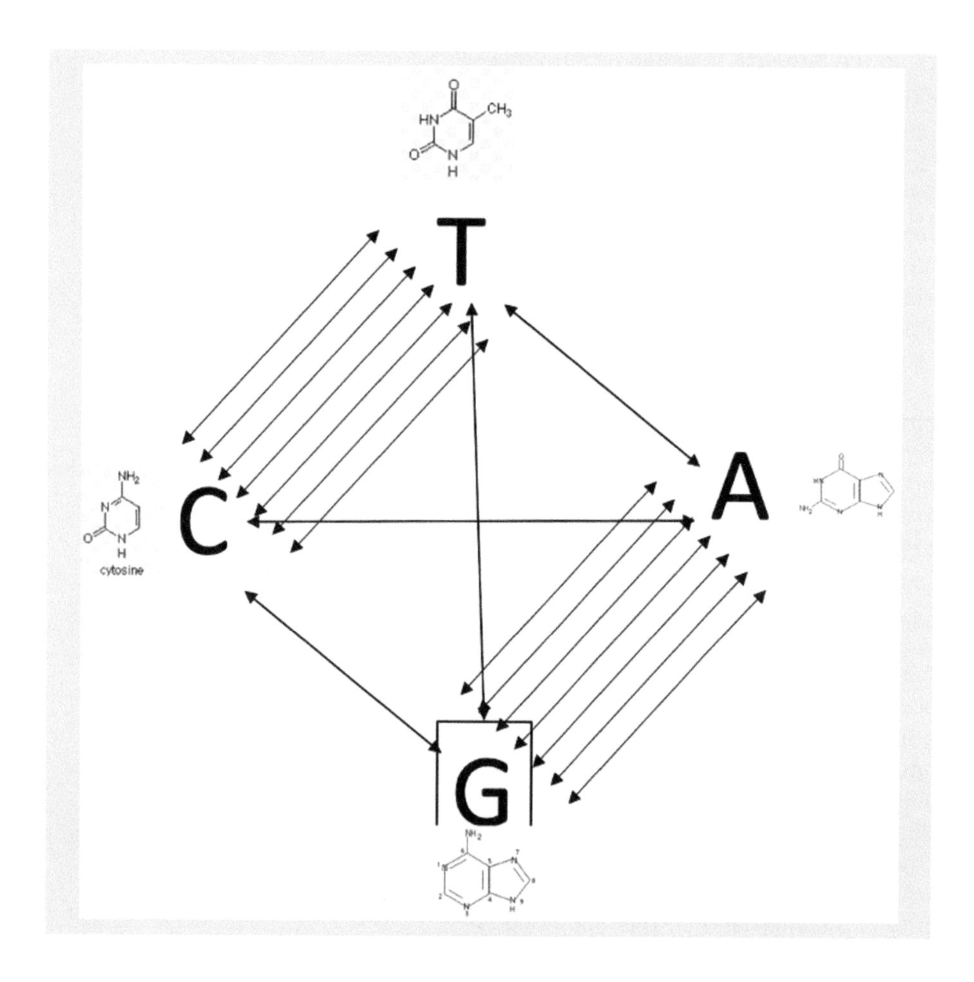

Figure 2.2 The four nucleotides showing the theoretical frequency (arrows) of change between them.

computer programs have been developed that are in use today to implement the phenetic method.

In 1956, the German entomologist Willi Hennig (1999) came up with a completely different way of thinking about phylogenetic classification that he called **phylogenetic systematics**. His approach was developed irrespective and in spite of overall similarity. Hennig realized that similarity might not indicate the complexity of evolution and that two taxa with the highest similarity might not be each other's closest relatives. He based his reasoning on the idea that branching occurred as the result of change in discrete characters, and that by analyzing the complexity of character state change using the principle of parsimony an accurate representation of the branching patterns of organisms could be reached. This approach is called **cladistics** and has grown into the modern criterion in tree building called **maximum parsimony** (**Sidebar 2.3**).

Up until the late 1960s, researchers had available to them mostly morphological data for systematic analysis. Sokal and Sneath outlined detailed ways how the morphological data could be compressed into pairwise distances and therefore used to generate a tree using dendrography approaches for summarizing distances. Hennig reasoned that since character states change over evolution and are the essence of the branching process, these character state changes should be used to generate a tree based on the **character state** information.

The characters that morphologists generated to use the maximum parsimony approach can be placed in several different kinds of categories. Morphological characters can be divided into two major kinds. The first is whether the character is discrete or continuous. Can the character states be coded as discrete morphological states or are they measured with quantitative meaning? A morphological example of a discrete character is whether the organism has wings or not. Typically, having wings would be coded as 1 and not having wings would be

Sidebar 2.3. Maximum parsimony

Maximum Parsimony (MP) analysis with morphological characters starts by scoring each taxon in the analysis for character states. There are several ways that morphological (and behavioral or other non-molecular traits) characters can be scored. The most common is presence (1)/ absence (0) or binary discrete characters. Such binary characters can be unordered meaning that transformation from 1 → 0 is as likely as from 0 → 1, or ordered where one of the directions of character change is allowed over the other direction (Dollo parsimony is a case of ordered characters where 0 → 1 is disallowed). Traits can also be scored as more than just two discrete states, in which case they can be either unordered or ordered. Finally, traits can be continuous, meaning that they span a range of values (like meristic characters).

Molecular maximum parsimony analysis starts with aligned DNA or amino acid sequences (see Chapter 3 for description of alignment procedures) and is performed by examining each position in the sequence or in the data matrix. The positions are examined for the number of steps they require to be placed on a specific tree topology. The number of steps each character takes is determined for non-additive characters by a process called the Fitch algorithm. Since many morphological characters and molecular characters are non-additive, this optimization approach has become the preferred one in MP. Ideally, every possible tree topology for the taxa in an analysis is examined. For instance, for three taxa (A, B and C) there are three ways to arrange these taxa: A with B, B with C and A with C. Figure 2.3

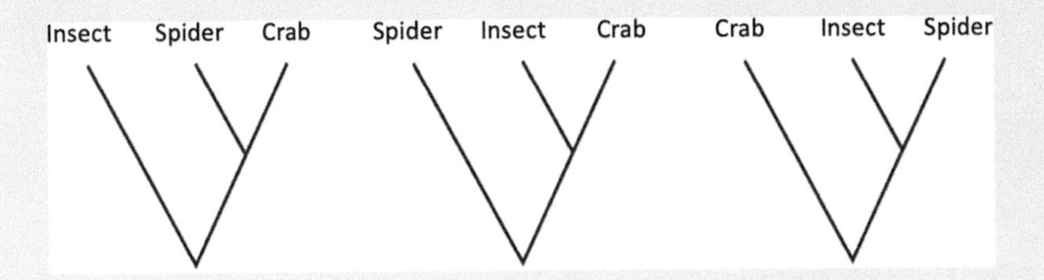

Figure 2.3 The three bifurcating topologies for Crab, Insect and Spider. In Newick notation these trees would be from left to right (insect, (spider, crab)); (spider, (crab, insect)); (crab, (spider, insect)). Theoretically only one of these trees can explain the divergence of these taxa. The tree is rooted with an outgroup (see text for definition and utility of outgroups) that is not shown.

shows the three arrangements for spider, insect and crab). As the number of taxa increases the number of possible bifurcating trees increases approximately logarithmically. By the time there are 20 taxa in an analysis there are more than 10^{28} possible bifurcating trees.

Once a tree is found to be most parsimonious the characters in the starting matrix can be classified into four categories (Figure 2.4):

Autapomorphies: Characters that change only in a single lineage on the tree.

Synapomorphies: Characters that are shared by two or more taxa to the exclusion of all others.

Symplesiomorphies: Characters shared by two or more taxa – but also with other taxa linked earlier in the clade. Such characters imply that the taxa have an earlier last common ancestor, with them, than theirs.

Homoplasies (Convergences): Characters that change in multiple lineages and are not synapomorphies or symplesiomorphies.

The large number of trees needed to evaluate even for 20 taxa is a problem in computational science called NP completeness, which means that the problem has a solution, but computing it is incredibly difficult with current algorithms. Hence, computer

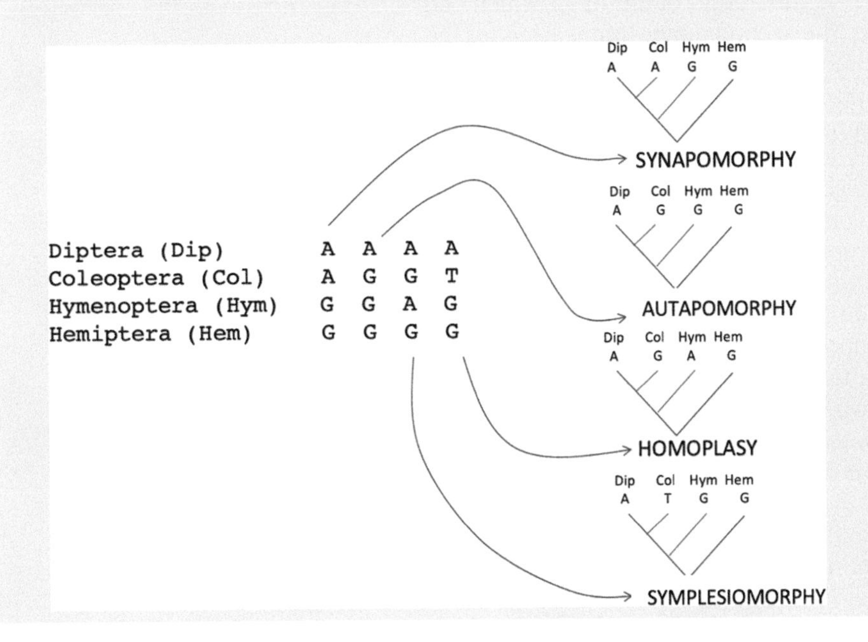

Figure 2.4 Examples of Synapomorphy, Autapomorphy, Homoplasy and Symplesiomorphy. For this example we focus on the relationships of three ingroup orders of insects – Dip = Diptera, Col = Coleoptera, Hym = Hymenoptera – and one outgroup order – Hem = Hemiptera.

programs that compute phylogenetic trees using MP (and maximum likelihood, ML) use some algorithmic tricks to increase the probability that an adequate sample of trees have been evaluated. The collection of trees that exist for a particular set of taxa is called the **tree space** for those taxa.

Tree searching algorithms begin with a starting tree which is evaluated and then a second tree is generated by swapping branches in the starting tree which is then evaluated. If the second tree is more parsimonious than the first, then it is saved and swapped upon. The process goes on until more parsimonious trees are no longer found. There are several ways to swap branches in these kinds of searches (which again are also done in ML analysis). Ensuring that the tree space is adequately examined is a difficult task and several approaches have been developed to streamline searches. These include doing multiple searches from randomly generated starting tree and a procedure called the "ratchet". The latter is the most-used approach and it simply allows for very efficient starting trees to be generated for evaluation.

The approach also requires that the tree be rooted so as to polarize the character state information and determine what is ancestral and what is derived. There are several ways to root a phylogenetic tree that include midpoint rooting, paralog rooting and **outgroup** rooting. The last of these (outgroup) is the preferred method in modern systematics of organisms. An outgroup is simply a taxon (or taxa) that is (are) not members of the ingroup. The ingroup includes all of the taxa a researcher is interested in analyzing.

A controversial aspect of parsimony analysis concerns how much weight characters should have in the analysis. It is reasonable to think that some characters are more reliable than others and so the strategy is to give those characters stronger weight in the analysis. Parsimony analysis can also include character transformation information as we saw in computing distances. Weights can be given to specific kinds of character transformations and applied in parsimony searches. Below we show a transformation matrix for DNA sequences implying that C to T, T to C, A to G and G to A transitions are less reliable than all of the transversions by a factor of 10. This is based on the dynamics of sequence change that we discussed in Sidebar 2.3 where transitions occur at a higher rate than transversions. Because transitions have very frequent occurrence, they are more prone to converging for a character. Such homoplasy complicates parsimony analysis. In essence, any transversion is weighted as 10 times more reliable as a transition.

	G	A	T	C
G	–	10	1	1
A	10	–	1	1
T	1	1	–	10
C	1	1	10	–

coded as 0. An example of a molecular character is DNA which has four discrete character states G, A, T and C (some systematists include the character state "gap" which results when the sequences are aligned. An example of a continuous trait might be the number of bristles on the notum of an insect which can vary across say 10, 11, 12, 13, 14, 15, etc. The second dimension is whether the character is binary or multistate. We already have an example of a binary discrete morphological character given above [wings (1) versus no wings (0)]. But we can code this kind of character even further by noting that we could have multiple character states for wings – no wings (0), 2 wings (1) and 4 wings (2). DNA and amino acid sequences are good examples of multistate characters with DNA having four states and amino acids having 20. By far the most common kinds of characters that are used in morphological phylogenetic analysis are called binary discrete characters. And as we point out molecular sequence characters are almost always multistate. With multistate characters, whether they are ordered or unordered becomes an important issue to address. For instance, if we have strong biological information that no wings (0) leads to four wings (2) and then to two wings (1) then we would want to order the character as $0 \rightarrow 2 \rightarrow 1$. Such an ordered character would behave very differently than if we allowed the character states to change randomly (or as unordered characters states). We will discuss the ramifications of the different kinds of character states in morphological analysis throughout this text, and the ability to make inferences about how molecular sequences change might depart from unordered is at the heart of modeling sequence change.

In the 1960s molecular information began to be used to build phylogenetic trees. Two methods that were developed then were preadapted, so to speak, to phenetic analysis. The method of measuring **immunological distances** was developed as a way to get an idea of the relatedness of members of a group of organisms. The method involves challenging the serum of one organism with the antibodies of another. The ensuing reaction could be assayed for the intensity of the interaction of the serum with the antibodies; the more intense the reaction, the closer two species were inferred to be. This method was employed to generate each pairwise immunological reaction of taxa in an analysis and in so doing generate a matrix of distances that could generate a **dendrogram**. The technique of **DNA-DNA hybridization** was also developed around this time and it was used to estimate the overall DNA sequence distance for generating dendrograms. The technique worked by mixing together the DNA from two taxa creating hybrid DNA molecules and then measuring the temperature at which the hybrid DNA strands dissociated. The better the match of DNA in a hybrid, the higher the dissociation or melting temperature. The melting temperature then serves as a proxy for genetic distances which are used to generate a dendrogram. A third method – protein electrophoresis – was also developed in the late 1960s. This method allowed researchers to characterize the charge characteristics of proteins and used electrophoretic gels to assay differences in charge. The idea with this approach was that if there are differences in protein charge, then this must lie in changes of the amino acids in the protein which are ultimately linked to changes in the gene sequences that coded for the protein. This approach could yield character state data, but by far the most common manipulation of the data was to compress the character states into an overall measure of genetic similarity or distance. In addition, during this period researchers were able to obtain sequence data from proteins through the laborious procedure of protein sequencing that gave amino acid sequences of isolated proteins. Again, the amino acid sequences could be used as characters in phylogenetic analysis but were more commonly compressed into overall measures of sequence similarity or distance and used to construct dendrograms. By the mid-1970s, several molecular methods had been utilized in phylogenetics. The field quietly awaited the development of DNA sequencing methods, which would occur in the next decade.

Edwards and Cavalli-Sforza (1967) used the idea that each residue in a DNA sequence or a protein sequence evolved independently (also an assumption of maximum parsimony) and suggested that likelihood estimation could be used to analyze character data in molecular sequence analysis. They developed the **maximum likelihood** (**Sidebar 2.4**) approach that will be discussed in more detail below. By 1968, then, the three major methods of phylogenetic inference had been developed. A great amount of debate erupted over which of these three methods was more appropriate for generating phylogenies, that is again beyond the scope of this chapter. Whatever way the debate is presented, the matter of the fact is that all three methods exist today in the systematist's toolbox. Despite their existence as tools though, the researcher is advised to understand the differences between the methods and to be prepared to defend the choice of analytical method or methods.

In addition, since a measure of likelihood can be obtained for the overall tree and for branch lengths as well as other parameters, several interesting statistical approaches can be implemented in the likelihood framework. As **Figure 2.5** shows, many of the models that are used in likelihood analysis are "embedded" within each other. In other words, K2P is a restricted version of K3ST, which in turn is a restricted version of GTR. This unique characteristic of sequence modeling allows for statistical tests of the adequacy of one model for another for a given dataset. Approaches to assess the "best" model for a particular dataset exist such as ModelTest (for DNA sequences) and ProTest (for protein sequences). In addition to this unique aspect of likelihood models,

Sidebar 2.4. Maximum likelihood

Maximum Likelihood (ML) approaches in systematics compute a probability of a dataset given a model and a tree topology. The tree with the maximum likelihood is then chosen as the best solution for the dataset. ML approaches are most commonly used in molecular systematic analyses where researchers claim to be able to model the substitution process. Likelihood approaches for morphological characters have also been developed using simple models of morphological character change.

As with MP analysis, likelihood estimates probabilities from position to position in aligned sequences. A probability for each position is computed for a given tree topology and model. Once the probability for each position is computed for a particular topology, an overall probability is computed from the individual position probabilities. The process is repeated for all topologies using topologies that are generated from algorithms that explore the tree space adequately (for a discussion see Sidebar 2.3). The tree with the greatest overall likelihood is chosen as the best tree for that dataset. The tree space problem also exists for ML as it does for MP so tree searching is accomplished with similar or identical tree search methods as with MP.

There are a large number of models that can be used in ML analysis so researchers have developed algorithms that can assist in choosing the best model for a given dataset. Models can include transformation probabilities of the residues in each alignment position (DNA sequence transition probabilities or amino acid transition probabilities), branch lengths and site-specific rate differences. Transformation probability matrices are an important aspect of modeling and several possible ways of doing this exist. The simplest is the Jukes Cantor model (JC) that assumes equal transition probabilities for all positions in DNA sequences. Slightly more complex models exist that take into consideration transition/transversion imbalance (Kimura 2 parameter model or K2P) and a model that takes into consideration base frequencies (Felsenstein 81 model or F81; Felsenstein, 1981). The most comprehensive model is called the Generalizable Time Reversal matrix or GTR, where different probabilities are allowed for each of the 12 possible transformations (or six if reversals are considered equally probable) for DNA sequences. The important thing about these models is that they are nested within each other, meaning that they are simply more complex due to the addition of parameters as you go from a simpler model (JC) to more complex (K2P) to most complex (GTR), and this allows interesting statistical tests to be performed with ease. Figure 2.5 shows several of these models and their nesting properties.

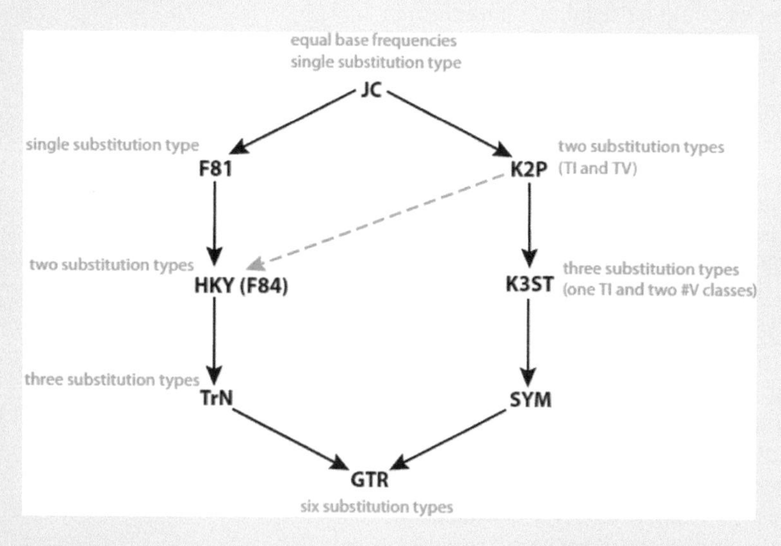

Figure 2.5 Diagram showing the relationships of general time reversible (GTR) family of substitution models.

Modeling site-specific rate differences allows for different rates of evolution in the sites of DNA or proteins and is represented by a gamma distribution (Figure 2.6). This distribution is quantified by the shape parameter of the distributions called alpha which is used in estimating the probability of base transformation in the likelihood estimation procedure. Each of the three curves below has different shapes and hence different alpha parameters that imply different dynamics of sequence change. DNA and protein sequences evolve more like the middle curve (many sites with little change and few sites with a lot of change) and so an alpha parameter that characterizes this kind of curve is most appropriate in a likelihood analysis. Most often though, likelihood analyses compute the alpha parameter from the sequences used in the analysis.

Figure 2.6 Three gamma distributions for site-specific differences. The top gamma distribution shows a system where the sites are distributed normally with respect to the amount of change they have incurred. The middle (which approximates real sequence change) shows a system where there are many sites with little change and few sites with a lot of change. The bottom gamma distribution shows a system where the number of sites is the same for any amount of change.

the overall likelihood of trees with the same dataset but different topologies can be computed. These likelihood values can then be used in statistical tests of the better fit of the model and different trees to the data. Approaches such as the SH (**Shimodaira-Hasegawa**) and KH (**Kishino-Hasegawa**) tests are therefore used to explore the statistical significance of different topologies of trees.

The development of modern systematics was also accompanied by the development of techniques to assess the reliability or robustness of inferences made using the various optimality criteria. Decay indices (DI) or Bremer support (BS) measures were developed for MP analyses to indicate the number of extra steps that an analysis would need to lose a particular branch. It is a branch-by-branch measure of the strength of phylogenetic signal found in a dataset. Decay indices can range from 0 (the node is not supported by the dataset) to 1 (the addition of a one character that disagrees with the node will collapse it) to larger numbers that indicate the number of characters needed to be added to collapse the node. Larger DIs are obviously preferred over smaller ones.

Felsenstein (1985) utilized a resampling technique called bootstrapping that gives bootstrap proportion for each node in the tree. Farris et al. (1996) utilized another resampling method called jackknifing to also give a proportional measure for each node of the tree. Bootstrap methods resample (i.e., choose certain sites to throw out) with replacement. Jackknife approaches resample without replacement. Bootstrap approaches can be applied to likelihood, distance and parsimony analyses to render some sense of robustness to the likelihood analysis. Bootstrap and Jackknife proportions range from 0.0 (no support) to 1.0 (highest support) and are used in most analyses of protein and DNA sequences in phylogenetics. Today researchers use all three of the major optimality criteria – distance, parsimony and likelihood – in phylogenetic research. One problem all of these methods have is how to asses optimality when there are such a large number of solutions (**Sidebar 2.5**).

Sidebar 2.5. Tree searches

Phylogenetic methods use what are called optimality criteria. For example, the criterion for maximum parsimony is to find the tree or trees with the fewest number of steps needed to explain the data on the tree topology. This criterion as well as the criteria for likelihood and distance analysis require that all possible trees be evaluated and the one with the fewest number of steps is reported as the maximum parsimony tree. For small numbers of taxa this is a simple task and it is computationally possible to evaluate all possible trees. For three ingroups and one outgroup in an analysis only three strictly bifurcating trees need to be evaluated. For four ingroups, 15 trees exist; for five ingroups, 105 trees exist; for six ingroups, 945 trees need to be evaluated and so on. Basically, with the addition of a taxon to analysis the number of trees increases by an order of magnitude. For instance, a matrix with 70 taxa will have well over 10^{65} trees to evaluate. The problem then becomes how to explore and evaluate this huge number of trees.

The number of trees is a difficult computational problem, but one that researchers have solved with clever tree-searching algorithms. One of the most important steps in tree searching is the starting tree. The initial tree can be a user-supplied one, or it can be generated by other rules. One rule that is default in many programs is called "as is", that is, in the order that the taxa are listed in the matrix. Other rules can be applied such as random addition of the taxa. One can perform a large number of different orders of addition of the taxa to make the initial tree. All of the trees that are found to fit the optimality criterion of the search from each replicate are then evaluated and the trees with the optimal statistics are chosen as optimal.

Once the starting tree is evaluated, the branches on the starting tree are moved around (swapped) and new evaluations are made. If the swapped trees are better, then they are kept and further swapping is accomplished. A specific number of swapping rounds (called replicates or reps) are set at the outset, and when the total number of reps are completed the search is stopped and the shortest tree during the process is reported as the maximum parsimony tree (or trees). There are several ways that the branches of a tree can be swapped and these are, in order of computational intensity: Nearest Neighbor Interchange (NNI – a process where branches very close to each other are swapped), Subtree Pruning and Regrafting (SPR – a process that expands exchanges branches on a preferred tree) and Tree Bisection and Reconnection (TBR – a process where the tree is split in half and the split part is reconnected in various places on the remaining half). The computation time between NNI, SPR and TBR can differ by orders of magnitude of computational time with large numbers of taxa. However, the most efficient searches are done with TBR and we recommend tree searches be accomplished with this branch-swapping method.

A clever tree search option called the "ratchet" is also available in many of the phylogenetic analysis programs (Nixon, 1999). In this approach, a starting tree is generated and a randomly chosen subset of characters is given additional weight. A new tree (or a few trees) is generated using the optimality criterion of the method. The weights of the characters are reset to be equal and a set of trees is obtained using the optimality criterion of the method. These trees are evaluated together and the shortest is retained. The process is repeated, in most cases 100 to 200 times (it can be more), and the pool of trees that fit the original criterion are evaluated to find the trees that fit the criterion best. This method more efficiently explores the wide range of starting trees that are possible and has been shown to be more efficient than randomly generating starting trees.

Bayesian Phylogenetic Inference has also been a popular approach employed by researchers. Since the 1990s, researchers have incorporated the principles of Bayesian reasoning into phylogenetic inference. The approach takes advantage of prior knowledge of parameters involved in phylogenetic analysis, models of evolution of sequences and Markov chain simulation to produce phylogenetic trees where each of the nodes in a tree could be assigned a posterior probability (**Sidebar 2.6**).

Sidebar 2.6. Bayesian Phylogenetic Inference

The advantage of Bayesian Phylogenetic Inference is its ability to incorporate prior probabilities and likelihood models with a unique method of exploring tree space to generate posterior probabilities for the existence of nodes in a phylogenetic tree. It is based on the famous Bayes equation, which we discuss below.

Bayes' equation:

$$P(H|D) = \left[P(D|H)P(H)\right]/P(D)$$

$P(H|D)$ is read as the *posterior probability* of the hypothesis given the data because it is computed after, or posterior to, knowing anything about the data. This posterior probability is in essence what we want to know about an experiment or about some scientific dataset. $P(H)$ is called the *prior probability* and it incorporates prior knowledge about the data (D). The quantity $P(D|H)$ is called the *likelihood operator or function* of D given H, and it can be approximated by using models. $P(D)$ is called the marginal probability of the data. This probability is typically computed from the law of total probability.

Note that once $P(H|D)$ is computed, there is an updated version of knowledge about the hypothesis and new probabilities can be estimated recursively. This procedure is very important for employing Bayesian methods in evolutionary biology with a basic similarity to the Markov chain model (MCM). It turns out that Markov chains have become an important part of Bayesian analysis.

Transposing this Bayesian framework onto phylogenetics is done in the following way. Let's say we have a DNA sequence dataset with 10 Cnidaria and 10 Bilateria as ingroups and 10 Porifera as an outgroup. In this example the mammals would be one clade and the reptiles would be another clade. The equation that would represent a Bayesian statement about the probability of a clade existing (that is, what is the probability that the group mammals exist?) would be written as

$$P(\text{clade}_{\text{Cnidaria}} \mid \text{data}) = \sum\nolimits_{(\text{any tree with Cnidaria})} \left[P(\text{data} \mid \text{clade}_{\text{Cnidaria}})P(\text{clade}_{\text{Cnidaria}})\right]/P(\text{data})$$

With this formulation of Bayes' theorem for phylogenetic analysis, there are several components that require further explanation: (1) how models are used and manipulated in Bayesian analysis, (2) how the computation of the posterior probability is accomplished and controlled to give results that are reliable, and (3) how prior probabilities impact the outcome of a Bayesian analysis.

The equation above means that the likelihood $P(\text{data}|\text{clade}_{\text{Cnidaria}})P(\text{clade}_{\text{Cnidaria}})$ needs to be estimated. But to do this, we also need to include other parameters like branch lengths, model of nucleotide (or amino acid) change, site-specific variation, and nucleotide frequency into the $P(\text{data}|\ \text{clade}_{\text{Cnidaria}})$ term; hence, a model of evolution needs to be applied. We also need to include the prior probability $P(\text{clade}_{\text{Cnidaria}})$. In general, realistic prior probabilities are hard to estimate due to a lack of information, so most analyses will substitute "safe" priors into the equations. These safe priors are usually what are called "flat" or "vague" priors. A flat prior for $P(\text{clade}_{\text{Cnidaria}})$ is simply where all trees have equal probability, and a vague prior for $P(\text{clade}_{\text{Cnidaria}})$ would be where the posterior probabilities being calculated have smaller variance than the prior probability of trees. A flat prior [when $P(\text{clade}_{\text{Cnidaria}})$ and $P(\text{data})$ are equal for all estimates in the sum above] means that the posterior probability distribution will then be proportional to the likelihood. So, in essence, when you do a Bayesian analysis in phylogenetics, it is approximately a lot like doing a likelihood analysis. In fact, Huelsenbeck has pointed out that "This [Bayesian inference] is roughly equivalent to performing a maximum likelihood analysis with bootstrap resampling, but much faster."

What do we need from a Bayesian phylogenetic analysis? The left side of the Bayes' rule equation above tips us off to this. We are after a distribution of trees in order to calculate the probability distribution of different topologies.

In essence, the MCMC (Markov chain Monte Carlo) simulation approach is used to generate trees to form a probability distribution by use of a Markov chain simulated under specific sets of rules. The MCMC is a simulation trick that is used to explore the distribution of data relevant to a specific question. The MCMC is implemented by using the Markov chain (MC) algorithm, which is a mathematical system that simulates change, as determined by specific rules set out at the beginning of the generation of the chain. Since the end product is a distribution, statistical methods can be applied to interpret the distribution. Generating a distribution of the trees, therefore, is the essence of the Bayesian approach as it is implemented in phylogenetic analysis programs. The distribution of trees can then be used to determine two properties. First, the single topology with the largest occurrence in the distribution can be found. Second, data can be represented by summarizing the frequency with which particular nodes appear in the posterior distribution of trees. For instance, with our example above, if the 10 Cnidaria are sister group to the group of 10 Bilateria in every tree in the posterior distribution, then the posterior probability of that node connecting Cnidaria with Bilateria would be 1.0. If the Cnidaria as a group is sister taxon to Bilateria in 95% of the trees in the posterior distribution of trees, then the posterior probability of that node would be 0.95. Thus, a Bayesian phylogenetic analysis can result in a tree, with branch lengths and posterior probabilities assigned to each node in the tree. Figure 2.7 shows the throughput of a typical Bayesian Phylogenetic Inference analysis (Huelsenbeck et al., 2001).

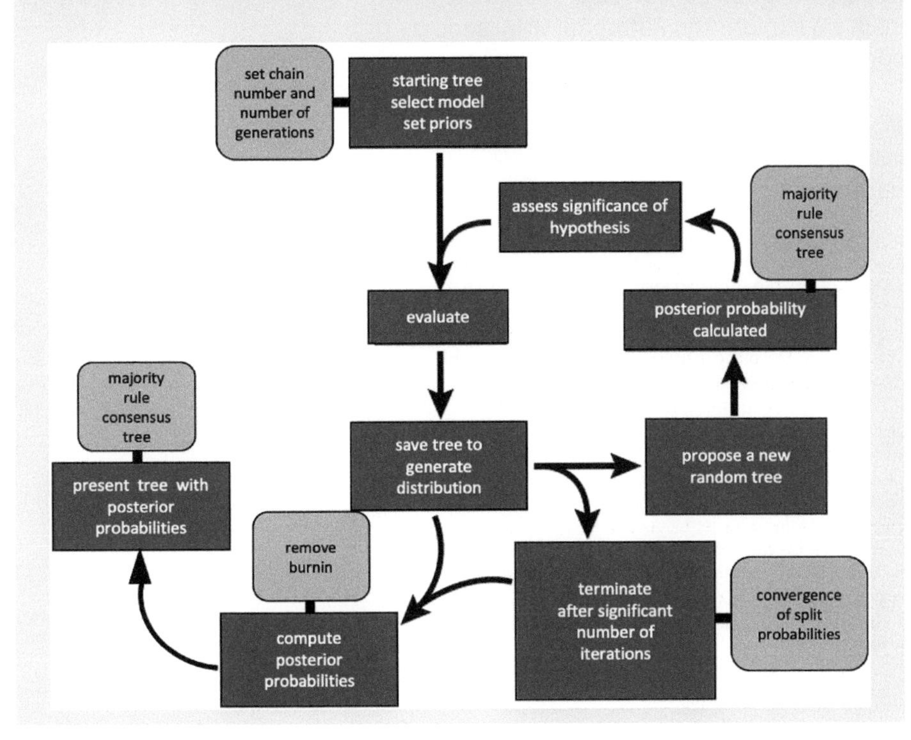

Figure 2.7 Flow diagram of the Bayesian phylogenetic approach.

PROGRAMS FOR PHYLOGENETIC ANALYSIS

Rather than list our favorites we refer the reader to the amazing compilation of programs for phylogenetic analysis compiled and maintained by Felsenstein (1993) and his colleagues (http://evolution.genetics.washington.edu/phylip/software.html#methods). This list includes general-purpose programs that apply multiple optimality criteria; programs that apply parsimony, likelihood and distance; and programs that use Bayesian approaches. Of these programs we will use TNT (Goloboff et al., 2008), RaxML (Stamatakis, 2014), MrBayes and PAUP in generating many of the examples in this text. The manuals for all of these computer programs should be consulted (see **Table 2.1** for website information for several phylogenetic packages) for brief descriptions of the optimality criteria

TABLE 2.1	Phylogenetic programs for modern systematic analysis	
PAUP*	http://paup.phylosolutions.com/get-paup/	MP, ML
TNT	http://www.lillo.org.ar/phylogeny/tnt	MP
IQTree	http://www.iqtree.org/	ML
RaxML	https://cme.h-its.org/exelixis/web/software/raxml/	ML
PHYLIP	http://evolution.genetics.washington.edu/phylip.html	MP, ML, NJ
MrBayes	http://nbisweden.github.io/MrBayes/download.html	BP
PhyloBayes	http://www.atgc-montpellier.fr/phylobayes/	BP
ASTRAL	https://github.com/smirarab/ASTRAL	BST
MEGA	https://www.megasoftware.net/	MP, NJ, ML, BP

MP = maximum parsimony; ML = maxiumum likelihood; NJ = neighbor joining; BP = Bayesian phylogenetics; BST = Bayesian species tree analysis. PAUP* (Swofford, 2001), TN T (Goloboff et al., 2008), IQTREE (Nguyen et al., 2015; Trifinopoulos et al., 2016), RaxML (Stamatakis, 2014), PHYLIP (Felsenstein, 1993), MrBayes (Ronquist et al., 2012), PhyloBayes (Lartillot et al., 2013), ASTRAL (Zhang et al., 2018), MEGA (Kumar et al., 1994).

used as well as for the operations of the programs. The programs all have different installation procedures and operate in different environments. As an example we show how to install and run a popular program called PAUP (Swofford, 2001) in **Sidebar 2.7**. In the next chapter we explore the nature of DNA and protein sequence data that have changed the face of the Invertebrate tree of life.

Sidebar 2.7. Installing phylogenetic programs on your computer

This chapter gives very basic attention to phylogenetic analysis using programs that are freely available to researchers. The program we will use here is PAUP, but there are other programs that are equally easy to use (see Table 2.1 for other programs). Many of the programs are used through a terminal which should be readily available on any laptop or desktop computer. We will start the sidebar by showing how to install a terminal-based program. We will use the Aristotle matrix we developed in Chapter 1. Next we will show how morphological data are analyzed using phylogenetic programs and then how molecular data are analyzed. We provide the student with over 80 phylogenetic matrices (both morphological and molecular) relevant to the taxa discussed in this text so that a detailed exploration of the relationships of these taxa is possible.

The different programs have specific installation instructions, so it is best to follow the instructions for the specific programs. Here we will show how PAUP is installed because we will use PAUP in this chapter. The PAUP version we will work with for this text is a command line version. The program can be downloaded from http://phylosolutions.com/paup-test/. Once downloaded (you should move the program called "**paup4a166_osx**" from your download folder to your desktop) it can be run in your terminal by resetting the access to the program in your terminal by typing "**chmod a + x paup4a166_osx**" into the command line and hitting return. The program icon on your desktop should turn into a terminal executable. You can simply double click on this terminal executable PAUP icon on your desktop to get it to open in your terminal (**Figure 2.8**). The program should now be accessible for analyzing phylogenetic matrices.

```
robdesalle — paup4a166_osx — paup4a166_osx — 80×24
Last login: Thu Jan 16 13:52:23 on ttys000
robdesalle@Robs-MacBook-Pro-2 ~ % /Users/robdesalle/Desktop/paup4a166_osx ; exit
;

P A U P *
Version 4.0a (build 166) for macOS (built on Sep  1 2019 at 23:02:50)
Thu Jan 16 14:01:46 2020

        -----------------------------NOTICE-----------------------------
          This is a test version that is still changing rapidly.
          Please report bugs to dave@phylosolutions.com
        ----------------------------------------------------------------

Running on Intel(R) 64 architecture
SSE vectorization enabled
SSSE3 instructions supported
Multithreading enabled for likelihood using Pthreads
Compiled using Intel compiler (icc) 11.1.0 (build 20091012)

paup>
```

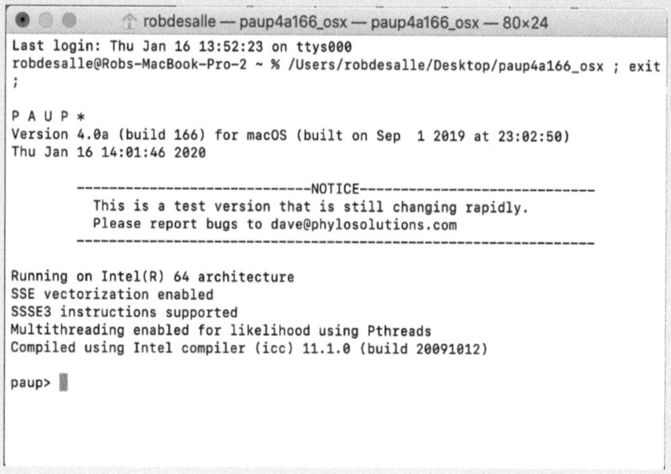

Figure 2.8 Screenshot of PAUP4a166_osx open in the terminal.

ANALYZING THE ARISTOTLE.NEX MATRIX

The matrix from Aristotle's writing in Chapter 1 (Table 1.7) can be analyzed in any phylogenetic program. There are 12 characters and 15 taxa (see Table 1.1). The Ys in Table 1.1 can be transformed to 1s and the Ns in Table 1.6 can be transformed into 0s. The format of the Aristotle matrix is discussed in **Sidebar 2.8**. The NEXUS formatted Aristotle matrix is shown in **Figure 2.9**. The steps for analysis of the Aristotle matrix are detailed in **Sidebar 2.9**.

The result of phylogenetic analysis of the Aristotle matrix is shown in **Figure 2.11**. The obvious root or outgroup of the tree from the Aristotle is inanimate matter and so the tree has been rooted (**Sidebar 2.10**) with "mineral_iron". Note that the tree gives a ladder like progression from inanimate matter to what *scala naturae* would suggest is more and more complex organisms. Even though no phylogenetic analysis was conducted by Aristotle, he was nonetheless very familiar with the data he had collected. He must have recognized this ladderlike trend in the data and it no doubt impacted how he interpreted his view of the world – as a ladder or a progression from less complex to more complex. The ladderlike progression is what he called the *Scala Naturae*.

The matrix can also be analysed for robustness using the bootstrap, and for character changes at the nodes in the tree (Sidebar 2.10).

Sidebar 2.8. Formatting a phylogenetic matrix

The different phylogenetic programs have different formats that are required for the program to execute the matrix. Format for PAUP requires specific items in the phylogenetic matrix file. The format of the file is in what is called nexus format and so first the "#NEXUS" statement has to appear in the first line of the file. The second line in this kind of nexus file is "Begin data;". The semicolon indicates that the end of line has occurred. The next line lists the dimensions of the matrix followed by a semicolon. Since there are 12 characters and 15 taxa the line reads "Dimensions nchar = 12 ntax = 15;". Next, the format of the matrix is listed. In this matrix we have presence/absence data in form of 0s and 1s; we also would have used a question mark (?) for missing data and we want to make sure that the program reads all of the lines in the file so we include a format statement for an interleaved matrix (this simply means that the matrix can occur in blocks) and so the format statement looks like this – "Format symbols = "01" Missing = ? interleave;". This last line is followed by a line with "Matrix" on it (note that there is no semicolon after the word) and then the matrix is typed in. Note that there are no white spaces in the taxa names and that underscores (_) are used between words. Do not use dashes in taxon names as these dashes will mean something when we talk about molecular sequences. The data are then followed by a line with a semicolon (to signify that the matrix has ended) and a final line with "end;" on it to signify that the file has ended (Figure 2.9). We can include other commands and assumptions after the "end;" statement that we will soon discuss. We prefer to use BBedit to do the editing, but the editing of these data can be done in any line editor such as textedit.

```
#NEXUS
Begin data;
Dimensions nchar=12 ntax=15;
Format symbols="01" Missing=? interleave;

Matrix
Man_man                                 110001111100
Live_bearing_tetrapods_Cat              101000111100
Cetaceans_Dolphin                       100000111100
Birds_Bee_eater                         110000111100
Egg_laying_tetrapods_Chameleon          101000110110
Snakes_Water_snake                      100000110110
Egg_laying_fishes_Sea_bass              100000110110
Placental_selachians_Shark              100000110111
Crustaceans_Shrimp                      000010110110
Cephalopods_Squid                       000010110110
Hard_shelled_animals_Cockle             000000110011
Larva_bearing_insects_Ant               000100110011
Spontaneous_generating_Sponge           000000110011
Plants_Fig                              000000010011
Minerals_Iron                           000000000011
;
end;
```

Figure 2.9 The Aristotle matrix transformed into a presence/absence matrix for phylogenetic analysis in PAUP. The format for the various lines in the file is described in the text.

Sidebar 2.9. Doing a phylogenetic analysis

Your PAUP executable should be on your desktop and you should also place the Aristotle.nex file on your desktop for easy access to the file. In order to make the PAUP executable able to access your data file, you need to change the directory of the executable by typing "cd desktop" (which is where your data file should be). Alternatively, you can create a folder anywhere on your computer for analyses with PAUP but when you do this you will need to change directories in the terminal to that folder for PAUP to execute your data files. Next open the PAUP executable and type "exe Aristotle.nex" into the command line and hit return. The file should execute, and the command line should now be available for issuing commands for the analysis of the file. This procedure is basically the same for any phylogenetic matrix you might want to analyze with PAUP (Figure 2.10).

The default setting for analyses is maximum parsimony (MP) and we will analyze the Aristotle matrix with MP so there is no need to change any of the settings for analysis. While PAUP can do exhaustive searches (meaning all possible trees for the number of taxa in the matrix will be evaluated and the shortest tree reported) and searches that are exact (branch and bound algorithm which promises to find the shortest tree), any matrix with more than 10 to 12 taxa cannot be analyzed exhaustively or exactly because of the huge number of trees that need to be evaluated. So we recommend that all analyses be accomplished with a heuristic search (a search that uses shortcuts to determine which tree or trees is the shortest). Heuristic searches (command "hsea" in PAUP) do not guarantee finding the shortest tree but they are good approximations when trillions of trees are involved in the search process. Hence, type "hsea" into the command line and the program will do the heuristic search.

After the search is completed the screen will show a lot of data about how the search proceeded. The last lines of the search will look like this:

Heuristic search completed

Total number of rearrangements tried = 9156
Score of best tree(s) found = 14
Number of trees retained = 8
Time used = 0.01 sec (CPU time = 0.01 sec)

Note that the search took 0.01 seconds and that eight trees were found that were of equal shortest length. Note also that the score of the shortest tree is 14 steps and that while there are billions of potential trees to examine for 15 taxa, only 9156 needed to be examined to find these shortest trees.

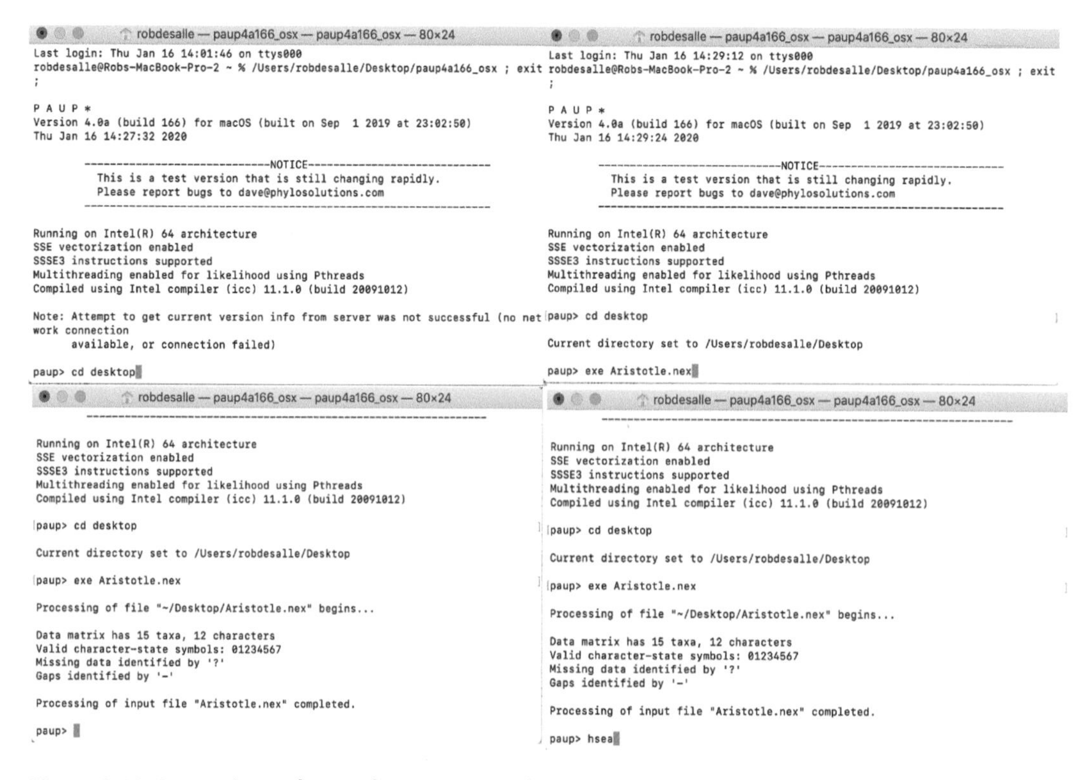

Figure 2.10 Screenshots of PAUP for executing a file. Top left – the "cd desktop" command is executed to make the desktop accessible (desktop is where the Aristotle.nex file is located.) Top right – the "exe Aristotle.nex" command to input the data matrrix into PAUP for further commands. Bottom left – screenshot showing that the Aristotle.nex file has been properly executed (note line "Processing of input file "Aristotle.nex" completed"). Bottom right – the "hsea" command typed into the command line.

Sidebar 2.10. Outgroups and the Aristotle matrix

As we described earlier in this chapter, phylogenetic trees generated in this way should be rooted with an outgroup. In the case of the Aristotle matrix, "minerals_iron" is the best outgroup for the collection of the taxa in this analysis. To set the outgroup type "outgroup minerals_iron" into the command line which will set the minerals_iron "taxon" as the outgroup and print out the following on the terminal screen.

Outgroup status changed:

1 taxon transferred to outgroup
Total number of taxa now in outgroup = 1
Number of ingroup taxa = 14

To view the tree results of the analysis type "describe" or "desc" into the command line and you should see the screen as in Figure 2.11. Other analyses can be accomplished while the Aristotle matrix is executed, such as a bootstrap analysis (discussed above) to explore the robustness (type "bootstrap" or "boot") and analyses to view which characters support which nodes (type "describe apolist" or "desc apolist"). These are the simplest commands for a phylogenetic analysis in PAUP. Other commands can be found in the PAUP command manual (available at http://phylosolutions.com/paup-documentation/paupmanual.pdf).

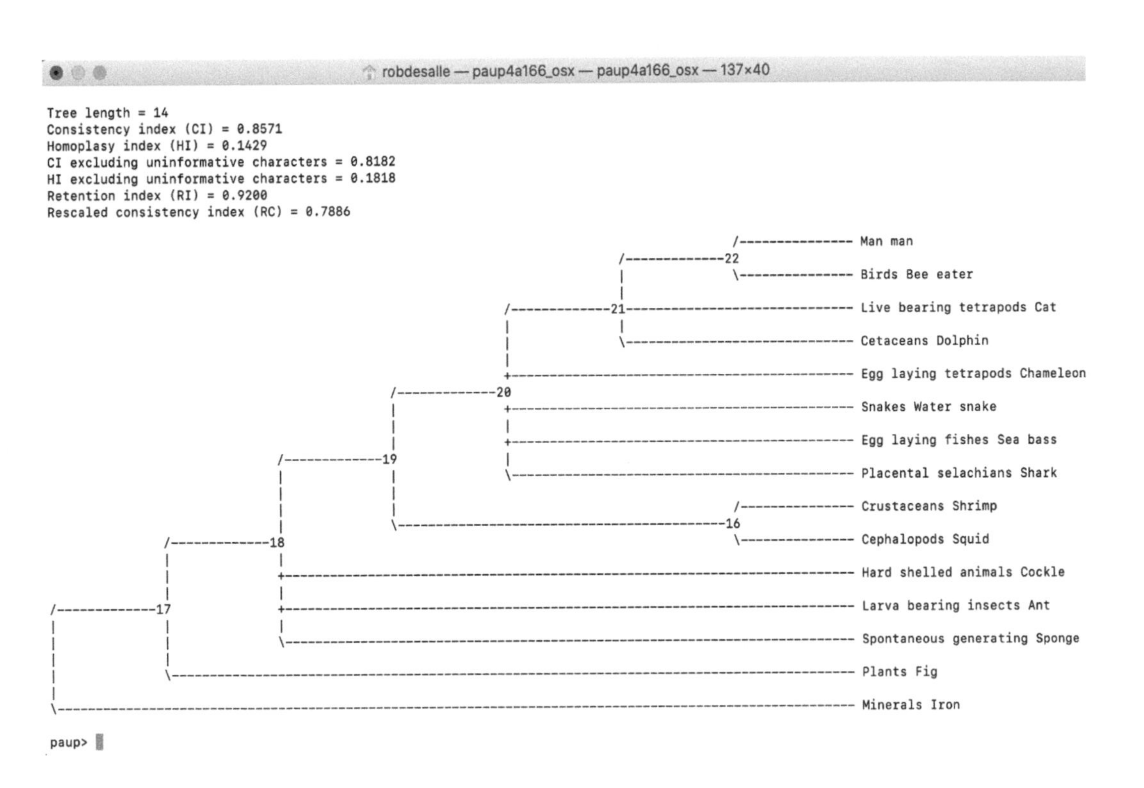

Figure 2.11 Screenshot of the most parsimonious solution for the Aristotle matrix. Note that all vertebrates appear in a clade defined by node number "28" and all animals defined by node numbered "18". Other groups are evident in the tree that may or may not make sense in our modern world. The information in the upper left of the screen refers to certain aspects of the analysis that are discussed in Chapter 3.

CHARACTERS USED IN MODERN INVERTEBRATE SYSTEMATICS

As we pointed out in Chapter 1, data for modern invertebrate systematics can in general be divided into morphological and molecular kinds. In other words, there are at least two major categories or partitions that systematic data can be organized into. But it was not until the 1980s, that molecular tools were widely applied to invertebrate phylogenetic questions. Also, at that time phylogenetic systematics and tree building were young disciplines with a substantial amount of background theory that had been developed. When molecular data started to be applied to phylogenetic questions, a broad array of new phylogenetic problems were met by systematists. First and foremost, it was recognized that morphological inferences were not always in line with molecular trees, meaning that the two major character partitions (Hillis, 1987; Brower et al., 1996; Baker et al., 1998) were oftentimes incongruent. Methods were developed to explore and attempt to correct for this incongruence of the two major categories of data over the following three decades. Some authors argued for total evidence (Kluge, 1998), simultaneous analysis (Nixon and Carpenter, 1996) or concatenation (Schierwater et al., 2009) of molecular with morphological data. More recently, because of the emergence of phylogenomic methods and applications, morphologists have given much thought to the problems involving incongruence of molecules and morphology. Some authors suggest that the role of morphology in modern systematics is "to understand how phenotypic diversity evolved" (Wanninger, 2015) and as important in "time-scaling phylogenies" (Lee and Palci, 2015). The authors of both of these works imply that while they recognize the importance of morphological characters, such characters are somewhat irrelevant to the actual construction of phylogenetic trees. Their reasoning is based on the idea that phylogenomic molecular characters simply swamp out the morphology. The argument does not dismiss morphology but rather relegates it to a dataset in need of explanation post tree construction.

Other researchers addressed this issue from the perspective of morphological cladistics and made the reasonable claim that morphological datasets needed to be expanded for questions involving metazoan (Jenner, 2003, 2004). He realized that many early morphological studies were based on a small number of characters, rendering them rather non-robust in comparison with phylogenomic datasets that were often times 1,000 to 10,000 times larger than morphological data with respect to number of characters. Jenner's (2003, 2004) solution was to use morphological characters as arbiters of phylogenetic accuracy in phylogenomics analyses. He argued more or less for a separation of molecular and morphological data and a larger focus on morphology for testing hypotheses about phylogenetic relationships. Jenner's (2003, 2004) papers appeared right on the cusp of phylogenomic data generation, but they are emblematic of an important idea about morphology in phylogenetics. In agreement, Scholtz (2010) suggested that "if morphological and molecular results clash, there is no logical necessity to dismiss morphological data". Giribet (2016) also made a cogent argument for the inclusion of morphological data in how we generate phylogenies. Neumann et al. (2020) demonstrated that addition of morphological analyses can influence the result of a phylogenetic analysis. As will be evident from the subsequent chapters we pay deep attention to both kinds of data and provide the student with data matrices with both kinds of data. Because a well-accepted method for how the two partitions should be treated we suggest, that the student be aware of the inferences from both.

CONCLUSION

Phylogenetic analysis has matured over the past 40 years. From the numerical taxonomy approaches of Sneath and Sokal that were developed to impart objectivity to phylogenies to the application of Bayesian reasoning in phylogenetics advances have been made at a dizzying pace. In this chapter we have described the major phylogenetic approaches available to the invertebrate biologist. We have also pointed the student to several programs that implement these modern phylogenetic approaches. Further, we have shown an example (Aristotle.nex) of how phylogenetic analysis is performed. In Chapter 4, we will go into detail about how to construct phylogenetic matrices for invertebrate taxa and discuss the ramifications of these approaches for understanding invertebrate systematics.

REFERENCES

Baker, Richard H., Xiaobo Yu and Rob DeSalle. "Assessing the relative contribution of molecular and morphological characters in simultaneous analysis trees." Molecular Phylogenetics and Evolution 9, no. 3 (1998): 427–436.

Brower, A. V. Z., R. DeSalle and A. Vogler. "Gene trees, species trees, and systematics: a cladistic perspective." Annual Review of Ecology and Systematics 27, no. 1 (1996): 423–450.

Cavalli-Sforza, Luigi L. and Anthony W. F. Edwards. "Phylogenetic analysis: models and estimation procedures." Evolution 21, no. 3 (1967): 550–570.

Farris, James S., Victor A. Albert, Mari Källersjö, Diana Lipscomb and Arnold G. Kluge. "Parsimony jackknifing outperforms neighbor-joining." Cladistics 12, no. 2 (1996): 99–124.

Felsenstein, Joseph. "Evolutionary trees from DNA sequences: a maximum likelihood approach." Journal of molecular evolution 17, no. 6 (1981): 368–376.

Felsenstein, Joseph. "Confidence limits on phylogenies: an approach using the bootstrap." Evolution 39, no. 4 (1985): 783–791.

Felsenstein, Joseph. PHYLIP (phylogeny inference package), version 3.5c. Joseph Felsenstein, 1993.

Giribet, Gonzalo. "New animal phylogeny: future challenges for animal phylogeny in the age of phylogenomics." Organisms Diversity & Evolution 16, no. 2 (2016): 419–426.

Goloboff, Pablo A., James S. Farris and Kevin C. Nixon. "TNT, a free program for phylogenetic analysis." Cladistics 24, no. 5 (2008): 774–786.

Hennig, Willi. Phylogenetic Systematics. University of Illinois Press, 1999.

Hillis, David M. "Molecular versus morphological approaches to systematics." Annual Review of Ecology and Systematics 18, no. 1 (1987): 23–42.

Huelsenbeck, John P., Fredrik Ronquist, Rasmus Nielsen and Jonathan P. Bollback. "Bayesian inference of phylogeny and its impact on evolutionary biology." Science 294, no. 5550 (2001): 2310–2314.

Jenner, Ronald A. "Unleashing the force of cladistics? Metazoan phylogenetics and hypothesis testing." Integrative and Comparative Biology 43, no. 1 (2003): 207–218.

Jenner, Ronald A. "The scientific status of metazoan cladistics: why current research practice must change." Zoologica Scripta 33, no. 4 (2004): 293–310.

Kluge, Arnold G. "Total evidence or taxonomic congruence: cladistics or consensus classification." Cladistics 14, no. 2 (1998): 151–158.

Kumar, Sudhir, Koichiro Tamura and Masatoshi Nei. "MEGA: molecular evolutionary genetics analysis software for microcomputers." Bioinformatics 10, no. 2 (1994): 189–191.

Lartillot, Nicolas, Nicolas Rodrigue, Daniel Stubbs and Jacques Richer. "PhyloBayes MPI: phylogenetic reconstruction with infinite mixtures of profiles in a parallel environment." Systematic Biology 62, no. 4 (2013): 611–615.

Lee, Michael S. Y. and Alessandro Palci. "Morphological phylogenetics in the genomic age." Current Biology 25, no. 19 (2015): R922–R929.

Neumann, Johannes S., Rob DeSalle, Apurva Narechania, Bernd Schierwater and Michael Tessler. "Morphological characters can strongly influence early animal relationships inferred from phylogenomic datasets." Systematic Biology 70, no. 2 (2021): 360–375. (2020).

Nguyen, Lam-Tung, Heiko A. Schmidt, Arndt Von Haeseler and Bui Quang Minh. "IQ-TREE: a fast and effective stochastic algorithm for estimating maximum-likelihood phylogenies." Molecular Biology and Evolution 32, no. 1 (2015): 268–274.

Nixon, Kevin C. "The parsimony ratchet, a new method for rapid parsimony analysis." Cladistics 15, no. 4 (1999): 407–414.

Nixon, Kevin C. and James M. Carpenter. "On simultaneous analysis." Cladistics 12, no. 3 (1996): 221–241.

Ronquist, Fredrik, Maxim Teslenko, Paul Van Der Mark, Daniel L. Ayres, Aaron Darling, Sebastian Höhna, Bret Larget, Liang Liu, Marc A. Suchard and John P. Huelsenbeck. "MrBayes 3.2: efficient Bayesian phylogenetic inference and model choice across a large model space." Systematic Biology 61, no. 3 (2012): 539–542.

Schierwater, Bernd, Michael Eitel, Wolfgang Jakob, Hans-Jürgen Osigus, Heike Hadrys, Stephen L. Dellaporta, Sergios-Orestis Kolokotronis and Rob DeSalle. "Concatenated analysis sheds light on early metazoan evolution and fuels a modern "urmetazoon" hypothesis." PLoS Biol 7, no. 1 (2009): e1000020.

Scholtz, Gerhard. "Deconstructing morphology." Acta Zoologica 91, no. 1 (2010): 44–63.

Sneath, Peter HA, and Robert R. Sokal. Numerical taxonomy. The Principles and Practice of Numerical Classification. 1973.

Stamatakis, Alexandros. "RAxML version 8: a tool for phylogenetic analysis and post-analysis of large phylogenies." Bioinformatics 30, no. 9 (2014): 1312–1313.

Swofford, David L. "Paup*: Phylogenetic analysis using parsimony (and other methods) 4.0. B5." 2001.

Trifinopoulos, Jana, Lam-Tung Nguyen, Arndt von Haeseler and Bui Quang Minh. "W-IQ-TREE: a fast online phylogenetic tool for maximum likelihood analysis." Nucleic Acids Research 44, no. W1 (2016): W232–W235.

Wanninger, Andreas. "Morphology is dead – long live morphology! Integrating MorphoEvoDevo into molecular EvoDevo and phylogenomics." Frontiers in Ecology and Evolution 3 (2015): 54.

Zhang, Chao, Maryam Rabiee, Erfan Sayyari and Siavash Mirarab. "ASTRAL-III: polynomial time species tree reconstruction from partially resolved gene trees." BMC Bioinformatics 19, no. 6 (2018): 153–156.

CHAPTER

3

INVERTEBRATE PHYLOGENOMICS

Rob DeSalle, Wiebke Feindt, and Bernd Schierwater

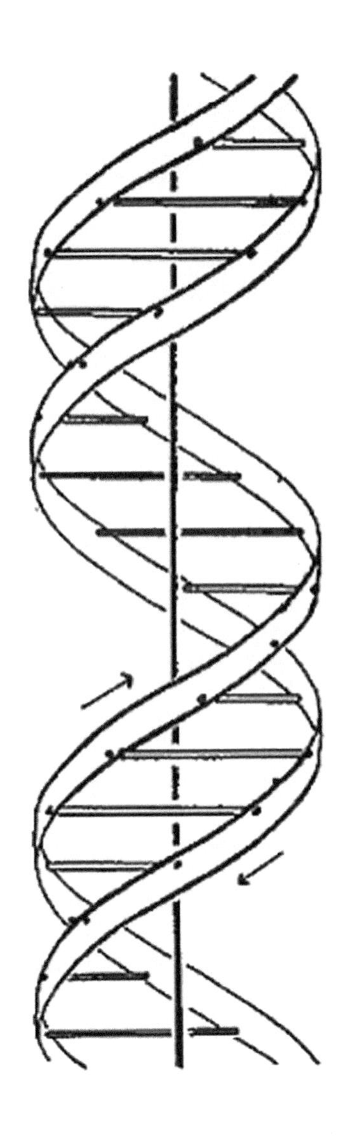

CONTENTS

THE DNA SEQUENCE REVOLUTION

While proteins were sequenced in the 1960s, rather laborious biochemical methods were used then. DNA sequencing methods were not developed until the early 1970s, and did not really gain prominence in systematics until the mid- 1980s. Specifically, two new methods – the polymerase chain reaction (PCR) and Sanger sequencing – were invented. A Nobel prize in chemistry was awarded in 1980 to Walter Gilbert and Fred Sanger for developing DNA sequencing techniques. Sanger's method, called chain termination or "sequencing by synthesis", turned out to be the method that most researchers would use from the 1990s and on. Its principles are still in use in many of the modern sequencing procedures. Another Nobel prize was awarded to Kerry Mullis for his timely invention of PCR. The details of DNA sequencing by synthesis and PCR are beyond the scope of this chapter, but we encourage students to use the internet or other textbooks to understand the biochemistry, because the two methods are at the heart of all modern biology including systematics and evolution (for a brief history of DNA sequencing see Shendure et al., 2017 and Heather and Chain, 2016)

As mentioned, one big advance in being able to manipulate DNA sequencing for biology in general and systematics in particular was the development of the polymerase chain reaction (PCR) by Kerry Mullis, who received the Nobel prize in chemistry in 1993. Mullis used basic DNA biochemistry to develop this method in 1984, synthesizing millions of copies of a targeted stretch of DNA from only a small amount of template. The perfecting of these so-called first-generation sequencing methods led to the generation of the whole genomes of several organisms in the 1990s, including *Haemophilus influenzae* (Feischmann et al., 1995), *Drosophila melanogaster* (Adams et al., 2000), *Saccharomyces cerevisiae* (Kim et al., 1998), *Caenorhabditis elegans* (The *C. elegans* Sequencing Consortium, 1998) and *Arabidopsis thaliana* (*Arabidopsis* Genome Initiative, 2000), and culminated in the sequencing of the first draft (meaning a preliminary but nearly complete version) of the human genome in 2001. Needless to say, DNA sequencing and PCR were manual and laborious processes prior to the development of what researchers called **high-throughput sequencing**.

BENCHMARKS AND SIGNPOSTS IN INVERTEBRATE MOLECULAR SYSTEMATICS

Prior to whole genome sequencing, systematists used PCR coupled with first-generation sequencing approaches (Sanger sequencing) to generate data for small stretches of specific genes to obtain DNA sequence data. This approach required a large amount of starting material for DNA isolation and focused on genes like 18s rDNA and mitochondrial genes like cytochrome oxidase subunit I (COI, cox 1). The generation of DNA sequences this way was somewhat laborious and expensive. The critical development in the modern methods we see today was the creation around 2005 of Next-Generation Sequencing (NGS; **Sidebar 3.1**) technology. NGS techniques highly parallelized the sequencing reactions so that thousands, if not millions or billions, of reactions could be done on a single small device. The miniaturization of single template reactions and of machines that could read the information from the highly or massively parallel reactions was a key in this development. Approaches such as 454, SOLID sequencing, Ion Torrent (MinION), Illumina, PacBio and Nanopre are some that were developed during this period (see Martin and Kelso, 2010). These NGS methods led to a logarithmic increase in sequence generation from the mid- 2000s to the present. Some of them have fallen by the roadside and some have survived, as we will see below.

Sidebar 3.1. Three Next-Generation Sequencing platforms

Here we describe three of the more widely used NGS platforms. Newer platforms are continually being developed, but we focus on these since they are three of the more popular ones currently available.

1. ***Illumina:*** This platform was originally developed by Solexa Inc. and introduced in 2006; shortly after its introduction its name was changed to Illumina. This technique uses a flow cell surface that allows for the detection of over millions of individual reads. This flow surface has eight channels on it, so eight separate sequencing experiments can be run per cell. The process uses the four nucleotides labeled with four different fluorescent molecules and involves sequencing by synthesis. Target DNA is fragmented and two small linkers are attached to the ends of each randomly fragmented piece of DNA. Each fragment is then bound to the surface of one of the eight channels mentioned above via small oligonucleotides. PCR is performed on all the samples bound to the cell by immersing the cell in the reaction mixture. This step will create a large number of paired end repeats for DNA sequencing, all of which are bound to the channels. The synthesized double-stranded DNA fragments bound to the flow cell surface are denatured. This step creates a large number of single-stranded paired end repeats that are suitable for another round of PCR. This second PCR step produces several million templates for DNA sequencing. The first base in each fragment is sequenced by using special nucleotides that are labeled with specific fluorescent dyes for each of the four bases. The base added to each of the millions of fragments is identified and recorded by exciting the sequencing cell with a laser beam. In this way, the first base of each fragment is recorded. The process is repeated and extended for about 150 rounds. Currently the length of fragments that can be sequenced using this platform is a little over 150 bases. This number of bases can also be enhanced by noting that the paired ends of fragments can be easily identified. By combining sequences from paired ends, the total length of fragments can be pushed to over 400 bases. The error rate (i.e. rate of incorporation of an erroneous base in the sequence read) is about 0.1%. This rate is among the best of all sequencing platforms. For a review of the Illumina technology see Mardis (2008).
2. ***PacBio:*** This platform was developed at Cornell University, and in 2004 Pacific Biosciences of California, Inc (PacBio) obtained the technology. The platform is described as "single-molecule real-time sequencing" (or SMRT). The platform uses a device called a SMRT cell where single very long molecules of DNA are sequenced from tissues of target organisms. The technology is called real time because sequences are generated in the SMRT cell in real time without pausing for each base to be washed over the cell as in Illumina sequencing. The steps of SMRT sequencing are now described: A typical PacBio run will generate about 150,000 reads ranging from 1,000 bases to 80,000 bases in length, with most of the fragments being about 16,000–30,000 bases in length. The error rate of this platform is quite high (over 10%), but because the reads are so long and overlap to such a high degree the error rate can be accommodated in most projects (for a review of PacBio technology, see Rhoads and Au (2015).
3. ***MinION:*** Originally developed as part of the Ion Torrent platform, which is a nanopore product. The MinION is another real-time sequencing approach using a flow cell that can generate millions of reads for each sample that are "ultra-long" (hundreds of kilobases per read). The major novel development for the MinION is its portability, as it is small (about the size of a large cell phone) and can be used in the field. For a review of the MinION system see Mikheyev and Tin (2014).

Amplicon versus shotgun sequencing: Amplicon sequencing refers to sequencing approaches where a PCR reaction is performed to generate a template for NGS sequencing. As we will see, some of the genome reduction approaches use amplicons as sequencing template. Shotgun sequencing refers to approaches where the DNA to be sequenced is randomly sheared by sonication to specific sizes and then the randomly sheared DNA is sequenced using NGS approaches. Depending on the size of the sheared fragments, different kinds of sequencing are used. For small sheared fragments (<250 bp) the Illumina platform can be used. For larger fragments (>2 kb) the PacBio platform is most appropriate. Some of the newer approaches (Nanopore, etc.) allow for real-time sequencing of extremely large fragments.

Whole genome sequencing: By far the most exhaustive approach to obtaining genome sequences is to sequence the whole genome. Organellar genomes like chloroplasts and mitochondrial genomes can be sequenced with ease due to their small size and higher concentration in cells. Usually the most informative aspect of mtDNA sequences across large phylogenetic distances is the gene order of the 13 protein coding genes, 2 RNA genes and 20 or so tRNA genes. Because the genes in the mtDNA genome change rather rapidly, the actual DNA or protein sequences of genes from this organelle are saturated with convergences that make phylogenetic analysis difficult at large phylogenetic distances. The mtDNA genome may however be useful in phylogenetic studies of closely related taxa over shorter evolutionary distances. Currently sequencing whole nuclear genomes is a very difficult task for studies with large numbers of taxa because of the sheer amount of sequence needed. When organellar genomes are sequenced using shotgun approaches, a high degree of coverage is attained. Coverage simply refers to the amount of sequence obtained for a target genome. So if one obtains 100 × coverage of a mitochondrial genome of length 16 kb, it means that 1600 kb were generated to give the mtDNA genome. Likewise, it is recommended to do at least 20 × or more coverage of nuclear genomes. Invertebrate genomes vary greatly in size, but let's take the *Drosophila* nuclear genome at a size of 200 mb (or 200 million bases). To obtain 20 × coverage of this genome we would need to sequence 4,000 Mb or 4 billion base pairs.

Genome assembly and post-sequencing bioinformatics: Whole genome sequencing (WGS; **Figure 3.1**) and genome reduction approaches are intricately connected to post-sequencing informatics. WGS results in billions of base pairs and hundreds of millions if not billions of short sequences called "reads" or "short reads". These short reads are practically useless if they are not assembled into longer stretches of DNA. So, the ultimate goal of genome assembly is to take the necessarily short fragments generated by WGS platforms and produce longer contiguous pieces (called contigs) of sequence, that can then be scaffolded onto chromosomes or that can simply be used to search for gene regions and other genomic phenomena.

A detailed treatment of the assembly approach is beyond the scope of this textbook, but it is straightforward and involves: (1) the sequencing phase, (2) an inspection and cleanup phase, (3) an assembly phase and (4) the draft genome phase (for a review of assembly methods see Nagarajan and Pop, 2013). With this platform in mind. Reads generated by the Illumina platform are between 36 and 150 bases in length (the lengths of Illumina reads have only gotten longer and longer as the technology has improved, so even longer read lengths are possible). Due to the design of the Illumina approach, there are two kinds of reads that the Illumina platform can produce. The first is called single ended sequencing and simply generates a sequence from one fragment

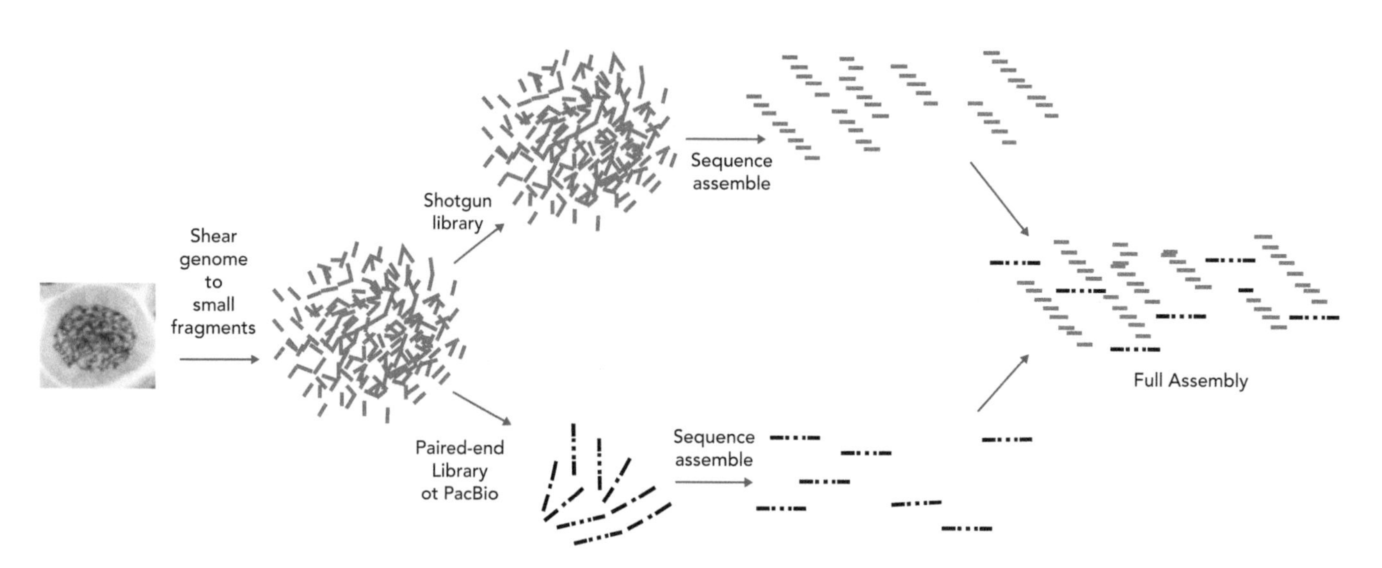

Figure 3.1 Scheme for shotgun sequencing. A diagrammatic representation of how the WGS approach would proceed with shotgun sequencing. (Drawn by RD.)

>SNP_3229
AATCATCAAAATTGACTCATATAAAATTGTCTATAATTTCTCCCACTTTGGACTTCAGT

>prot_222
SDDHSAGFUDYGFSUDGFGFYSDUGGDA

Figure 3.2 Nucleic acid and protein sequences in FASTA format.

in one direction. The second is called a paired end sequence, which is a pair of sequences from the same fragment of DNA read from the two ends of the fragment. Since the fragment length is known, paired end sequences can add extra information in the assembly process.

As we will soon see, refined, quality-controlled DNA sequences that are stored in the database are in a specific format called FASTA (pronounced 'fast aye'). The FASTA file has a defined format which starts on the first line of the file with a leftward caret (>), followed by the name of the sequence and any other information needed also placed on the first line. After this first line a carriage return is placed and then on the second and succeeding lines of the file the sequences are placed. The file ends when the sequences end. Examples of a DNA and a protein sequence in FASTA format are shown in **Figure 3.2**.

Raw short read DNA sequence files are called FASTQ files which not only contain the sequences but also a base-by-base assessment of the quality of the base pair in a position that is generated by the sequencing apparatus. The FASTQ files are useful in the quality-control step of genome sequencing and programs exist that take the FASTQ information and convert them to FASTA files. Large amounts of sequences from short read sequencing are stored in the Short Read Archives (SRA, https://www.ncbi.nlm.nih.gov/sra/).

The assembly phase is the most important as it can be manipulated by changing parameters in the algorithms used to find similarities among the short reads that allow for the linking together of short reads. If one is working with an organism that has its or a close relative's genome sequenced, then assembly is a relatively easy task because the known genome can be used as a scaffold on which assembly can be accomplished. Such assembly is called *ab initio* assembly. If one is sequencing the genome of a novel organism then the assembly procedure is much more difficult, because a reliable scaffold is not available. This kind of assembly is called *de novo* assembly. Another way to enhance *de novo* assembly process is to combine sequences from different approaches. Since the limit on sequence length of Illumina is between 100 and 200 bp, adding sequences from a PacBio run with lengths greater than 1,000 bp will enhance the assembly process.

Many systematists avoid WGS approaches in favor of ones that reduce the amount of sequence and assembly that is needed to produce high-quality data. This approach has taken hold because much of the genome of organisms does not code for proteins, and since researchers prefer protein coding sequences for phylogenetic analysis, avoiding non-coding sequences streamlines the workflow of phylogenomic projects. We discuss three of these approaches below.

Transcriptomics (Expressed Sequence Tags [EST] and RNAseq): Cells are continually expressing (transcribing RNA) a great many of the genes in the genome of an organism. EST sequencing takes advantage of this fact by using cellular RNA as templates for sequencing in systematic studies. In this approach the mRNA is separated from other cellular RNAs like ribosomal or tRNAs by taking advantage of the fact that most mRNAs are capped at the 3 prime end with a poly-A tract (AAAAA...). A ploy-T oligonucleotide is used as a primer to amplify any mRNA in a cellular nucleic acid extract and the amplified mRNA is sequenced using NGS approaches and is called RNAseq (Stark et al., 2019). This approach generates large amounts of sequence information specific to the organism being analyzed called a transcriptome. Similar to WGS, RNA is mostly sequenced on an Illumina platform, which produces small reads. Hence, the short reads have to be assembled into full-length transcripts.

There are drawbacks to this method though. First, not all genes in the genome are sequenced using this approach and so the overlap of genes across large numbers of taxa is sometimes low. Second, fresh tissue is required in order for there to be a good representation of mRNA in the sample. Oftentimes, this is a difficult task, especially with more exotic and rarer specimens. There are methods that are used to quantify the completeness of transcriptomes generated using RNAseq. A good example of one of these methods is BUSCO (Benchmarking sets of Universal Single-Copy Orthologs; https://busco.ezlab.org/) which "provides quantitative measures for the assessment of genome assembly, gene set, and transcriptome completeness, based on evolutionarily-informed expectations of gene content from near-universal single-copy orthologs" selected from an ortholog database (usually OrthoDB, a gene orthology [see below] database; https://www.orthodb.org/v9/index.html). The sequences obtained using RNAseq are also usually annotated so that researchers have some idea of the protein the sequences produce.

Genome Reduction Approaches: Several approaches have been developed that focus in genomic sequences but use methods that reduce the amount of DNA that is sequenced. One method, called Ultra-Conserved Elements (UCEs); for a review see Zhang et al., 2019), has been used to generate systematic data. UCEs refer to short, highly conserved sequences in the genomes across large phylogenetic distances. These short ultra-conserved sequences are used as primers in miniaturized PCR reactions to generate templates for sequencing. The number of primer pairs used in this approach is usually greater than a thousand so that DNA sequence information is obtained for over a thousand genomic regions. A related approach called exon-capture has also been used to generate systematic data. This approach uses highly conserved sequences in the genomes of organisms to obtain exon regions by hybridization and subsequent removal of any DNA sequences that do not hybridize (and hence are not part of the targeted capture system).

Another approach that generates large amounts of sequence for systematic studies is called RADseq (Andrews et al., 2016; Davey and Blaxter, 2010) which stands for Randomly Amplified DNA sequencing. This approach uses restriction enzymes that cut DNA at specific sequences. Some restriction enzymes will cut at a specific four-base sequence such as GAAG and others that cut at longer recognition sequences. Short linker sequences are ligated to the restriction cut DNA. The linkers have specific sequences that allow for amplification using PCR and these amplified fragments are sequenced via Illumina methods (for a more detailed account of the approach see Andrews et al., 2016; Davey and Blaxter, 2010). The trick with this approach is that closely related organisms will have similar restriction sites and so if the method is used on such organisms the same fragments will be generated creating maximum overlap. The whole genome is reduced to a subset of restriction fragments.

As with RNAseq, there are drawbacks to these genome reduction methods, the most prevalent being the overlap problem described above. Tissues used as source of DNA for these methods need to be in pretty good shape (i.e. have DNA of relatively long length >300 bp) as the primers used in the approaches are usually placed relatively far apart (200 to 500 bp apart). For further reading on targeted gene sequencing see McGinn et al., 2016.

Metagenomic or Environmental DNA (eDNA) Approaches: Recently, approaches that sequence whole environments have been developed. These approaches can be used to assay large numbers of small invertebrates, as well as to explore the microbial communities that live on and inside of invertebrates. eDNA approaches have evolved into two basic kinds – amplicon eDNA and shotgun eDNA. Both techniques take advantage of the fact that in a spoonful of dirt, water from the ocean, a lake or river and tissue from the gut of an organism contains microbes and the leftover tissues from animals and plants. The DNA is usually there in minuscule amounts but can be sequenced with modern next-generation technology. The amplicon approach takes DNA isolated from an environmental sample and amplifies it with a primer pair that amplifies a stretch of DNA that can be used to identify any organism that might have originally been a part of the sample. Once the DNA has been amplified, a library is prepared of the amplicons and the library is sequenced. The resulting archive of short reads (SRA) is then scanned with bioinformatic techniques that allow for the identification of the amplified fragments using a database of archived and identified sequences.

Shotgun eDNA has the same steps minus the amplicon step. This approach is a bit more difficult because instead of focusing on a well-characterized identification sequence (also known as a DNA barcode) every part of the genomes of every organism in the original sample is sequenced. Identifying these sequences to a single species depends on the completeness of the database used to do the identification. While both of these techniques have been used initially and primarily for examination of microbial communities, it has been gaining popularity among other specialists, including invertebrate zoologists who work with small specimens.

Whether a dataset is constructed with genome reduction, RNAseq, or even eDNA approaches, a whole suite of bioinformatic programs are available to facilitate the rapid processing of raw sequences to give high-quality FASTA files. Other bioinformatic platforms result in the processing of cleaned FASTA files into matrices for phylogenetic analysis. The first step toward matrix construction concerns the establishment of homology of the sequences and is followed by the alignment of the sequences.

Homology – Gene Orthology and Paralogy: In constructing a data matrix for phylogenetic analysis, two steps are required. First, traits of organisms need to be homologized. For instance, any trait concerning the front legs of a vertebrate needs to be compared to a trait on the front legs of other vertebrates, and not, say, to a trait on the tail of a vertebrate. This process is called establishing topological similarity or establishing primary homology. Once the traits have been homologized in this way they can then be used to construct a phylogenetic tree by establishing what is called secondary homology. A morphological matrix lists the taxon followed by the states of the traits for that taxon. Each taxon should have the same number of trait states and each column in the matrix should refer to the same trait.

The process of generating a matrix for molecular sequences is a bit different. Evolution of genes occurs in two "dimensions". First genes change with the speciation process. So, for instance when two species diverge, two "copies" of the gene exist, one in each of the two species. If another speciation event occurs then there are three "copies" of the gene, one in each of the three species. But genes can also duplicate within a species. So, if in an ancestral species a gene duplicates, two "copies" of the genes exist. If this ancestor diverges into two daughter species then four "copies" of the gene exist, two in each of the two daughter species. If another speciation even occurs, then six copies of the gene exist, two in each of the three species produced by the two divergence events. Genes that are produced as a result of speciation are called orthologs. Genes that are produced via duplication are called paralogs. For phylogenetic analysis, we have to be very careful that we are placing genes that are orthologous in the matrices. If not, we are prone to doing the proverbial comparison of "apples and oranges". To ensure this error is avoided, researchers have developed methods to determine orthology. **Figure 3.3** shows a series of duplications and speciation events.

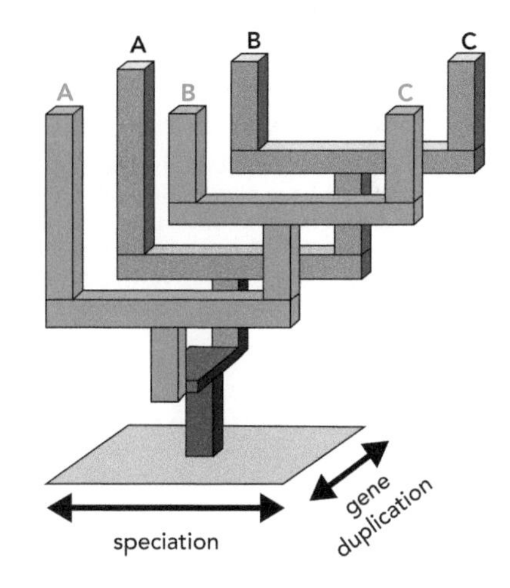

Figure 3.3 Speciation and gene duplication in a gene family. In this diagram an ancestral gene exists at the very bottom of the picture on the surface of the square. A duplication event occurs producing two genes, a green one and a gray one. Next, two speciation events occur that produce three new species. The green genes are said to be paralogous to the gray genes. Gray A, B and C are said to be orthologous to one another. Green A, B and C are said to be orthologous to one another. In constructing a matrix for phylogenetic analysis of these genes the green genes (green A, B and C) would be placed in one matrix and the gray genes (gray A, B and C) would be placed in another.

Sequence Databases: A major development in molecular biology was the prescient move to archive and make available electronically all of the sequences generated by researchers. On November 4, 1988, the United States Congress legislated that a database focused on the collection of molecular data should be established. In this way, the National Center for Biotechnology Information (NCBI) was established, placed under the aegis of the National Library of Medicine (NLM) at the National Institutes of Health (NIH). The National Institutes of Health, established in 1930 as an arm of the federal government of the United States, administers research and health advances for the country. The DNA sequence portions of the NCBI data took the form of a database known as GenBank which is linked to and coordinated with similar databases across the globe; in Europe, the European Molecular Biology Laboratories (EMBL), and in Japan at the DNA Databank of Japan (DDBJ). These three centers collect, archive and disburse data daily and provide the most up-to-date sequence information for researchers across the globe. The files that are available for access from GenBank vary with the targeted use of the information.

A researcher can easily download as many sequences as needed from these databases to construct a phylogenetic matrix. A lot of care must be given to making sure that gene orthologs are used in any phylogenetic matrix, and in documenting the sequences used in a phylogenetic analysis. Compiling a phylogenetic matrix might be easier using BLAST (Basic Local Alignment Search Tool) searches. One can search GenBank with a DNA or protein sequence under a large number of possible fields that include gene, DNA, protein, genome and taxonomy as an abbreviated list of possibilities. The output is dependent on the field searched. Information based on gene name using the "gene" field returns the most extensive set of information and both the protein and DNA sequences of the gene can be obtained from this search. Searching using the "DNA" or "protein" field gives simpler output and allows for the download of FASTA formatted information for incorporation into phylogenomic matrices. In addition, information is available in the XML format that is readily accessible to computational parsing. One of these XML formats is the GenBank (.gb) format (**Sidebar 3.2**). In addition to searching

Sidebar 3.2. GenBank accessions

A full description of bioinformatic tools is beyond the scope of this text. However, the student should have some principles in hand in order to understand how the phylogenomic matrices are constructed. Here we discuss the basic way that databases store and archive sequences. The top half of a typical GenBank formatted file includes the LOCUS, DEFINITION, ACCESSION, VERSION, KEYWORDS, SOURCE, and REFERENCE fields and these are listed in Table 3.1. These lines give all of the information appended and archived with the sequences.

The bottom half of the GenBank (.gb) file for the *Apis melifera* EF1a gene contains the detailed information for the accessioned sequence, listing both the protein sequence and the DNA sequence of the gene. Since FASTA files are the desired format for constructing phylogenetic matrices one can easily click on the FASTA format buttons for the DNA and protein sequences and easily obtain the FASTA formatted sequence (Figure 3.4) shown below.

TABLE 3.1 Typical GenBank accession fields
LOCUS
Locus name
Sequence length
Molecule type
GenBank division
Modification date
DEFINITION
ACCESSION
VERSION "GenInfo identifier"
KEYWORDS
SOURCE (organism name)
REFERENCE (or direct submission)
AUTHORS
TITLE (title of published work or tentative title of unpublished work)
JOURNAL
PUBMED
FEATURES

FASTA ▾

Apis mellifera elongation factor 1-alpha F2 (EF1a-F2), mRNA

NCBI Reference Sequence: NM_001014993.1

GenBank Graphics

```
>NM_001014993.1 Apis mellifera elongation factor 1-alpha F2 (EF1a-F2), mRNA
ATGGGTAAAGAAAAGATTCATATTAATATTGTCGTTATTGGACACGTCGACTCTGGCAAGTCTACCACCA
CTGGTCATTTGATCTACAAATGTGGTGGTATTGATAAACGTACCATTGAAAAATTCGAGAAGGAAGCCCA
GGAAATGGGCAAAGGATCCTTCAAATATGCCTGGGTATTGGATAAGTTAAAAGCTGAACGTGAACGTGGT
ATTACGATTGATATTGCTTTGTGGAAATTCGAAACGTCAAAATACTATGTTACTATTATTGATGCTCCTG
GACACAGAGATTTCATCAAAAACATGATTACTGGTACCTCTCAGGCTGATTGTGCTGTATTAATTGTTGC
TGCTGGTACTGGAGAGTTCGAAGCAGGCATTTCAAAGAATGGACAAACTCGTGAGCATGCTTTGCTCGCT
TTTACTCTTGGTGTGAAACAATTGATTGTTGGTGTTAATAAGATGGACTCCACTGAACCACCGTATTCTG
AAACCCGATTTGAAGAAATTAAAAAAGAAGTGTCATCTTACATTAAAAAAATTGGTTATAATCCAGCTGC
AGTTGCATTTGTGCCAATTTCTGGTTGGCATGGAGATAATATGTTGGAAGTTTCTTCAAAAATGCCTTGG
TTTAAGGGATGGACGGTTGAACGTAAAGAAGGAAAAGTTGAAGGAAAATGTCTCATTGAAGCGCTTGATG
CTATTCTTCCACCTACTAGACCTACAGACAAGGCTCTCCGTCTTCCTCTTCAGGACGTATATAAAATCGG
TGGTATTGGAACAGTACCAGTTGGTCGTGTCGAAACTGGTGTGTTGAAACCAGGTATGGTTGTCACATTC
GCTCCTGCTGGTTTGACTACTGAAGTCAAATCTGTTGAAATGCATCACGAAGCTTTGCAAGAGGCTGTTC
CTGGTGATAATGTTGGTTTTAACGTCAAGAACGTATCTGTCAAAGAATTACGTCGTGGTTATGTTGCTGG
TGATTCAAAGAATAATCCACCTAAAGGTGCTGCTGATTTTACTGCACAAGTTATTGTATTGAATCACCCT
GGTCAAATCAGCAATGGTTATACACCAGTATTGGATTGTCATACTGCACATATTGCATGTAAATTCGCTG
ATATCAAAGAGAAATGCGATCGTCGTAATGGAAAGACAACTGAAGAAAATCCGAAAAGCATCAAATCTGG
AGATGCTGCCATCGTTATGCTTGTGCCAAGCAAGCCTATGTGCGCTGAGGCTTTCCAAGAATTTCCGCCT
TTGGGGCGTTTCGCTGTTCGTGACATGCGTCAAACGGTAGCTGTTGGTGTTATCAAAGCTGTAACTTTCA
AGGACGCTGCTGGCAAGGTCACCAAGGCTGCCGAGAAGGCTCAGAAAAAGAAATAA
```

FASTA ▾

elongation factor 1-alpha [Apis mellifera]

NCBI Reference Sequence: NP_001014993.1

GenPept Identical Proteins Graphics

```
>NP_001014993.1 elongation factor 1-alpha [Apis mellifera]
MGKEKIHINIVVIGHVDSGKSTTTGHLIYKCGGIDKRTIEKFEKEAQEMGKGSFKYAWVLDKLKAERERG
ITIDIALWKFETSKYYVTIIDAPGHRDFIKNMITGTSQADCAVLIVAAGTGEFEAGISKNGQTREHALLA
FTLGVKQLIVGVNKMDSTEPPYSETRFEEIKKEVSSYIKKIGYNPAAVAFVPISGWHGDNMLEVSSKMPW
FKGWTVERKEGKVEGKCLIEALDAILPPTRPTDKALRLPLQDVYKIGGIGTVPVGRVETGVLKPGMVVTF
APAGLTTEVKSVEMHHEALQEAVPGDNVGFNVKNVSVKELRRGYVAGDSKNNPPKGAADFTAQVIVLNHP
GQISNGYTPVLDCHTAHIACKFADIKEKCDRRNGKTTEENPKSIKSGDAAIVMLVPSKPMCAEAFQEFPP
LGRFAVRDMRQTVAVGVIKAVTFKDAAGKVTKAAEKAQKKK
```

Figure 3.4 *Apis melifera* ef1a DNA (top) and translated protein sequence (bottom), both in FASTA format.

GenBank with word queries (like ef1a), the database can be searched using the protein or DNA sequence that the researcher desires to use in the phylogenetic matrix (**Sidebar 3.3**) as a query. The algorithm that accomplishes such searches is called BLAST.

Phylogenomic Matrices: Once the sequence data (and morphological data) to be used in a phylogenetic analysis have either been collated from the database or experimentally generated, they need to be placed into a format that phylogenetic programs can utilize. Phylogenetic analysis using parsimony, distance or likelihood (including Bayesian approaches) uses matrices of characters that come from and describe the taxa in the analysis. The dimensions of the matrix should have the number of taxa in one dimension (usually the rows) and the number of characters in the other (usually the columns). All of the taxa in the matrix must have the same number of characters. Raw molecular sequences (DNA and protein) can vary in length because of insertions or deletions during the evolutionary process. In order for molecular sequences with different lengths to be used in phylogenetic analysis the difference of numbers of positions in the sequences need to be accommodated in some way. The sequences need to be aligned (which we describe in **Sidebar 3.4**) or handled in some other way such as optimization alignment or implied alignment. The alignment process

Sidebar 3.3. Database searching via pairwise alignments: BLAST, the Basic Local Alignment Search Tool

The Basic Local Alignment Search Tool (BLAST) was developed to allow for efficient and rapid retrieval of sequences similar to a query sequence. The BLAST program was developed to balance computational speed with sensitivity of the search. To access BLAST, go to http://blast.ncbi.nlm.nih.gov/Blast.cgi. Figure 3.5 shows the home page and shows the four basic kinds of searches BLAST will do. There are several variations of BLAST for nucleotide and amino acid sequences. When using BLAST there are two kinds of query sequences – DNA and protein. But for each kind of sequence there are different ways to perform a search. The first method we discuss involves searching for nucleic acid sequence matches using a DNA sequence query called a "blastn". This kind of search will always return nucleic acid sequence accessions that fit the criteria of the search. The simplest search that can be performed using protein sequences as a query is called the "blastp" search. The blastp search returns protein sequences from the protein database targeted by the BLAST search.

A third type of search involves querying a protein sequence against the nucleic acid database. This search is called a "tblastn" search (for translated blastn). This kind of search uses protein translation of the entire nucleotide database in all potential reading frames (see below). Since the DNA has two strands that are complementary to each other and there are three reading frames for each strand, there are $2 \times 3 = 6$ different potential protein translations of each nucleotide sequence that need to be queried.

The next type of search is called blastx, a search of a nucleotide sequence against a protein database. In this case, the nucleotide query sequence is converted into its six potential amino acid translations, and all six are queried against the database. The most computationally complex search is called the "tblastx" search and it is NOT shown in Figure 3.5. This search queries a nucleotide sequence against a nucleotide database, but with the comparisons being done on a protein translation of both the query and the database. Thus, there are five ways of BLASTing your way through the NCBI database. Within each of the five types of BLAST (blastp, blastn, blastx, tblastn and tblastx) variations of how the search is accomplished exist. These variations on a theme are tailored to focus on specific tasks involved in BLASTing and one can explore their utility in searching the database.

One of the most important things to consider is that BLAST searches can be applied to the entire database or can be restricted to specific subsets of the database. The highest-quality database to search is the Reference Sequence (RefSeq) database. This collection of sequences is compiled by the International Nucleotide Sequence Database Collaboration (INSDC) made up of the US database (NCBI), the European sequence database (EMBL) and the Japanese database (DDBJ). It is updated daily, is a taxonomically diverse database and is by far the best database to search for comparative biology projects.

When a BLAST search report is generated, it contains sequence accessions that match the criteria of the search called "hits". Each BLAST report can contain a large number of hits, arranged from best to worst. Occasionally, a search will return few or no hits, indicating that the query sequence is not in the database in other organisms. Each hit includes an accession number, the name of the accession, the quality of the hit (given as an E-value, discussed below), selected metrics on the hit, and the alignment of the "query" sequence with the hit sequence, called the "subject" sequence. The coordinates within the gene or genome of the organism of the hit region are given for both the query and the subject sequence. Individual BLAST searches can be limited to specific groups of organisms if there is a focus on a specific group of organisms. Below we show part of a blastp table summarizing a search of ef1a protein sequence focused on only non-insect invertebrate accessions. An example results table is given in Figure 3.6. The quality of the alignment is represented by the E-value column, which gives a statistic that is similar to the statistical concept of a *p*-value. The E-value determines how likely it is to get a match of a certain level of significance given a database of a certain size. An E-value of 0.01 indicates that there is a 1% chance of randomly finding a match of that significance in the database. The main point to remember is that a smaller E-value (closer to 0) is better.

Below the list of "hits" the website will show several alignments of the query (in this case *Apis melifera* ef1a) to the "subject" hits. We show an example of the alignments for *Apis melifera* ef1a to some of the subject hits in Figure 3.7.

for molecular sequences is identical to the establishment of primary homology with morphological data as described above. However, instead of relying on observational topological similarity, sequence alignment operates through algorithms that use optimality criteria. Each gene or protein slated for inclusion into the data matrix should be aligned separately for collation or concatenation into the matrix. During the alignment process positions where a deletion has occurred are marked with dashes (–) and are called "gaps". How gaps are treated in phylogenetic analysis is an interesting subject, but most studies treat them as "missing" data (in other words they are ambiguous with respect to the search criterion).

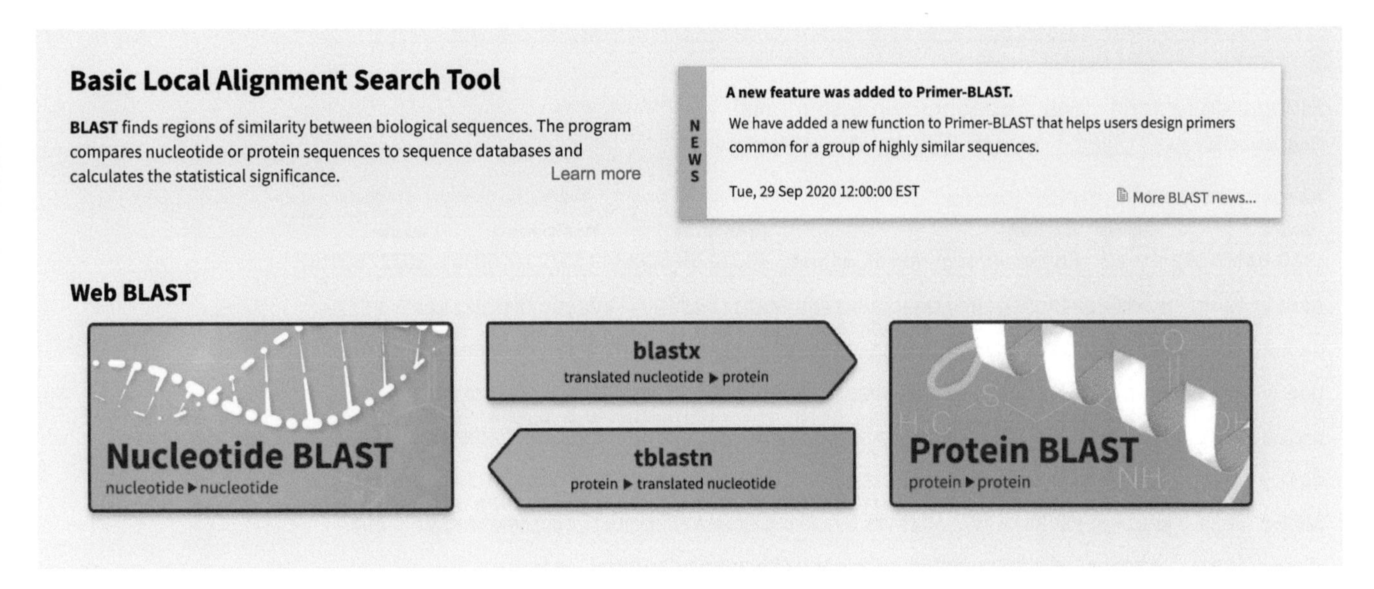

Figure 3.5 The home page for BLAST showing the various kinds of searches that can be accomplished.

Sequences producing significant alignments:

Select: All None Selected:0

Alignments Download GenPept Graphics Distance tree of results Multiple alignment

Description	Max Score	Total Score	Query Cover	E value	Per. Ident	Accession
LOW QUALITY PROTEIN: elongation factor 1-delta-like [Limulus polyphemus]	229	229	52%	6e-71	53.36%	XP_013775317.1
probable elongation factor 1-delta [Limulus polyphemus]	216	216	52%	6e-66	52.72%	XP_013779043.1
Elongation factor 1-delta [Lepeophtheirus salmonis]	210	210	51%	1e-62	50.00%	ACO12465.1
PREDICTED: elongation factor 1-delta-like isoform X3 [Crassostrea gigas]	203	203	53%	8e-61	50.21%	XP_019929767.1
probable elongation factor 1-delta [Parasteatoda tepidariorum]	205	205	41%	1e-60	59.34%	XP_015915345.1
PREDICTED: elongation factor 1-delta-like isoform X1 [Crassostrea gigas]	203	203	53%	1e-60	50.21%	XP_019929766.1
elongation factor 1 beta/delta chain, putative [Ixodes scapularis]	202	202	57%	2e-60	47.69%	XP_002400012.1
PREDICTED: elongation factor 1-delta-like isoform X4 [Crassostrea gigas]	201	201	53%	4e-60	51.28%	XP_011452062.1
PREDICTED: probable elongation factor 1-beta isoform X1 [Galendromus occidentalis]	201	201	57%	4e-60	44.83%	XP_003748381.1
PREDICTED: elongation factor 1-delta-like isoform X2 [Crassostrea gigas]	201	201	53%	7e-60	51.28%	XP_011452061.1
RecName: Full=Elongation factor 1-delta; Short=EF-1-delta	198	198	32%	3e-59	70.50%	P32192.2
elongation factor 1-beta [Parasteatoda tepidariorum]	196	196	29%	1e-58	73.60%	XP_015908446.1
Elongation factor 1-delta [Armadillidium vulgare]	196	231	68%	2e-58	51.30%	RXG52598.1
PREDICTED: elongation factor 1-delta-like [Hyalella azteca]	196	196	55%	4e-58	47.64%	XP_018020135.1
eukaryotic translation elongation factor 1 [Lycosa singoriensis]	195	195	28%	8e-58	75.61%	ABX75434.1
elongation factor 1-delta-like isoform X1 [Centruroides sculpturatus]	195	229	59%	9e-58	47.06%	XP_023235692.1
Elongation factor 1-beta [Stegodyphus mimosarum]	193	193	30%	1e-57	70.99%	KFM65912.1
putative Elongation factor 1-delta [Daphnia magna]	192	192	57%	9e-57	53.70%	KZS07147.1

Figure 3.6 **The hits list for a BLAST run of *Apis melifera* ef1a, restricted to non-insect invertebrates.** The name of the subject hit is in the first column, the next two columns give a score for the hit and the fourth column an estimate of the amount of subject sequence covered by the query. The fifth column gives the E-value (as discussed above) and the sixth column is the percent sequence identity. The final column gives the accession number of the subject hit and by clicking this button the sequence can be downloaded. There are methods described in the BLAST user manual that will allow one to download all of the hits simultaneously.

Download ∨ GenPept Graphics

Elongation factor 1-delta [Lepeophtheirus salmonis]

Sequence ID: ACO12465.1 **Length:** 334 **Number of Matches:** 1

Range 1: 122 to 334 GenPept Graphics ▼ Next Match ▲ Previous Matc

Score	Expect	Method	Identities	Positives	Gaps
210 bits(534)	1e-62	Compositional matrix adjust.	113/226(50%)	155/226(68%)	20/226(8%)

```
Query  208  EVAKARQHIKQSLQCMDDIAAVAGFATPNENKKDILDSS------VFQELKNTVEKLEER  261
            E+A+ARQHIK SL+ +D +AA+AG  +    K   L+S          +LKNTV  L+++
Sbjct  122  EIARARQHIKSSLEYVDGLAALAG-GSDVSGKIAKLESDNAKLHKAVADLKNTVLTLQDK  180

Query  262  VKALEIKIRTFVPADPIAVCPAKPQPASKPTQEKADDDEDVDLFGSDS-EGEDAEAAKLR  320
            VK LE              C +    A+ P  ++ ++D+DVDLFGS S E EDA+ A++R
Sbjct  181  VKILE------------KTCGSGKSVAAAPATKQVEEDDDVDLFGSSSDEEEDAQKARVR  228

Query  321  EERLAAYAAKKAKKPALIAKSNIILDVKPWDDETDMKAMEEEVRKIETDGLLWGASKLVP  380
            EERL AY  KK+KKP LIAK++++LDVKPWDDETDM A+ E  + I+ +GL+WGA KLVP
Sbjct  229  EERLKAYHEKKSKKPTLIAKTSVLLDVKPWDDETDMNAILENCKTIQKEGLVWGAHKLVP  288

Query  381  LAFGIHKLQISCVVEDDKVSVDWLTEQIQDIEDYVQSVDIAAFNKV  426
            + +GI KLQ+ CVVED+KVS+D L EQI + ED+VQSVD+AA +K+
Sbjct  289  IGYGIKKLQVMCVVEDEKVSIDELCEQIAEFEDFVQSVDVAAMSKI  334
```

Download ∨ GenPept Graphics

PREDICTED: elongation factor 1-delta-like isoform X3 [Crassostrea gigas]

Sequence ID: XP_019929767.1 **Length:** 269 **Number of Matches:** 1

Range 1: 39 to 269 GenPept Graphics ▼ Next Match ▲ Previous Matc

Score	Expect	Method	Identities	Positives	Gaps
203 bits(516)	8e-61	Compositional matrix adjust.	121/241(50%)	156/241(64%)	24/241(9%)

```
Query  200  SAGGSLANEVAKARQHIKQSLQ-----CMDDIAAVAGFATPNENKKDILDS---------  245
            S G ++ NE+A+ARQ I++ L      C+       G    N    + + S
Sbjct  39   SGGSTIKNEIAEARQQIQKVLNSHKGNCIQIFGGSGGAPADNSQVMNRVSSLEKENRDLK  98

Query  246  SVFQELKNTVEKLEERVKALEIKIRTFVPADPIAVCPAKPQPASKPTQEKADDDEDVDLF  305
             V +++K+ V+KLE RV  LE            +  A+      KP     DDD+D+DLF
Sbjct  99   KVVEDMKSLVQKLENRVTKLE---------GGSSAPAAQAPAPKKPAPADDDDDDDIDLF  149

Query  306  GSDSEGEDAEAAKLREERLAAYAAKKAKKPALIAKSNIILDVKPWDDETDMKAMEEEVRK  365
            GSD E  D EA K+R+ERLAAY AKK+KKPALIAKS+++LDVKPWDDETDMK ME+EVRK
Sbjct  150  GSDDE-VDEEAEKIRQERLAAYEAKKSKKPALIAKSSLLLDVKPWDDETDMKKMEQEVRK  208

Query  366  IETDGLLWGASKLVPLAFGIHKLQISCVVEDDKVSVDWLTEQIQDIEDYVQSVDIAAFNK  425
            I  DGLLWG +KLVP+ +GI KLQI+CV+EDDK+S D+L E+I  IED VQS+DIAAFNK
Sbjct  209  ITADGLLWGQAKLVPIGYGIKKLQINCVIEDDKISTDFLEEEITAIEDLVQSMDIAAFNK  268

Query  426  V  426
            +
Sbjct  269  I  269
```

Figure 3.7 Alignments of two of the top subject hits for *Apis melifera* ef1a. The accession number of the sequence is given after the name of the hit, with species name in parentheses. The alignment shows where the two sequence are identical (the amino acid letters in between the two sequences) and where there are differences (blank or +, where the latter indicates a change to an amino acid with similar physical properties).

If morphological characters are to be incorporated into the matrix, then data for the same taxa with sequence information should be scored for morphology. Such a matrix more than likely will be made up of presence/absence data for the morphological traits. The different gene, protein sequence submatrices and the morphological matrix (and any other kinds of characters scored for the taxa in the analysis) are then concatenated into a single matrix where the different gene/protein information is retained in the form of coordinates in the matrix. These coordinates allow a researcher to partition the matrix into its component genes/proteins for more refined analyses using the partitioned information. The

Sidebar 3.4. Sequence alignment

Here we describe the process of alignment of two sequences; the process for aligning multiple sequence alignment are similar but become very difficult computationally because of the large number of possible solutions to the alignment problem. Below we see two DNA sequences that we want to align

Sequence 1 ATCTGCATTCGATGCATCATGCATATGCATC

Sequence 2 ATCGATGTCGATGCATCATGCAATCTATGCATC

The alignment procedure uses an algorithm called Needleman-Wunsch algorithm or through dynamic programming. The process is beyond the scope of this chapter but setting up how it is done can be informative. In order to align the two sequences above we will need to insert some "gaps" into the sequences. Dynamic programming tells us where the gaps should go given costs for inserting a gap and for changing a base. Most alignment programs place the cost if inserting a gap as more than changing a base and the ratio of gap to change costs becomes an important part of the alignment procedure. Figure 3.8 shows two DNA sequence alignments for the two sequences shown above.

```
ATCTGCATTCGATGCATCATGCA---TATGCATC      Alignment 1
ATC-GATGTCGATGCATCATGCAATCTATGCATC

ATCTGCAT-TCGATGCATCATGCA---TATGCATC     Alignment 2
ATC-G-ATGTCGATGCATCATGCAATCTATGCATC
```

Figure 3.8 Two of the many possible alignments of the given sequences.

If we go through the comparisons of the two sequences in both alignments and count gaps (represented by dashes) and changes, we get the result that there are 4 gaps and 3 changes in Alignment 1 and 5 gaps and 0 changes for Alignment 2. If gaps cost 1 and changes cost 1 then Alignment 1 costs 7 units and Alignment 2 costs 5 units. Alignment 2 would be considered the better of the two alignments. However, if we change the gap cost to 10 and the change cost to 1 then Alignment 1 will cost 43 units and Alignment 2 will cost 50 units. In the latter case Alignment 1 would be considered the better of the two alignments. Dynamic programming uses the gap and change costs to settle on the optimal alignment.

We do not, of course, align sequences by hand as we have in the example above as there are dozens of programs and websites that can do alignments by computer. The following is a list of alignment programs that are commonly used by systematists: ClustalW, MUSCLE, MAFFT, T-Coffee. When using these programs always be aware of the input assumptions as the default settings may not be appropriate for your needs. All of these programs allow for alteration of default parameter values before running the alignment program. Also be careful when setting the parameters how the output format is set. Many of the alignment programs allow for direct output to the format that phylogenetic analysis programs use (see Table 3.2).

concatenated matrix is then placed into a format that the phylogenetic analysis program can recognize.

There are two widely used formats for phylogenetic analysis – NEXUS (.nex; Maddison, Swofford and Maddison, 1997) and PHYLIP (.phy; Retief, 2000). Most programs will allow one the use of both and to interchange both readily. NEXUS files start out with the "#NEXUS" statement to signify that the file is a NEXUS file. Next comes the "begin data;" statement followed by a line with the dimensions' statement where the number of characters (nchar) and the number of taxa (ntax) are given. A "format" line follows where the datatype, how gaps are treated, what designates missing data, how the data are arranged in the file and other important format information is listed. The last line before the actual matrix is the "matrix" line. All lines in this data block end with a semicolon (;)

TABLE 3.2 Commonly used alignment programs and links to download and webservers

Program	Download website	Online webserver
ClustalW	http://www.clustal.org/clustal2/	https://www.ebi.ac.uk/Tools/msa/clustalw2/
MUSCLE	https://www.drive5.com/muscle/downloads.htm	https://www.ebi.ac.uk/Tools/msa/muscle/
MAFFT	https://mafft.cbrc.jp/alignment/software/	https://mafft.cbrc.jp/alignment/server/
T-Coffee	http://www.tcoffee.org/Projects/tcoffee/	http://tcoffee.crg.cat/apps/tcoffee/all.html

except for the "matrix line". An example data block for a NEXUS formatted file is shown in **Figure 3.9**. The data are then given in matrix format and followed by a semicolon and then "end;". The data matrix can then be followed by any number of assumptions including how the data are partitioned, what weights to use, how to group taxa and other information that systematists use during the analysis phase. PHYLIP formatted matrices start with a single line with the dimensions of the matrix given with the number of taxa first followed by a space then the number of characters followed by a carriage return where the data matrix is then listed. Character partitions, weights and other information can also be appended to the PHYLIP matrix. See Table 2.1 in Chapter 2 for a list of phylogenetic analysis programs.

```
#NEXUS
Begin data;
Dimensions ntax=90 nchar=102481;
Format datatype=protein missing=? Gap=-;
Matrix
...
;
end;
```

Figure 3.9 The NEXUS format. "..." stands for the data matrix itself.

In Chapter 4 we include phylogenetic matrices that can be used by the student for direct analysis. The matrices are given in NEXUS format for consistency but can be transformed into PHYLIP format or used in PHYLIP format-based programs directly. We have attempted to include a matrix for each phylum and for several important phylogenetic questions (like the metazoan taxa that is sister to all other metazoan). These matrices should be executable in one of the more commonly used phylogenetic programs like TNT, PAUP, MmrBayes, phyloBayes, RaxML etc. User manuals for all of these programs are available and the student can use these manuals to get started. There are also a large number of online phylogenetic analysis programs (see Table 2.1).

CONSTRUCTING A PHYLOGENOMIC MATRIX WITH mtDNA GENOMES

Here we demonstrate how a phylogenomic matrix can be constructed from mitochondrial DNA genomes of invertebrate animals. To our knowledge very few pipelines exist that will do this automatically. One program that has the potential to do these operations is AGALMA from Dunn et al. (2013). Rather than run through this complicated pipeline we demonstrate here the individual steps using downloaded data and text editor programs. We use the mtDNA genome species list of Popova et al. (2016) and reconstruct a matrix from this study (**Table 3.3**).

The first step in constructing this matrix is to download the sequences using the accession numbers provided in Table 3.3. For this exercise we will construct a matrix with amino acid sequences, so we need to download the amino acid (protein) sequences for each species of each of the 13 mitochondrial proteins. The steps used to download the sequences are outlined in **Sidebar 3.5**.

Note that the FASTA sequence entries obtained by the approach in Sidebar 3.5 do not have the taxon name appended to the sequence. For this reason, prior to going further, it is recommended that the names be edited into the FASTA file's first line. If the reader is familiar with scripting or programming, then this step can be relatively simple. A cumbersome manual way to deal with the editing is through spreadsheeting. Finally, there are ortholog-finding programs that will find the orthologs of the 13 protein coding genes among the downloaded sequences and these can be used to group a large number of sequences into putative orthologous groups. For our example the annotation of the 70 mtDNA genomes is precise enough that we will use the GenBank annotations to organize the data by gene/protein.

So far, we have constructed the 13 individual protein FASTA files for the Popova et al. (2016) dataset in Table 3.3; these files are available on the textbook website. The next step is to align the raw amino acid sequences. We have discussed alignment in this chapter (mentioning the programs ClustalW, MUSCLE, MAFFT, T-Coffee), but point out here that there are many alignment programs that can be downloaded and that can be used online. In **Sidebar 3.6** we show the steps for alignment of sequences using MAFFT.

Sidebar 3.5. Downloading large molecular sequence datasets

Get online and go to the NIH NCBI search site (https://www.ncbi.nlm.nih.gov/). Next pull down the search data base to "Protein". Copy the numbers in the Accession Number column of Table 3.3 and paste this list into the NIH NCBI query box. The accession numbers should have a space or a tab between them. Press the "Search" button and if the search goes correctly you should see a page with the results of 70 mitochondrial proteomes (Figure 3.10).

The resulting accessions are in the same order as the input accession list with 70 accession numbers in it. Next go to the "Send To" pull-down just below the query box and drag down to "Coding Sequences" and click. This step should reveal a "Create File" button for the download to a file of all of the sequences in the query list. Click the "Create File" button and a file will download. This file is a text file and should be edited so that all of the say ND1 genes from the 70 taxa can be put into a new file. Make files for all 13 of the protein coding genes (NAD1, NAD2, NAD3, NAD4, NAD4L, NAD5, NAD6, ATP6, ATP8, COXI, COX2, COX3 and CYTB). The end product for this step should be 13 files with FASTA entries for each of the 70 taxa in the dataset. If a taxon does not have a gene sequence for a particular gene, don't worry. The sequence for this gene for this taxon will be replaced with Xs or? to signify the data are missing for that entry.

TABLE 3.3 The Popova et al. (2016) dataset

Species	Taxonomy 1	Taxonomy 2	Accession number
Tethya actinia	Porifera	Porifera	AY_320033
Sipunculus nudus	Sipunculida	Sipunculida	FJ_422961
Urechis caupo	Echiura	Echiura	NC_006379
Platynereis dumerilii	Annelida	Polychaeta	NC_000931
Phoronis architecta (syn. psammophila)	Phoronida	Phoronida	AY368231
Terebratalia transversa	Brachiopoda	Brachiopoda	NC_003086
Terebratulin a retusa	Brachiopoda	Brachiopoda	NC_000941
Laqueus rubellus	Brachiopoda	Brachiopoda	NC_002322
Lingula anatina	Brachiopoda	Brachiopoda	AB178773
Gyrodactylus derjavinoides	Platyhelminthes	Trematoda	NC_010976
Schistosoma mansoni	Platyhelminthes	Trematoda	NC_002545
Katharina tunicata	Mollusca	Polyplacophora	NC_001636
Biomphalaria glabrata	Mollusca	Gastropoda	NC_005439
Cepaea nemoralis	Mollusca	Gastropoda	NC_001816
Albinaria caerulea	Mollusca	Gastropoda	NC_001761
Nautilus macromphalus	Mollusca	Cephalopoda	NC_007980
Loligo bleekeri	Mollusca	Cephalopoda	NC_002507
Siphonodentalium lobatum	Mollusca	Scaphopoda	NC_005840
Graptacme eborea	Mollusca	Scaphopoda	NC_006162
Venerupis philippinarum	Mollusca	Bivalvia	NC_003354
Mytilus edulis	Mollusca	Bivalvia	NC_006161
Lampsilis ornata	Mollusca	Bivalvia	NC_005335
Inversidens japanensis	Mollusca	Bivalvia	AB055624
Loxocorone allax	Entoprocta	Entoprocta	NC_010431
Flustrellidra hispida	Bryozoa	Ectoprocta	NC_008192
Watersipora subtorquata	Bryozoa	Ectoprocta	NC_011820
Bugula neritina	Bryozoa	Ectoprocta	NC_010197
Paraspadella gotoi	Chaetognatha	Chaetognatha	NC_006083
Spadella cephaloptera	Chaetognatha	Chaetognatha	NC_006386
Sagitta enflata	Chaetognatha	Chaetognatha	NC_013814

(Continued)

TABLE 3.3 (Continued)

Species	Taxonomy 1	Taxonomy 2	Accession number
Sagitta nagae	Chaetognatha	Chaetognatha	NC_013810
Leptorhynchoides thecatus	Rotifera	Acanthocephala	NC_006892
Caenorhabditis elegans	Nematoda	Nematoda	NC_001328
Trichinella spiralis	Nematoda	Nematoda	NC_002681
Priapulus caudatus	Priapulida	Priapulida	NC_008557
Epiperipatus biolleyi	Onychophora	Onychophora	NC_009082
Limulus polyphemus	Arthropoda	Xiphosura	NC_003057
Steganacarus magnus	Arthropoda	Arachnida	NC_011574
Dermatophagoides pteronyssinus	Arthropoda	Arachnida	NC_012218
Leptotrombidium akamushi	Arthropoda	Arachnida	NC_007601
Nymphon gracile	Arthropoda	Pycnogonida	NC_008572
Narceus annularis	Arthropoda	Myriapoda	NC_003343
Ligia oceanica	Arthropoda	Crustacea	NC_008412
Argulus americanus	Arthropoda	Crustacea	NC_005935
Speleonectes tulumensis	Arthropoda	Crustacea	NC_005938
Tigriopus japonicus	Arthropoda	Crustacea	NC_003979
Vargula hilgendorfii	Arthropoda	Crustacea	NC_005306
Eriocheir sinensis	Arthropoda	Crustacea	NC_006992
Megabalanus volcano	Arthropoda	Crustacea	NC_006293
Cherax destructor	Arthropoda	Crustacea	NC_011243
Pagurus longicarpus	Arthropoda	Crustacea	NC_003058
Chinkia crosnieri	Arthropoda	Crustacea	NC_011013
Balanoglossus carnosus	Enteropneusta	Enteropneusta	NC_001887
Xenoturbella bocki	Xenoturbellida	Xenoturbellida	NC_008556
Florometra serratissima	Echinodermata	Crinoidea	NC_001878
Antedon mediterranea	Echinodermata	Crinoidea	NC_010692
Gymnocrinus richeri	Echinodermata	Crinoidea	NC_007689
Asterina pectinifera	Echinodermata	Asteroidea	NC_001627
Ophiura lukteni	Echinodermata	Ophiuroidea	NC_005930
Ophiopholis aculeata	Echinodermata	Ophiuroidea	NC_005334
Cucumaria miniata	Echinodermata	Holothuroidea	NC_005929
Strongylocentrotus purpuratus	Echinodermata	Echinoidea	NC_001453
Doliolum nationalis	Chordata	Tunicata	NC_006627
Phallusia fumigata	Chordata	Tunicata	NC_009834
Phallusia mammillata	Chordata	Tunicata	NC_009833
Ciona savignyi	Chordata	Tunicata	NC_004570
Ciona intestinalis	Chordata	Tunicata	NC_004447
Halocynthia roretzi	Chordata	Tunicata	NC_002177
Asymmetron inferum	Chordata	Cephalochordata	NC_009774
Homo sapiens	Chordata	Craniata	NC_012920

The four columns show, from left to right: the species name, two taxonomic schemes and the accession numbers.

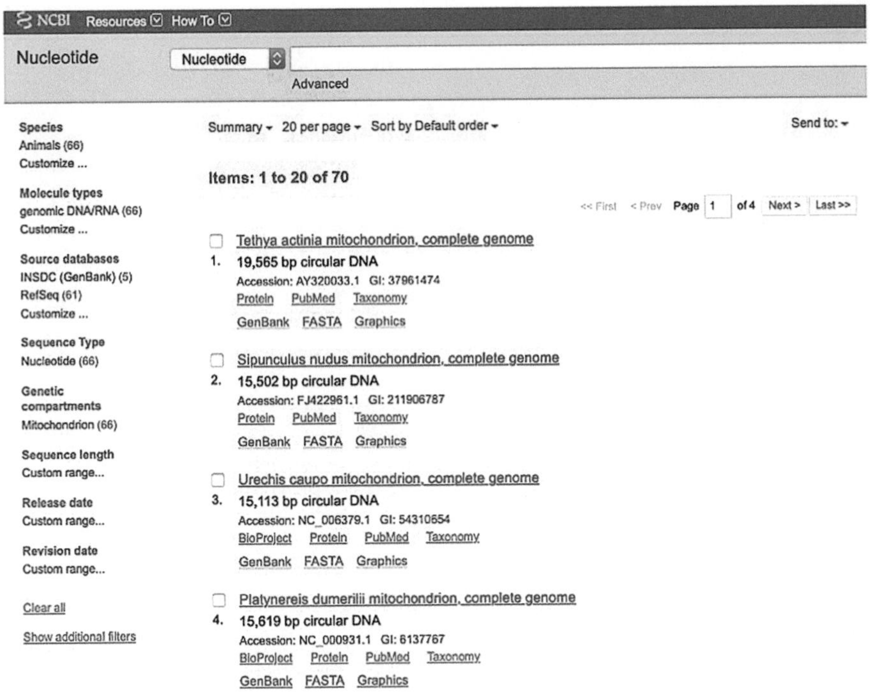

NCBI Resources How To
Nucleotide Nucleotide
Advanced

Species
Animals (66)
Customize ...
Molecule types
genomic DNA/RNA (66)
Customize ...
Source databases
INSDC (GenBank) (5)
RefSeq (61)
Customize ...
Sequence Type
Nucleotide (66)
Genetic compartments
Mitochondrion (66)
Sequence length
Custom range...
Release date
Custom range...
Revision date
Custom range...
Clear all
Show additional filters

Summary 20 per page Sort by Default order Send to:

Items: 1 to 20 of 70

<< First < Prev Page 1 of 4 Next > Last >>

1. Tethya actinia mitochondrion, complete genome
19,565 bp circular DNA
Accession: AY320033.1 GI: 37961474
Protein PubMed Taxonomy
GenBank FASTA Graphics

2. Sipunculus nudus mitochondrion, complete genome
15,502 bp circular DNA
Accession: FJ422961.1 GI: 211906787
Protein PubMed Taxonomy
GenBank FASTA Graphics

3. Urechis caupo mitochondrion, complete genome
15,113 bp circular DNA
Accession: NC_006379.1 GI: 54310654
BioProject Protein PubMed Taxonomy
GenBank FASTA Graphics

4. Platynereis dumerilii mitochondrion, complete genome
15,619 bp circular DNA
Accession: NC_000931.1 GI: 6137767
BioProject Protein PubMed Taxonomy
GenBank FASTA Graphics

Figure 3.10 Top four return accessions when searching with accession list from Popova et al. (2016).

The next step is to concatenate the 13 aligned genes for the 70 taxa into a single file and to edit the names of the taxa. Again, these steps can be done in a text editor. Alternatively, the "cat" command (**Sidebar 3.7**) can be used in the terminal to concatenate the 13 alignments.

Sidebar 3.6. Alignment of the mitogenomic data from Table 3.3

This step requires that you download and install MAFFT on your computer. Next, place the files you want to align in a folder or on your desktop. Open a terminal window and before opening the MAFFT program, change directory ("cd desktop" if the files you want to align are on the desktop or "cd directory" if they are in a different directory) so that MAFFT will be able to access the data files. Now simply type in "MAFFT" on the command line. If the program has been installed properly you should see the MAFFT shell appear in your terminal window (Figure 3.11).

```
---------------------------------------------------------------------

   MAFFT v7.312 (2017/Oct/16)

        Copyright (c) 2016 Kazutaka Katoh
        MBE 30:772-780 (2013), NAR 30:3059-3066 (2002)
        http://mafft.cbrc.jp/alignment/software/
---------------------------------------------------------------------


Input file? (fasta format)
@ 
```

Figure 3.11. MAFFT terminal shell. The program asks for an input file in FASTA format. Since we have 13 files (one each for the proteins in the mitochondrial genome) the analysis should be set to go.

The program asks for an input file; we start the 13 alignments with NAD6 (the file name is "IZ_NAD6.txt"). Type in the file name and hit return. The program will now ask you to give a name for the output file. We recommend that you simply append the word out to the initial file name to designate the output file. Type the designated name for the output file and hit return and you should see some options for format of the output file (Figure 3.12). Depending on which phylogenetic program you will use for the analysis you can choose one of the formats. We recommend outputting the file in PHYLIP format, as this format can be easily converted to other phylogenetic program formats. Since the input FASTA files for all of the 13 proteins have the same order of taxa in the file, you will want to force the output to be in the same order of taxa as the input. This step will save you a lot of editing after the alignments have been done.

The search strategy for the optimal alignment is set next and we recommend using the default setting (in this case a strategy called FFT-NS-2 which is a progressive alignment method with realignment). Once that option has been set you will see the following: "Additional arguments? (–ep # –op # –kappa # etc)", and you can either change the default parameters or proceed with the defaults by hitting return; we recommend at first to use the default parameters. The following line will appear: "command="/usr/local/bin/mafft"—localpair –maxiterate 16 –inputorder "iz_nd6.txt" > "iz_nd6.out" OK?". This line is just the command line for MAFFT with the parameters you have set so far. By hitting return you will start the alignment. When the run is complete the alignments will appear in the format you set in the terminal (Figure 3.13). The formatted output file should be in the directory where the input file sits.

To exit the alignment program type "q" into the command line and press "enter" or "return" and the terminal returns to the directory where your files are sitting. You should repeat these instructions for each of the 13 FASTA files you have. Note that in the PHYLIP format the name of the sequence is truncated to only 10 symbols; this problem can be fixed as we show below.

```
Input file? (fasta format)
@ iz_nd6.txt
OK. infile = iz_nd6.txt


Output file?
@ IZ_ND6.out
OK. outfile = IZ_ND6.out


Output format?
  1. Clustal format / Sorted
  2. Clustal format / Input order
  3. Fasta format   / Sorted
  4. Fasta format   / Input order
  5. Phylip format  / Sorted
  6. Phylip format  / Input order
@
```

Figure 3.12. MAFFT request to designate the output file format. If you have done a lot of work ordering your taxa in the alignment files then you want to choose/input order.

```
 70 392
AB055624.1 ---------- ---------- -----MAATV VVALVGMAFL TLLERKSLGY
AB178773.1 M-------FP WVFVIILL-- PIYLYMFSFQ VNLLVIIAFY SLSERAILSL
AY320033.1 ---------- ---MGIII-- ---IKVLIIL VSLLISIAYL TLAERKVLGY
AY368231.1 ---------- ----MTHT-- ---INYILLA VCILLAVAFF TLLERKVLGY
FJ422961.1 ---------- --MSVSFS-- ---LTLLITV LMAAVAMAFY TLLERKILGY
NC_000931. ---------- --MLLKSP-- ---ITILITF ICILLSMAFF TLLERKILGY
NC_000941. ---------- ---MPLFL-- ---TYSLLTI VPILVSVAFF TLMERKILSY
NC_001328. ---------- ---MILVL-- ---LMVILMM IFIVQSIAFI TLYERHLLGS
NC_001453. ---------- -MVYVFSI-- ---LELISFL IPILLSVAFL TLVERKVLGY
NC_001627. ---------- -MDWFVFF-- ---VNSILFI VPVLLAVALL TLVERKVLGY
NC_001636. ---------- ---MFLSF-- ---SWMVIAY ISILLAVAFF TLLERKGLGY
NC_001761. ---------- -----MVV-- ---FKSLLLN LCILLSVAFY TLLERKVLSS
NC_001816. ---------- ---------- ---------M LCVLLAVAYM TLLERKILSY
NC_001878. M--------- -NNNLIML-- ---INIVNII IPILIAVAFI VLIERKILGY
NC_001887. ---------- -MSWILII-- ---IHALILI VPVLLAVAFI TLGERKIIGY
NC_002177. MY-------- ---------- ---LFSLVFV LFLLLLVAFL VLLERKIFGL
NC_002322. ---------- -----MGV-- ---VYSVMVM VPILLAMAFF TLTERKILGY
NC_002507. ---------- --MMLVEL-- ---LSGVISC VCALLAVAFF TLLERKGLGY
NC_002545. ---------- --MLIYEL-- ---VVLIEGL IMILLLVSFY ILGERKILSY
NC_002681. ---------- ---MLLWI-- ---TNLLTII ITILLSIAFV TLMERQYIGI
NC_003057. ---------- --MLMSFI-- ---VCYVVIL ICVLVGVAFL TLLERSILGY
NC_003058. M--------- -FIFIMML-- ---LNYLMLI ICVLVGVAFV TLLERKILGY
NC_003086. ---------- ----MVSV-- ---TYTLFTL VPILLSMAFF TLLERKILGY
NC_003343. ---------- --MEWLLV-- ---MVLVLEF VCVLVGVAFF TLLERKILGY
NC_003354. ---------- ---------- ---MSVFLVS LVMLMSVAFF IVTERKGLGM
NC_003979. ---------- ---MSMDL-- ---VFSTYVL VLLLVNVAFV TLFERKILGC
NC_004447. ---------- ---MMLLI-- ---LIYLFFI LILLLMVAFL VLLERKVLGL
NC_004570. ---------- ---MILVI-- ---VFYFIFI LFLLLMVALL VLLERKVLGL
NC_005306. MNGTFCCINK LMEYVVCI-- ---VGYLMQW VGVLMGVAFF TLLERKILGY
NC_005334. M--YESVSGV YLAPVLII-- ---ISFLGLI VPVLLAVALL TLLERKVIGY
NC_005335. ---------- ---MIPHM-- ---ISTLITY LLILLGVAFF TLLERKALGY
NC_005439. ---------- -----MLW-- ---VTSIFTI VCVLIAVAFY TLYERKILGY
NC_005840. ---------- --MDIFGY-- ---IELLIMA VMVLLTVSFY TIVERKCLSY
NC_005929. ---------- -MSTALFT-- ---IQALIYI VPILLSVAFL TLVERKVLGY
NC_005930. MINFVAIINS YYTSILFV-- ---FNLILFI IPVLLAVALL TLLERKTLGY
NC_005935. ---------- -MSMMLFV-- ---FQYLVTM ILGLVGLAFV TLMERKVMSY
NC_005938. ---------- ----MIPF-- ---LLYIVLI IFVLVGVAFF TLLERSVLGY
NC_006083. MLS------- -FFFAESM-- ---FNLVLNI IIILIGVAFF TLMERKILSL
NC_006161. ---------- --MDWVAV-- ---IVSIIPF VGVLLAVGFY TLLERKILAI
NC_006162. ---------- -----MKL-- ---LNILVTF LMIMLSVAFF TLLERKGLGY
NC_006293. ---------- ---MVMLT-- ---IYYILLL ICVLISVAFL TLLERKVLGY
NC_006379. ---------- --MSISFS-- ---IAMILNM VGAMLAMAFF TLLERKTLGY
NC_006386. ---------- ------MM-- ---MNFVIQV VCVMLATALF TLLERKVLGY
NC_006627. MW-----YLI WGSFIVLLVQ SKTIWLLLVA LGLLLMVAFL VLIERNILGL
NC_006892. ME-------- -KTVLRLV-- ---LNSLIIV IIVLVLLSFF TLLERKVLGL
NC_006992. ---------- ---MIVEM-- ---VNFLVLM VCVLIGVAFV TLLERKILGY
NC_007601. ---------- ---------- ---MFGLLNI FPILLAVMFF TLFEVQLLGK
NC_007689. M--------- -NTPTTTI-- ---INTINFI IPVLIAVALI VLVERKILGY
NC_007980. ---------- ----MVSV-- ---LCVLVTC VCVLLGVAFF TLFERKGLGY
NC_008192. ---------- ---------- ---MKGAILS VVVALGVGFY TLLERKTLSY
NC_008412. ---------- -MLLAALT-- ---VKLALLV LCVLVSIAFI TLLERKILGY
NC_008556. MI-------- -TKYFSFT-- ---MLPLINL IPLLVGVAFL VLVERKLLGY
NC_008557. ---------- ------ML-- ---VNYIILT ICVLIAVAFF TLLERKVLGY
NC_008572. ---------- ----MDFI-- ---LVSVVLF VSVLLGVGFF TLFERKILGY
NC_009082. ---------- ----MITF-- ---IFYLVMI ISILISVAFL TLLERKILGY
NC_009774. ---------- --MWCINV-- ---IHLFLYF VPVLLAVAFL TLTERKVIGY
```

Figure 3.13 Example of an alignment of ND1 gene with MAFFT. The output is in PHYLIP format. By hitting return the terminal will scroll down by line and by hitting the space bar the alignment will scroll down a page.

Sidebar 3.7. Using the "cat" command on your terminal

Place all of the alignment output files in a folder and open your terminal. Change the directory (cd) to the folder with the output files in it. Next type cat "list of files" > "output file name" which for the output files for this exercise (also available on the textbook website) we have generated. The cat command that you should type into the command is – "cat IZ_ATP6.out IZ_ATP8.out IZ_COX1.out IZ_COX2.out IZ_COX3.out IZ_CYTB.out IZ_ND1.out IZ_ND2.out IZ_ND3.out IZ_ND4.out IZ_ND4L.out IZ_ND5.out IZ_ND6.out > IZ_cat.out". A file will appear in the folder with the output files with the name "IZ_cat.out" and it will be the file with all 13 IZ output files concatenated into a single file. This concatenated file can then be edited to a PAUP, TNT, MEGA, PHYLIP or other format. We will edit it here to the PAUP format. The number of characters concatenated into the final file are shown in Table 3.4.

TABLE 3.4 Number of taxa and characters in the individual mtDNA gene matrices

GENE	NTax	NChar
ATP6	70	349
ATP8	70	113
COX1	70	614
COX2	70	697
COX3	70	334
CYTB	70	465
ND1	70	392
ND2	70	549
ND3	70	202
ND4	70	568
ND4L	70	184
ND5	70	787
ND6	70	296
Total	70	5550

The final step is to format the file so that a phylogenetic analysis program can execute it and analyze it. We will use the NEXUS format for this example, but the reader should explore other formats for phylogenetic programs like the PHYLIP format which can be used in a variety of programs for phylogenetic analysis. PAUP NEXUS files have a series of required statements before the data matrix. The first is a line with "#NEXUS" on it to tell the program that the file is a NEXUX file. Next comes the data block introduced by a line with "begin data;" on it. The matrix dimensions come next and in this case they are ntax=70 and nchar=5550, so this line will have "dimensions nchar=5550 ntax=70;". The next line is the format line that tells the program what format the matrix is in. For this dataset the datatype is protein, the matrix is interleaved, gaps are represented by "- "and missing data are represented by "?", so the format line will look like "Format datatype=protein interleave gap=- missing=?;". Remember that each line so far needs a semicolon (;) at the end of it. The format line is then followed by a line with "matrix" on it, but note that there is no semicolon at the end of the line. The beginning block should look like **Figure 3.14**.

The matrix comes next and after the matrix are two lines, the first a semicolon and the second "end;". The Popova et al. dataset is now ready for a phylogenetic analysis and is available as "IZ_POPOVA.txt" on the textbook website. This file can then be converted to any of the phylogenetic analysis program formats. Once the matrix formatting is complete the file can be executed and analyzed as we discuss in Chapters 2 and 4.

```
#NEXUS
Begin data;
Dimensions ntax=70 nchar=5550;
Format datatype=protein interleave gap=- missing=?;
Matrix
```

Figure 3.14 Beginning NEXUS block of the mtDNA phylogenomic matrix discussed in the text.

CONCLUSION

This chapter has given a brief and condensed summary of the molecular approaches that are used in invertebrate zoology. Most early studies in invertebrate zoology used single genes like the structural RNA 18S ribosomal RNA gene or a protein like elongation factor 1 alpha (ef1a) for characters. If the student wants to construct a matrix with just a single gene in it, then all of the tools to do this are summarized in this chapter. Presently though, most invertebrate studies are done with phylogenomic approaches and use a large number of gene or protein sequences. We have also given the student rudimentary tools to construct these larger phylogenomic matrices. Further, we have included a large number of phylogenetic matrices for the student to use without having to fall back on constructing their own matrices. We discuss the use of phylogenetic programs in the next chapter.

REFERENCES

Adams, Mark D., Susan E. Celniker, Robert A. Holt, Cheryl A. Evans, Jeannine D. Gocayne, Peter G. Amanatides, Steven E. Scherer et al. "The genome sequence of Drosophila melanogaster." Science 287, no. 5461 (2000): 2185–2195.

Arabidopsis Genome Initiative, 2000. "Analysis of the genome sequence of the flowering plant Arabidopsis thaliana." Nature 408, no. 6814 (2000): 796.

Davey, John W., and Mark L. Blaxter. "RADSeq: next-generation population genetics." Briefings in Functional Genomics 9, no. 5–6 (2010): 416–423.

Dunn, Casey W., Mark Howison, and Felipe Zapata. "Agalma: an automated phylogenomics workflow." BMC Bioinformatics 14, no. 1 (2013): 330.

Fleischmann, Robert D., Mark D. Adams, Owen White, Rebecca A. Clayton, Ewen F. Kirkness, Anthony R. Kerlavage, Carol J. Bult, Jean-Francois Tomb, Brian A. Dougherty, and Joseph M. Merrick. "Whole-genome random sequencing and assembly of Haemophilus influenzae Rd." Science 269, no. 5223 (1995): 496–512.

Heather, James M., and Benjamin Chain. "The sequence of sequencers: The history of sequencing DNA." Genomics 107, no. 1 (2016): 1–8.

Kim, Jin M., Swathi Vanguri, Jef D. Boeke, Abram Gabriel, and Daniel F. Voytas. "Transposable elements and genome organization: a comprehensive survey of retrotransposons revealed by the complete Saccharomyces cerevisiae genome sequence." Genome Research 8, no. 5 (1998): 464–478.

Kircher, Martin, and Janet Kelso. "High-throughput DNA sequencing–concepts and limitations." Bioessays 32, no. 6 (2010): 524–536.

Maddison, David R., David L. Swofford, and Wayne P. Maddison. "NEXUS: an extensible file format for systematic information." Systematic Biology 46, no. 4 (1997): 590–621.

Mardis, Elaine R. "Next-generation DNA sequencing methods." Annu. Rev. Genomics Hum. Genet. 9 (2008): 387–402.

McGinn, Steven, David Bauer, Thomas Brefort, Liqin Dong, Afaf El-Sagheer, Abdou Elsharawy, Geraint Evans et al. "New technologies for DNA analysis – a review of the READNA Project." New Biotechnology 33, no. 3 (2016): 311–330.

Mikheyev, Alexander S., and Mandy MY Tin. "A first look at the Oxford Nanopore MinION sequencer." Molecular Ecology Resources 14, no. 6 (2014): 1097–1102.

Nagarajan, Niranjan, and Mihai Pop. "Sequence assembly demystified." Nature Reviews Genetics 14, no. 3 (2013): 157–167.

Popova, Olga V., Kirill V. Mikhailov, Mikhail A. Nikitin, Maria D. Logacheva, Aleksey A. Penin, Maria S. Muntyan, Olga S. Kedrova, Nikolai B. Petrov, Yuri V. Panchin, and Vladimir V. Aleoshin. "Mitochondrial genomes of Kinorhyncha: trnM duplication and new gene orders within animals." PloS One 11, no. 10 (2016).

Retief, Jacques D. "Phylogenetic analysis using PHYLIP." In Bioinformatics Methods and Protocols, pp. 243–258. Totowa, NJ: Humana Press, 2000.

Rhoads, Anthony, and Kin Fai Au. "PacBio sequencing and its applications." Genomics, Proteomics & Bioinformatics 13, no. 5 (2015): 278–289.

Shendure, Jay, Shankar Balasubramanian, George M. Church, Walter Gilbert, Jane Rogers, Jeffery A. Schloss, and Robert H. Waterston. "DNA sequencing at 40: past, present and future." Nature 550, no. 7676 (2017): 345–353.

Stark, Rory, Marta Grzelak, and James Hadfield. "RNA sequencing: the teenage years." Nature Reviews Genetics 20, no. 11 (2019): 631–656.

The *C. elegans* Sequencing Consortium, 1998. "Genome sequence of the nematode *C. elegans*: a platform for investigating biology". Science. 282 (5396): 2012–8. Bibcode: 1998Sci...282.2012. doi:10.1126/science.282.5396.2012. PMID 9851916.

Zhang, Yuanmeng, Jason Williams, and Andrea Lucky. "Understanding UCEs: A comprehensive primer on using Ultraconserved Elements for arthropod phylogenomics." (2019).

CHAPTER

4

MODERN INVERTEBRATE SYSTEMATICS

THE PHYLOGENETICS OF EARLY METAZOA

Johannes S. Neumann, Michael Tessler, Rob DeSalle, and Bernd Schierwater

CONTENTS

HYPOTHESES ON THE EARLIEST DIVERGING ANIMAL LINEAGE: WHO IS THE SISTER TO ALL OTHER METAZOANS?

When thinking about the origin of animals (Metazoa), many questions arise. Most fundamentally: What were the first animal lineages to evolve? Many of the early invertebrate lineages are soft-bodied, with unreliable to non-existent fossil records. So, to help answer this question, most researchers review animals that are alive today. Formally speaking, this scientific inquiry is usually focused on finding the sister to all other Metazoa (SOM), determining which lineage branched off first from the rest of the animal evolutionary tree. This remains a field of highly active research, with diverse and contentious results being produced constantly. The data we have to answer this question using extant animals are still limited. Anatomical and genetic clues are often obscured due to the hundreds of millions of years of evolution since the first animal lineages originated. Despite great difficulty in untangling the relationships among the earliest diverging animals, the implications for animal evolution cannot be overstated. Answering the SOM question informs the evolution of fundamental features, such as body symmetry, nervous systems and muscles, and allows us to search for "natural classification" of the animal kingdom.

In early treatises, biologists studying the SOM question were mainly limited to anatomical and morphological characters of the relevant animals, which works well for taxonomy (i.e. formally classifying organisms into groups) but often may not provide enough information about the evolutionary history. With the advent of molecular characters, which are now generated at constantly increasing speed, producing phylogenetic or evolutionary trees has become daily routine. However, many traditionally trained biologists argue that comparatively small numbers of morphological characters can hold strong evolutionary signal of a different quality. This regularly sparks debate when both data types give different clues. The SOM is just one example. Here, the question focuses on only five well-separated lineages: Bilateria (animals with bilateral symmetry as a result of two body axes: dorso-ventral plus left-right), Cnidaria (corals, anemones and jellyfish, radial symmetric with unique cnidocyte cells for feeding and defense), Ctenophora ("comb jellies", radial symmetric with unique colloblast cells for feeding), Placozoa (non-symmetric "plate animals" with the simplest bauplan) and Porifera (sponges; non-symmetric with lots of pores).

One of the earliest and most famous trees was drawn by Haeckel in 1874 and is shown in **Figure 4.1** (Haeckel, 1874). The first branch in the invertebrate section of his tree (i.e. the SOM in modern shorthand) unites sponges and cnidarians as Zoophyta, translated as "plant animals", thus placing the non-bilaterian groups together. Ctenophores were included within cnidarians, and placozoans were not in there as they were not discovered until a few years later, in 1883. Upon their discovery, placozoans were immediately hypothesized to represent the earliest level of animal evolution (i.e. the SOM) because they were so simple, so tiny, so different and so ideal as a model from which to derive all the other, more complex bauplans (Schulze, 1883; Bütschli, 1884).

When studying the SOM question, exceedingly few morphological features are phylogenetically informative. Datasets used for morphological phylogenetics of the groups have accordingly been fairly small, varying from just over ten to a few hundred phylogenetically informative characters (Neumann et al., 2020). This is in part because the relevant lineages are fairly simple and diverged many millions of years ago.

Cnidarians have a very simple nervous system and musculature, and a single opening that functions as both mouth and anus. Ctenophores have a comparably simple – but structurally very different – nervous system and muscle cells. Both

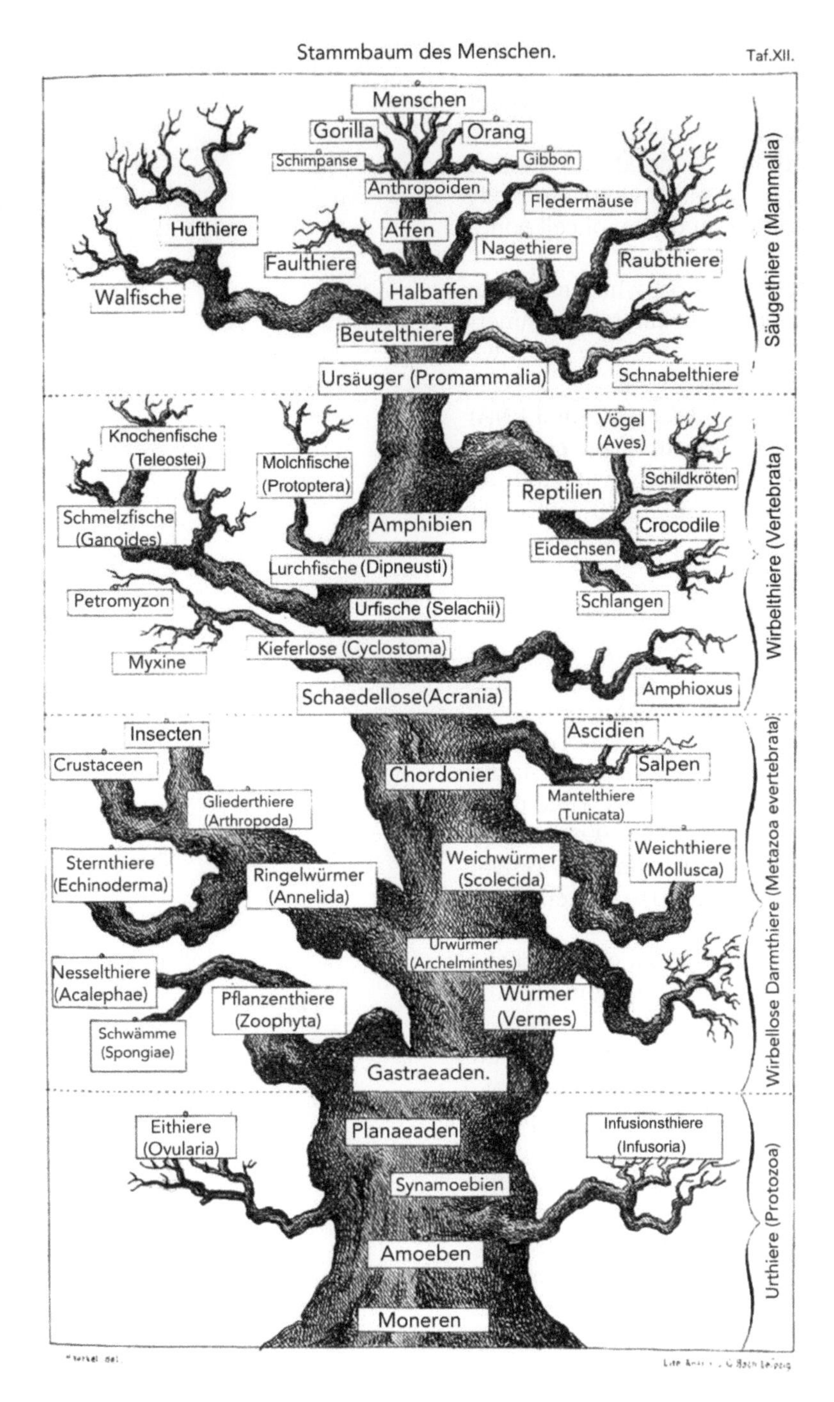

Figure 4.1 Ernst Haeckel's 1874 "Stammbaum der Tiere", which is German for "stem [i.e. family] tree of the animals". This image represents the relationships within the animal kingdom as branching patterns on a tree, with a thick stem leading from protists directly up to humans in the top center. While this is the most famous of Haeckel's trees, he did not draw a central stem and he also placed humans on side branches in earlier and later trees, which is more congruent with modern evolutionary thinking.

Cnidaria and Ctenophora show radial symmetry and a hollow gastral cavity, which is one reason why morphologists have often grouped them together as "Coelenterata". However, ctenophores are structurally more complex and can be said to sometimes have two radial symmetries (at times called "biradial"): one circle of eight hair combs, and another one that organizes two tentacles and two anal pores (Ryan and Baxevanis, 2007).

Sponges are much more simply built and they are not radially symmetric. They do not have a nervous system, muscles or a digestive system. Sponges also lack a mouth, an anus or mouth-anus; instead they bear pores that allow water to flow through chambers lined with filtering cells called choanocytes.

Placozoans are *extremely* simple. They not only lack a mouth, an anus, a mouth-anus and pores, but they also lack muscles, nerves, symmetry and body cavity; they do not even have the basal membrane, the foundation of cell adhesion in all other animal phyla.

The most complex lineage in our SOM approach is Bilateria, which show the two axes of symmetry allowing the complex architecture of bilaterally symmetric bauplans with ventral-dorsal and left-right body sides. It is the only group we discuss that has a mouth and an anus connected by a through-gut.

In addition to morphological comparisons, molecular data (DNA and RNA) have been applied extensively to help resolve the SOM question, since molecular phylogenetics has been particularly useful for comparing groups that do not share many morphological features (e.g. Foox and Siddall, 2015; Tessler et al., 2018).

SISTER OF OTHER METAZOA STUDIES ACROSS THE LAST THREE DECADES

In the 1990s, phylogenetic analyses mostly relied on single genes, especially the gene for the small subunit 18S ribosomal rRNA. Some of these studies placed Ctenophora on the same branch as Porifera, which was sometimes found to be paraphyletic (e.g. Winnepenninckx, Van de Peer, and Backeljau, 1998). Others inferred Placozoa as the SOM (Backeljau, Winnepenninckx, and De Bruyn, 1993), or joined all "lower" animals together, as the sister group to Bilateria (e.g. Peer and De Wachter, 1997).

Toward the end of the 2000s, the number of analyzed genes grew quickly, and alternative hypotheses began to seriously contend with the Porifera hypothesis, which was originally built on morphological similarities to the closest relatives of animals (**Sidebar 4.1**). In 2006 and 2007, two studies each again inferred Placozoa (Signorovitch, Buss, and Dellaporta, 2007; Dellaporta et al., 2006) or Bilateria (Erpenbeck et al., 2007; Signorovitch, Buss, and Dellaporta, 2007) as the SOM.

Phylogenomic studies now incorporate many hundreds or many thousands of loci. The first phylogenomic study and subsequent studies of Metazoa did not yet include data for all of the organisms that are important for the SOM question, such as Placozoa (Dunn et al. 2008; Telford 2008) or, in other cases, Ctenophora (Gissi, Iannelli, and Pesole, 2008; Srivastava et al., 2008).

A number of phylogenomic studies have found ctenophores as the SOM (first Dunn et al., 2008), sparking fierce scientific debate on potential tree-building errors, such as long-branch attraction and modeling issues. That is to say, ctenophores may have pushed into the SOM position in some analyses simply because they are genomically distant or derived compared to other animals, but not necessarily because they were the earliest lineage to split off from other animals. This debate is still prominent in the literature (Telford, 2008; Miller and Ball, 2008; DeSalle and Schierwater, 2008, Whelan et al., 2015, 2017, Fernández and Gabaldón, 2020). Most other recent phylogenomic studies have retrieved Porifera as the SOM (e.g., Laumer et al., 2018, 2019, Simion et al., 2017, Pett et al., 2019).

One morphologist recently summarized the implications of both the Porifera and the Ctenophora as the SOM hypotheses (Nielsen, 2019). Apart from citing cells with collar complex and intracellular digestion as morphological characters uniting Choanoflagellata and Porifera, Nielsen compared the number of shared derived characters (synapomorphies) supporting each hypothesis.

Sidebar 4.1. The closest relative of animals provides clues about the sister to all other metazoa

In phylogenetic analysis, it is always important to expand our perspective to the outgroup of the organisms in question. What is the closest relative of animals? Phylogenomics seems to confirm the 19th-century hypothesis that choanoflagellates are our closest non-animal relatives (Carr et al., 2008; Ruiz-Trillo et al., 2008). These protists have a flagellum surrounded by a collar of microvilli that they use for propulsion and to filter water for nutritious particles. Some species form colonies where one cell type is specialized for reproduction, while a second cell type feeds the colony. (This feature has evolved multiple times in different groups of "unicellular" protists.) A recently discovered colonial choanoflagellate species shows impressive collective behavior, contracting in unison to direct their flagella inward to feed or outward to swim (Brunet et al., 2019).

To evolve from a colonial protist to a truly "multicellular" organism requires somatic cell differentiation, which means to have more than one type of body cells. This happened three times independently and produced the multicellular kingdoms of plants, fungi, and animals. Perhaps the most common hypothesis for the evolution of the first animals is that a lineage of colonial protists that is sister to extant choanoflagellates developed additional somatic cell types with specialized functions. This hypothesis has often been linked to the similarly collared flagellated cells that sponges use to filter-feed, which are thus called choanocytes (Brunet and King, 2017).

Sponges have three types of cells that exhibit a collar around a flagellum – choanocytes, monoflagellated larval cells and sperm cells. Recent ultrastructure analyses show that there are considerable structural differences between the three cell types, and that the structure of these collared cells is mostly dependent on their function (Gonobobleva and Maldonado, 2009; Vasconcellos et al., 2019). They further suggest that the structure of sponge choanocytes may not be homologous to the structures in choanoflagellates. If truly non-homologous, this would be a dangerous character to use as the foundation of a hypothesis relating to the emergence of the animal kingdom. Still, other ultrastructure studies have indeed found compelling similarities (see Brunet and King, 2017 for a review). Accordingly, several authors continue to see this as a strong way to link Porifera to the closest relatives of animals, and thus as support for Porifera as the earliest-diverging animal phylum (Nielsen, 2019).

If ctenophores branched off first, Nielsen concludes, there are two options: (1) Porifera lost multiple morphological features that Ctenophora share with Placozoa and Cnidaria, or (2) these features evolved independently (i.e. they are homoplasies). In both cases, the resulting group comprised of Porifera, Cnidaria and Placozoa would show no synapomorphies at all, whereas all groups on the Porifera-first tree show synapomorphies. More specifically, inferring Ctenophora as the SOM would mean that fundamental features as the ectoderm, endoderm, extracellular digestion and critical eumetazoan genes were either secondarily lost or have evolved twice (Nielsen, 2019). The reader should be aware that phylogenomics is very different from traditional invertebrate zoology. The former relies on large numbers of genetic characters while the latter focuses a great deal of time and effort into examining physical structures.

Ctenophora are so complex that morphological discussions around their proposed SOM position usually revolve around how it may be possible that this complexity exists on two isolated branches of the animal kingdom (Nielsen, 2019). Scientists have looked for hints for secondary reduction in Placozoa and Porifera, but many have struggled to explain the loss of traits that most consider to be highly advantageous, such as a nervous system, muscles and a basal membrane for free-living animals (Ender and Schierwater, 2003; Raible and Steinmetz, 2010). Other researchers have instead hypothesized that the genetic toolkit to evolve these complex features was already present in the last common ancestor of all animals, which could have allowed their parallel evolution (see for example Moroz, 2012). Either way, the morphological data is at odds with the inferences from many phylogenomic reconstructions (Neumann et al., 2020).

An alternative hypothesis to sponges as the SOM first emerged based on morphological data and is called the plakula hypothesis (Bütschli, 1884). This hypothesis proposes that the animal kingdom began as a colony of single-celled organisms that received the initial stimulus for differentiation from gravity: cells at the bottom of the colony would have been in contact with the hard substrate, while the cells on top would have faced a different environment. According to this hypothesis, these separate stimuli would have led to the evolution of the first two somatic cell types and subsequently to the formation of an upper and a lower epithelium, as seen today in Placozoa. This hypothesis and the facts that placozoans are the most simple free-living extant animals and have the smallest nuclear (and the largest mitochondrial) genomes of all animals, have been brought forward as support for Placozoa as the SOM (Syed and Schierwater, 2002; Dellaporta et al., 2006).

Phylogenetics has resolved many evolutionary questions. However, in many cases it has also led to the publication of numerous contrasting evolutionary hypotheses, such as the different SOM inferences reviewed here. It has been proposed that the multitude of alternative phylogenetic methodologies available has created the shortest-lived hypotheses in biology (Schierwater et al., 2009a, 2016).

PHYLOGENETIC ANALYSIS OF THE BASE OF THE ANIMAL ToL – MORPHOLOGY

In this and the following section of this chapter, we will discuss six morphological matrices and five molecular datasets that can address the order of the five major taxa at the base of the animal tree of life. **Table 4.1** shows the morphological studies and **Table 4.2** shows the molecular studies that we include here. We have reduced the matrices to include only a single representative of each of the five major taxa (Bilateria, Ctenophora, Porifera, Cnidaria and Placozoa) as well as an outgroup taxon. We also include the full matrices from each of these six studies. These matrices are available on the textbook website. For this exercise, we will generate trees for all six reduced data matrices.

Analyzing small taxon matrices with morphology: You can easily access and execute each of the six morphological matrices with six taxa (Table 4.1) and perform a phylogenetic analysis with them (**Figure 4.2**) as well as a bootstrap analysis for each of the morphological datasets (**Figure 4.3**). Read **Sidebar 4.2** for instructions on how to run these analyses.

The results of these very simple analyses are summarized in **Figure 4.4** and indicate that four of the morphological matrices (BAK6, EER6, GLE6 and ZRZ6) give Porifera as the sister to all other metazoans at greater than 90% bootstrap support. Two matrices (SCH6, and BRU6) give Placozoa as the sister to all other metazoans. Note that all characters are equally weighted and unordered to obtain these results.

TABLE 4.1 The six morphological matrices discussed in this chapter

Study	#Taxa	All Ch Full	PI Ch Full	PI Ch 6 taxa
Sch	9	17	13	13
Eer	40	138	130	21
Gle	58	94	94	14
Zrz	56	276	252	20
Bru	35	96	50	11
Bak	38	78	62	9

Abbreviated names for these datasets are as follows: Sch = Schierwater et al., 2009b; Eer = Peterson and Eernisse, 2001; Gle = Glenner et al., 2004; Zrz = Zrzavý et al., 1998; Bru = Brusca and Brusca, 2003; Bak = Backeljau et al., 1993. PI stands for phylogenetically informative, Ch for characters. Table adapted from Neumann et al., 2020..

Sidebar 4.2. Using the morphological matrices in PAUP

Find the six matrices in Table 4.1 (BAK6.nex, BRU6.nex, EER6.nex, GLE6.nex, SCH6.nex and ZRZ6.nex) from the textbook website and place them on your desktop along with the PAUP program (Swofford et al., 2001). Open PAUP and change directory (cd) to "desktop". Next type "exe SCH6.nex" into the command line, then hit return. These steps should execute the six taxa file with the Schierwater et al. (2009b) dataset. The 17 characters used for this dataset are listed at the end of the file. The brackets ([]) indicate to PAUP that what is between them are the names, and not data to compute. You can do a tree search using an exhaustive search (command = alltrees), a branch and bound search (command = bandb) and a heuristic search (command = hsearch). All three searches will give the same result, so we will look at the result for a simple hsearch (Figure 4.2).

The search finds three trees. We can examine all three trees or we can construct what is called a strict consensus tree, which shows those nodes that are present in all of the shortest trees. Another option would be to construct a majority rule consensus tree which is a consensus tree where more than 50% of the trees show a specific node (any node that is present in all of the trees will be shown too). To get the strict consensus tree type "contree" into the command line. Since the outgroup (OG_Choanoflagellate) is listed first in the matrix and the default root for the analysis is the first taxon listed in the matrix, OG_Choanoflagellate is automatically used as the outgroup. If your outgroup is not listed first in the matrix, you can simply put a command into the command line telling the program which taxon is the outgroup.

If you are curious about the robustness of this inference you can run a bootstrap analysis; this is shown in Figure 4.3. Simply type "boot" into the command line and PAUP will generate bootstrap matrices, analyze them and report the bootstrap values. Note that Placozoa is placed as the most basal metazoan with 100% bootstrap values. In addition, sponges (Porifera) are the sister to the rest at 100% bootstrap, and Bilateria, Cnidaria, and Ctenophora are an unresolved derived group. Figure 4.4 shows a summary of these bootstrap analyses for all six morphological datasets.

```
Heuristic search completed
   Total number of rearrangements tried = 102
   Score of best tree (s) found = 23
   Number of trees retained = 3
   Time used = 0.00 sec (CPU time = 0.00 sec)

[paup> contree

Strict consensus of 3 trees:
```

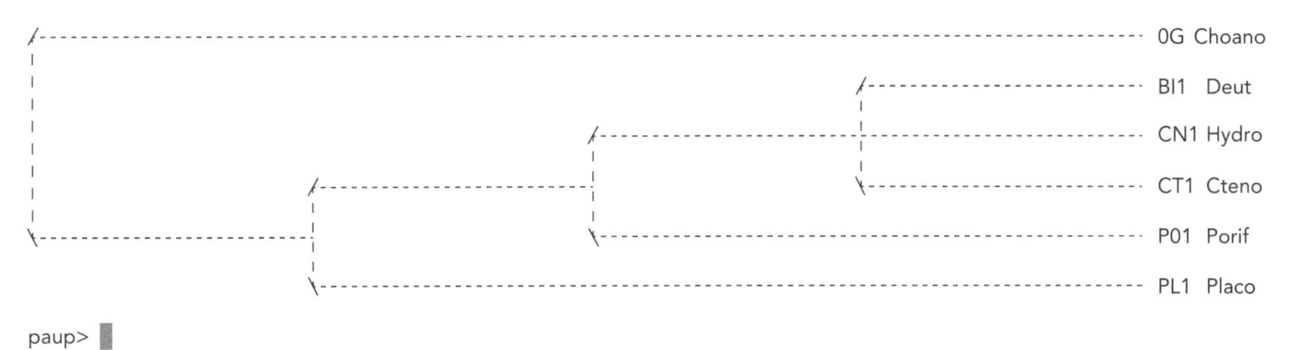

Figure 4.2 Morphological tree: Strict consensus tree of the phylogenetic analysis of the Schierwater et al. (2009b) morphological dataset. Three trees were found and a strict consensus tree was constructed. Some of the tree statistics, such as the tree length (23) and the number of trees found are shown in the upper left of the terminal. PL1 = Placozoa, PO1 = Porifera, CT1 = Ctenophora, CN1 = Cnidaria, BI1 = Bilateria and OG = outgroup (Choanoflagellate).

Analyzing morphological matrices with fuller taxonomic representation: Most of the studies listed in Table 4.1 have relatively large samples of metazoan phyla. Analysis of datasets with more taxa can oftentimes be more informative or precise as the addition of more taxa tends to break up long branches, which are problematic in phylogenetic analysis (Bergsten, 2005; Nosenko et al., 2013; Laumer et al., 2019). **Sidebar 4.3** summarizes how these analyses with fuller taxonomic representation are accomplished.

100 bootstrap replicates completed
Time used = 0.01 sec (CPU time = 0.01 sec)

Bootstrap 50% majority–rule consensus tree

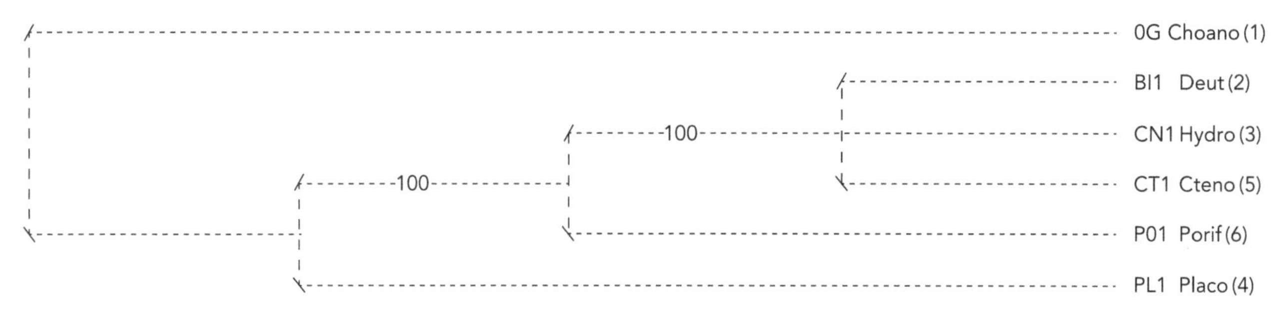

Bipartitions found in one or more trees and frequency of occurrence (bootstrap support values) :

```
123456          Freq
------------------------
.**.*.          99.50
.**.**          99.50
.*..*.          33.83
.**...          23.33
```

1 group at (relative) frequency less than 5% not shown

paup>

Figure 4.3 Morphological tree with bootstrap analysis of the Schierwater et al., (2009b) morphological dataset. PL1 = Placozoa, PO1 = Porifera, CT1 = Ctenophora, CN1 = Cnidaria, BI1 = Bilateria and OG = outgroup (Choanoflagellate). The bottom left hand shows the frequency of different hypothetical groups of metazoan.

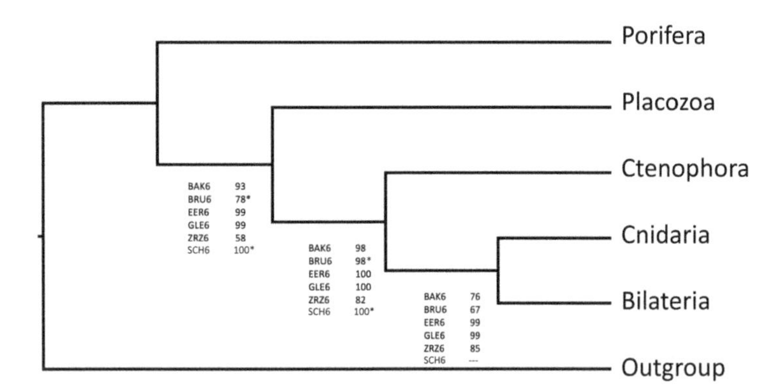

Figure 4.4 Morphological tree showing the results of analyzing the six taxa metazoan (and outgroup) morphological datasets (Table 4.1). Four of the morphological analyses give Porifera as the SOM (BAK6, EER6, GLE6 and ZRZ6) while two of the analyses give Placozoa as the SOM (BRU6 and SCH6). The numbers below the branches indicate the bootstrap support each matrix has for the monophyly of the branch. The asterisks next to BRU6 and SCH6 indicate the bootstrap support for Placozoa as the SOM, flipped with Porifera in the second branching position. The dashes in the figure indicate less than 50% bootstrap support for the monophyly of that branch.

PHYLOGENETIC ANALYSIS OF THE BASE OF THE ANIMAL ToL – MOLECULES

One of the most important aspects of phylogenetic analysis of molecular data for any difficult phylogenetic problem is how the data matrix is constructed. We discussed this problem in Chapters 2 and 3, and specifically the establishment of orthology and alignment issues are important. Examination of three recent

Sidebar 4.3. Analysis of morphological matrices with fuller taxonomic representation

We include six matrices with fuller taxonomic representation on the textbook website (BAK.nex, BRU.nex, EER.nex, GLE.nex, SCH.nex and ZRZ.nex). We will analyze two of the larger matrices here. Execute ZRZ.nex in PAUP and do a heuristic search by typing "hsearch" into the command line. The program rapidly analyzes the 56 taxa, 273 character matrix and results in 86 most parsimonious trees that are 649 steps long. When more and more taxa are added to a phylogenetic analysis the tree searches oftentimes need more involved searches. For instance, the default tree search that we just did that gave us 86 most parsimonious trees used a single search with very aggressive branch swapping and simple addition of taxa to the starting tree (i.e. addition of taxa in the order they appear in the matrix). These default parameters can oftentimes miss most parsimonious trees because they get "stuck" on less optimal inferences. Hence, we recommend when doing searches with large numbers of taxa to optimize the search by doing multiple rounds of searches with random addition of taxa to produce the starting tree. There is also a method called the ratchet that can be used (it is easily accessible in programs like RAxML and TNT). Here we simply alter the number of replicate searches and the addition sequence of taxa to produce starting trees. Execute ZRZ.nex and then type "hsearch addseq=random nreps=100" and hit return. The most aggressive and time-consuming branch-swapping algorithm called tree bisection and reconnection (TBR) is the default. The program will now perform 100 replicates of searching by changing the addition of taxa to the starting tree in each replicate randomly. When this second analysis is accomplished 80 shorter trees are found at 648 steps. A consensus tree for this search is shown in Figure 4.5.

To get the tree for the BRU full matrix, execute the BRU.nex file and type "boot" into the command line. The bootstrap analysis should take only a minute or so. How do the trees for the six matrices compare to each other?

We also show the tree obtained from a bootstrap analysis of the BRU.nex file for comparison. Note that both trees result in Porifera as the sister to all other metazoans. In fact, for all analyses with the full datasets (including BAC.nex, EER.nex, GLE.nex and SCH.nex), Placozoa or Porifera are obtained as the sister to all other metazoans and with relatively high bootstrap proportions.

studies on the metazoan problem demonstrate this importance: Simion et al. (2017; their Figure 4.1), Laumer et al. (2018; their Figure 4.1), and Whelan et al. (2015; their Figure 4.2) all use elaborate data filtering processes to generate molecular matrices. Within a study this procedure results in a large number of matrices based on the same primary data but comprised of different information because of the filtering. While the filtering processes are sometimes elaborate, it is a good idea to utilize this aspect of matrix construction to explore the behavior of the data.

Analyzing the molecular six taxon matrices with parsimony: We include five molecular matrices on the textbook website to demonstrate the following

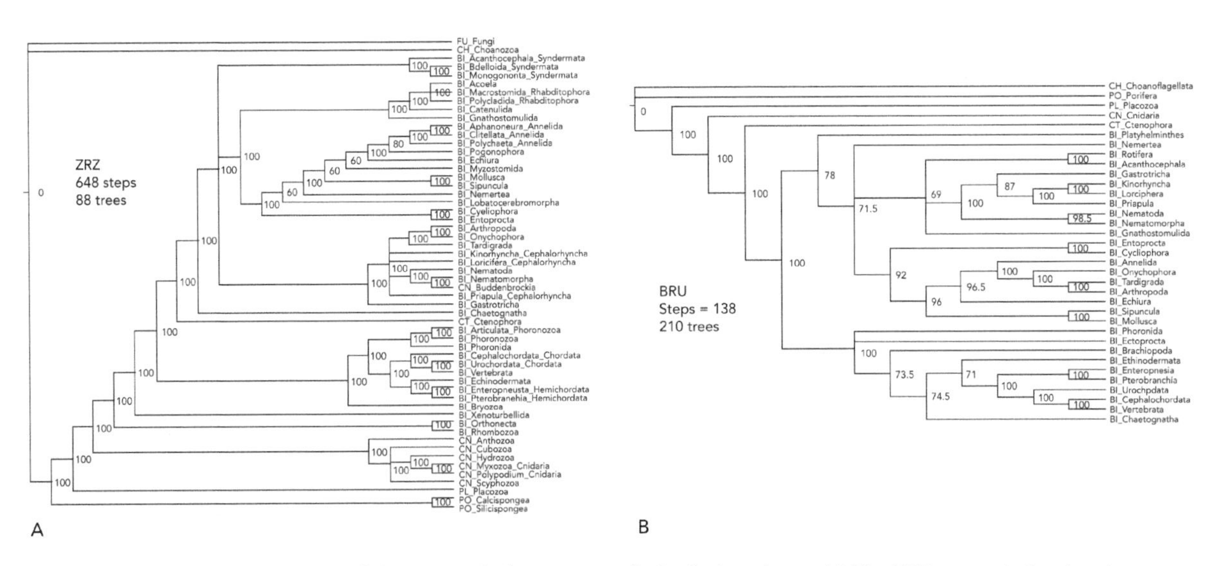

Figure 4.5 **Morphological trees of the expanded taxon morphological analyses.** (A) The ZRZ.nex analysis using the random addition, multiple replicate approach with 80 trees at 148 steps. The tree shown here is a majority rule consensus tree. (B) The BRU.nex dataset analyzed with parsimony and bootstrapped.

TABLE 4.2 The five molecular matrices discussed in this chapter

		#All Ch	#PI Ch	#PI Ch
Study	**#Taxa**	**Full**	**Full**	**6 taxa**
CHA	77	51,940	35,010	8,681
RYG	15	104,840	54,142	18,215
SIM	97	401,632	310,886	47,803
WHE	70	46,542	36,525	4,122
LAU	195	43,011	31,440	3,815

Abbreviated names for these datasets are as follows: RYG prefix = Ryan et al., 2013; SIM prefix = Simion et al., 2017; WHE prefix = Whelan et al., 2015; CHA prefix = Chang et al., 2015 and LAU = Laumer et al., 2018. PI stands for phylogenetically informative, Ch for characters. Table adapted from Neumann et al., 2020.

section. **Table 4.2** shows five matrices and as with the morphology we include a matrix for each study with six taxa (to represent the five major groups of metazoans (Bilateria, Cnidaria, Ctenophora, Porifera and Placozoa) as well as an outgroup (choanoflagellates). Each of these matrices has two character weighting matrices appended to the end of the file that can be used to weight the amino acid characters (WAG and LGM are cost matrices that are used in protein phylogenetics). Note that all of the trees have Ctenophora as the sister to all other metazoans using weighting schemes as suggested in **Sidebar 4.4**.

We have also included full taxonomic versions of the CHA, RYG, SIM, LAU and WHE matrices (Table 4.2). We recommend that the reader take the time to do analyses on these matrices too. PAUP (Swofford et al., 2001) can be used for all of the parsimony analyses we describe above and IQ-TREE (Ngyuen et al., 2015; Minh et al., 2020; for other likelihood and parsimony approaches, **Chapter 2**) can be used to do analyses for likelihood. One caveat is that parsimony analyses may not be ideal for analyzing molecular data at this level of phylogenetic divergence.

Likelihood analyses of molecular data: We now turn to analyzing molecular matrices with likelihood approaches. PAUP has the capacity to do likelihood analyses with amino acids and with DNA sequences. Since most of the analyses of higher-level invertebrate phylogenies use amino acid sequences, we focus on these. PAUP is not the most efficient program for likelihood analyses (try RAxML, IQ-TREE or GARLI) but here we use it to be consistent with the demonstrations in this chapter. The default optimality criterion in PAUP is parsimony. To change the criterion, type "set criterion=like" and hit return for likelihood and "set criterion=dist" for distance. There are many settings that one can use for a likelihood analysis. Most researchers first run programs to find the most appropriate models for an analysis (MODELTEST; ProTest etc.; Posada and Crandall,

Sidebar 4.4. Parsimony analysis of small molecular matrices with equal weights versus weighted schemes

This sidebar gives instructions for analyzing the molecular matrices in **Table 4.2**. The default optimality criterion for PAUP is maximum parsimony. If this parameter is not set to a different optimality approach then the analysis will be performed using parsimony. You can first analyze each matrix separately for the most parsimonious solution by executing each and typing "hsearch" into the terminal command line. For this sidebar we will also perform bootstrap analyses by typing "boot" into the command line. If we do this for all five of the datasets then the bootstrap analyses will result in the relationships represented in Figure 4.6 (black data below branches). Note that all of the datasets give Ctenophora as the sister to all other Metazoa. Now execute CHA6.nex and type in "ctype LGX2: all". This command tells PAUP to treat all of the characters as weighted according to the LGX2 matrix at the end of the CHA6.nex file. This matrix and the WAG matrix included in the file can correct for patterns of amino acid change that equal weighted parsimony does not account for. Next type in "boot" to do an analysis on the CHA6.nex matrix. We suggest that the student do bootstrap analyses using both LGX2 as a scoring matrix and WAG as a scoring matrix for all five molecular datasets. How do they compare?

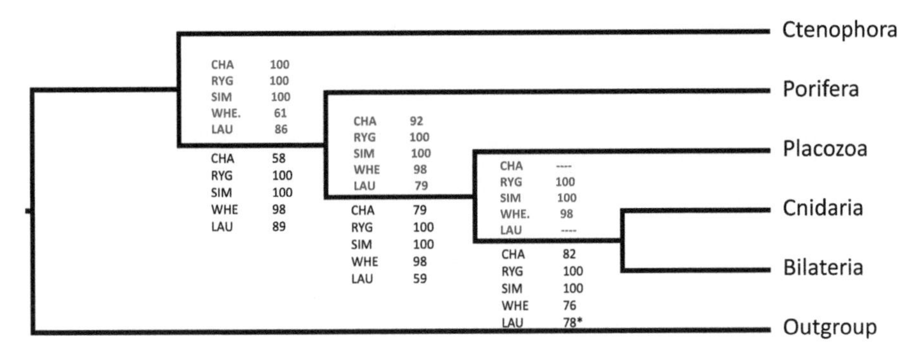

Figure 4.6 Molecular likelihood and parsimony tree based on the six taxa molecular datasets. This summarizes parsimony as well as likelihood analysis of the five molecular datasets listed in Table 4.2. They all produce almost identical tree topologies, with the following exception: CHA and LAU do not resolve the relationships between Bilateria, Cnidaria, and Placozoa. The numbers at the branches indicate the bootstrap support each matrix has for the monophyly of the branch after 1,000 bootstrap replicates: Red data above the branch are the respective likelihood bootstrap values, black data below the branch are parsimony bootstrap values. The values on the left show support for Ctenophora as the sister of all other metazoans. Note that this is 100% for three datasets under likelihood, and for two under parsimony; while two datasets each are ambiguous under likelihood (WHE: 61%, LAU: 86%), as well as parsimony (CHA: 58%, LAU: 89%).

1998; Posada, 2006; Darriba et al., 2020), but here we will use the WAG model for amino acids. We will also model the rates in the analysis with a distribution known as a gamma distribution. It is a good idea to become familiar with the models available and the other parameters that are included in likelihood runs by reading closely the manuals provided by the programs. Once the model has been selected a standard tree search is needed to find the tree or trees with the highest likelihood. This search is like the heuristic search we described for maximum parsimony (i.e. type "hsearch" on command line and PAUP will do the search). For these large molecular datasets and likelihood settings, the searches will be much longer even for searches with small numbers of taxa. Bootstrap (and jackknife) analyses in likelihood also use the same commands as in maximum parsimony.

We recommend that the reader execute and analyze the five molecular matrices we provide (CHA6.nex, RYG6.nex, SIM6.nex, WHE6.nex and LAU6.nex) using the following protocol which we show for CHA6.nex (**Sidebar 4.5**).

There are several issues that have been raised following the results from such studies. It is clear that the choice of model and analytical approach is critical in what results are obtained (Feuda et al., 2017), so is the choice of genes and validation of their orthology. It appears that most molecular systematists interested in invertebrates shy away from parsimony analysis for a variety of reasons, the most important of which is fear of long-branch attraction (LBA).

Sidebar 4.5. Likelihood analysis of small molecular matrices

This sidebar gives instructions for likelihood analysis of the small molecular matrices mentioned in the text. First, download all of the molecular matrices onto the computer desktop. Next execute CHA6.nex (type "exe CHA6.nex" on command line) and then change criterion to likelihood by typing "set criterion=like". Next set the model and parameters of the analysis with the following command "lset rates=gamma model=WAG". Next you can either do a search for the most likely tree or a bootstrap (jackknife) analysis. We have done bootstraps on all five of the matrices listed in Table 4.2 with the likelihood setting mentioned above (Figure 4.6, red data above branches), so we recommend the reader type "boot" into the command line and hit return which will tell PAUP to run the bootstrap. These bootstrap runs will be much longer than the parsimony runs of the same matrices, because likelihood is computationally more intense. We provide matrices with fuller taxon representation for the reader to explore using maximum likelihood. We recommend that the reader explore using these matrices with RAxML and IQ-TREE as for likelihood, as well as further exploration with TNT and MEGA.

TABLE 4.3 Amino acid recoding schemes

Dayhoff6	S&R-6	KGB-6
(AGPST) = A	(APST) = A	(AGPS) = A
(DENQ) = D	(DENG) = D	(DENQHKRT) = D
(HKR) = H	(QKR) = Q	(MIL) = M
(ILMV) = I	(MIVL) = M	(W) = W
(FWY) = F	(WC) = W	(FY) = F
(C) = C	(FYH) = F	(CV) = C

Each scheme reduces an original amino acid matrix to six character states as indicated in the table.

This problem occurs when two or more taxa have overly long branches. What happens in this case is that the long branch results in homoplasies being interpreted as synapomorphies (convergence of character states) and this results in inaccurate phylogenetic inference. Hence most molecular systematists use likelihood or Bayesian methods when performing phylogenetic analyses of molecular sequences. The model chosen for any molecular analysis is critical and we have discussed some of the nuances of model selection above.

It appears that invertebrate molecular systematists have settled on distinct strategies. Specifically, researchers have settled on Bayesian analysis with the CAT+GTR model (for a description of GTR – general time reversal –see **Chapter 3**) and sometimes also with a gamma distribution (Chapter 3) added to account for site heterogeneity. Some of the studies also use maximum likelihood but the models vary from study to study. Other strategies are to recode the 20 amino acid states into six functional categories using what is called the Dayhoff-6, the S&R-6 or the KGB-6 recoding schemes (**Table 4.3**). It is necessary to be aware of the model choice and the other parameters that researchers use when doing phylogenetic analysis with molecular data. Recently too, morphological data have been analyzed using Bayesian and likelihood approaches. The major obstacle to this approach with morphology is the choice of model. Simply put, there aren't many models to rely on for morphology and the ones that do exist appear to have shortcomings.

ANALYSIS OF MOLECULAR MATRICES WITH LARGE NUMBERS OF TAXA

For the previous section we trimmed the number of taxa down to six so as to represent each of the five major groups of animals and a sixth species for an outgroup. To fully reanalyze the studies in Table 4.2 would be a major undertaking, so here we will simply summarize the Laumer et al. (2018) and Simion et al. (2017) data matrices in detail. The phylogenetic result from both studies is shown in **Figure 4.7**. The Simion et al. (2017) study has Porifera as the SOM and is consistent with other published studies such as Feuda et al. (2017). The Laumer et al. (2018) study has Ctenophora as the SOM and is consistent with other studies such as Whelan et al. (2015; 2017).

For now and for the purposes of this textbook, we suggest that we cannot answer the fundamental question of what group is the SOM with certainty. Even when using the same data to address an evolutionary question, different studies often reach opposing conclusions. For instance, applying improved models to previously published genomic data has been found to flip the hypothesis from Ctenophora back to Porifera (Feuda et al., 2017). Recent work further suggests that even modestly weighted morphological data will often change phylogenomic inferences from a Ctenophora SOM hypothesis to a Porifera SOM

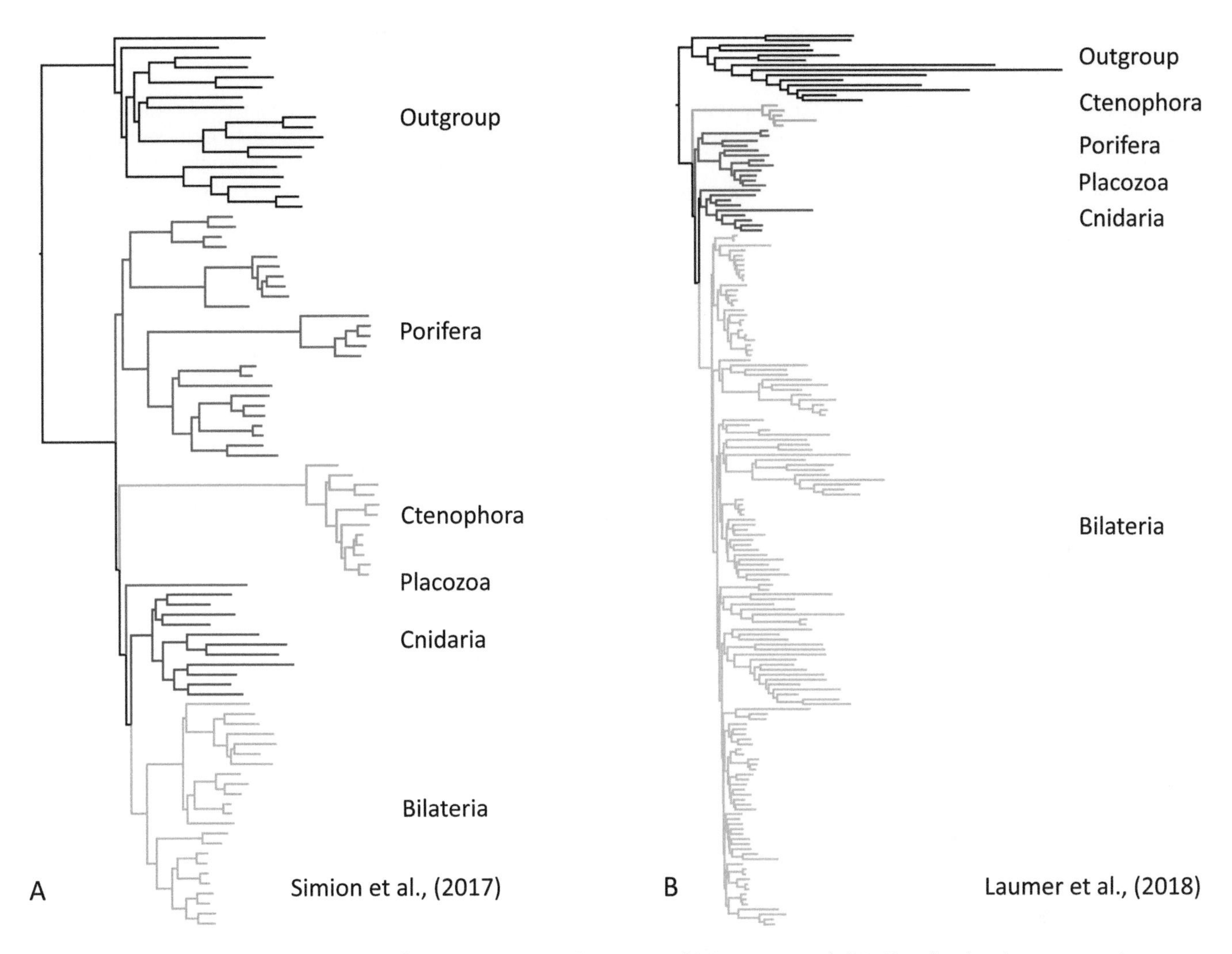

Figure 4.7 Molecular phylogenetic trees from (A) Simion et al. (2017) and (B) Laumer et al. (2018) molecular datasets. The five major metazoan groups are Bilateria = orange, Cnidaria = blue, Placozoa = purple, Ctenophora = green, Porifera = red and outgroups = black. For the Simion et al. study the analysis was a Bayesian CAT + Γ4 model. The nodes defining the five major metazoan lineages all have 1.0 posterior probabilities. The Laumer et al. analysis used a CAT + GTR + G4 model and Bayesian analysis. Trees based on recoded six amino acid matrices are similar.

hypothesis (Neumann et al., 2020), implying that morphological characters are especially useful or that the SOM node is flimsily supported (despite high published bootstrap and posterior probability support values).

So, what do we think? We suggest that a consideration of all hypotheses at this point is important for evolutionary questions about the invertebrates. As Schierwater et al. (2016) point out:

> In order to understand the dynamics of the current situation, we need to examine what the relevant data are and how different datasets and different analytical approaches can lead to conflicting inferences. We also need to discuss if there ever will be either (a) an unquestioned dataset and analysis, or (b) a middle ground where researchers will settle. Addressing these questions requires that we examine multiple sources of data for two reasons. First, such an examination might tell us why the controversy here is so pointed and sensitive to data input, and secondly to understand why the outcome of major studies addressing this problem are so prone to data handling and analytical approaches.

REFERENCES

Backeljau, Thierry, Birgitta Winnepenninckx, and Luc De Bruyn. "Cladistic analysis of metazoan relationships: A reappraisal." Cladistics 9(2) (1993): 167–81.

Bergsten, Johannes. "A review of long-branch attraction." Cladistics 21(2) (2005): 163–93.

Brunet, Thibaut, and Nicole King. "The origin of animal multicellularity and cell differentiation." Developmental Cell 43(2) (2017): 124–40.

Brunet, Thibaut, Ben T. Larson, Tess A. Linden, Mark J. A. Vermeij, Kent McDonald, and Nicole King. "Light-regulated collective contractility in a multicellular choanoflagellate." Science 366 (6463) (2019): 326–34.

Brusca, Richard C., and Gary J. Brusca. "Invertebrates" (2nd edn). Sunderland, MA: Sinauer Associates, 2003.

Bütschli, O. "Bemerkungen Zur Gastraeatheorie." Morphologisches Jahrblatt, no. 9 (1884): 415–27.

Carr, M., B. S. C. Leadbeater, R. Hassan, M. Nelson, and S. L. Baldauf. "Molecular phylogeny of choanoflagellates, the sister group to metazoa." Proceedings of the National Academy of Sciences 105 (43) (2008): 16641–46.

Chang, E. Sally, Moran Neuhof, Nimrod D. Rubinstein, Arik Diamant, Hervé Philippe, Dorothée Huchon, and Paulyn Cartwright. "Genomic insights into the evolutionary origin of Myxozoa within Cnidaria." Proceedings of the National Academy of Sciences 112, no. 48 (2015): 14912–17.

Collins A.G., Cartwright P., McFadden C.S., Schierwater B. "Phylogenetic context and basal metazoan model systems." Integr. Comp. Biol. 45 (2005): 585–594.

Darriba, Diego, David Posada, Alexey M. Kozlov, Alexandros Stamatakis, Benoit Morel, and Tomas Flouri. "ModelTest-NG: a new and scalable tool for the selection of DNA and protein evolutionary models." Molecular Biology and Evolution 37(1) (2020): 291–94.

Dellaporta, Stephen L., Anthony Xu, Sven Sagasser, Wolfgang Jakob, Maria A. Moreno, Leo W. Buss, and Bernd Schierwater. "Mitochondrial genome of trichoplax adhaerens supports placozoa as the basal lower metazoan phylum." Proceedings of the National Academy of Sciences of the United States of America 103 (23) (2006): 8751–56.

DeSalle, Rob, and Bernd Schierwater. "An Even 'newer' Animal Phylogeny." BioEssays 30 (11–12) (2008): 1043–47.

Dunn, Casey W., Andreas Hejnol, David Q. Matus, Kevin Pang, William E. Browne, Stephen A. Smith, Elaine Seaver, et al. "Broad phylogenomic sampling improves resolution of the animal tree of life." Nature 452 (7188) (2008): 745–49.

Ender, Andrea, and Bernd Schierwater. "Placozoa are not derived cnidarians: evidence from molecular morphology." Molecular Biology and Evolution 20 (1) (2003): 130–34.

Erpenbeck, D., O. Voigt, M. Adamski, M. Adamska, J. N. A. Hooper, G. Wörheide, and B. M. Degnan. "Mitochondrial diversity of early-branching metazoa is revealed by the complete Mt Genome of a haplosclerid demosponge." Molecular Biology and Evolution 24 (1) (2007): 19–22.

Fernández, Rosa, and Toni Gabaldón. "Gene gain and loss across the metazoan tree of life." *Nature ecology & evolution* 4(4) (2020): 524–533. doi:10.1038/s41559-019-1069-x

Feuda, Roberto, Martin Dohrmann, Walker Pett, Hervé Philippe, Omar Rota-Stabelli, Nicolas Lartillot, Gert Wörheide, and Davide Pisani. "Improved modeling of compositional heterogeneity supports sponges as sister to all other animals." Current Biology: CB 27 (24) (2017): 3864–70.e4.

Foox, Jonathan, and Mark E. Siddall. "The Road to Cnidaria: History of Phylogeny of the Myxozoa." The Journal of Parasitology 101 (3) (2015): 269–74.

Gissi, C., F. Iannelli, and G. Pesole. "Evolution of the mitochondrial genome of Metazoa as exemplified by comparison of congeneric species." Heredity 101 (4) (2008): 301–20.

Glenner, Henrik, Anders J. Hansen, Martin V. Sørensen, Frederik Ronquist, John P. Huelsenbeck, and Eske Willerslev. "Bayesian inference of the metazoan phylogeny: a combined molecular and morphological approach." *Current Biology* 14, no. 18 (2004): 1644–49.

Gonobobleva, Elisaveta, and Manuel Maldonado. "Choanocyte ultrastructure in Halisarca Dujardini (Demospongiae, Halisarcida)." Journal of Morphology 270 (5) (2009): 615–27.

Haeckel, Ernst. "Anthropogenie: Oder, Entwickelungsgeschichte des Menschen: Gemeinverständlich Wissenschaftliche Vorträge über die Grundzüge der Menschlichen Keimes- und Stammes-Geschichte." Leipzig: Engelmann, 1874.

Laumer CE, Gruber-Vodicka H, Hadfield MG, Pearse VB, Riesgo A, Marioni JC, Giribet G. "Support for a clade of Placozoa and Cnidaria in genes with minimal compositional bias." eLife 7, (2018) e36278. (doi:10.7554/eLife.36278)

Laumer, Christopher E., Rosa Fernández, Sarah Lemer, David Combosch, Kevin M. Kocot, Ana Riesgo, Sónia CS Andrade, Wolfgang Sterrer, Martin V. Sørensen, and Gonzalo Giribet. "Revisiting metazoan phylogeny with genomic sampling of all phyla." Proceedings of the Royal Society B 286, no. 1906 (2019): 20190831.

Littlewood, D. T. J. "Animal evolution: last word on sponges-first?" Current Biology 27(1) (2017): R259–R261.

Miller, David J., and Eldon E. Ball. "Animal evolution: trichoplax, trees, and taxonomic turmoil." Current Biology 18 (21) (2008): R1003–5.

Minh, Bui Quang, Heiko A. Schmidt, Olga Chernomor, Dominik Schrempf, Michael D. Woodhams, Arndt Von Haeseler, and Robert Lanfear. "IQ-TREE 2: New models and efficient methods for phylogenetic inference in the genomic era." Molecular Biology and Evolution 37 (5) (2020): 1530–34.

Moroz, L. L. "Phylogenomics meets neuroscience: How many times might complex brains have evolved?" Acta Biologica Hungarica 63 (Suppl 2) (2012): 3–19.

Neumann, Johannes S., Rob DeSalle, Apurva Narechania, Bernd Schierwater, and Michael Tessler. "Morphological characters can strongly influence early animal relationships inferred from phylogenomic datasets." Systematic Biology, (May 2020): https://doi.org/10.1093/sysbio/syaa038.

Nguyen, Lam-Tung, Heiko A. Schmidt, Arndt Von Haeseler, and Bui Quang Minh. "IQ-TREE: a fast and effective stochastic algorithm for estimating maximum-likelihood phylogenies." Molecular Biology and Evolution 32 (1) (2015): 268–74.

Nielsen, Claus. "Early Animal Evolution: A Morphologist's View." Royal Society Open Science 6 (7) (2019): 190638.

Nosenko, Tetyana, Fabian Schreiber, Maja Adamska, Marcin Adamski, Michael Eitel, Jörg Hammel, Manuel Maldonado et al. "Deep metazoan phylogeny: when different genes tell different stories." Molecular Phylogenetics and Evolution 67 (1) (2013): 223–33.

Peer, Yves Van de, and Rupert De Wachter. "Evolutionary relationships among the eukaryotic crown taxa taking into account site-to-site rate variation in 18S rRNA." Journal of Molecular Evolution 45 (6) (1997): 619–30.

Peterson, Kevin J., and Douglas J. Eernisse. "Animal phylogeny and the ancestry of bilaterians: inferences from morphology and 18S rDNA gene sequences." Evolution & Development 3 (3) (2001): 170–205.

Pett, Walker, Marcin Adamski, Maja Adamska, Warren R Francis, Michael Eitel, Davide Pisani, Gert Wörheide, "The Role of Homology and Orthology in the Phylogenomic Analysis of Metazoan Gene Conten." Molecular Biology and Evolution, 36(4) (April 2019): 643–49, https://doi.org/10.1093/molbev/msz013

Posada, David, and Keith A. Crandall. "Modeltest: testing the model of DNA substitution." Bioinformatics (Oxford, England) 14 (9) (1998): 817–18.

Posada, David. "ModelTest Server: a web-based tool for the statistical selection of models of nucleotide substitution online." Nucleic Acids Research 34 (no. suppl_2) (2006): W700–03.

Raible, Florian, and Patrick R. H. Steinmetz. "Metazoan Complexity." In Introduction to Marine Genomics, edited by J. Mark Cock, Kristin Tessmar-Raible, Catherine Boyen, and Frédérique Viard. Dordrecht: Springer Science & Business Media, 2010, 143–78.

Ruiz-Trillo, Iñaki, Andrew J. Roger, Gertraud Burger, Michael W. Gray, and B. Franz Lang. "A Phylogenomic Investigation into the Origin of Metazoa." Molecular Biology and Evolution 25 (4) (2008): 664–72.

Ryan, Joseph F., and Andreas D. Baxevanis. "Hox, Wnt, and the evolution of the primary body axis: insights from the early-divergent phyla." Biology Direct 2 (December 2007): 37.

Ryan, Joseph F., Kevin Pang, Christine E. Schnitzler, Anh-Dao Nguyen, R. Travis Moreland, David K. Simmons, Bernard J. Koch et al. "The genome of the ctenophore Mnemiopsis leidyi and its implications for cell type evolution." Science 342 (6164) (2013).

Schierwater, B., de Jong, D., DeSalle, R. "Placozoa, and the evolution of Metazoa and intrasomatic cell differentiation." The International Journal of Biochemistry & Cell Biology 41 (2009a): 370–79. doi:10.1016/j.biocel.2008.09.023.

Schierwater B., Eitel M., Jakob W., Osigus H.-J., Hadrys H., Dellaporta S.L., Kolokotronis S.-O., DeSalle R. "Concatenated analysis sheds light on early metazoan evolution and fuels a modern "urmetazoon" hypothesis." PLoS Biol. 7(1) (2009b): e1000020. doi:10.1371/journal.pbio.1000020.

Schierwater, B., Kamm, K., Srivastava, M., Rokhsar, D., Rosengarten, R. D., and Dellaporta, S. L. "The early ANTP Gene repertoire: insights from the placozoan genome." PLoS One 3: e2457 (2008): 109–10.

Schierwater, B., Stadler, P., DeSalle, R., and Podsiadlowski, L. "Mitogenomics and metazoan evolution." Molecular Phylogenetics and Evolution. 69 (2013): 311–12

Schierwater, Bernd, Michael Eitel, Wolfgang Jakob, Hans-Jürgen Osigus, Heike Hadrys, Stephen L. Dellaporta, Sergios-Orestis Kolokotronis, and Rob DeSalle. "Concatenated analysis sheds light on early metazoan evolution and fuels a modern 'urmetazoon' hypothesis." PLoS Biology 7 (1) (2009): e1000020.

Schierwater, Bernd, Peter W. H. Holland, David J. Miller, Peter F. Stadler, Brian M. Wiegmann, Gert Wörheide, Gregory A. Wray, and Rob DeSalle. "Never ending analysis of a century old evolutionary debate: 'unringing' the urmetazoon bell." Frontiers in Ecology and Evolution 4 (2016): https://doi.org/10.3389/fevo.2016.00005.

Signorovitch, Ana Y., Leo W. Buss, and Stephen L. Dellaporta. "Comparative genomics of large mitochondria in placozoans." PLoS Genetics 3 (1) (2007): e13.

Simion, P., H. Philippe, D. Baurain, M. Jager, D. J. Richter, A. Di Franco, B. Roure, N. Satoh, E. Queinnec, A. Ereskovsky, and P. Lapebie. "A large and consistent phylogenomic dataset supports sponges as the sister group to all other animals." Current Biology, 27(7) (2017): 958–67.

Srivastava, Mansi, Emina Begovic, Jarrod Chapman, Nicholas H. Putnam, Uffe Hellsten, Takeshi Kawashima, Alan Kuo, et al. "The trichoplax genome and the nature of placozoans." Nature 454 (7207) (2008): 955–60.

Swofford, David L. "Paup*: Phylogenetic analysis using parsimony (and other methods) 4.0. B5." (2001).

Syed, Tareq, and Bernd Schierwater. "The evolution of the placozoa: A new morphological model." Senckenbergiana Lethaea 82 (1) (2002): 315–24.

Syed T., and Schierwater, B. "Trichoplax adhaerens: discovered as a missing link, forgotten as a hydrozoan, re discovered as a key to metazoan evolution." Vie Milieu 52 (2002): 177–87.

Telford, Maximilian J. "Resolving animal phylogeny: A sledgehammer for a tough nut?" Developmental Cell 14 (4) (2008): 457–59.

Tessler, Michael, Danielle de Carle, Madeleine L. Voiklis, Olivia A. Gresham, Johannes S. Neumann, Stanisław Cios, and Mark E. Siddall. "Worms That suck: Phylogenetic Analysis of hirudinea solidifies the position of acanthobdellida and necessitates the dissolution of rhynchobdellida." Molecular Phylogenetics and Evolution 127 (October 2018): 129–34.

Vasconcellos, Vivian, Philippe Willenz, Alexander Ereskovsky, and Emilio Lanna. "Comparative Ultrastructure of the Spermatogenesis of Three Species of Poecilosclerida (Porifera, Demospongiae)." Zoomorphology 138 (1) (2019): 1–12.

Winnepenninckx, Birgitta M. H., Yves Van de Peer, and Thierry Backeljau. "Metazoan Relationships on the Basis of 18S rRNA Sequences: A Few years later...." American Zoologist 38 (6) (1998): 888–906.

Whelan, Nathan V., Kevin M. Kocot, Leonid L. Moroz, and Kenneth M. Halanych. "Error, signal, and the placement of Ctenophora sister to all other animals." Proceedings of the National Academy of Sciences 112 (18) (2015): 5773–78.

Whelan, Nathan V., Kevin M. Kocot, Tatiana P. Moroz, Krishanu Mukherjee, Peter Williams, Gustav Paulay, Leonid L. Moroz, and Kenneth M. Halanych. "Ctenophore relationships and their placement as the sister group to all other animals." Nature Ecology & Evolution 1, no. 11 (2017): 1737–46.

Zrzavý, Jan, Stanislav Mihulka, Pavel Kepka, Aleš Bezděk, and David Tietz. "Phylogeny of the Metazoa based on morphological and 18S ribosomal DNA evidence." Cladistics 14 (3) (1998): 249–285.

CHAPTER 5

ORGANIZING INVERTEBRATES

Bernd Schierwater and Rob DeSalle

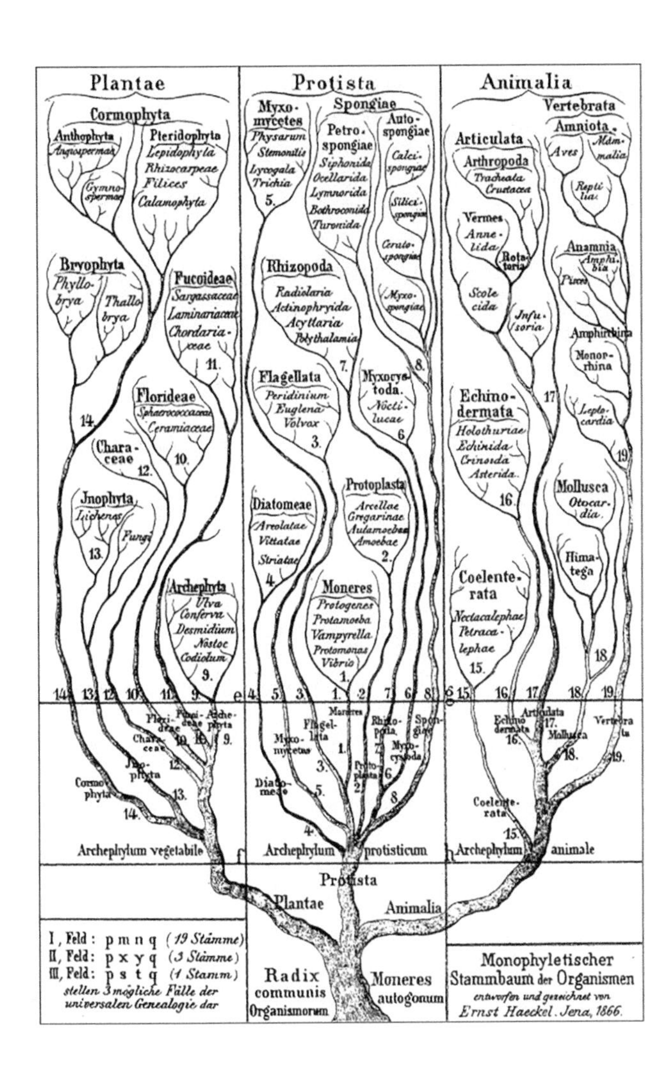

CONTENTS

This chapter describes the various phylogenetic hypotheses and taxonomies that have been posited over the past century to describe invertebrates. Before we go into too much detail here, we want to remind the student that one of the guiding principles of systematics is monophyly. We also remind the reader of the definition of monophyly we gave in Chapter 2:

The term "monophyletic" refers to a group of organisms with a single common ancestor to the exclusion of all others being considered. It is a relative term so that when something is referred to as monophyletic it needs to be in reference to something else. In **Figure 5.1** (Darwin's "I think" tree) "B", "C" and "D" are members of a monophyletic group relative to "A" because they have a single common ancestor to the exclusion of "A".

This definition is essential in evaluating a hypothesis of grouping for organisms at any level. It is used regularly in invertebrate zoology to name taxa and to understand higher-order taxonomy, i.e. the hierarchical arrangement of taxa within the Invertebrates. This principle, however, is not followed as a guiding principle by all authors, as we explain in what follows.

Through over two centuries of systematic interest in invertebrates many different (and unfortunately often ignorant) hypotheses have been offered to organize the taxa in this large group of organisms. Invertebrates should be defined as metazoan animals without vertebrae. This causes no problems with the definition of the group, if one accepts that this group is not a monophyletic taxon rather than a subgroup of Metazoa picked as such for educational/teaching reasons. The Vertebrates, which are terminal additions to one of the Invertebrate lineages. It has been a long-standing tradition in Animal Systematics to treat the most derived and most complex vertebrate bauplans in a separate volume. Similarly, single-cell ancestors of Metazoa, the Protozoa, are treated separately from the Invertebrates.

We would like to mention that the two main eukaryote groups, Protozoa (single-cell organisms) and Metazoa (multicellular animals) are clearly defined:

Def. Protozoa: Lack of intrasomatic variation.

Def. Metazoa: Intrasomatic variation, i.e. at least two different types of somatic cells are formed.

Please note that a protist specimen may well have more than one cell; many protists have more than 100 cells, but these are all the same type of somatic cell. If a protist has more than one cell type, then the second cell type are gamete (i.e. the opposite of somatic) cells. The definition given above has been valid and 100% clear for more than two centuries and any attempts to create confusing complications by searching for microscopic hints for differentiations between cells in a protist are unhelpful and unnecessary.

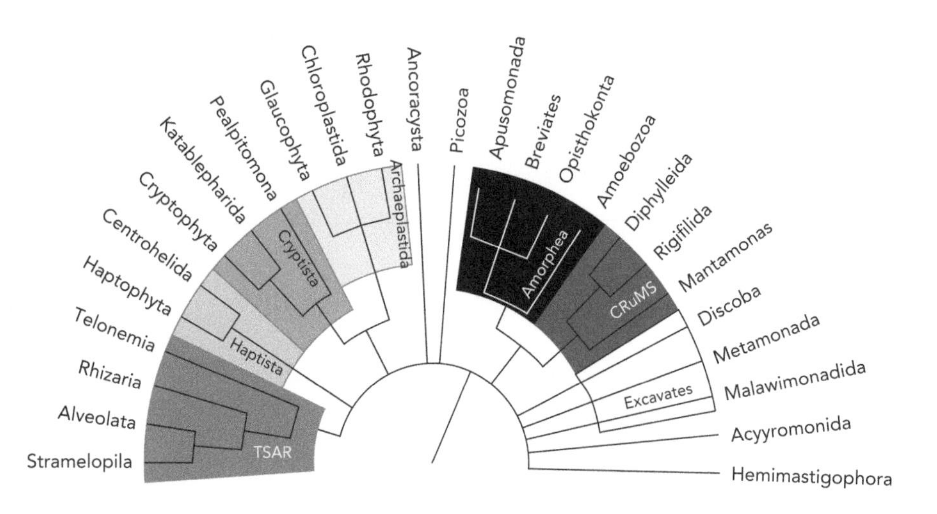

Figure 5.1 Eukaryote tree of life (after Burki et al., 2020). Within the Amorphea the three taxa – Breviates, Apusomonada and Opisthokonta make up the group Obazoa. Note that Excavates although indicated in the figure as a group is not a monophyletic group.

EUKARYOTES

Our goal in this chapter is to examine Invertebrate animals. Since all animals are eukaryotes, we will focus mostly on a discussion of relationships above the common ancestor of animals, except for those single-celled eukaryotes that are the closest relatives of animals – choanoflagellates. We have discussed in detail the phylogenetic approach to understand invertebrate zoology in Chapters 3, and 4, but we were more interested in mechanics than in results. In this chapter we will examine in detail some of the previous hypotheses about relationships of invertebrate taxa.

LOADS OF SINGLE-CELLED EUKARYOTES

While we discussed taxonomic hierarchy in Chapter 1 with great detail below the level of phylum, biologists who work on the overall relationships of eukaryotes originally settled on a "supergroup" concept for organization within the domain Eukrayota. Burki et al. (2020) present a cogent and logical picture of eukaryote phylogeny and suggest that most of the major supergroups are now subsumed into other groups or ignored altogether. Their scheme includes seven monophyletic supergroups – TSAR (the acronym uses the first letters of the major groups in the supergroup), Haptista, Cryptista, Archaeplastida (includes plant and algae), Amorphea (includes animals and fungi), CRuMs (again an acronym from the first letter of some of the major groups within) and Hemimastigophora. It also retains an older supergroup called Excavata which is more than likely polyphyletic, and several orphan taxa (Figure 5.1).

Of interest here is the supergroup Amorphea in which animals are a member. The Amorphea is comprised of four major lineages – Ameobozoa, Apusomonads, Breviates and Opisthokonta, of which the latter three make up a sub-supergroup called Obazoa. Animals are a member of the Opisthokonta and the major subgroups most closely related to opisthokonts are all single-celled organisms. While researchers have suggested several morphological and physiological characteristics that can define Opisthokonta, Burki et al. (2020) point out that this supergroup like others lacks clear defining characteristics. It is clear though from phylogenetic analysis that Obazoa is a monophyletic group and Opisthokonta is a monophyletic group within Obazoa.

Opisthokonta are comprised of the fungi (the Holomycota) and the animals plus several single-celled eukaryotes (collectively known as the Holozoa) (Keeling et al., 2019). The Holozoa contain several major kinds of single-celled eukaryotes, the Ichthyosporea, the Pluriformea (includes Corallochytrium and Syssomonas), Filasteria and the Choanozoa which includes Animalia and Choanoflagellata. The Filasteria and Choanozoa are combined into a higher group called the Filozoa. This scheme suggests that Pluriformes, Filasteria and Choanoflagellatea can all serve as the closest outgroups to Animalia and these organisms are discussed in detail in Chapter 6 in this textbook. The position of the Choanoflagellatea as the sister group to animals is supported by the existence of choanocytres in both the Choanoflagellatea and one major group of animals, the sponges. Choanocytes are a specialized kind of cell with a flagellum that has a collar of protoplasm at the base of the flagellum. This specialized kind of cell appears to have been lost in all other major groups of animals, but its occurrence in choanoflagellates and sponges suggests a close affinity of the single-celled choanoflagellates with animals.

Metazoan Animals

Traditionally, "animal" is NOT a taxonomic or systematic term, since "animals" include Metazoa and heterophagous protists (i.e. the border between animals and plants runs through the protists). However, some researchers prefer to say animals when they mean "Metazoa". So in this context we will use the term "metazoan animals" or "Metazoa". Furthermore, in the following description of metazoan animals and throughout this text we will use a simplification and sometimes write "animal" instead of "metazoan".

Animals as we point out above are most likely derived from single-celled eukaryotes resembling choanoflagellates. As we have also discussed at length the branching order of the five major kinds of Metazoa – Bilateria, Cnidaria, Placozoa, Ctenophora and Porifera – is contentious. Because our description of the major phyla in these five major groups is not dependent on a phylogeny, we settle on simply discussing the controversies and describing the groups as separate problems. We do not want to minimize the importance of solving the branching order of these groups, but a full discussion would be counterproductive to the overall goal of the chapters that follow this. While we would argue that there is little evidence to refute the sister relationship of Bilateria with Cnidaria, we point out that recent work suggesting that Placozoa are the sister group to Cnidaria conflicts with this notion. Hence, we treat the branching orders of the five major groups for the purposes of description in Chapters 6 through 36 as unresolved. This unresolved treatment poses some problems with respect to evolutionary scenarios for the evolution of certain organ systems in animals, so we attempt to order the major taxa in the last chapter of this book along a scheme that we feel is most logical. But first we discuss the taxonomic and grouping schemes that have appeared in the literature as of 2019.

DIFFERENT VIEWPOINTS: MANY SCHEMES FOR THE ORGANIZATION OF INVERTEBRATE ANIMAL ORGANISMS

Many treatments of the organization of invertebrate animals have appeared in the last two decades and here we present a summary of fifteen of these schemes (Dunn et al., 2014; Golaconda Ramalu, et al., 2012; Halanych et al., 2004; Moore 2001; Pechenik, 2004; Ruppert, 2004; Wallace and Taylor, 2003; Telford et al., 2015; Glenner at al., 2004.). This is not meant to be an exhaustive review of the schemes that have been generated since 2000, but the studies we summarize below should offer a representative view of the shifting landscape of taxonomic organization of invertebrate animals. **Figure 5.2** shows a summary of the fifteen studies and the phyla (sometimes including subphyla) of invertebrate animals. If a row is fully black, then this means all fifteen schemes included that taxon. So, for instance Annelida, Arthropoda and Brachiopoda are phyla that are included in all fifteen schemes. Taxa that are represented infrequently across the schemes are ones like Cephalochordata, Echiura and others with large amounts of white space in a row. Figure 5.2 should give the student an idea of the complexity and fluid nature of the schemes in the literature.

We have also summarized the above schemes at the level of phyla for the fifteen studies in **Figure 5.3**, where we include the phyla Chordata, Porifera, Placozoa, Cnidaria and Ctenophora. The figure demonstrates a fluid use of higher taxonomic schemes for some naming conventions and a rather static state for others. For instance, both Figures 5.2 and 5.3 show that the phyla Porifera, Cnidaria, Ctenophora and Placozoa are static entities in almost all schemes. In addition to these phylum level designations, the higher-level groups of Bilateria, Protostomia, Deuterostomia, and Ecdysozoa are also static. Names above the level of phyla that show intermediate stability are Spiralia, Lophotrochozoa, Scalidorpha, Nematodita, Trochozoa, Lophophorata, Panarthropoda and Ambulacraria. Note that Spiralia and Lophotrochozoa do not overlap. That is, when a scheme includes Spiralia it does not also include Lophotrochozoa and vice versa, except for the Giribet and Edgecombe scheme where the name

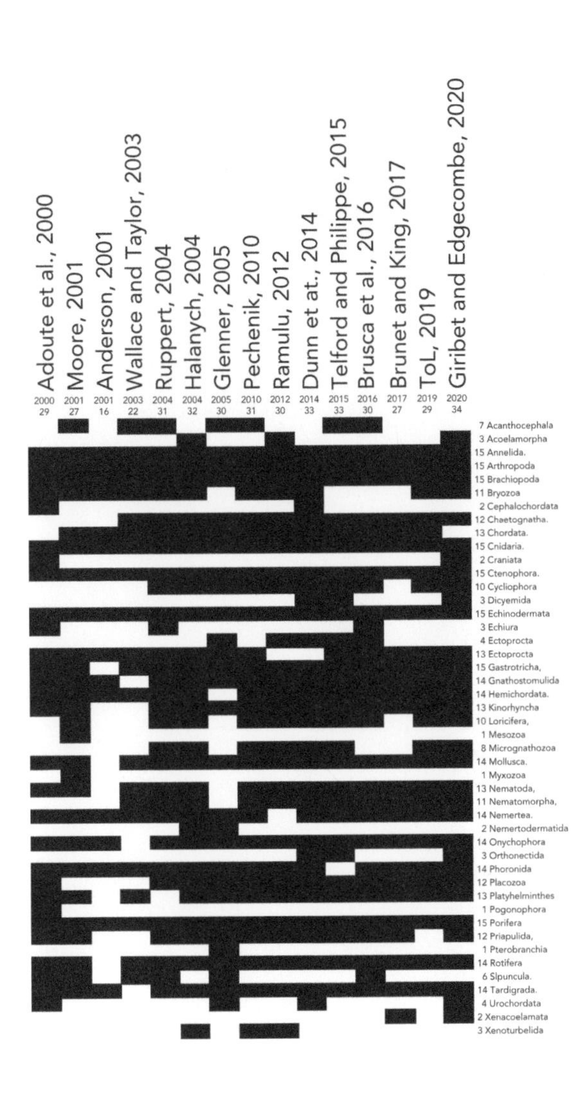

Figure 5.2 Distribution of phyla in the fifteen taxonomic schemes since 2000 treated here. The studies are listed left to right from least recent to most recent. Phyla are listed in the column to the right of the figure. The numbers across the top of the diagram indicate the number of higher taxa that are included in the specified study. The numbers in the column to the right of the diagram indicate the number of studies that include the indicated higher taxon.

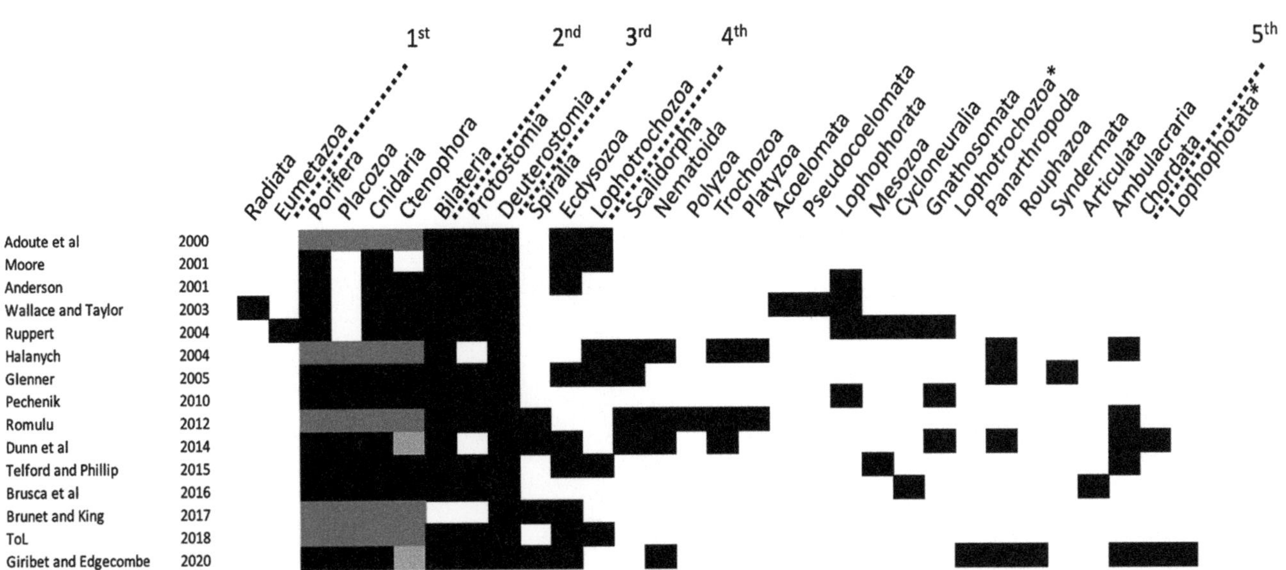

Figure 5.3 Survey of fifteen animal taxonomic schemes since 2000. The taxa are listed across the top of the diagram with the branching level indicated as 1st, 2nd, 3rd, 4th or 5th. The source of the scheme is given on the left and these are ordered by year from least recent (top) to most recent (bottom). The black squares indicate the presence of the taxon in the scheme on the left. The light-gray boxes indicate schemes that posit Ctenophora as the sister to all other metazoans. The dark-gray present the relationships of Porifera, Placozoa and Ctenophora as unresolved. All other schemes posit Porifera as the sister to all other Metazoa. Lophotrochozoa* indicates the use of this taxon by Giribet and Edgecombe (2020) as a non-sister group to Ecdysozoa and indicates a usage of this taxon name and generate at a level generally not used in other studies.

Lophotrochozoa is embedded within the Spiralia. Another trend that can be observed from the figure is that schemes agree after the first animal lineage is established and on to the 4th branching level in the figure. After the 4th branching level the names are dynamic and vary from study to study. This variability in naming below the 4th branching level in the figure is due to lack of agreement on phylogenetic affinity of the various bilaterian phyla with each other.

Of the 46 taxonomic names at the level of phylum Xenoturbelida, Xenacoelmata, Urochordata, Sipuncula, Pterobranchia, Pogonophora, Orthonectida, Nemertodermatida, Myxozoa, Ectoprocta, Echiura, Dicyemida, Craniata, Acoelamorpha and Cephalochordata are used rather sparingly by previous taxonomies. This leaves 31 phyla that are used in the majority of schemes by previous authors. In our treatment of phyla in the chapters in this text we cover all of these groups (**Tables 5.1–5.4**).

Bilateria and the Phyla Porifera, Cnidaria, Ctenophora, Placozoa

The first branching events within the Metazoa involve the establishment of the five major animal groups, four of which are given the status of phyla. While there is little doubt as to the reality of these five groups, we have discussed the controversies inherent in the branching order of these five taxa in several chapters so far and will return to our hypothesis of the relationships in Chapter 38.

Radiata and Coelenterata – Doomed Higher-Order Groups

Approaches to imparting hierarchy to these five higher taxa involve the use of body plan or common developmental benchmarks among the taxa. For instance, the Radiata, a now rejected taxonomic grouping, used the anatomical character of radial symmetry as a diagnostic. This approach erroneously grouped the radially symmetrical organisms Echinodermata, Cnidaria and Ctenophora. Later, Radiata was expanded by Cavalier-Smith (1998) to include Cnidaria, Placozoa, Myxozoa and Ctenophora and constricted by Margulis et al. (1999) to include only Cnidaria and Ctenophora. The inclusion of Echinodermata into the Radiata is a completely wrong classification and should be ignored by all invertebrate zoologists.

Another taxonomic epithet that is identical to the Radiata is the Coelenterata, so here were equate Radiata with Coelentrata. The diagnostic anatomical

TABLE 5.1 Orphan taxa used in some hierarchies shown in Figures 5.2 and 5.3

Name	Rank	Chapter no.	Higher
Ectoprocta	Phylum	26	Lophotrochozoa
Xenacoelomorpha	Phylum	36	Bilateria
Myxozoa	Subphylum	12	Cnidaria
Xenoturbelida	Subphylum	36	Xenacoelomorpha
Acoelamorpha	Subphylum	36	Xenacoelomorpha
Nemertodermatida	Class	19	Acoelamorpha
Sipuncula	Subphylum	15	Annelida
Pogonophora	Class	15	Annelida
Echiura	Subclass	15	Annelida
Pterobranchia	Class	37	Hemichordata
Craniata	Class	38	Chordata
Cephalochordata	Subphylum	38	Chordata
Urochordata	Subphylum	38	Chordata
Orthonectida	Phylum	*	Mesozoa
Dicyemida	Phylum	*	Mesozoa
Mesozoa	Superphylum	*	Protostomia

* Discussed below in "The Mesozoa Problem".

TABLE 5.2 Orphan taxa used in some hierarchies shown in Figures 5.2 and 5.3

Name	Rank	Chapter no.	Higher
Xenoturbelida	Subphylum	36	Xenacoelomorpha
Xenacoelomorpha	Phylum	36	Bilateria
Urochordata	Subphylum	38	Chordata
Sipuncula	Subphylum	15	Annelida
Pterobranchia	Class	37	Hemichordata
Pogonophora	Class	15	Annelida
Orthonectida	Phylum	*	Mesozoa
Nemertodermatida	Class	19	Acoelamorpha
Myxozoa	Subphylum	12	Cnidaria
Ectoprocta	Phylum	26	Lophotrochozoa
Echiura	Subclass	15	Annelida
Dicyemida	Phylum	*	Mesozoa
Craniata	Class	38	Chordata
Acoelamorpha	Subphylum	*	Xenacoelomorpha
Cephalochordata	Subphylum	38	Chordata
Mesozoa	Superphylum	*	Protostomia

* Discussed below in "The Mesozoa Problem".

character for Coelentrta and therefore Radiata in the context of our definition is the existence of a hollow body cavity with very simple tissue organization of only two cell layers (inner and outer) and radial symmetry. A second characteristic of coelenterates is the lack of a circulatory system and reliance on movement of nutrients and other substances throughout the body by diffusion across tissue layers. Ceolenterata has also been used to describe a group with both Cnidaria and Ctenophora. These attempts to organize phyla into superphyla like Coelenterata/Radiata have failed because they do not circumscribe phyla into monophyletic groups. The original Radiata is very problematic as it combines lower animals in two phyla – Ctenophora and Cnidaria – with bilateral animals in the phylum Echinodermata; as we mentioned above this is an absolutely incorrect grouping. Coelenterata, while focusing on a basic body organization

TABLE 5.3 Ecdysozoan phyla in Figures 5.2 and 5.3 and chapters covering them

Arthropoda	
Crustacea	Chapter 27
Insecta	Chapter 28
Chelicerata	Chapter 29
Myriapoda	Chapter 30
Onychophora	Chapter 31
Nematoda	Chapter 32
Nematomorpha	Chapter 32
Tardigrada	Chapter 33
Kinorhyncha	Chapter 34
Nematomorpha	Chapter 34
Priapulida	Chapter 34
Loricifera	Chapter 34

TABLE 5.4 Lophotrochozoa (Spiralia) phyla in Figures 5.2 and 5.3 and chapters covering them

Mollusca	Chapter 14
Polyplacophora	Chapter 14
Annelida	Chapter 15
Brachiopoda	Chapter 16
Bryozoa	Chapter 17
Phoronida	Chapter 17
Ectoprocta	Chapter 17
Nemertea	Chapter 18
Platyhelminthes	Chapter 19
Chaetognatha	Chapter 20
Gastrotricha	Chapter 21
Rotifera	Chapter 22
Cycliophora	Chapter 23
Gnathostomulida	Chapter 23
Micrognathozoa	Chapter 23
Acanthocephala	Chapter 24
Entoprocta	Chapter 25

trait (simple hollow bodies without circulatory systems), fails at all attempts to place phyla in that proposed group. Again, the problem is monophyly. In earlier schemes a group with Cnidaria and Ctenophora was tenable, but recent data on the position of Ctenophorans as more basal in the animal phylogeny, the possibility that Ctenophora and Cnidaria are sister taxa and the fact that Bilateria and Cnidaria are more than likely sister taxa all refute the idea of Coelenterata being a valid taxonomic group. Hence, Radiata/Coelenterata are doomed higher-order names. That some animals have coela (i.e. are hollow) is not the question here.

Subphyla, Superphyla and Higher Classification

The second major branching events that concern us here occur within the Bilateria. This does not mean though that major branching did not occur in Cnidaria, Ctenophora, Porifera or Placozoa. In fact, we recognize the four major cnidarian subphyla – Scyphozoa (Chapter 9), Anthozoa (Chapter 10), Hydrozoa (Chapter 11) and Cubozoa (Chapter 9) by giving them separate treatments in this text. We also expand on the Myxozoa (Chapter 12) due to their highly diverged anatomy and lifestyle. We also note that the authors of the Ctenophora (Chapter 13), Placozoa (Chapter 7) and Porifera (Chapter 8) chapters present well-established hierarchies within these groups. The bilaterian invertebrate taxonomy presents a challenge though at this level of branching with respect to phyla.

Developmental patterns in the different bilaterian phyla have been used to organize this group to finer levels. For instance, the developmental formation of the mouth and the anus (**Figure 5.4**) in bilaterians strikes a clear boundary between the higher taxonomic groups of Deuterostomea and Protostomea. Specifically, protostomes develop their mouths at the end of the embryo where the blastopore first develops and deuterostomes develop their mouths at the opposite end of the embryo where the blastopore forms. This basic developmental difference is a very precise indicator of membership in these two higher taxonomic entities. Within the protostomes some researchers recognize a major split between protostome animals that molt and animals that don't.

All Bilateria start with normal division. The ancestral protostome developed via spiral cleavage at the eight-cell stage, but this mode of embryonic

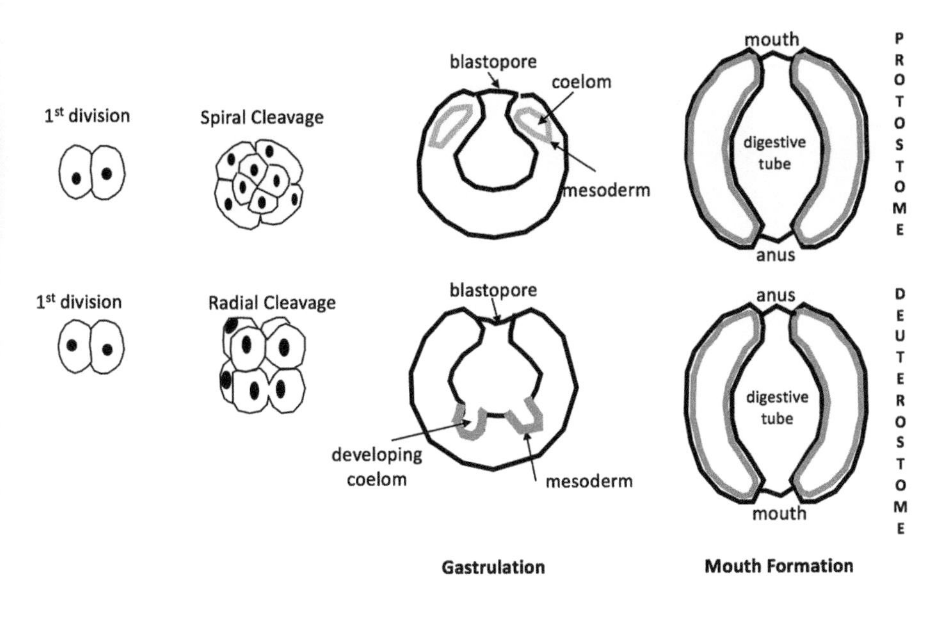

Figure 5.4 Diagram showing the difference between the developmental patterns of protostomes and deuterostomes.

development was lost in some lineages of protostomes (like the Ecdysozoa; see below). Deuterostomes, on the other hand, undergo radial cleavage at the same stage. As the embryo goes through gastrulation, the blastopore, mesoderm and developing coelom develop as shown in the diagram. For protostomes the coelom develops nearer the blastopore and in deuterostomes the coelom develops distant from the blastopore. This developmental pattern results in the mouth forming where the blastopore originated in protostomes and the mouth developing at the opposite end of the embryo from where the blastopore originated.

The next basic division occurs within protostomes and in some classifications it splits the protostomes into the Ecdysozoa and Lophotrochozoa (with several "orphan" taxa like Acanthocephala). Ecdysozoa are those animals that molt or have a molting stage. This higher group includes the phyla Priapulida, Kinorhyncha, Nematoda, Nematomprpha, Lorciphera, Onychophora, Tardigrada and Arthropoda. Several higher groupings are also discussed in the literature and these are mentioned in the various chapters on Ecdysozoa (Table 5.3).

Our organization of invertebrate groups follows the suggestions of the authors of chapters in the text. Visually we organize the metazoans based on the phylogenetic tree from Laumer et al. (2019) with some reservations. Whereas that publication resolves the order of branching for the Sister of all other Metazoans (SOM), we deresolve Porifera, Ctenophora and Placozoa. We suggest that a resolved topology of these groups is still in contention and we discuss the ramifications of one topology over another in Chapter 4. But we recognize the Laumer et al. (2019) study as a good organizing tool as they have included representatives from the grand majority of phyla that we describe in the text. **Figure 5.5** shows this topology with some tweaks.

PROBLEMATICA

The Ctenophore Problem

We have discussed the placement of Ctenophora in the animal tree of life in detail in Chapter 4. Briefly, morphological evidence places the phylum as sister to Cnidaria. Some molecular analyses place it in a more basal position. For molecular data (Chapter 4) depending on the model used either Porifera or Ctenophora appear to be the sister to all other metazoans (SOM). Feuda et al. (2017) argue that the models that produce Ctenophora as SOM suffer from inadequate accounting for the evolutionary process. When the adequacy of the model is increased then Porifera are the SOM. The topology of the five major

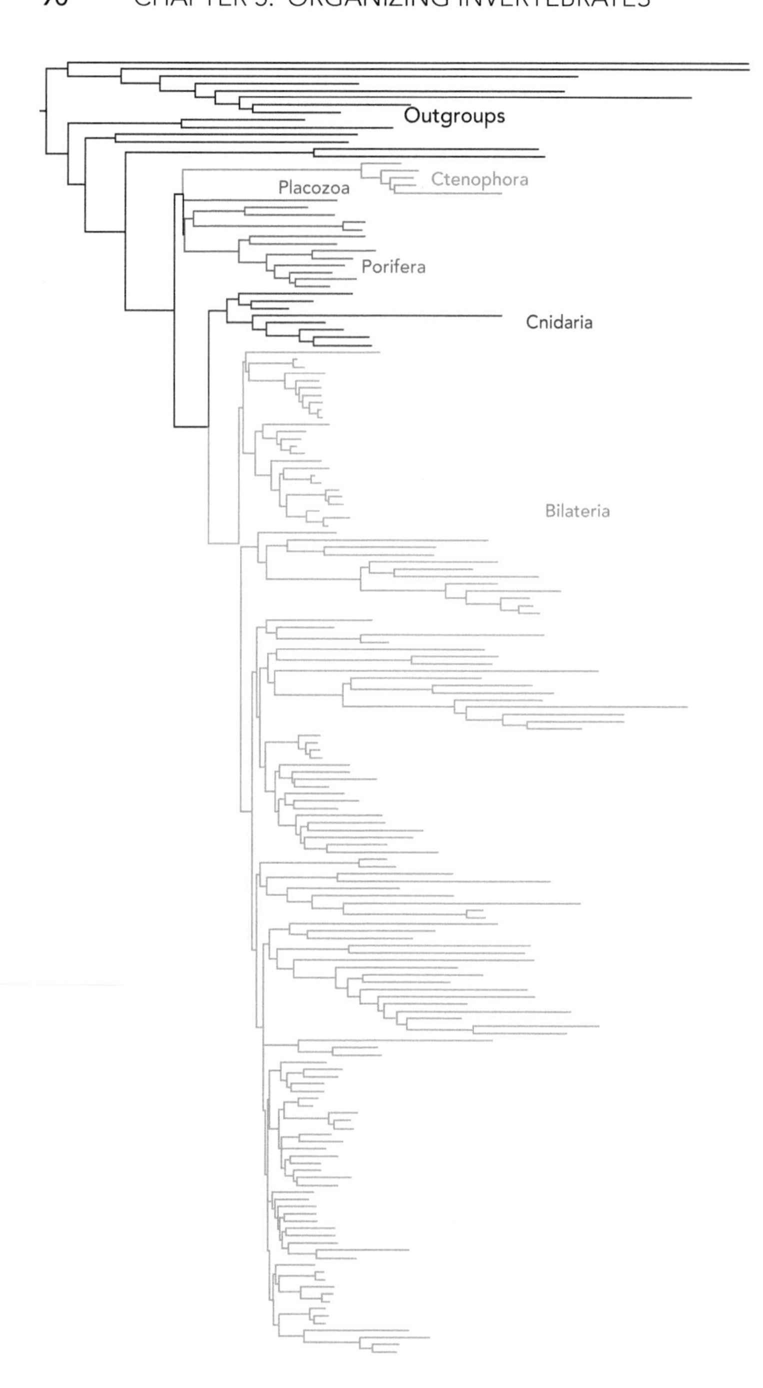

Figure 5.5 The topology we use is shown in the following tree where the five major groups of metazoans (Porifera = red, Ctenophora = green, Placozoa = purple, Cnidaria = blue and Bilateria = orange) are colored (the closest relatives to metazoan are used as outgroups here and are colored black. Since the majority of phyla approached in this textbook are bilaterians, the tree has taxa mostly colored orange. The arrangement of taxa within the Bilateria is highly resolved by Laumer et al. (2019). The reader should consult the chapters in this book for possible alternative arrangements of taxa.

taxa at the base of the Metazoa is essential in understanding certain aspects of morphological evolution, like the evolution of nervous systems and muscles. At this point in time, though, we have to conclude that the placement of this phylum is enigmatic.

The Cnidaria-Placozoa Problem: Recently, Laumer et al. (2019) have suggested that Cnidaria and Placozoa are sister taxa. They based this inference on phylogenomic analysis of several placozoan lineages and cnidarian lineages. The putative Placozoa-Cnidaria group is then suggested to be the sister to Bilateria. Because this inference is based on a single study with only a few placozoan lineages we will treat Placozoa as its own lineage sister to a Bilateria-Cnidaria clade or placed even more basally.

The Mesozoa Problem: The Mesozoa contain two potential phyla – the Orthonectida and the Rhombozoa (Dicyemida). While the name Mesozoa occurs frequently in the literature and on the web, some systematists have abandoned the name and now refer to the phyla Dicyemida and Orthonectida

(**Table 5.2**). We do not treat these potential phyla in any detail in this textbook, other than the short treatment below. Part of our reason for not treating them fully in this text is that they are a small group (about 50 known species) and their biology is poorly understood. As we noted above, molecular phylogenetic analysis suggests that they are on their own separate lineages, members of the Spiralia as defined by Giribet and Edgecombe (2020) and basal within the Spiralia. In general, they have very simple reduced bodies (about 50 cells). The two major groups Orthonectida (rare parasitic wormlike marine organisms) and Dicyemida (sometimes classified with the Orthonectids) can be distinguished by looking at their asexual forms where the former is amoeboid in shape and lacks cilia and the latter has cilia and is long and thin. Both groups are parasites on marine invertebrates (including platyhelminthes, echinoderms, molluscs and annelids).

The Xenacoelomorpha Problem: Basal to all other Bilateria are some two groups of animals – the Acoelomorpha and the Xenoturbellida in Chapter 36. Due to the small size of these two groups we focus only on one, the Xenoturbellida. Here we briefly treat the Acoelomorpha.

Acoels. Phylogenetic organization of this group is

Supergroup	Bilateria
Phylum	Xenacoelomorpha
Subphylum	Acoelomorpha
Order	Acoela

Acoels lack a fluid-filled body cavity (hence their name a [absence] of coelum [cavity]). More specifically they lack a conventional gut and so the mouth of acoels opens into the mesenchyme. Acoels digest food by means of a vacuolar system that facilitates the ingestion of food. However, there are no epithelial cells lining the acoel digestive system. They are small wormlike animals that appear flattened. They range in size from 2 mm to 15 mm. Acoels have a worldwide distribution and are found mostly in marine or brackish environments. Acoels used to be classified with Platyhelminthes, but molecular data clearly removes the acoels (as well as the Xenoturbellida). It is clear from both morphological and molecular information that Acoelomorpha and Xenoturbellida are sister taxa forming the phylum Xenacoelomorpha and branch from all other Bilateria. Together they are sister to another enigmatic group, the Nephrozoa. The position in the animal tree of life and relationships within the Xenacoelomorpha is a target of vibrant research by animal systematists.

The Spiralia/Lophotrochozoa Problem: Some researchers use Spiralia as the counter-group to Ecdysozoa because all of the so-called Lophotrochozoa undergo spiral cleavage. Which of these two names is used at this point in time depends on the definition of Lophotrochozoa. Most researchers consider the Cycliophora, Annelida, Mollusca, Brachiopoda, Phoronida, Entoprocta, Ectoprocta and Nemertea as members of the Lophotorochozoa (Table 5.4). The Gnathifera are problematic here because they do not have lophophore structures (one diagnostic for Lophotrochozoa) and are hence not considered Lophotrochozoa. The Gasrotricha and Platyhelminthes (grouped together in Rouphozoa) also lack lophophores and so they are usually excluded from the Lophotrochozoa. So, the name for the group Lophotrochozoa + Rouphozoa + Gnathifera has been amended to Spiralia. The name Platytrophozoa is also used at times and it refers to Rouphozoa + Lophotrochozoa. We recognize this higher-order naming problem, but point out that the organization is based entirely on molecular phylogenies.

There are many other problematic questions in invertebrate systematics, and these are treated throughout this book in the individual chapters.

CONCLUSION

A full treatment of invertebrates requires an understanding of phylogenetic relationships of the various phyla and the groups within these phyla. Because phylogenetic analysis is a revisionary approach, several controversies or problems have arisen in the development of a systematic treatment of invertebrate animals. These problems or controversies do not detract from the validity of the study of invertebrates. While some students might find these problems and controversies distracting and frustrating, the problems ultimately give vibrancy and excitement to the field of invertebrate zoology. The beginner should simply ignore the controversies and study group by group and pick an evolutionary scenario that helps him best to learn and memorize. Understanding the different bauplans is much more challenging than making different trees from the same dataset.

REFERENCES

Adl, Sina M., David Bass, Christopher E. Lane, Julius Lukeš, Conrad L. Schoch, Alexey Smirnov, Sabine Agatha et al. "Revisions to the classification, nomenclature, and diversity of eukaryotes." Journal of Eukaryotic Microbiology 66, no. 1 (2019): 4–119.

Adoutte, André, Guillaume Balavoine, Nicolas Lartillot, Olivier Lespinet, Benjamin Prud'homme, and Renaud De Rosa. "The new animal phylogeny: reliability and implications." Proceedings of the National Academy of Sciences 97 no. 9 (2000): 4453–4456.

Anderson, Donald Thomas. *Invertebrate Zoology*. ISBN 9780195513684. 2001.

Brunet T., King N. "The origin of animal multicellularity and cell differentiation." Dev. Cell 43, (2017): 124–140. 10.1016/j.devcel.2017.09.016, PMID 29065305.

Brusca, Richard C., and Gary J. Brusca. *Invertebrates*. No. QL 362. B78 2003. Basingstoke, 2003.

Burki, Fabien, Andrew J. Roger, Matthew W. Brown, and Alastair G. B. Simpson. "The new tree of eukaryotes." Trends in Ecology & Evolution 35, no. 1 (2020): 43–55.

Cavalier-Smith, Thomas. "A revised six-kingdom system of life." Biological Reviews 73, no. 3 (1998): 203–266.

Dunn, Casey W., Gonzalo Giribet, Gregory D. Edgecombe, and Andreas Hejnol. "Animal phylogeny and its evolutionary implications." Annual Review of Ecology, Evolution, and Systematics 45 (2014): 371–395.

Feuda, Roberto, Martin Dohrmann, Walker Pett, Hervé Philippe, Omar Rota-Stabelli, Nicolas Lartillot, Gert Wörheide, and Davide Pisani. "Improved modeling of compositional heterogeneity supports sponges as sister to all other animals." Current Biology 27, no. 24 (2017): 3864–3870.

Glenner, Henrik, Anders J. Hansen, Martin V. Sørensen, Frederik Ronquist, John P. Huelsenbeck, and Eske Willerslev. "Bayesian inference of the metazoan phylogeny: a combined molecular and morphological approach." Current Biology 14, no. 18 (2004): 1644–1649.

Giribet, Gonzalo, and Gregory D. Edgecombe. *The Invertebrate Tree of Life*. Princeton, NJ: Princeton University Press, 2020.

Golaconda Ramulu, Hemalatha, Pierre Pontarotti, and Didier Raoult. "The rhizome of life: what about metazoa?" Frontiers in Cellular and Infection Microbiology 2 (2012): 50.

Halanych, Kenneth M. "The new view of animal phylogeny." Annu. Rev. Ecol. Evol. Syst. 35 (2004): 229–256.

Keeling, Patrick J., and Fabien Burki. "Progress towards the Tree of Eukaryotes." Current Biology 29, no. 16 (2019): R808–R817.

Laumer, Christopher E., Rosa Fernández, Sarah Lemer, David Combosch, Kevin M. Kocot, Ana Riesgo, Sónia CS Andrade, Wolfgang Sterrer, Martin V. Sørensen, and Gonzalo Giribet. "Revisiting metazoan phylogeny with genomic sampling of all phyla." Proceedings of the Royal Society B 286, no. 1906 (2019): 20190831.

Margulis, Lynn, Karlene V. Schwartz, and Michael Dolan. *Diversity of Life: The Illustrated Guide to the Five Kingdoms*. Burlington, MA: Jones & Bartlett Learning, 1999.

Moore, Janet. *An Introduction to the Invertebrates*. Cambridge: Cambridge University Press, 2001.

Pechenik, J. A. *Biology of the Invertebrates*, 5th edition. Boston, MA: McGraw-Hill Higher Education, 2004.

Ruppert, Edward E., Richard S. Fox, and Robert D. Barnes. *Invertebrate Zoology: A Functional Evolutionary Approach*. Belmont, CA: Brooks/Cole (Thomson), 2004.

Telford, Maximilian J., Graham E. Budd, and Hervé Philippe. "Phylogenomic insights into animal evolution." Current Biology 25, no. 19 (2015): R876–R887.

Tree of life website: http://tolweb.org/tree/.

Wallace, Robert L., and Walter Kingsleycoaut Taylor. *Invertebrate Zoology:* A Laboratory Manual. No. 592 W3. 2003.

HIGHER TAXA

1

INTRODUCTION TO THE INVERTEBRATE TREE OF LIFE

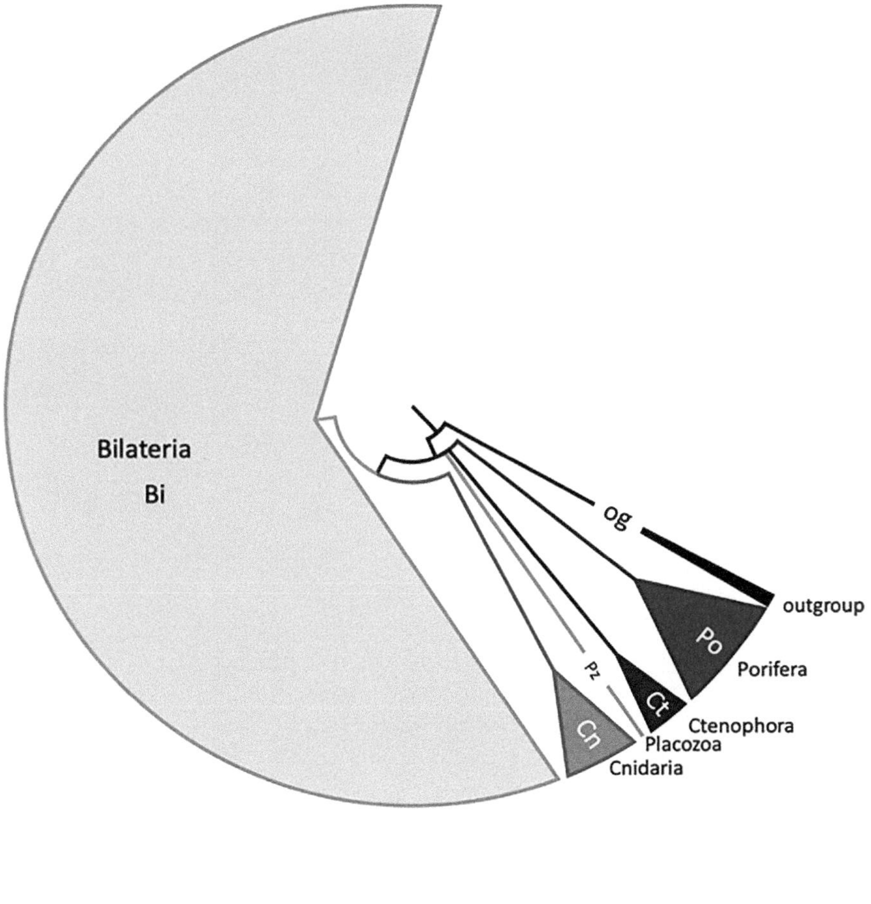

We loosely organize this textbook on a phylogenetic tree from Laumer et al. (2019) with some reservations. Whereas that publication resolves the order of branching for the Sister of all other Metazoans (SOM), we de-resolve Porifera (PO), Ctenophora (Ct) and Placozoa (Pz). Two other groups are often involved in discussing the invetebrate animals – the phylum Cnidaria (Cn) and the larger supergroup called Bilateria (Bi). We suggest that a resolved topology of these groups is still in contention and we discuss the ramifications of one topology over another in Chapter 4. But we recognize the Laumer et al. (2019) study as a good organizing tool as they have included representatives from the grand majority of phyla that we describe in the text.

The topology we use is shown in the following tree where the five major groups of metazoans (Porifera, Ctenophora, Placozoa, Cnidaria and Bilateria) are shown. The closest relatives to metazoans are used as outgroups here and are colored **black**. Since the majority of phyla approached in this textbook are bilaterians, the tree has taxa mostly in that group. In progressing through the major phyla described in this textbook and website for the textbook we will use blue to highlight the phylum being discussed.

General Characteristics

Metazoans have a complex multicellular structural organization that includes the tissues, organs and organ systems

Metazoans are relatively larger in size than unicellular protozoans

Locomotion in metazoans is in general highly developed resulting in contractile muscular elements and nervous structures

Most metazoans show differentiation of the anterior end or head (cephalization) resulting in a centralization of the nervous system in the head region

CHAPTER

6

THE CLOSEST UNICELLULAR RELATIVES OF ANIMALS

Michelle M. Leger and Iñaki Ruiz-Trillo

Holozoa

Classes: Choanoflagellatea, Filasterea, Ichthyosporea

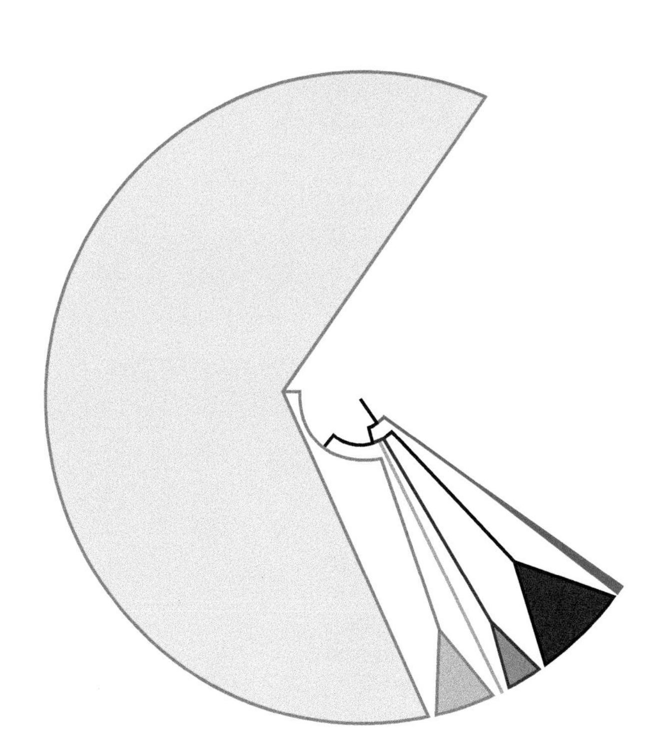

CONTENTS

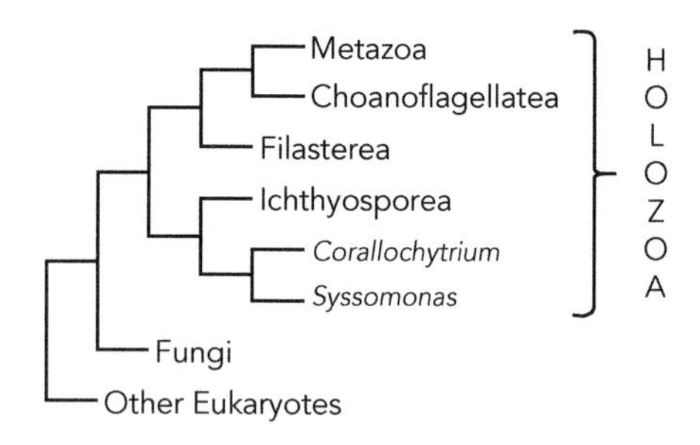

Figure 6.1 Schematic phylogeny highlighting the major groups within Holozoa. Data from Hehenberger et al. (2017); Torruella et al. (2015). Please note that some uncertainty remains with respect to the position of Pluriformea (*Corallochytrium* + *Syssomonas*). 'Metazoa' includes Porifera, Ctenophora, Placozoa, Cnidaria and Bilateria.

Animals (Metazoa) are by definition multicellular, and understanding how their multicellularity emerged is a major evolutionary question. Crucial to shedding light on this question, and to reconstructing the nature of the last metazoan common ancestor, are the close extant unicellular relatives of animals. Three major lineages of these organisms have been described: Choanoflagellatea, Filasterea and Ichthyosporea (**Figure 6.1**), as have additional species of uncertain phylogenetic placement. Together with animals themselves, these protist (also known as protozoan) lineages form the group Holozoa. Unlike the other organisms discussed in this book, relatively little is known about the unicellular holozoans; furthermore, there is considerable genetic and phenotypic variety within each group. Accordingly, there is little to say about the physiology and behaviour of these organisms. Hence we dispense with discussions of these two topics in this chapter.

HISTORY OF TAXONOMY AND CLASSIFICATION

Choanoflagellatea (Cavalier-Smith, 1998b emend. Nitsche et al., 2011)

The most closely related unicellular relatives of animals, forming the best-described group, are the choanoflagellates (Cavalier-Smith, 1997; Nitsche et al., 2011). Reported as far back as the early 19th century (reviewed in Leadbeater, 2015), around 360 extant species have been identified to date (Richter and Nitsche, 2017).

Cells are ovoid or rounded, with a cell body 1.2–10 μm in length, and possess a single apical flagellum surrounded by a collar of 20–50 microvilli (reviewed in Leadbeater, 2015; Richter and Nitsche, 2017). Cells may be naked, or covered in a mucilaginous glycocalyx, a rigid organic theca exhibiting a variety of morphologies, or a silicate, basket-like lorica (reviewed in Leadbeater, 2015; Richter and Nitsche, 2017). Clonal multicellularity in the form of colonies has been described in numerous species; colonies may take the form of chains or rosettes of varying sizes, and appear to be a transient life stage (Dayel et al., 2011) (reviewed in Leadbeater, 2015). Unicellular fast-swimming, slow-swimming and sessile stages have also been described (Dayel et al., 2011).

Two major groups of choanoflagellates have been recognized based on morphological characteristics, and confirmed by molecular data (Carr et al., 2008; Carr et al., 2017; Cavalier-Smith, 1997; Nitsche et al., 2011). The Craspedida comprise those choanoflagellates that only possess a glycocalyx or an organic theca, while those possessing a silicate lorica are placed in the Acanthoecida (Cavalier-Smith, 1997). The Acanthoecida are further subdivided based on molecular data, cell division morphology and lorica assembly into the Acanthoecidae and the Stephanoecidae (Carr et al., 2008; Carr et al., 2017; Leadbeater, 2015; Nitsche et al., 2011).

The morphological similarity between choanoflagellates and the choanocytes of sponges (Dujardin, 1841; James-Clark, 1867) led to speculation of a close relationship between choanoflagellates and animals (see e.g. Cavalier-Smith, 1981; James-Clark, 1867; Saville-Kent, 1878), and molecular data have confirmed this relationship (Carr et al., 2008; Lang et al., 2002; Shalchian-Tabrizi et al., 2008; Torruella et al., 2015).

Filasterea (Shalchian-Tabrizi et al., 2008)

Filasterea (Hehenberger et al., 2017; Shalchian-Tabrizi et al., 2008; Torruella et al., 2015) comprises three genera: Capsaspora (Hertel et al., 2002), Ministeria (Patterson et al., 1993), and Pigoraptor (Hehenberger et al., 2017). All are naked, uninucleate cells that appear to possess filopodia in at least some life stages (Shalchian-Tabrizi et al., 2008), and at least some members have been reported to exhibit aggregative multicellularity (Hehenberger et al., 2017; Sebé-Pedrós, Irimia et al., 2013).

Capsaspora owczarzaki was isolated from the mantle, pericardial tissue and haemolymph of the freshwater pulmonate snail *Biomphalaria glabrata* (Hertel et al., 2002; Owczarzak et al., 1980; Stibbs et al., 1979). *Capsaspora* cells are rounded, around 3–7 µm in diameter (Hertel et al., 2002), and exhibit temporally differentiated life stages (Sebé-Pedrós, Irimia et al., 2013): a filopodial amoebic stage during which cells proliferate and adhere to the substrate; an aggregative stage, during which cells produce an extracellular matrix and lose their filopodia in the course of cell–cell adhesion; and a cystic stage devoid of filopodia. These life stages are characterized by differences in gene expression (Sebé-Pedrós, Irimia et al., 2013), protein abundance and phosphorylation patterns (Sebé-Pedrós, Pena et al., 2016), and epigenomic changes (Sebé-Pedrós, Ballare et al., 2016).

In contrast to the commensal *Capsaspora*, members of the two other filasterean genera, *Ministeria* and *Pigoraptor*, are free-living heterotrophs. Two species have been described in each genus. *Ministeria vibrans* (Tong, 1997) and the poorly studied *Ministeria marisola* (Patterson et al., 1993) are bacterivores isolated from coastal waters off the United Kingdom and Cuba, respectively. The cells are small (1–4 µm), spherical and possess 14–30 radiating filopodia up to 9 µm (Patterson et al., 1993; Tong, 1997). *M. vibrans* possesses a flagellum (Torruella et al., 2015) initially described as a stalk (Tong, 1997).

Pigoraptor chileana and *P. vietnamica* (Hehenberger et al., 2017) are larger (5–14 µm), freshwater eukaryovores, feeding on bodonid flagellates, isolated from a submerged meadow in Chile and a lake in Vietnam respectively. The cells are uniflagellate, ovoid or rounded, and sometimes exhibit short, thin pseudopodia (Hehenberger et al., 2017). The cells have been observed to form cysts and clusters (Hehenberger et al., 2017).

Ichthyosporea (Cavalier-Smith, 1998b)

The earliest branching group of Holozoa are the Ichthyosporea (Cavalier-Smith, 1998b) (or Mesomycetozoea (Herr et al., 1999). Most of the more than 40 described species are parasites or commensals of fish or arthropods (Glockling et al., 2013), but ichthyosporeans are also found associated with mammals or birds (Pal et al., 2016); three free-living saprotrophic species have also been reported (Hassett et al., 2015; van Hannen et al., 1999).

Ichthyosporeans exhibit coenocytic life stages: the formation of new daughter cells is preceded by several rounds of nuclear division that result in large, multinucleate cells known as coenocytes, sporocytes or sporangia (Mendoza et al., 2002). Within a single species of ichthyosporean, these coenocytes may contain anywhere between 1–2 (in newborn cells) and hundreds of nuclei (Marshall and Berbee, 2011; Mendoza and Vilela, 2013; Ondracka et al., 2018). Cell size within a population is highly variable (Hassett et al., 2015; Marshall and Berbee, 2011; Marshall et al., 2008), and may be dependent on nutrient availability (Ondracka et al., 2018). Some species (Dyková and Lom, 1992; Marshall and Berbee, 2011; Okamoto et al., 1985) make elongated, coenocytic structures similar to fungal hyphae.

Ichthyosporeans were initially variously described as fungi, algae or protists (reviewed in Mendoza et al., 2002); phylogenetic analyses showed the need for these organisms to be placed separately (Ragan et al., 1996; Spanggaard et al., 1996), in what was briefly known as the DRIP clade (Ragan et al., 1996), an acronym of the original members (*Dermocystidium*, the rosette agent [now known as *Sphaerothecum destruens*], *Ichthyophonus* and *Psorospermium*) before Ichthyosporea was coined for the group (Cavalier-Smith, 1998a). Two major clades are recognized and supported by phylogenetic analyses: the Ichthyophonida and the Dermocystida (Cavalier-Smith, 1998b; Grau-Bové et al., 2017; Mendoza et al., 2002; Torruella et al., 2015). In general, members of the Ichthyophonida possess rounded or elongated coenocytes, and their endospores give rise to amoeboid zoospores, while Dermocystida possess rounded coenocytes, and uniflagellated zoospores (Mendoza et al., 2002).

Controversy surrounds the phylogenetic position of three other holozoan species, *Corallochytrium limacisporum* (Raghu-kumar, 1987), *Syssomonas multiformis* (Hehenberger et al., 2017), and *Tunicaraptor unikontum* (Tikhonenkov et al., 2020). *C. limacisporum* (Raghu-kumar, 1987) was isolated from coral reef lagoons (Raghu-kumar, 1987; Torruella et al., 2015); cells are usually small (4.5–7.5 μm) and with a thin cell wall, but may grow to a diameter of 20 μm (Raghu-kumar, 1987). It divides by binary fission, resulting in dyads or tetrads of adhered cells, and additionally possesses rounded single cell and elongated, motile amoebae (Mendoza et al., 2002). *S. multiformis* is a eukaryovore isolated from freshwater sediments (Hehenberger et al., 2017). Cells are naked, 7–14 μm in length and may occur as rounded uniflagellate cells, amoeboflagellates, unflagellated amoeboid cells, or rounded cysts. Torruella and colleagues confirmed earlier work (Mendoza et al., 2002) suggesting phylogenetic affinity between *Corallochytrium* and ichthyosporeans, and coined the term Teretosporea for the group (Grau-Bové et al., 2017; Torruella et al., 2015). A subsequent analysis including *Syssomonas*, but not including as many ichthyosporean taxa, failed to recover Teretosporea. Instead, it placed *Corallochytrium* sister to *Syssomonas*, at the base of animals, choanoflagellates and Filasterea; this grouping was named Pluriformea (Hehenberger et al., 2017). *T. unikontum* is a small (3–5 μm) eukaryovorous flagellate; while it is clearly a holozoan, its precise position within the group remains unclear (Tikhonenkov et al., 2020).

REPRODUCTION AND DEVELOPMENT

All of the organisms described above are known to reproduce clonally. In choanoflagellates and filastereans, this takes the form of binary fission.

In ichthyosporeans and *Corallochytrium* the process is more complex, with cells undergoing successive rounds of (clonal) nuclear division to form multinucleate coenocytes (Mendoza et al., 2002; Raghu-kumar, 1987). New cell walls form around daughter nuclei (a process known as cellularization), followed by rupture of the wall of the mother cell and the release of daughter cells, known as endospores.

Based on population genetics (Marshall and Berbee, 2010) and the presence of meiotic genes (Carr et al., 2010; Hofstatter and Lahr, 2019; Suga et al., 2013), at least some of these organisms are also capable of sexual reproduction, but the conditions under which this might occur are not known in most cases. An exception is the choanoflagellate *Salpingoeca rosetta*, in which sexual reproduction can be induced by starvation (Levin and King, 2013) or the presence of EroS, a chondroitinase produced by some of its bacterial prey species (Woznica et al., 2017).

Choanoflagellates undergo rapid clonal division to form chain- or 'rosette'-like colonies (Dayel et al., 2011). Intriguingly, colony formation in *Salpingoeca rosetta* (**Figure 6.2**) can be triggered, enhanced or inhibited by the presence of lipids produced by prey bacteria (Alegado et al., 2012; Woznica et al., 2016);

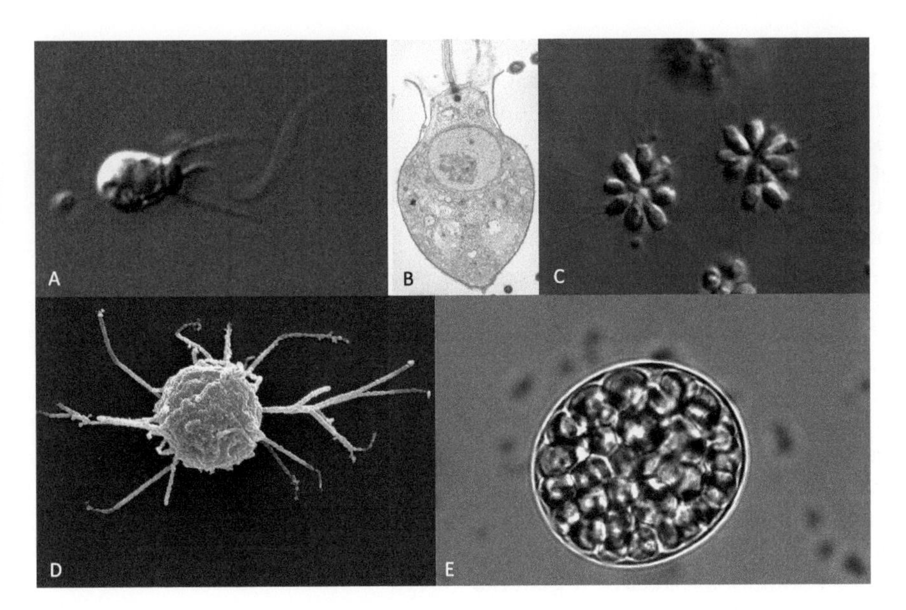

Figure 6.2 Micrographs of unicellular holozoans. (A) Phase contrast microscopy of *Salpingoeca rosetta* (Choanoflagellatea). (B) Transmission electron micrograph of *Salpingoeca* sp. (Choanoflagellatea) (C) Phase contrast microscopy of colonies of *Salpingoeca rosetta*. (D) Scanning electron micrograph of an adherent, filopodiated *Capsaspora owczarzaki* (Filasterea). (E) Cell phase contrast light microscopy of a *Pirum gemmata* (Ichthyosporea) coenocyte, showing cellularization. (Photos A, B, C: Adapted from Wikimedia Commons – Creative Commons Attribution-Share Alike 2.5 Generic license. A. Stephen Fairclough; B. Sergey Karpov; C. Mark J. Dayel. Photos in D and E: Courtesy of authors and from Flickr account under a Creative Commons license.)

possibly this allows *Salpingoeca* to detect conditions under which the benefits of rosette formation (increased prey capture rate) (Roper et al., 2013) outweigh its disadvantages (reduced motility) (Woznica et al., 2016). These colonies demonstrate simple multicellular features: there are indications that individual cells within rosette colonies may undergo cell differentiation (Laundon et al., 2019; Naumann and Burkhardt, 2019), and in some cases the colony's overall morphology changes, adopting shapes suited to predation or swimming (Brunet et al., 2019).

DISTRIBUTION AND ECOLOGY

Choanoflagellates are ubiquitous in a range of marine and freshwater habitats, including hypoxic and hypersaline waters, and in sea ice and waterlogged soils (reviewed in Leadbeater, 2015). All choanoflagellates identified to date are free-living heterotrophs, filter-feeding on bacteria and small eukaryotes (Pettitt et al., 2002; Richter and Nitsche, 2017).

Filastereans are far less abundant; they are notable for their scarcity in, or absence from, environmental sequence data (Arroyo et al., 2018; del Campo et al., 2015; del Campo and Ruiz-Trillo, 2013; Heger et al., 2018). Nevertheless, the five known species were isolated from a variety of environments: Cuban and British coastal waters (Patterson et al., 1993; Tong, 1997), freshwater environments in Chile and Vietnam (Hehenberger et al., 2017) and from the freshwater snail *Biomphalaria glabrata* (Owczarzak et al., 1980).

Ichthyosporeans have mostly been found as parasites or commensals of, or associated with, arthropods, fish, birds and mammals (Glockling et al., 2013; Pal et al., 2016), but free-living saprotrophic species have also been described (Hassett et al., 2015; van Hannen et al., 1999). Their abundance in low size fraction freshwater and marine environmental sequence data might reflect free-living life stages (Arroyo et al., 2018; del Campo and Ruiz-Trillo, 2013). In contrast, Pluriformea are much less abundant: they are rare in environmental sequence surveys (del Campo et al., 2015; del Campo and Ruiz-Trillo, 2013; Heger et al., 2018), an exception being data from the freshwater Paraná river (Arroyo et al., 2018). The two known species were isolated from coral reef lagoons in Hawaii (Raghu-kumar, 1987) and India, and from a lake in Vietnam (Hehenberger et al., 2017).

GENETICS AND GENOMICS

Relatively few genomes and transcriptomes are publicly available for unicellular holozoans to date (**Table 6.1**). These include four choanoflagellate genomes (two from single cells), and transcriptomes of 20 additional species; genomes and transcriptomes of two filastereans, and transcriptome-based predicted peptides for two additional species (Hehenberger et al., 2017); seven ichthyosporean genomes, and transcriptomes of two additional species; a genome and transcriptome of *Corallochytrium*, with transcriptome-based proteome data available for *Syssomonas* (Hehenberger et al., 2017); and a transcriptome of *Tunicaraptor* (Tikhonenkov et al., 2020).

Genetic manipulation in these lineages is in its infancy, but transfection and (in the first case) gene silencing protocols have been developed in *Creolimax fragrantissima* (Suga and Ruiz-Trillo, 2013), in *Abeoforma whisleri* (Faktorová et al., 2020), in *Corallochytrium limacisporum* (Kożyczkowska et al., 2018), in *Capsaspora owczarzaki* (Parra-Acero et al., 2018) and in *Salpingoeca rosetta* (Booth et al., 2018). Efforts to develop further genetic tools in these and other unicellular holozoans are underway.

POSITION IN THE ToL

Animals and their closest unicellular relatives together form the group Holozoa. The grouping of Holozoa with Holomycota (the group comprising fungi and their closest unicellular relatives) is known as the Opisthokonta (Cavalier-Smith, 1987), only one of the major eukaryotic groups currently recognized (Adl et al., 2019). It is important to note that unicellular holozoans are not monophyletic – that is, they do not form a single clade, but rather at least three separate clades (Ichthyosporea, Filasterea and Choanoflagellatea), with the precise position of the Pluriformea and *Tunicaraptor* remaining unclear (Grau-Bové et al., 2017; Hehenberger et al., 2017; Torruella et al., 2015; Tikhonenkov et al. 2020). These clades diverged successively from the lineage leading to animals (Figure 6.1).

DATABASES AND COLLECTIONS

Publicly available sequence data of unicellular holozoans are shown in Table 6.1; these are generally available through the International Nucleotide Sequence Database Collaboration, consisting of the National Institutes of Health's genetic sequence repository, GenBank® (http://www.ncbi.nlm.nih.gov/genbank/), the DNA Data Bank of Japan (http://www.ddbj.nig.ac.jp) and the European Nucleotide Archive (http://www.ebi.ac.uk). Some of these data are also available through the Ensembl genome database project (http://www.ensembl.org), and its specialized protist section Ensembl Protists (http://www.protist.ensembl.org).

TABLE 6.1 Publicly available genomes and transcriptomes of unicellular holozoans

				Genome	Transcriptome
Choanoflagellatea	Acanthoecida		Single-cell amplified genome UC4	ERS3941092[1]	-
		Acanthoecidae	*Acanthoeca spectabilis*	-	GGPA00000000.1[2]
			Helgoeca nana	-	GGOR00000000.1[2]
			Savillea parva	-	GGOL00000000.1[2]
		Stephanoecidae	*Diaphanoeca grandis*	-	GGPB00000000.1[2]
			Didymoeca costata	-	GGOQ00000000.1[2]
			Stephanoeca diplocostata	-	GGOS00000000.1[2]
	Craspedida		*Choanoeca perplexa*	-	GGOP00000000.1[2]
			Choanoeca flexa	-	SAMN11533889[3]
			Codosiga hollandica	-	GGOV00000000.1[2]
			Hartaetosiga balthica	-	GGOO00000000.1[2]
			Hartaetosiga gracilis	-	GGOU00000000.1[2]
			Microstomoeca roanoka	-	GGON00000000.1[2]
			Monosiga brevicollis	GCA_000002865.1[4]	-
			Mylnosiga fluctuans	-	GGOI00000000.1[2]
			Salpingoeca dolichothecata	-	GGOK00000000.1[2]
			Salpingoeca helianthica	-	GGOJ00000000.1[2]
			Salpingoeca infusionum	-	GGOW00000000.1[2]
			Salpingoeca kvevrii	-	GGOX00000000.1[2]
			Salpingoeca macrocollata	-	GGOT00000000.1[2]
			Salpingoeca punica	-	GGOZ00000000.1[2]
			Salpingoeca rosetta	GCA_000188695.1[5]	SRP005692[5]
			Salpingoeca urceolata	-	GGOY00000000.1[2]
			Single-cell amplified genome UC1	ERS3941091[1]	-
Filasterea			*Capsaspora owczarzaki*	GCA_000151315.2[6]	SRP061354[7]
			Ministeria vibrans	SRS213700[8]	SRX096927[8]
			Pigoraptor chileana	-	-
			Pigoraptor vietnamica	-	-
Teretosporea	Ichthyosporea	Ichthyophonida	*Sphaeroforma arctica*	GCA_008580545.1[9,10]	ERP115662[10]
			Sphaeroforma sirkka	GCA_001586965.2[11]	-
			Creolimax fragrantissima	GCA_002024145.1[8]	SRP058061[12]
			Abeoforma whisleri	GCA_002812265.1[9]	SRS502376[8]
			Pirum gemmata	GCA_002812295.1[9]	SRS502375[8]
			Amoebidium parasiticum	-	SRX179384[8]
			Ichthyophonus hoferi	GCA_002751075.1[8]	SRS726091[8]
		Dermocystida	*Sphaerothecum destruens*	-	SRS725801[8]
			Chromosphaera perkinsii	SRX2510795[9]	SRX2508208[9]
	Pluriformea		*Corallochytrium limacisporum (Hawaii strain)*	GCA_002811645.1[9]	SRS725979[8]
			Syssomonas multiformis	-	-
	Incertae sedis		*Tunicaraptor unikontum*	-	GIQG00000000.1[13]

[1]López-Escardó et al., 2019; [2]Richter et al., 2018; [3]Brunet et al., 2019; [4]King et al., 2008; [5]Fairclough et al., 2013; [6]Suga et al., 2013; [7]Sebé-Pedrós, Ballare et al., 2016; [8]Torruella et al., 2015; [9]Grau-Bové et al., 2017; [10]Dudin et al., 2019; [11]Ducluzeau et al., 2018; [12]de Mendoza et al., 2015; [13]Tikhonenkov et al., 2020.

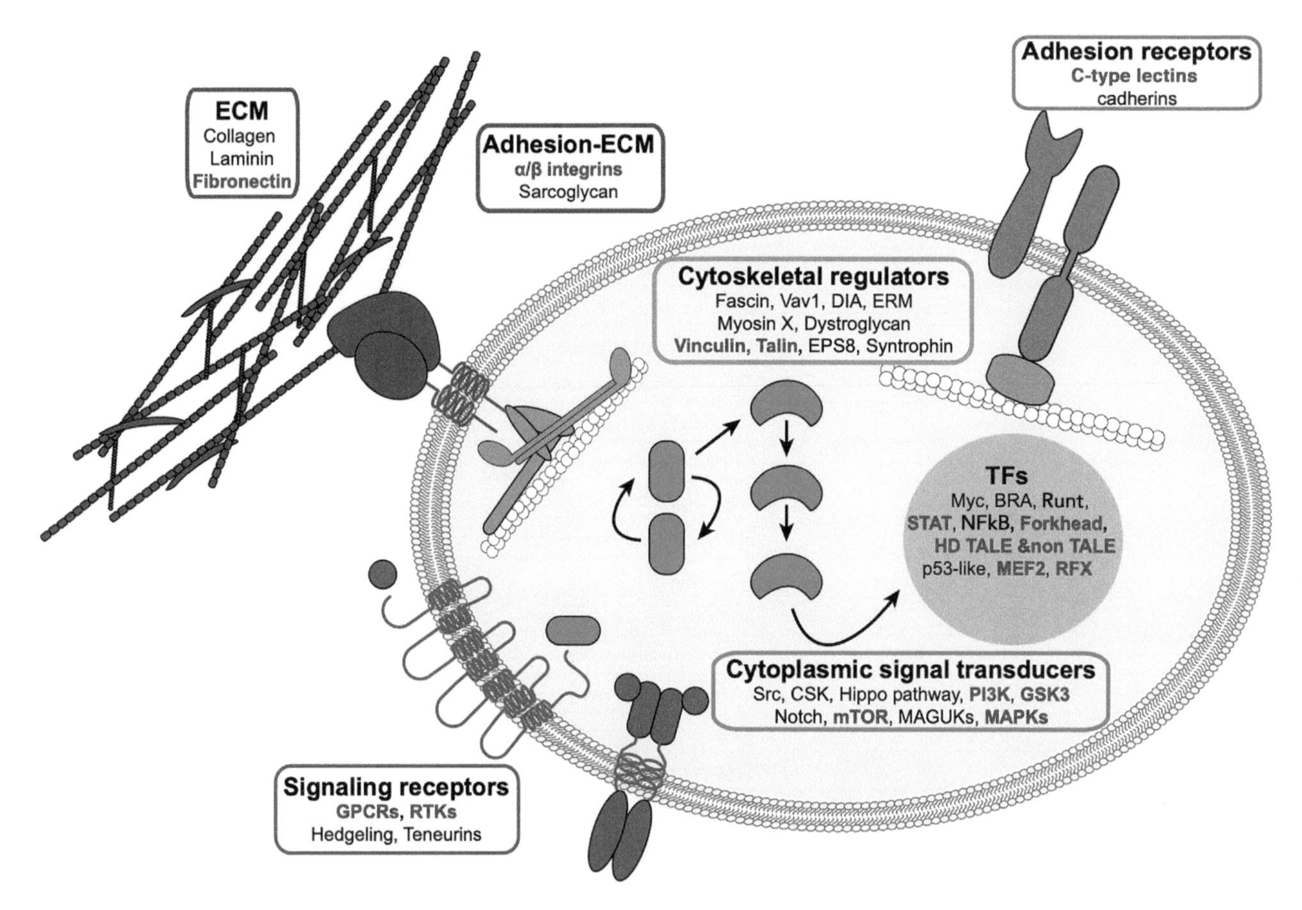

Figure 6.3 Key metazoan multicellularity-related genes that are also present in unicellular holozoans. Red = genes that are also inferred to have been present in the last common ancestor of all holozoans. (BRA, Brachyury; CSK, C-terminal Src kinase; DIA, diaphanous; EPS8, epidermal growth factor receptor kinase substrate 8; ERM, Ezrin–Radixin–Moesin proteins; GPCRs, G protein-coupled receptors; GSK3, glycogen synthase kinase 3; HD, homeodomain; MAGUKs, membrane-associated guanylate kinases; MAPKs, mitogen-activated protein kinases; MEF2, myocyte-specific enhancer factor 2; mTOR, mechanistic target of rapamycin; NFκB, nuclear factor-κB; PI3K, phosphatidylinositol 3-kinase; RTKs, receptor tyrosine kinases; STAT, signal transducer and activator of transcription; TALEs, three amino acid loop extensions; TF, transcription factor). Figure amended from Figure 3 of Sebé-Pedrós et al. (2017), with credit to Meritxell Antó. Data from Brown et al. (2013), de Mendoza et al. (2014), de Mendoza et al. (2013), de Mendoza et al. (2015), de Mendoza et al. (2010), Grau-Bové et al. (2017), Hehenberger et al. (2017), Philippon et al. (2015), Sebé-Pedrós, Burkhardt, et al. (2013), Suga et al. (2012, 2013).

CONCLUSION

Each of the unicellular holozoan groups discussed above possesses cellular and genomic features that are critical to multicellularity in animals (**Figure 6.3**; reviewed in (Richter and King, 2013; Sebé-Pedrós et al., 2017; see also Table 6.1). These include transcription factors such as members of the Homeobox or Forkhead domain families (Sebé-Pedrós et al., 2011); signaling pathways such as the Hippo (Sebé-Pedrós et al., 2012) or Notch (Richter et al., 2018) pathways; adhesion molecules such as integrins (Sebé-Pedrós and Ruiz-Trillo, 2010) or cadherins (Abedin and King, 2008; King et al., 2003); extracellular matrix-related domains (King et al., 2008); and structural remodeling proteins such as fascin or Ezrin–Radixin–Moesin proteins (Sebé-Pedrós, Burkhardt et al., 2013). As a result, efforts are underway to develop genetic tools in several unicellular holozoans in order to better understand these features, and how they might contribute to the emergence of animal multicellularity.

REFERENCES

Abedin, M., and King, N. (2008). The premetazoan ancestry of cadherins. *Science, 319*(5865), 946–948. doi:10.1126/science.1151084.

Adl, S. M., Bass, D., Lane, C. E., Lukes, J., Schoch, C. L., Smirnov, A., ... Zhang, Q. (2019). Revisions to the classification, nomenclature, and diversity of eukaryotes. *J. Eukaryot Microbiol 66*(1), 4–119. doi:10.1111/jeu.12691.

Alegado, R. A., Brown, L. W., Cao, S., Dermenjian, R. K., Zuzow, R., Fairclough, S. R., ... King, N. (2012). A bacterial sulfonolipid triggers multicellular development in the closest living relatives of animals. *Elife, 1*, e00013. doi:10.7554/eLife.00013.

Arroyo, A. S., López-Escardó, D., Kim, E., Ruiz-Trillo, I., and Najle, S. R. (2018). Novel diversity of deeply branching Holomycota and unicellular holozoans revealed by metabarcoding in Middle Paraná River, Argentina. *Front Ecol Evol.* doi:10.3389/fevo.2018.00099.

Booth, D. S., Szmidt-Middleton, H., and King, N. (2018). Choanoflagellate transfection illuminates their cell biology and the ancestry of animal septins. *Molecular Biology of the Cell,* Epublished ahead of print. doi:10.1091/mbc.E18-08-0514.

Brown, M. W., Sharpe, S. C., Silberman, J. D., Heiss, A. A., Lang, B. F., Simpson, A. G., and Roger, A. J. (2013). Phylogenomics demonstrates that breviate flagellates are related to opisthokonts and apusomonads. *Proc Biol Sci, 280*(1769), 20131755. doi:10.1098/rspb.2013.1755.

Brunet, T., Larson, B. T., Linden, T. A., Vermeij, M. J. A., McDonald, K., and King, N. (2019). Light-regulated collective contractility in a multicellular choanoflagellate. *Science, 366*(6463), 326–334. doi:10.1126/science.aay2346.

Carr, M., Leadbeater, B. S., and Baldauf, S. L. (2010). Conserved meiotic genes point to sex in the choanoflagellates. *J Eukaryot Microbiol, 57*(1), 56–62. doi:10.1111/j.1550-7408.2009.00450.x.

Carr, M., Leadbeater, B. S., Hassan, R., Nelson, M., and Baldauf, S. L. (2008). Molecular phylogeny of choanoflagellates, the sister group to Metazoa. *Proc Natl Acad Sci U S A, 105*(43), 16641–16646. doi:10.1073/pnas.0801667105.

Carr, M., Richter, D. J., Fozouni, P., Smith, T. J., Jeuck, A., Leadbeater, B. S. C., and Nitsche, F. (2017). A six-gene phylogeny provides new insights into choanoflagellate evolution. *Mol Phylogenet Evol, 107*, 166–178. doi:10.1016/j.ympev.2016.10.011.

Cavalier-Smith, T. (1981). Eukaryote kingdoms: seven or nine? *Biosystems, 14*(3–4), 461–481.

Cavalier-Smith, T. (1987). The origin of fungi and pseudofungi. In A. D. M. Rayner, B. C. M., and D. Moore (eds.), *Evolutionary Biology of the Fungi* (pp. 339–353): Cambridge University Press.

Cavalier-Smith, T. (1997). Amoeboflagellates and mitochondrial cristae in eukaryote evolution. *Archiv Fuer Protistenkunde, 147*, 237–258.

Cavalier-Smith, T. (1998a). A revised six-kingdom system of life. *Biol Rev Camb Philos Soc, 73*(3), 203–266.

Cavalier-Smith, T. (1998b). Neomonada and the origin of animals and fungi. In G. Coombs, K. Vickerman, M. Sleigh, and A. Warren (Eds.), *Evolutionary relationships among protozoa* (pp. 375–407). London: Chapman & Hall.

Dayel, M. J., Alegado, R. A., Fairclough, S. R., Levin, T. C., Nichols, S. A., McDonald, K., and King, N. (2011). Cell differentiation and morphogenesis in the colony-forming choanoflagellate *Salpingoeca rosetta. Dev Biol, 357*(1), 73–82. doi:10.1016/j.ydbio.2011.06.003.

de Mendoza, A., Sebé-Pedrós, A., and Ruiz-Trillo, I. (2014). The evolution of the GPCR signaling system in eukaryotes: modularity, conservation, and the transition to metazoan multicellularity. *Genome Biol Evol, 6*(3), 606–619. doi:10.1093/gbe/evu038.

de Mendoza, A., Sebé-Pedrós, A., Šestak, M. S., Matejčić, M., Torruella, G., Domazet-Loso, T., and Ruiz-Trillo, I. (2013). Transcription factor evolution in eukaryotes and the assembly of the regulatory toolkit in multicellular lineages. *Proc Natl Acad Sci U S A, 110*(50), E4858–4866. doi:10.1073/pnas.1311818110.

de Mendoza, A., Suga, H., Permanyer, J., Irimia, M., and Ruiz-Trillo, I. (2015). Complex transcriptional regulation and independent evolution of fungal-like traits in a relative of animals. *Elife, 4*, e08904. doi:10.7554/eLife.08904.

de Mendoza, A., Suga, H., and Ruiz-Trillo, I. (2010). Evolution of the MAGUK protein gene family in premetazoan lineages. *BMC Evol Biol, 10*, 93. doi:10.1186/1471-2148-10-93.

del Campo, J., Mallo, D., Massana, R., de Vargas, C., Richards, T. A., and Ruiz-Trillo, I. (2015). Diversity and distribution of unicellular opisthokonts along the European coast analysed using high-throughput sequencing. *Environ Microbiol, 17*(9), 3195–3207. doi:10.1111/1462-2920.12759.

del Campo, J., and Ruiz-Trillo, I. (2013). Environmental survey meta-analysis reveals hidden diversity among unicellular opisthokonts. *Mol Biol Evol, 30*(4), 802–805. doi:10.1093/molbev/mst006.

Ducluzeau, A. L., Tyson, J. R., Collins, R. E., Snutch, T. P., and Hassett, B. T. (2018). Genome Sequencing of Sub-Arctic Mesomycetozoean Sphaeroforma sirkka Strain B5, Performed with the Oxford Nanopore minION and Illumina HiSeq Systems. *Microbiol Resour Announc, 7*(15). doi:10.1128/MRA.00848-18.

Dudin, O., Ondracka, A., Grau-Bové, X., Haraldsen, A. A., Toyoda, A., Suga, H., ... Ruiz-Trillo, I. (2019). A unicellular relative of animals generates a layer of polarized cells by actomyosin-dependent cellularization. *Elife, 8*. doi:10.7554/eLife.49801.

Dujardin, F. (1841). *Histoire naturelle des zoophytes. Infusoires, comprenant la physiologie et la classification de ces animaux, et la manière de les étudier à l'aide du microscope.* Paris: Imprimerie de Fain et Thunot.

Dyková, I., and Lom, J. (1992). New evidence of fungal nature of *Dermocystidium koi* Hoshina and Sahara, 1950. *Journal of Applied Ichthyology, 8*, 180–185.

Fairclough, S. R., Chen, Z., Kramer, E., Zeng, Q., Young, S., Robertson, H. M., ... King, N. (2013). Premetazoan genome evolution and the regulation of cell differentiation in the choanoflagellate *Salpingoeca rosetta. Genome Biol, 14*(2), R15. doi:10.1186/gb-2013-14-2-r15.

Faktorová, D., R. Nisbet, E. R., Fernández Robledo, J. A., Casacuberta, E., Sudek, L., Allen, A. E., ... Lukeš, J. (2020). Genetic tool development in marine protists: emerging model organisms for experimental cell biology. *Nature Methods, In press.*

Glockling, S. L., Marshall, W. L., and Gleason, F. H. (2013). Phylogenetic interpretations and ecological potentials of the Mesomycetozoea (Ichthyosporea). *Fungal Ecology*, *6*(4), 237–247.

Grau-Bové, X., Torruella, G., Donachie, S., Suga, H., Leonard, G., Richards, T. A., and Ruiz-Trillo, I. (2017). Dynamics of genomic innovation in the unicellular ancestry of animals. *Elife*, *6*. doi:10.7554/eLife.26036.

Hassett, B. T., López, J. A., and Gradinger, R. (2015). Two new species of marine saprotrophic sphaeroformids in the Mesomycetozoea isolated from the sub-Arctic Bering Sea. *Protist*, *166*(3), 310–322. doi:10.1016/j.protis.2015.04.004.

Heger, T. J., Giesbrecht, I. J. W., Gustavsen, J., Del Campo, J., Kellogg, C. T. E., Hoffman, K. M., ... Keeling, P. (2018). High-throughput environmental sequencingreveals high diversity of litter and moss associated protist communities along a gradient of drainage and tree productivity. *Env Microbiol*, *20*(3), 1185–1203. doi:10.1111/1462-2920.14061.

Hehenberger, E., Tikhonenkov, D. V., Kolísko, M., Del Campo, J., Esaulov, A. S., Mylnikov, A. P., and Keeling, P. J. (2017). Novel predators reshape Holozoan phylogeny and reveal the presence of a two-component signaling system in the ancestor of animals. *Curr Biol*, *27*(13), 2043–2050 e2046. doi:10.1016/j.cub.2017.06.006.

Herr, R. A., Ajello, L., Taylor, J. W., Arseculeratne, S. N., and Mendoza, L. (1999). Phylogenetic analysis of *Rhinosporidium seeberi*'s 18S small-subunit ribosomal DNA groups this pathogen among members of the protoctistan Mesomycetozoa clade. *J Clin Microbiol*, *37*(9), 2750–2754.

Hertel, L. A., Bayne, C. J., and Loker, E. S. (2002). The symbiont *Capsaspora owczarzaki, nov. gen. nov. sp.*, isolated from three strains of the pulmonate snail *Biomphalaria glabrata* is related to members of the Mesomycetozoea. *Int J Parasitol*, *32*(9), 1183–1191.

Hofstatter, P. G., and Lahr, D. J. G. (2019). All eukaryotes are sexual, unless proven otherwise. *BioEssays*, *41*, 1800246. doi:10.1002/bies.201800246.

James-Clark, H. (1867). Conclusive proofs of the animality of the ciliate sponges, and of their affinities with the *Infusoria flagellata*. *The Annals and Magazine of Natural History; Zoology, Botany, and Geology*, *19*(3), 13–18.

King, N., Hittinger, C. T., and Carroll, S. B. (2003). Evolution of key cell signaling and adhesion protein families predates animal origins. *Science*, *301*(5631), 361–363. doi:10.1126/science.1083853.

King, N., Westbrook, M. J., Young, S. L., Kuo, A., Abedin, M., Chapman, J., and Rokhsar, D. (2008). The genome of the choanoflagellate *Monosiga brevicollis* and the origin of metazoans. *Nature*, *451*(7180), 783–788. doi:10.1038/nature06617.

Kożyczkowska, A., Najle, S. R., Ocaña-Pallarès, E. Aresté, C., Ruiz-Trillo, I. and Casacuberta, E. (2018). *Stable transfection in the protist Corallochytrium limacisporum allows identification of novel cellular features among unicellular relatives of animals*, bioRxiv https://doi.org/10.1101/2020.11.12.379420

Lang, B. F., O'Kelly, C., Nerad, T., Gray, M. W., and Burger, G. (2002). The closest unicellular relatives of animals. *Curr Biol*, *12*(20), 1773–1778.

Laundon, D., Larson, B. T., McDonald, K., King, N., and Burkhardt, P. (2019). The architecture of cell differentiation in choanoflagellates and sponge choanocytes. *PLoS Biol*, *17*(4), e3000226. doi:10.1371/journal.pbio.3000226.

Leadbeater, B. S. C. (2015). *The choanoflagellates: evolution, biology, and ecology*. Cambridge, UK: Cambridge University Press.

Levin, T. C., and King, N. (2013). Evidence for sex and recombination in the choanoflagellate *Salpingoeca rosetta*. *Curr Biol*, *23*(21), 2176–2180. doi:10.1016/j.cub.2013.08.061.

López-Escardó, D., Grau-Bové, X., Guillaumet-Adkins, A., Gut, M., Sieracki, M. E., and Ruiz-Trillo, I. (2019). Reconstruction of protein domain evolution using single-cell amplified genomes of uncultured choanoflagellates sheds light on the origin of animals. *Philos Trans R Soc Lond B Biol Sci*, *374*(1786), 20190088. doi:10.1098/rstb.2019.0088.

Marshall, W. L., and Berbee, M. L. (2010). Population-level analyses indirectly reveal cryptic sex and life history traits of Pseudoperkinsus tapetis (Ichthyosporea, Opisthokonta): a unicellular relative of the animals. *Mol Biol Evol*, *27*(9), 2014–2026. doi:10.1093/molbev/msq078.

Marshall, W. L., and Berbee, M. L. (2011). Facing unknowns: living cultures (*Pirum gemmata* gen. nov., sp. nov., and *Abeoforma whisleri*, gen. nov., sp. nov.) from invertebrate digestive tracts represent an undescribed clade within the unicellular Opisthokont lineage ichthyosporea (Mesomycetozoea). *Protist*, *162*(1), 33–57. doi:10.1016/j.protis.2010.06.002.

Marshall, W. L., Celio, G., McLaughlin, D. J., and Berbee, M. L. (2008). Multiple isolations of a culturable, motile Ichthyosporean (Mesomycetozoa, Opisthokonta), *Creolimax fragrantissima* n. gen., n. sp., from marine invertebrate digestive tracts. *Protist*, *159*(3), 415–433. doi:10.1016/j.protis.2008.03.003.

Mendoza, L., Taylor, J. W., and Ajello, L. (2002). The class mesomycetozoea: a heterogeneous group of microorganisms at the animal-fungal boundary. *Annu Rev Microbiol*, *56*, 315–344. doi:10.1146/annurev.micro.56.012302.160950.

Mendoza, L., and Vilela, R. (2013). Presumptive synchronized nuclear divisions without cytokinesis in the *Rhinosporidium seeberi* parasitic life cycle. *Microbiology*, *159*(Pt 8), 1545–1551. doi:10.1099/mic.0.068627-0.

Naumann, B., and Burkhardt, P. (2019). Spatial cell disparity in the colonial choanoflagellate *Salpingoeca rosetta*. *Front Cell Dev Biol*, *7*, 231. doi:10.3389/fcell.2019.00231.

Nitsche, F., Carr, M., Arndt, H., and Leadbeater, B. S. (2011). Higher level taxonomy and molecular phylogenetics of the Choanoflagellatea. *J Eukaryot Microbiol*, *58*(5), 452–462. doi:10.1111/j.1550-7408.2011.00572.x.

Okamoto, N., Nakase, K., Siuzuki, H., Nakai, Y., Fujii, K., and Sano, T. (1985). Life history and morphology of *Ichthyophonus hoferi* in vitro. *Fish Pathology*, *20*(2/3), 273–285.

Ondracka, A., Dudin, O., and Ruiz-Trillo, I. (2018). Decoupling of nuclear division cycles and cell size during the coenocytic growth of the Ichthyosporean *Sphaeroforma arctica*. *Curr Biol*. doi:10.1016/j.cub.2018.04.074

Owczarzak, A., Stibbs, H. H., and Bayne, C. J. (1980). The destruction of *Schistosoma mansoni* mother sporocysts in vitro by amoebae isolated from *Biomphalaria glabrata*: an ultrastructural study. *J Invertebr Pathol*, *35*(1), 26–33.

Pal, M., Shimelis, S., Rao, P. V. R., Samajpati, N., and Manna, A. K. (2016). Rhinosporidiosis: An enigmatic pseudofungal disease of humans and animals. *Journal of Mycological Research*, *54*(1), 49–54.

Parra-Acero, H., Ros-Rocher, N., Perez-Posada, A., Kożyczkowska, A., Sanchez-Pons, N., Nakata, A., ... Ruiz-Trillo, I. (2018). Transfection of *Capsaspora owczarzaki*, a close unicellular relative of animals. *Development*, 145(10). doi:10.1242/dev.162107.

Patterson, D. J., Nygaard, K., Steinberg, G., and Turley, C. M. (1993). Heterotrophic flagellates and other protists associated with oceanic detritus throughout the water column in the mid North Atlantic. *Journal of the Marine Biological Association of the United Kingdom*, *73*(1), 67–95.

Pettitt, M. E., Orme, B. A. A., Blake, J. R., and Leadbeater, B. S. C. (2002). The hydrodynamics of filter feeding in choanoflagellates. *European Journal of Protistology*, *38*(4), 313–332.

Philippon, H., Brochier-Armanet, C., and Perriere, G. (2015). Evolutionary history of phosphatidylinositol- 3-kinases: ancestral origin in eukaryotes and complex duplication patterns. *BMC Evol Biol*, *15*, 226. doi:10.1186/s12862-015-0498-7.

Ragan, M. A., Goggin, C. L., Cawthorn, R. J., Cerenius, L., Jamieson, A. V., Plourde, S. M., ... Gutell, R. R. (1996). A novel clade of protistan parasites near the animal-fungal divergence. *Proc Natl Acad Sci U S A*, *93*(21), 11907–11912.

Raghu-kumar, S. (1987). Occurrence of the thraustochytrid, *Corallochytrium limacisporum* gen. et sp. nov. in the coral reef lagoons of the Lakshadweep islands in the Arabian Sea. *Botanica Marina*, *30*(1), 83–90. doi:https://doi.org/10.1515/botm.1987.30.1.83.

Richter, D. J., Fozouni, P., Eisen, M., and King, N. (2018). Gene family innovation, conservation and loss on the animal stem lineage. *Elife*, *7*. doi:10.7554/eLife.34226.

Richter, D. J., and King, N. (2013). The genomic and cellular foundations of animal origins. *Annu Rev Genet*, *47*, 509–537. doi:10.1146/annurev-genet-111212-133456.

Richter, D. J., and Nitsche, F. (2017). Choanoflagellatea. In *Handbook of the Protists* (pp. 1479–1496). New York, NY: Springer Berlin Heidelberg.

Roper, M., Dayel, M. J., Pepper, R. E., and Koehl, M. A. (2013). Cooperatively generated stresslet flows supply fresh fluid to multicellular choanoflagellate colonies. *Phys Rev Lett*, *110*(22), 228104. doi:10.1103/PhysRevLett.110.228104

Saville-Kent, W. (1878). I.—Observations upon Professor Ernst Haeckel's group of the "Physemaria," and on the affinity of the sponges. *Annals and Magazine of Natural History*, *1*(1), 1–17.

Sebé-Pedrós, A., Ballare, C., Parra-Acero, H., Chiva, C., Tena, J. J., Sabido, E., ... Ruiz-Trillo, I. (2016). The dynamic regulatory genome of *Capsaspora* and the origin of animal multicellularity. *Cell*, *165*(5), 1224–1237. doi:10.1016/j.cell.2016.03.034.

Sebé-Pedrós, A., Burkhardt, P., Sanchez-Pons, N., Fairclough, S. R., Lang, B. F., King, N., and Ruiz-Trillo, I. (2013). Insights into the origin of metazoan filopodia and microvilli. *Mol Biol Evol*, *30*(9), 2013–2023. doi:10.1093/molbev/mst110.

Sebé-Pedrós, A., de Mendoza, A., Lang, B. F., Degnan, B. M., and Ruiz-Trillo, I. (2011). Unexpected repertoire of metazoan transcription factors in the unicellular holozoan *Capsaspora owczarzaki*. *Mol Biol Evol*, *28*(3), 1241–1254. doi:10.1093/molbev/msq309.

Sebé-Pedrós, A., Degnan, B. M., and Ruiz-Trillo, I. (2017). The origin of Metazoa: a unicellular perspective. *Nat Rev Genet*, *18*(8), 498–512. doi:10.1038/nrg.2017.21.

Sebé-Pedrós, A., Irimia, M., Del Campo, J., Parra-Acero, H., Russ, C., Nusbaum, C., ... Ruiz-Trillo, I. (2013). Regulated aggregative multicellularity in a close unicellular relative of metazoa. *Elife*, *2*, e01287. doi:10.7554/eLife.01287.

Sebé-Pedrós, A., Pena, M. I., Capella-Gutierrez, S., Antó, M., Gabaldón, T., Ruiz-Trillo, I., and Sabido, E. (2016). High-throughput proteomics reveals the unicellular roots of animal phosphosignaling and cell differentiation. *Dev Cell*, *39*(2), 186–197. doi:10.1016/j.devcel.2016.09.019.

Sebé-Pedrós, A., and Ruiz-Trillo, I. (2010). Integrin-mediated adhesion complex: Cooption of signaling systems at the dawn of Metazoa. *Commun Integr Biol*, *3*(5), 475–477. doi:10.4161/cib.3.5.12603.

Sebé-Pedrós, A., Zheng, Y., Ruiz-Trillo, I., and Pan, D. (2012). Premetazoan origin of the Hippo signaling pathway. *Cell Rep*, *1*(1), 13–20. doi:10.1016/j.celrep.2011.11.004.

Shalchian-Tabrizi, K., Minge, M. A., Espelund, M., Orr, R., Ruden, T., Jakobsen, K. S., and Cavalier-Smith, T. (2008). Multigene phylogeny of choanozoa and the origin of animals. *PLoS One*, *3*(5), e2098. doi:10.1371/journal.pone.0002098.

Spanggaard, B., Skouboe, P., Rossen, L., and Taylor, J. W. (1996). Phylogenetic relationships of the intercellular fish pathogen *Ichthyophonus hoferi*, and fungi, choanoflagellates and the rosette agent. *Marine Biology*, *126*(1), 109–115.

Stibbs, H. H., Owczarzak, A., Bayne, C. J., and DeWan, P. (1979). Schistosome sporocyst-killing Amoebae isolated from *Biomphalaria glabrata*. *J Invertebr Pathol*, *33*(2), 159–170.

Suga, H., Chen, Z., de Mendoza, A., Sebé-Pedrós, A., Brown, M. W., Kramer, E., ... Ruiz-Trillo, I. (2013). The *Capsaspora* genome reveals a complex unicellular prehistory of animals. *Nat Commun*, *4*, 2325. doi:10.1038/ncomms3325

Suga, H., Dacre, M., de Mendoza, A., Shalchian-Tabrizi, K., Manning, G., and Ruiz-Trillo, I. (2012). Genomic survey of premetazoans shows deep conservation of cytoplasmic tyrosine kinases and multiple radiations of receptor tyrosine kinases. *Sci Signal*, *5*(222), ra35. doi:10.1126/scisignal.2002733.

Suga, H., and Ruiz-Trillo, I. (2013). Development of ichthyosporeans sheds light on the origin of metazoan multicellularity. *Dev Biol*, *377*(1), 284–292. doi:10.1016/j.ydbio.2013.01.009.

Tikhonenkov, D. V., Mikhailov, K. V., Hehenberger, E., Karpov, S. A., Prokina, K. I., Esaulov, A. S., ... Keeling, P. J. (2020). New Lineage of Microbial Predators Adds Complexity to Reconstructing the Evolutionary Origin of Animals. *Curr Biol*, *30*(22), 4500–4509.e5. doi: 10.1016/j.cub.2020.08.061

Tong, S. M. (1997). Heterotrophic flagellates and other protists from Southampton Water, U.K. *Ophelia*, *47*(2), 71–131. doi:10.1080/00785236.1997.10427291.

Torruella, G., de Mendoza, A., Grau-Bové, X., Antó, M., Chaplin, M. A., del Campo, J., ... Ruiz-Trillo, I. (2015). Phylogenomics reveals convergent evolution of lifestyles in close relatives of animals and fungi. *Curr Biol, 25*(18), 2404–2410. doi:10.1016/j.cub.2015.07.053.

van Hannen, E. J., Mooij, W., van Agterveld, M. P., Gons, H. J., and Laanbroek, H. J. (1999). Detritus-dependent development of the microbial community in an experimental system: qualitative analysis by denaturing gradient gel electrophoresis. *Appl Environ Microbiol, 65*(6), 2478–2484.

Waller, R. F., Cleves, P. A., Rubio-Brotons, M., Woods, A., Bender, S. J., Edgcomb, V., ... Collier, J. L. (2018). Strength in numbers: Collaborative science for new experimental model systems. *PLoS Biol, 16*(7), e2006333. doi:10.1371/journal.pbio.2006333.

Woznica, A., Gerdt, J. P., Hulett, R. E., Clardy, J., and King, N. (2017). Mating in the closest living relatives of animals Is induced by a bacterial chondroitinase. *Cell, 170*(6), 1175–1183, e1111. doi:10.1016/j.cell.2017.08.005.

PHYLUM PLACOZOA

CHAPTER
7

Bernd Schierwater, Kai Kamm, Kathrin Wysocki, and Hans-Jürgen Osigus

Subregnum: Metazoa
1st Division: Diploblastic Animals
Phylum Placozoa

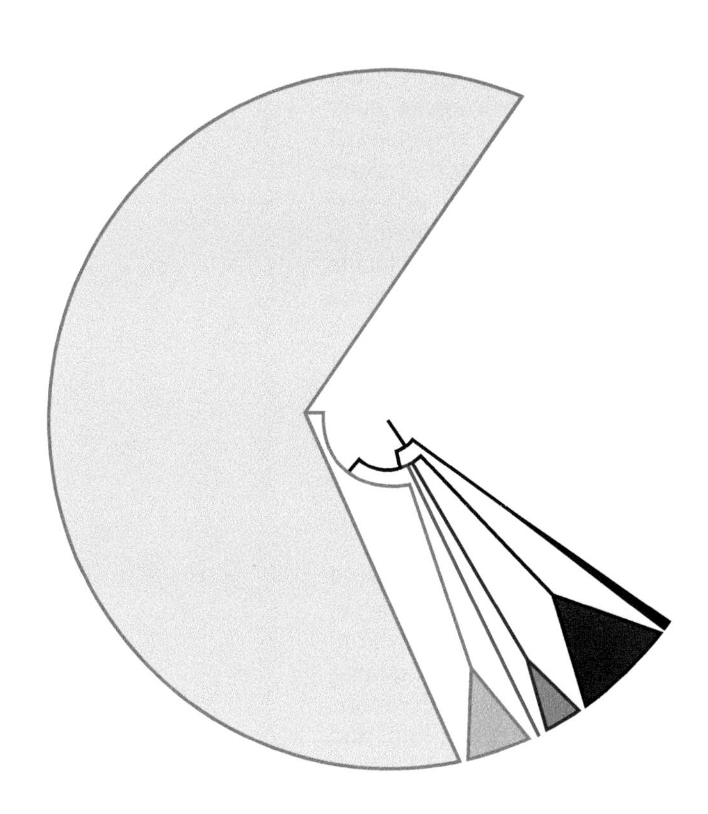

CONTENTS

General Characteristics

Three valid species: *Trichoplax adhaerens* Schulze 1883 (Schulze, 1883), *Hoilungia hongkongensis* Eitel, Schierwater and Wörheide 2018 (Eitel et al., 2018) and *Polyplacotoma mediterranea* Osigus and Schierwater 2019 (Osigus et al., 2019)

Microscopic non-symmetric animals, abundant in warm ocean waters

Most primitive metazoan bauplan with eight somatic cell types only

Ecological importance is probably small

Autapomorphies

- lack of ECM
- lack of basal membrane
- contractile fiber cells between epithelia
- no body axis
- single ParaHox gene

HISTORY OF TAXONOMY AND CLASSIFICATION

In 1883, the German zoologist Franz Eilhard Schulze discovered this microscopic marine animal on the glass walls of a seawater aquarium at the University of Graz, Austria (Schulze, 1883). The animal, usually measuring less than 3 mm in diameter and less than 20 µm in thickness, looked like an irregular hairy plate sticking to the glass surface and was thus named *Trichoplax adhaerens* (lat. tricho = hair, plax = plate) (**Figure 7.1A**). Later claims of another placozoan species, *Treptoplax reptans*, remained unconfirmed (Monticelli, 1893). The perfectly basic *Trichoplax* bauplan prompted the German zoologist Otto Bütschli to publish his famous "Placula hypothesis" (Bütschli, 1884). In Bütschli's Placula hypothesis, the "urmetazoon" had adapted a benthic life form, which derived from a multiple-cell protist colony switching from a planktonic to a benthic lifestyle by developing functionally and morphologically different epithelia for the side facing the substrate and the opposite side facing the open water. This hypothetical urmetazoan bauplan pretty much resembles the recent *Trichoplax* bauplan (if we ignore the later developed/evolved central fiber cells in the body center) and Bütschli called this hypothetical stage the "Placula". Thus, from the very beginning it has been believed that *Trichoplax* must be placed in an early branching position in the metazoan Tree of Life (ToL).

Despite its possible importance for reconstructing the metazoan Tree of Life, the interests in *Trichoplax* vanished for almost a century. In the 1960s, i.e. after about 60 years of silence on Placozoa, the German cell biologist Willy Kuhl (University of Frankfurt) found placozoans in a seawater aquarium containing samples from the Mediterranean Sea, and Kuhl reported some first observations on locomotion and regeneration in *Trichoplax* (Kuhl and Kuhl, 1963; Kuhl and Kuhl, 1966). A few years later, it was the German protozoologist Karl Gottlieb Grell (University of Tübingen) who restored the uniqueness and importance of this enigmatic animal. Grell found his *Trichoplax* in an algal sample from the Red Sea in 1969 and his meticulous research provided sufficient support for placing *Trichoplax adhaerens* in a new phylum. He named the new phylum "Placozoa" (Grell, 1971) following Bütschli's Placula hypothesis.

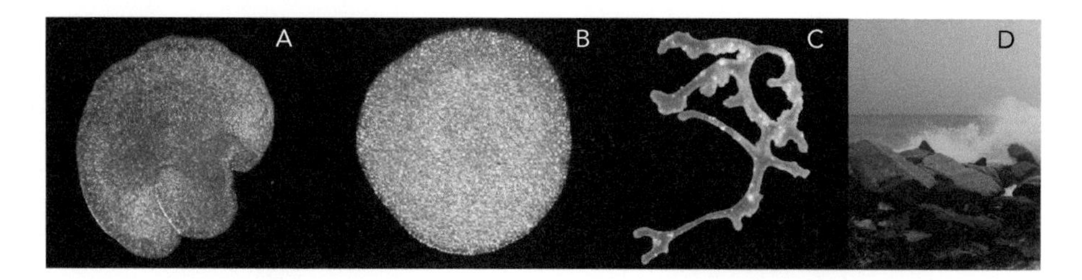

Figure 7.1 Light microscopy images of the three known placozoan species *Trichoplax adhaerens* H1 (A), *Hoilungia hongkongensis* H13 (B) and *Polyplacotoma mediterranea* H0 (C). The recently found *P. mediterranea* (C) is the only placozoon that does not show the placozoan typical "plate" habitus. This species lives in strong breaking wave zones (D) and has adapted a highly ramified body habitus. (Images are not to scale. Images A and B are taken from Eitel et al. (2018); Image C by Hans-Jürgen Osigus Image D courtesy of Nicole Bergjürgen.)

For more than a hundred years, the phylum Placozoa has been monotypic, harboring just the type species, *Trichoplax adhaerens* (Schulze, 1883) in the family Trichoplacidae (Bütschli and Hatschek, 1905). However, more recent genetic analyses of placozoan specimens from different ocean waters around the world revealed the presence of a large number of cryptic species, i.e. species which have been morphologically undistinguishable (reviewed in e.g. Eitel et al., 2013). The real placozoan biodiversity has been estimated to include several dozen genetically and ecologically distinguishable species (Eitel and Schierwater, 2010).

Due to the morphological uniformity of placozoans, a provisional classification of placozoan lineages has been based on a diagnostic sequence of the mitochondrial 16S ribosomal RNA gene (see e.g. Voigt et al., 2004; Eitel and Schierwater; 2010, Osigus et al., 2019). At present, some 20 placozoan lineages (so-called haplotypes) have been described (Voigt and Eitel, 2018, Osigus et al., 2019). Each so-called haplotype likely represents a separate, yet hard to describe species (**Figure 7.2**). The species description of *Trichoplax adhaerens* (Schulze, 1883) was solely based on morphological data, the description of the second placozoan species (*Hoilungia hongkongensis*, family Trichoplacidae, Figure 7.1B) was based on a taxogenomic approach mainly using nuclear genome data (Eitel et al., 2018). The third placozoan species, *Polyplacotoma mediterranea* (Figure 7.1C), has been described based on its unique morphology and highly divergent mitochondrial genome (Osigus et al., 2019). The genus *Polyplacotoma* has not been

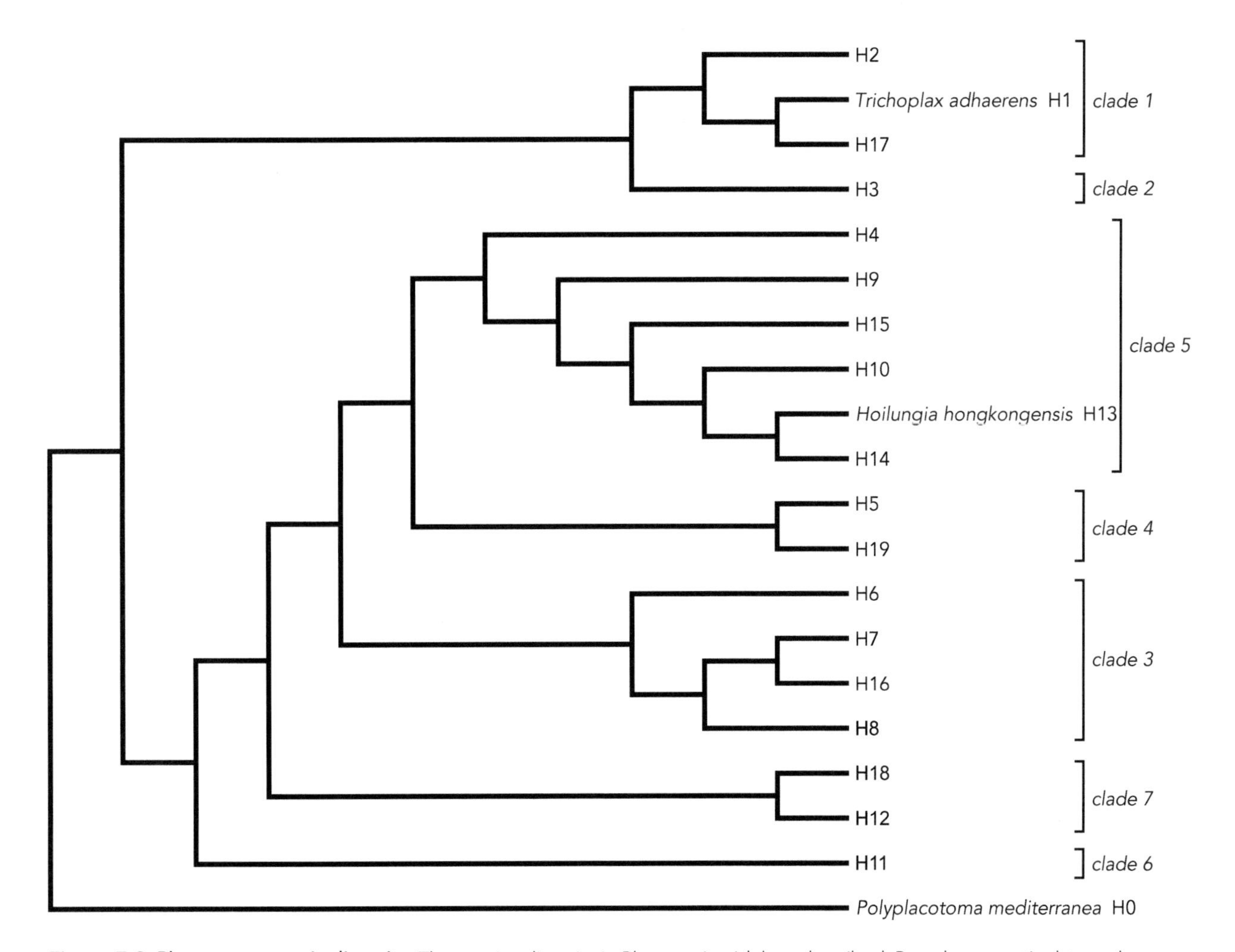

Figure 7.2 Placozoan genetic diversity. The species diversity in Placozoa is widely undescribed. Based on genetic data and the shown mitochondrial 16S rRNA gene phylogeny, at least 20 species are present in the different oceans (each haplotype likely represents a separate species). The numbering of haplotypes and clades corresponds to the numbering used in different studies (cf. Voigt and Eitel, 2018; Osigus et al., 2019 and references therein).

assigned to the family Trichoplacidae as it likely represents a member of a (yet undescribed) new placozoan family. The three accepted species are the systematic framework to assign the remaining known placozoan genetic lineages to Linnean ranks (cf. Osigus et al., 2019).

ANATOMY, HISTOLOGY AND MORPHOLOGY

When Franz Eilhard Schulze published his short communication on the description of a new species, *Trichoplax adhaerens*, he described several new features for metazoan animals (Schulze, 1883; 1891). Histological analyses reveal a three-layered sandwich organization with morphologically different upper and lower epithelia and an inner, fiber-like union of cells. These cells have been named "Faserzellen" ("fiber cells"), which are contractile. Contractions of these cells cause shape-changes of the animal (Schulze, 1883; 1891).

The upper epithelium consists of a thin squamous layer of upper epithelial cells while the lower epithelium consists of five different cell types: three different gland cells, lipophil cells and lower epithelium cells (Figure 7.3A) (following Schulze, 1883; Smith et al., 2014; Mayorova et al., 2019). Between the two epithelia the fiber cells and crystal cells (the latter at the edge of the animal) are found (Smith et al., 2014), which do not form a cell layer and thus are not comparable to the mesoderm of higher, triploblastic animals (Schierwater, 2005). The larger number of cell types in the lower epithelium relates to its function for extracellular digestion (see below). The only obvious specialized structures of the upper epithelium are large lipid droplets, which were named "Glanzkugeln" ("shiny spheres") by Schulze (1891). It is noteworthy that one type of gland cells in the lower epithelium can sporadically also be present in the upper epithelium (Mayorova et al., 2019; not shown in **Figure 7.3**). In contrast to the "typical" multicellular animal, placozoans do not show anything like an oral-aboral axis, nor do they possess any organs, nerve or muscle cells, basal lamina or extracellular matrix (for review see e.g. Schierwater, 2005; Schierwater et al., 2010; Eitel et al., 2013; Schierwater and Eitel, 2015; Schierwater and DeSalle, 2018). Lacking any body axis, placozoans also lack any type of symmetry, but show a clear "polarity" between the lower and upper epithelium. This may be seen as the precursor of symmetry, since pulling the placozoan body in the center and upward leads anatomically to a radial symmetric bauplan, as this is typical for the more derived Cnidaria and Ctenophora (Schierwater et al., 2009) (**Figure 7.4**).

According to the morphological description above, Placozoa possess eight defined somatic cell types: lower epithelia cells, upper epithelia cells, three different types of gland cells, lipophil cells, fiber and crystal cells (Figure 7.3A) (see Smith et al., 2014; Mayorova et al., 2019). If genetic and/or histochemical data are used as criteria, additional cell types could be defined (e.g. Jakob et al., 2004; Martinelli and Spring, 2004, 2005; Varoqueaux et al., 2018). For example, unique expression patterns of a ParaHox gene could define another functional cell type, i.e. small, potentially "omnipotent" cells in the fiber cell layer which likely serve as stem cells (Figure 7.3B; Jakob et al., 2004; Guidi et al., 2011). These are located in the region where the upper and lower epithelium meet, which led to the idea that these cells are involved in setting up "polarity" by differentially forming cells for the lower and upper epithelium (Jakob et al., 2004; Schierwater et al., 2009).

REPRODUCTION AND DEVELOPMENT

Vegetative reproduction is the first – and most often the only – mode of reproduction seen in the laboratory. When Grell (see Grell, 1971, 1972; Grell and Benwitz, 1974) discovered oogenesis and cleavage in *Trichoplax*, the embryos regularly

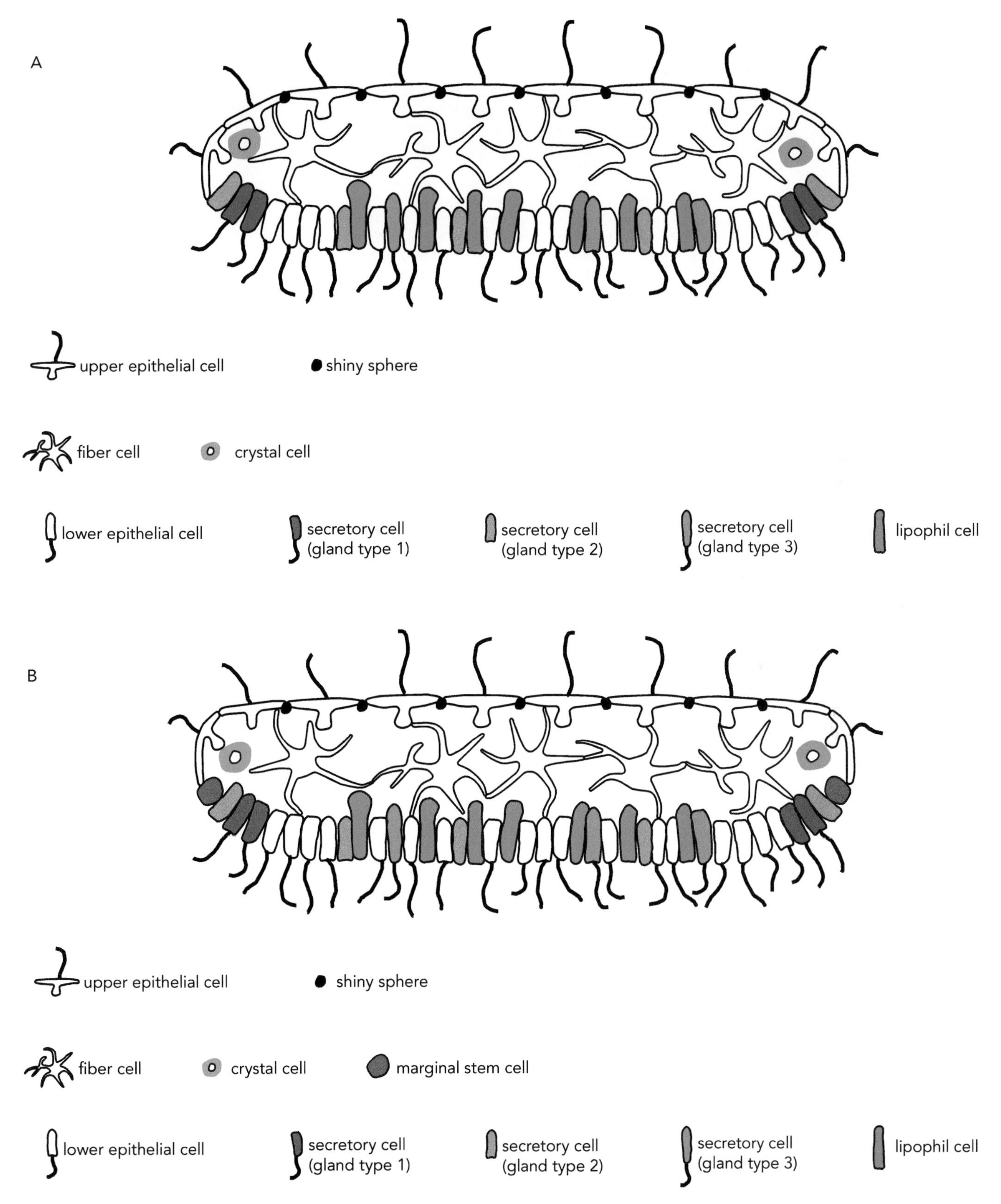

Figure 7.3 Schematic cross section of *Trichoplax adhaerens* H1. (A) Based on morphology, eight somatic cell types can be distinguished. The only cell type in the upper epithelium (UE) is upper epithelial cells, while the lower epithelium (LE) harbors three different types of gland cells as well as lower epithelial and lipophil cells. The interconnecting fiber cell layer mainly consists of fiber cells, and a smaller number of crystal cells is found near the margin. (B) Based on gene expression and histochemical data additional cell types may be defined, like the marginal stem cells at the border between the LE and UE. (Cross-sections following Jakob et al., 2004; Guidi et al., 2011; Smith et al., 2014; Mayorova et al., 2019.)

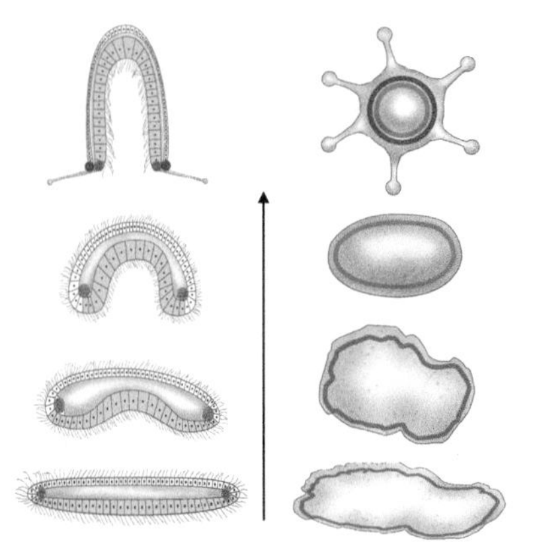

Figure 7.4 The "New Placula Hypothesis". As described in Schierwater et al. (2009), the non-symmetric placozoan bauplan can be transformed into a symmetric metazoan bauplan by exaggerating the forming of an extracorporal feeding cavity (**Figure 7.5**) to an extent that a larger and permanent feeding cavity is formed. During this process the bauplan turns from non-symmetric into radially symmetric, as shown here for a hydrozoan medusa. As a speculation, the process starts with a single regularity gene (gene expression in red) which later duplicates to adapt new functions for the more sophisticated morphology. In the cnidarian bauplan additional morphological features would include tentacles to assist in feeding.

died in early stages (cf. Ruthmann et al., 1981, Grell, 1984). While later studies could follow cleavage to the 128-cell stage (Eitel et al., 2011), a complete life cycle has never been observed.

Sexual reproduction

The life cycle of Placozoa must be suspected to be a simple life cycle with direct development and without any larval stage. Most likely, the adult placozoon – after a series of vegetative reproductions – becomes sexually mature as a simultaneous hermaphrodite. Whether outcrossing and/or selfing occurs is unknown. A switch to sexual reproduction is favored by high population densities, food scarceness and temperatures above 23°C (see Eitel et al., 2011).

Female gametocytes (oocytes) are produced in the lower epithelium (Grell and Benwitz, 1974) and maturation and fertilization occur somewhere in the center of the body. For maturation, the oocyte incorporates extensions from nursing fiber cells through pores on its surface (Grell and Benwitz, 1974, 1981; Eitel et al., 2011). In addition to yolk droplets and cortical granules, the oocyte stores lipid droplets and glycogen granules and a mature oocyte reaches a size of 90–120 μm (Grell, 1972). Vast amounts of bacteria, vertically transmitted from the nursing fiber cells (Grell and Benwitz 1974, 1981; Eitel et al., 2011), are transferred into the oocyte during maturation.

Although actual sperm cells have not been seen yet, the expression of sperm-associated marker genes strongly suggests spermatogenesis and sperm maturation in placozoans (Eitel et al., 2011). Even markers known to encode proteins for functional sperm flagella and sperm–oocyte recognition proteins used in fertilization were identified. Mature oocytes are fertilized internally and a "fertilization membrane" is built by fusion of accumulated granulae (Grell and Benwitz, 1974; Eitel et al., 2011). Early embryos grow inside the mother animal until the latter degenerates and releases the embryo.

Vegetative reproduction

Two modes of vegetative reproduction are known from Placozoa: (i) fission (normal type of vegetative reproduction in Placozoa) and (ii) swarmer formation (occasional type of vegetative reproduction). In fission, animals grow to a certain size and divide into two approximately equally sized daughter individuals,

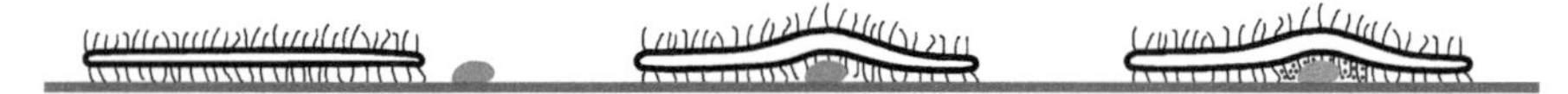

Figure 7.5 Placozoan feeding behavior. The animal surrounds the food particle (e.g. algae) and forms an extracorporal feeding cavity for digestion. The food digested outside the body is taken up by phagocytosis through the lower epithelium.

which then re-grow to the "normal" size (Schulze, 1883, 1891). This mode of vegetative reproduction can go on *ad infinitum*. The second mode of vegetative reproduction has only been observed in the laboratory when environmental conditions become unfavorable. Under such conditions, placozoans may develop small spherical swarmers, which are planktonic (free-floating), and thus are taken by water currents to new habitats (cf. Thiemann and Ruthmann 1988). These swarmers are made up of an outer layer of upper epithelial cells, a fiber cell layer and an inner layer of lower epithelial cells. After opening, the swarmers will fully rebuild the normal adult habitus within a day. In addition to swarmers, swarmer-like spheres (either hollow or solid) have also been reported, but their origin remains unclear (Thiemann and Ruthmann, 1990, 1991).

Developmental genes

In vivo studies on developmental genes in placozoans are very limited (e.g. Jakob et al., 2004; Martinelli and Spring, 2004, 2005; DuBuc et al., 2019). In a current view, a single ParaHox gene is present in placozoans and coordinates an ancestral symmetry pattern, i.e. "polarity" as the precursor of symmetry (Jakob et al., 2004). Since for a comparative morphologist polarity is the first step for creating symmetry, the "new Placula hypothesis" (Schierwater et al., 2009), i.e. the origin of all higher animal phyla from an ancestral "Placula" bauplan, derives as naturally as a baby's smile (Figure 7.4). Animals which give up symmetry and strict polarity, like sponges, no longer need Hox-like genes. More derived animals, which develop polarity into symmetry possibly have multiplied these genes or further developed them into genuine Hox genes. In placozoans, representatives of all important signaling pathways are present, including BMB/TGF beta, Wnt and Notch signaling pathways (Srivastava et al., 2008).

DISTRIBUTION AND ECOLOGY

Although discovered in the 1880s, very little has been known for more than a century about the distribution of placozoans in the oceans. The original assumption was that placozoans only live in tropical and subtropical waters on hard substrates (Eitel et al., 2013). We now know that the north–south distribution range is much wider and certainly reaches out at least to the Atlantic coast of Northern France (**Figure 7.6A**), and that placozoans can also been found in sandy habitats (e.g. Signorovitch et al., 2006). A map derived from modeling suggests a very broad distribution range and sympatry of several euryoecious as well as the existence of a few stenoecious and even endemic species (Paknia and Schierwater, 2015; Figure 7.6A and B). Different placozoan species respond sensitively and differentially to global warming and ocean acidification stress and regular seasonal fluctuations have been reported from placozoan populations, e.g. in Japan (Maruyama, 2004; Schleicherová et al., 2017). Very little is known about placozoans and their interaction with other organisms in the field. We only know that placozoans are feeding on algae and that placozoans themselves serve as food, e.g. for gastropods (Pearse and Voigt, 2007; Cuervo-Gonzáles, 2017).

PHYSIOLOGY AND BEHAVIOR

In culture, placozoans show undirected movements, sometimes interspersed with phases of rest or rotation on a specific spot. While smaller specimen glide on their cilia without major changes of their shape, larger specimens exhibit rather amoeboid-like movements combined with ciliary movements, which results in constant shape changes. If an animal is cut in half, the two halves quickly move away from each other in opposite directions (Kuhl and Kuhl, 1966; Schwartz, 1984). Surprisingly, placozoans do not show any signs of escape behavior while predators are feeding on them (Cuervo-Gonzáles, 2017). Placozoans

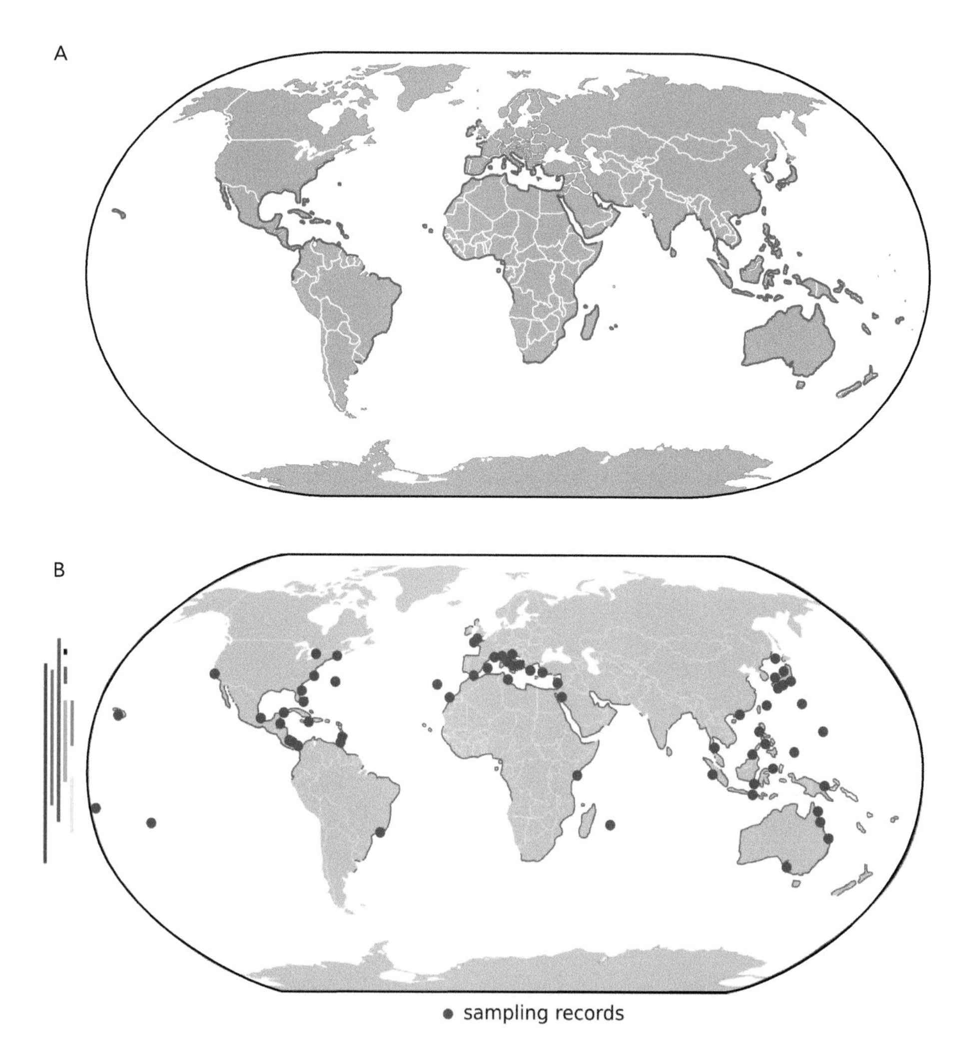

Figure 7.6 Global placozoan distribution. Placozoans are found worldwide in all warm and also many temperate ocean waters. (A) Predictions from ecological modeling approaches (Paknia and Schierwater, 2015) are shown in red. (B) Published sampling records by 05/2020 (blue dots) show that large regions of expected placozoan distribution have not yet been sampled. Especially the long coasts of Africa, Australia and South America have very poorly been examined for the presence of placozoans. Color bars on the left side of the map illustrate the latitudinal distribution of respective 16S clades, i.e. higher taxonomic placozoan orders (pink = clade 1, orange = clade 2, green = clade 3, brown = clade 4, blue = clade 5, purple = clade 6, yellow = clade 7, black = H0). A single blue dot might summarize multiple sampling sites.

show complex chemotactic behavior and also perceive light (Heyland et al., 2014), and coordinated feeding behavior of groups of animals has been reported (Smith et al., 2015; Fortunato and Aktipis, 2019; Smith et al., 2019). Most likely, placozoans are also capable of perceiving gravity (Mayorova et al., 2018).

With the aid of the cilia of the lower epithelium the animal can crawl. During feeding, the animal lifts up the center region of its body to form an external digestive cavity between the substrate and lower epithelium (e.g. Smith et al., 2015 and references therein). For this the animal creeps over small food particles, surrounds them and releases digestive enzymes into this extracorporal digestive cavity (Figure 7.5). Interestingly, the upper epithelium is also capable of feeding. Algae and other food particles are trapped in a slime layer coating the upper epithelium and are subsequently taken up (phagocytized) by the inner fiber cells; this unique mode of feeding has been called "transepithelial cytophagy" (Wenderoth, 1986). Placozoans harbor different endosymbiotic bacteria of unknown function in the endoplasmic reticulum of the fiber cells as well as in lower epithelial cells (e.g. Gruber-Vodicka et al., 2019; Kamm et al., 2019 and references therein).

Physiological studies on locomotion, feeding and the neuro-secretory system of placozoans have led to the identification of different types of gland cells (Varoqueaux et al., 2018; Mayorova et al., 2019) playing roles in animal attachment, movement and feeding processes. Studying the role of different neuropeptides for animal locomotion has revealed complex and uneven distribution of different cell populations, which are not yet understood (Varoqueaux et al., 2018). Functional studies on the role of shiny spheres (located in the upper epithelium) have suggested a role of these lipid droplets in chemical defense (Jackson and Buss, 2009).

GENETICS AND GENOMICS

The placozoan nuclear genome contains basic representatives of most major gene families present in humans (Srivastava et al., 2008). The very compact genome shows conserved gene content, gene structure and synteny in relation to the human and other complex eumetazoan genomes (Srivastava et al., 2008). With some 12,000 gene predictions and despite the apparent cellular and organismal simplicity of placozoans, the placozoan genome encodes a rich array of transcription factor and signaling pathway genes that are typically associated with diverse cell types and developmental processes in eumetazoans (Srivastava et al., 2008; Schierwater et al., 2008; Alie and Manuel, 2010; Varoqueaux et al., 2018; Eitel et al., 2018; Kamm et al., 2019).

So far three near complete placozoan nuclear genomes have been sequenced, up to scaffold levels in the megabase range (Srivastava et al., 2008; Kamm et al., 2018; Eitel et al., 2018). The placozoan genomes range in size from 87 to 95 megabases and all contain around 12,000 protein coding genes (Srivastava et al., 2008; Kamm et al., 2018; Eitel et al., 2018). Thus, placozoans harbor the smallest not secondarily reduced genomes among metazoan animals. The karyotype of placozoans has been investigated only once, showing six chromosome pairs without heterochromosomes (Birstein, 1989). We do not know if chromosome numbers vary in the phylum.

The genome of the well-investigated type species of the phylum, *Trichoplax adhaerens*, shows a significant amount of conserved synteny to eumetazoans from vertebrates to anthozoans, which is in contrast to derived and small invertebrate genomes like those of flies and nematodes (Srivastava et al., 2008). Within the phylum, the amount of genomic rearrangement varies from little (Kamm et al., 2018) to medium (Eitel et al., 2018) (comparable to different mammalian orders), while gene, exon and intron size show no deviation between the investigated genomes (Srivastava et al., 2008; Kamm et al., 2018; Eitel et al., 2018).

Although placozoans possess most of the gene families present in Bilateria, the complexity of their gene sets clearly represents a pre-cnidarian stage. For example, developmental signaling pathways present in Placozoa include Wnt- and TGF-β signaling while a complete Hedgehog pathway is absent (Srivastava et al., 2008). Up to 38 homeobox genes have been found in a single placozoan species, covering all major classes present in higher animals (Schierwater et al., 2008; Kamm et al., 2018). The amount and further diversification of homeobox genes, however, predates the expansion of this gene family that started with Cnidaria (Kamm et al., 2006; Kamm and Schierwater, 2006; Ryan et al., 2006), and led to the wealth of these important developmental genes in bilaterian phyla. Other examples are partial genetic toolkits for neuroendocrine signaling in the absence of a nervous system (Srivastava et al., 2008; Alie and Manuel, 2010; Varoqueaux et al., 2018) and for innate immunity (Kamm et al., 2019). Difficulties for homology assignments arise from observations that the domain architecture of some placozoan genes is difficult to predict because of substantial deviation from the bilaterian consensus (Kamm et al., 2019).

The placozoan mitochondrial genomes span sizes from 23 kb to 43 kb (Dellaporta et al., 2006; Signorovitch et al., 2007; Miyazawa et al., 2012; Osigus et al., 2019) and thus are the largest circular mitochondrial genomes found in metazoan animals if one excludes some aberrant molluscan mt genomes with secondary amplifications (e.g. Liu et al., 2013). Other unusual features include the presence of multiple introns and open-reading frames of unknown function (Dellaporta et al., 2006; Signorovitch et al., 2007; Miyazawa et al., 2012; Osigus et al., 2019). Altogether these features suggest that the placozoan mitochondrial genomes possess complex patterns of early evolutionary stages. The recent characterization of the compact mitochondrial genome of the aberrant placozoan, *Polyplacotoma mediterranea*, has raised more questions than answers yet (Osigus et al., 2019).

POSITION IN THE ToL

The phylogenetic position of Placozoa is controversial (cf. Schierwater et al., 2016). While traditional and parsimony approaches either support placozoans to be the first emerging metazoan animal group or a sister group to Porifera and Coelenterata within a non-Bilateria clade (e.g. Schierwater, 2005; Schierwater and DeSalle, 2007; Schierwater et al., 2009), recent molecular systematics approaches place Placozoa anywhere within the diploblast phyla (depending on the study) (e.g. Philippe et al., 2009; Pick et al., 2010; Simion et al., 2017; Laumer et al., 2019). All so far conducted phylogenetic analyses based on nuclear proteins only included data from members of the *Trichoplax*- and *Hoilungia*-group, but not from the more distantly related *Polyplacotoma*-group. Including these data might reduce a major problem of previous phylogenetic approaches, i.e. artifacts resulting from too long branches. As of today, putting all data together (from morphology to genome characteristics), assuming a basal position for Placozoa in the ToL remains the most parsimonious interpretation.

DATABASES AND COLLECTIONS

The state-of-the-art taxonomy of placozoans (although still far away from being complete) is curated by the Editors of the World Register of Marine Species (WoRMS) (http://www.marinespecies.org). However, there are currently no comprehensive collections of placozoans and preserved specimen can only sporadically be found in some museums. Any kind of confirmed fossil records are missing.

CONCLUSION

Placozoa are emerging and highly unique model organisms with tremendous potential for bridging crucial gaps between genetics, genomics, development and ecological adaptation. Independent of the phylogenetic position, placozoans represent the closest living surrogate for a simple urmetazoon bauplan and thus the simplest available metazoan model system.

ACKNOWLEDGMENTS

We would like to thank Frédérique Varoqueaux (University of Lausanne, Switzerland) for helpful comments on the morphology of *Trichoplax adhaerens*. We also would like to thank Kristin Fenske (Institute of Animal Ecology, Stiftung Tierärztliche Hochschule Hannover, Germany) for continuous help at all stages of the manuscript preparation.

REFERENCES

Alie, A., Manuel, M., The backbone of the post-synaptic density originated in a unicellular ancestor of choanoflagellates and metazoans, BMC Evol Biol 10 (2010) 34.

Birstein, V. J., On the karyotype of *Trichoplax* sp, (Placozoa), Biologisches Zentralblatt 108(1) (1989) 63–67.

Bütschli, O., Bemerkungen zur Gastraea-Theorie., Morphol. Jahrb. 9, (1884) 415–427.

Bütschli, O., Hatschek, B., Zoologisches Zentralblatt 12 (1905).

Cuervo-González, R., Rhodope placozophagus (Heterobranchia) a new species of turbellarian-like Gastropoda that preys on placozoans, Zoologischer Anzeiger – A Journal of Comparative Zoology 270 (Supplement C) (2017) 43–48.

Dellaporta, S. L., Xu, A., Sagasser, S., Jakob, W., Moreno, M. A., Buss, L. W., Schierwater, B., Mitochondrial genome of Trichoplax adhaerens supports placozoa as the basal lower metazoan phylum, Proc Natl Acad Sci USA 103(23) (2006) 8751–8756.

DuBuc, T. Q., Ryan, J. F., Martindale, M. Q., "Dorsal-Ventral" genes are part of an ancient axial patterning system: evidence from Trichoplax adhaerens (Placozoa), Mol Biol Evol 36(5) (2019) 966–973.

Eitel, M., Schierwater, B., The phylogeography of the Placozoa suggests a taxon-rich phylum in tropical and subtropical waters, Mol Ecol 19(11) (2010) 2315–2327.

Eitel, M., Guidi, L., Hadrys, H., Balsamo, M., Schierwater, B., New insights into placozoan sexual reproduction and development, PLoS One 6(5) (2011) e19639.

Eitel, M., Osigus, H. J., DeSalle, R., Schierwater, B., Global diversity of the Placozoa, PLoS One 8(4) (2013) e57131.

Eitel, M., Francis, W. R., Varoqueaux, F., Daraspe, J., Osigus, H. J., Krebs, S., Vargas, S., Blum, H., Williams, G. A., Schierwater, B., Worheide, G., Comparative genomics and the nature of placozoan species, PLoS Biol 16(7) (2018) e2005359.

Fortunato, A., Aktipis, A., Social feeding behavior of Trichoplax adhaerens, Front Ecol Evol 7 (2019).

Grell, K. G., Trichoplax adhaerens F.E. Schulze und die Entstehung der Metazoan, Naturwissenschaftliche Rundschau 24(4) (1971) 160–161.

Grell, K. G., Embryonalentwicklung bei Trichoplax adhaerens F. E. Schulze, Naturwissenschaften 58 (1971) 570.

Grell, K. G., Eibildung und Furchung von Trichoplax adhaerens F.E.Schulze (Placozoa), Zeitschrift für Morphologie der Tiere 73 (1972) 297–314.

Grell, K. G., Benwitz, G., Elektronenmikroskopische Beobachtungen über das Wachstum der Eizelle und die Bildung der "Befruchtungsmembran" von Trichoplax adhaerens F.E. Schulze (Placozoa). Zeitschrift für Morphologie der Tiere 79 (1974) 295–310.

Grell, K. G., Benwitz, G., Spezifische Verbindungsstrukturen der Faserzellen von Trichoplax adhaerens F.E. Schulze., Zeitschrift für Naturforschung C 29(11–12) (1974) 790.

Grell, K. G., Benwitz, G., Ergänzende Untersuchungen zur Ultrastruktur von Trichoplax adhaerens F. E. Schulze (Placozoa). Zoomorphology 98(1) (1981) 47–67.

Grell, K. G., Reproduction of Placozoa, in: Engels, W. (ed.), Advances in Invertebrate Reproduction, Elsevier, 1984, pp. 541–546.

Gruber-Vodicka, H. R., Leisch, N., Kleiner, M., Hinzke, T., Liebeke, M., McFall-Ngai, M., Hadfield, M. G., Dubilier, N., Two intracellular and cell type-specific bacterial symbionts in the placozoan Trichoplax H2, Nat Microbiol 4(9) (2019) 1465–1474.

Guidi, L., Eitel, M., Cesarini, E., Schierwater, B., Balsamo, M., Ultrastructural analyses support different morphological lineages in the phylum Placozoa Grell, 1971, J Morphol 272(3) (2011) 371–378.

Heyland, A., Croll, R., Goodall, S., Kranyak, J., Wyeth, R., Trichoplax adhaerens, an enigmatic basal metazoan with potential, Methods Mol Biol 1128 (2014) 45–61.

Jackson, A. M., Buss, L. W., Shiny spheres of placozoans (*Trichoplax*) function in anti-predator defense, Invertebrate Biology 128(3) (2009) 205–212.

Jakob, W., Sagasser, S., Dellaporta, S., Holland, P., Kuhn, K., Schierwater, B., The Trox-2 Hox/ParaHox gene of Trichoplax (Placozoa) marks an epithelial boundary, Dev Genes Evol 214(4) (2004) 170–175.

Kamm, K., Schierwater, B., Ancient complexity of the non-Hox ANTP gene complement in the anthozoan Nematostella vectensis: implications for the evolution of the ANTP superclass, J Exp Zool B Mol Dev Evol 306(6) (2006) 589–596.

Kamm, K., Schierwater, B., Jakob, W., Dellaporta, S. L., Miller, D. J., Axial patterning and diversification in the cnidaria predate the Hox system, Curr Biol 16(9) (2006) 920–926.

Kamm, K., Osigus, H. J., Stadler, P. F., DeSalle, R., Schierwater, B., Trichoplax genomes reveal profound admixture and suggest stable wild populations without bisexual reproduction, Sci Rep 8(1) (2018) 11168.

Kamm, K., Osigus, H. J., Stadler, P. F., DeSalle, R., Schierwater, B., Genome analyses of a placozoan rickettsial endosymbiont show a combination of mutualistic and parasitic traits, Sci Rep 9(1) (2019) 17561.

Kamm, K., Schierwater, B., DeSalle, R., Innate immunity in the simplest animals – placozoans, BMC Genomics 20(1) (2019) 5.

Kuhl, W., Kuhl, G., Bewegungsphysiologische Untersuchungen an Trichoplax adhaerens F. E. Schulze, Zoologischer Anzeiger Suppl. 26 (1963) 460–469.

Kuhl, W., Kuhl, G., Untersuchungen über das Bewegungsverhalten von Trichoplax adhaerens F. E. Schulze., Zeitschrift für Ökologie und Morphologie der Tiere 56 (1966) 417–435.

Laumer, C. E., Fernandez, R., Lemer, S., Combosch, D., Kocot, K. M., Riesgo, A., Andrade, S. C. S., Sterrer, W., Sorensen, M. V., Giribet, G., Revisiting metazoan phylogeny with genomic sampling of all phyla, Proc Biol Sci 286(1906) (2019) 20190831.

Liu, Y. G., Kurokawa, T., Sekino, M., Tanabe, T., Watanabe, K., Complete mitochondrial DNA sequence of the ark shell Scapharca broughtonii: an ultra-large metazoan mitochondrial genome, Comp Biochem Physiol Part D Genomics Proteomics 8(1) (2013) 72–81.

Martinelli, C., Spring, J., Expression pattern of the homeobox gene Not in the basal metazoan Trichoplax adhaerens, Gene Expression Patterns 4(4) (2004) 443–447.

Martinelli, C., Spring, J., T-box and homeobox genes from the ctenophore Pleurobrachia pileus: comparison of Brachyury, Tbx2/3 and Tlx in basal metazoans and bilaterians, FEBS Letters 579(22) (2005) 5024–5028.

Maruyama, Y. K., Occurrence in the field of a long-term, year-round, stable population of placozoans, Biological Bulletin 206(1) (2004) 55–60.

Mayorova, T. D., Smith, C. L., Hammar, K., Winters, C. A., Pivovarova, N. B., Aronova, M. A., Leapman, R. D., Reese, T. S., Cells containing aragonite crystals mediate responses to gravity in Trichoplax adhaerens (Placozoa), an animal lacking neurons and synapses, PLoS One 13(1) (2018) e0190905.

Mayorova, T. D., Hammar, K., Winters, C. A., Reese, T. S., Smith, C. L., The ventral epithelium of Trichoplax adhaerens deploys in distinct patterns cells that secrete digestive enzymes, mucus or diverse neuropeptides, Biol Open 8(8) (2019).

Miyazawa, H., Yoshida, M., Tsuneki, K., Furuya, H., Mitochondrial Genome of a Japanese Placozoan, Zool Sci 29(4) (2012) 223–228.

Monticelli, F. S., Treptoplax reptans n.g., n.sp., Atti dell'Academia dei Lincei, Rendiconti (5)II (1893) 39–40.

Osigus, H. J., Rolfes, S., Herzog, R., Kamm, K., Schierwater, B., Polyplacotoma mediterranea is a new ramified placozoan species, Curr Biol 29(5) (2019) R148–R149.

Paknia, O., Schierwater, B., Global habitat suitability and ecological niche separation in the Phylum Placozoa, PloS One 10(11) (2015).

Pearse, V. B., Voigt, O., Field biology of placozoans (Trichoplax): distribution, diversity, biotic interactions, Integr Comp Biol 47(5) (2007) 677–692.

Philippe, H., Derelle, R., Lopez, P., Pick, K., Borchiellini, C., Boury-Esnault, N., Vacelet, J., Renard, E., Houliston, E., Queinnec, E., Da Silva, C., Wincker, P., Le Guyader, H., Leys, S., Jackson, D. J., Schreiber, F., Erpenbeck, D., Morgenstern, B., Worheide, G., Manuel, M., Phylogenomics revives traditional views on deep animal relationships, Current Biology 19(8) (2009) 706–712.

Pick, K. S., Philippe, H., Schreiber, F., Erpenbeck, D., Jackson, D. J., Wrede, P., Wiens, M., Alie, A., Morgenstern, B., Manuel, M., Worheide, G., Improved phylogenomic taxon sampling noticeably affects nonbilaterian relationships, Molecular Biology Evolution 27(9) (2010) 1983–1987.

Ruthmann, A., Grell, K. G., Benwitz, B., DNA-content and fragmentation of the egg-nucleus of Trichoplax adhaerens., Zeitschrift für Naturforschung C 60 (1981) 564–567.

Ryan, J. F., Burton, P. M., Mazza, M. E., Kwong, G. K., Mullikin, J. C., Finnerty, J. R., The cnidarian-bilaterian ancestor possessed at least 56 homeoboxes: evidence from the starlet sea anemone, Nematostella vectensis, Genome Biol 7(7) (2006) R64.

Schierwater, B., My favorite animal, Trichoplax adhaerens, Bioessays 27(12) (2005) 1294–1302.

Schierwater, B., DeSalle, R., Can we ever identify the Urmetazoan?, Integr Comp Biol 47(5) (2007) 670–676.

Schierwater, B., Kamm, K., Srivastava, M., Rokhsar, D., Rosengarten, R. D., Dellaporta, S. L., The early ANTP gene repertoire: insights from the placozoan genome, PLoS One 3(8) (2008) e2457.

Schierwater, B., de Jong, D., Desalle, R., Placozoa and the evolution of Metazoa and intrasomatic cell differentiation, Int J Biochem Cell Biol 41(2) (2009) 370–379.

Schierwater, B., Eitel, M., Jakob, W., Osigus, H. J., Hadrys, H., Dellaporta, S. L., Kolokotronis, S. O., DeSalle, R., Concatenated analysis sheds light on early metazoan evolution and fuels a modern "urmetazoon" hypothesis, PLoS Biol 7(1) (2009) e20.

Schierwater, B., Kolokotronis, S. O., Eitel, M., DeSalle, R., The Diploblast-Bilateria sister hypothesis: parallel evolution of a nervous systems in animals, Cummunicative and Integrative Biology 2(5) (2009) 1–3.

Schierwater, B., Eitel, M., Osigus, H. J., von der Chevallerie, K., Bergmann, T., Hadrys, H., Cramm, M., Heck., L., M.R., L., DeSalle, R., Trichoplax and Placozoa: one of the crucial keys to understanding metazoan evolution, Key transitions in animal evolution (2010) 289–326.

Schierwater, B., Eitel, M., Placozoa, in: Wanninger, A. (ed.), Evolutionary Developmental Biology of Invertebrates 1: Introduction, Non-Bilateria, Acoelomorpha, Xenoturbellida, Chaetognatha, Vienna: Springer, 2015, pp. 107–114.

Schierwater, B., Holland, P. W. H., Miller, D. J., Stadler, P. F., Wiegmann, B. M., Wörheide, G., Wray, G. A., DeSalle, R., Never ending analysis of a century old evolutionary debate: "Unringing" the Urmetazoon Bell, Frontiers in Ecology and Evolution 4(5) (2016).

Schierwater, B., DeSalle, R., Placozoa, Curr Biol 28(3) (2018) R97–R98.

Schleicherová, D., Dulias, K., Osigus, H. J., Paknia, O., Hadrys, H., Schierwater, B., The most primitive metazoan animals, the placozoans, show high sensitivity to increasing ocean temperatures and acidities, Ecol Evol 7(3) (2017) 895–904.

Schulze, F. E., Trichoplax adhaerens, nov. gen., nov. spec., Zoologischer Anzeiger 6 (1883) 92–97.

Schulze, F. E., Über Trichoplax adhaerens, in: Reimer, G. (ed.), Abhandlungen der Königlichen Preuss. Akademie der Wissenschaften zu Berlin., Verlag der königlichen Akademie der Wissenschaften, Berlin, 1891, pp. 1–23.

Schwartz, V., The radial polar pattern of differentiation in Trichoplax adhaerens F. E. Schulze (Placozoa), Zeitschrift für Naturforschung C 39 (1984) 818–832.

Signorovitch, A. Y., Dellaporta, S. L., Buss, L. W., Caribbean placozoan phylogeography, Biological Bulletin 211(2) (2006) 149–156.

Signorovitch, A. Y., Buss, L. W., Dellaporta, S. L., Comparative genomics of large mitochondria in placozoans, PLoS Genet 3(1) (2007) e13.

Simion, P., Philippe, H., Baurain, D., Jager, M., Richter, D. J., Di Franco, A., Roure, B., Satoh, N., Queinnec, E., Ereskovsky, A., Lapebie, P., Corre, E., Delsuc, F., King, N., Worheide, G., Manuel, M., A Large and Consistent Phylogenomic Dataset Supports Sponges as the Sister Group to All Other Animals, Current Biology 27(7) (2017) 958–967.

Smith, C. L., Varoqueaux, F., Kittelmann, M., Azzam, R. N., Cooper, B., Winters, C. A., Eitel, M., Fasshauer, D., Reese, T. S., Novel cell types, neurosecretory cells, and body plan of the early-diverging metazoan Trichoplax adhaerens, Curr Biol 24(14) (2014) 1565–1572.

Smith, C. L., Pivovarova, N., Reese, T. S., Coordinated Feeding Behavior in Trichoplax, an Animal without Synapses, PLoS One 10(9) (2015) e0136098.

Smith, C. L., Reese, T. S., Govezensky, T., Barrio, R. A., Coherent directed movement toward food modeled in Trichoplax, a ciliated animal lacking a nervous system, Proc Natl Acad Sci USA 116 (18) (2019) 8901–8908.

Srivastava, M., Begovic, E., Chapman, J., Putnam, N. H., Hellsten, U., Kawashima, T., Kuo, A., Mitros, T., Salamov, A., Carpenter, M. L., Signorovitch, A. Y., Moreno, M. A., Kamm, K., Grimwood, J., Schmutz, J., Shapiro, H., Grigoriev, I.V., Buss, L. W., Schierwater, B., Dellaporta, S. L., Rokhsar, D. S., The Trichoplax genome and the nature of placozoans, Nature 454(7207) (2008) 955–960.

Thiemann, M., Ruthmann, A., Trichoplax adhaerens Schulze, F. E. (Placozoa) – The formation of swarmers, Zeitschrift für Naturforschung C 43(11–12) (1988) 955–957.

Thiemann, M., Ruthmann, A., Spherical forms of Trichoplax adhaerens., Zoomorphology 110(1) (1990) 37–45.

Thiemann, M., Ruthmann, A., Alternative modes of asexual reproduction in trichoplax-adhaerens (Placozoa), Zoomorphology 110(3) (1991) 165–174.

Varoqueaux, F., Williams, E.A., Grandemange, S., Truscello, L., Kamm, K., Schierwater, B., Jekely, G., Fasshauer, D., High Cell Diversity and Complex Peptidergic Signaling Underlie Placozoan Behavior, Curr Biol 28(21) (2018) 3495–3501 e2.

Voigt, O., Collins, A. G., Pearse, V. B., Pearse, J. S., Ender, A., Hadrys, H., Schierwater, B., Placozoa – no longer a phylum of one, Curr Biol 14(22) (2004) R944–R945.

Voigt, O., Eitel, M., Placozoa, in: Schmidt-Rhaesa, A. (ed.), Miscellaneous Invertebrates. Berlin, Boston: De Gruyter, 2018, pp. 41–54.

Wenderoth, H., Transepithelial cytophagy by Trichoplax adhaerens F. E.Schulze (Placozoa) feeding on yeast., Zeitschrift für Naturforschung C 41(3) (1986) 343–347.

PHYLUM PORIFERA

Maja Adamska

CHAPTER

8

Subregnum: Metazoa

1st Division: Diploblastic Animals

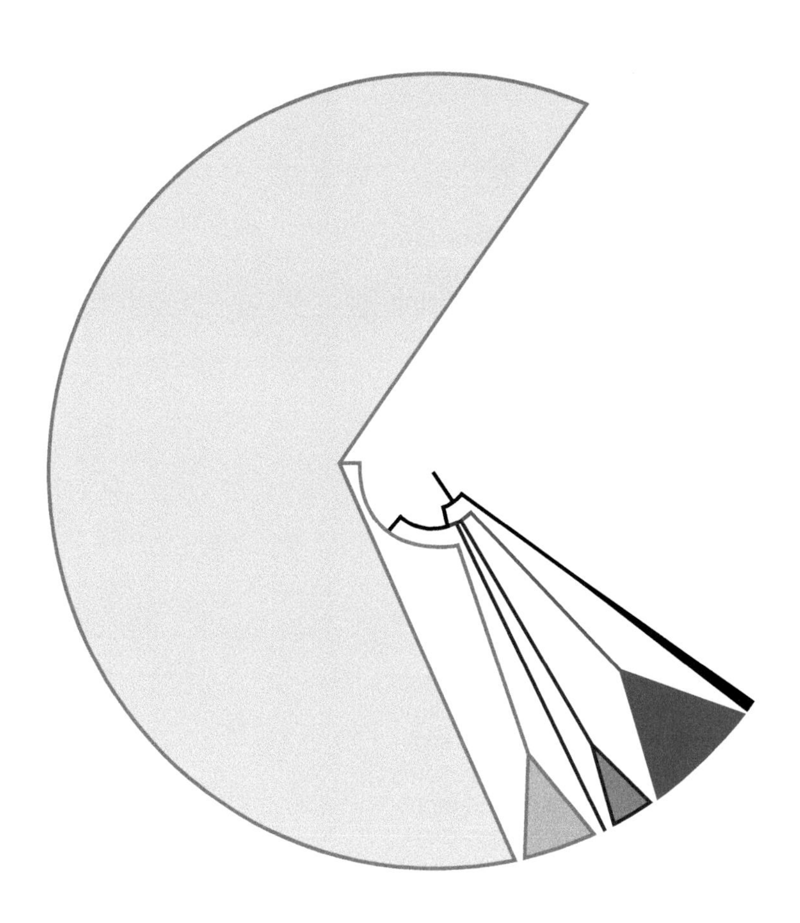

CONTENTS

General Characteristics

8,553 valid species

Number of "cryptic species" estimated to be up to 17,000

Sessile as adults, exclusively aquatic, mostly marine animals ranging in size from minute to approximately 200–500 μl in volume

Cell types: approximately 12

Mitochondrial genome: diverse, from "standard metazoan" to fragmented into linear segments containing one gene per segment

Nuclear genome size: diverse, from 125 to 350 Mb, possibly larger

Commercial value: bath sponges, bioprospecting, providing habitat for juvenile fish

Ecological importance: massive – multiple marine ecosystems are created or heavily influenced by sponges

Autapomorphies

- aquiferous system with flagellated chambers composed of choanocytes (lost in carnivorous sponges)

HISTORY OF TAXONOMY AND CLASSIFICATION

Sponges have been known and used by humans for thousands of years (Pronzato and Manconi, 2008), with the natural bath sponges – which are dried organic skeletons of sponges from the genera *Spongia* and *Hippospongia* – remaining a popular souvenir from the Mediterranean region (**Figure 8.1A**). Despite this familiarity, sponges have proven particularly difficult to classify.

In the words of Ernst Haeckel (1870), "No other class of the animal or vegetable kingdom, containing an equal number of abundant, large, and multifarious forms, has left naturalists, even up to the most recent times, so much in doubt as to its true nature, or called forth such a number of contradictory opinions." Early naturalists could in fact not decide whether sponges are plants or animals, with Aristotle (who must have known the Mediterranean bath sponges well) referring to them alternatively as animals or plants, and Linnaeus originally considering sponges "vegetables", before grouping them with corals into clade Zoophyta (Haeckel, 1870). The name Porifera (pore-bearing) was given to the phylum by Robert Grant (1825, 1826, 1836), who referred to the small incurrent openings on the surface of sponges as pores, the larger exhalant opening as the osculum, and placed sponges firmly among the animals. However, the discovery of striking similarity between sponge collar cells – choanocytes – and choanoflagellates (single-cell and colonial protists), along with observations of sponge cell motility, resulted in relegation of sponges to Protozoa by some authors (James-Clark, 1868). The subsequent "upgrade" to Parazoa (animals of cellular, rather than tissue grade of construction, Sollas, 1884) remained in use through much of the 20th century (Hyman, 1940). On the other hand, in Haeckel's view, sponges were not only clearly animals, but they shared body plan with cnidarians, with the major apical opening (mouth in polyps, osculum in sponges) and the digestive layers (gastrodermis in polyps, choanoderm in sponges) being homologous (Haeckel, 1870). While the views on homology of the body plans and their exact position in the tree of life remain hotly disputed, it is now broadly accepted that sponges form a monophyletic clade within metazoans.

All sponges are aquatic, and all are sessile in the adult form, but the shapes and sizes of sponges vary dramatically between and within the four currently recognized classes, which are defined by a combination of types of skeletal elements and larvae (Figure 8.1), and are in agreement with the recent molecular taxonomies (Van Soest et al., 2012). Homoscleromorpha, Hexactinellida and Demospongiae are sometimes referred to collectively as Silicispongia, as (in the species which contain inorganic skeletal elements) their spicules are built of silica. However, the homoscleromorphs, which are a small (approximately 100 species) class of exclusively marine, relatively small and inconspicuous species with simple spicules (Figure 8.1B,G), have recently been shown to be more closely related to the class Calcarea (Philippe et al., 2009; Gazave et al., 2010).

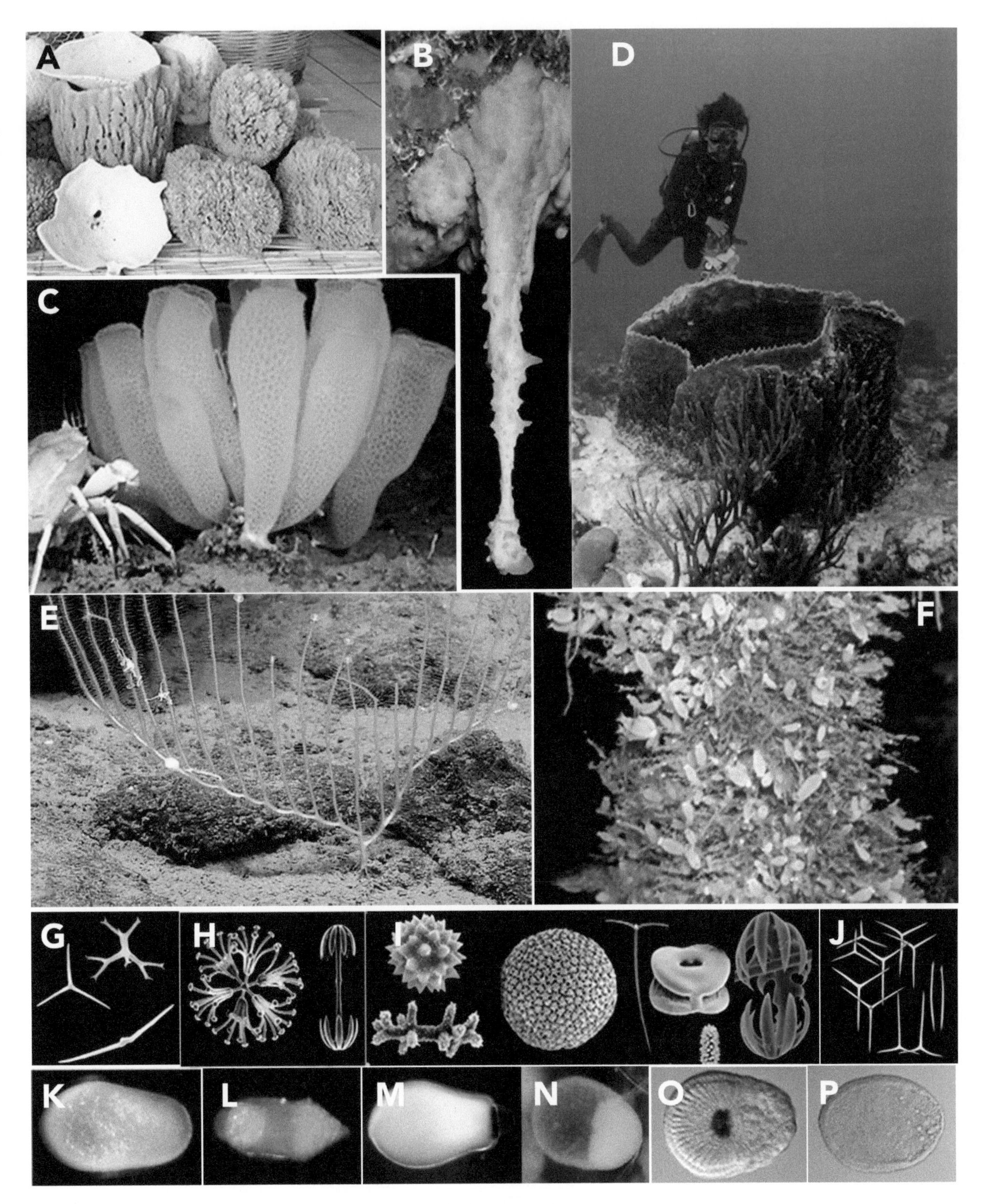

Figure 8.1 Diversity of sponge adult body types, spicule shapes and larvae. (A) Bath sponges (*Spongia* and *Hippospongia* species; *Demospongiae*) on display in Cretan tourist shop; (B) *Plakortis simplex* (*Homoscleromorpha*); (C) Venus flower basket (*Euplectella aspergillum, Hexactinellida*); (D) giant barrel sponge (*Xestospongia muta; Demospongiae*); (E) Cladorizhid carnivorous sponge (*Demospongiae*); (F) numerous specimens of *Sycon ciliatum* (*Calcarea*) on *Laminaria* kelp; (G–J) examples of spicules isolated from *Homoscleromorpha, Hexactinellida, Demospongiae* and *Calcarea* respectively; (K) cinctoblastula of *Oscarella lobularis* (*Homoscleromorpha*); (L) trichimella of *Oopsacas minuta* (*Hexactinellida*); (M, N) parenchymellas of *Amphimedon queenslandica* and *Ephydatia fluviatilis* (both *Demospongiae*); (O) amphiblastula of *Sycon ciliatum* (Calcarea, Calcaronea); (P) calciblastula of *Clathrina lacunosa* (Calcarea, Calcinea). (Photo credits: B, G–J: Van Soest et al., 2012; C, E: NOAA Okeanos Explorer Program; D: J. R. Pawlik, UNCW; K: A. Ereskovsky; L: Leys et al., 2016; N: N. Funayama.)

Hexactinellid (glass) sponges are characterized by unique syncytial body plan organization, which arises early during embryonic development of the trichimella larva (Boury-Esnault et al., 1999; Leys et al., 2016), and are generally found in deep sea environments (Figure 8.1C,H,L). Their spicules have triaxonic symmetry (Figure 8.1H); in some species the massive elongated spicules reach almost 3 m in length (Wang et al., 2011). Over 80% of the extant sponge species belong to the class Demospongiae, which occupy all marine environments, and which have also colonized freshwater habitats, with sizes ranging from minute to spectacular, as in case of the giant barrel sponges (Figure 8.1D). Demosponges have diverse body forms, spicule shapes and larval types (Figure 8.1E,I,L,M and see below). Some demosponges have abandoned filter-feeding lifestyle and trap prey with Velcro-style spicules (Vacelet and Duport, 2004) (Figure 8.1E,I far right).

The fourth class of sponges, *Calcarea*, is characterized by spicules built of calcium carbonate in the form of calcite. While not particularly species rich (only approximately 680 species have been described so far), *Calcarea* are characterized by variety of body types (see below), two unique types of larvae (calciblastula in *Calcinea*, and amphiblastula in *Calcaronea*), and since the time of Haeckel have provided important model species for evolutionary and developmental biology studies (Figure 8.1F,J,O,P; reviewed by Adamska, 2016a).

ANATOMY, HISTOLOGY AND MORPHOLOGY

With the notable exception of species from the carnivorous family *Cladorhizidae*, sponges are characterized by presence of flagellated aquiferous system, which can be considered the only autoapomorphy of the phylum. Sponge bodies are constructed from two epithelial layers – the choanoderm, built of squat, flagellated and collared cells (choanocytes) and the pinacoderm, built of flat pinacocytes, with non-epithelial layer of varied thickness, called mesohyl, sandwiched between (**Figure 8.2A,B**). Choanocytes are equipped with flagella, beating of which propels water entering the choanocyte chamber through the multiple pores and exiting through the osculum (or oscula); the microvillar collars of the choanocytes are responsible for capturing of food particles, predominantly bacteria. In its simplest form – asconoid grade of organization – the body is composed of the inner choanoderm and the outer pinacoderm, connected by thin-walled, well-shaped cells (porocytes) forming ostia; the thin mesohyl layer contains sclerocytes (spicule producing cells) and a small number of amoeboid cells (Figure 8.2C–E). In adult asconoid sponges this basic body plan becomes complicated by elongation, budding, and sometimes anastomosing of the bilayered tubes (Figure 8.2F,F'). A recently described solenoid organization (Cavalcanti and Klautau, 2011) resembles a complex, anastomosing asconoid system embedded in a thick mesohyl layer and emptying into a central, endopinacocyte-lined atrial cavity (Figure 8.2G,G'). The syconoid grade of organization can be viewed as collection of simple asconoid-grade units connected by the central atrium, lined with endopinacocytes (Figure 8.2H,H'). In some syconoid sponges (e.g. *Sycon ciliatum*, H'), the distal ends of the radial chambers are free, with porocytes directly connecting chambers to the external environment; in others (e.g. *Sycon capricorn*) they are embedded in the mesohyl layer, with water flowing in through pinacocyte-lined canals. Asconoid, solenoid and syconoid organization are only found among the *Calcarea*. In the sylleibid grade, characteristic for some homoscleromorph species, syconoid-like units are connected by a larger chamber leading to the osculum (Figure 8.2I,I'). The most common sponge body plan, found in all filter-feeding demosponges and many sponges from the remaining sponge classes, is called leuconoid. Here, ostia (pores) lead to canals lined with endopinacocytes, which in turn lead to round or ellipsoid choanocyte chambers, and then again from the chambers to atrium (Figure 8.2J,J').

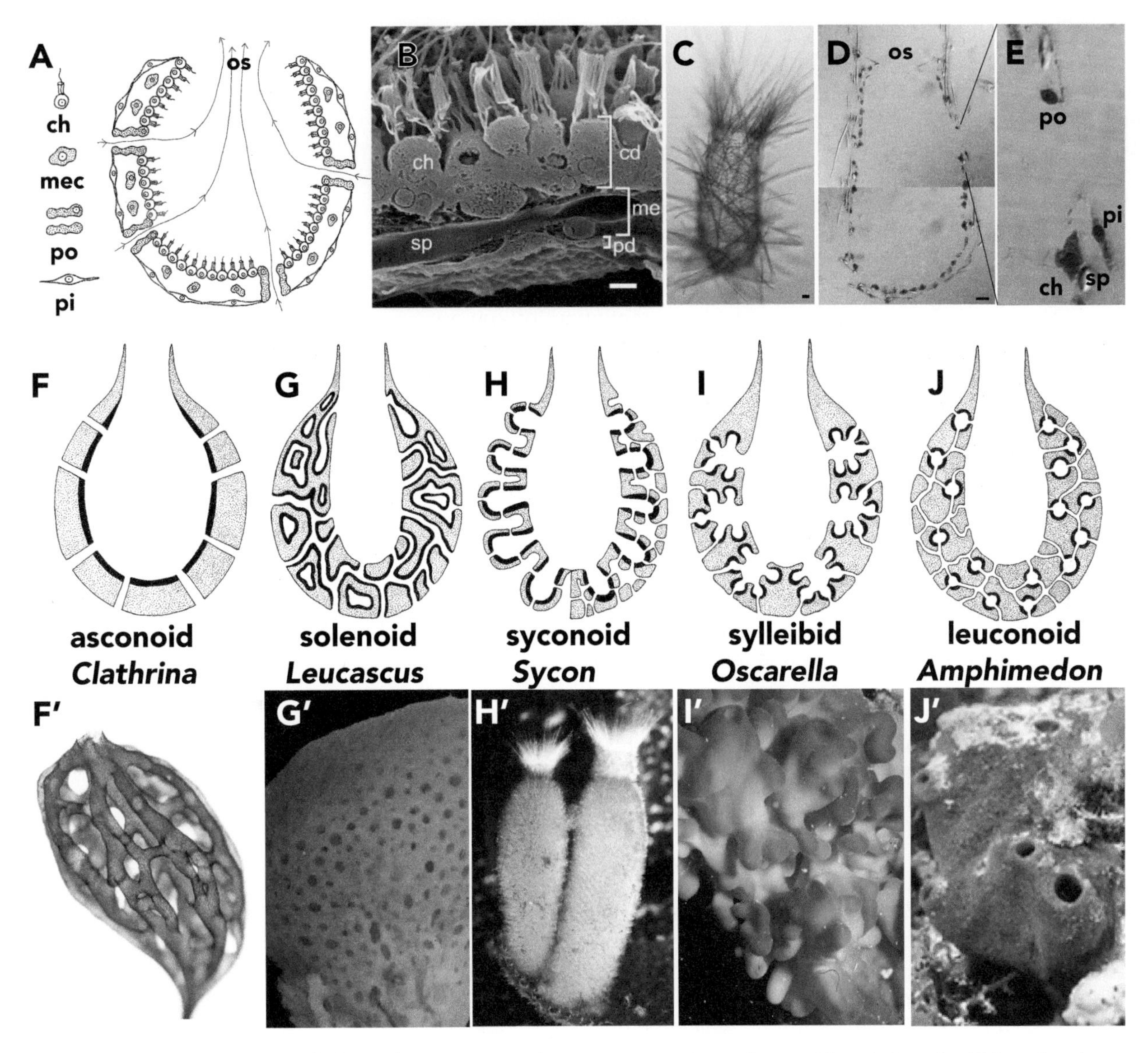

Figure 8.2 **Histology and morphology of sponges.** (A) Schematic representation of asconoid-grade juvenile. (B) Scanning electron microscopy image of body wall of *Sycon coactum* (*Calcaronea*); (C) live asconoid grade juvenile of *S. ciliatum*; (D–E) hematoxylin and eosin staining of section of a similar juvenile; (F–J) illustrated representations and (F'–J') live examples of sponge body plans. ch: choanocyte; mec: mesohyl cell; po: porocyte; pi: pinacocyte; os: osculum; sp: spicule; pd: pinacoderm; cd: choanoderm; me: mesohyl. (Image credits: A, F–J: Drawings by Ayla Manwaring; F -J: Modified after Hyman, 1940 and Cavalcanti and Klautau, 2011); B: Eerkes-Medrano and Leys, 2006; C–E: Leininger et al., 2014; G': Cavalcanti and Klautau, 2011; I': A. Ereskovsky; J': M. Adamski.)

The bodies are reinforced by skeletal elements, which can be organic (e.g. spongin as in bath sponges), or inorganic, either silica-based or calcium carbonate based (reviewed by Uriz, 2006). Spicules are produced by sclerocytes in the mesohyl, and at least in some sponges transported by specialized cells to their final location within the sponge body (Nakayama et al., 2015). The extremely diverse shapes of spicules provide foundation for the classical taxonomy of sponges (van Soest et al., 2012).

REPRODUCTION AND DEVELOPMENT

Sponges display a vast array of reproductive modes, both vegetative (asexual) and sexual. Asexual reproduction can include simple budding (as visible in Figure 8.1B) or formation of highly specialized structures called gemmules.

Gemmule formation appears to have independently evolved multiple times among demosponges (Simpson and Fell, 1974). One of the best studied examples is a freshwater sponge *Ephydatia fluviatilis* (Langenbruch, 1981; Mohri et al., 2008; Nakayama et al., 2015). In these sponges, gemmule production starts by aggregation of stem cells, archaeocytes, in the mesohyl. The aggregate is surrounded by spongocytes secreting a collagenous coat, into which gemmoscleres (dumbbell-shaped spicules) are incorporated. Within the coat, archaeocytes transform into resting, binuclear stem cells, called thesocytes, which remain dormant until gemmule hatching (spring in nature, after the remains of the adult sponge have disintegrated). During hatching, the thesocytes complete division, migrate out of the coat, and start proliferating and differentiating into pinacocytes, choanocytes and sclerocytes (**Figure 8.3A–D**). Within a few days, a complete new sponge with prominent oscular chimney is formed, demonstrating totipotency of the archeocytes.

Fertilization in sponges can be internal or external, with at least six major types of larvae (plus secondarily acquired direct development in some species) described to date (Maldonado, 2006; see also Figure 8.1K–P). Sponge embryogenesis involves the majority of the diverse cleavage and morphogenesis patterns known in the animal kingdom (for in-depth coverage of the subject, see Leys and Ereskovsky, 2006; Ereskovsky 2010; Degnan et al., 2015). Embryonic and postembryonic development have been particularly well studied in *Amphimedon queenslandica* (Demospongiae, Haplosclerida) (e.g. Leys and Degnan, 2002; Adamska et al., 2010) and in several species of Calcaronea (e.g. Franzen, 1988).

In *Amphimedon*, which reproduces through most of the year in the warm Great Barrier Reef environment, embryos of different stages develop in brood chambers within the mother tissue. Chaotic cleavage of the relatively large (500μm) yolk-rich eggs produces as solid spherical embryo of beige color (Figure 8.3E). The color is derived from small pigmented cells uniformly distributed throughout the outer layer of embryos at this stage. The pigment cells subsequently migrate to coalesce at the posterior (defined be swimming direction of the larva) pole of the embryo, forming a spot and then a ring (Figure 8.3F,G). The ring will become the photosensory and steering organ of the parenchymella larva (Leys and Degnan, 2001). The larvae "hatch" from surrounding follicle membranes, and – after a period of swimming – settle down on their anterior pole and begin metamorphosis, during which the pigment ring degenerates (Figure 8.3H–L). Within days, the leuconoid aquiferous system and a single oscular chimney form, completing the metamorphosis (Leys and Degnan, 2002) (Figure 8.3M,N).

Calcaronean sponges from cold and temperate climates, including *Leucosolenia complicata*, *Sycon coactum* and *S. ciliatum*, reproduce in summer (Eerkes-Medrano and Leys, 2006; Leininger et al., 2014). Synchronous development occurs in the mesohyl layer throughout the body of the sponge (Figure 8.3O,P). The initially equal cleavage of the small (approximately 50μm) fertilized eggs generates cup-shaped embryos with cilia pointing inward; inversion of the embryos results in translocation of the embryos to the choanocyte chambers (Figure 8.3P,Q; see also **Figure 8.4I–M**). The simple amphiblastula larva is composed of ciliated micromeres at the anterior pole and non-ciliated macromeres at the posterior pole (Figure 8.1O). During metamorphosis, the micromeres differentiate into choanocytes and (likely) sclerocytes, while macromeres become pinacocytes. In *S. ciliatum*, the asconoid-grade juvenile forms within four days after settlement (Figure 8.3R–V).

Analysis of developmental gene expression in *Amphimedon* and *Sycon* provides additional insights into cell behavior during embryogenesis, highlighting differences between these two model systems. For example, in contrast to calcaroneans, embryogenesis in *Amphimedon* is characterized by extensive individual movements of diverse cell types.

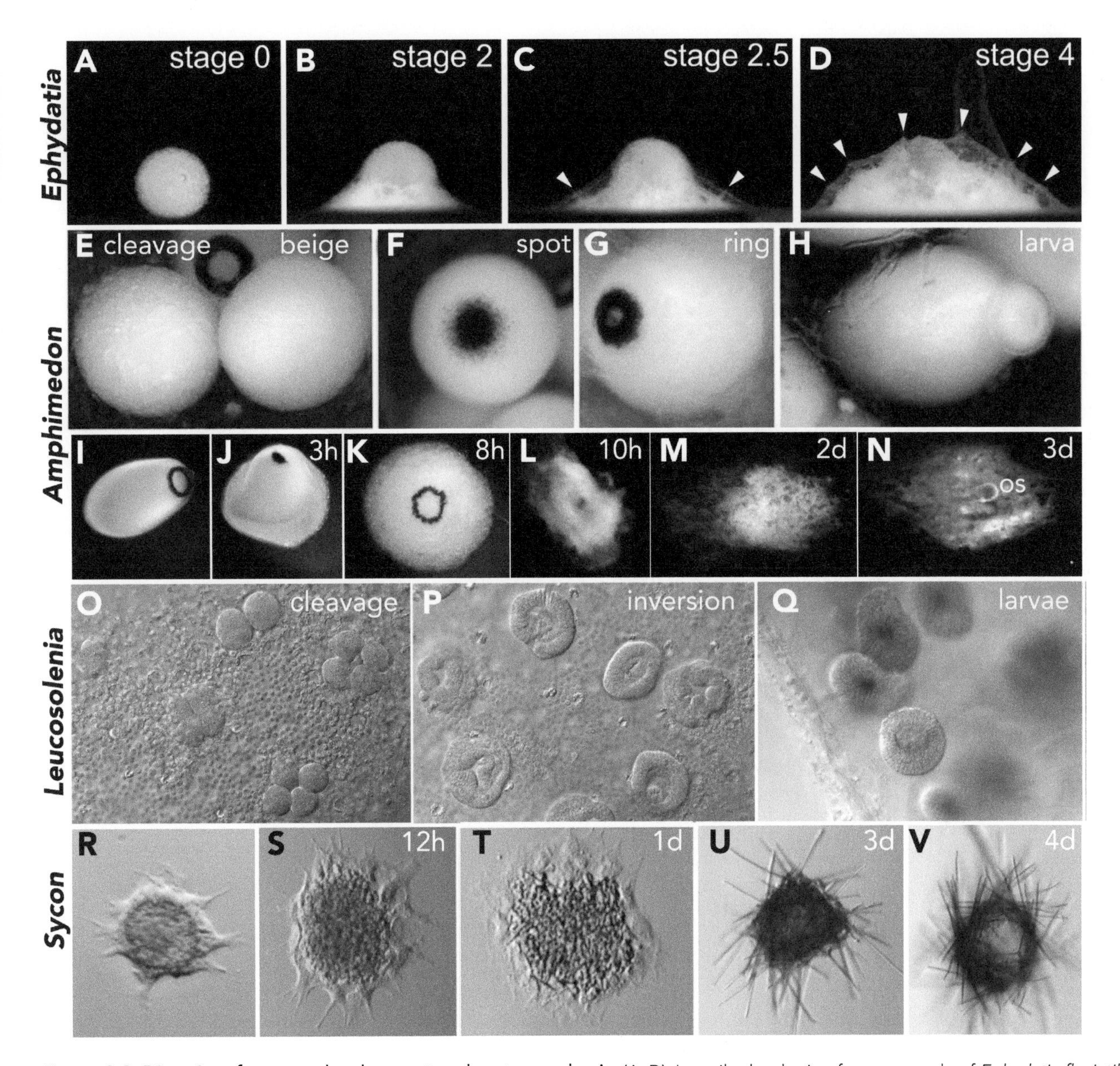

Figure 8.3 **Diversity of sponge development and metamorphosis.** (A–D) Juvenile developing from gemmule of *Ephydatia fluviatilis* (freshwater demosponge); arrowheads indicate spicule "poles"; (E–H) embryonic development and (I–N) metamorphosis in *Amphimedon queenslandica* (marine demosponge). (O–Q) embryonic development in *Leucosolenia complicata* (*Calcarea*); (R–V) metamorphosis in *Sycon ciliatum* (*Calcarea*). O–Q are images of adult tissue which has been fixed, despiculated and cleared; all other images are of live specimens. (Credits: A–D: Nakayama et al., 2015; E–H: Adamska et al., 2010; S–V: Leininger et al., 2014.)

One of these types is globular cells, which are distributed among the epithelial layer of the larvae and which likely have sensory functions. These cells, expressing transcripts of pro-neural bhlh gene related to atonal in addition to multiple Notch-Delta pathway components, can first be detected throughout the outer layer of the embryos at the spot stage (Richards et al., 2008). Globular cells temporarily localize to the sub-epithelial layer at the early ring stage, before migrating to the epithelial layer at the later (elongated) ring stage (Figure 8.4A–H).

In contrast, the putative four sensory cells of the calcaronean amphiblastula larvae – the cross cells – do not change position throughout the embryonic development (Ereskovsky, 2010; Fortunato et al., 2014a,b, 2016; Leininger et al., 2014). Instead, their future position can be visualized already at the four-cell stage as condensations of transcripts encoding *Nanos* gene in the outer "corners" of the blastomeres (Figure 8.4I,N). As the cleavage and cell differentiation continue, the cross cells express a variety of sensory cell markers

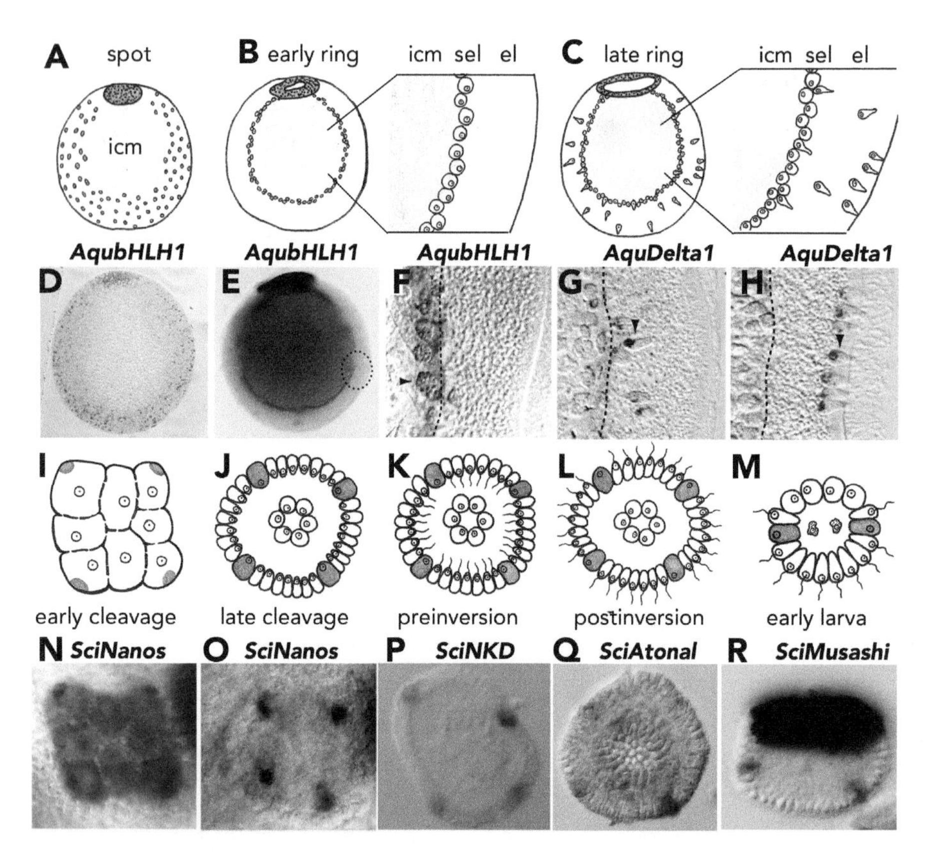

Figure 8.4 Expression of sensory cell markers in late embryogenesis of *Amphimedon queenslandica* (A–H) and *Sycon ciliatum* (I–R). Note migration of globular cells in *Amphimedon* and invariant position of cross cells (represented as the four darker cells) in *Sycon*. (Drawings by Ayla Manwaring. A–C: Modified after Richards et al., 2008); D–H: Richards et al., 2008; N–P: Fortunato et al., 2014a. Q: Fortunato et al., 2016; R: Fortunato et al., 2014b.)

including the atonal gene (Figure 8.4I–R). It appears that the relative position of all embryonic cells – micromeres, macromeres and cross cells – remains stable throughout larval formation, despite the major morphogenetic event of inversion. The macromeres maintain their epithelial character during metamorphosis, while micromeres undergo epithelial-to-mesenchymal transition during settlement. The cross cells, along with maternal cells which have been carried in the larval cavity, degenerate without contributing to the adult body plan (Amano and Hori, 1993).

The fact that putative sensory cells of sponge larvae express a number of genes involved in specification and differentiation of sensory cells and neurons in eumetaozoans, suggests that these cell types might be homologous. The potential homology of not only individual cell types, but also global body plans between sponges and eumetazoans, as proposed by Haeckel, is also receiving support from analysis of developmental gene expression in sponge larvae and adults. In particular, Wnt ligands, expression of which is almost universally associated with posterior pole of the bilaterian embryos and the oral end of cnidarians, have been found to be expressed in the posterior pole of diverse sponge larvae (parenchymella in *Amphimedon*, disphaerula in *Halisarca*, and amphiblastula in *Sycon*), as well as around the osculum in both demosponges and calcareous sponges (**Figure 8.5A–E**) (Adamska et al., 2007; Borisenko et al., 2016; Leininger et al., 2014). Similarly, the transcription factor GATA, which is a key element of the endomesoderm regulatory network in bilaterians, is expressed in the choanoderm lineage of *Amphimedon* and *Sycon* (Figure 8.5F–H) (Nakanishi et al., 2014; Leininger et al., 2014).

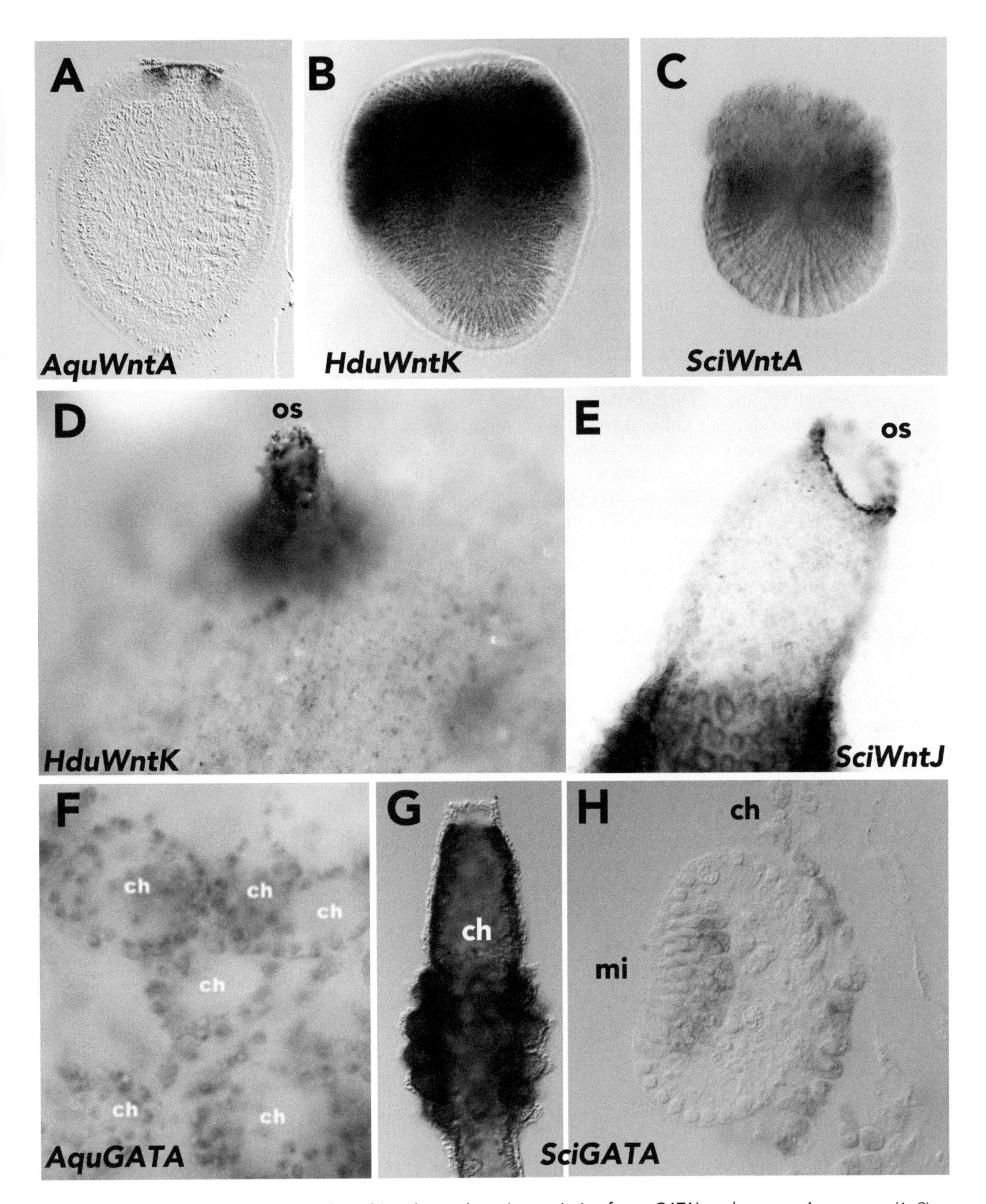

Figure 8.5 Expression of polarity (Wnt ligands) and germ layer (transcription factor GATA) marker genes in sponges. (A–C) Posterior Wnt expression in larvae: *Amphimedon* parenchymella, *Halisarca* disphaerula, *Sycon* amphiblastula; (D–E) oscular Wnt expression in *Halisarca* and *Sycon* adults; (F–H) GATA expression in choanocytes of juvenile and adult *Amphimedon* and *Sycon*, and in embryonic micromeres (which give rise to choanocytes) in *Sycon*. ch: choanocytes; mi: micromeres; os: osculum. (A: Adamska et al., 2007; B, D: Borisenko et al., 2016; C, E, G, H: Leininger et al., 2014; F: Nakanishi et al., 2014.)

DISTRIBUTION AND ECOLOGY

Sponges are broadly distributed in marine and freshwater environments, from the abyssal depths to intertidal zones (**Figure 8.6**) (Manconi M. and Pronzato, 2002; Van Soest et al., 2012). Especially in deeper environments, siliceous sponges can form spectacular aggregations, such as glass (hexactinellid) sponge reefs and carnivorous sponge grounds (Maldonado et al., 2016 and **Figure 8.7A**). Within those aggregations, and in mixed communities, sponges provide habitat and/or food for other organisms. Some zoanthid polyp species are only found living on surfaces of reef sponges (Crocker and Reiswig, 1981), while the sponge known as "candy cane sponge" among aquarists is in fact a red sponge (*Trikentrion flabelliforme*) covered in white zoanthids (Figure 8.7B). Worms, brittle stars and crustaceans often inhabit the aquiferous systems of sponges (Figure 8.7C). The Venus flower basket (*Euplectella* sp., see Figure 8.1C) deep sea hexactinellids often host pairs of *Spongicola* shrimps, whose progeny leaves the sponge through openings in the oscular mesh, which is too small for the adults to pass through (Saito and Konishi, 1999). Many animals, including angelfish and nudibranchs, have evolved morphological adaptations allowing them not only to efficiently use sponges as food source (Konow and Bellwood, 2005), but in extreme cases to look almost exactly like their prey (Figure 8.7D).

Even if not immediately visible, sponges play key roles in the coral reefs. Investigations into cryptic reef environments revealed that sponges can cover more than 50% of the surface in crevices and are capable of removing over 50% of passing phytoplankton and picoplankton (Richter et al., 2001). In addition, sponges are extremely effective in removing dissolved organic matter from the water column. As choanocytes undergo rapid turnover and are shed into surrounding water, these otherwise inaccessible nutrients, often brought by water currents from outside of the reef, are concentrated and made readily available to other reef animals (de Goeij et al., 2008, 2009). Thus, the ability of sponges to efficiently assimilate organic matter and pass it to other reef dwellers can explain Darwin's paradox – the ability of complex reef environments to exist in nutrient-poor waters (de Goeij et al., 2013).

However, the role of sponges in ecosystems, especially when disturbed by warming or pollution, can take on a darker side (Schoenberg et al., 2017). The bioeroding (boring) Clionid sponges excavate caverns within calcareous substrates, including live shells and coral skeletons, often to the detriment or complete demise of their hosts (Pomponi, 1980; de Bakker, 2018) (Figure 8.7E).

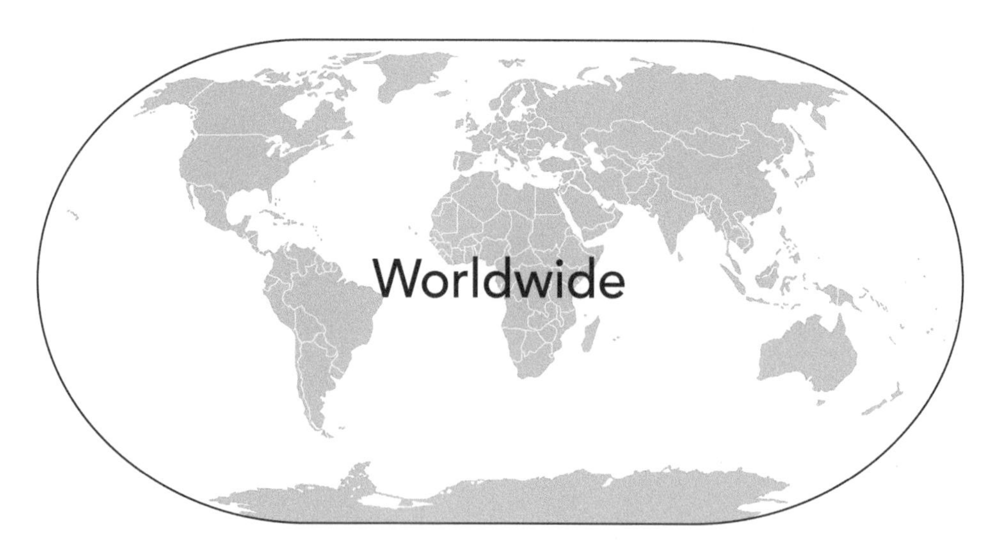

Figure 8.6 Global distribution of sponges.

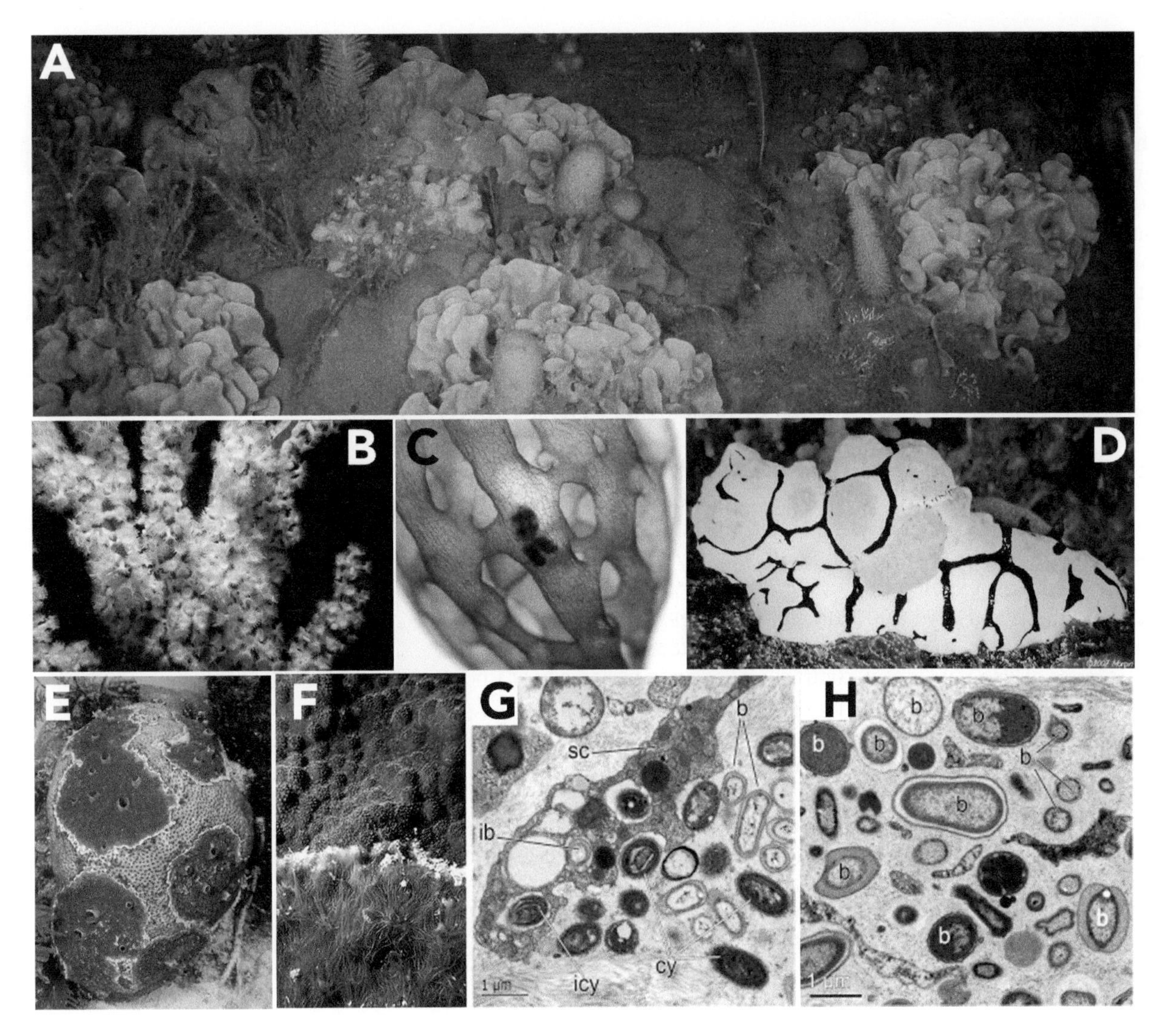

Figure 8.7 Ecology of sponges. (A) Sponges dominate benthic invertebrate assembly on the Antarctic sea floor; (B) Red *Trikentrion flabelliforme* covered in white zoanthids; (C) copepod inside of *Clathrina lacunosa*; (D) nudibranch *Notodoris minor* closely resembling its sponge prey, *Leucetta*; (E) excavating sponge *Cliona delitrix* on coral; (F) *Chalinula nematifera* on coral; (G, H) transmission electron microscopy images of the mesohyl of *Chondrilla nucula* and *Aplysina cavernicola*. b: intercellular bacteria; cy: intercellular cyanobacteria; sc: sponge cell; icy: intracellular cyanobacteria; ib: intracellular bacteria. (Photo credits: A: C. Jones, NOAA NMFS SWFSC Antarctic Marine Living Resources Program; B: J. Fromont, Western Australian Museum Collections; D: M. Adamski; E: J. R. Pawlik, UNCW; F: Rossi et al., 2015; G, H: Maldonado et al., 2012.)

Recently, reports of non-excavating sponges overgrowing and killing corals in the tropical regions have also been published (Rossi et al., 2015; Figure 8.7F).

Like many organisms, sponges live in tight association with microbes. Association with algae, including zooxanthellae, appears to have evolved multiple times, in both freshwater and marine environments, and provides the obvious advantage of photosynthesis (Van Trigt, 1919; Sara and Liaci, 1964). Sponges live in symbiosis with diverse – belonging to 52 phyla – bacteria, which can be transmitted horizontally (from environment) and/or vertically (from mother), and sometimes take up to 35% of the holobiont biomass (Reiswig, 1981; Schmitt et al., 2008; Webster et al., 2010). Sponge-inhabiting microbes are involved in nutrient (C, N, P, S) cycling as well as production of vitamins and secondary metabolites (Maldonado et al., 2012; Figure 8.7G,H). Some of these metabolites turned out to be extremely promising pharmaceuticals; these include cytarabine, eribulin mesylate and vidarabine, approved for cancer treatment and as anti-viral drugs (Brinkmann et al., 2017; Gerwick and Fenner, 2013).

PHYSIOLOGY AND BEHAVIOR

Despite not having nerves or muscles, sponges exhibit a range of behaviors not unlike those observed in other animals (recently reviewed by Leys et al., 2019). The most obvious are those of the motile larvae, which actively seek out surfaces suitable for metamorphosis (Maldonado, 2006). Larval behavior has been extensively studied in *Amphimedon queenslandica*, which has parenchymella type larvae, present also in many other demosponges. In this species, the long cilia at the posterior pole bend in response to light, effectively steering the larvae away from the light, allowing settlement in shaded areas under the coral rubble (Leys and Degnan, 2001). Once larvae reach competence, contact of the anterior pole with coralline algae (often growing on the coral rubble) induces metamorphosis, which appears to be regulated by nitric oxide signaling (Ueda et al., 2016).

The adult sponges are well known to regulate pumping rates in response to changes of environmental conditions, such as surrounding water current and presence of sediment (Tompkins-MacDonald and Leys, 2008). In at least some species, highly regulated contractions of the pinacocyte-lined canals and the osculum can effectively expel water, along with detritus which entered the aquiferous system (Elliott and Leys, 2007). In some species, the contractions and expansions of the canals lead to rhythmic changes in overall body volume. For example, in specimens of *Tethya wilhelma*, contractions occur in periodicity of 1 hour to 10 hours, decreasing the body volume by up to 70% (Nickel, 2004).

Perhaps most surprisingly, some adult sponges can move, albeit slowly, from their location. In many cases, the movement depends on complete reorganization of the body as a response to changes in the surrounding water current or light conditions. Freshwater sponges have been shown to crawl across glass aquarium walls with speeds up to 160 μm/hr (4 mm/day), leaving behind a trail of spicules (Bond and Harris, 1988). Similar speed, up to 186 μm/hr, has been recorded for *Tethya wilhelma* (Nickel, 2006). In both cases, the movement of the whole body appears to depend on concerted movement of amoeboid cells in contact with the substrate.

GENETICS

Sponge nuclear genome sizes estimated by flow cytometry and Feulgen image analysis densitometry range from 0.04–0.63 pg, which translates to approximately 40–600 Mb of haploid genome size (Jeffery et al., 2013). Complete nuclear genome sequence is available for only a handful of species, including *Amphimedon queenslandica* (167 Mb, Srivastava et al., 2010), *Sycon ciliatum* (350 Mb, Fortunato et al., 2014) and *Tethya wilhelma* (125 Mb, Francis et al., 2017). Gene content comparisons between the sequenced species, so far mainly limited to developmental regulatory genes, suggest that the last common ancestor of sponges had a much more complex gene repertoire than any of the extant sponges, indicating extensive independent gene loss among sponge lineages (Fortunato et al., 2015).

Sponge mitochondrial genomes are even more diverse in terms of structure and gene content (Lavrov and Pett, 2016). For example, the circular mitochondrial genome of *Amphimedon queenslandica* contains 13 protein coding genes; ATP9, which has been found in other demosponge mitochondrial genomes has been moved (apparently by a transposition event) to the nuclear genome (Erpenbeck et al., 2007). In contrast, mitochondrial genomes of calcaronean sponges, including *Sycon ciliatum*, are composed of linear segments containing one gene each, and are heavily edited (Lavrov et al., 2016).

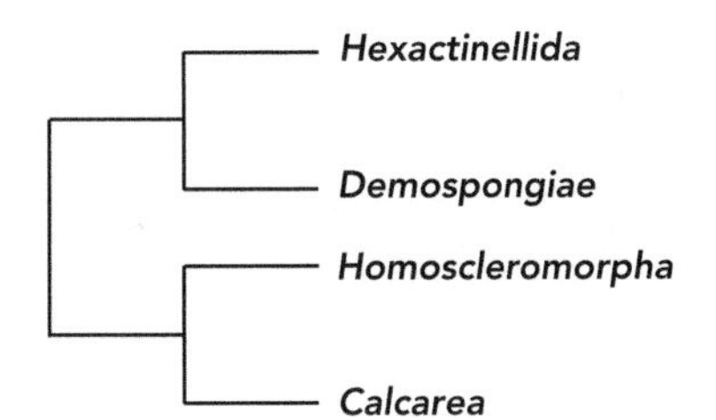

Figure 8.8 Phylogenetic relationships between sponge classes.

POSITION IN THE ToL

Haeckel's lament about the difficulty of placing sponges in the tree of life rings as true today as it did 150 years ago. While biologists currently agree that sponges are metazoans, and that choanoflagellates are the sister group to all metazoa including sponges, several scenarios of relationships among non-bilaterian animals have been proposed and remain hotly debated. In particular, it has been recently proposed that ctenophores, rather than sponges, might be the sister group to all remaining animals, with sponges-first and ctenophores-first hypotheses appearing to have a similar level of support in analyses of large phylogenomic datasets (see Whelan et al., 2017 vs Simion et al., 2017). The ctenophores-first scenario would imply either loss of morphological complexity (nerves, muscle, gut) in sponges, or independent evolution of these features in ctenophores, both of which seem unlikely, but are entertained in the current literature, with the most unbiased view advocating "embracing the uncertainty" (King and Rokas, 2017). With all of that said, the relationships within Porifera are well understood (**Figure 8.8**).

DATABASES AND COLLECTIONS

Extensive collections of sponge specimens can be found in the Queensland Museum (Brisbane), the Western Australian Museum (Perth) and the Natural History Museum (London). There are a number of accessible databases dedicated to sponges or including sponges among other phyla, focusing on distribution, phylogeny, and molecular resources or tools available. These include: The World Porifera Database (http://www.marinespecies.org/porifera/); Sponge Barcoding Project (https://www.spongebarcoding.org/); The Sponge Guide (https://spongeguide.uncw.edu/); Porifera Tree of Life (https://poriferatreeoflife.org/).

CONCLUSION

Similarities between sponge and choanoflagellate cells on the one hand, and between the bilayered, often radially symmetrical sponge and cnidarian body structures on the other, appear to firmly place sponge body plan between choanoflagellates and cnidarians (Adamska, 2016a,b). These features – combined with a large number of phylogenetic studies supporting basal position of sponges – make sponges key model systems to elucidate transitions between colonial protists and the Eumetazoans (Brunet and King, 2017). At the same time, sponges have a spectacular ability to regenerate, including re-aggregation of functional sponges from dissociated cells. This capacity, apparently driven by a combination of cell motility and the lability of cell fate specification allowing transdifferentiation, makes sponges ideal models to study cell differentiation mechanisms (Adamska, 2018).

REFERENCES

Adamska M. 2016. Sponges as models to study emergence of complex animals. Current Opinion in Genetics & Development, 39: 21–8.

Adamska, M. 2016. "Sponges as the Rosetta Stone of Colonial-to-Multicellular Transition" in: Multicellularity. Origins and Evolution. K., J. Niklas, S. A. Newman (Eds.), The MIT Press, Cambridge, Massachusetts; London, England. ISBN: 978-0-262-03415-9.

Adamska M. 2018. "Differentiation and transdifferentiation of sponge cells" in: Marine Organisms as Model Systems in Biology and Medicine. M. Kloc, J. Z. Kubiak (Eds.), Springer-Nature Series "Results and Problems in Cell Differentiation".

Adamska M., Degnan S. M., Green K. M., Adamski M., Craigie A., Larroux C., Degnan B. M. 2007. Wnt and Tgfβ expression in the sponge *Amphimedon queenslandica* and the origin of metazoan embryonic patterning. PLoS One, 10;2(10):e1031.

Adamska M., Larroux C., Adamski M., Green K., Lovas E., Koop D., Richards G. S., Zwafink C., Degnan, B. M., 2010. Structure and expression of conserved wnt pathway components in the demosponge *Amphimedon queenslandica*. Evolution & Development 12(5): 494–518.

Amano S., Hori I., 1993. Metamorphosis of calcareous sponges. 2. Cell rearrangement and differentiation in metamorphosis. Invert Reprod Dev, 24: 13–26.

Bond C., Harris A. K. 1988. Locomotion of sponges and its physical mechanism. J Exp Zool. 246(3): 271–284.

Borisenko I., Adamski M., Ereskovsky A., Adamska M. 2016. Surprisingly rich repertoire of Wnt genes in the demosponge Halisarca dujardini. BMC Evol Biol. 16: 123.

Boury-Esnault N., Efremova S., Bézac C., Vacelet J. 1999. Reproduction of a hexactinellid sponge: first description of gastrulation by cellular delamination in the Porifera, Invertebrate Reproduction & Development, 35: 3, 187–201. doi: 10.1080/07924259.1999.9652385.

Brinkmann C. M., Marker A., Kurtböke D. I. 2017. An Overview on Marine Sponge-Symbiotic Bacteria as Unexhausted Sources for Natural Product Discovery Diversity. 9, 40. doi:10.3390/d9040040.

Brunet T., King N. 2017. The Origin of Animal Multicellularity and Cell Differentiation. Dev Cell., 43 (2):124–140. doi: 10.1016/j.devcel.2017.09.016.

Cavalcanti F. F., Klautau M. 2011. Solenoid: a new aquiferous system to Porifera. Zoomorphology 130: 255–260. doi 10.1007/s00435-011-0139-7.

Crocker L. A., Reiswig H. M. 1981. Host Specificity in Sponge-Encrusting Zoanthidea (Anthozoa: Zoantharia) of Barbados, West Indies. Marine Biology 65: 231–236.

de Bakker D. M., Webb A. E., van den Bogaart LA, van Heuven S.M.A.C, Meesters E.H, van Duyl F.C. 2018. Quantification of chemical and mechanical bioerosion rates of six Caribbean excavating sponge species found on the coral reefs of Curacao. PLoS One 13(5): e0197824.

de Goeij J. M., van Oevelen D., Vermeij M. J. A., Osinga R., Middelburg J. J., de Goeij A. F. P. M., Admiraal W. 2013. Surviving in a marine desert: The Sponge Loop retains resources within Coral Reefs. Science 342(6154): 108–110. doi: 10.1126/science.1241981.

de Goeij J.M, van den Berg H, van Oostveen M.M, Epping E.H.G, van Duyl F.C. 2008. Major bulk dissolved organic carbon (DOC) removal by encrusting coral reef cavity sponges MEPS 357:139–151

de Goeij J. M., De Kluijver A., Van Duyl F. C., Vacelet J., Wijffels R. H., De Goeij A. F., Cleutjens J. P., Schutte B. 2009. Cell kinetics of the marine sponge Halisarca caerulea reveal rapid cell turnover and shedding. Journal of Experimental Biology. 212(Pt 23): 3892–3900. doi: 10.1242/jeb.034561.

Degnan B. M., Adamska M., Richards G. R., Larroux C., Leininger S., Bergum B., Calcino A., Maritz K., Nakanishi N., Degnan S. M. 2015. "Porifera" in: "Evolutionary Developmental Biology of Invertebrates", Vol. 1. Wanninger, A. (Ed.) pp. 65–106, Springer, ISBN 978-3-7091-1861-0.

Eerkes-Medrano D. I., Leys, S. P. 2006. Ultrastructure and embryonic development of a syconoid calcareous sponge. Invertebrate Biology 125(3): 177–194.

Elliott G. R. D, Leys S. P. 2007. Coordinated contractions effectively expel water from the aquiferous system of a freshwater sponge. Journal of Experimental Biology 210: 3736–3748.

Ereskovsky A. V. 2010. The comparative embryology of sponges. Springer, New York.

Erpenbeck D., Voigt O., Adamski M., Adamska M., Hooper J. N., Wörheide G., Degnan B. M. 2007. Mitochondrial diversity of early-branching metazoa is revealed by the complete mt genome of a haplosclerid demosponge. Mol Biol Evol. 24(1): 19–22. Epub 2006 Oct 19.

Fortunato S., Adamski M., Mendivil O., Leininger S., Liu J., Ferrier D. E. K., Adamska M. 2014. Calcisponges have a ParaHox gene and dynamic expression of dispersed NK homeobox genes. Nature 514(7524): 620–623.

Fortunato S., Leininger S., Adamska M. 2014. Evolution of the Pax-Six-Eya-Dach network: the calcisponge case study. EvoDevo 5: 23.

Fortunato S. A. V., Adamski M., Adamska M. 2015. Comparative analyses of developmental transcription factor repertoires in sponges reveal unexpected complexity of the earliest animals. Marine Genomics 2: 121–129.

Fortunato S. A., Vervoort M., Adamski M., Adamska M. 2016. Conservation and divergence of bHLH genes in the calcisponge *Sycon ciliatum*. Evodevo 7: 23.

Francis W. R, Eitel M., Vargas S., Adamski M., Haddock S. H. D., Krebs S., Blum H., Erpenbeck D., Wörheide G. 2017. The genome of the contractile demosponge *Tethya wilhelma* and the evolution of metazoan neural signalling pathways. bioRxiv 120998; doi: https://doi.org/10.1101/120998.

Franzen, W. 1988. Oogenesis and larval development of Scypha ciliata (Porifera, Calcarea) Zoomorphology 107: 349.

Gazave E., Lapébie P., Renard E., Vacelet J., Rocher C., et al. 2010. Molecular Phylogeny Restores the Supra-Generic Subdivision of Homoscleromorph Sponges (Porifera, Homoscleromorpha). PLoS One 5(12): e14290. doi:10.1371/journal.pone.0014290.

Gerwick W. H., Fenner A. M. 2013. Drug discovery from marine microbes. Microbial Ecology. 65(4): 800–806. doi:10.1007/s00248-012-0169-9.

Grant R. E. 1836. Animal Kingdom. in: The Cyclopaedia Of Anatomy And Physiology, Volume 1. Todd, R.B. (Ed.) pp. 107–118, Sherwood, Gilbert, and Piper, London.

Grant R. E. 1825. Observations and experiments on the structure and functions of the Sponge. Edinburgh Phil Journ xiii 94, 343; xiv, 113–124.

Haeckel E. 1874. Die Gastrae Theorie, die phylogenetische Classification des Thierreichs und die Homologie der Keimblatter. Jena Zeitschr Naturwiss, 8: 1–55.

Haeckel E. 1870. On the organization of sponges and their relationship to the corals. Annals and Magazine of natural History 5, 1–13: 107–120.

Hyman L. H. 1940. The Invertebrates: Protozoa through Ctenophora. McGraw-Hill Book Company, Inc.

James-Clark H. 1868. On the spongiae ciliatae as infusoria flagellata; or observations on the structure, animality, and relationship of Leucosolenia botryoides Annals and Magazine of Natural History 1, pp. 133–142; 188–215; 250–264

Jeffery N. W, Jardine C. B., Gregory T. R. 2013. A first exploration of genome size diversity in sponges. Genome.56(8): 451–456. doi: 10.1139/gen-2012-0122. Epub 2013 Mar 8.

King, N., 2004. The Unicellular Ancestry of Animal Development. Developmental Cell 7, 313–325.

King N., Rokas A.. 2017. Embracing uncertainty in reconstructing early animal evolution. Current Biology. 27(19): R1081–R1088. doi: 10.1016/j.cub.2017.08.054.

Konow N., Bellwood D. R. 2005. Prey-capture in Pomacanthus semicirculatus (Teleostei, Pomacanthidae): functional implications of intramandibular joints in marine angelfishes. The Journal of Experimental Biology 208: 1421–1433

Langenbruch P.-F. 1981. Zur entstehung der gemmulae bei Ephydatia fluviatilis L. (porifera). Zoolomorphology 97: 263–284.

Lavrov D. V, Adamski M., Chevaldonné P., Adamska M. 2016. Extensive Mitochondrial mRNA editing and unusual Mitochondrial Genome Organization in Calcaronean Sponges. Current Biology. 26: 86–92.

Lavrov D. V., Pett W. 2016. Animal Mitochondrial DNA as We Do Not Know It: mt-Genome Organization and Evolution in Nonbilaterian Lineages. Genome Biology and Evolution. 8(9): 2896–2913.

Leininger S., Adamski M., Bergum B., Guder C., Liu J., Laplante M., Bråte J., Hoffmann F., Fortunato S., Jordal S., Rapp H. T., Adamska M. 2014. Developmental gene expression provides clues to relationships between sponge and eumetazoan body plans. Nature Communications. 5: 3905.

Leys S. P., Mah J. L., McGill P. R., Hamonic L., De Leo F. C., Kahn A. S. 2019. Sponge behavior and the chemical basis of responses: A post-genomic view. Integr Comp Biol. 59(4): 751–764.

Leys S. P, Degnan B. M. 2001. Cytological basis of photoresponsive behavior in a sponge larva. Biol Bull. 201(3): 323–338.

Leys S. P, Ereskovsky A. V. 2006 Embryogenesis and larval differentiation in sponges. Canadian Journal of Zoology, 84(2): 262–287, https://doi.org/10.1139/z05-170.

Leys, S. P., Degnan, B. M. 2002. Embryogenesis and metamorphosis in a haplosclerid demosponge: gastrulation and transdifferentiation of larval ciliated cells to choanocytes. Inv. Biol 121:171–189.

Leys, S. P., Kamarul Zaman, A., Boury-Esnault, N. (2016), Three-dimensional fate mapping of larval tissues through metamorphosis in the glass sponge Oopsacas minuta. Inv. Biol 135: 259–272. doi:10.1111/ivb.12142.

Maldonado M. 2006. The ecology of the sponge larva. Canadian Journal of Zoology, 84(2):175–194, https://doi.org/10.1139/z05-177.

Maldonado, M., Aguilar R., Bannister R. J., Bell D., Geijzendorffer I. R. 2016. Sponge grounds as key marine habitats: a synthetic review of types, structure, functional roles, and conservation concerns. Sergio Rossi; Lorenzo Bramanti; Andrea Gori; Covadonga Orejas Marine Animal Forests: The Ecology of Benthic Biodiversity Hotspots, ISBN: 978-3-319-17001-5 (Print) 978-3-319-17001-5.

Maldonado M., Ribes M., van Duyl F. C. 2012. Nutrient fluxes through sponges: biology, budgets, and ecological implications. Advances in Marine Biology, Volume 62, pp 113–182. Elsevier Ltd ISSN 0065-2881, DOI: 10.1016/B978-0-12-394283-8.00003-5.

Manconi M. and Pronzato R. 2002. Suborder Spongillina subord. nov.: Freshwater sponges in systema porifera: A guide to the classification of sponges. Edited by John N.A. Hooper and Rob W.M. Van Soest. pp. 921–1021. Kluwer Academic/Plenum Publishers, New York.

Mohri K., Nakatsukasa M., Masuda Y., Agata K., Funayama N. 2008Toward understanding the morphogenesis of siliceous spicules in freshwater sponge: differential mRNA expression of spicule-type-specific silicatein genes in Ephydatia fluviatilis. Dev. Dyn.; 237: 3024–3039.

Nakanishi N., Sogabe S., Degnan B. M. 2014 Evolutionary origin of gastrulation: insights from sponge development. BMC Biol, 12: 26.

Nakayama S., Arima K., Kawai K., Mohri K., Inui C., Sugano W., Koba H., Tamada K., Nakata Y. J., Kishimoto K., Arai-Shindo M., Kojima C., Matsumoto T., Fujimori T., Agata K., Funayama N.. 2015. Dynamic transport and cementation of skeletal elements build up the pole-and-beam structured skeleton of sponges. Curr Biol. 25(19): 2549–2554. doi: 10.1016/j.cub.2015.08.023. Epub 2015 Sep 17.

Nickel M. 2004. Kinetics and rhythm of body contractions in the sponge Tethya wilhelma (Porifera: Demospongiae). J Exp Biol. 207(Pt 26): 4515–4524.

Nickel M. 2006. Like a 'rolling stone': quantitative analysis of the body movement and skeletal dynamics of the sponge Tethya wilhelma. J Exp Biol. 209(Pt 15): 2839–2846.

Philippe H., Derelle R., Lopez P., Pick K., Borchiellini C., Boury-Esnault N.Vacelet J., Renard E., Houliston E., Quéinnec E., Da Silva C., Wincker P., Le Guyader H., Leys S., Jackson D. J., Schreiber F., Erpenbeck D., Morgenstern B., Wörheide G., Manuel M. 2009 Apr 28. Phylogenomics revives traditional views on deep animal relationships. Curr Biol. 19(8): 706–12. doi: 10.1016/j.cub.2009.02.052. Epub 2009 Apr 2.

Pomponi S. A. 1980. Cytological mechanisms of calcium carbonate excavation by boring sponges. International Review of Cytology, 65: 301–319.

Pronzato R., Manconi R. 2008. Mediterranean commercial sponges: over 5000 years of natural history and cultural heritage. Marine Ecology 29: 146–166

Reiswig, H. M. 1981. Partial carbon and energy budgets of the bacteriosponge Verongia fistularis (Porifera: Demospongiae) in Barbados West-Indies. Mar. Biol. 2: 273–294.

Richards, G.S., Simionato, E., Perron, M., Adamska, M., Vervoort, M., Degnan, B. M. 2008. Sponge genes provide new insight into the evolutionary origin of the neurogenic circuit. Curr. Biol. 18: 1156–1161.

Richter C., Wunsch M., Rasheed M., Kötter I., Badran M. I. 2001. Endoscopic exploration of Red Sea coral reefs reveals dense populations of cavity-dwelling sponges. Nature 413(6857): 726–730.

Riesgo A., Taylor C., Leys S. P. 2007. Reproduction in a carnivorous sponge: the significance of the absence of an aquiferous system to the sponge body plan. Evol Dev. 9(6): 618–631.

Rossi G, Montori S, Cerrano C, Calcinai B. 2015. The coral killing sponge Chalinula nematifera (Porifera: Haplosclerida) along the eastern coast of Sulawesi Island (Indonesia), Italian Journal of Zoology, 82:1: 143–148. doi: 10.1080/11250003.2014.994046.

Saito T and Konishi K. 1999. Direct development in the sponge-associated deep-sea shrimp *Spongicola japonica* (Decapoda: Spongicolidae). Journal of Crustacean Biology 19: 46–52

Sara M and Liaci L. 1964. Symbiotic association between Zooxanthellae and two marine sponges of the genus cliona. Nature, Lond., 203: 32.

Schmitt S., Angermeier H., Schiller R., Lindquist N., Hentschel U. 2008. Molecular microbial diversity survey of sponge reproductive stages and mechanistic insights into vertical transmission of microbial symbionts. Appl Environ Microbiol. 74(24): 7694–708. doi: 10.1128/AEM.00878-08.

Schoenberg C. H., Fang J. K., Carreiro-Silva M., Tribollet A., Wisshak M. 2017 Bioerosion: the other ocean acidification problem. ICES Journal of Marine Science. 74(4): 895–925.

Simion P., Philippe H., Baurain D., Jager M., Richter D. J., Di Franco A., Roure B., Satoh N., Quéinnec É., Ereskovsky A., Lapébie P., Corre E., Delsuc F., King N., Wörheide G., Manuel M. 2017 Apr 3. A large and consistent phylogenomic dataset supports sponges as the sister group to all other animals. Curr Biol. 27(7): 958–967. doi: 10.1016/j.cub.2017.02.031. Epub 2017 Mar 16. PubMed PMID: 28318975.

Simpson T. L, Fell P. E. 1974. Dormancy among the porifera: gemmule formation and germination in fresh-water and marine sponges. Transactions of the American Microscopical Society, 93(4): 544–577.

Simpson T. L. 1984. The Cell Biology of Sponges. Springer Verlag.

Sollas, W. J. 1884. On the development of Halisarca lobularis (O. Schmidt). Quart. J. Microsc. Sci. 24: 603–621.

Srivastava, M., Simakov, O., Chapman, J., Fahey, B., Gauthier, M. E. A., Mitros, T., Richards, G.S., Conaco, C., Dacre, M., Hellsten, U., Larroux, C., Putnam, N. H., Stanke, M., Adamska, M., Darling, A., Degnan, S. M., Oakley, T. H., Plachetzki, D. C., Zhai, Y., Adamski, M., Calcino, A., Cummins, S. F., Goodstein, D. M., Harris, C., Jackson, D. J., Leys, S. P., Shu, S., Woodcroft, B. J., Vervoort, M., Kosik, K. S., Manning, G., Degnan, B. M., and Rokhsar, D. S. 2010. The *Amphimedon queenslandica* genome and the evolution of animal complexity. Nature 466(7307): 720–726.

Tompkins-MacDonald G. J., Leys S. P. 2008. Glass sponges arrest pumping in response to sediment: implications for the physiology of the hexactinellid conduction system. Marine Biology 154: 973–984

Van Trigt, H. 1919. A Contribution to the Physiology of the Freshwater Sponges (Spongillidae). (E. Brill, Leiden).

Ueda, N., Richards, G., Degnan, B. Kranz A, Adamska M, Croll R. P., and Degnan SM. 2016. An ancient role for nitric oxide in regulating the animal pelagobenthic life cycle: evidence from a marine sponge. Sci Rep 6: 37546.

Uriz M. J. 2006. Mineral skeletogenesis in sponges. Can. J. Zool. 84: 322–356.

Vacelet J., Duport E. 2004. Prey capture and digestion in the carnivorous sponge Asbestopluma hypogea (Porifera: Demospongiae). Zoomorphology 123: 179–190 doi: 10.1007/s00435-004-0100-0.

Vacelet, J., Boury-Esnault, N. 1995. Carnivorous sponges. Nature 373: 333–335.

Van Soest R. W. M., Boury-Esnault N., Vacelet J, Dohrmann M., Erpenbeck D., et al. 2012 Global diversity of Sponges (Porifera). PLoS One 7(4): e35105.doi:10.1371/journal.pone.0035105.

Voigt O., Adamska M., Adamski, Kittelmann A., Wencker L., Wörheide G. 2017. Spicule formation in calcareous sponges: Coordinated expression of biomineralization genes and spicule-type specific genes. Sci Rep, 7: 45658.

Voigt O., Adamski M., Sluzek K., Adamska M. 2014. Calcareous sponge genomes reveal complex evolution of alpha-carbonic anhydrases and two key biomineralization enzymes. BMC Evol Biol 14: 230.

Wang X., Wiens M., Schröder H. C., Jochum K. P., Schlossmacher U., Götz H., Duschner H., Müller W. E. 2011. Circumferential spicule growth by pericellular silica deposition in the hexactinellid sponge Monorhaphis chuni. J Exp Biol. 214(Pt 12): 2047–2056. doi: 10.1242/jeb.056275.

Webster N. S., Taylor M. W., Behnam F., Lücker S., Rattei T., Whalan S., Horn M., Wagner M. 2010. Deep sequencing reveals exceptional diversity and modes of transmission for bacterial sponge symbionts. Environ Microbiol. 12(8): 2070–2082. doi: 10.1111/j.1462-2920.2009.02065.x.

Whelan N. V., Kocot K. M., Moroz T. P., Mukherjee K., Williams P., Paulay G., Moroz L. L., Halanych K. M. Ctenophore relationships and their placement as the sister group to all other animals. Nat Ecol Evol. 2017 v;1(11): 1737–1746. doi: 10.1038/s41559-017-0331-3.

PHYLUM CTENOPHORA

Otto M. P. Oliveira and Sanna Majaneva

CHAPTER

9

Infrakingdom	Bilateria
1st Division	Diploblastic Animals
Phylum	Ctenophora

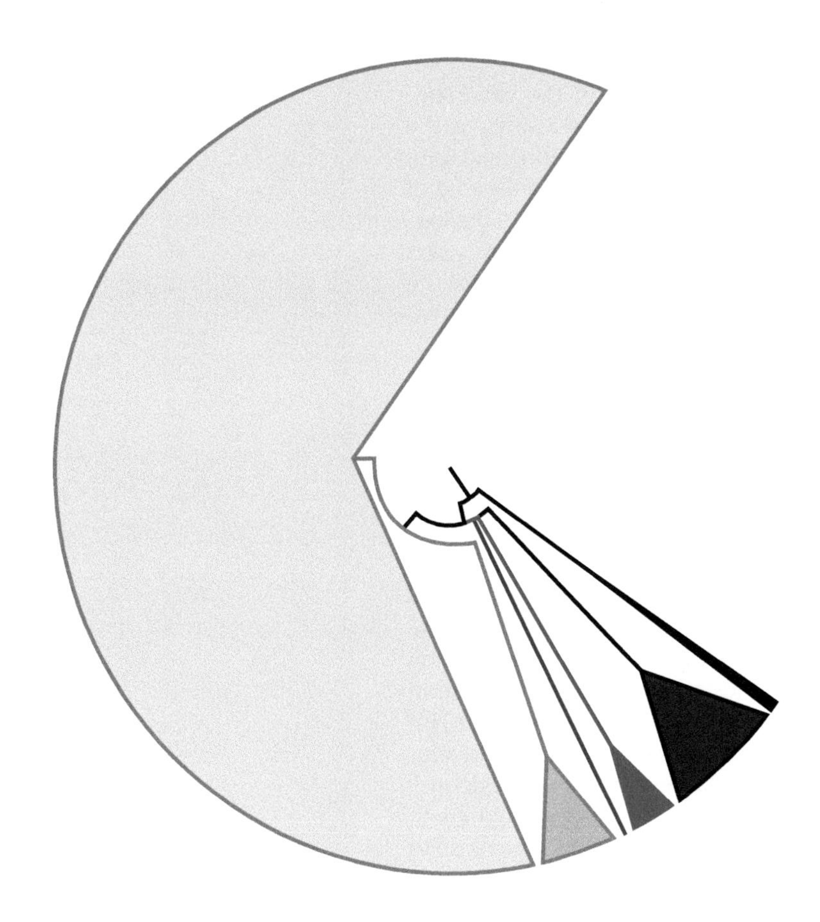

CONTENTS

General Characteristics

About 150–200 valid species

Number of "cryptic species" estimated up to 50

Animals with rotational symmetry, exclusively marine, abundant in all oceans

Animals with unique embryological cleavage pattern

Indirect commercial importance

Great importance in the pelagic trophic nets

Autapomorphies

- eight ctene rows distributed longitudinally along the body
- cydippid larva with long tentacles
- coloblasts

HISTORY OF TAXONOMY AND CLASSIFICATION

The Phylum Ctenophora was originally described in 1829 by the German zoologist J. F. Eschscholtz as a sibling group to cnidarians. Most of the valid species known today were described in the 19th century, in a period known as the "Golden Age of Gelata" (Haddock, 2004). The taxonomic classification used until now was created in that period. However, it has been known, since the end of the 20th century, that the classification of the group does not reflect its phylogeny (Harbison, 1985). Many molecular studies published in the last 20 years corroborated the polyphyletism of the Tentaculata, one of the two classes of the Phylum (Podar et al., 2001; Simion et al., 2015; Whelan et al., 2017). At least two orders, Cyddipida and Lobata, are also indicated by morphological and molecular studies to be non-monophyletical lineages. So, the classification of the group needs to be reformulated, with new taxonomic groups officially described to attend the phylogenetic findings.

ANATOMY, HISTOLOGY AND MORPHOLOGY

Ctenophores also known as comb jellies, have translucent body, with gelatinous consistence. Externally, the main body axis goes from the mouth opening to the apical organ (**Figure 9.1A**). Around this axis, the rotational symmetry, dividing body in two equal (not mirrored) portions, becomes evident when the position of the anal pores is detected adradially, nearby the apical organ (Figure 9.1B). The body can be cut in two main planes: the tentacular (cutting both tentacular sheaths), and the stomodeal (following the natural flatness of the stomodeum). Each quadrant between tentacular and stomodeal planes presents two ctene (comb plate) rows, named substomodeal and subtentacular, when more close to the stomodeal or tentacular plane, respectively. Internally, there are meridional canals above each ctene row, eight at all, connected aborally with the base of the stomodeum and the canals that end in the anal pores. Some species also present a radial canal, connecting the oral end of the meridional canals near the

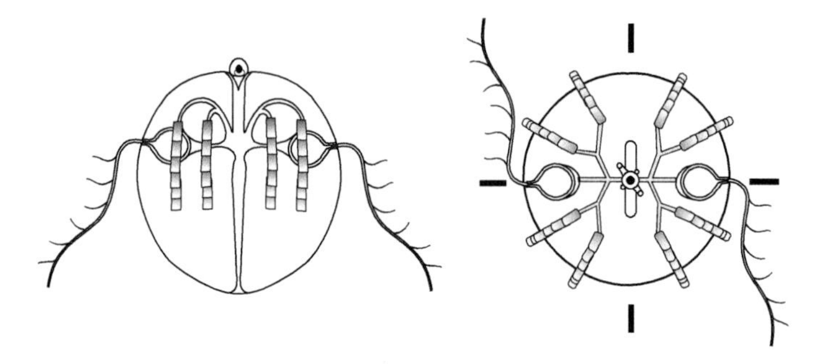

Figure 9.1 Anatomic scheme of a ctenophore, in (A) lateral view, showing the oral–aboral axis and in (B) aboral view, showing the stomodeal and tentacular planes. (Adapted from Martindale and Henry, 1999.)

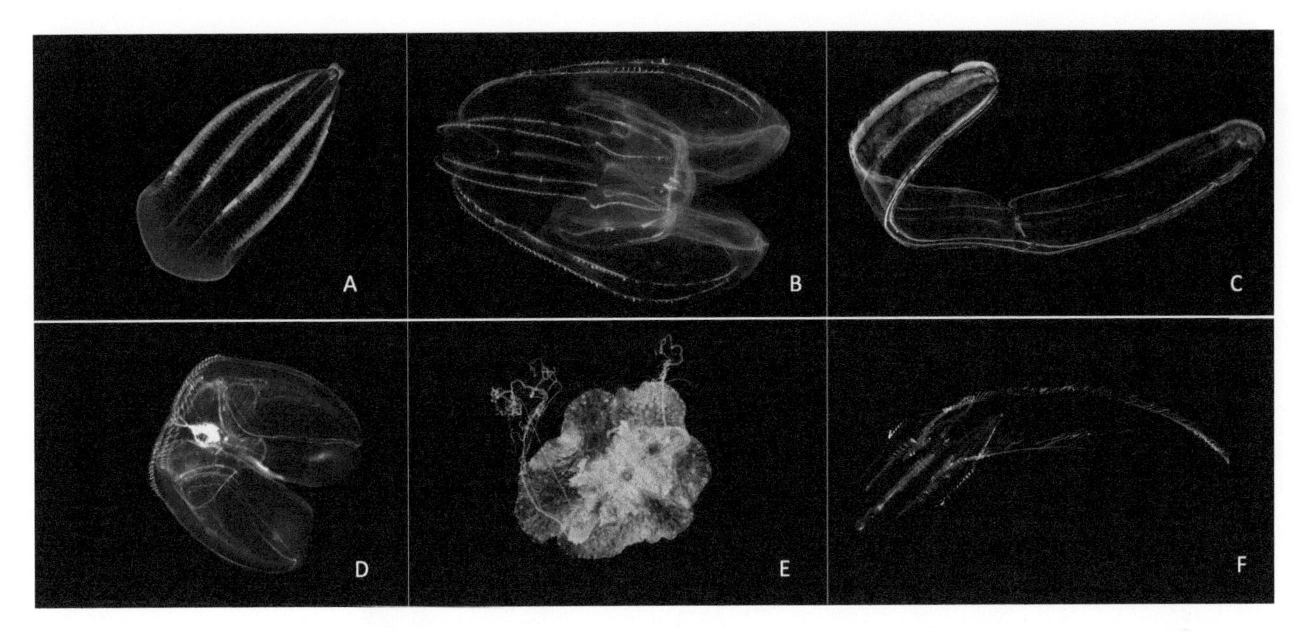

Figure 9.2 Plate showing ctenophore diversity. (A) Beroe forskalli; (B) *Bolinopsis vitrea*; (C) *Cestum* sp.; (D) *Ocyropsis crystallina*; (E) *Vallicula multiformis*; (F) *Callianira bialata*.

mouth. From each tentacular sheath a long, sometimes ramified, axial tentacle goes out. However, some species lack axial tentacles when adult. Beroids totally lack tentacular sheaths and tentacles. The appearance of comb jellies' bodies can be very distinct between the groups (**Figure 9.2**), changing from spherical to flattened in the oral–aboral axis (e.g. platyctenids, thalassocalycids) or very flattened in the tentacular plane (e.g. cestids). Others present oral projections in the stomodeal plane (e.g. lobated comb jellies) or have wide mouths (e.g. beroids, ganeshids).

Special structures present in the ctenophores are the colloblasts (**Figure 9.3**), adhesive cells present in tentacles, used to capture prey. Other structures only

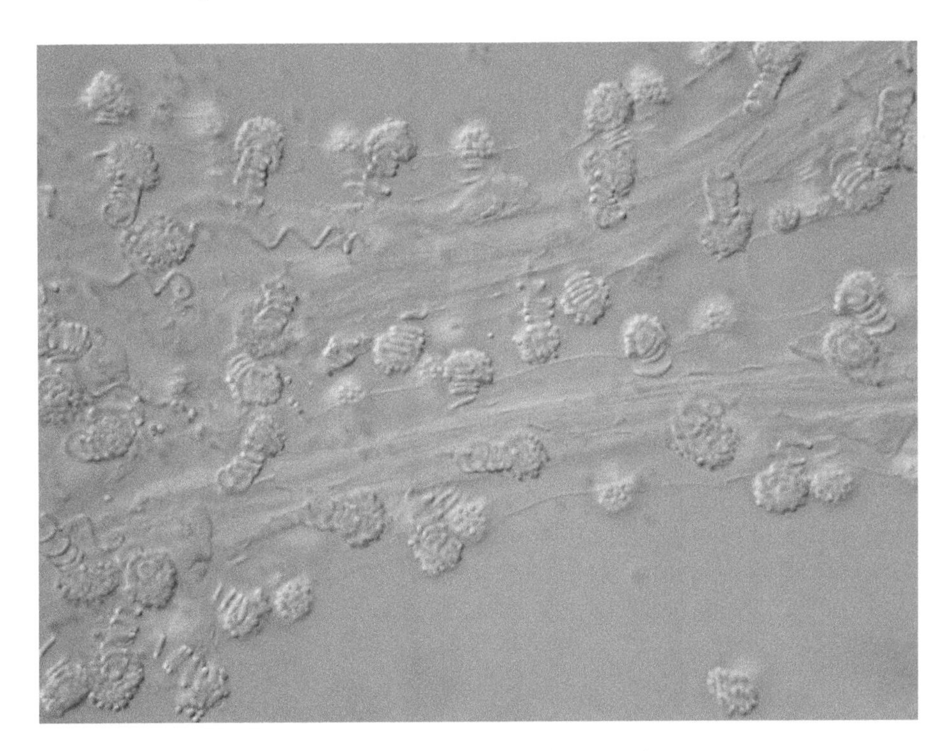

Figure 9.3 Colloblasts dispersed along the tentacles of a ctenophore.

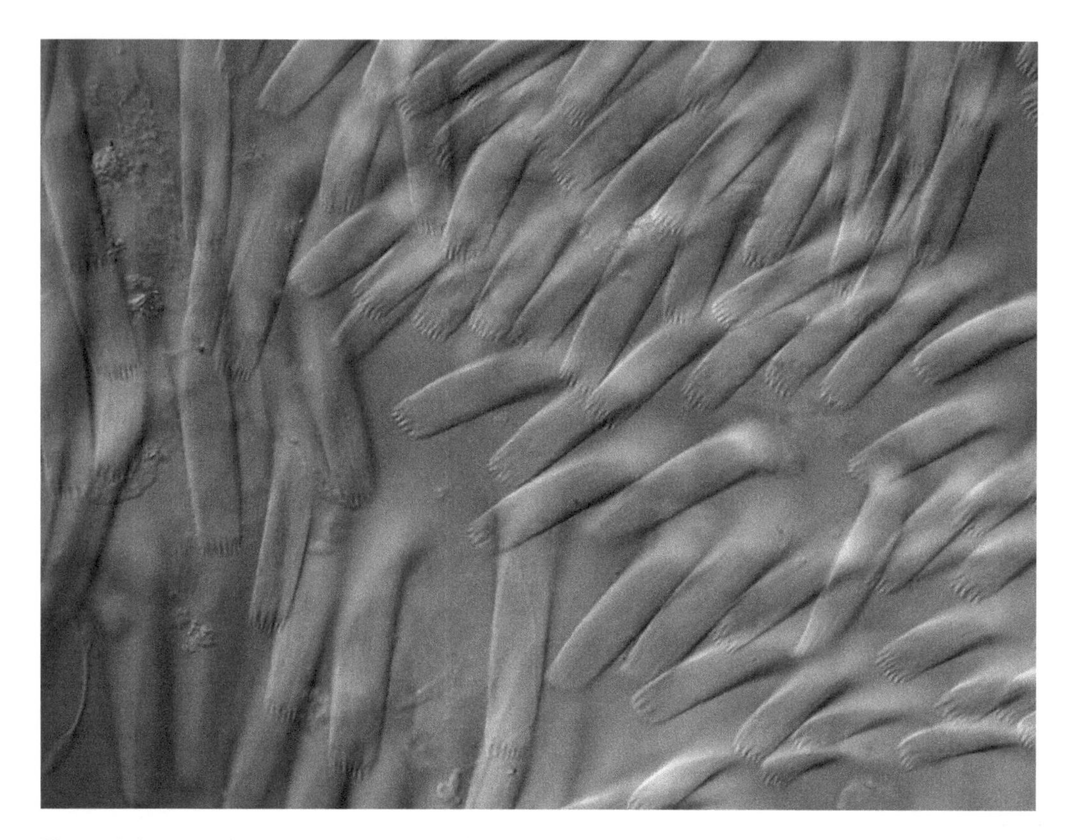

Figure 9.4 Macrocilia present in the mouth of a beroid.

known for ctenophores are the macrocilia (**Figure 9.4**) located inside beroids' mouths. Macrocilia are believed to help disrupt gelatinous prey tissues, working like small teeth (Tamm and Tamm, 1993).

REPRODUCTION AND DEVELOPMENT

Most of the pelagic ctenophores are hermaphrodites, producing both sex gametes simultaneously. The genus *Ocyropsis* is the only known exception between the pelagic species (Harbison and Miller, 1986). The gonads are arranged along the meridional canals. Gametes are liberated through small temporary pores that open between the ctenes. Spermatozoids are liberated firstly, followed by the ovules. Although fertilization is generally external, cases of internal self-fecundation inside the meridional canals have been reported for *Beroe ovata* (Oliveira and Migotto, 2006). The embryo develops from holoblastic and unequal cleavages to a cyddipid larvae. Larva are spherical, already presenting the eight ctene rows, an apical organ and mouth (**Figure 9.5**). Two long tentacles grow from the tentacular sheaths, determining the tentacular plane of symmetry. Exceptions are the beroid larvae, which lack the tentacles and even before hatching from the gelatinous egg already resemble a miniature adult. For *Mnemiopsis leidyi* the pedogenesis has been described in which an initial reproductive period occurs along the larval stage (Martindale, 1987). In this group, the reproductive stage is paused during the lobes grown and the individual's migration from estuarine to coastal waters.

In benthic ctenophores (platyctenids), the protandric hermaphroditism is common. Embryos are incubated externally, in egg masses attached to the parental body, or internally inside reproductive chambers. Most of the Platyctenida are also known to be capable of asexual reproduction through body fission and the regeneration of lost parts.

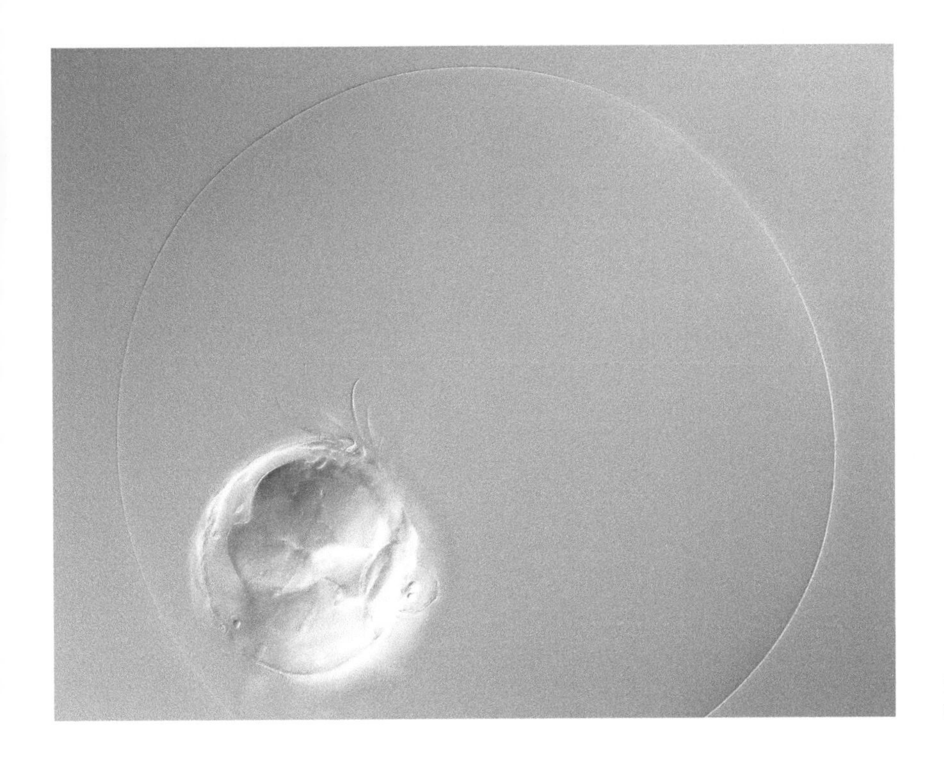

Figure 9.5 Cyddipid larva inside a gelatinous egg.

DISTRIBUTION AND ECOLOGY

Comb jellies occur in all marine environments, from coastal to deep-sea waters (**Figure 9.7**). In general, the larval stage lives among plankton. Most of the species are also pelagic along adult life, with large distributional ranges. The benthic species (platyctenids) present more restricted distribution patterns, most of them occurring in limited areas of the Indian and Western Pacific oceans.

Figure 9.6 Aboral portion of a beroid, showing the formation of temporary anal pore releasing digestion waste.

These animals are secondary to tertiary consumers. Benthic species feed mostly on epibenthic crustaceans. Pelagic species feed mostly on copepods, representing an important trophic chain between the microzooplankton and the nekton. During blooms and aggregative events, they can harvest prey, potentially cleaning the area from planktonic crustaceans. However, beroids and their relatives feed on other gelatinous organisms, including other comb jellies. To illustrate the importance of comb jellies in pelagic food webs, the invasion of *Mnemiopsis leidyi* in the Black Sea, in the mid- 1980s, associated with previous events of overfishing and eutrophication, resulted in the crash of the anchovy fisheries in this area (Kideys, 1994). *Mnemiopsis leidyi* is an efficient harvesting competitor with that fish, and also feeds on the anchovies' larvae. The blooms of *M. leidyi* in the Black Sea were eventually controlled after another comb jelly, *Beroe ovata*, invaded the area at the end of the 1990s. *Beroe ovata* is a natural predator of *M. leidyi* in its native area of occurrence.

PHYSIOLOGY AND BEHAVIOR

The main sensorial structure present in the comb jellies is the apical organ. It consists in four long tufts of cilia (balancers), arranged in a dome, supporting a statolith. This structure functions in the spatial orientation and balance of the animal with each balancer controlling the movement of a pair of ctene rows (Tamm, 2014). Furthermore, the polar field (the region around the apical organ) can also present photosensitive structures and, at least in the beroid, papillae with chemical sensors.

Most of the pelagic comb jellies are able to produce bioluminescence. The light-producing cells are distributed along the meridional canals and are connected by gap junctions, streaming the luminescent excitation (Anctil, 1985). Some species are able to release luminescent particles in escape behavior.

Ctenophores have no independent organs for circulation, gas exchange, excretion or osmoregulation, so that all these functions are made by direct diffusion to the water medium. Digestion occurs inside the stomodeum and digested particles are distributed to the body through the meridional canals. Although bigger indigestible parts of the prey can be discarded by the mouth, the anal pores are active in discarding small indigestible particles (Tamm, 2019) (**Figure 9.6**).

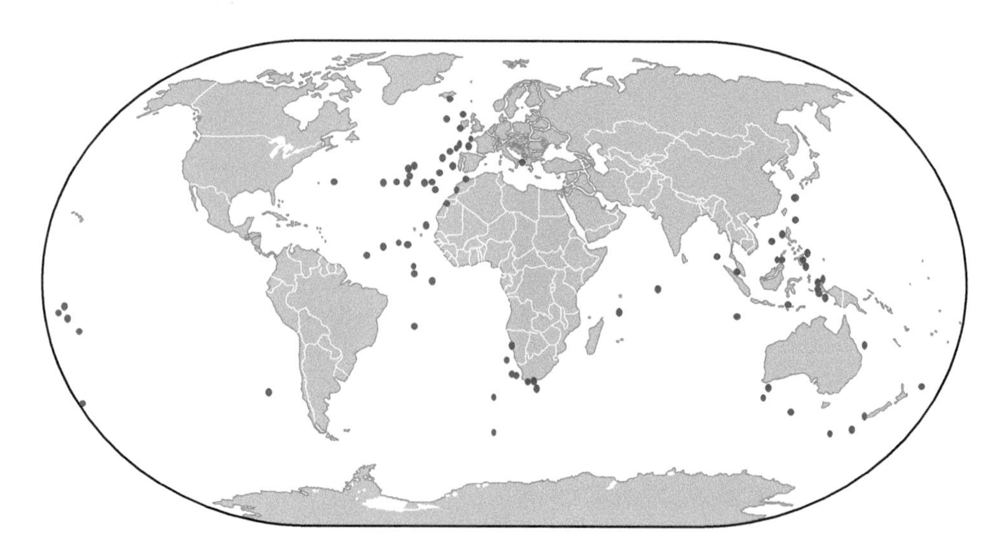

Figure 9.7 Distribution map of ctenophore. The phylum has a worldwide marine distribution.

GENETICS

Ctenophores have one of the smallest genomes within the Metazoa (Ryan et al., 2013; Gregory, 2019). For example, *Mnemiopsis leidyi* has 150 Mb genome tightly packed with gene sequences (Ryan et al., 2013) and the estimated genome of *Pleurobrachia bachei* is ~160 Mb (Moroz et al., 2014). However, when screening genome size of various ctenophore species using flow cytometry and densitometry, wider ranges have been detected with *Haeckelia rubra* having 3.1 Gb genome (Gregory et al., 2007). A highly reduced complement of pan-animal genes seems to be a feature shared in the Ctenophora phylum (e.g. apparent absence of canonical microRNA machinery and HOX genes; Moroz et al., 2014). On the other hand, collagens, RNA editing enzymes and RNA-binding proteins are examples of known gene families that are quite extensive among ctenophores (Moroz et al., 2014).

Unusually compact mitochondrial genomes (around 10 kb) in ctenophores represent one of the smallest animal mtDNA (Pett et al., 2011; Kohn et al., 2012; Arafat et al., 2018). Ctenophore mitochondrial genomes are also among the most derived suggesting that ctenophore mt genomes are strongly divergent from the typical animal mt genome (Arafar et al., 2018). The mt genes are characterized by high rates of sequence evolution, which is demonstrated in comparisons that yield low similarity between homologous coding sequences in ctenophores and in other animals, as well as unusual rRNA structures (Pett et al., 2011).

POSITION IN THE ToL

The position of Ctenophora in the metazoan phylogeny is still doubtful. Morphology-based phylogenies placed the ctenophores in several different positions within non-Bilateria animals. Recent phylogenomic studies diverge on the position of ctenophores as a sister group of all other animals (Dunn et al., 2008; Whelan et al., 2015) or a sister group of all metazoan groups, except the Porifera (Simion et al., 2017). The main discussion is the suggestion that the first hypothesis just happened due to methodological failures in the computational analysis (caused by, for instance, Long-Branching Attraction). Until there is a consensus, the most conservative position for the ctenophores is to consider the base of the metazoan phylogeny as a polytomy, with the three lineages: Ctenophora, Porifera and all other metazoans, rising at the same time (**Figure 9.8**).

DATABASES AND COLLECTIONS

Due to specimen fragility, there are few scientific collections with samples of ctenophores. A search for the phylum samples in the Smithsonian Museum of Natural History, for example, returns only 28 records, including DNA and image samples. A search in the Harvard Museum of Comparative Zoology returns 112 samples, but most of them contain secular specimens, impossible to identify correctly due to negative effects of the preservative medium in the specimens' tissues. GBIF records are scarce and need specific identification confirmation.

CONCLUSION

Ctenophora is a polemical group in Metazoan Tree of Life. The lineage position within the basal phyla is still to be confirmed with further studies. On the other hand, internal relationships are far to be correctly expressed in present taxonomical arrange. The consensus tree (Figure 9.8) suggests that many new taxa names need to be created and all taxonomical arrangement needs to be changed in order to better express the phylum internal phylogeny.

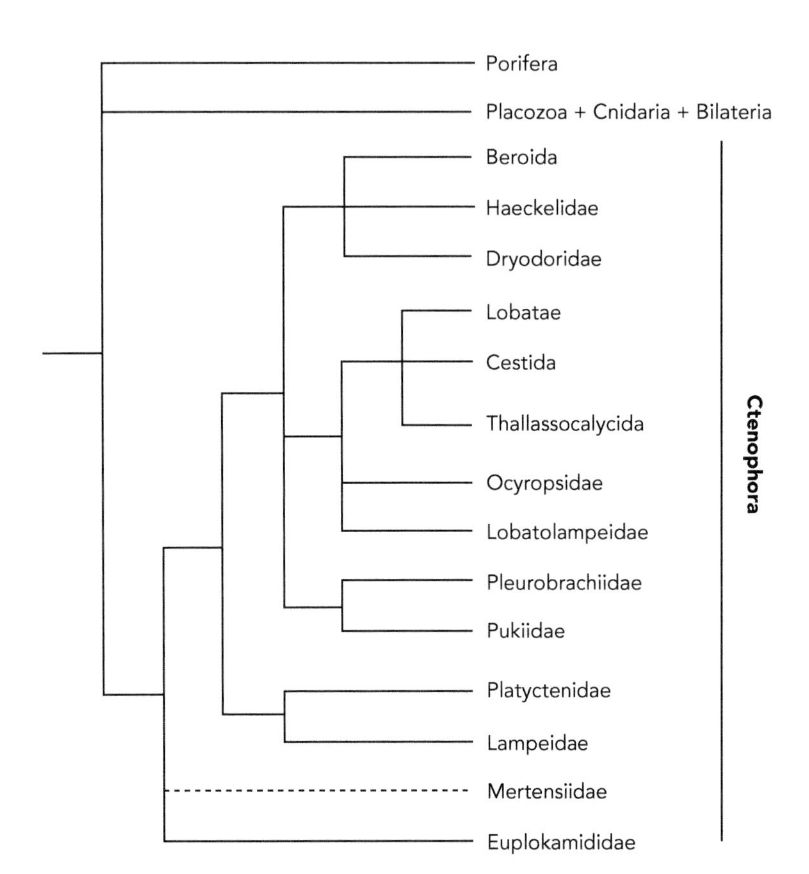

Figure 9.8 Most-conservative tree, showing the position of Ctenophora in metazoan phylogeny and a consensus of internal relationships. The position of Mertensiidae is shown in dotted line because there is no previous consensus on its position.

REFERENCES

Anctil M., 1985. Ultrastructure of the luminescent system of the ctenophore *Mnemiopsis leidyi*. Cell and Tissue Research 242, 333–340.

Arafat H., Alamaru A., Gissi C., Huchon D., 2018. Extensive mitochondrial gene rearrangements in Ctenophora: insights from benthic Platyctenida. BMC Evolutionary Biology 18, 65.

Dunn C. W., Hejnol A., Matus D. Q., Pang K, Browne W. E., Smith S. A., Seaver E., Rouse G. W., Obst M, Edgecombe G. D., Sørensen M. V., Haddock S. H. D., Schmidt-Rhaesa A., Okusu A., Kristensen R. M., Wheeler W. C., Martindale M. Q., Giribet G., 2008. Broad phylogenomic sampling improves resolution of the animal tree of life. Nature 452(7188), 745–749.

Eschscholtz J. F., 1829. System der Acalephen. F. Dümmler, Berlin, 190 pp.

Gregory T.R., 2019. Animal Genome Size Database. http://www.genomesize.com

Gregory, T. R., Nicol, J. A., Tamm H., Kullman B., Kullman K., Leitch I. J., Murray B. G., Kapraun D. F., Greilhuber J., Bennett M. D., 2007. Eukaryotic genome size databases. Nucleic Acids Research. 35, D332–D338.

Haddock S. H. D., 2004. A golden age of gelata: past and future research on planktonic ctenophores and cnidarians. Hydrobiologia 530(1–3), 549–556.

Harbison G. R., 1985. On the classification and evolution of the Ctenophora. In: Morris S. C., George J. D., Gibson R., Platt H. M. (Eds.), The origin and relationships of lower invertebrates. The Systematic Association, 28, Claredon Press, Oxford, pp. 78–100.

Kideys A. E., 1994. Recent dramatic changes in the Black Sea ecosystem: The reason for the sharp decline in Turkish anchovy fisheries. Journal of Marine Systems 5(2), 171–181.

Kohn A. B., Citarella M. R., Kocot K. M., Bobkova Y. V., Halanych K. M., Moroz L. L., 2012. Rapid evolution of the compact and unusual mitochondrial genome in the ctenophore, *Pleurobrachia bachei*. Molecular Phylogenetics and Evolution 63, 203–207.

Martindale M. Q., 1987. Larval reproduction in the ctenophore *Mnemiopsis mccradyi* (order Lobata). Marine Biology 94, 409–414.

Martindale M. Q., Henry J. Q., 1999. Intracellular fate mapping in a basal metazoan, the ctenophore *Mnemiopsis leidyi*, reveals the origins of mesoderm and the existence of indeterminate cell lineages. Developmental Biology 214(2), 243–257.

Moroz L. L., Kocot K. M., Citarella M. R., Dosung S., Norekian T. P., et al., 2014. The ctenophore genome and the evolutionary origins of neural systems. Nature 510(7503), 109–114.

Oliveira O. M. P., Migotto A. E., 2006. Pelagic ctenophores from the São Sebastião Channel, southeastern Brazil. Zootaxa 1183, 1–26.

Pett W., Ryan J. F., Pang K., Mullikin J. C., Martindale M. Q., et al., 2011. Extreme mitochondrial evolution in the ctenophore *Mnemiopsis leidyi*: Insight from mtDNA and the nuclear genome. Mitochondrial DNA 22(4), 130–142.

Podar M., Haddock S. H. D., Sogin M. L., Harbison G. R., 2001. A molecular phylogenetic framework for the phylum Ctenophora using 18S rRNA genes. Molecular Phylogenetics and Evolution 21(2), 218–230.

Ryan J. F., Pang K., Schnitzler C. E., Nguyen A.-D., Moreland R. T., et al., 2013. The genome of the ctenophore *Mnemiopsis leidyi* and its implications for cell type evolution. Science 342, 1242592.

Simion P., Bekkouche N., Jager M., Quéinnec E., Manuel M., 2015. Exploring the potential of small RNA subunit and ITS sequences for resolving phylogenetic relationships within the phylum Ctenophora. Zoology 118(2), 102–114.

Simion P., Philippe H., Baurain D., Jager M., Richter D. J., Di Franco A., Roure B., Satoh N., Queinnec E., Ereskovsky A., Lapebie P., Corre E., Delsuc F., King N., Wörheide G., Manuel M., 2017. A large and consistent phylogenomic dataset supports sponges as the sister group to all other animals. Current Biology 27(7): 958–967.

Tamm S. L., 2014. Cilia and the life of ctenophores. Invertebrate Biology 133(1), 1–46.

Tamm S. L., 2019. Defecation by the ctenophore *Mnemiopsis leidyi* occurs with an ultradian rhythm through a single transient anal pore. Invertebrate Biology 138(1), 3–16.

Tamm S. L. and Tamm S., 1993. Diversity of macrociliary size, tooth patterns, and distribution in *Beroë* (Ctenophora). Zoomorphology 113, 79–89.

Whelan N. V., Kocot K. M., Moroz, L. L., Halanych K. M., 2015. Error, signal, and the placement of Ctenophora sister to all other animals. PNAS 112(18), 5773–5778.

Whelan N. V., Kocot K. M., Moroz T. P., Mukherjee K., Williams P., et al., 2017. Ctenophore relationships and their placement as the sister group to all other animals. Nature Ecology & Evolution 1, 1737–1746.

CNIDARIA

HIGHER TAXON

2

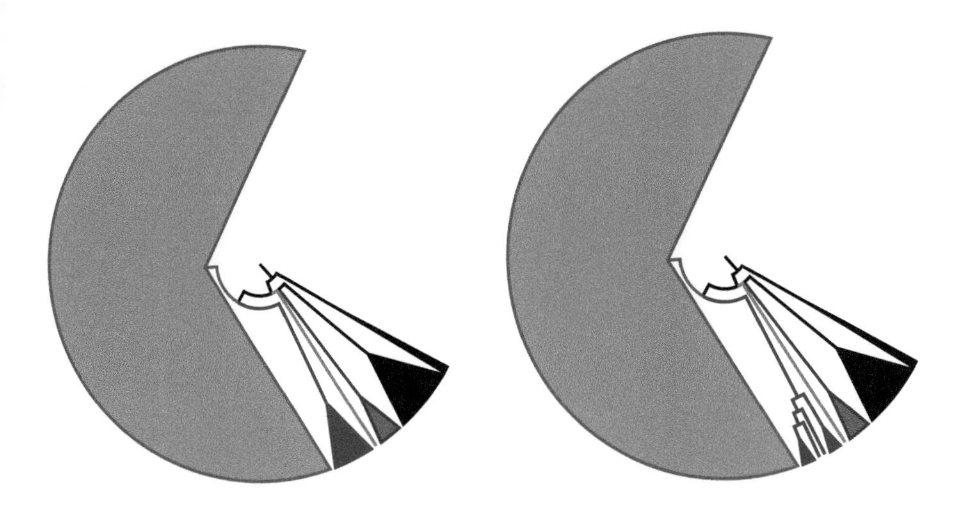

Phylum Cnidaria in the animal tree of life (left); the phylum opened showing the four major taxa we discuss here – Anthozoa, Scyphozoa, Hydrozoa and Myxozoa.

The phylum has over 11,000 species of aquatic animals (both freshwater and marine environments), but they are mostly marine. The group is defined by the presence of cnidocytes. These are specialized cells that are used by the cnidarian for prey capture. The cnidocytes are found in the mouth region where the radially symmetric cnidarian has tentacles bearing the cnidocytes. In addition, they have a distinctive body design which consists of an inert gelatinous substance that makes up the mesoglea. The mesoglea is sandwiched between very thin layers of epithelial cells. There are two basic forms that the typical cnidarian takes – a swimming medusa and a sessile polyp. Both the medusa and the polyp have a single orifice and a single body cavity where digestion and respiration occur. Their nervous system is interesting because it is decentralized and netlike (neural net). Specific details on the body plans of cnidarians can be found in the several of the following chapters.

The phylum is separated into five major classes – Anthozoa (sea anemones, corals, sea pens), Scyphozoa (jellyfish), Cubozoa (box jellies), Hydrozoa (hydroids) and Staurozoa. There are a couple of highly derived and reduced groups of Cnidarians embedded in this system such as Myxozoa and Polypodiozoa. Their phylogenetic relationships are indicated in the following figure. We treat three of the major classes of Cnidaria in this text – Scyphozoa, Hydrozoa and Anthozoa – and one of the minor clades of Cnidaria called Myxozoa.

Distinguishing Features

- More complex than sponges and less complex than bilaterians,
- Cells bound by inter-cell connections and carpet-like basement membranes
- Muscles
- Nervous systems
- Sensory organs
- Cnidocytes

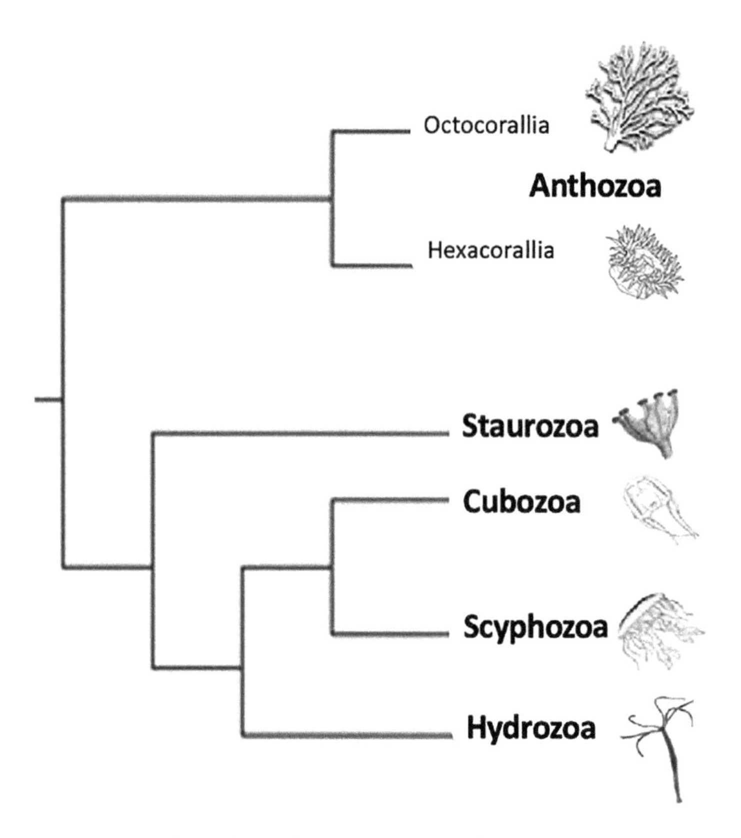

Phylogenetic relationships of the major classes of Cnidaria. The two subclasses of Anthozoa are also shown. (Drawn by R. DeSalle.)

- Only one opening in their body for ingestion and excretion
- Two main layers of cells that sandwich a middle layer of jelly-like material, which is called the mesoglea

CHAPTER

10

PHYLUM CNIDARIA: CLASSES SCYPHOZOA, CUBOZOA, AND STAUROZOA

Massimo Avian and Andreja Ramšak

Infrakingdom	Bilateria
1st Division	Diploblastic Animals
Phylum	Cnidaria
Subphylum	Medusozoa
Class	Scyphozoa, Cubozoa and Staurozoa

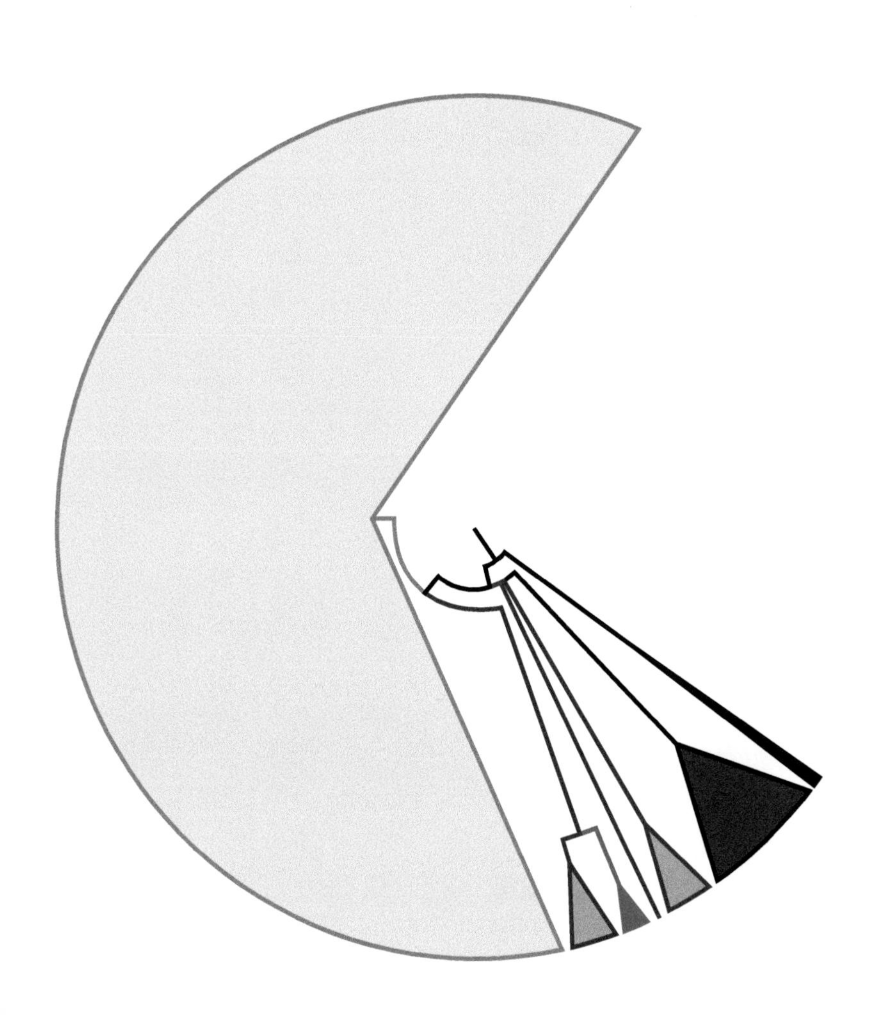

CONTENTS

CLASS SCYPHOZOA

General Characteristics

About 230 species described

Estimated number of species still to be described or cryptic forms: unknown

Marine, macroscopic jellyfish commonly (scyphomedusae or true jellyfish) ranging in size from less than 1 cm to about 2 m (Martellos et al., 2016; Jarms and Morandini, 2019). Found throughout the world's oceans, and at all latitudes, from the surface to great depths

Some scyphozoans (order Rhizostomeae) are fished commercially for food. An international trade based on jellyfish fishing and export is currently mostly limited to Southeast Asia, notably China

Their ecological importance is surely to be considered more than a trophic dead end (around 124 fish species and 34 other species prey on jellyfish including scyphozoans, see Pauly et al., 2009), even if their contribution to marine trophic chains is certainly still underestimated and needs to be further investigated

Synapomorphies

- umbrella saucer-shaped, hemispherical, flat-dish or high conical
- umbrella margins cleft into marginal lappets
- at the edge of the umbrella are located the marginal sense organs called rhopalia that contain statocysts, sensory niches and sometimes ocelli
- subumbrellar surface without velum (Hydrozoa) or velarium (Cubozoa)
- the mouth is surrounded by four very developed extensions of the manubrium, called oral arms
- the polyp possesses a peculiar budding mode that gives rise to the medusoid generation, called strobilation. The polyp is defined scyphistoma
- as a consequence, one or more free-swimming larvae are released, defined as ephyrae. Each of these will grow into a medusa

HISTORY OF TAXONOMY AND CLASSIFICATION

This class, established for a long time (1883), normally included several orders, such as Cubomedusae and Stauromedusae (Mayer, 1910; Kramp, 1961). However, this class has recently been the subject of several revisions, which led to the separation of the two taxa above, both elevated to the rank of class. Even at lower hierarchical level, the revisions are quite frequent; Bayha and Dawson (2010) introduced a new family, Drymonematidae, with one genus, *Drymonema* on the basis of genetic markers 18S, 28S, ITS1, COI and 16S, plus 55 morphological features. Again, in 2011, Straehler-Pohl et al. (2011) proposed the separation of the genus *Phacellophora* with a creation of a new family, Phacellophoridae, on the basis of morphological features either at polyp, ephyra and medusa stages.

At present, Scyphozoa can be so organized (**Figure 10.1**):

Subclass Coronamedusae, **Order** Coronatae (Fam. Atollidae, Atorellidae, Linuchidae, Nausithoidae, Paraphyllinidae, Periphyllidae).

Subclass Discomedusae, **Order** Semaeostomeae (Fam. Cyaneidae, Drymonematidae, Pelagiidae, Phacellophoridae, Ulmaridae); **Order** Rhizostomeae, **Suborder** Daktyliophorae (Fam. Catostylidae, Lobonematidae, Lychnorhizidae, Rhizostomatidae, Stomolophidae); **Suborder** Kolpophorae (Fam. Cassiopeidae, Cepheidae, Mastigiidae, Thysanostomatidae, Versurigidae), **Suborder** Ptychophorae? (Fam. Baginzidae? either the Suborder and the Family were created for a single species, *Bazinga rieki*, at present considered a doubtful species).

Another question concerns the continuous discovery of cryptic species, a phenomenon that seems to involve several species with large geographical areas. The worldwide-distributed species *Aurelia aurita* is a classic example; it has proved to be a case of cryptic species, when analyzed on molecular bases; at present it can be split into 28 species. Only in the Adriatic Sea, east of Italy, three different *Aurelia* spp. were defined, now denominated *A. solida*, *A. relicta*

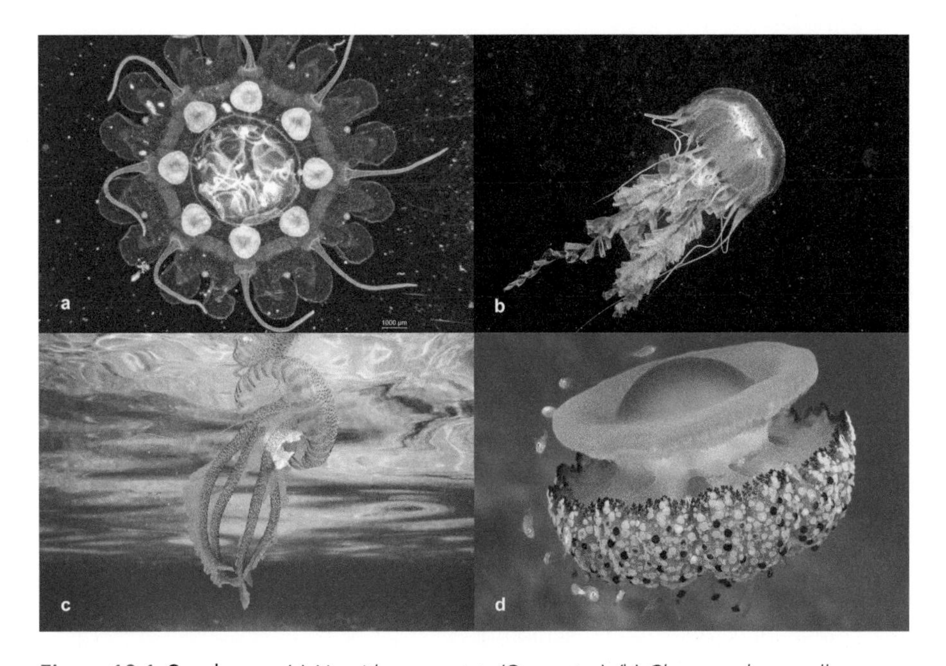

Figure 10.1 Scyphozoa. (a) *Nausithoe punctata* (Coronatae); (b) *Chrysaora hysoscella* (Semaeostomeae), young specimen; (c) *Pelagia noctiluca* (Semaeostomeae); (d) *Cotylorhiza tuberculata* (Rhizostomeae). (From: (a) Valentina Tirelli photo, Istituto Nazionale di Oceanografia e di Geofisica Sperimentale - OGS, Trieste Italy; (b) Tihomir Makovec photo, Marine Biology Station Piran, Slovenia; (c) Rocco Auriemma photo, Istituto Nazionale di Oceanografia e di Geofisica Sperimentale - OGS, Trieste Italy; (d) Tihomir Makovec photo, Marine Biology Station Piran, Slovenia).

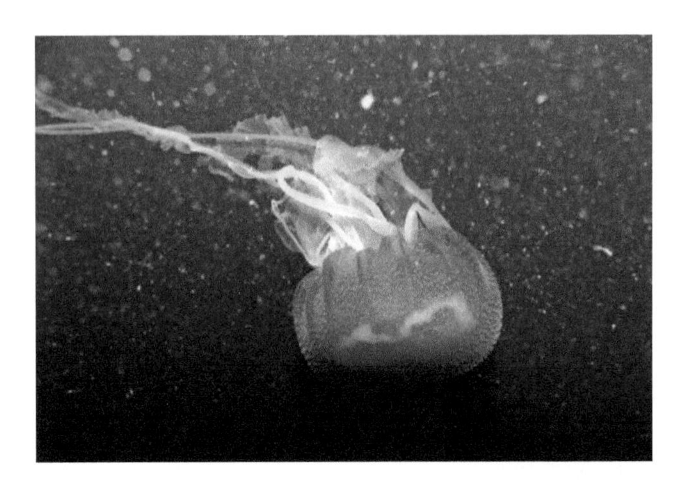

Figure 10.2 *Mawia benovici*, **lateral view of a jellyfish in the Bay of Piran, 5 February 2005.** (From Avian M., Ramšak A., Tirelli V., D'Ambra I., Malej A., 2016. Redescription of *Pelagia benovici* into a new jellyfish genus, *Mawia*, gen. nov., and its phylogenetic position within Pelagiidae (Cnidaria: Scyphozoa: Semaeostomeae). Invertebrate Systematics, 30: 523–546, *Figure 10.2c*.)

and *A. coerulea* (Dawson and Jacobs, 2001; Dawson, 2003; Ramšak et al., 2012; Scorrano et al., 2017).

Even today, an intriguing fact is the discovery of new jellyfish never previously observed, which can stimulate taxonomic revisions. As an example, in one of the most studied seas for at least a century and a half, the Northern Adriatic Sea, an unknown jellyfish was observed in 2013. Initially described as species nova, *Pelagia benovici* (Piraino et al., 2014), it was later redescribed as genus novum, *Mawia* (**Figure 10.2**) (Avian et al., 2016; Gómez Daglio and Dawson, 2019).

ANATOMY, HISTOLOGY AND MORPHOLOGY

Medusa (**Figures 10.3**, **10.4**): umbrella generically hemispherical with the mesogleal layer thick and solid, tapering to the umbrella margins. In Coronatae, the marginal zone of the umbrella is separated from a central disk by a circular furrow, the coronal groove. The marginal zone is also divided by radial furrows which divide the marginal area into thickened areas defined pedalia. Upper surface is defined exumbrella, inner surface is the subumbrella. The umbrella can be divided, like the wind rose, in quadrants by two orthogonal axes, the perradia, in turn divided by two interradia, subsequently divided by the adradia.

Margins interrupted by clefts, so forming separate lobes, the marginal lappets. Along the margin are housed in pits the sensory organs (rhopalia), in number of 6 up to 16, generally 8, containing a statocyst, some sensory pits ex- and subumbrellar, and may contain eyespots (**Figure 10.5**).

Adjacent marginal lappets are defined rhopaliar lappets. At the border of the umbrella may be present a variable number of marginal tentacles, solid (Coronatae) or hollow (Semaeostomeae), with their own ectodermal longitudinal musculature. Adjacent marginal lappets are defined tentacular lappets. Some semaeostomean jellyfish like *Drymonema* or *Cyanea* have many tentacles grouped in subumbrellar areas. Subumbrella presents ectodermal musculature organized in radial and circular (coronal muscle) muscle systems. The contraction of the coronal muscle, mainly, determines the typical pulsating swimming by alternately contracting and relaxing these muscles. In the center of the subumbrella there is a hollow, cruciform structure called manubrium, at whose apex a mouth opens. At the perradial mouth corners the manubrium continues in four elongated structures, the oral arms, often folded and frilled. The manubrium

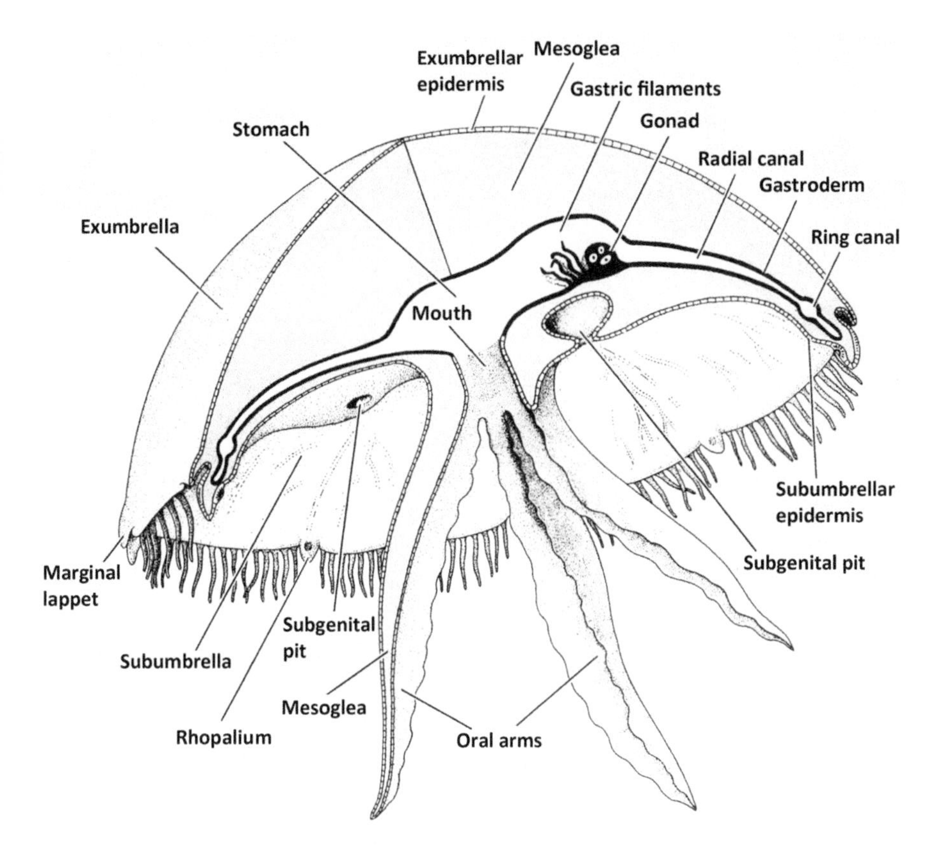

Figure 10.3 General anatomy of a scyphomedusa. (From Brusca and Brusca, 1990. Invertebrates, Sinauer Ass., 1008, *Figure 8.14a.*)

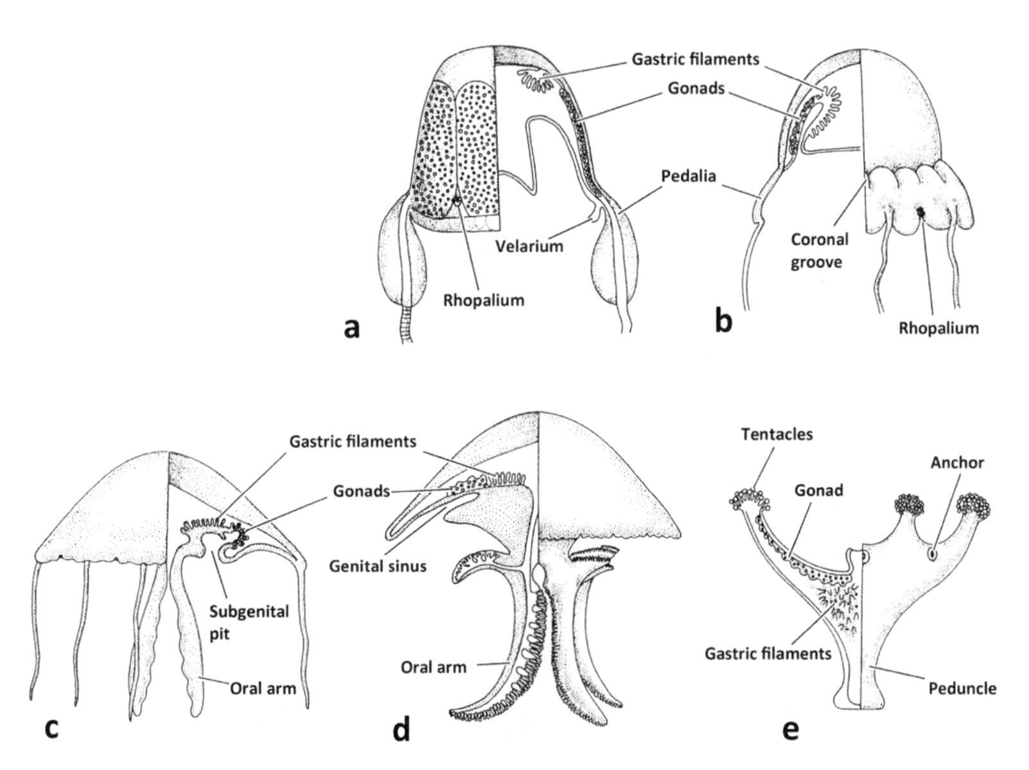

Figure 10.4 Comparison between Scyphozoa, Cubozoa and Staurozoa. (A) Cubozoan (B) scyphozoan of the order Coronatae (C) scyphozoan of the order Semaeostomeae (D) scyphozoan of the order Rhizostomeae (E) stauromedusa. (From Larson R. J., 1976. Marine Flora and Fauna of the Northeastern United States. Cnidaria: Scyphozoa. NOAA Technical Report NMFS Circular 397:17, *Figures 1–4, 8.*)

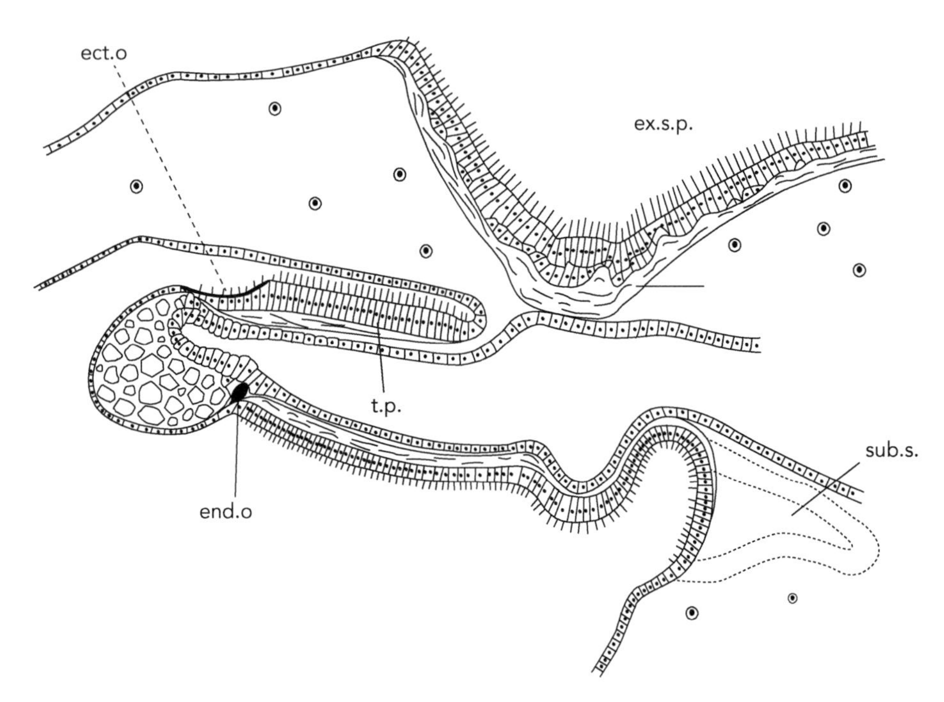

Figure 10.5 *Aurelia aurita* (Semaeostomeae), rhopalium radial section. ***ect.o.***, ectodermal ocellus; ***ex.s.p.***, exumbrella sensory pit; ***sub.s.p.***, subumbrellar sensory pit (more lateral); ***t.p.***, mechanoreceptor plate. (From Russell F. R. S., 1970. The Medusae of the British Isles. II. Pelagic Scyphozoa with a supplement to the first volume on Hydromedusae. Cambridge Univ. Press: 1–284. *Figure 10.3a*.)

communicates with a central stomach containing the gastric filaments, or cirri, and the gonads. From the stomach originate pouches, or numerous canals (the radial canal system), or the highly complicated network of anastomosing canals in Rhizostomeae, that reach the umbrella margin, giving origin to the gastrovascular system. In some medusae, the canals or pouches can communicate with a continuous ring canal just near the umbrella margin (in Coronatae the pouches communicate with each other through a festoon canal) (**Figures 10.6**, **10.7**).

The gonads are eight, adradial, peripheral to the central stomach, with a shape of round-oval-U-W in Coronatae (Figure 10.1a), or four subumbrellar interradial ribbon-like, just peripheral to the gastric filaments in the stomach in Semaeostomeae (Figure 10.2) and Rhizostomeae. These ribbons can be highly

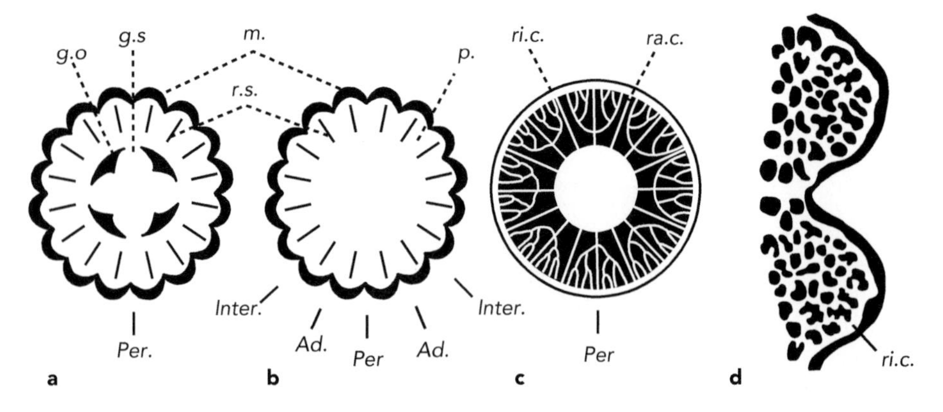

Figure 10.6 Drawings illustrating the endodermal fusion areas (black) that shape the gastrovascular system. (a) Coronatae (b and c) Semaeostomeae (d) Rhizostomeae. **Per.**, perradius; **Ad.**, adradius; **Inter.**, interradius; **g.o.**, gastric ostium; **g.s.**, gastric septum; **m.**, umbrellar margin; **p.**, pouch; **r.s.**, radial septum; **ra.c.**, radial canal; **ri.c.**, ring canal. (a,b,c: Modified from Russell F. R. S., 1970. The Medusae of the British Isles. II. Pelagic Scyphozoa with a supplement to the first volume on Hydromedusae. Cambridge Univ. Press: 1–284. *Figure 1;* d: ibid, *Figure 90b.*)

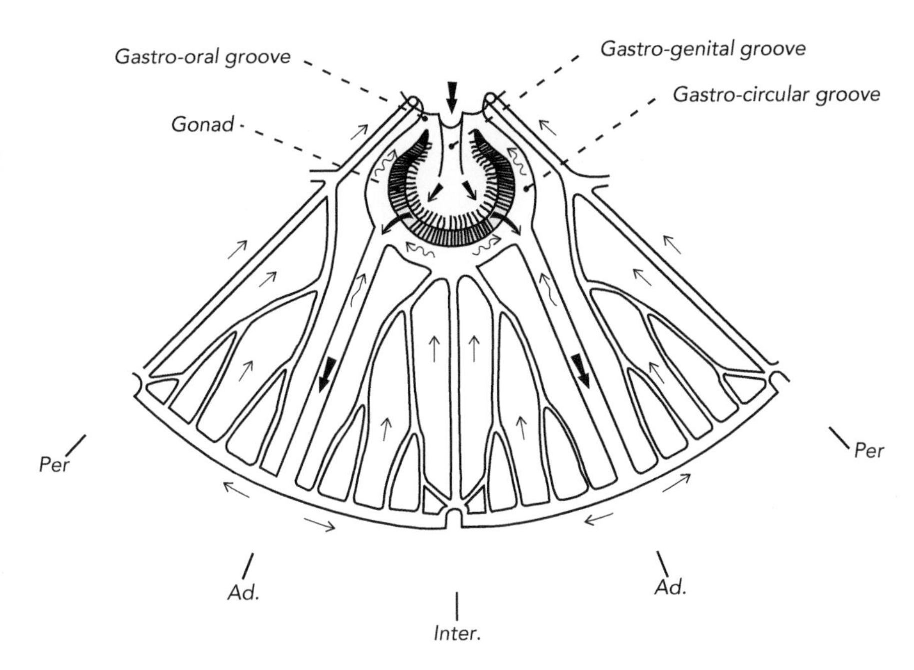

Figure 10.7 ***Aurelia aurita*** **circulation pattern in gastrovascular canal system (simplified by omission of branches of per- and interradial canals).** Wavy arrows indicate passage along bottoms of canals. *Per.*, perradial canals; *Ad.*, adradial canals; *Inter.*, interradial canals. (From Russell F. R. S., 1970. The Medusae of the British Isles. II. Pelagic Scyphozoa with a supplement to the first volume on Hydromedusae. Cambridge Univ. Press: 1–284. *Figure 86.*)

folded, and define with their subumbrellar surface a flat cavity, called genital sinus. Genital sinus opening is facing toward the stomach in Coronatae, and toward the periphery in Semaeostomeae and Rhizostomeae. Scyphozoans, where known, are mainly gonochoric, except the semaeostomean genus *Chrysaora*, which is protherandric/protandrous hermaphrodite. All Scyphozoa, like all cnidarians, have several types of nematocysts, mainly concentrated to the tentacles, but present also in other epithelia, like epidermis of manubrium and oral arms, exumbrella, and within the gastrovascular system in the gastroderm of gastric filaments. Many scyphozoans, both bathypelagic and epipelagic are able to emit bioluminescence (Herring and Widder, 2004) (**Figure 10.8**). The light-emitting molecule found in this class is coelenterazine, a type of luciferin.

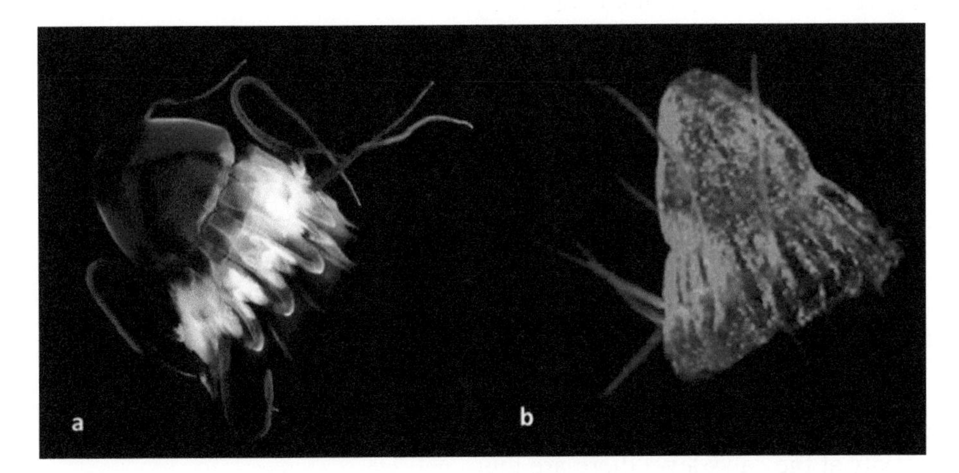

Figure 10.8 ***Periphylla periphylla*** **displaying bioluminescence.** (a) Artificially illuminated specimen, with the onset of bioluminescence flashes at the base of the tentacles (b) non-illuminated specimen displaying the wave of bioluminescence. (From: (a) https://www.wrobelphoto.com/deepseamarinelife/h1b491801#h1b491801; (b) https://66.media.tumblr.com/tumblr_lttellPF9U1qeeqk5o1_500.jpg, E. Widder photo, Ocean Research & Conservation Association, Fort Pierce, FL.)

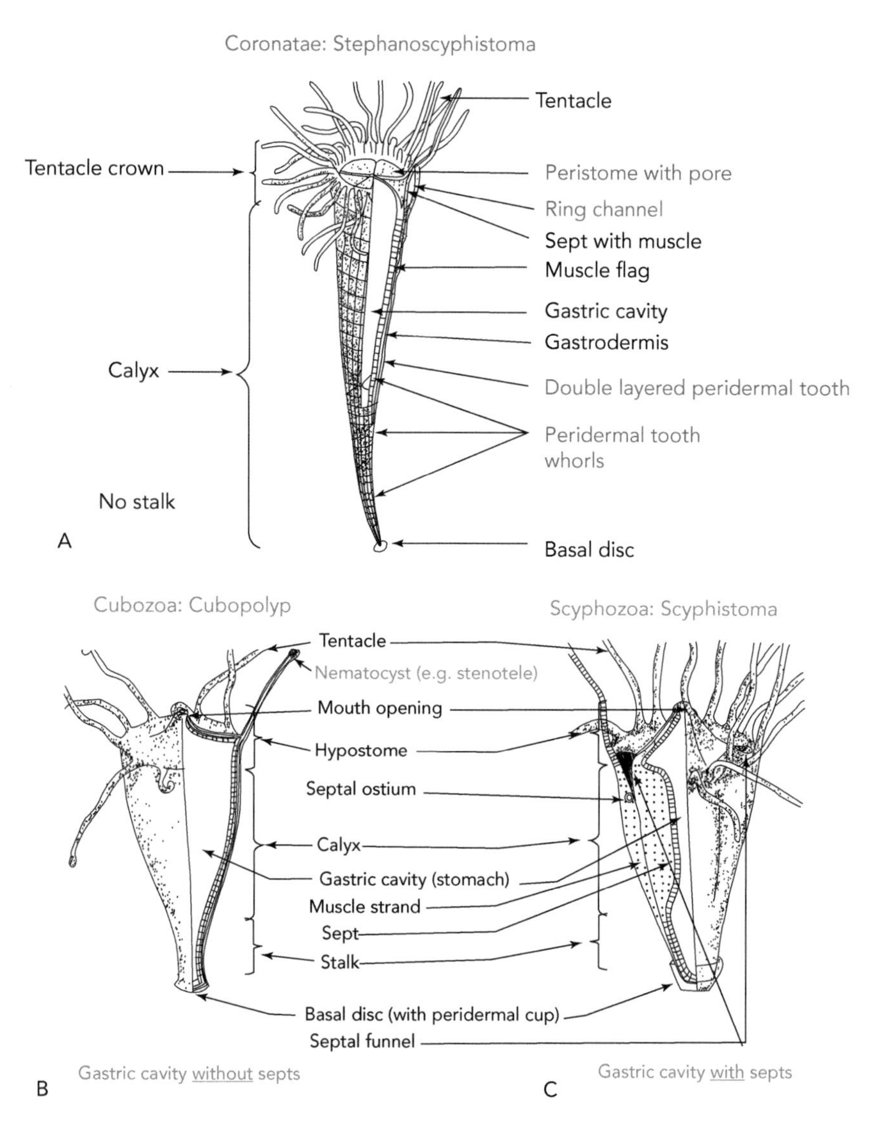

Figure 10.9 Anatomy of sessile stage of Scyphozoa and Cubozoa. (A) Coronatae: *Stephanoscyphistoma* (B) Cubozoa (C) Scyphozoa. (From Straehler-Pohl, I., 2017. Cubozoa and Scyphozoa: The results of 20 years of scyphozoan life cycle research with new results on cubozoan life cycles to suggest a new nomenclature referring to both classes. In: Frontiers in ecological studies of jellyfish (eds. Toyokawa, M., Miyake, H., Nishikawa, J.), Seibutsu Kenkyu Sha Co. Ltd. (Organisms Research Co. Ltd.), Tokyo, 17–29, *Figure 10.2, 10.3.*)

Despite its presence within cnidarians, jellyfish do not synthesize coelenterazine; rather they obtain it through their diet, largely from crustaceans and copepods (Haddock et al., 2001).

Polyp stage: called scyphistoma, it consists of a portion called calyx which continues into a cylindrical stalk, with a pedal disk through which it adheres to the substrate. At its oral end there is an enlarged, circular oral disk with in the center a protruding cruciform mouth, and the corners are perradial. At the border of the oral disk there are a number of tentacles, from 16 to 24 and more. Within the gastric cavity at the calyx level there are four gastric interradial longitudinal septa (tetramerous symmetry). At the basis of the septa there are four longitudinal muscles that run from the oral disk up to the pedal disk. Coronatae polyps are protected by a complete chitinous tube, while the other scyphopolyps possess a peridermal cup just around the basal disc (**Figure 10.9**).

REPRODUCTION AND DEVELOPMENT

Jellyfish are sexed, mostly gonochoric without a marked sexual dimorphism. Gametes are mostly released by both sexes followed by external fertilization, although there are species in which eggs are fertilized inside the stomach and/ or genital sinus, like *Rhizostoma* spp., or *Aurelia* spp., where fertilization is internal, then the embryos when reach the oral arms continue their development in special brood pouches (*Aurelia* sp.). Fully-grown planulae then will be released outside.

Planulae are typically freely swimming, two layered, ciliated, without eyespots. When settled onto the substrate, they metamorphose into a scyphistoma. The scyphistoma stage, solitary, is capable of carrying out a wide variety of budding (asexual reproduction), when environmental conditions are favorable with a particular asexual reproduction, called strobilation (transverse fission, form of asexual reproduction consisting of the spontaneous transverse segmentation of the calyx at oral disk level). The strobilation process produces a free-swimming larva called ephyra that quickly grows into a medusa. Strobilation may be monodisk (one ephyra per process) or polydisk (several ephyrae per process). Normally, within Scyphozoa strobilation is polydisk in Coronatae and Semaeostomeae (with the exception of *Sanderia* genus), and monodisk in some Rhizostomeae (Cepheidae). Most scyphozoans have a meroplanktonic biological cycle, which involves both polypoid and medusoid stages (**Figure 10.10**),

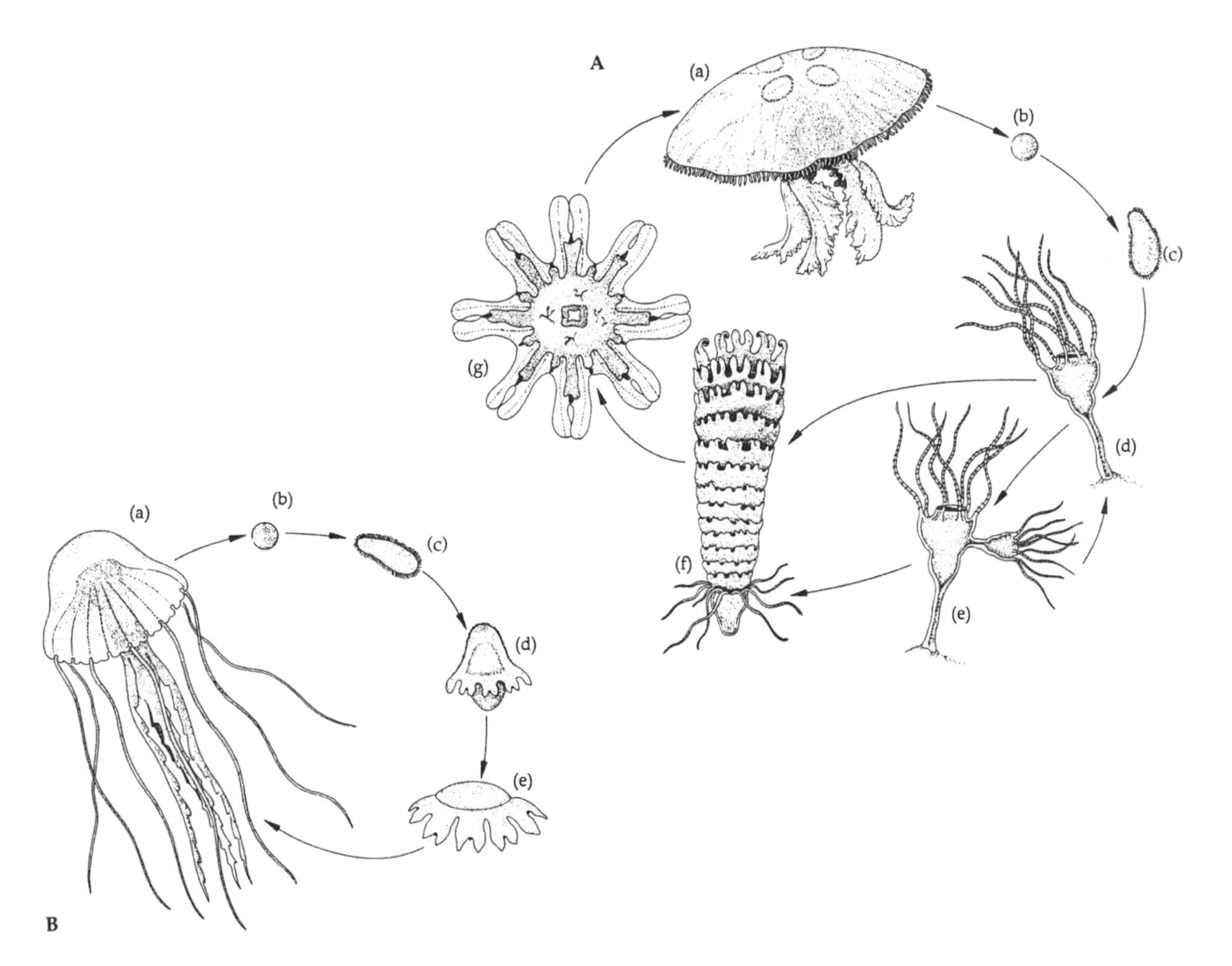

Figure 10.10 (A) Meroplanktonic life cycle of *Aurelia* sp. (a) adult jellyfish (b) fertilized oocyte (c) planula (d) scyphistoma (e) budding scyphistoma (f) strobila (g) ephyra. (B) Holoplanktonic life cycle of *Pelagia noctiluca*. (a) Adult jellyfish (b) fertilized oocyte (c) planula (d) developing ephyra (e) ephyra. (From Brusca and Brusca, 1990. Invertebrates, Sinauer Ass., 1008, *Figure 8.41 a, b.*)

but there are some exceptions, generally related to oceanic or deep sea environments. In these cases the polypoid stage has been suppressed, and the planula differentiates directly in an ephyra (holoplanktonic cycle) as in *Pelagia noctiluca* (Rottini Sandrini and Avian, 1983), or even with the suppression of the planuloid stage, like *Periphylla periphylla* (Jarms et al., 1999).

DISTRIBUTION AND ECOLOGY

Marine, worldwide distributed, epipelagic to neritic to bathypelagic and more (Coronatae were observed to at least 5,000 m depth), mainly within the temperate area, but well known both in cold temperate and polar waters. An alien species, *Aurelia* sp., is present even in the Caspian Sea, previously devoid of endemic scyphozoans, since 1999 (Korsun et al., 2012; see Figure 10.16 for the distribution map).

Jellyfish can aggregate actively into large blooms (or can be collected altogether by currents) in certain environmental conditions. Blooms can even reach an extension of several nautical miles (Arai, 1997). Coronatae and Semaeostomeae, with a central mouth, are opportunistic predators and can eat any prey of appropriate size; Rhizostomeae, with many little mouths, are planktofagous.

PHYSIOLOGY AND BEHAVIOR

The neuronal system enables scyphozoans to receive a stream of sensory inputs from the environment by several receptors, such as light receptors, mechanoreceptors, chemoreceptors, gravity receptors, and hydrostatic pressure receptors. Scyphozoans are sensitive to salinity and *Chrysaora quinquecirrha* can withstand salinity up to 13 ppt. They are sensitive to touch and able to escape and avoid turbulent water by diving; moreover they are not easily stranded, because they can avoid shallow water (Albert, 2011). Scyphozoan species are voracious planctivorous feeders, continuously swimming in order to cause flow that helps to catch the prey (e.g. *Aurelia*, *Rhizostoma*) and follow diel nocturnal migration of prey. Food is detected by chemoreceptors if it has a chemical signature of protein, and is otherwise eliminated and released back to water. Food is rapidly digested in the gastral cavity and then transported through canals. Functional differentiation of the canals has been observed (in *Aurelia*, *Cassiopea*); some canals transport fluids in only one direction, and others have bidirectional flow via the generation of ciliary currents (Russell, 1970). Moreover, some genera (*Cassiopea*, *Linuche*, *Linantha*, *Nausithoe*, *Cephea*, *Cotylorhiza*, *Netrostoma*, *Mastigias*, *Phyllorhiza*, and *Catostylus*) distributed in more oligotrophic waters establish symbiosis with diverse group of *Symbiodinium* algae (generally called Zooxanthellae). Scyphozoa has a great regenerative potential to regenerate polyp, ephyra or adult medusa and are able to regenerate sensory organs rhopalia as well, which are present in Scyphozoa and Cubozoa. Rhopalia sense light and gravity (statoliths), and control the rhythm of swimming-muscle contraction.

Jellyfish are often abundant in estuaries and coastal seas which are seasonally dysoxic (≤ 2 ml L^{-1} O_2) or anoxic, often as a result of eutrophication. Dysoxia and anoxia can directly or indirectly affect feeding, growth, reproduction and survival. Nevertheless, several of the mass-occurring Cnidaria (e.g. *Aurelia*, *Chrysaora quinquecirrha*, *Cyanea capillata*, etc.) are able to tolerate dissolved oxygen levels as low as 0.5 mg L^{-1}, in both the polyp and medusa phases of the life cycle in the case of scyphozoans. Their tolerance of exposure to low oxygen, and their high feeding rates under dysoxic conditions, may be partly responsible for the high abundance of these gelatinous species in coastal ecosystems where dysoxia is common. Scyphomedusae can be considered as cruising predators as well as ambush predators. They can be observed either dispersed in the sea, or in aggregations (blooms, swarms) sometimes of immense size and density. Within Semaeostomeae, which have tentacles, a swimming-search pattern can show a wide oblique sinusoidal movement with released tentacles, or by swimming vertically, ascending in a slow spiral and then sinking with umbrella upward and tentacles spreaded outward. When the tentacles come in contact with a prey, they sting-stick the prey, then they shorten and tilt toward the manubrium. Simultaneously, the oral arms begin to move and take the prey from the tentacles. Ciliary currents then transport the prey through the mouth and into the stomach. Rhizostomeae, lacking tentacles, equally are mainly active swimmers, forcing the water to be pumped toward their massive and complicated manubrium with numerous mouth openings. Coronatae were most often observed swimming, and feeding, with their tentacles in aboral to lateral position, suggesting that they behaved as "ramming" stealth predators. Various epi- mesopelagic scyphozoans undergo daily (diel) vertical migration (from a few meters to several hundreds) reaching shallower waters during the night (Arai, 1997).

In at least one species, *Pelagia noctiluca*, in western Mediterranean waters, a seasonal vertical migration was hypothesized (Canepa et al., 2014), with jellyfish that overcome the warmer months at colder, mid-water levels; by mid-autumn or early winter, jellyfish migrate upward for sexual reproduction; throughout spring to early summer, at shallow levels jellyfish feed on the seasonal spring plankton bloom, with rapid somatic growth; by the end of summer, jellyfish migrate downward to escape shortage of plankton food and warmer temperatures.

GENETICS AND GENOMICS

The size of genome in Scyphozoa is known only for a few species, the most investigated of which is *Aurelia aurita*, its genome size is estimated to C-value = 0.73 pg and chromosome number is in diploid stage $2n = 44$ (Goldberg et al., 1975; Diupotex-Chong et al., 2009). The genome of *Aurelia aurita* is the largest so far known, while the other genomes are in range of C-value = 0.26 pg (*Sanderia malayensis*) to C-value = 0.40 pg in *Cassiopea ornata* (Adachi et al., 2017) (genome size database http://www.genomesize.com). The assembly of genome in *Aurelia aurita* revealed high percentage of repetitive DNA (49.5% transposable elements and 0.8% simple tandem repeats) (Gold et al., 2019). *Hox* genes are present in all of the investigated scyphozoans and are involved in development and the regeneration processes (Ferrier and Holland, 2001). The whole genomes of *Aurelia aurita*, *Nemopilema nomurai* and *Cassiopea xamachana* are available in the Genbank.

The medusozoan mitochondrial genome is linear (Kayal et al., 2012), on single chromosome (split into eight linear chromosomes in Cubozoa and into two chromosomes in Hydrozoa). Mitochondrial DNA of several species of scyphozoan was sequenced among first from *Aurelia aurita* (Shao et al., 2006). The list of sequences and organization of the mitochondrial genome is deposited in Organelle Genome Resources database in Genbank, and currently five species have a sequenced mitochondrial genome (*Aurelia aurita*, *Cassiopea frondosa*, *Chrysaora quinquecirrha*, *Nemopilema nomurai* and *Rhopilema esculentum*). The linear mitochondrial genome poses several unique features in comparison with circular, such as maintenance of stability due to their susceptibility to exonuclease activity, replication mechanism and expression of genes. The ends of a linear chromosome should be protected by repetitive sequences (telomeres) to protect the chromosome from degradation and enable faithful replication.

Several scyphozoan species (*Aurelia* spp., *Cyanea capillata*, *C. lamarckii*, *Mastigias* spp. etc) were investigated in order to reveal their phylogenetic relationships, speciation processes and phylogeographic structure based on informative DNA sequences. Informative genetic markers are from mitochondrial DNA (COI and 16S), together with markers on nuclear chromosomes such as genes for ribosomal RNA (ITS regions, 28S and 18S) and microsatellites. The differences between populations *Aurelia aurita* are up to 40% divergence in ITS-1 and 23% in COI (Dawson and Jacobs, 2001).

CLASS CUBOZOA

General Characteristics

About 47 species described

Estimated number of species still to be described or cryptic forms: unknown

Marine, commonly called cubomedusae or box jellyfish (max 20–30 cm in diameter), mostly known in tropical-subtropical to temperate regions, especially subtropical and tropical Indo-Pacific waters, near shore habitats above the continental shelves (i.e. the neritic zone)

Some species, like *Chironex fleckeri*, the sea wasp, produce extremely potent toxins. Stings from this and a few other species are extremely painful and can be fatal to humans

No commercial value

Ecological importance is probably relevant, still to be investigated

Synapomorphies

- umbrella cube-shaped
- umbrella margins not cleft into lappets
- with four, single or groups, interradial tentacles emerging from gelatinous wing-shaped structures called pedalia
- subumbrellar cavity partially limited by an annular diaphragm called velarium
- with four perradial sense organs (rhopalia) localized into median umbrellar niches
- ocelli not only shaped as simple pigment-cup, but also with true eyes, complete with retinas, corneas and lenses
- box jellyfish have a central nervous system (CNS), with two major components: the rhopalial nervous system and the ring nerve
- the polyp generally does not strobilate, but is metamorphosed into a jellyfish (at least in some species)
- at least in some species, there is a complex sex mating behavior, with internal fecundation and ovoviviparity
- planula monostratified, may have eyespots

HISTORY OF TAXONOMY AND CLASSIFICATION

This taxon, previously considered an Order within the Class Scyphozoa, was elevated to the rank of Class by Werner (1973), on the basis of morphological characters and their peculiar biological cycle, a metamorphic event instead of a strobilation. These statements have given rise to a whole series of debates, as some authors consider the formation of the jellyfish plus a special case of monodisk strobilation (the basal part of the polyp remains attached to the substrate, then degenerates, or may regenerate the apical portion). Recently there have been suggestions both to reconsider the cubozoans as an order of Scyphozoa, and on the contrary to support validity of class.

A revision of the cubozoan systematics and phylogeny (Bentlage et al., 2009) identified several taxa as either para- or polyphyletic; designated a new family, Carukiidae; a new genus, *Copula*; and redefined a couple of families. At present, the class contains two orders, Carybdeida (with five families, Carybdeidae, Tripedaliidae, Tamoyidae, Carukiidae and Alatinidae) and Chirodropida (with three families, Chirodropidae, Chiropsalmidae and Chiropsellidae) (**Figure 10.11**).

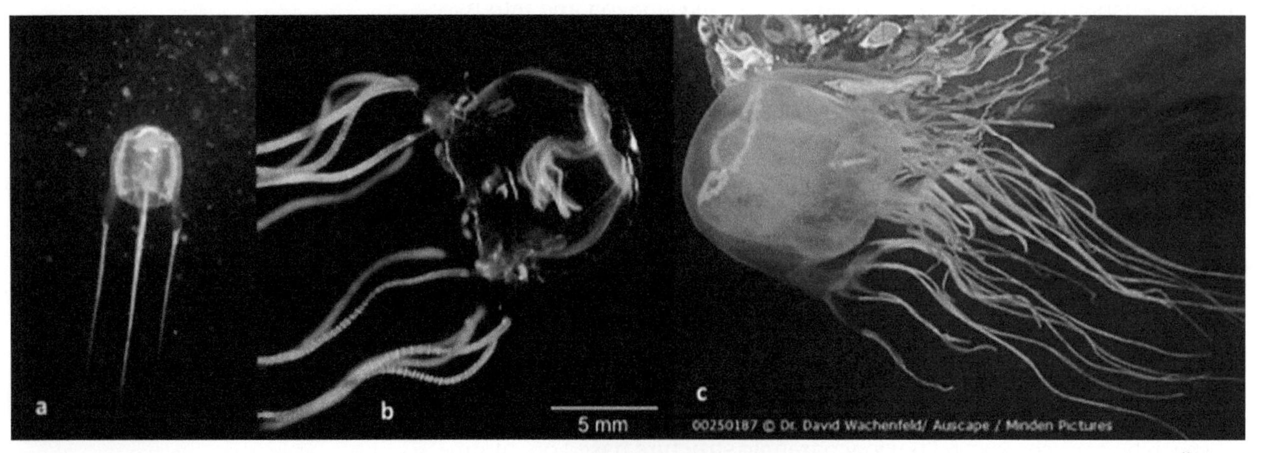

Figure 10.11 Cubozoans; *carybdeid cubomedusae.* (a) *Carybdea marsupialis* (b) *Tripedalia cystophora* (c) A chirodropidÜubomedusa, ***Chironex fleckeri.*** (From: (a) M. Rinaldi photo, Cesenatico, Italy; (b) J. Bieleki photo, Kiel, Germany; (c) https://www.mindenpictures.com/gallery/preview/1386/1972/1493/0/sea-wasp-chironex-fleckeri-highly-venomous-jellyfish-swimming-oceansurface/0_00250187.html.)

ANATOMY, HISTOLOGY AND MORPHOLOGY

Medusa (Figures 10.4a, 10.11): with a box-like umbrella. From each of the four umbrellar interradial lower corners hangs a short wing-shaped pedalia or stalk which bears one or more long, slender, hollow tentacle. The inner marginal side of the umbrella is folded inward to form a shelf known as velarium, which restricts the bell's aperture and creates a powerful jet during the pulsation, so the box jellyfish can move more rapidly than other jellyfish (observed speeds of up to 6 meters/min). At the center of the subumbrella is present a short manubrium, with a cruciform mouth at the tip, with small or without oral arms. Over the manubrium there is a central stomach, which communicates with four lateral, perradial, gastric pouches, partially separated interradially (in correspondence with the edges of the bell) by four septa. Each of these septa carries a pair of leaf-shaped gonads. The upper margins of the septa bear bundles of small gastric cirri. Sensory input largely comes from 24 eyes situated on four club-shaped sensory structures, the rhopalia (generally 6, two ocelli with lenses, one directed upward and the other downward and inward toward the manubrium, two simple pit and two-slit ocelli), (**Figure 10.12b–d**). To process the sensory input and convert it into the appropriate behavior, the box jellyfish have a central

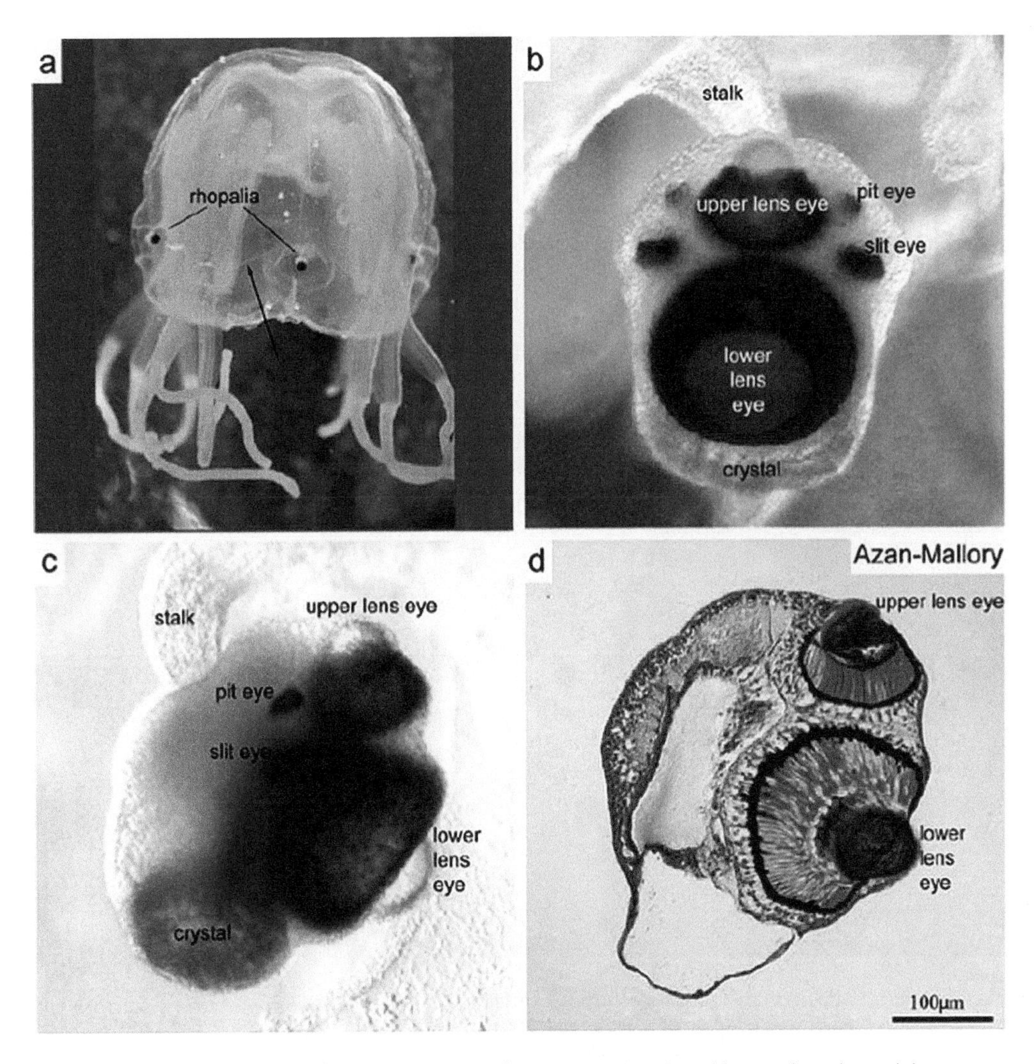

Figure 10.12 The rhopalia of *Tripedalia cystophora*. (a) Adult with visible two rhopalia and the nerve ring (arrow) (b) horizontal view from the subumbrellar cavity (c) lateral view (d) median section. (From Skogh C., Garm A., Nilsson D.-E., and Ekström P., 2006. Bilaterally symmetrical rhopalial nervous system of the box jellyfish *Tripedalia cystophora*. J. Morphol. 267, 1391–1405, *Figure 10.1*.)

nervous system (CNS). The CNS has two major components: the rhopalial nervous system and the ring nerve (Figure 10.12a). The rhopalar nervous system is situated within the rhopalia in close connection with the eyes, whereas the ring nerve encircles the bell. Nematocysts are mainly concentrated in rings on the tentacles. In some cubozoans, such as *Chironex fleckeri*, nematocysts are absent from the bell.

Polyp: apparently similar to that of the scyphopolyp, but without septa and without the peridermal cup around the basal disk (Figure 10.9b).

REPRODUCTION AND DEVELOPMENT

All gonochoric. Their reproduction seems generally based on an annual life cycle. Chirodropida have external fertilization, whereas Carybdeida have internal fertilization and ovoviviparity, sperms can be transferred through spermatophores called spermatozeugmata (**Figure 10.13**). The mating behavior and spermatophore transfer in *C. sivickisi* has been described in detail (Garm et al., 2015). Females then release embryo strands or planulae. Planulae swim in the water column for a few days and then settle on to the substrate. After settlement, the planulae grow into polyps. The polyps can move around, and they frequently bud off additional polyps. After a few months of feeding, the polyps are mature, undergoing a: monodisc strobilation-like metamorphosis (1 medusa + small polyp, Carukiidae); incomplete metamorphosis (1 medusa +

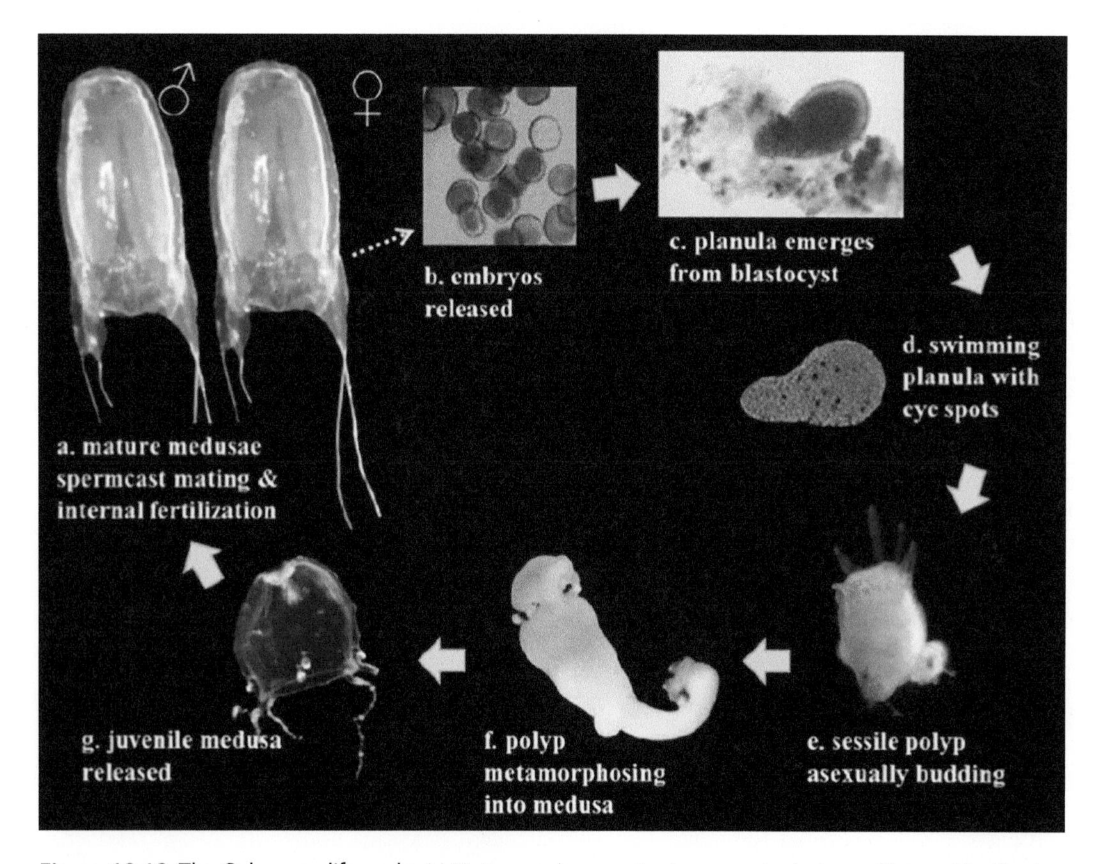

Figure 10.13 The Cubozoan life cycle. (a) Mature medusae mate via spermatophores and internal fertilization (b) embryos released into the water (c) planulae hatch from blastocysts with perisarc (d) sessile polyps adhere to the substrate and bud additional polyps asexually (e) polyp tentacles retract and the apical portion metamorphosis into a medusa (f) juvenile medusa is released. This figure is a composite of the life stages of two different cubozoan taxa: *Alatina alata* in Figures a–d, and cf. Alatinidae in Figures e–g. (Photos by Cheryl Lewis Ames and Allen G. Collins. (From Ames C. L., 2016. Taxonomy, Morphology, and RNA-Seq Transcriptomics of the Cubozoan Alatina alata. An Emerging Model Cnidarian.) Thesis dissertation, Smithsonian Institution, 144, *Figure 10.3*.)

polypoid residuum, *Carybdea xaymacana*); complete metamorphosis (1 medusa, Alatinidae, Tripedaliidae) (Straehler-Pohl, 2017).

DISTRIBUTION AND ECOLOGY

Cubozoans are primarily tropical jellyfishes with few species found outside of tropical latitudes. However, a few taxa have been recorded at higher latitudes, to 42°N and 42°S. The highest diversity of species is found in the Indo- East Pacific and within Coral Sea. Several species have been documented from the Philippine Sea, Caribbean Sea, Gulf of Mexico and both boundaries of the Atlantic Ocean. One species found in the Mediterranean Sea. Cubozoans are not only restricted to continental coastlines; they have also been found in waters of islands, some isolated by nearly 4,000 km of oceanic waters, including Polynesian islands and New Zealand in the Pacific basin as well as Bermuda and Saint Helena in the Atlantic basin. Box jellyfishes, therefore, are almost pantropical in distribution (Figure 10.16) (Kingsford and Mooney, 2014).

PHYSIOLOGY AND BEHAVIOR

Orientation based on eyes is a typical characteristic of cubomedusae when compared to scyphozoans. They are capable of strong directional swimming combined with rapid turns which can be up to 180° in a couple of bell contractions. Obstacle avoidance has also been documented in several cubozoan species (Kingsford and Mooney, 2014).

Most species exhibit a typical feeding behavior. With tentacles extended a cubomedusa will increase swimming speed vertically upward for a short period, perform a 180° turn then stop bell pulsation. The negatively buoyant medusa drifts down apex first with tentacles extended. When prey is entangled in tentacles, the pedalia fold inward, and the tentacles with attached prey are moved into the bell where the manubrium locates and removes the prey from the tentacles (Kingsford and Mooney, 2014). A specimen of *Carybdea marsupialis* was observed by one of the authors (M. Avian) in the harbor of Cesenatico (Italy) which was normally swimming parallel to the surface of the sea with the tentacles completely released. Casually approached by a small shrimp, it immediately blocked itself and performed a 180° rotation exposing the sub-umbrella and the partially contracted and curved tentacles toward the shrimp.

In at least in one species, *Copula sivickisi*, a sexual reproduction system including mating and internal fertilization has been described. When a mature male and female meet, they entangle their tentacles and the male transfers a sperm package, the spermatozeugma, which is ingested by the female fertilizing her eggs internally (Garm et al., 2015).

GENETICS AND GENOMICS

Among Cubozoa the size of the genome is only known for *Carybdea brevipedalia*, which has a C-value of 0.77 pg (Adachi et al., 2017), and until now among Cubozoa only *Morbakka virulenta* has a whole sequenced genome (sequences deposited in GeneBank).

Cubozoa are interesting because they have a mitochondrial chromosome which is segmented into linear molecules. The mitochondrial genome of *Alatina moseri* is highly subdivided and consists of 18 genes on eight linear chromosomes and the gene complement is similar to that in other medusozoans: 13 proteins, two ribosomal RNAs and one transfer RNA and putative beta DNA polymerase and region for DNA binding properties (orf 314). The linear chromosome has a stem-loop sequence in the telomere regions and probably serves as a control region (Smith et al., 2012).

CLASS STAUROZOA

General Characteristics

About 50 species described

Estimated number of species still to be described: unknown

Marine small, macroscopic, mostly known in anti-tropical regions. The greater part of the known species is present in the boreal hemisphere, including the arctic sea, but this may be due to a lack of studies performed in the southern hemisphere, rather than a real biogeographic distribution pattern

No commercial value

Ecological importance is still to be investigated

Synapomorphies

- larval form (planula) not ciliated
- polyp metamorphoses in an adult stauromedusa (stalked jellyfish)
- body with a calyx carrying hollow tentacles with a terminal knob
- aboral peduncle attached to substratum through an adhesive disk
- ovaries with follicle cells

HISTORY OF TAXONOMY AND CLASSIFICATION

This taxon was elevated to the rank of Class by Marques and Collins (2004), on the basis of cladistical analyses considering morphological and genetic (18S) characters. Further investigations (Collins et al., 2006; Miranda et al., 2016a,b) confirmed that Staurozoa are the sister group of all other medusozoans (Subphylum Medusozoa which comprises all cnidarians with a medusoid stage).

There is a single order, Stauromedusae (**Figure 10.14**), with two traditional suborders, Cleistocarpida and Eleutherocarpida. A revision (Miranda et al., 2016a) based on morphological and a combined set of molecular data (COI, 16S, ITS, 18S, and 28S), evidenced that these suborders were not monophyletic, now replaced by two new suborders, Amyostaurida (families Craterolophidae, Kishinouyeidae) and Myostaurida (families Haliclystidae, Kyopodiidae, Lipkeidae, Lucernariidae), characterized by the absence or presence of four interradial longitudinal muscles in the peduncle.

ANATOMY, HISTOLOGY AND MORPHOLOGY

Their general anatomy (Figure 10.4e, **10.15**) includes an upper part, the calyx, with tetraradial symmetry, and an aboral part called pedunculus (the stalk) that adheres to the substratum with an adhesive disk. Centrally in the upper surface of the calyx there is a small manubrium with a cruciform mouth. Four peristomial pits, or infundibula, are placed interradially. There are four interradial, intramesogleal longitudinal muscles associated with peristomial pits (infundibula) and are symplesiomorphic in Staurozoa, and shared by the ancestral staurozoan with some (but not all) other medusozoans (Collins et al., 2006). From the calyx margin eight groups of (secondary) tentacles emerge, with claviform endings, rich in nematocytes. In some staurozoans these clusters of tentacles originate from calyx extensions called "arms", that can be grouped in pairs, or symmetrically distributed, causing an apparent eight-fold symmetry. Between these clusters of tentacles the primary tentacles may be present (are residual of the polypoid stage), or reabsorbed, or modified into particular adhesive structures called "anchors" or "rhopalioids". Although the stauromedusae are sessile, they may occasionally detach from the substrate, and use these adhesive structures to anchor themselves temporarily. Gonads may be contained in the gastric cavity of the calyx or in the peduncle.

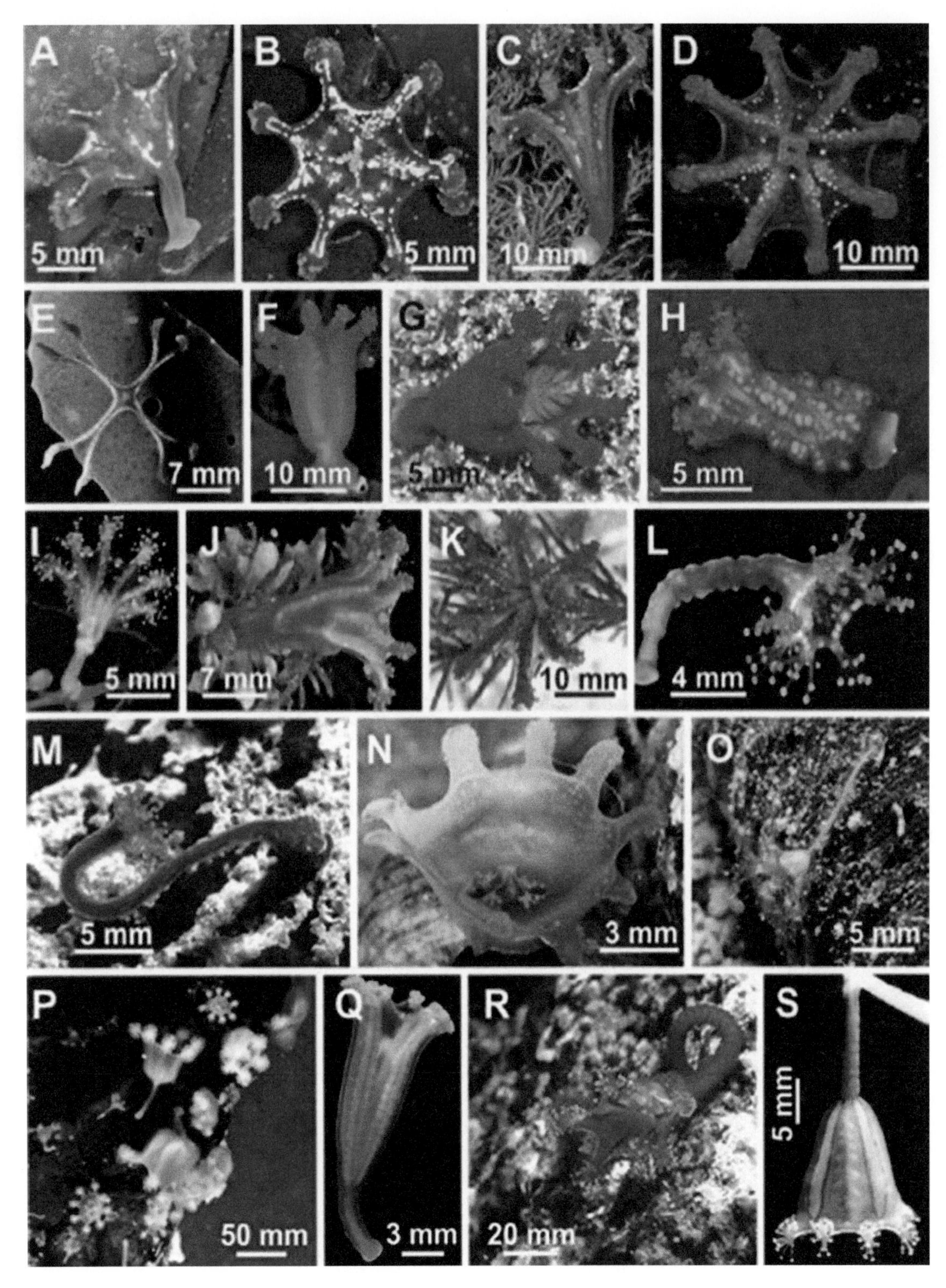

Figure 10.14 Diversity of stalked jellyfishes. *Calvadosia cruxmelitensis*: (A) lateral view, (B) oral view (photo: David Fenwick); *Calvadosia campanulata*: (C) lateral view, (D) oral view (photo: David Fenwick); *Calvadosia nagatensis*: (E) oral view (photo: Yayoi Hirano); *Craterolophus convolvulus*: (F, G) lateral view (photo: David Fenwick); *Depastromorpha africana*: (H) lateral view (photo: Yayoi Hirano); *Haliclystus tenuis*: (I) lateral view (photo: Yayoi Hirano); *Haliclystus borealis*: (J) lateral view (photo: Yayoi Hirano); *Haliclystus octoradiatus*: (K) oral view (photo: David Fenwick); *Haliclystus inabai*: (L) lateral view (photo: Yayoi Hirano); *Kyopoda lamberti*: (M) lateral view (photo: Ronald Shimek); *Lipkea* sp. Japan: (N) oral view (photo: Yayoi Hirano); *Stylocoronella riedli*: (O) lateral view (proto: Mat Vestjens and Anne Frijsinger); *Lucernaria janetae*: (P) lateral and oral views (photo: Richard Lutz); *Manania uchidai*: (Q) lateral view (photo: Yayoi Hirano); *Manania gwilliami*: (R) oral view (photo: Ronald Shimek); *Manania handi*: (S) lateral view (photo: Claudia Mills). (From Miranda L. S., Hirano Y. M., Mills C. E., Falconer A., Fenwick D., Marques A. C., Collins A. G., 2016a. Systematics of stalked jellyfishes (Cnidaria: Staurozoa). PeerJ 4:e1951, *Figure 10.1*.)

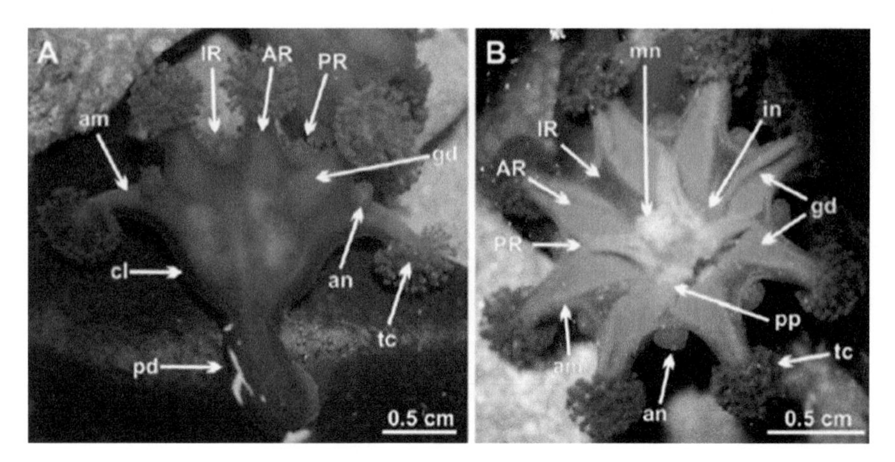

Figure 10.15 (A) Lateral view of living specimen of *Haliclystus antarcticus*. (B) Oral view of living specimen of *H. antarcticus*. am, arm; an, anchor; AR, adradii; cl, calyx; gd, gonad; in, infundibulum; IR, interradii; mn, manubrium; pd, peduncle; pp, perradial pocket; PR, perradii; tc, tentacles. (Photo: AC Morandini. From Miranda L. S., Collins A. G., Marques A. C., 2016b. Internal Anatomy of *Haliclystus antarcticus* (Cnidaria, Staurozoa) with a Discussion on Histological Features Used in Staurozoan Taxonomy. Journal of Morphology 274(12):1365–1383, *Figure 10.1*.)

REPRODUCTION AND DEVELOPMENT

It seems generally to be based on an annual life cycle. The population density increase from spring to fall. The growth of stauromedusae is rapid, gonochoric, and with external fertilization. Eggs are generally negatively buoyant, so they sink rapidly, thus staying near the point of release/emission. After fecundation, an elongated creeping planula larva develops, completely devoid of cilia, incapable of swimming. When settled it develops into a polyp with eight primary tentacles. At the beginning of the metamorphosis the polyp begins to develop the secondary tentacles, gonads and gastric cirri into the gastric cavity, and can maintain (or not) the primary tentacles, often rearranging them into anchors. Some species can be observed in patchy groups, while others seem to be solitary. In at least one species an early larval stage has been identified, capable of repeatedly budding, thus giving rise to many creeping planulae.

DISTRIBUTION AND ECOLOGY

Most stalked jellyfish species occur at mid-latitudes (but also in cold waters, both Arctic and Antarctic), especially at latitudes ranging from 30 to 80°N (Miranda et al., 2018), with a peculiar distribution pattern, not in conformity with the classical one with an increase of biodiversity at the equator (distribution map). Specimens are frequently found on algae, but they can attach to rocks, seagrasses, shells, mud, sand, coral/gorgonian, sea cucumber and serpulid tubes. Most of the species are found in the intertidal and shallow subtidal regions (**Figure 10.16** for distribution map), but at least one genus, *Lucernaria*, has been observed at a depth of more than 3,000 m. They prey mainly on amphipods and copepods, and their predators include nudibranch mollusks, pycnogonids and fishes.

PHYSIOLOGY AND BEHAVIOR

Stalked jellyfishes use mainly the secondary tentacles (with higher concentration of nematocysts), to capture their prey. Stalked jellyfishes are benthic animals, normally attached to substrate for most of their lives. They use their peduncle to

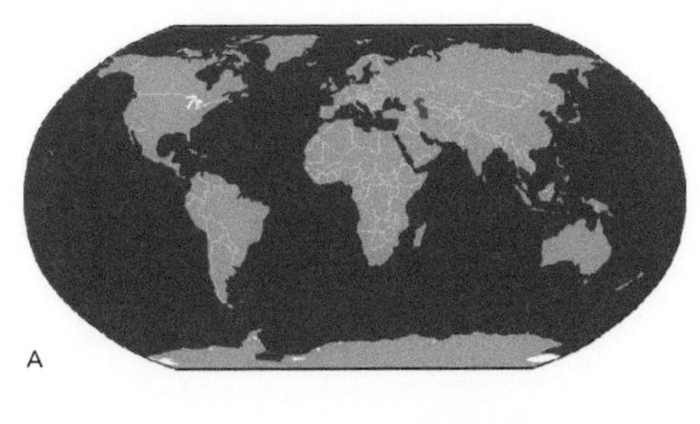

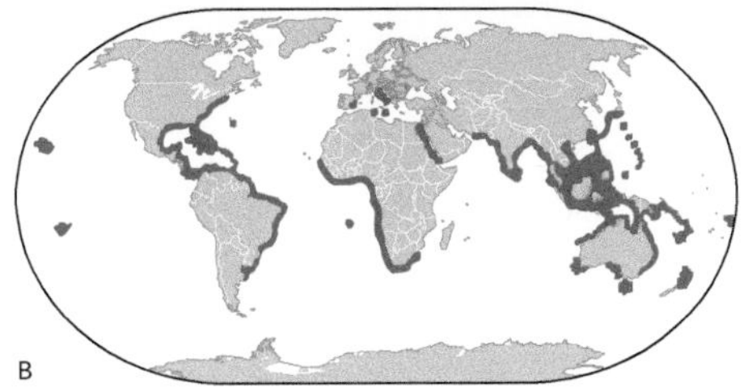

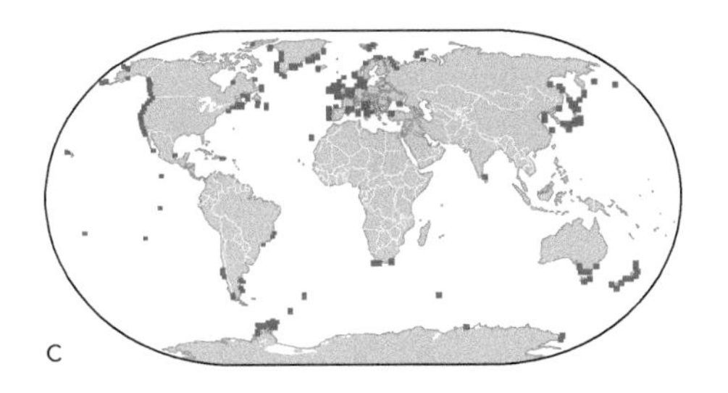

Figure 10.16 Distribution maps. (A) Scyphozoa, the presence in the Caspian Sea since 1999 evidenced in blue. (B) Cubozoa (C) Staurozoa. Global distributions of (A), (B), and (C). (http://marinespecies.org/aphia.php?p=taxdetails&id=1267)

attach to seaweeds, seagrasses, rocks, or other substrates, but they can detach themselves for short periods, floating freely in the water, or moving with the help of the tentacles, anchors, and pad-like structures, moving with a kind of somersaults, eventually reattaching with the peduncle (Miranda et al., 2018).

Their diet consists mainly of copepods, amphipods, isopods, juvenile decapods and ostracods.

GENETICS AND GENOMICS

The only staurozoan species with whole genome sequences is *Calvadosia cruxmelitensis* and its genome size is C-value = 0.21 pg (calculated from 209.4 Mb), while *Haliclystus antarcticus* is the only one staurozoan with a sequenced mitochondrial genome. The complete genome consists of 13 protein coding genes, seven transfer RNAs (tRNA) and two ribosomal RNAs genes (rRNA, Li et al., 2016). Mitochondrial markers (COI and 16S) and nuclear markers (ITS, 18S and 28S) were recently utilized in phylogenetic inference of staurozoan systematics (Miranda et al., 2016a). Recent analysis of mitochondrial genome data supports a clade of Staurozoa, Cubozoa and Scyphozoa (Zapata et al., 2015).

POSITION IN THE ToL

The positions of all three of these classes are at present under debate concerning their location within the clade Medusozoa (all cnidarians with a medusa stage), generally considered monophyletic (Kayal et al., 2018, reviving the term Acraspeda = Scyphozoa + Cubozoa + Staurozoa), but still with uncertainties at lower-level clades. Scyphozoa: Coronatae are considered a homogeneous taxon; not so Semaeostomeae, which seem to be paraphyletic (Collins, 2002; Daly et al.,

2007; Bayha et al., 2014) in respect of Rhizostomeae. Concerning the Rhizostomeae, Thiel (1970) proposed to split them, Rhizostomida and Cepheida (Straehler-Pohl, 2017), but to date the traditional Rhizostomeae order is still in use.

While traditional approaches support cubozoans as belonging within Scyphozoa, observations on the development of the jellyfish from the polyp have pushed us to consider Cubozoa as a separate class. Further research on the mode of formation of jellyfish led to the proposition to reinsert this taxon within the Scyphozoa class (Straehler-Pohl, 2017), but phylogenomic analyses of relationships within Cnidaria (Zapata et al., 2015; Kayal et al., 2018), tend on the contrary to support a clade composed of Staurozoa as sister group of Cubozoa and Scyphozoa.

Again, recent molecular and morphological systematics analyses see Staurozoa as a separate, independent sister group of the other medusozoans, that probably arose earlier than the taxa with swimming stage (Zapata et al., 2015; Kayal et al., 2018).

The synthetic phylogenetic tree at the end of this section (**Figure 10.17**) illustrates the possible phylogenetic relationships of these taxa, at least based on the current available knowledge (the asterisk indicate doubtful taxa).

DATABASES AND COLLECTIONS

Normally the specimens used to establish new species are deposited in museums around the world; however, establishing a holotype and depositing it in a museum was not yet common practice in the 18th and 19th centuries.

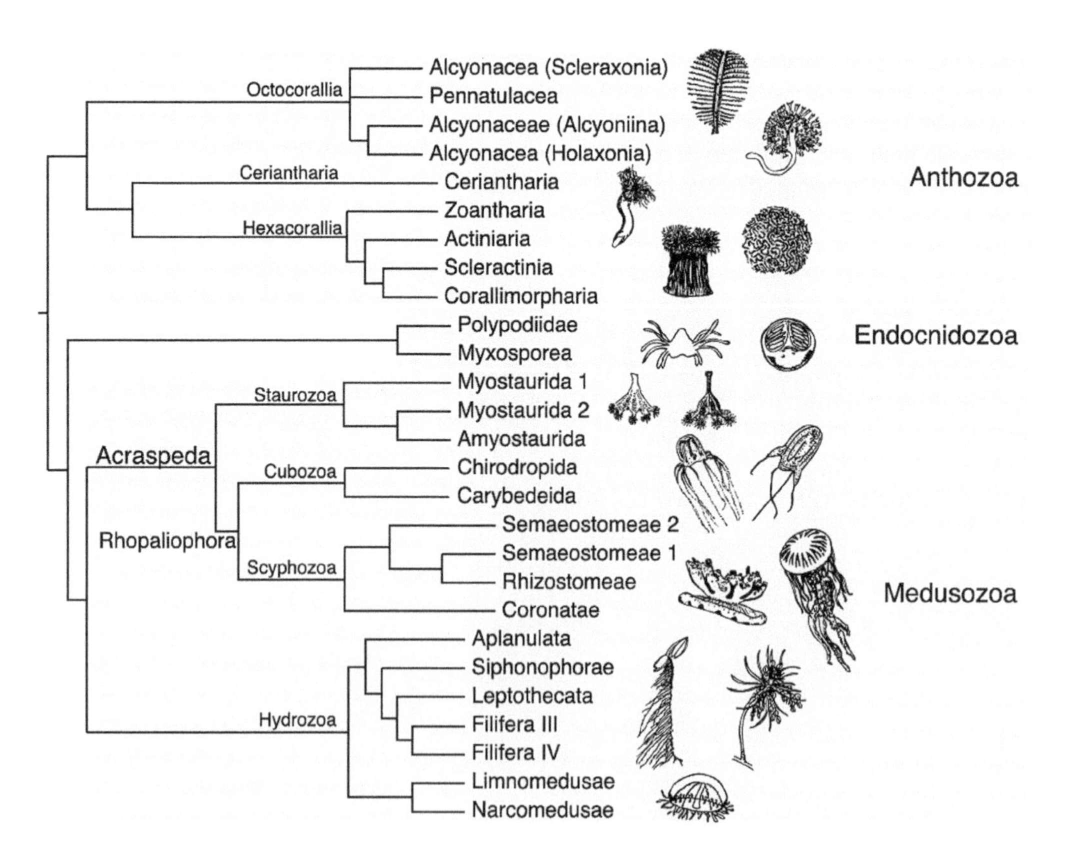

Figure 10.17 Hypothesis for cnidarian phylogeny based on phylogenomic analyses of new and existing genome-scale data that includes representatives of all cnidarian classes. (From Kayal E., Bentlage B., Pankey M.S., Ohdera A. H., Medina M., Plachetzki D. C., Collins A. G., Ryan J. F., 2018. Phylogenomics provides a robust topology of the major cnidarian lineages and insights on the origins of key organismal traits. BMC Evolutionary Biology 18:68, *Figure 10.7*.)

TABLE 10.1 Resources for Scyphozoa, Cubozoa and Staurozoa
Web Sites
Animal genome size database http://www.genomesize.com
https://boxjellies.weebly.com/index.html
http://dryades.units.it/jelly/index.php
GenBank https://www.ncbi.nlm.nih.gov/genbank/
http://www.marinespecies.org/aphia.php?p=taxdetails&id=265044
Organelle Genome Resources https://www.ncbi.nlm.nih.gov/genome/organelle/
https://www.reed.edu/biology/professors/srenn/pages/teaching/web_2010/mi_site/index.html
http://staurozoa.myspecies.info/
http://www.ucmp.berkeley.edu/cnidaria/cubozoalh.html

Furthermore, jellyfish, even if fixed in formalin or in ethanol, are not particularly durable samples, and tend sometimes to deteriorate over time. Moreover, several specimens have been destroyed from the devastation of the world wars. Consequently, several species established in past centuries no longer have their holotypes. Luckily, there are various synopses on scyphozoan, cubozoan, etc., both on paper and in digital format, some all-inclusive (at least at the time of printing). Noteworthy is the very recent print release of a world atlas of jellyfish, a monumental work that covers Scyphozoa and Cubozoa (Jarms and Morandini, 2019), a reliable reference work for future years. A pleasant introductory reading on jellyfish, popular and scientifically correct, can be found in Gershwin (2013, 2016). In this chapter's Bibliography some links to specific websites are given. We also list web resources for these groups of animals in **Table 10.1**.

CONCLUSION

Overall, the clade Scyphozoa + Cubozoa + Staurozoa appears to be monophyletic. However, the polarity of these taxa remains to be clearly defined (Staurozoa sister group of Scyphozoa + Cubozoa? or Hydrozoa + Scyphozoa + Cubozoa? etc). The discrepancies present in the phylogenetic studies of the last 20 years are probably linked to the still poor or absent phylogenetic assignment of many species. Most commonly used are mitochondrial genetic markers (COI and 16S) and nuclear markers ITS, 18S, and 28S. Moreover, there is a certain concern on quality control on sequences deposited in the available databases which is still low and without strict taxonomic attributions. The morphological-developmental features can be lent to misinterpretation, confusing symplesiomorphies with synapomorphies, i.e. is the process of strobilation symplesiomorphic with the cubozoan metamorphosis process considered as an apomorphism (Straehler-Pohl, 2017), or is the opposite true? Polydisc strobilation is likely ancestral within Scyphozoa, and monodisc strobilation is derived in at least two independent moments, once in *Sanderia* and once in Kolpophorae (Helm, 2018). The most recent revision (Kayal et al., 2018) considers Staurozoa as member of a monophyletic group containing Cubozoa and Scyphozoa (**Figure 10.18**), historically grouped in a clade, the Gegenbaur's Acraspeda, and sister group of Cubozoa + Scyphozoa, a clade named Rhopaliophora. If this interpretation is correct, then the (more or less) complete metamorphosis of the Staurozoa and the Cubozoa (complete-partial-monodisk like), could be considered as a symplesiomorphism, with the true strobilation *with the ephyra stage* a synapomorphism exclusive to Scyphozoa *sensu stricto*.

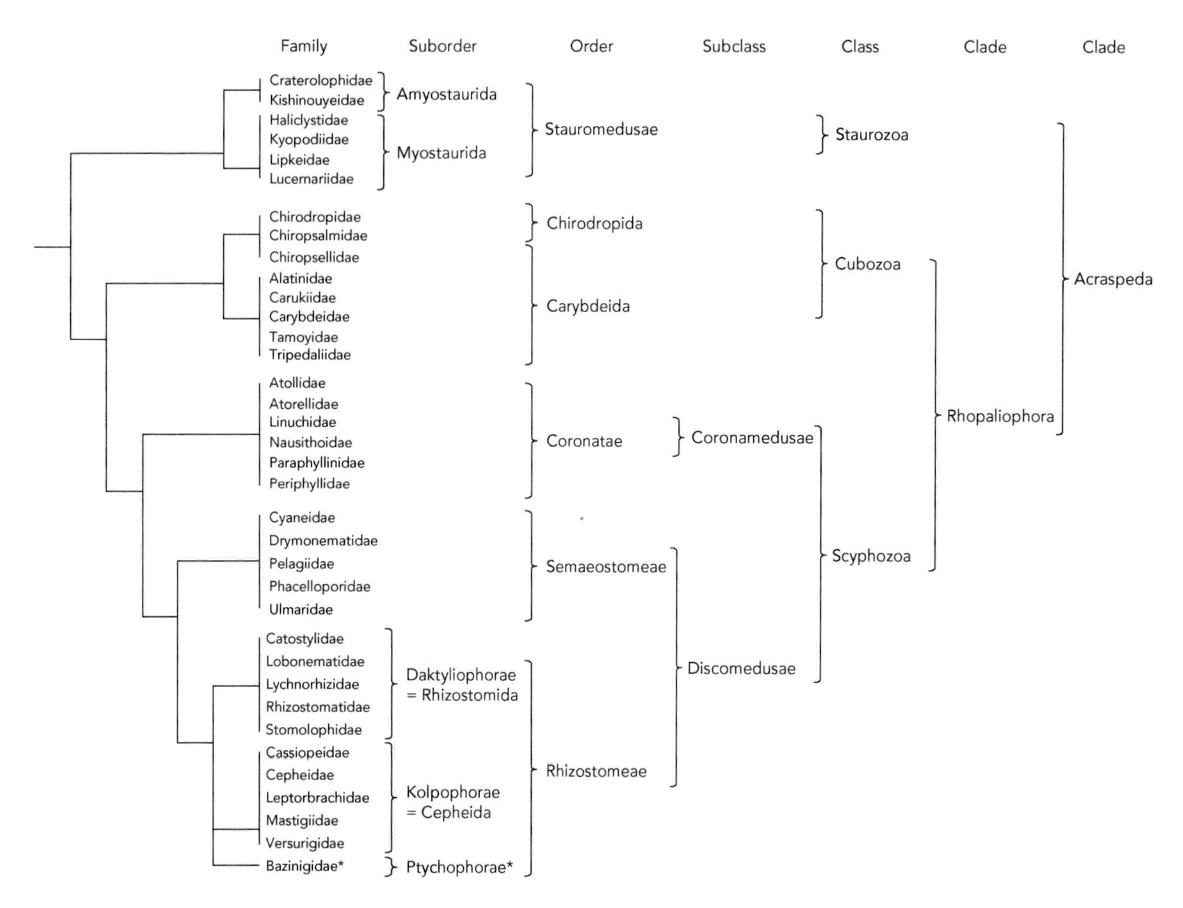

Figure 10.18 Classes Scyphozoa, Cubozoa and Staurozoa.

REFERENCES

Albert, J. D., 2011. What's on the mind of a jellyfish? A review of behavioural observations on Aurelia sp. jellyfish. Neuroscience and Behavioral Reviews 35, 474–482.

Adachi K., Miyake H., Kuramochi T., Mizusawa K., Okomura S., 2017. Genome size distribution in phylum Cnidaria. Fish Sci 83, 107–112.

Arai M.N., 1997. A Functional Biology of Scyphozoa. Chapman & Hall, 316 pp.

Avian M., Ramšak A., Tirelli V., D'Ambra I., Malej A., 2016. Redescription of *Pelagia benovici* into a new jellyfish genus, *Mawia*, gen. nov., and its phylogenetic position within Pelagiidae (Cnidaria: Scyphozoa: Semaeostomeae). Invertebrate Systematics 30, 523–546.

Bayha, K. M., Dawson M. N., 2010. New Family of Allomorphic Jellyfishes, Drymonematidae (Scyphozoa, Discomedusae), Emphasizes Evolution in the Functional Morphology and Trophic Ecology of Gelatinous Zooplankton. Biol Bull 219: 249–267.

Bayha K. M., Dawson M. N., Collins A. G., Barbeitos M. S., Haddock S. H. D., 2014. Evolutionary relationships among scyphozoan jellyfish families based on complete taxon sampling and phylogenetic analyses of 18S and 28S ribosomal DNA. Integrative and Comparative Biology, 50(3): 436–455.

Bentlage B., Peterson A. T., Cartwright, P., 2009. Inferring distributions of chirodropid box-jellyfishes (Cnidaria: Cubozoa) in geographic and ecological space using ecological niche modeling. Mar. Ecol. Prog. Ser. 384, 121–133. DOI:10.3354/meps08012.

Brusca R.C., and Brusca G. J., 1990. Invertebrates. Sinauer Associates, Sunderland, Massachusetts, 1008.

Canepa A., Fuentes V., Sabatés A., Piraino S., Boero F., Gili J.-M., 2014. *Pelagia noctiluca* in the Mediterranean Sea. In: Pitt K. A., Lucas C. H. (eds.) Jellyfish Blooms, Springer Netherlands, Dordrecht, Chapter 11, 237–266.

Collins A. G., 2002. Phylogeny of Medusozoa and the evolution of cnidarian life cycles. J Evol Biol 15, 418–432.

Collins A. G., Schuchert P., Marques A. C., Jankowski T., Medina M., Schierwater B., 2006. Medusozoan phylogeny and character evolution clarified by large and small subunit rDNA data and an assessment of the utility of phylogenetic mixture models. Syst Biol 55, 97– 115.

Daly M., Brugler M. R., Cartwright P., Collins A. G., Dawson M. N., Fautin D. G., France S. C., McFadden C. S., Opresko D. M., Rodriguez E., Romano S.L., Stake J. L., 2007. The phylum Cnidaria: a review of phylogenetic patterns and diversity 300 years after Linnaeus. Zootaxa 1668, 127–182.

Dawson M. N., 2003. Macro-morphological variation among cryptic species of the moon jellyfish, Aurelia (Cnidaria: Scyphozoa). Mar Biol 143, 369–379.

Dawson M. N., 2005. Renaissance taxonomy: integrative evolutionary analyses in the classification of Scyphozoa J Mar Biol Ass UK 85, 733–739.

Dawson M. N., Jacobs D. K., 2001. Molecular evidence for cryptic species of *Aurelia aurita* (Cnidaria, Scyphozoa). Biol Bull 200, 92–6.

Diupotex-Chong, M. E., Ocaña-Luna, A., Sánchez-Ramírez, M., 2009. Chromosome analysis of *Aurelia aurita* Linné, 1758 (Scyphozoa: Ulmaridae), southern Gulf of Mexico. Marine Biology Research 5 (4), 399–403.

Ferrier, D. E. K., Holland, P. W. H., 2001. Ancient origin of the Hox gene cluster. Nat. Rev. Genet. 2, 33–38.

García-Rodríguez J., Lewis Ames C., Marian J. E. A. R., Marques A. C., 2018. Gonadal histology of box jellyfish (Cnidaria: Cubozoa) reveals variation between internal fertilizing species *Alatina alata* (Alatinidae) and *Copula sivickisi* (Tripedaliidae). J Morphol 2018, 1–16.

Garm A., Poussart Y., Parkefelt L., Ekström P., Nilsson D.-E., 2007. The ring nerve of the box jellyfish *Tripedalia cystophora*. Cell Tissue Res 329, 147–157.

Garm A., Lebouvier M., Tolunay D., 2015. Mating in the box jellyfish *Copula sivickisi*—Novel function of cnidocytes. J Morphol. 276, 1055–1064.

Gold, D. A., Katsuki, T., Li, Y., Yan, X., Regulski, M., Ibberson, D., Holstein, T., Steele, R. E., Jacobs, D. K., Greenspan, R. J., 2019. The genome of the jellyfish *Aurelia* and the evolution of animal complexity. Nature Ecology & Evolution 3, 96–104.

Goldberg, R. B., Crain, W. R., Ruderman, J. V., Moore, G. P., Barnett, T. R., Higgins, R. C., Gelfand, R.A., Galau, G. A., Britten, R. J., and Davidson E. H., 1975. DNA sequence organization in the genomes of five marine invertebrates. Chromosoma 51, 225–251.

Gómez Daglio L., Dawson M. N., 2019. Integrative taxonomy: ghosts of past, present and future. Journal of the Marine Biological Association of the United Kingdom 1–10. DOI: 10.1017/S0025315419000201.

Haddock S. H. D., Moline M. A., Case J. F., 2001. Bioluminescence in the Sea. Annu Rev Mar Sci 2, 443–493.

Helm R. R., 2018. Evolution and development of scyphozoan jellyfish. Biol Rev 93, 1228–1250.

Herring, P. J., Widder E. A., 2004. Bioluminescence of deep-sea coronate medusae (Cnidaria: Scyphozoa). Mar Biol 146, 39–51.

Jarms G., Bamstedt U., Tiemann H., Martinussen M. B., Fossa J. H., 1999. The holopelagic life cycle of the deep-sea medusa *Periphylla periphylla* (Scyphozoa, Coronatae). Sarsia 84, 55–65.

Kayal E., Bentlage B., Collins A. G., Kayal M., Pirro S., Lavorov D. V., 2012. Evolution of linear mitochondrial genomes in medusozoan cnidarians. Genome Biol Evol 4, 1–12.

Kayal E., Bentlage B., Pankey M. S., Ohdera A. H., Medina M., Plachetzki D. C., Collins A. G., Ryan J. F., 2018. Phylogenomics provides a robust topology of the major cnidarian lineages and insights on the origins of key organismal traits. BMC Evolutionary Biology 18, 68.

Kingsford M. J., Mooney C. J., 2014. The Ecology of Box Jellyfishes (Cubozoa). In Pitt K. A., Lucas C. H. (eds.), Jellyfish Blooms, Springer Netherlands, Dordrecht, 267–302.

Korsun S., Fahrni J. F., Pawlowski J., 2012. Invading *Aurelia aurita* has established scyphistoma populations in the Caspian Sea. Mar Biol 159, 1061–1069.

Kramp, P. L., 1961. Synopsis of the medusae of the world. J Mar Biol Assoc UK 40, 292–303.

Li H.-.H., Sung P.-J.., Ho H.-C., 2016. The complete mitochondrial genome of the Antarctic stalked jellyfish, *Haliclystus antarcticus* Pfeffer, 1889 (Staurozoa: Stauromedusae). Genomics Data 8, 113–114.

Marques A. C., Collins A. G., 2004. Cladistic analysis of Medusozoa and cnidarian evolution. Inv. Biol 123, 32–42.

Martellos S., Ukosich L., Avian M., 2016. JellyWeb: an interactive information system on Scyphozoa, Cubozoa and Staurozoa. Zookeys 554, 1–25.

Mayer A. G., 1910. Medusae of the world. Volume III. Scyphomedusae. Carnegie Institution publishing, Washington, 109: 499–735.

Miranda L. S., Hirano Y. M., Mills C. E., Falconer A., Fenwick D., Marques A. C., Collins A. G., 2016a. Systematics of stalked jellyfishes (Cnidaria: Staurozoa). PeerJ 4, e1951.

Miranda L. S., Collins A. G., Hirano Y. M., Mills C.E., Marques A. C., 2016b. Comparative internal anatomy of Staurozoa (Cnidaria), with functional and evolutionary inferences. PeerJ 4, e2594.

Miranda L. S., Mills C. E., Hirano Y. M., Collins A. G., Marques A. C., 2018. A review of the global diversity and natural history of stalked jellyfishes (Cnidaria, Staurozoa). Mar Biodiv 48, 1695–1714.

Pauly D., Graham W., Libralato S., Morissette L., Deng Palomares M. L., 2009. Jellyfish in ecosystems, online databases and ecosystem models. Hydrobiologia 616, 67–85.

Piraino S., Aglieri G., Martell L., Mazzoldi C., Melli V., Milisenda G., Scorrano S., Boero F. 2014. *Pelagia benovici* sp. nov. (Cnidaria, Scyphozoa): a new jellyfish in the Mediterranean Sea. Zootaxa 3794 (3), 455–468.

Ramšak A., Stopar K.; Malej A., 2012. Comparative phylogeography of meroplanktonic species, *Aurelia* spp. and *Rhizostoma* pulmo (Cnidaria: Scyphozoa) in European Seas. Hydrobiologia 690, 69–80. DOI: 10.1007/s10750-012-1053-9.

Rottini Sandrini L., Avian M., 1983. Biological cycle of *Pelagia noctiluca*: morphological aspects of the development from planula to ephyra. Mar Biol 74, 169–174.

Russell F. R. S., 1970. The Medusae of the British Isles. II. Pelagic Scyphozoa with a supplement to the first volume on Hydromedusae. Cambridge Univ. Press: 1–284.

Skogh C., Garm A., Nilsson D. E., Ekström P., 2006. The bilateral symmetric rhopalial nervous system of box jellyfish. J Morphol,1391–1405. DOI: 10.1002/jmor.10472.

Scorrano S., Aglieri G., Boero F., Dawson M. N., Piraino S., 2017. Unmasking *Aurelia* species in the Mediterranean Sea: an integrative morphometric and molecular approach. Zoological Journal of the Linnean Society 180, 243–267. DOI: 10.1111/zoj.12494.

Shao Z., Graf S., Chaga O. Y., Lavrov D. V., 2006. Mitochondrial genome of the moon jelly *Aurelia aurita* (Cnidaria, Scyphozoa): a linear DNA molecule encoding a putative DNA-dependent DNA polymerase. Gene 381, 92–101.

Smith, D. R., Kayal, E., Yagihara, A. A., Collins, A. G., Pirro, S., Keeling, P. J., 2012. First complete mitochondrial genome sequence from a box jellyfish reveals a highly fragmented linear architecture and insights into telomere evolution. Genome Biology and Evolution 4(1), 52–58.

Straehler-Pohl I., Widmer C. L., Morandini A. C., 2011. Characterizations of juvenile stages of some semaeostome Scyphozoa (Cnidaria), with recognition of a new family (Phacellophoridae). Zootaxa 2741, 1–37.

Straehler-Pohl, I., 2017. Cubozoa and Scyphozoa: The results of 20 years of scyphozoan life cycle research with new results on cubozoan life cycles to suggest a new nomenclature referring to both classes. In: Toyokawa, M., Miyake, H., Nishikawa, J. (eds.), Frontiers in Ecological Studies of Jellyfish. Seibutsu Kenkyu Sha Co. Ltd. (Organisms Research Co. Ltd.), Tokyo, 17–29.

Zapata F., Goetz F. E., Smith S. A., Howison M., Siebert S., Church S. H., et al., 2015. Phylogenomic Analyses Support Traditional Relationships within Cnidaria. PLoS One 10(10), e0139068.

Additional Reading

Gershwin L. A., 2013. Stung! On Jellyfish Blooms and the Future of the Ocean. Univ. Chicago Press, 1–456.

Gershwin L. A., 2016. Jellyfish a Natural History. Univ. Chicago Press, 1–224.

Jarms G., Morandini A. C., 2019. World Atlas of Jellyfish. Scyphozoa except Stauromedusae. Dölling und Galitz Verlag, Hamburg, 815 pp.

PHYLUM CNIDARIA

CLASS ANTHOZOA

Andrea M. Quattrini, Danielle M. DeLeo, and Mercer R. Brugler

CHAPTER

11

Infrakingdom	Bilateria
1st Division	Diploblastic Animals
Phylum	Cnidaria
Class	Anthozoa

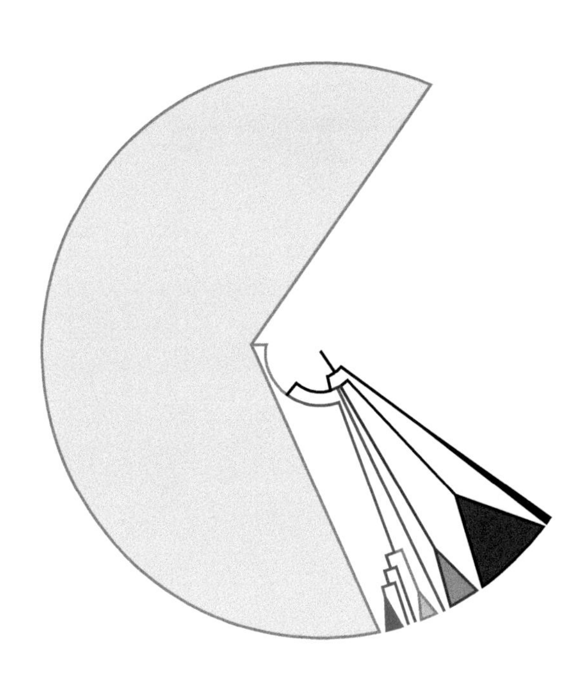

CONTENTS

General Characteristics

Number of valid extant species/estimated number of "cryptic species"

Hexacorallia

Actiniaria[1]: 1,000/300–500

Antipatharia[2]: 250/75–100

Ceriantharia[3]: 55/30–40

Corallimorpharia[4]: 46/~20

Scleractinia[5,A]: 735/200 azooxanthellate; 865/100 zooxanthellate

Zoantharia[6]: 450-500/700–1,000

Octocorallia

Alcyonacea[7,9]: 2,700/Unknown

Helioporacea[7,9]: 6/Unknown

Pennatulacea[8]: 200/Unknown

Estimates per: [1]Meg Daly (The Ohio State University); [2]Dennis Opresko (Smithsonian Institution National Museum of Natural History), Tina Molodtsova (P.P. Shirshov Institute of Oceanology), Marzia Bo (Università degli Studi di Genova) and Mercer R. Brugler (University of South Carolina Beaufort); [3]Sergio Stampar (São Paulo State University); [4]Daphne Fautin (University of Kansas); [5]Stephen Cairns (Smithsonian Institution National Museum of Natural History) and Bert Hoeksema (Naturalis Biodiversity Center); [6]James Reimer (University of the Ryukyus); [7]Catherine McFadden (Harvey Mudd College); [8]Gary Williams (California Academy of Sciences); [9]Andrea Quattrini (Smithsonian Institution National Museum of Natural History), [A]The family Dendrophyllidae contains both zooxanthellate and azooxanthellate (apozooxanthellate) species.

- Anthozoans occupy shallow waters to the abyss in tropical, temperate, and polar environments worldwide.
- Ecologically important foundation species, supporting most of the oceans' biodiversity.
- Some corals may reach more than 1,000 years old.
- Skeletons of some corals are turned into jewelry and living bone implants.
- Excretion of secondary metabolites with antimicrobial activity.
- Many anthozoans have photosynthetic algae, known as zooxanthellae.
- The deep-sea octocoral *Metallogorgia melanotrichos* has a "partner for life", the obligate symbiont *Ophiocreas oedipus.*
- Slow rates of mitochondrial genome evolution, 50–100× slower than most other animals.
- Mitochondrial genomes in octocorals include a putative mismatch repair gene and several gene order rearrangements.
- Multiple gains and losses of homing endonucleases as group I introns in the mitochondrial genomes of many hexacorals.

Synapomorphies

- actinopharynx – tube that projects into gastrovascular cavity
- siphonoglyph – ciliated groove in the gastrovascular cavity
- mesenteries – radially arranged sheets of tissue that extend from the body wall into the actinopharynx

HISTORY OF TAXONOMY AND CLASSIFICATION

The class Anthozoa (i.e. the flower animals), which contains approximately 7,500 extant species (Daly et al. 2007), is divided into the subclasses Hexacorallia (animals with six, or multiples of six, tentacles; ~4,300 species) and Octocorallia (animals with eight pinnate tentacles; ~3,000 species). Traditionally, classification of Anthozoa has been based on morphologies such as skeletal morphology, colony organization and soft-tissue anatomy of the polyps (Daly et al. 2007), including

Figure 11.1 Photo of a solitary anemone in the order Actiniaria attached to a deep-sea coral in the sub-class Octocorallia. Photo courtesy of the NOAA Okeanos Explorer Program.

the arrangement of internal mesenteries (Fautin and Mariscal 1991). The two subclasses, Octocorallia and Hexacorallia, are reciprocally monophyletic (Daly et al. 2007), and classifications are supported by morphological characters and phylogenomic datasets (Zapata et al. 2015; Pratlong et al. 2017; Quattrini et al. 2018; Kayal et al. 2018, Quattrini et al. 2020). The Octocorallia is comprised of three orders, including the Alcyonacea (soft corals and sea fans), Helioporacea (blue corals), and Pennatulacea (sea pens). The Hexacorallia is comprised of six orders, including the Actiniaria (sea anemones, **Figure 11.1**), Antipatharia (black corals), Ceriantharia (tube anemones), Corallimorpharia (corallimorphs), Scleractinia (stony corals), and Zoantharia (zoanthids). Black corals and tube anemones were formerly classified together in the subclass Ceriantipatharia based on the similarity of the cerianthid cerinula larva to the antipatharian adult polyp, and weak and indefinite musculature of the mesenteries. Molecular work by Brugler and France (2007), however, provided evidence that each group was unique enough to warrant two separate orders.

With the advent of NGS technologies and phylogenomic analyses of 100s to 1,000s of genes, the monophyly of each hexacoral order has been confirmed (Zapata et al., 2015; Kayal et al., 2018; Quattrini et al. 2018, 2020), with the exception of the Actiniaria. The monophyly of sea anemones was recently challenged with the discovery of *Relicanthus daphneae* (Rodriguez et al., 2014); however, this animal was originally described as an anemone (*Boloceroides daphneae*) by Daly (2006) and in a recent whole mitochondrial genome study (Xiao et al. 2019). A recent phylogenomic study by Quattrini et al. (2020), however, did not recover *R. daphneae* as monophyletic with other actiniarians. Molecular datasets have produced different results depending on gene region used, and many studies have encompassed limited taxonomic sampling (with exception of Quattrini et al. 2020) and many of these studies have encompassed limited taxonomic sampling. For example, the placement of Ceriantharia within the Anthozoa has been the subject of much debate. Stampar et al. (2014) suggested that cerianthids may represent either a new anthozoan subclass or a sister group to the Anthozoa based on mitochondrial 16S and nuclear 18S and 28S data. The recent discovery of multipartite linear mitochondrial genomes in select cerianthids may support that Ceriantharia is sister to the Anthozoa (Stampar et al. 2019). Recent phylogenomic analyses, however, indicated that Ceriantharia is in fact sister to the remaining hexacorals (Kayal et al. 2018; Quattrini et al. 2020), although other phylogenomic studies noted that the position of Ceriantharia remained unstable as different datasets produced different topologies (Zapata et al. 2015; Quattrini et al. 2018). Based on complete mitochondrial genomic data, corallimorphs were once thought to be "naked corals" within the Scleractinia (Medina et al. 2006), which suggested that the hard skeleton had simply been lost; however, subsequent work by Lin et al. (2016) using ~300 nuclear protein-coding genes showed that the non-calcifying corallimorphs form a monophyletic clade sister to the Scleractinia. Using a target-capture approach of 100s of ultraconserved element (UCE) and exon loci, Quattrini et al. (2020) corroborated that corallimorphs do not appear to be "naked corals". The position of Zoantharia has also been debated, but recently Quattrini et al. (2020) showed that zoantharians were sister to all hexacoral orders, with the exception of Ceriantharia.

Phylogenetic relationships within the Octocorallia remain perhaps the least resolved among all anthozoans. Molecular phylogenies indicate that both Pennatulacea and Helioporacea orders are nested within the order Alcyonacea (McFadden et al. 2006, 2010; Quattrini et al. 2020). Within the morphologically diverse Alcyonacea, six subordinal groups have been defined on the basis of skeletal composition and colony architecture (see Daly et al. 2007; McFadden et al. 2010), but molecular phylogenies indicate that none of these subordinal groups are monophyletic. Molecular phylogenies suggest instead the presence of two well-supported lineages: the Holaxonia–Alcyoniina clade and the Calcaxonia–Pennatulacea clade (McFadden et al. 2006, 2010; Brockman and McFadden 2012; Quattrini et al. 2018, 2020). Widespread homoplasy in skeletal characters has led to a lack of morphological synapomorphies that currently define these groupings.

ANATOMY, HISTOLOGY AND MORPHOLOGY

Anthozoa are considered 'simple' animals as they contain only two layers of cells, an inner layer called endoderm and an outer layer called ectoderm (**Figure 11.2**). These two cell layers are separated by an acellular jelly-like substance called mesoglea. Cnidarians are known for their unique dimorphic lifestyle (i.e., alternating between a benthic polyp and pelagic medusae); however, anthozoans spend their entire lives in the polypoid stage. A polyp is a tubelike sac (the gastrovascular cavity) with a single opening at the top of the sac that serves as both the mouth and anus and thus is called the mouth-anus (Figures 11.1, 11. 2). Surrounding the oral disc is a ring of tentacles that is armed with stinging cells (cnidocytes) to capture prey or repel predators or competitors. Octocorals have eight tentacles that are pinnate or subdivided into smaller perpendicular ramifications that give a feather-like appearance. Three types of cnidae are found within the Anthozoa: nematocysts (venomous; all anthozoans), spirocysts (agglutinant and non-venomous; Hexacorallia only), and ptychocysts (enveloping; used to construct the tube; Ceriantharia only). In general, anthozoans are predators but may obtain supplemental nutrition through a symbiotic relationship with intracellular photosynthetic dinoflagellates (Symbiodiniaceae) called zooxanthellae (LaJeunesse et al. 2018). Black corals, which have a preference for low-light environments and are considered a deep-water group, were not known to harbor zooxanthellae until a study by Wagner et al. (2011)

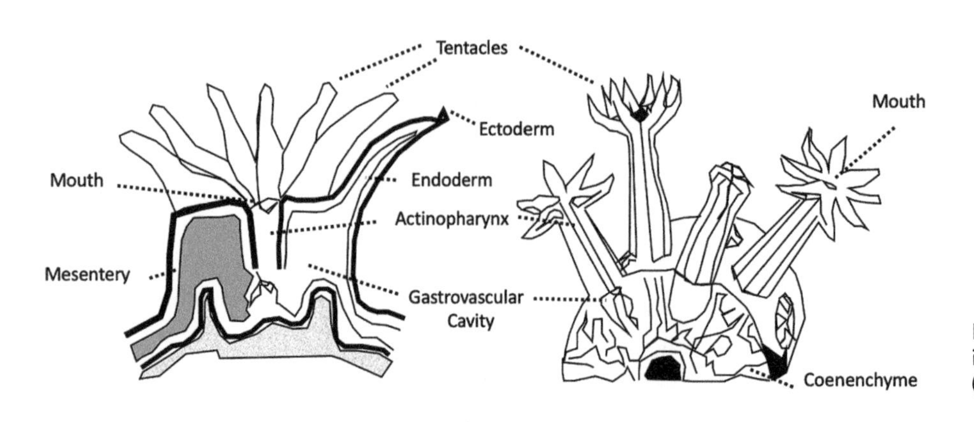

Figure 11.2 Schematic drawing of an idealized hexacoral (left) and octocoral (right). Body parts are labeled in the figure.

elucidated zooxanthellae in several species from Hawaii (including the deepest record for Symbiodiniaceae to date at 396 meters depth). There have been published examples of select anthozoans swimming to escape predators (anemones; Robson 1961; Lawn and Ross 1982), somersaulting to change locations (similar to the hydrozoan *Hydra*; Passano and McCullough 1964) or catching a ride on a crab's claws (Schnytzer et al. 2017); however, the polyp is nonetheless considered benthic and sedentary.

Anthozoans are either colonial, clonal, or solitary and, in certain groups, secrete a skeleton composed of calcium carbonate (select Octocorallia [High-Mg calcite, aragonite]; Scleractinia [aragonite]) or protein (Antipatharia [antipathin + chitin]; select Octocorallia [gorgonin]) (Daly et al 2007). Those groups that do not secrete a hard skeleton use a hydrostatic skeleton instead. Closing the sphincter and contracting muscles allows the water contained in the gastrovascular cavity to serve as a hydro-skeleton. The tissue of octocorals can be supported with microscopic calcareous sclerites. The shape and size of sclerites oftentimes aid in identification at the generic and/or species level (Lewis and Wallis 1991). Morphologically, there are three autapomorphies that are unique to the Anthozoa: the actinopharynx (analogous to an esophagus), siphonoglyph (ciliated groove in the actinopharynx), and mesenteries (sheets of tissue that partition the gastrovascular cavity and extend to varying degrees from the body wall to the actinopharynx) (Daly et al. 2007). Although anthozoans are externally radially symmetrical animals, the siphonoglyph imparts bilateral symmetry on the polyps. The mesenteries contain musculature (including retractor muscles to assist with contraction) and gametogenic tissue, and are arranged in cycles (Daly et al. 2003, 2007).

Several studies have revealed why using external morphology alone can be misleading when elucidating a species' taxonomy and phylogeny. Brugler et al. (2013) found that multiple species within the black coral genus *Stichopathes* group in to two separate families based on genetic analysis, although they are morphologically similar (unbranched whips). Molodtsova and Budaeva (2007) demonstrated that the presence of a symbiotic polychaete altered the morphology of the skeleton and species-specific skeletal spines of several black corals. Paz-Garcia et al. (2015) documented a switching between described morphospecies due to changing environmental conditions in the reef-building coral *Pocillopora*. Cairns et al. (2009) found that a cryptic shallow-water anemone (*Mimetridium cryptum*) is morphologically highly similar to *Metridium senile*, although it is not closely related.

REPRODUCTION AND DEVELOPMENT

Reproduction in Anthozoa can occur either vegetatively ("asexually") or sexually, with most species capable of employing both methods (reviewed in Fautin 2002). Asexual reproduction occurs "vegetatively" involving the complete detachment of body segments from the animal to form genetically identical individuals or genets. Alternatively, sexual reproduction involves the fusion of gametes – sperm and egg – to form a distinct fertilized embryo which develops into a free-swimming larva. Completely lacking a medusa stage, larvae subsequently settle and develop into the polyp body form (see Ivanova-Kazas 1975).

Asexual ("Vegetative") Reproduction

All anthozoans are able to reproduce "asexually", with various species capable of achieving this mode of reproduction through several mechanisms. This often occurs through fission, or division of the body segments either transversely or longitudinally, or through pedal laceration in which parts of the pedal disc (foot) detach from the animal to form a new genet or clone (reviewed in Grassle and Shick 1979; Shick 1991). Among some species, like those in order Scleractinia, partial fragmentation of the animal due to storms or wave action can also result in "asexual" replication by fission if subsequent fragment recruitment is successful (e.g., Highsmith 1982; Wallace 1985; Smith and Hughes 1999; Lirman; 2000). Moreover, Sammarco (1982) described a unique escape response in the stony coral *Seriatopora hystrix* in which a polyp isolates itself, detaches from the colony, disperses passively using negative buoyancy, resettles, and initiates reformation of the skeleton.

Sexual Reproduction

Sexual reproduction allows for increased genetic variation via meiotic crossover events during the formation of gametes and through the combination of genetically distinct sperm and egg during the fertilization process. In Anthozoa, both the sperm and egg are produced in the gametogenic tissue of the mesenteries which extends into the gastrovascular cavity (Dunn 1975; Larkman 1983, 1984; Wedi and Dunn 1983). Unlike the spermatocytes (sperm cells) found within male gonads, female gonads typically contain oocytes (egg cells) at different stages of development (reviewed in Boscharova and Kozevich 2011), which can range greatly in size from approximately 60 μm for the actinarian *Aiptasia pulchella* (Chen, Soong and Chen 2008) to as large as 1,500 μm in the antipatharian *Dendrobathypathes grandis* (Lauretta and Penchaszadeh 2017). These differences are linked to alternative reproductive strategies and the tradeoff between the energy invested in egg size and number; egg size influences the overall number of offspring that an animal can produce, as well as the overall fitness of the offspring (Levitan 2006).

Sexually reproducing anthozoans may have separate sexes – male and female – that produce either sperm or egg and are considered gonochoristic. Alternatively, they can be hermaphroditic in which a single individual is capable of producing both types of gametes. In simultaneous hermaphroditism, an organism is able to produce both sperm and egg at the same time, potentially leading to self-fertilization (reviewed in Richmond 1997). In sequential hermaphroditism, the organism is functionally male first and later develops into a reproductive female (protandry) (e.g. the scleractinian *Stylophora pistillata*, Rinkevich and Loya 1979). Anthozoans may also exhibit two different modes of sexual reproduction, primarily differing in respect to how the gametes come into contact. The eggs of brooding species are fertilized internally, and the embryo typically develops into the larval stage within the polyp or on the surface of the polyp prior to release (Coma et al. 1995; Dahan and Benayahu 1997; Larson 2017. Conversely, broadcast spawning species release their gametes into the water column where fertilization and larval development occur externally (Lasker et al. 1996; Richmond 1997). Some hermaphroditic species may also employ a combination of these strategies, retaining eggs to be fertilized internally, and releasing sperm during synchronized mass spawning events (e.g., *Goniastrea aspera*; Babcock et al. 1986, Sakai 1997).

Although less common, larvae of some anthozoans may also develop directly from unfertilized eggs, in a process called parthenogenesis. This phenomenon was first observed among actiniarians (Ottaway and Kirby 1975; Black and Johnson 1979; Gashout and Ormond 1979), octocorals (Hartnoll 1975; Brazeau and Lasker 1989) and scleractinians (Stoddart 1983; Ayre and Resing 1986), but has since been reported in other anthozoans (see Fautin 2002).

Each egg develops into a ciliated, lipid-rich planula larva (both in bisexual or monosexual reproduction). The larvae of brooding species tend to be relatively large upon release and more readily capable of settling and recruiting nearby, though they also have more energetic resources to disperse long distances (Richmond 1987, 1997). Regardless of which reproductive mode is employed, the anthozoan planktonic larvae can employ chemoreceptors to find suitable settlement areas before undergoing metamorphosis. During this developmental transformation into their benthic, juvenile polyp form, larvae undergo various morphological and biochemical changes (reviewed in Richmond 1997).

Developmental Genes

The study of developmental genes in anthozoans has primarily utilized the starlet sea anemone *Nematostella vectensis*. Even though anthozoans are considered simple animals (i.e., a single body axis, two germ layers [referred to as diploblastic] and either two or three cell lineages [two epithelial and an interstitial-like cell called an amoebocyte]; Tucker et al. 2011; Gold and Jacobs 2013), a diverse array of metazoan developmentally regulated signaling pathways, as well as antagonists, are present in anthozoans, including Wnt, TGFβ (including Smad subfamilies), Hedgehog, Ras-MAPK, and Notch (Technau et al. 2005). Finnerty et al. (2004) showed that the bilateral symmetry observed in *Nematostella* is obtained by expressing a combination of five *Hox* genes (which determine patterning of the anterior–posterior axis) and a single *decapentaplegic* (*dpp*) ortholog (determines patterning of the dorsal–ventral axis) across the primary and secondary body axes (respectively). The 12 *Wnt* genes found in *Nematostella* had expression patterns in a planula larva that suggests *Wnt* plays a role in gastrulation and axial differentiation (Kusserow et al. 2005). DuBuc et al. (2018) confirmed the interaction of select *Hox* genes (*NvAx6* [a putative ortholog of *Hox1*] and *NvAx1* [a central/posterior *Hox* gene]) with *Wnt* genes to pattern the oral–aboral axis in *Nematostella*. *Nematostella Hox-Gbx* genes are also responsible for segmentation of larval endoderm that ultimately determines the patterning of tentacles and mesenteries (*Gastrulation brain homeobox*, or *Gbx*, is a *Hox*-linked subfamily gene, He et al. 2018). Although anthozoans lack a mesoderm, Martindale et al. (2004) found the expression of homologous genes in *Nematostella*, six expressed in the endodermal layer (*twist*, *snailA*, *snailB*, *forkhead*, and GATA and LIM transcription factors) and one expressed in the ectoderm (*mef2*), all of which determine the specification and differentiation of mesodermal cell types in bilaterians and are predominantly expressed in the mesoderm. These endodermal expression patterns in *Nematostella* suggest that the mesodermal layer of more derived animals may have arisen from the endoderm of a diploblastic ancestor. Supporting this conclusion is work by Steinmetz et al. (2017) who constructed a fate map of germ layers, localized gut cell types, and analyzed the expression of germ layer-specific transcription factors, and found that in *Nematostella* the endoderm is homologous to bilaterian mesoderm. These studies demonstrate that the genetic complexity thought to have arisen in more derived animals has its origin in more ancestral animals, and for some gene families, anthozoans display greater diversity than more derived lineages such as *Drosophila melanogaster* and *Caenorhabditis elegans* (Kusserow et al. 2005).

DISTRIBUTION AND ECOLOGY

Anthozoans inhabit all oceans and live from the shoreline down to the hadal zone (**Figure 11.3**). The black coral *Schizopathes affinis* was collected from the Kuril–Kamchatka trench at 8,460 m depth (Molodtsova and Opresko 2017) and the sea anemone *Galatheanthemum profundale* was collected from the Marianas Trench at 10,710 m depth (Cairns et al. 2007). Corals form dense aggregations on shallow water reefs but are generally found as singletons in the deep sea; however, dense monotypic aggregations (i.e., coral gardens or forests) that extend over large areas (e.g., *Antipathella subpinnata* [de Matos et al. 2014]; *Stichopathes spiessi* [Opresko and Genin 1990], *Callogorgia* spp. [Quattrini et al. 2013]) and extensive cold-water reefs (*Lophelia pertusa* [Mortensen et al. 1995]) are known from deep waters. Anthozoans serve as foundation species, and often as ecosystem engineers, by creating biogenic structures (reviewed in Bruno and Bertness 2001) which provide structurally complex habitat, thereby supporting biodiversity (Wendt et al. 1985; Knowlton and Jackson 2001; Buhl-Mortensen and Mortensen 2004; Coma et al. 2006; Cordes et al. 2008, 2010; Buhl-Mortensen et al. 2010). Anthozoans host both obligate and facultative

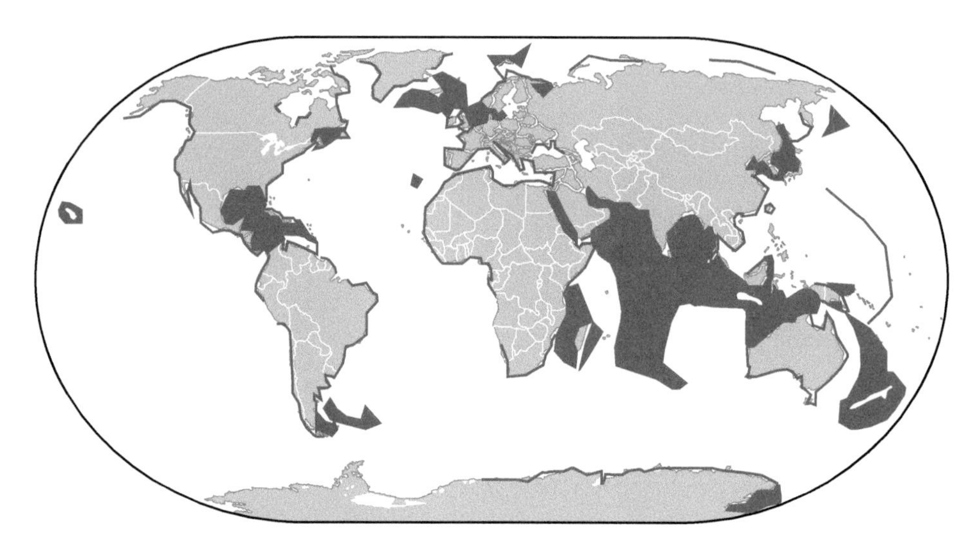

Figure 11.3 Distribution map of anthozoans.

symbionts from across the tree of life, including a diversity of invertebrates (e.g., brittle-star octocoral symbioses, [Mosher and Watling 2009] and crustacean–octocoral associations [Bracken-Grissom et al. 2018] and fishes [e.g., clown-fish–anemone symbioses, (Fautin 1991)]). Goffredi et al. (2020) showed that an actiniarian anemone (*Ostiactis pearseae*) from a hydrothermal vent is in a symbiosis with chemosynthetic bacteria. Many anthozoans also have one of the most well-known symbioses on earth – their partnership with the photosynthesizing Symbiodiniaceae (zooxanthellae, LaJeunesse et al. 2018).

Where zooxanthellae are not providing photosynthates to the host, zooplankton, including amphipods, copepods, chaetognaths, and in some instances, polychaetes (Lewis 1978; Tazioli et al. 2007), appear to be the major component of anthozoans' diets. Prey items are captured by suspension feeding, using a combination of mucus nets and strands, tentacles and nematocysts, and mesenterial filaments that extend out of the mouth or other openings in the body wall. Bacteria that are trapped in mucus are also consumed. Marine snow (sinking organic matter) composed of eukaryotic microalgae, prokaryotic cyanobacteria and heterotrophic bacteria (McMahon et al. 2018) is also captured and delivered to the mouth with the help of ciliary currents (dissolved organic compounds can also be absorbed through the body wall (Lewis 1978). Because many deep-sea anthozoans feed upon animals in the water column that previously grazed on phytoplankton from shallow, light-rich waters, anthozoans play a crucial role in the functioning of deep-sea benthic ecosystems through benthic–pelagic coupling (Gili et al. 2006; Fautin and Fitt 1991).

Notably, anthozoans can live from 10s to 1,000s of years. The slowest-growing species is the black coral in the genus *Leiopathes*. *Leiopathes* has an estimated radial growth rate of <5 μm/yr and age of 4,265 years (Roark et al. 2009). In terms of size, the black coral whip *Stichopathes* can be upward of 5 meters in length (Grigg 1964). Because of these long life spans, scientists use coral skeletons (in particular, scleractinians) to reconstruct past ocean chemistry and circulation patterns, as well as long-term changes in the carbon cycle (Druffel 1997; Williams et al. 2006; Robinson et al. 2014; Prouty et al. 2015).

PHYSIOLOGY AND BEHAVIOR

Anthozoans protect themselves from predators and competitors by producing copious amounts of mucus and/or discharging nematocysts that may contain a diversity of venoms (Fautin 2009). The bamboo coral *Isidella* has elongated polyps at the base of the colony that are modified into "sweeper tentacles" that have been hypothesized to serve as defense to predators (Etnoyer 2008), although these sweeping tentacles observed in other octocoral species have been documented to be used in feeding (Lopez-Gonzalez et al. 2018). In addition, many anthozoans produce chemical compounds that function in various physiological and ecological processes including predation deterrence, symbiosis, calcification/skeletal development and reproduction (reviewed in Tarrant 2005). For example, some octocorals produce toxins (see e.g. Hooper and Davies-Coleman 1995) to deter predation, though these same chemicals can function as sperm attractants for certain coral species (see e.g. Coll et al. 1995). Other chemical compounds (indoleamines) are thought to play a regulatory role in coral–zooxanthellae symbiosis (McCauley 1997). Moreover, corals are a rich source of secondary metabolites (Rocha et al. 2011) that have shown HIV-inhibitory (e.g. Rashid et al. 2000), cytotoxic, anticancer (e.g. Li et al. 2005), and anti-inflammatory/antimicrobial (e.g. Mayer et al. 1998) activities.

GENETICS AND GENOMICS

With the exception of the Ceriantharia (Stampar et al. 2014), anthozoans are characterized by exceptionally slow rates of mtDNA sequence evolution (the synonymous substitution rate is 50–100 times slower than most animals; Hellberg 2006). As an example, Fukami and Knowlton (2005) sequenced six mitogenomes (16,134 bp each) of the *Montastraea annularis* species complex (two individuals each of *M. annularis*, *M. faveolata*, and *M. franksi*) and found 25 variable sites, corresponding to a rate of evolution of 0.03–0.04% per million years. Additionally, variation is almost non-existent at the intraspecific level in numerous species of Anthozoa (Shearer et al. 2002; Brugler et al. 2013; but see Chen et al. 2008a, 2008b,2009; McFadden et al. 2011; Brugler et al. 2018). Several hypotheses have been proposed to account for this lack of variation, including the presence of a mismatch repair gene (to date only found in octocoral mitogenomes), extremely accurate replication machinery, molecular convergence, significant bottleneck, low metabolic rate, generation time, and the effective population size of mitochondria (Shearer et al. 2002; Hellberg 2006; Chen et al. 2009; Thomas et al. 2010), but the exact mechanism or combination of mechanisms remains elusive.

Octocorals are unique among all metazoans in that they have a putative mismatch repair gene, mtMutS, in their mitochondrial genomes. The octocoral mtMutS gene is ~3,000 bp and contains all coding necessary for a functional protein (Bilewitch and Dengan 2011). Although originally thought to have a euykaryotic origin, this gene does not share an immediate common ancestor with any eukaryotic *MSH* family (Bilewitch and Dengan 2011). Rather, this gene has an affinity to the *MutS7* gene family found in epsilonproteobacteria and a nucleocytoplasmic large DNA virus, and thus supports the hypothesis of a horizontal gene transfer (HGT) event from a microbial origin (see Bilewitch and Dengan 2011). However, the exact type of microbial vector remains unknown. Regardless, this mtMutS gene represents the only instance recorded to date of a HGT event into a metazoan mitochondrial genome (Bilewitch and Dengan 2011).

Numerous octocoral mitochondrial genomes (18,616 bp in *Acanella eburnea* to 20,246 bp in *Calicogorgia granulosa*) and hexacoral mitogenomes (16,137 bp in *Montastraea franksi* to 20,764 bp in *Savalia savaglia*) have been sequenced to date (169 total mitogenomes available on NCBI RefSeq, Jan 2021), however, more mitogenomes are likely to be published in coming years as the cost of NGS methods continues to decrease. Anthozoan mitogenomes encode 13 protein coding genes and two ribosomal RNA genes and they are circular, with the exception of at least some cerianthid mitogenomes that have multipartite linear mitochondrial genomes (Stampar et al. 2019). Interestingly, gene order rearrangements are common within the Octocorallia, with five gene orders found to date in the Calcaxonia-Pennatulacea clade (Brockman and McFadden 2012; Figueroa and Baco 2015; Poliseno et al. 2017). In octocorals, gene order rearrangements appear to have evolved by inversions of evolutionarily conserved blocks of protein-coding genes. This pattern differs from that observed within the Hexacorallia, in which extensive gene shuffling has occurred (Brockman and McFadden 2012). Hexacoral mitogenomes also differ from octocorals as all genes are encoded on the same strand (Brugler and France 2007; Emblem et al. 2011, 2014; Brockman and McFadden 2012). In addition, hexacoral mitogenomes encode two transfer RNAs, compared to one in Octocorallia. Group I introns are also found in hexacorals, in either one or two genes; one of these includes a homing endonuclease inserted as a group I intron into COI (Fukami et al. 2007; Emblem et al. 2011, 2014; Chi and Johansen 2017; Celis et al. 2017).

Nematostella vectensis was the first anthozoan genome assembled, and was assembled using a whole-genome shotgun approach followed by expressed-sequence tags for genome annotation (Putnam et al. 2007). Since then, numerous anthozoan genomes have been published. Currently, genome sizes of anthozoans range from ca. 172 Mb in the octocoral *Renilla muelleri* (Jiang et al. 2019) to 486 Mb in the hexacoral *Orbicella faveolata* (Prada et al. 2016). The number of protein-coding genes ranges from 21,372 in the corallimorpharian *Amplexidiscus fenestrafer* to 30,360 in *Montastraea cavernosa* (Wang et al. 2017; Jiang et al. 2019). These genomic resources have provided a wealth of insights into anthozoan evolution, including developmental and metabolic pathways, genomic architecture, and host–symbiont interactions (Putnam et al. 2007; Shinzato et al. 2011; Baumgarten et al. 2015). Genomic and transcriptomic data are available at http://reefgenomics.org (Liew et al. 2016), through the Global Invertebrate Genomics Alliance (GIGA Community of Scientists 2013), and on GenBank at NCBI (https://www.ncbi.nlm.nih.gov).

While genomes have revealed highly similar gene repertoires among anthozoans and more derived vertebrate lineages (reviewed in Technau and Schwaiger 2015), transcriptomic data for this group are becoming increasingly available. High-throughput RNA sequencing and subsequent analyses of select anthozoan transcriptomes in recent years have addressed a range of topics, including: evolution and development (Tulin et al. 2013; Helm et al. 2013), innate immunity (Libro et al. 2013; Burge et al. 2013; Pinzon et al. 2015), toxin/venom diversity (e.g. Macrander, Brugler and Daly 2015; Macrander, Broe and Daly 2016), symbiosis (e.g. Mohamed et al. 2016), and cellular-level responses to environmental change – both natural (e.g. diurnal fluctuations, Ruiz-Jones and Palumbi 2015; Leach et al. 2018) and anthropogenic (e.g. ocean acidification, Moya et al. 2012, Vidal-Dupiol et al. 2013; heat stress, Kenkel et al. 2013, Davies et al. 2016; pollutant exposure, DeLeo et al. 2018). Studies by Barshis et al. (2013) and Kenkel and Matz (2017) also used genome-wide expression profiling in calcifying corals to explore the molecular mechanisms behind resilience and adaptation to environmental change. RNA-Seq efforts are being progressively employed to elucidate the fate of these organisms in an era of global climate change and increased anthropogenic disturbance.

POSITION IN THE ToL

Cnidarians, including the Medusozoa and Anthozoa, are one of the earliest branching groups of metazoans in the animal tree of life, likely arising in the pre-Cambrian (Erwin et al. 2011; Quattrini et al. 2020). In phylogenomic studies, the Cnidaria has been resolved sister to the Bilateria, with the Placozoa sister to the Cnidaria + Bilateria clade (Zapata et al. 2015; Feuda et al. 2017; Kayal et al. 2018). After sequencing the genome of a second placozoan species (i.e.,

Hoilungia hongkongensis), Eitel et al. (2018) constructed a phylogeny based on 194 genes that confirmed the Placozoa groups sister to the Cnidaria + Bilateria.

The reciprocal monophyly of Anthozoa and Medusozoa has been long recognized, based on anatomy, life history, and a handful of DNA sequences (see Daly et al. 2007). The monophyly of the Anthozoa was, however, brought into question with the publication of two cnidarian phylogenies based on complete mitogenomes (Park et al. 2012; Kayal et al. 2013). Both studies recovered the Anthozoa as paraphyletic, with the Octocorallia sister to the Medusozoa (which includes the Cubozoa [box jellies], Hydrozoa [hydroids hydromedusae and siphonophores], Scyphozoa [true jellies], and Staurozoa [stalked jellies]). Subsequent phylogenomic analyses based on 10s to 100s of nuclear genes recovered a monophyletic Anthozoa that groups sister to a clade containing Medusozoa (Zapata et al. 2015; Pratlong et al. 2017) plus Endocnidozoa (the latter includes the parasitic Myxozoa and Polypodiozoa [Kayal et al. 2018]). Difficulty resolving relationships within the Anthozoa is not surprising given the ancient divergence estimates within the group and the slow rates of mitochondrial genome evolution. Pratlong et al. (2017) noted that high substitution saturation in mitochondrial genomes could be responsible for the incongruence between phylogenies constructed with mitochondrial and nuclear data. A recent phylogenomic study, however, found that the sister relationship between Octocorallia and Hexacorallia and the relationships among (mostly monophyletic) orders within each subclass were highly supported (**Figure 11.4**, Quattrini et al. 2018, 2020).

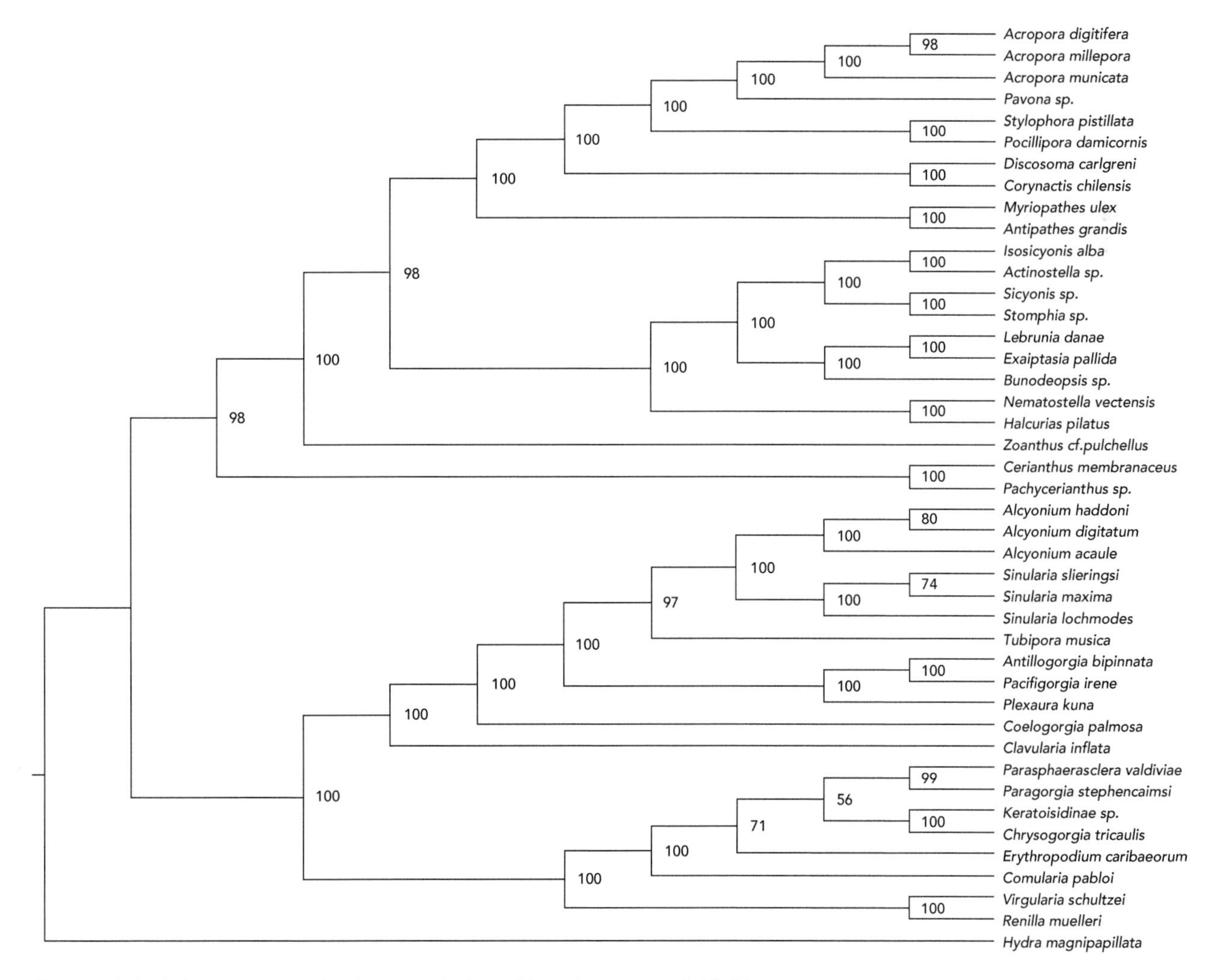

Figure 11.4 Phylogenetic tree of anthozoans. (Adapted from Quattrini et al. 2018.)

In addition, the fossil-calibrated phylogeny revealed that the most recent common ancestor (MRCA) of the Hexacorallia was estimated to be 711 million years old (MYO; 95% credible region: 599–828), while the MRCA of the Octocorallia was estimated at 578 MYO (95% CR: 483–685) (Quattrini et al. 2020).

DATABASES AND COLLECTIONS

WoRMS-World Register of Marine Species

http://www.marinespecies.org

NCBI's GenBank

https://www.ncbi.nlm.nih.gov

NCBI's RefSeq

https://www.ncbi.nlm.nih.gov/refseq/

NOAA Deep-Sea Coral Portal

https://deepseacoraldata.noaa.gov/

Ocean Biogeographic Information System

https://obis.org

National Museum of Natural History Invertebrate Zoology Collection, Smithsonian Institution

https://collections.nmnh.si.edu/search/iz/

American Museum of Natural History Invertebrate Zoology Collection

http://sci-web-001.amnh.org/imulive/iz.html

The National Institute of Water and Atmospheric Research (NIWA) Invertebrate Collection

https://obis.org/dataset/2c6db58f-ae91-4a17-9f0f-7db28506b94f

The Yale Peabody Museum of Natural History Invertebrate Zoology Collection

https://collections.peabody.yale.edu/search/Search/Advanced?collection=Invertebrate%20Zoology&searchText2=IZ&searchField2=CatalogNumber&matchTerms=AND

The Field Museum Invertebrate Collection

http://emuweb.fieldmuseum.org/iz/Query.php

California Academy of Sciences Invertebrate Zoology Collection

http://researcharchive.calacademy.org/research/izg/iz_coll_db/index.asp

Florida Museum Invertebrate Zoology Collection

http://specifyportal.flmnh.ufl.edu/iz/

Harvard University's Museum of Comparative Zoology

https://mczbase.mcz.harvard.edu/SpecimenSearch.cfm

Texas A&M Biodiversity Research and Teaching Collection

https://science.mnhn.fr/institution/mnhn/collection/ik/item/search

The French National Museum of Natural History (Muséum national d'histoire naturelle) Cnidaria Collection

https://science.mnhn.fr/institution/mnhn/collection/ik/item/search

RECOMMENDATIONS FOR NGS SUCCESS

Anthozoans produce copious amounts of mucus (polysaccharides), excrete a variety of secondary metabolites, are rich in collagen, and their tissue is typically preserved in ethanol for downstream molecular analysis, all of which may serve to inhibit PCR reactions (Wilson 1997; Schrader et al. 2012). Thus, for successful sequencing on next-generation instruments, PCR-free DNA libraries may be warranted. However, successful DNA libraries can be obtained by using commercial clean-up kits to remove inhibitors from genomic DNA prior to library generation. Repeated efforts to conduct NGS on black corals have been largely unsuccessful, including creating Illumina DNA libraries (this is also true for some species of stony corals) and a complete inability to obtain data on a PacBio RS II using freshly collected tissue of the Hawaiian black coral *Antipathes griggi*. It should be noted that 2XCTAB, which contains a number of potentially inhibitory detergents, is typically used during the DNA extraction process to rid black corals of polysaccharides. It is recommended to use commercial clean-up kits for black corals prior to NGS.

Due to the sensitive and labile nature of RNA, careful attention needs to be placed on sample preservation methods. While high salinity buffers (i.e, RNAlater) have proved successful in preserving high-quality RNA in soft-bodied anthozoans like octocorals and antipatharians, the hard exoskeletons of calcifying anthozoans (scleractinians) may limit tissue permeability and impede proper preservation. This issue can be avoided by flash-freezing specimens, though samples will be at higher risk of temperature-related nucleic acid degradation. While RNA can be successfully extracted with column based kits (i.e., Qiagen RNeasy), in some cases, like those with highly pigmented, scleritic or damaged/limited tissues, a modified protocol may be necessary for successful extraction (i.e., Trizol/Qiagen RNeasy protocol described in Burge et al. 2013; Polato, Vera, and Baums, 2011). When designing an RNA-Seq experiment, species preservation methods should be kept consistent, when possible as both preservation method – RNAlater or flash freezing (Passow et al. 2018) – and extraction/library preparation protocols (see Sultan et al. 2014) may influence data and gene expression results across taxa.

CONCLUSION

Anthozoan cnidarians are arguably some of the most important metazoans in the marine environment. They create habitat for 1,000s of other species and provide valuable ecosystem services. This class of Cnidaria has unique morphological characters; many species can live to be 1,000s of years old; they are in unique symbioses with other organisms; and they face extinction from anthropogenic global ocean change (Carpenter et al. 2008). Anthozoans arose prior to the Cambrian explosion, and have faced numerous ocean climate and chemistry changes since that time (Quattrini et al. 2020). With the advances of high-throughput sequencing and novel molecular techniques, new approaches to understand evolutionary history and genomic architecture in Anthozoa are emergent.

REFERENCES

Auster, P. J., 2007. Linking deep-water corals and fish populations. Bulletin of Marine Science, 81(3), pp. 93–99.

Ayre, D. J. and Resing, J. M., 1986. Sexual and asexual production of planulae in reef corals. Marine Biology, 90(2), pp. 187–190.

Babcock, R. C., Bull, G. D., Harrison, P. L., Heyward, A. J., Oliver, J. K., Wallace, C. C. and Willis, B. L., 1986. Synchronous spawnings of 105 scleractinian coral species on the Great Barrier Reef. Marine Biology, 90(3), pp. 379–394.

Baillon, S., Hamel, J. F., Wareham, V. E. and Mercier, A., 2012. Deep cold-water corals as nurseries for fish larvae. Frontiers in Ecology and the Environment, 10(7), pp. 351–356.

Barshis, D. J., Ladner, J. T., Oliver, T. A., Seneca, F. O., Traylor-Knowles, N. and Palumbi, S. R., 2013. Genomic basis for coral resilience to climate change. Proceedings of the National Academy of Sciences, 110(4), pp. 1387–1392.

Baumgarten, S., Simakov, O., Esherick, L. Y., Liew, Y. J., Lehnert, E. M., Michell, C. T., ... and Voolstra, C. R. 2015. The genome of Aiptasia, a sea anemone model for coral symbiosis. Proceedings of the National Academy of Sciences, 112(38) pp., 11893–11898.

Bertness, M. D., Gaines, S. D. and Hay, M. E. (eds.), 2001. Marine Community Ecology (No. QH 541.5. S3. M369 2001). Sinauer Associates, Sunderland, MA.

Bilewitch, J. P., and Degnan, S. M. 2011. A unique horizontal gene transfer event has provided the octocoral mitochondrial genome with an active mismatch repair gene that has potential for an unusual self-contained function. BMC Evolutionary Biology, 11(1), pp. 1–15.

Black, R. and Johnson, M. S., 1979. Asexual viviparity and population genetics of Actinia tenebrosa. Marine Biology, 53(1), pp. 27–31.

Bocharova, E. S. and Kozevich, I. A., 2011. Modes of reproduction in sea anemones (Cnidaria, Anthozoa). Biology Bulletin, 38(9), pp. 849–860.

Bracken-Grissom, H., Widder, E., Johnsen, S., Messing, C., and Frank, T. (2018). Decapod diversity associated with deep-sea octocorals in the Gulf of Mexico. Crustaceana, 91(10), pp. 1267– 1275.

Brazeau, D. A. and Lasker, H. R., 1989. The reproductive cycle and spawning in a Caribbean gorgonian. Biological Bulletin, 176(1), pp. 1–7.

Brockman, S. A., and McFadden, C. S. (2012). The mitochondrial genome of *Paraminabea aldersladei* (Cnidaria: Anthozoa: Octocorallia) supports intramolecular recombination as the primary mechanism of gene rearrangement in octocoral mitochondrial genomes. Genome Biology and Evolution, 4(9), pp. 994–1006.

Brugler, M. R., González-Muñoz, R. E., Tessler, M. and Rodríguez, E., 2018. An EPIC journey to locate single-copy nuclear markers in sea anemones. Zoologica Scripta.

Brugler, M. R. and France, S. C., 2007. The complete mitochondrial genome of the black coral Chrysopathes formosa (Cnidaria: Anthozoa: Antipatharia) supports classification of antipatharians within the subclass Hexacorallia. Molecular Phylogenetics and Evolution, 42(3), pp. 776–788.

Brugler, M. R., Opresko, D. M. and France, S. C., 2013. The evolutionary history of the order Antipatharia (Cnidaria: Anthozoa: Hexacorallia) as inferred from mitochondrial and nuclear DNA: implications for black coral taxonomy and systematics. Zoological Journal of the Linnean Society, 169(2), pp. 312–361.

Bruno, J. and Bertness, M. D., 2001. Positive interactions, facilitations and foundation species. Marine Community Ecology. Sinauer Associates, Inc Publishers, Sunderland, MA, pp. 201–218.

Buhl-Mortensen, L. E. N. E. and Mortensen, P. B., 2004. Symbiosis in deep-water corals. Symbiosis, 37(1), pp. 33–61.

Buhl-Mortensen, L. and Mortensen, P. B., 2005. Distribution and diversity of species associated with deep-sea gorgonian corals off Atlantic Canada. In Cold-water Corals and Ecosystems (pp. 849–879). Springer, Berlin, Heidelberg.

Buhl-Mortensen, L., Vanreusel, A., Gooday, A. J., Levin, L. A., Priede, I. G., Buhl-Mortensen, P., Gheerardyn, H., King, N. J. and Raes, M., 2010. Biological structures as a source of habitat heterogeneity and biodiversity on the deep ocean margins. Marine Ecology, 31(1), pp. 21–50.

Burge, C. A., Mouchka, M. E., Harvell, C. D. and Roberts, S., 2013. Immune response of the Caribbean sea fan, Gorgonia ventalina, exposed to an Aplanochytrium parasite as revealed by transcriptome sequencing. Frontiers in Physiology, 4, p. 180.

Cairns, S. D., Bayer, F. M. and Fautin, D. G., 2007. Galatheanthemum profundale (Anthozoa: Actiniaria) in the western Atlantic. Bulletin of Marine Science, 80(1), pp. 191–200.

Cairns, S. D., Gerhswin, L. A., Brook, F., Pugh, P. R., Dawson, E. W., Ocaña, V., Vervoort, W., Williams, G., Watson, J., Opresko, D. M. and Schuchert, P., 2009. Phylum Cnidaria; Corals, Medusae, hydroids, myxozoa. New Zealand Inventory of Biodiversity. *Volume 1. Kingdom Animalia: Radiata, Lophotrochozoa, Deuterostomia.*

Celis, J. S., Edgell, D. R., Stelbrink, B., Wibberg, D., Hauffe, T., Blom, J., ... & Wilke, T. (2017). Evolutionary and biogeographical implications of degraded LAGLIDADG endonuclease functionality and group I intron occurrence in stony corals (Scleractinia) and mushroom corals (Corallimorpharia). PloS One, *12*(3), e0173734.

Chen, C., Chiou, C. Y., Dai, C. F. and Chen, C. A., 2008a. Unique mitogenomic features in the scleractinian family Pocilloporidae (Scleractinia: Astrocoeniina). Marine Biotechnology, 10(5), p. 538.

Chen, C., Dai, C. F., Plathong, S., Chiou, C. Y. and Chen, C. A., 2008b. The complete mitochondrial genomes of needle corals, Seriatopora spp.(Scleractinia: Pocilloporidae): an idiosyncratic atp8, duplicated trnW gene, and hypervariable regions used to determine species phylogenies and recently diverged populations. Molecular Phylogenetics and Evolution, 46(1), pp. 19–33.

Chen, C., Soong, K., Chen, C. A., 2008. The smallest oocytes among broadcast- spawning actiniarians and a unique lunar reproductive cycle in a unisexual population of the sea anemone, *Aiptasia pulchella* (Anthozoa: Actiniaria). Zoological Studies 47(1), pp. 37–45.

Chen, I. P., Tang, C. Y., Chiou, C. Y., Hsu, J. H., Wei, N. V., Wallace, C. C., Muir, P., Wu, H. and Chen, C. A., 2009. Comparative analyses of coding and noncoding DNA regions indicate that Acropora (Anthozoa: Scleractina) possesses a similar evolutionary tempo of nuclear vs. mitochondrial genomes as in plants. Marine Biotechnology, 11(1), p. 141.

Chi, S. I., & Johansen, S. D., 2017. Zoantharian mitochondrial genomes contain unique complex group I introns and highly conserved intergenic regions. Gene, 628, pp. 24–31.

Coll, J., Leone, P., Bowden, B., Carroll, A., König, G., Heaton, A., De Nys, R., Maida, M., Alino, P. and Willis, R. 1995. Chemical aspects of mass spawning in corals. II.(-)-Epi-thunbergol, the sperm attractant in the eggs of the soft coral Lobophytum crassum (Cnidaria: Octocorallia). Marine Biology, 123, pp. 137–143.

Coma, R., Linares, C., Ribes, M., Diaz, D., Garrabou, J. and Ballesteros, E., 2006. Consequences of a mass mortality in populations of Eunicella singularis (Cnidaria: Octocorallia) in Menorca (NW Mediterranean). Marine Ecology Progress Series, 327, pp. 51–60.

Coma, R., Zabala, M. and Gili, J.M., 1995. Sexual reproductive effort in the Mediterranean gorgonian *Paramuricea clavata*. Marine Ecology Progress Series, pp. 185–192.

Cordes, E. E., Cunha, M. R., Galeron, J., Mora, C., Olu-Le Roy, K., Sibuet, M., Van Gaever, S., Vanreusel, A. and Levin, L. A., 2010. The influence of geological, geochemical, and biogenic habitat heterogeneity on seep biodiversity. Marine Ecology, 31(1), pp. 51–65.

Cordes, E. E., McGinley, M. P., Podowski, E. L., Becker, E. L., Lessard-Pilon, S., Viada, S. T. and Fisher, C. R., 2008. Coral communities of the deep Gulf of Mexico. Deep Sea Research Part I: Oceanographic Research Papers, 55(6), pp. 777–787.

Dahan, M. and Benayahu, Y., 1997. Reproduction of Dendronephthya hemprichi (Cnidaria: Octocorallia): year-round spawning in an azooxanthellate soft coral. Marine Biology, 129(4), pp. 573–579.

Daly, M., 2006. Boloceroides daphneae, a new species of giant sea anemone (Cnidaria: Actiniaria: Boloceroididae) from the deep Pacific. Marine Biology, 148(6), pp. 1241–1247.

Daly, M., Brugler, M. R., Cartwright, P., Collins, A. G., Dawson, M. N., Fautin, D. G., France, S. C., McFadden, C. S., Opresko, D. M., Rodriguez, E. and Romano, S. L., 2007. The phylum Cnidaria: a review of phylogenetic patterns and diversity 300 years after Linnaeus.

Daly, M., Fautin, D. G. and Cappola, V. A., 2003. Systematics of the hexacorallia (Cnidaria: Anthozoa). Zoological Journal of the Linnean Society, 139(3), pp. 419–437.

Davies, S. W., Marchetti, A., Ries, J. B. and Castillo, K. D., 2016. Thermal and pCO2 stress elicit divergent transcriptomic responses in a resilient coral. Frontiers in Marine Science, 3, p. 112.

de Matos, V., Gomes-Pereira, J. N., Tempera, F., Ribeiro, P. A., Braga-Henriques, A. and Porteiro, F., 2014. First record of Antipathella subpinnata (Anthozoa, Antipatharia) in the Azores (NE Atlantic), with description of the first monotypic garden for this species. Deep Sea Research Part II: Topical Studies in Oceanography, 99, pp. 113–121.

DeLeo, D. M., Herrera, S., Lengyel, S. D., Quattrini, A. M., Kulathinal, R. J., Cordes, E. E., 2018. Gene expression profiling reveals deep-sea coral response to the Deepwater Horizon oil spill. Molecular Ecology, 27(20), pp. 4066–4077. https://doi.org/10.1111/mec.14847

Dohrmann, M. and Wörheide, G., 2013. Novel scenarios of early animal evolution—is it time to rewrite textbooks? Integrative and Comparative Biology, 53(3), pp. 503–511.

Druffel, E. R., 1997. Geochemistry of corals: Proxies of past ocean chemistry, ocean circulation, and climate. Proceedings of the National Academy of Sciences, 94(16), pp. 8354–8361.

DuBuc, T. Q., Stephenson, T. B., Rock, A. Q. and Martindale, M. Q., 2018. Hox and Wnt pattern the primary body axis of an anthozoan cnidarian before gastrulation. Nature Communications, 9(1), p. 2007.

Dunn, C. W., 2017. Ctenophore trees. Nature Ecology & Evolution, 1(11), p. 1600.

Dunn, D. F., 1975. Reproduction of the externally brooding sea anemone *Epiactis prolifera* Verrill, 1869. The Biological Bulletin, 148(2), pp. 199–218.

Eitel, M., Francis, W. R., Varoqueaux, F., Daraspe, J., Osigus, H. J., Krebs, S., Vargas, S., Blum, H., Williams, G. A., Schierwater, B. and Wörheide, G., 2018. Comparative genomics and the nature of placozoan species. PLoS biology, 16(7), p. e2005359.

Ehrlich, H., Etnoyer, P., Litvinov, S. D., Olennikova, M. M., Domaschke, H., Hanke, T., Born, R., Meissner, H. and Worch, H., 2006. Biomaterial structure in deep-sea bamboo coral (Anthozoa: Gorgonacea: Isididae): perspectives for the development of bone implants and templates for tissue engineering. Materialwissenschaft und Werkstofftechnik: Entwicklung, Fertigung, Prüfung, Eigenschaften und Anwendungen technischer Werkstoffe, 37(6), pp. 552–557.

Emblem, Å., Karlsen, B. O., Evertsen, J., Johansen, S. D., 2011. Mitogenome rearrangement in the cold-water scleractinian coral Lophelia pertusa (Cnidaria, Anthozoa) involves a long-term evolving group I intron. *Molecular phylogenetics and evolution*, 61(2), pp. 495–503.

Emblem, Å., Okkenhaug, S., Weiss, E.S., Denver, D.R., Karlsen, B.O., Moum, T. and Johansen, S.D., 2014. Sea anemones possess dynamic mitogenome structures. *Molecular phylogenetics and evolution*, 75, pp.184–193.

Erwin, D. H., Laflamme, M., Tweedt, S. M., Sperling, E. A., Pisani, D., & Peterson, K. J. (2011). The Cambrian conundrum: early divergence and later ecological success in the early history of animals. Science, 334(6059), 1091–1097.

Etnoyer, P. J., 2008. A new species of Isidella bamboo coral (Octocorallia: Alcyonacea: Isididae) from northeast Pacific seamounts. Proceedings of the Biological Society of Washington, 121(4), pp. 541–553.

Etnoyer, P. and Warrenchuk, J., 2007. A catshark nursery in a deep gorgonian field in the Mississippi Canyon, Gulf of Mexico. Bulletin of Marine Science, 81(3), pp. 553–559.

Fautin, D. G., 1991. The anemonefishes: what is known and what is not. Symbiosis, 10, pp. 23–46.

Fautin, D. G., 2002. Reproduction of cnidaria. Canadian Journal of Zoology, 80(10), pp. 1735–1754.

Fautin, D. G., 2009. Structural diversity, systematics, and evolution of cnidae. Toxicon, 54(8), pp. 1054–1064.

Fautin, D. G. and Fitt, W. K., 1991, June. A jellyfish-eating sea anemone (Cnidaria, Actiniaria) from Palau: *Entacmaea medusivora* sp. Hydrobiologia, 216(1), pp. 453–461.

Fautin, D. G., and Mariscal, R. N., 1991. Cnidaria: anthozoa (Vol. 2, pp. 267–358). New York: Wiley-Liss.

Feuda, R., Dohrmann, M., Pett, W., Philippe, H., Rota-Stabelli, O., Lartillot, N., Wörheide, G. and Pisani, D., 2017. Improved modeling of compositional heterogeneity supports sponges as sister to all other animals. Current Biology, 27(24), pp. 3864–3870.

Figueroa, D. F., and Baco, A. R., 2015. Octocoral mitochondrial genomes provide insights into the phylogenetic history of gene order rearrangements, order reversals, and cnidarian phylogenetics. Genome Biology and Evolution, 7(1), pp. 391–409

Finnerty, J. R., Pang, K., Burton, P., Paulson, D. and Martindale, M. Q., 2004. Origins of bilateral symmetry: Hox and dpp expression in a sea anemone. Science, 304(5675), pp. 1335–1337.

Fukami, H., Chen, C. A., Chiou, C. Y., & Knowlton, N. (2007). Novel group I introns encoding a putative homing endonuclease in the mitochondrial cox1 gene of Scleractinian corals. Journal of molecular evolution, *64*(5), 591–600.

Fukami, H. and Knowlton, N., 2005. Analysis of complete mitochondrial DNA sequences of three members of the Montastraea annularis coral species complex (Cnidaria, Anthozoa, Scleractinia). Coral Reefs, 24(3), pp. 410–417.

Gashout, S. E. and Ormond, R. F., 1979. Evidence for parthenogenetic reproduction in the sea anemone Actinia equina L. Journal of the Marine Biological Association of the United Kingdom, 59(4), pp. 975–987.

GIGA Community of Scientists, 2013. The Global Invertebrate Genomics Alliance (GIGA): developing community resources to study diverse invertebrate genomes. Journal of Heredity, 105(1), pp. 1–18.

Gili, J.M., Rossi, S., Pagès, F., Orejas, C., Teixidó, N., López-González, P.J. and Arntz, W.E., 2006. A new trophic link between the pelagic and benthic systems on the Antarctic shelf. Marine Ecology Progress Series, 322, pp. 43–49.

Goffredi, S. K., Motooka, C., Fike, D. A., Gusmão, L. C., Tilic, E., Rouse, G. W., and Rodríguez, E., 2021. Mixotrophic chemosynthesis in a deep-sea anemone from hydrothermal vents in the Pescadero Basin, Gulf of California. BMC Biology, 19(1), pp. 1–18.

Gold, D. A. and Jacobs, D. K., 2013. Stem cell dynamics in Cnidaria: are there unifying principles?. Development genes and evolution, 223(1–2), pp. 53–66.

Grassle, J. F. and Shick, J. M., 1979. Introduction to the symposium: Ecology of asexual reproduction in animals. American Zoologist, 19(3), pp. 667–668.

Greco, G. R. and Cinquegrani, M., 2018. The Global Market for Marine Biotechnology: The Underwater World of Marine Biotech Firms. In Grand Challenges in Marine Biotechnology (pp. 261–316). Springer, Cham.

Grigg, R. W., 1964. A Contribution to the Biology and Ecology of the Black Coral, Antipathes Grandis in Hawai'i. MS thesis, University of Hawai'i: USA.

Hartnoll, R. G., 1975. The annual cycle of Alcyonium digitatum. Estuarine and Coastal Marine Science, 3(1), pp. 71–78.

He, S., del Viso, F., Chen, C. Y., Ikmi, A., Kroesen, A. E. and Gibson, M. C., 2018. An axial Hox code controls tissue segmentation and body patterning in Nematostella vectensis. Science, 361(6409), pp. 1377–1380.

Helm, R. R., Siebert, S., Tulin, S., Smith, J. and Dunn, C. W., 2013. Characterization of differential transcript abundance through time during Nematostella vectensis development. BMC Genomics, 14(1), p. 266.

Hellberg, M. E., 2006. No variation and low synonymous substitution rates in coral mtDNA despite high nuclear variation. BMC Evolutionary Biology, 6(1), p. 24.

Highsmith, R. C., 1982. Reproduction by fragmentation in corals. Marine Ecology Progress Series. Oldendorf, 7(2), pp. 207–226.

Hooper, G. J., and Davies-Coleman, M. T., 1995. New metabolites from the South African soft coral Capnella thyrsoidea. Tetrahedron, 51(36), pp. 9973–9984.

Ivanova-Kazas, O. M., 1975. Class Anthozoa–coral polyps. Comparable embryology of invertebrates: Protozoa and lower multicellular animals. Nauka, Novosibirsk, pp.190–205.

Jiang, J., Quattrini, A. M., Francis, W. R., Ryan, J. F., Rodriguez, E. and McFadden, C. S., 2019. A yybrid de novo assembly of the sea pansy (Renilla muelleri) genome. GigaScience, *8*(4), giz026.

Kayal, E., Bentlage, B., Pankey, M. S., Ohdera, A. H., Medina, M., Plachetzki, D. C., Collins, A. G. and Ryan, J. F., 2018. Phylogenomics provides a robust topology of the major cnidarian lineages and insights on the origins of key organismal traits. BMC Evolutionary Biology, 18(1), p. 68.

Kayal, E., Roure, B., Philippe, H., Collins, A. G. and Lavrov, D. V., 2013. Cnidarian phylogenetic relationships as revealed by mitogenomics. BMC Evolutionary Biology, 13(1), p. 5.

Kenkel, C. D. and Matz, M. V., 2017. Gene expression plasticity as a mechanism of coral adaptation to a variable environment. Nature Ecology & Evolution, 1(1), p. 0014.

Kenkel, C. D., Meyer, E. and Matz, M. V., 2013. Gene expression under chronic heat stress in populations of the mustard hill coral (P orites astreoides) from different thermal environments. Molecular Ecology, 22(16), pp. 4322–4334.

King, N. and Rokas, A., 2017. Embracing uncertainty in reconstructing early animal evolution. Current Biology, 27(19), pp. R1081–R1088.

Knowlton, N., Jackson, J. B. C., 2001. The ecology of coral reefs. In Bertness, M. D., Gaines, S., Hay, M. E., eds. Marine Community Ecology, pp. 395 – 422. Sinauer.

Kusserow, A., Pang, K., Sturm, C., Hrouda, M., Lentfer, J., Schmidt, H. A., Technau, U., von Haeseler, A., Hobmayer, B., Martindale, M. Q. and Holstein, T. W., 2005. Unexpected complexity of the Wnt gene family in a sea anemone. Nature, 433(7022), p. 156.

LaJeunesse, T.C., Parkinson, J. E., Gabrielson, P. W., Jeong, H. J., Reimer, J. D., Voolstra, C. R. and Santos, S. R., 2018. Systematic revision of Symbiodiniaceae highlights the antiquity and diversity of coral endosymbionts. Current Biology, 28(16), pp. 2570–2580.

Larkman, A. U., 1983. An ultrastructural study of oocyte growth within the endoderm and entry into the mesoglea in Actinia fragacea (Cnidaria, Anthozoa). Journal of Morphology, 178(2), pp. 155–177.

Larkman, A. U., 1984. An ultrastructural study of the establishment of the testicular cysts during spermatogenesis in the sea anemone Actinia fragacea (Cnidaria: Anthozoa). Gamete Research, 9(3), pp. 303–327.

Lasker, H. R. and Kim, K., 1996. Larval development and settlement behavior of the gorgonian coral Plexaura kuna (Lasker, Kim and Coffroth). Journal of Experimental Marine Biology and Ecology, 207(1–2), pp. 161–175.

Lasker, H. R., Brazeau, D. A., Calderon, J., Coffroth, M. A., Coma, R., and Kim, K., 1996. In situ rates of fertilization among broadcast spawning gorgonian corals. The Biological Bulletin, 190(1), pp. 45–55.

Lauretta, D. and Penchaszadeh, P. E., 2017. Gigantic oocytes in the deep sea black coral *Dendrobathypathes grandis* (Antipatharia) from the Mar del Plata submarine canyon area (southwestern Atlantic). Deep Sea Research Part I: Oceanographic Research Papers, 128, pp. 109–114.

Lawn, I. D. and Ross, D. M., 1982. The behavioural physiology of the swimming sea anemone Boloceroides mcmurrichi. Proc. R. Soc. Lond. B, 216(1204), pp. 315–334.

Leach, W. B., Macrander, J., Peres, R. and Reitzel, A. M., 2018. Transcriptome-wide analysis of differential gene expression in response to light: dark cycles in a model cnidarian. Comparative Biochemistry and Physiology Part D: Genomics and Proteomics, 26, pp. 40–49.

Levitan, D. R., 2006. The relationship between egg size and fertilization success in broadcast-spawning marine invertebrates. Integrative and Comparative Biology, 46(3), pp. 298–311.

Lewis, J. B., 1978. Feeding mechanisms in black corals (Antipatharia). Journal of Zoology, 186(3), pp. 393–396.

Lewis, J. C. and Wallis, E. V., 1991. The function of surface sclerites in gorgonians (Coelenterata, Octocorallia). Biological Bulletin, 181(2), pp. 275–288.

Li, G., Zhang, Y., Deng, Z., van Ofwegen, L., Proksch, P., and Lin, W., 2005. Cytotoxic Cembranoid Diterpenes from a Soft Coral Sinularia gibberosa. Journal of Natural Products, 68(5), pp. 649–652.

Libro, S., Kaluziak, S. T. and Vollmer, S. V., 2013. RNA-seq profiles of immune related genes in the staghorn coral Acropora cervicornis infected with white band disease. PLoS One, 8(11), p. e81821.

Liew, Y. J., Aranda, M. and Voolstra, C. R., 2016. Reefgenomics. Org – a repository for marine genomics data. Database, 2016.

Lin, M. F., Chou, W. H., Kitahara, M. V., Chen, C. L. A., Miller, D. J. and Forêt, S., 2016. Corallimorpharians are not "naked corals": insights into relationships between Scleractinia and Corallimorpharia from phylogenomic analyses. PeerJ, 4, p. e2463.

Lirman, D., 2000. Fragmentation in the branching coral Acropora palmata (Lamarck): growth, survivorship, and reproduction of colonies and fragments. Journal of Experimental Marine Biology and Ecology, 251(1), pp. 41–57.

López González, P. J., Bramanti, L., Escribano Álvarez, P., Benedetti, M. C., Martínez Baraldés, I., & Megina Martínez, C., 2018. Thread-like tentacles in the Mediterranean corals *Paramuricea clavata* and *Corallium rubrum*. Mediterranean Marine Science, 19, pp. 394–397.

Macrander, J., Broe, M. and Daly, M., 2016. Tissue-specific venom composition and differential gene expression in sea anemones. Genome Biology and Evolution, 8(8), pp. 2358–2375.

Macrander, J., Brugler, M. R. and Daly, M., 2015. A RNA-seq approach to identify putative toxins from acrorhagi in aggressive and non-aggressive Anthopleura elegantissima polyps. BMC Genomics, 16(1), p. 221.

Martell, L., Piraino, S., Gravili, C. and Boero, F., 2016. Life cycle, morphology and medusa ontogenesis of *Turritopsis dohrnii* (Cnidaria: Hydrozoa). Italian Journal of Zoology, 83(3), pp. 390–399.

Martindale, M. Q., Pang, K. and Finnerty, J. R., 2004. Investigating the origins of triploblasty: 'mesodermal' gene expression in a diploblastic animal, the sea anemone Nematostella vectensis (phylum, Cnidaria; class, Anthozoa). Development, 131(10), pp. 2463–2474.

Mayer, A. M., Jacobson, P. B., Fenical, W., Jacobs, R. S., and Glaser, K. B., 1998. Pharmacological characterization of the pseudopterosins: novel anti-inflammatory natural products isolated from the Caribbean soft coral, Pseudopterogorgia elisabethae. Life Sciences, 62(26), PL401–PL407.

McCauley, D. W. 1997. Serotonin Plays an Early Role in the Metamorphosis of the Hydrozoan *Phialidium gregarium*. Developmental Biology, 190, pp. 229–240.

McFadden, C. S., France, S. C., Sánchez, J. A., and Alderslade, P., 2006. A molecular phylogenetic analysis of the Octocorallia (Cnidaria: Anthozoa) based on mitochondrial protein-coding sequences. Molecular Phylogenetics and Evolution, 41(3), pp. 513–527.

McFadden, C. S., Benayahu, Y., Pante, E., Thoma, J. N., Nevarez, P. A. and France, S. C., 2011. Limitations of mitochondrial gene barcoding in Octocorallia. Molecular Ecology Resources, 11(1), pp. 19–31.

McFadden, C. S., Sánchez, J. A. and France, S. C., 2010. Molecular phylogenetic insights into the evolution of Octocorallia: a review. Integrative and Comparative Biology, 50(3), pp. 389–410

McMahon, K. W., Williams, B., Guilderson, T. P., Glynn, D. S. and McCarthy, M. D., 2018. Calibrating amino acid δ13C and δ15N offsets between polyp and protein skeleton to develop proteinaceous deep-sea corals as paleoceanographic archives. Geochimica et Cosmochimica Acta, 220, pp. 261–275.

Medina, M., Collins, A. G., Takaoka, T. L., Kuehl, J. V. and Boore, J. L., 2006. Naked corals: skeleton loss in Scleractinia. Proceedings of the National Academy of Sciences, 103(24), pp. 9096–9100.

Meyer, E., Aglyamova, G. V. and Matz, M. V., 2011. Profiling gene expression responses of coral larvae (Acropora millepora) to elevated temperature and settlement inducers using a novel RNA-Seq procedure. Molecular Ecology, 20(17), pp. 3599–3616.

Mohamed, A. R., Cumbo, V., Harii, S., Shinzato, C., Chan, C. X., Ragan, M. A., Bourne, D. G., Willis, B. L., Ball, E. E., Satoh, N. and Miller, D. J., 2016. The transcriptomic response of the coral Acropora digitifera to a competent Symbiodinium strain: the symbiosome as an arrested early phagosome. Molecular Ecology, 25(13), pp. 3127–3141.

Molodtsova, T. and Budaeva, N., 2007. Modifications of corallum morphology in black corals as an effect of associated fauna. Bulletin of Marine Science, 81(3), pp. 469–480.

Molodtsova, T. N. and Opresko, D. M., 2017. Black corals (Anthozoa: Antipatharia) of the Clarion-Clipperton Fracture Zone. Marine Biodiversity, 47(2), pp. 349–365.

Mortensen, P. B., Hovland, M., Brattegard, T., and Farestveit, R., 1995. Deep water bioherms of the scleractinian coral Lophelia pertusa (L.) at 64 N on the Norwegian shelf: structure and associated megafauna. Sarsia, 80(2), pp. 145–158.

Mosher, C. V., and Watling, L., 2009. Partners for life: a brittle star and its octocoral host. Marine Ecology Progress Series, *397*, pp. 81–88.

Moya, A., Huisman, L., Ball, E. E., Hayward, D. C., Grasso, L.., Chua, C. M., Woo, H. N., Gattuso, J. P., Foret, S. and Miller, D. J., 2012. Whole transcriptome analysis of the coral Acropora millepora reveals complex responses to CO_2-driven acidification during the initiation of calcification. Molecular Ecology, 21(10), pp. 2440–2454.

NCBI RefSeq. https://www.ncbi.nlm.nih.gov/refseq/, Accessed 26 Jan 2021.

Opresko, D. M., 1972. Biological results of the University of Miami Deep-Sea Expeditions. 97. Redescriptions and reevaluations of the antipatharians described by LF de Pourtales. Bulletin of Marine Science, 22(4), pp. 950–1017.

Opresko, D. M. and Genin, A., 1990. A new species of antipatharian (Cnidaria: Anthozoa) from seamounts in the eastern North Pacific. Bulletin of Marine Science, 46(2), pp. 301–310.

Ottaway, J. R. and Kirby, G.C., 1975. Genetic relationships between brooding and brooded Actinia tenebrosa. Nature, 255(5505), p. 221.

Park, E., Hwang, D. S., Lee, J. S., Song, J. I., Seo, T. K. and Won, Y. J., 2012. Estimation of divergence times in cnidarian evolution based on mitochondrial protein-coding genes and the fossil record. Molecular Phylogenetics and Evolution, 62(1), pp. 329–345.

Passow, C. N., Kono, T. J., Stahl, B. A., Jaggard, J. B., Keene, A. C. and McGaugh, S. E., 2018. RNAlater and flash freezing storage methods nonrandomly influence observed gene expression in RNAseq experiments. bioRxiv, p. 379834.

Paz-García, D. A., Hellberg, M. E., García-de-León, F. J. and Balart, E. F., 2015. Switch between morphospecies of Pocillopora corals. American Naturalist, 186(3), pp. 434–440.

Pinzón, J. H., Kamel, B., Burge, C. A., Harvell, C. D., Medina, M., Weil, E. and Mydlarz, L. D., 2015. Whole transcriptome analysis reveals changes in expression of immune-related genes during and after bleaching in a reef-building coral. Royal Society Open Science, 2(4), p. 140214.

Polato, N. R., Vera, J. C. and Baums, I. B., 2011. Gene discovery in the threatened elkhorn coral: 454 sequencing of the Acropora palmata transcriptome. PLoS One, 6(12), p. e28634.

Poliseno, A., Feregrino, C., Sartoretto, S., Aurelle, D., Wörheide, G., McFadden, C. S., and Vargas, S. (2017). Comparative mitogenomics, phylogeny and evolutionary history of Leptogorgia (Gorgoniidae). Molecular Phylogenetics and Evolution, 115, 181–189.

Prada, C., Hanna, B., Budd, A. F., Woodley, C. M., Schmutz, J., Grimwood, J., Iglesias-Prieto, R., Pandolfi, J. M., Levitan, D., Johnson, K. G. and Knowlton, N., 2016. Empty niches after extinctions increase population sizes of modern corals. Current Biology, 26(23), pp. 3190–3194.

Pratlong, M., Rancurel, C., Pontarotti, P. and Aurelle, D., 2017. Monophyly of Anthozoa (Cnidaria): why do nuclear and mitochondrial phylogenies disagree?. Zoologica Scripta, 46(3), pp. 363–371.

Prouty, N. G., Roark, E. B., Andrews, A., Robinson, L., Hill, T., Sherwood, O., Williams, B., Guilderson, T. P. and Fallon, S., 2015. Age, Growth Rates, and Paleoclimate Studies of Deep Sea Corals. NOAA.

Putnam, N. H., Srivastava, M., Hellsten, U., Dirks, B., Chapman, J., Salamov, A., ... Rokhsar, D. S. (2007). Sea anemone genome reveals ancestral eumetazoan gene repertoire and genomic organization. Science, 317(5834), 86–94.

Quattrini, A. M., Faircloth, B. C., Dueñas, L. F., Bridge, T. C., Brugler, M. R., Calixto-Botía, I. F., DeLeo, D. M., Foret, S., Herrera, S., Lee, S. M. and Miller, D. J., 2018. Universal target-enrichment baits for anthozoan (Cnidaria) phylogenomics: New approaches to long-standing problems. Molecular Ecology Resources, 18(2), pp. 281–295.

Quattrini, A. M., Georgian, S. E., Byrnes, L., Stevens, A., Falco, R., & Cordes, E. E., 2013. Niche divergence by deep-sea octocorals in the genus Callogorgia across the continental slope of the Gulf of Mexico. Molecular ecology, 22(15), pp. 4123–4140.

Quattrini, A. M., Rodríguez, E., Faircloth, B. C., Cowman, P. F., Brugler, M. R., Farfan, G. A., ... and McFadden, C. S., 2020. Palaeoclimate ocean conditions shaped the evolution of corals and their skeletons through deep time. Nature Ecology & Evolution, 4(11), pp. 1531–1538.

Rashid, M. A., Gustafson, K. R., and Boyd, M. R., 2000. HIV-inhibitory cembrane derivatives from a Philippines collection of the soft coral Lobophytum species. Journal of Natural Products, 63(4), pp. 531–533.

Reft, A. J., 2012. Understanding the morphology and distribution of nematocysts in sea anemones and their relatives (Doctoral dissertation, The Ohio State University).

Richmond, R. H., 1987. Energetics, competency, and long-distance dispersal of planula larvae of the coral Pocillopora damicornis. Marine Biology, 93(4), pp. 527–533.

Richmond, R. H., 1997. Reproduction and recruitment in corals: critical links in the persistence of reefs. Life and Death of Coral Reefs. Chapman & Hall, New York, pp. 175–197.

Rinkevich, B. and Loya, Y., 1979. The reproduction of the Red Sea coral Stylophora pistillata. I. Gonads and planulae. Marine Ecology Progress Series, pp. 133–144.

Roark, E.B., Guilderson, T.P., Dunbar, R.B., Fallon, S.J. and Mucciarone, D.A., 2009. Extreme longevity in proteinaceous deep-sea corals. Proceedings of the National Academy of Sciences, 106(13), pp. 5204–5208.

Robinson, L. F., Adkins, J. F., Frank, N., Gagnon, A. C., Prouty, N. G., Roark, E. B. and van de Flierdt, T., 2014. The geochemistry of deep-sea coral skeletons: a review of vital effects and applications for palaeoceanography. Deep Sea Research Part II: Topical Studies in Oceanography, 99, pp. 184–198.

Robson, E. A., 1961. The swimming response and its pacemaker system in the anemone Stomphia coccinea. Journal of Experimental Biology, 38(3), pp. 685–694.

Rocha, J., Peixe, L., Gomes, N., and Calado, R. (2011). Cnidarians as a source of new marine bioactive compounds—An overview of the last decade and future steps for bioprospecting. Marine Drugs, 9(10), 1860–1886.

Rodríguez, E., Barbeitos, M. S., Brugler, M. R., Crowley, L. M., Grajales, A., Gusmão, L., Häussermann, V., Reft, A. and Daly, M., 2014. Hidden among sea anemones: the first comprehensive phylogenetic reconstruction of the order Actiniaria (Cnidaria, Anthozoa, Hexacorallia) reveals a novel group of hexacorals. PLoS One, 9(5), p. e96998.

Ruiz-Jones, L. J. and Palumbi, S. R., 2015. Transcriptome-wide changes in coral gene expression at noon and midnight under field conditions. The Biological Bulletin, 228(3), pp. 227–241.

Sakai, K., 1997. Gametogenesis, spawning, and planula brooding by the reef coral Goniastrea aspera (Scleractinia) in Okinawa, Japan. Marine Ecology Progress Series, 151, pp. 67–72.

Sammarco, P. W., 1982. Polyp bail-out: an escape response to environmental stress and a new means of reproduction in corals. Marine ecology progress series. Oldendorf, 10(1), pp. 57–65.

Schnytzer, Y., Giman, Y., Karplus, I. and Achituv, Y., 2017. Boxer crabs induce asexual reproduction of their associated sea anemones by splitting and intraspecific theft. PeerJ, 5, p. e2954.

Schrader, C., Schielke, A., Ellerbroek, L. and Johne, R., 2012. PCR inhibitors–occurrence, properties and removal. Journal of Applied Microbiology, 113(5), pp. 1014–1026.

Shea, E. K., Ziegler, A., Faber, C. and Shank, T. M., 2018. Dumbo octopod hatchling provides insight into early cirrate life cycle. Current Biology, 28(4), pp. R144–R145.

Shearer, T. L., Van Oppen, M. J. H., Romano, S. L. and Wörheide, G., 2002. Slow mitochondrial DNA sequence evolution in the Anthozoa (Cnidaria). Molecular Ecology, 11(12), pp. 2475–2487.

Shick, J. M., 1991. A Functional Biology of Sea Anemones. Chapman & Hall, London. p. 395.

Shinzato, C., Shoguchi, E., Kawashima, T., Hamada, M., Hisata, K., Tanaka, M., ... and Satoh, N., 2011. Using the Acropora digitifera genome to understand coral responses to environmental change. *Nature*, *476*(7360), 320–323.

Simion, P., Philippe, H., Baurain, D., Jager, M., Richter, D. J., Di Franco, A., Roure, B., Satoh, N., Queinnec, E., Ereskovsky, A. and Lapebie, P., 2017. A large and consistent phylogenomic dataset supports sponges as the sister group to all other animals. Current Biology, 27(7), pp. 958–967.

Smith, L. D. and Hughes, T. P., 1999. An experimental assessment of survival, re-attachment and fecundity of coral fragments. Journal of Experimental Marine Biology and Ecology, *235*(1), pp. 147–164.

Stampar, S. N., Maronna, M. M., Kitahara, M. V., Reimer, J. D. and Morandini, A. C., 2014. Fast-evolving mitochondrial DNA in Ceriantharia: a reflection of Hexacorallia paraphyly?. PLoS One, 9(1), p. e86612.

Stampar, S. N., Broe, M. B., Macrander, J., Reitzel, A. M., Brugler, M. R., and Daly, M., 2019. Linear mitochondrial genome in Anthozoa (Cnidaria): a case study in Ceriantharia. Scientific reports, 9(1), pp. 1–12.

Steinmetz, P. R., Aman, A., Kraus, J. E. and Technau, U., 2017. Gut-like ectodermal tissue in a sea anemone challenges germ layer homology. Nature Ecology & Evolution, 1(10), p. 1535.

Stoddart, J. A., 1983. Asexual production of planulae in the coral Pocillopora damicornis. Marine Biology, 76(3), pp. 279–284.

Sultan, M., Amstislavskiy, V., Risch, T., Schuette, M., Dökel, S., Ralser, M., Balzereit, D., Lehrach, H. and Yaspo, M. L., 2014. Influence of RNA extraction methods and library selection schemes on RNA-seq data. BMC Genomics, 15(1), p. 675.

Tarrant, A. M. 2005. Endocrine-like signaling in cnidarians: current understanding and implications for ecophysiology. Integrative and Comparative Biology, 45, 201–214.

Tazioli, S., Bo, M., Boyer, M., Rotinsulu, H. and Bavestrello, G., 2007. Ecological observations of some common antipatharian corals in the marine park of Bunaken (North Sulawesi, Indonesia). Zoological Studies, 46(2), pp. 227–241.

Technau, U., Rudd, S., Maxwell, P., Gordon, P. M., Saina, M., Grasso, L. C., Hayward, D. C., Sensen, C. W., Saint, R., Holstein, T. W. and Ball, E. E., 2005. Maintenance of ancestral complexity and non-metazoan genes in two basal cnidarians. TRENDS in Genetics, 21(12), pp. 633–639.

Technau, U. and Schwaiger, M., 2015. Recent advances in genomics and transcriptomics of cnidarians. Marine Genomics, 24, pp. 131–138.

Thomas, J. A., Welch, J. J., Lanfear, R. and Bromham, L., 2010. A generation time effect on the rate of molecular evolution in invertebrates. Molecular Biology and Evolution, 27(5), pp. 1173–1180.

Tucker, R. P., Shibata, B. and Blankenship, T. N., 2011. Ultrastructure of the mesoglea of the sea anemone Nematostella vectensis (Edwardsiidae). Invertebrate Biology, 130(1), pp. 11–24.

Tulin, S., Aguiar, D., Istrail, S. and Smith, J., 2013. A quantitative reference transcriptome for Nematostella vectensis early embryonic development: a pipeline for de novo assembly in emerging model systems. EvoDevo, 4(1), p. 16.

Vidal-Dupiol, J., Zoccola, D., Tambutté, E., Grunau, C., Cosseau, C., Smith, K. M., Freitag, M., Dheilly, N. M., Allemand, D. and Tambutté, S., 2013. Genes related to ion-transport and energy production are upregulated in response to CO2-driven pH decrease in corals: new insights from transcriptome analysis. PloS One, 8(3), p. e58652.

Wagner, D., Pochon, X., Irwin, L., Toonen, R. J. and Gates, R. D., 2011. Azooxanthellate? Most Hawaiian black corals contain Symbiodinium. Proceedings of the Royal Society of London B: Biological Sciences, 278(1710), pp. 1323–1328.

Wallace, C. C., 1985. Reproduction, recruitment and fragmentation in nine sympatric species of the coral genus Acropora. Marine Biology, 88(3), pp. 217–233.

Wang, X., Liew, Y. J., Li, Y., Zoccola, D., Tambutte, S., and Aranda, M., 2017. Draft genomes of the corallimorpharians *Amplexidiscus fenestrafer* and *Discosoma* sp. Molecular Ecology Resources, 17(6), e187–e195.

Wedi, S. E. and Dunn, D. F., 1983. Gametogenesis and reproductive periodicity of the subtidal sea anemone *Urticina lofotensis* (Coelenterata: Actiniaria) in California. The Biological Bulletin, 165(2), pp. 458–472.

Wendt, P. H., Van Dolah, R. F. and O'Rourke, C. B., 1985. A comparative study of the invertebrate macrofauna associated with seven sponge and coral species collected from the South Atlantic Bight. Journal of the Elisha Mitchell Scientific Society, pp. 187–203.

Williams, B., Risk, M. J., Ross, S. W. and Sulak, K. J., 2006. Deep-water antipatharians: proxies of environmental change. Geology, 34(9), pp. 773–776.

Wilson, I. G., 1997. Inhibition and facilitation of nucleic acid amplification. Applied and Environmental Microbiology, 63(10), p. 3741.

Xiao, M., Brugler, M. R., Broe, M. B., Gusmao, L. C., Daly, M. and Rodriguez, E., in press. Mitogenomics suggests a sister relationship of Relicanthus daphneae (Cnidaria: Anthozoa: Hexacorallia: incerti ordinis) with Actiniaria. Scientific Reports.

Zapata, F., Goetz, F. E., Smith, S. A., Howison, M., Siebert, S., Church, S. H., ... and Cartwright, P., 2015. Phylogenomic analyses support traditional relationships within Cnidaria. PLoS One, 10(10), e0139068.

Additional Reading

Appeltans, W., Ahyong, S. T., Anderson, G., Angel, M. V., Artois, T., Bailly, N., Bamber, R., Barber, A., Bartsch, I., Berta, A. and Błażewicz-Paszkowycz, M., 2012. The magnitude of global marine species diversity. Current Biology, 22(23), pp. 2189–2202.

Arrigoni, R., Kitano, Y. F., Stolarski, J., Hoeksema, B. W., Fukami, H., Stefani, F., Galli, P., Montano, S., Castoldi, E. and Benzoni, F., 2014. A phylogeny reconstruction of the Dendrophylliidae (Cnidaria, Scleractinia) based on molecular and micromorphological criteria, and its ecological implications. Zoologica Scripta, 43(6), pp. 661–688.

Cairns, S. D., 2001. A generic revision and phylogenetic analysis of the Dendrophylliidae (Cnidaria: Scleractinia).

Cleves, P. A., Strader, M. E., Bay, L. K., Pringle, J. R. and Matz, M. V., 2018. CRISPR/Cas9-mediated genome editing in a reef-building coral. Proceedings of the National Academy of Sciences, 115(20), pp. 5235–5240.

DeLeo, D. M., Ruiz-Ramos, D. V., Baums, I.B. and Cordes, E. E., 2016. Response of deep-water corals to oil and chemical dispersant exposure. Deep Sea Research Part II: Topical Studies in Oceanography, 129, pp. 137–147.

Fautin, D. G., 2016. Catalog to families, genera, and species of orders Actiniaria and Corallimorpharia (Cnidaria: Anthozoa). Zootaxa, 4145(1), pp. 1–449.

Fautin, D. G., Malarky, L. and Soberon, J., 2013. Latitudinal diversity of sea anemones (Cnidaria: Actiniaria). The Biological Bulletin, 224(2), pp. 89–98.

Fine, M. and Tchernov, D., 2007. Scleractinian coral species survive and recover from decalcification. Science, 315(5820), p. 1811.

Grupstra, C. G., Coma, R., Ribes, M., Leydet, K. P., Parkinson, J. E., McDonald, K., Catlla, M., Voolstra, C. R., Hellberg, M. E. and Coffroth, M. A., 2017. Evidence for coral range expansion accompanied by reduced diversity of Symbiodinium genotypes. Coral Reefs, 36(3), pp. 981–985.

Hoeksema B., Cairns S., 2018. World List of Scleractinia. Accessed through: World Register of Marine Species at: http://www.marinespecies.org/index.php.

Matz, M. V., Treml, E. A., Aglyamova, G. V. and Bay, L. K., 2018. Potential and limits for rapid genetic adaptation to warming in a Great Barrier Reef coral. PLoS Genetics, 14(4), p. e1007220.

Roberts, J. M., Wheeler, A., Freiwald, A. and Cairns, S., 2009. Cold-water corals: the biology and geology of deep-sea coral habitats. Cambridge University Press.

Stampar, S. N., Maronna, M. M., Kitahara, M. V., Reimer, J. D., Beneti, J. S. and Morandini, A. C., 2016. Ceriantharia in current systematics: life cycles, morphology and genetics. In The Cnidaria, Past, Present and Future (pp. 61–72). Springer, Cham.

Titus, B. M. and Daly, M., 2018. Reduced representation sequencing for symbiotic anthozoans: are reference genomes necessary to eliminate endosymbiont contamination and make robust phylogeographic inference?. bioRxiv, p. 440289.

Williams, G. C., 2011. The global diversity of sea pens (Cnidaria: Octocorallia: Pennatulacea). PLoS One, 6(7), p. e22747.

PHYLUM CNIDARIA: *CLASS HYDROZOA*

Neil W. Blackstone and Paulyn Cartwright

CHAPTER **12**

Infrakingdom	Bilateria
1st Division	Diploblastic Animals
Phylum	Cnidaria
Class	Hydrozoa

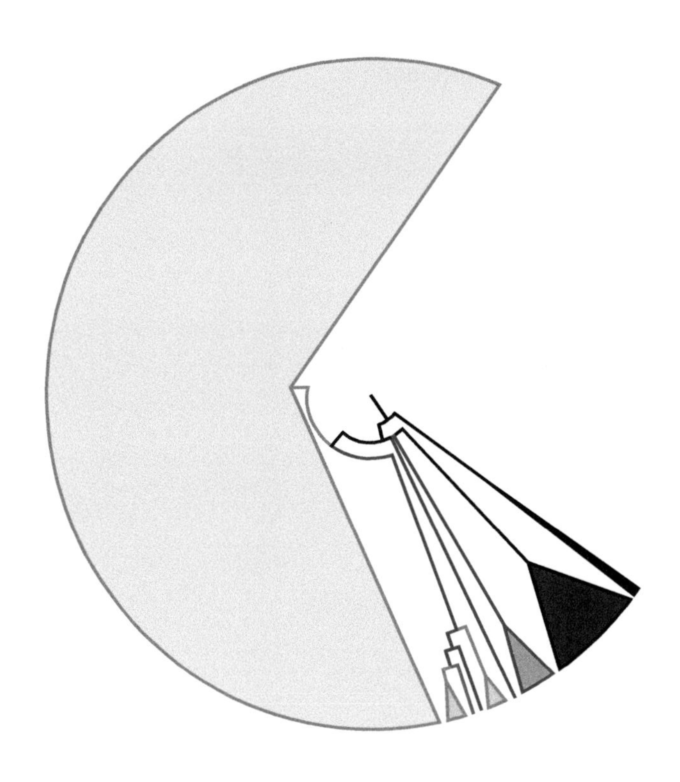

CONTENTS

General Characteristics

Roughly 3,800 species

Most diverse group of cnidarians

Marine and freshwater representatives

Polyps or medusae, or both, sometimes even included in the same life-cycle stage

Polyp stages frequently colonial

Usually carnivorous, sometimes symbiotic

A few species of commercial value

Moderate ecological importance

Autapomorphies

- gap junctions (unique within Cnidaria)
- medusa budded laterally
- medusa with velum and two nerve rings
- polyp lacks septae
- ectodermal origin and location of gametes

TAXONOMY AND CLASSIFICATION

A broad consensus for the classification of the Hydrozoa remains elusive. The classification below (to the subordinal level) has much to recommend it, although some taxa are generally acknowledged to be paraphyletic.

Hydrozoa
- Hydroidolina
  - Anthoathecata
    - Aplanulata
    - Capitata
    - Filifera
  - Leptothecata
    - Conica
    - Proboscoida
  - Siphonophora
    - Calycophorae
    - Cystonectae
    - Physonectae
- Trachylina
  - Limnomedusae
  - Narcomedusae
  - Trachymedusae
  - Actinulida

Aspects of this classification are strongly supported. Both molecular and morphological analyses support the monophyly of Hydrozoa, Hydroidolina, and Trachylina. On the other hand, Anthoathecata is not supported as monophyletic by molecular analyses. Features that putatively unite anthoathecates, e.g., the lack of a skeletal covering of the polyps, are likely shared primitive characters. Loss of the ciliated planula stage and molecular characters unite the monophyletic Aplanulata. Rounded or capitate tentacles at some stage of the life cycle and stenotele nematocysts unite Capitata. Filifera is clearly paraphyletic and includes several groups with strong to moderate support. Filiferan tentacles thus appear to be a shared primitive character for members of this group. Leptothecates, traditionally united by skeletal covering of the polyps, are also strongly united by molecular characters. The monophyly of Conica and Probscoida, however, is not yet clear. Similar to Leptothecata, molecular analyses recover a monophyletic Siphonophora that is derived from within anthoathecate lineages. Relationships among the major clades of the Hydroidolina remain obscure. Within Trachylina, molecular analyses support a monophyletic Narcomedusae, a paraphyletic Limnomedusae, and Actinulida derived within a paraphyletic Trachymedusae. Representatives of the latter are characterized by ecto-endodermal statocysts. Trachylina comprises holopelagic taxa (Narcomedusae, Trachymedusae), which lack a polyp stage, and Limnomedusae, which has a polyp stage. The medusae have gonads on their radial canals. The representatives of the enigmatic Actinulida are interstitial. The presence of statocysts leads to the interpretation that these are diminutive medusae. Their derivation within the holopelagic Tracheymedusae (that lacks a polyp stage) further supports their medusoid origin. A clade of freshwater Limnomedusae was recovered with molecular data, suggesting a single origin of the freshwater habit.

ANATOMY, HISTOLOGY AND MORPHOLOGY

Many features of hydrozoans are found in other cnidarians as well. They are diploblastic with two embryonic tissue layers and a primarily ectodermal derived acellular mesoglea, which separates the epidermis and gastrodermis (**Figure 12.1**). These tissues are largely made up of myoepithelial cells, sometimes termed epitheliomuscular cells. The gastrodermis encloses the gastrovascular system, which serves both nutritive and transport functions. In addition to epithelial cells, hydrozoans have a population of interstitial stem cells (i-cells), which produce various somatic cells including cnidocytes, the stinging cells that characterize all cnidarians (**Figure 12.2**). I-cells also produce gametes; thus, a germ line that is separate from these stem cells is lacking. Other cell types include sensory, secretory and nerve cells. The last typically form arrays or nerve nets. Small and transparent medusae usually exhibit a velum, a ring canal, and mouth on a pendant manubrium (**Figure 12.3**). Rhophalia are lacking. Polyps may be solitary or colonial, with a hypostome surrounding the mouth, a body column or gastric region, a peduncle and a basal region (Figure 12.3). Both oral and marginal tentacles may be present. Colonial polyps share a common coenosarc and gastrovascular system and may be encrusting or exhibit arborescent growth forms. Some hydrozoans display a division of labor with morphologically

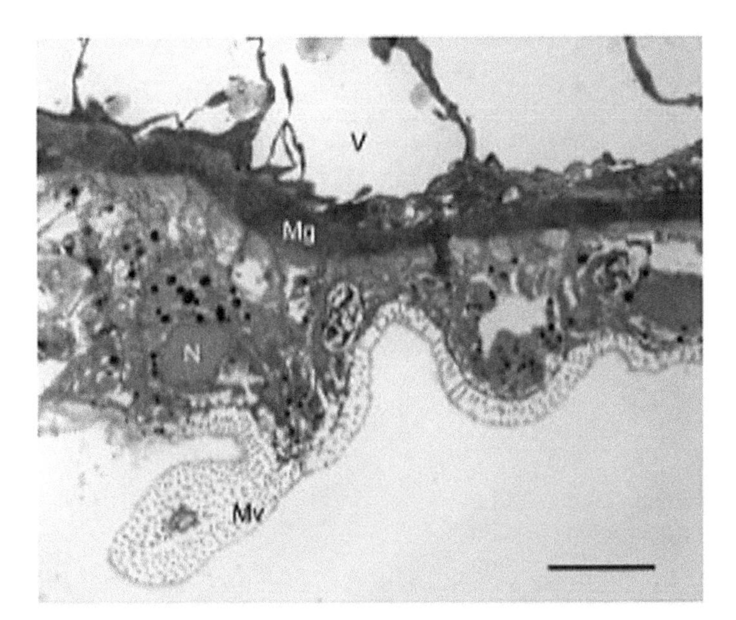

Figure 12.1 Transmission electron micrograph of the polyp of *Podocoryna carnea*, illustrating the tissue layers of hydrozoans. Above the acellular mesoglea (Mg) lie the cells of the gastrodermis; below lie the cells of the epidermis, separated from the external environment by a layer of microvilli (Mv). (N = nucleus; V = vacuole; scale = 5 µm.)

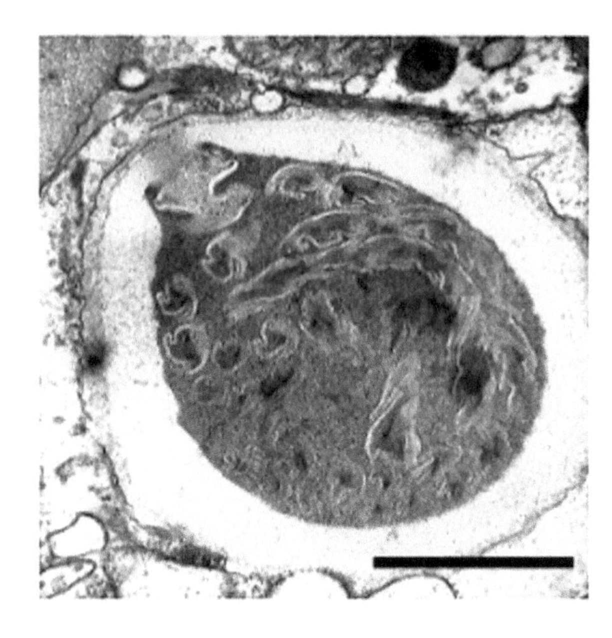

Figure 12.2 **As with other cnidarians, hydrozoans possess cnidae, which are produced by cells such as this cnidoblast.** Cnidae function in prey capture, defense, attachment and locomotion. Scale = 2 μm.

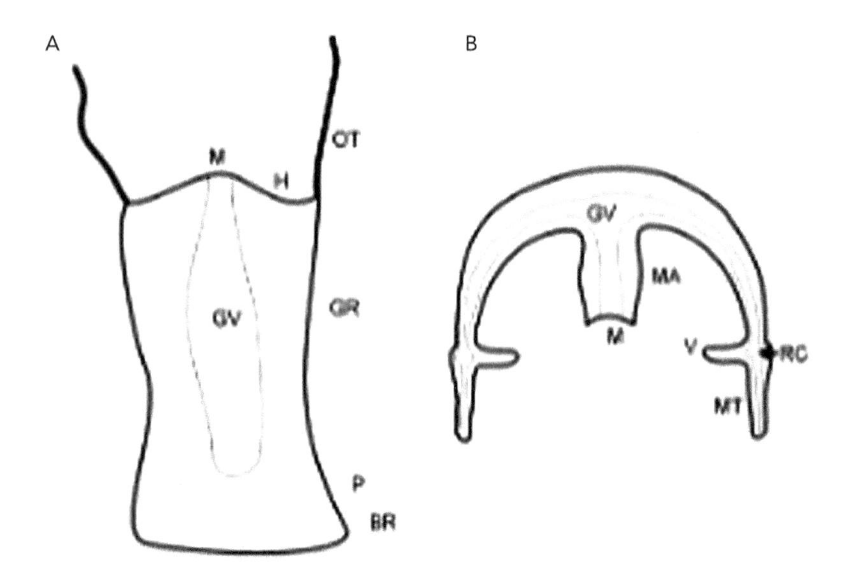

Figure 12.3 **Schemata of transverse sections of (A) polyp and (B) medusa.** (BR = basal region; GR = gastric region; GV = gastrovascular cavity; H = hypostome; M = mouth; MA = manubrium; MT = marginal tentacle; OT = oral tentacle; P = peduncle; RC = ring canal; V = velum.)

distinct polyp types such as feeding gastrozooids, reproductive gonozooids and defensive dactylozooids. Both polyps and medusae lack a pharynx and septa. Structural material, when present, is usually chitinous, but occasionally calcium carbonate.

REPRODUCTION AND LIFE CYCLE

Hydrozoans exhibit complex and diverse life cycles. What may be considered a typical hydrozoan life cycle alternates between asexual, benthic polyps, which may be colonial, and sexual planktonic medusae, produced by lateral budding of the entocodon (**Figure 12.4**). In many species, however, polyps or medusae may be suppressed, e.g. with vestigial medusae retained by the polyps, or entirely absent. For instance, a number of trachyline species develop from a planula into a medusa, while polyps of hydras lack gonophores and directly produce gametes within their epithelia. Numerous variant life cycles are also found,

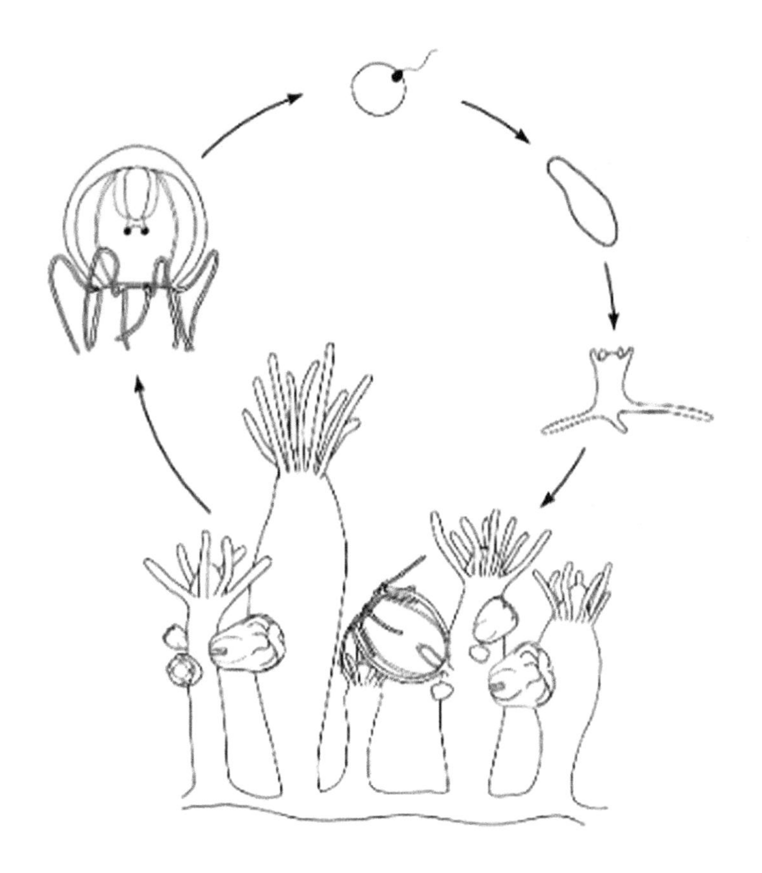

Figure 12.4 Schemata of an idealized life cycle of a hydrozoan, represented here by *Podocoryna carnea*. The colonial benthic polyps bud and release pelagic medusae, which produce gametes. Gametes fuse with those of medusae from another colony, forming a zygote, which develops into a ciliated planula larva, eventually giving rise to a new colony. Many hydrozoans deviate from this idealized life cycle, particularly in the development of free-swimming medusae.

e.g. the vegetative walking medusa of *Eleutheria dichotoma* and the development of a polyp from a medusa in *Turritopsis nutricula*. Remarkably, polyps and medusae are physically connected and physiologically integrated into single colonies in siphonophores. The diversity and abundance of hydrozoans derives in large part from the evolutionarily flexible deployment of the characteristics of the life cycle in relation to particular ecological niches.

Hydrozoan phylogeny reveals complex patterns of character evolution. Related to the evolution of the medusa is the position of the polyp's gonophore bud, which can produce medusae or gametes. Functionally specialized or polymorphic polyps may be found in some colonial forms, with distinct gonozooids the most common polymorphism (note that this is also related to the placement of the gonophore). While inferences are somewhat limited due to the ambiguous relationships among the major clades, some findings are nevertheless apparent. A medusa stage appears to be ancestral for Hydrozoa. Medusae were lost several times and likely regained at least twice. Polymorphism, particularly as represented by the presence of reproductive polyps, is widespread in colonial forms. The latter appears to be related to the evolution of gonozooids, and both accomplish a "division of labor" between feeding and reproductive activities. Developmental regulatory genes and pathways (e.g. the Hox and Wnt) have been implicated in major evolutionary changes in hydrozoan patterning and life-cycle changes. A solitary polyp stage appears to be ancestral for hydrozoans. After the divergence of Trachylina and Hydroidolina, coloniality was derived independently in the latter and in the trachyline clade Limnomedusae.

DISTRIBUTION AND ECOLOGY

Hydrozoans are widely distributed in all marine habitats and are found in freshwater as well (**Figure 12.5**). Temperature and salinity are likely major determinants of the distribution of individual taxa. The ecological effects of

Figure 12.5 Major areas of hydrozoan distribution are indicated in red. Hydromedusae, of course, may be more widely distributed in the ocean, and hydras are found in many freshwater lakes and rivers that are not indicated.

hydromedusae may be overshadowed by other carnivorous jellyfish, e.g. ctenophores and scyphozoans. Nevertheless, the latter may be more sensitive to low salinities, allowing hydromedusae to have a significant impact in some areas. The impact of "white weed" beds of colonial hydroids (*Sertularia* spp.) on benthic ecology is no doubt significant as well, although this remains largely unexplored. In parallel, *Hydra* (**Figure 12.6**) may be a significant predator in some freshwater lakes and ponds. Similar to several other species from the Caspian Sea, the freshwater hydrozoan *Cordylophora caspia* is an invasive species to North America and poses a threat to freshwater ecosystems there.

While perhaps not as noteworthy as those in anthozoans, symbioses are nonetheless important to the ecology of some hydrozoans. For instance, the tropical hydroids *Millepora* and *Myrionema* host dinoflagellates (*Symbiodinium*) much like tropical corals. Green hydra host symbionts from the green algae clade (*Chlorella*). The factors that underlie the greater frequency of photosymbioses in anthozoans relative to hydrozoans remain unexplored. As with many organisms, in cnidarians there is an increasing appreciation of the ecological role of symbiotic microbes and of the evolutionary interactions underlying these symbioses.

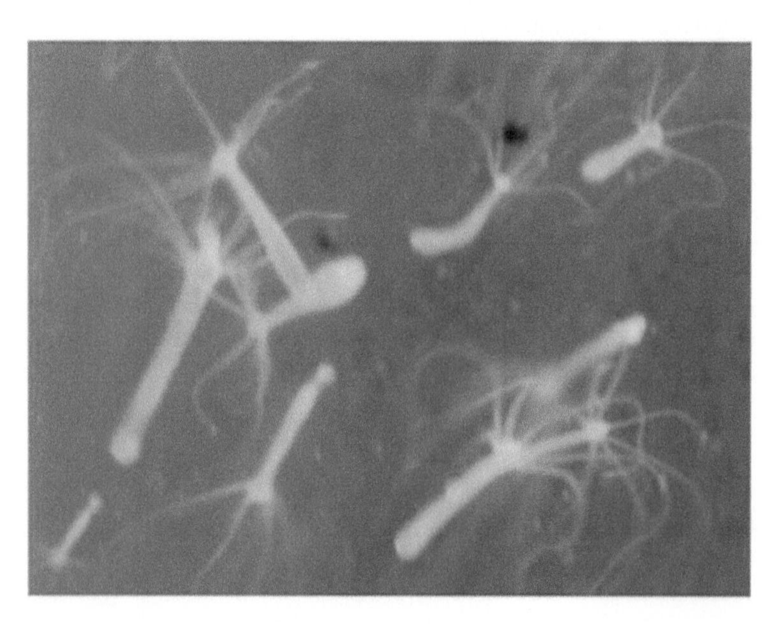

Figure 12.6 Hydras, such as the group shown here, are solitary, freshwater hydroids that can be important predators in some ecosystems. Adult polyps are about 1 cm in height.

PHYSIOLOGY AND BEHAVIOR

The mobile stages of the hydrozoan life cycle (i.e. planula, medusa) are most likely to exhibit complex behavior. Although the cues that trigger the metamorphosis of planulae have attracted some attention, other aspects of behavior (e.g. with regard to mating) remain largely unexplored. In polyps and perhaps in medusae as well, glutathione triggers a feeding response (e.g. tentacle movements and an opening of the mouth). Glutathione may also have complex effects on some hydroid colonies.

The centrality of the gastrovascular system to physiology has drawn considerable attention to this system. A hydroid colony exhibits a continuous gastrovascular cavity connecting polyps via the coenosarc. The gastrovascular system circulates food and other nutrients and removes waste from the colony. Patterns of pressure, shear stress and surface tension can vary throughout the system depending on various environmental inputs. In this way, the gastrovascular system integrates information from the environment with the metabolic state of the colony, providing an effective mediator of environmental signaling (see also the section below on transport systems).

GENETICS AND GENOMICS

Whole genome and transcriptome sequencing in species of Hydrozoa has revealed that hydrozoans possess complex genomes with evidence of extensive transposal element (TE) expansion, life-cycle-specific gene loss, gene duplication and horizontal gene transfer (HGT). Within the genus *Hydra*, genome size ranges from 300 Mb to 1 Gb. These size differences are thought to be primarily through extensive differential TE expansion. A number of surprising insights into hydrozoan evolution have been revealed through comparing genomes and transcriptomes of hydrozoan model systems. For example, hydrozoan genomes contain a UDP-glucose 6-dehydrogenase-like (UGDH) gene that was acquired through HGT by a virus of the Mimiviridae family. In the hydrozoan *Clytia hemisphaerica* the expression of this UGDH is restricted during the period of medusa formation. Interestingly, this gene is absent in *Hydra*, which lacks a medusa stage, and instead *Hydra* possesses a bacterial derived UGDH gene. Comparing genomes of *Hydra* and *Clytia* also reveals a number of transcription factors that are specific to *Clytia* medusa development and have been lost in *Hydra*. Medusa-specific genes have also been identified in the transcriptome of *Podocoryna carnea* and are found to be lost or downregulated in its non-medusa-bearing relative *Hydractinia symbiolongicarpus* (**Figure 12.7**).

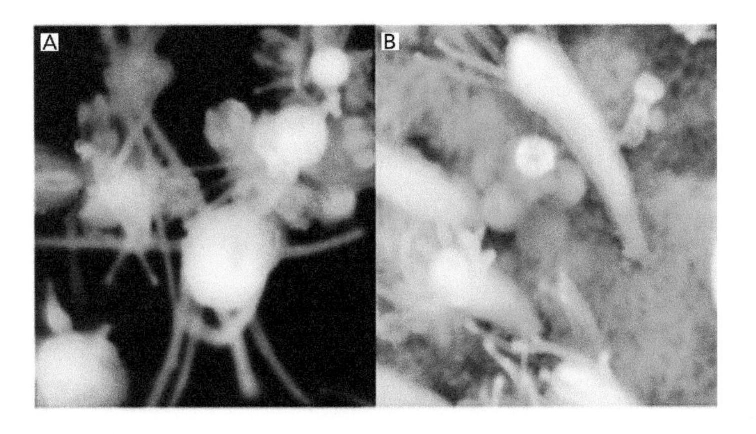

Figure 12.7 Colonies of two species of hydractiniid hydroid. (A) *Podocoryna carnea* exhibits a runner-like growth form and includes a medusa stage (several polyps have medusa buds). (B) *Hydractinia symbiolongicarpus* exhibits a sheet-like growth form and does not include a medusa stage (a reproductive polyp bearing gametes is in the center of the image).

POSITION IN THE ToL

The discovery of linear mitochondrial chromosomes in *Hydra* and subsequent work indicating that this is a shared derived character for Medusozoa (Staurozoa + Cubozoa + Scyphozoa + Hydrozoa) has greatly clarified relationships within Cnidaria. On the basis of this and other characters, it is now widely accepted that Cnidaria comprises two reciprocally monophyletic clades, Anthozoa and Medusozoa. Within the Medusozoa, some analyses suggest that Staurozoa may be sister to all other medusozoans, with Cubozoa and Scyphozoa forming a clade that is sister to the Hydrozoa. Alternatively, Staurozoa + Cubozoa + Scyphozoa may be sister to the Hydrozoa (**Figure 12.8**). In addition, Cnidaria has been well established as a sister group to Bilateria.

DATABASES AND COLLECTIONS

The World Hydrozoa Database (www.marinespecies.org/hydrozoa) is linked to the World Register of Marine Species (marinespecies.org), so they are duplicates of each other. They are curated by Peter Schuchert and are very up to date on the latest taxonomy. The Yale Peabody Museum and the National Museum of Natural History probably have the most extensive hydrozoan collection in the United States. Other institutions notable in this regard would be Royal Ontario Museum (curator Dale Calder) and Natural History Museum of Geneva (curator Peter Schuchert).

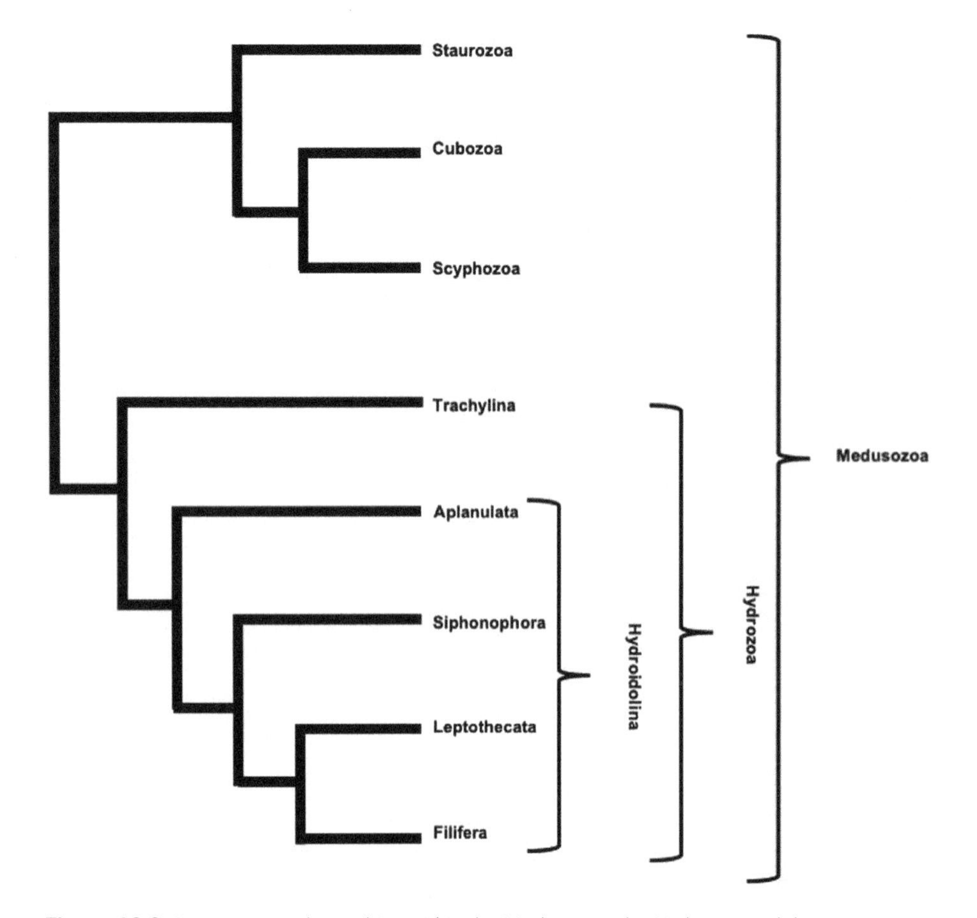

Figure 12.8 Sister-group relationships within the Medusozoa, the Hydrozoa and the Hydroidolina.

CONCLUSION

Character evolution. As discussed above, improving phylogenetic resolution has been followed by a greater understanding of character evolution in Hydrozoa. The impact of this research has been particularly pronounced in regard to putative homologies between hydrozoan and bilaterian features. For instance, based on evolutionary developmental studies of hydromedusae, striated muscle in hydroids appears to be homologous to that of vertebrates. However, if medusae were derived within the medusozoans, it is unlikely that a feature of these medusae would be shared with vertebrates. Research delving into the molecular toolkit of muscle development reveals an even more complex picture. A core set of contractile proteins characteristic of striated muscles in vertebrates was found to be conserved not only in sponges, which lack muscles, but even in unicellular organisms, which clearly lack any specialized cells. Similar challenges confront attempts to draw homologies between other hydrozoan and bilaterian characteristics (e.g. the entocodon and mesoderm). These insights demonstrate the power of the comparative approach.

Model systems. Given their phylogenetic position as early-diverging metazoans, studies of hydroids and other cnidarians have the potential to bridge gaps in the "yeast-worms-flies-mice" paradigms of modern biology. *Hydra* has a long history of study as a model system. Studies of regeneration dating back over a century led to more general studies of pattern formation, stem cells, transgenics and the first cnidarian genome. Given that *Hydra* is a representative of the highly derived Aplanulata, a case could be made for studies of more typical hydrozoans, e.g. marine, colonial forms that either include a medusoid state in their life cycle or have close relatives that do so. In addition, from a practical standpoint, modern genome editing techniques, such as CRISPR, rely on a consistent abundant supply of embryos. Hence, study of emerging models systems such *Clytia*, *Hydractinia* and *Podocoryna*, which display more complex life cycles and produce abundant gametes, promises to rapidly augment our current understanding of cellular and developmental processes in Cnidaria.

Stem cells. The persistence of a multipotent stem cell lineage (i-cells) throughout the adult polyp stage of hydrozoans is a feature that enables them to undergo clonal (asexual) reproduction. I-cells in *Hydra* and *Hydractinia* have been shown to give rise to nerve cells, gland cells, nematocysts and germ cells. In addition, i-cells have been shown to give rise to epithelial cells in *Hydractinia*, but not in *Hydra*. Frank and colleagues have used *Hydractinia* as a model for studying stem cell biology to better understand the molecular mechanisms responsible for the persistence of a stem cell line and its role in regeneration and development. They have identified several genes, including *Vasa* and *Piwi* (conserved markers for germ line cells in bilaterians) that appear responsible for the ability to maintain stem cell properties and self-renewal and play a key role in regeneration.

Allorecognition. A striking feature of colonial hydrozoans is the aggressive interactions that sometimes occur between colonies (**Figure 12.9**). As part of the larger phenomenon of histocompatibility or allorecognition, these interactions have attracted considerable study, particularly by Buss and colleagues. As colonies encrust a surface, they may encounter other colonies. While interspecific colonies necessarily compete, conspecific colonies may compete, or they may fuse. The latter option, however, does not eliminate competition, but rather shifts the level of competition from that of the colony to that of the cell, i.e. multipotent stem cells from one colony can enter the other colony and potentially replace the resident cells. Considerable effort has focused on elucidating the molecular genetic underpinnings of these interactions and recent genome sequencing methods promise to provide much insight into this complex mechanism. Similarities to the vertebrate immune system appear to be another fascinating case of parallelism, since the colonial habit appears to be derived within the Hydrozoa.

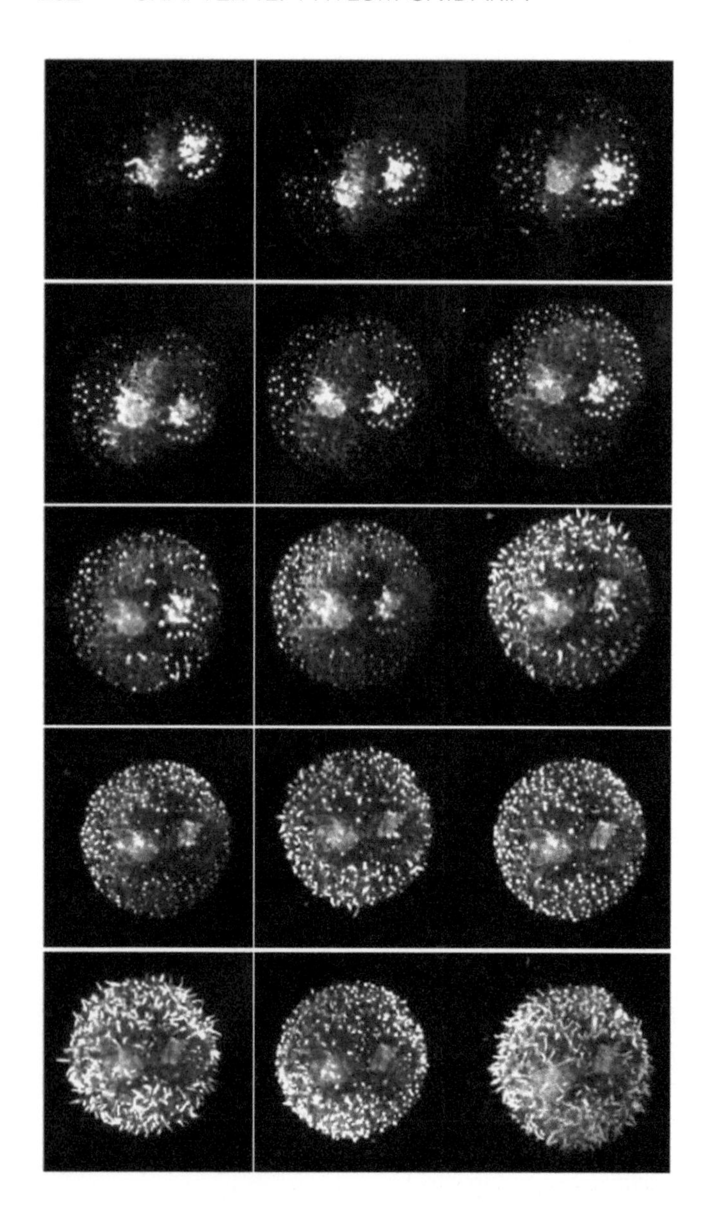

Figure 12.9 An aggressive encounter between two species of encrusting hydractiniid hydroids, growing on 15 mm diameter cover glass. Images were taken every several days for 50 days. Initially, the colony on the right (*Hydractinia symbiolongicarpus*) mounts a powerful attack with hyperplastic stolons. The colony on the left (*Podocoryna carnea*) is nearly overwhelmed but after flanking the colony of *H. symbiolongicarpus*, it eventually prevails.

Transport systems. In hydrozoan colonies, it is the common gastrovascular system that mediates many emergent features. In this regard, several derived features of hydrozoan colonies relative to anthozoans have been suggested, e.g. muscle-driven flow and bands of mitochondrion-rich myoepithelial at polyp–stolon junctions that function as valves (**Figure 12.10**). These and other features may lead to greater colony integration in hydrozoans, although more study is needed.

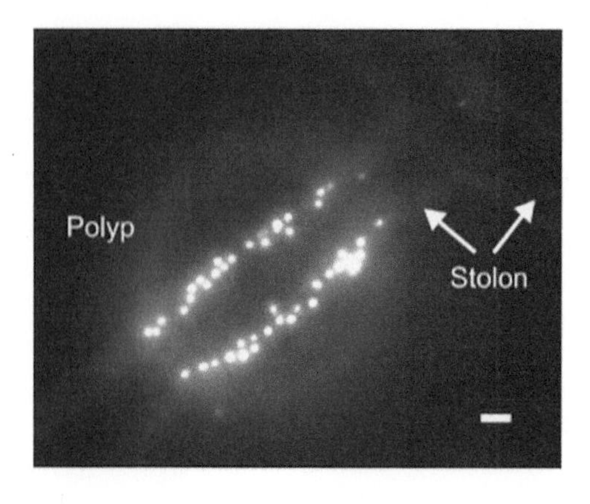

Figure 12.10 Fluorescent micrograph of a section of a living colony of *Podocoryna carnea*, visualized with an inverted microscope. Colony was treated for 1 h with a probe that detects reactive oxygen species (ROS). Tissue of the polyp and stolon dimly fluoresce, while the skeletal material of the perisarc fluoresces to a slightly greater extent. Two rows of cells at the polyp–stolon junction strongly fluoresce. Negative controls and transmission electron microscopy indicate that this fluorescence is due to high levels of mitochondrial ROS, which likely functions in signaling. Scale = 10 μm.

REFERENCES

Bang, C., Dagan, T., Deines, P., Dubilier, N., Duschl, W. J., Fraune, S., Hentschel, U., Hirt, H., Hülter, N., Lachnit, T., Picazo, D., Pita, L., Pogoreutz, C., Rädecker, N., Saad, M. M., Schmitz, R. A., Schulenburg, H., Voolstra, C. R., Weiland-Bräuer, N., Ziegler, M., Bosch, T. C. G., 2018. Metaorganisms in extreme environments: do microbes play a role in organismal adaptation. Zoology, 127, 1–19.

Blackstone, N. W., Golladay, J. M., 2018. Why do corals bleach? Conflict and conflict mediation in a host/symbiont community. BioEssays, 40, 1800021.

Blackstone, N. W., Steele, R. E., 2005. Introduction to the symposium. Integrative and Comparative Biology, 45, 583–584.

Bode, H. R., 2003. Head regeneration in *Hydra*. Developmental Dynamics, 226, 225–236.

Bosch, T. C. G., 2009. Hydra and the evolution of stem cells. BioEssays, 31, 478–486.

Bouillon, J., Boero, F., Cicogna, F., Cornelius, P. F. S. (eds.) 1987. Modern Trends in the Systematics, Ecology, and Evolution of Hydroids and Hydromedusae. Oxford, Clarendon Press.

Brusca, R., Moore, W., Shuster, S. M. 2016. Invertebrates. 3rd edn. Sinauer: Sunderland, MA.

Buss, L. W., Grosberg, R. K., 1990. Morphogenetic basis for phenotypic differences in hydroid competitive behavior. Nature, 343, 63–66.

Cartwright, P., Bowsher, J., Buss, L. W., 1999. Expression of a Hox gene, *Cnox-2*, and the division of labor in a colonial hydroid. Proceedings of the National Academy of Sciences USA, 96, 2183–2186.

Cartwright, P., Evans, N. M., Dunn, C. W., Marques, A. C., Miglietta, M. P., Schuchert, P., Collins, A. G., 2008. Phylogenetics of Hydroidolina (Hydrozoa: Cnidaria). Journal of the Marine Biological Association of the United Kingdom, 88, 1663–1672.

Cartwright, P., Nawrocki, A. M., 2010. Character evolution in the Hydrozoa (phylum Cnidaria). Integrative and Comparative Biology, 50, 456–472.

Collins, A. G., Schuchert, P., Marques, A. C., Jankowski, T., Medina, M., Schierwater, B. 2006. Medusozoan phylogeny and character evolution clarified by new large and small subunit rDNA data and an assessment of the utility of phylogenetic mixture models. Systematic Biology, 55, 97–115.

Collins, A. G., Bentlage, B., Lindner, A., Lindsay, D., Haddock, S. H. D., Jarms, G., Norenburg, J. L., Jankowski, T., Cartwright, P., 2008. Phylogenetics of Trachylina (Cnidaria: Hydrozoa) with new insights on the evolution of some problematical taxa. Journal of the Marine Biological Association of the United Kingdom, 88, 1673–1685.

Daly, M., Brugler, M. R., Cartwright, P., Collins, A. G., Dawson, M. N., Fautin, D. G., France, S. C., McFadden, C. S., Opresko, D. M., Rodriguez, E., Romano, S. L., Stake, J. L., 2007. The phylum Cnidaria: A review of phylogenetic patterns and diversity 300 years after Linnaeus. Zootaxa, 1668, 127–182.

Davy, S. K., Allemand, D., Weis, V. M., 2012. Cell biology of cnidarian-dinoflagellate symbiosis. Microbiology and Molecular Biology Reviews, 76, 229–261.

Denker, E., Manuel, M., Leclère, L., Le Guyader, H., Rabet, N. 2008. Ordered progression of nematogenesis from stem cells through differentiation stages in the tentacle bulb of *Clytia hemisphaerica* (Hydrozoa: Cnidaria). Developmental Biology, 315, 99–113.

Dunn, C. W., Pugh, P. R., Haddock, S. H., 2005. Molecular phylogenetics of the Siphonophora (Cnidaria), with implications for the evolution of functional specialization. Systematic Biology, 54, 916–935.

Folino-Rorem, N. C., Darling, J. A., D'Ausilio, C. A., 2008. Genetic analysis reveals multiple cryptic invasive species of the hydrozoan genus *Cordylophora*. Biological Invasions, 11, 1869–1882.

Frank, U., Leitz, T., Müller, W. A., 2001. The hydroid *Hydractinia*: a versatile, informative cnidarian representative. BioEssays, 23, 963–971.

Gahan, J. M., Bradshaw, B., Flici, H., Frank, U. 2017. The interstitial stem cells in *Hydractinia* and their role in regeneration. Current Opinions in Genetics and Development, 40, 65–73.

Hamada, M., Schröder, K., Bathia, J., Kürn, U., Fraune, S., Khalturina, M., Khalturina, K., Shinzato, C., Satoh, N., Bosch, T. C. G., 2018. Metabolic co-dependence drives the evolutionarily ancient *Hydra–Chlorella* symbiosis. eLife, 7, e35122.

Harmata, K. L., Parrin, A. P., Morrison, P. R., McConnell, K. K., Bross, L. S., Blackstone, N. W., 2013. Quantitative measures of gastrovascular flow in octocorals and hydroids: toward a comparative biology of transport systems in cnidarians. Invertebrate Biology, 132, 291–304.

Harmata, K. L., Somova, E. L., Parrin, A. P., Bross, L. S., Glockling, S. L., Blackstone, N. W., 2015. Structure and signaling at hydroid polyp-stolon junctions, revisited. Biology Open, 4, 1078–1093.

Hensel, K., Lotan, T., Sanders, S. M., Cartwright, P. and Frank, U. 2014. Lineage-specific evolution of cnidarian Wnt ligands. Evolution & Development 16, 259–269.

Kayal, E., Bentlage, B., Collins, A. G., Kayal, M., Pirro, S., Lavrov, D. V., 2011. Evolution of linear mitochondrial genomes in medusozoan cnidarians. Genome Biology and Evolution, 4, 1–12.

McFadden, C. S., McFarland, M. J., Buss, L. W., 1984. Biology of hydractiniid hydroids. I. Colony ontogeny in *Hydractinia echinata* (Flemming). Biological Bulletin, 166, 54–67.

Piraino, S., Boero, F., Aeschbach, B., Schmid, V., 1996. Reversing the life cycle: medusa transforming into polyps and cell transdifferentiation in *Turritopsis nutricula* (Cnidaria: Hydrozoa). Biological Bulletin, 190, 302–312.

Rosa, S. F. P., Powell, A. E., Rosengarten, R. D., Nicotra, M. L., Moreno, M. A., Grimwood, J., Lakkis, F. G., Dellaporta, S. L., Buss, L. W., 2010. *Hydractinia* allodeterminant *alr1* resides in an immunoglobulin superfamily-like gene complex. Current Biology, 20, 1122–1127.

Leclère, L., Horin, C., Chevalier, S., Lapébie, P., Dru, P., Peron, S., Jager, M., Condamine, T., Pottin, K., Romano, S., et al., 2019. The genome of the jellyfish *Clytia hemisphaerica* and the evolution of the cnidarian life-cycle. Nat Ecol Evol, 3, 801–810

Schierwater, B., Hadrys, H., 1998. Environmental factors and metagenesis in the hydroid *Eleutheria dichotoma*. Invertebrate Reproduction and Development, 34, 139–148.

Sanders, S. M. and Cartwright, P., 2015. Interspecific differential expression analysis of RNA-Seq data yields insight into medusae evolution in hydrozoans (Phylum Cnidaria). Genome Biology and Evolution, 7, 2417–2431.

Sanders, S. M., Ma, Z., Hughes, J. M., Riscoe, B. M., Gibson, G. A., Watson, A. M., Flici, H., Frank, U., Schnitzler, C. E., Baxevanis, A. D., Nicotra, M. L., 2018. CRISPR/Cas9-mediated gene knockin in the hydroid *Hydractinia symbiolongicarpus*. BMC Genomics, 19, 649.

Schuchert, P., 2018. World hydrozoan database. http://www.marinespecies.org/hydrozoa.

Thomas, M. B., Edwards, N. C., 1991. Cnidaria: Hydrozoa, In: Harrison, F. W., Westfall, J. A. (eds.), Microscopic Anatomy of Invertebrates, Placozoa, Porifera, Cnidaria, and Ctenophora. Wiley-Liss, New York, pp. 91–183.

Seipel, K., Schmid, V., 2006. Mesodermal anatomies in cnidarian polyps and medusae. International Journal of Developmental Biology, 50, 589–599.

Steinmetz, P. R. H., Kraus, J. E. M., Larroux, C., Hammel, J. U., Amon-Hassenzahl, A., Houliston, E., Wörheide, G., Nickel, M., Degnan, B. M., Technau, U., 2012. Independent evolution of striated muscles in cnidarians and bilaterians. Nature, 487, 231–234.

Tardent, P., 1995. The cnidarian cnidocyte, a high-tech cellular weaponry. BioEssays, 17, 963–971.

Wagler, H., Berghahn, R., Vorberg, R., 2009. The fishery for whiteweed, *Sertularia cupressina* (Cnidaria, Hydrozoa), in the Wadden Sea, Germany: history and anthropogenic effects. ICES Journal of Marine Science, 66, 2116–2120.

Wong, Y., Simakov, O., Bridge, D. M., Cartwright, P., Bellantuono, A. J., Kuhn, A., Holstein, T., David, C., Steele, R. E. and Martinez, D. E., 2019. Expansion of a single transposable element family is associated with genome size increase and radiation in the genus *Hydra*. Proc. Natl. Acad. Sci. USA.

PHYLUM CNIDARIA

LINEAGE MYXOZOA

Jonathan Foox

CHAPTER **13**

Infrakingdom	Diploblasta
1st Division	Radiata
Phylum	Cnidaria
Lineage	Myxozoa

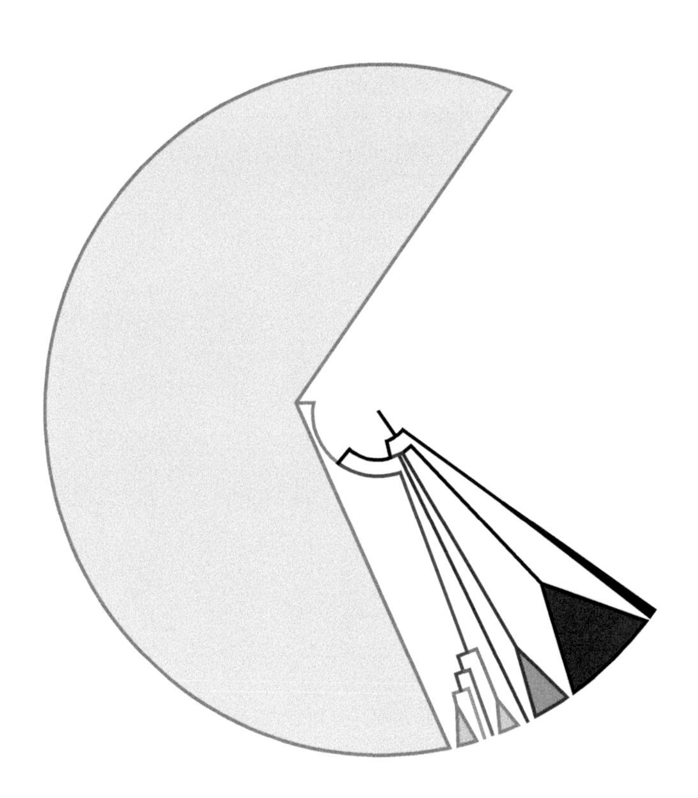

CONTENTS

HISTORY OF TAXONOMY AND CLASSIFICATION

Myxozoans were first observed by Jurine in 1825, who caught some whitefish from Lake Geneva in northern Switzerland and observed unusual cyst-like structures within the musculature (Jurine, 1825). Johannes Müller followed up on this work by publishing a detailed but informal description of these cysts in 1841, naming them "psorosperms," and grouping them with other microscopic organisms that would later be reclassified as coccidia (apicomplexans) and microsporidians. One of Müller's successors, Charles Robin, was the first to formally describe these microscopic forms (Robin, 1853), categorizing them as a tribe of diatoms in the phylum Protozoa. From here, myxozoan taxonomy underwent many overhauls as early milestones of understanding myxozoan biology were achieved (for a full review see Shulman, 1990; Foox and Siddall, 2015).

Otto Bütschli (1881), who observed the development of myxozoan spores and their role in transmission, reassigned the group to the class Sporozoa (reviewed in Lom, 1990). Štolc (1899) was among the first to remark upon the multicellular nature of myxospores and noted a potential metazoan origin for the group. As more species were described in comprehensive monographs (e.g. Kudo, 1933), researchers began to hypothesize a cnidarian origin for the group. Weill (1938) drew a comparison between myxozoan polar capsules and cnidarian nematocysts. Several researchers noted the resemblance between myxosporean pansporoblasts and narcomedusan larvae (Weill, 1938; Lom and de Puytorac, 1965; Mitchell, 1977; Shostak, 1993). Ontogenetic parallels were drawn between nematocysts and polar capsules (Lom and de Puytorac, 1965; Lom, 1969). Eventually, as profound differences in biology and ultrastructure were demonstrated between myxosporeans and other sporozoans, Myxozoa was reassigned to its own protozoan phylum by the International Commission on Protozoan Nomenclature (Levine et al., 1980).

Wolf and Markiw (1984) observed *Myxobolus cerebralis* (Hofer, 1903), a myxosporean parasite of salmonid fishes, transform during its life cycle into a so-called actinosporean parasite of tubificid worms, revealing that members of different classes of the phylum Myxozoa were in fact alternate stages of the same species. Soon after, the "demise of a class of protists" was declared (Kent et al., 1994), as the Actinosporea and Actinomyxidia were suppressed. At around the same time, molecular evidence supported a metazoan origin for Myxozoa (Smothers et al., 1994), and would soon be combined with morphological evidence to confirm a cnidarian origin (Siddall et al., 1995). Expanded molecular studies have since reaffirmed the cnidarian nature of Myxozoa (e.g. Jimenez-Guri et al., 2007; Nesnidal et al., 2013; Chang et al., 2015; Kayal et al., 2018).

ANATOMY, HISTOLOGY AND MORPHOLOGY

Myxozoans are a large and diverse radiation of microscopic spore-forming or cyst-forming endoparasites. As a result of their evolutionary transition to parasitism, myxozoans are extremely morphologically simplified, often consisting of as few as five cells within acellular proteinaceous shells. Myxozoans entirely lack organ systems and recognizable gametes or embryonic stages (Canning and Okamura, 2004). Their cells contain eukaryotic features including nuclei, endoplasmic reticulum and tubular mitochondria, though unlike other metazoans, myxozoans do not possess cilia or centrioles.

The primary clade, Myxosporea, comprises nearly 2,200 species (Lom and Dyková, 2006), united by their infective spore structure (**Figure 13.1**). Spores of the infective stage of the myxosporean life cycle are generally 10–15 µm in length, and in some cases may be as long as 45 µm. Myxospores are encased by a shell consisting of proteinaceous valves that fuse together and form a sutural ridge around their periphery, resembling the shell of a mollusc. The shell may be

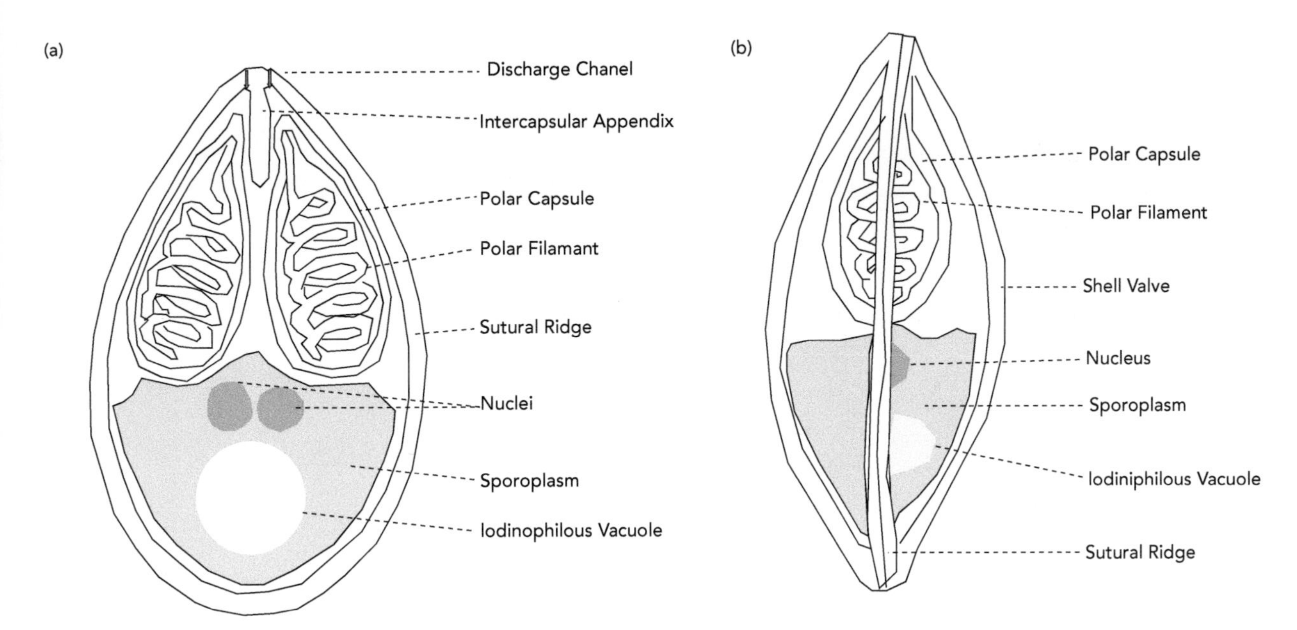

Figure 13.1 Diagrammatic representation of a typical myxospore. (a) Front view; (b) side view.

smooth, ridged or striated in texture; may be spherical, ovoid, pyriform or fusiform in shape; and may have elongated caudal features. These sporic characteristics are often the diagnostic features of a given species. Enveloped within the shell are capsulogenic cells. These cellular vessels harbor a complex post-Golgi apparatus known as the polar capsule (homologous to cnidarian cnidocytes), organelles containing a filament that is tightly coiled up in a helical fashion. Upon some chemical or mechanical stimulation of the spore, polar filaments are rapidly everted from the capsule like the finger of a glove through a channel in the valvogenic sutural ridge. The sticky filament latches onto the integument of the target animal like a grappling hook. Also encased within the spore is the infective agent: the binucleate germinative sporoplasm, which contains ribosomes, mitochondria, endoplasmic reticulum and glycogenic inclusions.

The secondary clade, Malacosporea, contains only four described species, and has retained more complex features such as musculature and epithelial cells. Their actinospores are tri-radially symmetrical, possessing three valves with one polar capsule in each, and a long caudal extension that enables buoyancy in the water column.

REPRODUCTION AND DEVELOPMENT

Myxozoans have complex life histories that involve radical physiological transformations. Most species exhibit a two-host life cycle that typically involves switching between an invertebrate as the definitive host and a vertebrate as the intermediate host (**Figure 13.2**). Spores achieve transmission in the water column by rapidly everting the filaments of their polar capsules and attaching to a new host. Once the spore penetratres the host, its infective sporoplasm is released and migrates to a specific target tissue. Sporoplasms will then enter a proliferative phase, in which dozens to hundreds of secondary cells will form within the sporoplasm via mitotic endogeny. These internal cells ("secondary cells") will then undergo sporogeny, in which they differentiate and develop into infective spores. At this time, the parasite as a whole is referred to as a plasmodium, and will eventually escape from the vertebrate host in search of its definitive invertebrate host. Meiotic development of both myxosporeans and malacosporeans occurs within fluid-filled stages in the invertebrate host (Janisewska, 1957).

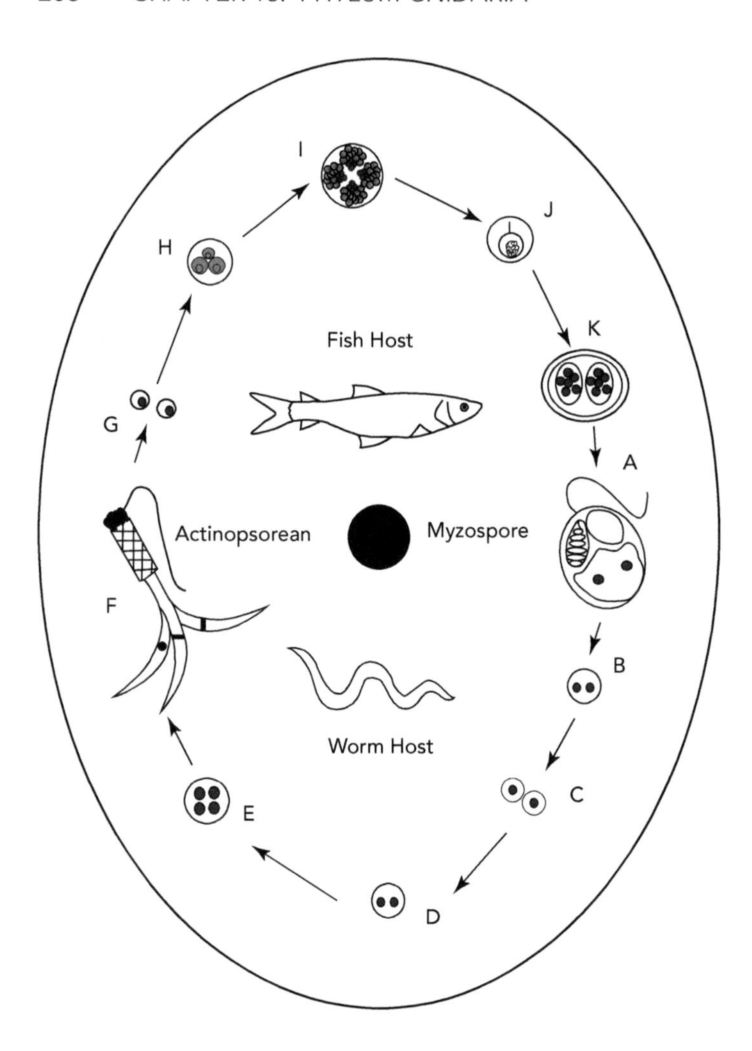

Figure 13.2 Generalized life cycle of the Myxozoa. A = Mature myzosporean from aquatic annelid. B through D = Schizigony resulting in the fusion of two uninucleate cells to form a binucleate cell. E = Gametogony. F = Mature actinopsorean. G = attachment to fish epidermis and release of uninucleate sporoplasm cells. H = Extrasporogenic development. I = Trophozoite stage. J = Sporogenesis. K = Formation of multicellular spores released to complete the life cycle. (Courtesy of R. DeSalle.)

During the trophic stage in which they develop internal secondary cells, myxosporeans may be organized as syncytial multinucleated plasmodia, uninucleated pseudoplasmodia or as hollow sacs or vermiform structures. Cells will rarely form into tissues, instead proliferating via endogeny. Daughter cells will quickly be surrounded by endoplasmic reticulum. Like myxosporeans, malacosporeans undergo nuclear division by meiosis in their invertebrate host, though diploidy through nuclear fusion is only observed in myxosporeans.

DISTRIBUTION AND ECOLOGY

Myxozoans are globally distributed parasites that can be found in marine and freshwater aquatic environments (**Figure 13.3**), with a few species exhibiting exclusively terrestrial life histories. Though they primarily alternate between an annelid definitive host and a teleost intermediate host, myxozoans have been observed in many other hosts including amphibians, birds, bryozoans, cephalopods, reptiles, shrews, and waterfowl. Myxozoans have been reported from nearly all tissue and organ sites within vertebrate hosts and most species exhibit tissue specificity, with coelozoic species infecting the lumen of hollow organs such as the gall bladder and histozoic species forming plasmodia in the intracellular space of tissues such as musculature and gill filaments (Mitchell, 1977) (**Figure 13.4**). A single host may harbor infections from multiple species in different organs. Myxozoa is an extremely diverse radiation comprising over 2,200 described species distributed among over 60 genera (Lom and Dyková, 2006), this likely representing a small fraction of the

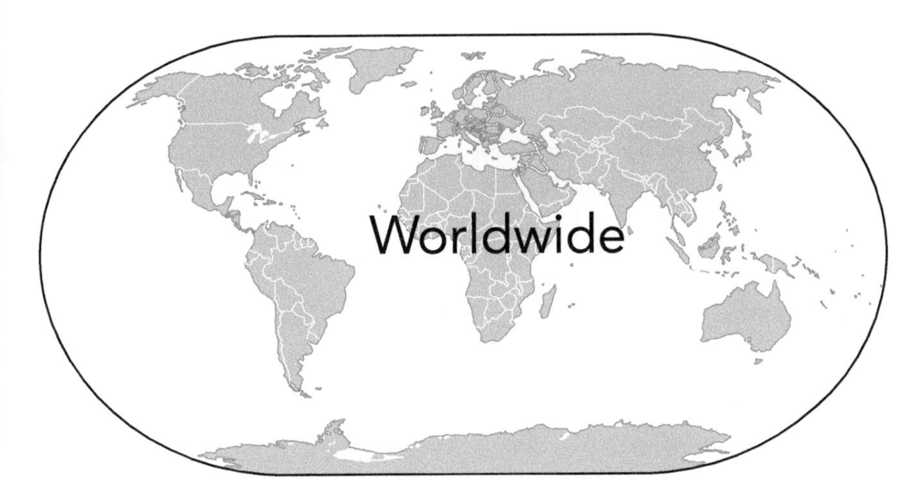

Figure 13.3 Distribution of Myxozoa. Myxozoans are globally distributed parasites that can be found in marine and freshwater aquatic environments.

total diversity, with some estimates of 16,000 species in the Neotropics alone (Naldonia et al., 2011).

Most research has concentrated upon infection of freshwater fishes because of their accessibility, and their ecological significance. While the vast majority of myxozoan infections appear to be innocuous, some species are well-known pathogens that cause fatal diseases in their hosts, such as *Myxobolus cerebralis*, which causes whirling disease in juvenile rainbow trout, or *Tetracapsuloides bryosalmonae*, the causative agent of proliferative kidney disease (PKD), which can wipe out 90% of infected salmonid populations. These infections have significant economic impact, particularly on netpen aquaculture (Kent et al., 2001). PKD alone has caused upward of £2.5 million per year in losses for the trout farming industry in the United Kingdom (Morris and Adams, 2008).

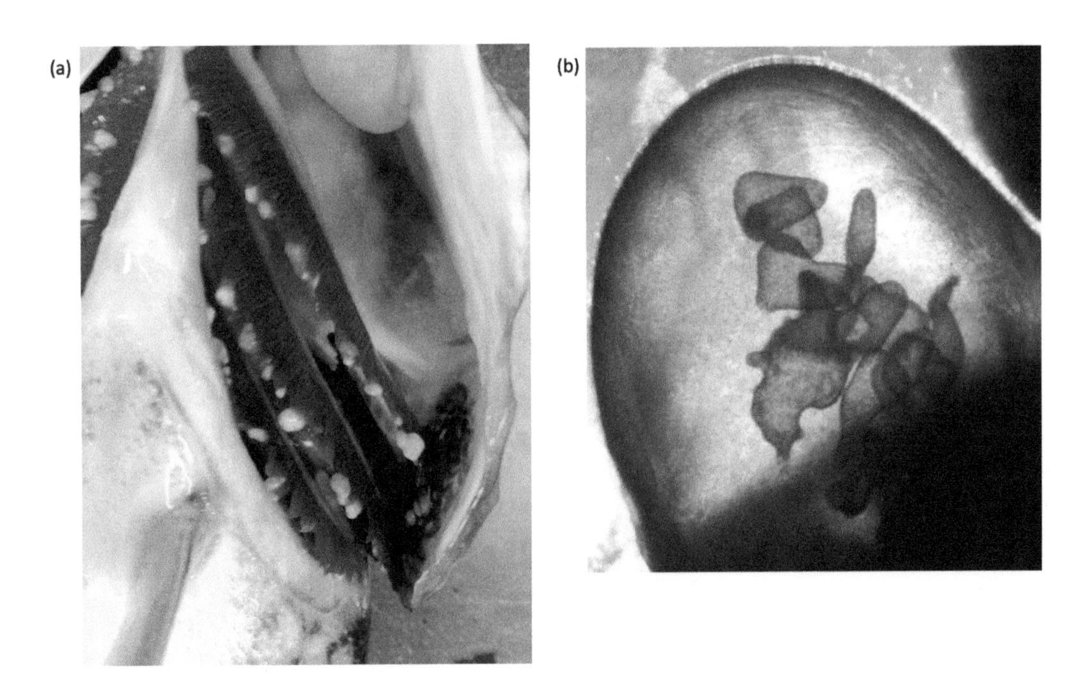

Figure 13.4 Typical myxozoan infection strategies. (a) histozoic infection of *Henneguya psorospermica* in the gill filaments of *Esox lucius*. (b) coelozoic infection of *Myxidium melleni* in the gall bladder of a western chorus frog (*Pseudacris triseriata*).

PHYSIOLOGY AND BEHAVIOR

Spores of the vertebrate infection stage of the myxozoan life cycle are able to chemically recognize potential hosts and extrude their polar filaments in response (Kallert et al., 2011). The myxosporean response to fish mucus has been well documented (e.g. Yokoyama et al., 1993). However, there is nearly no documentation about specific behavior of myxozoans, other than penetration into and migration within their host.

GENETICS AND GENOMICS

Sequence data derived from SSU rRNA was used to confirm the metazoan origin of Myxozoa (Smothers et al., 1994; Schlegel et al., 1996), and subsequently the cnidarian origin of Myxozoa (Siddall et al., 1995; Lom and Dyková, 1997). The majority of genetic studies on Myxozoa have relied on target loci sequencing (mostly 18S and 28S ribosomal subunits) to establish relationships within the clade (e.g. Holzer et al., 2004).

The first genome-level study utilized 50 protein-coding orthologs of the malacosporean *Buddenbrockia plumaetellae* (Jimenez-Guri et al., 2007) and established a myxozoan–medusozoan sister relationship. Another study, this time leveraging 128 genomic loci, verified these findings (Nesnidal et al., 2013). Since that time, several studies have employed RNAseq data to discover and describe cnidarian genes within the myxozoan genome (referred to as taxonomically restricted genes) related to cnidocyte formation and toxin secretion (Holland et al., 2010; Shpirer et al., 2014). Genomic investigations of Myxozoa forced the creation of novel methods to filter the relatively large proportion of host contamination among target myxozoan sequences (Foox et al., 2015). To date, four myxozoan genomes have been deposited, with some evidence that myxozoans possess the smallest metazoan genomes, as small as 23Mb (Yang et al., 2014; Chang et al., 2015).

Myxozoans also appear to have unusual mitochondrial genomes, partitioned into eight chromosomes, each harboring a large noncoding region (Yahalomi et al., 2017). The canonical mitochondrial genes are highly divergent from other cnidarians. Other remarkable observations included a total lack of tRNA genes, only five protein coding genes in total, and only two chromosomes in a second species, revealing high plasticity in myxozoan mitogenomes.

POSITION IN THE ToL

After many years of taxonomic uncertainty, Myxozoa is now definitively considered a highly derived cnidarian (**Figure 13.5**). Genomic studies (e.g. Chang et al., 2015) have supported a sister relationship to the Medusozoa. A sister relationship was long hypothesized (e.g. Weill, 1938; Siddall et al., 1995) between Myxozoa and *Polypodium hydriforme*, a species that begins its life cycle as an endoparasite of acipenseriform oocytes before emerging from its host and developing into a macroscopic, free-living stolon possessing tentacles and a gut (Raikova, 1994). Genomic data have affirmed this relationship, supporting monophyly of the Endocnidozoa (Zrzavý and Hypša, 2003), and therefore a single origin of endoparasitism within the Cnidaria.

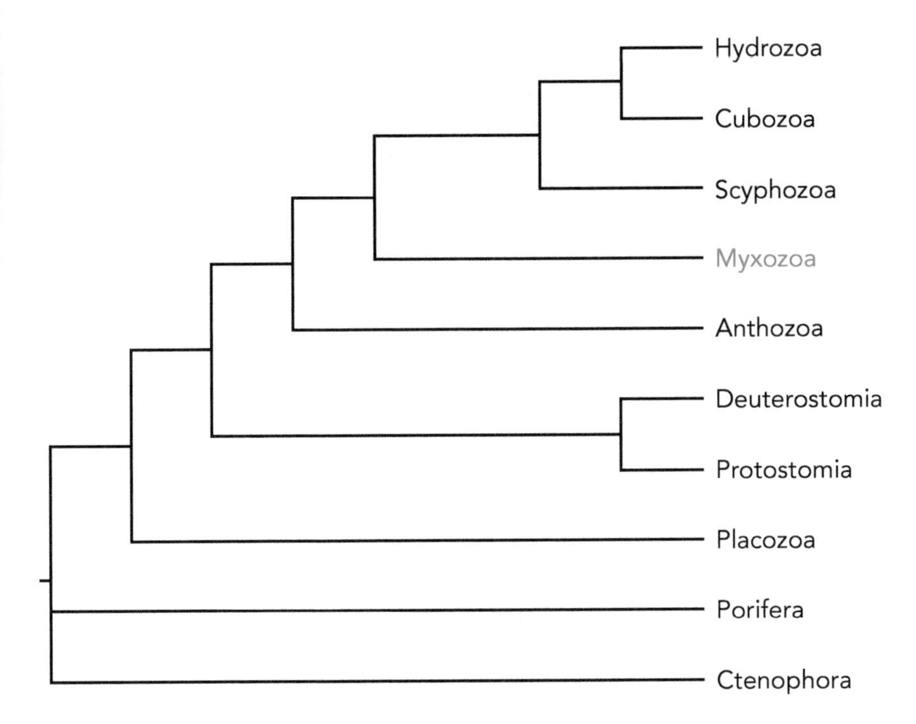

Figure 13.5 Phylogenetic tree showing the position of Myxozoa (in green) embedded within Cnidaria.

DATABASES AND COLLECTIONS

Myxozoans cannot be kept in culture and thus are not readily available in laboratory settings. No public or private collections of myxozoan specimens are available, either. Myxozoan researchers have a network for research communication called the Myxozoan Network (http://myxozoa.org/about.htm).

CONCLUSION

Myxozoa represent one of the most radical transformations over evolutionary time from free-living ancestor to microscopic obligate endoparasite. At the same time, they are an understudied group, with their life cycle only resolved within the last 30 years, and genomic data only becoming available in the last few years. Over 2,000 species have been described, but by many estimates this is a fraction of the extant diversity. Sampling additional sublineages, such as the marine coelozoic group, the marine chloromyxid clade and the basal sphaerosporid clade, will help to answer questions of internal phylogeny. Genomic data for malacosporeans will be critical to understanding this early-diverging lineage that has retained true musculature and epithelial tissues. Although highly unlikely, discovery of fossil evidence will go a long way toward reconstructing the timeframe of parasitic transitioning within Cnidaria. Myxozoans' trophic stages are entirely soft-bodied, but the proteinaceous external valve of some spores may have fossilized in a benthic deposit.

REFERENCES

Bütschli O. 1881. Myxosporidien. Zoologischer Jahrbericht für 1880(1): 162–164.

Canning E. U., Okamura B. 2004. Biodiversity and evolution of the Myxozoa. Advances in Parasitology 56: 43–131.

Chang E. S., Neuhof M, Rubinstein ND, Diamant A, Philippe H, Huchon D, Cartwright P. 2015. Genomic insights into the evolutionary origin of Myxozoa within Cnidaria. PNAS 112(48): 14912–14917.

Foox J., Siddall M. E. 2015. The Road to Cnidaria: History of Phylogeny of the Myxozoa. Journal of Parasitology 101(3): 269–274.

Foox J., Ringuette M., Desser S. S., Siddall M. E. 2015. In silico hybridization enables transcriptomic illumination of the nature and evolution of Myxozoa. BMC Genomics 16(1): 840.

Holland J. W., Okamura B., Hartikainen H., Secombes C. J. 2010. A novel minicollagen gene links cnidarians and myxozoans. Proceedings of the Royal Society B: Biological Sciences 278: 546–553.

Holzer A. S., Sommerville C., Wootten R. 2004. Molecular relationships and phylogeny in a community of myxosporeans and actinosporeans based on their 18S rDNA sequences. Int J Parasitol 34: 1099–1111.

Jiménez-Guri E., Philippe H., Okamura B., Holland P. W. 2007. *Buddenbrockia* is a cnidarian worm. Science 317(5834): 116–118.

Janisewska J. 1957. Actinomyxidia II; new systematics, sexual cycle, description of new genera and species. Zool Polon 8: 3–34.

Jurine L. L. 1825. Histoire des poissons du Lac Léman. Mém Soc Phys Hist Natg Geneve 3.

Kayal E., Bentlage B., Pankey M. S., Ohdera A. H., Medina M., Plachetzki D. C., Collins A. G., Ryan J. F. 2018. Phylogenomics provides a robust topology of the major cnidarian lineages and insights on the origins of key organismal traits. BMC Evolutionary Biology, 18(1): 68.

Kallert D. M., Bauer W., Haas W., El-Matbouli M. 2011. No shot in the dark: Myxozoans chemically detect fresh fish. International Journal for Parasitology, 41(3–4): 271–276.

Kent M. L., Margolis L., Corliss J. O. 1994. The demise of a class of protists: Taxonomic and nomenclatural revisions proposed for the protist phylum Myxozoa Grassé 1970. Canadian Journal of Zoology 72: 932–937.

Kent, M. L., Andree K., Bartholomew J. L., El-Matbouli M., Desser S. S., Devlin R. H., Feist S. W., Hedrick R. P., Hoffmann R W., Khattra J, et al. 2001. Recent advances in our knowledge of the Myxozoa. Journal of Eukaryotic Microbiology 48: 395–413.

Kudo R. R. 1933. A taxonomic consideration of Myxosporidia. Trans Am Microsc Soc 52: 195–216.

Levine N. D, Corliss J. O, Cox F. E, Deroux G, Grain J, Honigberg B. M, Leedale G. F, Loeblich A. R, Lom J, Lynn D, et al., 1980. A newly revised classification of the Protozoa. Journal of Protozoology 27: 37–58.

Lom J. 1969. Notes on the ultrastructure and sporoblast development in fish parasitizing myxosporidian of the genus *Sphaeromyxa* Zeitschrift für Zellforschung und Mikroskopische Anatomie 97: 416–437.

Lom J. 1990. Phylum Myxozoa. In Handbook of protoctista, L. Margolis, J. O. Corlis, M. Melkonian and D. J. Chapman (eds.), Jones and Bartlett Publishers, Boston, MA, pp. 36–52.

Lom J., de Puytorac P. 1965. Studies on the myxosporidian ultrastructure and polar capsule development. Protistologica 1: 53–65.

Lom J., Dyková I. 1997. Ultrastructural features of the actinosporean phase of Myxosporea (phylum Myxozoa): A comparative study. Acta Protozoologica 36: 83–103.

Lom J., Dykova I. 2006. Myxozoan genera: definition and notes on taxonomy, life-cycle terminology and pathogenic species. Folia Parasitology 53: 1–36.

Mitchell, L. G. 1977. Myxosporidia, in: Parasitic Protozoa, Vol. 4. pp. 115–154, Academic Press, London, UK.

Morris D. J., Adams A. 2008. Sporogony of *Tetracapsuloides bryosalmonae* in the brown trout *Salmo trutta* and the role of the tertiary cell during the vertebrate phase of myxozoan life cycles. Parasitology 135: 1075–1092.

Naldonia J., Aranab S., Maiac A. A. M., Silvac M. R. M., Carrieroc M M., Ceccarellid P. S., Tavarese L. E R., Adrianof E. A. 2011. Host-parasite-environment relationship, morphology and molecular analyses of *Henneguya eirasi* n. sp. parasite of two wild *Pseudoplatystoma* spp. in Pantanal Wetland, Brazil. Veterinary Parasitology 177: 247–255.

Nesnidal M P., Helmkampf M., Bruchhaus I., El-Matbouli M., Hausdorf B. 2013. Agent of whirling disease meets orphan worm: phylogenomic analyses firmly place Myxozoa in Cnidaria. PLoS One, 8(1): e54576.

Raikova E. V. 1994. Life cycle, cytology, and morphology of *Polypodium hydriforme*, a coelenterate parasite of the eggs of acipenseriform fishes. The Journal of Parasitology 1: 1–22.

Robin C. P. 1853. Histoire naturelle des végétaux parasites qui croissent sur l'homme et sur les animaux vivants. J.-B. Ballière, Paris, 734.

Schlegel M., Lom J., Stechmann A., Bernhard B., Leipe D., Dyková I., Sogin M. L. 1996. Phylogenetic analysis of complete small subunit ribosomal RNA coding region of *Myxidium lieberkuehni*: Evidence that Myxozoa are Metazoa and related to Bilateria. Archiv für Prosistenkunde 147: 1–9.

Shostak S., 1993. A symbiogenic theory for the origins of cnidocysts in Cnidaria. Biosystems 29: 49–58.

Shpirer E. Chang E. S., Diamant A., Rubinstein N., Cartwright P., Huchon D. 2014. Diversity and evolution of myxozoan minicollagens and nematogalectins. BMC Evolutionary Biology 14:205.

Shulman S. S. 1990. Myxosporidia of the USSR. Russian Translations Series 75. A.A. Balkema/Rotterdam.

Siddall M., Martin D. S., Bridge D., Desser S. S., Cone D. K. 1995. The demise of a phylum of protists: Phylogeny of Myxozoa and other parasitic Cnidaria. Journal of Parasitology 81: 961–967.

Smothers J, von Dohlen C, Smith L, Spall R. 1994. Molecular evidence that the myxozoan protists are metazoans. Science 265: 1719–1721.

Štolc A. 1899. Actinomyxidies, nouveau groupe de Mesozoaires parent des Myxosporidies. Bulletin International de l'Academie des Sciencesde Boheme 22: 1–12.

Weill R. 1938, L'interpretation des Cnidosporidies et la valeur taxonomique de leur cnidome. Leur cycle compare a la phase larvaire des Narcomeduses Cuninides. Travaux de la Station Zoologique de Wimereaux 13: 727–744.

Wolf K., Markiw M. E. 1984. Biology contravenes taxonomy in the Myxozoa: New discoveries show alternation of invertebrate and vertebrate hosts. Science 225: 1449–1452.

Yahalomi D., Haddas-Sasson M., Rubinstein N. D., Feldstein T., Diamant A., Huchon D. 2017. The multipartite mitochondrial genome of *Enteromyxum leei* (Myxozoa): eight fast-evolving megacircles. Molecular Biology and Evolution 34(7): 1551–1556.

Yang Y., Xiong J., Zhou Z., Huo F., Miao W., Ran C., Liu Y., Zhang J., Feng J., Wang M., Wang M., Wang L., Yao B. 2014. The Genome of the Myxosporean *Thelohanellus kitaeui* shows adaptations to nutrient acquisition within its fish host. Genome Biology and Evolution 6(12): 3182–3198.

Yokoyama H., Ogawa K., Wakabayashi H. 1993. Some biological characteristics of actinosporeans from the oligochaete *Branchiura sowerbyi*. Diseases of Aquatic Organisms 17(1993): 223–228.

Zrzavý, J., Hypša, V., 2003. Myxozoa, Polypodium, and the origin of the Bilateria: The phylogenetic position of "Endocnidozoa" in light of the rediscovery of Buddenbrockia. Cladistics 19(2): 164–169.

BILATERIA

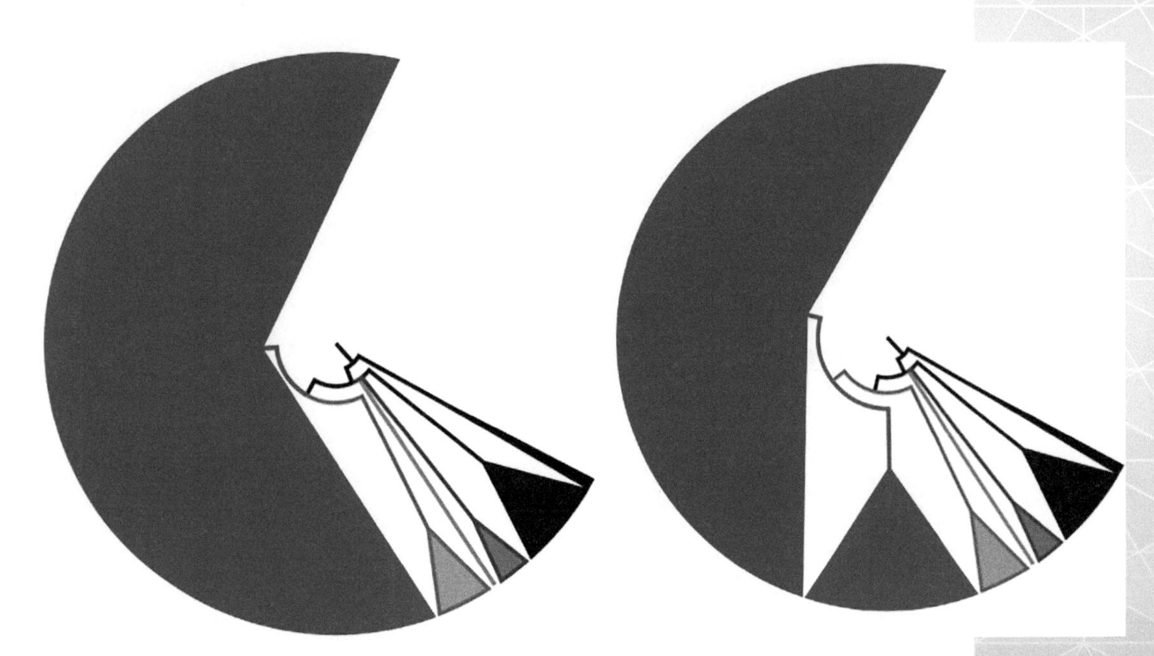

Left: Bilateria in blue. Right: Bilateria divided into Protostomea (larger clade) and Deuterostomea (smaller clade).

Bilateral symmetry is the hallmark character of this higher category of animals. The symmetry can occur in the adult or in the developing embryo. Bilaterians have a head and a tail (anterior – posterior axis). Echinodernata are the exception and these animals have pentaradial symmetry in the adult stage but the embryonic stages are bilaterally symmetrical.

The grand majority of animals are Bilateria and are what is called triploblastic (having three germ layers – endoderm, mesoderm and ectoderm. Bilateria have a mouth and an anus, the organization of which dictates whether they are protostome or deuterostomes (see these higher categories). Some bilaterians have simpler organization and lack body cavities (the so-called acoelomates, i.e. Platyhelminthes, Gastrotricha and Gnathostomulida). While most bilaterians have a digestive tract, some lack body cavities, some have primary body cavities and some have secondary cavities. The higher category Bilateria is separated into the Protostomea and Deuterostomea (see diagram above).

Synapomorphies

- bilateral symmetry of adult or larval stage (i.e. having a left and a right side that are, in general, mirror images of each other)
- a head and a tail (anterior–posterior axis) as well as a belly and a back

Protostomea

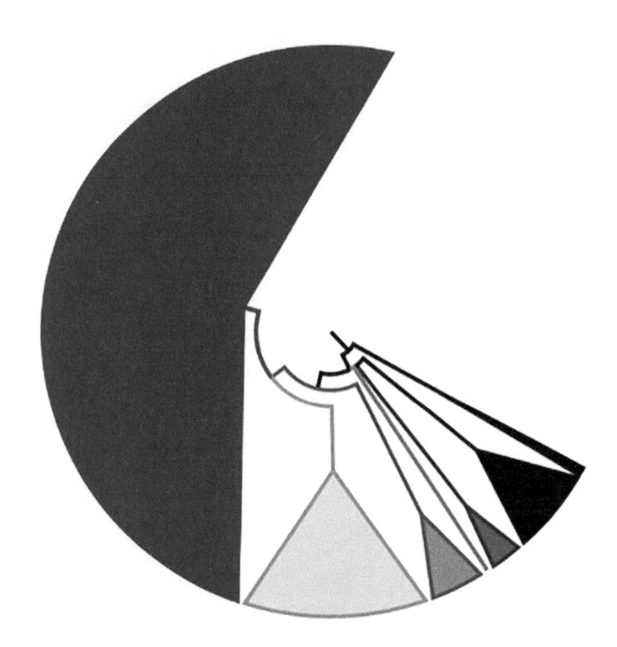

The Bilateria are divided into two major groups – Protostomea and Deuterostomea.

Synapomorphies

- mouth forms first, then the anus; the reverse is true in deuterostomes
- part of the mesoderm separates to form the coelom in a process called schizocoely
- determinate cleavage; in other words the developmental fate of each embryonic cell is pre-determined

The name Protostomea comes from the Greek; it is roughly translated as "mouth first". Protostomes develop differently than deuterostomes in that the blastopore develops into the mouth in protostomes and in deusterostomes the blastopore eventually develops into the anus.

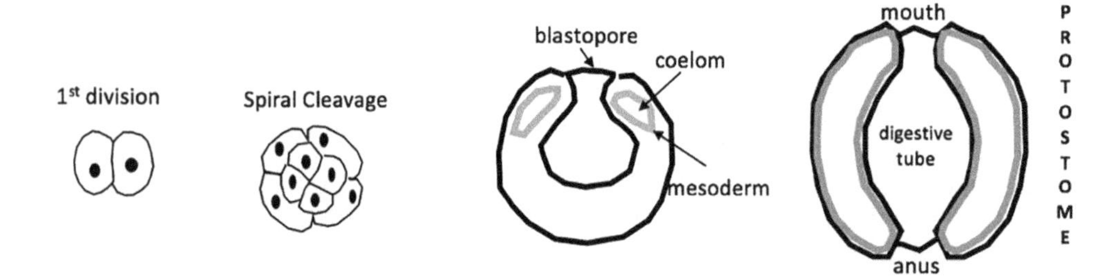

Protostomes are comprised of two larger groups, the Ecdysozoa and the Spiralia. Spiral cleavage during the early embryonic developmental process is the recognizable trait for the Spiralia. The process of molting is the defining character for the Ecdysozoa. The acoelomates are considered protostomes but are not included in the Ecdysozoa nor the Spiralia and hence here are considered as their own group.

Spiralia

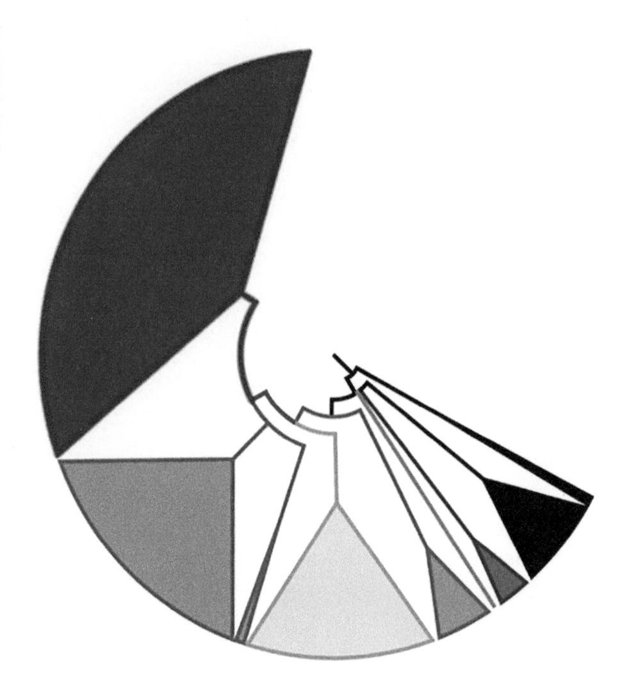

Synapomorphy

- Spiralia show canonical spiral cleavage (see above), a pattern of early development
- This character is also found in most (but not all) members of the Lophotrochozoa

Spiralia are a higher group within protostomes that are characterized by spiral cleavage during early embryogenesis (see figure below). There is a great deal of morphological variation within this higher group of animals. Spiralia includes annelids, platyhelminths, gastrotrichs, several small but interesting phyla (Rotifera, Gnathozoa and Micrognathozoa) and the larger group (superphylum) called Lophotrochozoa. Three names have arisen that complicate the higher taxonomy of the Spiralia – Platyzoa, Platytrochozoa and Rouphozoa. The Platytrochozoa are comprised of Gastrotricha, Platyhelminthes and Lophotrochozoa which would be consistent with the phylogeny pictured above, while the Rouphozoa are comprised of Platyhelminthes and Gastrotricha (also consistent with the phylogenetic tree above). The Platyzoa are said to consist of the Rouphozoa and the Gnathifera (Rotifera, Acanthocephala, Gnathostomulida, Micrognathozoa and Cycliophora), which is not consistent with the phylogeny shown above.

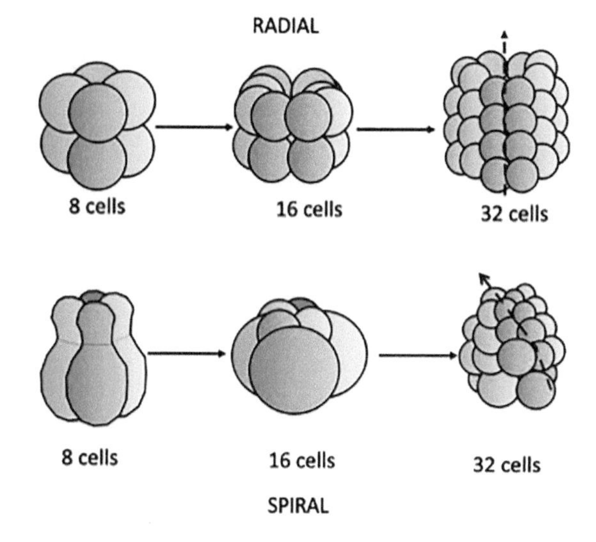

The difference between spiral (bottom) and radial (top) cleavage from the 8-cell to the 32-cell stage. The different colors represent distinct cell lineages. The arrow indicates the spiral nature of cleavage in the bottom scheme and the non-spiral cleavage in the top scheme. (Figure by R. DeSalle.)

Spiral cleavage is a complex developmental process so some discussion of it follows. Mollusca, Annelida, Platyhelminthes and Nemerteans all show the classic spiral cleavage schematically demonstrated in the figure. Rotifera, Brachiopoda, Phoronida, Gastrotricha and other spiralian phyla display spiral cleavage in only some of the species within these phyla and so this type of spiral cleavage is considered derived.

PHYLUM PLATYHELMINTHES

CHAPTER

14

Teresa Adell and Marta Riutort

Infrakingdom	Bilateria
Division	Protostomia
Subdivision	Lophotrochozoa
Phylum	Platyhelminthes

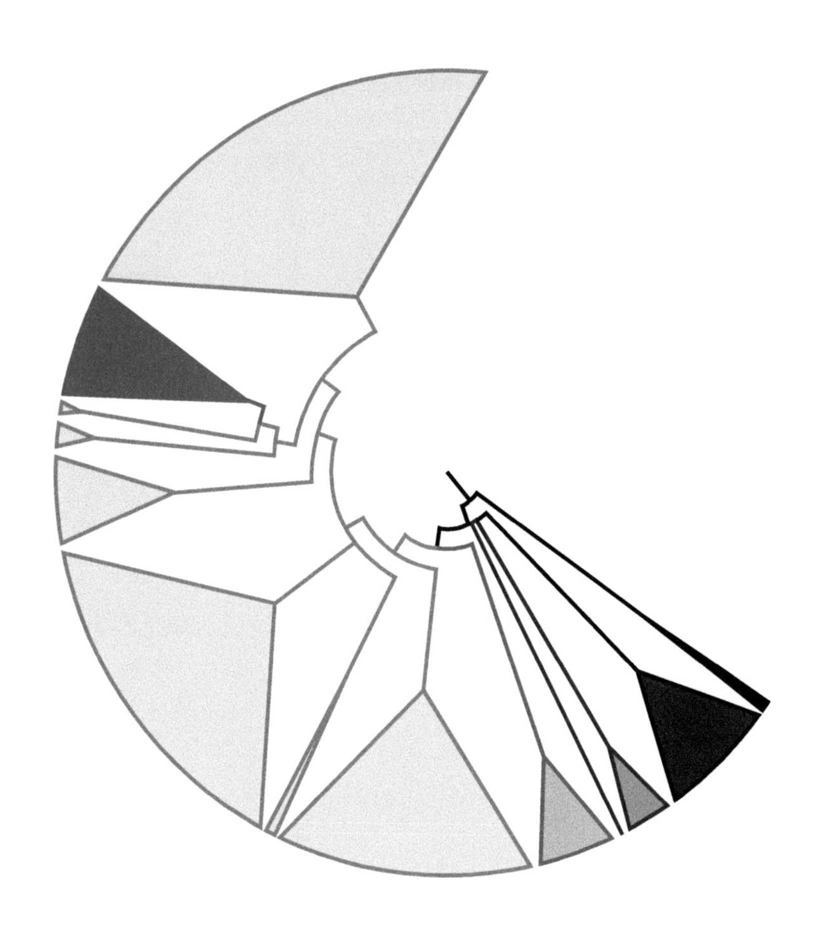

CONTENTS

The phylum **Platyhelminthes** comprises dorsoventrally flattened worms commonly known as **flatworms** (*platys*, flat; *helminthos*, worm). They are bilateral symmetric animals, triploblastic (with three embryonic layers) and non-segmented. Platyhelminthes is one of the largest animal phyla after arthropods, molluscs and chordates, and according to the Turbellaria database (turbellaria.umaine.edu) it includes more than 26,500 described species, of which around 20,000 are parasitic. **Neodermata** comprises all the parasitic flatworms, classified in three taxa: Monogenea, Cestoda and Trematoda (**Figure 14.1**). Monogenea are external parasites and have a single vertebrate host, most generally a fish; trematodes and cestodes are internal parasites of vertebrates with a very complex life cycle including several stages in invertebrate hosts; trematodes include such well-known species as *Fasciola hepatica* (liver fluke) and *Schistosoma mansoni* (Blood fluke), and cestodes include species as *Taenia solium* (pork tapeworm), all of these affecting either humans or our livestock. Free-living flatworms (classically referred to as "**Turbellaria**") include 11 orders, a few of which are popularly known, as for instance the Polycladida (beautiful marine worms usually confused with sea slugs, as the *Thysanozoon* genus) or the **Tricladida** or **planarians**, well known for the regenerative capabilities of many of its species, such as *Schmidtea mediterranea.* Macrostomida also comprise regenerative model species, like *Macrostomum lignano*. The "Turbellaria" live in a large variety of habitats, from freshwater springs, rivers, lakes and ponds (1,300 species, 426 belonging to the order Tricladida) to the ocean (4,340 species) and moist terrestrial habitats (860 species, most belonging to the Tricladida). These numbers may however increase as new studies, especially using molecular data, are unveiling many new and some cryptic species. Their size ranges from microscopic worms to the 30 m long tapeworms found in the sperm whale.

Synapomorphies

- they are acoelomates, since they do not have a secondary body cavity
- weak cephalization with well-defined cephalic ganglia
- real organs derived from the mesoderm (excretory system, copulatory apparatus ...), but no circulatory or respiratory system
- blind digestive system (anus and mouth are the same)
- they possess a special type of adult stem cells, called neoblasts
- a characteristic trait of this phylum is the presence of a large number and variety of adhesive secretions from specialized cells and organs

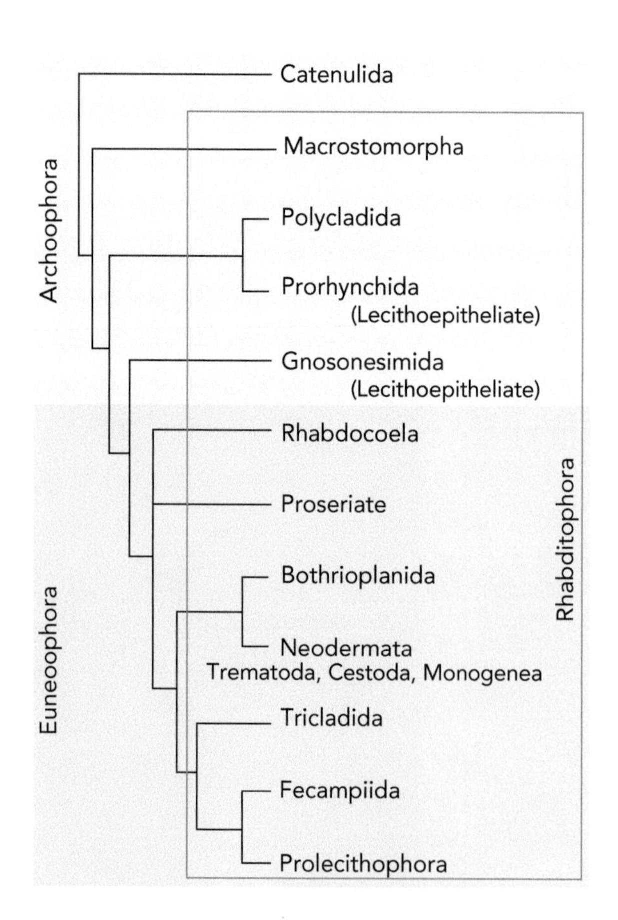

Figure 14.1 Phylogenetic hypothesis for the Platyhelminthes.

The adaptations to the parasitic life of most platyhelminths has resulted in profound modifications of their development and anatomy. This results in a lack of unique shared characteristics for the phylum that allow their diagnosis. Nonetheless they present a group of common features that may define them (although most are not exclusive of platyhelminths):

TAXONOMY AND SYSTEMATICS

The Platyhelminthes were originally divided in three classes, the free-living class "Turbellaria" and two parasitic classes, the Cestoda and the Trematoda (Gegenbaur, 1859; Haeckel, 1866; Hyman, 1951). The free-living were divided into around 11 orders, the number depending on the authors. These orders were classified based on a few characters, one of the basic characters being the structure of the oocyte: the **Archoophora** included those species with **endolecithal eggs** (that contain all the yolk needed for development); the **Neoophora** grouped **ectolecithal species** (with eggs that must take up yolk from outside). Ectolecithal eggs are considered a derived state since they required the invention of specialized yolk cell-producing organs called **vitellaria** (a detailed and recent revision can be found in Martín-Durán and Egger, 2012). However, through morphological analyses it was discovered that the parasitic Neodermata had evolved from free-living platyhelminths. The Turbellaria in consequence constituted a paraphyletic group (Ehlers, 1985), a situation indicated by the quotes around the name. The internal phylogenies based on molecular data supported the Neodermata as derived from free-living forms (and hence the paraphyly of the "Turbellaria") and the subdivision of the Platyhelminthes into two main groups: the earliest branching lineages, all free-living, belonged to the Archoophora, which was now a paraphyletic group since it included the more divergent and monophyletic Neoophora. However, in the most recent phylogenies based on transcriptomic data the Neoophora are neither monophyletic since ectolecithal eggs will have appeared independently in the Lecithoepitheliata (or one of its subclades, the Prorhynchida) that branch within the Arcoophora (Egger et al., 2015; Laumer et al., 2015) (Figure 14.1). The new clade, including all the ectolecithal species except the Lecithoepiteliata, has been named the **Euneoophora** (Figure 14.1). The molecular phylogenies have also confirmed or dismissed a series of previous proposed relationships:

1. Confirmed the Catenulida as sister group of the rest of Platyhelminthes, known as Rhabdipothora, which include the rest of the free-living orders and all the parasitic groups. All Rhabdipothora, except the Macrostomorhpha, are characterized by a biflagellate sperm.
2. The disappearance of the "Seriata" order, which grouped suborders Proseriata, Tricladida and Bothrioplanida. The Proseriata, now an order, disputes with the Rhabdocoela the basalmost position within the Euneoophora, while the order Tricladida has moved to a more derived position, and the Bothrioplanida (a monospecific order) is the sister group of the Neodermata.
3. The establishment of the Prolecitophora and the Fecampiida (groups overlooked in morphological studies) as sister to the Tricladida, although their relationships are still not fully resolved. This clade has been strongly supported by molecular data, but morphological characters that unite them are not known.
4. The Neodermata are no longer a group within the Rhabdocoela; instead they appear, together with Bothrioplana, as the sister group to the clade constituted by Tricladida + Prolecitophora + Fecampiida.

ANATOMY AND MORPHOLOGY

As already stated, platyhelminths are bilateral animals, flattened dorso-ventrally and non-segmented. They are acelomate, since the space between organs is filled with different cell types, called parenchyma. Due to the profound modifications suffered by parasitic platyhelminths it is difficult to establish a general morphological pattern for all of them (Adell et al., 2015; Cebrià et al., 2015). In general, they show a central nervous system consisting of two anterior cephalic ganglia from which a pair of longitudinal nerve cords run toward the tip of the tail (**Figures 14.2** and **14.3**). The longitudinal cords are interconnected by transversal commeasures. Eyes, absent in parasites and hypogean free-living species, are located in the anterior part of the head, usually a pair, and connected to the brain through the visual axons. Eyes allow them to detect the presence or absence of light, and they usually show a photophobic response. Free-living platyhelminths also show a digestive system, which in the planarians (Tricladida) is formed by three main branches, one anterior and two posterior, connected to a highly innervated pharynx (Figures 14.2 and 14.3). The pharynx can evaginate through the single opening, the mouth, which functions both as a mouth and as an anus. In most parasites there is a reduction of the digestive system, as an adaptation to their parasitic lifestyle. Platyhelminths also show a developed excretory system, which allows them to remove waste materials through filtration (Figure 14.2). Platyhelminths lack circulatory, skeletal and respiratory systems; the gases and nutrients reach the cells by diffusion, which explains why the animals are extremely flat. Their epidermis is a monolayer of multiciliated cells that provide locomotion to free-living platyhelminths. Several parasites have specific organs to attach to the host such as suckers and hooks.

REPRODUCTION AND DEVELOPMENT

Reproduction

Platyhelminths are hermaphrodites, possessing both testicles and ovaries (Figure 14.2). Some free-living species can also reproduce asexually either by parthenogenesis or by fission. Asexual reproduction by fission occurs in some catenulids and macrostomids, and it is common among triclads. In these animals, after the fission of the tail, new individuals are regenerated from both fragments (tail and head) in a few days (Figure 14.3). Fissiparous platyhelminths do not develop a mature reproductive system.

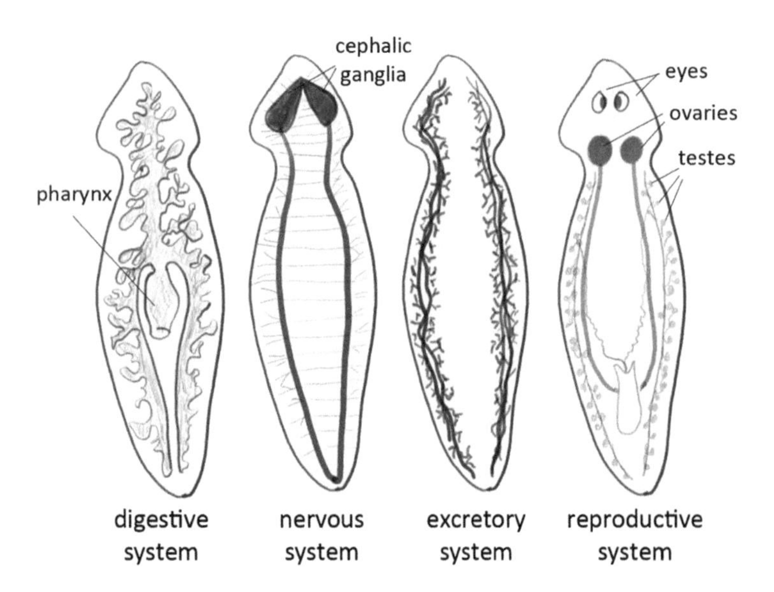

Figure 14.2 Diagrams showing the anatomical location of the digestive system, nervous system, excretory system and reproductive system.

Embryonic Development

In contrast to other animals, most platyhelminths have ectolecithal eggs (the Neoophora), which mean that they do not contain any yolk, which is supplied by specialized yolk cells called vitellaria, while the oocytes are produced in the germarium (Adell et al., 2015; Martín-Durán and Egger, 2012). All Neoophora embryos have developed mechanisms to engulf the external yolk cells, and this condition has probably determined the divergent mode of embryogenesis found within this group with respect to spiralians. In Archoophora a spiral cleavage pattern can be still recognized. However, Neoophora show a disperse cleavage, in which blastomeres are not kept attached one to the other, and develop an embryonic pharynx to engulf the yolk.

The most studied platyhelminths are triclads, due to the amazing regenerative abilities of the adults, and the parasitic species which represent a public health problem such as *Schistosoma*. In the following we describe the embryonic development of the triclad *Schmidtea* and the complex life cycle of *Schistosoma*.

Schmidtea lays ectolecithal eggs called cocoons, which contain several zygotes surrounded by a large number of extra-embryonic yolk cells (Cardona et al., 2006) (**Figure 14.4**). Cleavage shows no pattern and blastomeres remain detached from one another. After several divisions, a "cryptic larva" is formed, with an embryonic epidermis and the embryonic pharynx, to engulf the surrounding yolk (Figure 14.4). The rest of the blastomeres remain in the periphery until the larva has eaten the yolk. At this stage, a process of "metamorphosis" starts and the blastomeres start to migrate and differentiate into the definitive structures that will replace the embryonic ones (Cardona et al., 2006; Martín-Durán et al., 2012b) (Figure 14.3). During this stage the definitive adult body

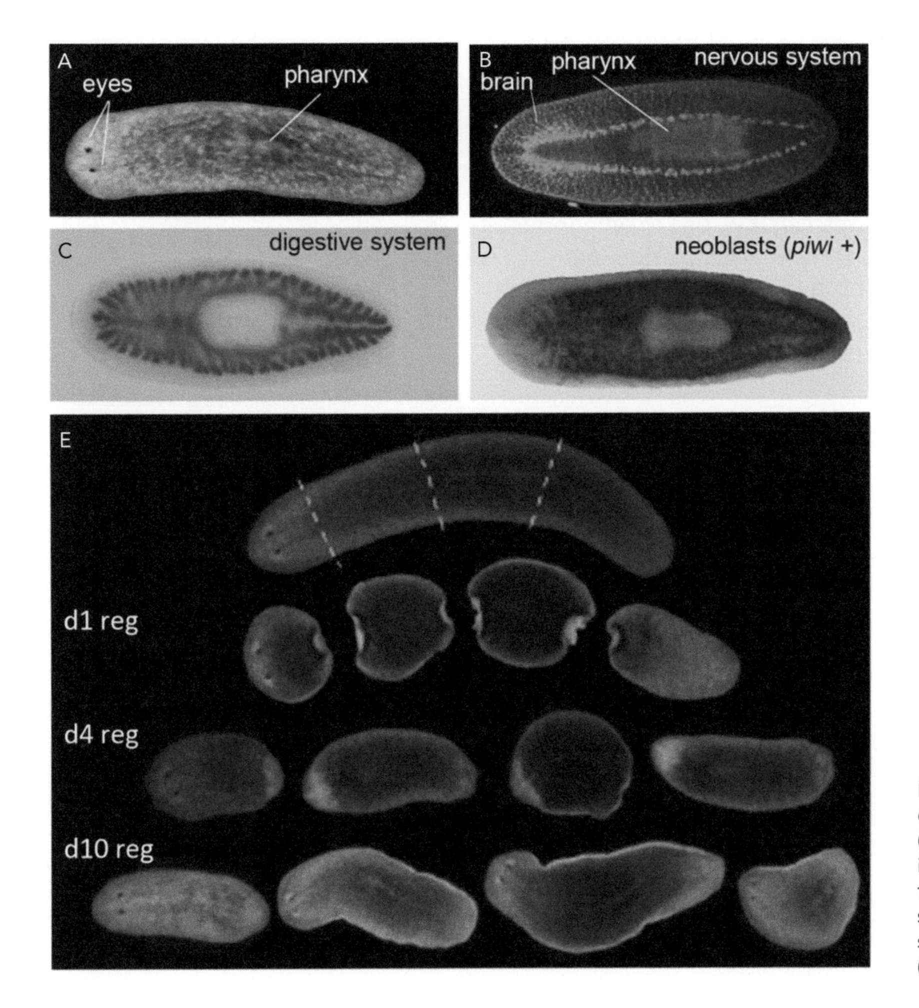

Figure 14.3 Micrographs showing some of the basic anatomy of platyhelminths. (A) The locations of the pharynx and eyes are indicated. (B) Flourescence microscopy showing the nervous system of this animal. (C) Antibody stain showing the digestive system. (D) Antibody stain showing the neoblasts of the organism. (E) Photos of regenerating organisms.

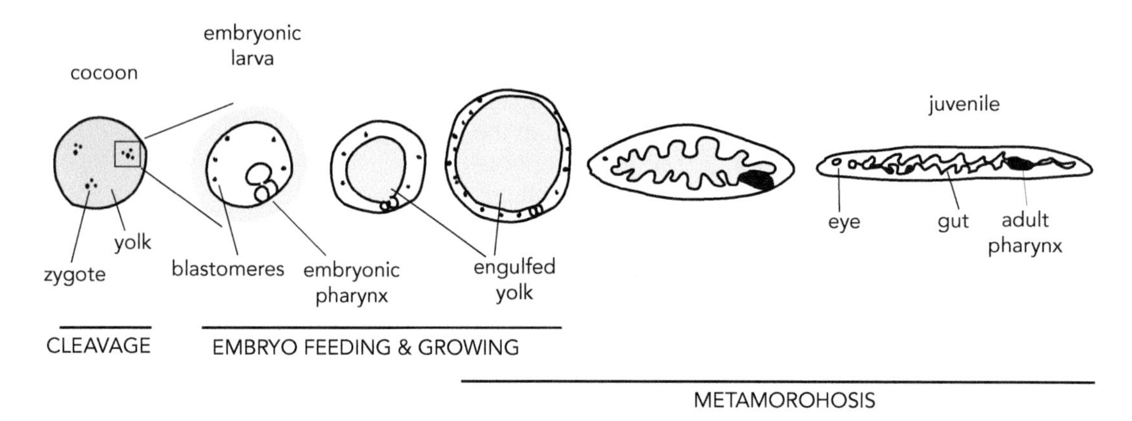

Figure 14.4 Diagram showing the developmental pathways for a typical platyhelminth.

plan is established, and patterning mechanisms are similar to those used during homeostasis and regeneration in the adult (see next section) (Martín-Durán et al., 2010).

Schistosoma mansoni has been the subject of several studies, since it is the causative agent of schistosomiasis, which ranks as one of the major infectious diseases in the world (Colley et al., 2014). *Schistosoma* infects more than 60 million people, most of them in Sub-Saharan Africa, causing intestinal and hepatosplenic schistosomiasis. As in all trematodes, the life cycle of *Schistosoma* is very complex. It has separate sexes that live for several years in blood vessels of vertebrates, their definitive host. Inside the host vasculature, each female can release about 300 eggs per day. When eggs are excreted into fresh water, a ciliated larva called a miracidium hatches in about five days. In snails of the genus *Biomphalaria*, miracidia undergo cycles of asexual reproduction, leading to the emergence of a second larval type, the cercaria. Cercariae swim actively and penetrate the definitive host again through the skin to enter blood vessels, where they develop into adults (Schwartz and Fallon, 2018). Through this complex cycle, a single zygote of parasitic flatworms can give rise to an enormous number of progeny. The disease is caused by chronic inflammation that occurs when eggs released by female worms get trapped in small blood vessels of the liver or other organs.

PHYSIOLOGY AND BEHAVIOR

Platyhelminths are well known for their extreme plasticity. Some planarians are capable of regenerating a whole organism from any small piece of their body (Figure 14.3) (Saló et al., 2009). Furthermore, they continuously change their size according to the environmental conditions (nutrients and temperature). This plasticity relies on the existence of a unique population of adult totipotent stem cells, termed neoblasts (Figure 14.3) (Baguñà, 2012).

All platyhelminths possess neoblasts. However, not all have the ability to regenerate (Cebrià et al., 2015). The most plastic ones are some triclad species, such as *Schmidtea mediterranea* and *Dugesia japonica*, as well as the macrostomid *Macrostomum lignano*. However, some groups, such as many rhabdocoels, cannot regenerate at all. Other groups can regenerate some organs or parts. For example, some species of triclads and macrostomids can regenerate the tail but not the head. In the parasitic species pluripotent stem cells are also fundamental during development. As expected, the species that reproduce asexually show more pronounced abilities of regeneration, since it is absolutely required for their survival.

Although the existence and the properties of neoblasts was proposed several years ago, it has only recently been shown that an individual neoblast is capable of differentiating into all cell types in the planarian *Schmidtea mediterranea* (Wagner et al., 2011). *Schmidtea mediterranea* has become the animal model for molecular and genetic approaches, and is offering valuable information about stem cell-based adult development. The transcriptomic study of this species at a single-cell level (SCseq) has allowed the identification of several differentiated cell types that compose those animals (epithelial, neural, secretory ...) as well as progenitor cells that are being specified to those differentiated cell types from the totipotent neoblasts (Fincher et al., 2018; Plass et al., 2018). In recent years, work from several groups has also demonstrated that planarians' plasticity does not only rely on neoblasts but also on the continuous activation of the intercellular molecular communication pathways that allow communication between cells (Reddien, 2018). The activation of those pathways controls the proliferation, differentiation and death of each planarian cell according to the cellular environment, thus allowing the proper maintenance and regeneration of structures in homeostatic and in regenerative context. As an example, it has been shown that the Wnt and BMP pathways are essential in specifying the antero-posterior and the dorso-ventral axis, respectively. The consequences of this are, for instance, that when the Wnt signaling is inhibited planarians in which the head and the tail have been amputated regenerate two heads. Importantly, even if there is no amputation, when the Wnt pathway is inhibited during several weeks of homeostasis planarians also develop a head at the posterior (Adell et al., 2010). These results demonstrate that, in their normal homeostasis, planarian neoblasts require the continuous signaling of those molecular pathways in order to properly specify their fate.

DISTRIBUTION AND ECOLOGY

The phylum Platyhelminthes originated during the early radiation of Metazoan phyla, most probably in a marine environment. Nowadays, however, we find that approximately a quarter of free-living species have colonized freshwater and terrestrial environments. In terms of their worldwide distribution, the data available are clearly biased depending on the area where the corresponding taxon's specialist worked. For instance, the distribution by biogeographical area of the number of known species and genera of freshwater "turbellaria" (**Table 14.1**) shows the Palearctic region as bearing the highest diversity of species – a situation difficult to believe, and which is more probably a reflection of that region having been more studied. In consequence we might foresee that most Platyhelminthes diversity, at least for free-living groups, is still to be discovered.

TABLE 14.1 Number of species and genera of freshwater "turbellaria" by biogeographic region (Modified from Schockaert et al., 2008)

Biogeographic area	Species	(%)	Genera	(%)
Palearctic	788	(56.2)	137	(46.4)
Nearctic	221	(15.8)	47	(15.9)
Neotropical	150	(10.7)	33	(11.2)
Afrotropical	85	(6.1)	28	(9.5)
Oriental	36	(2.6)	12	(4.1)
Australasia	116	(8.3)	30	(10.2)
Oceanic Islands	2	(0.1)	2	(0.7)
Antarctic	5	(0.4)	6	(2.0)

Again, given the high diversity of platyhelminths, their ecology is also very diverse. Most free-living forms are marine, although there are also many freshwater species. In the Tricladida there is a lineage that has adapted to live in terrestrial habitats, although these habitats have to be extremely humid (wet soils in forests). There are also platyhelminths that have adapted to hypogean life, losing their eyes and pigmentation. It has been found that some triclads have since long ago been transported by humans and introduced to new regions (for instance *Girardia tigrina* was introduceed from North America to Germany in the early 1920s and spread all over Europe). Unluckily, this human-mediated transport is becoming more and more frequent in recent years. The introduction of terrestrial planarians from neotropical areas (New Zealand, Australia and South America) to Europe and also North America, transported within the soil in pots of imported plants (Alvarez-Presas et al., 2014; Justine et al., 2018), is especially notorious. These introductions may have important economic consequences, as has been shown in the British Isles, where some of the introduced species prey on earthworms and have become a plague for earthworms farms and for composting worm cans (Murchie and Gordon, 2013). Presence of these species causes significant declines in earthworm populations, resulting in a reduction of ecological functions that earthworms provide, e.g. loss of soil drainage. Moreover, consequent flooding of the land can negatively affect crops and grassland productivity.

In the parasite groups we also find a great diversity. Monogeneans are ectoparasites of a wide variety of aquatic vertebrates, while trematodes and cestodes are endoparasites with complex life cycles that include one or more intermediate invertebrate hosts and a final vertebrate host, the intermediate hosts being either molluscs (for trematodes) or "crustaceans" (for cestodes). The parasites have an important impact on the economy, from affecting livestock (either vertebrates or invertebrates) to causing human diseases.

GENETICS

At present only four free-living platyhelminth genomes can be found in the public databases (*Schmidtea mediterranea*, *Dugesia japonica*, *Girardia tigrina* and *Macrostomum lignano*), while for parasitic species 29 genomes have been sequenced. For the free-living species, the genome size varies between 0.8 and 1.5 Gb, and all of them present a low GC content (around 30%). The best assembled and annotated genome, that of *S. mediterranea* (Grohme et al., 2018), has shown the presence of a high percentage of repetitive sequences (61.7%) including three LTR families with elements more than 30 Kb long. This genome also presents a high heterozygosity in spite of the fact that it was sequenced from a clonal lineage obtained through 17 successive sib-mating generations. A similar conclusion was reached for the genome of *D. japonica* basing on a comparison of the transcriptomes and genome data obtained from a lineage derived from a single individual that continued to undergo autonomous asexual reproduction for more than 20 years (Nishimura et al., 2015). All these characteristics may explain why the assembling of planarian genomes is such a nightmare. In the case of the parasitic platyhelminths, genome sizes range from around 150 Mb in some *Taenia* species (tapeworms) to around 1.2 Gb for *Fasciola*. Parasitic genomes also present low percentages of GC. *Schistosoma* (blood flukes) is the genus with most genomes sequenced, and the best studied is that of *Schistosoma mansoni*. The *S. mansoni* genome sequencing data was first published in 2009 and has been posteriorly improved (Protasio et al., 2012). It has a length of 364.5 Mb, and a total of 10,852 genes were identified, encoding over 11,000 proteins, 45% of which remain without known or predicted function. 81% of the genome was assembled onto the parasite's chromosomes, providing a partial genetic map. A phylogenetic study of their proteins (phylomes) has shown that besides the exploitation of host endocrine and immune signals, the parasite genome exhibits multiple events of gene duplication which may be, at least partially, an adaptive response related to the parasitic lifestyle. The availability of genomic data and its phylogenetic study therefore provides information for the identification of novel drug targets and vaccine candidates through a wide perspective.

For Platyhelminthes, the representation of mitochondrial genomes is better than for nuclear genomes; however, it is extremely biased toward the parasitic species. 97 mitochondrial genomes are available from the Neodermata, but only 14 from free-living flatworms, which represents only 4 of the 11 Platyhelminthes orders: one species of Catenulida, two species of Macrostomorpha (one incomplete), four species of Polycladida and eight species of Tricladida. The sizes of most free-living species' mitochondrial genomes range between 14,000 and 15,500 bp, with the exception of *Schmidtea mediterranea* (sexual biotype) with 27,133 bp. Cestodes have the smallest mitochondrial genomes, rarely exceeding 14,000 bp. Trematodes have values more similar to the free-living with an exception, the 16,901 bp of *Schistosoma spindale*. As in the nuclear genomes, the AT content for Platyhelminthes is extremely high, reaching values of 80% in *Obama nungara*. The gene order is diverse among groups; even within *Schistosoma* two distinct mitochondrial gene orders have been found. Although many studies had missed the presence of *atp8* gene in those mitochondrial genomes, the latest study (that of a Catenulida and a Macrostomorpha species [Egger et al., 2017]) has shown that this gene is also present in the free-living representatives of the phylum, although in the form of a very short version of the protein. This study has also shown that the mitochondrial genome of *Stenostomum sthenum* presents genes encoded in both strands, while all those sequenced before were encoded in the same strand. Finally, all the Rhabditophora present a unique genetic code shared with echinoderms (EMBL-NCBI genetic code 9), while the proposal of TAA coding for Tyr had not gained support, so the "alternative flatworm mitochondrial code", code 14 from EMBL-NCBI, proposed for some Platyhelminthes and Nematoda, is likely a feature exclusive to the latter (Solà et al., 2015).

POSITION IN THE ToL

Platyhelminthes had often been considered as representatives of the transition from diploblasts to triploblasts because they present a mix of simple (lack of coelom, anus, mitosis in somatic cells among others) and complex (spiral embryonic cleavage in many flatworms) features. Haeckel in his Gastraea theory placed flatworms in a key position in the animal tree, deriving an ancestral flatworm-like animal from the ancestral Gastraea (Haeckel, 1874). In these hypotheses the flatworm bodyplan would be simple and ancestral to bilaterian (**Figure 14.5A**). But later proposals, primarily based on cladistic principles, supported the idea that platyhelminths were derived protostomes with a secondarily simplified bodyplan (**Figure 14.5B**). Within the protostomes, multiple sister group relationships to other phyla had been proposed. For most of the traits used in those studies the platyhelminths shared with its proposed sister groups the primitive state for the protostomes. That is why, possibly, their similarity was not due to sharing a common ancestor but because they had retained the antique state. At the same time, the monophyly of the phylum was under debate. Due to the lack of shared derived characters, Platyhelminthes was divided into three groups: the Acoelomorpha (Acoela and Nemertodermatida), the Catenulida and the Rhabditophora (Smith et al., 1986).

Molecular phylogenetic studies have helped to solve some of these controversies (**Figure 14.5C**). They confirmed the lack of monophyly for Platyhelminthes, suggesting acoelomorphs as the first offshoot of the bilaterian stem or as a member of the Deuterostomia, whereas the Rhabditophora and the Catenulida constituted a monophyletic group situated within the Lophotrochozoa clade and not

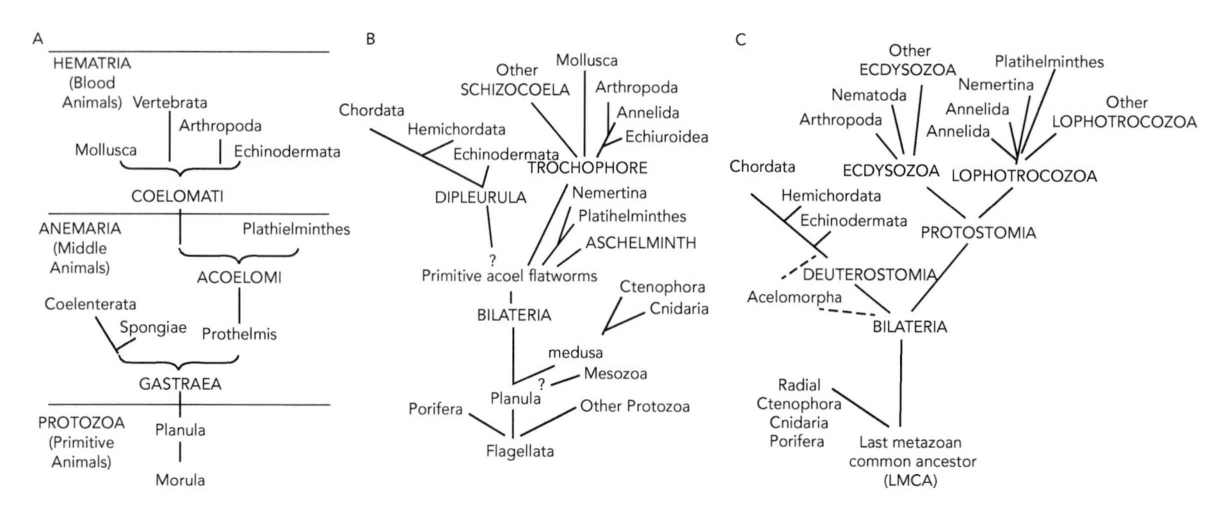

Figure 14.5 Three phylogenetic hypotheses for the position of Platyhelminthes in the animal tree of life. (A) The Gastrea hypothesis. (B) The hypothesis where platyhelminths are derived protostomes with a secondarily simplified bodyplan. (C) The hypothesis based on molecular data.

at the base of the Bilateria. Since the Acoelomorpha are not considered as platyhelminths any more, only the Catenulida and Rhabditophora remain as members of the phylum. The situation of Platyhelminthes within the Lophotrochozoa has posteriorly been well supported and widely accepted. However, their exact situation within this group remains elusive, in part due to their long branches that make them victims of the Long-Branch Attraction artifact in many analyses. Nonetheless each new contribution with more genes, more sophisticated methodologies and/or a better metazoan sampling is lending further support to their belonging to a clade including Annelida, Mollusca and Nemertina (see for instance, Marlétaz et al., 2018).

DATABASES AND COLLECTIONS

Data used to reconsruct phylogeny are from Laumer et al., Dryad Digital Repository under a CC0 Public Domain Dedication: 10.5061/dryad.622q4. https://datadryad.org/resource/doi:10.5061/dryad.622q4.

Two databases that archive information on Platyhelmithes are Wormbase-Parasite (https://parasite.wormbase.org/index.html) and Earthlife (https://www.earthlife.net/inverts/platyhelminthes.html). Collections can be found at most natural history museums. Two collections with detailed websites for Platyhelminthes are the National Museum of Natural History in Washington, DC (https://www.gbif.org/dataset/6c827f0d-9371-447b-b42b-199ad19bdb64) and the University of Copenhagen: https://samlinger.snm.ku.dk/en/dry-and-wet-collections/zoology/platyhelminthes-aschelminthes-proarthropoda-collections/. A limited number of formalin-preserved platyhelminths can be found in Ward's platyhelminth collection (https://us.vwr.com/store/product/8884189/ward-s-platyhelminth-collection).

CONCLUSION

As stated above, the identity of the closest relatives of the Platyhelminthes within this big clade that constitutes the Lophotrochozoa is still not totally clear. The hope is that adding data on more organisms, especially those rarer ones for which still little genomic data is available, may finally render better-resolved trees with higher supports for the basal nodes within the Lophotrochozoa. Although new methodologies and more sophisticated analyses may also appear and may assist better phylogenies, our opinion is that it will be difficult to get a better result only based in genetics, since many available genetic tools have already been used.

The same is true for the internal relationships of the Platyhelminthes. We have seen an impressive update of their phylogeny, thanks to the use of transcriptomic data and especially to the inclusion of many groups previously not considered. In this case, we do think that the addition of more representatives of some groups (Rhabdocoela, Proseriata and Lecitoepitheliata for instance) will help to solve open questions, such as the relative order of appearance of Rhabdocoela or Proseriata, the monophyly of Lecithoepitheliata, or the origin of endolecithality. Solving these questions will result in a better knowledge of the evolution of embryonic development in this interesting group of animals.

REFERENCES

Adell, T., Cebria, F., Salo, E., 2010. Gradients in Planarian Regeneration and Homeostasis. Cold Spring Harb. Perspect. Biol. 2: a000505.

Adell, T., Martín-durán, J. M., Saló, E., Cebrià, F., 2015. Plathyhelminthes. In: Wanninger, A. (ed.), Evolutionary Developmental Biology of Invertebrates 2: Lophotrochozoa (Spiralia). Springer-Verlag Wien, pp. 21–40.

Alvarez-Presas, M., Mateos, E., Tudó, A., Jones, H. D., Riutort, M., 2014. Diversity of introduced terrestrial flatworms in the Iberian Peninsula: a cautionary tale. PeerJ, e430.

Baguñà, J., 2012. The planarian neoblast : the rambling history of its origin and some current black boxes. Int. J. Dev. Biol. 56, 19–37.

Cardona, A., Hartenstein, V., Romero, R., 2006. Early embryogenesis of planaria : a cryptic larva feeding on maternal resources 667–681.

Cebrià, F., Saló, E., Adell, T., 2015. Regeneration and Growth as Modes of Adult Development: The Platyhelminthes as a Case Study. In: Evolutionary Developmental Biology of Invertebrates 2: Lophotrochozoa (Spiralia). Springer-Verlag Wien, pp. 41–78.

Colley, D. G., Bustinduy, A. L., Secor, W. E., King, C. H., 2014. Human schistosomiasis. The Lancet 383, 2253–2264.

Egger, B., Bachmann, L., Fromm, B., 2017. Atp8 is in the ground pattern of flatworm mitochondrial genomes. BMC Genomics 18, 414.

Egger, B., Lapraz, F., Tomiczek, B., Müller, S., Dessimoz, C., Girstmair, J., Skunca, N., Rawlinson, K. A., Cameron, C. B., Beli, E., Todaro, M. A., Gammoudi, M., Noreña, C., Telford, M. J., 2015. A transcriptomic-phylogenomic analysis of the evolutionary relationships of flatworms. Curr. Biol. 25, 1347–1353.

Ehlers, U., 1985. Das Phylogenetische System der Plathelminthes. Stuttgart and New York.

Fincher, C. T., Wurtzel, O., Hoog, T. De, Kravarik, K. M., Reddien, P. W., 2018. Cell type transcriptome atlas for the planarian Schmidtea mediterranea. Science 360, 874.

Gegenbaur, C., 1859. Grundzuge der vergleichenden Anatomie. Leipzig, Germany.

Grohme, M. A., Schloissnig, S., Rozanski, A., Pippel, M., Young, G. R., Winkler, S., Brandl, H., Henry, I., Dahl, A., Powell, S., Hiller, M., Myers, E., Rink, J. C., 2018. The genome of Schmidtea mediterranea and the evolution of core cellular mechanisms. Nature 554, 56–61..

Haeckel, E., 1866. Generelle Morphologie der Organismen: Allgemein Grunndzuge der Organischen Formen-Wissenschaft, Mechanisch Bergrundet durch die von Charles Darwin Reformite Descendenz-Theorie. Georg. Reimer, Berlin.

Haeckel, E., 1874. The Gastraea theory, the phylogenetic classification of the animal Animal kingdom and the homology of the germ-lamellae. Q. J. Microsc. Sci.

Hyman, L. H., 1951. The Invertebrates: Platyhelminthes and Rhynchocoela the acoelomate Bilateria. McGraw-Hill, London and New York.

Justine, J.-L., Winsor, L., Gey, D., Gros, P., Thévenot, J., 2018. Giant worms *chez moi!* Hammerhead flatworms (Platyhelminthes, Geoplanidae, *Bipalium* spp., *Diversibipalium* spp.) in metropolitan France and overseas French territories. PeerJ 6, e4672.

Laumer, C. E., Hejnol, A., Giribet, G., 2015. Nuclear genomic signals of the "microturbellarian" roots of platyhelminth evolutionary innovation. Elife 4, 1–31.

Marlétaz, F., T. C. A. Peijnenburg, K., Goto, T., Satoh, N., Rokhsar, D. S., 2018. A New Spiralian Phylogeny Places the Enigmatic Arrow Worms among Gnathiferans. Curr. Biol. 29, 1–7.

Martín-durán, J. M., Amaya, E., Romero, R., 2010. Germ layer specification and axial patterning in the embryonic development of the freshwater planarian Schmidtea polychroa. Dev. Biol. 340, 145–158.

Martín-durán, J. M., Egger, B., 2012. Developmental diversity in free-living flatworms Developmental diversity in free-living flatworms. Evodevo 3, 7.

Martín-Durán, J. M., Egger, B., 2012. Developmental diversity in free-living flatworms. Evodevo 3, 1–23.

Martín-durán, J. M., Monjo, F., Romero, R., 2012. Planarian embryology in the era of comparative developmental biology. Int J Dev Biol. 56, 39–48.

Murchie, A. K., Gordon, A. W., 2013. The impact of the "New Zealand flatworm", Arthurdendyus triangulatus, on earthworm populations in the field. Biol. Invasions 15, 569–586.

Nishimura, O., Hosoda, K., Kawaguchi, E., Yazawa, S., Hayashi, T., Inoue, T., Umesono, Y., Agata, K., 2015. Unusually Large Number of Mutations in Asexually Reproducing Clonal Planarian Dugesia japonica. PLoS One 10, e0143525.

Plass, M., Solana, J., Wolf, F. A., Ayoub, S., Misios, A., Glažar, P., Obermayer, B., Theis, F. J., Kocks, C., Rajewsky, N., 2018. Cell type atlas and lineage tree of a whole complex animal by single-cell transcriptomics. Science 360, eaaq1723.

Protasio, A. V., Tsai, I. J., Babbage, A., Nichol, S., Hunt, M., Aslett, M. A., De Silva, N., Velarde, G. S., Anderson, T. J. C., Clark, R. C., Davidson, C., Dillon, G. P., Holroyd, N. E., LoVerde, P. T., Lloyd, C., McQuillan, J., Oliveira, G., Otto, T.D., Parker-Manuel, S. J., Quail, M. A., Wilson, R. A., Zerlotini, A., Dunne, D. W., Berriman, M., 2012. A Systematically Improved High Quality Genome and Transcriptome of the Human Blood Fluke Schistosoma mansoni. PLoS Negl. Trop. Dis. 6, e1455.

Reddien, P. W., 2018. Review The Cellular and Molecular Basis for Planarian Regeneration. Cell 175, 327–345.

Saló, E., Abril, J. F., Adell, T., Cebrià, F., Eckelt, K. A. Y., Fernández-taboada, E., Handberg-thorsager, M., Iglesias, M., Molina, M. D., Rodríguez-Esteban, G., 2009. Planarian regeneration : achievements and future directions after 20 years of research 1327, 1317–1327.

Schockaert, E. R., Hooge, M., Sluys, R., Schilling, S., Tyler, S., Artois, T. J., 2008. Global diversity of free living flatworms (Platyhelminthes, "Turbellaria") in freshwater. Hydrobiologia 595, 41–48.

Schwartz, C., Fallon, P. G., 2018. Schistosoma "Eggs-Iting" the Host : Granuloma Formation and Egg Excretion. Front. Immunol. 9, 1–16.

Smith III, J. P. S., Tyler, S., Rieger, R. M., 1986. Is the Turbellaria polyphyletic? Hydrobiologia 132, 13–21.

Solà, E., Alvarez-Presas, M., Frías-López, C., Littlewood, D. T. J., Rozas, J., Riutort, M., 2015. Evolutionary Analysis of Mitogenomes from Parasitic and Free-Living Flatworms. PLoS One 1–20.

Wagner, D. E., Wang, I. E., Reddien, P. W., 2011. Clonogenic neoblasts are pluripotent adult stem cells that underlie planarian regeneration. Science 332, 811–6.

Additional Reading

Baguñà, J., Riutort, M., 2004. Molecular phylogeny of the Platyhelminthes. Can. J. Zool. 82, 168–193.

Riutort, M., Álvarez-Presas, M., Lazaro, E.M., Solà, E., Paps, J., 2012. Evolutionary history of the Tricladida and the Platyhelminthes: an up-to-date phylogenetic and systematic account. Int. J. Dev. Biol. 56, 5–17.

Ruiz-Trillo, I., Paps, J., 2016. Acoelomorpha: earliest branching bilaterians or deuterostomes? Org. Divers. Evol. 16, 391–399.

Valentine, J.W., 2004. On the Origin of Phyla. University Of Chicago Press.

PHYLUM CHAETOGNATHA

CHAPTER
15

Yvan Perez, Carsten H.G. Müller, and Steffen Harzsch

Infrakingdom	Bilateria
Division	Protostomia
Subdivision	Lophotrochozoa
Phylum	Chaetognatha

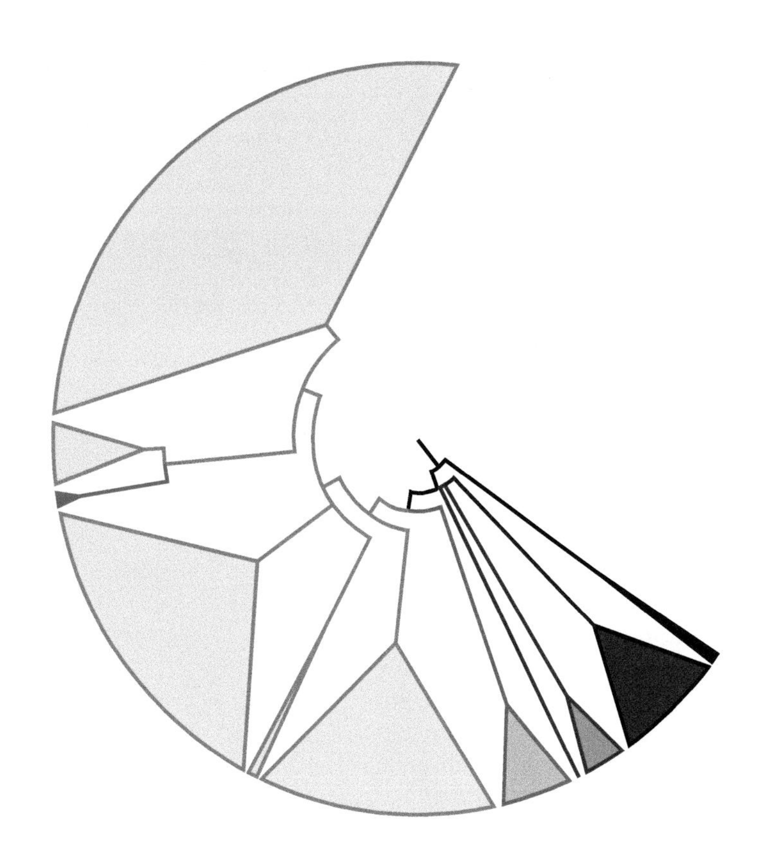

CONTENTS

General Characteristics

Vernacular name: arrow worm, referring to their torpedo-shaped and elongated body

Very ancient and peculiar bilaterian taxon with a long and isolated evolutionary pathway, as illustrated by the set of autapomorphies listed below. Oldest known fossils trace back to the lower Cambrian

Strictly marine. Very abundant in all oceanic environments. Meiobenthic, epibenthic, hyperbenthic and planktonic species. All chaetognaths are hermaphroditic

133 valid species allocated to 26 genera and eight families. Two or three orders are usually considered, but only one is validated by the molecular analyses

No economic value or direct significance to humans

Autapomorphies

- multilayered (pluristratified) epidermis constituted by two types of epidermal cells
- cuticle restricted to ventral part of the head
- secondary heterosarcomeric muscles
- parallel arrangement of perforated membranes in the distal segment of photoreceptor cells
- tactile receptor cells with unique bipartite ciliary rootlets (in ciliary fence and tuft organs)
- complex nerve plexus with both basi- and intraepidermal condensations (domains)
- neuromuscular innervations by diffusion of neurotransmitters through the epidermal ECM
- heterocoely (unique mode of gastrulation and coelomogenesis)
- fertilization via accessory fertilization cells
- expression of brachyury gene around the mouth opening of the hatchling
- duplicated ribosomal cluster

HISTORY OF TAXONOMY AND CLASSIFICATION

A generalized body plan of Chaetognatha (Leuckart, 1894), a taxon usually ranked as phylum, is illustrated in **Figures 15.1** and **15.2**. Recorded for the first time by the Dutch naturalist Martinus Slabber in 1775, chaetognaths are soft-bodied dimeric bilaterians, ranging in size from 1.3 mm for the smallest meiobenthic species up to 80 mm for the largest planktonic species (for a recent listing of valid species see Müller et al. 2019). The transparent body is fundamentally dimeric and divided into the cephalic region and an elongated trunk. A secondary division of the trunk occurs after hatching by the developmental emergence of a posterior septum and gives rise to a post-anal tail segment (Shinn 1997; Harzsch et al. 2015). The body shape is maintained by a hydroskeleton which functions based on the inner pressure applied by coelomic fluids associated with a well-developed musculature (Duvert and Salat 1990; Bone and Duvert 1991).

The systematics of chaetognaths is traditionally based on few key morphological characters, namely: the presence/absence and distribution of ventral transverse muscles and glandular structures on the body surface; the relative length of the tail (trunk/tail ratio); the number, shape and position of teeth, grasping spines and lateral fins; the grasping spine ornamentation; the type and position of eyes, seminal vesicles, and corona ciliata; the presence/absence of eyes, glandular neck channels, intestinal diverticula, tegumentary bridge between the lateral fins, vacuolated intestinal cells, collarette, and adhesive structures. However, the great homogeneity of the chaetognath's body organisation associated with high levels of phenotypic plasticity has hampered reaching consensus on their systematics.

Nuclear ribosomal DNA phylogenies, sometimes combined with morphological data, disentangled some aspects of the chaetognathan evolutionary history (**Figure 15.3**). Notably, molecular analyses invalidated the Monophragmophora/Biphragmophora hypothesis and showed the paraphyly of Phragmophora (Papillon et al. 2006; Miyamoto and Nishida 2011; Gasmi et al. 2014). Many of the morphological characters listed above are strongly subjected to homoplasy with numerous events of losses and reversions (Gasmi et al. 2014). According to molecular phylograms, the distribution of these morphological characters is neither related to phylogeny nor to ecology and suggests that the diversity in Chaetognatha was generated through a process of mosaic evolution, a conclusion previously supported by the analysis of their muscular diversity (Casanova and Duvert 2002). After the most recent discovery of new species, and according to the contributions of several molecular studies, 133 species allocated to 26 genera and eight families are considered as valid (Müller et al. 2019).

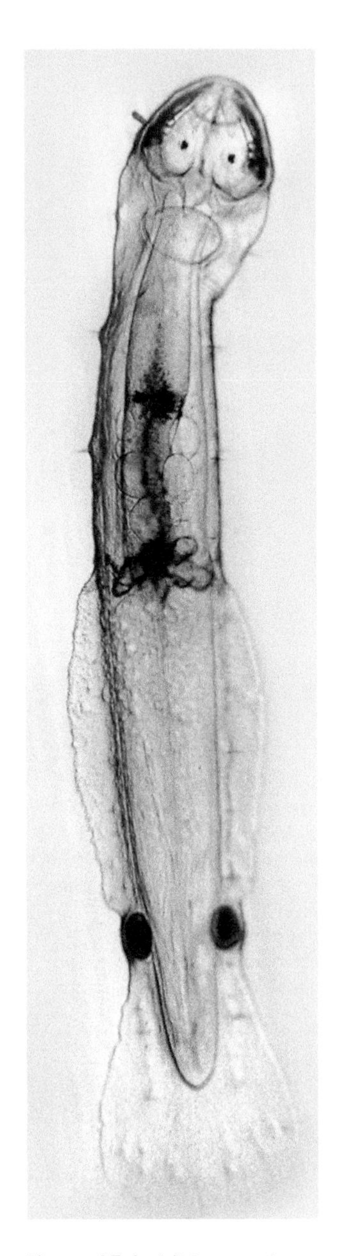

Figure 15.1 A living specimen of the benthic species *Spadella cephaloptera* (Phragmophora, Spadellidae) collected on a *Posidonia oceanica* seagrass bed near Marseilles (France).

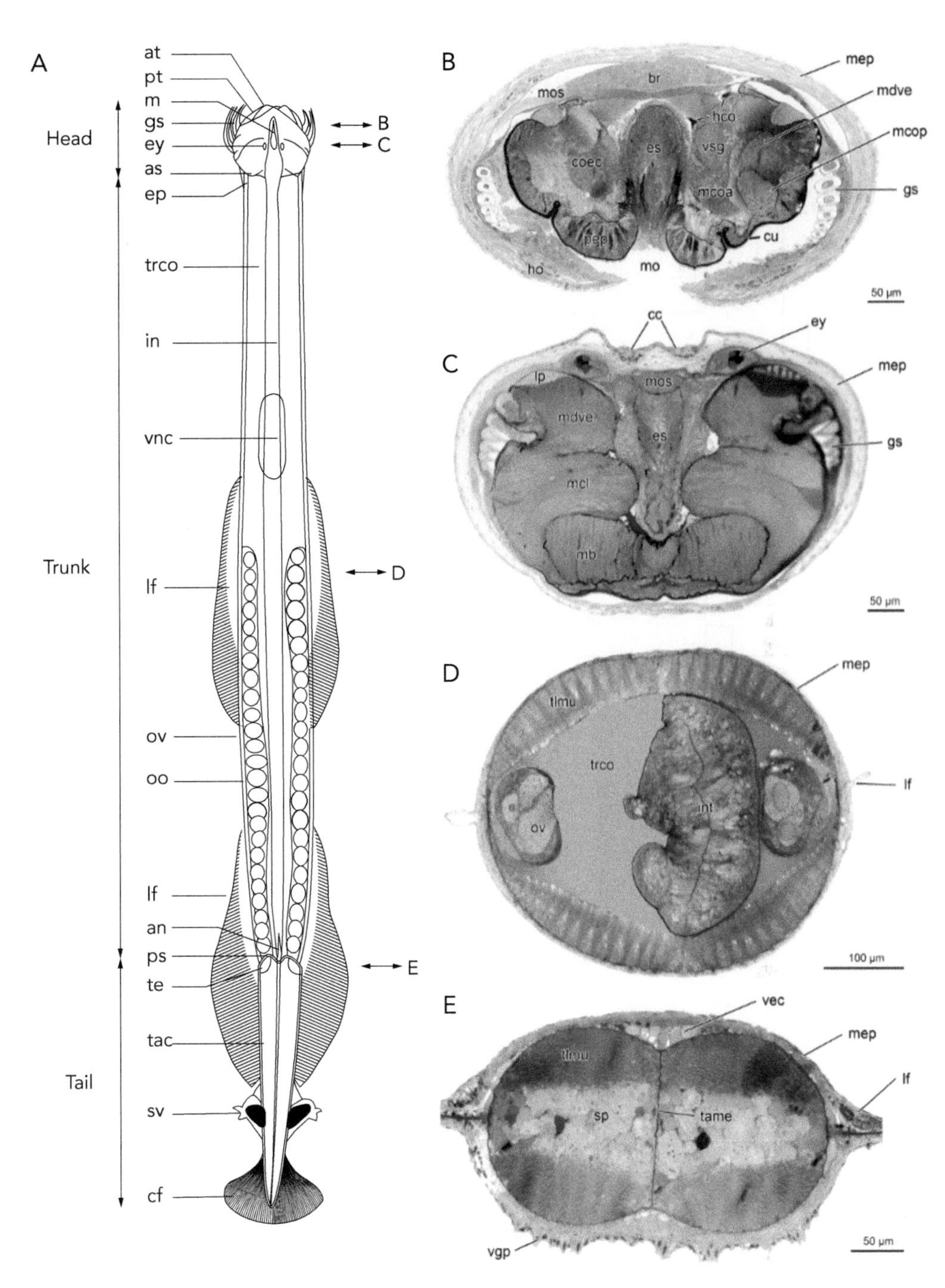

Figure 15.2 (A) Diagram showing the general morphology of a planktonic chaetognath, *Serratosagitta serratodentata* (Sagittidae, Aphragmophora). (B)–(E) Transverse semithin sections corresponding approximately to the section levels indicated on A in *Spadella cephaloptera* (B, E) and *Ferosagitta hispida* (C, D). lf, lateral fin; an, anus; as, anterior septum; at, anterior teeth; cf, caudal fin; m, ventral mouth; ep, epidermis; ey, eye; gs, grasping spines; in, intestine; oo, oocyte; ov, ovary; plf, posterior lateral fin; ps, posterior septum; pt, posterior teeth; tac, tail general cavity; te, testis; trc, trunk general cavity; sv, seminal vesicle; vnc, ventral nerve center. (A) Redrawn and modified from Alvariño (1969). (B)–(E) Modified from Müller et al. (2019).

ANATOMY, HISTOLOGY AND MORPHOLOGY

Chaetognaths usually have two eyes located on the dorsal side of the head (Goto and Yoshida 1984; Goto et al. 1989), but pigmented cells or the entire eye has been reported to be absent in few species living in troglobitic and deep

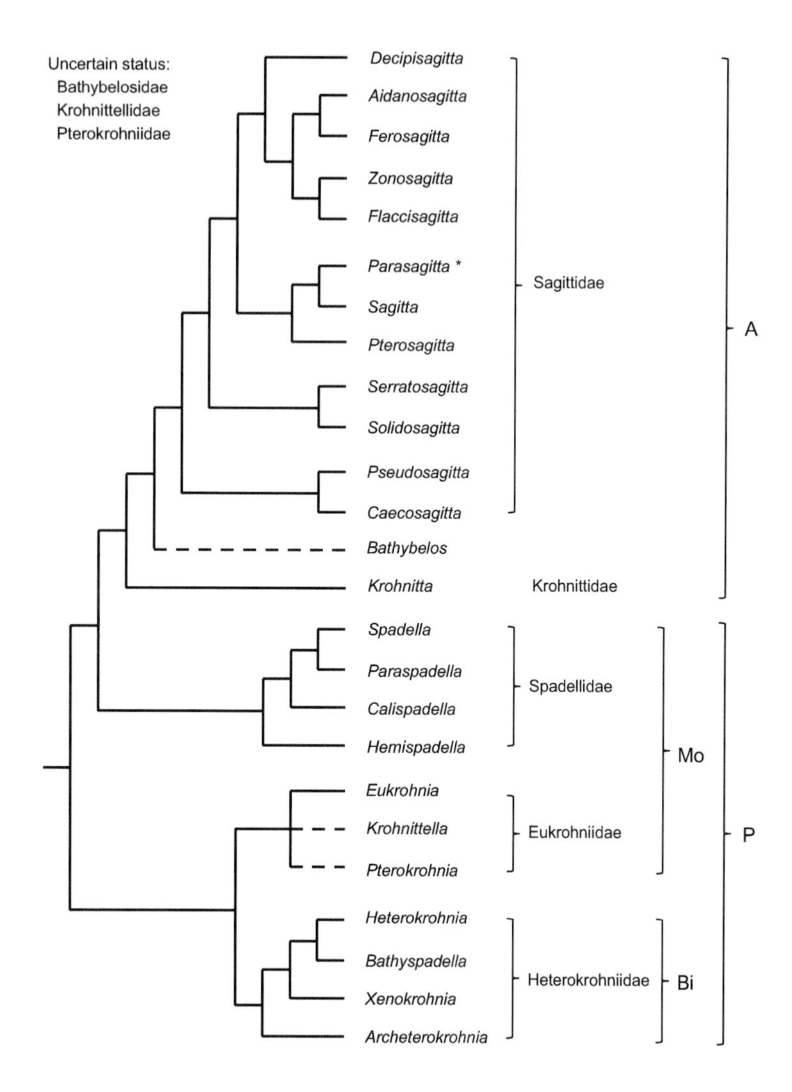

Figure 15.3 Consensus topology summarizing the evolutionary history of Chaetognatha based on morphological and molecular (rRNA) data. Molecular and morphological phylogenetic analyses invalidated the Biphragmophora (Bi) and Monophragmophora (Mo) orders which appear poly- and paraphyletic, respectively. In an alternative hypothesis, these clades are assembled in the Phragmophora (P) which are also paraphyletic in molecular trees. (A) Aphragmophora. Asterisk indicates the inclusion of *Mesosagitta minima* within Parasagitta in molecular phylogenies.

marine environments (Casanova 1986a, 1986b, 1986c, 1992, 1996; Bowman and Bieri 1989; Thuesen and Haddock 2013). The head is equipped with a chitinous cephalic armature. It is composed of two bilateral sets of occasionally serrated grasping spines (syn.: hooks) aligned along the lateral rim of the ventral mouth opening. The head also bears one or two teeth rows, which are located at the front of the mouth. The structure of the grasping spines and teeth is similar and consists of a central cavity with two concentric rings made of α-chitin (Atkins et al. 1979; Bone et al. 1983). A cuticle-covered area forming a ventral cephalic mask protects the epidermis from physical damage when capturing and swallowing prey. The mesoderm of the trunk and tail is occupied by large coelomic cavities fully lined by coelothelial cells (Shinn 1994a, 1997; Shinn and Roberts 1994, Müller et al. 2019). The cephalic coelom is residual as diminished in size by the massively elaborated head musculature. The developmental origin of an additional coelomic cavity, termed postovarian coelom, recently documented in a spadellid chaetognath (Müller et al. 2019) is still to be studied. While chaetognaths are usually devoid of pigmentation, the intestine of several deep-sea species is filled with orange-red carotenoid pigments (Thuesen et al. 2010; Miyamoto and Nishida 2011; Thuesen and Haddock 2013). Further, bioluminescence has also been demonstrated in the two deep-sea species *Caecosagitta macrocephala* and *Eukrohnia fowleri* (Haddock and Case 1994; Thuesen et al. 2010).

The multilayered epidermis consists of distal and proximal epidermal cells (**Figure 15.4**) and is recognize as an autapomorphy of chaetognaths. The thickness of the epidermis depends on the existence and number of layers of

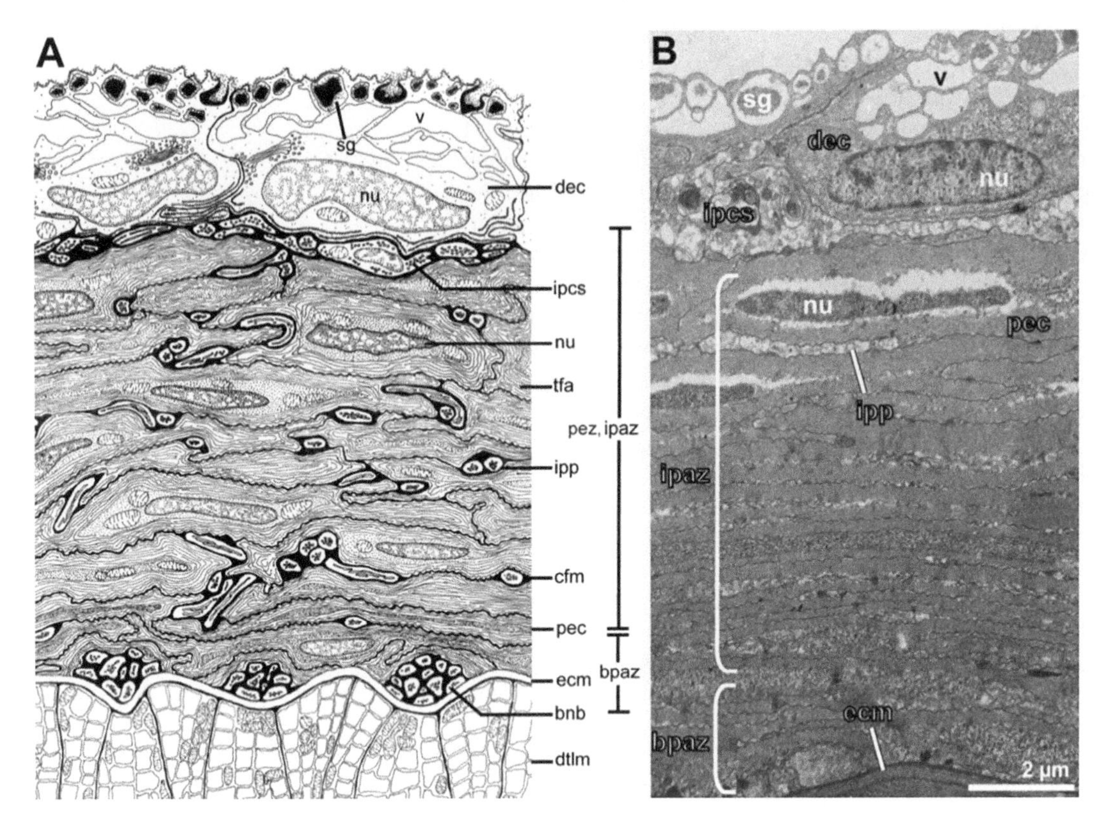

Figure 15.4 Cellular organization of the multilayered epidermis of chaetognaths. (A) Diagram of the epidermis from dorsolateral trunk region of *Spadella cephaloptera*. The epidermis comprises an upper unilayer of distal epidermal cells (dec) and a lower multilayer of flattened proximal epidermal cells (pec). The neuronal plexus integrates a basiepidermal domain (bpaz) with more-or-less orthogonally arranged neurite bundles (bnb) and an expanded intraepidermal domain (ipaz), including a distal layer of condensed neurites and neuronal somata (ipcs) as well as a network of numerous plexus profiles (ipp) making connection to basal neurite bundles. (B) TEM micrograph of a cross section of multilayered epidermis showing over 15 layers of proximal epidermal cells from ventrolateral region in *Spadella cephaloptera*. nu, nucleus; tfa, tonofilaments apparatus; pez, proximal epidermal zone; cfm, condensed fibrillous material; ecm, extracellular matrix; dtlm, dorsal longitudinal muscle. Adapted from Müller et al. (2019).

the proximal epidermal cells so that the epidermis may only be oligo- or even unilayered in specific body regions, as for example in the ventral and lateral cephalic areas (Shinn 1997; Müller et al. 2019). The body is laterally flanked by one or two pairs of lateral fins and a single tail fin, both types being supported by rays which consist of specialized epidermal cells. Another type of epidermal cell is concentrated at the level of the neck or arranged along the entire body and develops large vacuoles. These cells constitute an alveolar tissue named "collarette" the functional significance of which remains unclear, but it could participate in buoyancy and/or osmotic regulation (Kapp 1991; Perez et al. 2001). The integument also comprises receptor cells and neurons (see below 'Nervous and sensorial systems').

A specialized respiratory system is absent. Gas exchange likely occurs by diffusion through the body wall. While in the literature chaetognaths are commonly described as lacking a circulatory system, these animals nevertheless exhibit a rudimentary hemal system which consists of sinuses located in the extracellular matrix (ECM) between the basal laminae of the peritoneum and the lined tissues (Shinn 1993, 1997, Müller et al. 2019). The anatomical position of these sinuses resembles the organization of a simplified annelid circulatory system (Perez 2000). This hemal system is associated with podocyte-like cells anteriorly adjacent to the posterior septum, likely allowing the entry of ultrafiltered products into the tail coelom (Shinn 1997) and a presumably metanephridial system which consists of nephrostome-equipped coelomoducts, the distal portions of which open into the tail coelom via a funnel made up by multiciliated epithelial cells (Müller et al. 2019). However, the excretory function of these structures needs further investigations to be validated.

Muscular system: the muscular apparatus is rather complex and capable of delivering movements competing with the fastest yet studied in metazoans (Duvert 1989). It is composed of both locomotory and visceral musculature; the latter is made of myoepithelial cells that enwrap the entire gut system (Shinn 1997; Casanova and Duvert 2002; Müller et al. 2019). Regarding the locomotory system, the muscular pattern is highly complex in the head with a multidirectional arrangement of at least 16 distinct muscles that function to open the mouth and move the grasping spines and teeth. The great majority of locomotory muscles, however, is present in the trunk and tail and arranged in four longitudinal quadrants. Ventral transverse muscles, namely the phragms, can also be found in species belonging to the Phragmophora, in the trunk and tail (Heterokrohniidae) or in the trunk only (Eukrohniidae, Spadellidae). The presence of phragms has been associated with a benthic or hyperbenthic lifestyle (Tokioka 1965a, 1965b). There is not any circular muscle. The trunk and tail locomotory muscles can be classified as primary or secondary muscles (**Figure 15.5**). Primary longitudinal muscles occupy approximately 80% of the entire body. Two types of myocytes (type-A and -B) are distinguished in the primary musculature (Casanova and Duvert 2002). Furthermore, the ratio between type-A and -B myocytes is related to the ecology, with a higher proportion of type-B myocytes in large planktonic species. Type-B myocytes are absent in the benthic Spadellidae. The secondary longitudinal muscles constitute less than 1% of the trunk and tail musculature and constitute three bilateral band pairs: (1) the dorsomedian pair arranged at either side of the dorsal mesentery, (2) the ventromedian pair which mirrors the former, and (3) the lateral pair observed at the corner of the ventral primary muscles and coelothelial cells of the lateral fields. In some chaetognath taxa, these secondary muscles show a peculiar heterosarcomeric pattern (alternating of S1- and S2-sarcomeres) which represent a second autapomorphy of this phylum.

Digestive system: the gut is complete and consists of a ventral mouth opening (syn.: pharynx) followed by a pharynx and esophagus in the head as well as a straightly elongated intestine that extends along the trunk. The anus is situated ventrally just anterior to the posterior septum (John 1933; Parry 1944). In some species, the anterior part of the intestine exhibits a pair of intestinal diverticula. The pharyngeal and esophageal epitheliums consist of three types of secretory cells presumably responsible for secretion of mucus and digestive enzymes. The intestinal epithelium contains a fourth type of ciliated secretory cell type containing large, mucus-like granules and ciliated absorptive cells which predominate in the posterior half of the intestine. The absorptive cells have been demonstrated to be involved in secretion of glycoproteins and in endocytosis of macromolecules followed by intracellular digestion (Arnaud et al. 1996; Perez et al. 1999, 2000).

Nervous and sensorial systems: the architecture of the nervous system is typical for protostomes and comprises a chain of circumoral ganglia and a conspicuous ventral nerve center (**Figure 15.6**). Most of the constituents of the nervous system are basiepidermal, sitting on the epidermal ECM. An autapomorphic feature of the chaetognath nervous system is the peripheral, exclusively intra- and basiepidermal neuronal plexus domains which consist of fibers lacking direct synaptic contact and separated from underlying locomotory muscles by a thick ECM layer (see Harzsch et al. 2016 for a recent review).

The brain is divided into two functional domains (Rieger et al. 2010, 2011). The anterior neuropil domain is connected to the esophageal, vestibular, and the unpaired subesophageal ganglia via the frontal connectives and the esophageal commissure, which interconnects the vestibular ganglia as well as connects to the ventral nerve center via the paired main connectives. It is presumably involved in the control of the head muscles associated with the grasping spines and digestive system. The posterior neuropil domain is linked to the eyes

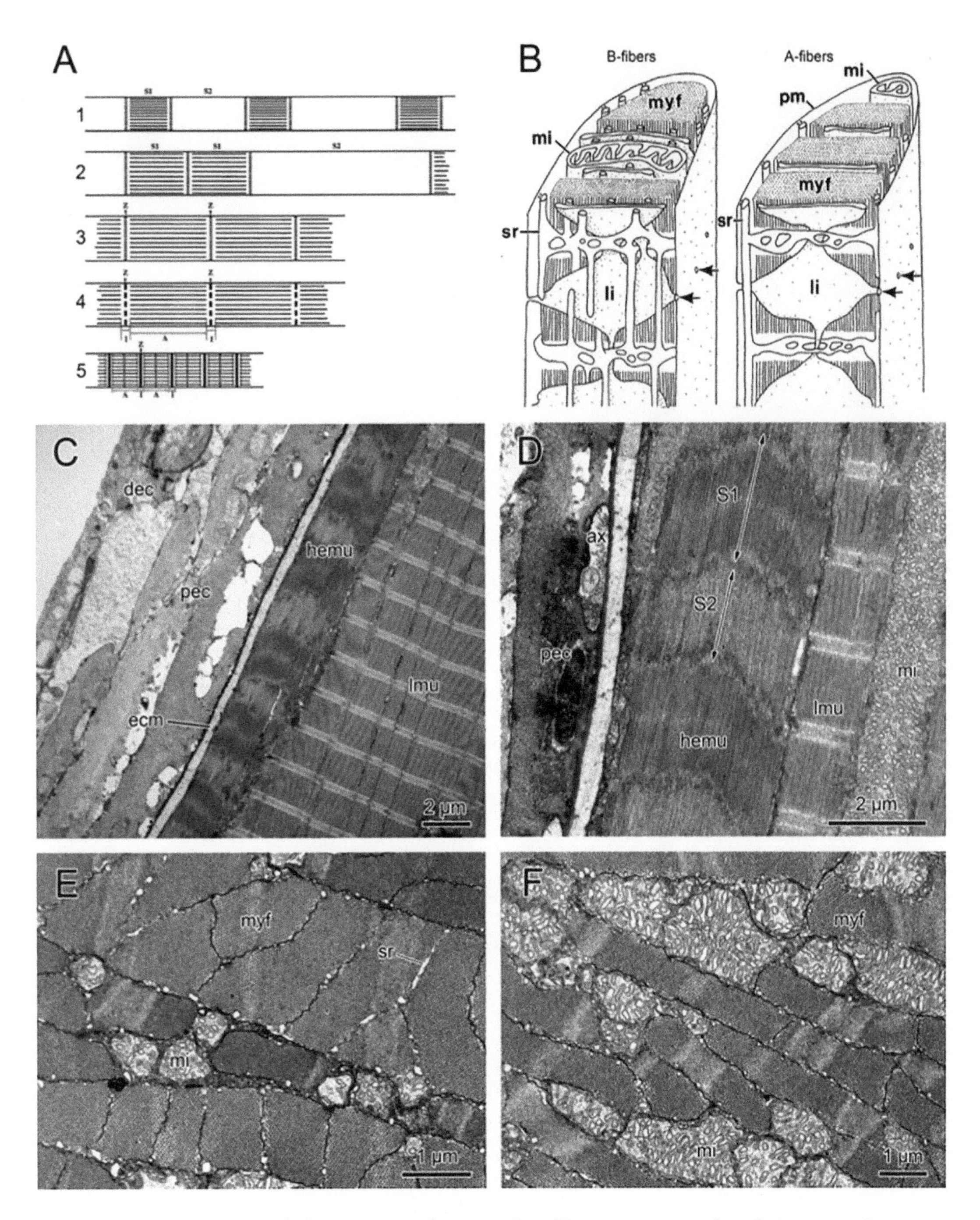

Figure 15.5 Organization and ultrastructure of cross-striated locomotory muscles of chaetognaths. (A) Schematic drawings of different sarcomeric organizations: (1) classical heterosarcomeric secondary muscle with aberrant S2-sarcomeres (S2) alternating with regular S1-sarcomeres (S1), e.g. Sagittidae; (2) irregular heterosarcomeric secondary muscle of *Archeterokrohnia rubra*; (3) homosarcomeric transverse muscle (only S1-sarcomeres present), e.g. Spadellidae; (4) super-contracting homosarcomeric secondary and super-contracting transverse muscles, *Spadella cephaloptera*; (5) primary muscle: all Chaetognatha. (B) Schematic reconstruction of sarcomeres and sarcoplasmic compartments in primary muscle (pm) of chaetognaths: (1) B-fibers with myofibrils (myf) alternating with sarcoplasmatic spaces mostly occupied by mitochondria (mi), sarcoplasmic reticulum (sr) shows longitudinal interconnections (li); (2) A-fibers contain more myofibrils and mitochondria are restricted to the periphery of the muscle cell. Arrows indicate openings of transverse tubules. (C) Parasagittal section through the tail of *Ferosagitta hispida* showing longitudinal primary musculature (lmu) and heterosarcomeric secondary musculature (hemu) with S1- and S2-sarcomeres alternating in irregular fashion. (D) Detail of heterosarcomeric secondary muscle; Axonal terminals (ax) indicate innervations of secondary muscle by basiepidermal neuronal plexus. (E) Detail of A-fibers in transverse section. (F) Detail of B-fibers in transverse section. dec, distal epidermal cells; ecm, extracellular matrix; proximal epidermal cells. (A) Adapted from Casanova and Duvert (2002). (B) Reprinted from Shinn (1997). (C)–(F) Adapted from Müller et al. (2019).

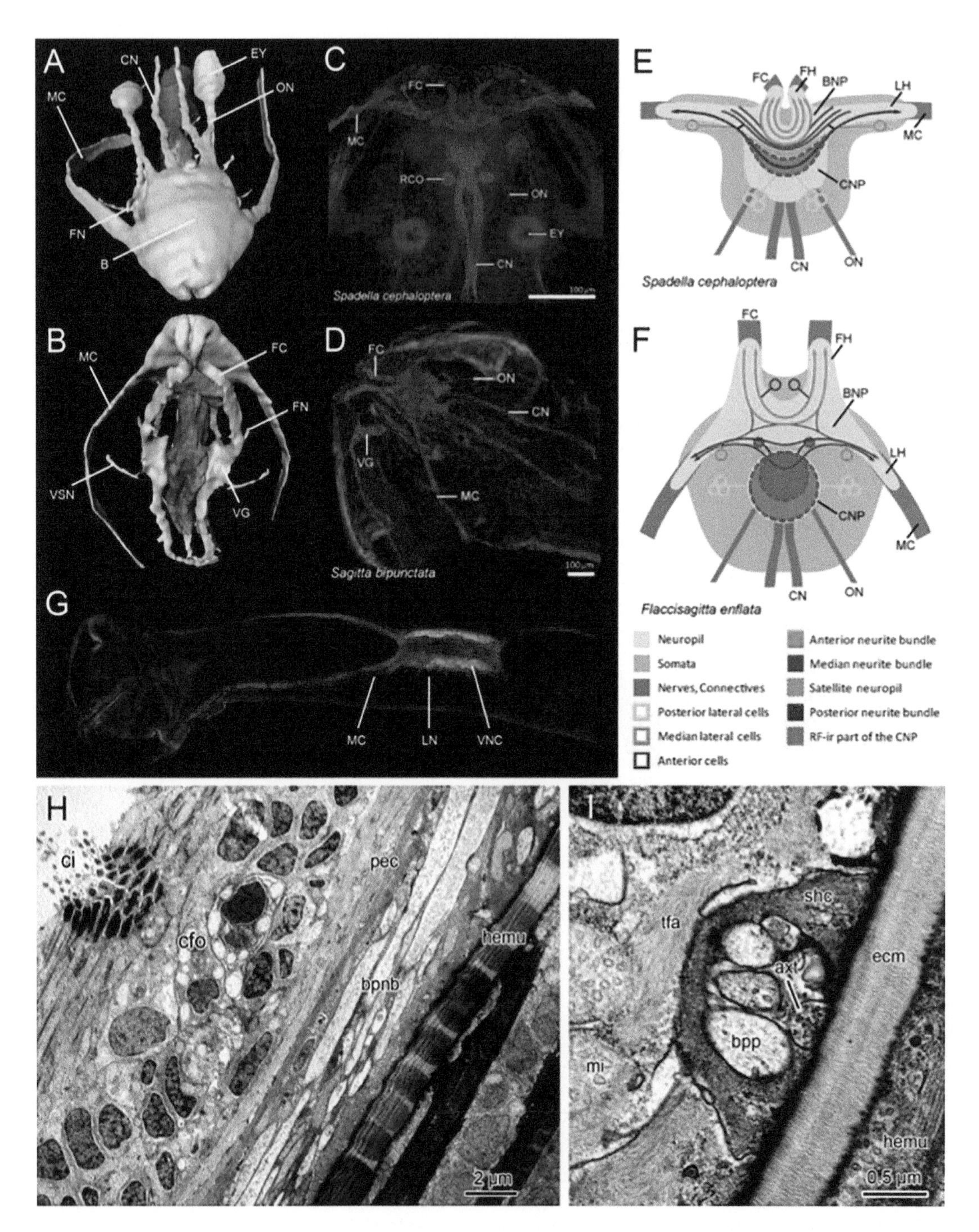

Figure 15.6 Organization and ultrastructure of the nervous system in various chaetognaths. (A, B) 3D reconstruction of the cephalic nervous system (yellow) in *Ferosagitta hispida*. The gut is stained in green. Fronto-dorsal (A) and fronto-ventral views (B). (C, D) Immunolocalization of tubulin (red) and histochemical staining of nuclei (blue) showing the organization of the cephalic nervous system. (E, F) Schematic drawings featuring the brains of *Spadella cephaloptera* and *Flaccisagitta enflata*. The retrocerebral organ is not included in these reconstructions. Neuropil, somata regions and main nerves of the brain (B) are shown in shades of gray. The RFamide-like (a neuropeptide) immunoreactive system within the posterior neuropil domain is colored in shades of red whereas those within the anterior neuropil domain are represented in shades of blue. (G) Immunolocalization of tubulin (red) and histochemical staining of nuclei (blue) showing the organization of nervous system in the trunk of *Sagitta bipunctata*. BNP, bridge neuropil; CN, coronal nerve; CNP, core neuropil; EY, eye; FC, frontal connective; FH, FN, frontal nerve; frontal horn; LH, lateral horn; Ln, lateral nerves; MC, main connective; ON, optic nerve; RCO, retrocerebral organ; VG, vestibular ganglion; VSN, vestibular nerve; VNC, ventral nerve center. (H, I) Ultrastructure of the basiepidermal neuronal plexus domain in dorsolateral trunk region of *Ferosagitta hispida*. Note the basiepidermal plexus neurite bundle (bpnb) with an axonal terminal (axt), encapsulated by glial sheath cell (shc) and sitting on the epidermal extracellular matrix (ecm). dec, distal epidermal cell; cfo, ciliary fence organ; ci, receptoral cilia; hemu, paramedian, heterosarcomeric, secondary tail musculature; mi, mitochondrion; nu, nucleus; pec, proximal epidermal cells; tfa, tonofilaments. (A)–(F) Reprinted from Rieger et al. (2010). (G) Reprinted from Perez et al. (2014). (H), (I) Reprinted from Müller et al. 2019.

and the corona ciliata by distinct nerves. It also accommodates a presumably sensory structure, the retrocerebral organ, the functional significance of which remains unknown. The posterior neuropil domain is connected to the ventral nerve center via paired main connectives. Immunolocalization revealed one serotonin immunoreactive neuron in the brain (Goto et al. 1992) and the presence of individually identifiable RFamide-immunoreactive neurons in the brain (Goto et al. 1992; Rieger et al. 2010).

The ventral nerve center occupies the middle or the first third of the trunk depending on the species considered. Similar to the brain, it consists of lateral clusters of neuronal somata and a central neuropil being composed of dozens of serially repeated synaptic microcompartments and is likely involved in the modulation of motor behaviors in response to changing the sensory input (Harzsch and Müller 2007). TEM analysis identified six different types of neurons in the ventral nerve center of *Spadella cephaloptera* (Harzsch et al. 2009). Similar to the brain, specific patterns of RFamide immunoreactive neurons are observed in the ventral nerve center (Goto et al. 1992; Harzsch and Müller 2007; Harzsch et al. 2009). These patterns are homologous between all species examined. Labeling of dividing neuronal progenitors with the s-phase mitosis marker (BrdU) disclosed serially organized domains of the developing ventral nerve center (Perez et al. 2013) as it has been previously documented based on serially iterated RFamide-like immunoreactive neurons.

As ambush predators, chaetognaths are equipped with a well-developed mechanosensory system to efficiently detect vibrations caused by swimming movements of their prey (Newbury 1972; Feigenbaum and Reeve 1977). It consists of a set of two types of ciliary receptor organs embedded in the multilayered epidermis, the transversally oriented ciliary fence organs and longitudinally oriented ciliary tuft organs (Malakhov et al. 2005; Müller et al. 2014). The neuronal epidermal plexus is particularly well-developed in the vicinity of these sensory organs. Immunolocalization of α-acetylated tubulin showed that mechanosensory inputs from both types of ciliary receptors feed into the plexus which acts as a semiautonomous system that mediates sensory–motor integration (Müller et al. 2014). The corona ciliata, a sensory organ presumably involved in chemoreception (Shinn 1997; Bleich et al. 2017), is situated dorsally on the neck region and connected by two coronal nerves to the posterior brain domain. The corona ciliata consists of two adjacent cell populations arranged in a ring-like formation. The external ring is exclusively composed of monociliated cells, their cilia are the only motile ones across the body surface. The internal ring includes both monociliated receptor cells and absorptive cells projecting one to several cilia. The cilia of receptor and absorptive cells remain hidden in a subsurface cavity (Müller et al. 2019). A water current is generated by the motile cilia and presumably serves to enhance olfactory sensitivity by recirculating seawater close to the coronal cells (Bleich et al. 2017).

Chaetognath eyes are classified according to the orientation of the photoreceptor cells as "indirect" (syn.: "inverted") or "direct" (syn.: "everted") eyes, with the latter being only found in Eukrohniidae species belonging to the hamata group (Eakin and Westfall 1964; Ducret 1975, 1978; Goto and Yoshida 1984; Goto et al. 1984). In both types, photoreceptor cells project their axons into the optic nerve towards the posterior compartment of the brain. In the well-studied "indirect" eyes, the numerous photoreceptor cells extend a receptive process, termed as distal segment, toward the single central pigment cell. The photoreceptor membranes in the distal segment are arranged as lamellar stacks. Apposing membranes of a lamella are continuous at numerous pores. The pores are 35–55 nm in diameter and arranged in a regular grid-like pattern with a center-to-center distance of 80–95 nm. Goto et al. (1984) considered this photoreceptive structure as unique and another autapomorphy of Chaetognatha (**Figure 15.7**).

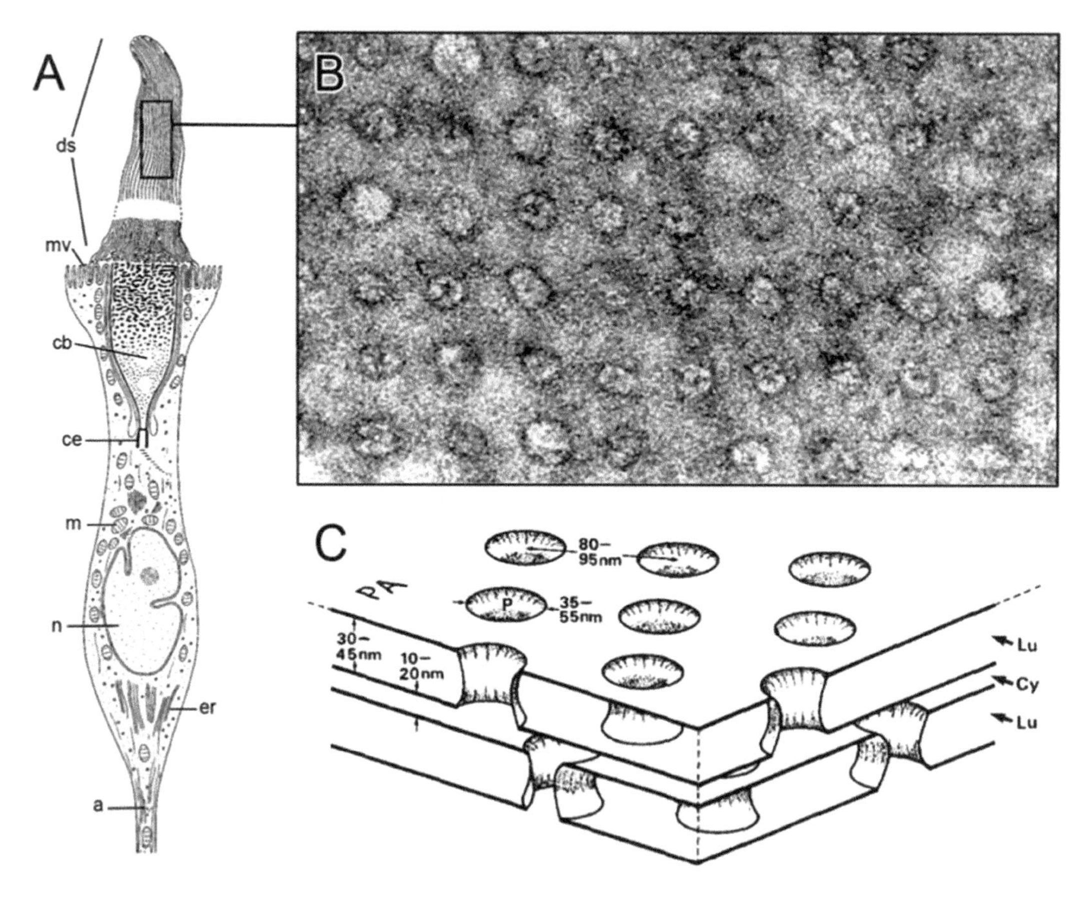

Figure 15.7 Photoreceptive system. (A) Diagram of a photoreceptor cell (indirect visual system) from the eye of *Pseudosagitta scrippsae*. (B) Distal segment of a photoreceptive cell in *Paraspadella gotoi* showing a regular arrangement of pores. (C) Schematic diagram of the lamellar structure in the distal segment. cb, conical body; ce, centriole (basal body); ds, distal segment; er, endoplasmic reticulum; m, mitochondrium; n, nucleus, mv, microvilli. (A) Adapted from Eakin and Westfall (1964). (B), (C) Adapted from Goto et al. (1984).

REPRODUCTION AND DEVELOPMENT

Reproductive system: the testes are arranged dorsolaterally in the tail coelomic cavities just posterior of the posterior septum. The testis wall consists of flattened specialized unciliated lateral field cells and medial mesodermal epithelial cells and is partly embedded in extensions of peritoneocytes (Bergey et al. 1994; Shinn 1997). Clusters of syncytial spermatogonia start their development in the testis space and are released into the tail coelomic cavities where mitosis and meiosis give rise to mature spermatozoa which are maintained in a circular movement by ciliated cells. Two sperm ducts (syn.: vasa deferentia), one on each posterior side of the tail, connect the coelomic cavities with the seminal vesicles gradually filled by spermatozoa. The anterior part of each duct resembles a ciliated nephrostome associated with metanephridia (Müller et al. 2019).

The ovaries are located dorsally in the posterior part of the trunk just anterior to the posterior septum. The ovarian space is delimited by a thin retrocoelothelial somatic wall that consists of squamous myoepithelial cells the contraction of which presumably facilitates egg deposition (Shinn 1992, 1997; Müller et al. 2019). The ovarian space contains differentiating oocytes including previtellogenic, vitellogenic and postvitellogenic oocytes. Each previtellogenic oocyte is associated with two oviducal specialized cells, the so-called accessory fertilization cells responsible for a very original and unique mode of fertilization (Shinn 1994b, 1997; Goto 1999, see below).

Reproduction and growth: chaetognaths are protandrous hermaphrodites and perform cross-fertilization (John, 1933; Ghirardelli 1968; Duvert et al. 2000; Goto and Suzuki 2001). During copulation, the mates exchange their spermatophores which might be species-specific and usually deposited dorsally on the neck. The transfer of spermatophores can be reciprocal or non-reciprocal. Then, the spermatozoa progress backwards and separate into two equal masses that enter each female genital opening. Spermatozoa accumulate within the seminal receptacles and the oviducal complex in which they are stored to perform several cycles of fertilization (Ghirardelli 1968). During the fertilization process, the spermatozoa pass through the oviducal complex to reach the oocyte in a canal formed by the accessory fertilization cells. After fecundation, the zygote moves into the oviducal syncytium through a pore formed by the detachment of the degenerative accessory fertilization cells (Shinn 1994b, 1997; Goto 1999).

Most planktonic species lay their eggs in the water column (Dallot 1968; Reeve 1970; Kotori 1975; Shimotori et al. 1997). One exception is known from *Ferosagitta hispida* which attach the fertilized eggs to the substrate (Reeve and Lester 1974). In the genus *Eukrohnia*, the adults carry two bilateral gelatinous sacs on their body in which they breed embryos and hatchlings. Benthic spadellids have been documented to perform elaborate courtships before copulating and attaching their eggs to the substrate (John 1933; Ghirardelli 1968; Goto and Yoshida 1997).

Depending on the species considered, eggs generally hatch after one to three days. Chaetognaths do not perform any metamorphosis; however, significant morphological changes occur in early life stages to unfold the adult body plan (Doncaster 1902; John 1933; Shinn 1994a, 1997; Rieger et al. 2011; Perez et al. 2013; Harzsch et al. 2015). Such changes concern, for instance, the completion of the muscular apparatus, the posterior septum and the neuro-sensorial system, the secondary opening of the gut, the elaboration of the chitinous structures, and the appearance of the second pair of lateral fins in Sagittidae. Chaetognaths are also subject to allometric changes during their life, with changes in the body proportions and the number of teeth and hooks.

Early and late development, coelomogenesis: cleavage is total and equal leading to a coeloblastula. Development is direct without any larval stage and accomplished quickly, from zygote to hatchling within less than 24 hours in *Spadella cephaloptera* (Harzsch et al. 2015). Over the past century, it has frequently been argued that chaetognaths display many embryological deuterostome-like features, for instance a radial cleavage of the embryo, a secondary opening of the mouth, a mesoderm originating from enterocoelic pouches and a trimeric body plan with a post-anal tail. However, and except for the deuterostomic gastrulation, the careful reappraisal of their developmental features refutes a deuterostome affinity:

- The four-cell embryo exhibits the tetrahedral configuration of spiral-cleaving embryos of some Lophotrochozoa (**Figure 15.8**). Cell fate analysis of the first four blastomeres with dye injections emphasized the link between the future body axes and the tetrahedral disposal of the blastomeres which is a further spiralian feature (Shimotori and Goto 2001).
- The mesoderm forms two entoderm infoldings which progresses towards the blastoporal site directly into the archenteron (inward pouches), while in deuterostomes the mesoderm arises by two evaginations which progress into the blastocoel towards the outside of the embryo (outward pouches, **Figure 15.9**). This peculiar developmental process, termed 'heterocoely' (proposed by Kapp 2000), is not homologous to that of enterocoelic deuterostomes and is in fact autapomorphic to Chaetognatha.

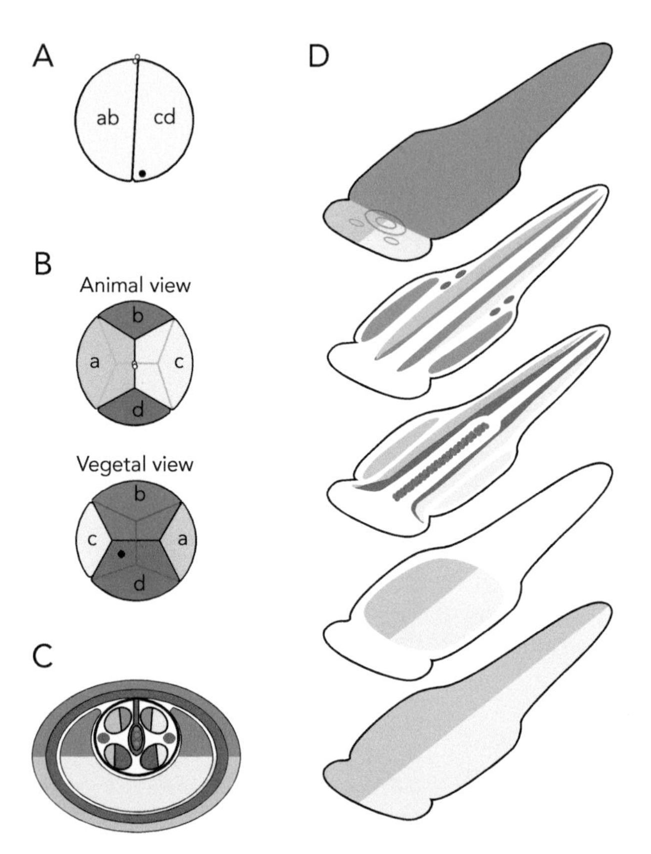

Figure 15.8 Diagrams summarizing the fate map of *Paraspadella gotoi* at the four-cell stage. The a, b, c and d blastomeres and the regions derived from each are colored in green, blue, yellow and red, respectively. (A) Lateral view of the two-cell embryo. (B) Animal and vegetal view of the four-cell embryo. Two open small circles represent the polar bodies. A dot indicates the germ plasm. (C) A transverse section through the trunk. (D) A series of horizontal sections. Hatched blue and light red indicates a mixture of clones derived from the b and d cells. Reprinted from Shimotori and Goto (2001).

- Regarding coelomogenesis, the late differentiation and cavitation of the coelothelium from two compact mesodermal bands occurring in the course of 48 hours after hatching can be considered as a derived, post-embryogenic variant of the schizocoely found in spiral-cleaving lophotrochozoans (Harzsch et al. 2015).
- Ultrastructural data showed that chaetognaths must be considered fundamentally dimeric bilaterians that become secondarily trimeric after the completion of the posterior septum. This step leads up to the segregation of the male and female primordial germ cells which implies that the posterior septum must be construed as a part of the reproductive system (Harzsch et al. 2015; Müller et al. 2019). Thus, the body cavities of chaetognaths are not homologous to the protocoel, mesocoel and metacoel found in archimeric lower deuterostomes.

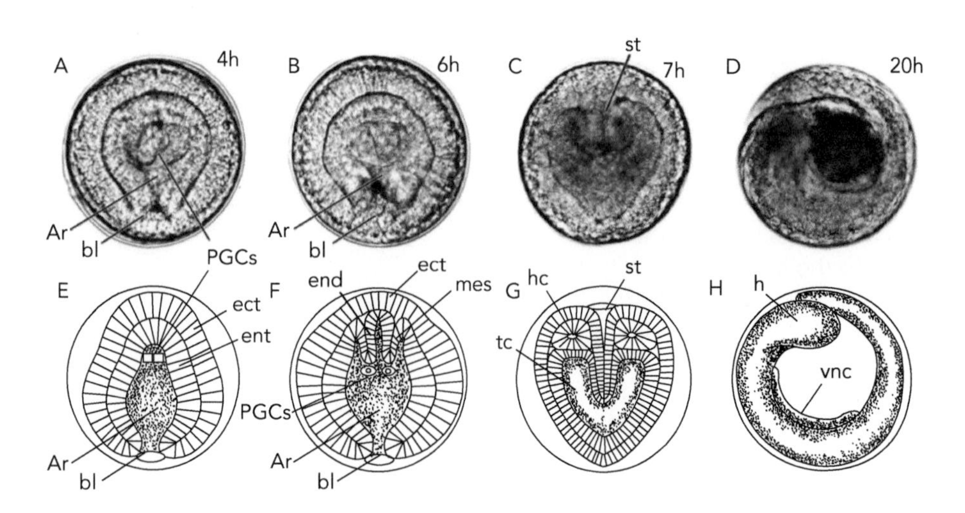

Figure 15.9 Gastrulation and coelomogenesis in *Spadella cephaloptera* (A)–(D) and *Sagitta bipunctata* (E)–(H). Ar, archenteron; bl, blastopore; ect, ectoderm; end, endoderm; ent, entoderm; h, head; hc, head coelom; mes, mesoderm; PGCs, primordial germ cells; st, stomodeum; tc, trunk coelom, vnc, ventral nerve center. (A–D) Adapted from Harzsch et al. (2015). (E–G) Adapted from Burfield (1927). (H) Modified from Doncaster (1903).

Developmental genes: overall, our knowledge of developmental genetics in chaetognaths is very limited and chiefly concerns Hox genes. The homeodomains of six Hox genes have been isolated from *Spadella cephaloptera*, one belonging to the paralogy group 3, four to the median class and one new homeodomain (SceMedPost) with a unique set of signature amino acid motifs shared by both median and posterior Hox proteins of protostomes and deuterostomes (Papillon et al. 2003). No sequence related to the posterior paralogy group has been isolated for this species. In *Flaccisagitta enflata*, in addition to the six previously identified Hox genes characterized in *Spadella cephaloptera*, two posterior genes have been isolated (Matus et al. 2007). The new SceMedPost mosaic gene discovered in the two species and firstly considered as an autapomorphy of chaetognaths has been subsequently isolated in rotifers, a finding that supports a sister-group relationship between chaetognaths and lophotrochozoans (Fröbius and Funch 2017). The posterior Hox genes of chaetognaths possess both ecdysozoan and lophotrochozoan signature amino acid motifs, while key amino acid signatures in Hox6/lox5 genes shared by rotifers and chaetognaths strengthen their close relationships.

The expression pattern of Hox genes has been investigated only for SceMed4, a putative orthologue to Scr/Hox5 or Antp paralogy groups, in the late embryo, hatchling and juvenile of *Spadella cephaloptera* (Papillon et al. 2005). The typical position-specific expression pattern found in the median region of the ventral nerve center suggests an implication of Hox genes in the regionalization of the central nervous system and in the diversification of neuronal populations and neuronal projection patterns (**Figure 15.10**). The expression pattern of Brachyury (Pg-Bra), a master developmental gene known as a key factor of the mesoderm determination, has been analyzed in *Paraspadella gotoi* (Takada et al. 2002). This gene is expressed in two specific domains; one designates the blastoporal region in the early embryo and the other the mouth opening region in the hatchling (Figure 15.10). The first specific expression domain is

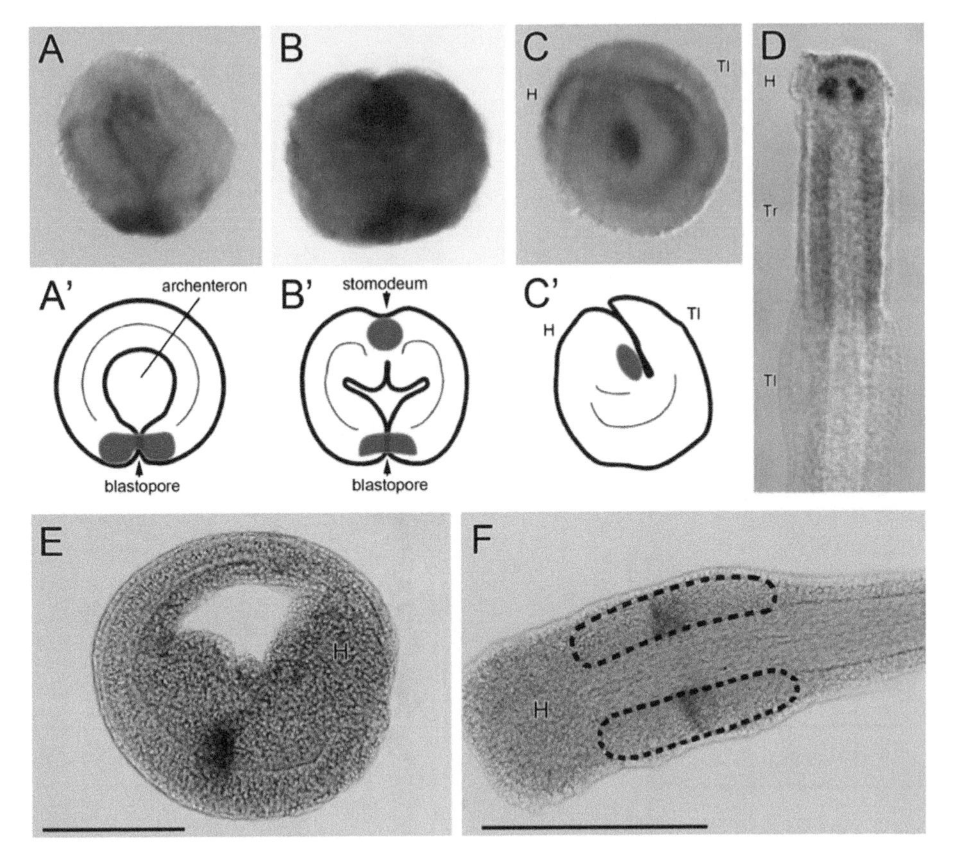

Figure 15.10 **(A)–(D) Spatial expression pattern of Pg-Bra (brachyury ortholog) in *Paraspadella gotoi*.** (A) Embryos at the early gastrula stage showing a Pg-Bra-positive circular region around the blastopore. (B) Embryos at the middle gastrula stage showing two Pg-Bra-positive regions observed around the blastopore and the stomodeum. (C) Embryos at the elongating stage and (D) in hatchling showing a Pg-Bra-positive region in the dorsal part of the head. (A')–(C') Diagrams showing Pg-Bra-positive regions (blue) in the embryonic stages showed in (A)–(C). (E), (F) Spatial expression pattern of the Hox gene SceMed4 (Scr/Hox5 orthology group) in *Spadella cephaloptera*. Embryo curved in late eggs (E) and hatchling (F) showing a Scemed4-positive region in the two bilateral somata clusters of the ventral nerve center (dashed line). H, head; Tr, trunk; Tl, tail. Scale bars: E, 0.15 mm; F, 0.4 mm. (A–D: Adapted from Takada et al. 2002. E, F: Adapted from Papillon et al. 2005.)

symplesiomorphic and resembles that of basal deuterostomes and cnidarians while the second represents once again a unique derived feature of chaetognaths.

DISTRIBUTION AND ECOLOGY

Despite limited species diversity, chaetognaths may be found in every oceanic region, from tropical to polar waters as well as brackish and estuarine waters (**Figure 15.11**). Most are planktonic, but about 30% of the species known are epi- and meiobenthic or live just above the substrate (hyperbenthic), in environments as diverse as seagrass beds, sediments, marine caves, hydrothermal vent sites and the deep ocean bottom (Bone et al. 1991; Müller et al. 2019). The highest individual and species densities are restricted to the photic zone of the shallow waters (above 200 m depth) but many species reach the bathypelagic layer (Alvariño 1964; Pierrot-Bults and Nair 1991). They usually rank as the second most dominant taxon in the zooplankton community and represent roughly 10 to 30% of the total biomass of copepods. All chaetognaths are carnivorous ambush predators, feeding essentially on copepods, cladocerans and, to a lesser extent, on euphausiids, appendicularians, various crustacean nauplii, larval fishes as well as smaller chaetognaths (see Müller et al. 2019 and references therein). They are capable of immobilizing their prey with tetrodotoxin-like venom. The toxin has been isolated from head extracts of chaetognaths and is presumably synthesized by a *Vibrio* bacterium (Thuesen et al. 1988; Thuesen and Kogure 1989). Consequently, chaetognaths are ecologically highly important since they constitute a significant proportion of the marine zooplankton and are a major food source for carnivores of higher trophic levels such as many fishes and other planktivores.

Chaetognaths are not able to actively swim against oceanic currents. However, as is commonly observed in many others planktonic organisms, they conduct diel vertical migrations, reaching the shallow waters at night where they encounter better feeding conditions and sinking downward during the day to avoid predators. Light intensity variation has been demonstrated to control upward vertical movement (Sweatt and Forward 1985a, b) but other parameters such as temperature, salinity, food supply and ontogenetic stage have been shown to significantly modulate the position in the water column and daily vertical migrations (see Müller et al. 2019 and references therein).

Many planktonic chaetognaths are excellent hydrological indicators associated with oceanic currents and are used for the delineation of specific water

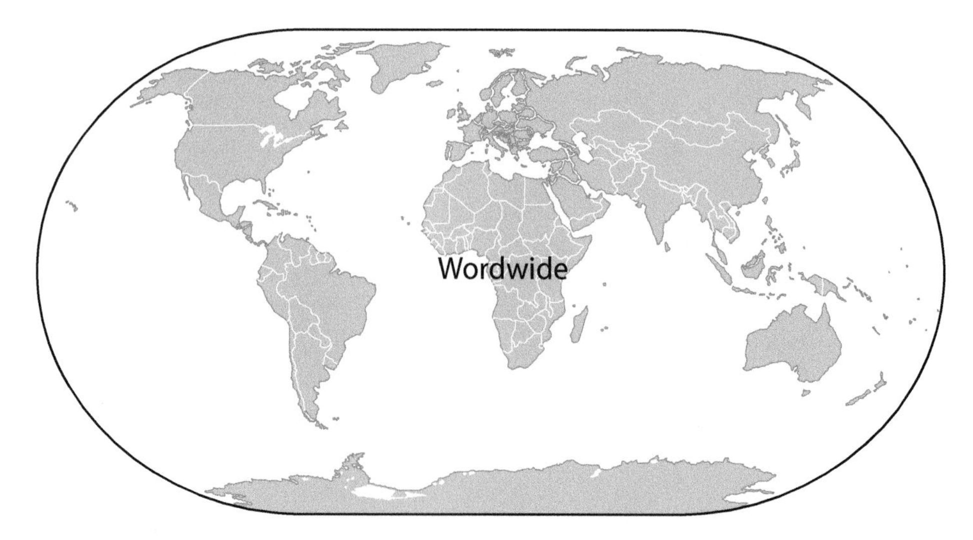

Figure 15.11 Distribution of chaetognaths in the world oceans.

masses (e.g. Resgalla 2008). The ongoing ocean warming will certainly have a deep impact on the geographical distribution and interaction of these species. A central issue, for instance, will be the impact on South Atlantic planktonic communities by the incoming of Indopacific chaetognaths from the Indian Ocean off southern South African coasts and carried by pockets of warm water pinched off from the Agulhas current.

PHYSIOLOGY AND BEHAVIOR

Few behavioral and physiological studies have been carried out on chaetognaths because of their small size, sampling issues (acquisition of healthy specimens), and difficulties in designing experimental settings which reproduce their natural living conditions. This is particularly true for the planktonic species, which have rarely been cultured with success. Most of the studies carried out so far have focused on development and growth (see section on reproductive development above), feeding frequency, energetic metabolism and locomotory activity. One behavioral aspect is the feeding frequency which is known to increase at higher temperatures. Food availability as well as prey category also significantly affect the hunting behavior and digestive metabolism.

Energetic and respiratory metabolism: a relatively high oxygen uptake (ca. 4–50μl^{-1}. h^{-1}. mg^{-1} dry weight) has been suggested by the studies of Reeve (1966), Reeve et al. (1970) and Duvert et al. (2000), an observation corroborated by the large volume of the mitochondrial compartment in primary locomotory muscles in planktonic (Dress and Duvert 1994; Shinn 1997) and benthic (Müller et al. 2019) species (**Figure 15.5F**). The oxygen/nitrogen ratio was found to be approximately 7:1, indicative of pure protein metabolism (Reeve et al. 1970). Using labeled-[14C] glucose and micro-radiorespirometric approach, Duvert et al. (2000) showed in *Parasagitta friderici* the prevalence of the Krebs–citric acid cycle rather than pentose cycle used for glycolysis. Radiographic data showed the incorporation of [^{14}C] 2-deoxyglucose in lateral fields and mesenteries where glycogen was also identified by cytochemistry. The storage of glycogen did not occur in muscles. Finally, these authors also demonstrated the occurrence of labeled-[14C] glucose as well as L-^{3}H-leucine taken up directly from seawater through the epidermis (osmotrophy). The intensity of L-^{3}H-leucine uptake was modified by osmotic shock.

Behavior and locomotory activity: benthic species spend most time of the day attached to various substrates, such as sea grass, macroalgae and rocks. Less frequently, benthic chaetognaths are found on the surface of or even in the interstitial of sandy sediments. All planktonic species adopt a slightly oblique position with the head pointing down, and most of them must swim continuously to maintain their position in the water layer (the so-called hop and sink swimming behavior, see Bone et al. 1987). A few planktonic chaetognaths display neutral buoyancy thanks to specific morphological (Kapp 1991; Perez et al. 2001) or physiological (Bone et al. 1987) adaptations. Unlike benthic species that behave like true ambush predators, motionless and fixed to their substrate waiting for prey to pass by, planktonic chaetognaths usually swim actively across short distances in order to catch their prey. When movement in their closest vicinity is detected by their ciliary fence and tuft organs, planktonic chaetognaths swiftly dart off toward their prey with a powerful vertical sweep of the tail fin (Newbury 1972; Feigenbaum and Reeve 1977). It has been shown that chaetognaths have phototactic abilities to follow their prey during diel vertical migrations (Goto et al. 1984; Foster 2006; Thuesen et al. 2010). Very few data are available on the physiology of the contractile apparatus (see Savineau and Duvert 1986, Duvert and Savineau 1986 and references therein). In a benchmark study, Duvert and Savineau (1986) demonstrated that locomotory muscles are directly excitable by electrical current, high K^+ solutions, and that acetylcholine is the main neuromuscular transmitter. Preincubation of the locomotory muscles with L-aspartate decreased the amplitude of the acetylcholine-induced contraction (Duvert et al. 1997). According to Duvert and Savineau (1986), the contractile response of chaetognath locomotory muscles is one of the fastest in the animal kingdom.

GENETICS AND GENOMICS

One nuclear genome has been sequenced (Carton 2017) but these data are not available in public databases yet. Using Feulgen Image Analysis Densitometry, the genome size has been estimated in two species, *Spadella cephaloptera* and *Flaccisagitta enflata*, at *c.*1.05 to 1.25 Gb (http://www.genomesize.com).

Complete mitochondrial genomes have been sequenced in ten species (Helfenbein et al. 2004; Papillon et al. 2004; Miyamoto et al. 2010a; Li et al. 2016; Wei et al. 2016; Marlétaz et al. 2017). Chaetognaths' mitogenome is the smallest among bilaterians; it has very short intergenic regions and misses two protein-coding (atp8 and atp6) and 21 tRNA genes (to date only tRNAmet gene has been identified with certainty).

Integrative and comparative analyses of nuclear transcriptomic and genomic data from cDNA and BAC libraries have enabled the building of broad molecular datasets and have convincingly established the phylogenetic position of chaetognaths within protostomes (Marlétaz et al. 2006, 2008, 2010). These studies also revealed one or several past genome duplication events followed by a high retention of duplicated genes in *Spadella cephaloptera*. Further, operonic transcription is subsequently resolved by trans-splicing maturation of transcripts. The chaetognaths' nuclear and mitochondrial genomes show evidence of a high evolutionary rate, and population genomics yielded evidence of an extremely high level of intraspecific genetic variation. Highly divergent mitochondrial lineages have been detected within interbreeding populations in *Parasagitta setosa*, *Parasagitta elegans*, and *Spadella cephaloptera* (Marlétaz et al. 2017). Recently, RNA-seq libraries have been prepared from all RNA of ten species and sequenced on Illumina instruments (Marlétaz et al. 2019).

Using mitochondrial (COI, COII) and/or nuclear (H3 and ITS1) molecular markers, spatial genetic structures have been detected in *Parasagitta setosa* (Peijnenburg et al. 2006), *Pseudosagitta maxima* (Kulagin and Neretina 2017), *Caecosagitta macrocephala* (Miyamoto et al. 2010b), *Eukrohnia bathypelagica* and *Eukrohnia hamata* (Kulagin et al. 2011, 2014; Miyamoto et al. 2012). Cryptic speciation has been documented in *Pseudosagitta maxima* and *Caecosagitta macrocephala*. Yet, because of the puzzling and unprecedented level of intraspecific polymorphism of the nuclear and mitochondrial genomes within interbreeding populations, markers traditionally used in population genetics should be interpreted cautiously (Marlétaz et al. 2017). In the future, single nucleotide polymorphisms (SNPs) or microsatellites will certainly help for a better understanding of chaetognath population genetics.

POSITION IN THE ToL

Due to a considerable number of conflicting studies, it has been virtually impossible to place chaetognaths in the ToL and their status is still hotly debated. Even though the analysis of broad phylogenomic datasets at least confirmed that Chaetognatha are protostomes, various affinities have been proposed, such as being the sister group to or a basal offshoot among protostomes (Marlétaz et al. 2006, 2008; Witek et al. 2009), an ingroup taxon of Ecdysozoa (Paps et al. 2009) or the sister group to Lophotrochozoa (Kocot et al. 2016), a clade equated with Spiralia in the most recent phylogenomic analyses. The synapomorphy of this assemblage (Protostomia plus Chaetognatha) is a circumoral arrangement of the cephalic nervous system elaborating a ganglionic ring chain connected with a ventral nerve center by two main connectives and may be named Hyponeuria (Nielsen 2012; Perez et al. 2014). Very recently, a phylogenomic analysis based on the most extensive and consistent dataset (1,174 genes belonging to 83 taxa, 416,663 amino acid supermatrix with 34.59% missing data) and using sophisticated methods for orthology selection and phylogenetic reconstruction found that chaetognaths cluster together with rotifers, gnathostomulids, and micrognathozoans within an expanded Gnathifera clade and that this clade is the sister group to Lophotrochozoa (Marlétaz et al. 2019). Such an inclusion within Gnathifera (or as sister group to all Gnathifera) has also recently been strengthened by paleontological data (Vinther and Parry 2019). However, morphological and developmental synapomorphies supporting the inclusion of Chaetognatha within Gnathifera are hard to find. Most of the character state interpretations used to corroborate such a hypothesis are contentious; for instance, the putative homology of the chitinous cephalic armatures, the nervous systems, or the tripartite body plan. In addition, Chaetognatha do not possess protonephridia, an excretory organ found in all ingroup taxa of Gnathifera. Finally, amino acid signatures of Hox homeodomains shared by Rotifera and Chaetognatha could represent plesiomorphic characters. As a cautious consensus, and according to evidence from new interpretation of morphological, developmental and molecular features (Harzsch et al. 2015; Müller et al. 2019), Chaetognatha are members of Spiralia and among them likely the sister group to Gnathifera (**Figure 15.12**). Nevertheless, it is not possible to refute that the character states shared by

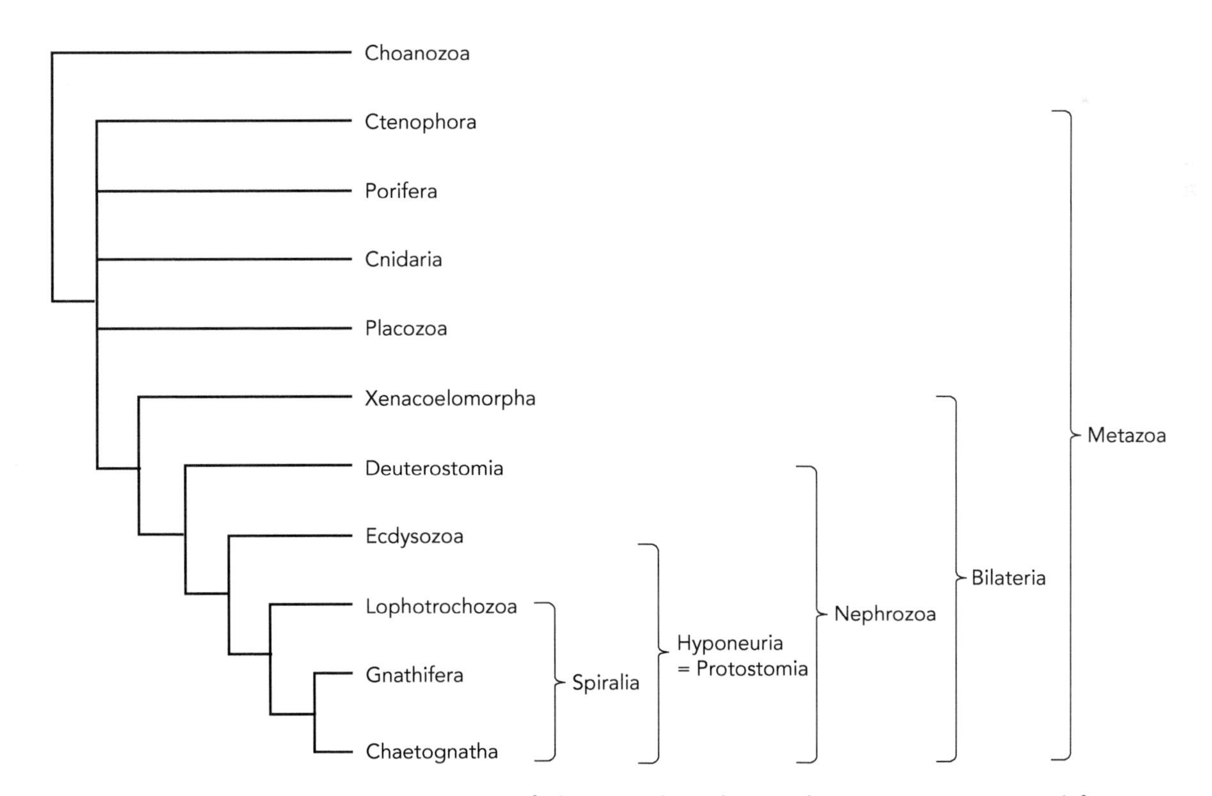

Figure 15.12 Preferred phylogenetic position of Chaetognatha within Spiralia as sister group to Gnathifera, a consensus inspired by the most recent phylogenomic study (Marlétaz et al. 2019), amino acid signatures of Hox homeodomains (Fröbius and Funch 2017), paleontological data (Vinther and Parry 2019) and the reappraisal of their developmental features (Harzsch et al. 2015; Müller et al. 2019). The basal nodes of the tree are shown as a polytomy. Xenacoelomorpha are a phylum whose monophyly has been discovered very recently. It has also been proposed that Xenacoelomorpha could be ingroup deuterostomes. The relationships of Chaetognatha represented here as sister group to Gnathifera are still under debate and previous conflicting analyses also support a sister group relationship to Spiralia or even to all Hyponeuria.

chaetognaths and spiralian ingroup taxa may represent symplesiomorphies. If this last view were to be confirmed, Chaetognatha might be basal for Spiralia or even for all Hyponeuria (= Protostomia *sensu lato*).

Wherever their final phylogenetic position along the stem lineages of Gnathifera, Spiralia or Hyponeuria may be, the mosaic of morphological and developmental characteristics of Chaetognatha has important implications for our understanding of the bilaterian evolutionary history. According to the phylogenetic distribution of the character states shared between Chaetognatha and other Bilateria, several hypotheses can be proposed:

1. The mouth opening develops separately from the embryonic blastopore, a situation which exactly fits with the deuterostomy definition. This contradicts two other hypothetical scenarios which predict the protostomic or amphistomic gastrulation to be plesiomorphic in Bilateria, the planuloid–acoeloid and trochaea theories, respectively.
2. Their early segmentation and coelomogenesis could be the first attempts toward the spiral cleavage and schizocoely as present in some Spiralia.
3. The neuroarchitecture and modes of neurogenesis suggest that a serially repeated organization of the central nervous system may be one essential step toward true metamery found in Annelida and Arthropoda.
4. Hox genes may primarily play a role in the regionalization of such an iterated organization of the central nervous system rather than in the usual expression patterns along the anterior–posterior axis.
5. The centralization of the nervous system in the ventral domain of the body represents a crucial step in bilaterian evolution and precedes the wide diversification of Hyponeuria.

DATABASES AND COLLECTIONS

Chaetognatha of the world (by A. Pierrot-Bults): http://species-identification.org/. This site offers morphological description and geographical distribution on all known species of chaetognaths.

Official species list (by E. Thuesen): http://www.marinespecies.org/.

Genome size: http://www.genomesize.com/.

EMBRC platform provides an exhaustive collection of chaetognaths at the Villefranche Oceanographic Laboratory (France) based on the large number of plankton samples collected aboard the schooner *Tara* during her 4-year expedition (2009–2011) on all the world's oceans.

CONCLUSION

Despite being highly important at the ecological level, chaetognaths are still mavericks in the ToL and zoosystematics, in general. Therefore, it is astonishing that only a few studies have been conducted in the past years (even including our own ones) targeting their evolutionary and functional morphology. Chaetognaths also seem more and more to be disappearing from basic lectures on Zoology or courses on Animal Anatomy. We are truly convinced that future studies on their adult morphology or developmental features will reveal the synapomorphies to whatever protostome ingroup we all so desperately seek to find. This expectation is not least based on the fact that for instance TEM data on the microanatomy of essential deep benthoplanktonic taxa such as Heterokrohniidae are still lacking.

REFERENCES

Alvariño A. (1964): Bathymetric distribution of chaetognaths. Pacific Science 8: 64–82.

Arnaud J., Brunet M., Casanova J. P., Mazza J., Pascalini V. (1996): Morphology and ultrastructure of the gut in *Spadella cephaloptera* (Chaetognatha). Journal of Morphology 228: 27–44.

Atkins E. D. T., Długosz J., Foord S. (1979): Electron diffraction and electron microscopy of crystalline ∞–chitin from the grasping spines of the marine worm *Sagitta*. International Journal of Biological Macromolecules 1: 29–32.

Bergey M. A., Crowder R. J., Shinn G. L. (1994): Morphology of the male system and spermatophores of the arrowworm *Ferosagitta hispida* (Chaetognatha). Journal of Morphology 221: 321–341.

Bone Q., Duvert M. (1991): Locomotion and Buoyancy. In: Bone Q., Kapp H., Pierrot-Bults A.C. (eds.). The Biology of Chaetognaths. Oxford University Press, New York: 32–44.

Bone Q., Ryan K. P., Pulsford A. L. (1983): The structure and composition of the teeth and grasping spines of chaetognaths. Journal of the Marine Biological Association of the United Kingdom 63: 929–939.

Bone Q., Brownlee C., Bryan G. W., Burt G. R., Dando P. R., Liddicoat M. I., Pulsford A. L., Ryan K. P. (1987): On the differences between the two 'indicator' species of chaetognath, Sagitta setosa and Sagitta elegans. Journal of the Marine Biological Association of the United Kingdom 67: 545–560.

Bone Q., Kapp H., Pierrot-Bults A. C. (1991): The Biology of Chaetognaths. Oxford University Press, New York.

Bowman T. E., Bieri R. (1989): *Paraspadella anops*, new species, from *Sagittarius* Cave, Grand Bahama Island, the second troglobitic chaetognath. Proceedings of the Biological Society of Washington 102: 586–589.

Bleich S., Müller C. H., Graf G., Hanke W. (2017): Flow generation by the corona ciliata in Chaetognatha–quantification and implications for current functional hypotheses. Zoology 125: 79–86.

Carton, R. (2017): Next generation sequencing technology and phylogenomics enhance the resolution of deep node phylogenies: A study of the Protostomia (Doctoral dissertation, National University of Ireland Maynooth).

Casanova J. P. (1986a): Quatre nouveaux Chaetognathes atlantiques abyssaux (genre *Heterokrohnia*): description, remarques éthologiques et biogéographiques. Oceanologia Acta 9: 469–477.

Casanova J. P. (1986b): *Archeterokrohnia rubra*, n. gen., n. sp., nouveau Chaetognathe abyssal de l'Atlantique nord-africain: description et position systématique, hypothèse phylogénétique. Bulletin du Muséum National d'Histoire Naturelle, Paris 8: 185–194.

Casanova J. P. (1986c): Deux nouvelles espèces d'*Eukrohnia* (Chaetognathes) de l'Atlantique sud-tropical africain. Bulletin du Muséum national d'Histoire Naturelle, Paris 8: 819–833.

Casanova J. P. (1992): Chaetognaths from Alvin dives in the Santa Catalina basin (California), with description of two new *Heterokrohnia* species. Journal of Natural History 26: 663–674.

Casanova J. P. (1996): A new genus and species of deep-benthic chaetognath from the Atlantic: a probable link between the families *Heterokrohniidae* and *Spadellidae*. Journal of Natural History 30: 1239–1245.

Casanova J. P., Duvert M. (2002): Comparative studies and evolution of muscles in chaetognaths. Marine Biology 141: 925–938.

Dallot S. (1968): Observations préliminaires sur la reproduction en élevage du Chaetognathe planctonique *Sagitta setosa* Müller. Rapport Commission Internationale Mer Méditerranée 19: 521–523.

Doncaster L. (1902): On the development of *Sagitta*, with notes on the anatomy of adult. Quarterly Journal of Microscopical Science 46: 351–398.

Ducret F. (1975): Structure et ultrastructure de l'œil chez les Chaetognathes (genres *Sagitta* et *Eukrohnia*). Cahiers de Biologie Marine 16: 287–300.

Ducret F. (1978): Particularités structurales du système optique chez deux Chaetognathes (*Sagitta tasmanica* et *Eukrohnia hamata*) et incidences phylogénétiques. Zoomorphologie 91: 201–215.

Dunn C. W., Hejnol A., Matus D. Q., Pang K., Browne W. E., Smith S. A., Seaver E., Rouse G. W., Obst M., Edgecombe G. D., Sorensen M. V., Haddock S. H., Schmidt-Rhaesa A., Okusu A., Kristensen R. M., Wheeler W. C., Martindale M. Q., Giribet G. (2008): Broad phylogenomic sampling improves resolution of the animal tree of life. Nature 452: 745–749.

Duvert M. (1989): Étude de la structure et de la fonction de la musculature locomotrice d'un invertébré. Apport de la biologie cellulaire à l'histoire naturelle des chaetognathes. Cuad Investigation Biologia Bilbao 15: 1–30.

Duvert M. (1991): A very singular muscle: the secondary muscle of chaetognaths. Philosophical Transactions of the Royal Society of London B 332: 245–260.

Duvert M., Savineau J. P. (1986): Ultrastructural and physiological studies of the contraction of the trunk musculature of Sagitta setosa (Chaetognath). Tissue and Cell 18: 937–952.

Duvert M., Salat C. (1990): Ultrastructural and cytochemical studies on the connective tissue of chaetognaths. Tissue and Cell 22: 865–878.

Duvert M., Savineau J. P., Campistron G., Onteniente B. (1997): Distribution and role of aspartate in the nervous system of the chaetognath Sagitta. Journal of Comparative Neurology 380: 485–494.

Duvert M., Perez Y., Casanova J.P. (2000): Wound healing and survival of beheaded chaetognaths. Journal of the Marine Biological Association of the United Kingdom 80: 891–898.

Eakin R. M., Westfall J. A. (1964): Fine structures of the eye of a chaetognath. Journal of Cell Biology 21: 115–132.

Edgecombe G. D., Giribet G., Dunn C. W., Hejnol A., Kristensen R. M., Neves R. C., Rouse G. W., Worsaae K., Sørensen M. V. (2011): Higher-level metazoan relationships: recent progress and remaining questions. Organisms, Diversity and Evolution 11: 151–172.

Feigenbaum D. L., Reeve M. R. (1977): Prey detection in the Chaetognatha: response to a vibrating probe and experimental determination of attack distance in large aquaria. Limnology and Oceanography 22: 1052–1058.

Feigenbaum D. L., Marris R. C. (1984): Feeding in the Chaetognatha. Annual Review of Oceanography and Marine Biology 22: 343–392.

Fröbius A. C., Funch P. (2017): Rotiferan Hox genes give new insights into the evolution of metazoan body plans. Nature Communications 8: 9.

Gasmi S., Nève G., Pech N., Tekaya S., Gilles A., Perez Y. (2014): Evolutionary history of Chaetognatha inferred from molecular and morphological data: a case study for body plan simplification. Frontiers in Zoology 11: 84.

Ghirardelli E. (1968): Some aspects of the biology of the chaetognaths. In: Russell F. S., Yonge M. (eds.). Advances in Marine Biology. Academic Press, New York: 271–375.

Goto T. (1995): Occurrence of *Spadella cephaloptera* during one year at Ischia Island (Gulf of Naples). Marine Ecology 16: 251–258.

Goto T. (1999): Fertilization process in the arrow worm *Spadella cephaloptera* (Chaetognatha). Zoological Science 16: 109–114.

Goto T., Yoshida M. (1981): Oriented light reactions of the arrow worm *Sagitta crassa* Tokioka. Biological Bulletin 160: 419–430.

Goto T., Yoshida M. (1984): Photoreception in Chaetognatha. In: Ali M. A. (ed.). Photoreception and Vision in Invertebrates. Plenum Publishing Corporation, New York: 727–742.

Goto T., Yoshida M. (1985): The mating sequence of the benthic arrow worm *Spadella schizoptera*. The Biological Bulletin 169: 328–333.

Goto T., Yoshida M. (1997): Growth and reproduction of the benthic arrowworm *Paraspadella gotoi* (Chaetognatha) in laboratory culture. Invertebrate Reproduction & Development 32: 201–207.

Goto T., Suzuki A. (2001): Variation of the mating behavior in the benthic arrow worm, *Spadella cephaloptera* (Chaetognatha). Zoological Science 18: 57.

Goto T., Takasu N., Yoshida M. (1984): A unique photoreceptive structure in the arrowworms *Sagitta crassa* and *Spadella schizoptera* (Chaetognatha). Cell and Tissue Research 235: 471–478.

Goto T., Terazaki M., Yoshida M. (1989): Comparative morphology of the eyes of *Sagitta* (Chaetognatha) in relation to depth of habitat. Experimental biology 48: 95–105.

Goto T., Katayama-Kumoi Y., Tohyama M., Yoshida M. (1992): Distribution and development of the serotonin-and RFamide-like immunoreactive neurons in the arrowworm, *Paraspadella gotoi* (Chaetognatha). Cell and Tissue Research 267: 215–222.

Haddock S. H., Case J. F. (1994): A bioluminescent chaetognath. Nature 367: 225–226.

Harzsch S., Müller C. H. G. (2007): A new look at the ventral nerve center of *Sagitta*: Implications for the phylogenetic position of Chaetognatha (arrow worms) and the evolution of the bilaterian nervous system. Frontiers in Zoology 4:14.

Harzsch S., Wanninger A. (2009): Evolution of invertebrate nervous systems: The Chaetognatha as a case study. Acta Zoologica (Stockholm) 91: 35–41.

Helfenbein K. G., Fourcade H. M., Vanjani R. G., Boore J. L. (2004): The mitochondrial genome of *Paraspadella gotoi* is highly reduced and reveals that chaetognaths are a sister group to protostomes. Proceedings of the National Academy of Sciences, USA 101: 10639–10643.

Harzsch S., Müller C. H. G., Rieger V., Perez Y., Sintoni S., Sardet C., Hansson B. (2009): Fine structure of the ventral nerve center and interspecific identification of individual neurons in the enigmatic Chaetognatha. Zoomorphology 128: 53–73.

Harzsch S., Müller C. H. G., Perez Y. (2015): Chaetognatha. In: Wanninger A. (ed.). Evolutionary Developmental Biology of Invertebrates 1. Introduction, Non-Bilateria, Acoelomorpha, Xenoturbellida, Chaetognatha. Springer, Vienna: 215–240.

Harzsch S., Perez Y., Müller C. H. G. (2016): Chaetognatha. In: Schmidt-Rhaesa A., Harzsch S., Purschke G. (eds.). Structure and Evolution of Invertebrate Nervous Systems. Oxford University Press, Oxford: 652–664.

John C. C. (1933): Habits, structure and development of *Spadella cephaloptera*. Quarterly Journal of Microscopical Science 75: 625–696.

Kapp H. (1991): Some aspects of buoyancy adaptations of chaetognaths. Helgoländer Meeresuntersuchungen 45: 263–267.

Kapp H. (2000): The unique embryology of Chaetognatha. Zoologischer Anzeiger 239: 263–266.

Kocot K. M., Struck T. H., Merkel J., Waits D. S., Todt C., Brannock P. M., Weese D. A., Cannon J. T., Moroz L. L., Lieb B., Halanych K. M. (2016): Phylogenomics of Lophotrochozoa with consideration of systematic error. Systematic Biology 66: 256–282.

Kotori M. (1975): Morphology of *Sagitta elegans* (Chaetognatha) in early larval stages. Journal of the Oceanographical Society of Japan 31: 139–144.

Kulagin D. N., Neretina T. V. (2017): Genetic and morphological diversity of the cosmopolitan chaetognath *Pseudosagitta maxima* (Conant, 1896) in the Atlantic Ocean and its relationship with the congeneric species. ICES Journal of Marine Science 74: 1875–1884.

Kulagin D. N., Stupnikova A. N., Neretina T. V., Mugue N. S. (2011): Genetic diversity of *Eukrohnia hamata* (Chaetognatha) in the South Atlantic: analysis of gene mtCO1. Invertebrate Zoology 8: 127–136.

Kulagin D. N., Stupnikova A. N., Neretina T. V., Mugue N. S. (2014): Spatial genetic heterogeneity of the cosmopolitan chaetognath *Eukrohnia hamata* (Möbius, 1875) revealed by mitochondrial DNA. Hydrobiologia 721: 197–207.

Li P., Yang M., Ni S., Zhou L., Wang Z., Wei S., Qin Q. (2016): Complete mitochondrial genome sequence of the pelagic chaetognath, *Sagitta ferox*. Mitochondrial DNA Part A 27(6): 4699–4700.

Malakhov V. V., Berezinskaya T. L., Solovyev K. A. (2005): Fine structure of sensory organs in the chaetognaths. 1. Ciliary fence receptors, ciliary tuft receptors and ciliary loop. Invertebrate Zoology 2: 67–77.

Marlétaz F., Martin E., Perez Y., Papillon D., Caubit X., Lowe C. J., Freeman B., Fasano L., Dossat C., Wincker P., Weissenbach J., Le Parco Y. (2006): Chaetognath phylogenomics: a protostome with deuterostome-like development. Current Biology 16: R577–R5778.

Marlétaz F., Gilles A., Caubit X., Perez Y., Dossat C., Samain S., Gyapay G., Wincker P., Le Parco Y. (2008): Chaetognath transcriptome reveals ancestral and unique features among bilaterians. Genome Biology 9: R94.

Marlétaz F., Gyapay G., Le Parco Y. (2010): High level of structural polymorphism driven by mobile elements in the Hox genomic region of the Chaetognath *Spadella cephaloptera*. Genome Biology and Evolution 2: 665–677.

Marlétaz F., Le Parco Y., Liu S., Peijnenburg K. T. (2017): Extreme Mitogenomic Variation in Natural Populations of Chaetognaths. Genome Biology and Evolution 9: 1374–1384.

Marlétaz F., Peijnenburg K. T., Goto T., Satoh N., Rokhsar D. S. (2019): A new Spiralian phylogeny places the enigmatic arrow Worms among Gnathiferans. Current Biology 29(2): 312–318.

Matus D. Q., Copley R. R., Dunn C. W., Hejnol A., Eccleston H., Halanych K. M., Martindale M. Q., Telford M. J. (2006): Broad taxon and gene sampling indicate that chaetognaths are protostomes. Current Biology 16: R575–R576.

Matus D. Q., Halanych K. M., Martindale M. Q. (2007): The Hox gene complement of a pelagic chaetognath, *Flaccisagitta enflata*. Integrative and Comparative Biology 47: 854–864.

Miyamoto H., Nishida S. (2011): New deep-sea benthopelagic chaetognath of the genus *Bathyspadella* (Chaetognatha) with ecological and molecular phylogenetic remarks, Journal of Natural History 45: 2785–2794.

Miyamoto H., Machida R. J., Nishida S. (2010a): Complete mitochondrial genome sequences of the three pelagic chaetognaths *Sagitta nagae, Sagitta decipiens* and *Sagitta enflata*. Comparative Biochemistry and Physiology - Part D Genomics and Proteomics 5: 65–72.

Miyamoto H., Machida R. J., Nishida S. (2010b): Genetic diversity and cryptic speciation of the deep sea chaetognath *Caecosagitta macrocephala* (Fowler, 1904). Deep Sea Research Part II: Topical Studies in Oceanography 57: 2211–2219.

Miyamoto H., Machida R.J., Nishida S. (2012): Global phylogeography of the deep-sea pelagic chaetognath *Eukrohnia hamata*. Progress in Oceanography 104: 99–109.

Müller C. H. G., Rieger V., Perez Y., Harzsch S. (2014): Immunohistochemical and ultrastructural studies on ciliary sense organs of arrow worms (Chaetognatha). Zoomorphology 133: 167–189.

Müller C. H. G., Harzsch S. Perez Y. (2019): Chaetognatha. In: Schmidt-Rhaesa, A. (Ed.) Handbook of Zoology. Miscellaneous Invertebrates. Walter de Gruyter GmbH & Co KG: 163–282.

Newbury T. K. (1972): Vibration perception by chaetognaths. Nature 236: 459–460.

Nielsen C. (2012): Animal Evolution: Interrelationships of the Living Phyla. 3rd edition. Oxford University Press, Oxford.

Papillon D., Perez Y., Fasano L., Le Parco Y., Caubit X. (2003): Hox gene survey in the chaetognath *Spadella cephaloptera*: Evolutionary implications. Development Genes and Evolution 213: 142–148.

Papillon D., Perez Y., Caubit X., Le Parco Y. (2004): Identification of chaetognaths as protostomes is supported by the analysis of their mitochondrial genome. Molecular Biology and Evolution 21: 2122–2129.

Papillon D., Perez Y., Fasano L., Le Parco Y, Caubit X. (2005): Restricted expression of a median Hox gene in the central nervous system of chaetognaths. Development Genes and Evolution 215: 369–373.

Papillon D., Perez Y., Caubit X., Le Parco Y. (2006): Systematics of Chaetognatha under the light of molecular data, using duplicated ribosomal 18S DNA sequences. Molecular Phylogenetics and Evolution 38: 621–634.

Paps J., Baguñà J., Riutort M. (2009): Bilaterian phylogeny: A broad sampling of 13 nuclear genes provides a new Lophotrochozoa phylogeny and supports a paraphyletic basal Acoelomorpha. Molecular Biology and Evolution 26: 2397–2406.

Parry D. A. (1944): Habits, structure and development of *Spadella cephaloptera* and *Sagitta setosa*. Journal of the Marine Biological Association of the United Kingdom 26: 16–36.

Peijnenburg K., Pierrot-Bults A. C. (2004): Quantitative morphological variation in *Sagitta setosa* Müller, 1847 (Chaetognatha) and two closely related taxa. Contributions to Zoology 73: 305–315.

Peijnenburg K., van Haastrecht E. K., Fauvelot C. (2005): Present day genetic composition suggests contrasting demographic histories of two dominant chaetognaths of the North-East Atlantic, *Sagitta elegans* and *S. setosa*. Marine Biology 147: 1279–1289.

Peijnenburg K., Fauvelot C., Breeuwer A. J., Menken S. (2006): Spatial and temporal genetic structure of the planktonic *Sagitta setosa* (Chaetognatha) in European seas as revealed by mitochondrial and nuclear DNA markers. Molecular Ecology 15: 3319–3338.

Perez Y. (2000): Structure and ultrastructure of the gut in chaetognaths. Functional and ecophysiological aspects. PhD thesis. Université d'Aix-Marseille 1, Marseille: 1–134.

Perez Y., Arnaud J., Brunet M., Casanova J. P., Mazza J. (1999): Morphological study of the gut in *Sagitta setosa, S. serratodentata* and *S. pacifica*. Functional implications in digestive processes. Journal of the Marine Biological Association of the United Kingdom 79: 1097–1109.

Perez, Y., Casanova, J. P., Mazza, J. (2000): Changes in the structure and ultrastructure of the intestine of *Spadella cephaloptera* (Chaetognatha) during feeding and starvation experiments. Journal of Experimental Marine Biology and Ecology 253: 1–15.

Perez Y., Casanova J. P., Mazza J. (2001): Degrees of vacuolation of the absorptive cells of five *Sagitta* (Chaetognatha) species: possible ecophysiological implications. Marine Biology 138: 125–133.

Perez Y., Rieger V., Martin E., Müller C. H. G., Harzsch S. (2013): Neurogenesis in an early protostome relative: progenitor cells in the ventral nerve center of chaetognath hatchlings are arranged in a highly organized geometrical pattern. Journal of Experimental Zoology 320: 179–193.

Perez Y., Müller C. H. G., Harzsch S. (2014): The Chaetognath: an anarchistic taxon between Protostomia and Deuterostomia. In: Wägele J. W., Bartholomaeus T. (eds.). Deep metazoan Phylogeny: the backbone of the tree of life. Walter De Gruyter GmbH, Berlin: 49–74.

Pierrot-Bults A. C., Nair V. (1991): The distribution patterns of chaetognaths. In: Q. Bone, H. Kapp, A.C. Pierrot-Bults (eds.). The Biology of Chaetognaths. Oxford University Press, New York: 86–116.

Reeve, M. R. (1966). Observations on the biology of a chaetognath. In: H. Barens (ed.). Some contemporary studies in marine science. Geroge Allen and Unwin Ltd., London: 613–630.

Reeve M. R. (1970): Complete cycle of development of a pelagic chaetognath in culture. Nature 227: 381.

Reeve M. R., Raymont J. E. G., Raymont J. K. B. (1970): Seasonal biochemical composition and energy sources of *Sagitta hispida*. Marine Biology 6: 357–364.

Reeve M. R., Lester B. (1974): The process of egg-laying in the chaetognath *Sagitta hispida*. The Biological Bulletin 147: 247–256.

Resgalla Jr. C. (2008): Pteropoda, Cladocera, and Chaetognatha associations as hydrological indicators in the southern Brazilian Shelf. Latin American Journal of Aquatic Research 36: 271–282.

Rieger V., Perez Y., Müller C. H. G., Lipke E., Sombke A., Hansson B. S., Harzsch S. (2010): Immunohistochemical analysis and 3D reconstruction of the cephalic nervous system in Chaetognatha: Insights into an early bilaterian brain? Invertebrate Biology 129: 77–104.

Rieger V., Perez Y., Müller C. H. G., Lacalli T., Hansson B. S., Harzsch S. (2011): Development of the nervous system in hatchlings of *Spadella cephaloptera* (Chaetognatha), and implications for nervous system evolution in Bilateria. Development Growth and Differentiation 53: 740–759.

Savineau J. P., Duvert M. (1986): Physiological and cytochemical studies of Ca in the primary muscle of the trunk of Sagitta setosa (Chaetognath). Tissue and Cell 18(6): 953–966.

Shimotori T., Goto T. (2001): Developmental fates of the first four blastomeres of the chaetognath *Paraspadella gotoi*: Relationship to protostomes. Development Growth and Differentiation 43: 371–382.

Shimotori T., Goto T., Terazaki M. (1997): Egg colony and early development of *Pterosagitta draco* (Chaetognatha) collected. Plankton Biology and Ecology 44: 71–80.

Shinn G. L. (1992): Ultrastructure of somatic tissues in the ovaries of a chaetognath (*Ferosagitta hispida*). Journal of Morphology 211: 221–241.

Shinn G. L. (1993): The existence of a hemal system in chaetognaths. In: Moreno I. (ed.). Proceedings of the second international workshop on Chaetognatha. Universitat de les Illes Balears, Palma: 17–18.

Shinn G. L. (1994a): Epithelial origin of mesodermal structures in arrowworms (Phylum Chaetognatha). American Zoologist 34: 523–532.

Shinn G. L. (1994b): Ultrastructural evidence that somatic "accessory cells" participate in chaetognath fertilization. In: Wilson W. H. Jr, Stricker S.A., Shinn G. L. (eds.). Reproduction and development of marine invertebrates. The Johns Hopkins University Press, London: 96–105.

Shinn G. L. (1997): Chaetognatha. In: Harrison F. W., Ruppert E. E. (eds.). Microscopic Anatomy of Invertebrates, Vol. 15: Hemichordata, Chaetognatha, and the Invertebrate Chordates. Wiley-Liss, New York: 103–220.

Shinn G. L., Roberts M. E. (1994): Ultrastructure of hatchling chaetognaths (*Ferosagitta hispida*): epithelial arrangement of mesoderm and its phylogenetic implications. Journal of Morphology 219: 143–163.

Sweatt A. J., Forward R. B. (1985a): Diel vertical migration and photoresponses of the chaetognath *Sagitta hispida* Conant. Biological Bulletin, Marine Biological Laboratory, Woods Hole 168: 18–31.

Sweatt A. J., Forward R. B. (1985b): Spectral sensitivity of the chaetognath *Sagitta hispida* Conant. Biological Bulletin, Marine Biological Laboratory, Woods Hole 168: 32–38.

Takada N., Goto T., Satoh N. (2002): Expression pattern of the Brachyury gene in the arrow worm *Paraspadella gotoi* (Chaetognatha). Genesis 32: 240–245.

Thuesen E. V., Kogure K. (1989): Bacterial Production of Tetrodotoxin in Four Species of Chaetognatha. Biological Bulletin 176: 191–194.

Thuesen E. V., Haddock S. H. (2013): *Archeterokrohnia docrickettsae* (Chaetognatha: Phragmophora: Heterokrohniidae), a new species of deep-sea arrow worm from the Gulf of California. Zootaxa 3717: 320–328.

Thuesen E. V., Kogure K., Hashimoto K., Nemoto T. (1988): Poison arrow worms: a tetrodotoxin venom in the marine phylum Chaetognatha. Journal of Experimental Marine Biology and Ecology 116: 249–256.

Thuesen E. V., Goetz F. E., Haddock S. H. (2010): Bioluminescent Organs of Two Deep-Sea Arrow Worms, *Eukrohnia fowleri* and *Caecosagitta macrocephala*, With Further Observations on Bioluminescence in Chaetognaths. Biological Bulletin 219: 100–111.

Tokioka T. (1965a): The taxonomical outline of chaetognaths. Publications of the Seto Marine Biological Laboratory 12: 335–357.

Tokioka T. (1965b): Supplementary notes on the systematics of Chaetognatha. Publications of the Seto Marine Biological Laboratory 13: 231–242.

Vannier J., Steiner M., Renvoisé E., Hu S. X., Casanova J. P. (2007): Early Cambrian origin of modern food webs: evidence from predator arrow worms. Proceedings of the Royal Society B, Biological Sciences 274: 627–633.

Vinther J., Parry L. A. (2019): Bilateral jaw elements in *Amiskwia sagittiformis* bridge the morphological gap between gnathiferans and chaetognaths. Current Biology 29(5): 881–888.

Wei S., Li P., Yang M., Zhou L., Yu Y., Ni S., Wang Z., Qin Q. (2016): The mitochondrial genome of the pelagic chaetognath, *Pterosagitta draco*. Mitochondrial DNA Part B 1(1): 515–516.

Witek A., Herlyn H., Ebersberger I., Welch D. B. M., Hankeln T. (2009): Support for the monophyletic origin of Gnathifera from phylogenomics. Molecular Phylogenetics and Evolution, 53(3): 1037–1041.

PHYLUM GASTROTRICHA

CHAPTER **16**

Maria Balsamo and Paolo Grilli

Infrakingdom	Bilateria
Division	Protostomia
Subdivision	Lophotrochozoa
Phylum	Gastrotricha

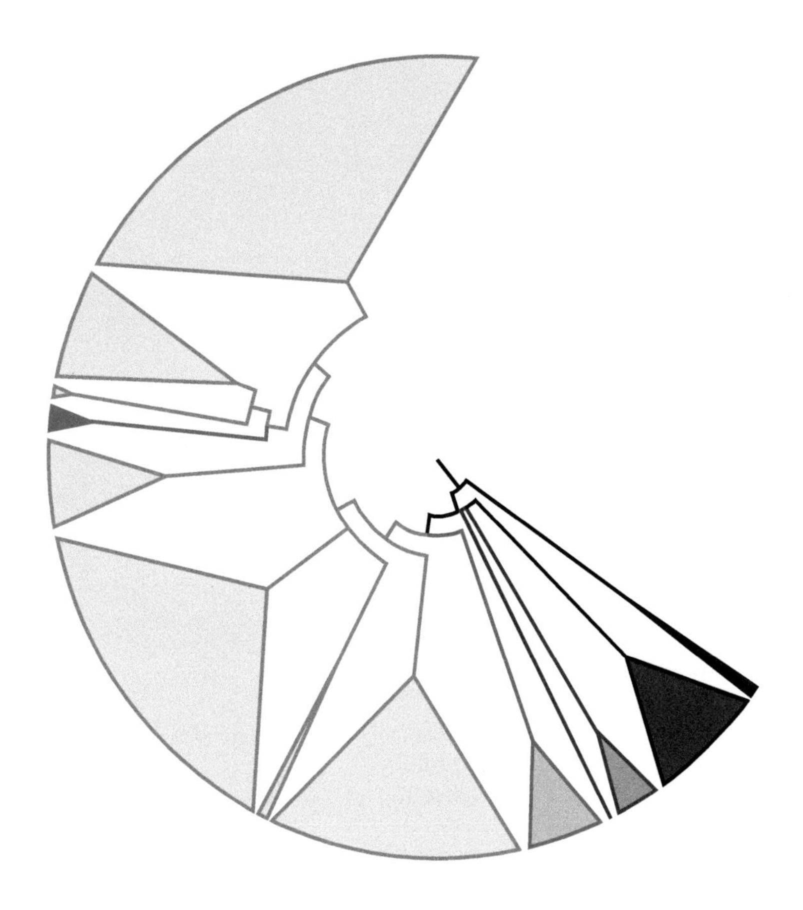

CONTENTS

General Characteristics

Over 860 valid species

Over 1,000 cryptic species estimated

Microscopic bilateral free-living aquatic animals with extensive ventral locomotory ciliature

Epibenthic/interstitial, widespread in marine and fresh waters, in sandy and/or detritus-rich habitats

Hermaphrodites or parthenogenic

Acoelomates with uncertain phylogenetic relationships among Protostomes

No commercial value

Ecological importance in the benthic microbial loop

Autapomorphies

- multi-layered body cuticle even covering sensory and locomotory cilia
- helicoidal muscles partially surrounding the pharynx and intestine

HISTORY OF TAXONOMY AND CLASSIFICATION

The first information on Gastrotricha dates back to the late 19th century: Joblot (1718) drew a *Chaetonotus* species from fresh waters, and Müller in 1786 described *Chaetonotus larus*, the first gastrotrich to receive a Latin binomial name. Remane (1925) started the interest for marine species considered at that time as 'aberrant Gastrotricha' and introduced the division of the phylum into two main orders, which were subsequently accepted as Macrodasyida and Chaetonotida, since their differences in morphology, biology and ecology are evident. Considering these groups as orders rather than classes was logical because the entire phylum had been proposed as a class of Aschelminthes (Hyman 1951). Macrodasyida are almost exclusively marine: only three species of the genus *Redudasys* and *Marinellina flagellata* are strictly freshwater. Chaetonotida include freshwater species and about a third of the marine and brackish-water gastrotrich species known (Zhang 2013).

Taxonomy and systematics of Gastrotricha have traditionally been based on morphology, and diagnostic anatomical characters have been the ground for classification and identification at the level of both species and superspecific taxa (Zelinka 1889; Remane 1936; Schwank 1990; Kisielewski 1998).

Current systematic research is justly focused on detecting natural relationships among taxa, and it applies molecular techniques in addition to morphological and ultrastructural investigations. The search for valid characters useful in taxonomical work and phylogenetic reconstruction has partially changed or integrated the traditional classification. However, since the intra-phylum relationships as well as those with other phyla of lower Metazoa have been actively debated and are still uncertain, the taxonomy of the group still appears anything but stable, even considering that molecular evidence highlights the presence of cryptic species within a number of nominal species (see Kieneke and Schmidt-Rhaesa 2015 for a review and references therein).

Currently the phylum Gastrotricha numbers 852 species: 371 Macrodasyida (367 marine and 4 freshwater) in ten families, and 481 Chaetonotida (130 marine, 4 brackish-water, 347 freshwater) in eight families (WoRMS 2019) (**Figure 16.1**).

ANATOMY, HISTOLOGY AND MORPHOLOGY

Worm-like (Macrodasyida) or tenpin-like (Chaetonotida) Gastrotricha have a bilateral body flattened ventrally and convex dorsally, ranging in length from less than 100 μm to over 3 mm (**Figure 16.2A,B**). Peculiarly the body cuticle also lines locomotory and sensory cilia (**Figure 16.3**) and is not molted during development: it is multi-layered, mainly proteinaceous and does not contain

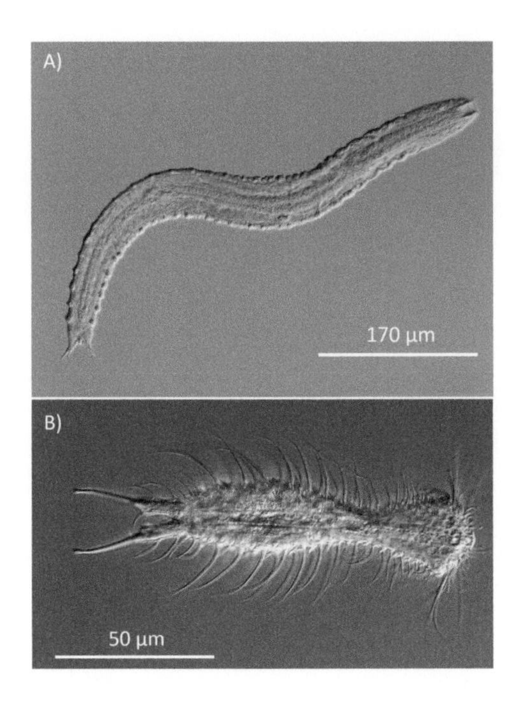

Figure 16.1 Representatives of the phylum Gastrotricha: (A) *Paraturbanella levantia* (order Macrodasyida); (B) *Chaetonotus* sp. (order Chaetonotida) (DIC microscopy). (Courtesy of M. Antonio Todaro.)

chitin. If it is thin and smooth the cuticle allows the animal to be very flexible and contractile, but mostly it shows external sculptures (scales, spines, plates) with protective function, that stiffen the body especially on the dorsal side (Figure 16.2C). Cuticular duo-gland adhesive tubes are generally numerous in Macrodasyida, arranged along the body often into paired groups, whereas Chaetonotida have only two tubes, rarely four, forming the caudal 'furca' (Figure 16.2B). These are especially important in the interstitial habitat, where they ensure a firm hold to the sediment grains, and the adhesive tubes are reduced or even absent in freshwater semiplanktonic species. Shape, size and arrangement of all the cuticular elements are important diagnostic characters at genus and species level.

The cellular epidermis is ventrally monociliated or multiciliated: ventral cells are cuboidal whereas dorsal cells are flat with no cilia. The extensive ventral locomotory ciliature accounts for the name of the phylum (Greek '*gastér*', venter, and '*thrix*', cilium) and allows active, smooth gliding on the substrate, on the submerged vegetation and in the interstitial habitat, and occasionally also active

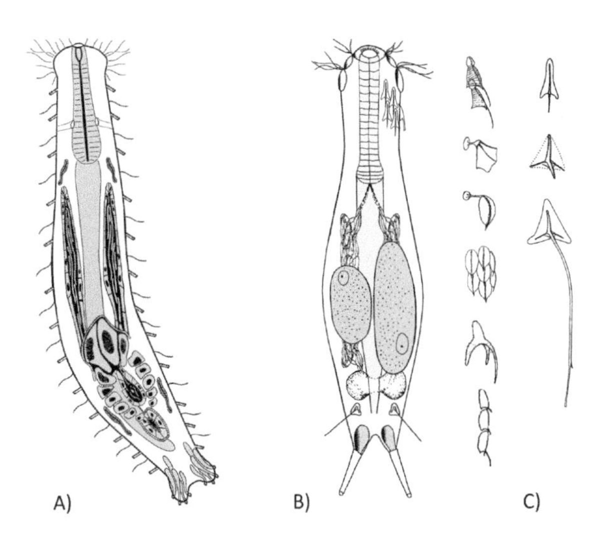

Figure 16.2 Schematic drawing of (A) a Macrodasyida, (B) a Chaetonotida and (C) examples of cuticular scales and spines of Chaetonotida. (Modified from various sources.)

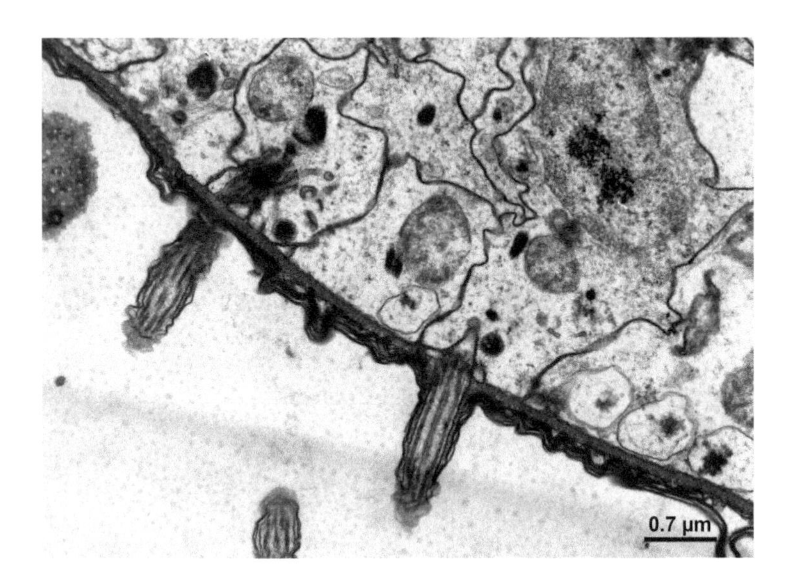

Figure 16.3 Detail of the body wall of a gastrotrich showing that the multilayered cuticle covers the cilia and the cellular epidermal layer (TEM). (Courtesy of Loretta Guidi.)

swimming. Locomotory ciliary patterns are various: usually arranged into two longitudinal bands, in semiplanktonic species ciliature may consist of only few paired ventral tufts (Balsamo et al. 2014).

Numerous paired large epidermal glands are often present and clearly visible in Macrodasyida.

Cross- and obliquely striated muscle bundles are variously arranged along the body: they are circular, longitudinal and helicoidal. Visceral longitudinal muscles surround the circular ones along the pharynx but this arrangement is reversed along the intestine. Both the circular and longitudinal muscles of the pharynx and the midgut are wrapped with a peculiar muscular double helix. Main longitudinal muscles are two large ventrolateral bundles that are directly involved in the great body contractility characteristic of Gastrotricha. Musculature arrangement, cytomorphology and ultrastructure have been intensively studied in recent years proving the phylogenetic value of muscle features of Gastrotricha (Hochberg and Litvaitis 2001; Kieneke and Schmidt-Rhaesa 2015).

The nervous system is composed of a dorsal bilateral symmetrical brain and a pair of ventrolateral longitudinal cords. The brain consists of a broad dorsal neuropil with a thin ventral commissure and lateral clusters of neuronal somata; thus it strongly differs from the 'Cycloneuralia' brain (Schmidt-Rhaesa et al. 2014). Numerous sensory organs are especially present on the head: cilia, often associated with adhesive tubes, palps or tentacles, 'pestle organs', with mechano- and chemoreceptive functions, in few cases pigmented photoreceptors, and gravity receptors (Kieneke and Schmidt-Rhaesa 2015).

The digestive system is complete. The terminal mouth opens into a cuticularized buccal cavity followed by a strong myoepithelial pharynx with a triradiate lumen, Y-oriented in cross-section in Macrodasyida and inverted-Y-oriented in Chaetonotida: the latter is a character shared with other invertebrate taxa (i.e., Nematoda, Tardigrada, Kinorhyncha), and is related to active suction feeding (**Figure 16.4**). Two thin lateral channels directly connect the pharyngeal lumen to the exterior ('pharyngeal pores') only in Macrodasyida, in which they have been interpreted as a device for allowing excess water ingested with the food to be expelled.

There are neither respiratory nor circulatory systems, due to small body size which allows these functions to be performed by diffusion. Some paired cyrtocytic protonephridia in Macrodasyida, and only one pair in Chaetonotida, lie laterally to the intestine and are the osmoregulatory-excretory system of the phylum. The 'pseudocoelom' of gastrotrichs is not evident because it is actually almost completely filled with organs, and in a few species it is partially lined with

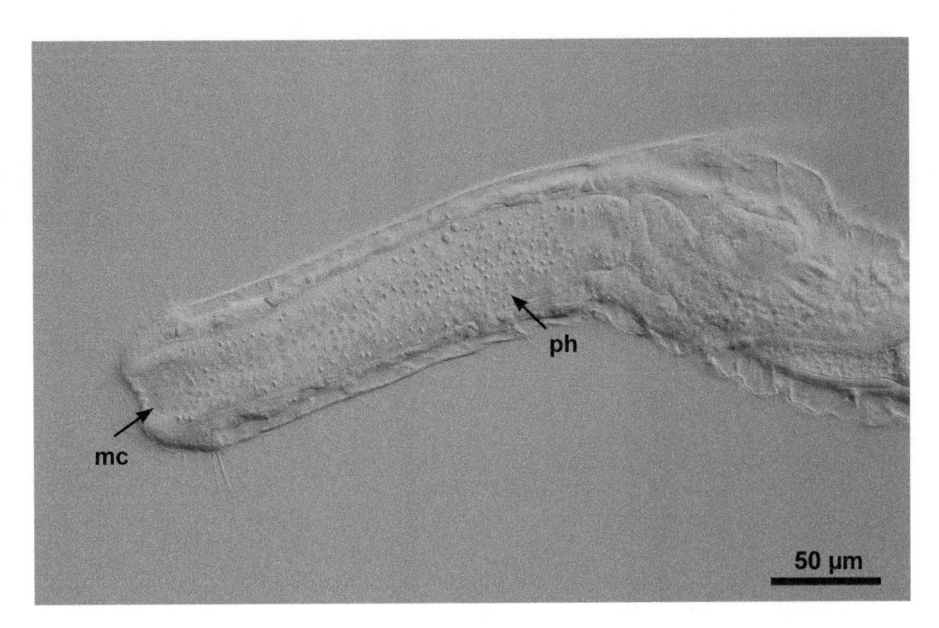

Figure 16.4 Detail of the anterior region of a Macrodasyida gastrotrich (*Macrodasys* sp.), showing the mouth cavity (mc) and the pharynx (ph) (DIC). (Courtesy of Andreas Schmidt-Rhäesa.)

mesothelium. The 'acoelomate' structure of gastrotrichs clearly distinguishes them from other 'pseudocoelomate' phyla, and this evidence caused a change of its position within the phylogenetic tree of lower invertebrates (Teuchert 1977).

REPRODUCTION AND DEVELOPMENT

Gastrotricha reproduce only sexually, by amphimixis or parthenogenesis. Hermaphroditism generally proterandric, then simultaneous or sequential is the rule in Macrodasyida. In a number of marine Chaetonotida, it is rarely proterogynic (Xenotrichulidae).

Two testes lateral to the intestine extend into sperm ducts opening ventrally into two or a single median male pore (in Thaumastodermatidae the sperm ducts open into the accessory caudal organ). Two or a single dorsal ovary mature very large oocytes generally in a caudo-cephalic direction.

Gastrotrich spermatozoa have been studied intensively in recent decades showing unexpected morphological variation and structural complexity which has also provided elements useful for phylogenetic purposes (Balsamo et al. 1999; Kieneke and Schmidt-Rhaesa 2015).

Fertilization is internal and crossed, which explains the complex reproductive system of Macrodasyida; they are equipped with 1–2 accessory organs for transmission and storage of sperm: the epithelial frontal organ, probably a seminal receptacle, and the strongly muscularized caudal organ that seems to act as a copulatory organ (Ruppert 1978; Kieneke and Schmidt-Rhaesa 2015; **Figure 16.5**).

A complex mating behavior has been reported only in two *Macrodasys* species by Ruppert (1978), but details of the transfer of reciprocal sperm or spermatophores are still poorly understood.

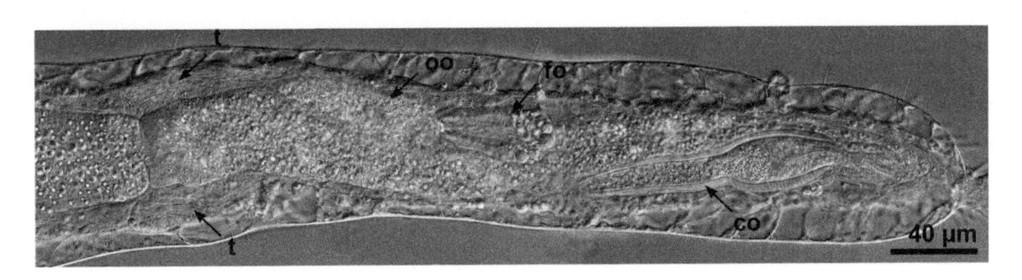

Figure 16.5 Reproductive system of a Macrodasyida gastrotrich (*Macrodasys* sp.): testes (t), mature oocyte (oo), frontal organ (fo) and caudal organ (co) (DIC). (Courtesy of M. Antonio Todaro.)

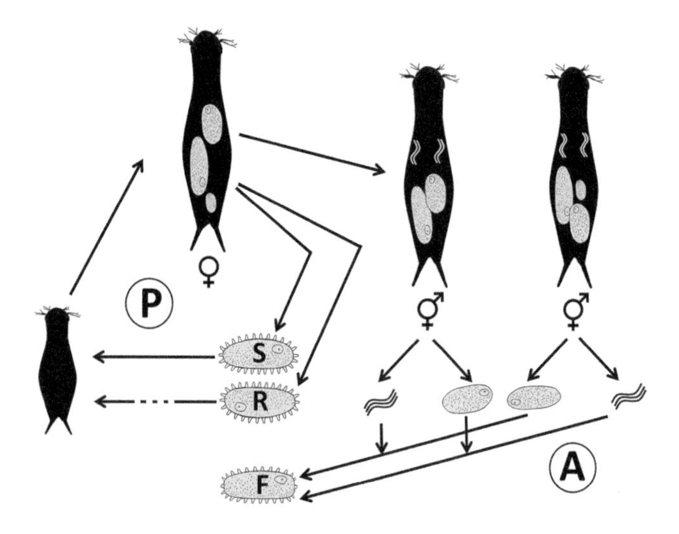

Figure 16.6 Biological cycle of freshwater Chaetonotida: parthenogenic phase (P) with production of subitaneous (S) or resting (R) eggs; postparthenogenic phase with production of few spermatozoa and hypothetical amphymixis (A). (Redrawn from various sources.)

Parthenogenesis is the rule in most Chaetonotida, very rare among Macrodasyida. The development of aberrant spermatozoa in postparthenogenic individuals of several Chaetonotida species has allowed researchers to hypothesize a biphasic life cycle with two different reproductive phases, the first parthenogenic and the second amphimictic within a single generation (Levy 1984; Balsamo 1992; **Figure 16.6**).

Single eggs are apparently released by breaking the body wall or perhaps through a temporary pore. Development is direct and usually starts just after fertilization following a modified spiral type pattern and completes in a few days, but it is affected by temperature and in marine species also by salinity (Balsamo and Todaro 1988; Hejnol 2015). Juveniles hatch from the eggs by active movements: they are much smaller than adults, with different body proportions, and in Macrodasyida they undergo significant changes in some morphological characters (i.e. number of adhesive tubes).

Freshwater parthenogenic Chaetonotida also produce resting eggs, morphologically different from the subitaneous egg and able to resist to extreme environmental conditions: they develop only after a 'dormancy' period (Figure 16.6).

DISTRIBUTION AND ECOLOGY

The phylum is generally cosmopolitan since gastrotrich species have been found all over the world except Antarctica, and at all marine depths. The geographic distribution and global diversity of marine gastrotrichs are well known from numerous world areas and many taxa show a broad distribution (Todaro et al. 2019) (**Figure 16.7**). The same cannot be stated about freshwater gastrotrichs, because available data are less numerous and extremely heterogeneous, and especially reported from Europe and North and South America (Balsamo et al. 2008) (**Figure 16.8**). The apparent wide distribution of a number of marine and freshwater species has proven to actually mask several cryptic species. Mitochondrial COI sequences analysis has recently been introduced in the study of gastrotrich populations and has highlighted significant genetic differences between separated populations, even at a small spatial scale (Todaro et al. 2006; Kieneke et al. 2012).

Gastrotricha are widely present in aquatic ecosystems where they are a significant component of the meiobenthic community. Marine species and some freshwater ones are well adapted to the psammic system, whereas most freshwater forms are epibenthic or periphytic, and a few have developed a semi-planktonic habitus mainly in standing waters. As meiobenthic organisms, gastrotrichs show a scattered and very variable spatial distribution, related to some

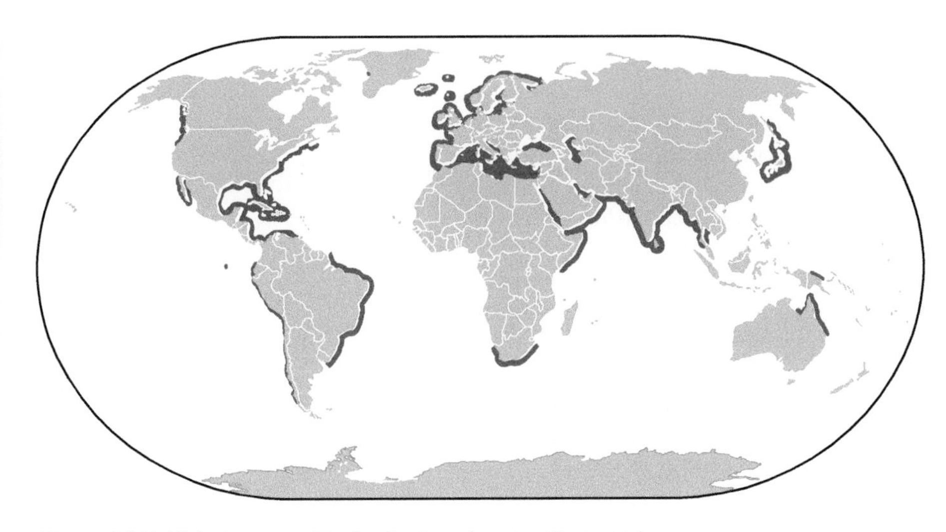

Figure 16.7 Global geographic distribution of marine Gastrotricha.

key environmental factors such as temperature, oxygen content, salinity, water flow and sediment grain size. That makes their effective sampling difficult, when technical problems in collecting and handling these soft-bodied and quite contractile animals are also considered (Balsamo et al. 2014). Marine species occasionally form abundant populations in fine sands of the shallow subtidal zone, up to 200–400 ind/cm^3 (Potel and Reise 1987). Freshwater species are mainly epibenthic on muddy sediments or periphytic on aquatic vegetation, and are more difficult to collect than marine ones, because they are usually very sporadic. Occasionally they can form large populations, but density values reported in the literature (es. 0.67×10^6 specimens per m^2, Muschiol and Traunspurger 2009) were generally calculated by extrapolation from small samples and have to be considered with caution.

Tolerance to low oxygen level allows some marine and freshwater species to colonize deeper sediments, or sands and muds rich in sulphide, as well as highly eutrophicated water bodies (see "Physiology and Behavior", below). The typical instability of freshwater habitats is faced by gastrotrichs through the ability to produce resting eggs (see "Reproduction", above).

Gastrotricha are mainly bacterivorous and are in turn predated on by other benthic animals: as a component of the microphagous and detritivorous benthic assemblage, they represent an important link between the microbial loop and higher trophic levels.

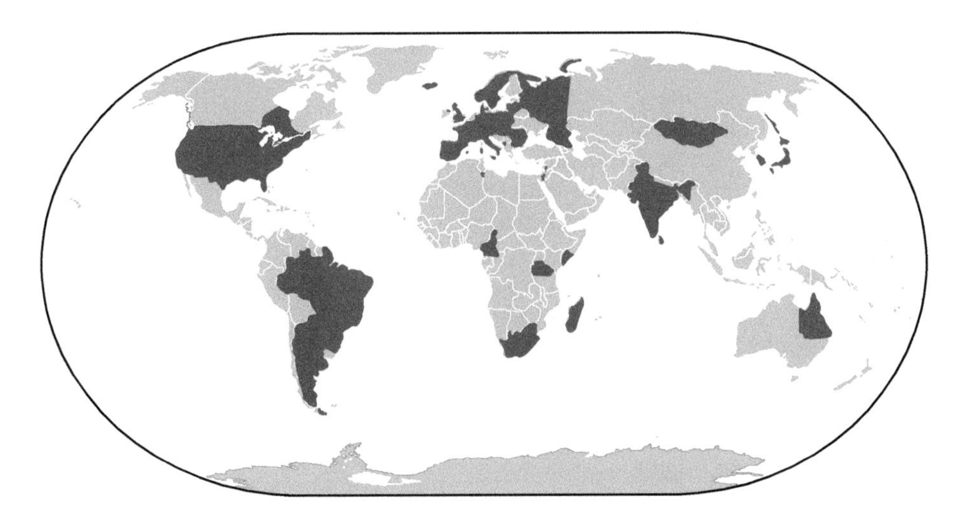

Figure 16.8 Global geographic distribution of freshwater Gastrotricha.

PHYSIOLOGY AND BEHAVIOR

A few studies have focused on physiological and behavioral aspects of Gastrotricha, mainly in laboratory conditions. Gastrotricha are free-living and very active in their habitat: the locomotory behavior mainly relies on the use of the ventral ciliature for active crawling or gliding on the substrate or briefly swimming over it, as well as of the ventral longitudinal muscles for sudden body contractions and flexions during locomotion. In the interstitial habitat the adhesive tubes, usually numerous in Macrodasyida, allow them to temporarily anchor to a substrate.

Gastrotricha are generally aerobic, but some marine species of Macrodasyida show an evident tolerance to anoxic habitats and also morphological adaptations like modifications to or even absence of mitochondria or of cells containing haemoglobin, suggesting that they may have evolved anaerobic metabolic pathways (Colacino and Kraus 1984; Balsamo et al. 2007). In fresh waters, where pH value is very variable, some chaetonotid species are able to tolerate low pH values down to 4 and others values up to 10. A negative phototaxis was observed in a chaetonotid with no photoreceptors and is to be ascribed to a generalized dermal photosensitivity (Balsamo 1980).

GENETICS

Nuclear DNA content was studied by cytofluorimetry in 15 species and highlighted a very small genome size (0.05–0.63 pg), similar in the two orders (Balsamo and Manicardi 1995).

Molecular analyses started recently on numerous species of both orders and focused on nuclear genes 18S RNA, 28S RNA and especially on COI1 mitochondrial DNA, with the aim of clarifying the degree of intraspecific variability, the genetic structure of populations and the phylogenetic relations of the phylum. Possible existence of cryptic species within a number of taxonomic species has been highlighted and new phylogenetic hypotheses have been advanced (see "Position in the ToL", below). Recent results of the first mitochondrial genome sequencing (Golombek et al. 2015) and of a 'multi-omics' approach to detect micro-RNA and piRNA complement in a freshwater species (Fromm et al. 2019) have led to support for a close relationship between gastrotrichs and flatworms.

POSITION IN THE ToL

The phylogenetic position of Gastrotricha is still unresolved. The multilayered body cuticle and the cuticle-lined myoepithelial sucking pharynx have been proposed as synapomorphies shared with Cycloneuralia, advancing the clade Nemathelminthes (Cycloneuralia + Gastrotricha). However, numerous cladistic analyses based on one to multiple genes loci as well as some recent data on transcriptomes suggested Gastrotricha are to be included in a large protostomian clade (Platyzoa), in which they appear to be the sister group of Platyhelminthes or Gnathostomulida. The phylum has also been proposed as the sister group of Ecdysozoa (Cycloneuralia + Panarthropoda), but molecular data do not support this hypothesis. Therefore, at present Gastrotricha appear as the potential sister group of Cycloneuralia, Platyhelminthes, Gnathostomulida or Ecdysozoa (Kieneke and Schmidt-Rhaesa 2015), even if the latest molecular findings suggest a closer relationship with Platyhelminthes (Fromm et al. 2019) supporting the monophyletic clade Rouphozoa within the Spiralia that had been proposed by Struck et al. (2014) and endorsed by two subsequent, independent phylogenomic studies (Egger et al. 2015; Laumer et al. 2015).

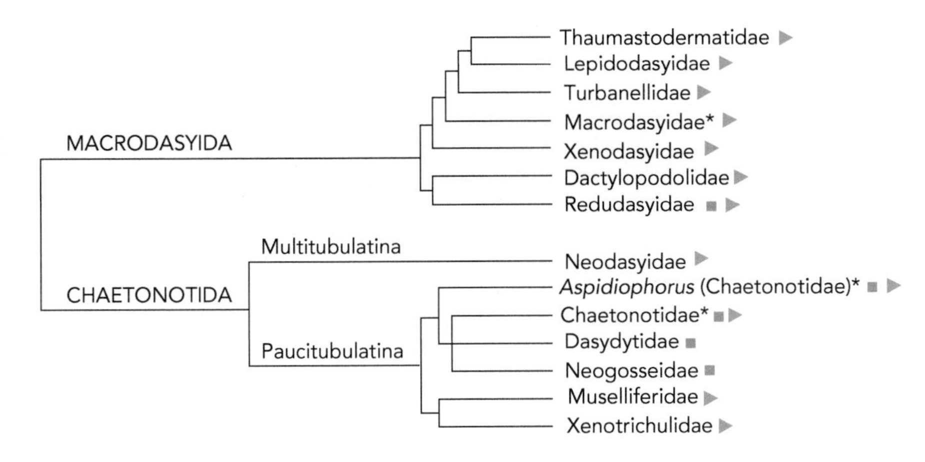

Figure 16.9 Intra-phylum relations of Gastrotricha: only the families whose relationships are sufficiently founded are reported. For each family the marine (▶) and/or freshwater (■) habitat is indicated.

The monophyly of the phylum is not generally questioned, whereas the intra-phylum phylogeny still shows many doubtful points, but it should be noted that DNA sequence-based investigations, though now numerous, still include only a fraction of the genera of the phylum (**Figure 16.9**).

DATABASES AND COLLECTIONS

Databases on Gastrotricha have been realized in the framework of the Projects: Fauna Europaea (https://fauna-eu.org), Freshwater Animal Diversity Assessment (http://fada.biodiversity.be), WoRMS (http://www.marinespecies.org), and on the website dedicated to the phylum (www.gastrotricha.unimore.it).

Collections of Gastrotricha are at various Museums worldwide, among which are: Museo Civico di Storia Naturale, Verona (IT), Museo Tridentino di Scienze Naturali, Trento (IT), British Museum of Natural History, London (UK), Universitets Zoologisk Museum of Copenhagen (DK), Zoologisk Museum Universitet I Bergen (NO), Muséum National d'Histoire Naturelle, Paris (FR), National Museum of Natural History, Smithsonian Institution, Washington, DC (US), American Museum of Natural History, New York (US), National Institute Biology Research, Incheon (KO), Museum of Osaka University (JP), Museu de Zoologia, Universidade Estadual Campinas (BR), Queensland Museum, Brisbane (Australia).

CONCLUSION

Gastrotrichs have had few followers over time because they are microscopic and often go unnoticed. However, their biology offers many cues to deal with, including the reproductive biology of freshwater Chaetonotida that seems to combine parthenogenesis and amphygony in a single life cycle, which is perhaps unique among the metazoans. Moreover, the ecology of the phylum points to an increasingly important role of the meiofauna within the trophic chains of the benthos and in particular the microbial loop. Last but not least, the long-debated position of the group among the basal Metazoa has not yet been definitely resolved. New technologies are allowing us to deepen morphological and molecular studies on microscopical organisms; thus they are and will be crucial in the study of Gastrotricha, which still has much to reveal.

REFERENCES

Balsamo, M., 1980. Spectral sensitivity in a fresh-water Gastrotrich (*Lepidodermella squamatum* Dujardin). Experientia, 36(7), 830–831.

Balsamo, M., 1992. Hermaphroditism and parthenogenesis in lower Bilateria: Gnathostomulida and Gastrotricha. In: Dallai, R. (ed.), Sex origin and evolution. Selected Symposia and monographs, 309–327. UZI, Mucchi, Modena.

Balsamo, M., Manicardi, G. C., 1995. Nuclear DNA content in Gastrotricha. Experientia, 51(4), 356–359.

Balsamo, M., Todaro, M. A. D., 1988. Life history traits of two chaetonotids (Gastrotricha) under different experimental conditions. International Journal of Invertebrate Reproduction and Development, 14(2–3), 161–176.

Balsamo, M., Fregni, E., Ferraguti, M., 1999. Gastrotricha. In: Adiyodi, K. G., Adiyodi, R. (eds.), Reproductive Biology of Invertebrates. vol. IX, Part A (Jamieson, B. G. M., ed.), Progress in male gamete ultrastructure and phylogeny, 171–191. Oxford & IBH Publishing, New Delhi.

Balsamo, M., Grilli, P., Guidi, L., d'Hondt, J. L., 2014. Gastrotricha: biology, ecology and systematics. Families Dasydytidae, Dichaeturidae, Neogosseidae, Proichthydiidae, Vol. 24, 187. Backhuys Publishers, The Netherlands.

Balsamo, M., d'Hondt, J. L., Kisielewski, J., Pierboni, L., 2008. Global diversity of gastrotrichs (Gastrotricha) in fresh waters. In: Balian, E. V., Lévêque, C., Segers, H. Martens, K. (eds.), Freshwater Animal Diversity Assessment, 85–91. Springer, Dordrecht.

Balsamo, M., Guidi, L., Pierboni, L., Marotta, R., Todaro, M. A., Ferraguti, M., 2007. Living without mitochondria: spermatozoa and spermatogenesis in two species of *Urodasys* (Gastrotricha, Macrodasyida) from dysoxic sediments. Invertebrate Biology, 126(1), 1–9.

Colacino, J. M., Kraus, D. W., 1984. Haemoglobin-containing cells of *Neodasys* (Gastrotricha, Chaetonotida)? II. Respiratory significance. Comparative Biochemistry and Physiology – Part A: Physiology, 79(3), 363–369.

Egger, B., Lapraz, F., Tomiczek, B., Müller, S., Dessimoz, C., Girstmair, J., Škunca, N., Rawlinson, K. A., Cameron, C. B., Beli, E., Todaro, M. A., Gammoudi, M., Noreña, C., Telford, M. J., 2015. A transcriptomic-phylogenomic analysis of the evolutionary relationships of flatworms. Current Biology, 25, 1347–1353.

Fromm, B., Tosar, J. P., Aguilera, F., Friedländer, M. R., Bachmann, L. Hejnol, A., 2019. Evolutionary Implications of the micro-RNA-and piRNA Complement of *Lepidodermella squamata* (Gastrotricha). Non-coding RNA, 5(1), 19.

Golombek, A., Tobergte, S. Struck, T. H., 2015. Elucidating the phylogenetic position of Gnathostomulida and first mitochondrial genomes of Gnathostomulida, Gastrotricha and Polycladida (Platyhelminthes). Molecular Phylogenetics and Evolution, 86, 49–63.

Hejnol, A., 2015. Gastrotricha. In: Wanninger, A. (ed.), Evolutionary Developmental Biology of Invertebrates, 2, 13–19. Springer, Vienna.

Hochberg, R., Litvaitis, M. K., 2001. Macrodasyida (Gastrotricha): a cladistic analysis of morphology. Invertebrate Biology, 120(2), 124–135.

Hyman, L. H., 1951. The invertebrates: Platyhelminthes and Rhynchocoela, the acoelomate Bilateria 3, 53–59. McGraw_ Hill, New York.

Joblot, L,. 1718. Descriptions et usages de plusieurs nouveaux microscopes: tant simples que composez; avec de nouvelles observations faites sur de multitude innombrable d'insectes, & d'autres animaux de diverses espèces, qui naissent dans des liqueurs préparées, et dans celles qui ne le sont point. Chez J. Collombat, imprimeur.

Kieneke, A., Martinez Arbizu, P. M., Fontaneto, D., 2012. Spatially structured populations with a low level of cryptic diversity in European marine Gastrotricha. Molecular Ecology, 21(5), 1239–1254.

Kieneke, A., Schmidt-Rhaesa, A., 2015. Gastrotricha. In: Schmidt-Rhaesa, A. (ed.), Handbook of Zoology, Gastrotricha, Cycloneuralia and Gnathifera. Vol. 3: Gastrotricha and Gnathifera, 1–134. De Gruyter, Berlin/Munich/Boston.

Kisielewski, J., 1998. Brzuchorzeski (Gastrotricha). Fauna Slodkowodna Polskie, 31, 156.

Levy, D. P., 1984. Obligate postparthenogenic hermaphroditism and other evidence for sexuality in the life cycles of freshwater Gastrotricha. PhD Dissertation, Rutgers University, New Brunswick, NJ. 257 pp.

Laumer, C. E., Hejnol, A., Giribet, G., 2015. Nuclear genomic signals of the 'microturbellarian' roots of platyhelminth evolutionary innovation. eLife 4, e05503.

Müller, O. F., 1786. Animalcula Infusoria fluviatilia et marina. Havniae, Nicholai Mölleri. 274 pp.

Muschiol, D., Traunspurger, W., 2009. Life at the extreme: meiofauna from three unexplored lakes in the caldera of the Cerro Azul volcano, Galápagos Islands, Ecuador. Aquatic Ecology, 43(2), 235–248.

Potel P., Reise, K., 1987. Gastrotricha Macrodasyida of intertidal and subtidal sandy sediments in the northern Wadden Sea. Microfauna Marina, 3, 363–376.

Remane, A., 1925. Organisation und systematische Stellung der aberranten Gastrotrichen. Verhandlungen der Deutschen Zoologischen Gesellschaft, 30, 121–128.

Remane, A., 1936. Gastrotricha. In: Bronns Klassen und Ordnungen des Tierreichs. Lieferung IV, Abt. 2, Buch 1, Teil 2: 1–142. Akademische Verlagsgesellschaft, Leipzig.

Ruppert, E. E., 1978. The reproductive system of gastrotrichs. II. Insemination in *Macrodasys*: A unique mode of sperm transfer in Metazoa. Zoomorphologie, 89 (3), 207–228.

Schmidt-Rhaesa, A., Rothe B. H., Wägele, J. W., Bartolomaeus, T., 2014. Brains in Gastrotricha and Cycloneuralia – a comparison. Deep metazoan phylogeny: the backbone of the Tree of Life. New insights from analyses of molecules, morphology, and theory of data analysis, 93–104. De Gruyter, Berlin.

Schwank, P., 1990. Gastrotricha. In: Schwoerbel, J., Zwick, P. (eds.), Gastrotricha und Nemertini. Süßwasserfauna von Mitteleuropa 3/1+21–252. Gustav Fischer, Stuttgart.

Struck, T. H., Wey-Fabrizius A. R., Golombek A., Hering L., Weigert A., Bleidorn C., Kück P., 2014. Platyzoan paraphyly based on phylogenomic data supports a noncoelomate ancestry of Spiralia. Molecular Biology and Evolution 31, 1833–1849.

Teuchert, G., 1977. The ultrastructure of the marine gastrotrich *Turbanella cornuta* Remane (Macrodasyoidea) and its functional and phylogenetical importance. Zoomorphologie, 88, 189–246.

Todaro, M. A., d'Hondt, J.-L., Hummon, W. D., 2019. WoRMS Gastrotricha: World Gastrotricha Database (version 2019-03-05). In: Roskov, Y., Orrell, T., Nicolson D., Bailly, N., Kirk, P. M., Bourgoin, T., DeWalt, R. E., Decock, W., De Wever, A., Nieukerken, E. van, Zarucchi, J., Penev, L. (eds.) Species 2000 & ITIS Catalogue of Life, 12 January 2020. Digital Resource at www.catalogueoflife/col. Species 2000: Naturalis, Leiden, the Netherlands. ISSN 2405-8858.

Todaro, M. A., Telford, M. J., Lockyer, A. E., Littlewood, D. T. J., 2006. Interrelationships of the Gastrotricha and their place among the Metazoa inferred from 18S rRNA genes. Zoologica Scripta, 35(3), 251–259.

WoRMS (2019). World Register of Marine Species. Available from http://www.marinespecies.org at VLIZ. Accessed 2019-11-30. doi:10.14284/170

Zelinka C., 1889. Die Gastrotrichen: eine monographische Darstellung ihrer Anatomie, Biologie und Systematik. Engelmann.

Zhang Z.-Q. ed. 2013. Animal biodiversity: An outline of higher-level classification and survey of taxonomic richness (Addenda 2013). Zootaxa, 3703 (1), 1–82.

Additional Reading

Ehlers, U., Ahlrichs, W., Lemburg, C. Schmidt-Rhaesa A., 1996. Phylogenetic systematization of the Nemathelminthes (Aschelminthes). Verhandlungen der Deutschen Zoologischen Gesellschaft, 89(8).

Giribet, G., Dunn, C. W., Edgecombe, G. D., Rouse, G. W., 2007. A modern look at the Animal Tree of Life. Zootaxa, 1668(1), 61–79.

Wägele, J. W., Bartolomaeus, T., 2014. Deep metazoan phylogeny: the backbone of the tree of life, 736 pp. De Gruyter GmbH, Berlin.

PHYLUM ROTIFERA

Diego Fontaneto and Michael Plewka

CHAPTER

17

Infrakingdom	Bilateria
Division	Protostoma
Subdivision	Lophotrochozoa
Phylum	Rotifera

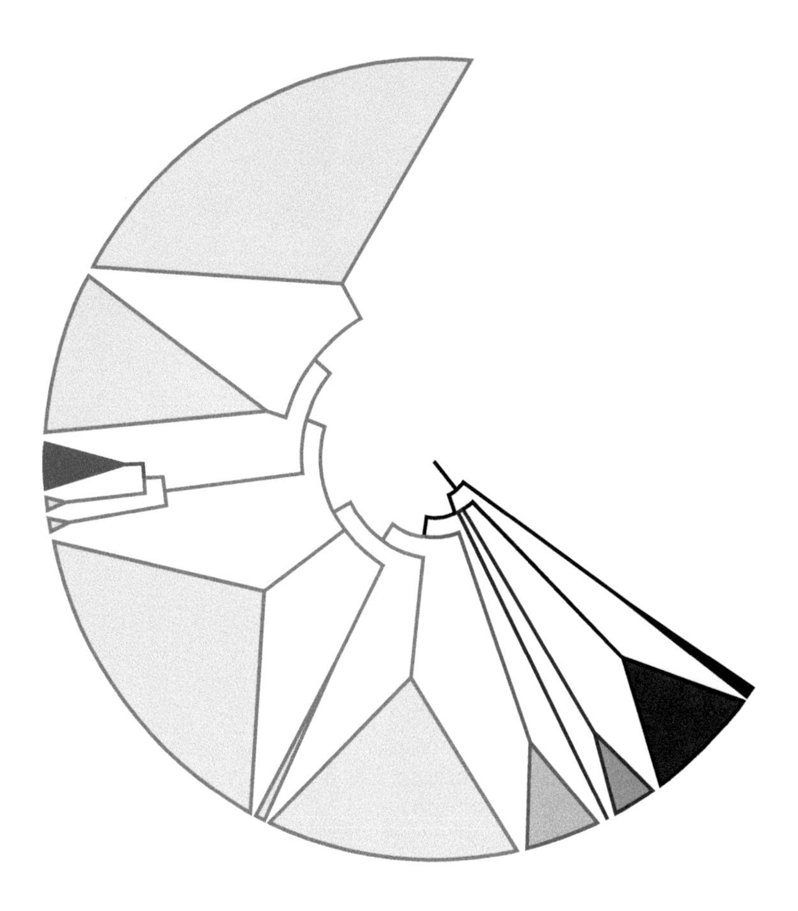

CONTENTS

General Characteristics

About 2,000 valid species, but the estimated number is higher, due also to the presence of cryptic species

Together with Acanthocephala, they belong to the group Syndermata

Microscopic bilateral animals, abundant in freshwater, limnoterrestrial and brackish habitats

Eutelic organisms with syncytial epithelia

Peculiar for desiccation capabilities, cyclical and obligate parthenogenesis, and massive horizontal gene transfer

Economic value for some species used as food in aquaculture

Synapomorphies

- corona of cilia
- masticatory apparatus (mastax) with hard parts (trophi)
- intracytoplasmic lamina

HISTORY OF TAXONOMY AND CLASSIFICATION

The first animals of the phylum Rotifera became known when the first microscopist, Anthony van Leeuwenhoek, described a rotifer collected in the debris of roof gutters, in a letter dated 1687. Three species of rotifers were already known by Linnaeus and described in the Systema Naturae of 1758: *Sinantherina socialis*, *Floscularia ringens*, and another one that is now considered a *species inquirenda* of uncertain taxonomic affiliation.

The close relationship between the three traditional groups of rotifers (Bdelloidea, Monogononta, and Seisonacea) and the Acanthocephala was initially suggested by morphological studies (Clément, 1993): rotifers and acanthocephalans share (1) a peculiar syncytial epidermis with an intracytoplasmic lamina (Ahlrichs, 1997), and (2) a similar morphology of the sperm (Ferraguti and Melone, 1999). Furthermore, molecular phylogeny assays using different genetic markers supported their close relationship, so that the four groups were brought together in the taxon Syndermata in the late 1990s.

The monophyly of each of the three major rotifer groups and of acanthocephalans is not in question (**Figure 17.1**). Nevertheless, the evolutionary relationships between them have not yet been clarified. Almost every possible relationship between Acanthocephala, Bdelloidea, Monogononta and Seisonacea has been suggested, without any final agreement on which hypotheses are more plausible (Sørensen and Giribet, 2006; Sielaff et al., 2016). However, two main constants emerge from all the analyses conducted so far: (1) the hypothesis of Acanthocephala as sister group of the other Rotifera, forming a monophyletic clade, has never been supported by molecular phylogeny, and (2) most

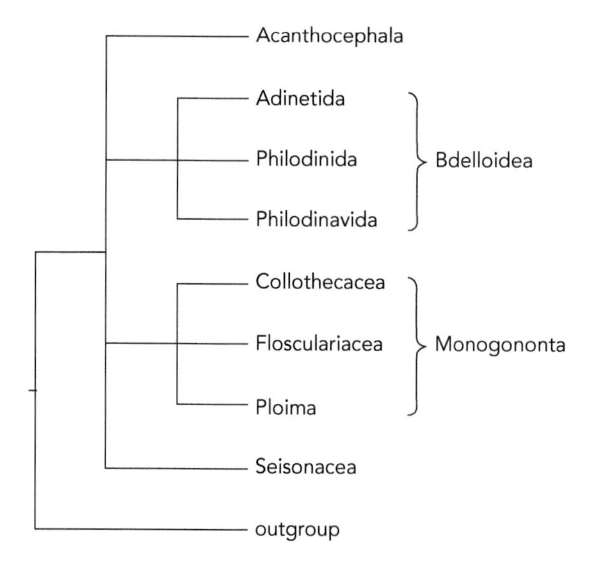

Figure 17.1 Phylogenetic relationships of Rotifera. Schematic cladogram of the relationships between the four major clades of rotifers (Acanthocephala, Bdelloidea, Monogononta, Seisonacea), with further internal subdivisions for bdelloids and monogononts.

recent analyses always keep Bdelloidea separate from Monogononta. However, Bdelloidea, Monogononta and Seisonacea share many morphological and ecological features, whereas there is no morphological support for the inclusion of Acanthocephala within them. This chapter will deal with Rotifera *sensu stricto*, without Acanthocephala.

Bdelloidea are a group of apparently obligate parthenogenetic animals that lack meiosis. They live mainly in fresh waters of all kinds, but also in soils, mosses, lichens, and any other environments that remain wet for only a short time, including deserts and glaciers (Donner, 1965; Kutikova, 2003).

Monogononta are characterized by cyclical parthenogenesis, alternating parthenogenesis with sexual reproduction (Koste, 1978). Males usually look like a reduced version of females with a large penis (Ricci and Melone, 1998). The outcome of sexual reproduction is a dormant embryo called resting egg. They live mainly in fresh waters of all kinds, but are also present in the sea.

Seisonacea strictly reproduce sexually. They live only as epibionts on marine crustaceans of the genus *Nebalia* (Ricci et al., 1993).

ANATOMY, HISTOLOGY AND MORPHOLOGY

The general body plan (**Figure 17.2**) is bilaterally symmetric, with a clear differentiation between the anterior and posterior extremities and between the ventral and dorsal sides. The body is subdivided into three parts: **head**, **trunk** and **foot** (Fontaneto and De Smet, 2015). The head bears a rostrum and a corona with cilia and sensory apparatuses. The trunk often appears to be pseudo-segmented due to the presence of transversal folds in the integument. The foot is ventral and posterior to the cloaca. A distinguishable pseudo-segment between the head and the trunk is often present and is called the neck, and is very evident in the seisonaceans. The head and the foot are usually retractable in the trunk (Wallace et al., 2015).

Regarding taxonomy and nomenclature, rotifers are the first animal group for which the official LAN (List of Available Names) has been established: https://www.iczn.org/list-of-available-names/rotifer-lan/ in 2018 (Segers et al., 2012).

The three groups recognized within the phylum Rotifera host about 2,000 known species (Segers, 2007): Bdelloidea with about 450 species in 19 genera and four families, Monogononta with about 1,600 species in 113 genera and 23 families, and Seisonacea with four species in two genera and one family.

The name Rotifera derives from the main distinctive feature of rotifers: the corona (**Figure 17.3**). The corona is formed by two bands of cilia (Melone, 1998), which in many species beat in a coordinated manner resulting in the optical appearance of the spokes of a turning wheel, hence the name "wheel bearers" (from the Latin *rota* = wheel and *fero* = to bear). The ciliary bands of the corona are extremely modified in certain genera with elongated cilia, presence of auricles, folds, reliefs, transverse ciliary fields, etc. Such modifications are related to the environment in which the species live, to their mode of movement and to the feeding type.

Another distinct feature of rotifers, visible internally through the transparent integument, is the masticatory apparatus, called **mastax**, with its muscles moving the hard jaws, called trophi (**Figure 17.4**). The trophi are formed by a series of hard extracellular elements, consisting of tubular sclerites (De Smet, 1998). The main elements of the trophi are: an unequal median fulcrum and three coupled elements called the rami, the unci and the manubira, connected by ligaments and moved by muscles that connect the different elements to each other and with the mastax wall. The shape and function of the trophi vary greatly between families and between the feeding styles of the different species (Figure 17.4).

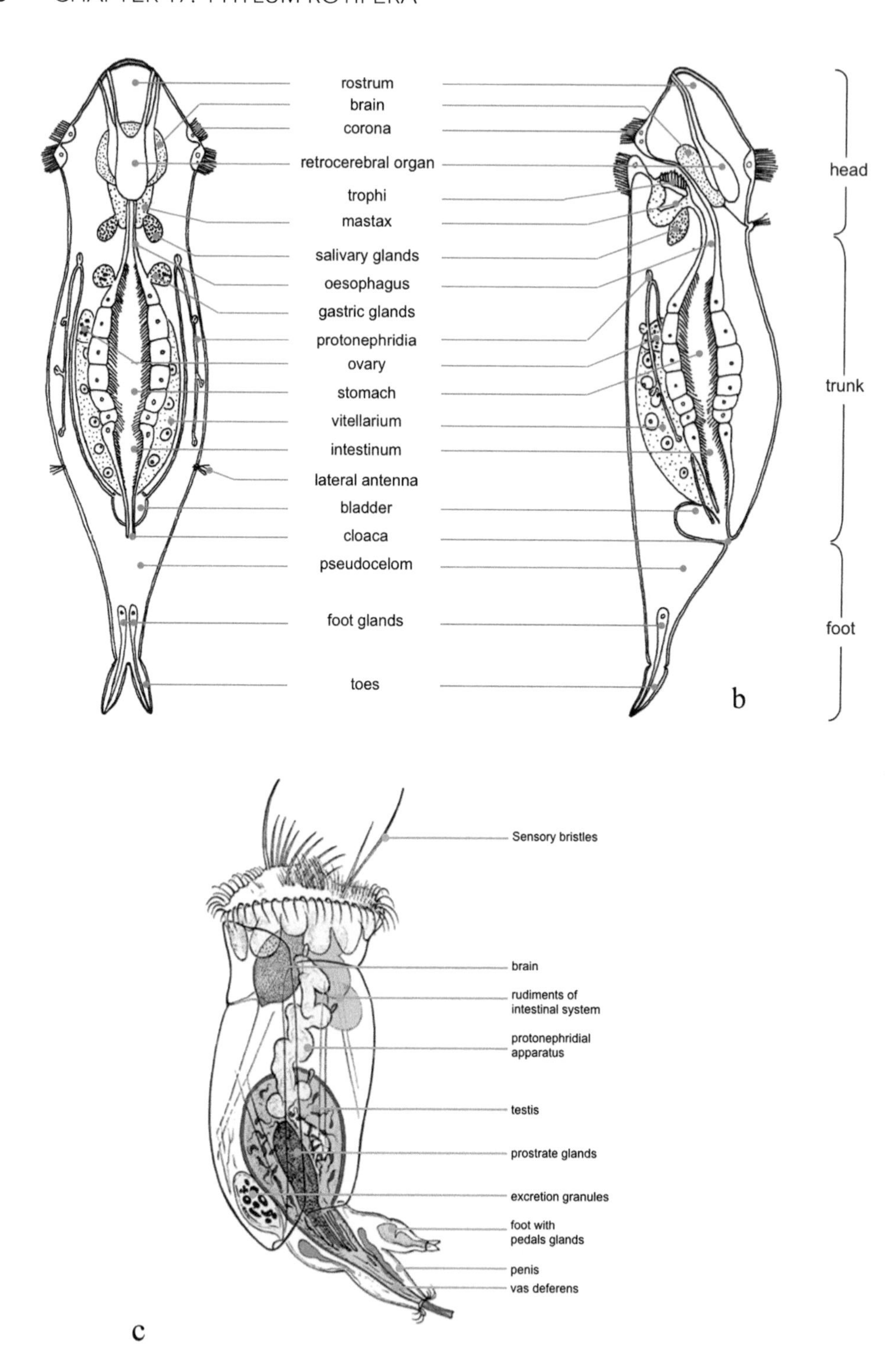

Figure 17.2 General body plan of a rotifer. Schematic drawings of a female in (a) dorsoventral and (b) lateral views, and of a male in lateral view (c). (Modified from Wallace et al., 2006; de Beauchamp ,1965.)

The **integument** is syncytial and, in addition to its function as an external protective coating, it has roles as hydraulic skeleton for the compression of internal fluids, as external skeleton for the attachment of muscles and as endocrine and exocrine system. The skeleton function is provided by a dense intracytoplasmic lamina, which is composed of keratin filamentous proteins, located just below the outer cell membrane of the syncytial integument, and is regularly perforated by pores (Ahlrichs, 1997). The secretions produced by the syncytial integument are expelled outward through secretory bulbs and pores

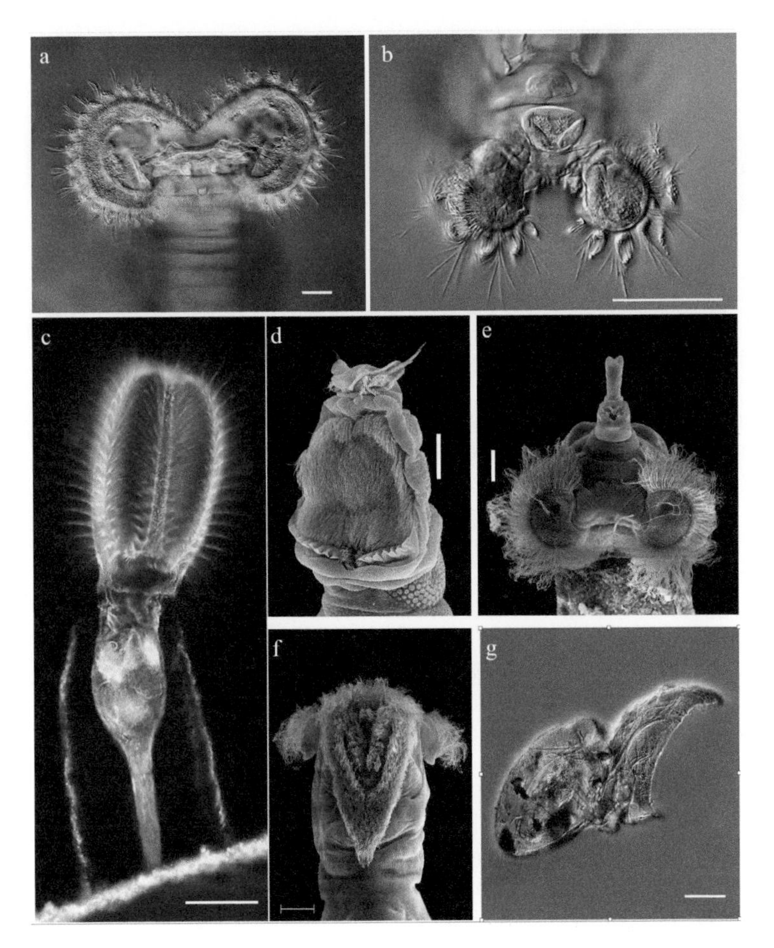

Figure 17.3 Corona. Light optical microscopy photographs of the corona of the monogonont *Limnias melicerta* (a) and the bdelloid *Pleuretra lineata* (b). Light microscopy photograph of the whole body of the monogononts *Stephanoceros fimbriatus* (c) and *Cupelopagis vorax* (g). Scanning electron microscopy photographs of the corona of the bdelloids *Adineta tuberculosa* (d) and *Dissotrocha aculeata* (e) and of the monogonont *Notommata* sp. (f). Rotifers or wheel animalcules got their name from the cilia of the corona, which in some species beat in such a metachronous way that the optical impression is of rotating gear wheels (a, b). While the corona of rotifers generally serves for food gathering and locomotion (a, b, d, e, f), it may also serve only for food gathering like a trap (c, g). Scale bars: a, b = 25 μm; c, g = 100 μm; d, e = 10 μm; f = 20 μm. (Light microcopy photographs: Michael Plewka; Scanning electron microscopy photographs: Diego Fontaneto and Giulio Melone.)

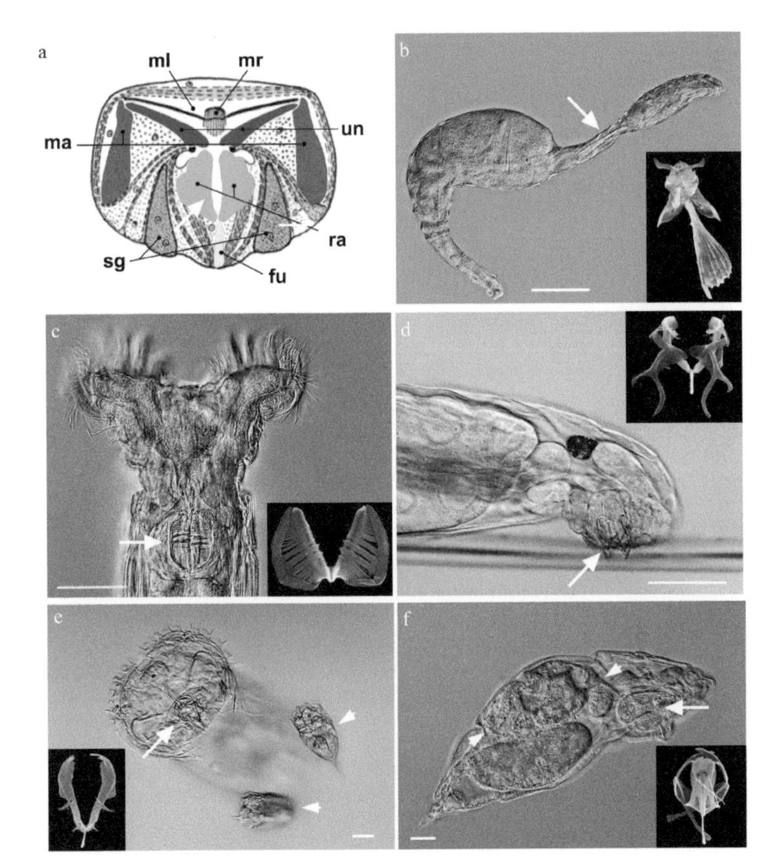

Figure 17.4 Mastax and trophi. General schematic drawing of the mastax (a). Abbrevitions are: fu: fulcrum; ma: manubrium; ml: mastax lumen; mr: mastax sensory receptor; ra: ramus; sg: salivary glands; un: uncus. The other images below show representative examples of species with some of the trophi types; the arrow marks the location of the trophi; an artificially colored SEM-image is also displayed to show the parts of the trophi in the same color scheme of the drawing in (a): fulcrate trophi of the seisonacean *Paraseison* sp. (b); ramate trophi of the bdelloid rotifer *Philodina* sp. (c); cardate trophi of the monogonont *Lindia* sp. (d) feeding on trichomes of Cyanobacteria; incudate trophi of the monogonont *Asplanchna* sp. (e) used to grab the prey (the monogonont *Keratella* sp., marked by arrowheads); forcipate trophi of the monogonont *Encentrum* sp. (f) used to pick up other rotifers (the monogonont *Cephalodella* sp., with head and toes marked by arrowheads). Scale bars: b = 100 μm; c, d, e, f = 25 μm. (Light microscopy photographs: Michael Plewka and Wilko Ahlrichs; Scanning electron microscopy photographs: Diego Fontaneto and Giulio Melone.)

that pass through the lamina. These secretions form an extracellular coating called glycocalyx, to which external particles can attach. The integument also has an endocrine function, discharging its secretion products into the liquid of the internal cavity through secretory granules.

The intracytoplasmic lamina also serves to fix the muscles of the body and the cutaneous-visceral muscles. Associated with the integument, there are several glands, including the pedal glands, the retrocerebral organ, and other glands present only in some groups of species. The foot contains the pedal glands with their reservoirs and ducts; part of these glands can also extend into the distal part of the trunk. They are unicellular or syncytial and polynucleate glands that secrete cement to temporarily or permanently attach the animal to the substrate (Wallace et al., 2006).

The retrocerebral organ is located dorsally to the brain and the mastax (Figure 17.2); it is involved in the production of various mucous secretions. An unpaired duct runs toward the corona and bifurcates anteriorly into two ducts. It is likely that this structure lubricates those cilia of the corona that are involved in movement on a substrate.

Many rotifers are able to secrete or construct sheaths or tubes of mucus around the body, adding also cemented debris, gelatinous material, pellets of bacteria and debris, fecal pellets and rigid material (**Figure 17.5a**).

Rotifers have smooth, crossed and oblique muscles (Kotikova et al., 2006). The common basic model consists of a system of external circular bands and internal longitudinal bands, coupled and bilaterally symmetrical (Hochberg and Ablak Gurbuz, 2008).

The digestive tract is similar in most rotifers and consists of a mouth, a buccal tube, a pharynx with a masticatory apparatus (mastax), an esophagus, a stomach, an intestine, a cloaca and the opening of the cloaca called the anus (Figure 17.2). In most species, the gastrointestinal system starts from the opening of the mouth and terminates at the anus as a relatively straight tube, which

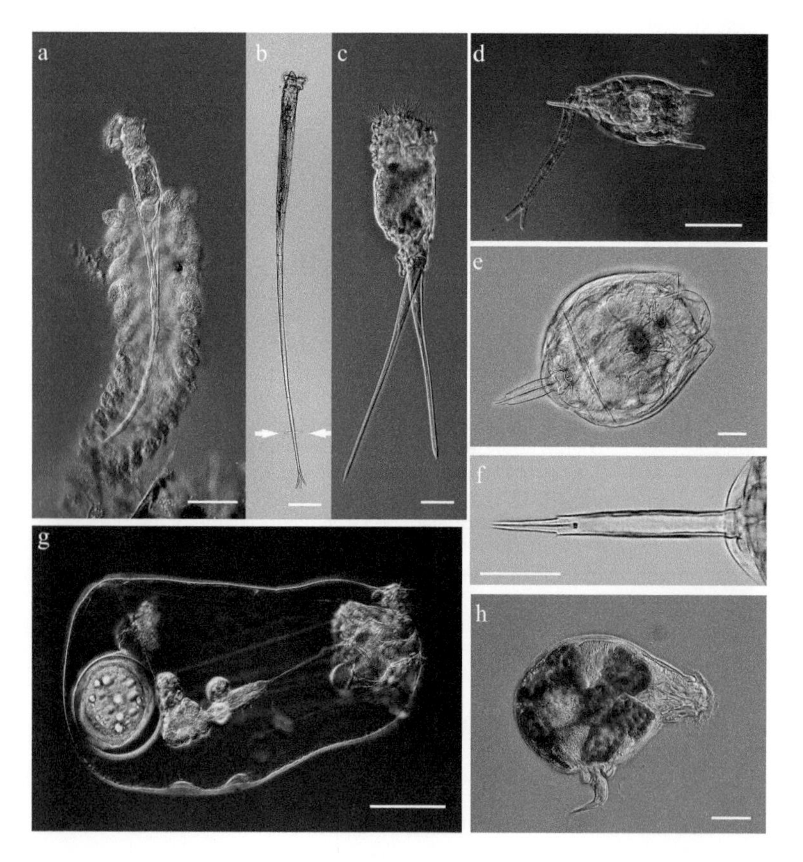

Figure 17.5 Foot. Light optical microscopy photographs of the monogononts *Ptygura pilula* (a), sessile and with long foot; *Monommata arndtii* (c), with short foot and long toes used for "jumping" in the water as an escape reaction; *Brachionus diversicornis* (d), with medium-size, very flexible foot; *Lecane luna* (e), with two separated toes; *Lecane bulla* (f), with almost completely fused toes; *Asplanchna* sp. (g), with totally reduced foot; *Gastropus minor* (h), with ventrally oriented foot; and the bdelloid *Rotaria neptunia* (b), with telescope-like contractable long foot with short toes (marked by arrowheads). Scale bars: a, b, d, g = 100 μm; c, e, f, h = 25 μm. (Photographs: Michael Plewka.)

can also be U-shaped; in some species, the intestine is blind-ended and has no anal opening.

The body cavity of rotifers is considered a pseudocelom, as it is not coated by an epithelium, but by an extracellular matrix. It is usually a spacious cavity, apparently devoid of fibrils or microfilaments. The pseudocelom often contains amoeboid cells, which, when present, form a highly dynamic three-dimensional polygonal network that facilitates internal particle transport. The pseudocelom works like a hydrostatic skeleton and presumably as a respiratory and circulatory system. The pseudocelomatic fluid is reintegrated from the water absorbed by the digestive tract and subsequently eliminated through the excretory system composed of protonephridia (Figure 17.2).

The organization of the nervous system is bilaterally symmetric, with a large cerebral ganglion, commonly called brain, ventral main nerves, caudal ganglia, secondary nerves and additional ganglia (Hochberg, 2009). The brain (Figure 17.2) is located behind the corona, dorsally to the mastax and surrounded by epithelial or muscle cells. The brain has various shapes and sizes: rounded, shaped like a bag, quadrangular, triangular, etc. The number of brain cells is constant for each species.

In bdelloids, all individuals are females. In monogononts, most of the individuals are females. Only under certain conditions (see reproduction) may males appear: they are generally smaller than the females (**Figure 17.6**), look different, and lack the trophi (Ricci and Melone, 1998). In seisonaceans, males and females are equally present and are not morphologically very different (Ricci et al., 1993).

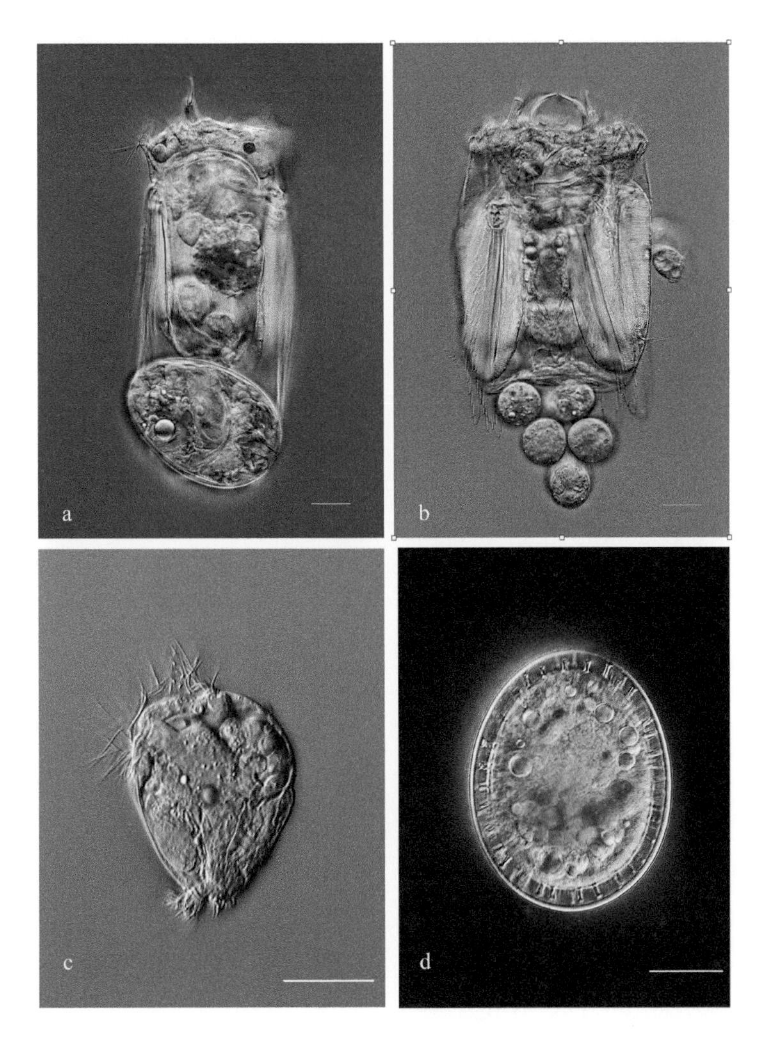

Figure 17.6 Reproduction in monogononts. Light optical microscopy photographs of *Polyarthra* sp.: female with amictic, diploid egg (a); female with unfertilized haploid eggs (b) from which males will hatch; haploid male (c); diploid resting egg (d). Scale bars = 25 µm. (Photographs: Michael Plewka.)

REPRODUCTION AND DEVELOPMENT

Rotifers have different types of reproduction (Gilbert, 1993). Seisonaceans have obligatory sexual reproduction, similar to what happens in most animals.

Bdelloids are obligate parthenogenetic animals and represent the most notorious animal group that has evolved and diversified in the absence of any known form of sexual reproduction (Mark Welch and Meselson, 2000). Because of their evolutionary success despite the absence of sexual reproduction, Maynard Smith in 1986 dubbed them as "something of an evolutionary scandal". Recent studies show that these animals are able to incorporate genes of other organisms into their genome, and about 10% of their genome is not of animal origin but derives from genes transferred horizontally from bacteria, algae, fungi, mosses, plants etc. (Gladyshev et al., 2008).

The reproductive cycle of monogononts is characterized by cyclic parthenogenesis: a combination of parthenogenetic and sexual reproduction (Hagiwara et al., 1995). The asexual females, called amictic (Figure 17.6a), produce female daughters through amictic or apomictic parthenogenesis, either by laying amictic eggs or by giving birth to fully developed daughters (Carmona et al., 1995). In response to certain environmental factors such as increase in population density (Stelzer and Snell, 2006), photoperiod, or chemical factors that signal the presence of predators or competitors in the environment (mixis stimulus), amictic females produce sexual daughters, called mictic females (Snell, 2011). Mictic females are morphologically and genetically indistinguishable from their mothers, but, through meiosis, they produce haploid eggs (Figure 17.6b), which are usually smaller than amictic eggs. These small haploid eggs, if not fertilized, develop into haploid males (Figure 17.6c). A male already present in the environment may fertilize the egg inside a mictic female. Fertilized eggs develop a very durable shell so that they are morphologically different from amictic eggs. Inside the shell, the embryo develops and arrests development at around 100 cells and becomes dormant. The dormant embryos are called resting eggs (Figure 17.6d) and can hatch even after months, years and over a century, giving rise to amictic females who restart the cycle (Gilbert, 1974).

Embryo development follows the three typical steps of segmentation, gastrulation and organogenesis (Gilbert, 1993). Rotifers have an embryonic development with deterministic segmentation following a modified spiral model. Cell fate is established very early in embryonic development. Moreover, rotifers are eutelic and cell division occurs only during embryogenesis, so that the newborn already has the same number of cells as the adult. Eggs are poor in yolk. During segmentation, subsequent divisions produce cells that are increasingly smaller and unequal in size and amount of cytoplasm, called blastomeres. The segmentation is holoblastic and the cell divisions are unequal. A typical 16-blastomere stage has four rows of four cells each, and gastrulation begins at this stage. Gastrulation occurs by epibolic movements and consequent involution; the resulting gastrula is a stereogastrula without a recognizable internal cavity. At the end of the gastrulation process the blastopore is formed by epithelial growth of the blastoderma coupled with its involution.

The mapping of cell lines has only been performed on a few species and the organogenesis is not well known. It seems that the stomodeum, the pharynx, the nervous system, the excretory system and the muscles come from the ectoderm; the reproductive system from the mesoderm; the digestive system from the endoderm.

DISTRIBUTION AND ECOLOGY

Rotifers are ubiquitous components of aquatic habitats: freshwater, limnoterrestrial and marine (**Figure 17.7**). More than 80% of the known species occur in continental environments (Fontaneto et al., 2006). Most species are free-living

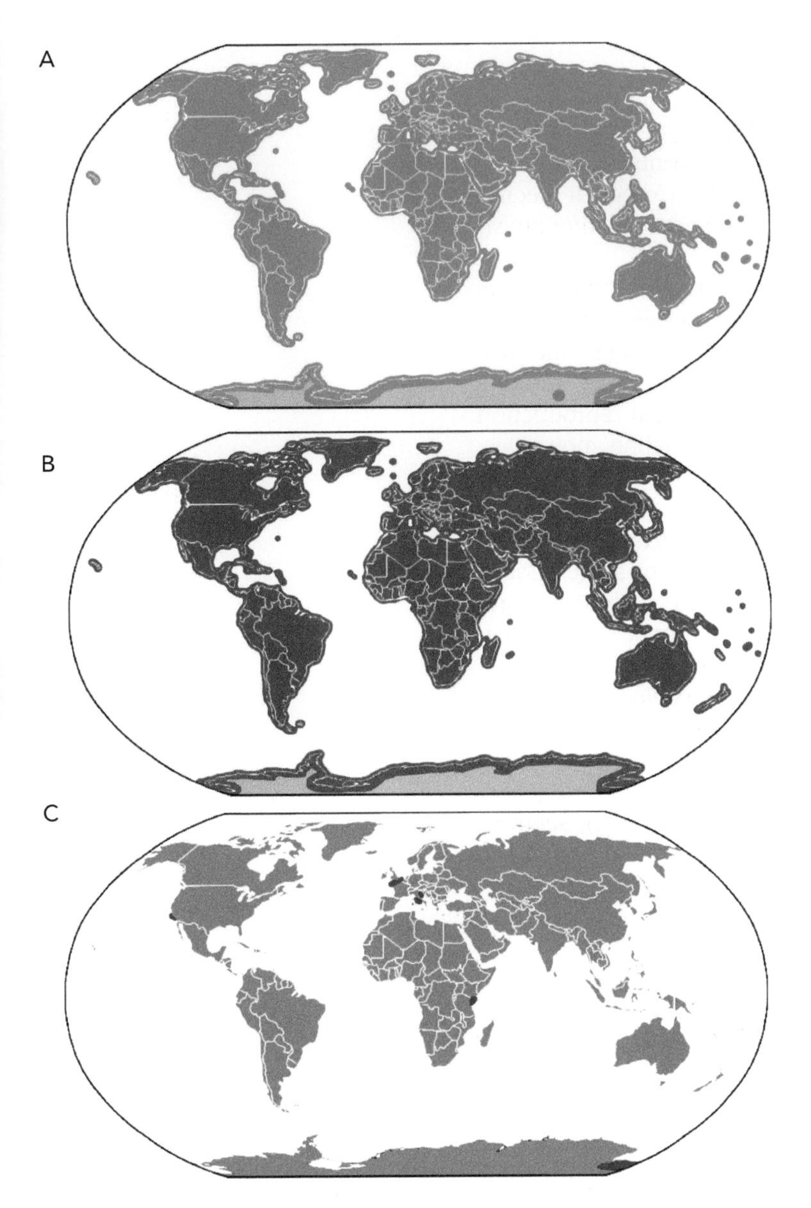

Figure 17.7 Distribution maps. Known and potential distribution of (A) Bdelloidea and (B) Monogononta; (C) known distribution of Seisonacea.

(Duggan, 2001), while others are sessile (Wallace, 1987), epibiotic, parasitic or symbiotic with algae or mosses (May, 1989). Species diversity of monogononts is lower in the polar regions (Green, 1972), whereas bdelloids are rarer in the subtropics and tropics (Fontaneto et al., 2015).

Due to the presence of easily dispersible resting stages, rotifers have the potential for cosmopolitan distribution. Information on their biogeography is scarce (Dumont, 1983). Yet, some species are known to have very limited distribution. Hotspots of diversity with high number of endemic species exist in Australia, China, North America, tropical South America and Antarctica, whereas few endemic species are present in Europe, Africa and the Indian subcontinent (Segers, 2008).

Biogeographical studies using phylogenetic information (= phylogeography) demonstrate that some species are indeed widely distributed, but they still experience the constraints of geography in their distribution (Gómez, 2005). Thus, the occurrence of refugia is present, biogeographical patterns exist, with evidence of enclave distribution, founder events and localized genetic differentiation (Gómez et al., 2007).

PHYSIOLOGY AND BEHAVIOR

The most studied physiological aspects of rotifers relate to the triggers of the switch between sexual and asexual reproduction in monogononts, dormancy in bdelloids, hormone-induced phenotypic plasticity and ecotoxicological mechanisms (Dahms et al., 2011). Very little is known on other aspects of rotifer physiology (Gilbert, 2017).

No rotifer hormones have been described, even given the fact that rotifers respond to hormones and neurotransmitters from vertebrates and insects. Moreover, rotifers indeed produce growth-promoting substances and use chemical signals as pheromones to regulate their reproduction (Rico-Martínez and Snell, 1996). The mechanisms that switch asexual to sexual reproduction in monogononts are controlled by the accumulation of signaling molecules, regulating gene expression in response to fluctuations in population density (Sarma et al., 2005).

Rotifers, similar to other microscopic aquatic animals such as tardigrades and nematodes, are able to survive desiccation by entering dormant stages (Ricci, 2001). Bdelloids can enter dormancy at any stage of their life cycle (Ricci and Caprioli, 2005), monogononts produce resting eggs (García-Roger et al., 2006), whereas seisonaceans do not have any dormant capability. The biochemical mechanisms used to survive during dormancy involve molecules that protect the intracellular and extracellular environment against damages induced by the lack of water: trehalose and sucrose, heat-shock proteins, late embryogenesis abundant proteins (LEA), chaperones, antioxidants, and others. Intriguingly, even foreign genes, acquired through horizontal gene transfer (HGT), seem to be involved in the desiccation response of bdelloids (Tunnacliffe and Lapinski, 2003). During the recovery phase from desiccation and after exposure to ionizing radiation, bdelloids are able to repair their broken chromosomes with a still unknown mechanism (Gladyshev and Meselson, 2008).

Rotifers are used in the risk assessment of pharmaceuticals, endocrine disruptors, and heavy metal pollution (Snell and Janssen, 1995), and in the effect of aging (King, 1969) through whole-animal bioassays and gene expression studies, especially using the laboratory models of the genus *Brachionus*.

GENETICS AND GENOMICS

The first sequenced genomes of bdelloid rotifers confirmed the high level of horizontally transferred genes (Flot et al., 2013), much higher than in other known eukaryotes (Boschetti et al., 2011, 2012), but the signature of asexuality was not clearly supported (Vakhrusheva et al., 2020). Only one genome is available for monogononts at the moment. Several more are on the way and projects on population genomics may shed light on diversification processes and adaptations in the group.

The reconstruction of mitochondrial genomes demonstrated that monogononts have two circular rings and not the usual single one (Suga et al., 2008).

Other studies using several sequences from single or few genetic markers (Gómez et al., 2002) revealed an extensive occurrence of cryptic species (Mills et al., 2017).

POSITION IN THE ToL

The homology between the hard pharyngeal parts of the masticatory apparatus of Gnathostomulida, Micrognathozoa and Rotifera supports a close relationship between these taxa within Spiralia (Laumer et al., 2015). The presence of an intracytoplasmic lamina also associates the rotifera with the Acanthocephala. These four taxa are usually united in the Gnathifera group.

Molecular phylogeny analyses clearly support the existence of Gnathiferans as a real group, while the relationships between the different taxa are not yet resolved. The consensus in the current state is that Gnathostomulida represent the sister taxon of the others, with Micrognathozoa the sister taxon of Syndermata (= Rotifera + Acanthocephala). In this chapter, we have kept Rotifera separated from Acanthocepahala only because they are morphologically and ecologically very different, notwithstanding their phylogenetic proximity.

DATABASES AND COLLECTIONS

The most relevant online reference is the Rotifer World Catalog (http://rotifera.hausdernatur.at, Jersabek and Leitner, 2013). This web page hosts photographs of the slides of the most important collection of rotifers, from the Academy of Natural Sciences of Drexel University, Philadelphia, USA. Live photographs of several species are present in Plingfactory, Life in Water (http://www.plingfactory.de/Science/Atlas/KennkartenTiere/Rotifers/01RotEng/E-TL/TL5Rotifera.html).

A biogeographical database for monogononts was published with all records from 1960 to 1992, and ongoing projects are trying to expand it (Fontaneto et al., 2012).

A database of all valid names is available on the web page of the International Code of Zoological Nomenclature (http://iczn.org) (Segers et al., 2012).

CONCLUSION

Rotifers are beautiful exquisite metazoans, common and abundant everywhere, and amenable for laboratory experiments and tests on different fundamental questions in biology (Declerck and Papakostas, 2017). Moreover, their peculiar features of desiccation resistance, obligate parthenogenesis, horizontally transferred genes etc make them excellent models to study the generality of biological processes (Fussmann, 2011), with applications in aquaculture (Lubzens, 1987).

REFERENCES

Ahlrichs W. H. 1997. Epidermal ultrastructure of *Seison nebaliae* and *Seison annulatus*, and a comparison of epidermal structures within the Gnathifera. Zoomorphology 117: 41–48.

Boschetti C., Pouchkina-Stantcheva N., Hoffmann P., Tunnacliffe A. 2011. Foreign genes and novel hydrophilic protein genes participate in the desiccation response of the bdelloid rotifer *Adineta ricciae*. Journal of Experimental Biology 214: 59–68.

Boschetti C., Carr A., Crisp A., Eyres I., Wang-Koh Y., Lubzens E., Barraclough T. G., Micklem G., Tunnacliffe A. 2012. Biochemical diversification through foreign gene expression in bdelloid rotifers. PLoS Genetics 8: e1003035.

Carmona M. J., Gómez A., Serra M. 1995. Mictic patterns of the rotifer *Brachionus plicatilis* Müller in small ponds. Hydrobiologia 313/314: 365–371.

Clément P. 1993. The phylogeny of rotifers: molecular, ultrastructural and behavioural data. Hydrobiologia 255/256: 527–544.

Dahms H. U., Hagiwara A., Lee J.-S. 2011. Ecotoxicology, ecophysiology, and mechanistic studies with rotifers. Aquatic Toxicology 101: 1–12.

de Beauchamp P. 1965. "Classe des rotifères". in: Traité de zoologie, T. 4, 3: Némathelmintes (Nématodes, Gordiacés). Grassé, P. P. (ed.). pp. 1225–1379, Rotifères, Gastrotriches, Kinorhynques, Masson et Cie, Paris.

Declerck S. A., Papakostas S. 2017. Monogonont rotifers as model systems for the study of micro-evolutionary adaptation and its eco-evolutionary implications. Hydrobiologia 796(1): 131–144.

De Smet W. H. 1998. Preparation of rotifer trophi for light and scanning electron microscopy. Hydrobiologia 387/388: 117–121.

Donner J. 1965. Ordnung Bdelloidea (Rotatoria, Rädertiere). Bestimmungsbüch. Bodefauna Eur. 6:1–267. Akademie Verlag, Berlin.

Duggan I. C. 2001. The ecology of periphytic rotifers. Hydrobiologia 446/447: 139–148.

Dumont H. J. 1983. Biogeography of rotifers. Hydrobiologia 104: 19–30.

Ferraguti M., Melone G. 1999. Spermiogenesis in *Seison nebaliae* (Rotifera, Seisonidae): further evidence of a rotifer-acanthocephalan relationship. Tissue Cell 31: 428–440.

Flot J. F., Hespeels B., Li X., Noel B., Arkhipova I., Danchin E. G., Hejnol A., Henrissat B., Koszul R., Aury J. M., Barbe V., et al. 2013. Genomic evidence for ameiotic evolution in the bdelloid rotifer *Adineta vaga*. Nature 500(7463): 453–457.

Fontaneto D., De Smet W. H. 2015. "Rotifera". in: Handbook of Zoology: Gastrotricha, Cycloneuralia and Gnathifera, Schmidt-Rhaesa, A. (ed.), pp. 217–300, Walter de Gruyter.

Fontaneto D., De Smet W. H., Ricci C. 2006. Rotifers in saltwater environments, re-evaluation of an inconspicuous taxon. Journal of the Marine Biological Association UK 86: 623–656.

Fontaneto D., Iakovenko N., De Smet W. H. 2015. Diversity gradients of rotifer species richness in Antarctica. Hydrobiologia 761(1): 235–248.

Fontaneto D., Barbosa A. M., Segers H., Pautasso M. 2012. The 'rotiferologist' effect and other global correlates of species richness in monogonont rotifers. Ecography 35: 174–182.

Fussmann G. F. 2011. Rotifers: excellent subjects for the study of macro- and microevolutionary change. Hydrobiologia 662: 11–18.

García-Roger E. M., Carmona M. J., Serra M. 2006. Patterns in rotifer diapausing egg banks: density and viability. Journal of Experimental Marine Biology and Ecology 336: 198–210.

Gilbert J. J. 1974. Dormancy in rotifers. Transactions of the American Microscopical Society 93: 490–513.

Gilbert J. J. 1993. Rotifera. in: Reproductive Biology of Invertebrates, Vol. VI, Part A. Jamieson, B. M. G. (ed.), pp. 231–263, Oxford & IBH Publishing, New Dehli.

Gilbert J. J. 2017. Non-genetic polymorphisms in rotifers: environmental and endogenous controls, development, and features for predictable or unpredictable environments. Biological Review 92: 964–992.

Gladyshev E., Meselson M. 2008. Extreme resistance of bdelloid rotifers to ionizing radiation. Proceedings of the National Academy of Sciences USA 105: 5139–5144.

Gladyshev E. A., Meselson M., Arkhipova I. R. 2008. Massive horizontal gene transfer in bdelloid rotifers. Science 320: 1210–1213.

Gómez A. 2005. Molecular ecology of rotifers: from population differentiation to speciation. Hydrobiologia 546: 83–99.

Gómez A., Serra M., Carvalho G. R., Lunt D. H. 2002. Speciation in ancient cryptic species complexes: evidence from the molecular phylogeny of *Brachionus plicatilis* (Rotifera). Evolution 56: 1431–1444.

Gómez A., Montero-Pau J., Lunt D. H., Serra M., Campillo S. 2007. Persistent genetic signatures of colonization in *Brachionus manjavacas* rotifers in the Iberian Peninsula. Molecular Ecology 16: 3228–3240.

Green J. J. 1972. Latitudinal variation in associations of planktonic Rotifera. Journal of Zoology 167: 31–39.

Hagiwara A., Kotani T., Snell T. W., Assava-Aree M., Hirayama K. 1995. Morphology, reproduction, genetics, and mating behavior of small, tropical marine *Brachionus* strains (Rotifera). Journal of Experimental Marine Biology and Ecology 194: 25–37.

Hochberg R. 2009. Three-dimensional reconstruction and neural map of the serotonergic brain of *Asplanchna brightwelii* (Rotifera, Monogononta). Journal of Morphology 270: 430–441.

Hochberg R., Ablak Gurbuz O. A. 2008. Comparative morphology of the somatic musculature in species of *Hexarthra* and *Polyarthra* (Rotifera: Monogononta): its function in appendage movement and escape behavior. Zoologischer Anzeiger 247: 233–248.

Jersabek C. D., Leitner M. F. 2013. The Rotifer World Catalog. World Wide Web electronic publication. http://www.rotifera.hausdernatur.at/

King C. E. 1969. Experimental studies of aging in rotifers. Experimental Gerontology 4: 63–79.

Koste W. 1978. Rotatoria. Die Rädertiere Mitteleuropas. Ein Bestimmungswerk, begründet von Max Voigt. Überordnung Monogononta. 1. Textband, 673 pp., 2. Tafelband, 234 pls. Gebrüder Borntraeger, Berlin.

Kotikova E. A., Raikova O. I., Flyatchinskaya L. P. 2006. Study of architectonics of rotifer musculature by the method of fluorescence with use of confocal microscopy. Journal of Evolutionary Biochemistry and Physiology 42: 89–97.

Kutikova L. A. 2003. Bdelloid rotifers (Rotifera, Bdelloidea) as a component of soil and land biocenoses. Biology Bulletin 30: 271–274.

Laumer C. E., Bekkouche N., Kerbl A., Goetz F., Neves R. C., Sørensen M. V., Kristensen R. M., Hejnol A., Dunn C. W., Giribet G., Worsaae K. 2015. Spiralian phylogeny informs the evolution of microscopic lineages. Current Biology 25: 2000–2006.

Lubzens E. 1987. Raising rotifers for use in aquaculture. Hydrobiologia 147: 45–255.

Mark Welch D. B., Meselson M. 2000. Evidence for the evolution of bdelloid rotifers without sexual reproduction and genetic exchange. Science 288: 1211–1215.

May L. 1989. Epizoic and parasitic rotifers. Hydrobiologia 186/187: 59–67.

Melone G. 1998. The rotifer corona by SEM. Hydrobiologia 387/388: 131–134.

Mills S., Alcántara-Rodríguez J. A., Ciros-Pérez J., Gómez A., Hagiwara A., Galindo K. H., Jersabek C. D., Malekzadeh-Viayeh R., Leasi F., Lee J. S., Welch D. B. et al., 2017. Fifteen species in one: deciphering the *Brachionus plicatilis* species complex (Rotifera, Monogononta) through DNA taxonomy. Hydrobiologia 796: 39–58.

Ricci C. 2001. Dormancy patterns in rotifers. Hydrobiologia 446/447: 1–11.

Ricci C., Melone G. 1998. Dwarf males in monogonont rotifers. Aquatic Ecology 32: 361–365.

Ricci C., Caprioli M. 2005. Anhydrobiosis in bdelloid species, populations and individuals. Integrative and Comparative Biology 45: 759–763.

Ricci C., Melone G., Sotgia C. 1993. Old and new data on Seisonidea (Rotifera). Hydrobiologia 255/256: 495–511.

Rico-Martínez R., Snell T. W. 1996. Mating behavior and mate recognition pheromone blocking of male receptors in *Brachionus plicatilis* Müller (Rotifera). Hydrobiologia 313/314: 105–110.

Sarma S. S. S., Gulati R. D., Nandini S. 2005. Factors affecting egg-ratio in planktonic rotifers. Hydrobiologia 546: 361–373.

Segers H. 2007. Annotated checklist of the rotifers (Phylum Rotifera), with notes on nomenclature, taxonomy and distribution. Zootaxa 1564: 1–104.

Segers H. 2008. Global diversity of rotifers (Rotifera) in freshwater. Hydrobiologia 595:49–59.

Segers H., De Smet W. H., Fisher C., Fontaneto D., Michaloudi E., Wallace R. L., Jersabek C. D. 2012. Towards a list of available names in zoology, partim Phylum Rotifera. Zootaxa 3179: 61–68.

Sielaff M., Schmidt H., Struck T. H., Rosenkranz D., Mark Welch D. B., Hankeln T., Herlyn H. 2016. Phylogeny of Syndermata (syn. Rotifera): Mitochondrial gene order verifies epizoic Seisonidea as sister to endoparasitic Acanthocephala within monophyletic Hemirotifera. Molecular Phylogenetics and Evolution 96: 79–92.

Snell T. W. 2011. A review of the molecular mechanisms of monogonont rotifer reproduction. Hydrobiologia 662: 89–97.

Snell T. W., Janssen C. R. 1995. Rotifers in ecotoxicology: a review. Hydrobiologia 313–314: 231–247.

Sørensen M. V., Giribet G. 2006. A modern approach to rotiferan phylogeny: combining morphological and molecular data. Molecular Phylogenetics and Evolution 40: 585–608.

Stelzer C. P., Snell T. W. 2006. Specificity of the crowding response in the *Brachionus plicatilis* species complex (Rotifera). Limnology and Oceanography 51: 125–130.

Suga K., Mark Welch D. B., Tanaka Y., Sakakura Y., Hagiwara A. 2008. Two circular chromosomes of unequal copy number make up the mitochondrial genome of the rotifer *Brachionus plicatilis*. Molecular Biology and Evolution 25: 1129–1137.

Tunnacliffe A., Lapinski J. 2003. Resurrecting Van Leeuwenhoek's rotifers: a reappraisal of the role of disaccharides in anhydrobiosis. Philosophical Transactions of the Royal Society London B 358: 1755–1771.

Vakhrusheva O. A., Mnatsakanova E. A., Galimov Y. R., Neretina T. V., Gerasimov E. S., Naumenko S. A., Ozerova S. G., Zalevsky A. O., Yushenova I. A., Rodriguez F., Arkhipova I. R. et al., 2020. Genomic signatures of recombination in a natural population of the bdelloid rotifer *Adineta vaga*. Nature Communications 11: 6421.

Wallace R. L. 1987. Coloniality in the phylum Rotifera. Hydrobiologia 147: 141–155.

Wallace R. L., Snell T. W., Ricci C., Nogrady T. 2006. "Rotifera. Vol. 1. Biology, ecology and systematics, 2nd edition". in: Guides to the Identification of the Microinvertebrates of the Continental Waters of the World, Vol. 23. Dumont, H. J. F. (ed.), pp. 1–299, Kenobi Productions, Ghen.

Wallace R. L., Snell T., Smith H. A. 2015. "Rotifer: ecology and general biology". in: Thorp and Covich's Freshwater Invertebrates, Thorp JH, Rogers DC. (eds.), pp. 225–271, Elsevier, Waltham.

PHYLA GNATHOSTOMULIDA, MICROGNATHOZOA, AND CYCLIOPHORA

Martin V. Sørensen

CHAPTER

18

Infrakingdom	Bilateria
Division	Protostomia
Subdivision	Lophotrochozoa
Phylum	Gnathostomulida, Micrognathozoa and Cycliophora

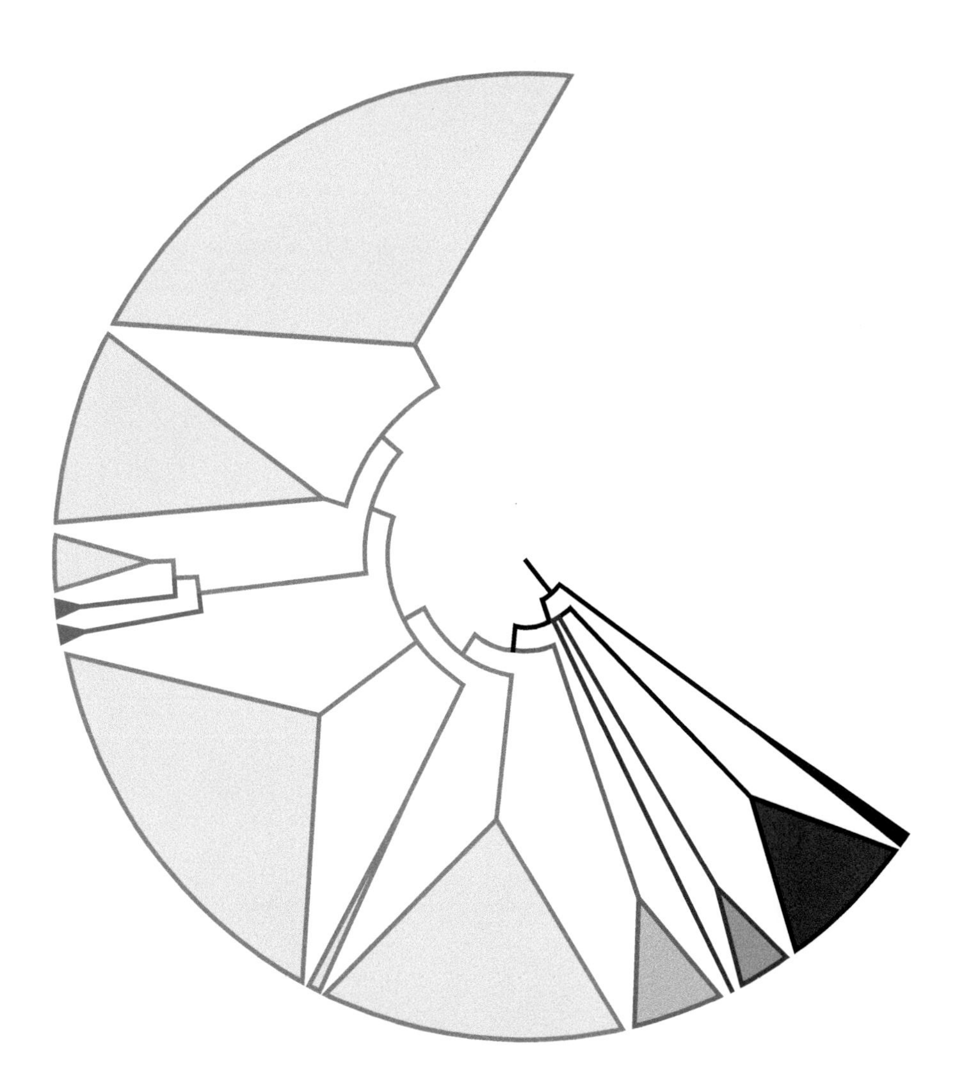

CONTENTS

GNATHOSTOMULIDA

General Characteristics

About 100 valid species

Microscopic worms with complex jaw apparatuses

Exclusively marine animals, found in shallow, sandy, detritus-rich habitats

Hermaphrodites

Synapomorphies

- monociliated epithelial cells
- complex paired jaws with unpaired basal plate

Closely related to the other gnathiferan taxa: Rotifera and Micrognathozoa
No commercial value
Ecological importance is probably small

HISTORY OF TAXONOMY AND CLASSIFICATION

The first gnathostomulid was collected in 1928 in the Baltic Sea at Kiel Bight, by the German meiofauna zoologist Adolf Remane. Remane suggested his Austrian colleague Josef Meixner to take a closer look at these intriguing worms, but the outbreak of World War II put an end to Meixner's studies. When Meixner passed away shortly after the war, all that was left of his work was some unpublished manuscript notes. These notes were passed to another Remane student, Peter Ax, who decided to continue where Meixner had stopped. Ax found the worms again at Kieler Bight, and also discovered them in large numbers at the north-west German Wadden Sea island Sylt, and near Banyuls on the French Mediterranean coast. Finally, in 1956 Ax presented the two first gnathostomulid species descriptions, *Gnathostomula paradoxa* and *Gnathostomaria lutheri*, but he considered them both to be flatworms and assigned them to a new platyhelminth order named Gnathostomulida (Ax, 1956). Thirteen years later, Riedl (1969) showed convincingly that gnathostomulids, despite their superficial resemblance to flatworms, should be considered a separate branch in the Tree of Life, and assigned the group phylum rank with uncertain phylogenetic affinities.

Since Ax's first descriptions, only a small number of scientists have worked with gnathostomulids, but the major contribution to our present-day knowledge of gnathostomulid systematics, biodiversity and biology comes from the nearly lifelong dedicated studies of Wolfgang Sterrer. Sterrer initiated his studies in the early 1960s (Sterrer, 1965, 1969), and found new species that differed so considerably from the gnathostomulids discovered by Ax and Remane that these taxa were later assigned to different orders and suborders of the phylum. In 1972 he published the most significant gnathostomulid contribution made so far, and established the framework for present-day gnathostomulid systematics (Sterrer, 1972). During the following decades he explored gnathostomulid biodiversity throughout the world, established 18 (out of 26) currently known genera, and described 72 (out of 101) species.

ANATOMY, HISTOLOGY AND MORPHOLOGY

Gnathostomulids are all worm-shaped animals, ranging in length from 270 μm to more than 3.5 mm in a few 'gigantic' species (**Figures 18.1** and **18.2**). The majority of the species have a body length between 500 μm and 1 mm. The largest species belong to the order Filospermoidea, and have a head with a long, pointed rostrum but otherwise a rather uniform and slender trunk (Figures 18.1A and 18.2A). The slightly shorter Bursovaginoidea have a shorter rostrum, and often a neck constriction and a tapered tail (Figures 18.1B, 18.2 B, C).

All gnathostomulids have a single-layered epidermis composed of monociliated cells. Locomotion is created by the beating of epidermal cilia, which gives the live animal a slow, continuous and gliding movement through the interstices in their sandy habitat. The musculature of the body is not directly involved in the locomotion of the animal but is used to turn and twist the body. All gnathostomulids have fairly strong, cross-striated longitudinal muscles that run through the entire trunk of the animals. Circular muscles are also found in many species, but they tend to be thinner, and are often strongest and more numerous in the posterior half of the trunk. Besides the trunk musculature,

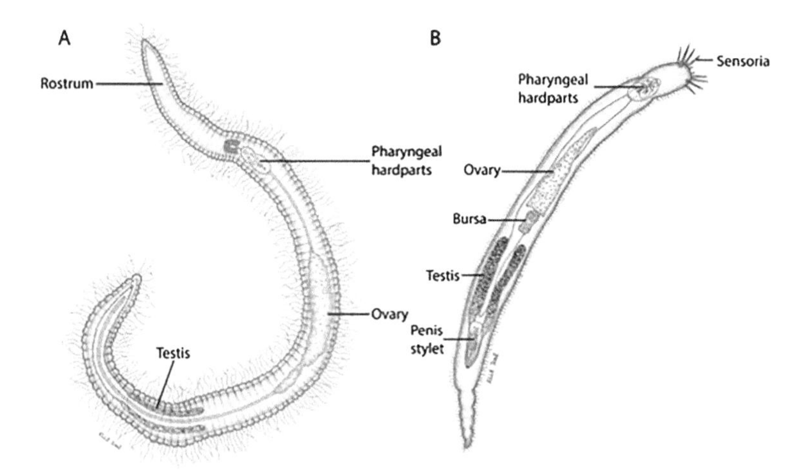

Figure 18.1 Two gnathostomulid species: (A) *Haplognathia simplex* (order: Filospermoidea) and (B) *Gnathostomula paradoxa* (order: Bursovaginoidea).

gnathostomulids have a very complex musculature associated with their jaw apparatuses.

The nervous system is intraepithelial and consists of a frontal ganglion (the brain), and a pair of ventral nerve cords that extend through the trunk and end in a terminal commissure. Additional commissures may be present in association with the pharynx and genital organs. All gnathostomulids have a single row of short, stiff, sensorial cephalic cilia. Other external sensory organs appear to be present in Bursovaginoidea only. They are restricted to the head region where they form bundles of stiff cilia, called sensoria. Other sensory organs include the depressed ciliary pits.

Gnathostomulids have a ventral mouth that leads through a pharynx to a blind gut. Studies have shown that some species might be able to form a temporary anal pore (Knauss, 1979), but generally most gnathostomulids appear to have a blind gut – a trait they share with free-living flatworms. The most prominent internal structure of gnathostomulids is found inside their muscular pharynx, where the pharyngeal hard parts are located. The hard parts form a set of pincer-like jaws, often accompanied by an unpaired basal plate located anterior to the jaws (**Figures 18.3** and **18.4**). The basal plate and movable parts of the jaws are all interconnected by minute muscles that all together make up the jaw apparatus. The morphology of jaws and basal plate varies greatly across the gnathostomulids (Sørensen and Sterrer, 2002), and is often the main key to recognizing species, genera and families.

The excretory system is composed of a series of protonephridia that individually open through nephridiopore cells in the epidermis. The terminal cells always have a single cilium only, whereas no cilia appear to be present in the canal cell.

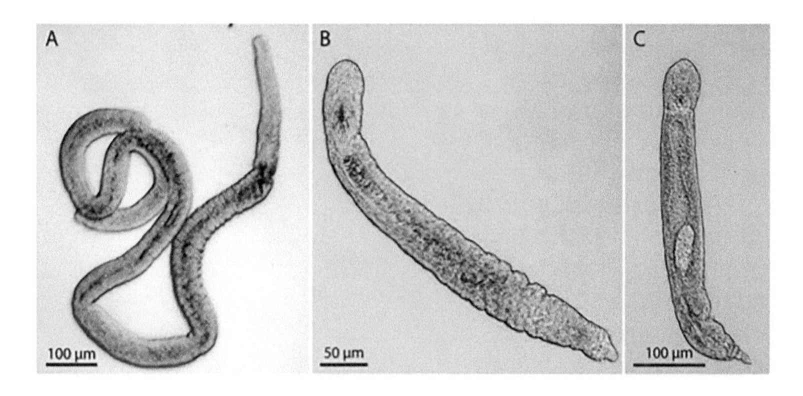

Figure 18.2 Light microscopical images gnathostomulids: (A) *Haplognathia* (order: Filospermoidea). (B) *Austrognathia* (suborder: Conophoralia). (C) *Gnathostomula* (suborder: Scleroperalia).

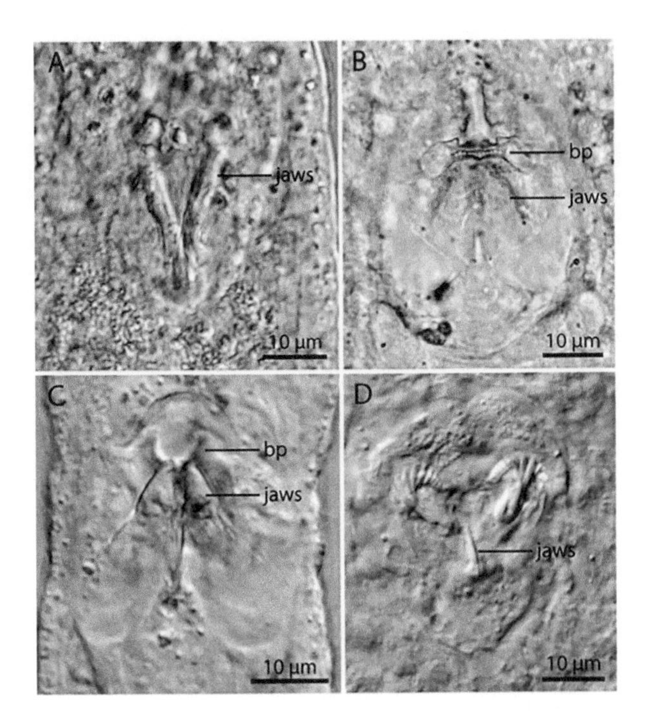

Figure 18.3 Light microscopical images showing gnathostomulid jaws and basal plates (bp): (A) Jaws in *Haplognathia* (order: Filospermoidea). (B) Jaws and basal plate in *Austrognathia* (suborder: Conophoralia). (C) Jaws and basal plate in *Gnathostomula* (suborder: Scleroperalia). (D) Jaws in *Valvognathia* (Suborder: Scleroperalia).

REPRODUCTION AND DEVELOPMENT

All gnathostomulids are hermaphrodites, and reproduction is always sexual. The worms only develop one fertilized egg at a time, and press it through a crack that appears in the body wall. The sticky egg will adhere to sediment grains, and a juvenile worm hatches after a few days. Development after hatching is direct, and the juvenile will increase in size and eventually reach sexual maturity.

Reproductive Organs and Spermatozoa

The morphology of the reproductive organs and spermatozoa differ between the two gnathostomulid orders, Filospermoidea and Bursovaginoidea, and between the bursovaginoid suborders Scleroperalia and Conophoralia.

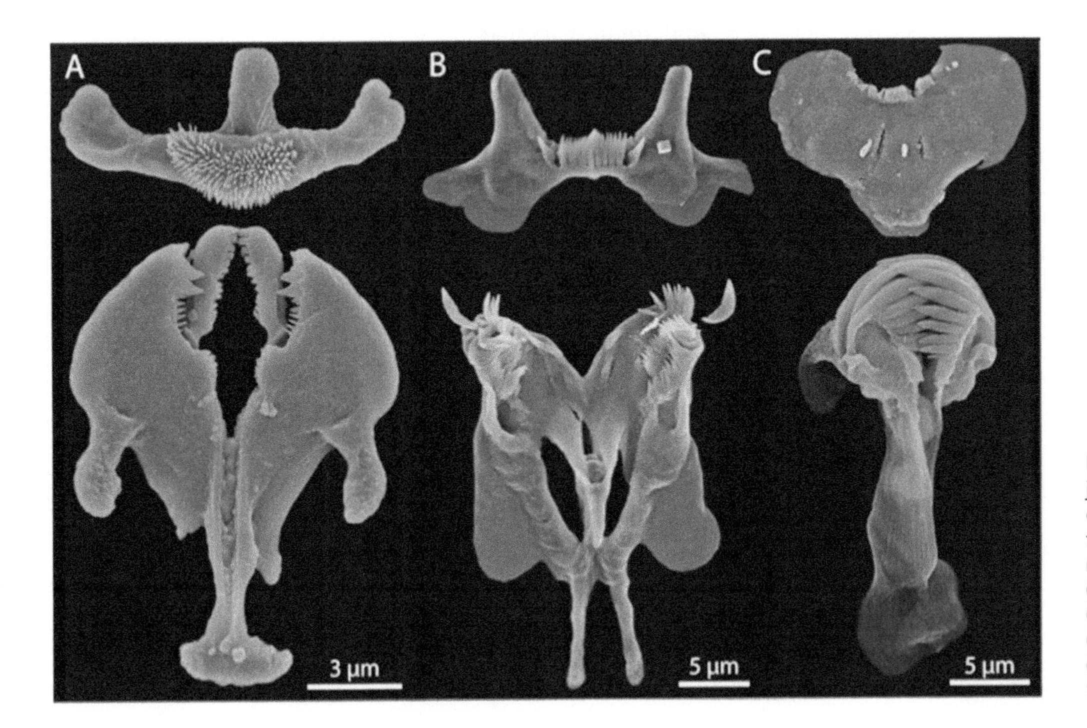

Figure 18.4 Gnathostomulid jaws and basal plates dissected and prepared for scanning electron microscopy: (A) *Cosmognathia* (order: Filospermoidea). (B) *Gnathostomula* (suborder: Scleroperalia). (C) *Valvognathia* (suborder: Scleroperalia).

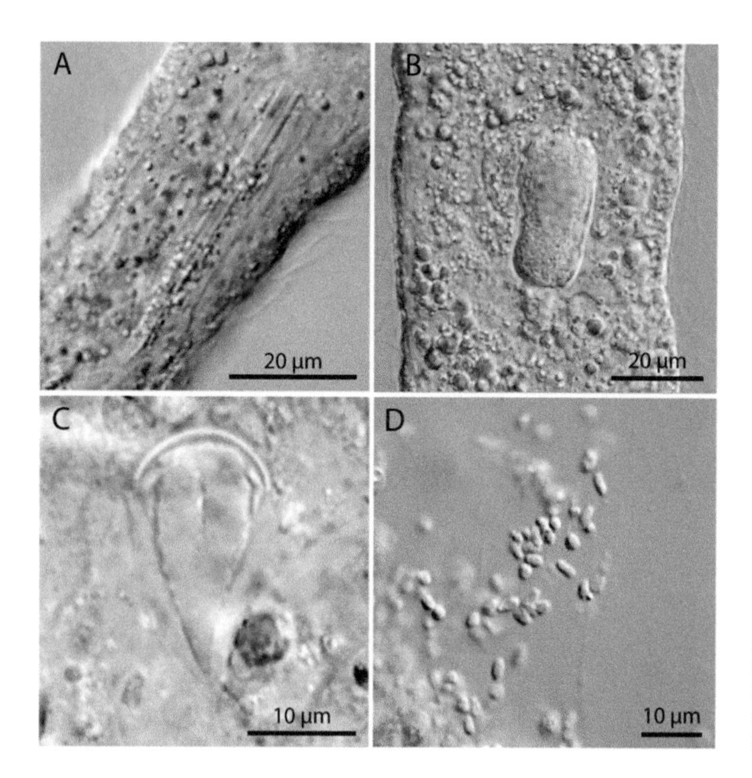

Figure 18.5 Light microscopical images gnathostomulid reproductive organs and spermatozoa: (A) Penis stylet and (B) Bursa in *Gnathostomula* (suborder: Scleroperalia). (C) Giant spermatozoa – conulus in *Austrognathia* (suborder: Conophoralia). (D) Dwarf sperm in *Valvognathia* (suborder: Scleroperalia).

The filospermoid worms have an unpaired ovary, but no vaginal opening, and paired or unpaired testes. The penis is soft and glandular, and not injectory. Spermatozoa are long and filiform (hence the name of the order) with pointed and corkscrew-shaped heads, and a terminal flagellum. It is hypothesized that the spermatozoa during mating are released directly on the integument of the partner, and actively drill their way through its epidermis until they reach the ovary and mature egg.

Species of Bursovaginoidea are characterized by the presence of a bursa – a special sac-like organ, used for sperm storage (Figure 18.5B). The bursa is located immediately behind the mature egg. The bursovaginoid suborder Scleroperalia has species with a rigid, sclerotized bursa, and an even harder, injectory penis stylet (**Figure 18.5A**). The spermatozoa are aflagellate and minute (hence referred to as dwarf sperm) (Figure 18.5D), and tightly packed in a testis follicle formed in the paired testes. The penis stylet injects the whole follicle through the partner's epidermis (or vagina if present), and after injection it merges with the bursa, if a bursa is present already. Alternatively, it forms a new bursa. The dwarf sperm inside the bursa can subsequently move through the bursa mouthpiece and fertilize the egg. Opposite to the scleroperalian dwarf sperm, the other bursovaginoid suborder, Conophoralia, has a few gigantic sperm cells called conuli (Figure 18.5C). A conulus is cone- (or droplet-) shaped and may measure up to 75 µm in length. Conophoralians have a single testis only, and the penile structure is softer than in scleroperalians, and without a penis stylet. Copulation among conophoralians is poorly understood, but even though their glandular penis appears to be non-injectory, it is still able to transfer a testis follicle with one or two conuli, that subsequently will form a bursa, and release the conuli that will fertilize the egg.

PHYSIOLOGY AND BEHAVIOR

Very little is known about gnathostomulid physiology, and behavioral studies are restricted to observations of live animals during morphological examinations. The animals respond negatively to light, and they will try to bury themselves when, for example, trapped in a net. Likewise, when trapped in a Petri dish, they will often seek the small detritus balls and try to hide inside them. Gnathostomulids move slowly by ciliary gliding, and some species are able to reverse the ciliary beats and hence move backward.

DISTRIBUTION AND ECOLOGY

Gnathostomulids can be found in sandy sediments throughout the world, but diversity and abundance appear to be higher in tropical and subtropical habitats. Only a few records are known from Arctic regions. They belong strictly to the interstitial fauna, and need spaces between the sand grains for their ciliary movement. Hence, gnathostomulids would rarely occur in silty or muddy sediments. They do, on the other hand, show a preference for detritus-rich sediments; thus, they would also be rare in sediments exposed to high currents. Their optimal habitat would be shallow, protected sandy areas without too much water movement. Coral sand with high contents of organic matter are often favorable for gnathostomulids, and they can also commonly be found in sand patches among corals. It has been shown that they may occur in high numbers around polychaete burrows in quartz sand (Reise, 1981). While some gnathostomulid species, genera or even families are known from single localities only, other species appear to be cosmopolitan, or at least represent mosaics of highly similar, potentially cryptic species.

GENETICS

Gnathostomulids have so far mostly been subjects for sequencing of selected target loci (18S rRNA, 28S rRNA, cytochrome *c* oxidase (COI), and Histone H3) used for phylogenetic analyses of relationships within the phylum (Sørensen et al., 2006). Transcriptomes for a species of *Austrognathia* (Conophoralia) and of Gnathostomulidae (Scleroperalia) were published by Laumer et al. (2015), and complete or nearly complete mitochondrial genomes are available for *Gnathostomula paradoxa* and *G. armata* (Golombek et al., 2015). The size of both mitochondrial genomes is around 14,000 bp. A complete, assembled nuclear genome is not yet available from any gnathostomulid.

POSITION IN THE ToL

After being considered as a flatworm ingroup, and subsequently as a separate phylum with uncertain phylogenetic affinities, gnathostomulids were finally matched up with Rotifera when Ahlrichs (1995) and Rieger and Tyler (1995) simultaneously suggested a close relationship between the two groups, based on the identical ultrastructure of their pharyngeal hard parts. When Micrognathozoa was described a few years later (see below), it was added as a close ally, and together the three groups make up the clade Gnathifera – 'those with jaws'. Gnathiferan monophyly has subsequently been confirmed by analysis of transcriptomic (Laumer et al., 2015) and mitogenomic (Golombek et al., 2015) data. Relationships within the Gnathostomulida are depicted in **Figure 18.6**.

DATABASES AND COLLECTIONS

Worms URL: http://www.marinespecies.org/aphia.php?p=taxdetails&id=14262

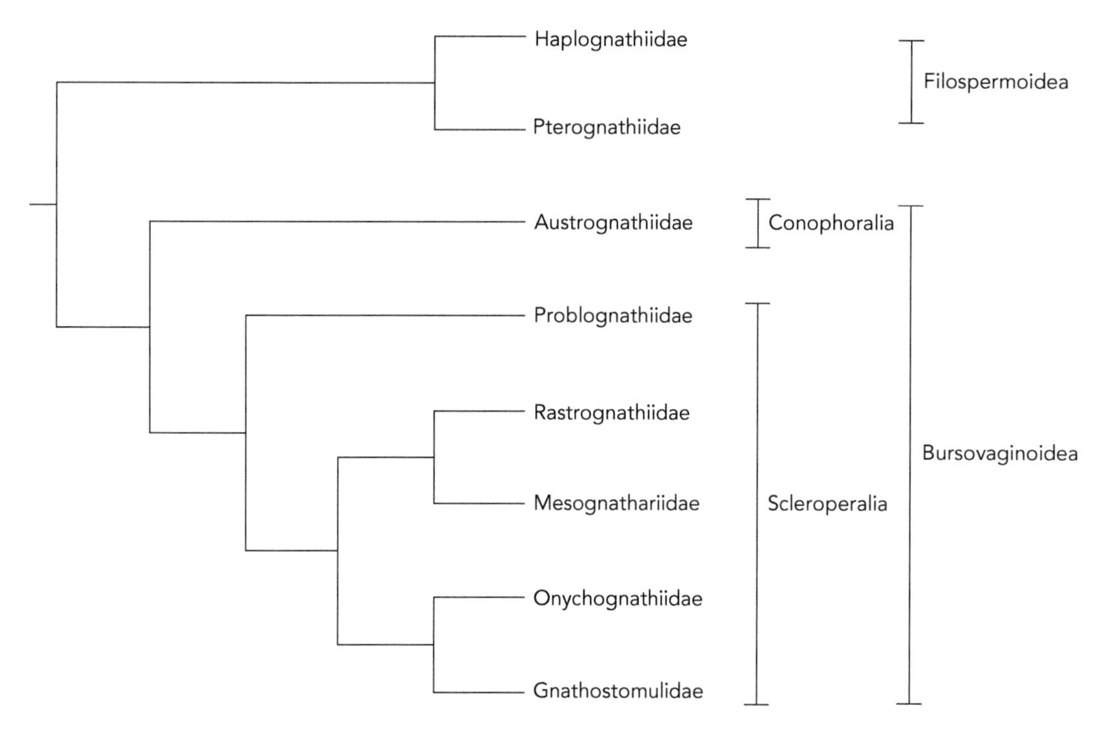

Figure 18.6 Cladogram based on analysis of four molecular loci (18S rRNA, 28S rRNA, Histone H3, COI) and morphology (Sørensen et al. 2006) showing the relationship of gnathostomulid families.

CONCLUSION

Gnathostomulids are one of the most neglected meiofaunal groups, and even though we have known about their existence for nearly a century, less than a handful of researchers have dedicated time to carry out in-depth studies on the group. They have obtained a reputation as "rare", "difficult to find and handle" and "taxonomically challenging", but this is not really the case. The truth is that the group has been neglected because they require observation while being alive, and thus are unsuitable for examination and identification after bulk fixation. They offer, however, great opportunities to study life in animals that are able to tolerate the extreme conditions close to the redox discontinuity layer, morphospecies that despite their seemingly low dispersal capabilities apparently have managed to become cosmopolitan, and morphologically an intriguing mix of simple ancient and highly complicated anatomical structures. When I started studying gnathostomulids, I had the chance to be the first to explore their morphologically amazing and disparate jaws with scanning electron microscopy, and it was like opening the door to a new microscopic world of shapes and complexity.

MICROGNATHOZOA

General Characteristics

One valid species, *Limnognathia maerski* (Kristensen and Funch, 2000)

Yet undescribed, but putatively new species have been observed at different localities in the United Kingdom, and on the Crozet Islands near Antarctica

Lives in small freshwater streams and ponds, among submerged vegetation

Microscopic animal consisting of a small head, an accordion-like thorax and a larger abdomen

Synapomorphies

- head and trunk with ventral rows of ciliophores
- dorsal side covered with epidermal plates, with intracellular protein lamina
- very complex jaw apparatus

One of the smallest metazoans, but with the most complex jaw apparatus
Closely related with Gnathostomulida and Rotifera
Males unknown

HISTORY OF TAXONOMY AND CLASSIFICATION

Micrognathozoa is the most recently discovered animal phylum. It was found for the first time in 1996 at Disko Island in West Greenland, during a student field course. The Danish researchers Peter Funch and Reinhardt M. Kristensen were collecting marine gastrotrichs and rotifers with a group of students on the northeast coast of Disko Island. They had set up a camp near a small freshwater stream that emerged from a cold spring, and they used water from the stream to freshwater shock the marine samples. In order to identify eventual cross-contaminating animals from the freshwater, they routinely examined samples from the stream, and while doing this, they noticed a small animal that moved gracefully in helical spirals through the water. The animal resembled a rotifer, but its movement was not rotifer-like, and a closer look under the light microscope revealed that it had a set of jaws that was much more complex than found in any rotifer. This intriguing finding prompted them to collect more animals from the mosses in the stream, and fix them for further light- and electron microscopical studies. After four additional years of intense studies, they named the animal *Limnognathia maerski*, and presented it as a new animal phylum, Micrognathozoa, closely related to gnathostomulids and rotifers (Kristensen and Funch, 2000).

ANATOMY, HISTOLOGY AND MORPHOLOGY

Micrognathozoans are strictly meiofaunal, and adults measure between 100 and 150 µm. Their bodies are divided into three conspicuous regions: a small, rounded head, a distinct accordion-like thorax and a bulbous abdomen (**Figure 18.7**). They do not have a cuticle, but the mostly unciliated epidermal cells on the dorsal and lateral sides form plates (each consisting of two to four cells). The presence of a rigid, intracellular protein lamina inside these plates compensates for the lack of cuticle. A similar internal protein lamina is found

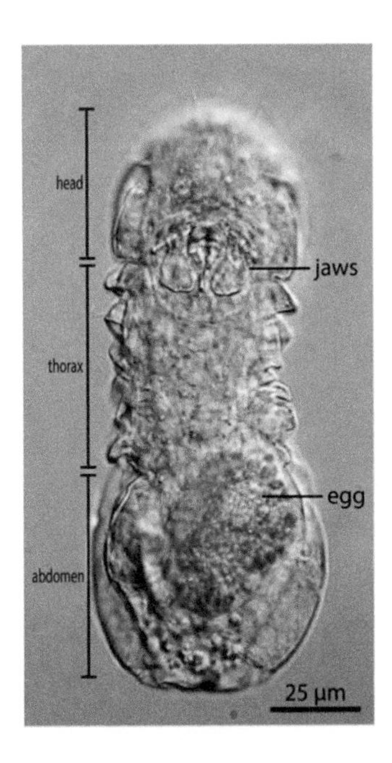

Figure 18.7 Light microscopical image of live *Limnognathia maerski* (Micrognathozoa).

inside the syncytial epidermis of rotifers, and it is considered to be synapomorphic for the two groups. Locomotory cilia are restricted to the head, and the ventral side of the trunk. The anterior part of the head has ciliary bands emerging from multiciliated epidermal cells. Even though the cilia are arranged in bands, comparison with the rotifer wheel-organ is not straightforward. Additional cilia are found laterally on the head, and along the ventral side of the trunk. These cilia emerge from distinct cells – so-called ciliophores – and the cilia on each ciliophore beat in unison. The lateral head ciliation consists of one row of small ciliophores on each side of it, and the ventral trunk ciliation of two distinct rows of large ciliophores. Ciliophores are also found in gastrotrichs and in the meiofaunal annelids *Diurodrilus* (Worsaae and Rouse, 2008).

The musculature of *Limnognathia maerski* includes, besides the extremely complex pharyngeal musculature, six pairs of longitudinal retractors that expand through the animal, and thirteen pairs of dorsoventral muscles in the trunk (Bekkouche et al., 2014). The nervous system is made up of a fairly large brain; auxiliary, pharyngeal and subpharyngeal ganglia; and two pairs of longitudinal nerves that expand through the trunk (Bekkouche and Worsaae, 2016). The ventrolateral nerve cords are connected with anterior and posterior commissures. Peripheral nerves connect the central nervous system with ciliary sensoria that are present on the head, and on the dorsal and lateral sides of the trunk.

The ventral mouth opening leads through the pharynx and a short esophagus to a blind midgut. A permanent anus is not present, but muscles attached to one of the posterior tegumental plates suggest that the animal might be able to form a temporary anus (Kristensen and Funch, 2000). The presence of a functional anus has never been observed though. The muscular pharynx encapsulates a very complex jaw apparatus (**Figure 18.8**). Sørensen (2003) identified six pairs of individual, paired, movable units, arranged around a central set of jaws. Some

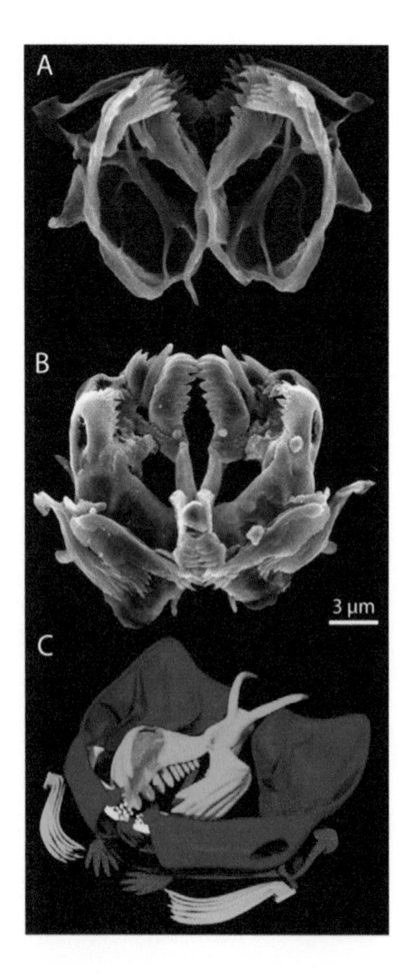

Figure 18.8 Dorsal (A) and ventral (B) side of *Limnognathia maerski* (Micrognathozoa) jaws dissected and prepared for scanning electron microscopy. (C) Three-dimensional reconstruction of jaws.

units are lamellar and supporting, whereas others are denticulated arms that are directly involved in grazing and handling of food items. The main jaws, and several of the accessory hard parts, are made up of the same rod-like elements that also form the jaws in gnathostomulids and rotifers.

Excretion is done via two or three pairs of protonephridia. The terminal cells are monociliated, which is similar to gnathostomulid terminal cells, but differs from the multiciliated terminal cells found in rotifers. The terminal cells connect with canal cells, and together they form a microvillar filter. The canal cells lead to larger collecting tubules, but it remains uncertain whether the tubules empty through nephridiopores or if they lead all the way to a cloaca.

REPRODUCTION AND DEVELOPMENT

Very little is known about micrognathozoan reproduction. So far, only females have been recorded, which suggests that they reproduce solely through parthenogenesis, as seen in bdelloid rotifers. However, this might be contradicted by the presence of two different kinds of eggs. The most common egg is a thin-shelled, non-sculptured kind that resembles the bdelloid egg, and eggs found during the amictic cycle of monogonont rotifers. But a second kind of egg has a thicker shell with a distinct ornamentation on the outside. Analogous with the rotifer life cycle, this could suggest that the thick-shelled eggs are results of mictic, and thus sexual, reproduction. However, this view of their reproduction remains obviously highly speculative until actual males have been observed.

Micrognathozoans lay one large egg at a time, and the embryo develops directly into a juvenile specimen that hatches from the egg. After hatching the juvenile increases slightly in size and gradually obtains reproductive maturity, but otherwise there are no conspicuous differences between juveniles and adults.

DISTRIBUTION AND ECOLOGY

The single described micrognathozoan species, *Limnognathia maerski*, is only known with certainty from the spring area in Isunngua on the northeast coast of Disko Island in Greenland (**Figure 18.9B**). Unlike many other springs on the island, that maintain the same temperature throughout the year due to underground thermal activity, the spring at Isunngua freezes by the end of the short, Arctic summer. This means that *L. maerski*'s active period is restricted to a short window from May to September. In the small streams that lead away from the spring area, *L. maerski* can be found in the submerged mosses, where it either swims among the vegetation or crawls on the moss leaves.

Outside Greenland, specimens of Micrognathozoa have been found at different localities in the United Kingdom, but the findings have never been documented, and it remains unclear whether they represent the same species as the one in Greenland. More surprisingly, micrognathozoans seem to be common on the Subantarctic Crozet Island, where they are found in ponds and lakes (De Smet, 2002). 18S rRNA sequences from the Subantarctic specimens differ from those of the Greenlandic population, but morphologically the Subantarctic specimens are very similar to *L. maerski*. Hence, despite the great distance, only sequence data suggests that the population on Crozet Island represents a second species of Micrognathozoa. It is expected that additional populations of micrognathozoans exist in cold subpolar and boreal freshwater bodies, but that they manage to remain concealed due to their minute body size.

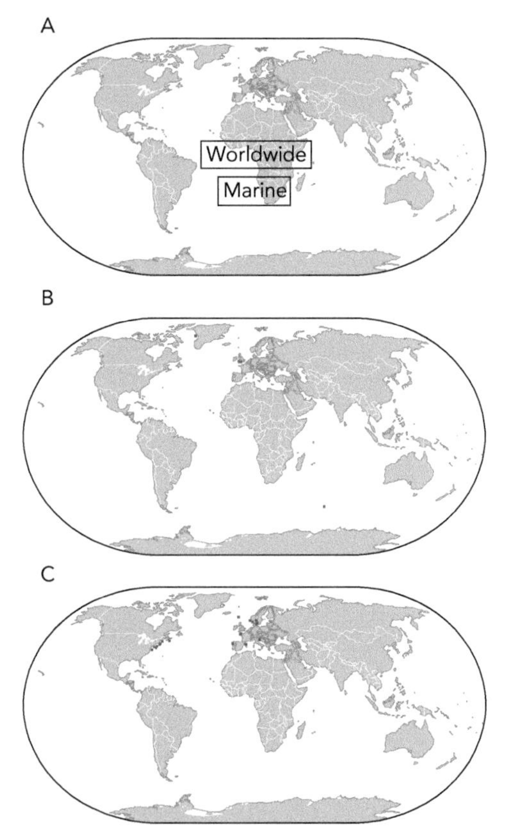

Figure 18.9 Distribution maps for Gnathostomulida (A), Micrognathozoa (B) and Cycliophora (C).

PHYSIOLOGY AND BEHAVIOR

Basically nothing is known about micrognathozoan physiology. Behavioral studies have shown that the animals like to crawl with their ventral ciliary bands along the vegetation or sand grains. If disturbed, for instance by a pipette tip, they will use their caudal adhesive ciliary pad to adhere strongly to the substratum. Instead of crawling, they may also enter the water and swim, using their preoral ciliary bands and ventral ciliophores. Swimming specimens move slowly in a characteristic alpha-helicoidal pattern that makes it very easy to distinguish them from the closely related rotifers.

GENETICS

Laumer et al. (2015) generated a transcriptome for *Limnognathia maerski*. Otherwise, genetic resources are limited to selected target loci: 18S rRNA, 28S rRNA, COI, and Histone H3 from the Greenlandic type locality (Giribet et al., 2004), and 18S rRNA from the Crozet Island population (Weekers et al., unpublished, GenBank AJ487046).

POSITION IN THE ToL

Morphological and ultrastructural similarities in the jaw apparatuses of Micrognathozoa, Gnathostomulida and Rotifera suggested from the first discovery of Micrognathozoa that a close relationship between the three groups exists, and that they are united in the clade Gnathifera (Kristensen and Funch, 2000; Sørensen, 2003). Initial attempts to confirm this relationship using molecular characters (four selected target loci) were inconclusive (Giribet et al., 2004), but recent analyses of transcriptomic data now support gnathiferan monophyly (Laumer et al., 2015).

DATABASES AND COLLECTIONS

The group does not appear in any major databases

CONCLUSION

Being known for less than twenty years, and with a distributional range that is restricted to remote, subpolar regions, our knowledge of Micrognathozoa is still extremely limited. One of the biggest questions that needs to be addressed regards their reproduction. Do they reproduce through parthenogenesis solely, or are males (dwarfish?) present for a very short period of their already shortened annual cycle? Is it perhaps possible for the specimens to change sex, and produce male gametes for a short period of time? These questions are indeed intriguing, but we are nowhere near answers. Another question regards their peculiar bipolar distribution. How did two morphologically similar and potentially cryptic species end up on Greenland and Crozet Island, basically as far away from each other as possible? Are these populations genetically related? And how many populations or as yet undiscovered species hide at remote freshwater bodies in between Disko and Crozet Island? Micrognathozoa is indeed among the smallest known metazoans, but it is also among the most fascinating and enigmatic.

CYCLIOPHORA

General Characteristics

Two described morphospecies, *Symbion pandora* and *S. americanus*

At least three cryptic species within *S. americanus*

Live as commensals on lobster mouthparts

Very complex life cycle

Synapomorphies

- asexual feeding stages with ciliated mouth funnel
- chordoid larva with internal chordoid support structure
- asexual, motile Pandora larvae
- Prometheus larva with internal dwarf males

HISTORY OF TAXONOMY AND CLASSIFICATION

Cycliophora is a microscopic animal that lives on the mouthparts of lobsters (**Figure 18.10**). It has an extremely complicated life cycle with various stages, inclusive females, dwarf males, Pandora larvae, Prometheus larvae, chordoid larvae, and feeding stages (**Figure 18.11**). The most prominent life cycle stage is the sessile feeding stage, and it was first observed in the 1960s by the Danish protozoologist Tom Fenchel, who found them on the mouthparts of the Norway lobster, *Nephrops norvegicus*. Fenchel passed the specimens to his colleague, Claus Nielsen, and they agreed that the animals looked like some sort of sessile, marine rotifer. Claus Nielsen kept the specimens in his drawer for several years, until the aspiring zoologist Peter Funch approached him in a search for a master thesis project. Peter Funch already had a background in rotifer research, so Claus Nielsen decided that Peter would be the right person to solve the mystery about the strange commensals on the lobster mouthparts. Over the next couple of years, Peter Funch studied the enigmatic animals, and concluded that they were not rotifers. Instead, they seemed to bear some superficial resemblance to Entoprocts and Bryozoans. After finishing his master's, Peter Funch

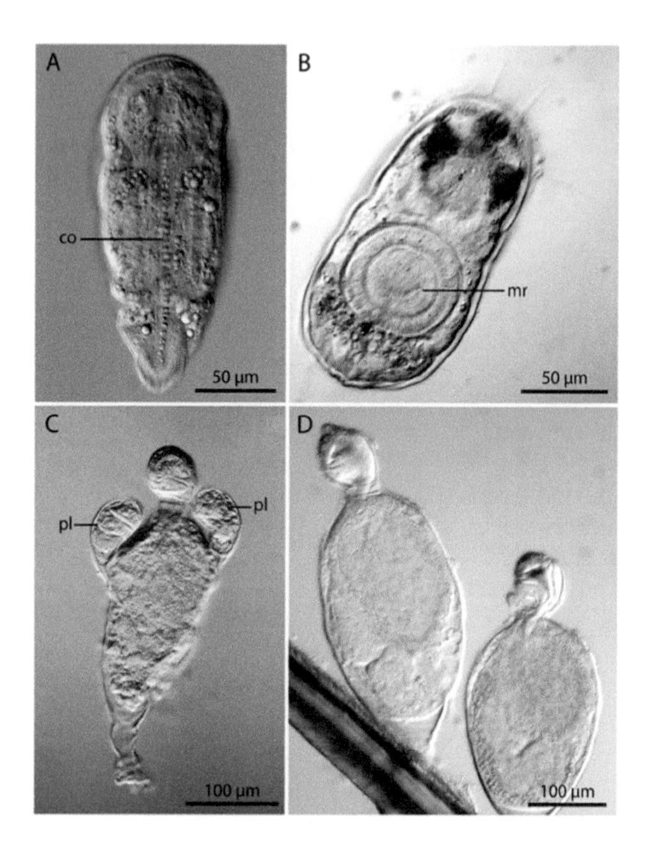

Figure 18.10 Light microscopical images of life cycle stages in Cycliophora. (A) Chordoid larva of *Symbion pandora* (note the chordoid organ, co). (B). Pandora larva of *Symbion americanus* (note the inner bud where the new developing mouth ring, mr, is already visible). (C) Feeding stage of *S. pandora* with two attached Prometheus larvae, pl. (D) Feeding stages of *S. americanus*. (Photos courtesy of Peter Funch.)

continued to explore the mystery animal of the lobster lips, and joined forces with Reinhardt M. Kristensen, who was well known for his insights into obscure meiofaunal organisms. Over the following years, with intense culturing experiments and studies of live animals, the complex life cycle of the cycliophorans unfolded for them, and in 1995 they described *Symbion pandora* as the first species of Cycliophora – 'a new phylum with affinities to Entoprocta and Ectoprocta [= Bryozoa]' (Funch and Kristensen, 1995).

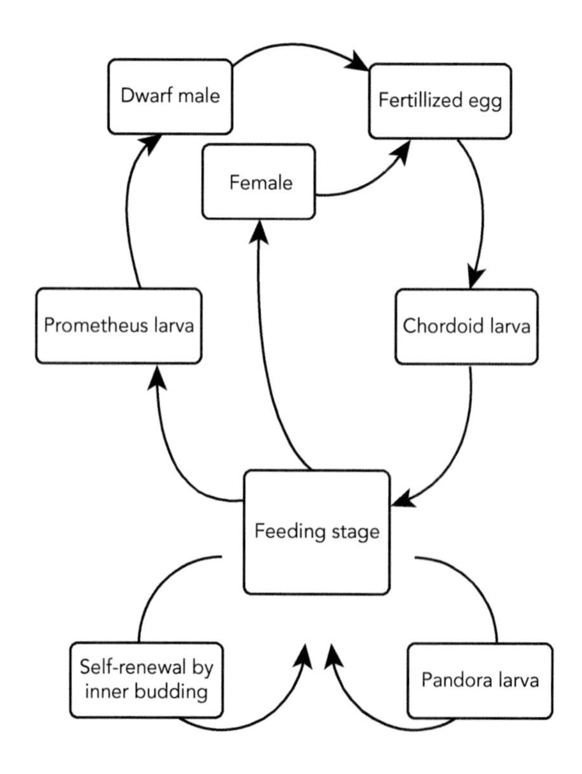

Figure 18.11 Life cycle of Cycliophora.

About ten years later, population genetic analyses showed that at least two additional species of Cycliophora existed: one on the European lobster, *Homarus gammarus*, and one on the American lobster, *H. americanus* (Obst et al., 2005). While the species of the European lobster remains undescribed, the American species was described shortly after, with the name *Symbion americanus* (Obst et al., 2006). Subsequent population genetic studies of *S. americanus* showed that this species is probably made of at least three cryptic species (Baker and Giribet, 2007; Baker et al., 2007). More recently, cycliophoran feeding stages have also been found on harpacticoid copepods (Neves et al., 2014). However, it remains uncertain whether the copepods can serve as a permanent host for a stable population, or if the *Symbion* larvae simply settled on the copepods by accident or because the copepods offered an alternative to densely populated lobster mouthparts.

LIFE CYCLE, ANATOMY, HISTOLOGY AND MORPHOLOGY

The cycliophoran life cycle includes a series of larval, as well as sexual and non-sexual stages, and the different stages are most easily understood when they are traced through this life cycle (Figure 18.11).

The primary cycliophoran larva is the chordoid larva (Figure 18.10A), which is a result of sexual reproduction. It is a motile, free-living stage, with two cephalic ciliary bands, a large ventral ciliated field, and a caudal ciliary foot. All cilia emerge from multiciliated epithelial cells. The larva has numerous longitudinal and incomplete circular muscles, but the most prominent part of its musculature is the so-called chordoid organ that expands ventrally, throughout the larva and acts as a supporting organ (Funch, 1996; Neves et al., 2009). Its nervous system includes a bilobed brain and paired, ventral nerve cords (Neves et al., 2010). Projections from the brain innervate different cephalic sensory organs, which probably play an important role when the larva is about to settle on the lobster. It has no mouth or digestive system, but is able to osmoregulate via a pair of protonephridia, each consisting of a single, multiciliated terminal cell, a canal cell and a nephridiopore. Its well-developed ciliation makes the chordoid larva a good swimmer, and it represents the primary dispersal stage in the life cycle. It can easily swim from one seta on a lobster mouthpart to another, but is also able to colonize new cuticle after the lobster has molted, or eventually leave the host and seek a new individual to settle on. When it has found a suitable spot on the lobster, it will settle on its head, and the larval tissue will degenerate. Inner buds are already present in the chordoid larva before it hatches, and these buds will now begin to develop into feeding stages.

The feeding stage (Figure 18.10C–D) represents the most prominent stage in the cycliophoran life cycle, and the lobster mouthparts may sometimes be densely covered with attached feeding stages. The feeding stage consists of a buccal funnel that via a narrow constriction attaches to an elongated trunk with a posterior stalk and attachment disc that adheres to the lobster. The rim of the buccal funnel is equipped with a ring of cilia that filter food items from the water, and direct them to the mouth opening at the bottom of the buccal funnel. The mouth leads to a u-shaped gut that terminates at the anus that is anterior on the trunk near the neck constriction. It does not have any excretory organs, and excretory waste products appear to be deposited as small crystals inside the trunk. The feeding stages only reproduce asexually, but they are capable of producing several new individuals including (1) new feeding stages via inner budding, (2) Pandora larvae, (3) females and (4) Prometheus larvae that will later develop into dwarf males.

Inner budding is a way for the feeding stage to renew itself. The inner budding take place inside the trunk, and in the beginning it resembles a new buccal funnel. The bud will gradually grow, and develop into a new feeding stage that will finally press the old feeding stage out of its cuticle. Hence, the whole feeding stage is renewed, except for the original cuticle that remains. This replacement of a feeding stage leaves a ring-shaped scar around the trunk, and observations of feeding stages with several scars suggest that a feeding stage is able to renew itself multiple times.

The feeding stage may also produce Pandora larvae (Figure 18.10B). The Pandora larva is a small, asexual life cycle stage, with ciliary cephalic sensory organs, and a ciliated apical field that enables it to creep, but not to swim. By the time it escapes the feeding stage, it already has an inner bud that develops into a new buccal funnel. It will only creep a short distance before it settles on the lobster mouthparts, and as soon as it is settled, the larval tissue will degenerate, and the inner bud will develop into a new feeding stage.

For sexual reproduction, the feeding stages may also produce females, and Prometheus larvae that will give rise to the males. The female in many ways resembles the Pandora larva, but it also carries a relatively large egg, and has a gonopore (Neves et al., 2012). The Prometheus larva has no reproductive capabilities by itself, but inside the larva one to three very minute (30–40 μm) dwarf males develop. After escaping the feeding stage, the Prometheus larvae move to a different feeding stage with a mature female inside and settle on it (Figure 18.10C). When the female escapes, the dwarf males are released, and they are assumed to fertilize the female just before she settles. Right after settling, the female tissue will degenerate, and the fertilized egg will develop into a new chordoid larva, which closes the life cycle.

REPRODUCTION AND DEVELOPMENT

Most of the cycliophoran life cycle stages – the feeding stages, the Pandora larvae, the Prometheus larvae and the females – develop from stem cells by inner budding in the feeding stages. Only the chordoid larva is a result of sexual reproduction. Very little is known about the chordoid larva's early embryology. Cleavage appears to be holoblastic, but detailed embryological studies are still required to confirm this.

DISTRIBUTION AND ECOLOGY

Except for a few records from harpacticoid copepods, Cycliophora appears to be restricted to specific lobster species – *Nephrops norvegicus*, *Homarus gammarus* and *H. americanus* – even though various other crustaceans have routinely been examined in search for cycliophoran commensals (Neves et al., 2014). On its three known host species, it is rather common, though. On *N. norvegicus* it has been found throughout Western Europe, and on the Faroe Islands. Cycliophorans living on *H. gammarus* are also present along the West European coast, as well as in the Mediterranean, and those living on the American lobster are found along the Canadian and US east coast. It is also quite commonly found on exported lobsters, raised for consumption.

The density of the feeding stages seems to increase with the lobster's age. Obst and Funch (2006) showed that only *N. norvegicus* with a carapace length >35 mm would be infected, and suggested that the younger and smaller lobsters would molt so frequently that the Cycliophora populations did not have time to establish. During molting, the Cycliophora population will collapse, and only the chordoid larvae are capable enough swimmers to escape from the shed cuticle and recolonize their host.

Cycliophorans will only attach to the lobster's mouthparts. They are occasionally found on mandibles and third maxillipeds, but they have a preference for maxillae and first to second maxillipeds, which are also the mouthparts with the highest concentration of suspended food particles when the lobster feeds (Obst and Funch, 2006).

PHYSIOLOGY AND BEHAVIOR

The cycliophoran feeding stages' ability to switch from asexual to sexual reproduction prior to the lobster's molting suggests that they are able to detect changes in their host's hormone concentrations (probably increased concentration of the molt-inducing hormone ecdysone). Likewise, behavorial studies show that the sexually produced chordoid larvae that after settling would develop into feeding stages display clear preferences when settling, and tend to aggregate on those parts of the mouth appendages where most food is suspended. On the other hand, the choice of settling site for the non-feeding chordoid cysts seem to be unaffected by food availability (Obst and Funch, 2006). This suggests that the different life-cycle stages of Cycliophora have very well-developed sensory organs, and that they behave in response to the sensorial inputs.

GENETICS

From *Symbion americanus* selected loci, 18S rRNA, 28S rRNA, COI and Histone H3, were sequenced for phylogenetic purposes (Giribet et al., 2004), and 16S rRNA, 18S rRNA, 28S rRNA and COI were sequenced from numerous specimens for population genetic studies (Baker and Giribet, 2007; Baker et al., 2007). Hejnol et al. (2009) produced Expressed Sequence Tags for *S. pandora*, and Laumer et al. (2015) published a transcriptome for *S. americanus*, though with low sequencing depth. Most recently, Neves et al. (2017) examined the differential gene expression in feeding stages from sexual and asexual generations and made a transcriptome available for *Symbion pandora*.

POSITION IN THE ToL

The first analyses of morphological data suggested a sister-group relationship with Entoprocta (Sørensen et al., 2000), which has subsequently been supported by analyses of ribosomal DNA and Expressed Sequence Tags (e.g., Baguñà et al., 2008; Hejnol et al., 2009; Nesnidal et al., 2013). Analyses of transcriptomic data failed to recover the Cycliophora–Entoprocta sister-group relationship, but Cycliophora was subsequently omitted from the final analyses, because the quality of its transcriptome was questionable (Laumer et al., 2015).

DATABASES AND COLLECTIONS

WoRMS URL: http://www.marinespecies.org/aphia.php?p=taxdetails&id=22586.

CONCLUSION

Cycliophorans still leave us with many open questions. Parts of their life cycle are still hypothetical, and it is not fully understood how the actual mating takes place. Also the apparently very narrow host preferences remain a bit of a riddle. Why have we only found stable populations of cycliophorans on three lobster species? And why have they never been found outside Europe or the American East Coast? And never on the southern hemisphere? A good and easily accessible spot to search for new cycliophoran species or new hosts is in fish markets. Hence, I encourage readers outside Europe and the Eastern US to visit their local fish markets, purchase a fresh decapod and examine its mouthparts – they might reveal new scientific discoveries.

REFERENCES

Ahlrichs, W. H., 1995. Ultrastruktur und Phylogenie von *Seison nebaliae* (Grube 1859) und *Seison annulatus* (Claus 1876). Cuvillier Verlag, Göttingen, p. 310.

Ax, P., 1956. Die Gnathostomulida, eine rätselhafte Wurmgruppe aus dem Meeressand. Akademie der Wissenschaften und der Literatur, 8, 1–32.

Baker, J., Funch, P., Giribet, G., 2007. Cryptic speciation in the recently discovered American *Symbion americanus*; genetic structure and population expansion. Marine Biology, 151, 2183–2193.

Baker, J., Giribet, G., 2007. A molecular phylogenetic approach to the phylum Cycliophora provides further evidence for cryptic speciation in *Symbion americanus*. Zoologica Scripta, 36, 353–359.

Baguñà, J., Martinez, P., Paps, J., Riutort, M., 2008. Back in time: a new systematic proposal for the Bilateria. Philosophic Transactions of the Royal Society B, 363, 1481–1491.

Bekkouche, N., Kristensen, R. M., Hejnol, A., Sørensen, M. V., Worsaae, K. 2014. Detailed reconstruction of the musculature in *Limnognathia maerski* (Micrognathozoa) and comparison with other Gnathifera. Frontiers in Zoology, 11, 71.

Bekkouche, N., Worsaae, K., 2016. Nervous system and ciliary structures of Micrognathozoa (Gnathifera): evolutionary insight from an early branch in Spiralia. Royal Society Open Science, 3, 160289.

De Smet, W. H., 2002. A new record of *Limnognathia maerski* Kristensen & Funch, 2000 (Micrognathozoa) from the subantarctic Crozet Islands, with description of the trophi. Journal of Zoology, London, 228, 381–393.

Funch, P. 1996. The chordoid larva of *Symbion pandora* (Cycliophora) is a modified trochophore. Journal of Morphology, 230, 231–263.

Funch, P., Kristensen, R. M., 1995. Cycliophora is a new phylum with affinities to Entoprocta and Ectoprocta. Nature, 378, 711–714.

Giribet, G., Sørensen, M. V., Funch, P, Kristensen, R. M., Sterrer, W., 2004. Investigations into the phylogenetic position of Micrognathozoa using four molecular loci. Cladistics, 20, 1–13.

Golombek, A., Tobergte, S., Struck, T. H., 2015. Elucidating the phylogenetic position of Gnathostomulida and first mitochondrial genomes of Gnathostomulida, Gastrotricha and Polycladida (Platyhelminthes). Molecular Phylogenetics and Evolution, 86, 49–63.

Hejnol, A., Obst., M., Stamatakis, A., Ott, M., Rouse, G. W., et al. 2009. Assessing the root of bilaterian animals with scalable phylogenomic methods. Proceedings of the Royal Society B, 276, 4261–4270.

Knauss, E. B., 1979. Indication of an anal pore in Gnathostomulida. Zoologica Scripta, 8, 181–186.

Kristensen, R. M., Funch, P., 2000. Micrognathozoa: A new class with complicated jaws like those of Rotifera and Gnathostomulida. Journal of Morphology, 246, 1–49.

Laumer, C. E., Bekkouche, N., Kerbl, A., Goetz, F., Neves, R. C., et al. 2015. Spiralian phylogeny informs the evolution of microscopic lineages. Current Biology, 25, 1–7.

Nesnidal, M. P., Helmkampf, M., Meyer, A., Witek, A., Bruchhaus, I., et al. 2013. New phylogenomic data support the monophyly of Lophophorata and an Ectoproct-Phoronid clade and indicate that Polyzoa and Kryptrochozoa are caused by systematic bias. BMC Evolutionary Biology, 13, 253.

Neves, R. C., Bailly, X., Reichert, H., 2014. Are copepods secondary hosts of Cycliophora? Organisms Diversity and Evolution, 14, 363–367.

Neves, R. C., da Cunha, M. R., Kristensen, R. M., Wanninger, A., 2010. Expression of synapsin and co-localization with serotonin, and RFamide-like immunoreactivity in the nervous system of the chordoid larva of *Symbion pandora* (Cycliophora). Invertebrate Biology, 129, 17–26.

Neves, R. C., Guimaraes, J. C., Strempel, S., Reichert, H., 2017. Transcriptome profiling of *Symbion pandora* (phylum Cycliophora): insights from a differential gene expression analysis. Organisms Diversity and Evolution, 17, 111–119.

Neves, R. C., Kristensen, R. M., Funch, P., 2012. Ultrastructure and morphology of the cycliophoran female. Journal of Morphology, 273, 850–869.

Neves, R. C., Kristensen, R. M., Wanninger, A., 2009. Three-dimensional reconstruction of the musculature of various life cycle stages of the Cycliophoran *Symbion americanus*. Journal of Morphology, 270, 257–270.

Obst, M., Funch, P., 2006. The microhabitat of *Symbion pandora* (Cycliophora) on the mouthparts of its host *Nephrops norvegicus* (Decapoda: Nephropidae). Marine Biology, 148, 945–951.

Obst, M., Funch, P., Giribet, G., 2005. Hidden diversity and host specificity in cycliophorans: a phylogeographic analysis along the North Atlantic and Mediterranean Sea. Molecular Ecology, 14, 4427–4440.

Obst, M., Funch, P., Kristensen, R. M. 2006. A new species of Cycliophora from the mouthparts of the American lobster, *Homarus americanus* (Nephropidae, Decapoda). Organisms Diversity and Evolution, 6, 83–97.

Reise, K., 1981. Gnathostomulida abundant alongside polychaete burrows. Marine Ecology Progress Series, 6, 329–333.

Riedl, R. J., 1969. Gnathostomulida from America – This is the first record of the new phylum from North America. Science, 163, 445–442.

Rieger, R. M., Tyler, S., 1995. Sister-group relationship of Gnathostomulida and Rotifera-Acanthocephala. Invertebrate Biology, 114, 186–188.

Sørensen, M. V. 2003. Further Structures in the Jaw Apparatus of *Limnognathia maerski* (Micrognathozoa), with Notes on the Phylogeny of the Gnathifera. Journal of Morphology, 255, 131–145.

Sørensen, M. V., Funch, P., Willerslev, E., Hansen, A. J., Olesen, J., 2000. On the phylogeny of the Metazoa in light of Cycliophora and Micrognathozoa. Zoologischer Anzeiger, 239, 297–318.

Sørensen, M. V., Sterrer, W., 2002. New characters in the gnathostomulid mouth parts revealed by scanning electron microscopy. Journal of Morphology, 253, 310–334.

Sørensen, M. V., Sterrer, W., Giribet, G., 2006. Gnathostomulid Phylogeny inferred from a combined approach of four molecular loci and morphology. Cladistics, 22, 32–58.

Sterrer, W., 1965. *Gnathostomula axi* Kirsteuer und *Austrognathia* (ein weiteres Gnathostomuliden-genus) aus der Nordadria. Zeitschrift für Morphologie und Ökologie der Tiere, 55, 783–795.

Sterrer, W., 1969. Beiträge zur Kenntnis der Gnathostomulida I. Anatomie und Morphologie des Genus *Pterognathia* Sterrer. Arkiv för Zoologi, Ser. 2, 22, 1–125.

Sterrer, W., 1972. Systematics and evolution within the Gnathostomulida. Systematic Zoology, 21, 151–173.

Worsaae, K., Rouse, G. 2008. Is *Diurodrilus* an Annelid? Journal of Morphology, 269, 1426–1455.

Additional Reading

Funch, P., Kristensen, R. M., 1997. Cycliophora. In: Harrison, F. W., Wollacott, R. M. (eds.), Microscopic Anatomy of Invertebrates. volume 13: Lophophorates, Entoprocta and Cycliophora, pp. 409–474. Wiley-Liss, New York.

Sørensen, M. V., Kristensen, R. M., 2015. Micrognathozoa. In: Schmidt-Rhaesa, A. (ed.), Handbook of Zoology, Gastrotricha, Cycloneuralia and Gnathifera. Volume 3: Gastrotricha and Gnathifera, pp. 197–216. De Gruyter, Berlin/Munich/Boston.

Sterrer, W., Sørensen, M. V., 2015. Gnathostomulida. In: Schmidt-Rhaesa, A. (ed.), Handbook of Zoology, Gastrotricha, Cycloneuralia and Gnathifera. Volume 3: Gastrotricha and Gnathifera, pp. 135–196. De Gruyter, Berlin/Munich/Boston.

LOPHOTROCHOZOA

HIGHER TAXON

4

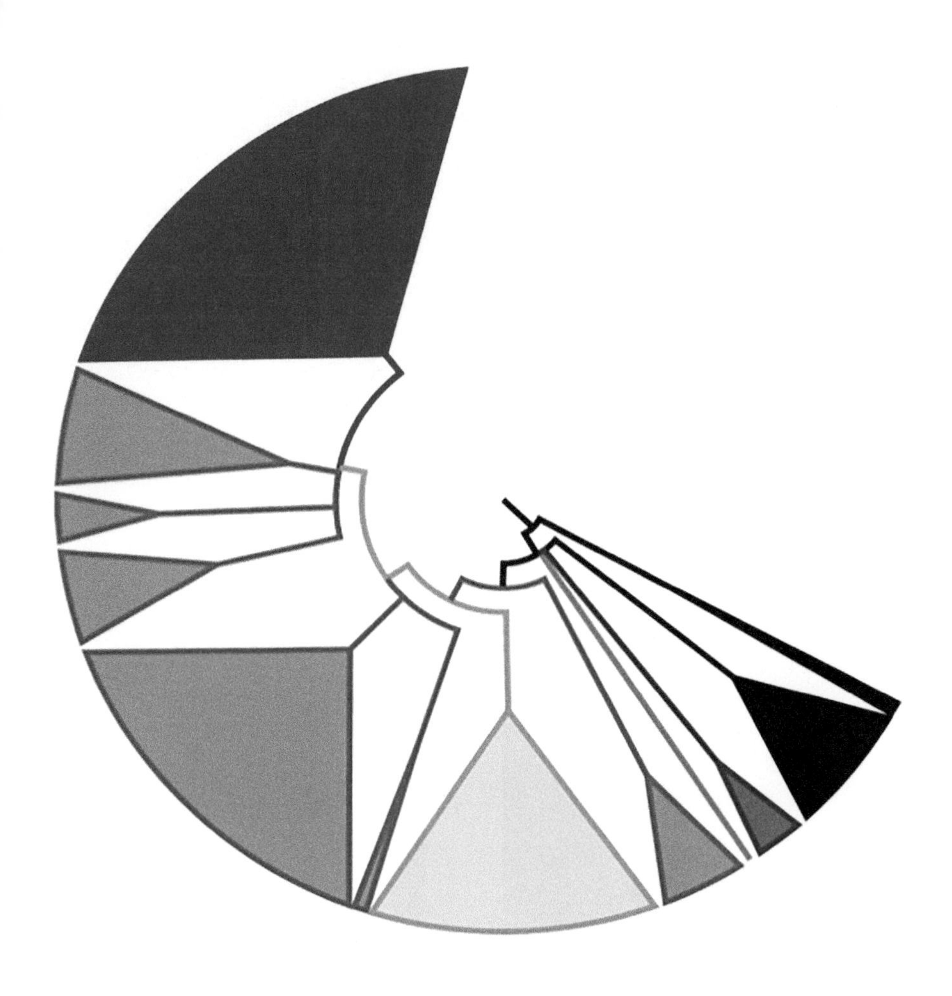

Lophotrochozoa (from Greek "crest/wheel animals") is a clade of protostome animals within the Spiralia. This group was first established using molecular data. The phyla within the Lophotrochozoa are Cycliophora, Entoprocta, Mollusca, Nemertea and the derived phyla Brachiopoda, Bryozoa, Phoronida, and Annelida. While the clade is well defined by molecular evidence, there are also two distinctive morphological characters that are in general consistent with Lophotrochozoa being monophyletic. The first is the feeding structure called the lophophore, best described as a ciliated crown of tentacles surrounding the mouth that facilitates feeding (Figure below). The second morphological trait is the occurrence of the larval developmental stage called the trochophore (Figure below). The two traits are not always found together in all lophotrochozoans. For instance, Brachiopoda and Bryozoa have lophophore feeding structures, while Mollusca and Annelida have trochophore larval stages.

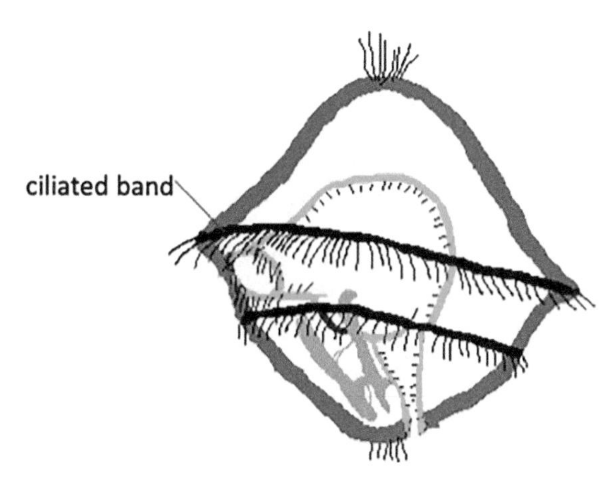

Planktotrophic trochophore larva with prototroch and metatroch. (Figure Public Domain and redrawn from Wlodzimierz.)

Defining Characteristics

Possess either a lophophore or trochophore larvae (see above)

Lophophore: a set of ciliated tentacles surrounding the mouth.

Possess an embryonic mesoderm sandwiched between the ectoderm and endoderm

Bilaterally symmetrical

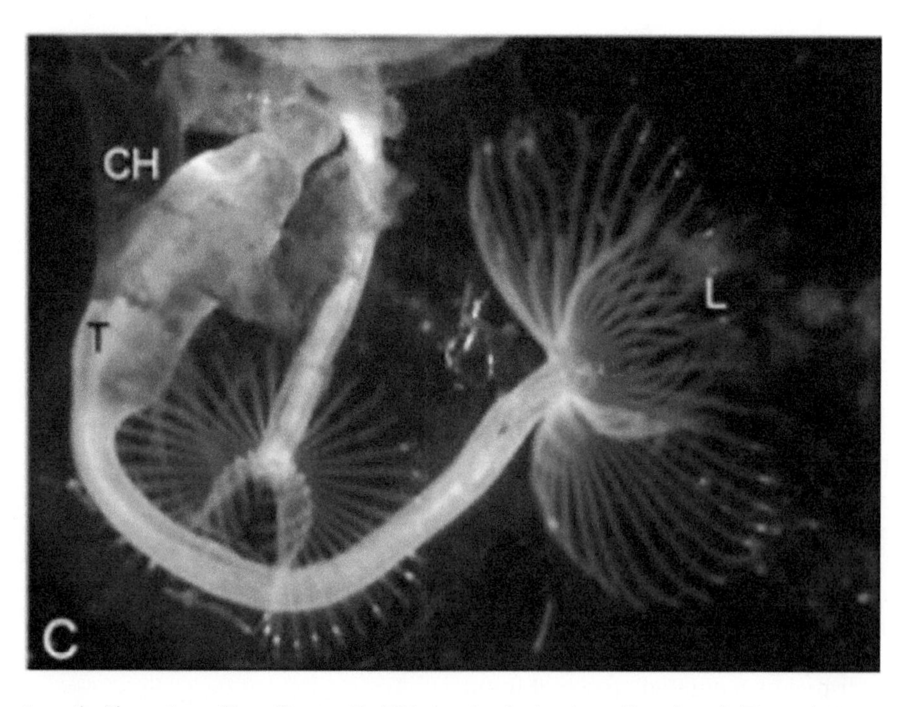

Juvenile *Phoronis* sp. (From Tampa, FL, USA showing lophophore (L) and trunk (T) in a chitinous tube (CH). Figure from Chapter 17.)

PHYLUM MOLLUSCA

Gerhard Haszprunar

CHAPTER

19

Infrakingdom	Bilateria
Division	Protostomia
Subdivision	Lophotrochozoa
Phylum	Mollusca

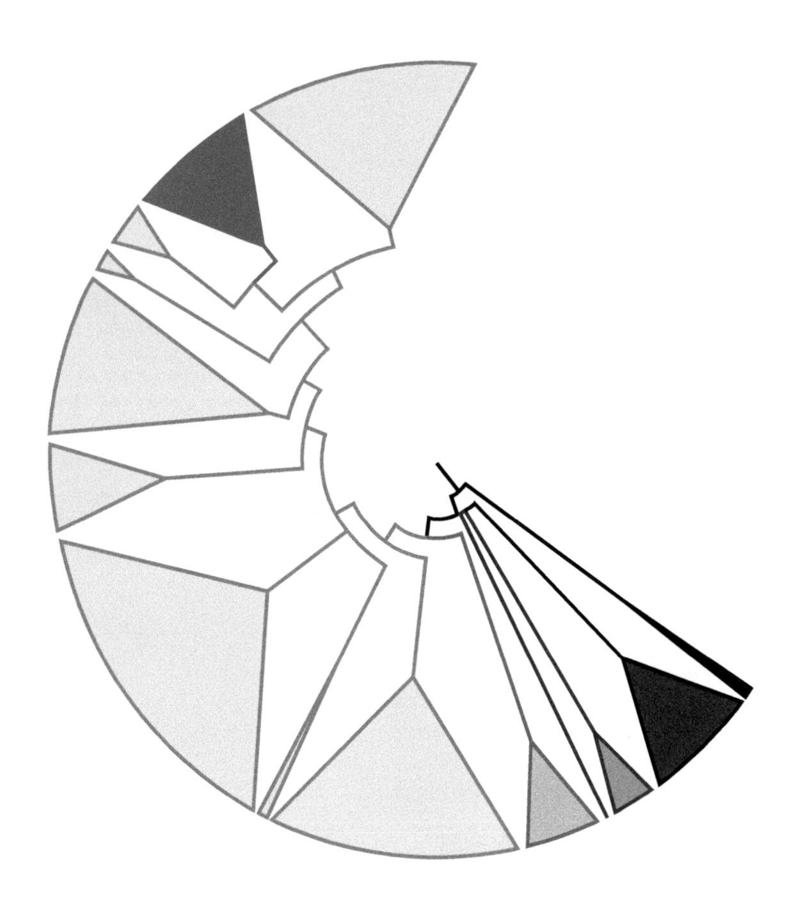

CONTENTS

General Characteristics

Less than 100,000 extant, valid species; 80% of these are Gastropoda, 15% are Bivalvia. About 100,000 fossils are named; "cryptic extant species" estimated of another 100,000

Animals range from microscopic (0.4 mm) to very large (6 m) size in all habitats except air

Quite distinct bilaterian bauplan, but there is enormous diversity in all respects (i.e., morphology, physiology, genetics, and ecology)

Numerous mitochondrial genomes available, <15 molluscan complete genomes available

High commercial value as food, artwork, and jewelry (nacre, pearls)

High ecological importance, many highly endangered species

Autapomorphies

- external aragonitic scales/spicules and/or dorsally located calcareous shells (plates) for protection
- ventral non-cuticularized, ciliated gliding sole with anterior foot gland
- dorsoventral muscles with ventral intercrossing
- mantle cavity with paired ctenidia (for ventilation) and release openings
- dorsal heart with paired atria and ventricle, sinusoid circulatory system
- rhogocytes (slit cells, pore cells) for production and recycling of hemocyanin
- podocytes at atrial wall of coelomic pericardium, the latter with releasing ducts
- coelomic system as gonopericardial system (no body coelom)
- chitinous radula consisting of serial teeth and being supplied by an odontophore
- central nervous system with anterior circumpharyngeal cerebro-pleural-pedal ring and four main longitudinal nerve cords (tetraneury: ventral/pedal and lateral/visceral cords)
- chemoreceptive osphradial sense organs
- decoupling of dorsal and ventral expression of Hox genes during development

HISTORY OF TAXONOMY AND CLASSIFICATION

The phylum Mollusca has a long and convoluted taxonomic history; indeed hundreds of suprageneric names have come and gone. Already Aristotle recognized molluscs with Malachia (Cephalopoda), and univalved and bivalved Ostrachodermata (the other conchiferan classes) as subgroups. Joannes Jonstonus (1650) created the name *Mollusca* for Cephalopoda and barnacles (cirripedian crustaceans). The taxon Mollusca was formally introduced as a subgroup of Vermes (worms) by Linnaeus (1758), who united shell-less gastropod taxa (*Limax, Doris, Tethys*) with annelids, priapulides, medusas, and echinoderms. Shelled taxa were united as *Testacea*, the latter included Polyplacophora, Bivalvia, Gastropoda, and Scaphopoda, but also included polychaetes with calcareous tubes like *Serpula*. Georg Cuvier (1817a,b) used Mollusca to include not only all conchiferan classes except Monoplacophora (not yet recognized at this time), but also Brachiopoda and Tunicata. Major progress toward our current understanding of Mollusca was seen during the 19th century: Von Ihering (1876) first recognized the aplacophoran classes as Mollusca and treated them together with Polyplacophora as Amphineura (now Aculifera). With the *Handbuch der Systematischen Weichthierkunde* by Johannes Thiele (1929–1935), the molluscan system seemed to be fixed, but the discovery of extant Monoplacophora (Tryblidia: Lemche, 1957) added another class of extant Mollusca.

Currently, there is broad agreement about the monophyly of each of the eight extant molluscan classes: The aplacophoran taxa Solenogastres (Neomeniomorpha) and Caudofoveata (Chaetodermomorpha) are usually united with Polyplacophora (chitons) as Aculifera; the Conchifera include Monoplacophora (Tryblidia), Bivalvia (Acephala, Lamellibranchia, Pelecypoda), Scaphopoda (Solenoconcha), Gastropoda, and Cephalopoda. In contrast, the phylogenetic relationships between these taxa are still debated: whereas morphological data suggest basal, paraphyletic Aculifera, molecular analyses, in particular, recent phylogenomic data call for a dichotomy of monophyletic Aculifera and Conchifera. The internal relationships of the latter are still fairly unknown, although the Tryblidia are probably the first extant offshoot (Kocot et al. 2020).

ANATOMY, HISTOLOGY, AND MORPHOLOGY

On the one hand, the basic bauplan of the primarily bilateral symmetric Mollusca is quite distinct but on the other hand shows extreme variability and diversity. Whereas a 7–8 fold seriality of dorsoventral (shell) muscles is considered a

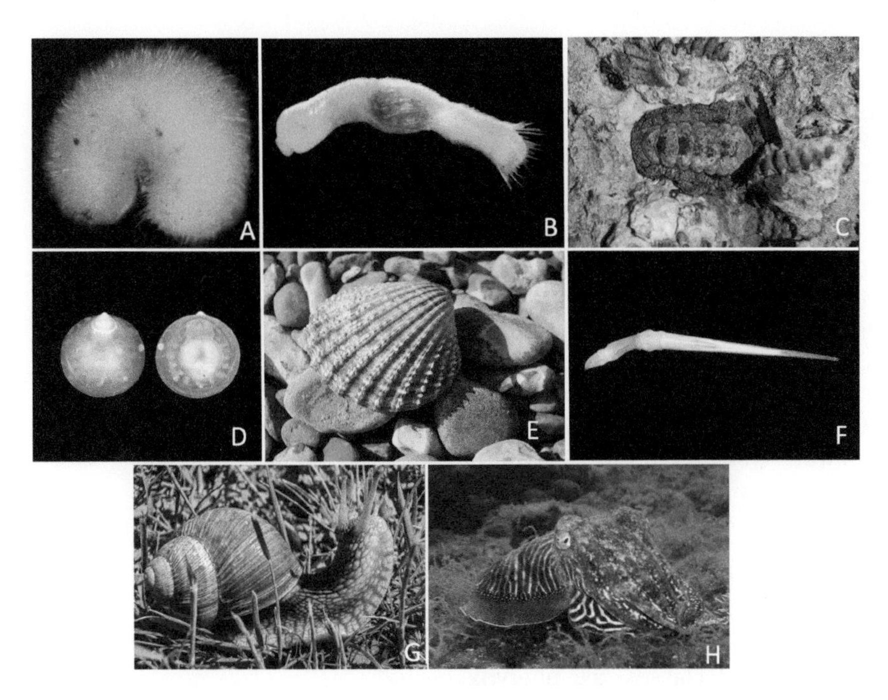

Figure 19.1 Live photos of representatives of molluscan classes.
A. Solenogastres – Simrothiellidae (courtesy Franziska Bergmeier).
B. Caudofoveata – Prochaetodermatidae (courtesy Franziska Bergmeier).
C. Polyplacophora – Chitonidae – *Acanthopleura* (pixabay: chiton-525167_1920).
D. Monoplacophora – Tryblidia – *Laevipilina theresae* (courtesy Michael Schrödl).
E. Bivalvia – Heterodonta – *Laevicardium* (cockle) (pixabay: Cardium-4199171_1920).
F. Scaphopoda – Dentaliida – *Fustiaria rubescens* (courtesy Bastian Brenzinger).
G. Gastropoda – Stylommmatophora – *Helix pomatia* (pixabay: Helix-1014715_1920).
H. Cephalopoda – Coleoidea – *Sepia officinalis* male (pixabay: Sepia-225423_1920).

plesiomorphic condition for Mollusca, seriality of several other organ systems (e.g., ctenidia, heart atria, nephridia, etc.) has been independently established in Polyplacophora, Tryblidia, and *Nautilus*. Scaphopoda, Cephalopoda, and, in particular, Gastropoda show (independently) a significant trend toward asymmetry of the reproductive system; asymmetric gut loops occur in Polyplacophora, Tryblidia, Bivalvia, and basal Gastropoda (**Figure 19.1**).

In general, molluscs show three main body regions: (1) A cerebrally innervated head (region) is always present. A buccal apparatus is entirely lacking in Bivalvia alone; however, a freely movable head equipped with eyes is restricted to Gastropoda and Cephalopoda. (2) A pedally supplied, ciliated or muscular foot sole is the main organ for locomotion in most species. Secondary loss of the foot occurs in Caudofoveata (few species retain a ventral suture); strong modifications of the foot are found in many bivalves, several groups of gastropods, and cephalopods. (3) The visceral body (region), with midgut, intestine, and the gonopericardial system, is innervated by the lateral/visceral nerve cords which may be concentrated to ganglia.

The epidermis generally comprises a monolayered epithelium with basal lamina. Cells may be covered by a microvillous border or cilia, may be cuticularized sometimes, or may produce calcareous hard parts such as aragonitic scales or spicules or shell (plates) of very variable mineralogy. The latter are usually covered by an organic layer, the periostracum, which is produced by a glandular ridge along the mantle border.

Since most of the epidermis is covered by armored cuticle or shell (plates) preventing diffusion, respiration mainly takes place at the inner, flat epithelium of the so-called mantle cavity. Except in Solenogastres and Scaphopoda, comb-like (bipectinate or monopectinate) organs called ctenidia are heavily ciliated and provide ventilation of the mantle cavity. In larger animals and in particular in Cephalopoda, the ctenidia may also contribute to respiration ("gills") as do secondary gills in Solenogastres and in many heterobranch gastropods. The mantle cavity also contains various ("hypobranchial "or "pallial") glands and release openings for germ cells, excretory products, and feces.

The mesenchymate primary body cavity (hemocoel) is mainly composed of muscle cells and connective tissue (collagen fibers) produced by colloblasts, leaving free spaces (sinuses and lacunae) which form the circulatory system with various hemocytes, rhogocytes with ultrafiltration sites (like podocytes) for production and recycling of respiratory proteins, and also amebocytes, various granulocytes, and occasionally even erythrocytes filled with hemoglobin. The coelomatic cavity (endothelial lining) is represented as a gonopericardial system: The heart consists of atria and ventricle that is surrounded by a pericardium. Ultrafiltration mostly takes place at the atrial wall of pericardium modified to podocytes, sometimes the outer pericardial wall ("pericardial glands") is also involved. The primary urine is further modified during passage through apericardioduct (usually elaborated as a nephridium); the latter often also release the gametes.

Molluscs are primarily equipped with a buccal apparatus (entirely lost in Bivalvia) with a radula (secondarily lost in many taxa) which consists of serially arranged chitinous teeth and is often supported by a cartilaginous odontophore. Jaws are restricted to the conchiferan classes. In most classes, the pharynx lumen is continued by an esophagus leading backward into a complicated stomach with associated midgut glands, followed by a usually looped intestine and straight rectum; but in some cases the gut is strongly modified or even entirely reduced (e.g., several solemyid bivalves).

The plesiomorphic central nervous system of molluscs is composed of an anterior circumpharyngeal ring (paired cerebral, pleural, and pedal ganglia), a buccal system, and two paired longitudinal nerve cords (tetraneury: ventral/pedal and lateral/visceral cords). True ganglia evolved several times independently. The foot sole additionally shows a serotonergic neural plexus at the base of the epidermis. In particular many Gastropoda and Cephalopoda, and also certain Bivalvia (e.g., Spondylidae) show "cerebralization" and true brains.

Only Gastropoda and Cephalopoda have cerebral eyes; further photoreceptive organs of different ultrastructure (open vs. closed; with or without lens; one or two retinae; evers vs. invers; rhabdomeric vs. ciliary) are found in the mantle cavity (larvae of Polyplacophora, larvae and juveniles of pteriomorph Bivalvia), along the mantle border (many Bivalvia), or on the dorsal body (many photoreceptive aesthetes or even shell eyes in Polyplacophora, dorsal eyes in onchidiid Gastropoda). True statocysts are restricted to Conchifera, being highly elaborate in the Cephalopoda. Most classes have chemoreceptive, so-called osphradial sense organs, but they are secondarily lacking in Tryblidia, Scaphopoda, coleoid Cephalopoda, and many slugs. There is a large number of further sensory organs that are characteristic for certain taxa.

REPRODUCTION AND DEVELOPMENT

Molluscs are vulnerable animals: The regeneration capacity of molluscs is generally poor, and asexual reproduction is unknown. Gonochorism and hermaphroditism are equally represented, parthenogenesis or self-fertilization occurs only occasionally. While most primitive taxa have entaquatic (in the mantle cavity) or ectaquatic (in the external environment) fertilization, Solenogastres, certain Bivalvia, most Gastropoda, and all Cephalopoda have repeatedly evolved internal fertilization by various copulatory structures (e.g., spermatophores, various kinds of penises, modified arm as hectocotylus in coleoid Cephalopoda).

Nearly all molluscs (no data on Tryblidia) show spiral-quartet cleavage, only Cephalopoda have superficial cleavage due to their large eggs. In marine

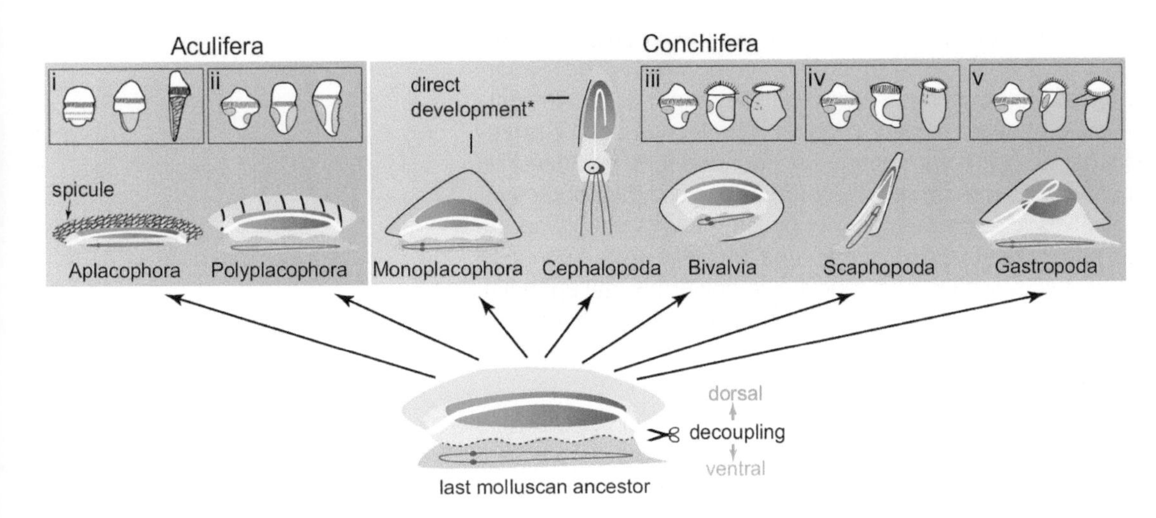

Figure 19.2 A comparison of Hox gene expression among molluscan classes (from Huam et al. 2020. PNAS, 117(1):503–512). Red: ventral nerve cords/ganglia. Green to violet: Hox gene expression from anterior to posterior. * Data on the development of monoplacophorans are still lacking.

taxa, except Cephalopoda, there is a trochophore larva with ciliary apical tuft and an apical organ with FMRF-amidergic "ciliary flask cells," one to three ciliary rings (prototroch, metatroch in case of planktotrophy, telotroch), and one pair of protonephridia. This (often intracapsular) stage is followed by a veliconcha (Bivalia) or veliger (Gastropoda) stage, which may be lecithotrophic or planktotrophic. Or the animal develops fully in the egg capsule and hatches as early juvenile. Certain marine, most freshwater, and all terrestrial molluscs have direct development as generally found among the Cephalopoda (**Figure 19.2**).

DISTRIBUTION AND ECOLOGY

While most molluscan classes solely comprise marine animals, bivalves and gastropods have repeatedly conquered freshwater habitats; gastropods may be found in all terrestrial habitats except air and permanent ice (**Figure 19.3**). Although certain pelagic taxa have ocean-wide distribution, freshwater or terrestrial taxa show narrow to extreme endemism and therefore are highly vulnerable to become endangered or extinct by anthropogenic activities. Indeed, terrestrial gastropods are the leading group in the Red Lists of many countries.

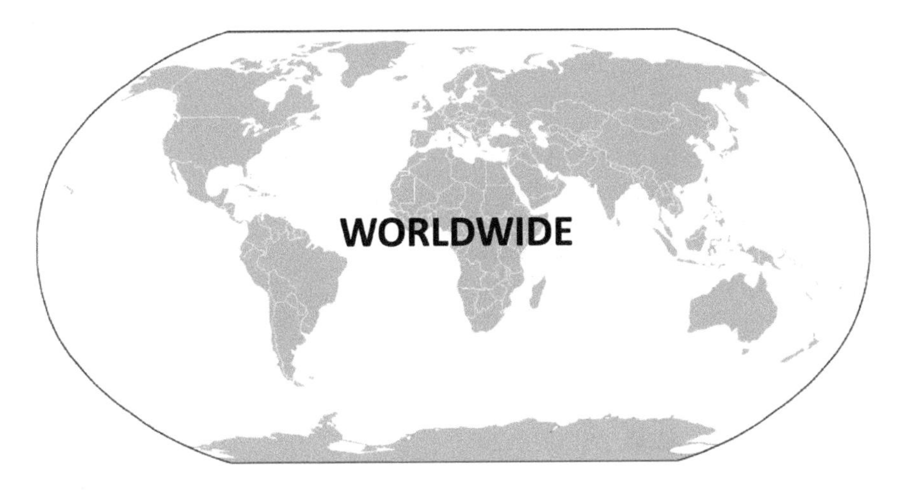

Figure 19.3 The worldwide distribution (except the continent Antarctica) of phylum Mollusca.

On the other hand, several species have been introduced (neozoans) and have become pests of agriculture, they also may predate on vulnerable island faunas.

Also, feeding ecology of molluscs varies in an outstanding matter. Not only detritovores and herbivores, but filter feeders (ctenidial or mucous trap), scavengers, predators, blood suckers, and parasites are also found among various taxa. Sacoglossa and certain nudibranch slugs show kleptoplasty to use their photosynthetic products as do bivalves (e.g., Tridacnidae) in housing zooxanthellae in the hemocoel of the mantle border. Several taxa in Tryblidia, Bivalvia, and Gastropoda have independently developed so-called bacteriocytes in epidermis, gill epithelia, or gut extensions to house endosymbiotic bacteria for living in sulfide- or methane-rich environments such as muddy bottoms, whale falls, cold seeps, or hot vents.

BEHAVIOR AND PHYSIOLOGY

Not only morphology and ecology, but behavior and physiology of molluscs also vary substantially. Most molluscs behave slowly, as reflected by the German word "Schneckentempo" (snail speed) for very slow movements in general. However, certain species and muscles are very fast: this is true for all kinds of muscular swimming behavior, the proboscis of cone snails, the closure of bivalve shells, digging of certain clams, and of cephalopods in general. Reproductive behavior ranges from simple shedding of gametes into the sea, simple to bizarre copulations with penises being derived from various parts of the male body (head, foot, mantle border) to very complicated and reciprocal (pseudo-)copulations with exchange of spermatophores.

Nearly all molluscs are aerobic animals who need at least a minimum of oxygen. The occasional existence of hemoglobin (instead of hemocyanin) as respiratory protein calls for very-low-oxygen aquatic microhabitats in these taxa. Recycling of these proteins happens in rhogocytes, solitary derivates of podocytes (again with ultrafiltration areas) lying in the haemocoel.

Excretion occurs in two steps as in most bilaterians: In adults, ultrafiltration takes place by podocytes which are situated in the atrial wall of the pericardium into the latter. The paired excretory ducts (nephridia) modify this primary filtrate to the final urine by absorption and secretion; in terrestrial gastropods also water is reabsorbed by passing through so-called urether.

Contrary to vertebrates and other phyla, the stomach of molluscs mainly is a sorting center rather than a digestion area. Uptake of food particles mainly takes place in the midgut (digestive) gland by means of endocytosis and (particularly in case of detritivory or herbivory) by the intestine.

GENETICS

Despite the long tradition of studying and the ecological importance of Mollusca, our genetic understanding of the phylum is still in its infancy. Only a dozen full genomes (three bivalves, five gastropods, and four cephalopods) have been elaborated up till now, mostly in species of high economic importance. A few more are in a draft stage or currently in the pipeline. Accordingly, there is still a long way for providing a general characteristic of the molluscan genome. What is known calls for caution to provide any generalization: For instance mitochondrial gene order may be very conservative even over classes (e.g., Polyplacophora, most basal Gastropoda) or may significantly change within genera (e.g., Vermetidae) or in heterobranch Gastropoda. Bivalvia are famous for their double uniparental inheritance of mitochondria (Doucet-Beaupré et al. 2010; Boyle and Etter 2012; Plazzi et al. 2014; Plazzi and Passamonti 2019), the sequences of which may differ up to 20% between sexes of the same species.

Concerning evo-devo genomics, the situation is somewhat better (De Oliveira et al. 2016; Huan et al. 2020: Figure 19.2). Mollusca uniquely show a decoupled dorsoventral Hox expression: Hox gene expression in the ventral ectoderm resembles that of other bilaterians and likely contributes to neurogenesis. In contrast, Hox expression on the dorsal side is strongly correlated with shell formation and exhibits lineage-specific characteristics in each class of molluscs. Certain Hox genes have lost their original function and were co-opted into the evolution of taxon-specific novelties. The bone morphogenetic protein (BMP)/decapentaplegic (Dpp) signaling pathway, that mediates dorsoventral axis formation, and molecular components that establish chirality appear to be more conserved in molluscs, nevertheless certain variations from the common scheme occur within molluscan sublineages (Wanninger and Wollesen 2019).

POSITION IN THE ToL AND FOSSIL RECORD

The tremendous fossil record of Mollusca is remarkable in several aspects: In general molluscs have been abundant throughout the Phanerozoic eon; however, distinct taxa are often characteristic for a particular time-frame. Systematic placement of Cambrian to Ordovician fossils suffers from the fact that mostly only a steinkern is retained, thus information on shell structure is lacking. This is particularly true for the quite simple limpet shell shape, which may represent a monoplacophoran, a gastropod, an extinct conchiferan taxon, or even a brachiopod (e.g., Craniida). On the other hand, the common preservation of muscle scars also provides some insights into soft-part anatomy even if only the shell is retained.

The correlation of protoconch morphology (embryonic shell only/sculptured or smooth larval shells) with the mode of development (lecithotrophic, planktotrophic, or secondarily intracapsular development) causes the rare possibility to infer reproductive biology in fossil invertebrates. It is still debatable, whether and which prominent fossil taxa (e.g., Helcionellida, Rostroconcha) should be given a class status proper.

Aside from the unsolved question of molluscan internal relationships (**Figure 19.4**), their position among the Lophotrochozoa is also doubtful and heavily debated. Also, recent phylogenomic analyses (Marlétaz et al.; Laumer et al. 2019; Bleidorn 2019) could not find a robust solution for the

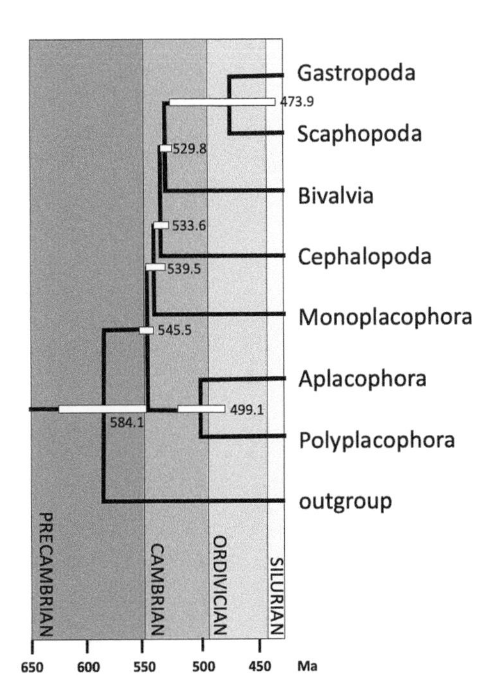

Figure 19.4 The most recent tree on molluscan relationships with an estimation of splitting times. (From Kocot et al. 2020. Drawing by Rob DeSalle.)

lophotrochozoan tree of life, the basal bifurcations of which certainly happened in Pre-Cambrian times, since the Lowest Cambrian already witnessed subgroups of e.g. Brachiopoda or Mollusca. At the phenotypic level (the level of selection) at least three other lophotrochozoan phyla show distinct similarities with Mollusca:

1. Polychaete Annelida and non-cephalopod Mollusca share a very similar cell-fate-pattern in their spiral-quartett cleavage. Also the trochophore larvae show significant similarities (apical tuft and organ, proto- and occasionally meta- and telotroch, one pair of anteriorly situated protonephridia) between the two phyla, although larval planktotrophy probably is a multiple secondary phenomenon (Haszprunar et al. 1995; Rouse 2000). These similarities are currently considered as lophotrochozoan plesiomorphies by most authors. To the contrary, annelid segmentation resp. seriality in Mollusca appear as independent steps in the evolution of each of the phyla (Haszprunar and Schaefer 1997).
2. The creeping larvae of solitary Kamptozoa (Entoprocta) show several remarkable similarities in particular with Solenogastres (Haszprunar and Wanninger 2008): A body being divided into a dorsal cuticularized episphere and a ventral foot sole; an anteriorly placed, large, subepidermal pedal gland and several bundles of cirri consisting of compound cilia and the anterior end of a pedal sole; ventrally intercrossing dorsoventral muscle fibers, a chitinous, but non-moulted cuticle, the sinusoid circulatory system, and tetraneury are shared characters. However, a close relationship between Mollusca and Kamptozoa has only been occasionally revealed as a result of the various molecular studies. It is interesting to note that the two most recent phylogenomic analyses showed Mollusca and Kamptozoa either as the most basal subsequent offshoots (Laumer et al. 2019) or as the basal clade (Sinusoida: Marlétaz et al. 2019; Bleidorn 2019) of Trochozoa.
3. Craniida and articulate Brachopoda and shelled Mollusca (incl. Polyplacophora) share a principally similar calcareous shell with an external periostracum. In addition, the structure and mode of development of brachiopod caeca and the aesthetes of Polyplacophora are very similar (Reindl and Haszprunar 1996; Reindl et al. 1995, 1997). Assumptions of a closer relationships between Brachiopoda and Mollusca are supported by the fact that extinct, shelled Hyolitha (formerly considered as Mollusca; e.g., Malinky and Yochelson 2007) had a lophophore (Moysiuk et al. 2017) and that the gastropod-like *Pelagiella* was equipped with annelid/brachiopod bristles (Thomas et al. 2020).

DATABASES AND COLLECTIONS

The high species numbers and the broad ecological and economic interest in molluscs result in a large number of molluscan databases: For taxonomic purposes, the World Register of Marine Species (WORMS: http://www.marinespecies.org/) still is the global standard and has been extended to MOLLUSCABASE (https://www.molluscabase.org/index.php) also including curated data on freshwater and terrestrial species. CEPHBASE (http://cephbase.eol.org/) informs about all aspects of Cephalopoda. Numerous further databases do exist which are specific to certain taxa or countries.

There are several world-famous molluscan collections, both for extant or fossil taxa, being housed by large, globally acting natural history museums, being supplied by numerous smaller collections, which may yet be important for certain taxa or geographic areas. Many of these are providing data on their collection via the WWW, although it will need many decades to have a global overview just of the location of type specimens.

CONCLUSION

Mollusca is one of the most fascinating phyla: Their tremendous variability, in particular among gastropods, is simply amazing and includes truly weird creatures such as certain heteropods, pteropods or nudibranchs, but also bizarre cephalopods do fascinate the observer. The aplacophoran classes are represented by worms as are certain secondarily modified bivalves (shipworms: Teredinidae) or gastropods (e.g., *Helminthope, Rhodope, Platyhedyle*). Molluscs created headlines even in most recent years, when new forms caused surprises, e.g., the "scaly-foot" gastropod *Chrysomallon squamiferum*, the only metazoan with iron sulfide scales, or the deep-sea slug *Bathyhedyle boucheti*, which is listed among the 10 most remarkable discovered species of the past decade. As a scientist I am deeply impressed by their enormous importance as models, in particular, for neurobiology, biomineralization, or evolutionary biology, but the human being in me pay attention also on their potential to develop new and promising therapies against cancer and pain. Last but not least their admirable beauty makes me happy to be a malacologist.

REFERENCES

Bleidorn, C., 2019. Recent progress in reconstructing lophotrochozoan (spiralian) phylogeny. Organisms, Diversity and Evolution, 19(4), 557–566.

Boyle, E.E., Etter, R.J., 2012. Heteroplasmy in a deep-sea protobranch bivalve suggests an ancient origin of doubly uniparental inheritance of mitochondria in Bivalvia. Marine Biology, 160, 413–422.

Cuvier, G., 1817a. Le régne animal, distribué d´aprés son organisation, pour servir de base à histoire naturelle des animaux et d´introduction à l´anatomie comparée. Tome II (xviii + 532 pp). Detervile, Paris.

Cuvier, G., 1817b. Mémoires pour servir à l'histoire et al´anatomie des Molluscques. (Viii +1073 pp). Detervile, Paris.

De Oliveira, A.L., Wollesen, T., Kristof, A., Scherholz, M., Redl, E., Todt, C., Bleidorn, C., Wanninger, A., 2016. Comparative transcriptomics enlarges the toolkit of known developmental genes in mollusks. BMC Genomics, 17(905), 1–23.

Doucet-Beaupré, H., Breton, S., Chapman, E.G., Blier, P.U., Bogan, A.E., Steward, D.T., Hoeh, A.R., 2010. Mitochondrial phylogenomics of the Bivalvia (Mollusca): searching for the origin and mitogenomic correlates of doubly uniparental inheritance of mtDNA. BMC Evolutionary Biology, 10(50), 1–19.

Haszprunar, G., 2000. Is the Aplacophora monophyletic? A cladistic point of view. American Malacologial Bulletin, 15(2), 115–130.

Haszprunar, G., Schaefer, K., 1997. Anatomy and phylogenetic significance of *Micropilina arntzi* (Mollusca, Monoplacophora, Micropilinidae fam.nov.). Acta Zoologica (Stockholm), 77(4), 315–334.

Haszprunar, G., Salvini-Plawen, L.V., Rieger, R.M., 1995. Larval planktotrophy – a primitive trait in the Bilateria? Acta Zoologica (Stockholm), 76(2), 141–154.

Haszprunar, G., Wanninger, A., 2008. On the fine structure of the creeping larva of *Loxosomella murmanica*: additional evidence for a clade of Kamptozoa (Entoprocta) and Mollusca. Acta Zoologica (Stockholm), 89(2), 137–148.

Huan, P., Wang, Q., Tan, S., Liu, B., 2020. Dorsoventral decoupling of Hox gene expression underpins the diversification of molluscs. Proceedings of the National Academy of Sciences of the USA, 117(1), 503–512.

Ihering, H.V., 1876. Versuch eines natürlichen Systemes der Mollusken. Jahrbuch der Deutschen Malakozoologischen Gesellschaft, 3, 97–148.

Jonstonus, J. 1650. Historiae naturalis de quadrupetibus libri: cum aeneis figuris. Illustrated and published by Matthaeus Merian. Francofurti ad Moenum.

Laumer, C.E., Fernández, R., Lemer, S., Combosch, D., Kocot, K.M., Riesgo, A., Andrade, S.C.S., Sterrer, W., Sørensen, M.V., Giribet, G., 2019. Revisiting metazoan phylogeny with genomic sampling of all phyla. Proceedings of the Royal Society of London, B286(1906), #20190831 (10 pp).

Lemche, H., 1957. A new living deep-sea mollusc of the Cambrio-Devonian class Monoplacophora. Nature, 179(4556), 413–416.

Linnaeus, C.V., 1758. Systema naturae per regna tria naturae, secundem classes, ordines, genera, species cum characteribus, differentis, synonymis. Holmiae, Laurentius Salvius, Stockholm (10th ed.), Vol. 1, 823.

Malinky, J.M., Yochelson, E.L., 2007. On the systematic position of the Hyolitha (Kingdom Animalia). Memoirs of the Association of Australasian Palaeontologists, 34, 521–536.

Marlétaz, F., Peijnenburg, K.T.C.A., Goto, T., Satoh, N., Rokhsar, D.S., 2019. A new spiralian phylogeny places the enigmatic arrow worms among gnathiferans. Current Biology, 29(2), 312–318.

Moysiuk, J., Smith, M.R., Caron, J.-B., 2017. Hyoliths are Palaeozoic lophophorates. Nature, 541(7637), 394–397, Supplements.

Plazzi, F., Cassano, A., Passamonti, M., 2014. The quest for doubly uniparental inheritance in heterodont bivalves and its detection in *Meretrix lamarckii* (Veneridae: Meretricinae). Journal of Zoological Systematics and Evolutionary Research, 53(1), 87–94.

Plazzi, F., Passamonti, M., 2019. Footprints of unconventional mitochondrial inheritance in bivalve phylogeny: Signatures of positive selection on clades with doubly uniparental inheritance. Journal of Zoological Systematics and Evolutionary Research, 57(2), 258–271.

Reindl, S., Haszprunar, G., 1996. Shell pores (caeca, aesthetes) of Mollusca: A case of polyphyly. In: Taylor J.D. (ed), Origin and Evolutionary Radiation of the Mollusca (pp. 115–118). Oxford University Press, Oxford.

Reindl, S., Salvenmoser, W., Haszprunar, G., 1995. Fine structural and immunocytochemical investigations of the caeca of *Argyrotheca cordata* and *Argyrotheca cuneata* (Brachiopoda, Terebratulida, Terbratellacea). Journal of Submicroscopical Cytology and Pathology, 27(4), 543–556.

Reindl, S., Salvenmoser, W., Haszprunar, G., 1997. Fine structure and immunocytochemical studies on the eyeless aesthetes of *Leptochiton algesirensis*, with comparison to *Leptochiton cancellatus* (Mollusca, Polyplacophora). Journal of Submicroscopical Cytology and Pathology, 29(1), 135–151.

Rouse, G.W. 2000. Polychaetes have evolved feeding larvae several times. Bulletin of Marine Science, 67(1), 391–409.

Thomas, R.D.K., Runnegar, B., Matt, K. 2020. *Pelagiella exigua*, an early Cambrian stem gastropod with chaetae: lophotrochozoan heritage and conchiferan novelty. Palaeontology, 63(4), 601–627.

Wanninger, A., Wollesen, T., 2015. Mollusca. In: Wanninger, A. (ed.), Evolutionary Developmental Biology of Invertebrates, Vol 2, (pp. 103–153). Lophotrochozoa (Spiralia). Springer Verlag, Wien.

Wanninger, A., Wollesen, T., 2019. The evolution of molluscs. Biological Review, 94(1), 102–115.

Additional Readings

Avila, C. 2007. Molluscan natural products as biological models: Chemical ecology, histology, and laboratory culture. In: Cimino, G., Gavagnin, M. (eds.), Molluscs – From Chemo-ecological Study to Biotechnological Application (pp 1–23). Springer Verlag, Berlin.

Beesley, P.L., Ross, G.J.B., Wells, A. (eds.) 1998. Mollusca: The Southern Synthesis. Fauna of Australia, Vol. 5A (xvi, 563 pp); Vol.5B (viii, pp. 565–1234). CSIRO Publishing, Melbourne.

Benkendorff, K., 2010. Molluscan biological and chemical diversity: secondary metabolites and medicinal resources produced by marine molluscs. Biological Reviews, 85(4), 757–775.

Brusca, R.C., Lindberg, D.L., Ponder, W.P. 2016. Mollusca. In: Brusca, R.C., Moore, W., Shuster, S.M. (eds.), Invertebrates (pp 453–530). 3rd Ed. Sinauer Association.

Glaubrecht, M., 2009. On "Darwinian Mysteries" or molluscs as models in evolutionary biology: From local speciation to global radiation. American Malacological Bulletin, 27(1–2), 3–23.

Kocot, K.M., Poustka, A.J., Stöger, I., Halanych, K.M., Schrödl, M., 2020. New data from Monoplacophora and a carefully-curated dataset resolve molluscan relationships. Scientific Reports, 10, 101 (8 pp).

Ponder, W.F., Lindberg, D.R. (eds.) 2008. Phylogeny and Evolution of the Mollusca (1019). University of California Press, Berkeley.

Ponder, W.F., Lindberg, D.R. (illustrations by J.M. Ponder) 2019/2020. Biology and Evolution of the Mollusca. 2 Vols. 864 + 870 pp, CRC Press.

Sturm, C.F., Pearce, T.A., Valdes, A., 2006. The Mollusks: A Guide to Their Study, Collection, & Preservation (xii+445 pp, 101 figure). University Publication, Boca Raton.

Williams, S.T., 2017. Molluscan shell colour. Biological Reviews, 92(2), 1039–1058.

PHYLUM ANNELIDA

CHAPTER

20

Sebastian Kvist and Alejandro Oceguera-Figueroa

Infrakingdom	Bilateria
Division	Protostome
Subdivision	Lophotrochozoa
Phylum	Annelida

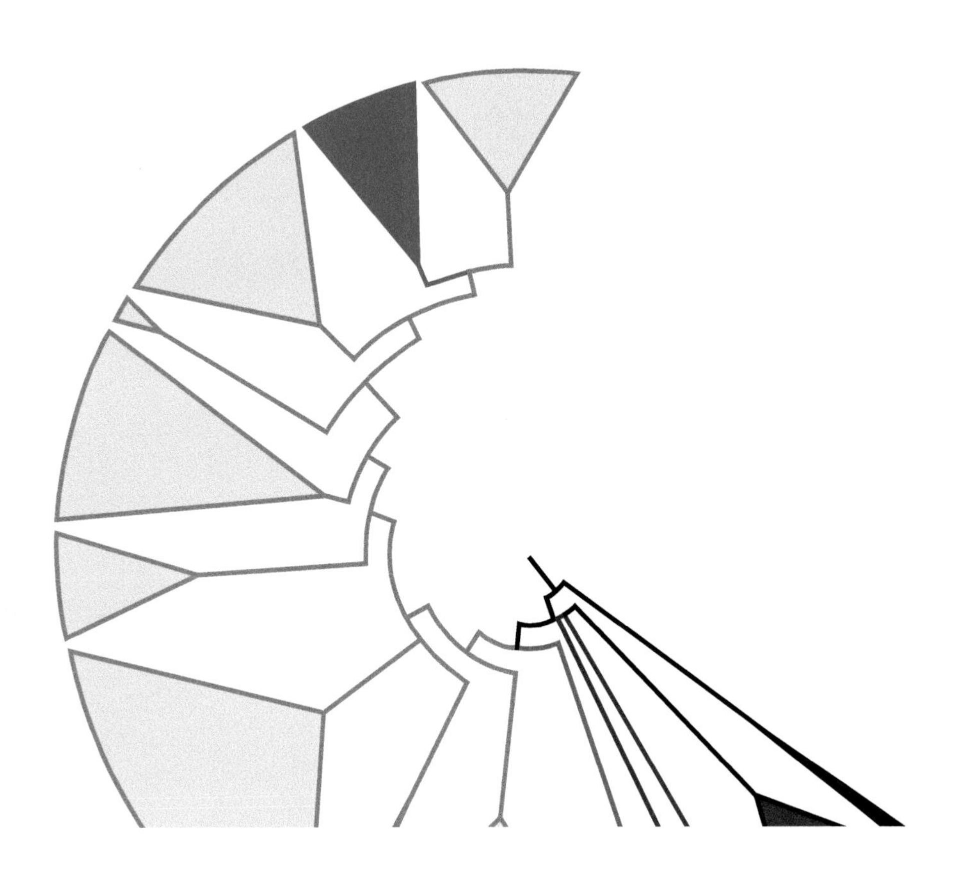

CONTENTS

The phylogenetic position of the phylum in the animal tree of life and the phylogenetic patterns within the phylum are discussed in detail.

General Characteristics

Over 22,000 valid species

Number of undescribed or "cryptic species" estimated between 5,000 and 20,000

HISTORY OF TAXONOMY AND CLASSIFICATION

The phylum Annelida (Greek *annelus* = small ring, *ida* = suffix), whose members are commonly known as segmented worms, includes bristle worms, earthworms and their relatives, leeches, crayfish worms and acanthobdellidands (Lamarck, 1818; Hatschek, 1878). The phylum is part of the superphylum Trochozoa, which also includes Nemertea, Mollusca, Brachiopoda, and Phoronida (see Dunn et al., 2014; Kocot, 2016; Kocot et al., 2017). Annelida was historically divided into two classes of exclusively segmented worms: Errantia and Sedentaria (de Quatrefages 1865; Fauvel 1923; Fauvel 1927). As the names imply, Errantia harbored highly vagile and errant worms, whereas Sedentaria included more sedentary, often sessile species; awkwardly, Sedentaria also included leeches, some of which are highly mobile. At that point in time, rather than being based on morphological synapomorphies, the classification scheme was probably a matter of convenience (Kvist & Siddall, 2013). As our understanding of the detailed morphology and evolutionary histories of annelids expanded, annelid taxonomists strayed from this traditional classification scheme, ultimately leading to Fauchald's (1977) elimination of these taxa and erection of 17 new groups. These groups were later expanded to include more orders (Rouse & Fauchald, 1997; Rouse & Pleijel, 2001; Bartolomaeus et al., 2005). For most of the 1900s, however, the two recognized classes within Annelida were Polychaeta (Greek *poly* = many, *khaitē* = long hair), which includes the bristle worms, and Clitellata (from the term clitellum, a glandular girdle that is used to secrete the casing for their eggs during cocoon-laying), which includes Oligochaeta (earthworms, sludgeworms, etc.) and Hirudinea (leeches, crayfish worms, and acanthobdellidans; Rouse and Fauchald 1995, 1998; McHugh 2000; Rousset et al., 2007); **Figure 20.1** shows some representative taxa from the different groups. The advent of DNA sequencing in biological sciences revolutionized the field of phylogenetics and allowed researchers to understand that Clitellata is nested within the paraphyletic Polychaeta (McHugh 2000; Rousset et al. 2007; Zrzavý et al. 2009; Struck et al. 2007, 2011; Kvist and Siddall 2013; Weigert et al. 2014; **Figure 20.2**). Due to the predicaments related to the use of non-monophyletic groups in taxonomic classification schemes and, in part, due to our improved understanding of annelid evolutionary relationships, some authors argued for the re-erection of the previous classes Errantia and Sedentaria, based on evidence from phylogenetic analysis (Struck et al., 2011; Weigert et al., 2014; but see also Kvist & Siddall, 2013). Although most contemporary systematists favor the use of Errantia and Sedentaria, there is currently no full consensus in the scientific community as to the nomenclature of higher taxonomic groups.

Errantia includes the subclasses Aciculata and Protodriliformia (Andrade et al., 2015; Weigert & Bleidorn, 2016); note that the taxonomic ranks of these are sometimes dubious. Aciculata, in turn, contains the orders Phyllodocida and Eunicida which include mainly macroscopic bristle worm species (Andrade et al., 2015). Protodriliformia, by contrast, includes mainly interstitial taxa (those that live in the spaces created between sand grains in marine sediments) (Struck et al., 2015). The second class, Sedentaria, includes most of the families that were previously placed within the orders Canalipalpata and Scolecida. Importantly, however, several taxa that have variably been referred

Autapomorphies

no clear morphological autapomorphies exist, but the following characters can help circumscribe the phylum:

- true metamerism, with body divided into "head" region (prostomium and peristomium), trunk, and pygidium
- chaetae that arise from the parapodia or body wall, generally paired on each segment (secondarily lost in Hirudinea)
- nuchal organs; chemoreceptors located behind the prostomium (secondarily lost in Clitellata)
- epidermis covered by cuticle with basal collagenous matrix

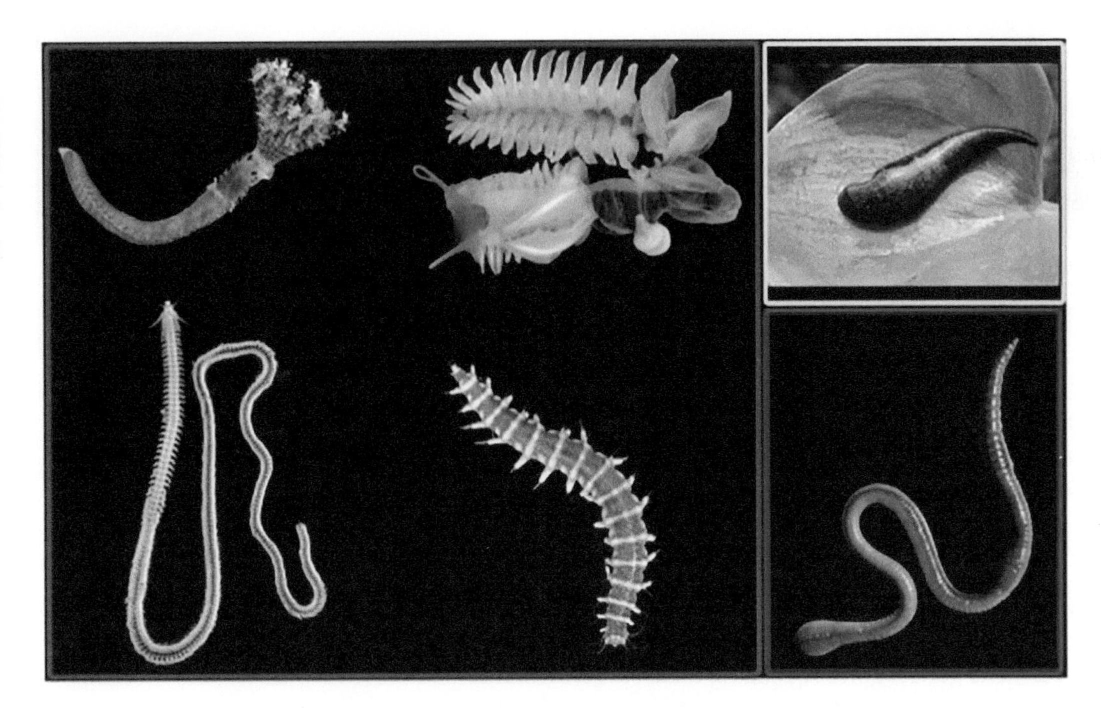

Figure 20.1 Live habitus of selected specimens of Polychaeta (red box), Oligochaeta (blue box), and Hirudinea (yellow box). (Photos of polychaetes are courtesy of Gonzalo Giribet (Harvard University), the photo of the oligochaete is adapted from National Public Radio (taken by Ed Marshall), and the photo of the hirudinean is courtesy of Charlotte Calmerfalk Kvist.)

to as annelids do not seem to find a place within either Errantia or Sedentaria. Due to challenges in identifying morphological characters that unify these groups and the fact that modern molecular phylogenetics places them outside of Errantia and Sedentaria, each of Amphinomida, Chaetopteridae, Magelonidae, Oweniidae, and Sipuncula is often represented outside of the major diffractions of Annelida (Weigert et al., 2014; Andrade et al., 2015).

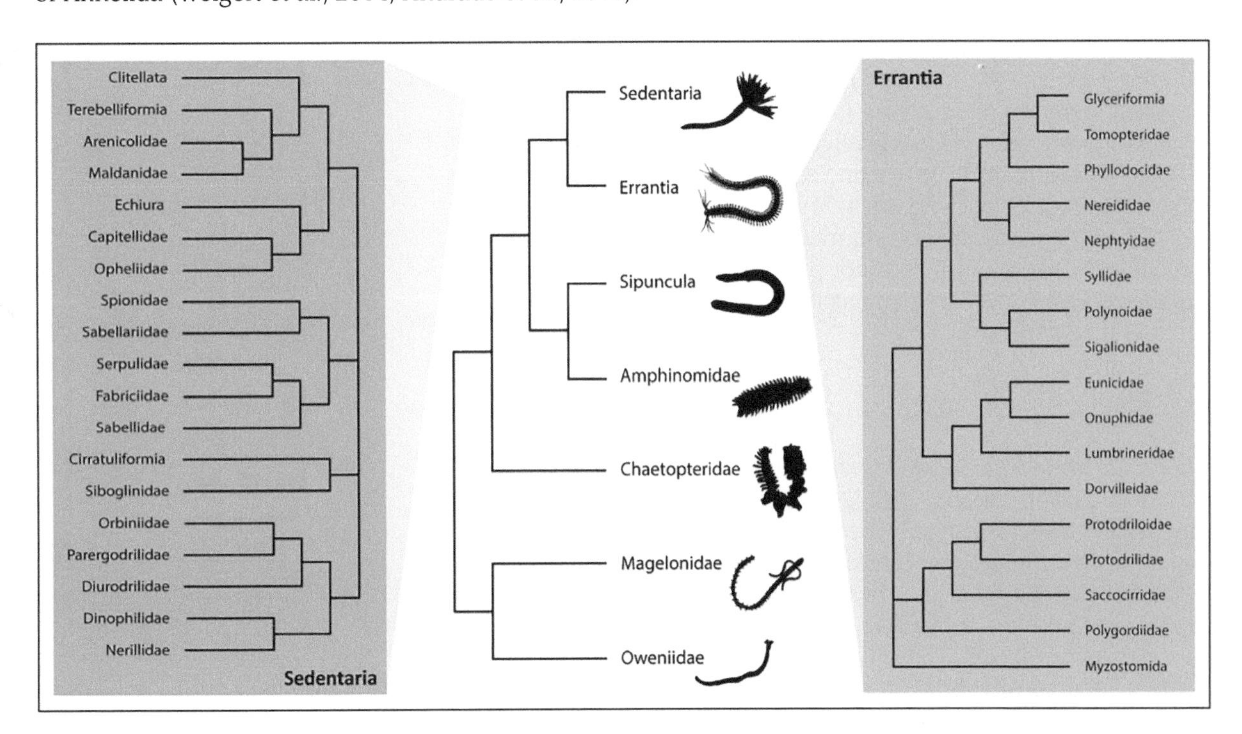

Figure 20.2 Phylogenetic hypothesis of Annelida, based on our current understanding. (The image is taken from Weigert and Bleidorn 2016.)

Although Oligochaeta is rendered paraphyletic in most modern phylogenetic analyses by virtue of Hirudinea nesting inside of it, it is useful to discuss taxonomic classification of Clitellata from the standpoint of separating Oligochaeta and Hirudinea. The separation of these taxa is largely historical and, as a result, the classification schemes within each of these have been held separate from the other. The higher-level classification within Oligochaeta, which includes some 5,000 species, is still contentious and taxonomic rearrangements are not infrequent, although the utility of molecular phylogenetics has begun to resolve some of the recalcitrant splits in the trees. Erséus (2005) and Timm & Martin (2015) provide excellent accounts of both historical and modern knowledge leading to the classification system of today. One of the first overarching classification schemes for Oligochaeta was proposed by Beddard (1895), who suggested a basic split between Microdrili (small, often microscopic, worms) and Megadrili (species with larger bodies, including earthworms) – the two taxa that had previously been suggested to encompass all oligochaete species (Benham, 1890). Later, Jamieson (1988) suggested that Megadrili be replaced with the name Metagynophora and, based on formal phylogenetic analyses, also erected a less inclusive taxon Crassiclitellata to encompass the 3,000 species of earthworms that possess multilayered clitellums. Although Crassiclitellata has been shown to be a natural group based on phylogenetic analyses using both standard Sanger sequencing as well as next-generation sequencing datasets (Jamieson et al., 2002; Anderson et al., 2017), the phylogenetic status of Metagynophora is yet to be tested. In general, oligochaete experts assume that Microdrili is not monophyletic and, as a result, taxonomy within this group is normally discussed from a family perspective.

The subclass Hirudinea includes Hirudinida ("true leeches"), Acanthobdellida, and Branchiobdellida (Odier, 1823; Livanow, 1931; Sawyer, 1986; Siddall et al., 2001; Tessler et al., 2018a). Traditional classification subdivided Hirudinida into two orders based on the morphology of feeding structures: Rhynchobdellida, which includes proboscis-bearing leeches, and Arhynchobdellida, which includes species that lack such a structure (Sawyer, 1986). However, despite concerted efforts, modern phylogenetics has failed to recover Rhynchobdellida as a monophyletic group (Siddall & Burreson, 1998; Apakupakul et al., 1999; Trontelj et al., 1999). Consequently, Tessler et al. (2018a) abolished these orders and erected and re-erected five new orders: Oceanobdelliformes (including the families Piscicolidae and Ozobranchidae, most members of which are marine), Glossiphoniformes (including the family Glossiphoniidae), Americobdelliformes (including the monotypic family Americobdellidae), Erpobdelliformes (including the families Erpobdellidae, Orobdellidae, Gastrostomobdellidae and Salifidae), and Hirudiniformes (including several families of both hematophagous and macrophagous leeches). It has not yet been convincingly inferred whether or not phylogenomics and the utility of genome-scale datasets will recover Rhynchobdellida as monophyletic.

ANATOMY, HISTOLOGY, AND MORPHOLOGY

Excellent overviews of the morphology and natural history of Annelida are available in Aguado et al. (2014), Verdonschot (2015), and Timm & Martin (2015). Annelida is one of only three phyla with species that show true segmentation, i.e., serially repeated structures along the anteroposterior body axis including repetition of features derived from the mesoderm and endoderm; the others being Arthropoda and Chordata. Almost all annelids are elongated and worm-like, often referred to as vermiform. These worms range in length from less than a millimeter to more than 6 m (Avel, 1959; Aguado et al., 2014) and even the longest species are generally less than 5–10 cm wide. The general body plan of polychaetes includes two anterior pre-segmental regions (prostomium and peristomium – together, these constitute the "head' region), a segmented trunk

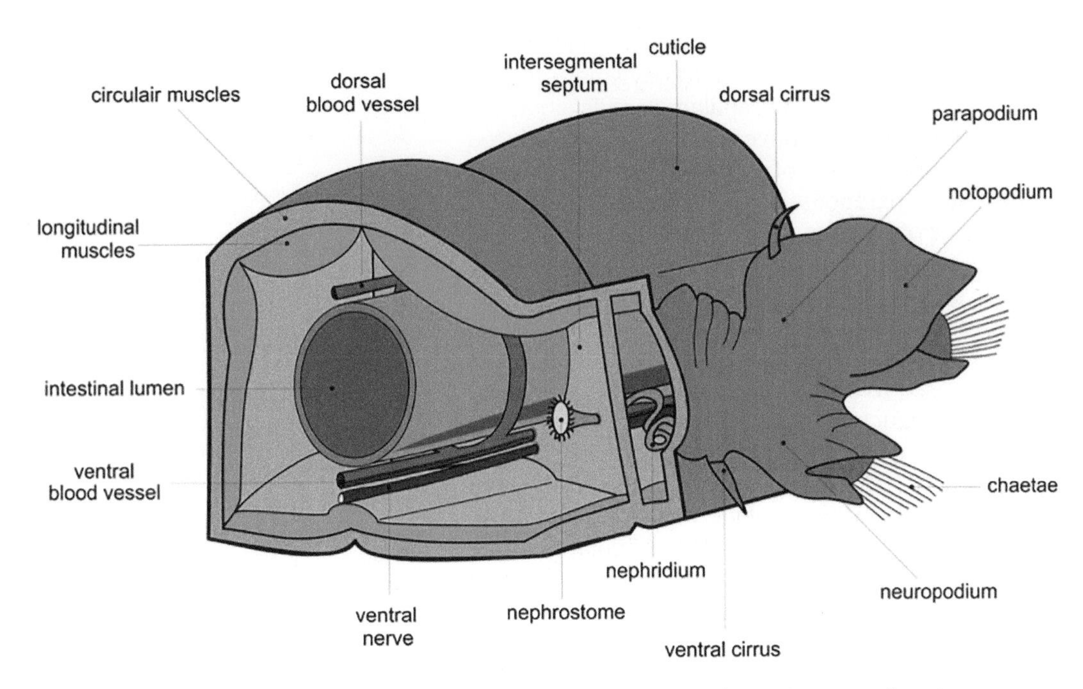

Figure 20.3 Cross-section of a polychaete showing the gross internal anatomy, as well as the parapodial composition. (The image is used here under the Creative Commons Attribution-Share Alike 4.0 International license and is attributed to Hans Hillewaert.)

or body, and a post-segmental tail region known as the pygidium. In extant polychaetes, the head region is devoid of parapodia (chaetae-bearing, paired, unjointed outgrowths on each segment) but extinct forms have been shown to carry these structures, as well as chaetae, in the head region (Nanglu & Caron, 2018). The parapodia in the segmented portion of the body are normally divided into a dorsal lobe (notopodium) and ventral lobe (neuropodium), with chaetal bundles extending from each lobe (**Figure 20.3**). Often, these lobes are supported by internal, thickened, bristle-like structures called aciculae (*pl.*). The pygidium does not possess parapodia but some species possess post-anal cirri extending from the posteriormost part of the body. The prostomium bears sets of bicellular eyes (with receptor cells and supporting cells; the number of eyes varies between species), palps (cushion-like structures arising from the sides of the head region, normally used as sensory structures or for feeding), nuchal organs (a pair of chemosensory structures on the lateral margins of the prostomium; this seems to be a synapomorphic feature in Polychaeta), and some species also bear paired antennae. The peristomium bears the mouth and normally also a set of tentacular cirri. These cirri are particularly evident in suspension-feeding taxa, which may bear them as a crown around the mouth. Some species also possess gills in the anterior portion of the body.

Polychaetes possess both circular and longitudinal musculature throughout their bodies, with an epithelium-derived, soft, cellular cuticle surrounding the body. A "brain," or mass of cerebral ganglia, is present in the head of polychaetes and this mass is connected to the ventral nerve cord, which runs the length of the body. The nerve cord dilates into a ganglion in each segment – this is true also for oligochaetes and hirudineans, although the latter also possesses two sets of enlarged ganglionic masses, one at each end of the body (Eckert, 1963). Each segment of the main body also possesses paired nephridia that are used to evacuate metabolic waste products. The digestive system consists of (i) a foregut, which includes the pharynx and anterior esophagus, (ii) a midgut specialized in breaking down and digesting nutrients, and (iii) a hindgut that connects the midgut to the anus. Polychaetes normally have two fluid systems: the coelomic system and the circulatory system – both are used to evacuate waste products – although circulatory systems can be absent in smaller taxa. The circulatory system (hemolymph system) is also used to oxygenate organs and is normally closed, consisting of both major and capillary vessels.

Members of Oligochaeta, as the name implies, have bristles that are reduced both in number and size. Moreover, the bristles of oligochaetes eject directly from pores in the body wall, instead of parapodia as in Polychaeta. Each segment normally possesses four bundles of chaetae, laterodorsally and lateroventrally on each side of the body. The prostomium is followed by chaetae-bearing segments and the oligochaete "head" lies within the first four segments of the body; note here that "head" refers to the grouping of cerebral ganglionic masses. The segments are commonly numbered using Roman numerals, beginning at I following the head region. As opposed to polychaetes, the oligochaete head lacks sensory structures in the form of nuchal organs. The first segment (after the prostomium) bears the mouth in oligochaetes. The main body follows the prostomium and the last segment, the pygidium, bears the anus. The body wall of oligochaetes consists of a thin cuticle covering the epidermis. Importantly, the epidermis is lined with mucus-secreting glands. The circular musculature layer is present directly underneath the epidermis and the longitudinal muscle layer lies underneath the circular layer.

The easiest way to picture the internal morphology of oligochaetes is by imagining a tube within a tube. The coelomic cavity is located between the central cylindrical gut and the outer dermal layer (**Figure 20.4**). Because of the segmentation, the coelomic cavity is divided into compartments. The mouth opens up into a buccal cavity, which, in turn, opens into a rather muscular pharynx. Posteriorly, the pharynx becomes the crop, which, in some species, has a specialized compartment known as the gizzard. The function of the gizzard is mainly to mechanically macerate food particles. The crop/gizzard opens up into the intestinal tract. The circulatory system is present both ventrally and dorsally and the two major vessels branch off into smaller capillaries to supply organs with oxygen and nutrition. Oligochaetes have several "hearts" and these are arranged as commissural vessels in the anterior half of the body. All clitellates are simultaneous hermaphrodites and, in oligochaetes, male and female reproductive

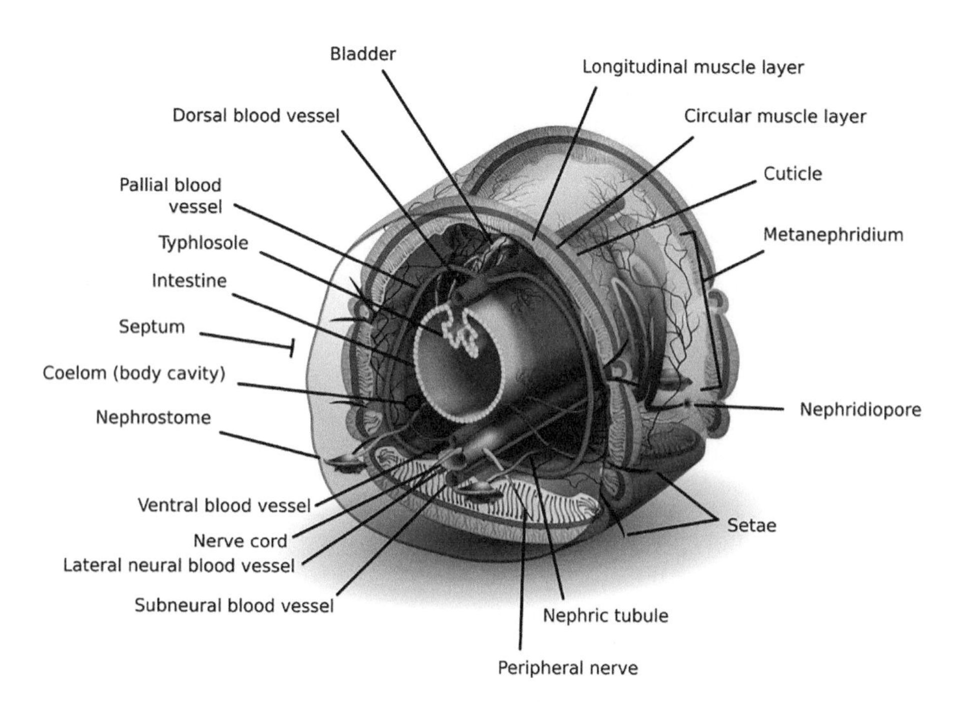

Figure 20.4 Cross-section of an oligochaete showing the gross internal anatomy, as well as the chaetal arrangement. (The image is used here under the Creative Commons Attribution-Share Alike 3.0 International license.)

structures are paired and, most commonly, lie close to each other in the anterior half of the body. As opposed to hirudineans (see below) the anteroposterior sequence of male and female organs varies between families and species.

In many ways, the overarching morphological characters of hirudineans resemble those of oligochaetes. However, some important differences do exist. First, hirudineans have secondarily lost their chaetae throughout the length of the body. Second, leeches possess attachment organs, or suckers, both at the anterior and posterior ends of the body. Third, as opposed to oligochaetes, hirudineans have a fixed number of 34 segments (referred to as somites in leech literature, and numbered with Roman numerals from I to XXXIV). However, each of the somites is superficially subdivided into annuli that can easily be mistaken for extra somites when internal morphology is not investigated. The annulation pattern varies between species although general patterns can be discerned for certain families. The prostomium is commonly referred to as the first somite and bears the mouth, which lies within the oral sucker, normally toward the posterior lip. Eyespots are normally present in one or a few consecutive annuli, beginning in somite II. The number and arrangement of eyespots is often diagnostic at the familial level. The anus is dorsal and anterior to the last somite. In contrast to most oligochaetes, hirudineans often possess ornate pigmentation on the dorsal and/or ventral surfaces, sometimes with elaborate coloration. At various levels of conspicuousness, leeches possess a series of sensory papillae on the dorsal surface of the body.

Phylogenetic systematics of Hirudinea suggests that proboscis-bearing represents the ancestral condition within the group. Several families have secondarily transformed the feeding structures into jaws or completely lost these structures all together. In at least one family, Salifidae, jaws seem to have evolved into stylets inside of the mouth. The general condition within medicinal leeches, which are not a monophyletic group, is the possession of three jaws, lined with 50–90 teeth or denticles. Regardless of feeding structures, these open up into a pharynx (often less muscular than in Oligochaeta), which is followed by the crop. As opposed to oligochaetes, the hirudinean crop is formed by a series of digitate processes (caeca) stemming from the alimentary canal. Designed to store blood or other food sources for extended periods, the crop consists of highly expandable tissues, where ingested meals are stored and from which water is removed. Although seemingly specialized for blood storage, the crop morphology is conserved in some non-bloodfeeding leeches, possibly attesting to the hematophagous nature of the ancestral leech (Siddall et al., 2011, 2016; Tessler et al., 2018a). Often, the last pair of processes is longer than the rest and positioned posteriorly toward the caudal sucker – these are commonly known as post caeca. Posterior to the post caeca lies the intestine, which ultimately leads to the anus. As in polychaetes and oligochaetes, the nerve cord lies ventrally throughout the length of the body and ganglionic masses of nerves allocated throughout its length, one in each mid-body somite. Each mid-body somite also harbors a pair of nephridia, which typically empty to the outside via nephridiopores. In hirudineans, the male gonopore always precedes the female gonopore anteroposteriorly.

ECOLOGY AND GEOGRAPHIC DISTRIBUTION

Out of the over 22,000 species of annelids that are currently recognized (Aguado et al., 2014), the vast majority dwell in the ocean. However, annelids are abundant in freshwater and terrestrial environments also; in fact, all three major groups have been recorded from all three major ecosystems (freshwater, marine, and terrestrial; see **Figure 20.5**). Notwithstanding the wide geographic distribution of marine polychaetes, leeches alone exist on every major continent (including the oceans of Antarctica), in every investigated major body of water (freshwater

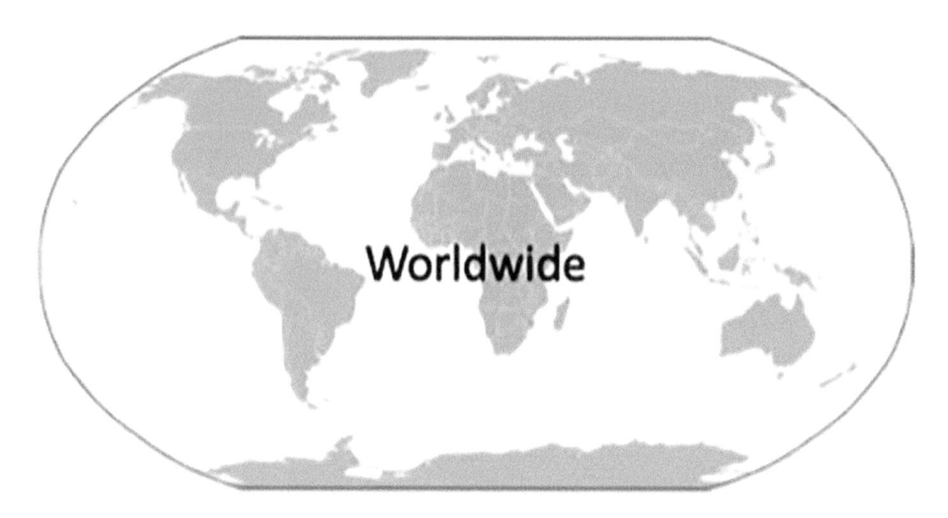

Figure 20.5 Distribution of annelid taxa across the world. Note that annelids are cosmopolitan, occurring in all three major ecosystems (land, freshwater, and marine) across Earth.

and marine ecosystems) and on land (Sket & Trontelj, 2008). Indeed, annelids have been found living in both temperate and extreme environments, from inside arctic ice sheets (Shain et al., 2001; Hartzell et al., 2005) and hydrothermal vents thousands of meters below the sea surface (Gaill et al., 1995; Black et al., 1997; Hurtado et al., 2004), along with hot springs and extremely high saline environments (Verdonschot, 2015). Members of this phylum are also often recovered from ephemeral habitats that are subject to heavy fluctuations in environmental conditions, such as intermittent streams, seasonal ponds, and mudflats. This phylum, like a few other phyla, truly is cosmopolitan. In addition to this astonishingly broad geographic distribution, annelids are often mega abundant, sometimes dominating ecosystems. Fossil annelids date back 508 million years to the Cambrian period (Nanglu and Caron, 2018), and seem to be relatively abundant throughout the fossil record with the pinnacle of their abundance (using the number of described taxa as proxy) falling within the Cretaceous period (Morris and Peel, 2008; Bracchi and Alessandrello, 2005; Vinn et al., 2013). Unsurprisingly, due to the soft-bodied nature of oligochaetes and leeches, the overwhelming majority of annelid fossils pertain to Polychaeta. Polychaetes have been recorded at varying depths in the ocean, from intertidal down to over 10,000 m (Paterson et al., 2009). Marine oligochaetes are perhaps less extreme in the choice of environment, existing from the surface to a few thousand meters in depth (Martin et al., 1999). Importantly, annelids have been heavily employed as bioindicators of ecosystem health. Their suitability for this purpose stems both from the fact that some species are sensitive to environmental change and will not occur in heavily polluted areas, and from other species being extremely tolerant to pollution and environmental change (Paoletti et al., 1996; Martins et al., 2008; Sharma and Rawat, 2009; Pérès et al., 2011).

Feeding

Although most errant polychaetes normally are predators or scavengers, the sedentary (or even sessile) forms are commonly suspension or deposit feeders. Predaceous polychaetes are often decent swimmers and actively hunt for their prey either on the ocean floor or the water column directly above the bottom. Some of the more notorious taxa are "sit-and-wait" predators, effectively ambushing their prey. Arguably, the most well-known polychaete to adhere to this system is the bobbit worm (family Eunicidae), which can reach lengths of up to 3 m; the jaws are armed at the interface between the benthos and the water column, and the remainder of the body lay hidden inside the sediment. Its infamy largely stems from the scissor-like jaws that snap shut like a trapdoor when hunting; large eunicids can easily capture and devour fish that are several times their width. Several species of predatory polychaetes forcefully and rapidly evert their proboscis, latch on to prey, and then pull it toward their mouth. In the family Glyceridae, some species have taken this approach one step further and have evolved large fangs at the end of the proboscis (these fangs are conspicuous through the body wall even when the proboscis is inverted). Moreover, several species of the family have evolved venom that are connected to the fangs and administered during attacks (Böggemann, 2002; von Reumont et al., 2014). For suspension-feeding taxa, the tentacles and cirri are normally used to create water currents that bring suspended food particles to the mouth. In some cases, the animal will pick particles from the benthos with their tentacles and deposit them in the mouth.

Oligochaetes are scavengers or detritivors, feeding mostly on dead organic matter, especially dead vegetation available in the soil. Aquatic oligochaetes are commonly positioned head down in the sediment, where they selectively feed on clay or silt particles, which travel against gravity to the digestive tract and are excreted as pellets that drop to the surface of the benthos. Such is their appetite and abundance that the amount of fecal pellets can effectively change the habitat (Verdonschot, 2015). This activity is called bioturbation and is often beneficial in oxygen-poor environments and terrestrial soils. Some taxa capture clay or silt particles using their chaetal bundles, whereafter these are wiped off inside the mouth (Verdonschot, 2015).

Hirudineans seem to have evolved three main modes of feeding throughout their evolutionary history. The most common mode, occurring in about two-thirds of the known species, is hematophagy or blood feeding. Blood-feeding leeches either insert their proboscis subcutaneously into the prey to extract the blood, or use jaws armed with numerous teeth in a serrating motion to create an incision wound on the skin surface. In both cases, the leeches can feed up to five times their body weight in blood and these prolonged periods of feeding are only permitted by a cocktail of anticoagulation factors that are injected into the wound during feeding (Min et al., 2010; Kvist et al., 2011, 2013, 2014, 2016; Siddall et al., 2016; Müller et al., 2016; Tessler et al., 2018b). Salzet (2001) provides an excellent overview of our knowledge of leech anticoagulants prior to 2001. The remaining leeches are either liquidosomatophagous (these feed mainly on the hemolymph of aquatic invertebrates) or macrophagous (swallowing their prey whole). It is worth noting that there is still much debate as to the reasoning and foundation behind the separation of liquidosomatophagy and hematophagy, seeing as hemolymph and blood are relatively similar in composition between several taxa. Some macrophagous leeches are particularly voracious and can swallow earthworms or other leeches of approximately their own size. Our knowledge regarding the sensory structures used by leeches to detect prey is still scarce, but it is known that they react to motion, temperature, amino acid concentrations in the water, and carbon dioxide emissions – the latter is thought to be constrained to terrestrial taxa.

Locomotion

At various levels for different taxa, polychaetes actively burrow, creep, crawl, and/or swim. Their locomotion is enabled both by the contraction and relaxation of circular and longitudinal musculature, as well as the extension and retraction of parapodia and chaetae. Some species, such as those in the genera *Nereis* and *Hediste*, are adept swimmers and use these skills to hunt prey.

Locomotion in oligochaetes happens through peristalsis which creates wave-like movements through symmetrical contraction and relaxation of the circular and longitudinal muscles (Gardner, 1976).

Although locomotion in hirudineans is far more complex, two modes of moving present themselves as most dominant among the known taxa: a dorsoventral wave-like motion that sweeps posteriad during swimming and inchworm-like locomotion during periods of attachment to the substrate. Most aspects of leech biology and behavior, including locomotion, are covered in Sawyer's (1986) seminal treatise.

REPRODUCTION AND DEVELOPMENT

Whereas reproduction and development is well-documented in ecdysozoans, mainly growing from the foundation permitted by the plethora of studies into the fruit fly (genus *Drosophila*) model system, model organisms are much scarcer and much shallower studied in Trochozoa. In addition, there are few studies that address development in Annelida in general, but a comparatively higher number of studies that focus particularly on Polychaeta (Agassiz, 1866; Clapérede & Mecznikow, 1869; Day, 1934; Nolte, 1938; Segrove, 1941; Giangrande, 1997). In particular, D. P. Wilson, S. Okuda, and J. Blake have provided great service to our knowledge of polychaete development, through their tedious work rearing eggs, larvae, and juveniles through to adulthood. Leeches of the genus *Helobdella* also have been used as model organisms to study development and other aspects of trochozoan biology; one of the pioneers of this model system is D. Weisblat.

In broad brushstrokes, annelid development can be divided into five ontogenetic stages: egg – trochophore larva – metamorphosis – juvenile – adult. The trochophore larva (**Figure 20.6**) is of special interest (see Rouse, 1999 and reference therein for an overview of the spiralian trochophore larva) because a similar larval type is shared by the other phyla within Trochozoa; this character is, in part, the namesake for the superphylum Lophotrochozoa. As such, much of the literature revolves around the transition from trochophore larva to juvenile worm, including metamorphosis. Due to limited space and the tedious nature of

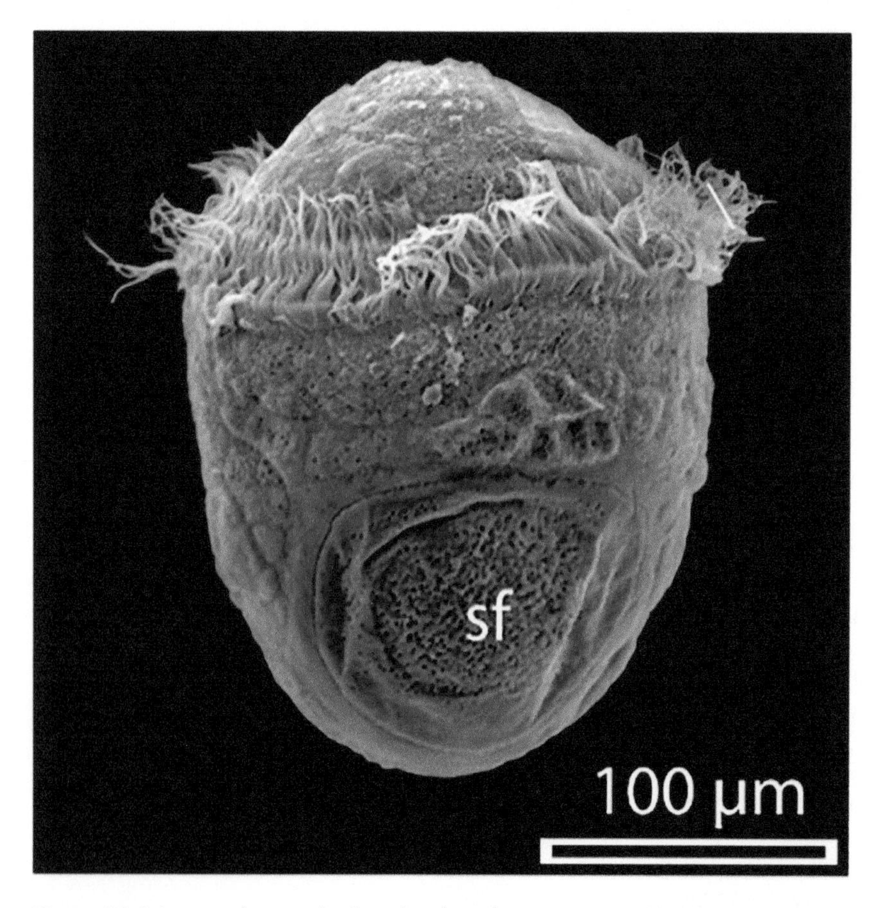

Figure 20.6 Image of a standard trochophore larva, possessed by the members of Trochozoa (Annelida, Nemertea, Mollusca, Brachiopoda, and Phoronida) and further discussed in the text. sf, shell field. (The image is taken from Jackson et al. (2007) and is used here under the Creative Commons Attribution License 2.0.)

exhaustively detailing the development of *all* polychaetes, the following account will, in most cases, be overly simplified and general. Although most polychaetes are dioecious (individuals possess either female or male reproductive structures), some are hermaphroditic (individuals possess both female and male reproductive structures) (Schroeder & Hermans, 1975; Fischer & Pfannenstiel, 1984). Polychaetes exhibit a dizzying diversity of reproductive strategies – both sexual and asexual – with some species employing internal fertilization and others using a broadcast spawning technique, whereby gametes are emitted into the water column and fertilization happens externally. In other cases, males will fertilize the eggs after the female has attached them to the substrate. In taxa that use internal fertilization, the gelatinous egg masses are commonly laid on a substrate such as rocks or vegetation. Typically, in these groups, the embryos will undergo indirect development via larval stages inside of the egg masses – in some species, the trochophore stage is bypassed and replaced by, e.g., a metatrochophore. In some tube-dwelling worms, eggs will be brooded inside of the tubes and some scale worms (family Polynoidae) will carry the eggs in specialized compartments under their scales (Rasmussen, 1956; Daly, 1972). By contrast, in broadcast spawners, the pelagic, fertilized eggs will develop into a trochophore larva, which bears distinct morphological features. In particular, trochophore larvae (Figure 20.6) are round or oval in shape, possess an apical organ (ornate with a tuft of cilia), pre-oral and post-oral ciliated bands that surround the body (mainly used for locomotion), and both a complete intestinal tract and two protonephridia (Hatschek, 1878; Rouse, 1999). The lifespan of the trochophore varies between taxa, from hours to weeks (Pleijel & Rouse, 2006), but the larvae ultimately settle on the benthos where they undergo metamorphosis and become juveniles. The time span of metamorphosis surely differs between species, but Okuda (1946) reports that it takes a minimum of 3 days in the investigated species.

Developmental Genes

Studies of annelid Hox genes have centered around the long-standing questions of whether or not the segmental patterns in Annelida are homologous to those of arthropods. We know now that the segmentation in annelids arose independently to that of arthropods and, although the genetic pathways that afford a segmented bauplan are often conserved or convergent, remnants of this independence can be traced in annelid developmental studies (Eernisse et al., 1992; Moore and Willmer, 1997; Kourakis et al., 1997). Kourakis et al. (1997 and references therein), state that leeches generate their segmentation through stereotyped cell lineages and that this cellular mechanism is rather different from their arthropod counterparts. Somewhat surprisingly, early studies of annelid development focused on leeches (pioneered by Marty Shankland and Mark Martindale) and much of our contemporary knowledge is based on our understanding of early development in the leech model organism *Helobdella robusta* (Shankland et al., 1992). So far, studies have found seven leech Hox genes (commonly referred to as Lox genes in the literature) with proven orthology to Hox genes known from *Drosophila melanogaster* (Kourakis and Martindale, 2001): *Lox 7* (orthologous to *lab*), *Lox 6* (orthologous to *Dfd*), *Lox 20* (orthologous to *Scr*), *Lox 5* (orthologous to *Antp*), *Lox 2* (orthologous to *abdA/Ubx*), *Lox 4* (orthologous to *abdA/Ubx*), and *Lox 18* (orthologous to *Dfd/Hox4*). Importantly, based on these orthology determinations, leech Hox genes, with the exception of *Lox 18*, are consistent with the hypothesis of collinearity, suggesting that the transcription domains are collinear in both time and space, along the anteroposterior axis. Hox genes and their role in developmental processes and segmentation have also been studied in some polychaetes, notably *Chaetopterus, Nereis virens* (Sars, 1835) and *Platynereis dumerilii* (Audouin and Milne-Edwards, 1833) (Irvine and Martindale, 2000; Kulakova et al., 2007; Steinmetz et al., 2011; Pfeifer et al., 2012).

PHYSIOLOGY AND BEHAVIOR

Respiration

Beyond locomotion, reproduction, and feeding (detailed above), annelid respiration has received some attention in the literature. Gas exchange normally occurs across the surface of the organisms, but can also occur through gill filaments in some members of each of Polychaeta, Oligochaeta, and Hirudinea. Oxygen is either transported free in the bloodstream or coelomic cavity (rare) or in respiratory pigments (more common), such as hemoglobin. The hemoglobin of annelids is, by and large, comparable to that of vertebrates, but differs in the details of the composition and molecular weight (Jouan et al., 2001; Roesner et al., 2005). The composition and expression of respiratory pigments in Annelida depends on the ecology of the species. For example, in members of the lugworm genus *Arenicola*, hemoglobin releases oxygen into the tissues only under anoxic or hypoxic conditions, ultimately allowing the organisms to withstand periods of respiratory stress (Toulmond, 1973). In some oligochaetes, oxygen is amassed from the atmosphere into the hemoglobin, but is only released when the oxygen level in the tissue reaches critically low levels – this may prevent the effects of oxygen poisoning in the worms.

Digestion

Most annelids have a complete digestive system with a mouth and an anus, and digestion is normally an entirely extracellular activity; although some polychaetes seem to also use intracellular digestion (Jeuniaux, 1969). The terminology for the various parts of the digestive tract differs between the groups, despite seemingly obvious homology in most cases. Generally, the digestive tract can be divided into esophagus, "stomach," or crop, intestine, and rectum. The stomach and intestines are normally divided into three regions with separate histological compositions; these are the foregut, midgut, and hindgut. Some species (mainly in Clitellata) have evolved a crop and/or gizzard as part of the digestive tract and this is likely as a result of the varying diet of the species. Jeuniaux (1969) provides a detailed overview of annelid digestion and nutrient acquisition.

Of particular note is the fact that some species of oligochates in the genera *Inanidrilus* and *Olavius* lack mouths, guts, and anuses entirely. Instead, these species have formed obligate, species-specific symbioses with a variety of extracellular bacterial symbionts that provide the worms with the needed nutrition. Often occurring in hypoxic or anoxic environments, the worms rely on the bacteria for the oxidizing of sulfur and reduction of sulfates (Woyke et al., 2006). Detailed overviews for these intriguing systems can be found in Dubilier et al. (1999, 2006) and Blazejak et al. (2005).

Behavior

Although most aspects of annelid behavior have been covered in sections above, a unique and, therefore, notable trait exists in the hirudinean order Glossiphoniformes: parental care (Sawyer, 1986). In members of the order, parental care is initiated by the deployment of eggs onto the ventral surface of the leech parent. The parent will then cover the eggs with its dorsoventrally flattened body as protection until the eggs hatch. After hatching, the offspring will attach to the venter of the parent who will carry them to their first bloodmeal. Depending on the circumstances and on the species, some juveniles will reattach to the parent after feeding. The act of brooding reaches its pinnacle in the genera *Marsupiobdella* and *Maiabdella* wherein the parents carry the eggs and juveniles in a specialized internal brood pouch or marsupium (Sawyer, 1971).

GENETICS AND GENOMICS

Annelid phylogenetics has focused heavily on commonly used nuclear and mitochondrial genes (18S rRNA, 28S rRNA, Histone H3, 12S rRNA, 16S rRNA, cytochrome c oxidase subunit I [COI], and, to some extent, elongation factor 1α [EF1α]) (Kim et al., 1996; Apakupakul et al., 1999; McHugh, 2000; Bleidorn et al., 2003; Rousset et al., 2007; Zrzavý et al., 2009). As a result, a plethora of genetic data for these loci exists on GenBank (https://www.ncbi.nlm.nih.gov/), yet these have not heretofore been leveraged collectively to encompass as much of the living diversity as possible in a phylogenetic analysis. Instead, most phylogenetic analyses regarding Annelida have focused on only a small subset of taxa. Our knowledge of higher-level taxonomy, as derived from robust phylogenetic analyses, was therefore based on a handful of studies with broader taxonomic sampling. Recent phylogenomic studies (Struck et al., 2011; Kvist and Siddall, 2013; Weigert et al., 2014; Andrade et al., 2015) have somewhat remedied this by including broader sampling from across the three groups of annelids, albeit with oligochaetes and hirudineans less completely sampled.

Two annelid genomes have so far been sequenced. These belong to the polychaete *Capitella teleta* (Blake et al., 2009) and the leech *Helobdella robusta.* Both of these genomes were sequenced using whole-genome shotgun sequencing with Sanger DNA reads and the assemblies are available on the Joint Genome Institute portal website (https://genome.jgi.doe.gov/portal/) or at GenBank (accession numbers AMQM00000000 and AMQN00000000). In describing these genomes (as well as that of the mollusc *Lottia gigantea* Gray in Sowerby, 1834), Simakov et al. (2013) asserted that the genome sizes were 324 mega basepairs (Mbp) for *C. teleta* and 228 Mbp for *H. robusta.* The genomes encode between 23,000 (*H. robusta*) and 32,000 (*C. teleta*) genes, approximately 50–60% of which found orthologues in other annotated bilaterian genomes. Interestingly, while the polychaete genome shows extensive macrosyntenic tendencies when compared to several metazoan genomes (including human), the lack of macrosynteny in the leech genome implies a high level of genomic rearrangement.

As for mitochondrial genomes, there exist over 50 fully sequenced genomes, the bulk of which belong to Polychaeta. For such an ancient group, mitochondrial gene order is fairly conserved within Annelida (Oceguera-Figueroa et al., 2016), although substitution rates have been shown to be elevated compared to other phyla (Seixas et al., 2017). Most, if not all, mitochondrial genomes of annelids contain 37 genes, 13 of which are protein coding genes. The largest mitochondrial genome sequenced thus far seems to be that of *Spirobranchus giganteus* (Pallas, 1766) which is 22,058 nucleotides in size. By contrast, the shortest mitogenome present on GenBank is that of *Myzostoma seymourcollegiorum* (Rouse & Grygier, 2005) at 11,505 nucleotides in size; note that this genome is linear and thus not fully sequenced.

Beyond genomes, there are numerous transcriptomes and expressed sequence tag libraries for annelids present on GenBank, the most complete of which belong to *Alvinella pompejana* (Desbruyères and Laubier, 1980) (Polychaeta, Alvinellidae), *Capitella teleta*, and *Hirudo medicinalis* (Linnaeus, 1758) (Hirudinea, Hirudinidae).

POSITION IN ToL

As mentioned above, Annelida nests within Trochozoa (Eernisse et al., 1992; Bleidorn, 2008; Giribet, 2008; Dunn et al., 2014), which also includes Nemertea, Mollusca, Brachiopoda, and Phoronida. Although the sister taxon to Annelida has been long debated, phylogenomic analyses seem to robustly infer that its sister taxon is Nemertea (Kocot et al., 2017). It is now fully accepted that within Annelida, Polychaeta is rendered paraphyletic by virtue of Clitellata nesting within this group (Zrzavý et al., 2009; Struck et al., 2011; Kvist & Siddall, 2013; Weigert et al., 2014; Aguado et al., 2015). Moreover, since the advent of molecular phylogenetics, almost all studies have found that Oligochaeta is rendered paraphyletic by virtue of Hirudinea nesting within. As such, and in order to maintain taxonomic stability based on monophyletic groupings, one might argue that Annelida and Polychaeta are synonyms, as they are inclusive of exactly the same taxa.

DATABASES AND COLLECTIONS

Resources and databases for annelid research can be broadly divided into two main categories: taxonomic databases and molecular resources. Some of the most commonly used taxonomic databases that handle vast amounts of data for annelid taxa include the World Register of Marine Species (WORMS; http://www.marinespecies.org/) and its affiliated websites such as the World Polychaeta Database (http://www.marinespecies.org/polychaeta/) as well as the Integrated Taxonomic Information System (ITIS; https://www.itis.gov/). More regionally defined databases also exist, such as NONATObase (http://nonatobase.ufsc.br/) for polychaetes from the Southwestern Atlantic Ocean and Fauna Europaea (https://fauna-eu.org/) for European taxa.

In terms of molecular resources, most DNA and RNA data are deposited in the National Center for Biotechnology Information (NCBI, sometimes also referred to as GenBank; https://www.ncbi.nlm.nih.gov/). In particular, next-generation sequencing data are often stored in NCBI's Short Read Archive (SRA; https://www.ncbi.nlm.nih.gov/sra). The European Nucleotide Archive (ENA; https://www.ebi.ac.uk/ena) also provides molecular sequence data across geographic and taxonomic boundaries. Molecular datasets for specific scientific articles (on Annelida and other groups) can also be found in Zenodo (https://zenodo.org/), Dryad (https://datadryad.org/), and TreeBASE (https://www.treebase.org/). Morphological data matrices that can be used for phylogenetic inference are also available from MorphoBank (https://morphobank.org/), although the data regarding annelid taxa are rather scarce.

CONCLUSION

Our understanding of annelid biology, like that of most other groups of organisms, is evolving rapidly due to the possibilities of generating and incorporating massive amounts of DNA sequences through innovative sequencing platforms. Phylogenomic studies have clarified long-standing questions regarding the evolution of annelids; furthermore, new sequencing technologies have also evinced clues about the detailed functions of specific tissues (e.g., salivary glands) and the differential patterns of gene expression during development. The trend is rather conspicuous: genomic-scale data will become more frequently used moving forward. Annelida is clearly a model group that can be leveraged to explore the complex relationship between evolution and development (evo-devo), in particular because of the extreme phenotypic diversity and the variety of life histories within this taxon. Notwithstanding the importance of new technologies for the study of Annelida, this group is still understudied in many areas of the world, opening the possibility to new discoveries that can change our view of the phylogeny. Finally, we think that ecological studies of Annelida have a promising future, of particular interest are those studies that incorporate the analyses of microbiota associated with worms and the key role that this association may play in the colonization of new environments that otherwise would be inaccessible for the worms alone; indeed, the ecological roles of most of the species remain largely unknown.

REFERENCES

Agassiz, A., 1866. On the young stages of a few Annelids. Annals of the Lyceum of Natural History, 8.

Aguado, M.T., Capa, M., Oceguera-Figueroa, A. and Rouse, G.W., 2014. Annelids: segmented worms. In: Vargas, P. and Zardoya, R (eds.), The Tree of Life (254–269). Sinauer Associates, Inc., Sunderland.

Anderson, F.E., Williams, B.W., Horn, K.M., Erséus, C., Halanych, K.M., Santos, S.R. and James, S.W., 2017. Phylogenomic analyses of Crassiclitellata support major northern and southern hemisphere clades and a Pangaean origin for earthworms. BMC Evolutionary Biology, 17: 123.

Andrade, S.C., Novo, M., Kawauchi, G.Y., Worsaae, K., Pleijel, F., Giribet, G. and Rouse, G.W., 2015. Articulating "archiannelids": phylogenomics and annelid relationships, with emphasis on meiofaunal taxa. Molecular Biology and Evolution, 32: 2860–2875.

Apakupakul, K., Siddall, M.E. and Burreson, E.M., 1999. Higher level relationships of leeches (Annelida: Clitellata: Euhirudinea) based on morphology and gene sequences. Molecular Phylogenetics and Evolution, 12: 350–359.

Avel, M., 1959. Classe des Annélides Oligochètes. In: Grassé, P.P. (ed.), Traité de Zoologie (V: 224–470). Masson et Cie, Paris.

Bartolomaeus, T., Purschke, G. and Hausen, H., 2005. Polychaete phylogeny based on morphological data – a comparison of current attempts. Hydrobiologia, 535/536: 341–356.

Blazejak, A., Erséus, C., Amann, R. and Dubilier, N., 2005. Coexistence of bacterial sulfide oxidizers, sulfate reducers, and spirochetes in a gutless worm (Oligochaeta) from the Peru margin. Applied Environmental Microbiology, 71: 1553–1561.

Beddard, F.E., 1895. A Monograph of the Order Oligochaeta. Clarendon Press, Oxford, 769.

Benham, W. B., 1890. An attempt to classify earthworms. Quarterly Journal of Microscopical Science (new series), 31: 201–315.

Black, M.B., Halanych, K.M., Maas, P.A.Y., Hoeh, W.R., Hashimoto, J., Desbruyeres, D., Lutz, R.A. and Vrijenhoek, R.C., 1997. Molecular systematics of vestimentiferan tubeworms from hydrothermal vents and cold-water seeps. Marine Biology, 130: 141–149.

Bleidorn, C., 2008. Lophotrochozoan relationships and parasites. A snap-shot. Parasite, 15, 329–332.

Bleidorn, C., Vogt, L. and Bartolomaeus, T., 2003. New insights into polychaete phylogeny (Annelida) inferred from 18S rDNA sequences. Molecular Phylogenetics and Evolution, 29: 279–288.

Böggemann, M., 2002. Revision of the Glyceridae Grube, 1850 (Annelida: Polychaeta). Abhandlungen der Senckenbergischen Naturforschenden Gesellschaft, 555: 1–249.

Bracchi, G. and Alessandrello, A., 2005. Paleodiversity of the free-living polychaetes (Annelida, Polychaeta) and description of new taxa from the Upper Cretaceous Lagerstätten of Haqel, Hadjula and Al-Namoura (Lebanon). Società Italiana di Scienze Naturali e Museo Civico di Storia Naturale di Milano, 32: 1–64.

Clapèrede, E.D. and Mecznikow, E., 1869. Beiträge zur Kenntnis der Entwicklung der Anneliden. Zeitschrift für wissenschaftliche Zoologie, 19.

Daly, J.M., 1972. The maturation and breeding biology of *Harmothoe imbricata* (Polychaeta: Polynoidae). Marine Biology, 12: 53–66.

Day, J.H., 1934. Development of *Scolecolepis fuliginosa* (Clapérède). Journal of the Marine Biological Association of the United Kingdom, 19: 633–654.

Desbruyéres, D. and Laubier, L., 1980. *Alvinella pompejana* gen. sp. nov., Ampharetidae aberrant des sources hydrothermales de la ride Est-Pacifique. Oceanologica Acta, 3, 267–274.

Dubilier, N., Amann, R., Erséus, C., Muyzer, G., Park, S., Giere, O. and Cavanaugh, C.M., 1999. Phylogenetic diversity of bacterial endosymbionts in the gutless marine oligochete *Olavius loisae* (Annelida). Marine Ecology Progress Series, 178: 271–280.

Dubilier, N., Blazejak, A. and Rühland, C., 2006. Molecular basis of symbiosis between bacteria and gutless marine oligochaetes. In: Overmann J (ed.), Progress in Molecular and Subcellular Biology (41: 251–275). Springer, Berlin.

Dunn, C.W., Giribet, G., Edgecombe, G.D. and Hejnol, A., 2014. Animal phylogeny and its evolutionary implications. Annual Review of Ecology, Evolution, and Systematics, 45: 371–395.

Eckert, R., 1963. Electrical interaction of paired ganglion cells in the leech. Journal of General Physiology, 46: 573–87.

Eernisse, D.J., Albert, J.S. and Anderson, F.E., 1992. Annelida and Arthropoda are not sister taxa: a phylogenetic analysis of spiralian metazoan morphology. Systematic Biology, 41, 305–330.

Erséus, C., 2005. Phylogeny of oligochaetous Clitellata. Hydrobiologia, 535: 357–372.

Fauchald, K., 1977. The polychaete worms: definitions and keys to the orders, families and genera. Natural History Museum of Los Angeles County, Science Series, 28, Los Angeles, CA.

Fauvel, P., 1923. Polychètes errantes. Faune de France, 5: 1–488.

Fauvel, P., 1927 Polycheètes sédentaires. Faune de France, 16: 1–494.

Fischer, A., Pfannenstiel, H.D., 1984. Polychaete reproduction. Progress in comparative reproductive biology. Fortschritte der Zoologie, 341.

Gaill, F., Mann, K., Wiedemann, H., Engel, J. and Timpl, R., 1995. Structural comparison of cuticle and interstitial collagens from annelids living in shallow sea-water and at deep-sea hydrothermal vents. Journal of Molecular Biology, 246: 284–294.

Gardner, C.R., 1976. The neuronal control of locomotion in the earthworm. Biological Reviews, 51: 25–52.

Giangrande, A., 1997. Polychaete reproductive patterns, life cycles and life histories: an overview. Oceanography and Marine Biology, 35: 323–386.

Giribet, G., 2008. Assembling the lophotrochozoan (= spiralian) tree of life. Philosophical Transactions of the Royal Society of London B: Biological Sciences, 363: 1513–1522.

Hartzell, P.L., Nghiem, J.V., Richio, K.J. and Shain, D.H., 2005. Distribution and phylogeny of glacier ice worms (*Mesenchytraeus solifugus* and *Mesenchytraeus solifugus rainierensis*). Canadian Journal of Zoology, 83: 1206–1213.

Hatschek, B., 1878. Studien über Entwicklungsgeschichte der Anneliden. Ein Beitrag zur Morphologie der Bilaterien. Arbeiten aus den Zoologischen Instituten der Universität Wien, 1: 277–404.

Hurtado, L.A., Lutz, R.A. and Vrijenhoek, R.C., 2004. Distinct patterns of genetic differentiation among annelids of eastern Pacific hydrothermal vents. Molecular Ecology, 13: 2603–2615.

Irvine, S.Q. and Martindale, M.Q., 2000. Expression patterns of anterior Hox genes in the polychaete *Chaetopterus*: correlation with morphological boundaries. Developmental Biology, 217: 333–351.

Jamieson, B.G.M., 1988. On the phylogeny and higher classification of the Oligochaeta. Cladistics, 4: 367–410.

Jamieson, B.G.M., Tillier, S., Tillier, A., Justine, J.-L., Ling, E., James, S. McDonald, K. and Hugall, A.F., 2002. Phylogeny of the Megascolecidae and Crassiclitellata (Oligochaeta, Annelida): combined versus partitioned analysis using nuclear (28S) and mitochondrial (12S, 16S) rDNA. Zoosystema, 24: 707–734.

Jeuniaux, C., 1969. Nutrition and digestion. In: Chemical Zoology. Volume IV: Annelida, Echiura, and Sipuncula, 69–91.

Jouan, L., Taveau, J.C., Marco, S., Lallier, F.H. and Lamy, J.N., 2001. Occurrence of two architectural types of hexagonal bilayer hemoglobin in annelids: comparison of 3D reconstruction volumes of *Arenicola marina* and *Lumbricus terrestris* hemoglobins. Journal of Molecular Biology, 305: 757–771.

Kim, C.B., Moon, S.Y., Gelder, S.R., Kim, W., 1996. Phylogenetic relationships of annelids, molluscs, and arthropods evidenced from molecules and morphology. Journal of Molecular Evolution, 43, 207–215.

Kocot, K.M., 2016. On 20 years of Lophotrochozoa. Organisms Diversity & Evolution, 16: 329–343.

Kocot, K.M., Struck, T.H., Merkel, J., Waits, D.S., Todt, C., Brannock, P.M., Weese, D.A., Cannon, J.T., Moroz, L.L., Lieb, B. and Halanych, K.M., 2017. Phylogenomics of Lophotrochozoa with consideration of systematic error. Systematic Biology, 66: 256–282.

Kourakis, M.J. and Martindale, M.Q., 2001. Hox gene duplication and deployment in the annelid leech *Helobdella*. Evolution & Development, 3: 145–153.

Kourakis, M.J., Master, V.A., Lokhorst, D.K., Nardelli-Haefliger, D., Wedeen, C.J., Martindale, M.Q. and Shankland, M., 1997. Conserved anterior boundaries of Hox gene expression in the central nervous system of the leech *Helobdella*. Developmental Biology, 190: 284–300.

Kulakova, M., Bakalenko, N., Novikova, E., Cook, C.E., Eliseeva, E., Steinmetz, P.R., Kostyuchenko, R.P., Dondua, A., Arendt, D., Akam, M. and Andreeva, T., 2007. Hox gene expression in larval development of the polychaetes *Nereis virens* and *Platynereis dumerilii* (Annelida, Lophotrochozoa). Development Genes and Evolution, 217: 39–54.

Kvist, S., Brugler, M.R., Goh, T.G., Giribet, G. and Siddall, M.E., 2014. Pyrosequencing the salivary transcriptome of *Haemadipsa interrupta* (Annelida: Clitellata: Haemadipsidae): anticoagulant diversity and insight into the evolution of anticoagulation capabilities in leeches. Invertebrate Biology, 133: 74–98.

Kvist, S., Min, G.S., Siddall, M.E., 2013. Diversity and selective pressures of anticoagulants in three medicinal leeches (Hirudinida: Hirudinidae, Macrobdellidae). Ecology and Evolution, 3: 918–933.

Kvist, S., Oceguera-Figueroa, A., Tessler, M., Jiménez-Armenta, J., Freeman Jr, R.M., Giribet, G. and Siddall, M.E., 2016. When predator becomes prey: investigating the salivary transcriptome of the shark-feeding leech *Pontobdella macrothela* (Hirudinea: Piscicolidae). Zoological Journal of the Linnean Society, 179: 725–737.

Kvist, S., Sarkar, I.N. and Siddall, M.E., 2011. Genome-wide search for leech antiplatelet proteins in the non-blood-feeding leech *Helobdella robusta* (Rhyncobdellida: Glossiphoniidae) reveals evidence of secreted anticoagulants. Invertebrate Biology, 130: 344–350.

Kvist, S., Siddall, M.E., 2013. Phylogenomics of Annelida revisited: a cladistic approach using genome-wide expressed sequence tag data mining and examining the effects of missing data. Cladistics 29: 435–448.

Lamarck, J.-B.M., 1818. Histoire naturelle des animaux sans vertebres. Deterville/Verdiere, Paris.

Livanow, N., 1931. Die organisation der Hirudineen und die Beziehungen dieser Gruppe zu den Oligochäten. Ergebnisse und Fortschritte der Zoolgie, 7: 378–484.

Martin, P., Martens, K. and Goddeeris, B., 1999. Oligochaeta from the abyssal zone of Lake Baikal (Siberia, Russia). Hydrobiologia, 406: 165–174.

Martins, R.T., Stephan, N.N.C. and Alves, R.G., 2008. Tubificidae (Annelida: Oligochaeta) as an indicator of water quality in an urban stream in southeast Brazil. Acta Limnologica Brasiliensia, 20: 221–226.

McHugh, D., 2000. Molecular phylogeny of the Annelida. Canadian Journal of Zoology, 78: 1873–1884.

Min, G.S., Sarkar, I.N. and Siddall, M.E., 2010. Salivary transcriptome of the North American medicinal leech, *Macrobdella decora*. Journal of Parasitology, 96: 1211–1221.

Moore, J. and Willmer, P., 1997. Convergent evolution in invertebrates. Biological Reviews, 72: 1–60.

Morris, S.C. and Peel, J.S., 2008. The earliest annelids: lower Cambrian polychaetes from the Sirius Passet Lagerstätte, Peary Land, North Greenland. Acta Palaeontologica Polonica, 53: 137–148.

Müller, C., Mescke, K., Liebig, S., Mahfoud, H., Lemke, S. and Hildebrandt, J.P., 2016. More than just one: multiplicity of hirudins and hirudin-like factors in the medicinal leech, *Hirudo medicinalis*. Molecular Genetics and Genomics, 291: 227–240.

Nanglu, K. and Caron, J.B., 2018. A new Burgess Shale polychaete and the origin of the annelid head revisited. Current Biology, 28, 319–326.

Nolte, W., 1938. Annelidlarven. Nordisches Plankton, 23: 59–169.

Oceguera-Figueroa, A., Manzano-Marín, A., Kvist, S., Moya, A., Siddall, M.E. and Latorre, A., 2016. Comparative mitogenomics of leeches (Annelida: Clitellata): genome conservation and *Placobdella*-specific trnD gene duplication. PLoS One, 11: e0155441.

Odier, A., 1823. Memoire sur le *Branchiobdella*, nouveau genere d'Annelides de la famille des Hirudinees. Mémoires de la Société d'historie naturelle de Paris, 1: 70–78.

Okuda S., 1946. Studies on the development of Annelida Polychaeta I. Journal of the Faculty of Science, Hokkaido Imperial University, 9: 115–219.

Paoletti, M.G., Bressan, M. and Edwards, C.A., 1996. Soil invertebrates as bioindicators of human disturbance. Critical Reviews in Plant Sciences, 15: 21–62.

Paterson, G.L., Glover, A.G., Froján, C.R.B., Whitaker, A., Budaeva, N., Chimonides, J. and Doner, S., 2009. A census of abyssal polychaetes. Deep Sea Research Part II: Topical Studies in Oceanography, 56: 1739–1746.

Pérès, G., Vandenbulcke, F., Guernion, M., Hedde, M., Beguiristain, T., Douay, F., Houot, S., Piron, D., Richard, A., Bispo, A. and Grand, C., 2011. Earthworm indicators as tools for soil monitoring, characterization and risk assessment. An example from the national Bioindicator programme (France). Pedobiologia, 54: S77–S87.

Pfeifer, K., Dorresteijn, A.W. and Fröbius, A.C., 2012. Activation of Hox genes during caudal regeneration of the polychaete annelid *Platynereis dumerilii*. Development Genes and Evolution, 222: 165–179.

Pleijel, F. and Rouse, G.W., 2006. Phyllodocida. In: Rouse, G.W. and Pleijel, F (eds.) Reproductive Biology and Phylogeny of Annelida (441–530). Science Publishers, Jersey & Plymouth.

de Quatrefages, A., 1865. Note sur la classification des Annélides. In: Comptes rendus hebdomadaires des séances de l'Académie des sciences (60: 586–600). Paris.

Rasmussen, E., 1956. The reproduction and larval development of some polychaetes from the Isefjord, with some faunnistic notes. Biologiske Meddelelser, 23: 1–84.

Roesner, A., Fuchs, C., Hankeln, T. and Burmester, T., 2005. A globin gene of ancient evolutionary origin in lower vertebrates: evidence for two distinct globin families in animals. Molecular Biology and Evolution, 22: 12–20.

Rouse, G.W., 1999. Trochophore concepts: ciliary bands and the evolution of larvae in spiralian Metazoa. Biological Journal of the Linnean Society, 66: 411–464.

Rouse, G.W. and Fauchald, K., 1995. The articulation of annelids. Zoologica Scripta, 24: 269–301.

Rouse, G.W. and Fauchald, K., 1997. Cladistics and polychaetes. Zoologica Scripta, 26: 139–204.

Rouse, G.W. and Fauchald, K., 1998. Recent views on the status, delineation and classification of the Annelida. American Zoologist, 38: 953–964.

Rouse, G.W. and Grygier, M.J., 2005. *Myzostoma seymourcollegiorum* n. sp. (Myzostomida) from southern Australia, with a description of its larval development. Zootaxa, 1010: 53–64.

Rouse, G. and Pleijel, F., 2001. Polychaetes. Oxford University Press.

Rousset, V., Pleijel, F., Rouse, G.W., Erséus, C. and Siddall, M.E., 2007. A molecular phylogeny of annelids. Cladistics, 23: 41–63.

Salzet, M., 2001. Anticoagulants and inhibitors of platelet aggregation derived from leeches. FEBS Letters, 492: 187–192.

Sawyer, R.T., 1986. Leech biology and behaviour. Clarendon Press, Oxford.

Schroeder, P.C. and Hermans, C.O., 1975. Annelida: Polychaeta. In: Giese A.C. and Pearse J.S. (eds.) Reproduction of marine invertebrates (1–213). Academic Press, New York.

Segrove, F., 1941. The development of the serpulid *Pomatoceros triqueter* L. Quarterly Journal of Microscopical Science, 82: 467–540.

Seixas, V.C., de Moraes Russo, C.A. and Paiva, P.C., 2017. Mitochondrial genome of the Christmas tree worm *Spirobranchus giganteus* (Annelida: Serpulidae) reveals a high substitution rate among annelids. Gene, 605: 43–53.

Shankland, M., Bissen, S.T. and Weisblat, D.A., 1992. Description of the Californian leech *Helobdella robusta* sp. nov., and comparison with *Helobdella triserialis* on the basis of morphology, embryology, and experimental breeding. Canadian Journal of Zoology, 70: 1258–1263.

Shain, D.H., Mason, T.A., Farrell, A.H. and Michalewicz, L.A., 2001. Distribution and behavior of ice worms (*Mesenchytraeus solifugus*) in south-central Alaska. Canadian Journal of Zoology, 79: 1813–1821.

Sharma, R.C. and Rawat, J.S., 2009. Monitoring of aquatic macroinvertebrates as bioindicator for assessing the health of wetlands: a case study in the Central Himalayas, India. Ecological Indicators, 9: 118–128.

Siddall, M.E. and Burreson, E.M., 1998. Phylogeny of leeches (Hirudinea) based on mitochondrial cytochrome c oxidase subunit I. Molecular Phylogenetics and Evolution, 9: 156–162.

Siddall, M.E., Apakupakul, K., Burreson, E.M., Coates, K.A., Erséus, C., Gelder, S.R., Källersjö, M. and Trapido-Rosenthal, H., 2001. Validating Livanow: molecular data agree that leeches, branchiobdellidans, and *Acanthobdella peledina* form a monophyletic group of oligochaetes. Molecular Phylogenetics and Evolution, 21: 346–351.

Siddall, M.E., Min, G.S., Fontanella, F.M., Phillips, A.J. and Watson, S.C., 2011. Bacterial symbiont and salivary peptide evolution in the context of leech phylogeny. Parasitology, 138: 1815–1827.

Siddall, M.E., Brugler, M.R. and Kvist, S., 2016. Comparative transcriptomic analyses of three species of *Placobdella* (Rhynchobdellida: Glossiphoniidae) confirms a single origin of blood feeding in leeches. Journal of Parasitology, 102, 143–150.

Simakov, O., Marletaz, F., Cho, S.J., Edsinger-Gonzales, E., Havlak, P., Hellsten, U., Kuo, D.H., Larsson, T., Lv, J., Arendt, D. and Savage, R., 2013. Insights into bilaterian evolution from three spiralian genomes. Nature, 493: 526–531.

Sket, B. and Trontelj, P., 2008. Global diversity of leeches (Hirudinea) in freshwater. In: Freshwater Animal Diversity Assessment (129–137). Springer, Dordrecht.

Steinmetz, P.R., Kostyuchenko, R.P., Fischer, A. and Arendt, D., 2011. The segmental pattern of otx, gbx, and Hox genes in the annelid *Platynereis dumerilii*. Evolution & Development, 13, 72–79.

Struck, T.H., Schult, N., Kusen, T., Hickman, E., Bleidorn, C., McHugh, D. and Halanych, K.M., 2007. Annelid phylogeny and the status of Sipuncula and Echiura. BMC Evolutionary Biology, 7: 57.

Struck, T.H., Paul, C., Hill, N., Hartmann, S., Hösel, C., Kube, M., Lieb, B., Meyer, A., Tiedemann, R., Purschke and G., Bleidorn, C., 2011. Phylogenomic analyses unravel annelid evolution. Nature, 471: 95–98.

Struck, T.H., Golombek, A., Weigert, A., Franke, F.A., Westheide, W., Purschke, G., Bleidorn, C. and Halanych, K.M., 2015. The evolution of annelids reveals two adaptive routes to the interstitial realm. Current Biology, 25, 1993–1999.

Tessler, M., de Carle, D., Voiklis, M.L., Gresham, O.A., Neumann, J.S., Cios, S., Siddall, M.E., 2018a. Worms that suck: phylogenetic analysis of Hirudinea solidifies the position of Acanthobdellida and necessitates the dissolution of Rhynchobdellida. Molecular Phylogenetics and Evolution, 127: 129–134.

Tessler, M., Marancik, D., Champagne, D., Dove, A., Camus, A., Siddall, M.E. and Kvist, S., 2018b. Marine Leech Anticoagulant Diversity and Evolution. Journal of Parasitology, 104: 210–220.

Timm, T., Martin, P.J., 2015. Clitellata: oligochaeta. In: Thorp and Covich's Freshwater Invertebrates, (Fourth Edition) 529–549.

Toulmond, A., 1973. Tide-related changes of blood respiratory variables in the lugworm *Arenicola marina* (L.). Respiration Physiology, 19: 130–144.

Trontelj, P., Sket, B. and Steinbrück, G., 1999. Molecular phylogeny of leeches: congruence of nuclear and mitochondrial rDNA data sets and the origin of bloodsucking. Journal of Zoological Systematics and Evolutionary Research, 37: 141–147.

Verdonschot, P.F.M., 2015. Introduction to Annelida and the class Polychaeta. In: Thorp and Covich's Freshwater Invertebrates, (Fourth Edition) 509–528.

Vinn, O., Kupriyanova, E.K. and Kiel, S., 2013. Serpulids (Annelida, Polychaeta) at Cretaceous to modern hydrocarbon seeps: ecological and evolutionary patterns. Palaeogeography, Palaeoclimatology, Palaeoecology, 390:35–41.

von Reumont, B.M., Campbell, L.I., Richter, S., Hering, L., Sykes, D., Hetmank, J., Jenner, R.A. and Bleidorn, C., 2014. A polychaete's powerful punch: venom gland transcriptomics of *Glycera* reveals a complex cocktail of toxin homologs. Genome Biology and Evolution, 6:2406–2423.

Weigert A., and Bleidorn, C., 2016. Current status of annelid phylogeny. Organisms Diversity & Evolution, 16:345–362.

Weigert, A., Helm, C., Meyer, M., Nickel, B., Arendt, D., Hausdorf, B., Santos, S.R., Halanych, K.M., Purschke, G., Bleidorn, C. and Struck, T.H., 2014. Illuminating the base of the annelid tree using transcriptomics. Molecular Biology and Evolution, 31:1391–1401.

Woyke, T., Teeling, H., Ivanova, N.N., Huntemann, M., Richter, M., Gloeckner, F.O., Boffelli, D., Anderson, I.J., Barry, K.W., Shapiro, H.J. and Szeto, E., 2006. Symbiosis insights through metagenomic analysis of a microbial consortium. Nature, 443:950–955.

Zrzavý, J., Říha, P., Piálek, L. and Janouškovec, J., 2009. Phylogeny of Annelida (Lophotrochozoa): total-evidence analysis of morphology and six genes. BMC Evolutionary Biology, 9, 189.

PHYLUM BRACHIOPODA

Masato Hirose and Kazuyoshi Endo

CHAPTER
21

Infrakingdom	Bilateria
Division	Protostomia
Subdivision	Lophotrochozoa
Phylum	Brachiopoda

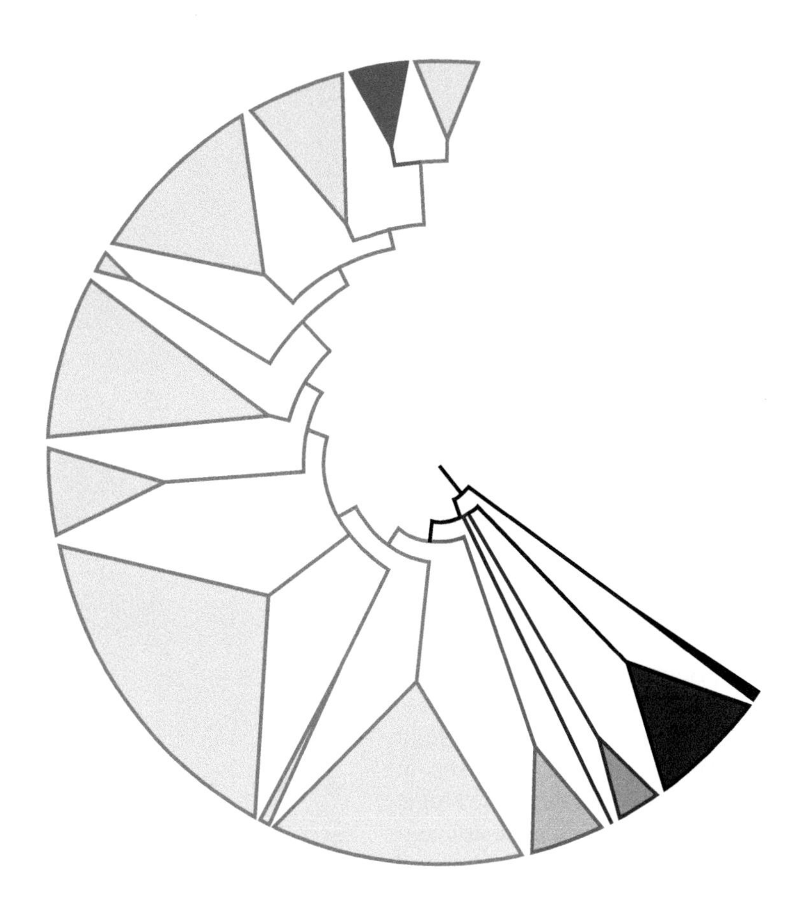

CONTENTS

General Characteristics

Approximately 380 valid living species
Solitary animal having two valves and a lophophore
Recent species have a size range of about 1 mm–10 cm in shell length
Abundant fossil record from early Cambrian
Found in all oceans
Mostly distributed in continental shelves and bathyal slopes
Known as lamp shells
Two different types of shell components: calcium carbonate and phosphate
Develop both deuterostome and protostome features

Synapomorphies

- possession of bilaterally symmetrical shells, consisting of calcium carbonates or calcium phosphates, covering the ventral and dorsal sides of the body
- U-shaped ciliated lophophore occupies most of the volume in the mantle cavity
- attached to the substrate by a pedicle, which is an outgrowth of the posterior body wall or continuous with the mantle rudiment

HISTORY OF CLASSIFICATION

Traditionally, Brachiopoda had been regarded as a class in the phylum Lophophorata based on the diagnostic ciliate lophophore with the other two taxa Phoronida and Bryozoa. Recent molecular phylogenetic work suggests a strong phylogenetic relationship between Brachiopoda and Phoronida, but the position of Bryozoa is still not certain (Cohen et al., 1998; Halanych, 2004; Dunn et al., 2008; Nesnidal et al., 2013; Dunn et al., 2014; Marlétaz et al., 2019). In some studies, Brachiopoda is inferred as a paraphyletic group including Phoronida as the sister group with the linguliform brachiopods; Brachiopoda and Phoronida are sometimes regarded to constitute a single phylum called Brachiozoa or Phoronozoa.

As with Bryozoa and Phoronida, phylum Brachiopoda is characterized by its U-shaped ciliated lophophore lacking the anus inside. Since it has several characters such as enterocoel, radial cleavage, and adult mouth not based on the blastopore, Brachiopoda was traditionally regarded as a deuterostome or the intermediate taxon between the deuterostomes and protostomes. Recent molecular phylogenetics, however, revealed that Brachiopoda is a member of protostomes, forming a clade called Lophotrochozoa with Mollusca, Annelida, and other phyla (Halanych, 2004; Dunn et al., 2008; Hejnol et al., 2009; Nesnidal et al., 2013; Dunn et al., 2014; Luo et al., 2018; Marlétaz et al., 2019).

Brachiopoda has a continuous fossil record from early Cambrian throughout the Phanerozoic eon. Traditionally, it had been classified into two groups based on the shell morphology: Articulate brachiopods have a pair of articulation (a hinge) at the posterior part of the valves, while inarticulate brachiopods lack the articulation. Recently, however, the phylum has been classified into three subphyla based on the development, anatomy, and the chemical component of the valves: inarticulate brachiopods are divided into two groups of Linguliformea and Craniiformea, and the articulate brachiopods are now called Rhynchonelliformea.

MORPHOLOGY AND ANATOMY

External morphology

The outer part of brachiopods consists of the epidermis and the biomineralized shell. The shell is normally bilaterally symmetrical and consists of two dissimilar valves. The valve that accommodates the pedicle is referred to as the ventral valve. The opposing dorsal valve is typically smaller than the ventral valve. Brachiopod shells often have a concentric ornamentation, so-called growth lines, on the outer shell surface, reflecting accretionary shell growth, and changes and breaks in the secretion of the shell.

The valves of the Rhynchonelliformea and the Craniiformea are composed of calcium carbonate in the form of low-magnesium calcite, while those of the Linguliformea are phosphatic in the form of apatite (carbonate fluorapatite), which is at least crystallochemically similar to francolite. The carbonate valves contain low levels of organic materials such as intracrystalline proteins, lipids, and carbohydrates, while phosphatic valves contain much higher levels of organic materials such as proteins including collagen, chitin, and glycosaminoglycans.

Most living brachiopods are attached to the substrate by a pedicle. The pedicles of brachiopods vary in their origin; the pedicle of linguliforms is an outgrowth of the posterior body wall, while that of rhynchonelliforms is continuous with the mantle rudiment.

The lophophore, a ciliated tentacular organ that occupies most of the volume in the mantle cavity, is composed of brachial groove and ciliated tentacles which generate an inhalant and exhalant water current through the mantle cavity. The lophophore functions as a feeding and respiratory organ by capturing food particles, removing the metabolic waste products, and facilitating uptake of oxygen. In some rhynchonelliform brachiopods, the lophophore also functions as a brooding organ.

Most rhynchonelliform brachiopods have a calcareous brachidium inside the dorsal valve as a support for their lophophore. The brachidia of rhynchonelliform brachiopods are connected with spirally coiled ribbons (spiralia) or loops; they are greatly variable in form depending on the taxa. Some brachiopods also have mesodermal calcareous spicules in the lophophore and mantle.

Internal anatomy

The body of brachiopods occupies the posterior part of the space inside the valves. The body wall extends forward and reaches the anterior periphery of the inner surface of both valves as a pair of mantles which enclose the mantle cavity (**Figure 21.1**). The mantle cavity is mostly occupied by the lophophore.

The coelom consists of the body cavity (perivisceral cavity) and some coelomic canals; the former occupies the posterior part of the shell and contains the digestive tract, excretory and reproductive organs, and main muscles, while the latter contains ganglia as the central nerve. Some part of the coelom also extends into the mantles, brachia, and tentacles, and, in linguliforms, into the pedicle.

All brachiopods have an open circulatory system containing a colorless or red-purple fluid, the latter containing hemerythrin as blood pigments. The circulatory system is composed of a series of blood vessels and communicating coelomic canals; therefore, the coelomic fluid and blood are mixing. Ciliated epithelium lining in the coelom (peritoneum) also circulates the coelomic fluids.

Several muscles cross between the valves; adductor and diductor muscles close and open the valves, respectively, and another two sets of muscles

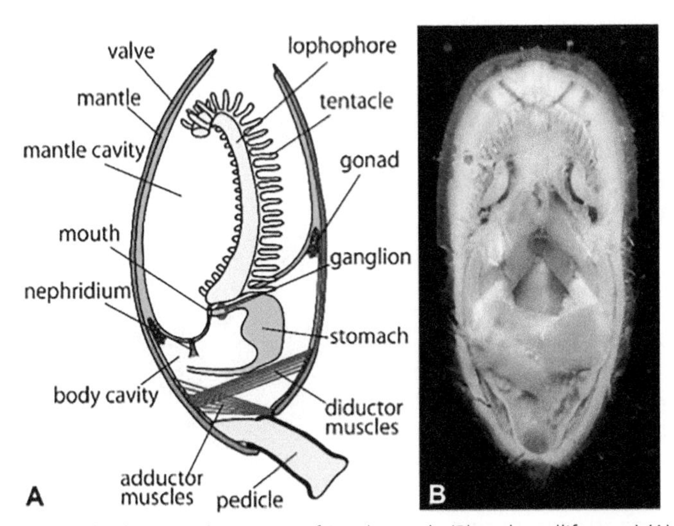

Figure 21.1 General anatomy of Brachiopoda (Rhynchonelliformea) (A); muscular system of *Lingula* (Lingulliformea) (B).

(adjustor muscles) control the pedicle and rotate the entire shell. The posterior hinge of rhynchonelliform brachiopods permits valve opening and closing in a single plane. The absence of a hinge in linguliform brachiopods permits rotation of the valves relative to one another. The muscle bases on the inner surface of the valves are commonly seen as distinct impressions, which are called muscle scars.

The opening of the digestive system at the mouth occurs medially in the brachial groove at the base of the lophophore. The digestive system, or the gut, of the Linguliformea and the Craniiformea terminates at the opened anus in the mantle cavity and in a posterior part, respectively, while that of the Rhynchonelliformea is blind with no anus, terminating in a cecum.

The excretory system consists of one or two pairs of metanephridia. Metabolic waste products are carried by the coelomocytes through the body to the metanephridia. The pairs of metanephridia are also used as gonoducts during spawning; gametes are discharged from the coelom into the mantle cavity. The gonads of the Linguliformea are usually confined to the visceral cavity, while those of the Rhynchonelliformea and the Craniiformea are mostly extended anteriorly into the mantle canals.

Brachiopods possess a central nervous system. Nerves emanate laterally from two ganglia, subenteric ganglion and supraenteric ganglion, which lie above and below the esophagus, respectively; the latter ganglion is absent in the Linguliformea and the Craniiformea.

Some species bear sensory chitinous setae along the margin of the mantle, projecting beyond the edges of the valves. Some brachiopod larvae have eyespots (ocelli) and a pair of statocysts. Both are assumed to be controlling and maintaining optimal position and behavior of the larvae.

REPRODUCTION AND DEVELOPMENT

Most living brachiopods are gonochoristic, but some micromorphic species tend to be hermaphroditic. The sexual dimorphism is quite limited. Most brachiopods release their gametes and fertilize in the surrounding waters. In the Linguliformea, *Lingula* spawns about 28,000 oocytes over a 6-month period, while *Glottidia* produces about 130,000 ova over a 4-month period. The spawning occurs in the morning or evening. The early development of the zygote is similar among all brachiopods; cleavage into blastomeres is holoblastic, equal, and radial.

Larvae of the Rhynchonelliformea and the Craniiformea are lecithotrophic, while the larvae of the Linguliformea are planktotrophic. The lecithotrophic larvae of the former two groups have only hours or a few days of motile planktonic phases. The later stages of larvae of rhynchonelliform brachiopods usually consist of mantle and pedicle lobes, but lack the functional gut. A bivalved protegulum is secreted during metamorphosis. Some species of rhynchonelliforms brood their larvae in the specialized pouches or within the female lophophore to advanced larval stages. The postmetamorphic juvenile of rhynchonelliforms externally resembles adult, but internally has only some muscles and gut rudiments. The larvae of craniiform brachiopods also lack a functional gut and the lophophore, and have about 4 to 6 days of swimming stage before settlement. The larvae of linguliform brachiopods have several weeks of swimming life. The larvae of linguliforms consist of a functional gut, a lophophore, some muscular systems, and a protegulum. The duration of the planktonic larval stage varies from 3 to 6 weeks. The pedicle and setae, both dictated by the availability of settlement, develop during the later stages in the larvae of linguliforms; the postlarval forms are almost the same as adults both externally and internally.

As in larvae of many other sessile organisms, the behavior of brachiopod larvae is influenced by gravity and/or light; positive phototaxis is seen during the early stages, which becomes negative prior to settlement.

PHYSIOLOGY AND BEHAVIOR

Physiology

The physiology and behavior of living brachiopods could help interpret the life-forms and habits of fossil brachiopods. Brachiopods are active suspension feeders which feed phytoplankton (e.g., diatoms, peridiniales, and filamentous algae), some meiobenthos (e.g., foraminifers, rotifers, and copepods), and other organic detritus. Brachiopods use metabolic energy to produce currents for feeding and respiration by the lophophore. Some brachiopods such as rhynchonelliforms and discinids also rotate their shells and orient themselves to the external water currents, augmenting the filtration efficiently and helping the removal of waste products and previously filtered water. The clearance rates of brachiopods are highly variable, but generally low relative to other filter feeders such as bivalves. Rates of oxygen consumption of brachiopods are also generally low and are known to be lower than those of similar sized bivalves.

Metabolic end products of brachiopods are excreted across the mantle and the lophophore by diffusion and through the pairs of metanephridia situated on either side of the intestine. Brachiopods are known to have low levels of physiological activities, which probably indicate energy-saving adaptations and/or low energy requirements. The lifespans of some lingulid brachiopods have been estimated 5–8 years.

Behavior

The free-living lingulids use their pedicle as a support and burrow into sediments by cyclical movements of both valves. The lateral setae and the flow through the setae also enable lingulids to maintain their position within the sediments and withdraw rapidly into the sediments. The burrowing behavior is based on the scissor-like shell movements achieved by contraction of the oblique muscles and the water jets from the margin of the shells. The burrowing speed reaches about 3 cm/hr. Lingulids can also rapidly retract into the burrow as an escape reflex by quick closure of the valves with an expulsion of water and contraction of the pedicle when the lingulids get tactile stimulations. In intertidal environments, the lingulids retract into the burrow during low tide. At the surface of the sediments, lingulids use anterior setae of the mantle for forming central exhalant and two lateral inhalant apertures for water circulation. The water currents are separated by the mantle crests and tentacle tips to prevent mixing of the flows.

Discinid brachiopods also have the tactile sensitive setae on the mantle edges, and the sensitivity results in closure of the valves by the adductor muscles and contraction of the pedicle drawing the shell near the substrate. The densely packed long setae on the anterior mantle margin function as an incurrent siphon. The craniiforms are, however, lacking any tactile setae on the mantle edges.

The lifestyle of the rhynchonelliform brachiopods is principally governed by the type of substrate. Most of the rhynchonelliforms use their pedicle for attaching to the substrate such as rocks and reefs; the pedicle functions as an attachment organ and the rotational movements of the valves around the pedicle prevent overgrowth by other sessile organisms such as colonial sponges and ascidians. Some larvae also settle on small substrates such as the shell of molluscs and fragments of bryozoans and lie on the substrate. In this case, the movement of valves and pedicle also prevent burial by the sediments of the substrate.

ECOLOGY AND DISTRIBUTION

All brachiopods are marine suspension feeders and require good circulation of the water. The rhynchonelliforms generally distribute at the continental shelves and bathyal slopes. Some species reach abyssal depths and a few species

Figure 21.2 Crowded population of *Terebratulina crossei* (Rhynchonelliformea) on rocky substrate in Otsuchi, Japan.

distribute intertidally. The shells of rhynchonelliform brachiopods are generally oriented with the ventral valve side up and the dorsal valve next to the substrate, and the anterior end opening above the substrate surface (**Figure 21.2**). In the population on the stable hard substrate, most individuals are consistently oriented actively to the currents.

The linguliforms and the craniiforms distribute from the littoral waters to bathyal zone. Lingulids live in vertical burrows in compact and stable soft sediments. The burrow has the transversally oval upper part and the cylindrical lower part. The length of the whole burrow is about ten times the length of the shell. The walls of the burrow are lined with mucus secreted by the edges of the mantle and the pedicle. The distal bulb of the pedicle, surrounded by sand, is anchored at the bottom of the burrow.

Discinids and craniiforms are epifaunal and are attached to the hard substrate such as rock and mollusc fragments. Most of the discinid brachiopods are attached to the substrate by a short highly muscular pedicle. The ventral valve is always facing closely toward the substrate. Although craniiform brachiopods are also epifaunal, they lack a pedicle and cement themselves by the entire surface of the ventral valve to the hard substrate. The craniiform larvae settle with the posterior end and secreting the ventral valve, which is cemented to the substrate. Discinids redirect the shells, and thus the lophophore, relative to the water current, while the orientation of craniiforms largely depends on the larval settlement.

As in many other organisms, the ranges of temperature and salinity tolerance of brachiopods might vary depending on the species and population. The strength of the current might be one of the abiotic factors determining the distribution of rhychonelliform brachiopods. Generally, lingulids are capable of osmotic response to stresses of strong salinity variations, particularly at low tide in the tidal flat habitat. Lingulids are also able to survive temporarily poor-oxygen water during low tide, because of the presence of hemerythrin within the coelomocytes.

Rhynchonelliform brachiopods are seasonal breeders and are presumed to spawn between spring and autumn. The breeding and spawning season of lingulids depends mainly on water temperature, with optimal conditions varying from 1.5 months in midsummer in temperate waters to year-round breeding in tropical waters (Hammond, 1982).

The geographical distribution of brachiopods is largely controlled by the capacity of larvae dispersal and environmental factors that may limit their distribution. The planktotrophic larvae of the linguliform brachiopods result in the broad geographic distributions; most of the linguliform genera have broad geographic distributions. The Lingulidae are abundant in tropical and subtropical areas, while the Discinidae occur in temperate to tropical areas (**Figure 21.3A,B**). The Linguliformea principally occur in littoral waters, most species being restricted to the continental shelf from intertidal to less than 100 m deep, except *Pelagodiscus atlanticus* that occurs at abyssal depths down to 6,000 m.

The larvae of rhynchonelliforms are lecithotrophic and short-lasting; the species composition of rhynchonelliforms have differences both within and between geographic areas. The common range of bathymetric distribution of rhynchonelliform brachiopods is down to 600 m, but they are also found in bathyal waters.

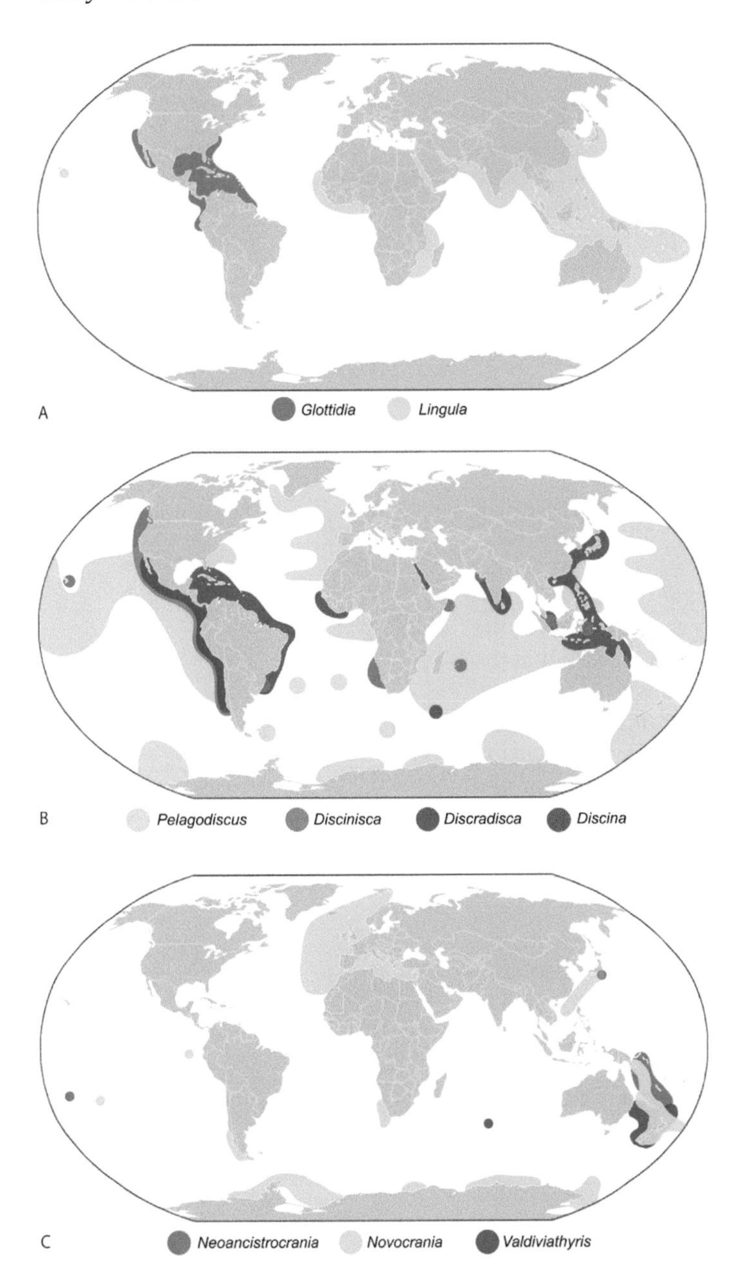

Figure 21.3 Global distributions of major genera in Linguliformea (A, B) and Craniiformea (C). (Altered from Emig 1997, 2017, 2018.)

Although larvae of the Craniiformea are known to be short-lasting and lecithotrophic, *Neocrania* has a worldwide distribution from northern to southern high latitudes (Figure 21.3C). The other two genera of Craniidae shows restricted distributions. Approximately half of the craniiform species occur between 20 to 420 m deep, while *Neocrania* has bathymetric distribution from shallow waters on the continental shelf to about 1,000 m deep on the bathyal slope.

Associations with other organisms: Several associations of brachiopods and other organisms have been reported. The principal impact of other organisms appears to be drilling by carnivorous gastropods and grazing by echinoderms and chitons. Some bottom-feeding fishes are also known to be predators for brachiopods. The competition between brachiopods and other organisms, such as tubicolous polychaetes and red algae, is also inferred as the factor responsible for shorter lifespan of brachiopods in rock-surface populations than in rock-wall populations. Some associations of amphipods and crabs with rhynchonelliform brachiopods have been recorded; the parasites are often found in the mantle cavity. Various epibionts, such as bryozoans, serpulids, barnacles, calcareous algae, sponges, and other brachiopods, are found on the rhynchonelliform brachiopod shells. Lingulids are eaten by various crustaceans (e.g., hermit crabs and amphipods), echinoderms (e.g., asteroids and ophiuroids), gastropods, and fishes. People also eat *Lingula* on almost all the western Pacific islands from Japan to New Caledonia. In the Linguliformea, unencysted metacercariae of trematodes of Gymnophallinae have been seasonally recorded around the nephrostomes and gonads of *Glottidia pyramidata*. Some poecilostomatoid copepods have also been reported as associating with *Lingula*. For the Craniiformea, monocystid protozoa have been reported in *Neocrania*.

GENETICS

There are several studies on the genetics of the Linguliformea (e.g., *Lingula anatina*), Rhynchonelliformea (e.g., *Terebratalia transversa*), and Craniiformea (e.g., *Novocrania anomala*).

The study on the expression of neural related genes during the larval development of *T. transversa* showed conservation and common origin of some of the genes within the larval apical organs of spiralians. The results suggested that the broader field of sensory neurons and the larger number of cells within the larval apical organ of *T. transversa* are mirrored in the broader spatial and temporal expression patterns of some neural related genes (Santagata et al., 2012).

Several studies have focused on brachiopod shell formation proteins and genes. In 2015, the shell proteome and shell-forming transcriptome of the rhynchonelliform brachiopod, *Magellania venosa*, was first surveyed and characterized (Jackson et al., 2015). Some proteins show biochemical and sequence similarities to other metazoan biomineralization proteins; this similarity suggests that some elements of the brachiopod shell-forming proteome are deeply evolutionarily conserved. The shell matrix proteins in *Laqueus rubellus* were also analyzed and 40 proteins were identified from the shell (Isowa et al., 2015). Most of the identified proteins have no homologues with previously reported proteins in the database at that time. Among these unknown proteins, one shell matrix protein was also identified with a domain architecture that includes a kind of region which had not been detected in other biominerals. The study suggests an independent origin and unique mechanisms for brachiopod shell formation. In 2015, the 425 Mb genome of *L. anatina* was decoded (Luo et al., 2015). The results indicate that the gene number in this species increased to 34,000 by extensive expansion. *Lingula* is phylogenetically closer to Mollusca than to Annelida, and the origin of phosphate biomineral formation in *Lingula* is independent from that of vertebrates. Several genes involved in *Lingula* shell formation are shared by molluscs, but *Lingula* has independently undergone domain combinations to produce shell matrix collagens. These features might characterize the unique biomineralization in *Lingula*.

Some molecular studies also focused on the homeobox (Hox) genes, which pattern the anteroposterior axis of bilateral animals in brachiopods. In 2016, investigation of the expression of the arthropod segment polarity genes *engrailed*, *wnt1*, and *hedgehog* in the development of brachiopods (*T. transversa* and *N. anomala*) revealed that these genes might be involved in the anterior patterning of brachiopod larvae, and bilaterians might share an ancestral non-segmental domain of *engrailed* expression during early embryogenesis (Vellutini and Hejnol, 2016). In 2017, genomic organization and embryonic expression of Hox genes in two brachiopod species *T. transversa* and *N. anomala* were also investigated (Schiemann et al., 2017). The results show that *T. transversa* has a split cluster with 10 genes as in *Lingula*, while *N. anomala* has 9 genes. The specific expression of the Hox genes in the chaetae and the shell fields in brachiopods shared with molluscs and annelids suggested conservation of the common molecular basis for those hard tissues characterizing these lophotrochozoans. Also in 2017, a study on gene expression of *engrailed* during shell development of *L. anatina* demonstrated that *engrailed* is involved in embryonic shell development in *Lingula* as in molluscs, while analyses of molecular phylogeny of *engrailed* paralogues and their genomic architectures suggested that the *engrailed* gene was independently co-opted for shell formation in each of those phyla (Shimizu et al., 2017).

Recently, the study on the draft genomes of the nemerteans and the phoronids inferred Nemertea as a sister group to the Phoronida and Brachiopoda, and revealed that the lophotrochozoans share many gene families with deuterostomes (Luo et al., 2018). The lophophores of phoronids and brachiopods are similar not only morphologically, but also at the molecular level; the results suggested a common origin of bilaterian head patterning, although different heads evolved independently in each lineage.

Study on the molecular components of the immune system based on the genome of *L. anatina* revealed that brachiopod immune gene repertories show significant overlap with those of molluscs but less with those of arthropods; the results confirmed the divergence of the defense system between Lophotrochozoa and Ecdysozoa (Gerdol et al., 2018).

POSITION IN THE ToL

Brachiopods are among the most significant group of fossil invertebrates in diversity and abundance. The brachiopod fossil record dates back to early Cambrian. Brachiopods were abundant in Paleozoic, however, were greatly reduced in diversity in the end-Permian mass extinction (Carlson, 2016).

In several studies on broad animal phylogeny, based on various gene sequences, brachiopods always appeared close to phoronids (Cohen et al., 1998), however, the group of Lophophorata is not supported in many of the recent studies, since the phylum Bryozoa always appears in a different place, and often distant from brachiopods and phoronids, among lophotrochozoans. Lophophorata, however, still appears as a monophyletic clade in some recent studies (Nesnidal et al., 2013; Luo et al., 2018).

Brachiopods also appeared as a close or even sister taxon to the phylum Nemertea in some recent molecular studies (Dunn et al., 2008; Hejnol et al., 2009). The recent genome-based study also suggests that Nemertea is sister group of Phoronida and Brachiopoda (Luo et al., 2018). Based on those molecular studies, it was suggested that the spiral cleavage in the mode of development was lost or at least has deviated from the original pattern in several lineages including brachiopods and phoronids (Hejnol, 2010).

Phylogenetic relationships among brachiopods and phoronids have also been studied based on several genes; Rhynchonelliformea is shown as a monophyletic group, while the other two subphyla often constitute a paraphyletic group with Phoronida (Santagata and Cohen, 2009). But recent phylogenomic analyses indicated that phoronids and brachiopods form a separate clade, respectively, and that within the brachiopod clade, the two inarticulate subphyla form a clade sister to the articulate subphylum (Luo et al., 2018).

DATABASES AND COLLECTIONS

The Brachiopoda Database (http://paleopolis.rediris.es/brachiopoda_database/index.html) is a part of Catalogue of Life (ITIS species 2000) and provides a list of recent brachiopod species, higher taxa names, references, as well as distribution maps of some species and genera (Emig et al., 2019). The database partly originated from SIBIC (Smithsonian International Brachiopod Information Center) and is closely related to the web portal BrachNet (http://paleopolis.rediris.es/BrachNet/). The World Register of Marine Species (WoRMS) (http://www.marinespecies.org/aphia.php?p=taxdetails&id=1803) also provides a list of recent brachiopod species and their global distribution maps for certain species. The Wallace Brachiopod Collection (http://www.jsg.utexas.edu/npl/databases/wallace-brachiopod-collection/) and the Type and Figured Collection (PaleoCentral) (http://paleocentral.org/home/) provide detailed information about the fossil brachiopod collections including images of some specimens.

CONCLUSION

Brachiopoda is unique in having species with calcium phosphate shells in addition to those with calcium carbonate shells; these shell mineralogies provide intriguing questions as to how biomineralized skeletons may have originated during the course of early animal evolution.

Brachiopods, which date back to early Cambrian, are arguably among the most diverse and abundant fossil invertebrates; they flourished and adapted to a range of paleoenvironments especially in Paleozoic seas, while they yielded some lineages which became to be known as "living fossils," such as *Lingula*. These lineages indicated little morphological changes for a long time, as evidenced by the continuous fossil record. Therefore, living and fossil brachiopods contribute to the understanding of adaptive evolution as well as evolutionary stasis. Their ecology, behaviors, associations with other organisms, and factors, such as water currents, that control their distribution patterns are yet to be studied in detail.

Phylogenetic relationships within the phylum are still unresolved in some groups of brachiopods. For instance, a study of the order Rhynchonellida based on the small subunit (SSU/18S) and large subunit (LSU/28S) ribosomal DNA (rDNA) sequences showed that none of the clades inferred from the molecular analysis corresponds to the four extant superfamilies represented in the morphological taxonomy (Cohen and Bitner, 2013) (**Figure 21.4**). Therefore, it is hoped that further reappraisal of the taxonomic characters as well as phylogenetic inferences based on much larger sequence datasets will reveal a more coherent view on the relationships within this order. The phylogenetic positions of Brachiopoda within the lophotrochozoan tree, especially relative to such taxa as the lophophorate Bryozoa and the ribbon worm Nemertea, are also still under discussion (Luo et al., 2018; Marlétaz et al., 2019). Further genetic studies may help make clear the phylogenetic position of brachiopods and contribute to the understanding of the whole history of metazoan evolution.

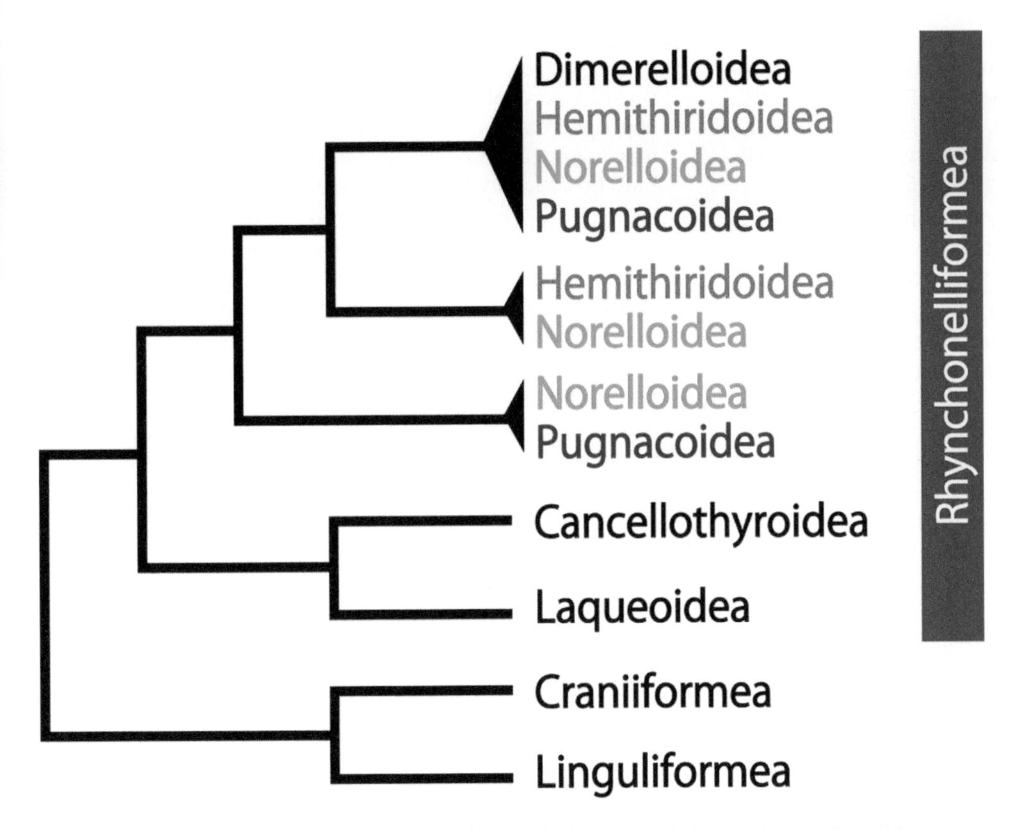

Figure 21.4 Molecular phylogeny of Rhynchonellida based on rDNA analysis. (Altered from Cohen and Bitner, 2013.)

REFERENCES

Cohen, B.L., Gawthrop, A., Cavalier-Smith, T., 1998. Molecular phylogeny of brachiopods and phoronids based on nuclear-encoded small subunit ribosomal RNA gene sequences. Philosophical Transactions of the Royal Society B Biological Sciences, 353, 2039–2061.

Cohen, B.L, Bitner, M.A., 2013. Molecular phylogeny of Rhynchonellide articulate brachiopods (Brachiopoda, Rhynchonellida). Journal of Paleontology, 87(2), 211–216.

Carlson, S.J., 2016. The evolution of Brachiopoda. Annual Review of Earth and Planetary Sciences, 44, 409–438.

Dunn, C.W., Hejnol, A., Matus, D.Q., Pang, K., Browne, W.E., Smith, S.A., Seaver, E., Rouse, G.W., Obst, M., Edgecombe, G.D., Sørensen, M.V., Haddock, S.H., Schmidt-Rhaesa, A., Okusu, A., Kristensen, R.M., Wheeler, W.C., Martindale, M.Q., Giribet, G., 2008. Broad phylogenomic sampling improves resolution of the animal tree of life. Nature, 452(7188), 745–749.

Dunn, C.W., Giribet, G., Edgecombe, G.D., Hejnol, A., 2014. Animal phylogeny and its evolutionary implications. Annual Review of Ecology, Evolution, and Systematics, 45, 371–395.

Emig, C.C., 1997. Biogeography of inarticulated brachiopods. In: Kaesler, R.L. (ed.), Treatise on Invertebrate Paleontology, Part H. Brachiopoda, 1 (pp. 497–502). The Geological Society of America and the University of Kansas, Colorado, and Lawrence, Kansas.

Emig, C.C., 2017. Atlas of Antarctic and sub-Antarctic Brachiopoda. Carnets de Géologie, Madrid, CG2017_B03.

Emig, C.C., 2018. Brachiopodes récoltés lors de campagnes (1976–2014) dans l'étage Bathyal des côtes françaises méditerranéennes. Redéfinition des limites du système phytal dans le domaine marin benthique, Carnets de Géologie, Madrid, CG2018_B01.

Emig, C.C., Bitner, M.A., Alvarez, F. 2019. Brachiopoda Database (version Feb 2016). In Roskov, Y., Ower, G., Orrell, T., Nicolson, D., Bailly, N., Kirk, P. M., Bourgoin, T., DeWalt, R. E., Decock, W., Nieukerken, E. van, Zarucchi, J., Penev, L. (eds.), Species 2000 & ITIS Catalogue of Life, 2019 Annual Checklist. Digital resource at www.catalogueoflife.org/annual-checklist/2019. Species 2000: Naturalis, Leiden, The Netherlands. ISSN 2405-884X.

Gerdol, M., Luo, Y-J., Satoh N., Pallavicini, A., 2018. Genetic and molecular basis of the immune system in the brachiopod *Lingula anatina*. Developmental and Comparative Immunology, 82, 7–30. doi: 10.1016/j.dci.2017.12.021

Halanych, K.M., 2004. The new view of animal phylogeny. Annual Review of Ecology, Evolution, and Systematics, 35, 229–256. doi: 10.1146/annurev.ecolsys.35.112202.130124

Hammond, L.S., 1982. Breeding season, larval development and dispersal of *Lingula anatine* (Brachiopoda, inarticulata) from Townsville, Australia. Journal of Zoology, London, 198, 183–196.

Hejnol, A., Obst, M., Stamatakis, A., Ott, M., Rouse, G.W., Edgecombe, G.D., Martinez, P., Baguñà, J., Bailly, X., Jondelius, U., Wiens, M., Müller, W.E.G., Seaver, E., Wheeler, W.C., Martindale, M.Q., Giribet, G., Dunn, C.W., 2009. Assessing the root of bilaterian animals with scalable phylogenomic methods. Proceedings of the Royal Society B, 276, 4261–4270.

Hejnol, A., 2010. A twist in time—the evolution of spiral cleavage in the light of animal phylogeny. Integrative and Comparative Biology, 50(5), 695–706. doi: 10.1093/icb/icq103

Isowa, Y., Sarashina, I., Oshima, K., Kito, K., Hattori, M., Endo, K., 2015. Proteome analysis of shell matrix proteins in the brachiopod *Laqueus rubellus*. Proteome Science, 13, 21. doi: 10.1186/s12953-015-0077-2

Jackson, D.J., Mann, K., Häussermann, V., Schilhabel, M.B., Lüter, C., Griesshaber, E., Schmahl, W., Wörheide, G., 2015. The *Magellania venosa* biomineralizing proteome: A Window into brachiopod shell evolution. Genome Biology and Evolution, 7(5), 1349–1362. doi: 10.1093/gbe/evv074

Luo, Y-J., Takeuchi, T., Koyanagi, R., Yamada, L., Kanda, M., Khalturina, M., Fujie, M., Yamasaki, S., Endo, K., Satoh, N., 2015. The *Lingula* genome provides insights into brachiopod evolution and the origin of phosphate biomineralization. Nature Communications, 6, 8301. doi: 10.1038/ncomms9301

Luo, Y-J., Kanda, M., Koyanagi, R., Hisata, K., Akiyama, T., Sakamoto, H., Sakamoto, T., Satoh, N., 2018. Nemertean and phoronid genomes reveal lophotrochozoan evolution and the origin of bilaterian heads. Nature Ecology and Evolution, 2, 141–151. doi: 10.1038/s41559-017-0389-y

Marlétaz, F., Peijnenburg, K.T.C.A., Goto, T., Satoh, N., Rokhsar, D.S., 2019. A new Spiralian phylogeny places the enigmatic arrow worms among Gnathiferans. Current Biology, 29(2), 312–318.

Nesnidal, M.P., Helmkampf, M., Meyer, A., Witek, A., Bruchhaus, I., Ebersberger, I., Hankeln, T., Lieb, B., Struck, T.H., Hausdorf, B., 2013. New phylogenomic data support the monophyly of Lophophorata and an ectoproct-phoronid clade and indicate that Polyzoa and Kryptrochozoa are caused by systematic bias. BMC Evolutionary Biology, 13, 253.

Santagata, S., Cohen, B.L., 2009. Phoronid phylogenetics (Brachiopoda; Phoronata): evidence from morphological cladistics, small and large subunit rDNA sequences, and mitochondrial *cox1*. Zoological Journal of the Linnean Society, 157, 34–50. doi: 10.1111/j.1096-3642.2009.00531.x

Santagata, S., Resh, C., Hejnol, A., Martindale, M.Q., Passamaneck, Y.J., 2012. Development of the larval anterior neurogenic domains of *Terebratalia transversa* (Brachiopoda) provides insights into the diversification of larval apical organs and the spiralian nervous system. EvoDevo, 3, 3. doi: 10.1186/2041-9139-3-3

Schiemann, S.M., Mart ín-Durán, J.M., Børve, A., Vellutini, B.C., Passamaneck, Y.J., Hejnol, A., 2017. Clustered brachiopod Hox genes are not expressed collinearly and are associated with lophotrochozoan novelties. Proceedings of National Academy of Sciences if the United States of America, 114(10), E1913–E1922. doi: 10.1073/pnas.1614501114

Shimizu, K., Luo, Y.-J., Satoh, N., Endo, K., 2017. Possible co-option of engrailed during brachiopod and mollusc shell development. Biology Letters, 13, 20170254. doi: 10.1098/rsbl.2017.0254

Vellutini, B.C., Hejnol, A., 2016. Expression of segment polarity genes in brachiopods supports a non-segmental ancestral role of *engrailed* for bilaterians. Scientific Reports, 6, 32387. doi: 10.1038/srep32387

PHYLA ECTOPROCTA AND PHORONIDA

CHAPTER

22

Scott Santagata

Infrakingdom	Bilateria
Division	Protostomia
Subdivision	Lophotrochozoa
Phyla	Ectoprocta and Phoronida

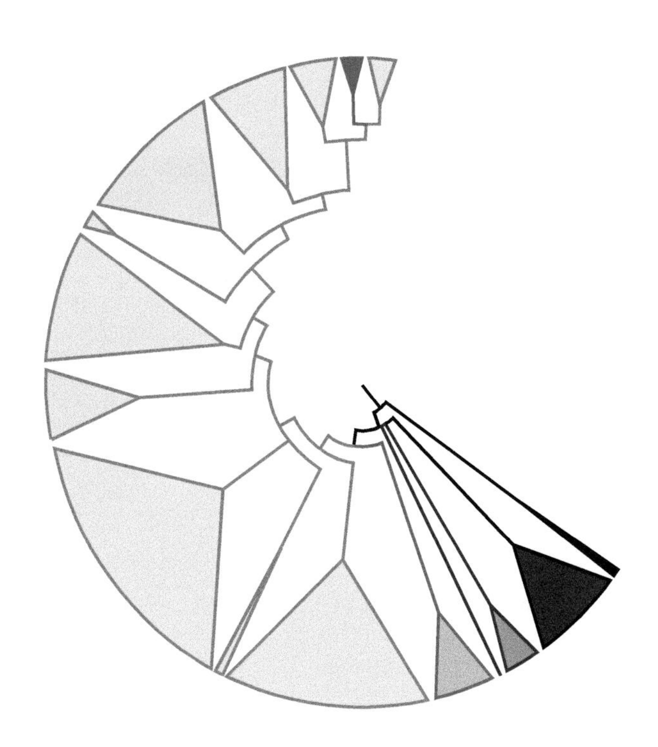

CONTENTS

General Characteristics

Ectoprocta

5869 valid extant species (Bock and Gordon, 2013)

With additional cryptic species the total species may be approximately 8,000 (Ryland, 2005)

Mostly colonial, bilaterally symmetrical animals found in freshwater, brackish, and marine environments worldwide

Six mitochondrial genomes available

One assembled genome for *Bugula neritina* (Rayko et al., 2020) and several transcriptomes

Commercial value related to isolated natural products (e.g., Bryostatin and other compounds)

Ecological importance as providing secondary habitats for numerous other animals

Phoronida

14 valid extant species (Temereva and Chichvarkhin, 2017)

Mostly solitary, bilaterally symmetrical animals found in brackish and marine environments worldwide

Two mitochondrial genomes available, *Phoronis architecta* and *Phoronopsis harmeri*

Genome for one species, *Phoronis australis*

No established commercial value

Ecological importance as suspension feeders

Synapomorphies—Ectoprocta

- colonies consist of polymorphic zooids with an outer cystid and inner polypide
- no circulatory or excretory organs
- nutrients may be transported through colony via a funicular cord system connected by communication pores between adjacent zooids
- distinctive cyphonautes (feeding) and coronate (non-feeding) larval forms

Synapomorphies—Phoronida

- one pair of metanephridia
- closed circulatory system
- distinctive actinotroch (feeding) and slug-like (non-feeding) larval forms

Synapomorphies—Potential Ectoprocta-Phoronida

- lophophore
- U-shaped gut
- trimeric body with coelomic cavities as ancestral state
- blastomere arrangements derived from a spiralian ancestor
- cerebral ganglion of ectoprocts and the inner ganglion of *Phoronis ovalis* (Temereva, 2017)

HISTORY OF TAXONOMIC RELATIONSHIPS AMONG ECTOPROCTS AND PHORONIDS

Both the ectoprocts and entoprocts were once grouped together in the Bryozoa (Ehrenberg, 1831), but were later separated later by Nitsche (1869) and then reorganized into different phyla by Cori (1929). There are 5869 extant ectoproct species represented by three classes (Brock and Gordon, 2013) occurring in freshwater (Phylactolaemata), brackish, and marine (Stenolaemata and Gymnolaemata) environments worldwide. More than 15,000 fossil species have been described going back to the Ordovician period (Xia et al., 2007; Ma et al., 2015).

Conversely, the phoronids are one of the least speciose groups in the animal kingdom. Based on adult characters at least 24 species have been described by various authors, however the majority of phoronid morphotypes have been synonymized under 14 cosmopolitan species (Emig, 1974; Temereva and Chichvarkhin, 2017) and two genera, *Phoronis* (Wright, 1856) and *Phoronopsis* (Gilchrist, 1907). The chitinous sandy tubes of phoronids have been associated with some trace fossils. A *Phoronis ovalis*-like boring pattern may have produced the Devonian ichnofossil, *Talpina* (Thomas, 1911). However, it is likely that the crown phoronid lineage is much older, and may be linked to vermiform filter-feeding forms from the lower Cambrian (Skovsted et al., 2008, 2011).

Based on embryonic cleavage patterns and the morphology of their feeding structures and body cavities, earlier work united ectoprocts with the phoronids and brachiopods under the super phyletic assemblage Lophophorata (Hyman, 1959), all of whose members were believed to share affinities with deuterostome animals. This hypothesis was initially refuted by 18S rDNA-based phylogenetic data (Halanych et al., 1995), placing the ectoprocts, phoronids, and brachiopods in a new superphyletic assemblage of protostome animals called the Lophotrochozoa. How more recent phylogenomic analyses have influenced the interpretations of the evolutionary origins and relationships among these lophophore-bearing animals is discussed later in this chapter.

MORPHOLOGY AND ANATOMY

Ectoprocts are aquatic invertebrates that can form elaborate and occasionally large colonies, composed of numerous individual zooids each typically no more than a millimeter in length (**Figure 22.1A**). Zooids in a colony may be one of several different polymorphic forms specialized for different functions such as feeding, reproduction, or defense. All ectoprocts feed with a lophophore which is a circular or U-shaped ring of ciliated tentacles surrounding the mouth. Ciliary currents direct suspended food particles toward the mouth, to be gathered in a muscular pharynx. Ingested food particles are then processed in a U-shaped gut. Ectoprocts do not have nephridia, but solid waste is expelled via the anus located outside the tentacle ring. Feeding zooids, called autozooids, can transfer nutrients to other non-feeding polymorphic zooids (e.g., *Avicularia*, kenozooids, etc.) in the colony by a network of tissue cords and communication pores called the funicular system (Best and Thorpe, 1985). Neuronal cell bodies of the intraepithelial nervous system are mainly concentrated in a single ganglion positioned between the mouth and anus (Schwaha et al., 2011) and can integrate sensory-motor information among zooids (Thorpe et al., 1975). Collectively, the lophophore, gut, nervous system, musculature, and funicular tissue are grouped together as the polypide. Surrounding the polypide is an outer stratified epithelial layer called the cystid having a tubular, ovoid, or box-like shape. Gymnolaemate bryozoans are comprised of two orders, the uncalcified Ctenostomata and the diversely calcified Cheilostomata (**Figure 22.1B**). Further subdivisions of the Ctenostoma have been based mainly on colony-level traits such as budding patterns or colony form. Family-level evolutionary relationships under these systematic categories differ among authors due to varying interpretational significances assigned to particular morphological characters (Jebram, 1992; Todd, 2000; Schwaha and Wanninger, 2018).

Different morphological grades of cheilostomes (Anasca, Cribimorpha, and Ascophora) were separated according to the structure of the frontal cystid membrane covering the polypide and its role in eversion of the lophophore. Among the Anasca, the frontal membrane may be a flexible structure that can be depressed by parietal muscles, thus increasing pressure in the coelom surrounding the polypide, and resulting in lophophore eversion (e.g., *Membranipora*). Cribimorph species enclose the frontal membrane with a series of fused calcified spines (e.g., *Cribrilina*). Numerous species of ascophoran cheilostomes have a calcified wall with pores of varying numbers and sizes beneath their frontal membranes. These species have a membranous sac called an ascus below the frontal wall that is open to the surrounding seawater. Parietal muscles retract to expand this sac causing the eversion of the lophophore. Living members of the class Stenolaemata are represented only by the order Cyclostomata, whose cylindrical calcified zooids lack an operculum or muscular constriction at their openings. Cyclostomes occur strictly in marine environments and their colonies can be relatively small in comparison to those of other ectoproct classes. Similar to the other orders, systematic classifications within the cyclostomes based solely on zooidal morphology have also been questioned (Taylor and Weedon, 2000). Some of the more divergent zooidal traits are found in the freshwater phylactolaemates whose colonies are often described as being either gelatinous or a series of chitinous, branched tubes (Wood and Lore, 2005). Unlike other bryozoans, phylactolaemate zooids have lophophores with a hollow neuronal ganglion (Gruhl and Bartolomaeus, 2008; Schwaha and Wanninger, 2012), bud from a fleshy mass, and have a U-shaped lophophore more similar to those of

Phoronis spp.

Adult phoronid bodies are divided by transverse septa into three regions: the epistome, tentacle crown, and trunk (**Figure 22.1C**). These body regions are lined by mesoderm, forming true coelomic cavities. Collectively the protocoel of the epistome, the mesocoel of the lophophore, and the metacoel of the trunk were believed to be similar to the tripartite coeloms found in echinoderms (Masterman, 1898). However, the epistomal lining in two species of *Phoronis* are composed of myoepithelial cells lacking adherent junctions surrounding a gel-like extracellular matrix (Bartolomaeus, 2001; Gruhl et al., 2005), but the epistomal lining of *Phoronopsis harmeri* does contain adherent junctions (Temereva and Malakhov, 2011). Further variation in this trait involves whether the adult epistome and protocoel are derived from some portion of the larval hood and protocoel, or formed *de novo* after metamorphosis (Zimmer, 1978; Santagata, 2002). As discussed previously, the ectoproct Phylactolaemate lophophore is more similar to that of some phoronids than to those of other bryozoans. Phylactolaemate epistomes do contain a true protocoel, however, their tripartite body cavities are confluent with each other rather than being separate (Schwaha, 2018). Whether adult phoronids have two or three true coelomic compartments, this distinction does not help resolve their phylogenetic position relative to other lophophore-bearing animals or to annelids, nemerteans, and molluscs.

In both ectoprocts and phoronids, the adult gut is U-shaped and the anus is positioned outside of the tentacles. In contrast to ectoprocts, each phoronid tentacle contains a blind capillary with red blood cells using a type of hemoglobin as an oxygen carrier (Garlick et al., 1979). Tentacular capillaries are connected to lophophoral ring vessels that are linked to the efferent and afferent blood vessel loop traversing the trunk region. The adult trunk is divided into a more posterior ampullary region and a more anterior and tapered muscular region containing some diagonal muscle fibers in the body wall (Chernyshev and Temereva, 2010). The main body wall musculature of the trunk epithelium consists of a layer of circular muscles underlying numerous feathery or bush-like longitudinal muscles (**Figure 22.1D**; Herrmann, 1997). The distribution of longitudinal muscles in the four mesenteric divisions of the trunk is the character used to separate most phoronid species (Emig, 1974). However, almost all phoronid species are distributed worldwide, and variation in longitudinal muscle patterns makes it difficult to identify cryptic species.

The anterior center of the adult phoronid nervous system is a group of basiepidermal neuronal cell bodies concentrated between the mouth and the anus (Fernández et al., 1996). Although this structure has often been called the dorsal ganglion (Silén, 1954a) or a dorsal neural plexus (Temereva and Malakhov, 2009), some aspects of its post-metamorphic development are not consistent with a completely dorsal origin (Santagata, 2002). Despite these interpretational differences, this anterior group of ciliated neuronal cells is the adult "brain," connected to a collar nerve ring with unciliated neuronal cells along its length at the base of the tentacles (Temereva and Malakhov, 2009). One potential synaptomorphy shared among the nervous systems of phoronids and ectoprocts may be found in the colonial-like phoronid species, *P. ovalis*, which has two anterior ganglions. One of these ganglions is similar to that of other phoronid species, with the other inner ganglion bearing similarities to the cerebral ganglion of ectoprocts (Temereva, 2017). Beyond anterior neuronal structures sporadically distributed neuronal cells and fibers are found throughout the surface of the trunk epithelium. A subset of these cells and fibers are serotonergic (Santagata, 2002), but the most centralized neuronal structure is the giant nerve fiber embedded in the anterior portion of the trunk epithelium (Temereva and Malakhov, 2009). Among various phoronid species, the single or multiple giant nerve fibers have different morphologies with respect to their position and posterior limit in the trunk region (Emig, 1974).

Figure 22.1 **Ectoprocts and phoronids.** (A) Colony of *Bugula turrita* collected from Orr's Island, ME, USA. (B) Colony of *Schizoporella floridana* collected from Fort Pierce, FL, USA, overgrowing a colony of *Electra*. (C) Juvenile *Phoronis* sp. from Tampa, FL. (D) Volume rendering of confocal z-series through the trunk region of *Phoronis architecta* stained for musculature (green) and nuclei (blue). Dashed lines show how the trunk region is divided into three internal zones by mesenteries. AF, afferent blood vessel; CH, chitinous tube; CM, circular muscle; EF, efferent blood vessel; GF, giant nerve fiber; L, lophophore; LM, longitudinal muscle; O, ovicell; T, trunk region.

REPRODUCTION

Ectoproct reproductive patterns are diverse but overall, zooids within ectoproct colonies may be hermaphroditic, or separate male and female zooids may be present in the same colony (Reed, 1991; Santagata and Banta, 1996; Temkin, 1996; Ostrovsky et al., 2009; Ostrovsky, 2013). Typically, mature sperm are released as a cloud from a coelomopore in the lophophore (Silén, 1966), or in packets of 32 or 64 called spermatozeugmata (Temkin and Bortolami, 2004). Anascan cheilostome ectoprocts, such as *Membranipora* spp., broadcast-spawn fertilized (60 μm) oocytes into the water column (**Figure 22.2A**) via an intertentacular organ in the lophophore (Temkin, 1994), which develop into a feeding larval form called the cyphonautes (**Figure 22.2B**; Kupelweiser, 1905). However, the vast majority of marine bryozoans brood embryos within maternal zooids or specialized chambers called ovicells, producing diverse forms of spheroid-shaped non-feeding larvae, the most common being the coronate larva (**Figure 22.2C**; Santagata, 2008a,b). The cell types and the anatomy of the nervous systems and musculature of larval ectoprocts have been better studied in selected species of gymnolaemates and, overall, these data support some broad-scale homology with the larval neuromuscular systems of spiralian protostomes (Santagata, 2008b; Gruhl, 2009; Wanninger, 2009), but differ from the larval morphologies of phoronids and brachiopods (Santagata and Zimmer, 2002; Altenburger and Wanninger, 2009; Santagata, 2011; Temereva and Wanninger, 2012).

Adult phoronids are either gonochoristic or hermaphroditic, with some species having a brief temporal bias toward protandry (Zimmer, 1991). Gonads develop from the peritoneal lining covering, the capillaries of the efferent blood vessel on the surface of the stomach (Ikeda, 1903; Rattenbury, 1953; Zimmer, 1991). The four known reproductive types among phoronid species are based largely on the sex of the adult, egg size, and how embryos are retained in some species (Zimmer, 1991). Group 1 includes members of both genera and is comprised almost entirely of gonochoristic species (except *Phoronis pallida*) that freely spawn 60 μm ova (**Figure 22.2D**). Members of group 2 mostly produce larger mature eggs (≥ approx. 100) and brood embryos to an early larval or competent larval stage on specialized nidamental glands at the base of the lophophore. Patterns within the latter group are unclear due to incomplete knowledge regarding the reproductive properties of species such as *Phoronopsis albomaculata* and other disputed species morphotypes (*Phoronis capensis* and *Phoronis bhadurii*). Group 3 has only one member, *P. ovalis*, whose sex type remains unconfirmed, produces 125 μm eggs, and broods embryos in the parental tube (Harmer, 1917; Silén, 1954b). A potential group 4 is represented by *Phoronis embryolabi* in which larval development begins in the metanephridia prior to release (Temereva and Chichvarkhin, 2017). Morphological and molecular characteristics as well as life history traits suggest that this species is closely related to *P. pallida* (Santagata, 2004a,b; Santagata and Cohen, 2009; Temereva and Chichvarkhin, 2017).

Fertilization is largely internal in phoronids. Primary oocytes are arrested at metaphase and released from the ovary to fuse with activated V-shaped sperm (Zimmer, 1964, 1991; Reunov and Klepal, 2004). Male pronuclei have been found in the coelomic oocytes of *P. harmeri* (Rattenbury, 1953) and *Phoronis ijimai* (Zimmer, 1991; Hirose et al., 2014). There is one largely undocumented report of external fertilization in *Phoronis muelleri* (Herrmann, 1986), but this observation requires further investigation. Among species that are simultaneous hermaphrodites, self-fertilization has not been reported, as mature sperm released from the testes in the trunk are not active (Zimmer, 1964; Reunov and Klepal, 2004). The ciliary currents of the paired metanephridial funnels eventually collect mature sperm in the trunk region. Although sperm can be released directly from the metanephridial ducts into the surrounding seawater (Rattenbury, 1953; Silén, 1954b), more often released sperm are enclosed by the paired lophophoral organs at the base of the lophophore (**Figure 22.2E**). Sperm are then packaged into spherical, bean- or club-shaped spermatophores, some of which have elaborate sail-like structures (Zimmer, 1967, 1991). Spermatophores drift away and activate when contacting another adult conspecific individual. One way for activated sperm to reach the ovary is through the metanephridial ducts (Brooks and Cowles, 1905; Rattenbury, 1953; Zimmer, 1967). Another route has been described in *Phoronopsis viridis* (= *P. harmeri*; Emig, 1974; Santagata and Cohen, 2009), in which the spherical portion of the spermatophore contacts the tip of a tentacle where sperm lyse through the epithelial tissue into the collar coelom (Zimmer, 1991). From there the activated sperm swim down and traverse another septum, traveling to the base of the trunk to fertilize oocytes from the ovary. Additionally, released spermatophores may be ingested by nearby adults.

The majority of phoronid species produce a feeding larval form called the actinotroch (**Figure 22.2F**). Competent actinotroch anatomy varies in maximum size, number of blood corpuscle masses, maximum number of larval tentacles, and how the juvenile tentacles are differentiated in the larva (Santagata and Zimmer, 2002). Only one species, *P. ovalis*, produces a non-feeding slug-like larva that lacks tentacles. Feeding actinotroch larvae have a complex complement of striated musculature for lifting and closing the preoral hood, extending or lowering the tentacles, and changing the angle of the teletrochal band of cilia while swimming (Zimmer, 1964; Santagata and Zimmer, 2002; Santagata, 2002, 2004b; Temereva and Tsitrin, 2014). Although many aspects of larval musculature are shared among several phoronid species, the morphology of particular larval muscles does differ, especially among the smaller actinotroch types (*P. pallida*, *P. ijimai*, and Phoronis *hippocrepia*) in contrast with the more elongate forms (*P. harmeri* and *P. muelleri*; Santagata and Zimmer, 2002; Temereva and Tsitrin, 2014).

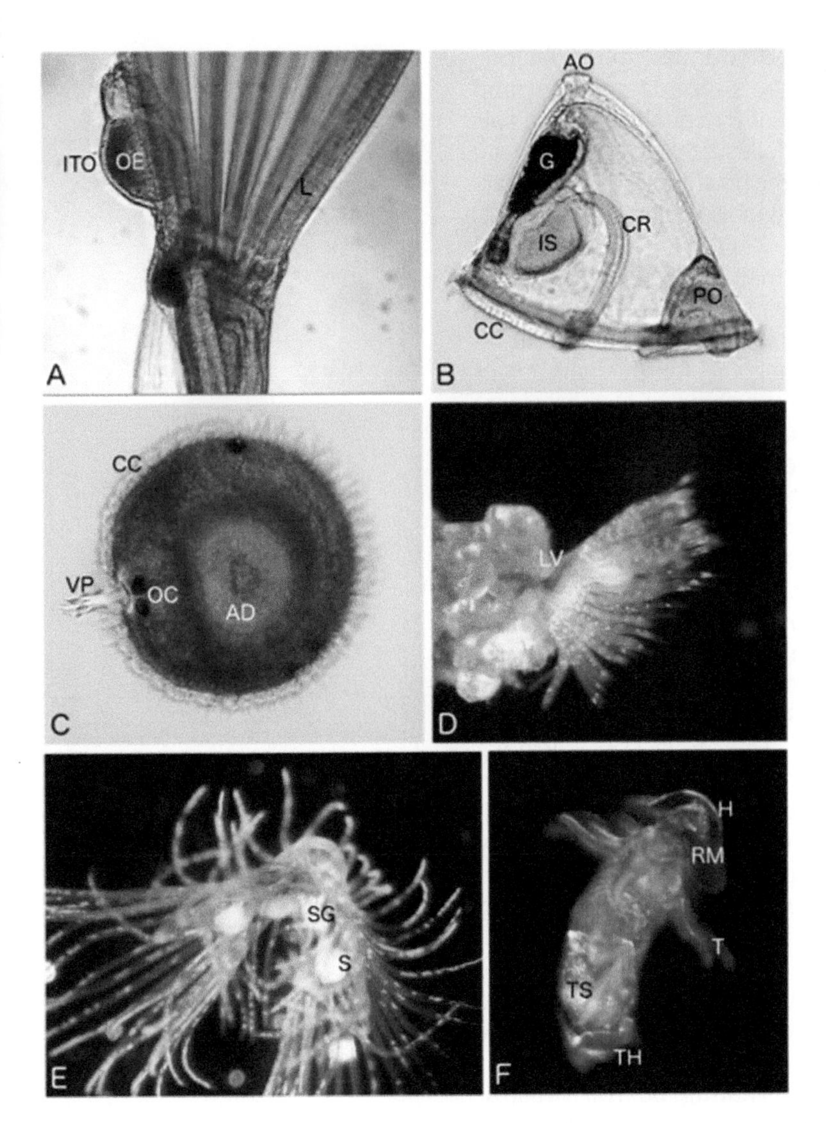

Figure 22.2 Reproductive traits of ectoprocts and phoronids. (A) Intertentacular organ of *Membranipora membranacea* (Friday Harbor, WA, USA) spawning a fertilized primary oocyte. (B) Late stage cyphonautes larva of *Membranipora membranacea* (Orr's Island, ME, USA). (C) Coronate larval form of *Bugula turrita*. (D) Lophophore of *Phoronis architecta* (Panacea, FL, USA). (E) Spermatophoral glands and spermatophores of *Phoronis architecta*. (F) Competent actinotroch larva of *Phoronis architecta* collected from the plankton of lower Chesapeake Bay, VA, USA. AD, apical disc; AO, apical sense organ; CC, coronal cells; CR, ciliated ridges; G, larval gut; H, larval hood; IS, internal sac; ITO, intertentacular organ; L, lophophore; LV, lophophoral ring blood vessels; PO, pyriform organ; OC, ocelli; OE, fertilized primary oocyte; RM, red blood corpuscle mass; S, spermatophore; SG, spermatophoral gland; T, larval tentacle; TH, telotroch; TS, trunk sac; VP, vibratile plume.

DEVELOPMENT

Much of the debate concerning lophophorate development has centered around early cleavage patterns, which have typically been described as being radial or biradial in their orientation (Zimmer, 1997). Blastomeres of ectoproct early embryos at the 32-cell stage and later are arranged in tiers reminiscent of similar stages in equally cleaving spiralian embryos (Nielsen, 2002). As there are examples of spiral cleavage being lost in some spiralians (Wadeson and Crawford, 2003; Martín-Durán and Marlétaz, 2020), the possibility exists that early cleavage patterns in ectoprocts are derived from a spiral-cleaving ancestor. Vellutini et al. (2017) tested this hypothesis using detailed cell lineage tracking and mapping the expression of key developmentally related transcription factors. Interestingly, in *Membranipora* embryos, specification of the D quadrant and mitogen-activated protein kinase activity resembles other equally cleaving spiralian embryos (**Figure 22.3A,B**; Freeman and Lundelius, 1992; Gonzales et al., 2006; Vellutini et al., 2017). As in other protostomes the blastopore forms the larval mouth, and the larval corona shares a cellular origin as well as gene expression patterns with some spiralians (Vellutini et al., 2017). Spiralians typically form mesoderm from two (ectodermal and endodermal) sources, and although Vellutini et al. (2017) only found evidence for endodermal sources of mesoderm, other accounts suggest additional ectodermal sources for larval mesoderm do exist (Gruhl, 2010). Endodermal patterning involving the fourth quadrant of blastomeres and the expression of gata456b in the larval gut of *Membranipora* also support that these features stem from a spiralian ancestor. However, nonfeeding coronate larval forms of other bryozoan species such as *Bugula neritina* lack a larval gut, and gata456 is expressed in some of the epidermal blastemal cells near the apical organ (Fuchs et al., 2011). Therefore, GATA456 expression may have undergone a heterochronic shift in this species since subsets of epidermal blastemal cells likely contribute to the adult gut post-metamorphosis.

Whether cleavage stages of various phoronid species exhibit aspects of radial cleavage or spiral cleavage has been debated since the first embryological observations were gathered. Radial (or biradial) cleavage patterns have been found in several species (Masterman, 1900; Ikeda, 1901), but spiral-like cleavage patterns have also been observed (Foettinger, 1882; Brooks and Cowles, 1905), and moreover, both patterns have been observed in the same species (Herrmann, 1986). A more typical, spiral cleavage pattern was described for *P. viridis* by Rattenbury (1954), but this was interpreted by Zimmer (1964) as an artifact introduced by compaction of the blastomeres by a tightly fitting vitelline envelope. All of the phoronid species having spiral-like cleavage patterns (or both spiral and radial) make smaller eggs, but other species that produce larger eggs such as *P. ijimai* have radially cleaving embryos (Zimmer, 1964; Wu et al., 1980; Malakhov and Temereva, 2000; Freeman and Martindale, 2002). Using four-dimensional microscopy techniques, Pennerstorfer and Scholtz (2012) determined that *P. muelleri* embryos exhibit oblique cell divisions with alternating dextral-sinistral twists. Third cleavage of *P. pallida* embryos produces some eight-cell stages that have a radial arrangement of blastomeres, and others that have animal cells slightly offset (with a clockwise twist) from those in the vegetal half of the embryo (**Figure 22.3C**; Santagata, 2015). Fourth cleavage is also variable among embryos; some 16-cell stages have the characteristic open radial arrangement of blastomeres, while other embryos have four tiers of cells offset from one another, arranged by oblique divisions with a counterclockwise twist. The embryos from the larger eggs of *Phoronis vancouverensis* have a radial arrangement of blastomeres, and although irregular early cleavage stages have been observed (Zimmer, 1964; Malakhov and Temereva, 2000), none of these patterns exhibit spiral-like cell divisions.

Further, embryonic development results in a blastula with a rounded blastopore that becomes slit-like as the embryo elongates. The slit-like blastopore is progressively closed from the posterior end, leaving only a small anterior remnant to form the larval mouth. Cell labeling and other embryological experiments show that much of the larval mesoderm is derived from ectodermal cells, especially where the ectoderm is in contact with the endoderm, with the remaining mesoderm derived from the endoderm (**Figure 22.3D**; Freeman and Martindale, 2002; Santagata, 2004b; Temereva and Malakhov, 2007). Information from the expression of developmentally related transcription factors is not known for phoronid embryos, however genomic studies have found that patterning of phoronid and brachiopod lophophores as well as the bilaterian heads all share several gene expression domains (Luo et al., 2018).

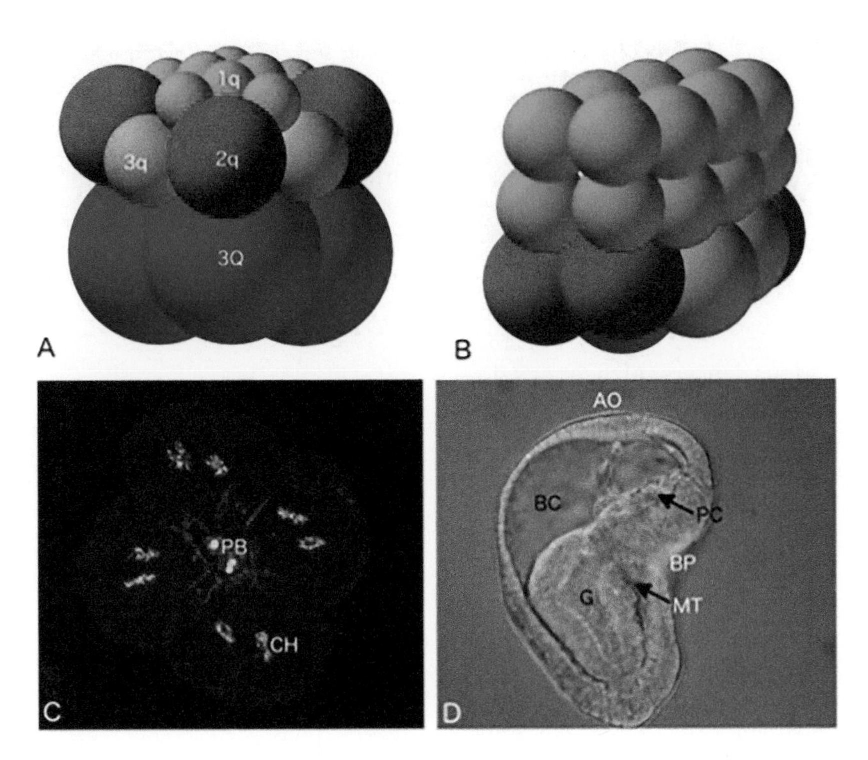

Figure 22.3 Embryonic stages of ectoprocts and phoronids. (A) Generalized spiralian embryo based on and modified from Vellutini et al. (2017) and Henry and Martindale, (1999). (B) Similar stage as A of *Membranipora membranacea* embryo based on and modified from Vellutini et al. (2017). (C) Confocal z-projection of the animal tier of an eight-cell *Phoronis pallida* embryo in the process of karyokinesis (anaphase to telophase) going to the 16-cell stage. The fibrous actin filaments of cell borders are stained red and the DNA is stained yellow. (D) Late gastrula of *Phoronis architecta*. 1q, first quartet; 2q, second quartet; 3q, third quartet; 3Q, third quartet macromeres; AO, apical sense organ; BC, remnant of the blastocoelic space; BP, blastopore; CH, chromosomes; G, larval gut; PB, polar body; MT, mesodermal cells that will line the trunk coelom; PC, mesodermal cells that will line the hood coelom.

ECOLOGY (INCLUDING ECONOMIC IMPORTANCE)

The ecology of both ectoprocts and phoronids is linked to their roles as suspension feeders that may, in particular habitats, occur at relatively high densities where they form secondary habitats for other species. Habitat-forming bryozoan communities have been documented worldwide (Wood et al., 2012). However, little is known about the species assemblages that create these unique habitats, or their importance to marine shelf and deeper marine communities. Among some Antarctic shelf habitats of the Ross, Scotia, and Weddell Seas, clade composition varies significantly in habitat-forming bryozoan communities (**Figure 22.4A**; Santagata et al., 2018). Although, phoronids can dominate the density and coverage of some benthic marine habitats (Larson and Stachowicz, 2009), and very little is known about their ecological role in such habitats. One aspect in common to both ectoprocts and phoronids is their production of secondary metabolites that can be effective chemical defenses against predators. At least for ectoprocts, chemical defenses are present in both adult colonies (Tian et al., 2017) and larvae (Lopanik et al., 2006). Both phyla have worldwide distributions (**Figure 22.5**).

Perhaps the greatest economic and health-related impact created by either animal group involves the discovery of Bryostatin, a polyketide produced by *Candidatus Endobugula sertula*, the bacterial symbionts found in the ectoproct, *B. neritina* (Woollacott, 1981; Haygood and Davidson, 1997; Linneman et al., 2014). Bryostatin (**Figure 22.4B**) is a protein kinase C activator potentially having

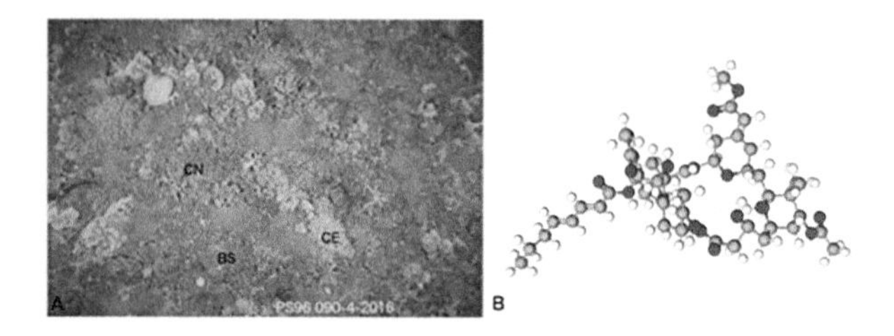

Figure 22.4 Ecological and economic significance of ectoprocts. (A) Habitat-forming ectoproct community located on the marine shelf of the Weddell Sea, Antarctica. Photograph taken of the seabed in the Weddell Sea using the AWI Ocean Floor Observation System as part of the DynAMO project during the PS96 cruise of *R/V Polarstern* in 2016 (Piepenburg et al., 2017). (B) Molecular model of the bryostatin compound in which carbon atoms are shaded gray, hydrogen atoms are shaded white, and oxygen atoms are shaded in red. BS, brittle star; CE, cellariiform colonies of ectoprocts; CN, crinoid.

multiple beneficial effects for the treatment of various cancers, Alzheimer's disease, and AIDS (Marsden et al., 2018; Sarajärvi et al., 2018; Song et al., 2018). Natural products isolated from ectoproct species remains an active line of research, the potential benefits of which are only now becoming known.

GENETICS AND GENOMICS

Evolutionary relationships among the ectoprocts, phoronids, and other bilaterian animals are not well resolved by currently available morphological or genomic information. A robust molecular phylogeny of all extant ectoprocts based on adequate sampling of taxa and genes is still forthcoming, but some macroevolutionary hypotheses about ectoproct evolution have been tested

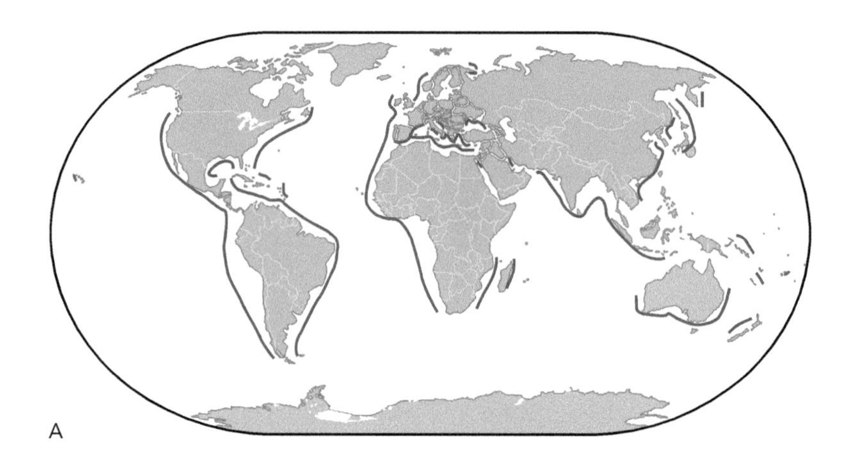

A

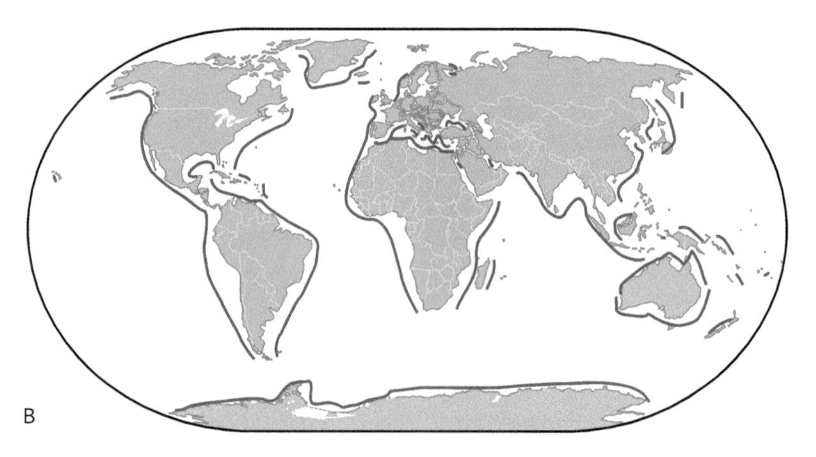

B

Figure 22.5 Species of phoronids and ectoprocts are generally found in circum-global brackish and marine habitats. Phylactolaemate ectoprocts, however, also occur in freshwater ecosystems. (A) Phoronids. (B) Ectoprocts.

(Waeschenbach et al., 2009, 2012). Strong support exists for the morphological characters separating the three main classes: Stenolaemata, Gymnolaemata, and Phylactolaemata. However, some family-level divisions based on zooidal frontal wall morphology in cheilostomes and colony budding types in ctenostomes and phylactolaemates are not congruent with molecular phylogenetic analyses (**Figure 22.6A**; Waeschenbach et al., 2012). Homoplasy among characters associated with skeletal features and colonial growth strategies is not unexpected, as predation and competition for space are convergent selective pressures acting on all forms of ectoprocts (McKinney, 1992, 1995). How ectoprocts are related to other bilaterian animals remains ambiguous. Based on 18S ribosomal DNA, the ectoprocts were grouped into a supraphyletic assemblage of protostome animals called the Lophotrochozoa, in which the evolutionary position of the ectoprocts relative to phoronids, brachiopods and spiralians was unresolved (Halanych et al., 1995). Increased taxon sampling bolstered with data from more rDNA genes, full mitochondrial genomes, and select nuclear genes continued to support the Lophotrochozoa hypothesis (Passamaneck and Halanych, 2004, 2006; Jang and Hwang, 2009). More recent phylogenomic approaches supported the resurrection of the Polyzoa concept in which ectoprocts are linked with entoprocts (Hausdorf et al., 2007, 2010; Helmkampf et al., 2008; Hejnol et al., 2009), or reuniting all the lophophorates in one monophyletic group where the phoronids are a sister taxon to the ectoprocts (Nesnidal et al., 2013). Considering the available genomes coupled with transcriptomic datasets, there is some support for phoronids and ectoprocts as sister taxa basal to all brachiopods (Luo et al., 2015; Luo et al., 2018; Laumer et al., 2019).

Species- and genus-level relationships among phoronids based on either morphological or molecular characters are fairly congruent (Santagata and Cohen, 2009), and support *P. ovalis* as a divergent lineage, a *Phoronopsis* spp. clade, and a *Phoronis* spp. clade (**Figure 22.6B**). Based on rDNA, mitochondrial genes, and microRNAs, it is not clear whether or not phoronids are an ancestral sister taxon to all brachiopods (Sperling et al., 2011) or instead as a subtaxon within the brachiopods as a whole (Cohen and Weydmann, 2005; Santagata and Cohen, 2009; Cohen, 2013). Phoronids are clearly related to brachiopods and share evolutionary affinities with spiralian protostomes

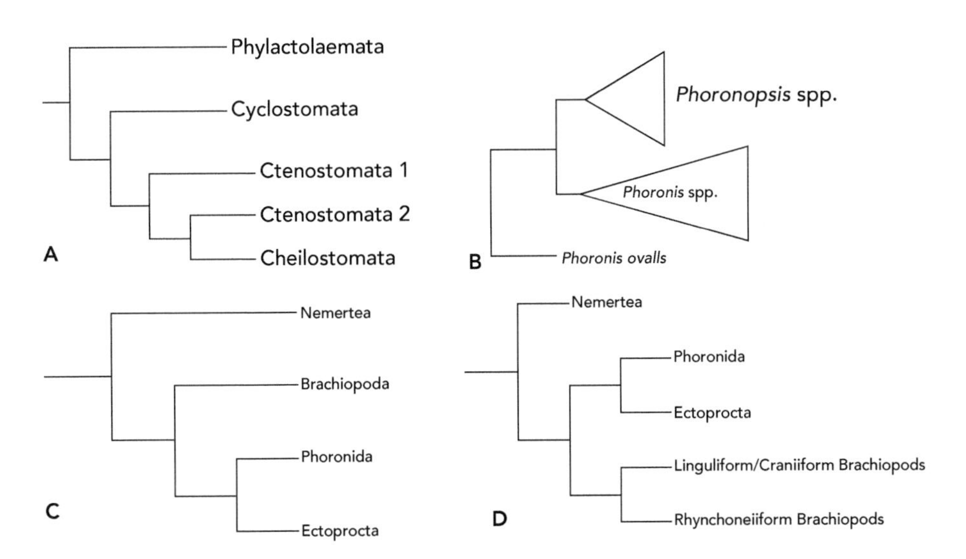

Figure 22.6 Phylogenetic hypotheses about ectoprocts and phoronids. (A) Molecular phylogeny of major extant ectoproct clades based on and modified from Waeschenbach et al. (2012) and Taylor and Waeschenbach (2015). (B) Molecular phylogeny of extant phoronids species based on and modified from Santagata and Cohen (2009) and Santagata (unpublished data). (C) Lophophorate evolutionary relationships based on and modified from Nesnidal et al. (2013) and Laumer et al. (2015). (D) Lophophorate evolutionary relationships based on and modified from Luo et al. (2018).

such as nemerteans, annelids, and molluscs, but their phylogenetic position relative to ectoprocts is not stable (Halanych et al., 1995; Dunn et al., 2008; Hejnol et al., 2009; Hausdorf et al., 2010; Mallatt et al., 2012; Kocot et al., 2017). However, more recent phylogenomic studies have recovered the lophophorate monphyly (**Figure 22.6C**; Nesnidal et al., 2013; Laumer et al., 2015). When considering the available genomes (one phoronid and one brachiopod; Luo et al., 2015, 2018) coupled with transcriptomic datasets, there is support for phoronids and ectoprocts as sister taxa basal to all brachiopods (**Figure 22.6D**; Luo et al., 2018). Further phylogenetic resolution among these lineages will require additional taxon sampling, especially with respect to potentially intermediate forms such as *P. ovalis* and various species of phylactolaemate and ctenostome ectoprocts. Despite numerous differences in anatomy, reproduction, and development among species of phoronids and ectoprocts, when weighed against the available paleontological and molecular evidence, it is reasonable to test the hypothesis that ectoprocts (and brachiopods) evolved from a phoronid-like ancestor. The derived features of ectoprocts may have been secondarily acquired along with a switch to colonial adult forms and reduced body (zooid) size.

POSITION IN THE ToL

Based on the available transcriptomic and genomic information, phoronids and ectoprocts (as well as brachiopods) are lophotrochozoan protostomes sharing evolutionary affinities with spiralians (Luo et al., 2015; Luo et al., 2018; Laumer et al., 2019). Discrepancies among dataset types and phylogenetic analyses exist with respect to whether phoronids are more closely related to brachiopods or to ectoprocts, the validity of the lophophorate monophyly, and the phylogenetic position of the entoprocts.

DATABASES AND COLLECTIONS

Taxonomic databases for both phoronids and ectoprocts (bryozoans) may be downloaded through the World Register of Marine Species (Worms) and the Integrated Taxonomic Information System (ITIS). Specimen collections for ectoprocts are much more common and widespread than for phoronids. Querying the invertebrate zoology collections at the Smithsonian National Museum of Natural History results in 4,000 ectoproct records, but only 503 phoronid records. Similar differences in taxonomic collections are apparent at the American Museum of Natural History (bryozoans = 295 records; phoronids = 0), the California Academy of Sciences (ectoprocts = 1,608 records; phoronids = 100), and the Natural History Museum of London, U.K. (bryozoans = 46,121 records; phoronids = 60).

CONCLUSION

The unofficial title submitted for this chapter was "Heads and Tales of Two 'Phyla,' the Ectoprocts and Phoronids" as play on words to describe the variation in (and losses of) novel head-like structures exhibited by larval and adult forms of these animals. The word phyla appears in single quotes due to the ambiguous status of phoronids being a separate phylum from ectoprocts and brachiopods. As someone who is inclined to study invertebrates having enigmatic evolutionary origins and relationships, it is my hope that this chapter will inspire a new generation of invertebrate zoologists and evolutionary biologists to investigate lophophore-bearing animals.

REFERENCES

Altenburger A, Wanninger A. 2009. Comparative larval myogenesis and adult myoanatomy of the rhynchonelliform (articulate) brachiopods *Argyrotheca cordata, A. cistellula*, and *Terebratalia transversa*. Frontiers in Zoology 6:3.

Bartolomaeus T. 2001. Ultrastructure and formation of the body cavity lining in *Phoronis muelleri* (Phoronida, Lophophorata). Zoomorphology 120:135–148.

Best MA, Thorpe JP. 1985. Autoradiographic study of feeding and the colonial transport of metabolites in the marine bryozoan *Membranipora membranacea*. Marine Biology 84:295–300.

Bock PE, Gordon DP. 2013. Phylum Bryozoa Ehrenberg, 1831. Zootaxa 3703:1–8.

Brooks W, Cowles R. 1905. *Phoronis architecta*: its life history, anatomy and breeding habits. Memoirs of the National Academy of Sciences 10:72–113.

Chernyshev AV, Temereva EN. 2010. First report of diagonal musculature in phoronids (Lophophorata: Phoronida). Doklady Biological Sciences 433:264–267.

Cohen BL. 2013. Rerooting the rDNA gene tree reveals phoronids to be "brachiopods without shells;" dangers of wide taxon samples in metazoan phylogenetics (Phoronida; Brachiopoda). Zoological Journal of the Linnean Society 167:82–92.

Cohen BL, Weydmann A. 2005. Molecular evidence that phoronids are a subtaxon of brachiopods (Brachiopoda: Phoronata) and that genetic divergence of metazoan phyla began long before the early Cambrian. Organism Diversity & Evolution 5:253–273.

Cori CI. 1929. "Kamptozoa, dritter Cladus der Vermes Amera" in: Handbuch der Zoologie. Kükenthal W and Krumbach T. (Eds.), pp. 1–64, Zweiter Band, Berlin: Walter de Gruyter and Co..

Dunn CW, Hejnol A, Matus DQ, Pang K, Browne WE, Smith SA, Seaver E, Rouse GW, Obst M, Edgecombe GD, Sørensen MV, Haddock SHD, Schmidt-Rhaesa A, Okusu A, Kristensen RM, Wheeler WC, Martindale MQ, Giribet G. 2008. Broad phylogenomic sampling improves resolution of the animal tree of life. Nature 452:745–749.

Ehrenberg CG. 1831. Symbolae physicae, seu Icones et Descriptiones Mammalium Avium, Insectorum et Animalium Evertebratorum. Berlin, Pars Zoologica No pagination.

Emig CC. 1974. The systematics and evolution of the phylum Phoronida. Zeitschrift fur Zoologische Systematik und Evolutionsforschung 12:128–151.

Fernández I, Pardos F, Benito J. 1996. Ultrastructural observations on the phoronid nervous system. Journal of Morphology 230:265–281.

Foettinger A. 1882. Note sur la formation du mesoderme dans la larve de *Phoronis hippocrepia*. Archives de Biologie. Paris 3:679–686.

Freeman G, Lundelius JW. 1992. Evolutionary implications of the mode of D quadrant specification in coelomates with spiral cleavage. Journal of Evolutionary Biology 5:205–247.

Freeman G, Martindale MQ. 2002. The origin of mesoderm in phoronids. Developmental Biology 252:301–311.

Fuchs J, Martindale MQ, Hejnol A. 2011. Gene expression in bryozoan larvae suggest a fundamental importance of pre-patterned blastemic cells in the bryozoan lifecycle. EvoDevo 2:1–13.

Garlick RL, Williams BJ, Riggs AF. 1979. The hemoglobins of *Phoronopsis viridis*, of the primitive invertebrate phylum Phoronida: Characterization and subunit structure. Archives of Biochemistry and Biophysics 194:13–23.

Gilchrist JDF. 1907. New forms of the Hemichordata from South Africa. Transactions of the South African Philosophical Society 17:151–176.

Gonzales EE, van der Zee M, Dictus WJAG, van den Biggelaar J. 2006. Brefeldin A or monensin inhibits the 3D organizer in gastropod, polyplacophoran, and scaphopod molluscs. Development Genes and Evolution 217:105–118.

Gruhl A, Bartolomaeus T. 2008. Ganglion ultrastructure in phylactolaemate Bryozoa: evidence for a neuroepithelium. Journal of Morphology 269:594–603.

Gruhl A. 2009. Serotonergic and FMRFamidergic nervous systems in gymnolaemate bryozoan larvae. Zoomorphology 128:135–156.

Gruhl A. 2010. Ultrastructure of mesoderm formation and development in *Membranipora membranacea* (Bryozoa: Gymnolaemata). Zoomorphology 129:45–60.

Gruhl A, Grobe P, Bartolomaeus T. 2005. Fine structure of the epistome in *Phoronis ovalis:* significance for the coelomic organization in Phoronida. Invertebrate Biology 124:332–343.

Halanych KM, Bacheller JD, Aguinaldo AM, Liva SM, Hillis DM, Lake JA. 1995. Evidence from 18S ribosomal DNA that the lophophorates are protostome animals. Science 267:1641–1643.

Harmer SF. 1917. Harmer: on *Phoronis ovalis*, Strethill Wright. Quarterly Journal of Microscopical Science 62:115–148.

Hausdorf B, Helmkampf M, Meyer A, Witek A, Herlyn H, Bruchhaus I, Hankeln T, Struck TH, Lieb B. 2007. Spiralian phylogenomics supports the resurrection of Bryozoa comprising Ectoprocta and Entoprocta. Molecular Biology and Evolution 24:2723–2729.

Hausdorf B, Helmkampf M, Nesnidal MP, Bruchhaus I. 2010. Phylogenetic relationships within the lophophorate lineages (Ectoprocta, Brachiopoda and Phoronida). Molecular Phylogenetics and Evolution 55:1121–1127.

Haygood MG, Davidson SK. 1997. Small-subunit rRNA genes and in situ hybridization with oligonucleotides specific for the bacterial symbionts in the larvae of the bryozoan *Bugula neritina* and proposal of "*Candidatus endobugula sertula*". Applied and Environmental Microbiology 63:4612–4616.

Hejnol A, Obst M, Stamatakis A, Ott M, Rouse GW, Edgecombe GD, Martinez P, Baguna J, Bailly X, Jondelius U, Wiens M, Muller WEG, Seaver E, Wheeler WC, Martindale MQ, Giribet G, Dunn CW. 2009. Assessing the root of bilaterian animals with scalable phylogenomic methods. Proceedings of the Royal Society B: Biological Sciences 276:4261–4270.

Helmkampf M, Bruchhaus I, Hausdorf B. 2008. Multigene analysis of lophophorate and chaetognath phylogenetic relationships. Molecular Phylogenetics and Evolution 46:206–214.

Henry J, Martindale MQ. 1999. Conservation and innovation in spiralian development. Hydrobiologia. 402:255–265.

Herrmann K. 1986. Die Ontogenese von *Phoronis mulleri* (Tentaculata) unter besonderer Berücksichtigung der Mesodermdifferenzierung und Phylogenese des Coeloms. Zoologische Jahrbücher Abteilung für Anatomie und Ontogenie der Tiere 114:441–463.

Herrmann K. 1997. "Phoronida" in: Microscopic Anatomy of Invertebrates, Volume 13. Harrison FW, and Woollacott RM. (Eds.), pp. 207–235, Lophophorates, Entoprocta, and Cyliophora Wiley-Liss, New York.

Hirose M, Fukiage R, Katoh T, Kajihara H. 2014. Description and molecular phylogeny of a new species of *Phoronis* (Phoronida) from Japan, with a redescription of topotypes of *P. ijimai* Oka, 1897. Zookeys 398:1–31.

Hyman LH. 1959. The Invertebrates: smaller coelomate groups: Chaetognatha, Hemichordata, Pogonophora, Phoronida, Ectoprocta, Brachipoda, Sipunculida, the coelomate Bilateria. Volume V. McGraw-Hill, New York, pp. 1–783

Ikeda I. 1901. Observations on the development: structure and metamorphosis of Actinotrocha. Journal of the College of Science, Imperial University of Tokyo 13:507–591.

Ikeda I. 1903. On the development of the sexual organs and of their products in *Phoronis*. Annotationes Zoologicae Japan 4:141–153.

Jang K, Hwang U. 2009. Complete mitochondrial genome of *Bugula neritina* (Bryozoa, Gymnolaemata, Cheilostomata): phylogenetic position of Bryozoa and phylogeny of lophophorates within the Lophotrochozoa. BMC Genomics 10:1–18.

Jebram DHA. 1992. The polyphyletic origin of the "Cheilostomata" (Bryszoa). Journal of Zoological System 30:46–52.

Kocot KM, Struck TH, Merkel J, Waits DS, Todt C, Brannock PM, Weese DA, Cannon JT, Moroz LL, Lieb B, Halanych KM. 2017. Phylogenomics of Lophotrochozoa with consideration of systematic error. Systematic Biology 66:256–282.

Kupelweiser H. 1905. Untersuchungen über den feineren Bau und die Metamorphose des Cyphonautes. Zoologica (Stuttgart) 19:1–50.

Larson AA, Stachowicz JJ. 2009. Chemical defense of a soft-sediment dwelling phoronid against local epibenthic predators. Marine Ecology Progress Series 374:101–111.

Laumer CE, Bekkouche N, Kerbl A, Goetz F, Neves RC, Sørensen MV, Kristensen RM, Hejnol A, Dunn CW, Giribet G, Worsaae K. 2015. Spiralian phylogeny informs the evolution of microscopic lineages. CURBIO 25:2000–2006.

Laumer CE, Fernández R, Lemer S, Combosch D, Kocot KM, Riesgo A, Andrade SCS, Sterrer W, Sørensen MV, Giribet G. 2019. Revisiting metazoan phylogeny with genomic sampling of all phyla. Proceedings of the Royal Society B: Biological Sciences 286:1–10.

Linneman J, Paulus D, Lim-Fong G, Lopanik NB. 2014. Latitudinal Variation of a Defensive Symbiosis in the *Bugula neritina* (Bryozoa) Sibling Species Complex. PLoS One 9:1–10.

Lopanik NB, Targett NM, Lindquist N. 2006. Ontogeny of a symbiont-produced chemical defense in *Bugula neritina* (Bryozoa). Marine Ecology Progress Series 327:183–191.

Luo Y-J, Kanda M, Koyanagi R, Hisata K, Akiyama T, Sakamoto H, Sakamoto T, Satoh N. 2018. Nemertean and phoronid genomes reveal lophotrochozoan evolution and the origin of bilaterian heads. Nature Ecology & Evolution 1–13.

Luo Y-J, Takeuchi T, Koyanagi R, Yamada L, Kanda M, Khalturina M, Fujie M, Yamasaki S-I, Endo K, Satoh N. 2015. The *Lingula* genome provides insights into brachiopod evolution and the origin of phosphate biomineralization. Nature Communications 6:1–10.

Ma J, Taylor PD, Xia F, Zhan R. 2015. The oldest known bryozoan: *Prophyllodictya* (Cryptostomata) from the lower Tremadocian (Lower Ordovician) of Liujiachang, south-western Hubei, central China. Palaeontology 58:925–934.

Malakhov VV, Temereva EN. 2000. Embryonic development of the phoronid *Phoronis ijimai*. Russian Journal of Marine Biology 26:412–421.

Mallatt J, Craig CW, Yoder MJ. 2012. Nearly complete rRNA genes from 371 Animalia: Updated structure-based alignment and detailed phylogenetic analysis. Molecular Phylogenetics and Evolution 64:603–617.

Marsden MD, Wu X, Navab SM, Loy BA, Schrier AJ, DeChristopher BA, Shimizu AJ, Hardman CT, Ho S, Ramirez CM, Wender PA, Zack JA. 2018. Characterization of designed, synthetically accessible bryostatin analog HIV latency reversing agents. Virology 520:83–93.

Martín-Durán JM, Marlétaz F. 2020. Unravelling spiral cleavage. Developmental Biology 147:1–7.

Masterman AT. 1898. On the theory of archimeric segmentation and its bearing upon the phyletic classification of the Coelomata. Proceedings of the Royal Society Edinburgh 270–310.

Masterman AT. 1900. Memoirs: On the Diplochorda III. The early development and anatomy of *Phoronis buskii*, McIntosh. Quarterly Journal of Microscopical Science 43:375–418.

McKinney FK. 1992. Competitive interactions between related clades: evolutionary implications of overgrowth interactions between encrusting cyclostome and cheilostome bryozoans. Marine Biology 114:645–652.

McKinney FK. 1995. One hundred million years of competitive interactions between bryozoan clades: asymmetrical but not escalating. Biological Journal of the Linnean Society 56: 465–481.

Nesnidal MP, Helmkampf M, Meyer A, Witek A, Bruchhaus I, Ebersberger I, Hankeln T, Lieb B, Struck TH, Hausdorf B. 2013. New phylogenomic data support the monophyly of Lophophorata and an Ectoproct-Phoronid clade and indicate that Polyzoa and Kryptrochozoa are caused by systematic bias. BMC Evolutionary Biology 13:253–13.

Nielsen C. 2002. The phylogenetic position of Entoprocta, Ectoprocta, Phoronida, and Brachiopoda. Integrative and Comparative Biology 42:685–691.

Nitsche H. 1869. Beiträge zur Erkenntnis der Bryozoen. I Beobachtungen ueber die Entwicklungsgeschichte einiger cheilostomen Bryozoen. Zeitschrift für Wissenschaftliche Zoologie 20:1–13.

Ostrovsky AN. 2013. From incipient to substantial: evolution of placentotrophy in a phylum of aquatic colonial invertebrates. Evolution 67:1368–1382.

Ostrovsky AN, Gordon DP, Lidgard S. 2009. Independent evolution of matrotrophy in the major classes of Bryozoa: transitions among reproductive patterns and their ecological background. Marine Ecology Progress Series 378:113–124.

Passamaneck YJ, Halanych KM. 2004. Evidence from Hox genes that bryozoans are lophotrochozoans. Evolution & Development 6:275–281.

Passamaneck YJ, Halanych KM. 2006. Lophotrochozoan phylogeny assessed with LSU and SSU data: Evidence of lophophorate polyphyly. 2006. Lophotrochozoan phylogeny assessed with LSU and SSU data: Evidence of lophophorate polyphyly. Molecular Phylogenetics and Evolution 40:20–28.

Pennerstorfer M, Scholtz G. 2012. Early cleavage in *Phoronis muelleri* (Phoronida) displays spiral features. Evolution & Development 14:484–500.

Piepenburg D., Buschmann A., System ADE. (2017). Seabed images from Southern Ocean shelf regions off the northern Antarctic Peninsula and in the southeastern Weddell Sea. Earth System Science Data 9: 461–469.

Rattenbury JC. 1953. Reproduction in *Phoronopsis viridis*. The annual cycle in the gonads, maturation and fertilization of the ovum. Biological Bulletin 104:182–196.

Rattenbury JC. 1954. The embryology of *Phoronopsis viridis*. Journal of Morphology 95:289–349.

Rayko M, Komissarov A, Kwan JC, Lim-Fong G, Rhodes AC, Kliver S, Kuchur P, O'Brien SJ, Lopez JV. 2020. Draft genome of *Bugula neritina*, a colonial animal packing powerful symbionts and potential medicines. Sci Data 7:1–5.

Reed CG. 1991. "Bryozoa" in: Reproduction of Marine Invertebrates, Volume VI: Echinoderms and Lophophorates, Giese AC, Pearse JS, Pearse V. (Eds.), pp. 85–245The Boxwood Press, Pacific Grove, CA.

Reunov A, Klepal W. 2004. Ultrastructural study of spermatogenesis in *Phoronopsis harmeri* (Lophophorata, Phoronida). Helgoländer Wissenschaftliche Meeresuntersuchungen 58:1–10.

Ryland JS. 2005. Bryozoa: an introductory overview. Denisia 16:9–20.

Santagata S, Ade V, Mahon AR, Wisocki PA, Halanych KM. 2018. Compositional differences in the habitat-forming bryozoan communities of the antarctic shelf. Frontiers in Ecology and Evolution 6:1–15.

Santagata S. 2004a. Larval development of *Phoronis pallida* (Phoronida): Implications for morphological convergence and divergence among larval body plans. Journal of Morphology 259:347–358.

Santagata S. 2004b. A waterborne behavioral cue for the actinotroch larva of *Phoronis pallida* (Phoronida) produced by *Upogebia pugettensis* (Decapoda: Thalassinidea). Biological Bulletin 207:103–115.

Santagata S. 2008a. Evolutionary and structural diversification of the larval nervous system among marine bryozoans. Biological Bulletin 215:3–23.

Santagata S. 2008b. The morphology and evolutionary significance of the ciliary fields and musculature among marine bryozoan larvae. Journal of Morphology 269:349–364.

Santagata S. 2011. Evaluating neurophylogenetic patterns in the larval nervous systems of brachiopods and their evolutionary significance to other bilaterian phyla. Journal of Morphology 272:1153–1169.

Santagata S. 2015. "Phoronida" in: Evolutionary Developmental Biology of Invertebrates, 2. Wanninger A. (Ed.), p. 231–245, Springer, Vienna.

Santagata S, Banta WC. 1996. Origin of brooding and ovicells in cheilostome bryozoans: interpretive morphology of *Scrupocellaria ferox*. Invertebrate Biology 170–180.

Santagata S, Cohen BL. 2009. Phoronid phylogenetics (Brachiopoda; Phoronata): evidence from morphological cladistics, small and large subunit rDNA sequences, and mitochondrial cox1. Zoological Journal of the Linnean Society 157:34–50.

Santagata S, Zimmer RL. 2002. Comparison of the neuromuscular systems among actinotroch larvae: systematic and evolutionary implications. Evolution and Devlopment 4:43–54.

Sarajärvi T, Jäntti M, Paldanius KMA, Natunen T, Wu JC, Mäkinen P, Tarvainen I, Tuominen RK, Talman V, Hiltunen M. 2018. Protein kinase C -activating isophthalate derivatives mitigate Alzheimer's disease-related cellular alterations. Neuropharmacology 141:76–88.

Schwaha T. 2018. Morphology and ontogeny of *Lophopus crystallinus* lophophore support the epistome as ancestral character of phylactolaemate bryozoans. Zoomorphology 137:355–366.

Schwaha T, Wanninger A. 2012. Myoanatomy and serotonergic nervous system of plumatellid and fredericellid Phylactolaemata (Lophotrochozoa, Ectoprocta). Journal of Morphology 273:57–67.

Schwaha TF, Wanninger A. 2018. Unity in diversity: a survey of muscular systems of ctenostome Gymnolaemata (Lophotrochozoa, Bryozoa). 1–18.

Schwaha T, Wood TS, Wanninger A. 2011. Myoanatomy and serotonergic nervous system of the ctenostome *Hislopia malayensis*: Evolutionary trends in body plan patterning of ectoprocta. Frontiers in Zoology 8:1–16.

Silén L. 1954a. On the nervous system of *Phoronis*. Archiv fur Zoologi 1–40.

Silén L. 1954b. Developmental biology of Phoronidea of the Gullmar Fiord area (west coast of Sweden). Acta Zoologica Stockholm. 35:215–257.

Silén L. 1966. On the fertilization problem in the gymnolaematous bryozoa. Ophelia 3:113–140.

Skovsted CB, Brock GA, Paterson JR, Holmer LE, Budd GE. 2008. The scleritome of *Eccentrotheca* from the Lower Cambrian of South Australia: lophophorate affinities and implications for tommotiid phylogeny. Geology 36:171.

Skovsted CB, Brock GA, Topper TP, Paterson JR, Holmer LE. 2011. Scleritome construction, biofacies, biostratigraphy and systematics of the tommotiid *Eccentrotheca helenia* sp. nov. from the Early Cambrian of South Australia. Palaeontology 54:253–286.

Song X, Xiong Y, Qi X, Tang W, Dai J, Gu Q, Li J. 2018. Molecular targets of active anticancer compounds derived from marine sources. Marine Drugs 16:175–22.

Sperling EA, Pisani D, Peterson KJ. 2011. Molecular paleobiological insights into the origin of the Brachiopoda. Evolution & Development 13:290–303.

Taylor PD, Waeschenbach A. 2015. Phylogeny and diversification of bryozoans. Paleontology. 58:585–599.

Taylor PD, Weedon MJ. 2000. Skeletal ultrastructure and phylogeny of cyclostome bryozoans. Zoological Journal of the Linnean Society 128:337–399.

Temereva EN. 2017. Innervation of the lophophore suggests that the phoronid *Phoronis ovalis* is a link between phoronids and bryozoans. Scientific Report 1–16.

Temereva EN, Chichvarkhin A. 2017. A new phoronid species, *Phoronis embryolabi*, with a novel type of development, and consideration of phoronid taxonomy and DNA barcoding. Invertebrate Systematics 31:65–84.

Temereva EN, Malakhov VV. 2007. Embryogenesis and larval development of *Phoronopsis harmeri* Pixell, 1912 (Phoronida): dual origin of the coelomic mesoderm. Invertebrate Reproduction & Development 50:57–66.

Temereva EN, Malakhov VV. 2009. On the organization of the lophophore in phoronids (Lophophorata: Phoronida). Russian Journal of Marine Biology 35:479–489.

Temereva EN, Malakhov VV. 2011. Organization of the epistome in *Phoronopsis harmeri* (Phoronida) and consideration of the coelomic organization in Phoronida. Zoomorphology 130:121–134.

Temereva EN, Tsitrin EB. 2014. Development and organization of the larval nervous system in *Phoronopsis harmeri* : new insights into phoronid phylogeny. Frontiers in Zoology 11:1–25

Temereva E, Wanninger A. 2012. Development of the nervous system in *Phoronopsis harmeri* (Lophotrochozoa, Phoronida) reveals both deuterostome- and trochozoan-like features. BMC Evolutionary Biology 12:1–27.

Temkin MH. 1994. Gamete spawning and fertilization in the gymnolaemate bryozoan *Membranipora membranacea*. Biological Bulletin 187:143–155.

Temkin MH. 1996. Comparative fertilization biology of gymnolaemate bryozoans. Marine Biology 127:329–339.

Temkin MH, Bortolami SB. 2004. Waveform dynamics of spermatozeugmata during the transfer from paternal to maternal individuals of *Membranipora membranacea*. Biological Bulletin 206:35–45.

Thomas AO. 1911. A fossil burrowing sponge from the Iowa Devonian. Bulletin of the Laboratory of Natural History, State University of Iowa, Iowa City. 6:165–166.

Thorpe J, Shelton G, Laverack M. 1975. Colonial nervous control of lophophore retraction in cheilostome Bryozoa. Science 189:60–61.

Tian XR, Gao YQ, Tian XL, Li J, Tang HF, Li YS, Lin HW, Ma ZQ. 2017. New cytotoxic secondary metabolites from marine Bryozoan *Cryptosula pallasiana*. Marine Drugs 15:1–10.

Todd JA. 2000. "The central role of ctenostomes in bryozoan phylogeny" in: Proceedings of the 11th International Bryozoology Association Conference Smithsonian Tropical Research Institute. Cubilla H, Jackson JBC. (Eds.), pp. 104–135, Balboa, Republic of Panama.

Vellutini BC, Martín-Durán JM, Hejnol A. 2017. Cleavage modification did not alter blastomere fates during bryozoan evolution. 1–28.

Wadeson PH, Crawford K. 2003. Formation of the blastoderm and yolk syncytial layer in early squid development. Biological Bulletin 205:179–180.

Waeschenbach A, Cox CJ, Littlewood DTJ, Porter JS, Taylor PD. 2009. First molecular estimate of cyclostome bryozoan phylogeny confirms extensive homoplasy among skeletal characters used in traditional taxonomy. Molecular Phylogenetics and Evolution 52:241–251.

Waeschenbach A, Taylor PD, Littlewood DTJ. 2012. A molecular phylogeny of bryozoans. Molecular Phylogenetics and Evolution 62:718–735.

Wanninger A. 2009. Shaping the things to come: ontogeny of lophotrochozoan neuromuscular systems and the Tetraneuralia concept. Biological Bulletin 216:293–306.

Wood ACL, Probert PK, Rowden AA, Smith AM. 2012. Complex habitat generated by marine bryozoans: a review of its, structure, diversity, threats and conservation. Aquatic Conservation: Marine and Freshwater Ecosystem 22:547–563.

Wood TS, Lore M. 2005. "The higher phylogeny of phylactolaemate bryozoans inferred from 18S ribosomal DNA sequences" in: Proceedings of the 13th International Bryozoology Association Conference. Wyse Jackson PN, Cancino JM, Moyano GHI, (Eds.), pp. 361–368, Taylor & Francis, Concepción/Chile.

Woollacott RM. 1981. Association of bacteria with bryozoan larvae. Marine Biology 65:155–158.

Wright TS. 1856. Description of two tubicolar animals. Proceedings of The Royal Society of Edinburgh 1:165–167.

Wu B, Chen M, Sun R. 1980. On the occurrence of *Phoronis ijimai* Oka in the Huang Hai, with notes on its larval development. Studia Marina Sinica. 16:101–122.

Xia FS, Zhang SG, Wang ZZ. 2007. The oldest bryozoans: new evidence from the late Tremadocian (early Ordovician) of east Yangtze Gorges in China. Journal of Paleontology 81:1308–1326.

Zafer KM, Seong-Geun K, Do DT, Chang-Bae K. 2016. Complete mitochondrial genome analysis of *Lingula anatina* from Korea (Brachiopoda, Lingulida, Lingulidae). Mitochondrial DNA Part B: Resources 2:829–830.

Zimmer RL. 1964. Reproductive biology and development of Phoronida. Ph.D. Thesis, p. 416, University of Washington, Seattle, WA.

Zimmer RL. 1967. The morphology and function of accessory reproductive glands in the lophophores of *Phoronis vancouverensis* and *Phoronopsis harmeri*. Journal of Morphology 121:159–178.

Zimmer RL. 1978. "The comparative structure of the preoral hood coelom" in: Settlement and Metamorphosis of Marine Invertebrate Larva. Chia FS Rice ME. (Eds.), pp. 23–40, Elsevier, New York.

Zimmer RL. 1991. "Phoronida" in: Reproduction of Marine Invertebrates, Volume VI: Echinoderms and Lophophorates. Pearse JS, Pearse VB, Giese AC. (Eds.), pp. 1–45, Boxwood Press, Pacific Grove, CA.

Zimmer RL. 1997. "Phoronids, brachiopods, and bryozoans, the lophophorates" in: Embryology: Constructing the Organism. Gilbert SF, Raunio AM. (Eds.), pp. 279–305, Sinauer Associates, Sunderland, MA.

CHAPTER
23

PHYLUM NEMERTEA

Hiroshi Kajihara

Infrakingdom	Bilateria
Division	Protostomia
Subdivision	Lophotrochozoa
Phylum	Nemertea

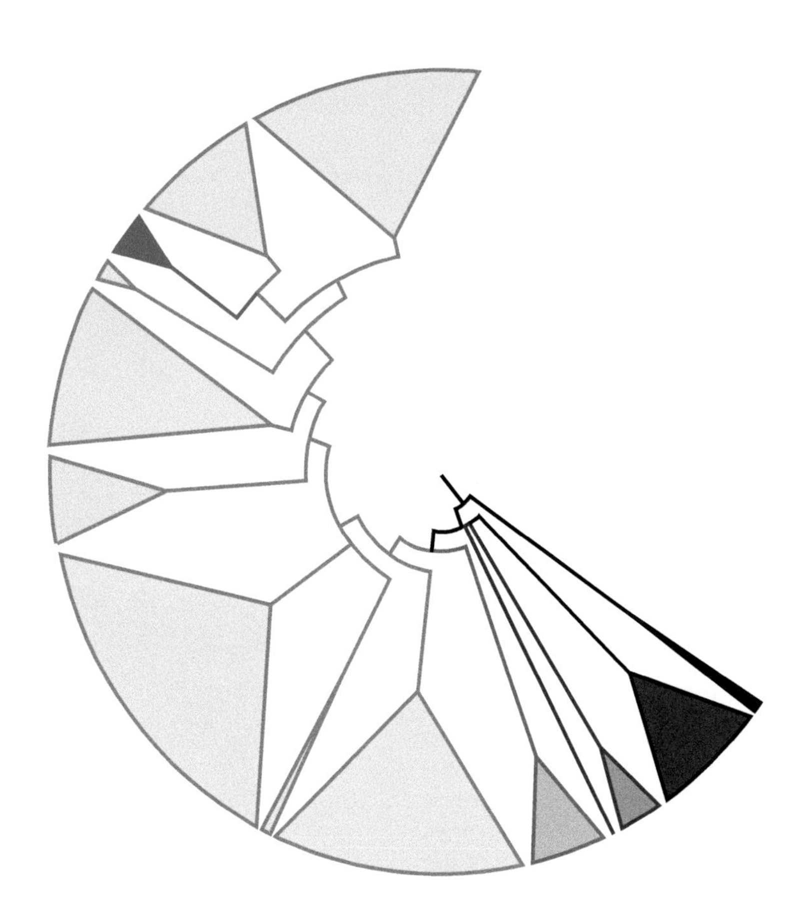

CONTENTS

General Characteristics

About 1300 species

Likely sister to a clade that comprises Phoronida and Brachiopoda (and possibly Bryozoa, Cycliophora, and Entoprocta as well), comprising Lophotrochozoa along with Annelida and Mollusca

Three major subgroups: Palaeonemertea, Pilidiophora, and Hoplonemertea

Basically carnivorous, preying upon crustaceans, annelids, and molluscs

Eversible proboscis housed in a body cavity, the rhynchocoel, used in attacking prey

Synapomorphies

- eversible proboscis housed in fluid-filled, mesoderm-derived rhynchocoel
- brain ring encircling the rhynchocoel
- closed blood vascular system lined with mesoderm-derived endothelium

HISTORY OF TAXONOMY AND CLASSIFICATION

The earliest mention to nemerteans in literature was made in 1555 as "*De vermibus longissimis*" (Magnus, 1555 [p. 496] according to Cedhagen and Sundberg, 1986). The worm was later described as *Ascaris longissima* in 1770 (transferred to *Lineus* in 1806; now *Lineus longissimus* [**Figure 23.1A**]), which represents the oldest-established, currently valid nemertean species. To the same species, the French comparative anatomist Georges Cuvier gave the name *Nemertes borlasii* in 1817. Although *Nemertes* (a sea nymph in Greek mythology) is a junior synonym of *Lineus*, its derivatives such as Nemertea, Nemertina, Nemertinea, and Nemertini were adopted by earlier zoologists to refer to nemerteans as a group. The common use among English literature until the 1980 was "nemertines." Rhynchocoela (Schultze, 1851) – a synonym of Nemertea – has not gained a wide acceptance. From 18th to early 19th centuries, nemerteans were grouped together with other vermiform invertebrates – for example, platyhelminths, nematodes, and annelids – under now-abandoned higher taxa such as Vermes or Zoophyta. Nemerteans were not mentioned in Carolus Linnaeus' *Systema Naturae*. Many zoologists in the 19th to early 20th centuries appear to have considered nemerteans as constituting a subgroup within Platyhelminthes (especially Turbellaria) or Annelida. The idea of Nemertea as an independent phylum was first proposed in the late 19th century (Sedgwick, 1898 according to Chernyshev, 2011) and gradually accepted among zoologists by the mid-20th century (Coe, 1943; Hyman, 1951). In 1853, the German microscopic anatomist Max J. S. Schultze grouped nemerteans with non-armed proboscides in Anopla and those with armed ones in Enopla. Different taxonomic schemes were proposed subsequently, variously relying on the presence/absence of the horizontal lateral cephalic slits (**Figure 23.1B**) or the position of the nervous system relative to the body wall musculature (**Figure 23.2** and **Table 23.1**). In the currently accepted system, the phylum is divided into Palaeonemertea and Neonemertea, with the latter comprising Pilidiophora and Hoplonemertea. A mid-dorsal vessel is absent in Palaeonemertea s.str., but is present in Neonemertea. Hoplonemertea is almost synonymous with Schultze's Enopla. Pilidiophora, with members having a characteristic pilidium larva, consists of Heteronemertea (with a dermal layer, 445 spp.) and *Hubrechtella* (without a dermal layer, 14 spp.); the latter was formerly placed in Palaeonemertea s.lat. "Palaeonemerteans" with a mid-dorsal vessel (*Hubrechtia*, *Sundbergia*, and *Tetramys*) may also belong to Pilidiophora. Classification in the family and genus levels is in state of flux. The present author's species enumeration in each major taxon as of 2018 resulted as follows: Palaeonemertea (114 spp.), Pilidiophora (462), and Hoplonemertea (593); the last consists of Monostilifera (593), Reptantia (43), and Pelagica (98). *Arhynchonemertes axi* lacks the proboscis apparatus, whose taxonomic placement in the phylum remains uncertain. Fossil taxa at least once treated as nemerteans include *Amiskwia* (alternative interpretation is Chaetognatha), *Archisymplectes*, *Legnodesmus*, *Lumbricaria* (listed as nemerteans by Diesing, 1850, but most likely represent cephalopod coprolites), and *Nemertites*.

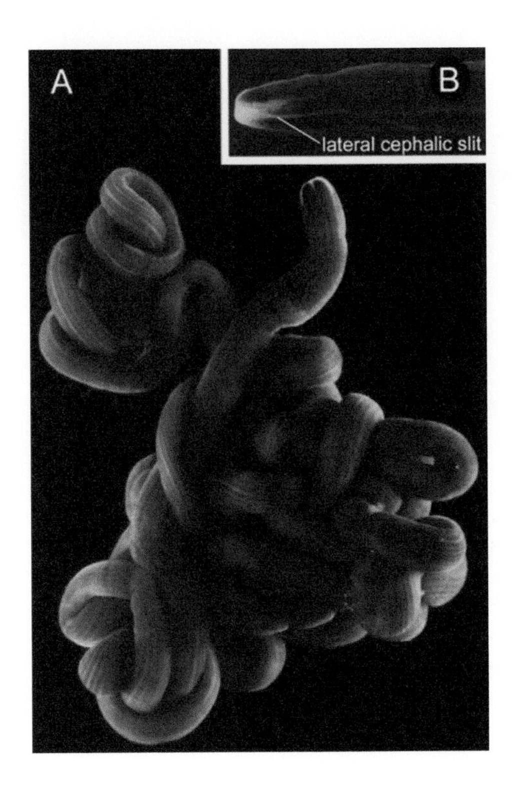

Figure 23.1 The longest worm. The bootlace worm *Lineus longissimus* is reported to have attained 55 m length. (A) A specimen (65 cm long) from Bergen, Norway. (B) Magnification of head, lateral view showing lateral cephalic slit.

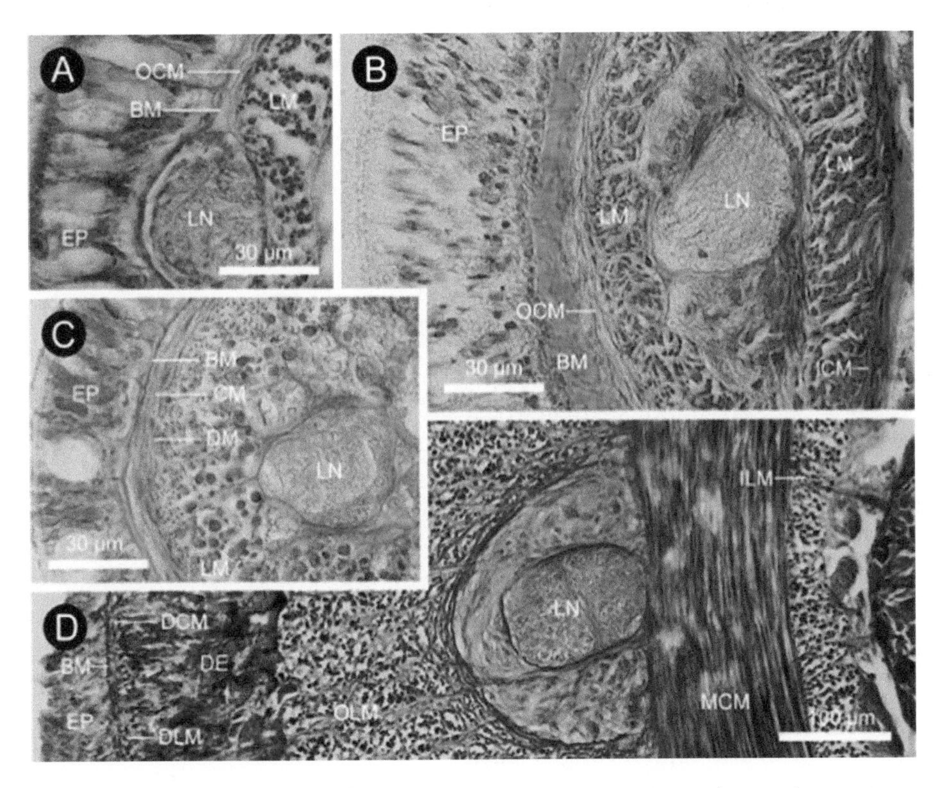

Figure 23.2 Relative position of nervous system. Transverse sections showing the position of lateral nerve cord within body wall: (A) *Tubulanus tamias* (Palaeonemertea); (B) *Cephalothrix simula* (Palaeonemertea); (C) *Nemertopsis mitellicola* (Hoplonemertea); (D) *Lineopselloides albilineus* (Heteronemertea). BM, basement membrane; CM, circular muscle; DE, dermis (cutis); DM, diagonal muscle; DCM, dermal circular muscle; DLM, dermal longitudinal muscle; EP, epidermis; ICM, inner circular muscle; ILM, inner longitudinal muscle; LM, longitudinal muscle; LN, lateral nerve; MCM, middle circular muscle; OCM, outer circular muscle; OLM, outer longitudinal muscle.

TABLE 23.1 Historical transition of major nemertean higher taxonomic scheme

Representative genus	Schultze (1853)	Hubrecht (1879)	Bürger (1892)	Coe (1905)	Iwata (1960)	Thollesson and Norenburg (2003)
Cephalothrix	Anopla	Palaeonemertini	Mesonemertini	Palaeonemertea	Archinemertea	Palaeonemertea
Tubulanus			Protonemertini		Palaeonemertea	
Hubrechtella[a]						Pilidiophora
Baseodiscus			Heteronemertini	Heteronemertea	Heteronemertea	
Lineus		Schizonemertini				
Amphiporus	Enopla	Hoplonemertini	Metanemertini	Hoplonemertea	Hoplonemertea	Hoplonemertea
Malacobdella[b]	—	—		Bdellonemertea	Bdellonemertea	

[a] Not considered until Thollesson and Norenburg (2003), although Bürger (1892) considered the similar form *Hubrechtia* as belonging to Palaeonemertini.
[b] Not considered by Schultze (1853) and Hubrecht (1879).

ANATOMY, HISTOLOGY, AND MORPHOLOGY

Nemerteans range in body length from 1–3 mm, in mesopsammic hoplonemerteans of the genus *Arenonemertes*, to >10 m in the bootlace worm *L. longissimus*. The nemertean body is covered with an epidermis containing multiciliated cells, glandular cells, and sensory cells, without cuticular layer. Underneath the epidermis lies the basement membrane – a layer of extracellular matrix. Below it, outer circular, middle diagonal, and inner longitudinal muscle layers are present in palaeo- and hoplonemerteans (**Figure 23.2A–C**). In heteronemerteans, a dermis (also called cutis) lines the epidermis. The heteronemertean dermal layer consists of outer muscular (circular and/or diagonal muscles and longitudinal muscles) and inner glandular- and connective-tissue components; beneath the dermis lie the body wall outer longitudinal, middle circular, and inner longitudinal muscle layers (**Figure 23.2D**). The heteronemertean middle circular and inner longitudinal muscle layers are likely homologous with the outer circular and inner longitudinal muscle layers in Palaeo- and Hoplonemertea, respectively.

The proboscis is a blind-ended tube-like structure connected to the body by the proboscis insertion mostly in front of the brain ring. The proboscis is usually introverted and housed in a fluid-filled cavity called the rhynchocoel; the latter has been hypothesized to be homologous to coeloms in annelids and molluscs. The proboscis consists of a glandular epithelium, nerve plexus, muscle layers, and endothelium; it is not vascularized. A few species of heteronemerteans possess branched proboscis. In Palaeonemertea and Pilidiophora, the proboscis epithelium may contain pseudocnides or rhabdites. In Hoplonemertea, the proboscis is compartmentalized into anterior, middle, and posterior chambers; the middle chamber contains the stylet apparatus, which contains a basis bearing a single, long stylet (Monostilifera) (**Figure 23.3A**) or multiple, short (Polystilifera) stylets (**Figure 23.3B**). By the increased hydraulic pressure as a result of contraction of the rhynchocoel-wall musculature, the proboscis can be extroverted inside-out via the rhynchodaeum that leads from the proboscis insertion to the exterior, with an opening called the rhynchostome or proboscis pore.

The central nervous system (CNS) comprises the brain and lateral (medullary) nerve cords. The brain consists of paired dorsal and ventral ganglia, each connected by a transverse commissure, forming a brain ring that encircles the rhynchocoel, not the esophagus as in other groups of animals. The lateral nerve is a posterior continuation of the ventral ganglion. Posteriorly, the lateral nerve cords are connected to each other by a commissure above the intestine near anus. The nemertean CNS is embedded with various depths in the body wall during ontogeny. In Palaeonemertea s.lat., it retains in the epidermis outside the body-wall musculature (e.g., *Carinina*, *Tubulanus*, *Hubrechtella*) or in

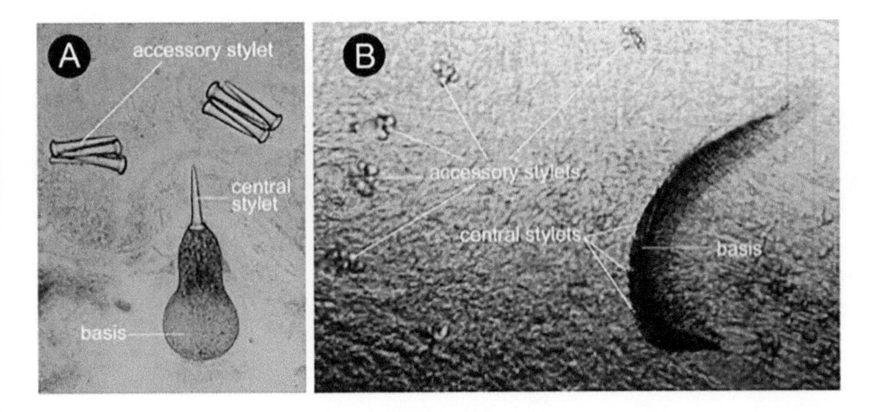

Figure 23.3 Stylet apparatus. (A) Monostilifera. (B) Polystilifera. (Photos courtesy of N. Hookabe.)

an intramuscular position (e.g., *Cephalothrix*). In Heteronemertea, it is situated between the body-wall middle circular and outer longitudinal muscle layers. In Hoplonemertea, the CNS is submuscular in position.

Nemertean sensory organs include the cerebral organ, ocelli, and statocyst, in addition to some putative sensory epidermal structures such as the frontal organ, lateral (side) organ, and sensory pits. The cerebral organ, situated on each side in connection with the brain lobes, usually takes a form of a pear-shaped, neuroglandular mass, leading to the exterior via a ciliated canal with an opening situated on the cephalic furrow. In Pilidiophora, the cerebral organ is partially bathed by blood lacuna. In Palaeonemertea (*Tubulanus*) and Hoplonemertea, the cerebral organ is closely attached by blood vessels. Certain endocrinological function is assumed in the cerebral organs. Lateral organs are present near nephridial openings in tubulanid palaeonemerteans and the heteronemertean *Zygeupolia*. Eyes in nemerteans are pigment-cup ocelli. Only a few species of Palaeonemertea possess eyes in the adult stage. Nothing is known about the ultrastructure of the adult palaeonemertean eyes; they are always embedded in the epidermis. Larvae of the palaeonemertean *Cephalothrix oestrymnica* possess ciliary photoreceptor cells (von Döhren and Bartolomaeus, 2018). Neonemertean ocelli are subepidermal, inverse pigment-cup type, associated with rhabdomeric photoreceptor cells. Statocysts are known among interstitial hoplonemerteans in the genus *Ototyphlonemertes*. A frontal organ is a ciliated pit situated at the tip of head above the proboscis pore. In general, heteronemerteans have three frontal organs, hoplonemerteans have one, and palaeonemerteans have none. Cephalic glands discharge via the frontal organ in hoplonemerteans. In life, heteronemertean frontal organs may appear sometimes protruded and can be retractable. Epidermal sensory pits are known in the palaeonemertean *Carinoma*.

The alimentary canal consists of the mouth, esophagus, foregut (stomach), midgut (intestine), and anus. In Palaeonemertea and Pilidiophora, the mouth opens separately from the rhynchodaeum. In Hoplonemertea (with some exceptions), the esophagus is united with the rhynchodaeum, leading anteriorly as the rhynchostomodaeum to open to the exterior.

Nemerteans have a closed circulatory system. The basic design consists of paired lateral vessels connected to each other at the anterior and posterior ends. Modifications to this basic design include a mid-dorsal vessel running between the rhynchocoel and alimentary tract (Neonemertea), transverse connectives metamerically arranged in the intestinal region (Heteronemertea and some Hoplonemertea), and various branches of blood vessels protruding into the rhynchocoel (Palaeonemertea). No heart is present. The blood circulates by body movements. It may contain corpuscles in some species. Nemertean blood vessels (along with the rhynchocoel) are mesodermal in origin, for which homology to schizocoels in Annelida has been proposed (Turbeville and Ruppert, 1985; Turbeville, 1986). All of this eventually led to the abolition of "Acoelomata" as a natural group (reviewed by Turbeville, 2002).

The excretory system in Nemertea is represented by protonephridia which consist of: (1) a terminal unit serving in ultrafiltration, (2) an excretory duct where secretion and resorption occur, and (3) nephridiopore cell. The terminal unit is protruded into the blood vessel in Palaeonemertea, while it is not in Neonemertea (Bartolomaeus and von Döhren, 2010).

Ovary (Stricker et al., 2001) and testis (Stricker and Folsom, 1997) are simple sacs arranged in a row on each side of intestine, often alternating with intestinal lateral diverticula. When matured, a duct is formed from each gonad leading to the exterior through the body wall. The origin of germ cells is not known with certainty.

REPRODUCTION AND DEVELOPMENT

Sexual reproduction predominates in the phylum, with only a few to some species having been confirmed to perform asexual reproduction by fragmentation/fission.

Sexual reproduction

Nemerteans are mostly dioecious. Hermaphroditism has been reported in ~40 species (mostly freshwater and terrestrial but some marine) in all the three major higher taxa. External fertilization involves free spawning and possibly mucus spawning as well (Thiel and Junoy, 2006). In the latter mode, one female and one or more males release gametes within a mucus matrix. Internal fertilization is confirmed in the mucus-spawning *Lineus viridis*; about 10 viviparous species, as well as bathypelagic species with pseudopenes, would also undergo internal fertilization. The marine monostiliferous hoplonemertean *Prosorhochmus americanus* is simultaneously hermaphroditic and viviparous, undergoing self-fertilization with high rates (Caplins and Turbeville, 2015).

Oocytes of the heteronemertean *Cerebratulus* show a similar mechanism to that of certain bivalves (*Crassostrea*, *Hiatella*, *Limaria*, and *Mytilus*), and possibly that of annelids in the genus *Chaetopterus* as well, in that (1) fertilization occurs in metaphase I, (2) fertilization induces external Ca^{2+} influx, called cortical flash, and then (3) Ca^{2+} are released repetitively from endoplasmic reticulum via inositol-1,4,5 trisphosphate receptor (Stricker, 2014). In general, cleavage is holoblastic, equal (homoquadrant), and spiral. Palaeonemerteans develop directly via planuliform larvae. In the palaeonemertean *Carinoma*, the larva is covered by a preoral belt of cleavage-arrested large squamous cells, which are derived from trochoblast lineage; the palaeonemertean planuliform larvae likely represent modified trochophores (Maslakova et al., 2004a,b). Hoplonemerteans develop more or less directly via decidula larvae. In hoplonemertean decidulae, larval epidermis is replaced by definitive one by either being shed or gradually resorbed (Maslakova and von Döhren, 2009; Hiebert et al., 2010). Pilidiophorans undergo metamorphosis. In pilidial indirect development, the juvenile forms inside the larva by fusion of a set of eight rudiments including three pairs of imaginal disks (cephalic, cerebral organ, and trunk disks) and unpaired proboscis and dorsal rudiments (**Figure 23.4**). Metamorphosis is completed when the larval epidermis is shed, which is then mostly swallowed by the juvenile (Maslakova, 2010, 2011).

Vegetative reproduction

Cyst formation has been reported in the freshwater monostiliferous hoplonemertean *Prostoma*. Asexual reproduction by fragmentation followed by anterior regeneration has been observed in some species (cf. Ikenaga et al., 2019; Kajihara and Hookabe, 2019; Zattara et al., 2019) including *L. pseudolacteus*, which is a triploid hybrid between *L. sanguineus* and the non-fissiparous *Lineus lacteus* (Ament-Velásquez et al., 2016).

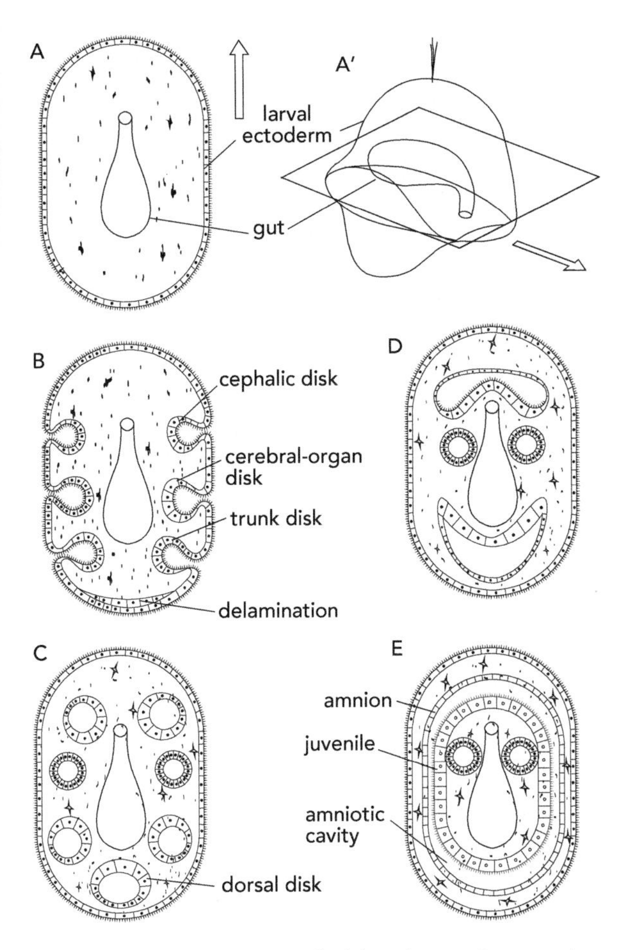

Figure 23.4 Development of pilidium larva. After gastrulation, a pair of flaps are formed on both sides of the mouth (A, A′); three pairs of imaginal disks are formed by invagination of larval ectoderm (B); unpaired dorsal disk and proboscis rudiment [not shown for clarity] are also formed (C); juvenile is formed by fusion of imaginal disks (D, E).

Developmental genes

The genome of the heteronemertean *Notospermus* contains 16 *Hox* genes and 2 *ParaHox* genes; the clusters of *Hox*, *Wnt*, and *NK* genes are not arranged in tandem but scattered either in different regions within a single chromosome or possibly even in different chromosomes. In adult *Notospermus*, *Hox1* and *Hox2* are expressed anteriorly, *Lox2* and *Lox4* mid-posteriorly, and *Post2* posteriorly. No *Hox* genes are expressed in the head and proboscis in *Notospermus* (Luo et al., 2018). In Pilidiophora, *Hox* genes pattern the anteroposterior axis of the juvenile but not the larva (Hiebert and Maslakova, 2015a); expression of *Hox*, *Cdx*, and *Six3/6* genes indicates the homology between the imaginal disks in pilidia (Pilidiophora) and paired invaginations in decidulae (Hoplonemertea; Hiebert and Maslakova, 2015b). Of the five transcription factors (*nkx2.2*, *nkx6*, *pax6*, *pax3/7*, and *msx*) involved in dorsoventral (DV) patterning in vertebrates, three (*nkx2.2*, *nkx6*, and *pax6*) are also employed in DV patterning in the heteronemertean *Lineus ruber* (Martín-Durán et al., 2018). Retinal determination gene network has been studied in the heteronemertean *L. viridis* and *L. sanguineus* (von Döhren, 2015). Developmental gene expression has not been investigated so far in Palaeonemertea.

DISTRIBUTION AND ECOLOGY

Nemerteans are distributed in the world oceans (**Figure 23.5**). The majority of the ~1,300 species are marine benthic (intertidal to hadal zones, with the deepest record being 8,339 m; *Nemertovema hadalis*, Chernyshev and Polyakova, 2018).

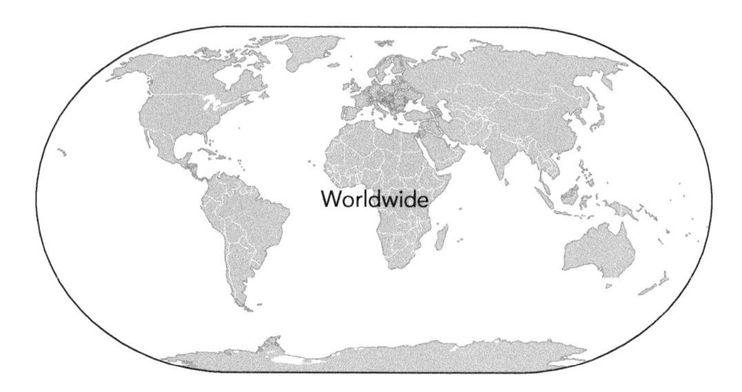

Figure 23.5 Worldwide distribution of Nemertea.

About 100 species (Pelagica and a few Monostilifera) live in water column, most abundant at 625–2,500 m. About 10 species dwell in freshwater and another 10 species in terrestrial environments, both in the tropics and subtropics.

Nemerteans occur in most marine environments although usually in low abundances. During specific time periods appropriate for hunting, nemerteans roam about in search of prey. Upon receiving a stimulus (usually chemical cues), many nemertean species actively pursue their prey and follow them into their dwellings or in their tracks. Other species (many hoplonemerteans) adopt a sit-and-wait strategy, awaiting prey items at strategic locations. Most nemertean species prey on live marine invertebrates, but some also gather on recently dead organisms to feed on them. Heteronemerteans preferentially feed on polychaetes, while most hoplonemerteans prey on small crustaceans (McDermott and Roe, 1985; Thiel and Kruse, 2001).

Nemerteans are eaten by fish, sea birds, and other invertebrates, though quite infrequently. An analysis (McDermott, 2001) of a Food Habits Data Base from the U.S. National Marine Fisheries Service, compiled for fishes collected in the Atlantic Ocean from the Canadian border to Cape Hatteras, North Carolina (1973–1990), showed that nemerteans were recovered from the stomachs of 27 species of fishes in 14 families. They were found in 223 of 26,642 (0.84%) fish stomachs examined in the laboratory, but only in 0.09% of 58,812 fish examined in the field. Among species in the former category, for which ≥1000 were examined, the winter flounder *Pseudopleuronectes americanus* and the yellowtail flounder *Limanda ferruginea* had the greatest frequency of nemerteans, 71 of 1545 (4.6%) and 33 of 1045 (3.2%), respectively. The black-bellied plover *Pluvialis squatarola*, the semipalmated plover *Charadrius semipalmatus*, and the herring gull *Larus argentatus* are known to feed on nemerteans. Several species of nemerteans are known to ingest other nemerteans. Other invertebrates known to feed on nemerteans include the horseshoe crab *Limulus polyphemus*; the shrimp *Sergia* sp.; the crabs *Cancer pagurus*, *Carcinus maenas*, and *Callinectes ornatus*; the squid *Illex illecebrosus*; the gastropod *Aglaja diomedea*; the polychaete *Aphrodite acuteata*; and the scyphozoan *Cyanea capillata*.

Life history is not known for many species; the monostiliferous hoplonemertean *Amphiporus lacteus* is likely iteroparous (Thiel and Dernedde, 1996); the monostiliferous hoplonemertean *Paranemertes peregrina* is also likely iteroparous, but its life span is 1.5–1.75 years (Roe, 1976); the heteronemertean *L. viridis* is perennial and iteroparous (von Döhren et al., 2012).

Decapod-egg predatory monostiliferans (*Alaxinus*, *Carcinonemertes*, *Ovicides*, and *Pseudocarcinonemertes*) may have had indirect negative impact on fisheries of economically important species such as the red king crab *Paralithodes camtschaticus*, the Dungeness crab *Metacarcinus magister*, and the blue crab *Callinectes sapidus* (Kuris, 1993). The estuarine heteronemertean *Yininemertes pratensis* is giving economic damage to local fishermen in Han River Estuary,

South Korea, by neurotoxins to the juveniles of Japanese eel *Anguilla japonica* migrating to fresh water (Kwon et al., 2017).

PHYSIOLOGY AND BEHAVIOR

Active prey are caught by the proboscis and killed or immobilized within few seconds by toxic substances secreted from the organ before being ingested. Inert or decaying organic material is swallowed directly without prior proboscis eversion. Food may be swallowed whole or extracellularly digested by the everted foregut (stomach) before ingestion. Fat and glycogen are stored as small globules in the intestinal columnar cells (Gibson, 1972).

Nemerteans use toxic substances for both defense and offence. Toxins can be contained in the body surface and the proboscis. Nemertean toxins range from low-molecular-weight substances, such as anabasein and tetrodotoxin, to polypeptides (Jacobsson et al., 2018 and references therein). Nemertean peptide toxins, including putative ones based on sequence similarities, are mostly neurotoxins (ion channel blockers) and cytotoxins (those with pore-forming activity and DNase activity; Whelan et al., 2014).

In small nemerteans, gaseous exchange is likely done through the body surface. Larger species (e.g., the heteronemertean *Cerebratulus*, *Baseodiscus*, and *Gastropion*) perform foregut respiration by rhythmic uptake and expulsion of seawater. A vascular network is developed around the foregut in such species; nephridial ducts open not only on the epidermis but also into the foregut lumen in *Baseodiscus* and *Gastropion* (Moretto, 1998).

Females of Antarctic monostiliferous hoplonemerteans in the genus *Antarctonemertes* oviposit in a cocoon (~140 eggs), showing a defensive behavior (proboscis eversion) when disturbed (Taboada et al., 2013).

GENETICS

The haploid nuclear genome size of the heteronemertean *Notospermus* is 859 Mb, comparable to those of other spiralians, such as the polychaete *Capitella teleta* (324 Mb), Pacific oyster *Crassostrea gigas* (558 Mb), the phoronid *Phoronis australis* (498 Mb), and the brachiopod *Lingula anatina* (406 Mb) (Luo et al., 2018). The genome size (estimated by flow cytometry) appears to be correlated to the body size, ranging from 0.28 pg in the monostiliferous hoplonemertean *P. peregrina* (8–11 cm in body length) to 1.17 pg in the heteronemertean *Cerebratulus marginatus* (20–30 cm in body length; Mulligan et al., 2014).

The mitochondrial genome (mitogenome) size tends to be larger in palaeonemerteans (16,296 bp in *Cephalothrix hongkongiensis*; *Cephalothrix simula* per Chen et al., 2009) than in pilidiophorans (15,333 bp in *Micrura ignea*; Gonzalez-Cueto et al., 2015); 15,388 bp in *L. viridis* (Podsiadlowski et al., 2009); 15,513 bp in *Zygeupolia rubens* (Chen et al., 2012) and in hoplonemerteans (14,558 bp in *Paranemertes* cf. *peregrina* (Chen et al., 2011); 15,365 bp in *Nectonemertes* cf. *mirabilis* (Gonzalez-Cueto et al., 2015); 14,580 bp in *Tetrastemma olgarum* (Sun et al., 2016), although the largest mitogenome so far known is 16,847 bp in the pilidiophoran *Micrura bella* (Shen and Sun, 2016). All the mitogenomes investigated so far contain 37 genes (13 protein-coding genes, 2 ribosomal RNAs, and 22 transfer RNAs).

POSITION IN THE ToL

Nemertea appear to be closely related to Brachiopoda and Phoronida (and possibly also Bryozoa, Cycliophora, and Entoprocta) within Lophotrochozoa, which also contain Annelida and Mollusca within Spiralia (e.g., Luo et al., 2018).

DATABASES AND COLLECTIONS

World Nemertea Database (http://www.marinespecies.org/nemertea/) maintains taxonomic information.

CONCLUSION

Nemertean systematic studies have long been hampered by the soft-bodied nature of the animals. Although information in the living state – including the body shape and color – is useful in species identification, it is lost in preserved material. Classification has been based on internal anatomy examined by time-consuming serial histological sectioning. These impediments are now overcome by recent technological advancement in high-quality digital imaging and molecular sequencing. There is at least two to three times more species diversity than currently known in the nature, awaiting formal description.

REFERENCES

Ament-Velásquez, S. L., Figuet, E., Ballenghien, M., Zattara, E. E., Norenburg, J. L., Fernández-Álvarez, F. A., Bierne, J., Bierne, N., and Galtier, N. 2016. Population genomics of sexual and asexual lineages in fissiparous ribbon worms (*Lineus*, Nemertea): hybridization, polyploidy and the Meselson effect. Molecular Ecology 25: 3356–3369.

Bartolomaeus, T. and von Döhren, J. 2010. Comparative morphology and evolution of the nephridia in Nemertea. Journal of Natural History 44: 2255–2286.

Bürger, O. 1892. Zur Systematik der Nemertinenfauna des Golfs von Neapel. Nachrichten von der Königlichen Gesellschaft der Wissenschaften und der Georg-Augusts-Universität zu Göttingen 5: 137–178.

Caplins, S. A. and Turbeville, J. M. 2015. High rates of self-fertilization in a marine ribbon worm (Nemertea). Biological Bulletin 229: 255–264.

Cedhagen, T. and Sundberg, P. 1986. A previously unrecognized report of a nemertean in the literature. Archives of Natural History 13: 7–8.

Chen, H.-X., Sun, S.-C., Sundberg, P., Ren, W.-C., and Norenburg, J. L. 2012. A comparative study of nemertean complete mitochondrial genomes, including two new ones for *Nectonemertes* cf. *mirabilis* and *Zygeupolia rubens*, may elucidate the fundamental pattern for the phylum Nemertea. BMC Genomics 13: 139.

Chen, H.-X., Sundberg, P., Norenburg, J. L., and Sun, S.-C. 2009. The complete mitochondrial genome of *Cephalothrix simula* (Iwata) (Nemertea: Palaeonemertea). Gene 442: 8–17.

Chen, H.-X., Sundberg, P., Wu, H.-Y., and Sun, S.-C. 2011. The mitochondrial genomes of two nemerteans, *Cephalothrix* sp. (Nemertea: Palaeonemertea) and *Paranemertes* cf. *peregrina* 9 (Nemertea: Hoplonemertea). Molecular Biology Reports 38: 4509–4525.

Chernyshev, A. V. 2011. Comparative Morphology, Systematics and Phylogeny of the Nemerteans. Vladivostok, Dalnauka, 309 pp. [In Russian]

Chernyshev, A. V. and Polyakova, N. E. 2018. Nemerteans of the Vema-TRANSIT expedition: first data on diversity with description of two new genera and species. Deep Sea Research Part II: Topical Studies in Oceanography 148: 64–73.

Coe, W. R. 1905. Nemerteans of the west and northwest coasts of North America. Bulletin of the Museum of Comparative Zoölogy at Harvard College 44: 1–318.

Coe, W. R. 1943. Biology of the nemerteans of the Atlantic coast of North America. Transactions of the Connecticut Academy of Arts and Sciences 35: 129–328.

Diesing, C. M. 1850. Systema Helminthum, Vol. I. Vienna, W. Braumüller, 679 pp.

Gibson, R. 1972. Nemerteans. London, Hutchinson University Library, 224 pp.

Gonzalez-Cueto, J., Escarraga-Fajardo, M. E., Lagos, A. M., Quiroga, S., and Castro, L. R. 2015. The complete mitochondrial genome of *Micrura ignea* Schwartz & Norenburg 2005 (Nemertea: Heteronemertea) and comparative analysis with other nemertean mitogenomes. Marine Genomics 20: 33–37.

Hiebert, L. S., Gavelis, G., von Dassow, G., and Maslakova, S. A. 2010. Five invaginations and shedding of the larval epidermis during development of the hoplonemertean *Pantinonemertes californiensis* (Nemertea: Hoplonemertea). Journal of Natural History 44: 2331–2347.

Hiebert, L. S. and Maslakova, S. A. 2015a. *Hox* genes pattern the anterior-posterior axis of the juvenile but not the larva in a maximally indirect developing invertebrate, *Micrura alaskensis* (Nemertea). BMC Biology 13: 23.

Hiebert, L. S. and Maslakova, S. A. 2015b. Expression of *Hox*, *Cdx*, and *Six3/6* genes in the hoplonemertean *Pantinonemertes californiensis* offers insight into the evolution of maximally indirect development in the phylum Nemertea. EvoDevo 6: 26.

Hubrecht, A. A. W. 1879. The genera of European nemerteans critically revised, with description of several new species. Notes from the Leyden Museum 1: 193–232.

Hyman, L. H. 1951. The Invertebrates. Vol. II: Platyhelminthes and Rhynchocoela. The Acoelomate Bilateria. New York, McGraw-Hill, 550 pp.

Ikenaga, J., Hookabe, N., Kohtsuka, H., Yoshida, M., and Kajihara, H. 2019. A population without females: males of *Baseodiscus delineatus* (Nemertea: Heteronemertea) reproduce asexually by fragmentation. Zoological Science 36: 348–353.

Iwata, F. 1960. Studies on the comparative embryology of nemerteans with special reference to their interrelationships. Publications from the Akkeshi Marine Biological Station 10: 1–51.

Jacobsson, E., Andersson, H. S., Strand, M., Peigneur, S., Eriksson, C., Lodén, H., Shariatgorji, M., Andrén, P. E., Lebbe, E. K. M., Rosengren, K. J., Tytgat, J., and Göransson, U. 2018. Peptide ion channel toxins from the bootlace worm, the longest animal on Earth. Scientific Reports 8: 4596.

Kajihara, H. and Hookabe, N. 2019. Anterior regeneration in *Baseodiscus hemprichii* (Nemertea: Heteronemertea). Tropical Natural History 19: 39–42.

Kuris, A. M. 1993. Life cycles of nemerteans that are symbiotic egg predators of decapod Crustacea: 10 adaptations to host life histories. Hydrobiologia 266: 1–14.

Kwon, Y.-S., Min, S.-K., Yeon, S.-J., Hwang, J.-H., Hong, J.-S., and Shin, H.-S. 2017. Assessment of neuronal cell-based cytotoxicity of neurotoxins from an estuarine nemertean in the Han River Estuary. Journal of Microbiology and Biotechnology 27: 725–730.

Luo, Y.-J., Kanda, M., Koyanagi, R., Hisata, K., Akiyama, T., Sakamoto, H., Sakamoto, T., and Satoh, N. 2018. Nemertean and phoronid genomes reveal lophotrochozoan evolution and the origin of bilaterian heads. Nature Ecology and Evolution 2: 141–151.

Magnus, O. 1555. Historia de gentibus septentrionalibus. Antverpiae, I. Bellerum, 512 pp.

Martín-Durán, J. M., Pang, K., Børve, A., Lê, H. S., Furu, A., Cannon, J. T., Jondelius, U., and Hejnol, A. 2018. Convergent evolution of bilaterian nerve cords. Nature 553: 45–50.

Maslakova, S. A. 2010. Development to metamorphosis of the nemertean pilidium larva. Frontiers in Zoology 7: 30.

Maslakova, S. A. 2011. The invention of the pilidium larva in an otherwise perfectly good spiralian phylum Nemertea. Integrative and Comparative Biology 50: 734–743.

Maslakova, S. A., Martindale, M. Q., and Norenburg, J. L. 2004a. Fundamental properties of the spiralian developmental program are displayed by the basal nemertean *Carinoma tremaphoros* (Palaeonemertea, Nemertea). Developmental Biology 267: 342–360.

Maslakova, S. A., Martindale, M. Q., and Norenburg, J. L. 2004b. Vestigial prototroch in a basal nemertean, *Carinoma tremaphoros* (Nemertea; Palaeonemertea). Evolution and Development 6: 219–226.

Maslakova, S. A. and von Döhren, J. 2009. Larval development with transitory epidermis in *Paranemertes peregrina* and other hoplonemerteans. Biological Bulletin 216: 273–292.

McDermott, J. J. 2001. Status of the Nemertea as prey in marine ecosystems. Hydrobiologia 456: 7– 20.

McDermott, J. J. and Roe, P. 1985. Food, feeding behavior and feeding ecology of nemerteans. American Zoologist 25: 113–125.

Moretto, H. J. 1998. A new heteronemertean from the Argentine coast of the Southern Atlantic. Hydrobiologia 365: 215–222.

Mulligan, K. L., Hiebert, T. C., Jeffery, N. W., and Gregory, T. R. 2014. First estimates of genome size in ribbon worms (phylum Nemertea) using flow cytometry and Feulgen image analysis densitometry. Canadian Journal of Zoology 92: 847–851.

Podsiadlowski, L., Braband, A., Struck, T. H., von Döhren, J., and Bartolomaeus, T. 2009. Phylogeny and mitochondrial gene order variation in Lophotrochozoa in the light of new mitogenomic data from Nemertea. BMC Genomics 10: 1–14.

Roe, P. 1976. Life history and predator-prey interactions of the nemertean *Paranemertes peregrina* (Coe). Biological Bulletin 150: 80–106.

Sedgwick, A. 1898. A Student's Text-Book of Zoology, Vol. I. London, Swan Sonnenschein and Co., 619 pp.

Schultze, M. S. 1851. Beiträge zur Naturgeschichte der Turbellarien. Greifswald, C. A. Koch, 78 pp.

Schultze, M. 1853. Zoologische Skizzen. Zeitschrift für wissenschaftliche Zoologie 4: 178–195.

Shen, C.-Y. and Sun S.-C. 2016. Mitochondrial genome of *Micrura bella* (Nemertea: Heteronemertea), the largest mitochondrial genome known to phylum Nemertea. Mitochondrial DNA Part A 27: 2899–2900.

Stricker, S. A. 2014. Calcium signalling and endoplasmic reticulum dynamics during fertilization in marine protostome worms belonging to the phylum Nemertea. Biochemical and Biophysical Research Communications 450: 1182–1187.

Stricker, S. A. and Folsom, M. W. 1997. A comparative ultrastructural analysis of spermatogenesis in nemertean worms. Hydrobiologia 365: 55–72.

Stricker, S. A., Smythe, T. L., Miller, L., and Norenburg, J. L. 2001. Comparative biology of oogenesis in nemertean worms. Acta Zoologica 82: 213–230.

Sun, W.-Y., Shen, C.-Y., and Sun, S.-C. 2016. The complete mitochondrial genome of *Tetrastemma olgarum* (Nemertea: Hoplonemertea). Mitochondrial DNA Part A 27: 1086–1087.

Taboada, S., Junoy, J., Andrade, S. C., Giribet, G., Cristobo, J., and Avila, C. 2013. On the identity of two Antarctic brooding nemerteans: redescription of *Antarctonemertes valida* (Bürger, 1893) and description of a new species in the genus *Antarctonemertes* Friedrich, 1955 (Nemertea, Hoplonemertea). Polar Biology 36: 1415–1430.

Thiel, M. and Dernedde, T. 1996. Reproduction of *Amphiporus lactifloreus* (Hoplonemertini) on tidal flats: implications for studies on the population biology of nemerines. Helgoländer Meeresuntersuchungen 50: 337–351.

Thiel, M. and Junoy, J. 2006. Mating behavior of nemerteans: present knowledge and future directions. Journal of Natural History 40: 1021–1034.

Thiel, M. and Kruse, I. 2001. Status of the Nemertea as predators in marine ecosystems. Hydrobiologia 456: 21–32.

Thollesson, M. and Norenburg, J. L. 2003. Ribbon worm relationships: a phylogeny of the phylum Nemertea. Proceedings of the Royal Society B 270: 407–415.

Turbeville, J. M. 1986. An ultrastructural analysis of coelomogenesis in the hoplonemertine *Prosorhochmus americanus* and the polychaete *Magelona* sp. Journal of Morphology 187: 51–60.

Turbeville, J. M. 2002. Progress in nemertean biology: development and phylogeny. Integrated and Comparative Biology 42: 692–703.

Turbeville, J. M. and Ruppert, E. E. 1985. Comparative ultrastructure and the evolution of nemertines. American Zoologist 25: 53–71.

von Döhren, J. 2015. Nemertea. In: Wanninger, A. (Ed.) Evolutionary Developmental Biology of Invertebrates 2: Lophotrochozoa (Spiralia). Wien, Springer-Verlag, pp. 155–192.

von Döhren, J. and Bartolomaeus, T. 2018. Unexpected ultrastructure of an eye in Spiralia: the larval ocelli of *Procephalothrix oestrymnicus* (Nemertea). Zoomorphology 137: 214–248.

von Döhren, J., Beckers, P., and Bartolomaeus, T. 2012. Life history of *Lineus viridis* (Müller, 1774) (Heteronemertea, Nemertea). Helgoland Marine Research 66: 266.

Whelan, N. V., Kocot, K. M., Santos, S. R., and Halanych, K. M. 2014. Nemertean toxin genes revealed through transcriptome sequencing. Genome Biology and Evolution 6: 3314–3325.

Zattara, E. E., Fernández-Álvarez, F. A., Hiebert, T. C., Bely, A. E., and Norenburg, J. L. 2019. A phylum-wide survey reveals multiple independent gains of head regeneration in Nemertea. Proceedings of the Royal Society B 286: 20182524.

PHYLUM ACANTHOCEPHALA

CHAPTER **24**

Bahram Sayyaf Dezfuli and Luisa Giari

Infrakingdom	Bilateria
Division	Protostome
Subdivision	Lophotrochozoa
Phylum	Acanthocephala

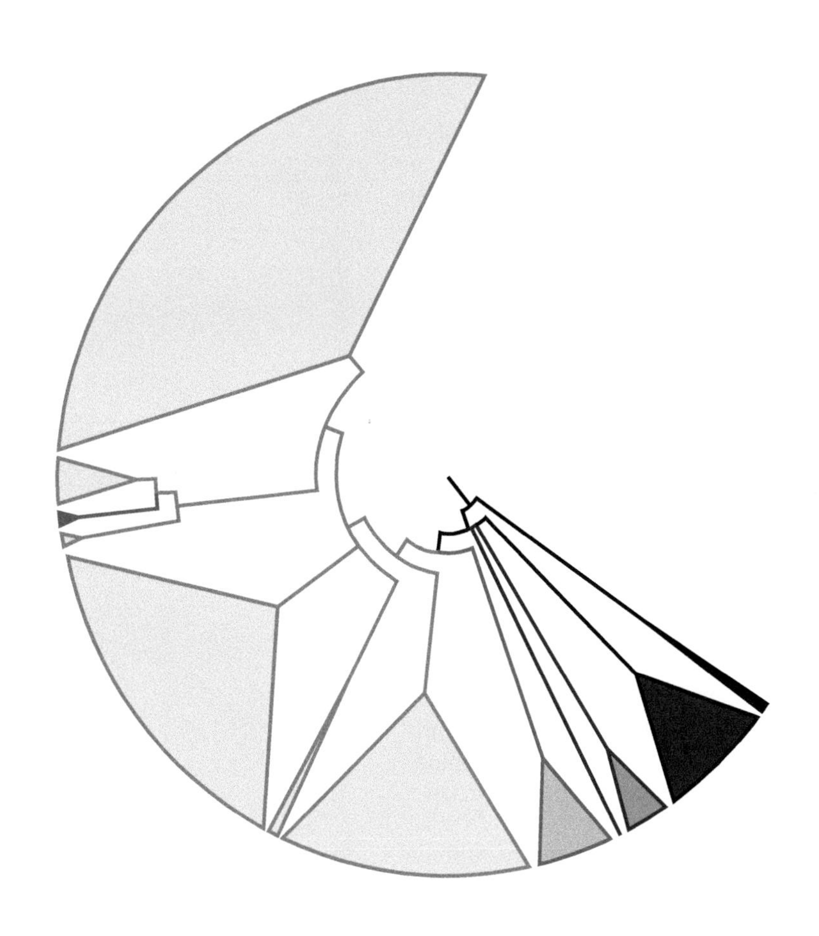

CONTENTS

General Characteristics

Over 1,300 valid species placed in 4 classes and 11 genera

Symmetric animals

They possess pseudocoelom

Endoparasites mainly of aquatic organisms

They have complex life cycles, involving at least two hosts – an invertebrate and a vertebrate

Are of no commercial value

Ecological importance is probably small

Synapomorphies

- eversible proboscis with rows of recurved hooks
- lack of digestive tract
- absence or presence of an excretory system is uncertain
- males with cement glands
- body is divided into two regions: praesoma and metasoma

INTRODUCTION: HISTORY AND SYSTEMATICS

Acanthocephala (from Greek *ákantha* "thorn", *kephalos* derivative of *kephalē* "head") is a phylum of endoparasitic metazoans commonly known as spiny-headed worms. The main characteristic of this phylum is that they possess an eversible proboscis which bears rows of recurved hooks used for their attachment to the intestinal wall in their definitive hosts. Acanthocephalans are extremely successful group of endoparasites which can be ubiquitously found in all freshwater and marine systems. During their life cycle, they can infect a vast range of definitive hosts (vertebrates) and commonly arthropods as intermediate hosts. Although Acanthocephala induce important diseases to farm and domestic animals, they are usually considered to be a small and rather insignificant taxon of parasites and represent a minor phylum of interest to specialists since these parasites seldom infect human. Acanthocephala are frequently neglected even in most parasitology textbooks and few people seem interested to work with them. Some pioneering authors who worked on this phylum are: Rudolphi, Lühe, Kaiser, Hamann, Meyer, Travassos, and Van Cleave, but the monumental work of Meyer (1932, 1933) still stands as one of the principal reference works for the Acanthocephala. In the last century, some books were devoted to them exclusively. Two monographs with systematic approaches are Petrochenko (1956, 1958) and Yamaguti (1963). Whereas, the monograph "An Ecological Approach to Acanthocephalan Physiology" (Crompton, 1970) and subsequently "Biology of the Acanthocephala" (Crompton and Nickol, 1985) provided fundamental insights and basics with respect to morphology, physiology, reproduction, taxonomy, and several other facets of biology of this taxon.

Amin (1985) provided a detailed historical account of the Acanthocephala since the first known reference to worms having a proboscis published by Italian scientist Francesco Redi in 1684 (Redi, 1684). About 1,300 acanthocephalan species are known to date (García-Varela and Perez-Ponce de Leon, 2015) but new species are continuously being discovered (Gomes et al., 2015). The phylum Acanthocephala comprises four classes (Archiacanthocephala, Palaeacanthocephala, Eoacanthocephala, and Polyacanthocephala). Phylogenetic interrelationships of these classes still remain unresolved (Gazi et al., 2016). Current classification schemes of the taxon are in transitional change, as more species are being subjected to molecular phylogentic analyses (Goater et al., 2014). A monophyletic origin of Acanthocephala and wheel-animals (Rotifera) is the widely accepted norm, nevertheless, the phylogeny inside the clade, be it called Syndermata or Rotifera, lacks validation by mitochondrial data (Sielaff et al., 2016).

MORPHOLOGY AND ANATOMY

Acanthocephala have a cylindrical or slightly flattened body (**Figure 24.1**). Most species are around 10 mm in length, but some, like *Macracanthorhynchus hirudinaceus* which infect the intestines of pigs, can measure up to 40–70 cm.

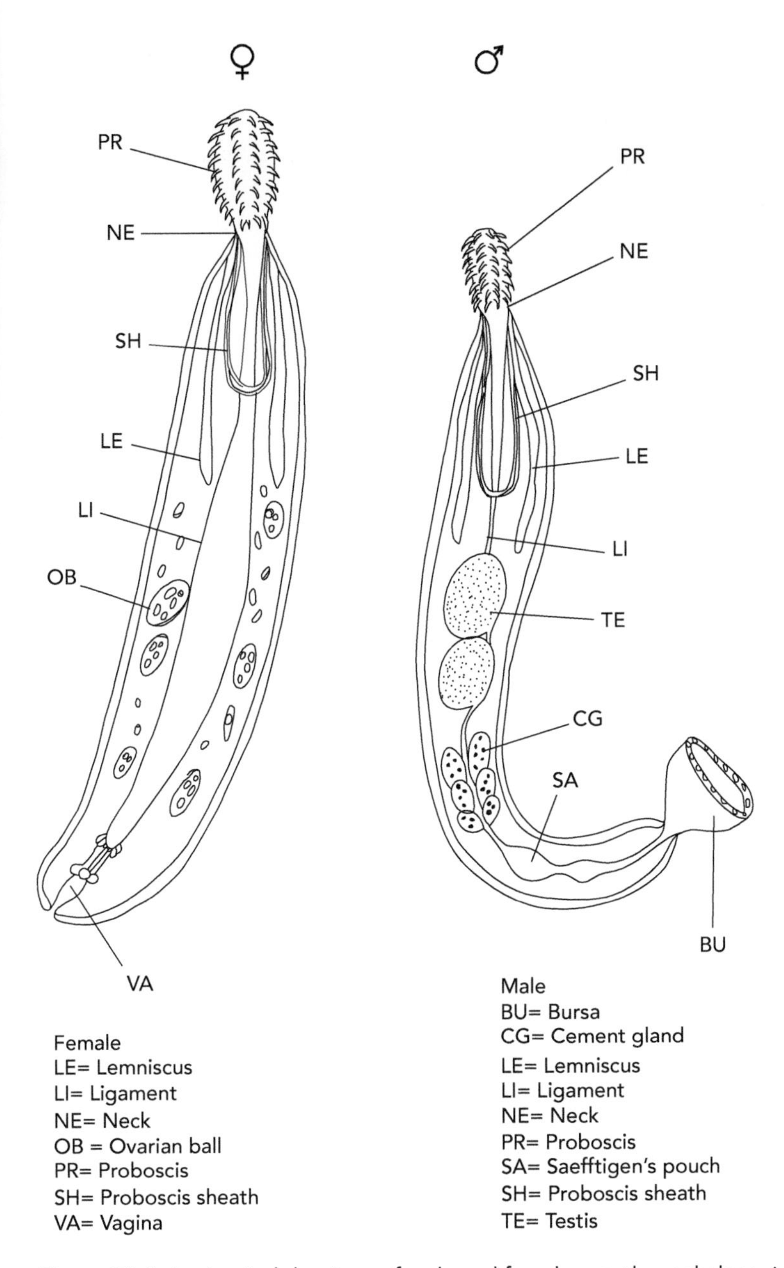

Figure 24.1 Anatomical drawings of male and female acanthocephalans. Anatomical parts are labeled as in the figure.

Acanthocephala are often white to cream in color, but some species are yellow and even orange (**Figure 24.2**). The body is divided into two regions: praesoma and metasoma or trunk (**Figure 24.3a**). The praesoma comprises a proboscis and unarmed neck (**Figure 24.3b**); the parasite attaches to the intestinal wall of the host with the proboscis, which is highly variable in shape and size, is retractile (Herlyn, 2017), and covered with curved hooks which are arranged in longitudinal rows (Figure 24.3b) or in a spiral pattern. Hooks are an important taxonomic feature for Acanthocephala; nevertheless, very limited records have been published on their chemical composition (Brázová et al., 2014; Taraschewski, 2000). Conversely, the ultrastructure of hooks and their X-ray microanalysis (EDXA) have been extensively described (Amin and Heckmann, 2014; Brázová et al., 2014; Dezfuli et al., 2008a). There are still several open questions on anatomy of the Acanthocephala and one of the reasons for this lack of information is the incorrect description of structure of an organ and in some cases even misinterpretation about functionality of these organs. For example, only recently

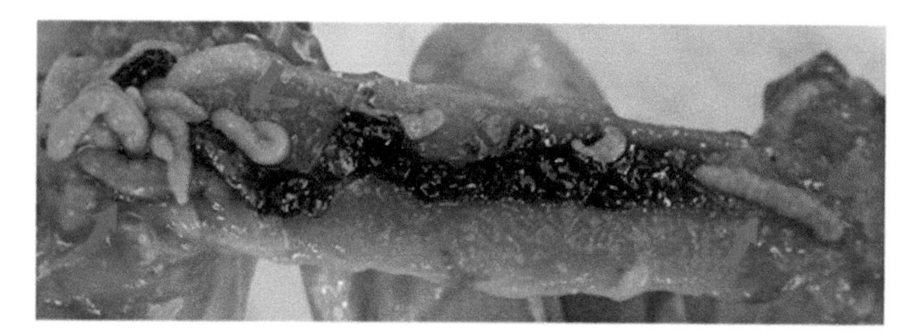

Figure 24.2 Live acanthocephalans (arrows) in the intestine of the fish *Squalius cephalus*.

Herlyn and Taraschewski (2017) described in detail that the literature spanning over 200 years on muscles involved in movement of the praesoma and the proboscis contains inaccuracies in interpretations by several authors.

Below the neck, tube-shaped metasoma encloses the pseudocoelom (body cavity), receptacle, lemnisci, ligament sacs, reproductive organs, excretory organs when present, genital ganglion in the male, and various muscles. Lemnisci are paired organs that extend into the body cavity from the neck region. Each lemniscus serves as a fluid reservoir when the proboscis is invaginated. In some species, the external surface of the trunk entirely or partially can be covered with spines (**Figure 24.3c**). In comparison to proboscis hooks, spines are smaller and without roots. There is lack of a hollow core in the trunk of these spines and it is believed that they are secondary holdfast structures.

The body wall of the Acanthocephala is formed by three distinct sections – an outer tegument, a middle group of circular muscles, and an inner group of longitudinal muscles (Miller and Dunagan, 1985). The term "tegument" is preferred for Acanthocephala over the term "cuticle," "epidermis," or "dermis" because it is advantageous in that it does not bear the connotation of nonliving structures (Miller and Dunagan, 1985) and does not imply a cellular structure. There are no indications of cell division and it is assumed that the entire tegument forms a syncytium. Pervading the entire tegument is a set of channels named the lacunar system which is organized differently in four classes of the taxon. It was reported that the tegument of the body wall of acanthocephalans has trophic, metabolic, protective, and supportive functions (Barabashova, 1971). Acanthocephala lack a digestive tract and all uptake of nutrients takes place across the tegument. The outer surface coat of the tegument is made up of carbohydrate rich glycocalyx and closely packed pores that lead to pore canals which branch and anastomose. There is lack of a directional-flow circulatory system in the Acanthocephala, but they possess a loose network of tube-like cavities with no lining or force for fluid flow. It is presumed that the trunk wall musculature acts as the motive force for fluid flow (Miller and Dunagan, 1985). These canal-like cavities connect the lacunar system and the pseudocoel. The function of this system is uncertain, it may play a role in the functioning of the body wall musculature or serve as some kind of "circulatory system."

The nervous system and sensory receptor system of Acanthocephala are poorly known. The cerebral ganglion or brain is situated inside the proboscis receptacle or sheath. The number of cells which constitute this ganglion and position of the brain are variable in different orders. From the brain, lateral ganglia extend posteriorly and anteriorly to other organs and tissues. In addition, in males, a pair of genital ganglia with nerves arises from the brain. Indeed, sense organs are found in the proboscis and in males, in the penis and copulatory bursa. Most of our current understanding of the nervous system of this phylum results from the work of Miller and Dunagan on select few species of Archiacanthocephala parasites of mammals. These studies have paved the way for studying Acanthocephala in other hosts (see Miller and Dunagan, 1985).

Unfortunately, most descriptive studies of Acanthocephala do not provide any definitive insight on the absence or presence of an excretory system, thus the vast majority of the acanthocephalans seem to lack this system. Nonetheless, excretory organs, when present often, are of the nature of protonephridia (see Miller and Dunagan, 1985). The same authors indicate that a protonephridial system is restricted to the family Oligacanthorhynchidae (Dunagan and Miller, 1986). It is assumed that metabolic waste products are removed by diffusion via numerous pores within the body wall.

LIFE CYCLE, DEVELOPMENT, AND REPRODUCTION

In the Acanthocephala life cycle, there are no free-living stages and at least two hosts are necessary – an intermediate and a definitive host. Certain species of insects have been identified to be intermediate hosts of acanthocephalans which infect terrestrial vertebrates. However, intermediate hosts for acanthocephalan parasites of fish or aquatic birds are generally benthic amphipods, isopods, or ostracods. An egg (**Figure 24.3d**) eaten by an arthropod, hatches into an acanthor (first larval stage) which leaves the host's gut and develops into an acanthella (second larval stage) within the hemocoel. After few weeks, the larva transforms into the cystacanth stage (third larval stage, **Figure 24.3e**); the parasite at this stage has all the structures of the adult acanthocephalan and is

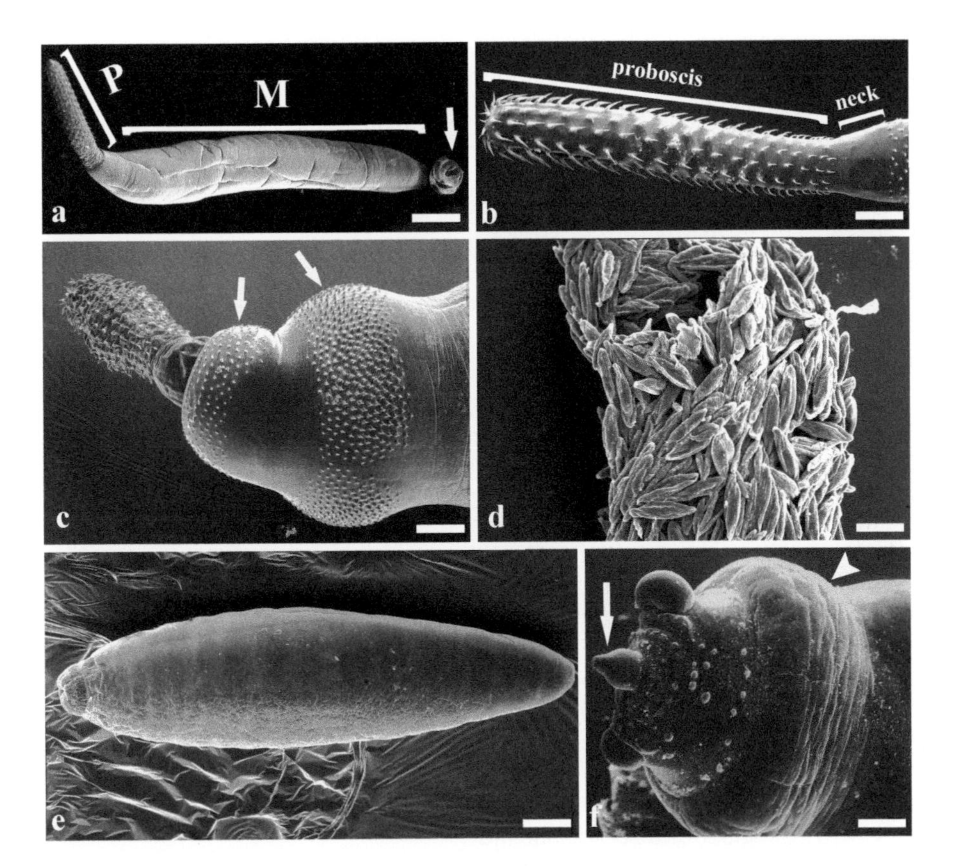

Figure 24.3 Scanning electron micrographs of acanthocephalans. (a) A male acanthocephalan with the copulatory bursa extruded (arrow), P, praesoma; M, metasoma; bar = 0.39 mm. (b) Higher magnification of praesoma of an acanthocephalan, arrangement of proboscis hooks and unarmed neck are evident, bar = 0.10 mm. (c) Micrograph shows distribution of trunk spines (arrows), bar = 0.18 mm. (d) Numerous eggs isolated from the body of a mature female *Pomphorhynchus laevis*, bar = 0.03 mm. (e) Cystacanth isolated from the haemocoel of crustacean intermediate host, bar = 0.01 mm. (f) Fully extruded copulatory bursa (arrow head) of a mature male and penis (arrow) are visible, bar = 0.09 mm.

able to infect its definitive host (vertebrate), in which it becomes an adult within few weeks (Figure 24.3a).

Reproduction in Acanthocephala is exclusively sexual and the most prominent interior organs of the phylum are the reproductive ones. The sexes are separate, wherein females commonly attain a larger size than male counterparts (Crompton, 1985). The female reproductive system consists of (i) gonads from which ovarian balls develop to produce oocytes and eggs, (ii) ligament sacs, (iii) efferent duct system which comprises a uterine bell, and (iv) uterus and vagina (Parshad and Crompton, 1981). The male acanthocephalan reproductive apparatus includes two testes, vasa efferentia, vas deferens, seminal vesicles, cement glands (variable in size, shape, and number), cement reservoir, Saeffigen's pouch, copulatory bursa, and penis (**Figure 24.3f**) (Miller and Dunagan, 1985; Parshad and Crompton, 1981). Variations in the above components can occur according to the species. Most of our knowledge on reproduction and spermatogenesis in Acanthocephala are the result of work done by Crompton and colleagues, Marchand and Mattei (see Crompton, 1985). Nevertheless, the last review on reproduction of the taxon was published by Carcupino and Dezfuli (1999).

Cement gland secretion of the male acanthocephalan is believed to contribute to the formation of the copulatory cap (mating plug) which is commonly observed on the posterior end of the female specimens (Parshad and Crompton, 1981). Until two decades ago, very little was known about the nature of cement gland secretions and different chemical origins were attributed to it. Based on electrophoretic separation of material from the cement gland of male *Pomphorhynchus laevis*, an acanthocephalan parasite of fish, Dezfuli and colleagues (1998) reported that a protein of 23 kDa probably was the main component of cement secretion. These authors produced a polyclonal antibody raised against purified *P. laevis* p23 (anti-p23PL) and applied it to sections of male and female *P. laevis.* In sum, p23 was present in cement glands and gland ducts. Indeed, this cement gland was identified at the posterior ends of females retaining the cap (Dezfuli et al., 1999). The ultrastructure of cement glands in other species of acanthocephalans has also been described (Dezfuli, 2000; Dezfuli et al., 2001). It is interesting that the first absolute photographic documentation of acanthocephalans during copulation appeared in Dezfuli and De Biaggi (2000).

DISTRIBUTION AND ECOLOGY

Infra-populations of marine acanthocephalans (**Figure 24.4**) are less known in comparison to freshwater counterparts which are often abundant in host populations (Taraschewski, 2005). Acanthocephalans, as other groups of enteric helminths, show preference for specific regions of vertebrate definitive hosts' digestive tract (Crompton, 1973). The preferred site may relate to the physicochemical conditions of that region (Crompton, 1973), to typical nutritional requirements, or to interspecific interactions (Holmes, 1973). The distribution of the same species of Acanthocephala may differ in different host species or even strains (Kennedy, 1985).

The book entitled "Ecology of the Acanthocephala" (Kennedy, 2006) appears to be an excellent monograph for ecologists, parasitologists, and zoologists and is the only detailed literature devoted exclusively to the ecology of this phylum. Moreover, the second edition of monograph "Evolutionary Ecology of Parasites" by Poulin (2007) presented a synthesis of the most exciting aspects of parasites' evolutionary ecology and is vastly useful for advanced students and practicing researchers in parasitology.

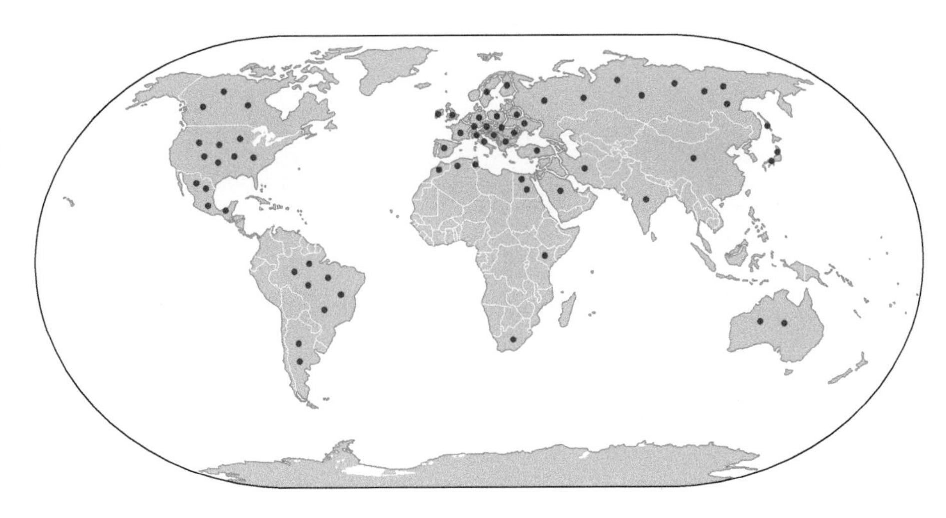

Figure 24.4 **The distribution of acanthocephalans is worldwide where hosts live.** In this distribution map we show their presence based on available publications.

PHYSIOLOGY AND BEHAVIOR

Acanthocephalans do not normally alter the behavior of their vertebrate definitive and paratenic hosts. There is only one single record of an acanthocephalan altering the swimming and diving behavior of a skink species (Daniels, 1985). Several studies have been done in field and laboratory settings on the crustacean intermediate host of acanthocephalan parasites of fish (see Maynard et al., 1998; Wellnitz et al., 2003). It was reported that one common response to infection is hyperactivity of intermediate host (Dezfuli et al., 2003; Maynard et al., 1998). There are examples of parasites that change their intermediate host's behavior and downstream drift in ways consistent with increased predation (Bethel and Holmes, 1973; Maynard et al., 1998; Moore, 2002; Wellnitz et al., 2003). Overactive intermediate hosts are likely to be more visible and exposed to predators; this could be a strategy adopted by the parasite to increase its transmission (Moore, 2002). Moreover, the failure of infected amphipods to reduce activity levels even in the presence of fish odor underlines this unique parasite strategy (Dezfuli et al., 2003). Effects of acanthocephalans on reproductive potential of crustacean intermediate hosts has been described in detail in Dezfuli and Giari (1999) and Dezfuli et al. (2008b).

GENETICS

Genetic information provides essential data on taxonomic relationship as well as the history of evolution of the parasite species and their variants (Huyse and Littlewood, 2007). Hence, several parasitology studies implement DNA-sequence based approaches. To mention a few, random amplified polymorphic DNA (RAPD) markers revealed to be essential for species recognition of congeneric acanthocephalans (Dezfuli and Tinti, 1998). Besides RAPD, polymerase chain reaction (PCR)-based assays, using oligonucleotide primers derived from species-specific sequences, have been utilized due to high sensitivity and reproducibility (Gasser, 2006). Furthermore, there are numerous accounts of DNA sequencing studies in acanthocephalans (García-Varela et al., 2013; García-Varela and Perez-Ponce de Leon, 2015; Herlyn et al., 2003).

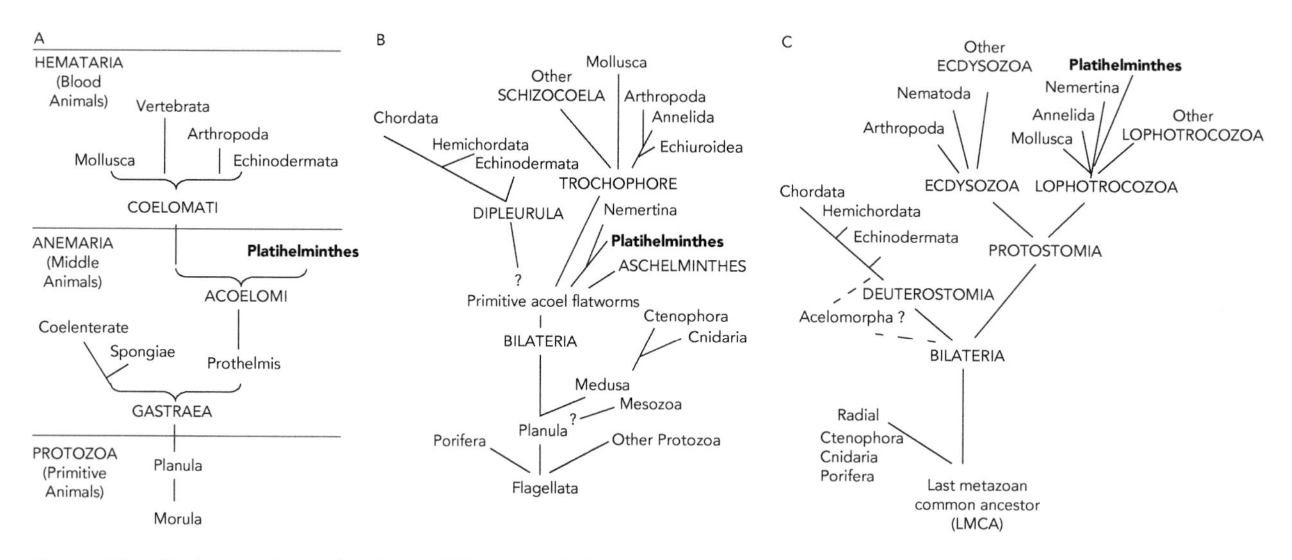

Figure 24.5 Phylogenetic relationships of Bilaterian phyla. The arrow points to the position of Acanthocephala in this tree of Bilaterians.

POSITION IN THE ToL

A close phylogenetic relationship of acanthocephalans and rotifers is widely accepted (**Figure 24.5**). Inside the clade, acanthocephalans are probably sisters of Seisonidea, as suggested by mitochondrial gene order (Sielaff et al., 2016) and part of nuclear sequence analyses (Herlyn et al., 2003; Wey-Fabrizius et al., 2014). Specificities in the organization of the epidermis and spermatozoon are in congruence with a sister-group relationship of Seisonidea and Acanthocephala (Ahlrichs, 1997). Alternative associations lack evidence of morphological data (Ricci, 1998). Additionally, taking into account the life styles, a first association with mandibulate arthropods presumably evolved in the stem line of the Seisonidea + Acanthocephala clade (Pararotatoria). Invasion of the mandibulate host and inclusion of gnathostome vertebrates as second hosts later occurred in the acanthocephalan stem line (Sielaff et al., 2016).

DATABASE AND COLLECTIONS

Some information on acanthocephalans can be found in "The Catalogue of Life" and WoRMS.

CONCLUSION

The phylum Acanthocephala is minor in comparison to platyhelminths and nematodes. Acanthocephalans have little diversity, but they are widespread and might be considered a successful group. The anatomical simplicity of the Acanthocephalans is not a sign of degeneracy but rather of high level of adaptation to their peculiar mode of life. Acanthocephala most likely evolved as an aquatic group and hence majority of the species are still aquatic. Acanthocephalans are often considered unimportant parasites because they do not affect humans. Study of evolutionary patterns at the molecular level promises to be an exciting and informative field for future research. There is paucity of data on many aspects of acanthocephalan biology and ecology, for example, several questions of functional morphology of the acanthocephalan reproductive system need convincing answers. We hope that information provided in this chapter will spur further research.

ACKNOWLEDGMENT

The authors would like to thank Dr. R.C. Shankaraiah, Vancouver, Canada for English correction of the draft. Thanks are due to Prof. H. Herlyn from the University of Mainz, Germany for help on the phylogeny paragraph.

REFERENCES

Ahlrichs, W.H., 1997. Epidermal ultrastructure of *Seison nebaliae* and *Seison annulatus*, and a comparison of epidermal structures within the Gnathifera. Zoomorphology, 117, 41–48.

Amin, O., 1985. Classification. In Crompton D.W.T., and Nickol B. B. (eds.), Biology of the Acanthocephala (pp. 27–71). Cambridge University Press, Cambridge.

Amin, O.M. and Heckmann, R.A., 2014. First description of *Pseudoacanthocephalus lutzi* from Peru using SEM. Scientia Parasitologica, 15, 19–26.

Barabashova, V.N., 1971. The structure of the integument and its role in the vital activity of Acanthocephala. Parazitologiya, 5, 446–454.

Bethel, W.M. and Holmes, J.C., 1973. Altered evasive behavior and responses to light in amphipods harboring acanthocephalan cystacanths. Journal of Parasitology, 59, 945–956.

Brázová, T., Poddubnaya, L.G., Ramírez Miss, N. and Hanzelová, V., 2014. Ultrastructure and chemical composition of the proboscis hooks of *Acanthocephalus lucii* (Müller, 1776) (Acanthocephala: Palaeacanthocephala) using X-ray elemental analysis. Folia Parasitologica, 61, 549–557.

Carcupino, M. and Dezfuli, B.S., 1999. Acanthocephala. In Jamieson B.G.M. (ed.), Progress in Male Gamete Biology (pp. 229–242). Oxford and IBH Publishing Co., New Delhi.

Crompton, D.W.T., 1970. An Ecological Approach to Acanthocephalan Physiology. Cambridge University Press, Cambridge.

Crompton, D.W.T., 1973. The sites occupied by some parasitic helminths in the alimentary tract of vertebrates. Biological Reviews, 48, 27–83.

Crompton, D.W.T, 1985. Reproduction. In Crompton D.WT. and Nickol B.B. (eds.), Biology of the Acanthocephala (pp. 213–271). Cambridge University Press, Cambridge.

Crompton, D.W.T and Nickol, B., 1985. Biology of the Acanthocephala. Cambridge University Press, Cambridge.

Daniels, C.B., 1985. The effect of infection by a parasitic worm on swimming and diving in the water skink, *Sphenomorphus quoyii*. Journal of Herpetology, 19, 160–162.

Dezfuli, B.S., 2000. Study of cement apparatus, cement production and transportation in adult male *Neoechinorhynchus rutili* (Acanthocephala: Eoacanthocephala). Parasitology Research, 86, 791–796.

Dezfuli, B.S., Capuano, S., Pironi, F. and Mischiati C., 1999. The origin and function of cement gland secretion in *Pomphorhynchus laevis* (Acanthocephala). Parasitology, 119, 649–653.

Dezfuli, B.S. and De Biaggi, S., 2000. Copulation of *Acanthocephalus anguillae* (Acanthocephala). Parasitology Research, 86, 524–526.

Dezfuli, B.S. and Giari, L., 1999. Amphipod intermediate host of *Polymorphus minutus* (Acanthocephala), parasite of water birds, with notes on ultrastructure of host-parasite interface. Folia Parasitologica, 46, 117–122.

Dezfuli, B.S., Lui, A., Giari, L., Boldrini, P. and Giovinazzo, G., 2008a. Ultrastructural study on the body surface of the acanthocephalan parasite *Dentitruncus truttae* in brown trout. Microscopy Research and Technique, 71, 230–235.

Dezfuli, B.S., Lui, A., Giovinazzo, G. and Giari, L., 2008b. Effects of Acanthocephala infection on the reproductive potential of crustacean intermediate hosts. Journal of Invertebrate Pathology, 98, 116–119.

Dezfuli, B.S., Maynard, B.J. and Wellnitz, T., 2003. Activity levels and predator detection by amphipods infected with an acanthocephalan parasite, *Pomphorhynchus laevis*. Folia Parasitologica, 50, 129–134.

Dezfuli, B.S., Onestini, S., Carcupino, M. and Mischiati C., 1998. The cement apparatus of larval and adult *Pomphorhynchus laevis* (Acanthocephala: Palaeacanthocephala). Parasitology, 116, 437–447.

Dezfuli, B.S., Simoni, E. and Mischiati, C., 2001. The cement apparatus of larval and adult *Acanthocephalus anguillae* (Acanthocephala), with notes on the copulatory cap and origin of gland secretion. Parasitology Research, 87, 299–305.

Dezfuli, B.S. and Tinti, F., 1998. Species recognition of congeneric acanthocephalans in slider turtle by random amplified polymorphic DNA (RAPD) markers. Journal of Parasitology, 84, 860–862.

Dunagan, T.T. and Miller, D.M., 1986. A review of protonephridial excretory systems in Acanthocephala. Journal of Parasitology, 72, 621–632.

García-Varela, M. and Perez-Ponce de Leon, G., 2015. Advances in the classification of acanthocephalans: evolutionary history and evolution of the parasitisms. In Morand S., Krasnov B.R., Littlewood D.T.J. (eds.), Parasite Diversity and Diversification: Evolutionary Ecology Meets Phylogenetics (pp. 182–201). Cambridge University Press, Cambridge.

García-Varela, M., Pérez-Ponce de León, G., Aznar, F.J. and Nadler, S.A., 2013. Phylogenetic relationship among genera of Polymorphidae (Acanthocephala), inferred from nuclear and mitochondrial gene sequences. Molecular Phylogenetics and Evolution, 68, 176–184.

Gasser, R.B., 2006. Molecular tools – advances, opportunities and prospects. Veterinary Parasitology, 136, 69–89.

Gazi, M., Kim, J., García-Varela, M., Park, C., Littlewood, D.T. and Park, J.K., 2016. Mitogenomic phylogeny of Acanthocephala reveals novel class relationships. Zoologica Scripta, 45, 437–454.

Goater, T.M., Goater, C.P. and Esch, G.W., 2014. Parasitism: The Diversity and Ecology of Animal Parasites. Cambridge University Press, Cambridge.

Gomes, A.P., Olifiers, N., Souza, J.G., Barbosa, H.S., D'Andrea, P.S. and Maldonado, A., 2015. A new acanthocephalan species (Archiacanthocephala: Oligacanthorhynchidae) from the crab-eating fox (*Cerdocyon thous*) in the Brazilian pantanal wetlands. Journal of Parasitology, 101, 74–79.

Herlyn, H., 2017. Organization and evolution of the proboscis musculature in avian parasites of the genus *Apororhynchus* (Acanthocephala: Apororhynchida). Parasitology Research, 116, 1801–1810.

Herlyn, H., Piskurek, O., Schmitz, J., Ehlers, U. and Zischler, H., 2003. The syndermatan phylogeny and the evolution of acanthocephalan endoparasitism as inferred from 18S rDNA sequences. Molecular Phylogenetics and Evolution, 26, 155–164.

Herlyn, H. and Taraschewski, H., 2017. Evolutionary anatomy of the muscular apparatus involved in the anchoring of Acanthocephala to the intestinal wall of their vertebrate hosts. Parasitology Research, 116, 1207–1225.

Holmes, J.C., 1973. Site selection by parasitic helminths: interspecific interactions, site segregation, and their importance to the development of helminth communities. Canadian Journal of Zoology, 51, 333–347.

Huyse, T. and Littlewood, D.T.J., 2007. Parasite species and speciation – tackling a host of problems. International Journal for Parasitology, 37, 825–828.

Kennedy, C.R., 1985. Site segregation by species of Acanthocephala in fish, with special reference to eels, *Anguilla anguilla*. Parasitology, 90, 375–390.

Kennedy, C.R., 2006. Ecology of the Acanthocephala. Cambridge University Press, Cambridge.

Maynard, B.J., Wellnitz, T., Zanini, N., Wright, B. and Dezfuli, B.S., 1998. Parasite-altered behaviour in a crustacean intermediate host: field and laboratory studies. Journal of Parasitology, 84, 1102–1106.

Meyer, A., 1932. Acanthocephala. In Dr. H. G. Bronn's Klassen und Ordnungen des Tier Reichs (Vol. 4, pp. 1–332). Akad. Verlagsgesellschaft MBH, Leipzig.

Meyer, A., 1933. Acanthocephala. In Dr. H. G. Bronn's Klassen und Ordnungen des Tier Reichs (Vol. 4, pp. 333–582). Akad. Verlagsgesellschaft MBH, Leipzig.

Miller, D.M. and Dunagan, T.T., 1985. Functional morphology (chapter 5). In Crompton D.W.T. and Nickol B.B. (eds.), Biology of the Acanthocephala (73–123). Cambridge University Press, Cambridge.

Moore, J., 2002. Parasites and the Behavior of Animals. Oxford University Press, Oxford.

Parshad, V.R. and Crompton, D.W.T., 1981. Aspects of acanthocephalan reproduction. Advances in Parasitology, 19, 73–138.

Petrochenko, V.I., 1956. Acanthocephala of Domestic and Wild Animals, vol. 1. Izdatel'stvo Akademii Nauk SSSR, Moscow (In Russian) (English translation by Israel Program for Scientific Translations, Ltd., Jerusalem, 1971, 465).

Petrochenko, V.I., 1958. Acanthocephala of Domestic and Wild Animals, vol. 2. Izdatel'stvo Akademii Nauk SSSR, Moscow (In Russian) (English translation by Israel Program for Scientific Translations, Ltd., Jerusalem, 1971, 478).

Poulin, R., 2007. Evolutionary Ecology of Parasites, Second edition. Princeton University Press, Princeton.

Redi, F., 1684. Osservazioni interna agli animali viventi che si trovano negli animali viventi. Firenze, 253.

Ricci, C., 1998. Are lemnisci and proboscis present in the Bdelloidea? Hydrobiologia, 387/388, 93–96.

Sielaff, M., Schmidt, H., Struck, TH., Rosenkranz, D., Mark Welch, D.B., Hankeln, T. and Herlyn, H., 2016. Phylogeny of Syndermata (syn. Rotifera): mitochondrial gene order verifies epizoic Seisonidea as sister to endoparasitic Acanthocephala within monophyletic Hemirotifera. Molecular phylogenetics and evolution, 96, 79–92.

Taraschewski, H., 2000. Host-parasite interactions in Acanthocephala: a morphological approach. Advances in Parasitology, 46, 1–179.

Taraschewski, H., 2005. Helminth parasites: Acanthocephala. In Rohde K. (ed.), Marine Parasitology (116–121). CABI Publishing, Wallingford.

Wellnitz, T., Giari, L., Maynard, B.J. and Dezfuli, B.S., 2003. A parasite spatially structures its host population. Oikos, 100, 263–268.

Wey-Fabrizius, A.R., Herlyn, H., Rieger, B., Rosenkranz, D., Witek, A., Welch, D.B., Ebersberger, I. and Hankeln, T., 2014. Transcriptome data reveal Syndermatan relationships and suggest the evolution of endoparasitism in Acanthocephala via an epizoic stage. PLoS One, 9, e88618.

Yamaguti, S., 1963. Acanthocephala. Systema Helminthum (vol. 5, 1–423). Wiley Intersci., New York, London.

PHYLUM ENTOPROCTA

Tohru Iseto and Hiroshi Kajihara

CHAPTER

25

Infrakingdom	Bilateria
Division	Protostomia
Subdivision	Lophotrochozoa
Phylum	Entoprocta

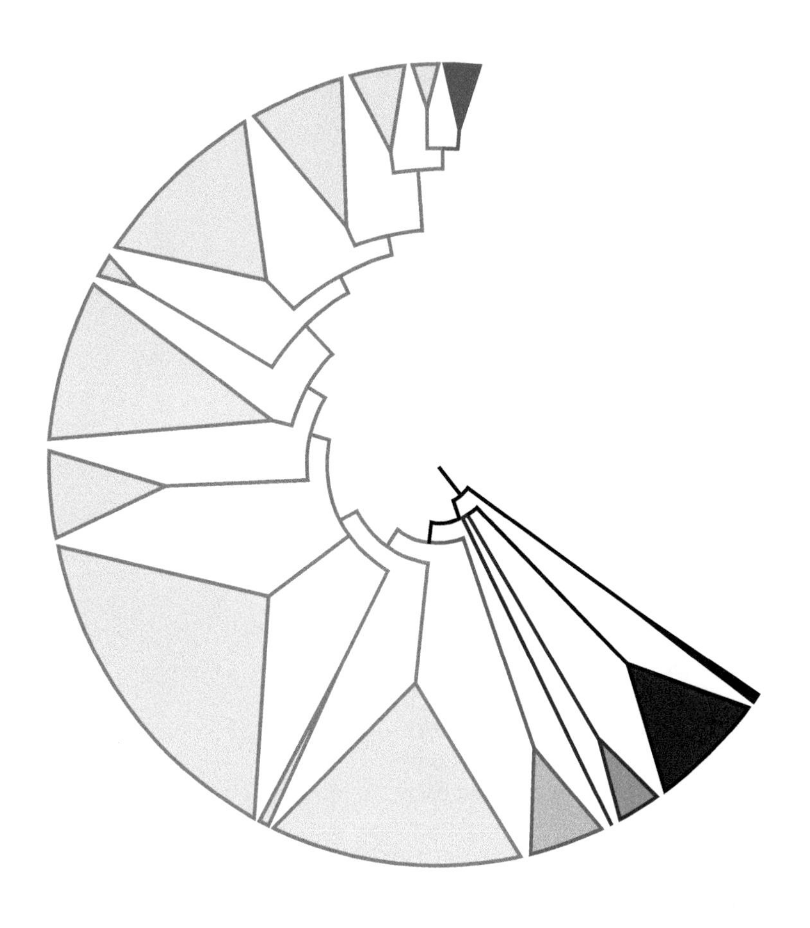

CONTENTS

General Characteristics

About 180 valid species

Benthic, mostly marine animals. Some species live in brackish waters and two in freshwaters

Distributed worldwide, ranging from shallow intertidal to more than 5,000 m deep waters

Bilaterally symmetric body, exist either solitary or in colonial conditions

No commercial value

Ecological importance is probably small

Synapomorphies

- U-shaped gut with anus opening inside the tentacle crown
- suspension feeding with downstream-collecting system
- no coelom. The cavity is filled by a loose fluid matrix with mesenchyme cells

HISTORY OF TAXONOMY AND CLASSIFICATION

The oldest-known entoproct species in the literature is the marine colonial *Pedicellina cernua*, which was found on a dead, filamentous colony of the bryozoan *Cellularia salicornia* (now *Cellaria fistulosa*). It was originally described and placed in the now-rotifer genus *Brachionus* by the Prussian zoologist and botanist Peter S. Pallas in 1774. The earliest entoproct genus *Pedicellina* was established by the Norwegian theologian and biologist Michael Sars (1835) for *Pedicellina echinata* and *P. gracilis* (now *Barentsia gracilis*); these were placed in Polypi along with hydroids, anthozoans, and bryozoans. Freshwater colonial[1] and marine solitary[2] taxa were first discovered in the mid-19th century. In 1870, the German zoologist Hinrich Nitsche correctly apprehended the anatomical peculiarity of entoprocts in comparison with bryozoans, and established Entoprocta[3] and Ectoprocta[4] within Bryozoa. In 1888, the Austrian zoologist Berthold Hatschek raised Entoprocta to the phylum level, for which other synonymous names Calyssozoa[5] and Kamptozoa[6] were also proposed later.

The phylum now comprises about 180 (>130 solitary and ~40 colonial) species in 13 genera and 4 families. Dichotomy of Entoprocta into solitary forms, Solitaria, and colonial forms, Coloniales,[7] seems to be consistent with recent molecular phylogenetic data.[8] However, inclusion of the colonial *Loxokalypus* in a molecular phylogenetic context is essential for this dichotomous classification to be validated, because *Loxokalypus* shares some important characters with members in Solitaria, but was placed in Coloniales because of the colonial body plan[9] (**Figure 25.1**).

Figure 25.1 Colonized individuals of solitary entoproct, *Loxosomella* sp., on a sponge (Izu Oshima Island, Japan, 27 m deep). (Photo by Osamu Hoshino.)

ANATOMY, HISTOLOGY, AND MORPHOLOGY

Entoprocts are either solitary or colonial (**Figure 25.2**). The body, 0.1–15 mm in size, consists of the calyx, stalk, and basal attaching structure. The calyx contains a U-shaped gut, ganglion (nerve), a pair of protonephridia, and one or two pairs of gonads. The mouth and anus open upward, inside the tentacle crown. The mouth opens anteriorly and anus posteriorly. The upper and bottom sides of the calyx are defined as ventral and dorsal, respectively (**Figure 25.3**). The tentacles have well-organized longitudinal rows of

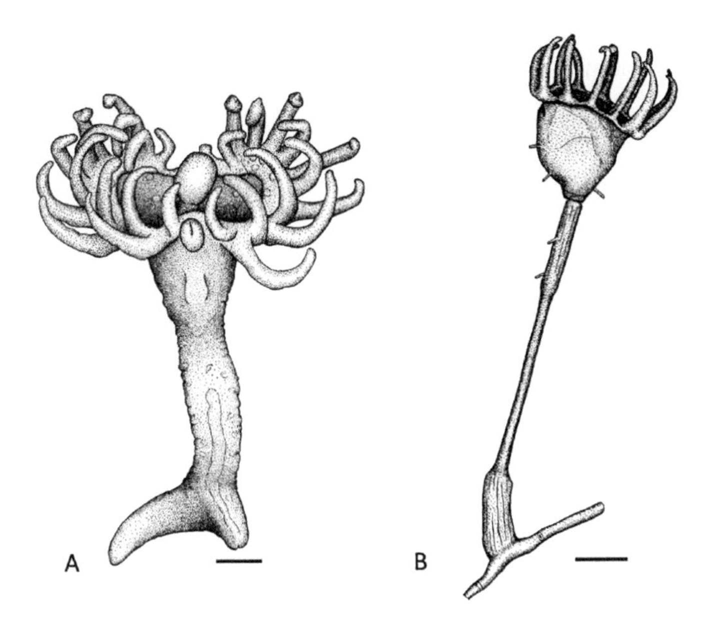

Figure 25.2 Two forms of Entoprocta. (A) Solitary form (*Loxosomella dicotyledonis*). (B) Colonial form (*Barentsia* sp.). Scale bar = 100 μm. (From Iseto, 2001.)

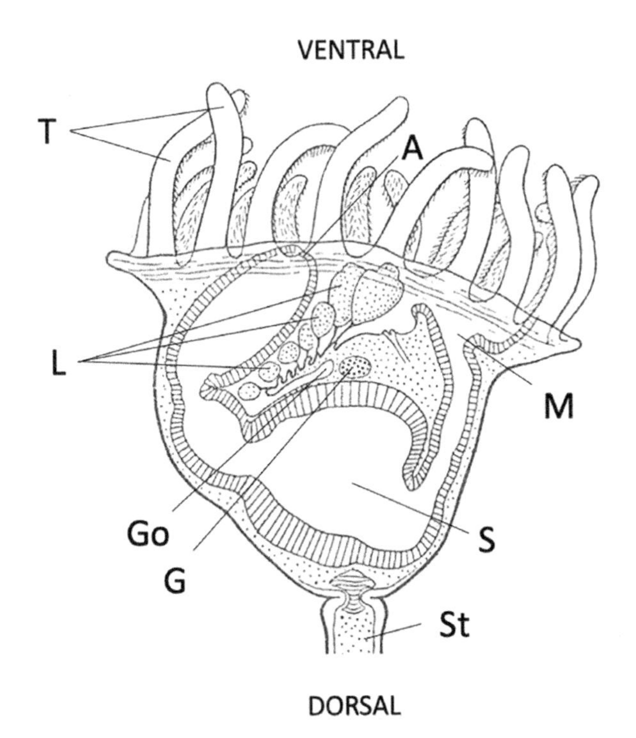

Figure 25.3 A schematic drawing of the calyx of an entoproct. A, anus; G, ganglion; Go, gonad; L, larvae; M, mouth; S, stomach; St, stalk; T, tentacles. (Modified from Brien, 1959.)

ciliated cells which create water current, capture particles, and transport the particles toward the mouth. Solitary species have a basal attaching organ, the foot. Some solitary species degenerate the foot and cement on the substrate. The Entoproct body wall comprises a single layer of epithelium and a cuticle layer covering the whole body. The cuticle structure resembles that of annelids.

REPRODUCTION AND DEVELOPMENT

Sexual reproduction

Solitary entoprocts are mostly protandric hermaphrodite. In colonial species, zooids are mostly gonochoric. Some species have only one sex in a colony but others have both. In a few cases, a single zooid is simultaneously hermaphroditic.

Sperm is apparently released directly to the water column and eggs are fertilized in the ovary.[10] Cleavage is spiral, equal, and asynchronous. After fertilization, embryos are transported to a deep depression at the atrium, called the brood chamber. Each zooid has a single brood chamber that may contain multiple embryos. In most species, each embryo is connected to the mother via a string, which does not involve nutrient transfer. In a few species the embryo connects with the mother by a placenta and enjoys nutrient supply. Fully grown larvae swim away from the brood chamber.

Entoproct larvae resemble trochophores in having a prototroch and an apical organ, but differ in having a frontal organ and a ciliated foot[11]. In many species, larvae can not only swim but also creep with the foot. They have a functional, inverted U-shaped gut, with the mouth and anus opening on the bottom side. Metamorphosis involves a rotation (a nose down, tail up pitching) of inner structures including the gut, so that the gut now takes an upright U shape, with the mouth and anus directing upward in adult stage. In some species, larvae do not metamorphose; instead, inner buds come out and become new individuals[12] (**Figure 25.4**).

Vegetative reproduction

All entoproct species are capable of reproducing asexually. Solitary species produce buds on the calyx, which will be detached from the parent. Colonial species also make buds, typically at the tip of the stolon, which become new members of the same colony. Budding is achieved by epidermal growth with a partial contribution from mesenchyme, but without any endoderm contribution. Hibernacula are known in some colonial species but not in solitary species.

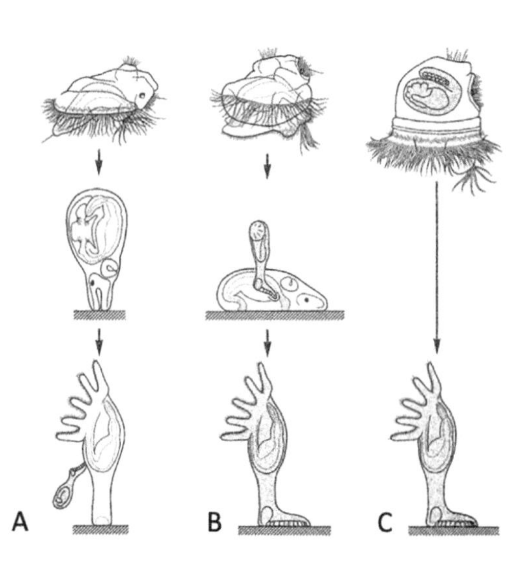

Figure 25.4 Formation of new adult individuals from larvae in the genus *Loxosomella*. (A) A new individual is formed through the metamorphoses of a larva (*Loxosomella harmeri*). (B and C) Larvae do not metamorphose but new individuals are formed by budding from the larvae (B) *L. leptoclini*; (C) *L. vivipara*. In the cases of B and C, larvae die after the liberation of the buds. (From Nielsen, 1971.)

Regeneration is common in most colonial species. They often replace old or injured calyxes. The colonial *Barentsia discreta* can generate complete bodies from a fragment of stalk or stolon.[13] Only one solitary species, *Loxosomella antarctica,* is known to be able to regenerate their calyx.[14]

Developmental genes

Developmental genes are not yet studied for entoprocts. One reason may be the difficulty in handling eggs and embryos, which are connected to the mother atrium until hatching.

DISTRIBUTION AND ECOLOGY

Entoprocts occur in all the world oceans, from tropical to polar regions and from shallow intertidal habitat to deep seas **Figure 25.5**. The deepest dwelling solitary species has been reported from 5,222 m.[15]

The freshwater colonial *Urnatella gracilis* is known to be dispersed as "propagation stolons," which are fragments of tiny colonies.[16] The worldwide occurrence of this species is considered to be anthropogenic.

Most solitary species live on other animals such as polychaetes, bryozoans, and sponges. By association to other animals, solitary entoprocts may enjoy a safe habitat to avoid being predated upon by natural enemies and being covered by other fouling organisms and/or sediments. Entoprocts seems not affect host food capture, indicating that they are not parasites but commensals.

Solitary entoprocts are often found with high density on a single host sponge. One might speculate that such a local population would comprise genetically identical clones as a result of a single migration event, because fewer chances are likely for entoproct larvae/buds to reach a particular host. Besides, asexual reproduction by budding is common among entoprocts. Contrary to expectations, however, a molecular genetic analysis demonstrated that individuals of *Loxosomella plakorticola* on the same demosponge *Plakortis* did not belong to a single clone.[17] Thus, the entoprocts probably migrate to sponges more frequently than expected.

All entoprocts are suspension feeders, feeding on phytoplankton and organic particles. Some entoprocts are known to trap larger organisms such as ciliates by contracting the tentacles. The Antarctic *L. antarctica* discharges sticky threads (lime-twig gland), which is assumed to catch larger foods like ciliates.

Turbellarians and nudibranchs prey on entoprocts.[18]

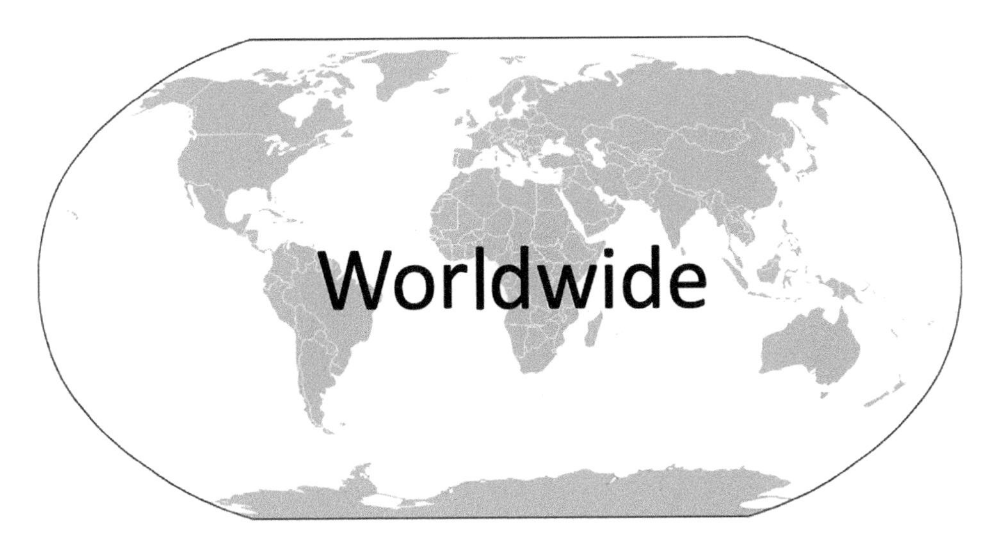

Figure 25.5 Distribution of entoprocts.

PHYSIOLOGY AND BEHAVIOR

Entoprocts usually have a pair of protonephridia. Both of the two freshwater entoprocts, *U. gracilis and Loxosomatoides sirindhornae*, have a considerably larger number of protonephridia not only in the calyx but also in the stalk. Protonephridia are thought to function mainly in osmoregulation. A brownish stomach "roof" is the place where the metabolites are excluded into stomach for the animal to expel.

Nerves radiate from a dumbbell-shaped ganglion in the calyx to other parts of the calyx, tentacles, and stalk. Unicellular tactile receptors are present on tentacles and calyx.[19] Nervous connections between zooids are not present within a colony. A pair of eyes is present in larvae of some species but is lost in the adult stage.

Because of their small body size, diffusion may be sufficient for circulation and respiration for most entoprocts. In solitary species, their active body movement may be enough to diffuse their body-fluid material. Most colonial species have a "star-cell complex" at the connection of the calyx and stalk, which functions as a circulation pump.

The behavior of solitary species is quite active (**Figure 25.6**). Some solitary species creep like caterpillars by alternatively attaching their basal foot and sticky frontal tentacles. Other species can somersault on the substratum. One peculiar species, *Loxosomella bifida*, can simulate a "walking" motion with a pair of leg-like expansions of the basal attaching organs. Colonial species show nonlocomotory bending motions.

GENETICS AND GENOMICS

Complete sequence of mitochondrial genome is determined in two solitary species.[20] The mitochondrial genome size is about 15,000 and it carries typical gene set of metazoans. Unusual initiation codon, ACG, instead of the usual AUG in a COI gene is suggested in a solitary species.

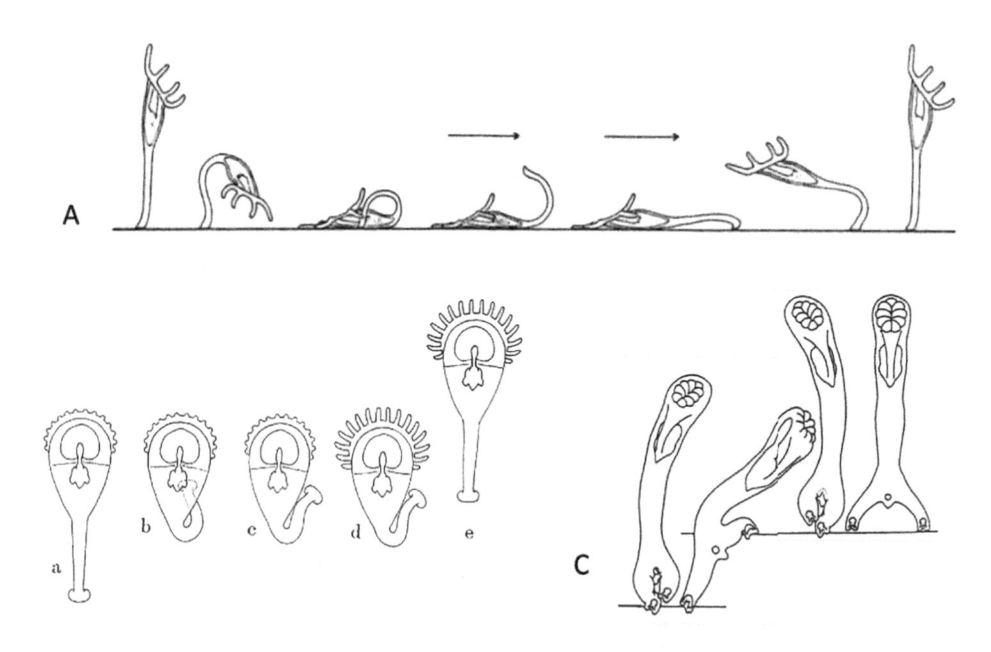

Figure 25.6 **Various active locomotion movements of solitary entoprocts.** (A) Somersault of *Loxosoma agile*. (B) Looper-like locomotion of *Loxosoma monilis*. (C) Walking of *Loxosomella bifida* by a pair of "legs". ((A) From Nielsen, 1964; (B) From Konno, 1973; (C) From Konno, 1972.)

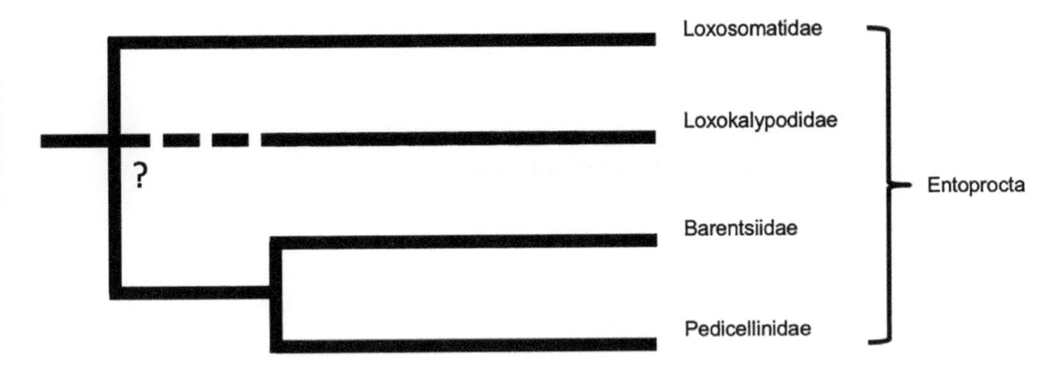

Figure 25.7 Systematic relationships of the four families of the phylum Entoprocta.

POSITION IN THE ToL

Entoprocta belong to Lophotrochozoa but the relationships among lophotrochozoan phyla are still unclear. Mitochondrial gene order and morphology of larvae suggest close relationship to Mollusca. Sequence-based analyses of mitochondrial genes, however, support close affinities of Entoprocta to Phoronida and Nemertea (**Figure 25.7**). Recent molecular phylogenomic studies support the sister relationship of Entoprocta and Cycliophora.[21]

DATABASES AND COLLECTION

World Register of Marine Species (WoRMS) maintains entoproct scientific names.

CONCLUSION

Our current knowledge of the Entoprocta is apparently only a fraction of the true diversity. Opportunities to find new species and new biological phenomena associated with them are still highly possible for this animal. In particular, studies on genomics, developmental genes, and evolutionary developmental biology are critically needed. Because of relatively poor species-level characters, misidentification and presence of cryptic species must be carefully studied for the progress of entoproct studies.

NOTES

1 *Urnatella gracilis* Laidy, 1851.
2 *Loxosoma singulare* Keferstein, 1862.
3 from the Ancient Greek ἐντός (*entos*, "inside") and πρωκτός (*proctos*, "anus"), indicating the position of the anal opening within the tentacle crown, as opposed to ectoprocts (= bryozoans), in which the anus opens outside the tentacle crown (or lophophore).
4 from the Ancient Greek ἐκτός (*ektos*, "outside") and πρωκτός (*proctos*, "anus").
5 by Clark (1921).
6 by Cori (1929), because Calyssozoa was already used for a group of cnidarians.
7 Emschermann (1972).
8 for example, Fuchs et al. (2010), Kajihara et al. (2015), Borisanova et al. (2018).
9 thus Emschermann's (1972) dichotomous system was criticized by Nielsen (1989); it is not employed in this chapter as well.
10 Mukai and Makioka (1980).

11 Nielsen (1971).

12 Nielsen (1971).

13 Mukai and Makioka (1978).

14 Emschermann (1993).

15 *Loxosomella profundorum* Borisanova, Chernyshev, Neretina, and Stupnikova, 2015 from the Kuril-Kamchatka trench area.

16 Emschermann (1987).

17 Sugiyama et al. (2010).

18 Nakano (2019)

19 Nielsen and Rostgaard (1976).

20 *Loxosomella allax* and *Loxosomella aloxiata*, Yokobori et al. (2008).

21 Hejnol et al. (2009), Kocot et al. (2017).

REFERENCES

Borisanova, A. O., Chernyshev, A. V., Neretina, T. V., and Stupnikova, A. N. 2015. Description and phylogenetic position of the first abyssal solitary kamptozoan species from the Kuril–Kamchatka trench area: *Loxosomella profundorum* sp. nov. (Kamptozoa: Loxosomatidae). Deep Sea Research Part II: Topical Studies in Oceanography 111: 351–356.

Borisanova, A. O., Chernyshev, A. V., and Ekimova, I. A. 2018. Deep-sea Entoprocta from the Sea of Okhotsk and the adjacent open Pacific abyssal area: New species and new taxa of host animals. Deep Sea Research Part II: Topical Studies in Oceanography 154: 87–98.

Brien, P. 1959. Classe des Endoproctes ou Kamptozoaires. In: Grassé, P. (Ed.) Traité de Zoologie 5: 927–1007. Paris, Masson et Cie.

Clark, A. H. 1921. A new classification of animals. Bulletin de l'Institut Océanographique de Monaco 400: 1–24.

Cori, C. 1929. Kamptozoa. In: Krumbach, T. and Kükenthal, W. (Eds.) Handbuch der Zoologie 2: 1–64. Berlin, Walter de Gruyter.

Emschermann, P. 1972. *Loxokalypus socialis* gen. et sp. nov. (Kamptozoa, Loxokalypodidae fam. nov.), ein neuer Kamptozoentyp aus dem nördlichen Pazifischen Ozean. Ein Vorschlag zur Neufassung der Kamptozoensystematik. Marine Biology 12: 237–254.

Emschermann, P. 1987. Creeping propagation stolons—an effective propagation system of the freshwater entoproct *Urnatella gracilis* Leidy (Barentsiidae). Archiv für Hydrobiologie 108: 439–448.

Emschermann, P. 1993. Lime-twig glands: A unique invention of an Antarctic entoproct. Biological Bulletin 185: 97–108.

Fuchs, J., Iseto, T., Hirose, M., Sundberg, P., and Obst, M. 2010. The first internal molecular phylogeny of the animal phylum Entoprocta (Kamptozoa). Molecular Phylogenetics and Evolution 56: 370–379.

Hatschek, B. 1888. Lehrbuch der Zoologie: eine morphologische Übersicht des Thierreiches zur Einführung in das Studium dieser Wissenschaft. Jena, G. Fischer.

Hejnol, A., Obst, M., Stamatakis, A., Ott, M., Rouse, G. W., Edgecombe, G. D., Martinez, P., Baguña, J., Bailly, X., Jondelius, U., Wiens, M., Müller, W. E. G., Seaver, E., Wheeler, W. C., Martindale, M. Q., Giribet, G., and Dunn, C. W. 2009. Assessing the root of bilaterian animals with scalable phylogenomic methods. Proceedings of the Royal Society B 276: 4261–4270.

Iseto, T. 2001. Okinawa no naikou-doubutsu/Okinawa de bunruigaku [Entoprocts in Okinawa/Taxonomy at Okinawa]. Umiushi-Tsushin 30: 6–8. In Japanese.

Kajihara, H., Tomioka, S., Kakui, K., and Iseto, T. 2015. Phylogenetic position of the queer, backward-bent entoproct *Loxosoma axisadversum* (Entoprocta: Solitaria: Loxosomatidae). Species Diversity 20(1): 83–88.

Keferstein, W. 1862. Ueber *Loxosoma singulare* gen. et sp. n., den Schmarotzer einer Annelide. Zeitschrift für wissenschaftliche Zoologie 12: 131–132.

Kocot, K. M., Struck, T., Merkel, J., Waits, D. S., Todt, C., Brannock, P. M., Weese, D. A., Cannon, J., Moroz, L. L., Leib, B., and Halanych, K. 2017. Phylogenomics of Lophotrochozoa with consideration of systematic error. Systematic Biology 66: 256–282.

Konno, K. 1972. Studies on Japanese Entoprocta. I. On four new species of Loxosomatids found at Fukaura, Aomori prefecture. Sci. Rep. Hirosaki Univ., 19:22–36.

Konno, K. 1973. Studies on Japanese Entoprocta. II. On a new species of *Loxosoma* and *Loxosomella akkesiensis* (YAMADA) from Fukaura. Sci. Rep. Hirosaki Univ., 20:79–87.

Leidy, J. 1851. On some American freshwater Polyzoa. Proceedings of the Academy of Natural Sciences of Philadelphia 5: 320–322.

Mukai, H. and Makioka, T. 1978. Studies on the regeneration of an entoproct, *Barentsia discreta*. Journal of Experimental Zoology 205: 261–276.

Mukai, H. and Makioka, T. 1980. Some observations on the sex differentiation of an entoproct, *Barentsia discreta* (Busk). Journal of Experimental Zoology 213: 45–59.

Nakano, R. 2019. Entoprocts and nudibranchs. Quarterly Umiushi 5(3-4): 7. (in Japanese)

Nielsen, C. 1964. Studies on Danish Entoprocta. Ophelia 1: 1–76.

Nielsen, C. 1971. Entoproct life-cycles and the entoproct/ectoproct relationship. Ophelia 9: 209–341.

Nielsen, C. 1989. Entoprocta. Synopses of the British Fauna, n.s. 41, 1–131.

Nielsen, C. and Rostgaard, J. 1976. Structure and function of an entoproct tentacle with a discussion of ciliary feeding types. Ophelia 15(2): 115–140.

Nitsche, H. 1870. Beiträge zur Kenntniss der Bryozoen. Zeitschrift für wissenschaftliche Zoologie 20: 1–36.

Pallas, P. S. 1774. Spicilegia zoologica quibus novae imprimis et obscurae animalium species iconibus, descriptionibus atque commentariis illustrantur. Fasiculus decimus. Berlin, G. A. Lange, 41 : pls I–IV.

Sars, M. 1835. Beskrivelser og iagttagelser over nogle mærkelige eller nye i havet ved den bergenske kyst levende dyr af polypernes, acalephernes, radiaternes, annelidernes, og molluskernes classer, med en kort oversigt over de hidtil af forfatteren sammesteds fundne arter og deres forekommen. Bergen, T. Hallagers, iv + 81:pls 1–15.

Sugiyama, N., Iseto, T., Hirose, M., and Hirose, E. 2010. Reproduction and population dynamics of the solitary entoproct *Loxosomella plakorticola* inhabiting a demosponge, *Plakortis* sp. Marine Ecology Progress Series 415: 73–82.

Yokobori, S., Iseto, T., Asakawa, S., Sasaki, T., Shimizu, N., Yamagishi, A., Oshima, T., and Hirose, E. 2008. Complete nucleotide sequences of mitochondrial genomes of two solitary entoprocts, *Loxocorone allax* and *Loxosomella aloxiata*: implications for lophotrochozoan phylogeny. Molecular Phylogenetics and Evolution 47: 612–628.

ECDYSOZOA AND ARTHROPODA

HIGHER TAXA 5

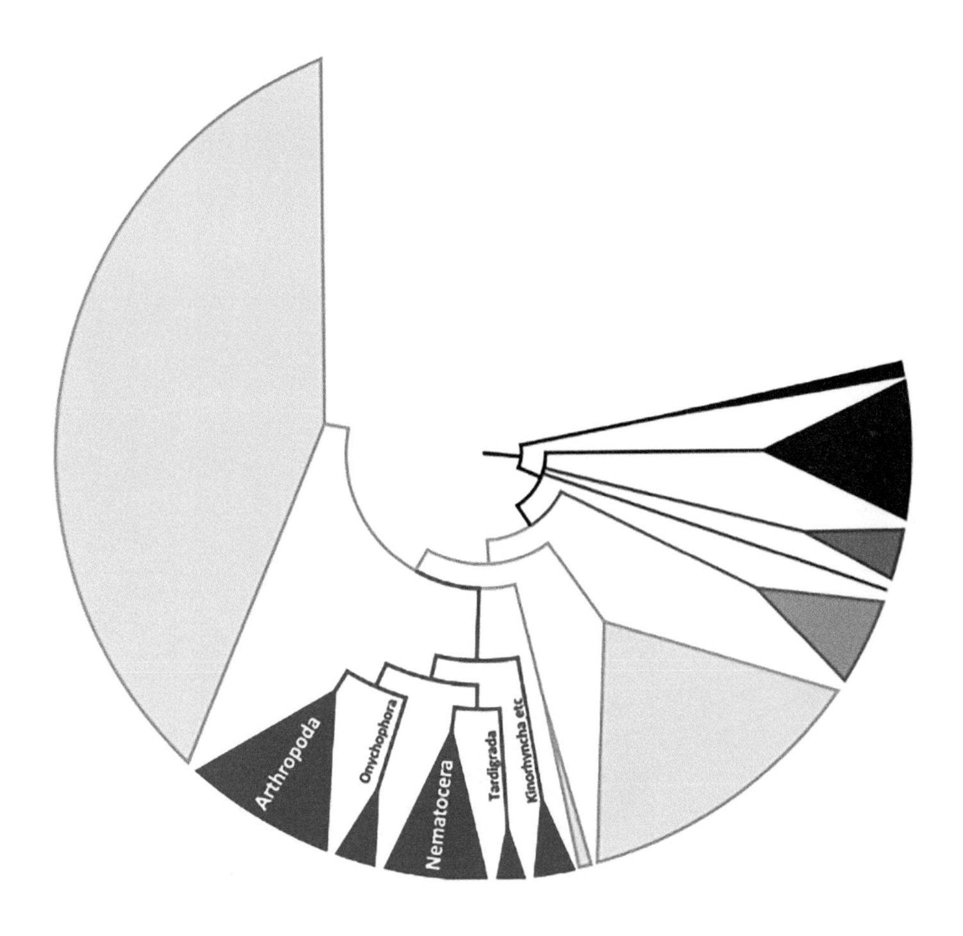

Ecdysozoa

The name Ecdysozoa comes from the Greek meaning "shedding animal," and indeed the defining morphological character of the group is that these animals grow by molting or ecdysis (**Figure 1**). The group contains the Kinorhyncha, Priapulida, Nematoda, Tardigrada, Nematomorpha, Lorciphera, Onychophora, and the large group named Arthropoda. The Ecdysozoa is also supported by extensive phylogenomic data. Arthropoda and Onychophora are considered sister groups. Nematoda are a group of round worms, together with Nematomorpha make up the group Nematoida which is in turn sister group to the Tardigrada.

Figure 1 A mayfly (*Cloeon* sp.) on the left and its molt on the right. (Licensed under the Creative Commons Attribution-Share Alike 2.5 Generic license.)

Several minor ecdysozoan phyla also exist – Kinorhyncha, Priapulida, and Loricifera. The Arthropoda are by far the most complex subdivision of the Ecdysozoa and they are further divided as below (**Figure 2**).

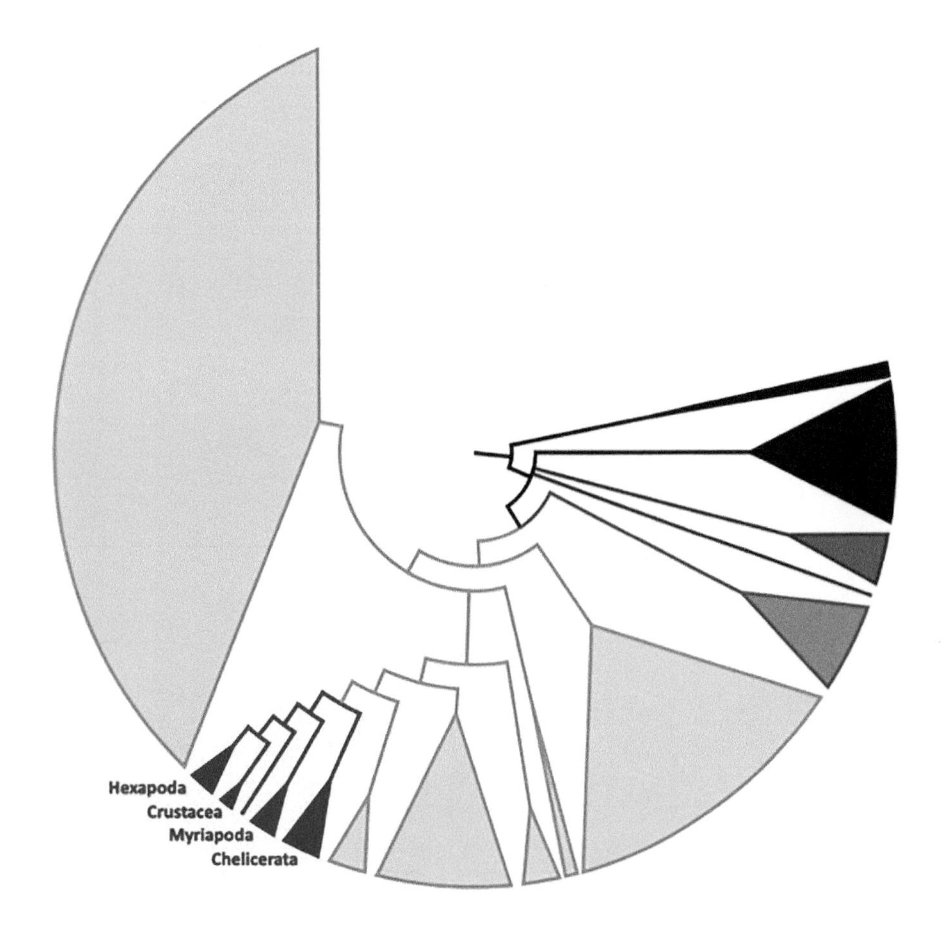

Arthropoda

Arthropoda is made up of Chelicerata, Myriapoda, Crustacea, and Hexapoda. Chelicerata and Myriapoda are monophyletic groups unto themselves. However, Crustacea is paraphyletic with regards to Hexapoda, and the hexapods plus crustaceans clade is called Pancrustacea.

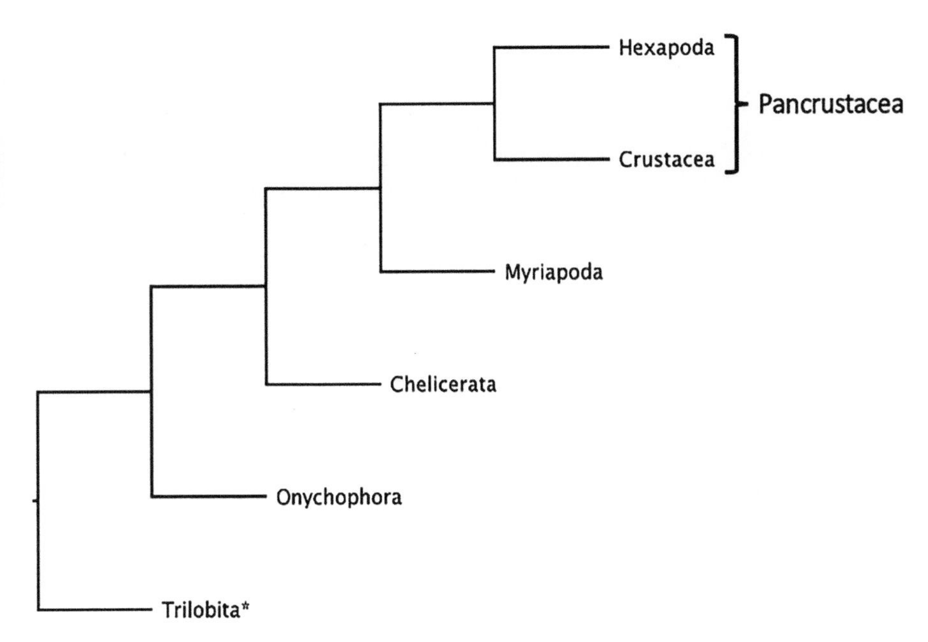

Figure 2 Phylogenetic relationships of the Arthropoda. The asterisk (*) indicates the fossil taxon Trilobita.

SUBPHYLUM CRUSTACEA

Nicolas Rabet

CHAPTER
26

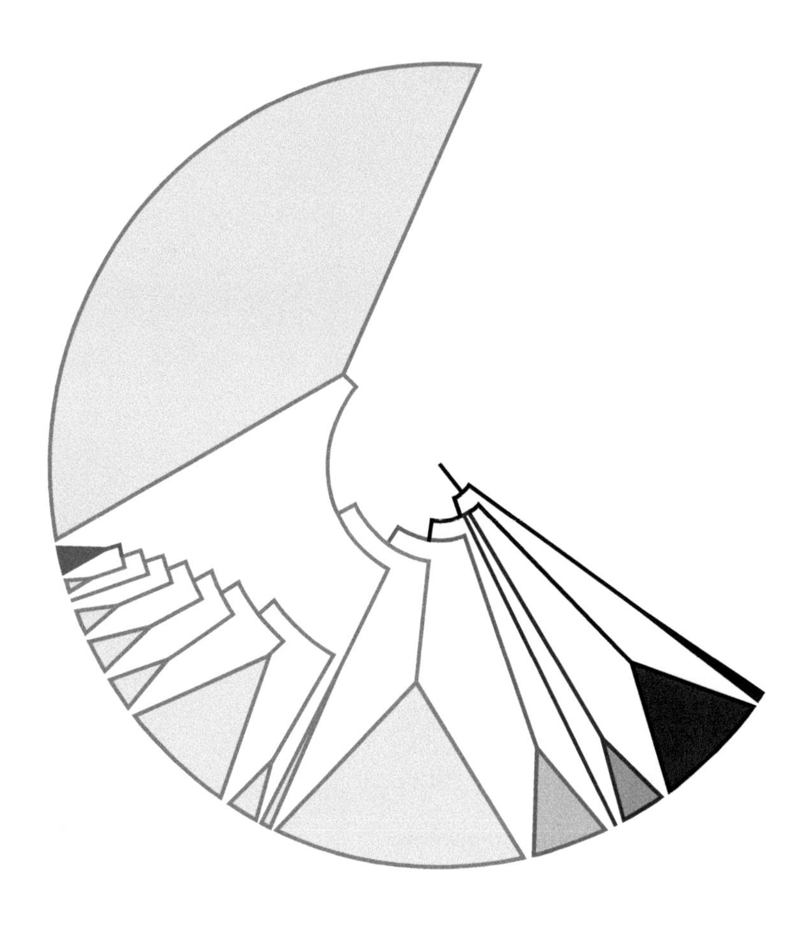

Infrakingdom	Bilateria
Division	Protostome
Subdivision	Ecdysozoa
Phylum	Arthropoda
Subphylum	Crustacea (paraphyletic)

CONTENTS

General Characteristics

Present in almost every ecosystem, from deep ocean environments to deserts

Segmented animals with large variations in the number and shape of the associated segments and appendages

Ecologically very important in aquatic environments

A large size range, with the largest terrestrial and aquatic arthropods and also some of the smallest animals with 0.1 mm length

Considerable biomass, particularly for planktonic species

Presence of a nauplius larva with three pairs of appendages and a single odd eye sometimes retained in the egg

Appendices fundamentally biramed

Great commercial importance due to the use of many species as human food

Synapomorphies

since Crustacea is paraphyletic within Arthropoda we list the following characters as potential synapomorphies:

- two pairs of antennae
- one pair on H1
- one pair on H2 (biramous ancestrally, but uniramous in some derived taxa)

HISTORY OF TAXONOMY AND CLASSIFICATION

The word *crustacea* is derived from the Latin *crusta* which means that the body is covered with a hard shell. Large crustaceans have probably always been known to man as a food source, since it is known that even Neanderthals ate crabs (Zilhão et al., 2020). It is therefore not surprising that in Aristotle's classification of some crustaceans was already listed under the name « μαλακόστρακα » (malakostraka), i.e., animals with soft (malakós) shell (óstrakon) (Zucker, 2005). For a long time, all Arthropods were gathered in the group of Insecta. Thus Linnaeus in 1758 classified certain crustaceans known at the time in the order of Aptera with only three genera *Cancer*, *Monoculus*, and *Oniscus*. Cirripeds (with the genus *Lepas*) and parasitic copepods (with the genus *Lernaea*) were not classified as Insecta but as Vermes, Testacea, and Mollusca, respectively.

The name Crustacea was proposed by Brünnich in 1772. However, it would be many decades before the different groups of crustaceans gradually came together in a more rational taxonomy. The scientific community also integrated the newly discovered groups of cephalocarids (Sanders, 1955) and remipedes (Yager, 1981). Bowman and Abele in 1982 proposed a classification of crustaceans into six classes (i.e., Cephalocarida, Branchiopoda, Remipedia, Maxillopoda, Ostracoda, and Malacostraca). The Maxillopoda brings together Mystacocarida, Cirripedia, Copepoda, and Branchiura.

Molecular data of the 18S gene allowed the confirmation and integration of the amazing Pentastomida with the Branchiura (Abele et al., 1989, Martin and Davis, 2001). Further analyses challenged the link between Hexapoda and Crustacea. Regier et al. (2010) profoundly modified the systematics of crustaceans: Hexapods must now be considered as a lineage of crustaceans that have become terrestrial and the group of crustaceans *sensu stricto* becomes paraphyletic. Furthermore, Regier et al. (2010) indicated that maxillopods are not monophyletic.

Within the pancrustaceans, relationships are not well understood (**Figure 26.1**). The group most closely related to Hexapods could be the Cephalocarid associated to the remipedes constituting the Xenocarida group (Regier et al., 2010, Lee et al., 2013) and together with Hexapods these form the Miracrustacea. But the position of the cephalocarids is still uncertain and some propose that the remipedes should be a sister group to the Hexapods to form the Labiocarida group (Schwentner et al., 2017). The other crustaceans are distributed in the Multicrustacea, which includes the malacostracans, copepods, thecostracans, and may form with the branchiopods the Vericrustacea even though these position are not certain. The group of Oligostraca of basal emergence includes the Ostracoda, Branchiura, Tantulocarida, and Pentastomida and is a sister group to all the other Pancrustacea called the Altocrustacea.

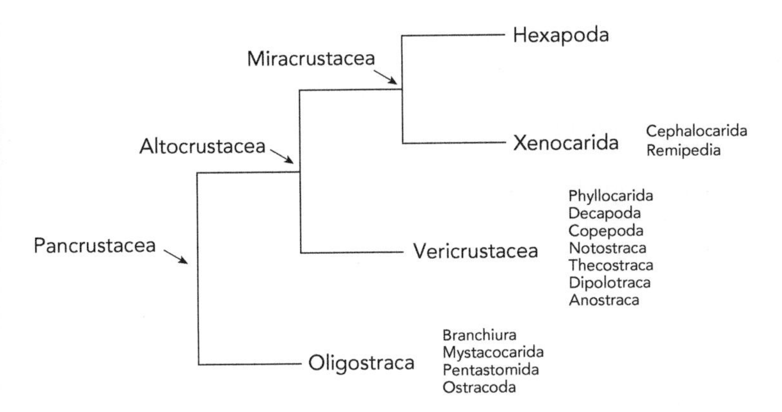

Figure 26.1 Major phylogenetic organization of the Pancrustacea based on Regier et al. (2011). The six classes established earlier and listed in the text are included in this tree but some of these are not monophyletic. The nuances of the discrepancies between the classical organization and the molecular one are discussed in the text.

ANATOMY, HISTOLOGY, AND MORPHOLOGY

Crustaceans are arthropods (**Figure 26.2**) with a well-individualized head, having an ocular region and a series of paired appendages that from front to back include two pairs of antennae (A1 and A2), the mandibles, and two pairs

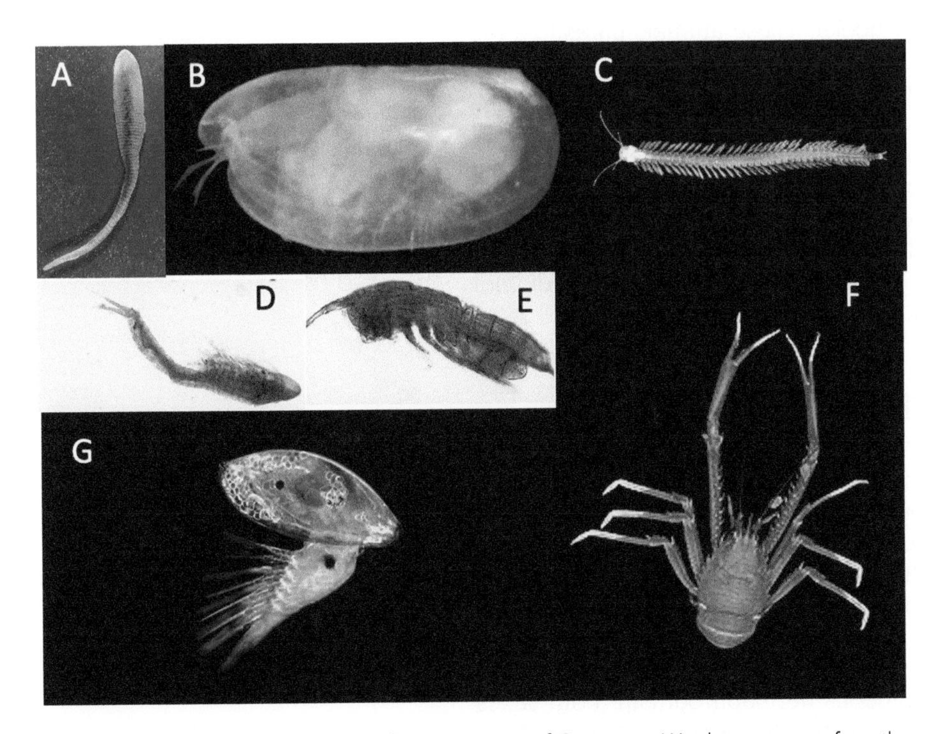

Figure 26.2 Photographs of some of major groups of Crustacea. We show two taxa from the Oligostraca (sensu Reier et al., 2010; A and B), two taxa from the Xenocarida (sensu Regier et al., 2010; C and D), and three taxa from the Vericrustacea (sensu Regier et al., 2010; E, F, and G). (A) Pentastomida: *Linguatula serrata*. (From Dennis Tappe and Dietrich W. Büttner license: Creative Commons Attribution 2.5 Generic license.); Major group Oligostraca. (B) Ostracoda; *Spelaeoecia* sp. NOAA; Major group Oligostraca. (C) Remipedia; Speleonectidae. (From Neiber, M. T., Hartke, T. R., Stemme, T., Bergmann, A., Rust, J., Iliffe, T. M., Koenemann, S. (2011). Global biodiversity and phylogenetic evaluation of Remipedia (Crustacea). *PLoS One* 6 (5): e19627, figure 1. doi:10.1371/journal.pone.0019627.); Major group Xenocarida). (D) Cephalocarida. (This file is made available under the Creative Commons CC0 1.0 Universal Public Domain Dedication.); Major group Xenocarida. (E) Copepoda; harpacticoid copepod. (NOAA.); Major group Vericrustacea. (F) Decapoda; *Eumunida picta* the squat lobster. (NOAA.); Major group Vericrustacea. (G) *Cirripedia* sp. (NOAA.); Major group Vericrustacea.

of maxillae diagnostic of the group (Scholtz and Edgecombe, 2006). All of these cephalic appendages may vary among taxa. For example, in cladocerans the second pair of antennae is locomotor and has lengthened considerably (Fryer, 1987) and in wood lice, terrestrialization has led to a strong reduction in A1 (Schmalfuss, 1998).

The rest of the body is sometimes made up of a trunk as in remipedes (Yager, 1981, Neiber, 2011) but in other lineages there are two distinct tagmata. They can be generically called as thorax and abdomen, although there are sometimes specific names used for the anatomy in some lineages. It is quite possible that these tagmata are not homologous between the different lineages (Averof and Akam, 1995).

The number of body segments is stable within a lineage such as Malacostraca (Richter and Scholtz, 2001). It is also stable in copepods and thecostracans. However, in the latter, the abdomen seems to have disappeared in barnacles (Darwin, 1851) and may only be rudimentary in the larvae (Gibert et al., 2000). On the other hand, in branchiopods or remipedes the number of body segments varies. In *Triops* (Branchiopoda) the number of segments changes even within a population (Korn and Hundsdoerfer, 2016).

Another diagnostic characteristic of crustaceans is the presence of biramous appendages. There is in fact an external branch or exopodite and an internal branch or endopodite. These appendages may also bear expansions on the external and internal side. The functions of its appendages are multiple and show high adaptive diversity (Boxshall, 2004). The morphological modifications are more pronounced in some parasites. Pentastomids have an elongated worm-like body ringed with two pairs of hooks. These animals parasitize the respiratory tracts of vertebrates and are really far from the "idea" of a crustacean. They have been classified in many groups such as Tardigrada, Annelida, Platyhelminthes, and Nematoda before finally being related to more classical ectoparasitic crustaceans Branchiura (Riley et al., 1978; Lavrov et al., 2004). In the case of the *Sacculina*, the internal part resembles a kind of root that penetrates the crab and profoundly modifies its behavior. The external part is a kind of sac that contains the ovaries and the living males in hyper-parasite. The integration with the other barnacles was possible with the larvae which did not evolve much (Delage, 1884; Høeg, 1995).

REPRODUCTION AND DEVELOPMENT

Reproduction is always performed through gametes but the modalities are highly variable in crustaceans. The majority of species are gonochoric but there are many cases of parthenogenesis in freshwater ostracods (Butlin et al., 1998), brine shrimp *Artemia* (Bowen et al., 1978), some wood lice (Christensen, 1983), and in at least one crayfish (Scholtz et al., 2003). Cyclic parthenogenesis (parthenogenesis alternating sexual reproduction) is a rule in Cyclestherida and Cladocerans (Olesen, 2014; Hebert, 1987).

There are also many cases of simultaneous hermaphroditism (both sexes are present simultaneously) especially in some Branchiopoda (Scanabissi and Mondini 2002; Weeks et al. 2014), Remipedia (Neiber et al., 2011), Cephalocarida (Addis et al., 2012), and Cirripedia (Charnov 1987). Among the Malacostraca, there is little unisexuality but many cases of sequential hermaphroditism (sex change during life; Benvenuto and Weeks, 2020). More rarely, androdioecia (hermaphroditic hermaphrodiosis cohabiting with males) is also reported in branchiopods (Sassaman and Weeks, 1993; Zierold et al., 2007) or cirripedes (Benvenuto, C., Weeks, S.C., 2020). *Eulimnadia*'s androdioecia has been generalized to a number of species and this has been interpreted as a key innovation that has allowed this group to colonize the world (Bellec and Rabet, 2016).

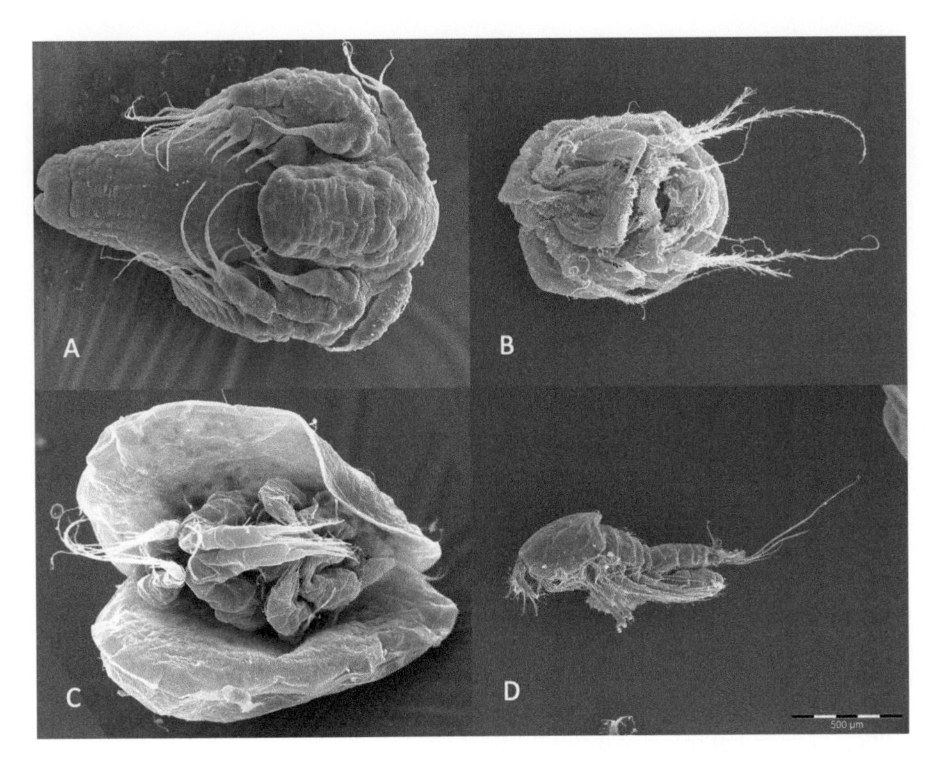

Figure 26.3 Free nauplius stage of three crustaceans: (A) Branchiopoda (*Artemia franciscana*); (B) Copepoda (*Tigriopus brevicornis*); (C) Ostracode (*Heterocyrpris incongruens*); (D) Adult stage of *Tigriopus brevicornis*.

The mating modalities are extremely varied and one can retain the originality of barnacles which have a very long penis compensating for the immobility of the adults and also the capacity to leave their sperm to the waves (Barazandeh et al., 2013). The modalities of embryonic development vary according to the different lineages. Currently, the emerging study model is *Parhyale hawaiensis* (Browne et al., 2005).

Larval development begins ancestrally with a free nauplius stage (**Figure 26.3**) typically consisting of three pairs of appendages and a nauplius eye. This larva is a synapomorph of Pancrustacea (Regier et al., 2010). In many groups, the addition of segments occurs progressively from a subterminal region (Copf et al., 2003). In many Malacostraca, an embryo with a nauplius-like appearance appears transiently in the embryo reminiscent of ancestral development (Scholtz, 2002; Jirikowski et al., 2013). Embryonated egg supporting dehydration (resting egg) have appeared in several lines of continental crustaceans (i.e., Copepoda, Ostracoda, and Branchiopoda). This key innovation in the success of crustaceans living in instable environment existed since at least the Devonian period (Brendonck, 2008; Gueriau et al., 2016).

DISTRIBUTION AND ECOLOGY

All the major lineages are present in marine environments and it is also there that the number of species is the highest (**Figure 26.4**). About 15% of all crustaceans live in fresh or brackish water. The main lineages are ostracods, malacostracans, copepods, and branchiopods (Thorp and Rogers, 2011). The land is less favorable to crustaceans but in the phylogenetic sense the divergent hexapods are a lineage belonging to the same clade and which dominates the aerial world. Among crustaceans *sensu stricto*, isopods with wood lice are the most frequent in land.

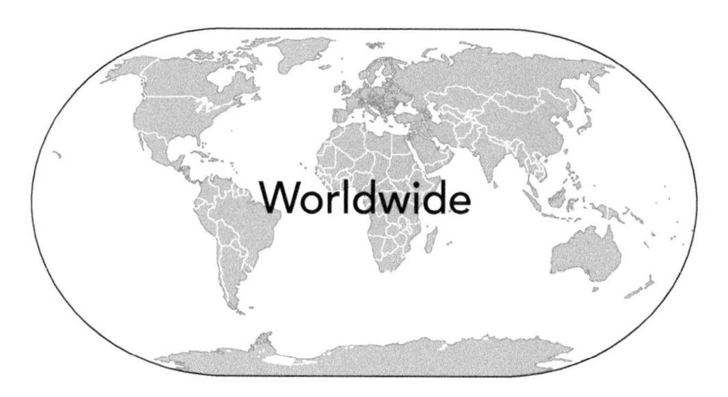

Figure 26.4 Map showing the worldwide distribution of Crustacea.

They live in coastal areas, in dark and humid environments but also in arid environments. For example, the genus *Hemilepistus* digs galleries of several tens of centimeters to escape the high temperatures (Edney, 1954). There are also many decapods within Brachyura and also Anomura, among which are the largest land-living arthropods like the coconut crab (*Birgus latro*) (Krieger, et al., 2010).

Terrestrial humid land has also been colonized by jumping Amphipods of the Talitridae family. They only occur in the Indo-Pacific region and the southern hemisphere (Friend and Richardson, 1986). Somewhat more marginally, the Ostracoda group has also put its appendages on land by living in the litter and moss of tropical rainforests in the southern hemisphere (Harding, 1953; Pinto et al., 2008).

Continental supersalty aquatic environments have also been colonized by ostracods and copepods but the extreme champion of the salinity remains the branchiopod *Artemia* which can survive in supersaturated salty environments (340 g/L; Gajardo and Beardmore, 2012). Except for the branchiopods, phylogenies indicate that colonization of the continental environment has been repeated many times within the great pancrustacean lineages. Cave environments and marine or continental sediments have also been colonized by all the main lineages with incredible diversity and are also refuges for groups such as remipedes and cephalocarids (Neiber et al., 2011; Sanders, 1955).

Crustaceans are ecologically dominant in the plankton of which they are a permanent component (copepods, krill in the sea, some cladocerans in fresh water) or temporary environs for the larvae. This last point is essential because the dispersion of species in the ocean environment is largely ensured by this planktonic phase for both marine and sometimes continental adults (terrestrial or amphidromous decapods).

PHYSIOLOGY AND BEHAVIOR

The number of studies on the physiology and behavior of crustaceans is considerable, but given the diversity of the group, much remains unknown. In small species, breathing takes place through the tegument, but the larger species have gills optimized for aquatic life. When comparing the respiratory patterns of crabs, the gills shorten when they move from aquatic to terrestrial environments, with an 85% reduction in surface area lost between *Callinectes* and *Geocarcinus*. In addition, finger-like projections between gill leaflets for support in air are present in aerial species (O'Mahoney and Full, 1984). In the wood lice, the exopodites of pleopods may have air-filled respiratory tubules called pseudotracheae that function as respiratory organ and are called pleopodal lung. The structure of these organs appears to be a determining factor in the ecology of different species (Unwin, 1931; Csonka et al., 2013). The physiology and behavior of crustaceans is extremely varied. In extreme environments, the shrimp *Rimicaris exoculata*, from 1,700–4,000 m deep in the Atlantic, form high-density populations on the wall of the hydrothermal vent. In order to feed, this species houses filamentous bacteria under its shell with which they establish symbiotic relationships transferring nutrient compounds directly through its tegument without passing through the digestive tract (Petersen et al., 2010; Ponsard et al., 2013). In temporary ponds or hyper-salted environments, the majority of crustaceans produce resistant eggs that survive for years and are used for the dispersion of species (Brendonck et al., 2008). During this suspended phase of life, no metabolism is recorded (Lavens and Sorgeloos, 1987).

Another originality among the crustaceans is that only the remipedes are venomous (von Reumont et al., 2014). The behavior of certain crustaceans also deserves our attention. Christmas Island knows crabs invasion due to the mass migration of animals during the egg-laying season (Adamczewska and Morris, 2001). It is also necessary to report the presence of surprising weapons in some Malacostraca. Due to extreme speeds of these strikes underwater, cavitation occurs between appendages in Alpheid shrimps and stomatopods causing localized phenomena of extreme violence (Patek and Caldwell, 2005; Lohse et al., 2001). Cases of eusociality have recently been reported in the Alpheid shrimp in addition to those already known in insects and vertebrates (Duffy, 1996).

GENETICS

Formal genetic experiments have been carried out on the branchiopod *Artemia* but despite interesting results (Bowen, 1965; Bowen et al., 1966), the species has never become a true model as simple and powerful as *Drosophila melanogaster*. The development of sequencing methods has made it possible to use pieces of DNA sequence to better identify species. In crustaceans the cytochrome c oxidase subunit I (COI) gene amplified by PCR is used as "DNA barcode." Since 2005, this method has been used specifically in the malacostracan decapods and also in the branchiopods. For the moment, the origin of the sequenced species and the groups represented are very uneven (Raupach and Radulovici, 2015; Weigand et al. 2019).

New sequencing methods (NGS) provide much more sequences at a lower cost and DNA fragment to reconstruct genome and transcriptome. These sequences are used to produce phylogenies or to deepen knowledge of species of interest. The mitochondrial genome is relatively easy to reconstitute and it is then possible to compare the mitogenomes with each other or to use the sequences to make phylogenies more interesting than that obtained with a single mitochondrial gene. As in other lineages, the distribution of mitogenome reorganization events in crustaceans appears to be random but may be very useful at times. For example, the reorganization brought Pentastomida closer to Branchiura (Lavrov et al., 2004). Until now, the control region of the polar copepods of the genus *Calanus* is known to be the longest of the crustaceans (Weydmann et al., 2017).

The first sequenced genome of crustaceans was *Daphnia pulex* (Colbourne et al., 2011). More and more genomes are becoming available, but there is a wide variation in the size of genomes in crustaceans and it appears that marine crustaceans living in cold and constant temperature waters have much larger genomes than others and will therefore be difficult to sequence (Alfsnes et al., 2017).

Developmental genetics were first developed in *Artemia* (Averof and Akam, 1995) and then in barnacle *Sacculina carcini* (Gibert et al., 2000; Rabet et al., 2001). Today Malacostraca *Parhyale hawaiensis* is a very promising model for understanding development, regeneration, and immunity. The genome has been sequenced and transcriptomes are available. The embryos are accessible and the development is mosaic-like allowing cell ablation. Tools such as gene knockdown, knockout, and transgenesis as well as temporal control of gene expression are available (Browne et al., 2005; Kao et al., 2016; Sun and Patel, 2019).

POSITION IN THE ToL

Today, among crustaceans, there are around 70,000 valid current species distributed in nine major lineages (i.e., Remipedia, Cephalocarida, Malacostraca, Copepoda, Thecostraca, Branchiopoda, Mystacocarida, Branchiura, and Ostracoda) and in nearly 1000 families (Ahyong et al., 2011 and Regier et al., 2010). The Cambrian fossils show that the group is already diverse, with representatives of the Malacostracea (Collette and Hagadorn, 2010) or Branchiopoda (Waloszek, 1993) already found, and that the evolutionary history is therefore really long and probably quite complex with a large number of extinct lineages.

The phylogenetic relationship between crustaceans and hexapods has long been controversial. It appears that hexapods are ultimately a lineage within crustaceans (Regier et al., 2010). As a result, crustaceans are paraphyletic but still Crustacea is a very popular and practical name. The Pancrustacea proposed by Zrzavý and Štys (1997) became the monophyletic group including all crustaceans and hexapods which has since been confirmed now by several studies (Lee et al., 2013; Schwentner et al. 2017).

Pancrustacea is the sister group of the myriapods (**Figure 26.5**) and together they form the clade of the mandibulates. It is characterized by the presence of gold jaws mandibles (Snodgrass, 1938). Mandibulates and chelicerates constitute the phylum Arthropoda which is the most important in number of animal species. They share many synapomorphies like articulated appendages and organization of the cuticle.

DATABASES AND COLLECTIONS

The sampling of crustaceans and their collection is old, and several processes have proceeded according to the times and the objectives. The preparation of the specimens varies if we want to present the animals to the public (dry or included

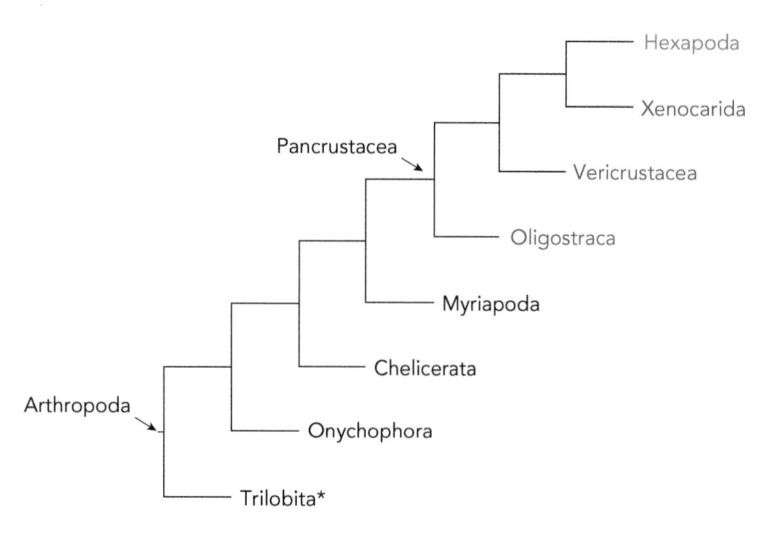

Figure 26.5 Tree showing the relationship of crustacean taxa (Xenocarida, Vericrustacea, and Oligostraca) to other Arthropod taxa. Note that Crustacea (red) are paraphyletic, with Pancrustacea begin defined by crustacean taxa with Hexapoda. Trilobita* an extinct taxon is shown as the most basal arthropod. (DOI: 10.1046/j.1420-9101.1997.10030353.x.)

* mean extinct taxon

in a resin) or if we want to preserve the colors and size (formalin) or even if we want to do molecular analyses (ethanol). Recent recommendations on sampling and conservation not only include an ethical and legal component (Martin, 2016) but also the specificities of each group to achieve a modern taxonomy (Martin et al., 2016; Watling, 2016; Yager, 2016; Ng, 2017).

Many museums around the world devote part of their collection to crustaceans. It is obviously not possible to list all of them here. The most important collections are probably that of the Smithsonian National Museum of Natural History (https://naturalhistory.si.edu/research/invertebrate-zoology), the British Natural History Museum (https://www.nhm.ac.uk/our-science/collections/zoology-collections/crustacean-collection.html), and the French Muséum national d'Histoire naturelle (https://www.mnhn.fr/en/collections/collection-groups/marine-invertebrates/decapod-and-non-decapod-crustaceans). Most of important institutions have databases associated with collections making it easier to find specimens and schedule visits or loans.

There are websites that are generalist and contain valuable information on species names, many of which relate to crustaceans such as the World Register of Marine Species (WoRMS; http://www.marinespecies.org/) or for gene sequences the American National Center for Biotechnology Information (NCBI; https://www.ncbi.nlm.nih.Gov/).

In addition, there are personal or "society" websites that provide information, sometimes coupled with a newsletter or/and a scientific journal. The Crustacean Society (TCS) is intended to be international and cover the whole world (http://www.thecrustaceansociety.org/). There are geographically specialized societies such as the Carcinological Society of Japan (https://csjenglish.webnode.jp/) or the Carcinological Society of Brazil (http://crustacea.org.br/en/). There is also a mass of information in specialized group websites such as for Ostracods irgo, International Research Group on Ostracods (https://www.ostracoda.net/index.php/irgo-home) or for copepods the World Association of Copepodologists (http://www.monoculus.org/).

Crustybase (Hyde et al., 2020; https://crustybase.org) and CAT (Nong et al., 2020; http://cat.sls.cuhk.edu.hk/) are two transcriptomics databases that summarize the genome-level tranascriptome data.

CONCLUSION

Darwin was once stunned by an unknown barnacle taken from the shores of southern Chile. These animals were so enigmatic that it took him years to start from this sample and study many species to build a proper systematics (Darwin, 1851; Stott, 2003)! So, yes, the crustacean group is fascinating. It presents a significant specific diversity but above all an incredible morphological and ecological diversity. This diversity translates into a real puzzle in systematics with a real difficulty in establishing strong parenting relationships. As a young student, I could already hear my teachers complaining about the changes that always had to be made in order to get the course on the famous crustaceans right!

What about today? One wonders what there is left to do in this group!

Already there is still work to clarify the systematics at a high level and it will probably take years and analyses with even more genes and species to solve this interesting problem!

On another scale, it is always necessary to collect and study crustaceans of all groups and sizes during naturalist expeditions to find out that there are many species still hiding in the wild. Many of these species are localized and others could be a good indicator of future changes and of course many may well disappear in the coming years.

REFERENCES

Abele, L.G., Kim, W. and Felgenhauer, B.E., 1989. Molecular evidence for inclusion of the phylum Pentastomida in the Crustacea. Molecular Biology and Evolution, 6, 685–691.

Adamczewska, A.M. and Morris, S., 2001, Ecology and behavior of *Gecarcoidea natalis*, the Christmas Island red crab, during the annual breeding migration. Ecology and Evolution. The Biological Bulletin, 200(3), 305–320.

Addis, A., Fabiano, F., Delogu, V. and Carcupino, M., 2012. Reproductive system morphology of *Lightiella magdalenina* (Crustacea, Cephalocarida): functional and adaptive implications, Invertebrate Reproduction and Development. doi: 10.1080/07924259.2012.704408.

Ahyong, S.T., Lowry, J.K. Alonso, M., Bamber, R.N., Boxshall, G.A., Castro, P., Gerken, S., et al., 2011. Subphylum Crustacea Brünnich, 1772. In: Zhang, Z.-Q. (ed.), Animal Biodiversity: An Outline of Higher-Level Classification and Survey of Taxonomic Richness.

Alfsnes, K., Leinaas, H.P. and Hessen, D.O., 2017. Genome size in arthropods; different roles of phylogeny, habitat and life history in insects and crustaceans. Ecology and Evolution, 7, 5939–5947. doi: https://doi.org/10.1002/ece3.3163.

Averof, M. and Akam, M., 1995. Hox genes and the diversification of insect and crustacean body plans. Nature, 376, 420–423.

Barazandeh, M., Davis, C.S., Neufeld, C.J., Coltman, D.W. and Palmer, A.R., 2013. Something Darwin didn't know about barnacles: spermcast mating in a common stalked species. Proceedings of the Royal Society B, 280, 20122919. doi: http://dx.doi.org/10.1098/rspb.2012.2919.

Bellec, L. and Rabet, N., 2016. Dating of the Limnadiidae family suggests an American origin of Eulimnadia. Hydrobiologia, 773(1), 149–161.

Benvenuto, C. and Weeks, S.C., 2020. Hermaphroditism and gonochorism. In: The Natural History of the Crustacea: Reproductive Biology. Oxford University Press.

Bowen, S.T., 1965. The genetics of *Artemia salina*. V. Crossing over between the X and Y chromosomes. Genetics, 52(3), 695–710.

Bowen, S.T., Durkin, J.P., Sterling, G. and Clark, L.S., 1978. Artemia hemoglobins: genetic variation in parthenogenetic and zygogenetic populations. Biological Bulletin, 155(2), 273–287. doi: 10.2307/1540952.

Bowen, S.T., Hanson, J., Dowling, P. and Poon, M.-C., 1966. The genetics of *Artemia salina*. VI. Summary of mutations. Biological Bulletin. doi: 10.2307/1539753.

Bowman, T.E. and Abele, L.G., 1982. Classification of the recent Crustacea. In: Abele, L.G. (ed.), Systematics, the Fossil Record, and Biogeography. Vol. I, The Biology of Crustacea. In: Bliss, D.E. (ed.). Academic Press, New York, pp. 1–27.

Boxshall, G.A., 2004. The evolution of arthropod limbs. Biological Reviews, 79(2), 253–300. doi: 10.1017/s1464793103006274.

Brendonck, L., Rogers, D.C., Olesen, J., Weeks, S. and Hoeh, W.R., 2008. Global diversity of large branchiopods (Crustacea: Branchiopoda) in freshwater. Hydrobiologia, 595(1), 167–176. doi: 10.1007/978-1-4020-8259-7_18.

Browne, W.E., Price, A.L., Gerberding, M. and Patel, N.H., 2005. Stages of embryonic development in the amphipod crustacean, *Parhyale hawaiensis*. Genesis, 42(3), 124–149. doi: 10.1002/gene.20145.

Brünnich, M.T., 1772. Zoologiae fundamenta praelectionibus academicis accommodata: Grunde i Dyrelaeren. Hafniae et Lipsiae: Apud Frider. Christ. Pelt.

Butlin, R., Schön, I. and Martens, K., 1998. Asexual reproduction in nonmarine ostracods. Heredity, 81, 473–480.

Charnov, E.L., 1987. Sexuality and hermaphroditism in barnacles: a natural selection approach. In: Southward, A.J. (ed.), Barnacle Biology, Crustacean Issues 5. A. A. Belkema, Rotterdam, pp. 89–103.

Christensen, B., 1983. Genetic variation in coexisting sexual diploid and parthenogenetic triploid *Trichoniscus pusillus*. Hereditas, 98, 201–207. doi: https://doi.org/10.1111/j.1601-5223.1983.tb00594.x.

Colbourne, J.K., Pfrender, M.E., Gilbert, D., Thomas, W.K., Tucker, A., Oakley, T.H., Tokishita, S., et al., 2011. The ecoresponsive genome of *Daphnia pulex*. Science, 331(6017), 555–561. doi: 10.1126/science.1197761.

Collette, J.H. and Hagadorn, J.W., 2010. Early evolution of Phyllocarid Arthropods: phylogeny and systematics of Cambrian-Devonian archaeostracans. Journal of Paleontology, 84(5), 795–820. doi: https://doi.org/10.1666/09-092.1.

Copf, T., Rabet, N., Celniker, S.E. and Averof, M., 2003. Posterior patterning genes and the identification of a unique body region in the brine shrimp *Artemia franciscana*. Development, 130(24), 5915–5927.

Csonka, D., Halasy, K., Szabó, P., Mrak, P., Strus, J. and Hornung, E., 2013. Eco-morphological studies on pleopodal lungs and cuticle in *Armadillidium* species (Crustacea, Isopoda, Oniscidea). Arthropod Structure and Development, 42(3), 229–235. doi: https://doi.org/10.1016/j.asd.2013.01.002.

Darwin, C., 1851. A Monograph on the Sub-class Cirripedia, the Lepadidae or Pedunculated Cirripedes. Ray Society, London.

Delage, Y., 1884. Evolution de la sacculine (Sacculina carcini Thompson), Crustacé endoparasite de l'ordre nouveau des Kentrogonides. Archives de zoologie expérimentale et générale, 2e série, II, 417–736.

Duffy, E., 1996. Eusociality in a coral-reef shrimp. Nature, 381(6582), 512–514. doi: 10.1038/381512a0.

Edney, E.B., 1954. Woodlice and the land habitat. Biological Reviews, 29(2), 185–219. doi: 10.1111/j.1469-185X.1954.tb00595.x.

Friend, J.A. and Richardson, A.M.M., 1986. Biology of terrestrial amphipods. Annual Review of Entomology, 31, 25–48. doi: 10.1146/annurev.en.31.010186.000325.

Fryer, G.F.L.S., 1987. A new classification of the branchiopod Crustacea. Zoological Journal of the Linnean Society, 91(4), 357–383. doi: https://doi.org/10.1111/j.1096-3642.1987.tb01420.x.

Gajardo, G. and Beardmore, J.A., 2012. The brine shrimp *Artemia*: adapted to critical life conditions. Frontiers in Physiology, 3, 185.

Gibert, J.M. and Mouchel-Vielh, E., Quéinnec, E. and Deutsch, J.S., 2000. Barnacle duplicate *engrailed* genes: divergent expression patterns and evidence for a vestigial abdomen. Evolution and Development, 2(4), 194–202.

Gueriau, P., Rabet, N., Clément, G., Lagebro, L., Vannier, J., Briggs, D.E.G., Charbonnier, S., Olive, S. and Béthoux, O., 2016. A 365-million-year-old freshwater community reveals morphological and ecological stasis in branchiopod crustaceans. Current Biology, 26(3), 383–390.

Harding, J.P., 1953. The first known example of a terrestrial ostracod, *Mesocypris terrestris* sp. Nov. Annals of the Natal Government Museum, 12, 359–365.

Hebert, P.D.N., 1987. Genotypic characteristics of the Cladocera. Hydrobiologia, 145, 183–193.

Høeg, J.T., 1995. The biology and life cycle of the Rhizocephala (Cirripedia). Journal of the Marine Biological Association of the UK, 75(3), 517–550.

Hyde, C.J., Fitzgibbon, Q.P., Elizur, A., Smith, G.G. and Ventura, T., 2020. CrustyBase: an interactive online database for crustacean transcriptomes. BMC Genomics, 21(1), 1–10.

Jirikowski, G.J., Richter, S. and Wolff, C., 2013. Myogenesis of Malacostraca – the "egg-nauplius" concept revisited. Frontiers in Zoology, 10, 76.

Kao, D., Lai, A.G., Stamataki, E., Rosic, S., Konstantinides, N., Jarvis, E., Di Donfrancesco, A., et al., 2016. The genome of the crustacean *Parhyale hawaiensis*, a model for animal development, regeneration, immunity and lignocellulose digestion. Elife, 5, e20062.

Koenemann, S. and Iliffe, T.M., 2013. Class Remipedia. In: von Vaupel Klein, J.C., Charmantier-Daures, M., Schram, F.R. (eds.), Treatise on Zoology – Anatomy, Taxonomy, Biology. The Crustacea. Brill.

Korn, M. and Hundsdoerfer, A.K., 2016. Molecular phylogeny, morphology and taxonomy of Moroccan *Triops granarius* (Lucas, 1864) (Crustacea: Notostraca), with the description of two new species. Zootaxa, 4178(3), 328–346.

Krieger, J., Sandeman, R.E., Sandeman, D.C., Hansson, B.S. and Harzschl, S., 2010. Brain architecture of the largest living land arthropod, the Giant Robber Crab *Birgus latro* (Crustacea, Anomura, Coenobitidae): evidence for a prominent central olfactory pathway? Frontiers in Zoology, 7, 25.

Lavens, P. and Sorgeloos, P., 1987. The cryptobiotic state of Artemia cysts, its diapause deactivation and hatching: a review. In: Sorgeloos, P., Bengtson, D.A., Decleir, W. and Jaspers, E. (eds.), Artemia Research and Its Applications (3). Ecology, Culturing, Use in Aquaculture, 556 p. Universa Press, Wetteren, Belgium, pp. 27–63.

Lavrov, D.V., Brown, W.M. and Boore, J.L., 2004. Phylogenetic position of the Pentastomida and (pan)crustacean relationships. Proceedings of the Royal Society B: Biological Sciences, 271, 537–544.

Lee, M.S.Y., Soubrier, J. and Edgecombe, G.D., 2013. Rates of phenotypic and genomic evolution during the Cambrian explosion. Current Biology, 23, 1889–1895.

Linné, C. von, 1758. Caroli Linnaei Equitis De Stella Polari, Archiatri Regii, Med. & Botan. Profess. Upsal.; Systema Naturae Per Regna Tria Naturae, Secundum Classes, Ordines, Genera, Species, Cum Characteribus, Differentiis, Synonymis. Locis, Holmiae.

Lohse, D., Schmitz, B. and Versluis, M., 2001. Snapping shrimp make flashing bubbles. Nature, 413(6855), 477–478. doi: 10.1038/35097152.

Martin, J.W., 2016. Collecting and processing Crustaceans: an introduction. Journal of Crustacean Biology, 36(3), 393–395. doi: https://doi.org/10.1163/1937240X-00002436.

Martin, J.W. and Davis, G.E., 2001. An updated classification of the recent Crustacea. Natural History Museum of Los Angeles County Contributions in Science, 39, 1–124.

Martin, J.W., Rogers, D.C. and Olesen J., 2016. Collecting and processing Branchiopods. Journal of Crustacean Biology, 36(3), 396–401. doi: https://doi.org/10.1163/1937240X-00002434

Neiber, M.T., Hartke, T.R., Stemme, T., Bergmann, A., Rust, J., Iliffe, T.M. and Koenemann S., 2011. Global biodiversity and phylogenetic evaluation of Remipedia (Crustacea). PLoS One, 6(5), e19627. doi: https://doi.org/10.1371/journal.pone.0019627.

Ng, P.K.L., 2017. Collecting and processing freshwater shrimps and crabs. Author Notes Journal of Crustacean Biology, 37(1), 115–122. doi: https://doi.org/10.1093/jcbiol/ruw004.

Nong, W., Chai, Z.Y.H., Jiang, X., Qin, J., Ma, K.Y., Chan, K.M., Chan, T.F., et al., 2020. A crustacean annotated transcriptome (CAT) database. BMC Genomics, 21(1), 1–5.

Olesen, J., 2014. Cyclestherida. In: Joel W. Martin, Jørgen Olesen, Jens T. Høeg (eds.), Atlas of Crustacean Larvae. Johns Hopkins University Press.

O'Mahoney, P.M. and Full, R.J., 1984. Respiration of crabs in air and water. Comparative Biochemistry and Physiology Part A: Physiology, 79(2), 275–282.

Patek, S.N. and Caldwell, R.L., 2005. Extreme impact and cavitation forces of a biological hammer: strike forces of the peacock mantis shrimp *Odontodactylus scyllarus*. Journal of Experimental Biology, 208, 3655–3664. doi: 10.1242/jeb.01831.

Petersen, J.M., Ramette, A., Lott, C., Cambon-Bonavita, M.-A., Zbinden, M. and Dubilier, N., 2010. Dual symbiosis of the vent shrimp *Rimicaris exoculata* with filamentous gamma- and epsilon proteobacteria at four Mid-Atlantic Ridge hydrothermal vent fields. Environmental Microbiology, 12(8), 2204–2218. doi: 10.1111/j.1462-2920.2009.02129.x.

Pinto, R.L., Rocha, C. and Martens, K., 2008. On the first terrestrial ostracod of the Superfamily Cytheroidea (Crustacea, Ostracoda): description of *Intrepidocythere ibipora* n. gen. n. sp. from forest leaf litter in São Paulo State, Brazil. Zootaxa, 828, 29–42.

Ponsard, J., Cambon-Bonavita, M.-A., Zbinden, M., Lepoint, G., Joassin, A., Corbari L., Shillito B., Durand L., Cueff-Gauchard, V. and Compère, P., 2013. Inorganic carbon fixation by chemosynthetic ectosymbionts and nutritional transfers to the hydrothermal vent host-shrimp *Rimicaris exoculata*. The International Society for Microbial Ecology Journal, 7, 96–109.

Rabet, N., Gibert, J.M.,Quéinnec, E., Deutsch, J.S. and Mouchel-Vielh, E., 2001. The *caudal* gene of the barnacle *Sacculina carcini* is not expressed in its vestigial abdomen. Development Genes and Evolution, 211(4), 172–178.

Raupach, M.J. and Radulovici, A.E., 2015. Looking back on a decade of barcoding crustaceans. Zookzeys, 539, 53–81.

Regier, J.C., Shultz, J.W., Zwick, A., Hussey, A., Ball, B., Wetzer R., Martin, J.W. and Cunningham C.W., 2010. Arthropod relationships revealed by phylogenomic analysis of nuclear protein-coding sequences. Nature, 463(7284), 1079–1083. doi: 10.1038/nature08742.

Richter, G. and Scholtz, G., 2001. Phylogenetic analyses of the Malacostraca (Crustacea). Journal of Zoological Systematics and Evolutionary Research, 39(3), 113–136. doi: 10.1046/j.1439-0469.2001.00164.x.

Riley, J., Banaja, A.A. and James, J.L., 1978. The phylogenetic relationships of the Pentastomida: The case for their inclusion within the crustacea. International Journal for Parasitology, 8(4), 245–254.

Sanders, H.L., 1955. The Cephalocarida, a new subclass of Crustacea from Long Island Sound. Proceedings of National Academy of Sciences of the United States of America, 41(1), 61–66. doi: 10.1073/pnas.41.1.61.

Scanabissi, F. and Mondini, C., 2002. A survey of the reproductive biology in Italian branchiopods. Hydrobiologia, 486(1), 263–272. doi: 10.1023/A:1021371306687.

Schmalfuss, H., 1998. Evolutionary strategies of the antennae in terrestrial Isopods. Journal of Crustacean Biology, 18(1), 10–24. doi: 10.1163/193724098X00025Corpus.

Sassaman, C. and Weeks, S., 1993. The genetic mechanism of sex determination in the Conchostracan shrimp *Eulimnadia texana*. American Naturalist, 141(2), 314–328. doi: 10.1086/285475.

Scholtz, G., 2002. Evolution of the nauplius stage in malacostracan crustaceans. Journal of Zoological Systematics and Evolutionary Research,38(3),175–187.doi:10.1046/j.1439-0469.2000.383151.x.

Scholtz, G., Braband, A., Tolley, L., Reimann, A., Mittmann, B., Lukhaup, C., Steuerwald F. and Vogt, G., 2003. Parthenogenesis in an outsider crayfish. Nature, 421, 806.

Scholtz, G. and Edgecombe, G.D., 2006. The evolution of arthropod heads: reconciling morphological,developmental and palaeontological evidence. Development Genes and Evolution, 216, 395–415.

Schwentner, M., Combosch, D.J., Pakes, J.N. and Giribet, G., 2017. A phylogenomic solution to the origin of insects by resolving Crustacean-Hexapod relationships. Current Biology, 27(12), 1818–1824. https://doi.org/10.1016/j.cub.2017.05.040.

Snodgrass, R.E., 1938. Evolution of the Annelida, Onychophora and Arthropoda. Smithsonian Miscellaneous Collections, 97(6), 1–159.

Stott, R., 2003. Darwin and the Barnacle – The story of one tiny creature and history's most spectacular scientific breakthrough, 309 p.

Sun, D.A., Patel, N.H., 2019. The amphipod crustacean *Parhyale hawaiensis*: An emerging comparative model of arthropod development, evolution, and regeneration. WIREs Developmental Biology 8:e355.

Thorp, J.H., Rogers, D.C., 2011 Introduction to Freshwater Invertebrates in the Phylum Arthropoda in « Field Guide to Freshwater Invertebrates of North America.

Unwin, E.E., 1931. On the structure of the respiratory organs of the terrestrial Isopoda. Papers and Proceedings of the Royal Society of Tasmania, 37–104.

von Reumont, B.M., Blanke, A., Richter, S., Alvarez, F., Bleidorn, C. and Jenner, R.A., 2014. The first venomous Crustacean revealed by transcriptomics and functional morphology: Remipede venom glands express a unique toxin cocktail dominated by enzymes and a neurotoxin. Molecular Biology and Evolution, 31(1), 48–58. doi: 10.1093/molbev/mst199.

Waloszek, D., 1993. The Upper Cambrian Rehbachiella and the phylogeny of Branchiopoda and Crustacea. Fossils and Strata, 32(4), 1–202. doi: 10.1111/j.1502-3931.1993.tb01537.x.

Watling, L., 2016. Collecting and processing Bathynellaceans, Anaspidaceans,Spelaeogriphaceans,andThermosbaenaceans. Journal of Crustacean Biology, 36(3), 402–404. https://doi.org/10.1163/1937240X-0000243.

Weeks, S.C., Brantner, J.S., Astrop, T.I., Ott, D.W., Rabet, N., 2014. The evolution of hermaphroditism from dioecy in Crustaceans: selfing hermaphroditism described in a fourth *Spinicaudatan* genus. Evolutionary Biology, 41, 251–261.

Weigand, H., Beermann, A.J., Čiampor, F., Costa, F.O., Csabai, Z., Duarte, S., Geiger, M.F., et al., 2019. DNA barcode reference libraries for the monitoring of aquatic biota in Europe: gap-analysis and recommendations for future work. Science of the Total Environment 678, 499–524. doi: 10.1016/j.scitotenv.2019.04.247.

Weydmann, A., Przyłucka, A., Lubośny, M., Walczyńska, K.S., Serrão E.A., Pearson, G.A. and Burzyński, A., 2017. Mitochondrial genomes of the key zooplankton copepods Arctic *Calanus glacialis* and North Atlantic *Calanus finmarchicus* with the longest crustacean non-coding regions. Scientific Reports, 7, 13702. doi: https://doi.org/10.1038/s41598-017-13807-0.

Yager, J., 1981. A new class of Crustacea from a marine cave in the Bahamas. Journal of Crustacean Biology, 1(3), 328–333.

Yager, J., 2016. Collecting and processing remipedes. Journal of Crustacean Biology, 36(3), 405–407. doi: https://doi.org/10.1163/1937240X-00002433.

Zierold, T., Hänfling, B. and Gómez, A., 2007. Recent evolution of alternative reproductive modes in the 'living fossil' Triops cancriformis. BMC Evolutionary Biology, 7(1), 161.

Zilhão, J., Angelucci, D.E., Araújo Igreja, M., Arnold, L.J., Badal, E., Callapez, P., Cardoso, J.L., et al., 2020. Last interglacial Iberian Neandertals as fisher-hunter-gatherers. Science, 367. doi: 10.1126/science.aaz7943.

Zrzavý, J. and Štys, P., 1997. The basic body plan of arthropods: insights from evolutionary morphology and developmental biology. Journal of Evolutionary Biology, 10(3), 353–367.

Zucker, A., 2005, Aristote et les classifications zoologiques. Peeters.

SUBPHYLUM HEXAPODA

Rob DeSalle

CHAPTER
27

Infrakingdom	Bilateria
Division	Protostome
Subdivision	Ecdysozoa
Phylum	Arthropoda
Subphylum	Hexapoda

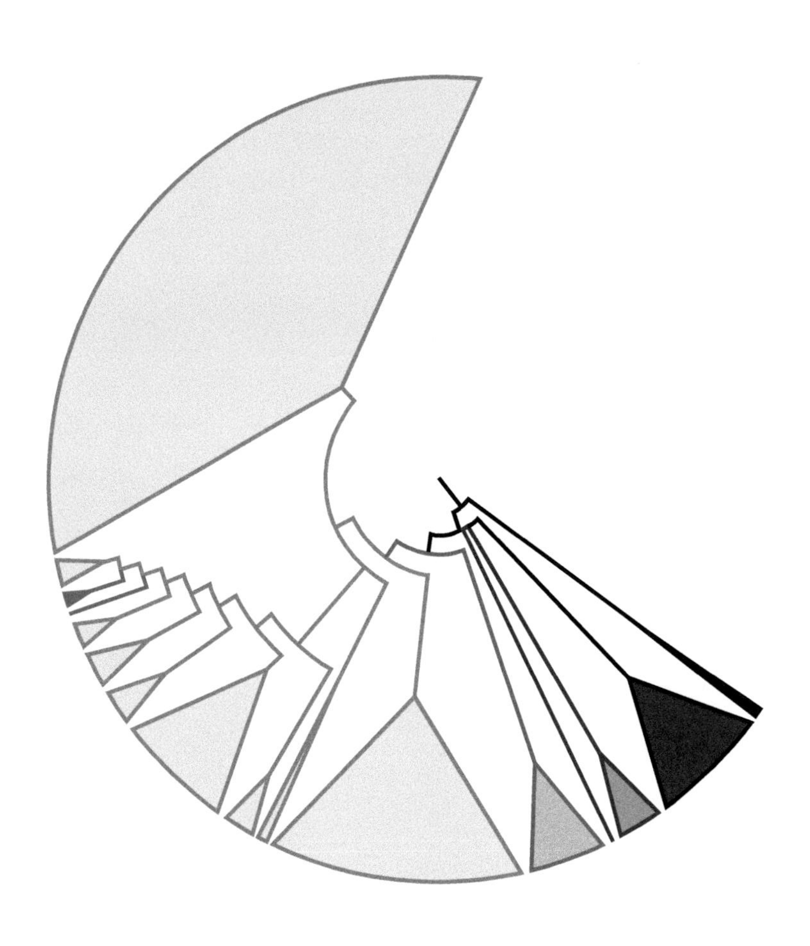

CONTENTS

General Characteristics

There are nearly 1 million named species, 85,000 genera, 1,600 families, and 29 orders in the subphylum

Bilaterally symmetric animals

Mostly terrestrial but also found in freshwater and marine environments

They have complex life cycles – with holometabolous and hemimetabolous cycles being the norm

Considerable economic value as many insects are pest species. Considerable health issues as many insects are vectors of human infectious disease. Insects are of considerable importance in forensics too

Ecological importance is high in that some species are used as biomonitors (caddisflies etc.)

Synapomorphies (After Kristensen, 1975; 1981)

- lack of musculature beyond the first segment of antenna
- Johnston's organ in pedicel (second segment) of antenna. This organ is a collection of sensory cells that detect movement of the flagellum.
- a transverse bar forming the posterior tentorium inside the head
- tarsi are subsegmented
- females with ovipositor formed by gonapophyses from segments 8 and 9
- annulated, terminal filament extending out from end of segment 11 of abdomen (subsequently lost in most groups of insects)

INTRODUCTION: HISTORY AND SYSTEMATICS

Subphylum Hexapoda

There are four groups of Hexapoda – Collembola, Protura, Diplura, and Insecta. The phylogenetic relationships (Sasaki et al., 2013) of the extant groups of Hexapoda are shown in **Figure 27.1** and examples of the four groups are shown in **Figure 27.2**.

Protura: These animals are on average 2 mm or less in length. They hide quite well in the soil and were mostly missed by zoologists up until recently. They were at once thought to be an order of insects, but due to phylogenetic analysis they have now been removed from Insecta. There are around 800 known species in three families.

Diplura: This group contains about 800 species in 11 families (2 of which are extinct). Some Diplura can be as long as 50 mm, but most are around 10 mm in length. They are wingless and as their name indicates, have two tails or caudal filaments at the posterior end of the body They were also once thought to be an order of Insecta, but are now considered as a different group.

Collembola: These are also known as springtails. There are over 6000 species in 16 families. They are in general tiny (less than 6 mm) and wingless.

Insecta

The group Insecta is a member of the subphylum Hexapoda (Figure 27.1), one of four recognized extant subphyla within the phylum Arthropoda (Myriapoda, Hexapoda, Chelicerata, and Crustacea), which also includes one extinct subphylum (Trilobita). Along with Diplura, Protura, and Collembola

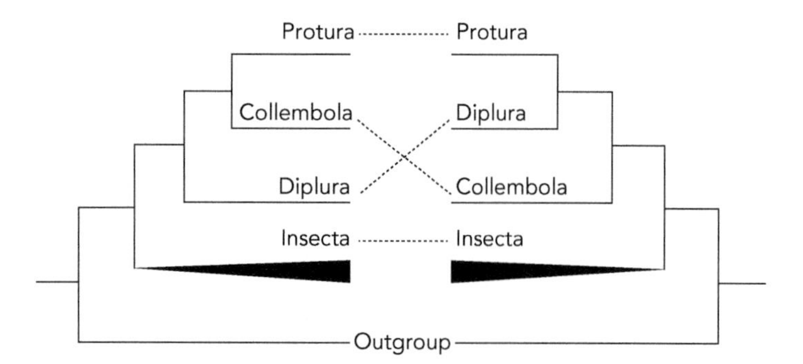

Figure 27.1 Phylogenetic hypotheses for the Hexapoda. The tree on the left is from classical morphological analysis (Hennig, 1981) and the one on the right is from nuclear DNA phylogenomic analysis (Meusemann et al., 2010). Another large phylogenomic dataset shows Protura and Collembola as sister taxa. (Misof et al., 2014.)

Figure 27.2 Photographs of the four groups within the Hexapoda. (A) Diplura (*Campodea staphylinus*, attribution by Michel Vuijlsteke); (B) Collembola (*Willowsia nigromaculata*, attribution by Cody Hough, college student and photographer in the Michgian area); (C) Insecta (*Antherophagus* sp., attribution by Sam Droege from Beltsville, USA; (D) Protura, attribution Sikander Kiani, Invertebrate Zoology Division, Yale Peabody Museum.

(springtails), the Insecta are one of the most speciose groups of animals on the planet with nearly 1 million named species and a large number of species yet to be discovered. The history of insect study is deep and goes all the way back to Aristotle who described dozens of insect species (Weiss, 1929). Insects have been important across the ages and their systematization really began with Rene-Antoine Ferchault de Reamur's six-volume tome entitled *Memoirs Serving as a History of Insects* (*Mémoires pour servir à l'histoire des insects*). This work was followed by Carolus Linnaeus who described about 1900 species (Usinger, 1964; Winsor, 1976; Cannings and Scudder, 2001) in the 10th edition of *Systema Naturae*. Post Linnaeus era saw an immense amount of work on natural history, taxonomy, and biology of insects. Today insects are the group with the most formally described species on this planet making up around half of all formally described and named organisms (Smith and Kennedy, 2009; Stork, 2018).

The intense work on insects has resulted in one of the most extensive lists of species for any group (**Table 27.1**; Stork, 2018). According to Stork (2018), there are 29 orders of insects (**Figure 27.3**) and their phylogenetic relationships have been examined over the past half century with the most prominent early work coming from Will Hennig (summarized in Dupuis, 1984; Hennig, 1981; Kristensen, 1995). Other significant resources for summaries of insect systematics come from Kristensen (1981; 1995), Grimaldi (2001), Grimaldi and Engel (2005), Carpenter (1992), Larink (1997), Willmann (2004), Wheeler et al. (2001), and more recently Stork (2018), Kjer et al. (2016a,b), Wipfler et al. (2016) Johnson (2019), Yeates et al. (2016), Beutel et al. (2017), Singh et al. (2017), Giribet and Edgecombe (2019), Chesters (2019), Johnson et al. (2018), Wipfler et al. (2019), and Misof et al. (2014). According to Misof et al. (2014), the orders can be grouped into four major higher groups Holometabola, Condylognatha, Polyneoptera, and Palaeoptera. This scheme leaves Psocodes, Zygentoma, and Archaeognatha as single orders. Other higher order schemes exist and the student is encouraged to examine the hypotheses in the references mentioned above (**Figure 27.4**).

TABLE 27.1 Number of extant species in the orders of Insecta

Order	Species
Archaeognatha	513
Zygentoma	560
Ephemeroptera	3240
Odonata	5899
Orthoptera	23,855
Neuroptera	5868
Phasmatodea	3014
Embioptera	463
Grylloblattodea	34
Mantophasmatodea	20
Plecoptera	3743
Dermaptera	1978
Zoraptera	37
Mantodea	2400
Blattodea	7314
Psocoptera	5720
Phthiraptera	5102
Thysanoptera	5864
Hemiptera	103,590
Hymenoptera	116,861
Strepsiptera	609
Coleoptera	386,500
Megaloptera	354
Raphidioptera	254
Trichoptera	14,391
Lepidoptera	157,338
Diptera	155,477
Siphonaptera	2075
Mecoptera	757

Figure 27.3 **Photographs of representatives of the 29 phyla of insects.** (A) Archaeognatha (*Petrobius brevistylis*; PD, JungleDragon); (B) Zygentoma (*Ctenolepisma longicaudatum*; PD, JungleDragon); (C) Ephemeropta (PD, needpix); (D) Odonata (*Helocordulia uhleri*; PD, USDA); (E) Orthoptera (*Anabrus simplex*; PD, USDA); (F) Neuroptera (*Chrysopa carnea*; PD, Peter Häger, CC0 Public Domain); (G) Phasmatodea (*Phasmatodea* sp., PD; Maky Orel, CC0 1.0 Universal Public Domain Dedication). (H) Plecoptera (Chloroperlidae, PD, EPA); (I) Embioptera (*Oligotoma saundersii*, PD; Free Art License, Dean Rider, Jr.); (I) Grylloblattodea (*Grylloblatta* sp., PD, Alex Wild, Creative Commons CC0 1.0 Universal Public Domain Dedication). (K) Mantodea (Unidentified Mantodea; PD, Emőke Dénes); (L) Blattodea (*Periplaneta americana*, PD, Insects Unlocked, Creative Commons CC0 1.0 Universal Public Domain Dedication); (M) Psocoptera (*Dorypteryx domestica*; Arp, Creative Commons CC0 1.0 Universal Public Domain Dedication); (N) Phthiraptera (*Pediculus humanus*; PD, CDC). (O) Zoraptera (*Zorotypus hubbardi*; David Maddison-CC BY 3.0); (P) Dermaptera (*Titanolabis colossea*, PD. Fritz Geller-Grimm); (Q) Thysanoptera (*Rohrthrips schizovenatus*, PD, Ulitzka, 2019, Creative Commons CC0 1.0 Universal Public Domain Dedication); (R) Hemiptera (*Podisus maculiventris*; PD, USDA); (S) Hymenoptera (*Euglossini* sp., PD, https://www.goodfreephotos.com); (T) Strepsiptera (*Xenos vesparum*, PD, https://www.dreamstime.com/photos-images/strepsiptera.html); (U) Coleoptera (Buprestidae sp., PD, Insects Unlocked, A. Santillana); (V) Megaloptera (*Sialis lutaria*, Attribution:©entomart); (W) Raphidioptera (*Raphidia ophiopsis* PD, Attribution: olaru andreea); (X) Lepidoptera (*Actiasl luna*, PD, NPS); (Y) Diptera (*Erebomyia exalloptera*, USFWS, public domain); (Z) Trichoptera (Caddisfly, USDA/Public Domain); (AA) Siphonaptera (Siphonaptera sp., PD, Daniel J. Drew); (BB) Mecoptera (*Panorpa rosemaria*, PD, USDA); (CC) Mantophasmatodea (*Mantophasma zephyra*, attribution: P.E. Bragg; licensed under the Creative Commons Attribution-Share Alike 3.0 Unported license: https://creativecommons.org/licenses/by-sa/3.0/deed.en).

MORPHOLOGY AND ANATOMY

Insects are a morphologically diverse group, but with many recognizable characteristics. They are bilaterally symmetric, segmented, and have an exoskeleton made of chitin (**Figure 27.5**). Their general body plan has three specialized tagmata: head, thorax, and abdomen. The head is very complex with much of its developmental biology worked out in the model organism *Drosophila melanogaster*. Also much of the development of the segmental organization of the thorax and the abdomen has been worked out for this model organism. Of course, what is true for *Drosophila* is commonly NOT true for other insects. The thorax has three partitions: prothorax, mesothorax, and metathorax from anterior to posterior. Each of these segments has a symmetrical pair of legs extending from the segment. The mesothorax and metathorax segments both have wings in all orders of winged insects except for Diptera (so named because they have only two wings coming from their mesothorax) and Strepsiptera (two wings extending from their metathorax). On the third thoracic segment in Diptera and the second segment in Strepsiptera, a balancing structure called a pair of halteres has derived from reduced wings. The terminology for these segments is as follows: the dorsal part of each segment is called the notum (or tergum), the regions on the side of the animal are called pleura, and the ventral side of the animal is called the sternum.

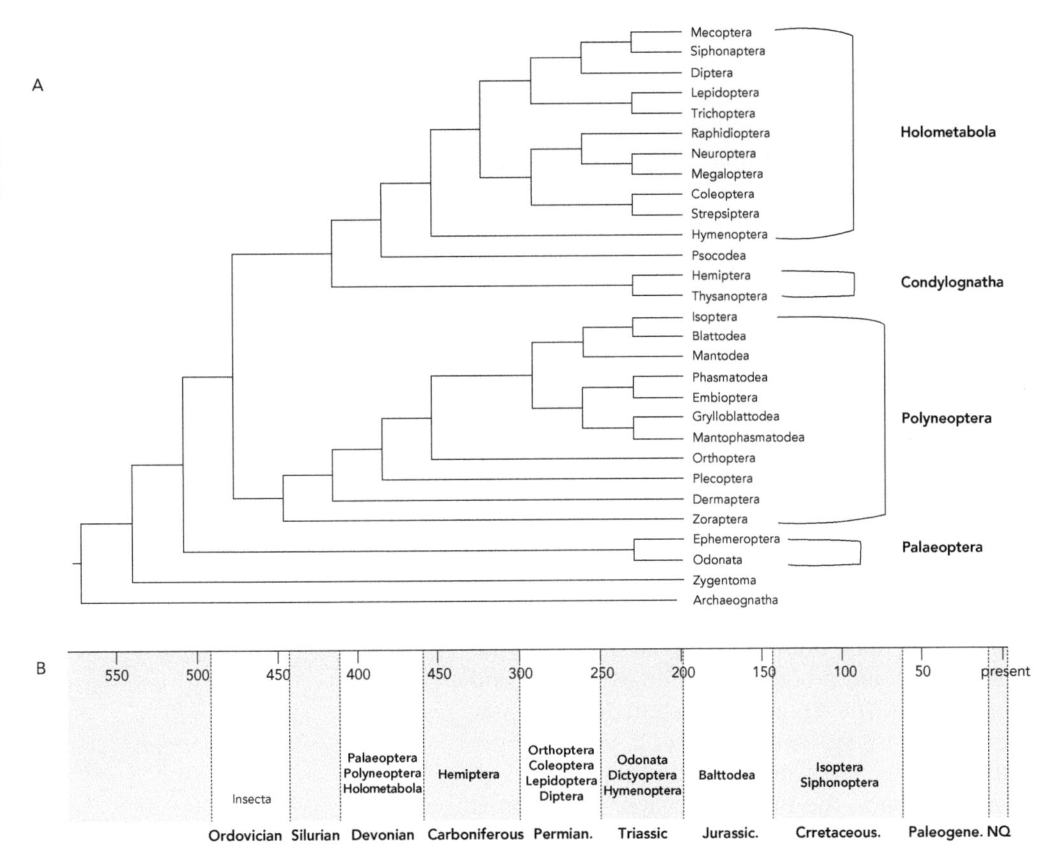

Figure 27.4 Phylogenetic tree of the 29 orders of insects from Misof et al., 2014. (A) The tree with higher groups Holometabola, Conylognatha, Polyneuroptera, and Palaeoptera marked on the right. (B) Time frame for the origin of 15 of the orders of insects. (From Misof et al., 2014.)

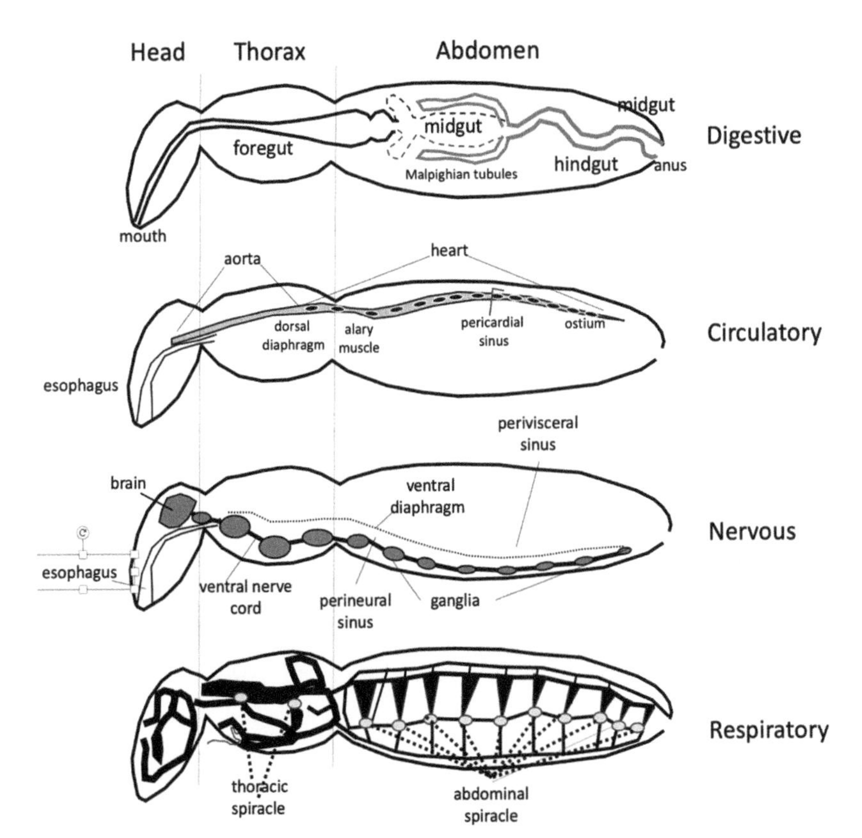

Figure 27.5 Generalized internal anatomy of digestive, circulatory, respiratory, and nervous systems of insects. The nervous system shows just the ventral ganglia; smaller ganglia emanate from the ganglia to enervate the typical insect body.

The abodomen is primitively 11 abdominal segments with varying degrees of reduction or fusion of segments as the group diverged. The abdomen harbors most of the major organ systems (digestive, excretory, and respiratory). The external structure of the insect abdomen is usually slightly sclerotized with spiracles for respiration located on the side of the animal. Variations on these generalities exist.

The basic internal anatomy of insects

The internal anatomy of insects is complex, but we can generalize some aspects of the body plan (Figure 27.5). The digestive system runs the length of the body with the mouth connected to a foregut. This digestive organ then joins the midgut, continues to the hindgut and finally the anus. Terrestrial insects have structures called Malphigian tubules that perform excretory and osmoregulatory functions. The nervous system usually consists of a brain in the anterior head region and a series of ganglia that run the length of the body (Smarandache-Wellmann, 2016). Note that the nervous system is ventral, and the circulatory and digestive systems are dorsal like in other protostomes. This is the typical protostome anatomy and opposite of the deuterostome body plan. The evolution of the rather opposite body plans is a vibrant subject (Holland, 2016; Arendt and Nübler-Jung, 1997; Peterson, 1995; Louryan and Vanmuylder, 2018). The idealized circulatory system of insects is open and also runs the length of the body with a large dorsal aorta anteriorly and a body-length heart (Hillyer and Pass, 2019) that circulates the hemolymph. The circulatory system does not play a role in oxygen transport. The respiratory system of insects is aerated by a series of thoracic and abdominal openings (spiracles), leading to a network of inner tubes (tracheae) that deliver oxygen to the internal organs. Excretion is modulated by the Malphigian tubules which remove nitrogenous waste from the hemolymph. In addition, waste is dumped directly into the alimentary canal at the beginning of the hindgut.

LIFE CYCLE, DEVELOPMENT, AND REPRODUCTION

Insects can directly develop or go through larval and pupal stages (a process called holometabolism; **Figure 27.6**). It appears that holometabolous insects are monophyletic. While hemimetabolism, or direct development through nymph stages, is ancestral in the insects and characterizes a larger number of the orders of insects.

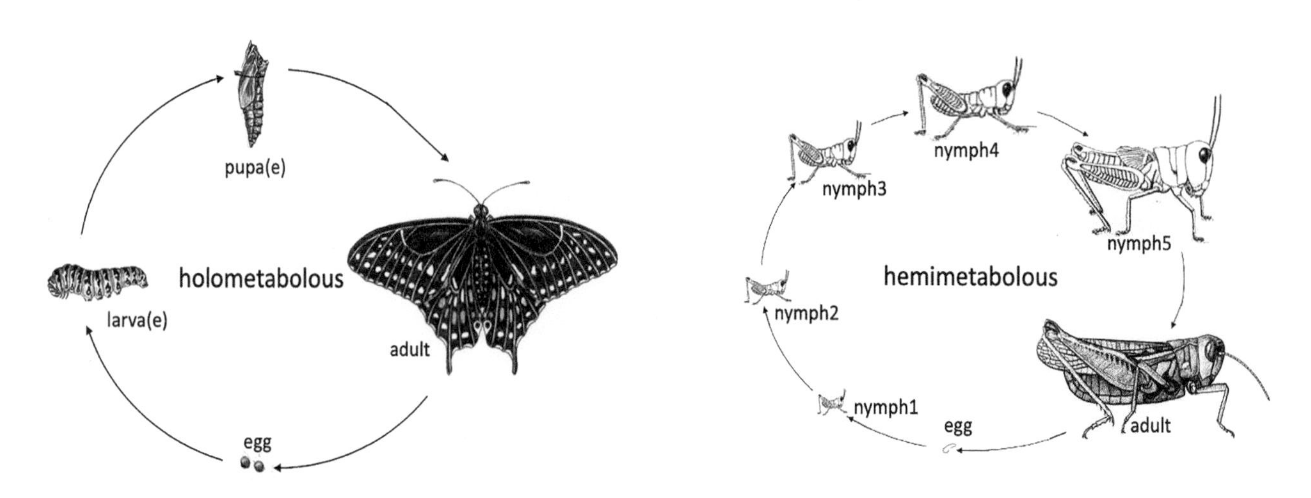

Figure 27.6 Holometabolous and hemimetabolous life cycles of insects. Butterfly larvae shown here are also known as caterpillars; butterfly pupae shown here are also known as chrysalis. Each arrow between the nymph stages and between nymph5 and adult are called molts.

The developmental biology of insects has been facilitated by intense work on the model organism – *D. melanogaster*. This dipteran has been used for over a century as a genetic, behavioral, neurobiological, and developmental study system. While *D. melanogaster* is a highly derived insect species, it nonetheless has been used to mine the genome for developmental genes and other important biological and biochemical pathways in insects (Perry et al., 2019). Sex determination is an important aspect of the developmental biology of insects and has been examined most intensively in the context of holometabolous insects (Mesa and De Mesa, 1967; Sánchez, 2004; Verhulst et al., 2010). Hemimetabolous insects have also been examined (Blackman et al., 1995; Ashman et al., 2014) and a wide array of genetic sex determination mechanisms exist across the insects. The most common mechanism of sex determination is that of male heterogamety XX/XY where the presence of a Y chromosome determines maleness in the organism. Species with an XX/XO system have lost their Y chromosome entirely so that males are determined by the presence of only one X chromosome (i.e., males are XO and females XX). Less abundant mechanisms are sex determination by haplodiploidy (males result from unfertilized haploid eggs) or female heterogamety, but also several other sex determining mechanisms are distributed across the insects (for a review; see Blackman et al., 1995; Beukeboom, 2005).

The list of reproductive strategies or sex-determining systems among insects is highly variable. The three major categories are systems with diploid males (diplodiploid), effective haploid males (haplodiploid), no males (thelytoky) and rarely hermaphroditism. The normal reproductive mode of insects is sexual between diploid males and females (diplodiploidy). In haplodiploid, reproduction is where the males are haploid (as a result of developing without fertilization) and the females are diploid (developed from eggs that were inseminated). This mode of reproduction occurs mostly in social hymenopterans. Parthenogenesis (a form of thelytoky) is a process where females reproduce without fertilization by a male and is common in aphids. While hermaphroditism is relatively common in groups closely related to Insecta like Crustacea, it is incredibly rare in insects (Normark, 2003). However, some reports of hermaphrodites, gynandromorphs (individuals with both male and female characteristics, sometimes one half of the individual will show male characteristics and the other half female) in insects do exist (Gardner and Ross, 2011; Gullan and Kosztarab, 1997; Langton, 2008). A very interesting and recent development in the study of sex determination and reproductive strategies is the discovery that endosymbionts may have a role in the process. (non-genetic, i.e., modificatory sex determination; Ma et al., 2014; Ma and Schwander, 2017; Duplouy and Hornett, 2018; Treanor et al., 2018).

In general, the reproductive system in female insects has a symmetrical pair of ovaries, an accessory gland, and one or a pair of spermathecae (tubes or sacs where sperm is stored). The ovaries consist of tubules known as ovarioles where the eggs sit. The number of ovarioles is highly variable depending on the species being examined. This system allows for some versatility and control by the female during the fertilization process as the accessory glands regulate the release of sperm through the spermathecae. The male insect reproductive system consists of a pair of testes, usually suspended in a fat body, where sperm are produced and stored. Male genitalia are often complex and bear a rich source of morphological characters for systematics.

DISTRIBUTION AND ECOLOGY

Insect distribution is worldwide (**Figure 27.7**). They are found on all seven continents and in terrestrial and freshwater and marine environments (Stork, 2018) as well as passively drifting animals in the air (Glick, 1939). Estimating

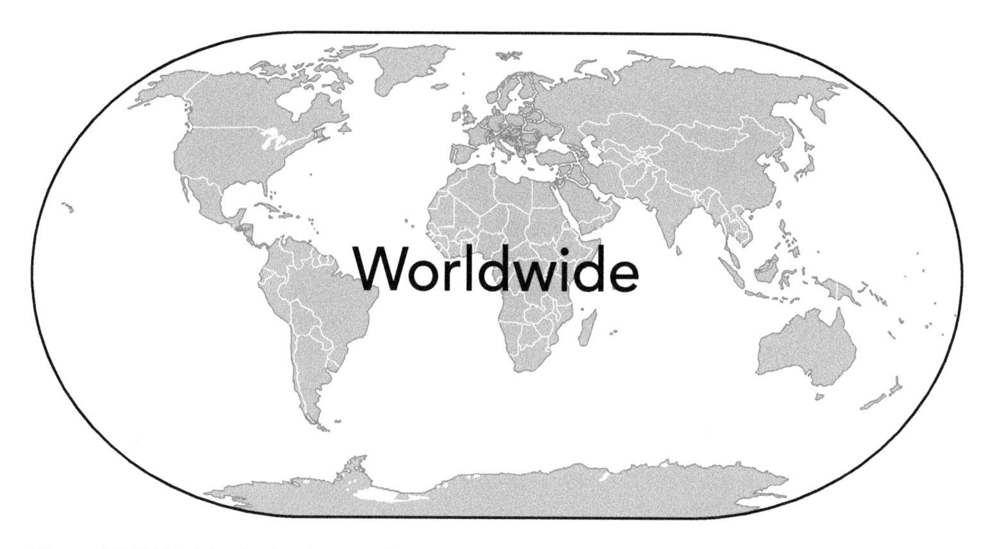

Figure 27.7 Worldwide distribution of Insecta.

the number of species of any group on the planet is a tricky business and many researchers have developed ways to make these estimates. Such estimates for insects can be made from the number of vascular plants in a biogeographic region as the insects are highly dependent on vascular host plants (**Table 27.2**). Because insects have such diverse distributions, ranging from the tropics to the polar regions, their ecological habits are also diverse. Many also endure varying ecological constraints and show a broad altitudinal distribution. They also endure varying ecological constraints because of their diverse altitudinal distribution (McCoy, 1990). Many insects are ecologically important organisms and play major roles (e.g., in pollinator systems; Ehrlich and Raven, 1964; Janz, 2011; Faheem et al., 2004). With respect to species interactions, insect fungal interactions have been the subject of recent work (Biedermann and Vega, 2020) as well as urban ecology (Brown 2018). Finally, global climate change (Sable and Rana, 2016) has been shown to impact the distribution of species in this group. Insects are well-investigated models in behavioral ecology (Burk, 1988; Kalinkat et al., 2015), understanding

TABLE 27.2 Estimates of the number of insect species based on number of vascular plants in a biogeographic region. Beetles are included as they are the most speciose order of insects. Antarctic has a single endemic species (Usher and Edwards, 1984) – *Belgica antarctica* (a chironomid) – along with some tick and mite species that migrate with birds

Region	Vascular plants	Insects	Beetles
Australasia	52,728	720,521	196,515
Afrotropical	71,363	975,179	265,971
Central America	45,000	614,918	167,713
Indo-Malayan	53,774	734,822	200,416
North America	8453	115,503	31,502
Neotropics	118,577	1,620,348	441,935
Oceanic	14,249	194,706	53,104
Palearctic	38,358	524,165	142,961
Total	402,500	5,500,163	1,500,118

From Stork (2018).

of invasiveness (Kenis et al., 2009) and are involved in many different kinds of species interactions including parasitism, commensalism, mutualism, and predator–prey interactions.

PHYSIOLOGY AND BEHAVIOR

Insects have been instrumental in the understanding of many physiological systems such as vision (Land, 1997; Warrant and Dacke, 2011; Briscoe and Chittka, 2001; Egelhaaf and Kern 2002), olfaction (Hallem et al., 2006; Hansson and Stensmyr, 2011; Fleischer et al., 2018), nociception (pain reception; Eisemann et al., 1984; Tiffin, 2016), hearing (Hoy and Robert, 1996; Goepfert and Hennig, 2016), excretion (Weiss, 2006), and many other physiological processes including pheromone biology (Shorey, 1973; Yew and Chung, 2015; Stökl and Steiger, 2017).

The degree of behavioral variation among insects is also staggering. Social behavior is a major subject in the behavioral realm (Matsuura, 2020; Dukas, 2008; Weitekamp et al., 2017; Fischman et al., 2011; Jandt and Gordon, 2016; Wheeler, 2015; Brian, 2012). Camouflage and mimicry can also be considered animal behavioral strategies and these have been studied extensively in insects. Mimicry systems are abound in the Lepidoptera (Zhang et al., 2017; Punnett, 2016; Turner, 1977; Shih et al., 2019), but it is also acute in other insects orders (Vanitha, 2019; Rettenmeyer, 1970; Sheikh et al., 2017). Mimicry is a strategy, often used for defense, whereby one species evolves to appear similar in appearance to another species (or object), thereby pretending to be (e.g.) poisonous, dangerous, or even harmless to gain an advantage. The best known example is that of a harmless species mimicking a harmful one (Batesian mimicry). Mimicry can be an interaction between closely related species, for example, in many butterfly mimicry systems or between very distant species such as mantis mimicking plant flowers. Examples of camouflage through coloration or morphological changes are abundant across insects and studied intensively in a wide range of insects (Pérez-de la Fuente et al., 2012; Dettner and Liepert, 1994; Guerra-Grenier 2019). This behavior is a defense strategy against predators as the camouflaged individual can blend into the background to avoid visual recognition by predators. Chemical defense in insects has also been examined in detail (Laurent et al., 2005; Howard and Blomquist, 1982; Bowers, 1992; Shorter and Rueppell, 2012). Chemical defenses of insects are generally discussed in the context of plant–insect interactions. In addition, some insect chemical defense is inter- and intraspecific. Some social insects use chemicals to influence other species or individuals within their species. The investigation of interaction of insects with microbes is also a more recent development (Biedermann and Vega, 2020; Boucias et al., 2018; Gurung et al., 2019; Blow and Douglas, 2019; Dillon and Dillon, 2004; Weiss and Aksoy, 2011; Lewis and Lizé, 2015; Gil and Latorre, 2019). This work has demonstrated intricate (mostly endosymbiotic) interactions between bacteria and fungi within the guts of insects. Depending on the insect and microbe species in the interaction, the systems range from mutualism, in which both partners complement each other's needs, to parasitism, in which the endosymbiont exploits the insect host and may have a deep influence on its metabolism or even its behavior.

GENETICS AND GENOMICS

As mentioned above, one of the most used model systems in biology is an insect *D. melanogaster*. This dipteran has been genetically mapped extensively and its genome, as well as those of nearly a hundred congeners, has been sequenced so far. Genome size has also been estimated for a broad range of insects. Some

Coleoptera have tiny genomes in the range of 150 megabases (million bases), while the biggest genomes are over 7,500 megabases in some Orthoptera (Hanrahan and Johnston, 2011). The genome of *D. melanogaster* has lent greatly to our understanding of genome evolution, gene expression, functional genomic elements, and comparative genomics. Other insect species with formal genetic systems exist but are rare.

Whole genome sequencing, genome reduction approaches, and RNA-seq have all been used to generate genomic information about insects. A recent review of insect whole genomes (Li et al., 2019) summarizes the information on numbers of genomes sequenced or assembled. This is an important number because it indicates the quantity and kinds of insect genomes that have been worked on. Probably more important than the number of assemblies is the number of annotations that exist for a single taxon. Annotated genomes are much more valuable than simple assemblies because they offer the researcher an immediate and readily analyzed dataset, while assemblies require a large amount of preparatory work. Perhaps of equal importance to understanding the scope of insect genomics is the number of BioProjects that exist for insect genomics. According to the NIH, a BioProject "is a collection of biological data related to a single initiative, originating from a single organization or from a consortium"; so this means that if a BioProject exists for a group, a genome assembly can most likely be added to that taxon. The number of BioProjects gives the upper boundary of number of genomes in a group that have either been assembled or in the pipeline for genome production. **Table 27.3** shows the distribution of BioProjects (BP), assemblies (As), and annotated (An) genomes in insects.

Transcriptomics or RNA-seq have been used extensively to generate genome level data (Oppenheim et al., 2015). A more recent summary of insect transcriptomics is the 1KITE (1K Insect Transcriptome Evolution; https://www.1kite.org/) consortium which has generated more than 1,000 transcriptomes for evolutionary and systematic studies.

POSITION IN THE ToL

The position of insects in the tree of life has been of particular interest to phylogenomicists (Whiting, 2004; Meusemann et al., 2010). Insects are considered a group within a subphylum we discussed in the beginning of this chapter called Hexapoda (consisting of Insecta, Protura, Diplura, and Collembola). Hexapods are one of five major groups that make up the phylum Arthropoda (the extant groups Hexapoda, Myriapoda, Chelicerata, Crustacea, and one extinct group Trilobita). Some authors (Giribet and Edgecombe, 2012; 2015; 2019) suggest that Chelicerata comprises a group called the Euchelicerata plus Pycnogonida. Arthropoda are an ecdysozoan phylum that are in a higher group called Panarthropoda (includes Arthropoda, Tardigrada, and Oncychophora).

In one view Hexapoda and Crustacea are sister taxa (**Figure 27.8**). Another scenario based on the fossil record is that Hexapoda originate within Crustacea (Giribet and Edgecombe 2012 and 2019), and together make up the (monophyletic) taxon Pancrustacea.

DATABASES AND COLLECTIONS

Vast valid insect collections can be found at any accredited museum and taxonomy websites for insects and other hexapods is given in **Table 27.4**. In essence, insect collections are most likely going to be the largest collection in any natural history museum with respect to specimens and species (SCAN; Symbiota

TABLE 27.3 BioProjects (BP), assembled genomes (As), and annotated genomes (An) for the 29 orders of Insecta. (Modified from Li et al., 2019.)

Order	BP	As	An
Archaeognatha	1	1	0
Zygentoma	3	0	0
Ephemeroptera	2	2	0
Odonata	5	2	0
Orthoptera	20	3	0
Neuroptera	5	0	0
Phasmatodea	6	4	0
Embioptera	7	0	0
Grylloblattodea	1	0	0
Mantophasmatodea	9	0	0
Plecoptera	4	3	0
Dermaptera	4	0	0
Zoraptera	4	1	0
Mantodea	78	20	0
Blattodea	16	4	3
Psocoptera	4	3	0
Phthiraptera	244	1	1
Thysanoptera	57	1	1
Hemiptera	192	30	16
Hymenoptera	137	30	47
Strepsiptera	2	1	0
Coleoptera	137	22	15
Megaloptera	28	4	0
Raphidioptera	24	1	0
Trichoptera	5	3	0
Lepidoptera	165	97	25
Diptera	168	132	45
Siphonaptera	1	1	1
Mecoptera	2	0	0

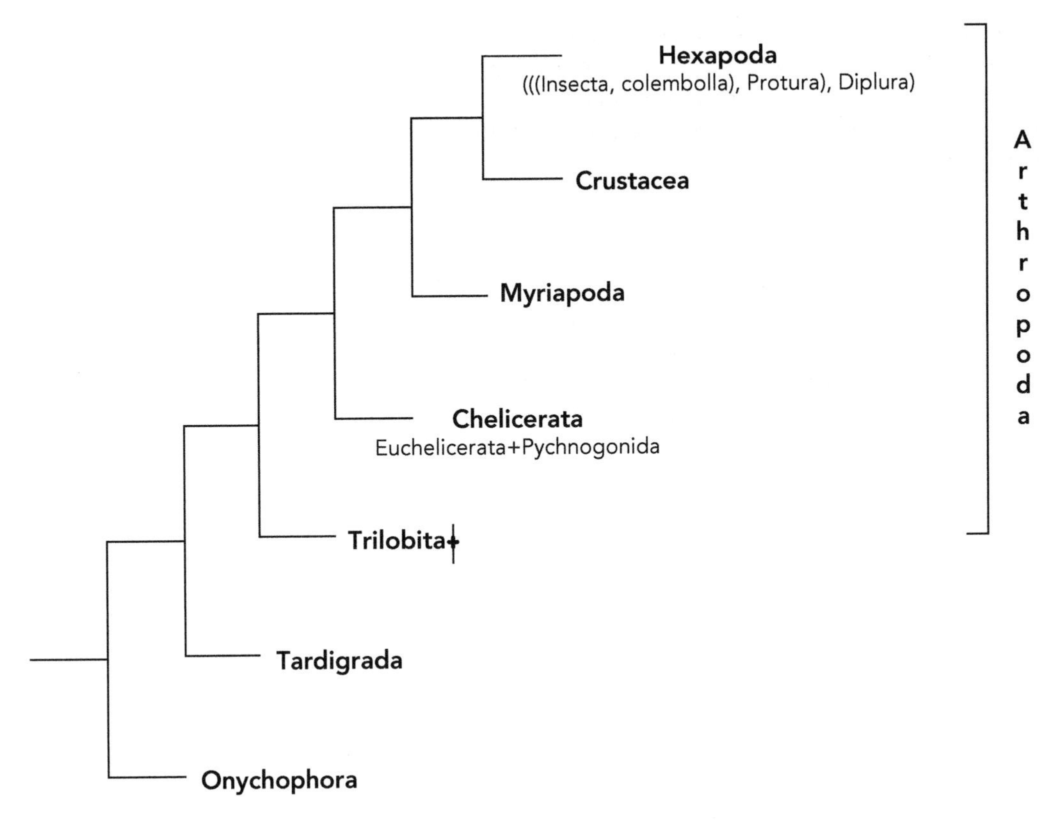

Figure 27.8 **Position of Insecta in the tree of life.** The cross indicates extinct group. Insecta is a group within the subphylum Hexapoda, which in turn is one of the five groups within the Arthropoda.

TABLE 27.4 A rather eclectic list of representative insect-based databases and the websites where they can be accessed

Database	Subject
https://data.world/datasets/insects	Insect datasets
https://instr.iastate.libguides.com/ento/id	Insect identification
https://www.insectidentification.org/insect-identification.asp	Insect identification
https://www.bugs.com/bug-database/	Insect identification
https://www.entsoc.org/common-names	Common names
https://entonation.com/entomological-societys-database-	Common names
https://scan-bugs.org/portal/collections/index.php	Collections
insect-common-names/	
https://www.insectimages.org/	Images
https://datadryad.org/stash/dataset/doi:10.5061/dryad.pv40d2r	Egg size
https://problemsolvedpest.com/bugs-database/	Pests
http://cdfd.org.in/INSATDB/home.php	Microsatellites
http://insect-plant.org/insectindb/insect/search	Chemical ecology and genomics
http://www.brc.ac.uk/dbif/	Host plants
https://irac-online.org/pests/	Insecticide resistance
http://www.insect.glfc.cfs.nrcan.gc.ca/prod/index.cfm?lang=eng	Cultures
http://www.insect-genome.com/	Genomes
http://www1.biologie.uni-hamburg.de/b-online/library/genomeweb/Genome	Genomes
Web/	

Collections of Arthropods Network; https://scan-bugs.org/portal/collections/index.php). Frozen tissue collections will also archive a vast number of specimens and species of insects. We have already mentioned BioProjects at NIH and this web tool can be consulted for the many insect genome and transcriptome projects finished and in progress.

CONCLUSION

The Insecta are probably the single most complex group of invertebrates discussed in this text with respect to systematics and phylogenetics because of their sheer numbers (Table 27.1). A 5,000 word chapter on the Insecta such as this can only cover so much and the student should consult the many textbooks in the literature on insects. Some good introductory texts are by Chapman (1998), Johnson and Triplehorn (2005), Resh and Cardé (2009), Burton and Burton (1975), but perhaps the gold standard is the text by Grimaldi and Engel (2005), *Evolution of the Insects*, which can be consulted for more details on many of the topics discussed in this chapter.

REFERENCES

Arendt, Detlev, and Katharina Nübler-Jung. "Dorsal or ventral: similarities in fate maps and gastrulation patterns in annelids, arthropods and chrodates." Mechanisms of Development 61, no. 1–2 (1997): 7–21.

Ashman, Tia-Lynn, Doris Bachtrog, Heath Blackmon, Emma E. Goldberg, Matthew W. Hahn, Mark Kirkpatrick, Jun Kitano et al. "Tree of sex: a database of sexual systems." Scientific Data 1 (2014): 140015.

Beukeboom, Leo. "Insects and sex." In Proceedings of the Netherlands Entomological Society Meeting, vol. 16, pp. 9–16. Centre for Ecological and Evolutionary Studies, University of Groningen, 2005.

Beutel, Rolf G., Margarita I. Yavorskaya, Yuta Mashimo, Makiko Fukui, and Karen Meusemann. "The phylogeny of Hexapoda (Arthropoda) and the evolution of megadiversity." Proceedings of Arthropodan Embryological Society of Japan 51 (2017): 1–15.

Biedermann, Peter H.W., and Fernando E. Vega. "Ecology and evolution of insect–fungus mutualisms." Annual Review of Entomology 65 (2020).

Blackman, Roger L., S.R. Leather, and J. Hardie. "Sex determination in insects." In Insect Reproduction, pp. 57–94. CRC Press, Boca Raton, 1995.

Blow, Frances, and Angela E. Douglas. "The hemolymph microbiome of insects." Journal of Insect Physiology 115 (2019): 33–39.

Boucias, Drion G., Yonghong Zhou, Shuaishuai Huang, and Nemat O. Keyhani. "Microbiota in insect fungal pathology." Applied Microbiology and Biotechnology 102, no. 14 (2018): 5873–5888.

Bowers, M. Deane. "The evolution of unpalatability and the cost of chemical defense in insects." In Insect Chemical Ecology: An Evolutionary Approach, pp. 216–244. Chapman & Hall, New York, (1992).

Brian, Michael Vaughan. *Social insects: ecology and behavioural biology*. Springer Science & Business Media, 2012.

Briscoe, Adriana D., and Lars Chittka. "The evolution of color vision in insects." Annual Review of Entomology 46, no. 1 (2001): 471–510.

Brown, Brian V. "After "the call": a review of urban insect ecology trends from 2000–2017." Zoosymposia 12, no. 1 (2018): 4–17.

Burk, Theodore. "Insect behavioral ecology: some future paths." Annual Review of Entomology 33, no. 1 (1988): 319–335.

Burton, Maurice, and Robert Burton. *Encyclopedia of insects & arachnids. No. C/595.7003 B8*. Octopus Books, 1975.

Cannings, Roeert A., and Geoffrey G.E. Scudder. "An overview of systematics studies concerning the insect." Journal of Entomological Society of British Columbia 98 (2001): 33.

Carpenter, F.M. "Superclass Hexapoda." In Volumes 3 and 4 of Part R, Arthropoda 4 of Treatise on Invertebrate Paleontology. Geological Society of America, Boulder, Colorado, 1992.

Chapman, Reginald Frederick. *The insects: structure and function*. Cambridge University Press, 1998.

Chesters, Douglas. "The phylogeny of insects in the data-driven era." Systematic Entomology 2019 (2019):1–12.

Dettner, Konrad, and Caroline Liepert. "Chemical mimicry and camouflage." Annual Review of Entomology 39, no. 1 (1994): 129–154.

Dillon, Rod J., and V.M. Dillon. "The gut bacteria of insects: nonpathogenic interactions." Annual Reviews in Entomology 49, no. 1 (2004): 71–92.

Dukas, Reuven. "Evolutionary biology of insect learning." Annual Review of Entomology 53 (2008): 145–160.

Duplouy, Anne, and Emily A. Hornett. "Uncovering the hidden players in Lepidoptera biology: the heritable microbial endosymbionts." PeerJ 6 (2018): e4629.

Dupuis, Claude. "Willi Henning's impact on taxonomic thought." Annual Review of Ecology and Systematics 15, no. 1 (1984): 1–25.

Helen, E. Tiffin. "Do insects feel pain?" Animal Studies Journal 5, no. 1 (2016): 80–96.

Egelhaaf, Martin, and Roland Kern. "Vision in flying insects." Current Opinion in Neurobiology 12, no. 6 (2002): 699–706.

Ehrlich, Paul R., and Peter H. Raven. "Butterflies and plants: a study in coevolution." Evolution 18, no. 4 (1964): 586–608.

Eisemann, C.H., W.K. Jorgensen, D.J. Merritt, M.J. Rice, B.W. Cribb, P.D. Webb, and M.P. Zalucki. "Do insects feel pain? A biological view." Experientia 40, no. 2 (1984): 164–167.

Faheem, Muhammad, Muhammad Aslam, and Muhammad Razaq. "Pollination ecology with special reference to insects a review." Journal of Research (Science) 4 (2004): 395–409.

Fischman, Brielle J., S. Hollis Woodard, and Gene E. Robinson. "Molecular evolutionary analyses of insect societies." Proceedings of the National Academy of Sciences 108, no. Supplement 2 (2011): 10847–10854.

Fleischer, Joerg, Pablo Pregitzer, Heinz Breer, and Jürgen Krieger. "Access to the odor world: olfactory receptors and their role for signal transduction in insects." Cellular and Molecular Life Sciences 75, no. 3 (2018): 485–508.

Gardner, Andy, and Laura Ross. "The evolution of hermaphroditism by an infectious male-derived cell lineage: an inclusive-fitness analysis." American Naturalist 178, no. 2 (2011): 191–201.

Gil, Rosario, and Amparo Latorre. "Unity makes strength: a review on mutualistic symbiosis in representative insect clades." Life 9, no. 1 (2019): 21.

Giribet, Gonzalo, and Gregory D. Edgecombe. "Reevaluating the arthropod tree of life." Annual Review of Entomology 57 (2012): 167–186.

Giribet, Gonzalo, and Gregory D. Edgecombe. "The phylogeny and evolutionary history of arthropods." Current Biology 29, no. 12 (2019): R592–R602.

Glick, Perry A. The distribution of insects, spiders, and mites in the air. No. 1488-2016-124024. 1939.

Goepfert, Martin C., and R. Matthias Hennig. "Hearing in insects." Annual Review of Entomology 61 (2016): 257–276.

Grimaldi, D. "Insect evolutionary history from Handlirsch to Hennig, and beyond." Journal of Paleontology 75 (2001): 1152–1160.

Grimaldi, D. and M. S. Engel. *Evolution of the insects*. Cambridge University Press, 2005.

Guerra-Grenier, Eric. "Evolutionary ecology of insect egg coloration: a review." Evolutionary Ecology 33, no. 1 (2019): 1–19.

Gullan, P. J., and Kosztarab, M. "Adaptations in scale insects." Annual Review of Entomology 42, (1997): 23–50.

Gurung, Kiran, Bregje Wertheim, and Joana Falcao Salles. "The microbiome of pest insects: it is not just bacteria." Entomologia Experimentalis et Applicata 167, no. 3 (2019): 156–170.

Hallem, Elissa A., Anupama Dahanukar, and John R. Carlson. "Insect odor and taste receptors." Annual Review of Entomology 51 (2006): 113–135.

Hanrahan, Shawn Jason, and J. Spencer Johnston. "New genome size estimates of 134 species of arthropods." Chromosome Research 19, no. 6 (2011): 809.

Hansson, Bill S., and Marcus C. Stensmyr. "Evolution of insect olfaction." Neuron 72, no. 5 (2011): 698–711.

Hennig, W. Phylogenetic Systematics. Urbana: Univ. Ill. Press, 1966, 263 pp. Reprinted 1979.

Hennig W. *Insect phylogeny*. John Wiley & Sons, New York, 1981.

Hillyer, Julián F., and Günther Pass. "The insect circulatory system: structure, function, and evolution." Annual Review of Entomology 65 (2020): 121–143.

Holland, Nicholas D. "Nervous systems and scenarios for the invertebrate-to-vertebrate transition." Philosophical Transactions of the Royal Society B: Biological Sciences 371, no. 1685 (2016): 20150047.

Howard, Ralph W., and Gary J. Blomquist. "Chemical ecology and biochemistry of insect hydrocarbons." Annual Review of Entomology 27, no. 1 (1982): 149–172.

Hoy, Ronald R., and D. Robert. "Tympanal hearing in insects." Annual Review of Entomology 41, no. 1 (1996): 433–450.

Jandt, J. M., and D. M. Gordon. "The behavioral ecology of variation in social insects." Current Opinion in Insect Science 15 (2016): 40–44.

Janz, Niklas. "Ehrlich and Raven revisited: mechanisms underlying codiversification of plants and enemies." Annual Review of Ecology, Evolution, and Systematics 42 (2011): 71–89.

Johnson, Kevin P. "Putting the genome in insect phylogenomics." Current Opinion in Insect Science 36 (2019): 111–117.

Johnson, Kevin P., Christopher H. Dietrich, Frank Friedrich, Rolf G. Beutel, Benjamin Wipfler, Ralph S. Peters, Julie M. Allen et al. "Phylogenomics and the evolution of hemipteroid insects." Proceedings of the National Academy of Sciences 115, no. 50 (2018): 12775–12780.

Johnson, Norman F., and Charles A. Triplehorn. *Borror and DeLong's Introduction to the Study of Insects*. Thompson Brooks/Cole, Belmont, CA, 2005.

Kalinkat, Gregor, Malte Jochum, Ulrich Brose, and Anthony I. Dell. "Body size and the behavioral ecology of insects: linking individuals to ecological communities." Current Opinion in Insect Science 9 (2015): 24–30.

Kenis, Marc, Marie-Anne Auger-Rozenberg, Alain Roques, Laura Timms, Christelle Péré, Matthew JW Cock, Josef Settele, Sylvie Augustin, and Carlos Lopez-Vaamonde. "Ecological effects of invasive alien insects." Biological Invasions 11, no. 1 (2009): 21–45.

Kjer, Karl, Marek L. Borowiec, Paul B. Frandsen, Jessica Ware, and Brian M. Wiegmann. "Advances using molecular data in insect systematics." Current Opinion in Insect Science 18 (2016a): 40–47.

Kjer, Karl M., Chris Simon, Margarita Yavorskaya, and Rolf G. Beutel. "Progress, pitfalls and parallel universes: a history of insect phylogenetics." Journal of the Royal Society Interface 13, no. 121 (2016b): 20160363.

Kristensen, N.P. "The phylogeny of hexapod "orders". A critical review of recent accounts." Zeitschrift für zoologische Systematik und Evolutionsforschung 13 (1975): 1–44.

Kristensen, N.P. "Phylogeny of insect orders." Annual Review of Entomology 26 (1981): 135–157.

Kristensen, N. P. "Forty years' insect phylogenetic systematics." Zoologische Beiträge NF 36, no. 1 (1995): 83–124.

Land, Michael F. "Visual acuity in insects." Annual Review of Entomology 42, no. 1 (1997): 147–177.

Langton, Peter H. "A structural hermaphrodite Micropsectra (Diptera: Chironomidae)." Bol. mus. mun. Funchal, sup. N.° 13 (2008): 213–216.

Larink, O. "Apomorphic and plesiomorphic characteristics in Archaeognatha, Monura, and Zygentoma." Pedobiologia 41 (1997): 3–8.

Laurent, Pascal, Jean-Claude Braekman, and Désiré Daloze. "Insect chemical defense." In The Chemistry of Pheromones and Other Semiochemicals II, pp. 167–229. Springer, Berlin, Heidelberg, 2005.

Lewis, Zenobia, and Anne Lizé. "Insect behaviour and the microbiome." Current Opinion in Insect Science 9 (2015): 86–90.

Li, Fei, Xianxin Zhao, Meizhen Li, Kang He, Cong Huang, Yuenan Zhou, Zekai Li, and James R. Walters. "Insect genomes: progress and challenges." Insect Molecular Biology 28, no. 6 (2019): 739–758.

Louryan, S., and N. Vanmuylder. "The dorsoventral inversion: an attempt of synthesis." Morphologie: bulletin de l'Association des anatomistes 102, no. 337 (2018): 122–131.

Ma, W-J., Fabrice Vavre, and Leo W. Beukeboom. "Manipulation of arthropod sex determination by endosymbionts: diversity and molecular mechanisms." Sexual Development 8, no. 1–3 (2014): 59–73.

Ma, W.-J., and T. Schwander. "Patterns and mechanisms in instances of endosymbiont-induced parthenogenesis." Journal of Evolutionary Biology 30, no. 5 (2017): 868–888.

Matsuura, Kenji. "Genomic imprinting and evolution of insect societies." *Population Ecology* 62, no. 1 (2020): 38–52.

McCoy, Earl D. "The distribution of insects along elevational gradients." Oikos (1990): 313–322.

Mesa, A., and R.S. De Mesa. "Complex sex-determining mechanisms in three species of South American grasshoppers (Orthoptera, Acridoidea)." Chromosoma 21, no. 2 (1967): 163–180.

Meusemann K, von Reumont BM, Simon S, Roeding F, Strauss S, Kück P, Ebersberger I, Walzl M, Pass G, Breuers S, Achter V, von Haeseler A, Burmester T, Hadrys H, Wägele JW, and Misof B. "A phylogenomic approach to resolve the arthropod tree of life." Molecular Biology and Evolution 27 (2010): 2451–2464. doi:10.1093/molbev/msq130

Misof, Bernhard, Shanlin Liu, Karen Meusemann, Ralph S. Peters, Alexander Donath, Christoph Mayer, Paul B. Frandsen et al. "Phylogenomics resolves the timing and pattern of insect evolution." Science 346, no. 6210 (2014): 763–767.

Normark, Benjamin B. "The evolution of alternative genetic systems in insects." Annual Review of Entomology 48, no. 1 (2003): 397–423.

Oppenheim, Sara J., Richard H. Baker, Sabrina Simon, and Rob DeSalle. "We can't all be supermodels: the value of comparative transcriptomics to the study of non-model insects." Insect Molecular Biology 24, no. 2 (2015): 139–154.

Pérez-de la Fuente, Ricardo, Xavier Delclòs, Enrique Peñalver, Mariela Speranza, Jacek Wierzchos, Carmen Ascaso, and Michael S. Engel. "Early evolution and ecology of camouflage in insects." Proceedings of the National Academy of Sciences 109, no. 52 (2012): 21414–21419.

Perry, Caitlyn, Jack Scanlan, and Charles Robin. "Mining insect genomes for functionally affiliated genes." Current Opinion in Insect Science 31 (2019): 114–122.

Peterson, K.J. "Dorsoventral axis inversion." Nature 373, no. 6510 (1995): 111–112.

Punnett, Reginald Crundall. *Mimicry in butterflies*. Cambridge University Press, 2016.

Resh, Vincent H., and Ring T. Cardé, eds. *Encyclopedia of insects*. Academic Press, 2009.

Rettenmeyer, Carl W. "Insect mimicry." Annual Review of Entomology 15, no. 1 (1970): 43–74.

Sable, M.G., and D.K. Rana. "Impact of global warming on insect behavior—A review." Agricultural Reviews 37, no. 1 (2016): 81–84.

Sánchez, Lucas. "Sex-determining mechanisms in insects." International Journal of Developmental Biology 52, no. 7 (2004): 837 856.

Sasaki, Go, Keisuke Ishiwata, Ryuichiro Machida, Takashi Miyata, and Zhi-Hui Su. "Molecular phylogenetic analyses support the monophyly of Hexapoda and suggest the paraphyly of Entognatha." BMC Evolutionary Biology 13, no. 1 (2013): 236.

Sheikh, Aijaz Ahmad, N. Z. Rehman, and Ritesh Kumar. "Diverse adaptations in insects: a review." Journal of Entomology and Zoology Studies 5 (2017): 343–350.

Shih, Chungkun, Yongjie Wang, and Dong Ren. "Camouflage, mimicry or eyespot warning." Rhythms of Insect Evolution: Evidence from the Jurassic and Cretaceous in Northern China (2019): 651–665.

Shorey, Harry H. "Behavioral responses to insect pheromones." Annual Review of Entomology 18, no. 1 (1973): 349–380.

Shorter, J. R., and Olav Rueppell. "A review on self-destructive defense behaviors in social insects." Insectes sociaux 59, no. 1 (2012): 1–10.

Singh, Satyapriya, V. K. Mishra, and Tanmaya Kumar Bhoi. "Insect molecular markers and its utility—a review." International Journal of Agriculture, Environment and Biotechnology 10, no. 4 (2017): 469–479.

Smarandache-Wellmann, Carmen Ramona. "Arthropod neurons and nervous system." Current Biology 26, no. 20 (2016): R960–R965.

Smith, Edward H., and George G. Kennedy. "History of entomology." In Encyclopedia of Insects, pp. 449–458. Academic Press, 2009.

Stökl, Johannes, and Sandra Steiger. "Evolutionary origin of insect pheromones." Current Opinion in Insect Science 24 (2017): 36–42.

Stork, Nigel E. "How many species of insects and other terrestrial arthropods are there on earth?" Annual Review of Entomology 63 no. 1, (2018): 31–45. doi:10.1146/annurev-ento-020117-043348.

Treanor, D., T. Pamminger, and W.O.H. Hughes. "The evolution of caste-biasing symbionts in the social hymenoptera." Insectes sociaux 65, no. 4 (2018): 513–519.

Turner, John RG. "Butterfly mimicry: the genetical evolution of an adaptation." *AGRIS* (1977).

Ulitzka, Manfred R. "Five new species of Rohrthrips (Thysanoptera: Rohrthripidae) from Burmese amber, and the evolution of Tubulifera wings." Zootaxa 4585, no. 1 (2019): 27–40.

Usher, Michael B., and Marion Edwards. "A dipteran from south of the Antarctic Circle: Belgica antarctica (Chironomidae) with a description of its larva." Biological Journal of the Linnean Society 23, no. 1 (1984): 19–31.

Usinger, Robert L. "The role of Linnaeus in the advancement of entomology." Annual Review of Entomology 9, no. 1 (1964): 1–17.

Vanitha, K. "Mimicry in insects and spiders." (2019).

Verhulst, Eveline C., Louis van de Zande, and Leo W. Beukeboom. "Insect sex determination: it all evolves around transformer." Current Opinion in Genetics and Development 20, no. 4 (2010): 376–383.

Warrant, Eric, and Marie Dacke. "Vision and visual navigation in nocturnal insects." Annual Review of Entomology 56 (2011): 239–254.

Weiss, Brian, and Serap Aksoy. "Microbiome influences on insect host vector competence." Trends in Parasitology 27, no. 11 (2011): 514–522.

Weiss, Harry B. "The entomology of Aristotle." Journal of the New York Entomological Society 37, no. 2 (1929): 101–109.

Weiss, Martha R. "Defecation behavior and ecology of insects." Annual Review of Entomology. 51 (2006): 635–661.

Weitekamp, Chelsea A., Romain Libbrecht, and Laurent Keller. "Genetics and evolution of social behavior in insects." Annual Review of Genetics 51 (2017): 219–239.

Wheeler, W.C., M. Whiting, Q.D. Wheeler, and J.M. Carpenter. "The phylogeny of the extant hexapod orders." Cladistics 17 (2001): 113–169.

Wheeler, William Morton. *The social insects: their origin and evolution*. Routledge, 2015.

Whiting, M. "Phylogeny of the holometabolous insects." In Assembling the Tree of Life, pp. 345–359. 2004.

Willmann, Rainer. "Phylogenetic relationships and evolution of insects." In Assembling the Tree of Life, pp. 330–344. 2004.

Winsor, Mary P. "The development of Linnaean insect classification." Taxon (1976): 57–67.

Wipfler, Benjamin, Hans Pohl, Margarita I. Yavorskaya, and Rolf G. Beutel. "A review of methods for analysing insect structures—the role of morphology in the age of phylogenomics." Current Opinion in Insect Science 18 (2016): 60–68.

Wipfler, Benjamin, Harald Letsch, Paul B. Frandsen, Paschalia Kapli, Christoph Mayer, Daniela Bartel, Thomas R. Buckley et al. "Evolutionary history of Polyneoptera and its implications for our understanding of early winged insects." Proceedings of the National Academy of Sciences 116, no. 8 (2019): 3024–3029.

Yeates, David K., Karen Meusemann, Michelle Trautwein, Brian Wiegmann, and Andreas Zwick. "Power, resolution and bias: recent advances in insect phylogeny driven by the genomic revolution." Current Opinion in Insect Science 13 (2016): 16–23.

Yew, Joanne Y., and Henry Chung. "Insect pheromones: an overview of function, form, and discovery." Progress in Lipid Research 59 (2015): 88–105.

Zhang, Wei, Erica Westerman, Eyal Nitzany, Stephanie Palmer, and Marcus R. Kronforst. "Tracing the origin and evolution of supergene mimicry in butterflies." Nature Communications 8, no. 1 (2017): 1–11.

PHYLUM ARTHROPODA

CHELICERATA

Shahan Derkarabetian

CHAPTER **28**

Infrakingdom	Bilateria
Division	Protostomia
Subdivision	Ecdysozoa
Phylum	Arthropoda
Class	Chelicerata

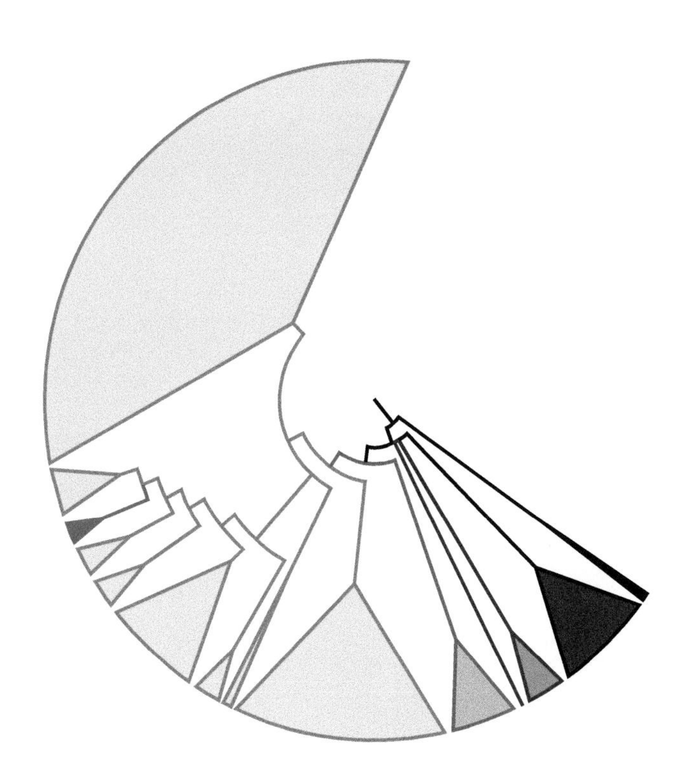

CONTENTS

General Characteristics

Approximately 120,000 valid species described

Published estimates of cryptic or undiscovered species range from ~400,000 to 1,000,000

Morphologically, ecologically, and behaviorally diverse; some relationships still uncertain

Arthropod body plan with varying patterns of segmentation

Two separate genome duplications and a diversity of silk, venom, and sensory genes

Medical, agricultural, manufacturing, and some pet trade value

High ecological importance

Synapomorphies

- presence of chelicerae/chelifores (raptorial appendage on somite II)
- two tagmata: prosoma (fused head and thorax) and opisthosoma (abdomen)

HISTORY OF TAXONOMY AND CLASSIFICATION

The Chelicerata (**Figure 28.1**) includes three major extant lineages: Pycnogonida (sea spiders, ~1,300 species), Xiphosura (horseshoe crabs, four species), and Arachnida (mites/ticks, spiders, harvestmen, scorpions, etc., >116,000 species), plus several extinct lineages. The first hypotheses of chelicerate relationships based on a cladistic analysis were proposed by Weygoldt and Paulus (1979), and

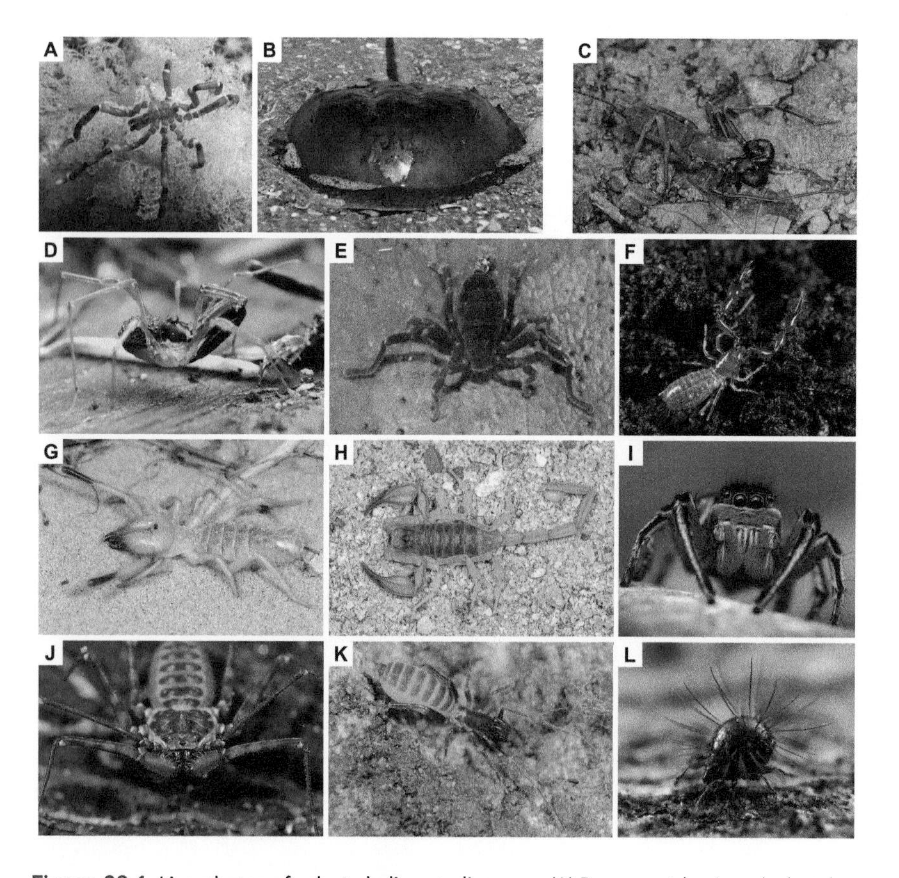

Figure 28.1 Live photos of select chelicerate lineages. (A) Pycnogonida: *Anoplodactylus evansi*. Photo: M. Harris. (B) Xiphosura: *Limulus polyphemus*. Photo: Marshal Hedin. (C) Thelyphonida: *Thelyphonus* sp. Photo: Marcus Ng. (D) Opiliones, *Taracus pallipes*. Photo: Shahan Derkarabetian. (E) Ricinulei: *Ricinoides karschii*. Photo: Gonzalo Giribet. (F) Pseudoscorpiones: *Pararoncus* sp. Photo: Gonzalo Giribet. (G) Solifugae: Eremobatinae sp. Photo: Marshal Hedin. (H) Scorpiones: *Smeringurus vachoni*. Photo: Marshal Hedin. (I) Araneae: *Habronattus americanus*. Photo: Thomas Shahan. (J) Amblypygi: *Heterophrynus* sp. Photo: Marshal Hedin. (K) Schzomida: *Hubbardia briggsi*. Photo: Marshal Hedin. (L) Acari: Oribatida, *Neotrichoribates* sp. Photo: Nicholas Porch.

since this publication, there have been a vast number of phylogenetic studies of chelicerate relationships using morphology, molecular, and fossil data (Shultz 1990; Wheeler and Hayashi 1998; Giribet et al. 2002; Shultz 2007; Regier et al. 2010; Sharma et al. 2014). There was some early controversy as to the placement of Pycnogonida within arthropods (summarized in Dunlop and Arango 2005), but it is now generally accepted that Pycnogonida are chelicerates. While some studies recovered Pycnogonida as sister to Arachnida (Garwood et al. 2017), the majority of recent morphological and molecular analyses (Dunn et al. 2008; Legg et al. 2013) have recovered Pycnogonida as the sister lineage to a clade including Xiphosura and Arachnida called Euchelicerata which also includes two extinct lineages Eurypterida (giant water scorpions) and Chasmataspidida and several extinct arachnid lineages.

The monophyly of Euchelicerata is generally highly supported with both morphological and molecular data. Xiphosura has been recovered within Arachnida in some genetic analyses, although this was attributed to accelerated molecular evolution in some arachnid orders as analyses only utilizing the slowest evolving loci recover a monophyletic Arachnida (Sharma et al. 2014). Analyses of mitochondrial genomes have also recovered a paraphyletic Arachnida recovering both Xiphosura and Pycnogonida within Arachnida (Masta et al. 2009; Masta et al. 2010; Ovchinnikov and Masta 2012); although these results alone are somewhat questionable considering some almost universally recovered arachnid lineages are not monophyletic based on mitochondrial genomes. A recent phylogenomic study consistently recovered Xiphosura within Arachnida even after accounting for common sources of phylogenetic error (Ballesteros and Sharma 2019), although shortly after, another study with increased sampling again recovered a monophyletic Arachnida (Lozano-Fernandez et al. 2019; **Figure 28.2**).

Arachnida includes 11 main extant lineages: the "Acari" (mites and ticks) with ~56,000 species total, Araneae (spiders, ~48,000 species), Opiliones (harvestmen or daddy-long-legs ~6,700 species), Pseudoscorpiones (false scorpions, ~3,500 species), Scorpiones (scorpions, ~2,400 species), Solifugae (camel spiders, ~1,100 species), Schizomida (short-tailed whip scorpions, ~300 species), Amblypygi (whip spiders, ~220 species), Thelyphonida (whip scorpions or vinegaroons ~110 species), Palpigradi (micro-whip scorpions, ~92 species), and Ricinulei (hooded tickspiders, ~76 species). The name "Acari" has historically been used to encompass all mites and ticks, but this is now generally split into three main lineages that may not form a monophyletic group: the Acariformes (mites), Parasitiformes (ticks), and the rare and primitive Opilioacariformes mites, which are sometimes included within Parasitiformes. In discussions of arachnids, these acarine lineages are often introduced at an equivalent level to the arachnid orders, but they are formally ranked as superorders that include six orders (Lindquist et al. 2009; Giribet and Hormiga 2016), resulting in a total of 16 extant arachnid orders. Currently, four extinct arachnid orders are known, including Haptopoda, Phalangiotarbida, Trigonotarbida, and Uraraneida, and significant progress had been made to include these taxa in phylogenetic analyses (Garwood and Dunlop 2014). The monophyly of arachnids has traditionally been well supported via morphology (Shultz 1990; Wheeler and Hayashi 1998; Shultz 2007; Garwood and Dunlop 2014) and morphology plus molecules (Wheeler and Hayashi 1998; Giribet et al. 2002), with morphological synapomorphies summarized in Shultz (2001) and more recently in Garwood and Dunlop (2014). The recent phylogenomic study of Ballesteros and Sharma (2019) does not recover a monophyletic Arachnida, instead recovering Xiphosura as the sister group to Ricinulei suggesting reinterpretation of morphology. However, a more recent study by Lozano-Fernandez et al. (2019) with additional samples recovers a monophyletic Arachnida. Putting aside the issue of monophyly, internal relationships of Arachnida have been highly unstable with many poorly supported nodes, particularly with molecular data. The placement of Scorpiones

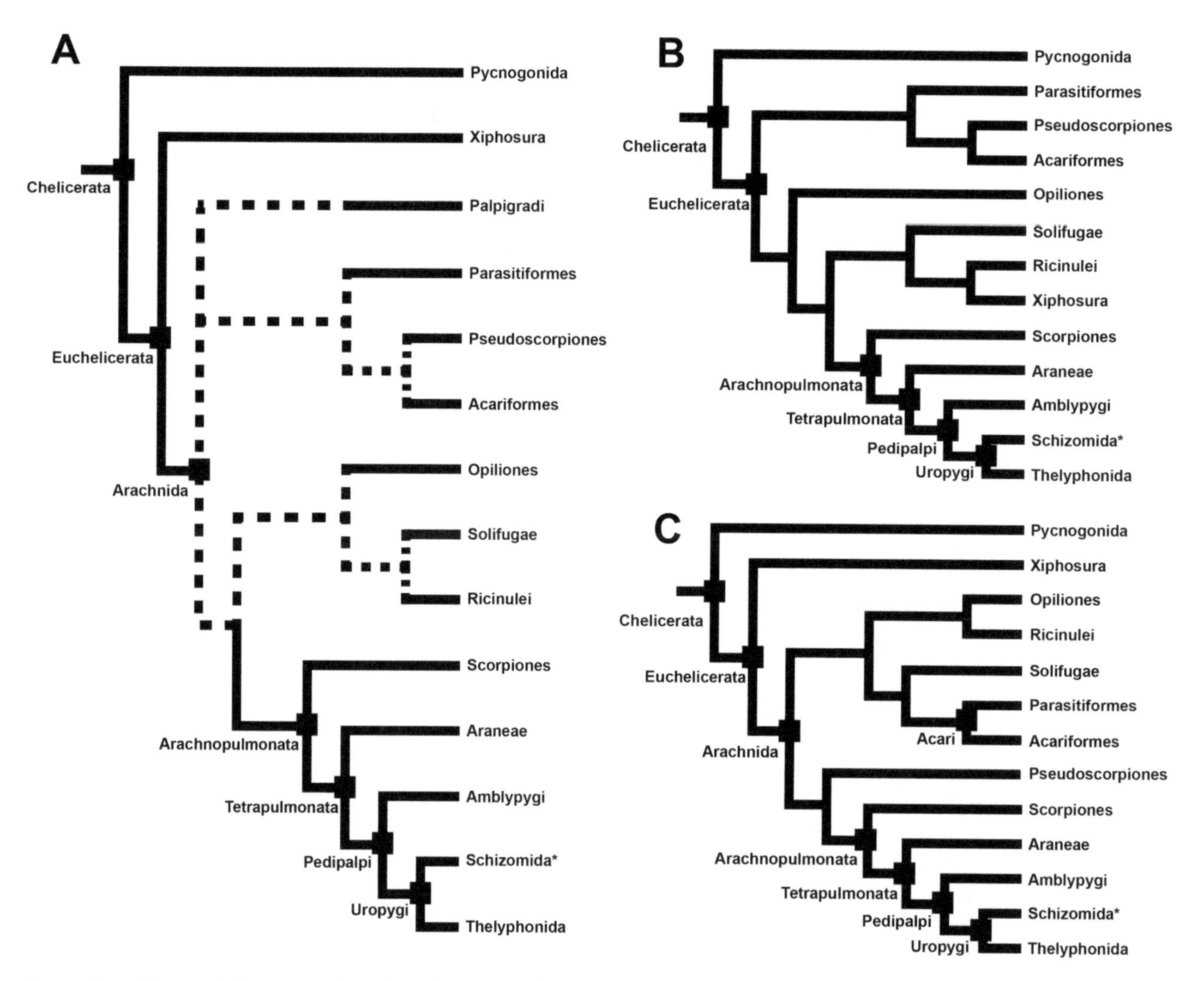

Figure 28.2 **Recent chelicerate phylogenies.** (A) Phylogeny based on Sharma et al. (2014) analysis of 500 slowest evolving loci. Dashed branches indicate relationships supported only with slowly evolving loci. Nodes with filled squares are generally highly supported and accepted across analyses and studies. Some analyses recovered Pseudoscorpiones as sister to either Arachnopulmonata or Scorpiones with weak support. (B) Phylogeny based on Ballesteros and Sharma (2019) analysis of the "decisive dataset" showing a paraphyletic Arachnida. (C) Phylogeny based on Lozano-Fernandez et al. (2019) analysis of "Matrix A" showing monophyletic Arachnida and Acari. Weakly supported relationships across analyses are not indicated. Schizomida and Palpigradi were not sampled in any analyses; here, placement of Schizomida is based on overwhelming support from other data (noted with an asterisk), and Palpigradi are left unresolved in A and not included in B or C.

has been highly variable, with the earliest hypotheses considering them as the earliest diverging arachnids (Weygoldt and Paulus 1979). Later morphological studies recovered Scorpiones in multiple places including close to Opiliones and Pseudoscorpiones (Shultz 2007; Garwood and Dunlop 2014), usually as an early-diverging lineage. Along with pseudoscorpions, the acarine lineages tend to be highly unstable across and within phylogenetic studies. Some of this uncertainty may be due to highly accelerated rates of molecular evolution in these groups, particularly the Acariformes (Sharma et al. 2014). Despite much debate and uncertainty regarding relationships among arachnids, some relationships have been consistently recovered through time in morphological and molecular datasets, for example, the Pedipalpi (Amblypygi + Schizomida + Thelyphonida) and Tetrapulmonata (Araneae + Pedipalpi). Recent molecular studies also generally support a sister relationship between Scorpiones and Tetrapulmonata, called Arachnopulmonata (Regier et al. 2010; Sharma et al. 2014; Starrett et al. 2017), a derived clade also supported by respiratory anatomy and a relatively complex circulatory system (Klußmann-Fricke and Wirkner 2016).

Since the first explicit phylogenetic hypotheses were made regarding arachnid relationships (Weygoldt and Paulus 1979), a great deal of effort has been exerted by numerous researchers to resolve the arachnid phylogeny using ever-increasing amounts and types of data (summarized in Giribet 2018). Despite a plethora of analyses and studies, including datasets with hundreds and thousands of genes and incorporation of fossils, we are still uncertain of many relationships within arachnids owing largely in part to a likely ancient and rapid radiation and highly variable rates of molecular evolution. Despite these difficulties, some consensus currently exists across all phylogenomic analyses (Figure 28.2). New and promising molecular approaches may yet help when fully applied to the inference of chelicerate phylogeny (Starrett et al 2017), particularly in combination with the most recent phylogenomic datasets (Ballesteros and Sharma 2019; Lozano-Fernandez et al. 2019).

ANATOMY, HISTOLOGY, AND MORPHOLOGY

The chelicerates are a relatively old lineage with the oldest confirmed fossils dating to the Cambrian, at least ~510 million years ago (Wolfe et al. 2016). Chelicerates are united by possessing modified forms of the first pair of appendages termed chelifores in Pycnogonida and chelicerae in Euchelicerata – structures shown to be homologous to each other based on Hox gene expression patterns and neuroanatomy (Jager et al. 2006; Brenneis et al. 2008), and homologous to the antennae of their sister group the mandibulate arthropods. The chelicerae are typically formed into pincers for feeding, however in spiders they are modified into fangs. The three-segmented chelicerae found in horseshoe crabs and some arachnids, such as scorpions and harvestmen, are considered the primitive condition, while all other arachnids have derived two-segmented chelicerae. The chelicerate ground body plan consists of two tagmata: the anterior portion called the prosoma or cephalothorax bearing the feeding, sensory, and locomotion structures, and the posterior portion called the opisthosoma or abdomen which bears the digestive, respiratory, and reproductive organs (**Figure 28.3**). There is variation in the number of somites across lineages, and chelicerate orders and higher-level groupings can be defined based on segmentation, structure of tagmata, and limb morphology (Dunlop and Lamsdell 2017). Sea spiders have a rather unusual morphology – they tend to have a small narrow body where the opisthosoma is highly reduced, and many of their organs including the gonads are found in the appendages. Most arachnid orders possess opisthosomal segmentation (e.g., scorpions, amblypygids). Araneae are interesting in this regard as the early-diverging Mesothelae are the only spiders to show external segmentation of the opisthosoma. Similarly, Opiliones show varying degrees of fusion of the opisthosomal segments.

Pycnogonida typically have four simple eyes, although some species have lost eyes completely. Xiphosura have a pair of compound eyes, the only chelicerates to possess them, and a pair of simple eyes. Arachnids have median and lateral ocelli that vary in number and presence: scorpions have one pair of median and up to five pairs of lateral eyes (Yang et al. 2013); Amblypygi and Thelyphonida have one pair of median and three pairs of lateral eyes; Opiliones have only a single pair of median eyes, and although the early-diverging Cyphophthalmi are mostly blind, some have median eyes which are laterally displaced; and Palpigradi and Schizomida do not possess eyes. Of course, many taxa have reduced numbers of eyes, for example, many spiders have less than four pairs of eyes and many cave-dwelling arachnids have reduced or lost eyes completely. Visual acuity varies across chelicerates from basic detection of movement to well-developed vision in jumping spiders which have a wide field of vision, very high spatial resolution, and the ability to detect multiple colors and UV-light (Zurek et al. 2015).

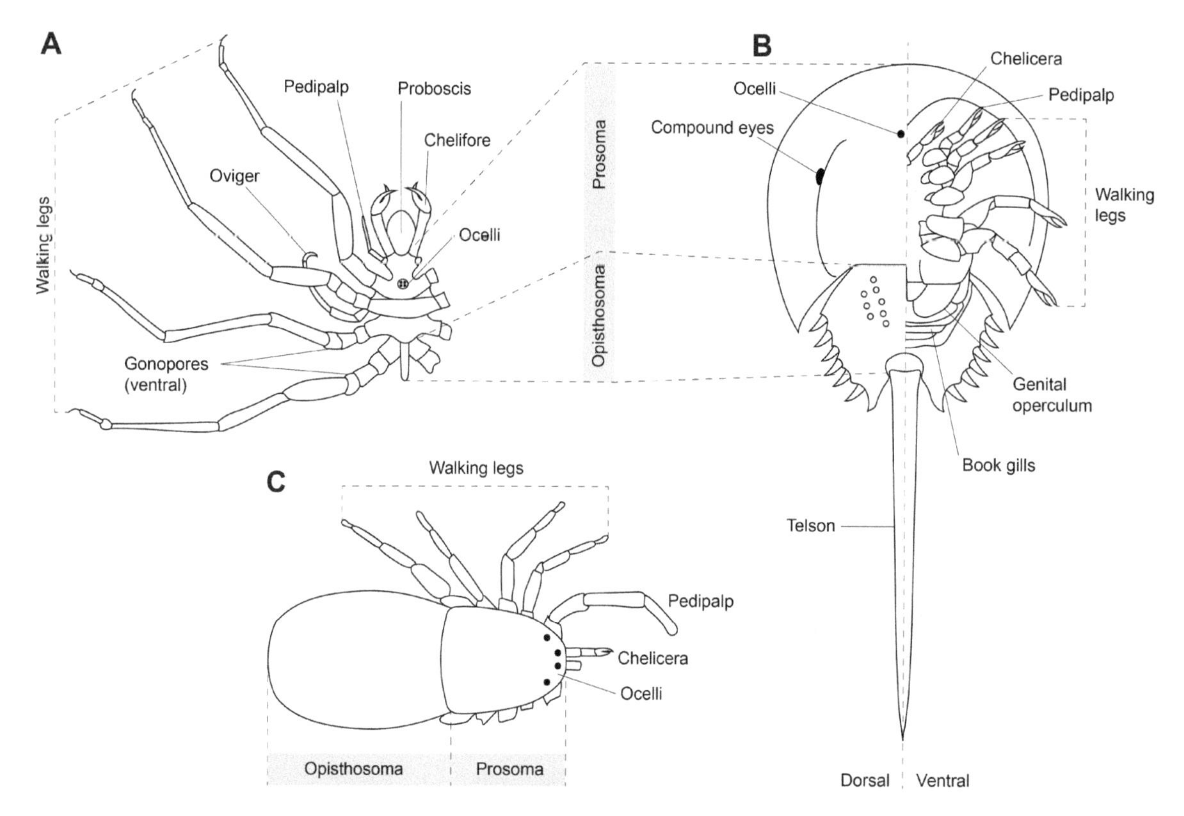

Figure 28.3 General body plan of chelicerate lineages. (A) Pycnogonida, (B) Xiphosura, and (C) a generalization of Arachnida.

Arachnids have evolved a diversity of unique morphological characters used for prey capture, defense, and sensory purposes **(Figure 28.4)**. The second pair of appendages (pedipalps) have been modified across lineages: they are formed into pincers for prey capture in scorpions, pseudoscorpions, and Ricinulei; spider pedipalps are largely sensory organs that become sexually dimorphic in adults as male palps are modified for sperm transfer; camel spider pedipalps are leg-like and used as sensory organs (Cushing and Casto 2012); and raptorial pedipalps used for prey capture are found in the Pedipalpi and most Opiliones. A segmented flagellum is present at the terminal end of the abdomen in palpigrades and is a key character supporting the Uropygi (schizomids and thelyphonids), and a flagelliform telson is found in extinct lineages leading to modern spiders. In addition to modified pedipalps in the form of pincers, the opisthosoma of scorpions terminates in a telson and aculeus (metasoma, commonly called a stinger) with venom glands. Interestingly, reports in some scorpions suggest that they possess photoreceptors in their metasoma allowing the detection of and response to light (Zwicky 1968; Rao and Rao 1973). Scorpions also possess a pair of chemosensory organs called pectines found ventrally on the second abdominal segment. Solifugae have similar organs called malleoli (or racquet organs) found ventrally on the fourth pair of legs that are used as chemoreceptors. In addition, Solifugae show extreme exaggeration of their chelicerae, which are often bigger than the prosoma, resulting in powerful pincers. Ricinulei are called hooded tickspiders due to their rather unique cucullus, a moveable hood-like projection that can cover their mouthparts. Ticks (Parasitiformes) possess a chemosensory structure on the first leg called Haller's organ which is used to find hosts by detecting odors, chemicals, and changes in temperature and humidity (Guerin et al. 2000). A pair of walking legs are modified and used as sensory structures in the phalangid harvestmen (all harvestmen except the early-diverging mite harvestmen, Cyphophthalmi), Ricinulei, and Pedipalpi.

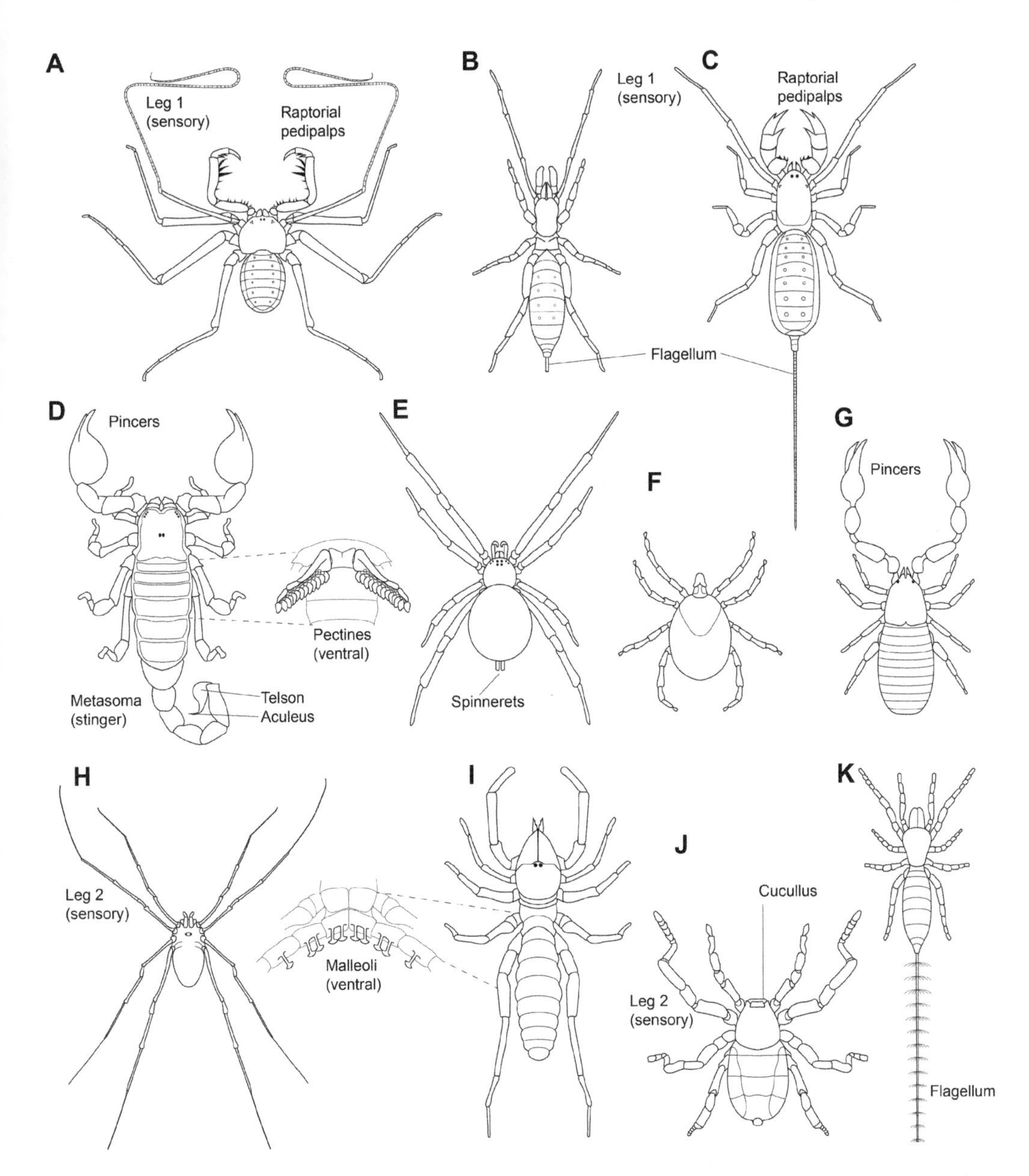

Figure 28.4 **General morphology of Arachnida lineages.** (A) Amblypygi, (B) Schizomida, (C) Thelyphonida, (D) Scorpiones, with inset of partial ventral view, (E) Araneae, (F) Acari, (G) Pseudoscorpiones, (H) Opiliones, (I) Solifugae, with inset of partial ventral view, (J) Ricinulei, (K) Palpigradi. Note: drawings not to scale.

Spiders are characterized by the presence of up to four pairs of silk-spinning organs (spinnerets) on their abdomen each connected to a silk-producing gland within the abdomen. Fossil evidence suggests that spinnerets evolved earlier than modern spiders with the extinct arachnid order Uraraneida and other recently discovered fossils that together may form a grade leading to extant spiders, also possessing spinnerets (Wang et al. 2018; Huang et al. 2018). Silk production is also known in some mites and pseudoscorpions. While spiders and scorpions may use their ability to envenomate as a defensive mechanism, other particularly interesting defensive strategies are found in some other lineages. Harvestmen possess a pair of ozopores on the prosoma leading to repugnatorial glands which produce diverse and unpleasant chemicals that are secreted as a defensive mechanism. Thelyphonida (also commonly called vinegaroons) similarly utilize a defensive spray via two anal glands producing a mixture of mainly acetic and caprylic acid that smells like vinegar.

REPRODUCTION AND DEVELOPMENT

Multiple reproductive strategies are used across chelicerates, including external and internal fertilization, with internal fertilization occurring through indirect and direct fertilization. All chelicerates lay eggs except scorpions and some mites. There is a great deal of variation across and within chelicerate orders in terms of number of eggs laid, development time, and time to maturity. In addition, parthenogenetic taxa have been reported in some arachnids including mites, scorpions, harvestmen, and amblypygids.

External fertilization

External fertilization occurs in both Pycnogonida and Xiphosura. Sea spiders show some courtship and precopulatory behavior, after which females release eggs that the male fertilizes (Bain and Govedich 2004). The mating and reproductive behavior of horseshoe crabs is easily observable on beaches during their mating season: the male mounts the female, the female digs in the sand and lays eggs, which she then buries after the male has fertilized and separated.

Internal fertilization

Two forms of internal fertilization exist in arachnids: indirect and direct fertilization.

Indirect fertilization: With the exception of phalangid harvestmen and some acariform mites, arachnids rely on indirect fertilization; and in most taxa, this involves the male transferring a sperm package (spermatophore) to the female. In many taxa, some courtship behavior is involved and the spermatophore is placed on the substrate where the female picks it up. In scorpions, for example, a complex courtship (sometimes aptly referred to as a "dance") is usually involved where the male grasps the female and guides her to a suitable place for spermatophore deposition, then guides the female directly over the spermatophore, which is picked up by the female through her genital opening. In spiders, sperm transfer does not involve a spermatophore, but instead sperm is transferred via highly modified pedipalpal tarsi formed into copulatory bulbs (secondary sexual characters) to the female's reproductive organs. The male typically produce a sperm web to hold their sperm, then use their palp to collect the sperm somewhat like a syringe, which is then inserted into the female's epigynum. Spiders may have very elaborate courtship rituals where the use of silk is an important component (Scott et al. 2018), like precise silk web plucking by males of web-building spiders to entice the female. Many male jumping spiders use very elaborate, complex, colorful, and enjoyable-to-observe courtship dances to impress females (Girard et al. 2011; Elias et al. 2012). Other means of indirect fertilization also occur, for example, Ricinulei use a highly modified third walking leg to transfer sperm to the female, akin to sperm transfer in spiders.

Direct fertilization: This type of fertilization is known to occur in some mites and phalangid Opiliones. Male phalangid harvestmen possess an eversible reproductive organ termed a penis, which is used to deposit sperm directly inside the female's ovipositor during copulation. The majority of harvestmen species copulate face-to-face, with males grasping the females with their pedipalps. Some well-studied taxa like sclerosomatids show evidence of significant precopulatory antagonism and courtship behavior including nuptial gifts that are secreted from paired gland-bearing sacs on the penis (Burns et al. 2013). Opiliones are also noted for their alternative reproductive tactics, which in many cases are associated with male dimorphism or polymorphism. These behavioral and morphological traits have evolved many times independently (Buzatto and Machado 2014). In taxa with male dimorphism, the major forms typically have some exaggerated features such as thickened palps, chelicerae or longer legs, and the minor forms possess unexaggerated features and can sometimes be morphologically similar to females. Major males will fight with each other for territory and mating opportunities, while minors may sneak by the major males to mate with females. The harvestman family Neopilionidae is well-known for the exaggerated and extremely elongated chelicerae of males, for example, *Pantopsalis cheliferoides*, with trimorphic male forms (Painting et al. 2015). In this species, the males include two exaggerated cheliceral forms, long-slender and short-broad, and a third form with reduced short-slender chelicerae. The exaggerated chelicerae are used as weapons in male–male combat in different ways: the long-slender forms waving their chelicerae before combat potentially assessing rivals based on cheliceral length, and the short-broad forms using the chelicerae to stab and pinch.

Parental care

Pycnogonids exclusively show paternal care. After fertilization the eggs are transferred to a special appendage often found in male sea spiders, called ovigers, where the male will carry and care for the eggs and young (Bain and Govedich 2004). Most arachnids show maternal care in some form, the most easily observable of which are the spiders which produce and guard silken egg cases. In some spiders (e.g., wolf spiders) and other arachnids, such as amblypygids and whip scorpions, the female will carry the egg sac, and even after hatching the female will carry the young on their body for some time. Scorpions are unique among arachnids in that they give birth to live young that also remain on the mother's back. Recently, and quite surprisingly, it was reported that mothers of the ant-mimicking jumping spider species *Toxeus magnus* will produce and feed their offspring "milk" for an extended period of time (Chen et al. 2018). All pseudoscorpions have matrotrophic development where females will carry a brood chamber housing the fertilized eggs which are provided with nutritive secretions from the ovary (Ostrovsky et al. 2016). Opiliones are rather interesting in that they also show paternal care, a behavior that evolved independently in several families. Most well-known are the mud-nest harvestmen (e.g., *Zygopachylus albomarginis*) of the Neotropics where the males build open circular nests on trees which are used as arenas to attract females, fight any invading males, and actively care for eggs once laid (Mora 1990).

Developmental genomics

Thorough summaries of development and evolutionary developmental genetics in chelicerates can be found in Schwager et al. (2015) and Sharma (2018). Most developmental studies have focused on the relatively derived spiders, but recent work is aiming to incorporate other arachnid lineages like Opiliones which retain some ancestral chelicerate characteristics. It is generally believed that the ancestral condition of euarthropods is a cluster of 10 Hox genes; all chelicerates studied to date show this same cluster with the exception of the mite *Tetranychus urticae*, which has lost *Abd-A* (Grbić et al. 2011; summarized in Leite and McGregor 2016). As a result of two separate whole genome duplication events inferred within chelicerates, multiple copies of Hox genes have been reported in horseshoe crabs with up to four copies per gene (Kenny et al. 2016), in scorpions with two copies for all but one Hox gene (Sharma et al. 2015; Di et al. 2015), and two copies for several genes in spiders (Schwager et al. 2007). Similarly, regulatory homeobox genes show much higher levels of duplication in Arachnopulmonata relative to other arachnids (Leite et al. 2018). Extensive duplication of Hox genes have not been found in the other lineages examined, again with the exception of *T. urticae* which has two copies of the *ftz* gene.

In the antennae of mandibulate arthropods, the *dachshund* gene is not expressed or not required for antennal development. Similarly, in spiders, which have the derived two-segmented chelicerae (homologous to the mandibulate antennae), the *dachshund* gene is absent. However, in Opiliones, which have the ancestral three segmented chelicerae, the expression domain of *dachshund* is present suggesting a potential mechanism of cheliceral modification in chelicerates (Sharma et al. 2013). The importance of *Distal-less* genes in development of chelicerates was demonstrated in several lineages using methods like RNA interference (Sharma et al. 2012). Silencing of *Distal-less* resulted in loss of the distal portion of particular appendage segments, as also reported in insects.

DISTRIBUTION AND ECOLOGY

Pycnogonids are benthic organisms with a cosmopolitan distribution found in habitats ranging from shallow estuaries to deep water and abyssal trenches, spanning oceans from pole to pole. Most of their diversity is found in shallow habitats, but sea spiders have been collected from the deep sea including collections from depths of over 7,000 m (Arnaud and Bamber 1987). Interestingly, the largest species of sea spiders tend to inhabit very cold waters. All four species of Xiphosura are found in coastal marine and/or brackish habitats. The well-studied *Limulus polyphemus* is native to the Americas along the Atlantic coastline while the other three species are restricted to Asia (**Figure 28.5**). Arachnida are terrestrial and, as a whole, are found in all continents in all types of habitats with the greatest diversity found in the tropical forests, including several independent secondary invasions of aquatic habitats in mites and spiders. The most widely distributed arachnids are mites which occur on all continents, including Antarctica, in essentially all terrestrial, freshwater, and marine habitats as either free-living or parasitic organisms. Mites have independently invaded aquatic habitats at least 10 times (Walter and Proctor 2013). The majority of marine mite species are found in the Halacaroidea with records approaching depths of 7,000 m and several species are associated with hydrothermal deep-sea vents (Bartsch 1994). A few species of spiders have associations with aquatic habitats (e.g., fishing spiders), but the species *Argyroneta aquatica* lives almost entirely underwater constructing a "diving bell" web and breathing oxygen trapped by hydrophobic abdominal hairs (Schütz and Taborsky 2003). Intertidal taxa are recorded in mites, spiders, palpigrades, and pseudoscorpions. While groups like pycnogonids and Acari are widespread, other arachnid lineages are more restricted in their distributions (Figure 28.5). Some arachnid lineages are restricted to particular habitat types, for example, camel spiders are mainly found in arid and semiarid regions, and Ricinulei and the non-spider tetrapulmonates are largely restricted to tropical habitats with many species found in caves and leaf litter. Spiders, scorpions, and harvestmen are widely distributed throughout temperate and tropical climates. Many arachnids tend to have high ecological constraints and low dispersal ability, biological characteristics that lead to a considerable number of species which may be considered short-range endemics (Harvey 2002) with very small geographic distributions, sometimes only known from one to several localities (Edward and Harvey 2008; Emata and Hedin 2016).

While pycnogonid adults are free-living, the larvae and juveniles are largely parasitic living in or on hosts like hydroids, bivalves, and echinoderms. They are carnivorous, preying or scavenging upon various invertebrates, using a proboscis to puncture the body of the prey. Xiphosurans are generalist feeders and may play important roles in coastal habitats by feeding on bivalves or as prey items for shorebirds and loggerhead turtles (Botton 2009). The majority of arachnids are carnivorous predators feeding mainly on smaller invertebrates, but some large spiders are known to catch and eat small vertebrates like rodents and birds. Although spiders are strictly carnivorous, an interesting case is known in the jumping spider *Bagheera kiplingi* that is predominantly herbivorous feeding on the Beltian food bodies of acacias (Meehan et al. 2009). Spiders play a significant ecological role as insect predators with estimates of their global prey kill at 400–800 million tons per year (Nyffeler and Birkholder 2017). Specialized predators exist in some arachnids, for example, some spiders and harvestmen are well-adapted gastropod predators. Gastropod-specialists are found in multiple

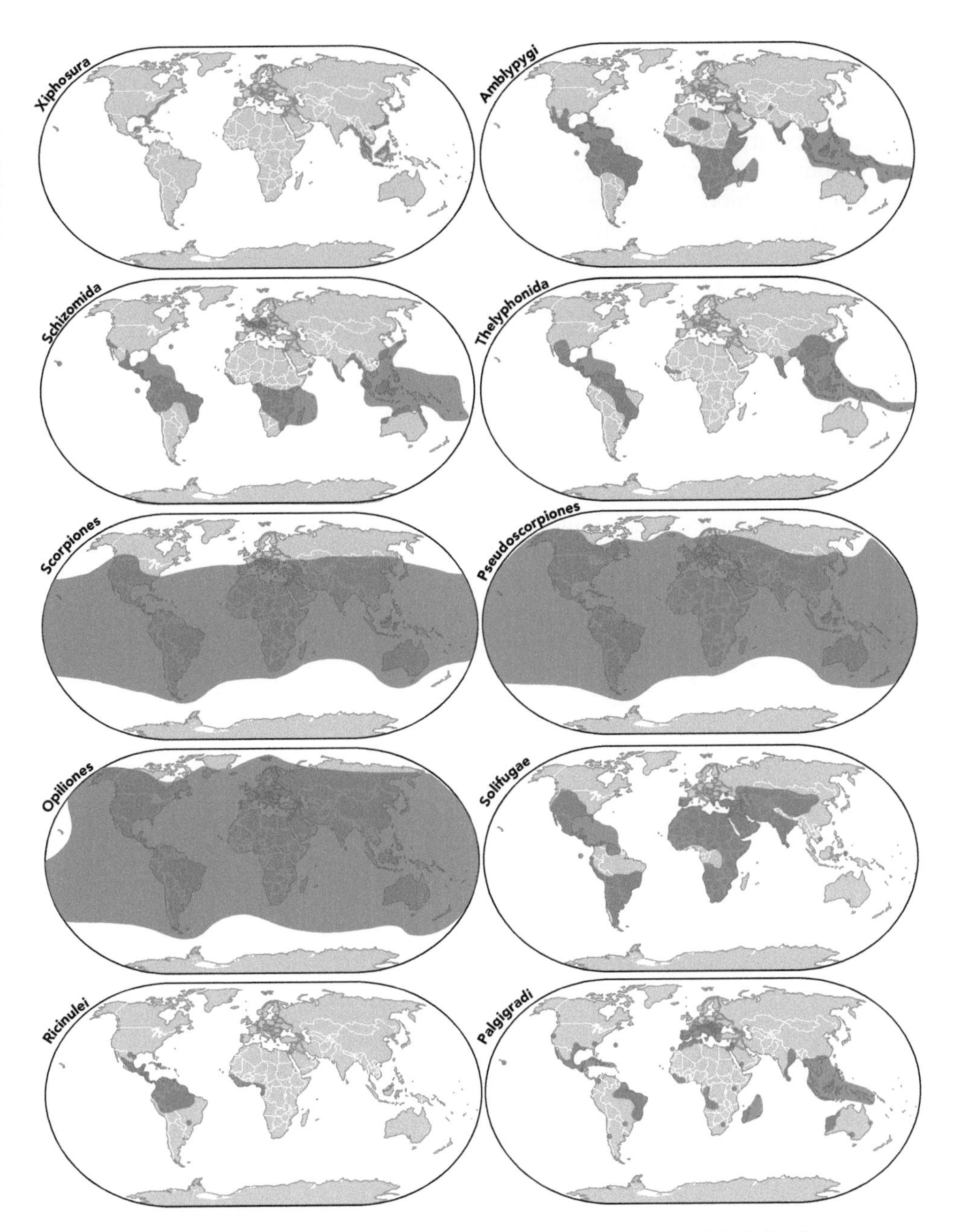

Figure 28.5 Distribution maps for most chelicerate lineages based on published distribution data. Distribution maps were not drawn for Pycnogonida (widespread in all marine habitats), Acari (widespread in all terrestrial and aquatic habitats), or Araneae (widespread in all terrestrial habitats except mainland Antarctica).

harvestmen lineages: *Ischyropsalis hellwigi* evolved powerful chelicerae to break open shells and devour the soft tissue, and species in Trogulidae are even reported to lay eggs in the empty shell after feeding (Nyffeler and Symondson 2001). The acarines as a whole are omnivorous, with certain taxa primarily predators, fungivores, microbivores, detritivores, saprophages, algivores, etc. Although primarily predators, some harvestmen have also been recorded to be scavengers, frugivores, and fungivores. Hard ticks, the most familiar of Parasitiformes, are well-adapted obligate blood-feeding ectoparasites of many terrestrial vertebrates. Given their blood-feeding habits, they are also important vectors of diseases like typhus and Lyme disease in humans and other animals, and species like *Ixodes scapularis* are rather well-studied (Gulia-Nuss et al. 2016; Eisen et al. 2016). Equally well-studied are spider mites (Acariformes) which live and feed on plants. *Tetranychus urticae* is considered a worldwide pest of many crops and vegetables, and much effort and research has been devoted to its biological control (Attia et al. 2013).

The ecological diversity and niche specificity among the acarines can be quite astonishing. Many mites are endo- or ectoparasites of other animals with varying degrees of host specificity ranging from strict monoxeny (one host) to pleioxeny (multiple hosts across multiple genera). For example, any given species of bird may have 2–25 species of feather-associated mites that coexist through niche partitioning, occupying different feathers or regions of a feather (Proctor and Owens 2000). Not all mite associations with hosts are parasitic; feather mites might indeed be beneficial for hosts benefitting them by cleaning their feathers (Doña et al. 2018).

When it comes to interactions with humans, spiders perhaps receive the most worldwide public attention, and unfortunately, nearly all of this attention is undeservedly negative. While all spiders have fangs and venom, bites are rare, and the vast majority of bites are not harmful to humans, and as such, the associated fear and blame of spiders is largely unsubstantiated with multiple causes (Vetter and Isbister 2008). Australia has perhaps some of the most dangerous spiders known, like the funnel-web spiders in the genera *Atrax* and *Hadronyche*. The Sydney funnel-web spider *Atrax robustus* is considered to be the most venomous spider species with potential to cause death in humans (Nicholson and Graudins 2002), although antivenom is highly effective and envenomings are relatively rare (Vetter and Isbister 2008). In North America, the brown recluse *Loxosceles reclusa* is perhaps the most notorious and feared spider due to its potent venom that causes necrotic lesions at the bite site. However, the species is not as widespread as commonly believed, confirmed bites with verified identifications are rare, and other medical issues are very often misdiagnosed as brown recluse spider bites (Vetter 2015). In some regions of the world like Mexico, scorpions receive more attention perhaps due to the higher actual threat they pose relative to spiders. Scorpion envenomation is actually more common and more dangerous with 1.5 million envenomings reported annually causing over 2000 deaths worldwide, most significantly in the Middle East (Chippaux 2012).

PHYSIOLOGY AND BEHAVIOR

In chelicerates, the digestive system generally consists of a preoral chamber where food is liquified with digestive enzymes or chewed, before entering the pharynx, which in most chelicerates acts as a pump. Food then enters the midgut where it is digested and nutrients are absorbed. In horseshoe crabs, the pharynx acts as more of a crop and gizzard. Two structures act as excretory organs: the coxal glands which are involved in water balance with fluid exiting from pores near the base of the legs, and the Malphagian tubules which concentrate and release solid nitrogenous waste through the anus. Due to their generally small size, pycnogonids do not possess a respiratory system, instead they rely on diffusion of oxygen absorbed through pores in their legs (Lane et al. 2018). Horseshoe crabs use book gills, while the arachnids have book lungs and/or a tracheal system. The Arachnopulmonata are united by the possession of book lungs; scorpions have four pairs, while the Tetrapulmonata possess two pairs, although derived spiders possess only a single pair, the second being replaced by a pair of tracheae. All other arachnids possess a tracheal system, except in palpigrades in which they are completely lacking. Typical of arthropods, chelicerates possess an open circulatory system where the body fluid (hemolymph) is pumped through the body cavity (hemocoel) by a dorsal tube-shaped heart that collects hemolymph and directs it from the opisthosoma to the prosoma. Like the standard arthropod model, the chelicerate central nervous system is located ventrally and consists of a set of ganglia for each body segment connected by paired nerve cords, and a brain composed of fused ganglia typically found in the prosoma.

Chelicerates are extremely diverse in behavior, as previous sections have discussed, in regards to reproduction. Behaviorally, spiders are perhaps the most charismatic and well-studied of the chelicerates. One of the most obvious (and visually appealing) behaviors exhibited is silk use and web-building in spiders, a behavior associated primarily with the purpose of food capture using diverse types of silk structures. For example, the early-diverging Mesothelae spiders build trap-door burrows with silken "trip-lines" radiating outward from the burrow entrance which alert the spider to any prey that come in contact with these lines. The most obvious silk structures are the commonly encountered orb-webs, a key innovation which led to the diversification of the lineages that employ them (Bond and Opell 1998). Some spiders have developed interesting alternative ways to use silk in food capture, like the bolas spiders that attach a droplet of glue at the end of a silk string which they swing at approaching moths. Remarkably, the droplet of glue includes chemicals that mimic the sexual pheromones of the moths, attracting them to the spider (Stowe et al. 1987).

The vast majority of arachnids are solitary, but social behavior is known in some taxa. In spiders, social behavior has evolved several times independently with individuals cooperating in prey capture, web maintenance, and brood care. Opiliones will sometimes form dense aggregations potentially numbering in the thousands on rocks, trees, or in caves. Amblypygi are becoming an excellent system to study fitness consequences and sensory control of behavior, due to their impressive homing ability in complex environments (Chapin and Hebets 2016).

GENETICS AND GENOMICS

A handful of chelicerate genomes have been sequenced including the horseshoe crab *Limulus polyphemus* and multiple arachnids representing Acariformes, Parasitiformes, Scorpiones, and several Araneae. Sizes of published whole genomes are highly variable ranging from 90 Mb and 18,000 genes in the highly reduced genome of *T. urticae* to >3,000 Mb and >30,000 genes in other taxa. Whole genome duplication events have been reported in horseshoe crabs, spiders, and scorpions, with analyses indicating two distinct whole genome duplication events in chelicerates: one (or possibly two) in Xiphosura and a separate duplication inferred for all Arachnopulmonata (Kenny et al. 2016; Schwager et al. 2017).

Due to its consideration as an agricultural pest, the spider mite *T. urticae* was the first chelicerate to have its genome sequenced revealing genomic signatures associated with its lifestyle, for example, unique and expanded composition of gene families associated with its polyphagous plant-feeding habits and detoxification of xenobiotics (Grbić et al. 2011). Genome sequencing of the parasitic tick *Ixodes scapularis* provided much insight into its unique parasitic lifestyle showing expanded genomic repertoires associated with feeding and detoxification of xenobiotic elements, and novel and unique mechanisms for host detection and blood digestion (Gulia-Nuss et al. 2016). The genome of the scorpion *Mesobuthus martensii* demonstrated a unique model in which morphological evolution is rather conservative while gene turnover is high, and interestingly, expression analyses recover a complete pathway for light sensitivity in the scorpion metasoma (Cao et al. 2013). Concomitant with the diverse chemosensory adaptations, processes, and mechanisms within chelicerates, comparative genomic analyses across chelicerates identified many new chemosensory genes and lineage-specific chemosensory repertoires and expansions (Vizueta et al. 2018).

Genetic and genomic studies are shedding light on various evolutionary aspects in arachnids, most notably, silk production and venom evolution (Sanggaard et al. 2014). With multiple genomes sequenced, spiders have received considerable attention from a genomics perspective in regard to silk production as spider silk is one of the toughest known biological materials with applications in industry and medicine. Spiders may have up to seven different types of silk glands, each producing a distinct silk type for specific needs. The silk is composed primarily of spidroin proteins; the spidroin gene family having undergone radiation and duplication producing silk with a variety of properties (Garb et al. 2010). Genome sequencing of the orb-weaver *Triconephila* (*Nephila*) *clavipes* demonstrated gland-specific patterns of spidroin expression and production, gene duplication, and high polymorphism, and also revealed many motifs shared across various silk types that, when differentially organized into larger ensembles, may account for the diversity of mechanical properties of the silk types (Babb et al. 2017). Venoms are of wide interest due to their effects on humans, their biochemical properties, and pharmaceutical potential including as an anticancer agent (Ghosh et al. 2018). In arachnids, venom is produced and used by spiders, scorpions, and the Iocheirata pseudoscorpions, and is present in the salivary secretions of some ticks. In addition, recent research identified the presence of venom homologues in palpigrades suggesting a deeper evolution of venoms within arachnids (Santibañez-López et al. 2019). Genome sequencing shows that the venom neurotoxin genes are the most expanded gene families in scorpions and, in both scorpions and spiders, venom genes evolved via tandem duplication events (Cao et al. 2013).

The mitochondrial genomes of chelicerates are generally around 15,000 base pairs, although larger mitogenomes approaching 25,000 bp have been reported in parasitiform mites. *Limulus* and a few other chelicerates retain the arthropod ground pattern for mitochondrial gene order, although most show some gene rearrangements, and some taxa show reduction in tRNA size and structure and variation in nucleotide composition bias (Fahrein et al. 2007; Masta and Boore 2008).

POSITION IN THE ToL

Much research has been conducted regarding the phylogenetic relationships among arthropod lineages and their relatives (Edgecombe 2010; and summarized in Giribet and Edgecombe 2019). Within the Panarthropoda, the Onychophora and Tardigrada are consistently recovered as early-diverging lineages that are the sister groups to the Arthropoda (also called Euarthropoda). Within the Arthropoda, it is agreed that two major lineages exist: the Chelicerata and its sister clade called Mandibulata, which contains the Myriapoda and Pancrustacea.

DATABASES AND COLLECTIONS

Several types of databases exist for chelicerates. Given the medical importance of some arachnids, several toxin databases are kept for spiders and scorpions, for example, a database specifically for toxins acting on potassium channels (http://https://kaliumdb.org), and another for all proteins derived from spider venoms (http://www.arachnoserver.org). An EST (expressed sequence tags) database exists for all chelicerates that have been sequenced in this way (http://www.nematodes.org/NeglectedGenomes/ARTHROPODA/Chelicerata.html). From the taxonomic and biological perspectives, there are multiple databases for specific lineages of chelicerates. For example, there is a world database for pycnogonids that includes taxonomic, distribution, biological, and ecological information, as well as taxonomic references for all species (http://www.marinespecies.org/pycnobase). There is a World Spider

Catalog (https://wsc.nmbe.ch) which maintains an up-to-date taxonomy for all extant and extinct species of Araneae including links to primary taxonomic literature, and a similar catalog exists for the Acari (http://www.miteresearch.org). The Western Australian Museum includes taxonomic, distribution, and fossil catalogs for the minor orders of arachnids (http://museum.wa.gov.au/catalogues-beta/).

CONCLUSION

The ongoing burst of phylogenomic analyses are rearranging what we know about the tree of life for many arachnid taxa while also providing high support for novel classifications and new higher level lineages (Hedin et al. 2018; Derkarabetian et al. 2018). These phylogenomic analyses are refining our understanding of chelicerate evolution and the morphological, behavioral, ecological, and developmental aspects that have led to their diversification. Of utmost importance will be robustly resolving the chelicerate phylogeny in order to provide the necessary phylogenetic and evolutionary context for future studies. Sequencing of additional chelicerate genomes not only has great potential for advancing our understanding of their evolution, but also offers promising and exciting avenues of research with broader applications across the sciences.

Chelicerates are appealing taxa for many reasons, most notably their incredible diversity in morphology, behavior, ecology, etc. but also in their biodiversity, particularly what yet remains to be discovered. Relative to their species-level diversity, chelicerates remain incredibly understudied and a large percentage of current research is still purely taxonomic in nature. New species are continuously being discovered and described, particularly so for spiders and mites. In 2018 alone, over 50 new genera and over 800 new species of spiders were described (World Spider Catalog). Compared to what is currently known, the actual biodiversity of chelicerates is vastly underestimated. For example, estimates of undescribed diversity are anywhere from 400,000 to 1,000,000 species for the acarines alone (Krantz 2009). Total estimated diversity (described and undescribed) is up to 170,000 species for spiders (Coddington and Levi 1991) and at least 10,000 species for harvestmen (Machado et al. 2007). Cryptic speciation is a commonly occurring phenomenon in arachnids with high ecological constraints and low dispersal ability, whereby highly conservative morphology obscures the actual number of species (Boyer et al. 2007; Satler et al. 2013) necessitating the incorporation of genetic data into modern integrative taxonomic revisions (Derkarabetian and Hedin 2014). The potential unknown diversity of mites is particularly astounding when you consider their ecological specificity and the diversity of hosts with which they associate (Young et al. 2019). As numerous taxonomic studies can attest, there is an extremely high probability that any focused taxonomic research in any chelicerate group, except horseshoe crabs, will lead to discovery of new taxa. Even relatively well-studied regions are turning up previously unknown higher level lineages (Griswold et al. 2012). While spiders and mites are incredibly well-studied in many regards, taxa like Ricinulei and Palpigradi are rather enigmatic, unknown, and neglected. From the perspectives of basic natural history, ecology, and behavior, there is still a great deal that remains unknown. As with all invertebrate lineages, much remains to be discovered in chelicerates.

ACKNOWLEDGMENTS

I thank Gonzalo Giribet for providing helpful comments that greatly improved this chapter.

REFERENCES

Arnaud, F. and Bamber, R.N., 1987. The biology of Pycnogonida. In: Blaxter J.H.S., Southward, A.J. (eds), Advances in Marine Biology (vol. 24, pp. 1–96). Academic Press, London.

Attia, S., Grissa, K.L., Lognay, G., Bitume, E., Hance, T. and Mailleux, A.C., 2013. A review of the major biological approaches to control the worldwide pest *Tetranychus urticae* (Acari: Tetranychidae) with special reference to natural pesticides. Journal of Pest Science, 86, 361–386.

Babb, P.L., Lahens, N.F., Correa-Garhwal, S.M., Nicholson, D.N., Kim, E.J., Hogenesch, J.B., Kuntner, M., Higgins, L., Hayashi, C.Y., Agnarsson, I. and Voight, B.F., 2017. The *Nephila clavipes* genome highlights the diversity of spider silk genes and their complex expression. Nature Genetics, 49, 895.

Bain, B.A. and Govedich, F.R., 2004. Courtship and mating behavior in the Pycnogonida (Chelicerata: Class Pycnogonida): a summary. Invertebrate Reproduction and Development, 46, 63–79.

Ballesteros, J.A. and Sharma, P.P., 2019. A critical appraisal of the placement of Xiphosura (Chelicerata) with account of known sources of phylogenetic error. Systematic Biology, 68, 896–917.

Bartsch, I. 1994. Halacarid mites (Acari) from hydrothermal deep-sea sites: new records. Cahiers de Biologie Marine, 35, 479–490.

Brenneis, G., Ungerer, P. and Scholtz, G., 2008. The chelifores of sea spiders (Arthropoda, Pycnogonida) are the appendages of the deutocerebral segment. Evolution and Development, 10, 717–724.

Bond, J.E. and Opell, B.D., 1998. Testing adaptive radiation and key innovation hypotheses in spiders. Evolution, 52, 403–414.

Botton, M.L., 2009. The ecological importance of horseshoe crabs in estuarine and coastal communities: a review and speculative summary. In Tanacredi J., Botton M., Smith D. (eds.) Biology and conservation of horseshoe crabs (pp. 45–63). Springer, Boston, MA.

Boyer, S.L., Baker, J.M. and Giribet, G., 2007. Deep genetic divergences in *Aoraki denticulata* (Arachnida, Opiliones, Cyphophthalmi): a widespread 'mite harvestman' defies DNA taxonomy. Molecular Ecology, 16, 4999–5016.

Burns, M.M., Hedin, M. and Shultz, J.W., 2013. Comparative analyses of reproductive structures in harvestmen (Opiliones) reveal multiple transitions from courtship to precopulatory antagonism. PLoS One, 8, e66767.

Buzatto, B.A. and Machado, G., 2014. Male dimorphism and alternative reproductive tactics in harvestmen (Arachnida: Opiliones). Behavioural Processes, 109, 2–13.

Cao, Z., Yu, Y., Wu, Y., Hao, P., Di, Z., He, Y., Chen, Z., Yang, W., Shen, Z., He, X., Sheng, J., Xu, X., Pan, B., Feng, J., Yang, X., Hong, W., Zhao, W., Li, Z., Huang, K., Li, T., Kong, Y., Liu, H., Jiang, D., Zhang, B., Hu, J., Hu, Y., Wang, B., Dai, J., Yuan, B., Feng, Y., Huang, W., Xing, X., Zhao, G., Li, X., Li, Y. and Li, W., 2013. The genome of *Mesobuthus martensii* reveals a unique adaptation model of arthropods. Nature Communications, 4, 2602.

Chapin, K.J. and Hebets, E.A., 2016. The behavioral ecology of amblypygids. The Journal of Arachnology, 44, 1–15.

Chen, Z., Corlett, R.T., Jiao, X., Liu, S.-J., Charles-Dominique, T., Zhang, S., Li, H., Lai, R., Long, C. and Quan, R.-C., 2018. Prolonged milk provisioning in a jumping spider. Science, 362, 1052–1055.

Chippaux, J.P., 2012. Emerging options for the management of scorpion stings. Drug Design, Development and Therapy, 6, 165–173.

Coddington, J.A. and Levi, H.W. 1991. Systematics and evolution of spiders (Araneae). Annual Review of Ecology and Systematics, 22, 565–592.

Cushing, P.E. and Casto, P., 2012. Preliminary survey of the setal and sensory structures on the pedipalps of camel spiders (Arachnida: Solifugae). Journal of Arachnology, 40, 123–127.

Derkarabetian, S. and Hedin, M., 2014. Integrative taxonomy and species delimitation in harvestmen: a revision of the western North American genus *Sclerobunus* (Opiliones: Laniatores: Travunioidea). PLoS One, 9, e104982.

Derkarabetian, S., Starrett, J., Tsurusaki, N., Ubick, D., Castillo, S. and Hedin, M., 2018. A stable phylogenomic classification of Travunioidea (Arachnida, Opiliones, Laniatores) based on sequence capture of ultraconserved elements. ZooKeys, 760, 1–36.

Di, Z., Yu, Y., Wu, Y., Hao, P., He, Y., Zhao, H., et al., 2015. Genome-wide analysis of homeobox genes from *Mesobuthus martensii* reveals Hox gene duplication in scorpions. Insect Biochemistry and Molecular Biology, 61, 25–33.

Doña, J., Proctor, H., Serrano, D., Johnson, K.P., Oploo, A.O.V., Huguet-Tapia, J. C., Ascunce, M.S. and Jovani, R., 2018. Feather mites play a role in cleaning host feathers: new insights from DNA metabarcoding and microscopy. Molecular Ecology, 28, 203–218.

Dunlop, J.A. and Arango, C.P., 2005. Pycnogonid affinities: a review. Journal of Zoological Systematics and Evolutionary Research, 43, 8–21.

Dunlop, J.A. and Lamsdell, J.C., 2017. Segmentation and tagmosis in Chelicerata. Arthropod Structure and Development, Evolution of Segmentation, 46, 395–418.

Dunn, C.W., Hejnol, A., Matus, D.Q., Pang, K., Browne, W.E., Smith, S.A., Seaver, E., Rouse, G.W., Obst, M., Edgecombe, G.D., Sørensen, M.V., Haddock, S.H.D., Schmidt-Rhaesa, A., Okusu, A., Kristensen, R.M., Wheeler, W.C., Martindale, M.Q. and Giribet, G., 2008. Broad phylogenomic sampling improves resolution of the animal tree of life. Nature, 452, 745–749.

Edgecombe, G.D., 2010. Arthropod phylogeny: an overview from the perspectives of morphology, molecular data and the fossil record. Arthropod Structure and Development, Fossil Record and Phylogeny of the Arthropoda, 39, 74–87.

Edward, K.L. and Harvey, M.S., 2008. Short-range endemism in hypogean environments: the pseudoscorpion genera *Tyrannochthonius* and *Lagynochthonius* (Pseudoscorpiones: Chthoniidae) in the semiarid zone of Western Australia. Invertebrate Systematics, 22, 259–293.

Eisen, R.J., Eisen, L. and Beard, C.B., 2016. County-scale distribution of *Ixodes scapularis* and *Ixodes pacificus* (Acari: Ixodidae) in the continental United States. Journal of Medical Entomology, 53, 349–386.

Elias, D.O., Maddison, W.P., Peckmezian, C., Girard, M.B. and Mason, A.C., 2012. Orchestrating the score: complex multimodal courtship in the *Habronattus coecatus* group of *Habronattus* jumping spiders (Araneae: Salticidae). Biological Journal of the Linnean Society, 105, 522–547.

Emata, K.N. and Hedin, M., 2016. From the mountains to the coast and back again: ancient biogeography in a radiation of short-range endemic harvestmen from California. Molecular Phylogenetics and Evolution, 98, 233–243.

Fahrein, K., Talarico, G., Braband, A. and Podsiadlowski, L., 2007. The complete mitochondrial genome of *Pseudocellus pearsei* (Chelicerata: Ricinulei) and a comparison of mitochondrial gene rearrangements in Arachnida. BMC Genomics, 8, 386.

Garb, J.E., Ayoub, N.A. and Hayashi, C.Y., 2010. Untangling spider silk evolution with spidroin terminal domains. BMC evolutionary Biology, 10, 243.

Garwood, R.J. and Dunlop, J., 2014. Three-dimensional reconstruction and the phylogeny of extinct chelicerate orders. PeerJ, 2, e641.

Garwood, R.J., Dunlop, J.A., Knecht, B.J. and Hegna, T.A., 2017. The phylogeny of fossil whip spiders. BMC Evolutionary Biology, 17, 105.

Ghosh, A., Roy, R., Nandi, M. and Mukhopadhyay, A., 2018. Scorpion venom – Toxins that aid in drug development: a review. International Journal of Peptide Research and Therapeutics, 25, 27–37.

Girard, M.B., Kasumovic, M.M. and Elias, D.O., 2011. Multimodal courtship in the peacock spider, *Maratus volans* (O.P.-Cambridge, 1874). PLoS One, 6, e25390.

Giribet, G., 2018. Current views on chelicerate phylogeny – A tribute to Peter Weygoldt. Zoologischer Anzeiger, 273, 7–13.

Giribet, G., Edgecombe, G.D., Wheeler, W.C. and Babbitt, C., 2002. Phylogeny and systematic position of Opiliones: a combined analysis of chelicerate relationships using morphological and molecular data. Cladistics, 18, 5–70.

Giribet, G. and Hormiga, G., 2016. Phylum Arthropoda. The Chelicerata. In Brusca, R.C., Moore, W., Shuster, S.M. (eds.) Invertebrates, 3rd edition (pp. 911–966). Sinauer Associates, Inc. Sunderland, MA.

Giribet, G. and Edgecombe, G.D., 2019. The phylogeny and evolutionary history of arthropods. Current Biology, 29, R592-R602.

Grbić, M., Van Leeuwen, T., Clark, R.M., Rombauts, S., Rouzé, P., Grbić, V., et al., 2011. The genome of *Tetranychus urticae* reveals herbivorous pest adaptations. Nature, 479, 487–492.

Griswold, C.E., Audisio, T. and Ledford, J.M., 2012. An extraordinary new family of spiders from caves in the Pacific Northwest (Araneae, Trogloraptoridae, new family). ZooKeys, 215, 77–102.

Guerin, P.M., Kröber, T., McMahon, C., Guerenstein, P., Grenacher, S., Vlimant, M., Diehl, P.A., Steullet, P. and Syed, Z., 2000. Chemosensory and behavioural adaptations of ectoparasitic arthropods. Nova Acta Leopoldina, 83, 213–229.

Gulia-Nuss, M., Nuss, A.B., Meyer, J.M., Sonenshine, D.E., Roe, R.M., Waterhouse, R. M., et al., 2016. Genomic insights into the *Ixodes scapularis* tick vector of Lyme disease. Nature Communications, 7, 10507.

Harvey, M.S., 2002. Short-range endemism amongst the Australian fauna: some examples from non-marine environments. Invertebrate Systematics, 16, 555–570.

Hedin, M., Derkarabetian, S., Ramírez, M. J., Vink, C. and Bond, J.E., 2018. Phylogenomic reclassification of the world's most venomous spiders (Mygalomorphae, Atracinae), with implications for venom evolution. Scientific Reports, 8, 1636.

Huang, D., Hormiga, G., Xia, F., Cai, C., Yin, Z., Su, Y. and Giribet, G., 2018. Origin of spiders and their spinning organs illuminated by mid-Cretaceous amber fossils. Nature Ecology and Evolution, 2, 623–627.

Jager, M., Murienne, J., Clabaut, C., Deutsch, J., Guyader, H.L. and Manuel, M., 2006. Homology of arthropod anterior appendages revealed by Hox gene expression in a sea spider. Nature, 441, 506–508.

Kenny, N.J., Chan, K.W., Nong, W., Qu, Z., Maeso, I., Yip, H.Y., Chan, T.F., Kwan, H.S., Holland, P.W.H., Chu, K.H. and Hui, J.H.L., 2016. Ancestral whole-genome duplication in the marine chelicerate horseshoe crabs. Heredity, 116, 190.

Klußmann-Fricke, B.J. and Wirkner, C.S., 2016. Comparative morphology of the hemolymph vascular system in Uropygi and Amblypygi (Arachnida): complex correspondences support Arachnopulmonata. Journal of Morphology, 277, 1084–1103.

Krantz, G.W., 2009. Introduction. In Krantz, G.W., Walter, D.E. (eds.) A Manual of Acarology, 3rd edition (pp. 1–2). Texas Tech University Press, Lubbock, TX.

Lane, S.J., Moran, A.L., Shishido, C.M., Tobalske, B.W. and Woods, H.A., 2018. Cuticular gas exchange by Antarctic sea spiders. Journal of Experimental Biology, 221, jeb177568.

Legg, D.A., Sutton, M.D. and Edgecombe, G.D., 2013. Arthropod fossil data increase congruence of morphological and molecular phylogenies. Nature Communications, 4, 2485.

Leite, D.J. and McGregor, A.P., 2016. Arthropod evolution and development: recent insights from chelicerates and myriapods. Current Opinion in Genetics and Development, 39, 93–100.

Leite, D.J., Baudouin-Gonzalez, L., Iwasaki-Yokozawa, S., Lozano-Fernandez, J., Turetzek, N., Akiyama-Oda, Y., Prpic, N.M., Pisani, D., Oda, H., Sharma, P.P. and McGregor A.P., 2018. Homeobox gene duplication and divergence in arachnids. Molecular Biology and Evolution, 35, 2240–2253.

Lindquist, E.E., Krantz, G.W. and Walter, D.E., 2009. Classification. In Krantz, G.W., Walter, D.E. (eds.), A Manual of Acarology, 3rd edition (pp. 97–106). Texas Tech University Press, Lubbock, TX.

Lozano-Fernandez, J., Tanner, A.R., Giacomelli, M., Carton, R., Vinther, J., Edgecombe, G.D. and Pisani, D. 2019. Increasing species sampling in chelicerate genomic-scale datasets provides support for monophyly of Acari and Arachnida. Nature communications, 10, 2295.

Machado, G., Pinto-da-Rocha, R. and Giribet, G., 2007. What are harvestmen? In Pinto-da-Rocha, R., Machado, G., Giribet, G. (eds.), Harvestmen: The Biology of Opiliones (pp. 1–13). Harvard University Press, Cambridge, MA.

Masta, S.E. and Boore, J.L., 2008. Parallel evolution of truncated transfer RNA genes in arachnid mitochondrial genomes. Molecular Biology and Evolution, 25, 949–959.

Masta, S.E., Longhorn, S.J. and Boore, J.L., 2009. Arachnid relationships based on mitochondrial genomes: asymmetric nucleotide and amino acid bias affects phylogenetic analyses. Molecular Phylogenetics and Evolution, 50, 117–128.

Masta, S.E., McCall, A. and Longhorn, S.J., 2010. Rare genomic changes and mitochondrial sequences provide independent support for congruent relationships among the sea spiders (Arthropoda, Pycnogonida). Molecular Phylogenetics and Evolution, 57, 59–70.

Meehan, C.J., Olson, E.J., Reudink, M.W., Kyser, T K. and Curry, R.L., 2009. Herbivory in a spider through exploitation of an ant–plant mutualism. Current Biology, 19, R892-R893.

Mora, G., 1990. Paternal care in a neotropical harvestman, *Zygopachylus albomarginis* (Arachnida, Opiliones: Gonyleptidae). Animal Behaviour, 39, 582–593.

Nicholson, G.M. and Graudins, A., 2002. Spiders of medical importance in the Asia–Pacific: atracotoxin, latrotoxin and related spider neurotoxins. Clinical and Experimental Pharmacology and Physiology, 29, 785–794.

Nyffeler, M. and Symondson, W.O. 2001. Spiders and harvestmen as gastropod predators. Ecological Entomology, 26, 617–628.

Nyffeler, M. and Birkhofer, K., 2017. An estimated 400–800 million tons of prey are annually killed by the global spider community. The Science of Nature, 104, 30.

Ovchinnikov, S. and Masta, S.E., 2012. Pseudoscorpion mitochondria show rearranged genes and genome-wide reductions of RNA gene sizes and inferred structures, yet typical nucleotide composition bias. BMC Evolutionary Biology, 12, 31.

Ostrovsky, A.N., Lidgard, S., Gordon, D.P., Schwaha, T., Genikhovich, G. and Ereskovsky, A.V., 2016. Matrotrophy and placentation in invertebrates: a new paradigm. Biological Reviews, 91, 673–711.

Painting, C.J., Probert, A.F., Townsend, D.J. and Holwell, G.I., 2015. Multiple exaggerated weapon morphs: a novel form of male polymorphism in harvestmen. Scientific Reports, 5, 16368.

Proctor, H. and Owens, I., 2000. Mites and birds: diversity, parasitism and coevolution. Trends in Ecology and Evolution, 15, 358–364.

Rao, G. and Rao, K.P., 1973. A metasomatic neural photoreceptor in the scorpion. Journal of Experimental Biology, 58, 189–196.

Regier, J.C., Shultz, J.W., Zwick, A., Hussey, A., Ball, B., Wetzer, R., Martin, J.W. and Cunningham, C.W., 2010. Arthropod relationships revealed by phylogenomic analysis of nuclear protein-coding sequences. Nature, 463, 1079.

Sanggaard, K.W., Bechsgaard, J.S., Fang, X., Duan, J., Dyrlund, T.F., Gupta, V., Jiang, X., Cheng, L., Fan, D., Feng, Y., Han, L., Huang, Z., Wu, Z., Liao, L., Settepani, V., Thøgersen, I.B., Vanthournout, B., Wang, T., Zhu, Y., Funch, P., Enghild, J.J., Schauser, L., Andersen, S.U., Villesen, P., Schierup, M.H., Bilde, T. and Wang, J., 2014. Spider genomes provide insight into composition and evolution of venom and silk. Nature Communications, 5, 3765.

Santibañez-López, C.E., Gavish-Regev, E., Ballesteros, J.A., Ontano, A.Z., Kováč L., Harvey, M. and Sharma, P.P., 2019. Neglected no longer: phylotranscriptomics and molecular modeling reveals venom homologs in Pseudoscorpiones and Palpigradi. Poster presented at International Congress of Arachnology, Christchurch, New Zealand.

Satler, J.D., Carstens, B.C. and Hedin, M., 2013. Multilocus species delimitation in a complex of morphologically conserved trapdoor spiders (Mygalomorphae, Antrodiaetidae, *Aliatypus*). Systematic Biology, 62, 805–823.

Schütz, D. and Taborsky, M., 2003. Adaptations to an aquatic life may be responsible for the reversed sexual size dimorphism in the water spider, *Argyroneta aquatica*. Evolutionary Ecology Research, 5, 105–117.

Schwager, E.E., Schoppmeier, M., Pechmann, M. and Damen, W.G., 2007. Duplicated Hox genes in the spider *Cupiennius salei*. Frontiers in Zoology, 4, 10.

Schwager, E.E., Schönauer, A., Leite, D.J., Sharma, P.P. and McGregor, A.P., 2015. Chelicerata. In Wanninger, A. (ed.) Evolutionary Developmental Biology of Invertebrates, 3 (pp. 99–139). Springer, Vienna.

Schwager, E.E., Sharma, P.P., Clarke, T., Leite, D.J., Wierschin, T., Pechmann, M., Akiyama-Oda, Y., Esposito, L., Bechsgaard, J., Bilde, T., Buffry, A.D., Chao, H., Dinh, H., Doddapaneni, H., Dugan, S., Eibner, C., Extavour, C.G., Funch, P., Garb, J., Gonzalez, L.B., Gonzalez, V.L., Griffiths-Jones, S., Han, Y., Hayashi, C., Hilbrant, M., Hughes, D.S.T., Janssen, R., Lee, S.L., Maeso, I., Murali, S.C., Muzny, D.M., Nunes da Fonseca, R., Paese, C.L.B., Qu, J., Ronshaugen, M., Schomburg, C., Schönauer, A., Stollewerk, A., Torres-Oliva, M., Turetzek, N., Vanthournout, B., Werren, J.H., Wolff, C., Worley, K.C., Bucher, G., Gibbs, R.A., Coddington, J., Oda, H., Stanke, M., Ayoub, N.A., Prpic, N.-M., Flot, J.-F., Posnien, N., Richards, S. and McGregor, A.P., 2017. The house spider genome reveals an ancient whole-genome duplication during arachnid evolution. BMC Biology, 15, 62.

Scott, C.E., Anderson, A.G., Andrade, M.C.B., 2018. A review of the mechanisms and functional roles of male silk use in spider courtship and mating. Journal of Arachnology, 46, 173–207.

Sharma, P.P. 2018. Chelicerates. Current Biology, 28, R774–R778.

Sharma, P.P., Schwager, E.E., Extavour, C.G. and Giribet, G., 2012. Evolution of the chelicera: a *dachshund* domain is retained in the deutocerebral appendage of Opiliones (Arthropoda, Chelicerata). Evolution and Development, 14, 522–533.

Sharma, P.P., Schwager, E.E., Giribet, G., Jockusch, E.L. and Extavour, C.G., 2013. Distal-less and dachshund pattern both plesiomorphic and apomorphic structures in chelicerates: RNA interference in the harvestman *Phalangium opilio* (Opiliones). Evolution and Development, 15, 228–242.

Sharma, P.P., Kaluziak, S.T., Pérez-Porro, A.R., González, V.L., Hormiga, G., Wheeler, W.C. and Giribet, G., 2014. Phylogenomic interrogation of arachnida reveals systemic conflicts in phylogenetic signal. Molecular Biology and Evolution, 31, 2963–2984.

Sharma, P.P., Santiago, M.A., González-Santillán, E., Monod, L. and Wheeler, W.C., 2015. Evidence of duplicated Hox genes in the most recent common ancestor of extant scorpions. Evolution and Development, 17, 347–355.

Shultz, J.W., 1990. Evolutionary morphology and phylogeny of Arachnida. Cladistics, 6, 1–38.

Shultz, J.W., 2001. Gross muscular anatomy of *Limulus polyphemus* (Xiphosura, Chelicerata) and its bearing on evolution in the Arachnida. The Journal of Arachnology, 29, 283–303.

Shultz, J.W., 2007. A phylogenetic analysis of the arachnid orders based on morphological characters. Zoological Journal of the Linnean Society, 150, 221–265.

Starrett, J., Derkarabetian, S., Hedin, M., Bryson Jr, R.W., McCormack, J.E. and Faircloth, B.C., 2017. High phylogenetic utility of an ultraconserved element probe set designed for Arachnida. Molecular Ecology Resources, 17, 812–823.

Stowe, M.K., Tumlinson, J.H. and Heath, R.R., 1987. Chemical mimicry: bolas spiders emit components of moth prey species sex pheromones. Science, 236, 964–967.

Vetter, R.S., 2015. The Brown Recluse Spider. Cornell University Press, Cornell, NY.

Vetter, R.S. and Isbister, G.K., 2008. Medical aspects of spider bites. Annual Review of Entomology, 53, 409–429.

Vizueta, J., Rozas, J. and Sánchez-Gracia, A., 2018. Comparative genomics reveals thousands of novel chemosensory genes and massive changes in chemoreceptor repertories across chelicerates. Genome Biology and Evolution, 10, 1221–1238.

Wang, B., Dunlop, J.A., Selden, P., Garwood, R.J., Shear, W.A., Müller, P. and Lie, X., 2018. Cretaceous arachnid *Chimerarachne yingi* gen. et sp. nov. illuminates spider origins. Nature Ecology and Evolution, 2, 614–622.

Walter, D.E. and Proctor, H.C., 2013. Acari underwater, or, why did mites take the plunge? In Walter, D.E., Proctor, H.C. (eds.), Mites: Ecology, Evolution & Behaviour (pp. 229–280). Springer, Dordrecht.

Weygoldt, P. and Paulus, H.F., 1979. Untersuchungen zur Morphologie, Taxonomie und Phylogenie der Chelicerata1 II. Cladogramme und die Entfaltung der Chelicerata. Journal of Zoological Systematics and Evolutionary Research, 17, 177–200.

Wheeler, W.C. and Hayashi, C.Y., 1998. The phylogeny of the extant chelicerate orders. Cladistics, 14, 173–192.

Wolfe, J.M., Daley, A.C., Legg, D.A. and Edgecombe, G.D., 2016. Fossil calibrations for the arthropod tree of life. Earth-Science Reviews, 160, 43–110.

World Spider Catalog, version 20.0. https://wsc.nmbe.ch/

Yang, X., Norma-Rashid, Y., Lourenço, W.R. and Zhu, M., 2013. True lateral eye numbers for extant buthids: a new iscovery on an old character. PLoS One, 8, e55125.

Young, M.R., Proctor, H.C., deWaard, J.R. and Hebert, P.D., 2019. DNA barcodes expose unexpected iversity in Canadian mites. Molecular Ecology. Doi:10.1111/mec.15292

Zurek, D.B., Cronin, T.W., Taylor, L.A., Byrne, K., Sullivan, M.L. and Morehouse, N.I., 2015. Spectral filtering enables trichromatic vision in colorful jumping spiders. Current Biology, 25, R403–R404.

Zwicky, K.T., 1968. A light response in the tail of *Urodacus*, a scorpion. Life sciences, 7, 257–262.

Additional reading

Beccaloni, J., 2009. Arachnids. University of California Press, 320 pages.

Dunlop, J.A., 2010. Geological history and phylogeny of Chelicerata. Arthropod Structure and Development, 39, 124–142.

Foelix, R., 2010. Biology of Spiders. Oxford University Press, 432 pages.

Garb, J.E., Sharma, P.P. and Ayoub, N.A., 2018. Recent progress and prospects for advancing arachnid genomics. Current Opinion in Insect Science, 25, 51–57.

Giribet, G. and Sharma, P.P., 2015. Evolutionary biology of harvestmen (Arachnida, Opiliones). Annual Review of Entomology, 60, 157–175.

Mammola, S., Michalik, P., Hebets, E.A. and Isaia, M., 2017. Record breaking achievements by spiders and the scientists who study them. PeerJ, 5, e3972.

Pinto-da-Rocha, R., Machado, G. and Giribet, G., 2007. Opiliones: Biology of Harvestmen. Harvard University Press, 608 pages.

Polis, G.A., 1990. The Biology of Scorpions. Stanford University Press, 587 pages.

Sharma, P.P., 2017. Chelicerates and the conquest of land: a view of arachnid origins through an evo-devo spyglass. Integrative and Comparative Biology, 57, 510–522.

Walter, D.E. and Proctor, H.C., 2013. Mites: Ecology, Evolution and Behavior, 2nd edition. Springer, Dordrecht.

Weygoldt, P., 1969. The Biology of Pseudoscorpions. Harvard University Press, 159 pages.

CHAPTER
29

SUBPHYLUM MYRIAPODA

José D. Gilgado

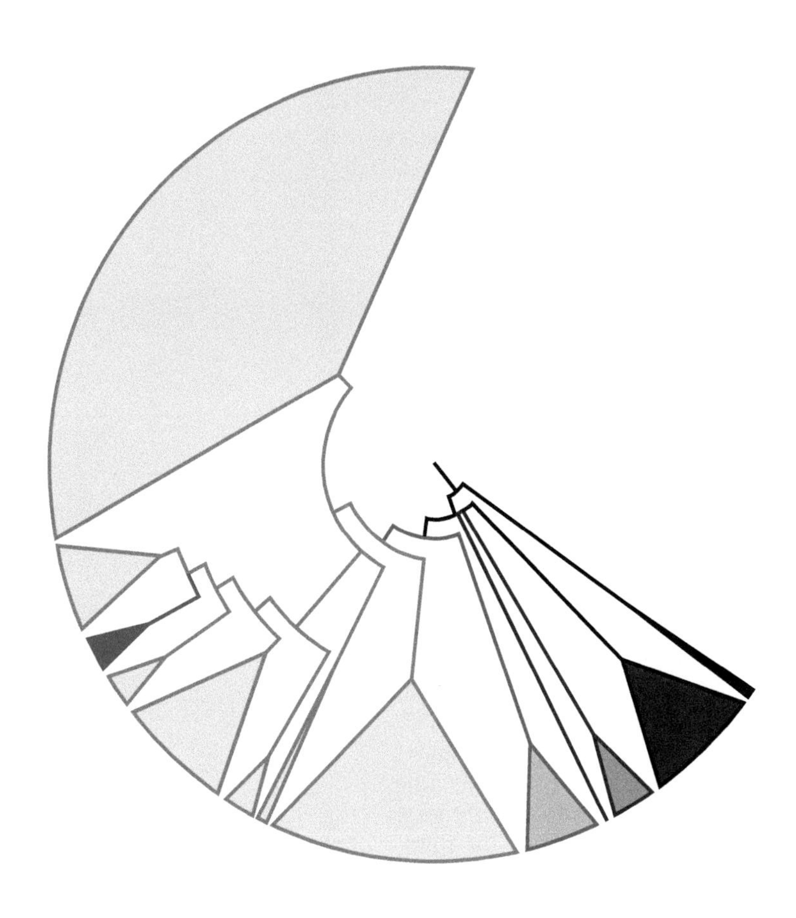

CONTENTS

General Characteristics

Myriapoda is divided in four Classes: Chilopoda, Diplopoda, Pauropoda and Symphyla

More than 16,000 described species

Between 60,000 and 93,000 estimated species

Predominantly soil animals, inhabiting all continents except Antarctica

Body divided into two tagmata (head and multisegmented trunk) with large number of legs (up to 750) and body length ranging from a couple of millimeters to 30 cm

Myriapoda together with the Pancrustacea ("Crustacea" + Hexapoda) form the clade Mandibulata

Nuclear genome size ranges from 290 to 538 Mb (so far only three species have been studied)

Of very limited economic importance, except for pet trading or damages by invasive and pest species, and for the potential pharmacological use of centipede venom

As detritivores, Diplopoda are of high ecological importance in soil ecosystems, while Chilopoda are important predators

Autapomorphies

- structure of tentorial endoskeleton (tentorium) and its functional relationship to the mandible
- groups of serotonin-immunoreactive neurons formed only by single cells or pairs of them
- eucone cells of ommatidium with nuclei positioned proximally to the cone compartments

HISTORY OF TAXONOMY AND CLASSIFICATION

Despite their relative minor economic importance, hazard or applications, the two more conspicuous classes of the Myriapoda, the Chilopoda (centipedes) and Diplopoda (millipedes) have been known to humans for a long time. The Greek natural scientist and philosopher Theophrastus reported millipede swarms more than 2,000 years ago (Sharples 1994, Enghoff and Kebapći 2008). Millipedes and centipedes have traditionally been used as food in some parts of Africa (Enghoff et al. 2014) and as medicine in some regions of Asia (Zimian et al. 1997). The first myriapod descriptions meeting modern scientific standards were made by Linnaeus in 1758. Linnaeus described a few millipede and centipede species in the genera *Julus* and *Scolopendra*, respectively. As soon as more naturalists started paying attention to this group, the complexity of Myriapoda became more evident (**Figure 29.1**), and new genera were described which were later assigned to new orders (Minelli 2015a). Since the early 19th century, the number of Myriapoda studies increased, with important works published by Attems, Brölemann, Cook, Latzel, Newport, Meinert, Silvestri, and Verhoeff among others. Extensive monographs and systematic revisions facilitated further studies of Myriapoda. At that time, myriapodologists usually worked with more than one class of Myriapoda. However, since the middle of the 20th century it became common to specialize in just one of the four classes of this subphylum.

Apart from the widely known centipedes and millipedes, the subphylum Myriapoda includes two less conspicuous classes: Symphyla and Pauropoda. The first described species of Symphyla was originally thought to belong to the genus *Scolopendra* (Scopoli 1763). As explained by Domínguez (1992), it was only Ryder (1880), more than 100 years later, who created the modern class Symphyla as a distinct group. In contrast, Lubbock (1867) clearly identified Pauropoda as a separate group from the outset (Scheller 2011). While Myriapoda as a whole is an understudied group, its four classes have received varying amounts of attention. Centipedes are probably slightly better known than millipedes regarding their systematics and biology, and these two groups with relatively large-sized animals are significantly better known than the smaller and more cryptic-living Symphyla and Pauropoda. The systematics of the subphylum has changed repeatedly over time and even its monophyly was

Figure 29.1 Representatives of the four main classes of Myriapoda. (A) A centipede (Chilopoda: Scolopendromorpha). (B) A millipede (Diplopoda: Julida). (C) A symphylan (Symphyla). (D) A pauropod (Pauropoda), photograph by Andy Murray (chaos of delight).

questioned until recently (see section Position in the ToL). Even today the relationships among the classes are not fully resolved. However, recent advances in phylogenetics have clarified that Diplopoda and Pauropoda are a monophyletic clade which seems to be closer to Symphyla (Fernández et al. 2018), with Chilopoda as a sister group of all them. However, whether the sister group of all is Symphyla or Chilopoda seems to depend on what groups are included in the analysis (Giribet and Edgecombe 2019).

ANATOMY, HISTOLOGY, AND MORPHOLOGY

The morphology and anatomy of the Myriapoda are similar to those of insects in several respects. Commonalities include the tracheal system for gas exchange, the Malpighian tubules for excretion, a single pair of antennae and head appendages, articulated uniramous legs, and the presence of a pair of mandibles and maxillae (although in Chilopoda and Symphyla there is a second additional pair of maxillae). Nevertheless, the Myriapoda body plan is very different from that of insects (**Figure 29.2**), with their body divided in just two tagmata: A cephalon and a multisegmented trunk showing some external and internal signs of metamerization (Brusca et al. 2016). The segments in Chilopoda are simple and their number ranges from 15 to 191. They are also simple in Symphyla, but species of this class have only 14 segments. However, in Diplopoda, the majority of the segments are not simple, but formed by the fusion of pairs of simple ones. These double segments are called diplosegments, while the simple ones are called haplosegments. In Chilopoda, the number of segments ranges from 9 to 192. Lastly, Pauropoda have 11 segments, mostly simple but some pairs are fused. The most conspicuous trait of the Myriapoda, and the source of their name, is their large number of legs. Diplopoda species *Illacme plenipes* has 750 legs which is highest number of legs (Marek et al. 2012). The number of legs per segment is originally one pair, but diplosegments have two pairs. Chilopoda have mostly one pair of legs per segment, with the first pair modified into forcipules used to inject venom into the prey, and the last pair of legs usually presenting sexual dimorphism. This venom is of interest to biomedicine (Undheim et al. 2014). Millipedes generally have two pairs of legs per (diplo)segment, except the first four (haplo)segments (a legless segment called Collum and three "thoracic"

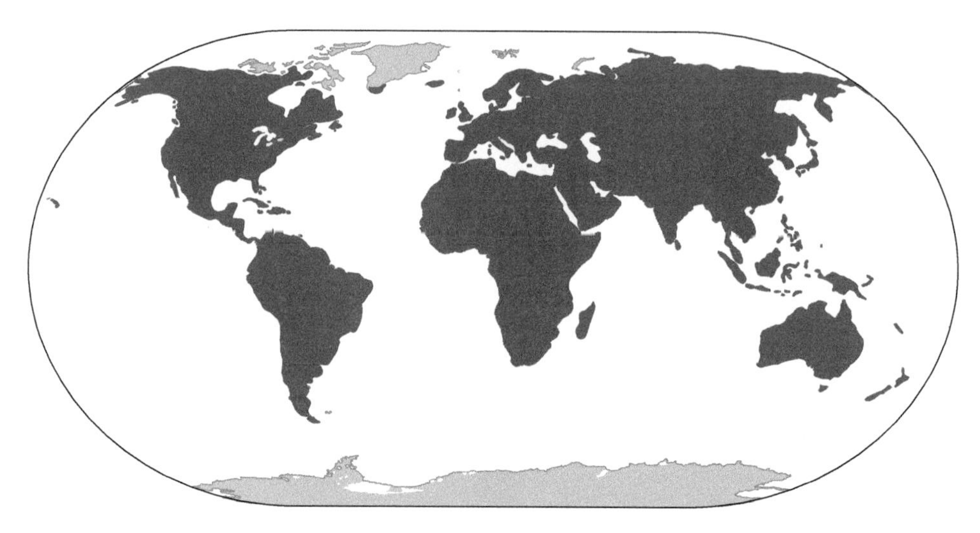

Figure 29.2 Distribution map of Myriapoda. They are distributed in all continental lands and most islands.

single-paired legged segments) and the last segment of the body (also legless) called telson. One or more pairs of legs are usually modified as reproductive organs in males. The modified legs are called gonopods or telopods, depending on the group. Symphylans have one pair of legs per segment, and while pauropods mostly have one pair of legs per segment, in those segments which are fused they have two pairs.

The cuticle of Myriapoda is sclerotized to a different extent, depending on the group or species. Symphylans are weakly sclerotized and this is also true for most pauropods, but it depends on the family, while centipedes have a normal sclerotized cuticle. The cuticle of millipedes is not only sclerotized but even calcified, except in penicillates (Makarov 2015), and has repugnatory glands. Myriapoda do not usually have a well-developed wax layer on their cuticle, except for some desert-adapted Diplopoda species (Brusca et al. 2016).

The sensory organs vary among classes and species. Many groups and species lack eyes. Species with eyes have usually simple groups of ocelli. An exception is the centipede order Scutigeromorpha whose individuals possess a particular type of compound eye. It seems that the plesiomorphic state of the character is homologous to the compound eye of crustaceans and insects (Müller et al. 2011, Müller and Sombke 2015). A relevant character of the myriapod eye, proposed as an apomorphy of the group, is the position of the nuclei in the eucone cells of the ommatidia, proximal to the cone compartments (Müller et al. 2007). Myriapoda have very sensitive antennae, with different shapes in each class. The most remarkable are those of Pauropoda, biramous and with three flagella and a unique candelabra-shaped or globular sense organ (Scheller 2011). The most characteristic and intriguing sensory organ of Myriapoda is the Tömösvary organ, whose function is not fully clear. It is located in the head and several studies have proposed different functions such as olfaction, gravity perception, hygroreception, perception of CO_2, or even thermoreception, although it probably evolved as a chemoreception organ (Müller and Sombke 2015).

The nervous system is generally similar to that of other arthropods. However, the presence of groups of serotonin-immunoreactive neurons formed by just single cells or pairs of cells is a character only present in Myriapoda and has been proposed to be an apomorphy of the group (Harzsch 2004; 2006). The circulatory system of the Myriapoda is open, with their organs surrounded by hemolymph, similar to other arthropods. It has a vascular system, including a tubular heart that pumps the hemolymph toward the head, and a lacunar system with spaces among the organs (Wirkner et al. 2011, Wirkner and Xylander

2015). The circulation as well as the pressure of the hemolymph are relatively low (Brusca et al. 2016). The digestive system is simple, with no ramifications, formed by a foregut, midgut, and hindgut (Hopkin and Read 1992; Koch et al. 2011). A very remarkable anatomical character of the Myriapoda is the structure of the cephalic tentorial endoskeleton (tentorium). It appears to be a very strong argument for the monophyly of the subphylum, especially, the hypopharyngeal bar that serves as support for the hypopharynx and which has no homologue within the Pancrustacea (Edgecombe 2011).

REPRODUCTION AND DEVELOPMENT

Myriapoda are gonochoric, although some species have parthenogenetic females and some populations of those species can even lack males. There are some external morphological differences between the two sexes, although they may not be present or evident in all groups. In general, Myriapoda transfer the sperm via a spermatophore. Exceptions include the order Geophilomorpha (Chilopoda) where the males release the sperm without spermatophore. The spermatophore transfer can be direct (most Diplopoda except Penicillata) or indirect (Symphyla, Pauropoda, and Chilopoda) (Hopkin and Read 1992; Minelli 2011; Scheller 2011; Szucsich and Scheller 2011). Chilopoda and Symphyla males produce dimorphic spermatozoa of two sizes (Minelli 2011; Szucsich and Scheller 2011), while Diplopoda and Pauropoda produce monomorphic spermatozoa, aflagellate in Diplopoda (Minelli and Michalik 2015) and flagellate in Pauropoda (Rosati et al. 1970). Females produce yolky eggs, although there is some secondary reduction in Pauropoda and Symphyla. They fertilize the eggs with the sperm received from the male, usually in the moment of laying them (Brusca et al. 2016). The number of eggs varies among groups and females may lay them individually or in clusters. In some groups of Chilopoda, the female stays with the eggs until hatching or sometimes even for longer (Minelli 2011).

The embryonic development of Myriapoda has been summarized by Brusca et al. (2016), and here their information is presented together with additional data from Minelli and Sombke (2011) and Minelli (2015b). During the embryonic development, the cleavage is meroblastic with early intralecithal nuclear divisions. The newly formed nuclei migrate to the peripheral cytoplasm and continue dividing, and only when they form a dense layer, new membranes start developing conforming a periblastula surrounding a yolky sphere with some scattered nuclei in it. In some Chilopoda and Symphyla groups, it has been observed that the cleavage of the yolky egg leads to the formation of yolk pyramids. A group of columnar cells forms a germinal disc, and in it, endodermal and mesodermal cells start differentiating as germinal centers as they proliferate, locating themselves beneath their parental cells that now form the ectoderm. A grove starts forming in the mesoderm, which will give place to the gut, and the cells start absorbing the yolky central mass of the embryo. Next, the segments start differentiating and the embryo develops appendage buds. In millipedes, the head and first three trunk segments are formed almost at the same time, while the next ones will be added sequentially. The mouth and anus, as in other arthropods, are formed via invaginations of the ectoderm, forming the fore- and hindgut. By the end of the embryonic development, in Diplopoda and Pauropoda a "pupoid" phase appears (Scheller 2011, Minelli 2015b). Although it breaks the chorion, the animal remains in the egg until the next molt (stage I), since it does not have articulated legs or antennae and has very little musculature.

There are two main types of postembryonic development in Myriapoda: epimorphosis (individuals are born with the final number of segments that they will have as adults) and anamorphosis (individuals are born with an incomplete number of segments). The anamorphic type of development is further divided in subtypes (Enghoff et al. 1993): euanamorphosis, where individuals continue adding segments until they die, even after reaching sexual maturity; hemianamorphosis, where individuals stop adding segments while still molting before reaching maturity; and teloanamorphosis, where individuals stop adding segments when they have the final number, reaching maturity at the same time. Centipedes comprise epimorphic and hemianamorphic groups (Minelli and Sombke 2011), while millipedes include groups with the three types of anamorphic developments (Enghoff et al. 1993). Symphyla are hemianamorphic (Szucsich and Scheller 2011) and Pauropoda are mostly teloanamorphic, although hemianamorphic cases are also known (Scheller 2011). There is a particularity observed in some groups of millipedes, where mature males can molt to an intercalary, nonreproductive stage and molt back to a mature stage. This process is called periodomorphosis (Minelli 2015b). In Myriapoda, depending on the group, there is a varying ability of regenerating lost limbs while molting.

DISTRIBUTION AND ECOLOGY

Myriapoda are predominantly soil organisms, but there are exceptions such as arboreal or semiaquatic species (Golovatch and Kime 2009, Dányi et al. 2019). With exceptions, each class has a different lifestyle. Centipedes are venomous soil-living predators, preferring inhabiting dark spaces and humid microhabitats, although some species have been observed to consume plant matter (Voigtländer 2011). The type of vegetation does not seem to be the most important factor for their communities, but vegetation may indirectly affect centipede communities by influencing the porosity or structure of the soil. Other variables seem to have a stronger effect, such as the soil pollution, the availability of prey, and the presence of other predators. The latter include many vertebrates, insects, and arachnids, with especially arachnids being a very important factor determining their abundance (Voigtländer 2011). Chilopoda feature three ecomorphotypes that correspond with different lifestyles (Manton 1977): A burrowing type as in Geophilomorpha; an intermediate type, as in Scolopendromorpha; and a running type, as in Lithobiomorpha and Scutigeromorpha. Millipedes are detritivores, usually living in decaying plant material in forests, but they have high range of habitats and microhabitats, with some species living even in relatively extreme environments such as deserts, high mountains, or deep caves (Golovatch 2009). The species richness tends to be higher in calcareous soils, presumably because of the higher mineral content of the plants living there but it could also be because of the microclimatic characteristics of those plant communities (David 2015). Although millipedes feed mostly on decaying plant material, many species also occasionally feed on fungus, feces, lichens, algae, and rarely, dead invertebrates (David 2015). They have an impact on litter decomposition, consuming up to 15% of the annual leaf fall in temperate forests (Golovatch and Kime 2009). Millipedes seem to play an important role in nitrogen mineralization despite their relatively low assimilation efficiency (Cárcamo et al. 2000).

Millipedes have defensive glands that are assumed to deter predators. Nevertheless, millipedes have many predators (vertebrates and arthropods). Indeed predation seems to be one of the main factors limiting the growth of millipede populations, which otherwise can increase exponentially under favorable conditions, even leading to population outbreaks and mass migrations (David 2015). However, swarming and migrations may also occur without a population outbreak due to adverse environment conditions.

Diplopoda can be divided into five morphotypes (Hopkin and Read 1992; Kime and Golovatch 2000): bulldozers or rammers (Julida, Spirobolida, and Spirostreptida), wedge types or litter-splitters (Polydesmida), borers (Chordeumatida, Polyzoniida, Platydesmida, and Siphonophorida), rollers (Glomerida and Sphaerotheriida), and bark dwellers (Polyxenida). In addition to the morphotypes, there are five lifestyles (stratobionts, pedobionts, troglobionts, subcorticoles, and epiphytobionts), and although the different morphotypes seem to provide an advantage for some lifestyles, there is not an absolute correspondence between them.

The ecology of the other two Myriapoda orders, Symphyla and Pauropoda, is less known. Symphylans live in different types of soil, but they seem to be less abundant in clay and sand (Szucsich and Scheller 2011). Their survival depends on the soil moisture and they stay in moist spots only leaving them for feeding (Edwards 1961). Symphylans are mostly saprophagous, although some of them may feed on live plant material, and gut contents show that some species could be predators (Walter et al. 1989). Like symphylans, pauropods usually inhabit humid (but not damp) soils but can also be found in decaying logs. Remarkably, some species have even been found in deserts (Scheller 2011). Pauropods cannot burrow, so their presence depends on soil pore size. Pauropods can migrate vertically, and they have been collected in soil at depths of up to 3 m (Price 1975). Although there is not much information on their feeding habits, the two main groups, Hexamerocerata and Tetramerocerata, seem to differ in mandible morphology. The first group chews and eats fungal hyphae and plant tissue, while the second sucks fluid from fungi or vegetables (Hüther 1959; Scheller 2011).

The four Myriapoda classes are present in all continents except Antarctica. The four main centipede orders (Scutigeromorpha, Lithobiomorpha, Geophilomorpha, and Scolopendromorpha) have very wide distributions, being present in most continents and most biotopes. In contrast, Craterostigomorpha, with only two known species, are only found in Tasmania and New Zealand (Bonato and Zapparoli 2011). Most millipede species have small distribution areas and high levels of endemicity (Golovatch and Kime 2009). Higher taxonomical levels also tend to have limited distribution ranges, making them a suitable group for biogeographical studies (Enghoff 2015). In both Chilopoda and Diplopoda, there are non-native and even invasive species that have been transported by humans and have nowadays a broad distribution (Stoev et al. 2010). In Symphyla, some genera have wide distributions. However, an insufficient sampling effort results in a lack of records for many regions (Szucsich and Scheller 2011). Very little is known about pauropod species distribution. Ten genera have a very wide distribution, and 13 genera, consisting of just one or two species, are only known from a single country (Scheller 2011). Overall though they are predominantly soil animals, inhabiting all continents except Antarctica (**Figure 29.3**).

PHYSIOLOGY AND BEHAVIOR

Most of the available information on the physiology and behavior of Myriapoda is derived from studies on centipedes and millipedes. Nevertheless, some of this information may equally apply to symphylans and pauropods, and comments on these classes will be made here when relevant. In Myriapoda the digestion occurs mainly in the midgut, and the food is surrounded by chitinous peritrophic membranes. Chilopoda process the food more strongly before ingestion, usually liquefying it inside the prey's body (Koch et al. 2011), while Diplopoda chew the plant material into small pieces. In this latter case, the digestion is carried out by enzymes and also by microorganisms (Hopkin and Read 1992).

Myriapoda breathe by means of tracheae similar to those of other arthropods, but there are some differences among groups. Chilopoda usually have a higher oxygen consumption rate than Diplopoda, and the oxygen travels through the tracheae mostly by active ventilation and not by passive diffusion, in contrast to the situation for Diplopoda. Interestingly, the fast-running Scutigeromorpha (Chilopoda) have a unique system of short tracheae to transfer the oxygen into the hemolymph where it is transported via hemocyanin. This pigment is also present in some large (and even some not so large) Diplopoda, although the

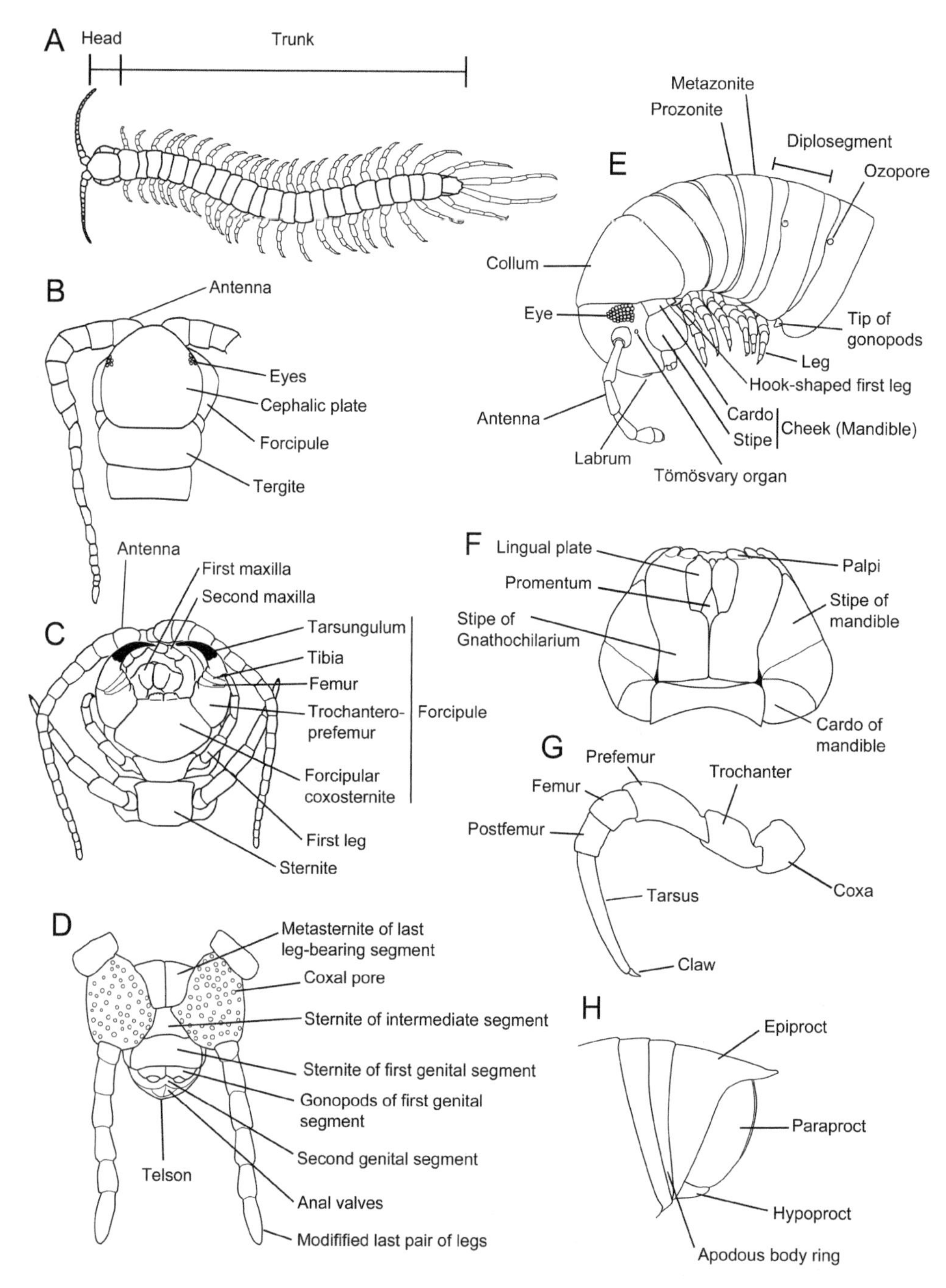

Figure 29.3 Morphology of Myriapoda. (A) General morphology based on a scolopendromorph centipede (Chilopoda). (B) Head and first segments of a scolopendromorph centipede in dorsal view (no walking legs drawn). (C) Head and first segments of a scolopendromorph centipede in ventral view. (D) Last segments of a female geophilomorph centipede in ventral view. (E) Head and first trunk segments of a male julid millipede (Diplopoda). (F) A millipede head in ventral view. (G) Leg of a millipede. (H) Telson of a millipede.

general rule is that the oxygen is delivered directly to the tissues or indirectly dissolved in the hemolymph (Hilken et al. 2011; Hilken et al. 2015). Some species of Myriapoda are adapted to survive flooding under water for a relatively long time, and it is known that usually Diplopoda do it by plastron respiration, and in some species also by adapting the metabolic rate to the available oxygen concentration (Hilken et al. 2015).

The regulation of body water content is important for Myriapoda, and this also applies to Symphyla and Pauropoda. For this purpose, Myriapoda use several behavioral and physiological strategies. Chilopoda burrow holes and even make vertical migrations between soil strata to survive in the dryer seasons (Voigtländer 2011). Most Diplopoda are more vulnerable to water loss than insects (Mantel 1979), but even so some species can recover after temporarily losing a third of their body water content (Meyer and Eisenbeis 1985). Diplopoda have different strategies to avoid desiccation, such as moving to more humid places, rolling up to minimize cuticular water loss, reducing respiratory metabolism, increase of osmotic pressure, or aggregating (Hopkin and Read 1992). For excretion, all Myriapoda have Malpighian tubules connected to the lumen of the digestive system, but some groups have additional organs or other strategies. For example, Scutigeromoprha and Lithobiomorpha (both Chilopoda) have maxillary nephridia, which are thought to reduce the loss of water (Rosenberg et al. 2011). Diplopoda have nephridia, while Symphyla and Pauropoda have different glands in the head for excretion (Scheller 2011; Szucsich and Scheller 2011). Mainly two types of nitrogenous waste have been detected in Myriapoda: ammonia and uric acid; but in some cases, the excretion of urea was also observed (Rosenberg et al. 2011; Hopkin and Read 1992).

The thermal limits of Myriapoda depend on the group and region. A temperature of 35°C may cause irreversible damage in European centipedes (Voigtländer 2011), while at least one species of desert-living millipede can reach this temperature when exposed to direct sunlight, supposedly to increase digestion efficiency (Wooten et al. 1975).

GENETICS

The genetics of Myriapoda is not so well studied as in some insect groups, but major advances have been made in recent years, such as the publication of the first complete nuclear genomes of representatives of Chilopoda and Diplopoda (Chipman et al. 2014; Kenny et al. 2015; Qu et al. 2020). Most of the studies about Myriapoda involving genetics have been made with the aim to solve their position within the Arthropoda, and the inner systematics of the taxon (see section Position in the ToL). Studies have also been made about the genes involved in the development of some Chilopoda and Diplopoda species (Minelli and Sombke 2011; Minelli 2015b).

The chromosome number varies among groups in Myriapoda, ranging from 2n = 14 to 2n = 54 in Chilopoda (Minelli 2011) and from 2n = 12 to 2n = 30 in Diplopoda (Minelli and Michalik 2015). Different species of millipedes within in the same genus can have different numbers of chromosomes. In Pauropoda, chromosome number ranges from 2n = 12 to 2n = 28 (Fratello and Sabatini 1990) and in Symphyla it may be from 2n = 11 to 2n = 18 (Fischer 1987). Sex determination by XX and XY chromosomes seems to be the most common and the ancestral condition, although it does not seem to be the same for all species (Green et al. 2016).

The first completely sequenced genome of a Myriapoda species that was published was that of the Chilopoda *Strigamia marina* (Leach 1817) (Chipman et al. 2014; Robertson et al. 2015), and the first genome of a Diplopoda published was that of *Trigoniulus corallinus* (Gervais 1847) (Kenny et al. 2015). A recent work by Qu et al. (2020) published the genome of an additional millipede species, *Helicorthomorpha holstii* (Pocock 1895), and also published a new genome of *T. corallinus*. The genome of *S. marina* is approximately 290 Mb long, while that of *T. corallinus* with 538 Mb is twice as long. The genome of *H. holstii* is more similar in length to *S. maritima*, and this difference among millipede species maybe due to the high number of repeated sequences in *T. corallinus* (Qu et al. 2020). The nuclear genome of *S. maritima* seems to be very conservative when compared to that of other arthropods. However, this fact contrasts with its derived mitochondrial genome showing an atypical gene order as well as a different tRNA secondary structure. The comparison of the information in the genome of Myriapoda with that of other land arthropods allows some conclusions concerning their evolution. For example, genes involved in the air chemoreception in insects, such as the odorant-binding protein or CheA/B families, are absent in Myriapoda. This means that since insects and myriapods colonized the land independently, they evolved different sets of genes for that function, and thus that family of genes may be an insect novelty (Chipman et al. 2014; Kenny et al. 2015). While none of the known arthropod light receptor or circadian clock genes was found in the blind centipede *S. maritima*, canonical opsin genes and circadian clock driving protein genes were found in the eyed millipede *T. corallinus*. Therefore, it seems that the Myriapoda share the same gene families for that function as other arthropods, while the anophthalmous and subterranean adapted lineages such the Geophilomorpha have lost these genes during evolution. There is still the question of what mechanism is involved in the active avoidance of open spaces by this species, as this should be due to another type of receptor (Chipman et al. 2014). The study of millipede genomes has also shown the cyanogenic pathway genes of *H. holstii*, and the antimicrobial function of the secretions of *T. corallinus* (Qu et al. 2020).

POSITION IN THE ToL

The monophyly of Myriapoda was discussed for more than a century. Despite some relatively recent publications pointing at its polyphyly (Kraus 2001), it is nowadays accepted that Myriapoda are a monophyletic group through accumulated morphological and molecular evidence (Giribet and Edgecombe 2019). For a long time and before molecular phylogenetics was available, Myriapoda was thought to form a clade with Hexapoda, called Atelocerata (Giribet and Edgecombe 2012). However, the first molecular analysis demonstrated that Hexapoda was a group within Crustacea, which is nowadays called Pancrustacea or Tetraconata (Dohle 2001; Richter 2002), leaving the myriapods "alone." There were two competing hypotheses on the systematic position of the Myriapoda. The first hypothesis states that Myriapoda form a clade together with Pancrustanceans, called Mandibulata. This is nowadays widely accepted, because of its molecular and morphological support (Giribet and Edgecombe, 2019). The second hypothesis states that Myriapoda form a clade together with Chelicerata, called Myriochelata or Paradoxopoda. Although this hypothesis received little morphological support (Rehm et al. 2014), restricted to some similarities in neurogenesis (Kadner and Stollewerk 2004), it was recurrently supported by some phylogenetic trees (Mallatt et al. 2004; Pisani et al. 2004; Dunn et al. 2008). This has been demonstrated to be a result of the high sensitivity of the position of Myriapoda to the taxa sampled, outgroups selected, and methods applied (Rota-Stabelli et al., 2011).

The phylogenetic relationship among the different classes within Myriapoda is not fully solved, but their monophyly is strongly supported by morphological and developmental characters (Edgecombe 2011). It seems that the most stable relationship is Diplopoda together with Pauropoda (Miyazawa et al. 2014; Fernández et al. 2018) forming a clade called Dignatha because the mandibles and first maxillae are the only functional mouthparts. There are two main hypotheses for the position of Symphyla and Chilopoda in relation to Dignatha (**Figure 29.4**). The first includes Chilopoda forming a clade together with Dignatha, with Symphyla as sister group (Miyazawa et al. 2014; Rehm et al. 2014). The second includes Symphyla and Dignatha forming a clade, called Progoneata because of the gonopore placed anteriorly in the body, that would have Chilopoda as a sister group (Fernández et al. 2018). Many other groupings have been proposed but have received little or no support at all (Edgecombe 2011). It seems that whether Symphyla or Chilopoda appear as closer to Dignatha depends on what groups are included in the analysis (Giribet and Edgecombe 2019).

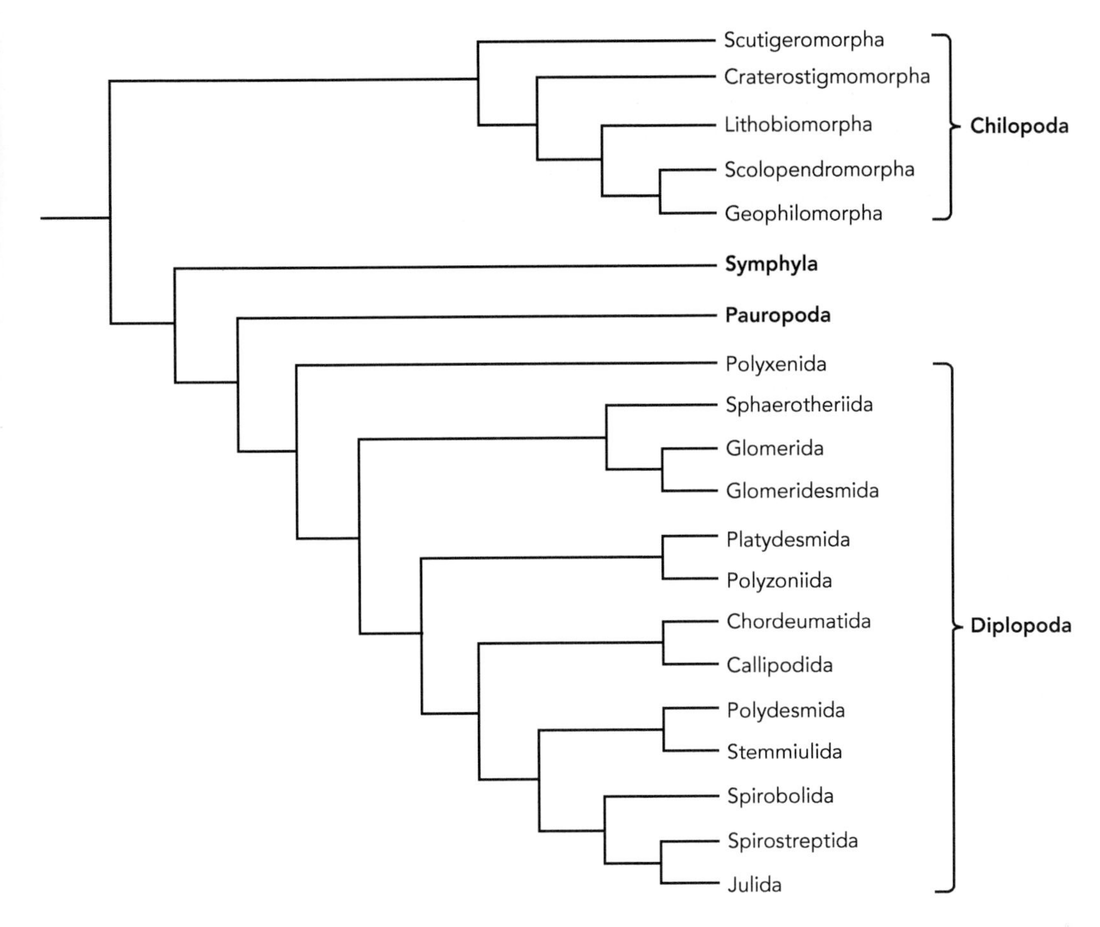

Figure 29.4 Phylogenetic tree of Myriapoda, including Diplopoda and Chilopoda orders. Based on Fernández et al. (2016) and Fernández et al. (2018).

The Myriapoda seem to have their origin in the early Cambrian, as suggested by molecular dating and the existence of "pancrustacean" fossils 518 million years ago, when they had already diverged (Miyazawa et al. 2014; Rehm et al. 2014; Fernández et al. 2018; Giribet and Edgecombe 2019). All extant Myriapoda are terrestrial, and despite the absence of clear fossil evidence, the divergence from the rest of Mandibulata was probably made by an aquatic lineage. This may have been similar to an Euthycarcinoid (Giribet and Edgecombe 2019), which is in a stem position in cladistic analysis together with Myriapoda (Vannier et al. 2018). Some authors suggest the possibility that Myriapoda performed multiple land colonizations (Rehm et al. 2014; Miyazawa et al. 2014), but this would be in contradiction with the fact that they all share the same terrestrial adaptations (e.g., tracheae, Malpighian tubules), suggesting a single event of terrestrial colonization (Fernández et al. 2018). The estimates for terrestrialization of Myriapoda are similar or somewhat previous to the other land arthropods (Lozano-Fernandez et al. 2016; Schwentner et al. 2017). Although the oldest air-breathing terrestrial animal was supposed to be a Myriapoda (Wilson and Anderson 2004), this has been challenged recently (Suarez et al. 2017).

DATABASES AND COLLECTIONS

Some of the worldwide most important taxonomic databases for Myriapoda are: ChiloBase 2.0 (Bonato et al. 2016) dedicated to Chilopoda; MilliBase (Sierwald and Spelda 2019) dedicated to Diplopoda, Pauropoda, and Symphyla. A recently created but quickly growing collaborative database is Myriatrix (2020), with information on Myriapoda and Onycophora. They gather exhaustive information on each species, such as the reference and author of the original description, synonymy, and higher classification. The databases are public and freely accessible on the internet. Additional information on taxonomical and bibliographical databases can be found on the webpage of the CIM (Centre International de Myriapodologie) part of the International Society for Myriapodology (myriapodology.org). This webpage provides additional relevant information on the field of Myriapodology. Sierwald and Reft (2004) made an exhaustive compendium of available Diplopoda collections in museums and other institutions around the world, listing 268 institutions with collections from 143 countries. Probably, most of those institutions include Chilopoda collections, but no similar listing effort seems to have been made for that group.

CONCLUSION

Myriapoda is a highly diverse subphylum of arthropods whose individuals play a key role in terrestrial ecosystems. However, myriapods have been by far less studied than other arthropods such as Hexapoda, despite its potential for biogeography, development, ecology, systematics, etc. In fact, for many species, little is known apart from how they look and where they were found. This is especially true for the two most cryptic Myriapoda classes, Pauropoda and Symphyla. The inner systematics of the Myriapoda is still a subject of debate, and questions about their evolutionary history are still not fully solved, even though great progress has been made in the last decades. Myriapoda are also of economic interest because of their potential applications for biomedicine, the trade with exotic species kept as pets, and also for the damage caused by some invasive species. But apart from this economic point of view, the Myriapoda are a fascinating and interesting group of arthropods, and I would encourage any amateur naturalist and entomologist in the early stage of his/her career to give a closer look to these many-legged animals.

REFERENCES

Bonato, L., Chagas Jr, A., Edgecombe, G.D., Lewis, J.G.E., Minelli, A., Pereira, L.A., Shelley, R.M., Stoev, P. and Zapparoli, M., 2016. ChiloBase 2.0 – A World Catalogue of Centipedes (Chilopoda). Available at http://chilobase.biologia.unipd.it.

Bonato, L. and Zapparoli, M., 2011. Chilopoda—geographical distribution. In: Minelli, A. (Ed.), Treatise on Zoology-Anatomy, Taxonomy, Biology. The Myriapoda, vol. 1. Brill, Leiden-Boston, pp. 327–337.

Brusca, R.C., Moore, W., Schuster, M., 2016. Invertebrates (3rd ed). Sinauer Associated. Inc, Publishers, Massachusetts.

Cárcamo, H.A., Abe, T.A., Prescott, C.E., Holl, F.B. and Chanway, C.P., 2000. Influence of millipedes on litter decomposition, N mineralization, and microbial communities in a coastal forest in British Columbia, Canada. Canadian Journal of Forest Research, 30(5), 817–826.

Chipman, A.D., Ferrier, D.E., Brena, C., Qu, J., Hughes, D.S., Schröder, R., (...) and Alonso, C.R., 2014. The first myriapod zhtent and genome organisation in the centipede *Strigamia maritima*. PLoS Biology, 12(11), e1002005.

Dányi, L., Balázs, G., and Tuf, I.H., 2019. Taxonomic status and behavioural documentation of the troglobiont Lithobiusmatulici (Myriapoda, Chilopoda) from the Dinaric Alps: are there semi-aquatic centipedes in caves? ZooKeys, 848, 1–20.

David, J.F., 2015. Diplopoda—Ecology. In: Minelli, A. (Ed.), Treatise on Zoology-Anatomy, Taxonomy, Biology. The Myriapoda, vol. 2. Brill, Leiden-Boston, pp. 303–327.

Dohle, W., 2001. Are the insects terrestrial crustaceans? A discussion of some new facts and arguments and the proposal of the proper name 'Tetraconata' for the monophyletic unit Crustacea + Hexapoda. Annales de la Société Entomologique de France, 37, 85–103.

Domínguez, M.T., 1992. Symphyla y Pauropoda (Myriapoda) de suelos de España II. PhD Thesis, Universidad Complutense de Madrid.

Dunn, C.W., Hejnol, A., Matus, D.Q., Pang, K., Browne, W.E., Smith, S.A., Seaver, E., Rouse, G.W., Obst, M., Edgecombe, G.D., Sorensen, M.V., Haddock, S.H., Schmidt-Rhaesa, A., Okusu, A., Kristensen, R.M., Wheeler, W.C., Martindale, M.Q. and Giribet, G., 2008. Broad phylogenomic sampling improves resolution of the animal tree of life. Nature, 452, 745–749.

Edgecombe, G.D., 2011. Phylogenetic relationships of Myriapoda. In: Minelli, A. (Ed.), Treatise on Zoology-Anatomy, Taxonomy, Biology. The Myriapoda, vol. 1. Brill, Leiden-Boston, pp. 1–20.

Edwards, C.A., 1961. The ecology of Symphyla part III. Factors controlling soil distributions. Entomologia Experimentalis et Applicata, 4(4), 239–256.

Enghoff, H., 2015. Diplopoda—Geographical distribution. Minelli, A. (Ed.), Treatise on Zoology-Anatomy, Taxonomy, Biology. The Myriapoda, vol. 2. Brill, Leiden-Boston, pp. 329–336.

Enghoff, H., Dohle, W. and Blower, J.G., 1993. Anamorphosis in millipedes (Diplopoda)—the present state of knowledge with some developmental and phylogenetic considerations. Zoological Journal of the Linnean Society, 109(2), 103–234.

Enghoff, H. and Kebapći, Ü., 2008. *Calyptophyllum longiventre* (Verhoeff, 1941) invading houses in Turkey, with the first description of the male (Diplopoda: Julida: Julidae). Journal of Natural History, 42, 2143–2150.

Enghoff, H., Manno, N., Tchibozo, S., List, M., Schwarzinger B., Schoefberger, W., Schwarzinger C. and Paoletti M.G., 2014. Millipedes as food for humans: their nutritional and possible antimalarial value—a first report. Evidence Based Complementry and Alternative Medicine, 651768. doi:10.1155/2014/651768

Fernández, R., Edgecombe, G.D. and Giribet, G. 2016. Exploring phylogenetic relationships within Myriapoda and the effects of matrix composition and occupancy on phylogenomic reconstruction. Systematic Biology, 65(5), 871–889.

Fernández, R., Edgecombe, G.D. and Giribet, G., 2018. Phylogenomics illuminates the backbone of the Myriapoda tree of life and reconciles morphological and molecular phylogenies. Scientific Reports, 8, 83.

Fischer, A., 1987. Chromosome studies in nine species of Austrian Symphyla (Myriapoda, Tracheata, Arthropoda). Genetica, 75(2), 109–116.

Fratello, B. and Sabatini, M.A., 1990. Chromosomes of Pauropoda. In: Minelli, A. (Ed.), Proceedings of the 7th International Congress of Myriapodology. Brill, Leiden, pp. 109–114.

Giribet, G. and Edgecombe, G.D., 2012. Reevaluating the arthropod tree of life. Annual Review of Entomology, 57, 167–186.

Giribet, G. and Edgecombe, G.D., 2019. The phylogeny and evolutionary history of arthropods. Current Biology, 29(12), R592–R602.

Golovatch, S.I., 2009. Millipedes (Diplopoda) in extreme environments. In Golovatch, S.I., Makarova, O.L., Babenko, A.B., Penev, L. (Eds.), Species and Communities in Extreme Environments. Festschrift towards the 75th Anniversary and a Laudatio in Honour of Academican Yuri Ivanovich Chernov (pp. 87–112). Pensoft & KMK Scientific Press, Sofia–Moscow.

Golovatch, S.I., and Kime, R.D., 2009. Millipede (Diplopoda) distributions: a review. Soil Organisms, 81(3), 565–597.

Green, J.E., Dalíková, M., Sahara, K., Marec, F. and Akam, M., 2016. XX/XY system of sex determination in the geophilomorph centipede *Strigamia maritima*. PLoS One, 11(2), e0150292.

Harzsch, S., 2004. Phylogenetic comparison of serotonin-immunoreactive neurons in representatives of the Chilopoda, Diplopoda, and Chelicerata: implications for arthropod relationships. Journal of Morphology, 259(2), 198–213.

Harzsch, S., 2006. Neurophylogeny: architecture of the nervous system and a fresh view on arthropod phyologeny. Integrative and Comparative Biology, 46(2), 162–194.

Hilken, G., Müller, C.H., Sombke, A., Wirkner, C.S., and Rosenberg, J., 2011. Chilopoda—Tracheal system. In: Minelli, A. (Ed.), Treatise on Zoology-Anatomy, Taxonomy, Biology. The Myriapoda, vol. 1. Brill, Leiden-Boston, pp. 137–155.

Hilken, G., Sombke, A., Müller, C. H., and Rosenberg, J., 2015. Diplopoda—Tracheal system In: Minelli, A. (Ed.), Treatise on Zoology-Anatomy, Taxonomy, Biology. The Myriapoda, vol. 2. Brill, Leiden-Boston, pp. 129–152.

Hopkin, S.P. and Read, H.J., 1992. The Biology of Millipedes. Oxford University Press, United Kingdom.

Hüther, W., 1959. Zur Ernährung der Pauropoden. Naturwissenschaften, 46(19), 563–564.

Kadner, D. and Stollewerk, A., 2004. Neurogenesis in the chilopod *Lithobius forficatus* suggests more similarities to chelicerates than to insects. Development Genes and Evolution, 214(8), 367–379.

Kenny, N.J., Shen, X., Chan, T.T., Wong, N.W., Chan, T.F., Chu, K.H., Lam, H.M. and Hui, J.H., 2015. Genome of the rusty millipede, *Trigoniulus corallinus*, illuminates diplopod, myriapod, and arthropod evolution. Genome Biology and Evolution, 7(5), 1280–1295.

Kime, R.D. and Golovatch, S.I., 2000. Trends in the ecological strategies and evolution of millipedes (Diplopoda). Biological Journal of the Linnean Society, 69(3), 333–349.

Koch, M., Müller, C.H., Hilken, G. and Rosenberg, J., 2011. Chilopoda—Digestive system. In: Minelli, A. (Ed.), Treatise on Zoology-Anatomy, Taxonomy, Biology. The Myriapoda, vol. 1. Brill, Leiden-Boston, pp. 121–136.

Kraus, O., 2001. 'Myriapoda' and the ancestry of the Hexapoda. Annales de la Société Entomologique de France, 37, 105–127.

Lozano-Fernandez, J., Carton, R., Tanner, A.R., Puttick, M.N., Blaxter, M., Vinther, J., Olesen, J., Giribet, G., Edgecombe G.D. and Pisani, D.A., 2016. A molecular palaeobiological exploration of arthropod terrestrialisation. Philosophical Transactions of the Royal Society B: Biological Sciences, 371, 20150133.

Lubbock, J., 1867. On *Pauropus*, a new type of centipede. Transactions of the Linnean Society of London, 26, 181–190.

Makarov, S.E., 2015. Diplopoda—Integument. In: Minelli, A. (Ed.), Treatise on Zoology-Anatomy, Taxonomy, Biology. The Myriapoda, vol. 2. Brill, Leiden-Boston, pp. 69–99.

Mallatt, J.M., Garey, J.R. and Shultz, J.W., 2004. Ecdysozoan phylogeny and Bayesian inference. First use of nearly complete 28S and 18S rRNA gene sequences to classify the arthropods and their kin. Molecular Phylogenetics and Evolution, 31, 178–191.

Mantel, L.H., 1979. Terrestrial invertebrates other than insects. In: Maloiy, G.M.O. (Ed.), Comparative Physiology of Osmoregulation in Animals, vol. 1. Academic Press, London-New York-San Francisco, pp. 75–218.

Manton, S.M., 1977. The Arthropoda. In Habits, Functional Morphology and Evolution. Clarendon Press, Oxford.

Marek, P.E., Shear, W.A. and Bond, J.E., 2012. A redescription of the leggiest animal, the millipede *Illacme plenipes*, with notes on its natural history and biogeography (Diplopoda, Siphonophorida, Siphonorhinidae). ZooKeys, 241, 77–112.

Meyer, E. and Eisenbeis, G., 1985. Water relations in millipedes from some alpine habitat types (Central Alps, Tyrol) (Diplopoda). Bijdragen tot de Dierkunde, 55(1), 131–142.

Minelli, A., 2011. Chilopoda—Reproduction. In: Minelli, A. (Ed.), Treatise on Zoology-Anatomy, Taxonomy, Biology. The Myriapoda, vol. 1. Brill, Leiden-Boston, pp. 279–294.

Minelli, A., 2015a. Diplopoda—An outline of research history. In: Minelli, A. (Ed.), Treatise on Zoology-Anatomy, Taxonomy, Biology. The Myriapoda, vol. 2. Brill, Leiden-Boston, pp. 1–6.

Minelli, A., 2015b. Diplopoda—Development. In: Minelli, A. (Ed.), Treatise on Zoology-Anatomy, Taxonomy, Biology. The Myriapoda, vol. 2. Brill, Leiden-Boston, pp. 267–302.

Minelli, A. and Michalik, P., 2015. Diplopoda—Reproduction. In: Minelli, A. (Ed.), Treatise on Zoology-Anatomy, Taxonomy, Biology. The Myriapoda, vol. 2. Brill, Leiden-Boston, pp. 237–265.

Minelli, A. and Sombke, A., 2011. Chilopoda—Development. In: Minelli, A. (Ed.), Treatise on Zoology-Anatomy, Taxonomy, Biology. The Myriapoda, vol. 1. Brill, Leiden-Boston, pp. 295–308.

Miyazawa, H., Ueda, C., Yahata, K. and Su, Z.H., 2014. Molecular phylogeny of Myriapoda provides insights into evolutionary patterns of the mode in post-embryonic development. Scientific Reports, 4, 4127.

Müller, C.H., Sombke, A., Hilken, G. and Rosenberg, J., 2011. Chilopoda—Sense organs. In: Minelli, A. (Ed.), Treatise on Zoology-Anatomy, Taxonomy, Biology. The Myriapoda, vol. 1. Brill, Leiden-Boston, pp. 327–337.

Müller, C.H. and Sombke, A., 2015. Diplopoda—Sense organs. In: Minelli, A. (Ed.), Treatise on Zoology-Anatomy, Taxonomy, Biology. The Myriapoda, vol. 2. Brill, Leiden-Boston, pp. 181–235.

Müller, C.H., Sombke, A. and Rosenberg, J., 2007. The fine structure of the eyes of some bristly millipedes (Penicillata, Diplopoda): additional support for the homology of mandibulate ommatidia. Arthropod Structure and Development, 36(4), 463–476.

Myriatrix 2020. The Fellegship of the Rings. Available at http://myriatrix.myspecies.info.

Pisani, D., Poling, L.L., Lyons-Weiler, M. and Hedges, S.B., 2004. The colonization of land by animals: molecular phylogeny and divergence times among arthropods. BMC Biology, 2, 1.

Price, D.W., 1975. Vertical distribution of small arthropods in a California pine forest soil. Annals of the Entomological Society of America, 68(1), 174–180.

Qu, Z., Nong, W., So, W.L., Barton-Owen, T., Li, Y., Leung, T.C.N., Li, C., Baril, T. Wong, A. Y. P., Swale, T., Chan, T.-F., Hayward, A., Ngai, S.-M., Hui, J.J.L. 2020. Millipede genomes reveal unique adaptations during myriapod evolution. PLoS Biology 18(9): e3000636. https://doi.org/10.1371/journal.pbio.3000636

Rehm, P., Meusemann, K., Borner, J., Misof, B. and Burmester, T., 2014. Phylogenetic position of Myriapoda revealed by 454 transcriptome sequencing. Molecular Phylogenetics and Evolution, 77, 25–33.

Richter, S., 2002. The Tetraconata concept: hexapod-crustacean relationships and the phylogeny of Crustacea. Organisms Diversity and Evolution, 2, 217–237.

Robertson, H.E., Lapraz, F., Rhodes, A.C. and Telford, M.J., 2015. The complete mitochondrial genome of the geophilomorph centipede *Strigamia maritima*. PLoS One, 10(3), e0121369.

Rosati, F., Baccetti, B. and Dallai, R., 1970. The spermatozoon of Arthropoda. X. Araneids and the lowest Myriapods. In: Baccetii (Ed.), Comparative Spermatology, Proceedings of the International Symposium held in Rome and Siena 1969, pp. 247–254.

Rosenberg, J., Sombke, A. and Hilken, G., 2011. Chilopoda—Excretory system. In: Minelli, A. (Ed.), Treatise on Zoology-Anatomy, Taxonomy, Biology. The Myriapoda, vol. 1. Brill, Leiden-Boston, pp. 177–193.

Rota-Stabelli, O., Campbell, L., Brinkmann, H., Edgecombe, G.D., Longhorn, S.J., Peterson, K.J., Pisani, D., Philippe, H. and Telford, M.J., 2011. A congruent solution to arthropod phylogeny: phylogenomics, microRNAs and morphology support monophyletic Mandibulata. Proceedings of the Royal Society of London B: Biological Sciences, 278, 298–306.

Ryder, J.A., 1880. *Scolopendrella* as the type of a new order of Articulates (Symphyla). American Naturalist, 14, 375–376.

Scheller, U., 2011. Pauropoda. In: Minelli, A. (Ed.), Treatise on Zoology-Anatomy, Taxonomy, Biology. The Myriapoda, vol. 1. Brill, Leiden-Boston, pp. 467–508.

Sharples, R.W., 1994. Theophrastus of Eresus. Sources for his life, writings, thought and influence. In Sources on Biology, vol. 5. Brill, Leiden.

Schwentner, M., Combosch, D.J., Nelson, J.P. and Giribet, G., 2017. A phylogenomic solution to the origin of insects by resolving crustacean hexapod relationships. Current Biology, 27, 1818–1824.

Scopoli, J.A., 1763. Entomologia carniolica, exhibiens insecta Carnioliae indigena et distributa in ordines, genera, species, varietates. Methodo Linnaeana, Vindobonae.

Sierwald, P. and Reft, A.J., 2004. The millipede collections of the world. Fieldiana N.S., 104 (1532), 1–100.

Sierwald, P. and Spelda, J. 2019. MilliBase. Accessed at http://www.millibase.org on 2019-11-04. doi:10.14284/370

Stoev, P., Zapparoli, M., Golovatch, S., Enghoff, H., Akkari, N. and Barber, A., 2010. Myriapods (Myriapoda). Chapter 7.2. BioRisk, 4(1), 97–130 doi: 10.3897/biorisk.4.51

Suarez, S.E., Brookfield, M.E., Catlos, E.J. and Stöckli, D.F., 2017. A U-Pb zircon age constraint on the oldest-recorded air-breathing land animal. PLoS One, 12(6), e0179262.

Szucsich, N. and Scheller, U., 2011. Symphyla. In: Minelli, A. (Ed.), Treatise on Zoology-Anatomy, Taxonomy, Biology. The Myriapoda, vol. 1. Brill, Leiden-Boston, pp. 445–466.

Undheim, E.A., Jones, A., Clauser, K.R., Holland, J.W., Pineda, S.S., King, G.F. and Fry, B.G., 2014. Clawing through evolution: toxin diversification and convergence in the ancient lineage Chilopoda (Centipedes). Molecular Biology and Evolution, 31(8), 2124–2148.

Vannier, J., Aria, C., Taylor, R.S. and Caron, J.B., 2018. *Waptia fieldensis* Walcott, a mandibulate arthropod from the middle Cambrian Burgess Shale. Royal Society Open Science, 5(6), 172206.

Voigtländer, K., 2011. Chilopoda—Ecology. In: Minelli, A. (Ed.), Treatise on Zoology-Anatomy, Taxonomy, Biology. The Myriapoda, vol. 1, 309–325. Brill, Leiden-Boston.

Walter, E.W., Moore, J.C. and Loring, S.J., 1989. *Symphylella* sp. (Symphyla: Scolopendrellidae) predators of arthropods and nematodes in grassland soils. Pedobiologia, 22, 113–116.

Wilson, H. and Anderson, L., 2004. Morphology and taxonomy of Paleozoic millipedes (Diplopoda: Chilognatha: Archipolypoda) from Scotland. Journal of Paleontology, 78, 169–184. doi:10.1666/0022-3360(2004)078.

Wirkner, C.S., Hilken, G. and Rosenberg, J., 2011. Chilopoda—Circulatory system. In: Minelli, A. (Ed.), Treatise on Zoology-Anatomy, Taxonomy, Biology. The Myriapoda, vol. 1. Brill, Leiden-Boston, pp. 157–176.

Wirkner, C.S. and Xylander, W.E., 2015. Diplopoda—Circulatory system. In: Minelli, A. (Ed.), Treatise on Zoology-Anatomy, Taxonomy, Biology. The Myriapoda, vol. 2. Brill, Leiden-Boston, pp. 153–160.

Wooten Jr, R.C., Crawford, C.S. and Riddle, W.A., 1975. Behavioural thermoregulation of Orthoporus ornatus (Diplopoda: Spirostreptidae) in three desert habitats. Zoological Journal of the Linnean Society, 57(1), 59–74.

Zimian D., Yonghua, Z. and Xiwu, G., 1997. Medicinal insects in China. Ecology of Food and Nutrition, 36(2–4), 209–220. doi: 10.1080/03670244.1997.9991516

CHAPTER

30

PHYLUM ONYCHOPHORA

Ivo de Sena Oliveira and Georg Mayer

Infrakingdom	Bilateria
Division	Protostomia
Subdivision	Ecdysozoa
Phylum	Onychophora

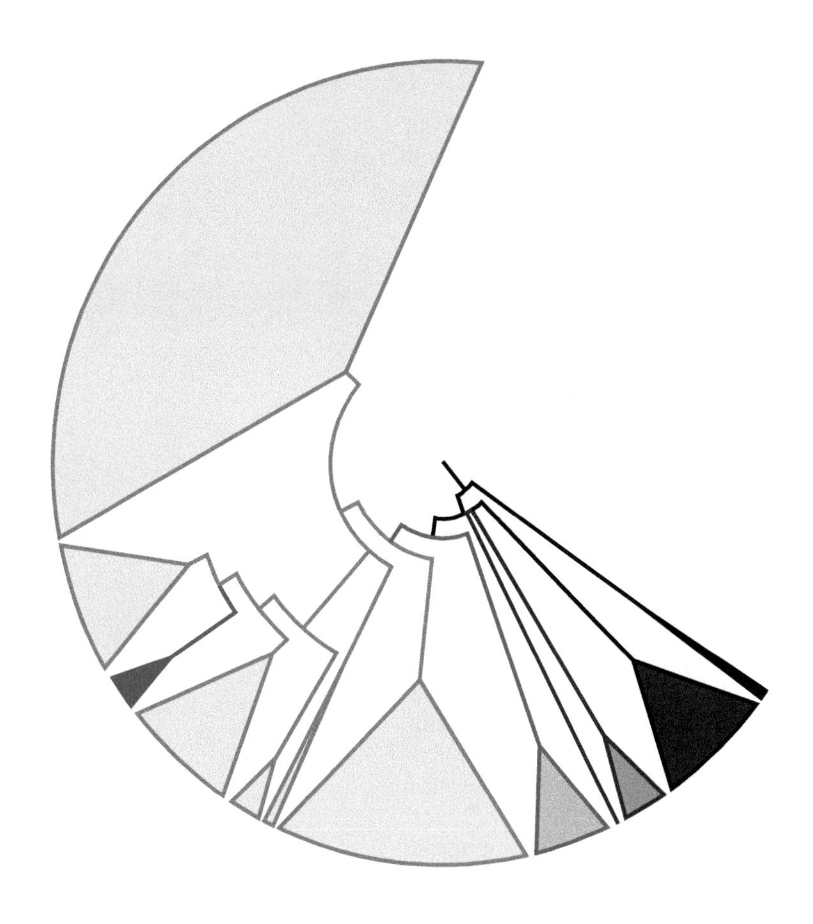

CONTENTS

General Characteristics

207 valid species (Peripatidae: 82; Peripatopsidae: 125) and 20 *nomina dubia*

Estimated number of "cryptic" species: several hundred or higher

Terrestrial animals, mainly found in leaf litter and decaying logs in temperate and tropical forests of the southern hemisphere and around the equator

Together with tardigrades, the only living lobopodians

No commercial value

Ecological importance is probably small

Potential flagship taxon for habitat conservation

Synapomorphies

- jaws (derivatives of second pair of limbs innervated by deutocerebrum)
- jaw apodemes
- slime papillae (limb derivatives of third body segment)
- slime glands (derivatives of crural glands of third body segment)
- salivary glands (derivatives of nephridia of third body segment)
- ultrahydrophobic body surface covered with microscales
- tracheae that develop after birth
- distal foot connected to leg via ventral bridge

HISTORY OF TAXONOMY AND CLASSIFICATION

Onychophorans, or "velvet worms" (**Figure 30.1**), comprise a small group of soft-bodied, terrestrial invertebrates first discovered on the island of Saint Vincent, Antilles, almost 200 years ago (Guilding 1826). The name Onychophora (= claw-bearers) was suggested a few years after the description of the first velvet worm species (*Peripatus juliformis*; Guilding, 1826), making reference to the paired claws on the distal leg portion of these animals (Grube 1853). Initially, the onychophoran affinities were exhaustively discussed and these animals were allocated to a range of different groups, including Mollusca (Guilding 1826), Annelida (Audouin and Milne-Edwards 1833), Malacopoda, i.e., a transitional form between annelids and arthropods (de Blainvillé *apud* Gervais 1837; Blanchard 1847), as well as Arthropoda (Sedgwick 1888). As additional velvet worm species were investigated, the evolutionary position of these animals was gradually clarified, and to date, onychophorans are classified as an independent protostome clade within the megadiverse taxon of Panarthropoda (Aguinaldo et al. 1997; Telford et al. 2008).

Initially, velvet worms from different parts of the globe were consistently assigned to a single taxon, namely *Peripatus*, until detailed taxonomic studies began to unravel the actual diversity within the group (Sedgwick 1888; Dendy 1894; Pocock 1894; Bouvier 1905, 1907; Clark 1913). Consequently, *Peripatus* became restricted to a few neotropical species, while numerous additional taxa have been raised to accommodate over 200 velvet worm species currently described (Oliveira et al. 2012a,b, 2015, 2018). The previous separation of onychophoran species into two major subgroups, the Peripatidae (Evans, 1901) and the Peripatopsidae (Bouvier, 1905), also became strongly supported by both morphological and molecular studies (Reid 1996; Murienne et al. 2014; Giribet et al. 2018). Nevertheless, the classification within each of these groups, especially inside Peripatidae, is still debated (Giribet et al. 2018). Peripatidae is arguably the least studied onychophoran subgroup, with the last monograph published over 100 years ago (Bouvier 1905), while studies on Peripatopsidae already resulted in several revisions and monographs (e.g., Bouvier 1907;

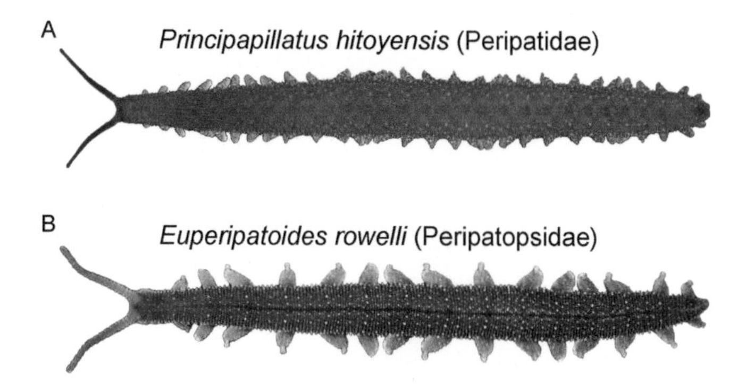

Figure 30.1 Representatives of the two major onychophoran subgroups. Photographs of (A) a peripatid from Costa Rica and (B) a peripatopsid from Australia. (Modified from Martin et al. 2017b; Creative Commons [CC].)

Ruhberg 1985; Reid 1996; Hamer et al. 1997). To date, 82 valid species are assigned to Peripatidae and 125 to Peripatopsidae (Oliveira et al. 2012a,b, 2013a, 2015, 2018; Daniels et al. 2013, 2016; Ruhberg and Daniels 2013; Oliveira and Mayer 2017), with an additional 20 species being regarded as *nomina dubia* (Oliveira et al. 2012b). This number clearly does not reflect the actual diversity of Onychophora, as a high number of undescribed and cryptic species may exist in both Peripatidae and Peripatopsidae (Oliveira et al. 2012b).

EXTERNAL AND INTERNAL MORPHOLOGY

Onychophorans exhibit a worm-like body composed of a head (**Figure 30.2A**), an elongated trunk with 13–43 leg-bearing segments, and an anal cone, the latter being a true limbless segment (Mayer et al. 2015a). The integument comprises numerous dermal papillae covered with microscales (**Figure 30.2B**) which provide onychophorans with their characteristic velvety appearance (hence, the name "velvet worms"). Each trunk segment bears a pair of unjointed limbs (= lobopods) equipped with a distal foot and a pair of sclerotized claws (**Figure 30.2C**; Oliveira and Mayer 2013). The head comprises three anteriormost body segments, each of which holds a pair of modified limbs: the antennae, the jaws, and the slime papillae (**Figure 30.2A,F**; Mayer et al. 2010). The annulated antennae are located anterodorsally and equipped with numerous mechano- and chemoreceptors (**Figure 30.2D**; Storch and Ruhberg 1977; Oliveira et al. 2012a). A pair of simple eyes, which are probably homologous to the median ocelli of arthropods, occurs at the antennal basis (**Figure 30.2E**; Eakin and Westfall 1965; Mayer 2006a). The eyes contain rhabdomeric photoreceptors but do not have high visual resolution and only serve for monochromatic vision (Hering et al. 2012; Kirwan et al. 2018). The jaws are serially homologous to claws, composed of two blades (Figure 30.2F) and lie inside the ventral mouth together with an unpaired, median tongue (Mayer et al. 2010, 2015b). Numerous oral lips surround the mouth opening (Figure 30.2A; Oliveira et al.

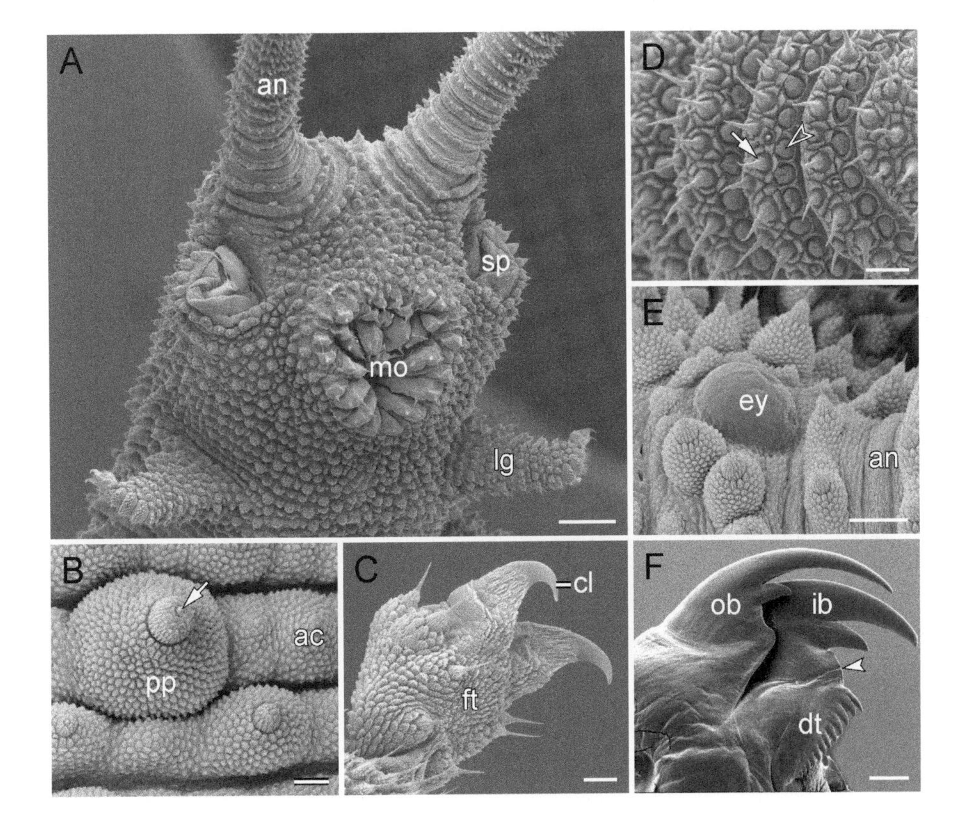

Figure 30.2 External morphology of onychophorans. Scanning electron micrographs of *Principapillatus hitoyensis* (Peripatidae). (A) Ventral view of the head region. (B) Dermal papillae of the dorsal integument. Note the numerous scales covering the dermal papillae. Arrow points to the sensory bristle of a primary papilla. (C) Ventral view of the foot with paired distal claws. (D) Detailed view of the antennal tip with rows of chemoreceptors (arrowhead) and mechanoreceptors (arrow). (E) Dorsolateral view of the eye. (F) Dissected jaw (modified from Oliveira *et al.* 2012a creative commons [CC]) composed of an outer and an inner blade, the latter being linked with the denticle blade via a soft diastemal membrane (arrowhead), which occurs only in representatives of Peripatidae. ac, accessory papilla; an, antenna; dt, denticle blade; ey, eye; ib, inner jaw blade; lg, leg; mo, mouth; ob, outer jaw blade; pp, primary papilla; sp, slime papilla. Scale bars 300 µm in (A); 500 µm in (B,F); 25 µm in (C,D) and 100 µm in (E).

Figure 30.3 Slime ejection and feeding behavior in onychophorans. High-speed camera footage (A) and photograph (B) from living specimens of *Euperipatoides rowelli* (Peripatopsidae). (A) Slime jet being ejected through the slime papillae (arrowheads)—a hunting/defense behavior characteristic to all onychophorans. (B) Specimen feeding on a cricket. (Image modified from Mayer et al. (2015b) copyright (2015) Oxford University Press on behalf of the Society for Integrative and Comparative Biology.)

2012a; Martin and Mayer 2014), while internally, the mouth cavity leads into a muscular pharynx followed posteriorly by a tube-like esophagus, a thick-walled midgut, a short hindgut, and a terminal anus (Storch et al. 1988; Mayer et al. 2013a, 2015a).

The slime papillae represent a unique feature of onychophorans and are used to eject a sticky secretion either for prey capture or defense (**Figure 30.3A**; Baer and Mayer 2012; Baer et al. 2014; Mayer et al. 2015a; von Byern et al. 2017). The fluid onychophoran slime becomes adhesive by mechanical stress (Baer et al. 2017; Baer et al. 2018), thus immobilizing the prey (such as arthropods) as it attempts to escape. The prey is then predigested by a cocktail of digestive saliva, which is injected into its body after the velvet worm has punctured the prey's cuticle using the jaws (**Figure 30.3B**; Mayer et al. 2015a). The liquefied contents are finally ingested using the suctorial pharynx (Manton and Heatley 1937; Nielsen 2012; Mayer et al. 2015a) and digested within a peritrophic membrane secreted by gut cells (Ruhberg and Mayer 2013). Digestion is complete after ~18 hours and non-digested parts are excreted together with the peritrophic membrane (Ruhberg and Mayer 2013).

The onychophoran soft body exhibits a conspicuous muscular system mainly composed of three muscle layers: circular, diagonal, and longitudinal (**Figure 30.4A**; Manton, 1973; Hoyle and Williams, 1980). Longitudinal muscles form the innermost layer and are organized into seven bundles of muscle fibers: two dorsal, two lateral, and three ventral (Figure 30.4A). Additional transverse muscles connect the ventral and dorsolateral body walls (Figure 30.4A). Since none of these layers exhibit a segmental arrangement, muscles associated with the legs constitute the only segmental elements of the muscular system (Hoyle and Williams 1980; Oliveira and Mayer 2013; Oliveira et al. 2013b; Müller et al. 2017). Each leg shows a complex myoanatomy formed by a dense mesh of muscle fibers (Müller et al. 2017; Oliveira et al. 2019). A total of 15 muscles have been identified in the leg of an Australian species of Peripatopsidae (Oliveira et al. 2019).

The onychophoran nervous system differs mainly from that of their closest relatives, the tardigrades and arthropods, in that it lacks segmental ganglia (**Figure 30.4B,C**; Whitington and Mayer 2011; Mayer et al. 2013a). A pair of longitudinal, ventral nerve cords widely separated from each other are connected anteriorly to a bilobed, ganglionic brain and interconnected transversally by numerous median (ventral) and ring (dorsal) commissures (Figure 30.4B,C). Each nerve cord is a medullary structure with neuronal somata distributed along its entire length (Mayer and Harzsch 2007, 2008; Mayer et al. 2013b; Martin et al. 2017a). Ring commissures further link the nerve cords to two dorsolateral nerves and the dorsal heart nerve, thus forming an orthogonal nervous system (Figure 30.4C; Mayer and Harzsch 2008). Like the musculature, the leg and nephridial nerves are the only segmental elements of the nervous system. The brain is bipartite and consists of a proto- and a deutocerebrum (Figure 30.4B; Mayer et al. 2010; Martin and Mayer 2014; Martin et al. 2017a). The protocerebrum innervates the antennae and the eyes, whereas the deutocerebrum supplies the

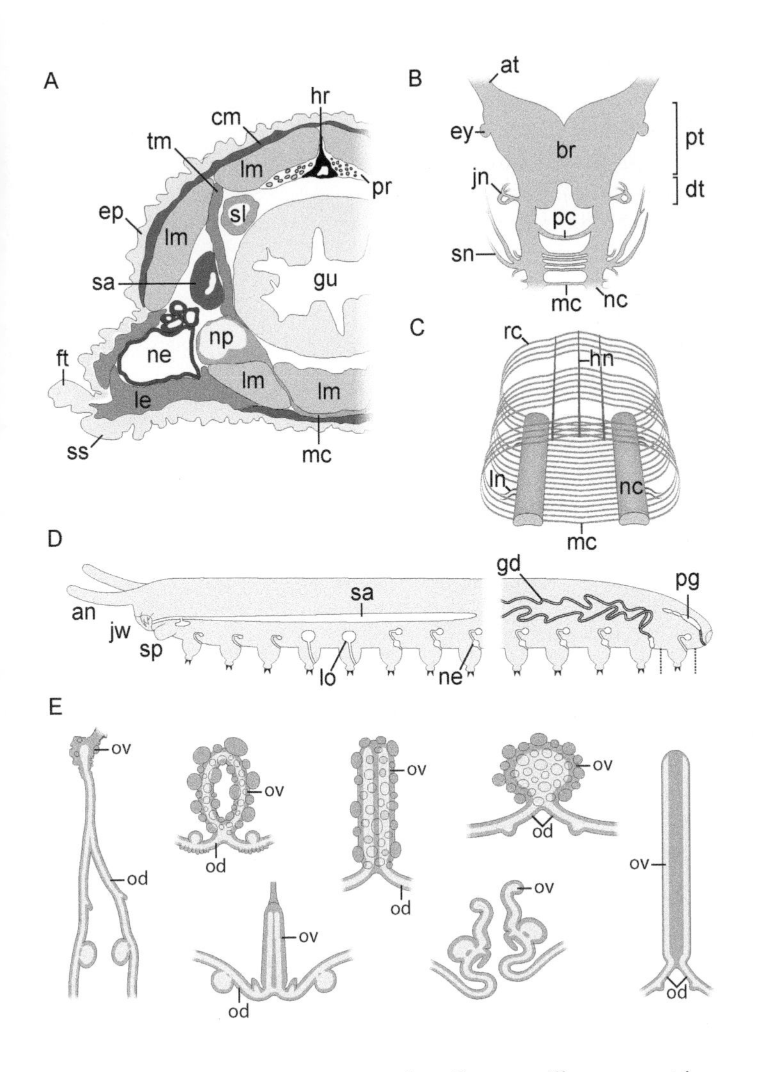

Figure 30.4 Internal anatomy of onychophorans. (A) Representation of the anterior trunk region based on a histological cross-section. (B) Anterior part of the nervous system showing the brain. (C) Orthogonal organization of the nervous system in the trunk. (D) Distribution of nephridia and their derivatives. Dotted lines indicate a body segment missing in peripatopsids. (E) Diversity of ovarian structures in different taxa. (Diagrams modified from Martin et al. (2017a) copyright 2017 Springer in B,C; Mayer (2006b) copyright 2006 Springer in D; and Mayer and Tait (2009) copyright 2009 The Linnean Society of London in E.) an, antenna; at, antennal tract; br, brain; cm, circular and diagonal musculature; dt, deutocerebrum; ep, epidermis with extracellular matrix; ey, eye; ft, foot; gd, gonoduct; gu, gut lumen; hn, heart nerve; hr, heart; jn, jaw nerve; jw, jaw; le, leg musculature; lm, longitudinal musculature; ln, leg nerve; lo, labyrinth organ; mc, median commissure; nc, ventral nerve cord; ne, nephridium; np, neuropil of nerve cord; od, oviduct; ov, ovary; pc, first post-oral commissure; pg, posterior accessory gland; pr, pericardial sinus with pericardial cells; pt, protocerebrum; rc, ring commissure; sa, salivary gland; sl, slime gland; sn, slime papilla nerve; sp, slime papilla; ss, spinous pad; tm, transverse musculature.

jaws (Martin and Mayer 2015). The slime papillae are not innervated by the brain but by the anterior region of the ventral nerve cords (Figure 30.4B; Mayer et al. 2010; Martin and Mayer 2015). The numerous oral lips surrounding the mouth are supplied by different parts of the central nervous system (Martin and Mayer 2014, 2015; Martin et al. 2017a).

Segmental nephridia play the role of renal organs and are associated with most trunk appendages, commonly opening to the exterior at the ventral basis of each leg (Figure 30.4A,D). Each nephridium shows a coelomic sacculus connected via a ciliated duct to a contractile bladder which opens to the outside through a short duct (Figure 30.4A; Ruhberg and Mayer 2013). The nephridia of the fourth and fifth leg pairs have been modified into labyrinth organs which end distally on specialized nephridial tubercles but their function remains unclear (**Figure 30.4D**; Mayer 2006b; Oliveira et al. 2012a). The salivary glands represent modified nephridia of the slime papilla segment, whereas the anal, uterine, and accessory genital glands as well as the gonoducts are generally regarded as derivatives of nephridia of their corresponding body segments (Figure 30.4D; Mayer 2006b).

Crural glands, which are mainly restricted to male onychophorans and believed to produce pheromones (Reid 1996; Barclay et al. 2000), also open at the ventral basis of the leg either in numerous trunk segments (in most peripatopsids and representatives of *Peripatus*) or in only 2–3 pre-genital segments (in most peripatids) (Reid 1996; Oliveira et al. 2012a, 2015). These glands may also have given rise to the slime glands of onychophorans (Figure 30.4A), as well as to the sexual anterior accessory glands of some male peripatopsids (Ruhberg and Storch 1977; Baer and Mayer 2012).

The reproductive system of onychophorans is variable among taxa but shows a simple overall construction. Male reproductive structures include a pair of testes, seminal vesicles, spermioducts, and an unpaired ejaculatory duct (Storch and Ruhberg 1990; Storch et al., 1995). The reproductive tract of females is comparatively more diverse (**Figure 30.4E**; Mayer and Tait 2009) and consists of different types of ovary, a pair of oviducts (associated or not with seminal receptacles, accessory pouches and/or ovarian funnels), paired uteri, and a common duct leading into the vagina (Herzberg et al. 1980; Brockmann et al. 1999, 2001; Walker et al. 2006; Mayer and Tait 2009). In both sexes, the genital opening is located ventrally in the posterior body region, either between the last (in peripatopsids) or penultimate leg pair (in peripatids). In peripatopsids, the genital opening is modified into an ovipositor in females of oviparous species and into a penis-like structure in males of *Paraperipatus* (Ruhberg 1985; Reid 1996).

Like many arthropods, onychophorans also exhibit tracheae as respiratory organs, which form a prominent and elaborate system of tubes and fascicles supplying oxygen to tissues and organs of the digestive, reproductive, nervous, and muscular systems (Storch and Ruhberg 1993; Hilken 1998; Mayer and Tait 2009; Baer and Mayer 2012; Oliveira et al. 2012a, 2013b; Ruhberg and Mayer 2013). Different from arthropods, the onychophoran tracheae develop postembryonically, are unbranched, and open to the outside via randomly distributed atria that lack a closing mechanism (Bicudo 1986; Ruhberg and Mayer 2013). The onychophoran tracheal tubes show a unique structure and their lining is not molted periodically together with the body cuticle. The periodic molting process of onychophorans is most likely mediated by ecdysteroid hormones, as in other representatives of Ecdysozoa (Manton 1938a; Holliday 1942; Hoffmann 1997), although the majority of genes responsible for the ecdysteroid biosynthesis remain unknown (Schumann et al. 2018).

REPRODUCTION AND DEVELOPMENT

Different sperm transfer modes have been described from onychophorans, ranging from dermal insemination (Manton 1938b; Mayer 2007) to direct transfer by either pairing their genital openings (Oliveira et al. 2012a) or using specialized head structures in males (Reid 1996; Tait and Norman 2001). Irrespective of the sperm transfer mode, internal fertilization is the rule among onychophorans. Embryonic nourishment modes vary from oviparity to lecithotrophic, matrotrophic or placentotrophic viviparity (Mayer et al. 2015a). Embryos of placentotrophic viviparous species are connected to the uterus wall via a hollow stalk (Mayer et al. 2015a); however, nourishment is not transferred through this structure but rather via the embryonic surface (Walker and Campiglia 1990). The term "ovoviviparity" used for several onychophoran species seems inappropriate, as it oversimplifies a range of distinct embryonic nutrition modes (Mayer et al. 2015a; Oliveira et al. 2018). For example, the peripatopsid *Euperipatoides rowelli* previously classified as "ovoviviparous", rather exhibits a combination of lecithotrophic and matrotrophic viviparity, in that embryos of this species develop from yolky eggs but also receive extra nourishment from the mother (Sunnucks et al. 2000). Such a "mixed" mode may also be present in other species previously regarded as "ovoviviparous" (Tutt et al. 2002; Oliveira et al. 2018).

Matrotrophic viviparous species produce yolkless eggs and their embryos usually exhibit large "trophic vesicles" or "trophic organs" believed to take up nourishment from the mother (Mayer et al. 2015a). This is the case in *Metaperipatus inae* (Chile), *Peripatopsis sedgwicki* (South Africa), and *Paraperipatus* spp. (Indonesia and Papua New Guinea) (Willey 1898; Bouvier 1905; Pflugfelder 1948; Manton 1949; Hofmann 1988; Mayer 2007), but it is unclear whether their trophic organs are homologous structures or evolved convergently. True placentotrophic viviparity might have evolved either in the neotropical Peripatidae or in the last common ancestor of the tropical African and the neotropical Peripatidae (Mayer et al. 2015a). Oviparity is also a derived condition that evolved at least twice independently among Australian/New Zealand peripatopsids (Reid 1996; Mayer et al. 2015a; Giribet et al. 2018), probably by heterochronic evolution from either lecithotrophic or combined lecithotrophic/matrotrophic viviparity, as at least some embryonic development takes place before egg deposition (Brockmann et al. 1997; Norman and Tait 2008).

The development, from cleavage to gastrulation, varies across Onychophora according to the nourishment type. Two types of cleavage patterns have been confirmed: discoidal meroblastic in most lecithotrophic/matrotrophic viviparous species and holoblastic (total) cleavage in placentotrophic viviparous species (Mayer et al. 2015a). The cleavage pattern of oviparous species is unknown. Development proceeds differently even in closely related species, in particular regarding the gastrulation modes (see Mayer et al. 2015a for further details). Remarkable developmental patterns include the putative lack of a true blastula in the matrotrophic viviparous *Peripatopsis capensis* (cf. Manton 1949) and the formation of a germ disk prior to gastrulation in non-placentotrophic viviparous species (Mayer and Whitington 2009; Mayer et al. 2015a). At the onset of gastrulation in *E. rowelli*, two openings corresponding to the stomodeum and proctodeum are observed inside the blastoporal pit. The anteroposterior body axis is already defined at this stage (Mayer et al. 2015a). As development proceeds, these two openings are then moved apart, forming a longitudinal slit (= blastoporal slit) which separates from the openings and eventually closes by amphistomy. The origin of the stomodeum in placentotrophic viviparous species is ambiguous but the occurrence of a blastoporal slit in late developmental stages suggests a similar process as in *E. rowelli*. The mechanism of dorsoventral axis determination, on the other hand, remains uncharacterized (Treffkorn and Mayer 2013). Further embryonic development, including growth, mesoderm differentiation from endomesodermal tissue, and organogenesis, proceeds similarly in placentotrophic and non-placentotrophic species (Mayer et al. 2015a).

DISTRIBUTION AND ECOLOGY

The two major onychophoran subgroups show a disjointed and mutually exclusive distribution range along the tropical and subtropical regions, mainly in the southern hemisphere (**Figures 30.5, 30.6**). Representatives of Peripatidae

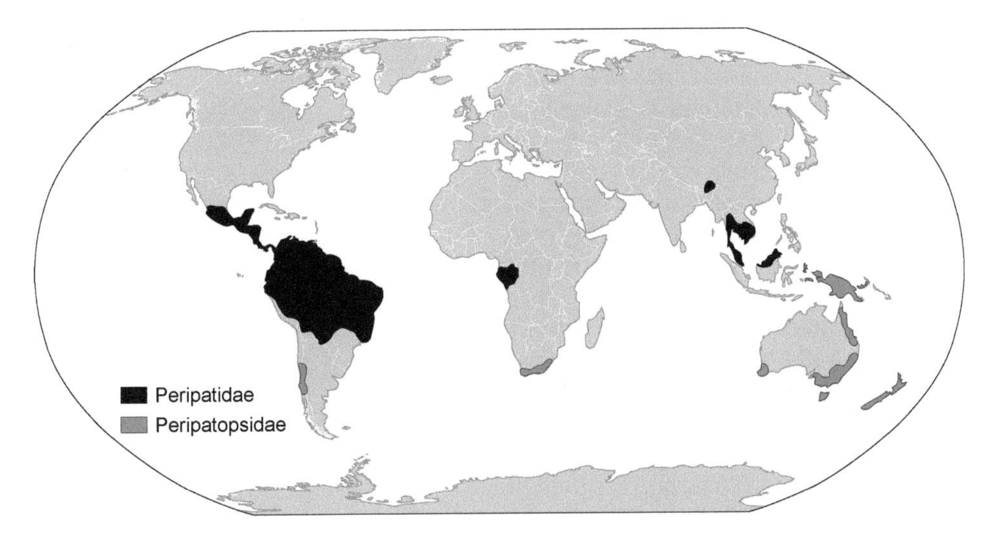

Figure 30.5 Map showing the geographical distribution of the two major onychophoran subgroups. (Modified from Oliveira et al. 2012b; Creative Commons [CC].)

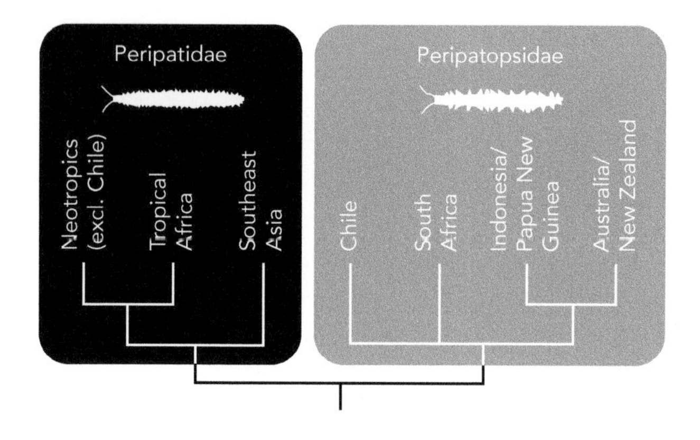

Figure 30.6 Phylogeography of onychophorans. Phylogenetic relationships after Murienne et al. (2014) and Giribet et al. (2018).

occupy the neotropical region of central Mexico to Brazil, as well as tropical Africa and Southeast Asia. The Peripatopsidae has a southern distribution in Chile, South Africa, Papua New Guinea, parts of Indonesia, Australia, and New Zealand. These taxa remain naturally isolated by arid belts in South America and Africa and by water masses in Southeast Asia and Oceania (Monge-Nájera 1995), whereas their co-occurrence in certain areas most likely resulted from recent human activities (van der Lande 1991). Despite the close association of onychophorans with landmasses derived from the break-up of Gondwana (Murienne et al. 2014), fossil records indicate that velvet worms had a wider distribution in the past (Garwood et al. 2016; Oliveira et al. 2016). The evolutionary relationships within each major onychophoran subgroup are relatively well investigated (**Figure 30.6**) and evidence suggests that recent lineages of Peripatidae and Peripatopsidae might have diverged from each other and dispersed prior to the continental drift, over 300 million years ago, subsequently separating by vicariance (Murienne et al. 2014; Oliveira et al. 2016; Giribet et al. 2018).

Velvet worms are exclusively terrestrial animals that show low dispersal capability, cryptic behavior, and small populations (New 1995), with most species being point endemic and restricted to small areas. Inside the forests, onychophorans are typically confined to moist microhabitats such as the leaf litter, decaying logs, and soil cavities (Brinck 1957; Ruhberg 1985; Reid 1996), given a dearth of mechanisms to control water loss (Ruhberg and Mayer 2013; Weldon et al. 2013). Nearly all species are solitary, although several individuals may be found together in microhabitats with suitable humidity and temperature (Lavallard et al. 1975; Ruhberg 1985). Social-like behavior has only been described for *E. rowelli*: small groups are putatively organized in a hierarchy based on female dominance and males play an important role in dispersion mediated by pheromones (Barclay et al. 2000; Reinhard an Rowell 2005). Tropical onychophorans are nocturnal predators that forage the forest ground for potential prey, such as crickets, spiders, cockroaches, and other arthropods (Read and Hughes 1987; González et al. 2018). *Macroperipatus torquatus* selects the prey based on its size, as small prey would scarcely repay the energy costs of slime production and too large prey is more prone to escape (Read and Hughes 1987). In this species, ingestion of the prey takes up to 9 hours and adults feed once every 22 days, while juveniles feed approximately once every 5 days (Read and Hughes 1987). Reproduction may or may not be seasonal depending on the species and females most likely have postcopulatory control of fertilization (Lavallard and Campiglia 1975a; Curach and Sunnucks 1999; Sunnucks et al. 2000; Oliveira et al. 2012a, 2015). In general, gestation time may exceed 12 months and reproduction rates are substantially lower in placentotrophic species than other onychophorans (Lavallard and Campiglia 1975a; Oliveira et al. 2012a; Ruhberg 1985; Sunnucks et al. 2000; Walker and Tait 2004).

PHYSIOLOGY AND BEHAVIOR

In contrast to arthropods, velvet worms are unable to close their tracheal openings (Lavallard and Campiglia-Reimann 1966; Bicudo 1986; Storch and Ruhberg 1993), thus they are highly prone to desiccation. Water loss and oxygen consumption measurements in several species indeed indicate that respiratory loss accounts for 34%–50% of total water loss, which contrasts with less than 20% in arthropods (Campiglia et al. 1978; Glens and Campiglia 1986; Clusella-Trullas and Chown 2008; Weldon et al. 2013). Interestingly, different from arthropods, the water loss in onychophorans does not seem to be associated with changes in metabolic rates or gas exchange but is rather affected by temperature and humidity (Clusella-Trullas and Chown 2008; Weldon et al. 2013), with lower rates observed in species inhabiting warmer and drier areas (Weldon et al. 2013). Experimental data from *Epiperipatus biolleyi* further indicate that onychophorans may endure long contact with freshwater but rapidly die when exposed to seawater (Monge-Nájera et al. 1993). Apparently, they do not tolerate the hyperosmotic stress.

Like most other ecdysozoans, onychophorans regularly molt their cuticle (Holliday 1942, 1944; Storch and Ruhberg 1993), a process known as ecdysis. Ecdysteroids have been detected in most tissues and might act as molting hormones, but beyond this only little is known about the molecular pathway of molting in onychophorans (Hoffmann 1997; Schumann et al. 2018). The two pigment-dispersing factor neuropeptides identified in onychophorans might also play a role as hormones involved in circadian control (Mayer et al. 2015c; Martin et al. 2020). The light input to the circadian clock might be provided by one of the three opsin proteins present in onychophorans: an r-opsin (onychopsin), a c-opsin, and an arthropsin (Hering et al. 2012; Eriksson et al. 2013; Beckmann et al. 2015; Schumann et al. 2016). The r-opsin has been shown to be expressed in the eye and responds to wavelengths ranging from ultraviolet to green light, with maximum sensitivity to blue light (peak sensitivity: ~480 nm) suggesting monochromatic vision (Beckmann et al. 2015). Pheromone-mediated dispersion is believed to occur in *E. rowelli* (Barclay et al. 2000), although no pheromone has been identified in onychophorans yet.

The blood–brain barrier does not seem to exist in velvet worms, as peripheral perineural glial cells are not connected by junctional complexes, such as septate junctions, allowing substances to enter the nervous system without restriction (Lane and Campiglia 1987). The blood contains various ions, including Na^{+}, K^{+}, Ca^{2+}, Mg^{+}, Cl^{-}, HCO_3^{-}, and $(PO_4)^{3-}$ and shows a pH of ~7.3 (Robson et al. 1966; Campiglia 1976). The ionic composition, however, may vary slightly between the species (Campiglia 1976). Hemocyanin may act as the only oxygen-binding pigment (Kusche et al. 2002). Six types of cells have been reported from the hemolymph, including the so-called nephrocytes and five types of hemocytes (Lavallard and Campiglia 1975b; Seifert and Rosenberg 1977). These cells are believed to be responsible for nutrition and excretory processes, with some of the hemocytes also playing a role in immune response (Silva and Coelho 2000). Induced inflammatory process and immune-inducible genes have been analyzed in two species of Peripatidae (Silva and Coelho 2000; Altincicek and Vilcinskas 2008). These studies revealed that the immune-related signaling is mediated by 36 genes and that the inflammatory reaction causes substantial migration of hemocytes into the connective tissue toward the artificially damaged area.

Only little information is available on the physiology of the excretory system in velvet worms. Evidence of uric acid as a potential excretory product has been found in the gut tissue of both peripatids and peripatopsids (Heatley 1936; Manton and Heatley 1937; Lavallard 1967). This might be a physiological adaptation to reduce water loss (Manton and Heatley 1937). The identified enzymes in the gut include invertase, maltase, lipase, esterase, amino- and carboxypolypeptidase, and dipeptidase (Heatley 1936). The nephridial fluid (urine) has a similar ionic composition to the blood but shows different ionic concentrations (Campiglia and Lavallard 1983). The urine has an osmolarity of 196.9 μOsm/L and a pH of 5.3 (Campiglia and Lavallard 1983). The salivary glands, which are derivatives of nephridia, possess an enlarged secretory part composed of mucus-, protein-, and enzyme-producing cells (Storch et al. 1979; Storch and Ruhberg 1993). The saliva produced by these glands is rich in acid phosphatase (Nelson et al. 1980) and contains amylase, glycogenase, protease, and carboxypolypeptidase (Heatley 1936). It is assumed to be involved in extraintestinal digestion of the prey (Mayer et al. 2015b).

The heart physiology has been studied in two species (Sundara Rajulu and Singh 1969; Sundara Rajulu and Santhana Krishnan 1973; Hertel et al. 2002). The rhythmicity of the heart beat is increased by adrenaline and acetylcholine and strongly decreased by histamine, whereas the presence of the neuropeptide proctolin and the head peptide has no significant effect on cardiac physiology (Sundara Rajulu and Singh 1969; Hertel et al. 2002). The somatic musculature is cholinoceptive and innervated by cholinergic motor neurons; the oblique and ring musculature reacts additionally to L-glutamate but no muscle is sensitive to γ-aminobutyric acid (GABA) (Florey and Florey 1965; Stern and Bicker 2008), although GABA-like immunoreactivity is widely distributed in the onychophoran nerve cords (Martin et al. 2017b). Like in many other animals, CB_1 cannabinoid receptors have been detected in neural membranes of the onychophoran *Peripatoides novaezealandiae* (McPartland et al. 2006), but the specific role of these G-protein-coupled receptors has not been clarified.

The composition and properties of the onychophoran slime have been investigated in several species (Benkendorff et al. 1999; Haritos et al. 2010; Baer et al. 2014, 2017, 2018, 2019a,b). The slime is a fluid, protein-rich secretion, which rapidly transforms into stiff fibers upon agitation. In the native slime, the proteins are organized into nanodroplets that are stabilized by electrostatic interactions (Baer et al. 2017). Mechanical stress leads to a spontaneous self-assembly of nanodroplets accompanied by partial unfolding of proteins, which form a strong network by solidifying into fibers. This process is totally reversible, as the β-crystalline structure of proteins is only partially unfolded during fibrillation (Baer et al. 2019).

A few behavioral features of velvet worms, in particular those dealing with their reproduction and ecology, have been pointed out above (see sections Reproduction and Development and Distribution and Ecology of this chapter). Beyond this, there are not many other reports of onychophoran behavior, which is probably due to their cryptic lifestyle. As nocturnal animals, onychophorans show negative phototaxis (Alexander 1957; Beckmann et al. 2015; Kirwan et al. 2018) and when exposed to light, most species tend to hide quickly, while some peripatopsids display a characteristic curling behavior (Ruhberg 1985; Reid 1996). Humidity also seems to affect the behavior of onychophorans. For example, a Costa Rican species of the Peripatidae has been reported to be more active during the driest and darkest nights of the year (Barquero-González et al. 2018), whereas some Australian peripatopsids have been observed to fall into a state of dormancy when desiccated (Hardie 1972). The locomotory behavior of velvet worms includes different stepping movements and gaits, including a bottom gear, a middle gear, and top gear gaits (Manton 1950; Oliveira et al. 2019). A metachronal rhythm of legs is not always observed during locomotion and the movement of each leg might be regulated independently (Oliveira et al. 2019).

Under laboratory conditions, the peripatid *E. biolleyi* preferred moss over other types of vegetal substrates and no aggression was observed among specimens of this species (Monge-Nájera et al. 1993). In contrast, aggressive interaction between individuals, such as biting and chasing, has been reported from the Australian peripatopsids *Phallocephale tallagandensis* and *E. rowelli* (Reinhard and Rowell 2005; Mayer et al. 2015b). Moreover, specimens of *E. rowelli* were recorded to tear off and swallow large pieces of prey (Mayer et al. 2015b). This and the fact that the animals are able to swallow other kind of solid matter, such as their own exuvia (Storch and Ruhberg 1993), suggest that extraintestinal digestion is not the only type of digestion in onychophorans.

Death is often preceded by contraction of the body and ejection of saliva, slime, feces, and sometimes embryos (Monge-Nájera et al. 1993). Interestingly, different tissues decay at different rates (Murdock et al. 2014). For example, while the alimentary tract and the epidermis are prone to very fast decay lasting only a few hours, decomposition of the eyes and the nerve cords is evident only after several days; no effects of decay were observed in sclerotized cuticular structures, such as jaws and claws, which may persist indefinitely after death (Murdock et al. 2014).

GENETICS

Onychophorans possess large, repetitive genomes that still challenge sequencing based on current techniques. Genome size estimates vary from 4.9 Gb up to 18.5 Gb, the latter being over a six-fold the size of the human genome (Jeffery et al. 2012). Despite recent attempts, an assembled complete genome sequence of Onychophora is yet unavailable. On the other hand, a wealth of molecular investigations based on a limited number of genes has been conducted on onychophorans in recent years. These studies mainly focused on either clarifying phylogenetic relationships among species at local or broader scales (Daniels et al. 2009, 2016, 2017; Lacorte et al. 2011; Oliveira et al. 2011, 2012a, 2013a, 2015, 2018; Murienne et al. 2014; Oliveira and Mayer 2017; Giribet et al. 2018) or assessing the evolution and/or expression pattern of different genes and protein groups in velvet worms (Hering et al. 2012; Janssen and Budd 2013, 2016, 2017; Franke and Mayer 2014; Franke et al. 2015; Schumann et al. 2018; Treffkorn et al. 2018; Vizueta et al. 2020).

POSITION IN THE ToL

Onychophorans have consistently been classified within Ecdysozoa (= molting animals) as a clade of the megadiverse taxon Panarthropoda which also includes the tardigrades (water bears) and the arthropods (chelicerates, myriapods, and pancrustaceans) (**Figure 30.7**; Nielsen 2012). Nevertheless, the phylogenetic position of Onychophora inside Panarthropoda is still a subject of discussion, mainly due to the uncertain position of tardigrades. Hence, onychophorans may be classified as either the sister group to Arthropoda or to Tardigrada plus Arthropoda (Martin et al. 2017a).

DATABASES AND COLLECTIONS

An updated checklist of all valid onychophoran taxa is available at www.onychophora.com and www.myriatrix.myspecies.info together with a list of fossil onychophorans and taxa regarded as *nomina dubia*.

CONCLUSION

A number of aspects make onychophorans relevant for studying biodiversity, biogeography, animal evolution, and conservation. These include their ancient

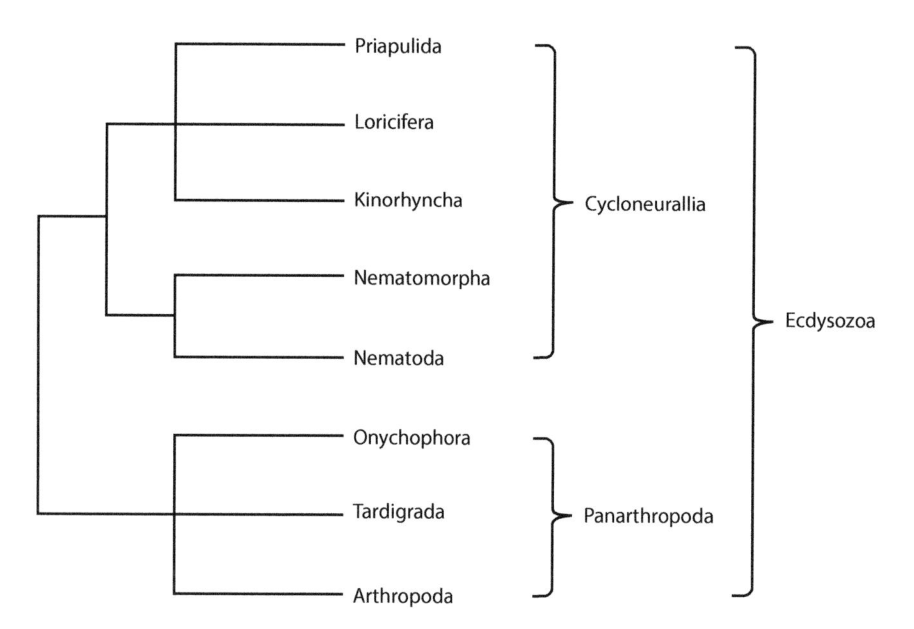

Figure 30.7 Position of Onychophora within Ecdysozoa.

origin, remarkable distribution pattern, conserved morphology, restriction to humid habitats, and key phylogenetic position as sister group to the world's most diverse animal group, the arthropods. However, Onychophora remains understudied and research efforts in many directions are still necessary to close the numerous gaps in our knowledge.

REFERENCES

Aguinaldo A.M.A., Turbeville J.M., Linford L.S., Rivera M.C., Garey J.R., Raff R.A. & Lake J.A. (1997) Evidence for a clade of nematodes, arthropods and other moulting animals. Nature 387: 489–493.

Alexander A.J. (1957) Notes on onychophoran behaviour. Annals of the Natal Museum 14: 35–43.

Altincicek B. & Vilcinskas A. (2008) Identification of immune inducible genes from the velvet worm *Epiperipatus biolleyi* (Onychophora). Developmental & Comparative Immunology 32: 1416–1421.

Audouin M. & Milne-Edwards H. (1833) Classification des Annélides, et description de celles qui habitent les côtes de la France - sixième Famille: Péripatiens. Annales des Sciences Naturelles 30: 411–414.

Baer A. & Mayer G. (2012) Comparative anatomy of slime glands in Onychophora (velvet worms). Journal of Morphology 273: 1079–1088. Doi: 10.1002/jmor.20044.

Baer A., Hänsch S., Mayer G., Harrington M.J. & Schmidt S. (2018) Reversible supramolecular assembly of velvet worm adhesive fibers via electrostatic interactions of charged phosphoproteins. Biomacromolecules 19: 4034–4043. Doi: 10.1021/acs.biomac.8b01017.

Baer A., Horbelt N., Nijemeisland M., Garcia S.J., Fratzl P., Schmidt S., Mayer G. & Harrington M.J. (2019a) Shear-induced β-crystallite unfolding in condensed phase nanodroplets promotes fiber formation in a biological adhesive. ACS Nano 13: 4992–5001. Doi: 10.1021/acsnano.9b00857.

Baer A., Oliveira I.S., Steinhagen M., Beck-Sickinger A.G. & Mayer G. (2014) Slime protein profiling: a non-invasive tool for species identification in Onychophora (velvet worms). Journal of Zoological Systematics and Evolutionary Research 52: 265–272. Doi: 10.1111/jzs.12070.

Baer A., Schmidt S., Haensch S., Eder M., Mayer G. & Harrington M.J. (2017) Mechanoresponsive lipid-protein nanoglobules facilitate reversible fibre formation in velvet worm slime. Nature Communications 8: 974. Doi: 10.1038/s41467-017-01142-x.

Baer A., Schmidt S., Mayer G. & Harrington M.J. (2019b) Fibers on the fly: multiscale mechanisms of fiber formation in the capture slime of velvet worms. Integrative and Comparative Biology 59: 1690–1699. Doi: 10.1093/icb/icz048.

Barclay S.D., Rowell D.M. & Ash J.E. (2000) Pheromonally mediated colonization patterns in the velvet worm *Euperipatoides rowelli* (Onychophora). Journal of Zoology 250: 437–446. Doi: 10.1111/j.1469-7998.2000.tb00787.x.

Barquero-González J.P., Morera-Brenes B. & Monge-Nájera J. (2018) The relationship between humidity, light and the activity pattern of a velvet worm, *Epiperipatus* sp. (Onychophora: Peripatidae), from Bahía Drake, South Pacific of Costa Rica. Brazilian Journal of Biology 78: 408–413. Doi: 10.1590/1519-6984.166495.

Beckmann H., Hering L., Henze M.J., Kelber A., Stevenson P.A. & Mayer G. (2015) Spectral sensitivity in Onychophora (velvet worms) revealed by electroretinograms, phototactic behaviour and opsin gene expression. Journal of Experimental Biology 218: 915–922. Doi: 10.1242/jeb.116780.

Benkendorff K., Beardmore K., Gooley A.A., Packer N.H. & Tait N.N. (1999) Characterisation of the slime gland secretion from the peripatus, *Euperipatoides kanangrensis* (Onychophora: Peripatopsidae). Comparative Biochemistry and Physiology, Part B 124: 457–465.

Bicudo J.E.P.W. (1986) Morphometric estimates of the surface areas of the tracheal and cuticular systems in *Peripatus acacioi* Marcus and Marcus (Onychophora). Journal of Morphology 188: 251–255.

Blanchard E. (1847) Recherches sur l'organisation des Vers. Annales des Sciences Naturelles [3e Série] 8: 119–149.

Bouvier E.L. (1905) Monographie des Onychophores. Annales des Sciences Naturelles Zoologie et Biologie Animale [9e Série] 2: 1–383.

Bouvier E.L. (1907) Monographie des Onychophores. Annales des Sciences Naturelles Zoologie et Biologie Animale [9e Série] 5: 61–318.

Brinck P. (1957) Onychophora, a review of South African species, with a discussion on the significance of the geographical distribution of the group. In: Hanström B., Brinck P. & Rudebeck G. (Eds.), South African Animal Life Results of the Lund University expedition 1950–1951. Almqvist & Wiksell, Stockholm, Sweden, 7–32.

Brockmann C., Mesibov R. & Ruhberg H. (1997) Observations on *Ooperipatellus decoratus*, an oviparous onychophoran from Tasmania (Onychophora: Peripatopsidae). Entomologica Scandinavica, Supplement 51: 319–329.

Brockmann C., Mummert R., Ruhberg H. & Storch V. (1999) Ultrastructural investigations of the female genital system of *Epiperipatus biolleyi* (Bouvier, 1902) (Onychophora, Peripatidae). Acta Zoologica 80: 339–349.

Brockmann C., Mummert R., Ruhberg H. & Storch V. (2001) The female genital system of *Ooperipatus decoratus* (Onychophora, Peripatopsidae): an ultrastructural study. Journal of Morphology 249: 77–88.

Campiglia S.S. (1976) The blood of *Peripatus acacioi* Marcus and Marcus (Onychophora) - III. The ionic composition of the hemolymph. Comparative Biochemistry and Physiology, Part A 54: 129–133.

Campiglia S.S. & Lavallard R. (1983) Extracellular space and nephridial function in *Peripatus acacioi*, Marcus and Marcus (Onychophora). Comparative Biochemistry and Physiology 76A: 167–171.

Campiglia S.S., Lavallard R. & Ribeiro M.d.G.C. (1978) Pression partielle de gaz carbonique et concentration des bicarbonates dans l'hemolymphe de *Peripatus acacioi* Marcus et Marcus (Onychophore). Boletim de Fisiologia Animal 2: 23–32.

Clark A.H. (1913) A revision of the American species of *Peripatus*. Proceedings of the Biological Society of Washington 26: 15–19.

Clusella-Trullas S. & Chown S.L. (2008) Investigating onychophoran gas exchange and water balance as a means to inform current controversies in arthropod physiology. Journal of Experimental Biology 211: 3139–3146. Doi: 10.1242/jeb.021907.

Curach N. & Sunnucks P. (1999) Molecular anatomy of an onychophoran: Compartmentalized sperm storage and heterogeneous paternity. Molecular Ecology 8: 1375–1385.

Daniels S.R., Dambire C., Klaus S. & Sharma P.P. (2016) Unmasking alpha diversity, cladogenesis and biogeographical patterning in an ancient panarthropod lineage (Onychophora: Peripatopsidae: *Opisthopatus cinctipes*) with the description of five novel species. Cladistics 32: 506–537. Doi: 10.1111/cla.12154.

Daniels S.R., Dreyer M. & Sharma P.P. (2017) Contrasting the population genetic structure of two velvet worm taxa (Onychophora: Peripatopsidae: *Peripatopsis*) in forest fragments along the south-eastern Cape, South Africa. Invertebrate Systematics 31: 781–796. Doi: 10.1071/IS16085.

Daniels S.R., McDonald D.E. & Picker M.D. (2013) Evolutionary insight into the *Peripatopsis balfouri* sensu lato species complex (Onychophora: Peripatopsidae) reveals novel lineages and zoogeographic patterning. Zoologica Scripta 42: 656–674. Doi: 10.1111/zsc.12025.

Daniels S.R., Picker M.D., Cowlin R.M. & Hamer M.L. (2009) Unravelling evolutionary lineages among South African velvet worms (Onychophora: *Peripatopsis*) provides evidence for widespread cryptic speciation. Biological Journal of the Linnean Society 97: 200–216. Doi: 10.1111/j.1095-8312.2009.01205.x.

Dendy A. (1894) Additions to the cryptozoic fauna of New Zealand. Annals and Magazine of Natural History [Series 6] 14: 393–401.

Eakin R.M. & Westfall J.A. (1965) Fine structure of the eye of *Peripatus* (Onychophora). Zeitschrift für Zellforschung und mikroskopische Anatomie 68: 278–300.

Eriksson B.J., Fredman D., Steiner G. & Schmid A. (2013) Characterisation and localisation of the opsin protein repertoire in the brain and retinas of a spider and an onychophoran. BMC Evolutionary Biology 13: 186. Doi: 10.1186/1471-2148-13-186.

Evans R. (1901) On two new species of Onychophora from the Siamese Malay States. Quarterly Journal of Microscopical Science 44: 473–538.

Florey E. & Florey E. (1965) Cholinergic neurons in the Onychophora: A comparative study. Comparative Biochemistry and Physiology 15: 125–136.

Franke F.A. & Mayer G. (2014) Controversies surrounding segments and parasegments in Onychophora: Insights from the expression patterns of four "segment polarity genes" in the peripatopsid *Euperipatoides rowelli*. PLoS One 9: e114383. Doi: 10.1371/journal.pone.0114383.

Franke F.A., Schumann I., Hering L. & Mayer G. (2015) Phylogenetic analysis and expression patterns of Pax genes in the onychophoran *Euperipatoides rowelli* reveal a novel bilaterian Pax subfamily. Evolution and Development 17: 3–20. Doi: 10.1111/ede.12110.

Garwood R.J., Edgecombe G.D., Charbonnier S., Chabard D., Sotty D. & Giribet G. (2016) Carboniferous Onychophora from Montceau-les-Mines, France, and onychophoran terrestrialization. Invertebrate Biology 135: 179–190. Doi: 10.1111/ivb.12130.

Gervais P. (1837) Etudes pour servir à l'historie naturelle des Myriapodes. Annales des Sciences Naturelles [2e Série] 7: 35–60.

Giribet G., Buckman-Young R.S., Costa C.S., Baker C.M., Benavides L.R., Branstetter M.G., Daniels S.R. & Pinto-da-Rocha R. (2018) The '*Peripatos*' in Eurogondwana? – Lack of evidence that south-east Asian onychophorans walked through Europe. Invertebrate Systematics 32: 842–865. Doi: doi.org/10.1071/IS18007.

Glens F. & Campiglia S. (1986) Temperature de la surface corporelle et taux d'evaporation chez *Peripatus acacioi* Marcus et Marcus (Onychophora: Peripatidae). Boletim de Fisiologia Animal 10: 103–113.

González M., Sosa-Bartuano A. & Monge-Nájera J. (2018) A velvet worm (Onychophora: Peripatidae) feeding on a free ranging spider in Sierra Llorona, Panama. Cuadernos de Investigación UNED 10: 283–284.

Grube E. (1853) Über den Bau von *Peripatus edwardsii*. Müller's Archives of Anatomy and Physiology 20: 322–360.

Guilding L. (1826) Mollusca Caribbaeana. Zoological Journal 2: 437–449.

Hamer M.L., Samways M.J. & Ruhberg H. (1997) A review of the Onychophora of South Africa, with discussion of their conservation. Annals of the Natal Museum 38: 283–312.

Hardie H.R. (1972) Studies on the Onychophora. Dissertation thesis, University of New England, Armidale, Australia. 1–122.

Haritos V.S., Niranjane A., Weisman S., Trueman H.E., Sriskantha A. & Sutherland T.D. (2010) Harnessing disorder: onychophorans use highly unstructured proteins, not silks, for prey capture. Proceedings of the Royal Society B: Biological Sciences 277: 3255–3263. Doi: 10.1098/rspb.2010.0604.

Heatley N.G. (1936) The digestive enzymes of the Onychophora (*Peripatopsis* spp.). Journal of Experimental Biology 13: 329–343.

Hering L., Henze M.J., Kohler M., Bleidorn C., Leschke M., Nickel B., Meyer M., Kircher M., Sunnucks P. & Mayer G. (2012) Opsins in Onychophora (velvet worms) suggest a single origin and subsequent diversification of visual pigments in arthropods. Molecular Biology and Evolution 29: 3451–3458. Doi: 10.1093/molbev/mss148.

Hertel W., Wirkner C.S. & Pass G. (2002) Studies on the cardiac physiology of Onychophora and Chilopoda. Comparative Biochemistry and Physiology, Part A 133: 605–609.

Herzberg A., Ruhberg H. & Storch V. (1980) Zur Ultrastruktur des weiblichen Genitaltraktes der Peripatopsidae (Onychophora). Zoologische Jahrbücher, Abteilung für Anatomie und Ontogenie der Tiere 104: 266–279.

Hilken G. (1998) Vergleich von Tracheensystemen unter phylogenetischem Aspekt. Verhandlungen des Naturwissenschaftlichen Vereins in Hamburg 37: 5–94.

Hoffmann K.H. (1997) Ecdysteroids in adult females of a "walking worm": *Euperipatoides leuckartii* (Onychophora, Peripatopsidae). Invertebrate Reproduction and Development 32: 27–30.

Hofmann K. (1988) Observations on *Peripatopsis clavigera* (Onychophora, Peripatopsidae). South African Journal of Zoology 23: 255–258.

Holliday R.A. (1942) Some observations on natal Onychophora. Annals of the Natal Museum 10: 237–244.

Holliday R.A. (1944) Further notes on natal Onychophora. Annals of the Natal Museum 10: 433–436.

Hoyle G. & Williams M. (1980) The musculature of *Peripatus* and its innervation. Philosophical Transactions of the Royal Society B, Biological Sciences 288: 481–510.

Janssen R. & Budd G.E. (2013) Deciphering the onychophoran 'segmentation gene cascade': Gene expression reveals limited involvement of pair rule gene orthologs in segmentation, but a highly conserved segment polarity gene network. Developmental Biology 382: 224–234. Doi: 10.1016/j.ydbio.2013.07.010.

Janssen R. & Budd G.E. (2016) Gene expression analysis reveals that *Delta/Notch* signalling is not involved in onychophoran segmentation. Development Genes and Evolution 226: 69–77. Doi: 10.1007/s00427-016-0529-4.

Janssen R. & Budd G.E. (2017) Investigation of endoderm marker-genes during gastrulation and gut-development in the velvet worm *Euperipatoides kanangrensis*. Developmental Biology 427: 155–164. Doi: 10.1016/j.ydbio.2017.04.014.

Jeffery N.W., Oliveira I.S., Gregory T.R., Rowell D.M. & Mayer G. (2012) Genome size and chromosome number in velvet worms (Onychophora). Genetica 140: 497–504. Doi: 10.1007/s10709-013-9698-5.

Kirwan J.D., Graf J., Smolka J., Mayer G., Henze M.J. & Nilsson D.E. (2018) Low-resolution vision in a velvet worm (Onychophora). Journal of Experimental Biology 221: jeb175802.

Kusche K., Ruhberg H. & Burmester T. (2002) A hemocyanin from the Onychophora and the emergence of respiratory proteins. Proceedings of the National Academy of Sciences of the United States of America 99: 10545–10548.

Lacorte G.A., Oliveira I.S. & Fonseca C.G. (2011) Phylogenetic relationships among the *Epiperipatus* lineages (Onychophora: Peripatidae) from the Minas Gerais State, Brazil. Zootaxa 2755: 57–65.

Lane N.J. & Campiglia S.S. (1987) The lack of a structured blood-brain barrier in the onychophoran *Peripatus acacioi*. Journal of Neurocytology 16: 93–104.

Lavallard R. (1967) Ultrastructure des cellules prismatiques de l'épithélium intestinal chez *Peripatus acacioi* Marcus et Marcus. Comptes Rendus Hebdomadaires des Seances Serie D: Sciences Naturelles 264: 929–932.

Lavallard R. & Campiglia S. (1975a) Contribution à la biologie de *Peripatus acacioi* Marcus et Marcus (Onychophore). V. Etude des Naissances dans un elevage de laboratiore. Zoologischer Anzeiger 195: 338–350.

Lavallard R. & Campiglia S. (1975b) Contribution a l'hématologie de *Peripatus acacioi* Marcus et Marcus (Onychophora) - I. Structure et ultrastructure des hémocytes. Annales des Sciences Naturelles, Zoologie et Biologie Animale [12e Série] 17: 67–92.

Lavallard R. & Campiglia-Reimann S. (1966) Structure et ultrastructure de l'appareil trachéen chez *Peripatus acacioi* Marcus et Marcus (Onychophore). Comptes Rendus Hebdomadaires des Seances Serie D: Sciences Naturelles 263: 1728–1731.

Lavallard R., Campiglia S., Parisi Alvares E. & Valle C.M.C. (1975) Contribution à la biologie de *Peripatus acacioi* Marcus et Marcus (Onychophore). III. Étude descriptive de l'habitat. Vie et Milieu 25: 87–118.

Manton S.M. (1938a) Studies on the Onychophora, VI. The life-history of *Peripatopsis*. Annals and Magazine of Natural History [Series 11] 1: 515–529.

Manton S.M. (1938b) Studies on the Onychophora, IV. The passage of spermatozoa into the ovary in *Peripatopsis* and the early development of the ova. Philosophical Transactions of the Royal Society B, Biological Sciences 228: 421–444.

Manton S.M. (1949) Studies on the Onychophora VII. The early embryonic stages of *Peripatopsis*, and some general considerations concerning the morphology and phylogeny of the Arthropoda. Philosophical Transactions of the Royal Society B, Biological Sciences 233: 483–580.

Manton S.M. (1950) The evolution of arthropodan locomotory mechanisms - Part I. The locomotion of *Peripatus*. Journal of the Linnean Society of London Zoology 41: 529–570.

Manton S.M. (1973) The evolution of arthropodan locomotory mechanisms. Part II: Habits, morphology and evolution of the Uniramia (Onychophora, Myriapoda and Hexapoda) and comparisons with the Arachnida, together with a functional review of uniramian musculature. Zoological Journal of the Linnean Society 53: 257–375.

Manton S.M. & Heatley N.G. (1937) Studies on the Onychophora. II. The feeding, digestion, excretion, and food storage of *Peripatopsis*, with biochemical estimations and analyses. Philosophical Transactions of the Royal Society B, Biological Sciences 227: 411–464.

Martin C. & Mayer G. (2014) Neuronal tracing of oral nerves in a velvet worm—Implications for the evolution of the ecdysozoan brain. Frontiers in Neuroanatomy 8: 1–13.

Martin C. & Mayer G. (2015) Insights into the segmental identity of post-oral commissures and pharyngeal nerves in Onychophora based on retrograde fills. BMC Neuroscience 16: 1–13. Doi: 10.1186/s12868-015-0191-1.

Martin C, Hering L, Metzendorf N, Hormann S, Kasten S, Fuhrmann S, Werckenthin A., Herberg FW, Stengl M, Mayer G (2020) Analysis of pigment-dispersing factor neuropeptides and their receptor in a velvet worm. Frontiers in Endocrinology 11:273.

Martin C., Gross V., Hering L., Tepper B., Jahn H., Oliveira I.S., Stevenson P.A. & Mayer G. (2017a) The nervous and visual systems of onychophorans and tardigrades: Learning about arthropod evolution from their closest relatives. Journal of Comparative Physiology A: Neuroethology, Sensory, Neural, and Behavioral Physiology 203: 565–590. Doi: 10.1007/s00359-017-1186-4.

Martin C., Gross V., Pflüger H.-J., Stevenson B.J. & Mayer G. (2017b) Assessing segmental versus non-segmental features in the ventral nervous system of onychophorans (velvet worms). BMC Evolutionary Biology 17: 3. Doi: 10.1186/s12862-016-0853-3.

Mayer G. (2006a) Structure and development of onychophoran eyes — what is the ancestral visual organ in arthropods? Arthropod Structure and Development 35: 231–245.

Mayer G. (2006b) Origin and differentiation of nephridia in the Onychophora provide no support for the Articulata. Zoomorphology 125: 1–12.

Mayer G. (2007) *Metaperipatus inae* sp. nov. (Onychophora: Peripatopsidae) from Chile with a novel ovarian type and dermal insemination. Zootaxa 1440: 21–37.

Mayer G. & Harzsch S. (2007) Immunolocalization of serotonin in Onychophora argues against segmental ganglia being an ancestral feature of arthropods. BMC Evolutionary Biology 7: 118.

Mayer G. & Harzsch S. (2008) Distribution of serotonin in the trunk of *Metaperipatus blainvillei* (Onychophora, Peripatopsidae): Implications for the evolution of the nervous system in Arthropoda. Journal of Comparative Neurology 507: 1196–1208.

Mayer G. & Tait N.N. (2009) Position and development of oocytes in velvet worms shed light on the evolution of the ovary in Onychophora and Arthropoda. Zoological Journal of the Linnean Society 157: 17–33.

Mayer, G. & Whitington, P.M. (2009): Velvet worm development links myriapods with chelicerates. Proceedings of the Royal Society B: Biological Sciences 276:3571–3579.

Mayer G., Whitington P.M., Sunnucks P. & Pflüger H.-J. (2010) A revision of brain composition in Onychophora (velvet worms) suggests that the tritocerebrum evolved in arthropods. BMC Evolutionary Biology 10: 1–10.

Mayer G., Kauschke S., Rüdiger J. & Stevenson P.A. (2013a) Neural Markers Reveal a One-Segmented Head in Tardigrades (Water Bears). PLoS One 8: e59090. Doi: 10.1371/journal.pone.0059090.

Mayer G., Martin C., Rüdiger J., Kauschke S., Stevenson P.A., Poprawa I., Hohberg K., Schill R.O., Pflüger H.-J. & Schlegel M. (2013b) Selective neuronal staining in tardigrades and onychophorans provides insights into the evolution of segmental ganglia in panarthropods. BMC Evolutionary Biology 13: 230. Doi: 10.1186/1471-2148-13-230

Mayer G., Franke F.A., Treffkorn S., Gross V. & Oliveira I.S. (2015a) Onychophora. In: Wanninger A. (Ed), Evolutionary Developmental Biology of Invertebrates. Springer, Berlin, 53–98.

Mayer G., Hering L., Stosch J.M., Stevenson P.A. & Dircksen H. (2015b) Evolution of pigment-dispersing factor neuropeptides in Panarthropoda: insights from Onychophora (velvet worms) and Tardigrada (water bears). Journal of Comparative Neurology 523: 1865–1885. Doi: 10.1002/cne.23767.

Mayer G., Oliveira I.S., Baer A., Hammel J.U., Gallant J. & Hochberg R. (2015c) Capture of prey, feeding, and functional anatomy of the jaws in velvet worms (Onychophora). Integrative and Comparative Biology 55: 217–227. Doi: 10.1093/icb/icv004.

McPartland J.M., Agraval J., Gleeson D., Heasman K. & Glass M. (2006) Cannabinoid receptors in invertebrates. Journal of Evolutionary Biology 19: 366–373. Doi: 10.1111/j.1420-9101.2005.01028.x.

Monge-Nájera J. (1995) Phylogeny, biogeography and reproductive trends in the Onychophora. Zoological Journal of the Linnean Society 114: 21–60. Doi: 10.1111/j.1096-3642.1995.tb00111.x.

Monge-Nájera J., Barrientos Z. & Aguilar F. (1993) Behavior of *Epiperipatus biolleyi* (Onychophora: Peripatidae) under laboratory conditions. Revista de Biologia Tropical 41: 689–696.

Müller M., Oliveira I.S., Allner S., Ferstl S., Bidola P., Mechlem K., Fehringer A., Hehn L., Dierolf M., Achterhold K., Gleich B., Hammel J.U., Jahn H., Mayer G. & Pfeiffer F. (2017) Myoanatomy of the velvet worm leg revealed by laboratory-based nanofocus X-ray source tomography. Proceedings of the National Academy of Sciences of the United States of America 114: 12378–12383. Doi: 10.1073/pnas.1710742114.

Murdock D.J.E., Gabbott S.E., Mayer G. & Purnell M.A. (2014) Decay of velvet worms (Onychophora), and bias in the fossil record of lobopodians. BMC Evolutionary Biology 14: 1–9. Doi: 10.1186/s12862-014-0222-z.

Murienne J., Daniels S.R., Buckley T.R., Mayer G. & Giribet G. (2014) A living fossil tale of Pangean biogeography. Proceedings of the Royal Society B: Biological Sciences 281: 20132648. Doi: 10.1098/rspb.2013.2648.

Nelson L., van der Lande V. & Robson E.A. (1980) Fine structural and histochemical studies on salivary glands of *Peripatoides novae-zealandiae* (Onychophora) with special reference to acid phosphatase distribution. Tissue and Cell 12: 405–418.

New T.R. (1995) Onychophora in invertebrate conservation: priorities, practice and prospects. Zoological Journal of the Linnean Society 114: 77–89. Doi: 10.1111/j.1096-3642.1995.tb00113.x.

Nielsen C. (2012) Animal Evolution: Interrelationships of the Living Phyla. Oxford University Press Inc., New York, 402 pp.

Norman J.M. & Tait N.N. (2008) Ultrastructure of the eggshell and its formation in *Planipapillus mundus* (Onychophora: Peripatopsidae). Journal of Morphology 269: 1263–1275.

Oliveira I.S. & Mayer G. (2013) Apodemes associated with limbs support serial homology of claws and jaws in Onychophora (velvet worms). Journal of Morphology 274: 1180–1190. Doi: 10.1002/jmor.20171.

Oliveira I.S. & Mayer G. (2017) A new giant egg-laying onychophoran (Peripatopsidae) reveals evolutionary and biogeographical aspects of Australian velvet worms. Organisms Diversity & Evolution 17: 375–391. Doi: 10.1007/s13127-016-0321-3.

Oliveira I.S., Lacorte G.A., Fonseca C.G., Wieloch A.H. & Mayer G. (2011) Cryptic speciation in Brazilian *Epiperipatus* (Onychophora: Peripatidae) reveals an underestimated diversity among the peripatid velvet worms. PLoS One 6: 1–13. Doi: 10.1371/journal.pone.0019973.

Oliveira I.S., Franke F.A., Hering L., Schaffer S., Rowell D.M., Weck-Heimann A., Monge-Nájera J., Morera-Brenes B. & Mayer G. (2012a) Unexplored character diversity in Onychophora (velvet worms): a comparative study of three peripatid species. PLoS One 7: e51220. Doi: 10.1371/journal.pone.0051220.

Oliveira I.S., Read V.M.S.J. & Mayer G. (2012b) A world checklist of Onychophora (velvet worms), with notes on nomenclature and status of names. ZooKeys 211: 1–70. Doi: 10.3897/zookeys.211.3463.

Oliveira I.S., Schaffer S., Kvartalnov P.V., Galoyan E.A., Palko I.V., Weck-Heimann A., Geissler P., Ruhberg H. & Mayer G. (2013a) A new species of *Eoperipatus* (Onychophora) from Vietnam reveals novel morphological characters for the South-East Asian Peripatidae. Zoologischer Anzeiger 252: 495–510. Doi: 10.1016/j.jcz.2013.01.001.

Oliveira I.S., Tait N.N., Strübing I. & Mayer G. (2013b) The role of ventral and preventral organs as attachment sites for segmental limb muscles in Onychophora. Frontiers in Zoology 10: 73. Doi: 10.1186/1742-9994-10-73.

Oliveira I.S., Lacorte G.A., Weck-Heimann A., Cordeiro L.M., Wieloch A.H. & Mayer G. (2015) A new and critically endangered species and genus of Onychophora (Peripatidae) from the Brazilian savannah—a vulnerable biodiversity hotspot. Systematics and Biodiversity 13: 211–233. Doi: 10.1080/14772000.2014.985621.

Oliveira I.S., Bai M., Jahn H., Gross V., Martin C., Hammel J.U., Zhang W. & Mayer G. (2016) Earliest onychophoran in amber reveals Gondwanan migration patterns. Current Biology 26: 2594–2601. Doi: 10.1016/j.cub.2016.07.023.

Oliveira I.S., Ruhberg H., Rowell M.D. & Mayer G. (2018) Revision of Tasmanian viviparous velvet worms (Onychophora: Peripatopsidae) with descriptions of two new species. Invertebrate Systematics 32: 909–932. Doi: doi.org/10.1071/IS17096.

Oliveira I.S., Kumerics A., Jahn H., Müller M., Pfeiffer F. & Mayer G. (2019) Functional morphology of a lobopod: Case study of an onychophoran leg. Royal Society Open Science 6: 191200. Doi: 10.1098/rsos.191200.

Pflugfelder O. (1948) Entwicklung von *Paraperipatus amboinensis* n. sp. Zoologische Jahrbuecher Abteilung fuer Anatomie und Ontogenie der Tiere 69: 443–492.

Pocock R.I. (1894) Contributions to our knowledge of the arthropod fauna of the West Indies - Part III. Diplopoda and Malacopoda, with a supplement on the Arachnida of the class Pedipalpi. The Journal of the Linnean Society of London, Zoology 24: 473–544.

Read V.M.S.J. & Hughes R.N. (1987) Feeding behaviour and prey choice in *Macroperipatus torquatus* (Onychophora). Proceedings of the Royal Society of London, Part B 230: 483–506.

Reid A.L. (1996) Review of the Peripatopsidae (Onychophora) in Australia, with comments on peripatopsid relationships. Invertebrate Taxonomy 10: 663–936.

Reinhard J. & Rowell D.M. (2005) Social behaviour in an Australian velvet worm, *Euperipatoides rowelli* (Onychophora: Peripatopsidae). Journal of Zoology 267: 1–7.

Robson E.A., Lockwood A.P.M. & Ralph R. (1966) Composition of the blood in Onychophora. Nature 209: 533.

Ruhberg H. (1985) Die Peripatopsidae (Onychophora). Systematik, Ökologie, Chorologie und phylogenetische Aspekte. Zoologica (Stuttgart) 137: 1–183.

Ruhberg H. & Daniels S.R. (2013) Morphological assessment supports the recognition of four novel species in the widely distributed velvet worm *Peripatopsis moseleyi sensu lato* (Onychophora: Peripatopsidae). Invertebrate Systematics 27: 131–145. Doi: 10.1071/IS12069.

Ruhberg H. & Mayer G. (2013) Onychophora, Stummelfüßer. In: Westheide W. & Rieger G. (Eds.), Spezielle Zoologie. Springer, Berlin, 457–464.

Ruhberg H. & Storch V. (1977) Über Wehrdrüsen und Wehrsekret von *Peripatopsis moseleyi* (Onychophora). Zoologischer Anzeiger 198: 9–19.

Schumann I., Hering L. & Mayer G. (2016) Immunolocalization of Arthropsin in the Onychophoran *Euperipatoides rowelli* (Peripatopsidae). Frontiers in Neuroanatomy 10: 80. Doi: 10.3389/fnana.2016.00080.

Schumann I., Kenny N., Hui J., Hering L. & Mayer G. (2018) Halloween genes in panarthropods and the evolution of the early moulting pathway in Ecdysozoa. Royal Society Open Science 5: 180888. Doi: 10.1098/rsos.180888.

Sedgwick A. (1888) A monograph on the species and distribution of the genus *Peripatus* (Guilding). Quarterly Journal of Microscopical Science 28: 431–493.

Seifert G. & Rosenberg J. (1977) Die Ultrastruktur der Nephrozyten von *Peripatoides leuckarti* (Saenger 1869) (Onychophora, Peripatopsidae). Zoomorphologie 86: 169–181.

Silva J.R.M.C. & Coelho M.P.D. (2000) Induced inflammatory process in *Peripatus acacioi* Marcus & Marcus (Onychophora). Journal of Invertebrate Pathology 75: 41–46.

Stern M. & Bicker G. (2008) Mixed cholinergic/glutamatergic neuromuscular innervation of Onychophora: a combined histochemical/electrophysiological study. Cell & Tissue Research 333: 333–338. Doi: 10.1007/s00441-008-0638-0.

Storch V. & Ruhberg H. (1977) Fine structure of the sensilla of *Peripatopsis moseleyi* (Onychophora). Cell & Tissue Research 177: 539–553.

Storch V. & Ruhberg H. (1990) Electron microscopic observations on the male genital tract and sperm development in *Peripatus sedgwicki* (Peripatidae, Onychophora). Invertebrate Reproduction and Development 17: 47–56.

Storch V. & Ruhberg H. (1993) Onychophora. In: Harrison F.W. & Rice M.E. (Eds), Microscopic Anatomy of Invertebrates. Wiley-Liss, New York, 11–56.

Storch V., Alberti G. & Ruhberg H. (1979) Light and electron microscopical investigations on the salivary glands of *Opisthopatus cinctipes* and *Peripatopsis moseleyi* (Onychophora: Peripatopsidae). Zoologischer Anzeiger 203: 35–47.

Storch V., Holm P. & Ruhberg H. (1988) Zur Ultrastruktur und Histochemie des Darmkanals verschiedener Onychophora. Zoologischer Anzeiger 221: 281–294.

Storch V., Mummert R. & Ruhberg H. (1995) Electron microscopic observations on the male genital tract, sperm development, spermatophore formation, and capacitation in *Epiperipatus biolleyi* (Bouvier) (Peripatidae, Onychophora). Mitteilungen aus dem Hamburgischen Zoologischen Museum und Institut 92: 365–379.

Sundara Rajulu G. & Santhana Krishnan G. (1973) Studies on the nature of the cardio-excitor neurohormone in *Eoperipatus weldoni* (Onychophora: Arthropoda), together with observations on its phylogenetic significance. Zeitschrift für zoologische Systematik und Evolutionsforschung 11: 104–110.

Sundara Rajulu G. & Singh M. (1969) Physiology of the heart of *Eoperipatus weldoni* (Onychophora). Naturwissenschaften 56: 38.

Sunnucks P., Curach N.C., Young A., French J., Cameron R., Briscoe D.A. & Tait N.N. (2000) Reproductive biology of the onychophoran *Euperipatoides rowelli*. Journal of Zoology 250: 447–460. Doi: 10.1111/j.1469-7998.2000.tb00788.x.

Tait N.N. & Norman J.M. (2001) Novel mating behaviour in *Florelliceps stutchburyae* gen. nov., sp. nov. (Onychophora: Peripatopsidae) from Australia. Journal of Zoology 253: 301–308.

Telford M.J., Bourlat S.J., Economou A., Papillon D. & Rota-Stabelli O. (2008) The evolution of the Ecdysozoa. Philosophical Transactions of the Royal Society B, Biological Sciences 363: 1529–1537.

Treffkorn S. & Mayer G. (2013) Expression of the *decapentaplegic* ortholog in embryos of the onychophoran *Euperipatoides rowelli*. Gene Expression Patterns 13: 384–394. Doi: 10.1016/j.gep.2013.07.004.

Treffkorn S., Kahnke L., Hering L. & Mayer G. (2018) Expression of NK cluster genes in the onychophoran *Euperipatoides rowelli*: implications for the evolution of NK family genes in nephrozoans. BMC EvoDevo 9: 17. Doi: 10.1186/s13227-018-0105-2.

Tutt K., Daugherty C.H. & Gibbs G.W. (2002) Differential life-history characteristics of male and female *Peripatoides novaezealandiae* (Onychophora: Peripatopsidae). Journal of Zoology 258: 257–267.

van der Lande V.M. (1991) Native and introduced Onychophora in Singapore. Zoological Journal of the Linnean Society 102: 101–114.

Vizueta, J., Escuer, P., Frías-López, C., Guirao-Rico, S., Hering, L., Mayer, G., Rozas, J. & Sánchez-Gracia, A. (2020): Evolutionary history of major chemosensory gene families across Panarthropoda. Molecular Biology and Evolution 37:3601–3615.

von Byern J., Müller C., Voigtländer K., Dorrer V., Marchetti-Deschmann M., Flammang P. & Mayer G. (2017) Examples of Bioadhesives for Defence and Predation. In: Gorb S. & Gorb E. (Eds.), Functional Surfaces in Biology III Biologically-Inspired Systems: Diversity of the Physical Phenomena. Springer, Cham, 141–191.

Walker M. & Campiglia S. (1990) Some observations on the placenta and embryonic cuticle during development in *Peripatus acacioi* Marcus & Marcus (Onychophora, Peripatidae). In: Minelli A. (Ed), Proceedings of the 7th International Congress of Myriapodology. E.J.Brill, Leiden, New York, 449–459.

Walker M.H. & Tait N.N. (2004) Studies on embryonic development and the reproductive cycle in ovoviviparous Australian Onychophora (Peripatopsidae). Journal of Zoology 264: 333–354.

Walker M.H., Roberts E.M., Roberts T., Spitteri G., Streubig M.J., Hartland J.L. & Tait N.N. (2006) Observations on the structure and function of the seminal receptacles and associated accessory pouches in ovoviviparous onychophorans from Australia (Peripatopsidae; Onychophora). Journal of Zoology 270: 531–542.

Weldon C.W., Daniels S.R., Clusella-Trullas S. & Chown S.L. (2013) Metabolic and water loss rates of two cryptic species in the African velvet worm genus *Opisthopatus* (Onychophora). Journal of Comparative Physiology B Biochemical Systemic and Environmental Physiology 183: 323–332. Doi: 10.1007/s00360-012-0715-2

Whitington P.M. & Mayer G. (2011) The origins of the arthropod nervous system: Insights from the Onychophora. Arthropod Structure & Development 40: 193–209. Doi: 10.1016/j.asd.2011.01.006.

Willey A. (1898) The anatomy and development of *Peripatus novae-britanniae*. 1–52.

CHAPTER

31

PHYLUM NEMATODA

Karin Kiontke, Mark Blaxter, and David H. A. Fitch

Infrakingdom	Bilateria
Division	Protostoma
Subdivision	Ecdysozoa
Phylum	Acanthocephala
Phylum	Nematoda

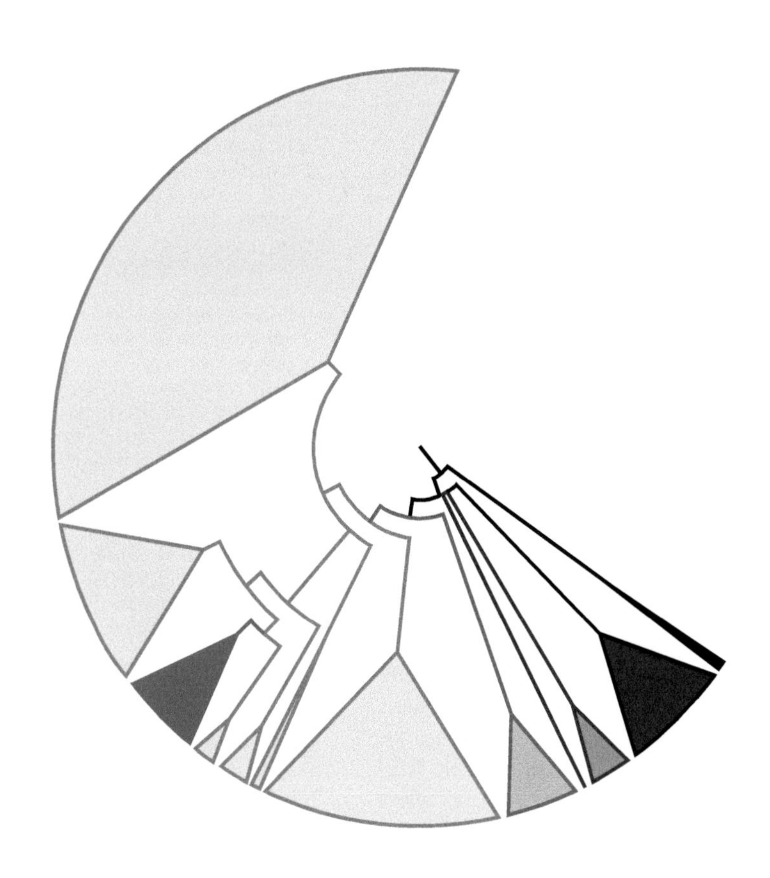

CONTENTS

General Characteristics

Consistent vermiform body plan

Speciose phylum, with >28,000 species described and perhaps millions undescribed

Numerous, among the most abundant metazoa

Distributed worldwide and adapted to nearly all habitats

Free-living and parasitic species, generalists and specialists

Many species of agricultural and medical importance

Origin of many scientific and technological breakthroughs

Poor fossil record

In this chapter, we present only a brief summary of what is known about nematodes. For more information, we refer readers to previously published books and reviews (Blaxter, 2011; Chitwood and Chitwood, 1974; De Ley and Blaxter, 2004; Haag et al., 2018; Kiontke and Fitch, 2013; Maggenti,1981; Malakhov, 1994; Nielsen, 2012; Schmidt-Rhaesa et al., 2013). In particular, we highlight only a few of the evolutionarily important morphological and developmental characters and life history strategies. For several characters, we have tried to infer the nematode ancestral state (i.e., stem species pattern). A caveat is that studies to date – especially of development, physiology, and ecology – have only sparsely sampled nematode phylogenetic diversity. We especially need more information from basally diverged groups.

Synapomorphies

Apomorphies for nematodes (a, bold), synapomorphies with Nematomorpha (s), and other plesiomorphies (p):

- bilaterally symmetric (p)
- no appendages (p)
- no segments (p or s?)
- no locomotory cilia (p)
- multilayered cuticle (p) with collagen (s)
- **four (a) molts** (p)
- **body musculature consists of four bands (a?) of obliquely striated longitudinal muscles (a)**; no ring muscles (s)
- muscle cells extend processes to neurons (s)
- **movement sinusoidal (a?)**
- lateral epidermal cords (s)
- terminal mouth (p) with **six lips that bear three circles of sensilla (a)**
- triradiate sucking pharynx with myoepithelial cells (a/p?)
- primary body cavity (p)
- no protonephridia (s), **excretory system consists of one to five cells (a)**
- nervous system with unpaired ventral cord and nerve ring around pharynx (p)
- male with cloaca (s); **female genital opening at midbody (a)**
- two ovaries (p), two testes (p)
- **male with complex copulatory structures (a)**
- **aflagellate and amoeboid sperm (a), using major sperm protein instead of actin for locomotion (a)**
- internal fertilization (s)
- **direct development (a?)**
- early embryogenesis with equal cleavage (p)
- early cell lineages largely indeterminate, endoderm produced by the anterior blastomere (p)
- gastrulation by invagination of archenteron (p)
- postembryonic cell proliferation is limited (?)
- aquatic habitat (p)

HISTORY OF TAXONOMY AND CLASSIFICATION

Possibly the most ancient reference to nematodes are human parasites mentioned in the *Huang Ti Nei Ching*, ca. 2700 BC (Maggenti, 1981). There is some controversy whether or not the "fiery serpent" mentioned by Moses in the Bible corresponds to the guinea worm (*Dracunculus medinensis*; Chitwood and Chitwood, 1974). The first discovery of a free-living nematode, the vinegar eel worm, was made in 1656 by P. Borelli, also noted by Robert Hooke in 1667 using his microscope. The first discovery of a plant parasitic nematode was made by Needham in 1743 (Chitwood and Chitwood, 1974). The subsequent development of nematology has been well documented by others (Chitwood and Chitwood, 1974; Maggenti, 1981; Malakhov, 1994). From the earliest days, nematode taxonomy was hampered by several problems. First, classifications were often incomplete or biased because researchers specialized in marine and terrestrial free-living or animal/plant-parasitic groups, but not both, precluding a coherent synthesis. Second, nematodes are superficially quite uniform morphologically and there may be considerable homoplasy. Nevertheless, several attempts have been made to systematize the phylum, resulting in confusingly diverse classifications (De Ley and Blaxter, 2004). For example, Chitwood treated "Nematoda" as a phylum that he split into two groups: Phasmidia and Aphasmidia, names later dropped in favor of Secernentea and Adenophorea (Chitwood and Chitwood, 1950). Later, Andrássy (1974, 1976) and Malakhov (1994) split nematodes into three (but different) groups. Early phylogenetic analysis with morphological and molecular data showed that all of these classification systems had problems with paraphyly and even polyphyly. For this reason, De Ley and Blaxter (2002) abandoned most of the older groupings and names in favor of a taxonomy with clades that are likely monophyletic based on molecular data (**Figure 31.1**; De Ley and Blaxter, 2004).

Indeed, molecular data have revolutionized phylogenetics within Nematoda, permitting assessment of relationships unbiased by assumptions of morphological homology. Analyses to date have primarily used the gene for small subunit (18S) ribosomal RNA (Blaxter, 2011; De Ley and Blaxter, 2002; Blaxter et al., 1998; Fitch et al., 1995; Sudhaus and Fitch, 2001) only, or with additional gene sets (Kiontke et al., 2007; Kiontke et al., 2011; Stevens et al., 2019). More recently, datasets derived from whole genomes or transcriptomes have been added (Blaxter and Koutsovoulos, 2015). Three major clades have been distinguished: Dorylaimia and Enoplia (Clades I and II, respectively) and Chromadoria (Figure 31.1). Within Chromadoria, a ladder-like phylogeny has yielded numerous groups. One of these, the Rhabditida, includes three major clades of terrestrial and parasitic species: Spirurina (Clade III), Tylenchina (Clade IV), and Rhabditina (Clade V). Because it is the most basally divergent, characteristics of nematodes in Enoplia that are shared with either of the other clades or the outgroup were likely present in the nematode stem species.

The fossil record of Nematoda is sparse, but an increasing number of discoveries include records from three kinds of deposits (Poinar, 2015). Nematodes have been described from amber deposits worldwide, often in association with presumed insect hosts. Coprolites have yielded microfossils ascribed to nematode eggs. Finally, the earliest terrestrial ecosystem, the Rhynie chert (419 million years ago) has yielded nematode fossils in association with the earliest land plants, suggesting a very early evolution of plant parasitism. These fossils have not, as of yet, been used to significantly revise or date nematode phylogeny, and most have been placed within extant groups.

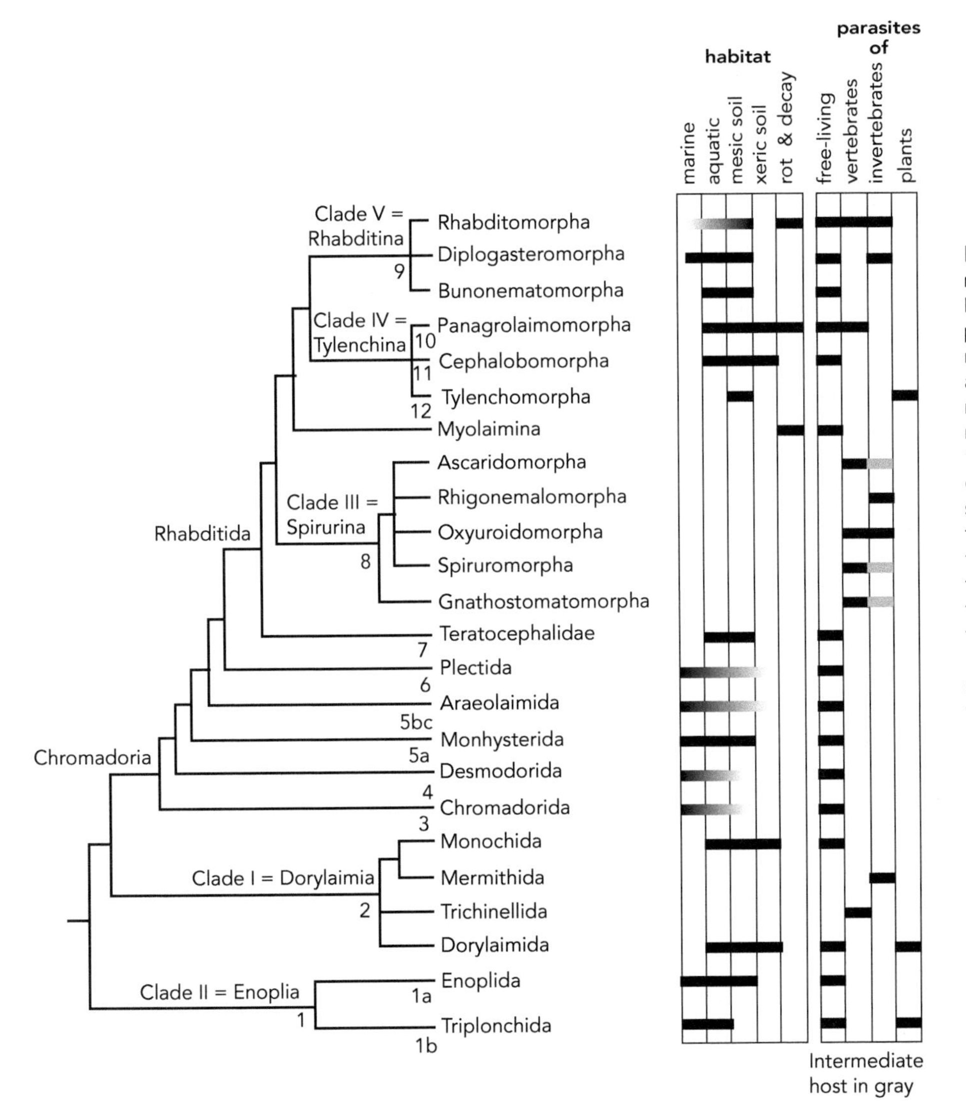

Figure 31.1 Phylogeny and ecological ranges of major groups in phylum Nematoda, based on molecular data. The phylogeny of Nematoda has been derived mainly from ribosomal RNA (rRNA) genes and contains several well-defined clades: major clades I–V, designated with roman numerals and taxon names, and clades 1–12 designated in Arabic numerals below corresponding branches, as found in studies incorporating a large number of taxa. Some taxa have been left out here for simplicity. Black bars in the columns to the right of the phylogeny indicate the ecological ranges of species in these taxa (grey bars in the parasites columns indicate intermediate hosts). Enoplia species are generally microbivores that inhabit marine environments (e.g., *Enoplus*), although some (e.g., *Tobrillus*) have adapted to freshwater habitats or soil. *Dorylaimia* species include free-living marine and terrestrial microbivores as well as plant and animal parasites. Chromadoria again includes marine and terrestrial microbivores and many radiations of parasites, both plant and animal. The distribution of parasitism across the phylum suggests multiple independent origins of both animal and plant parasitism, generally from terrestrial ancestors (Dorris et al., 1999). (Adapted with permission from previous reviews Blaxter, 2011; Kiontke and Fitch, 2013.)

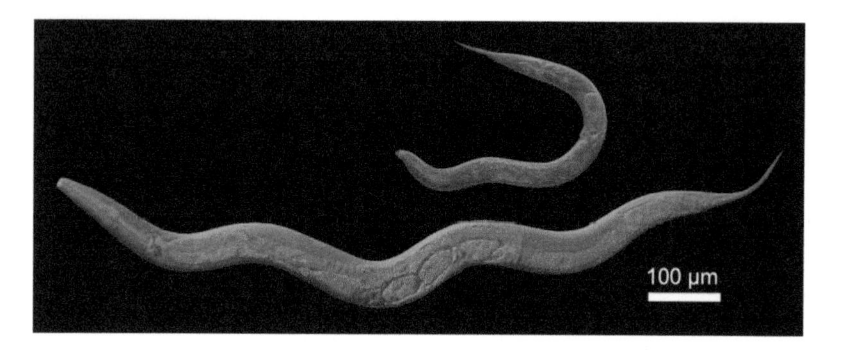

Figure 31.2 Whole animal photographs. Photomicrographs of females of *Diploscapter pachys* (parthenogenic, top) and *Caenorhabditis elegans* (hermaphroditic, bottom), both Rhabditina. Color enhancement of the otherwise colorless worms was obtained using a lambda plate placed in the illumination path with differential interference contrast optics.

ANATOMY, HISTOLOGY, AND MORPHOLOGY

The simple body plan of nematodes is often described as a tube (gut) inside a tube (body wall) separated by a fluid-filled body cavity that is under pressure, forming a hydrostatic skeleton (**Figures 31.2** and **31.3**). Indeed, all nematodes lack segmentation or appendages and their organization is fairly uniform across a very large number of species living in very diverse habitats and showing large differences in body size (from less than a millimeter to several meters in the extreme case of a parasite in the sperm whale placenta, *Placentanema gigantissima*). Most free-living nematodes are of microscopic size and transparent, their body is spindle-shaped and round in cross-section (which gives rise to the common name, roundworms). While this spindle-shape model fits most nematode species, some have evolved bodies with distinct anterior and posterior tagmata, or have clear dorsal and ventral differentiation. In some cyst-forming plant parasitic species in Tylenchina, the vermiform body form is absent in adult stages, which are instead globular. In some arthropod parasites (*Sphaerularia* and relatives, Tylenchina), the nematode builds relatively huge trophic organs (20 mm compared to body size of 1 mm).

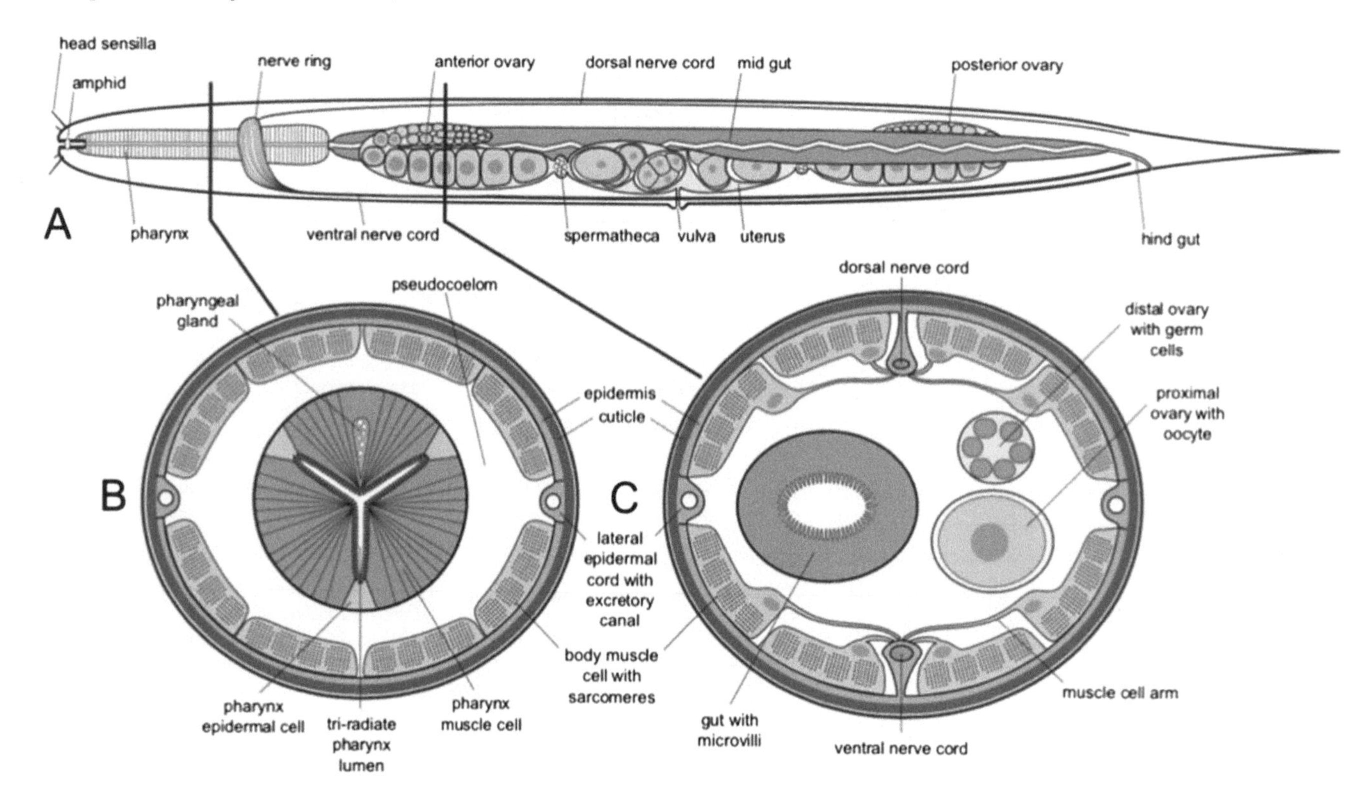

Figure 31.3 Schematic of general nematode morphology. (A) Lateral view of a typical female nematode. (B) Cross-section in the pharynx region. (C) Cross-section near the anterior gonad arm.

The main body of nematodes is surrounded by a tough and flexible extracellular **cuticle**, which, together with the turgor of the body cavity, is the antagonist for the longitudinal muscles and thus functions as an exoskeleton in small species (Figure 31.3). However, in the very large animal parasites, the mechanical bases of locomotion must be very different (Blaxter and Robertson, 1998; Wharton, 1986). Cuticle also lines the pharynx, vagina, rectum/cloaca and, when present, the excretory duct. Even though the cuticle protects the nematode from the external environment, it is semipermeable and is the site of excretion and uptake of substances (including nutrients in some parasitic taxa). The body cuticle has multiple layers or zones (possibly four ancestrally) and is primarily composed of collagens. Chitin is absent from the cuticle, though present in nematode eggshells. Its surface can be smooth or bear rings, striations, wings ("alae"), and other ornaments. Cuticle is overall covered by a glycocalyx which is continually secreted by various glands and plays a crucial role as an interface between the nematode and its environment. The cuticle is molted a fixed number of four times and can differ in its fine structure between larval stages. Cuticle is synthesized by the underlying **epidermis** (often called hypodermis) that was ancestrally cellular but is syncytial in many derived species. The epidermis forms lateral cords containing the excretory canals, and dorsal and ventral cords containing nerve cords.

The somatic **musculature** of nematodes consists of two dorsal and two ventral longitudinal bundles that are anchored through the epidermis to the cuticle (Figure 31.3). Ring muscles are absent. Each somatic muscle cell is spindle-shaped and has three parts: a contractile part containing the sarcomere, a non-contractile part containing the nucleus, and a thin "innervation arm" that projects to the ventral or dorsal nerve cord or to the nerve ring. Another apomorphic trait of the nematode body wall musculature is its oblique striation. In addition to the body wall muscles, nematodes possess a muscular pharynx (see section below on "digestive system") and specialized muscles, e.g., at the female vulva and the copulatory organs of males.

The **nervous system** of nematodes is best known from extensive studies of larger parasites (e.g., ascarids as early as 1874 by Buetschli and *Ancylostoma* by Loos in 1905) and in the free-living *Caenorhabditis elegans*. Here, the number of neurons is small and fixed (302 in a *C. elegans* hermaphrodite, 298 in an *Ascaris suum* female). The nervous system of these distantly related species of very different size is similar enough to allow for identification of homologous neurons and connections. These features are likely apomorphic for the Chromadoria. The nervous system of earlier-branching nematodes is much less well known but likely more complex with many more neurons (and a lateral plexus). However, all nematodes have a nerve ring looping around the pharynx which is composed of axons and dendrites of neurons whose cell bodies lie in bundles (ganglia) (Figure 31.3). In addition, there is a major ventral nerve cord, which contains cell bodies and processes, and a dorsal and two sublateral nerve cords that consist only of processes. Further ganglia are found in the posterior near the rectum.

Like the nervous system as a whole, nematode **sensory organs** are best known from Rhabditida, especially with respect to their function. Generally, most nematode sense organs are sensilla composed of one or more neurons with ciliated sensory processes that are surrounded by two non-neuronal cells. The sensory processes can be exposed to the exterior via a pore or embedded in the cuticle. A set of 2× 6 labial and 4 cephalic sensilla at the anterior body end is likely ancestral for nematodes (**Figure 31.4**). In addition, numerous sensilla that can take the shape of very long setae are found along the body of many marine (Clade I, Clade II and basally arising Chromadoria) nematodes, whereas Rhabditida have only deirids and posterior deirids (mechanoreceptors) and phasmids (chemoreceptors). The main sense organs of all nematodes are a pair of lateral amphids near the front end of the animal (Figures 31.3, 31.4). The

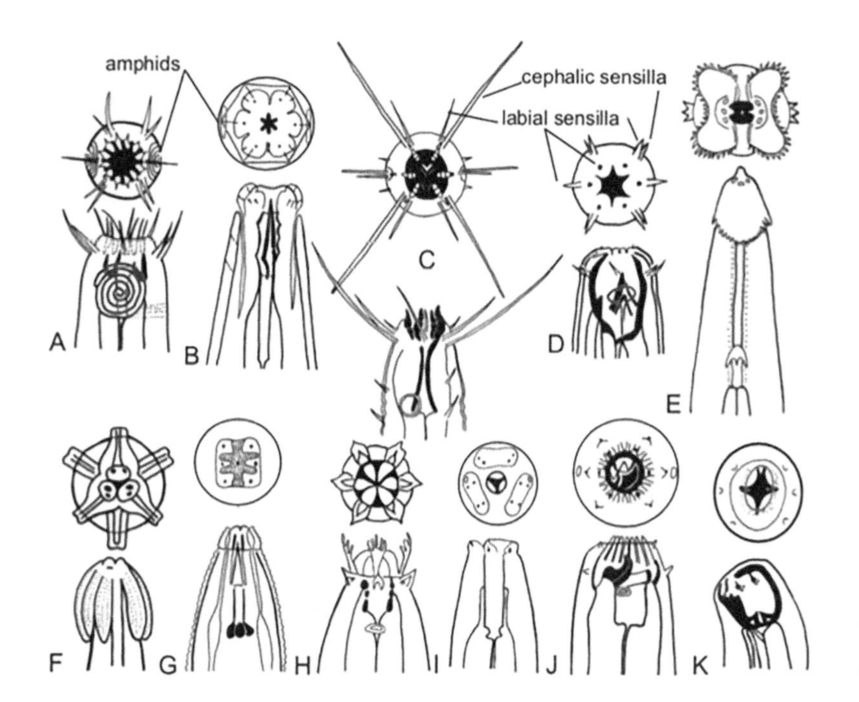

Figure 31.4 Anterior body regions of selected nematodes. Top: *en-face* view with dorsal side up. Bottom: lateral or ventral view. (A) *Pomponema mirabile* (Chromadorida), (B) *Actinolaimus* sp. (Dorylaimida), (C) *Scaptrella cineta* (Monhysterida), (D) *Metoncholaimus pristuris* (Oncholaimina), (E) *Stegophorus stellaepolaris* (Spiruromorpha), (F) *Pulchrocephala* sp. (Oxyuridomorpha), (G) *Hoplolaimus bradys* (Tylenchomorpha), (H) *Chambersiella rodens* (Cephalobomorpha), (I) *Rhabditis* sp. (Rhabditidomorpha), (J) *Mononchoides* sp. (Diplogasteromorpha), *Necator americanus* (Rhabditidomorpha, Strongyloidea). (Images modified from Chitwood and Chitwood, 1950.)

amphids contain numerous sensilla inside a pouch with a duct and an opening to the outside that can take clade- or species-specific shapes from a pore (in Rhabditida) to a pocket (e.g., Enoplia and Dorylaimia) or a complex spiral (in marine Chromadorida).

From *C. elegans* it is known that amphid neurons are chemosensory for taste and smell, thermosensory, nociceptory, oxygen sensors and osmoreceptors. Males have sex-specific genital sensory organs in the tail in a ventral row and/or paired around the cloaca and in each spicule (**Figure 31.5**). In addition, nematodes possess internal sense organs that are proprioreceptors, thermoreceptors, oxygen sensors and in some species photoreceptors. "Eyespots" or ocelli in the form of paired melanin pigmented cells and associated neurons are common in marine Chromadoria, Clade I and Clade II species, and likely form shadowing photoreceptors (Vanfleteren and Coomans, 1976). In the Mermithida, the photoshadowing pigment is a hemoglobin in pharyngeal cells, and this confers directional photosensitivity (Burr et al., 2000).

The **digestive system** is a simple tube (Figure 31.3). The terminal mouth leads to the foregut with a cuticular lining that is differentiated into a buccal cavity and a muscular pharynx. The **buccal cavity** experienced the most dramatic evolutionary changes within nematodes, associated with diversification of feeding habits and food sources: A needle-like stylet evolved three times, associated with feeding on plant cells (Tylenchidae, Triplonchida) or fungal and animal cells (Dorylaimidae). Large teeth evolved in predatory species (e.g., Monochida and Diplogastridae) and in some animal gut parasites (Rhigonematomorpha, Strongylida). The **pharynx** can have multiple parts. Its cross section is triradiate with three blocks of myoepithelial cells interspersed by epidermal cells. The cuticular lining of the pharynx can have differentiations that facilitate filtering or crushing. The pharynx also contains 5 (or 3) unicellular glands (one dorsal, two on each subventral side). Innervation of the pharynx is largely independent of the rest of the animal. The **midgut** is a single-layered epithelium with microvilli. It is followed by a **hindgut** with a valve and sphincter muscle. In males, the hindgut connects to the vas deferens and forms a cloaca. Three or six unicellular rectal glands are present.

Nematodes have a primary **body cavity** that for historical reasons is called a pseudocoelom (Figure 31.3). It contains **pseudocoelomocytes** (6 in *C. elegans* hermaphrodites) that are surrounded by basal lamina and are thought to be

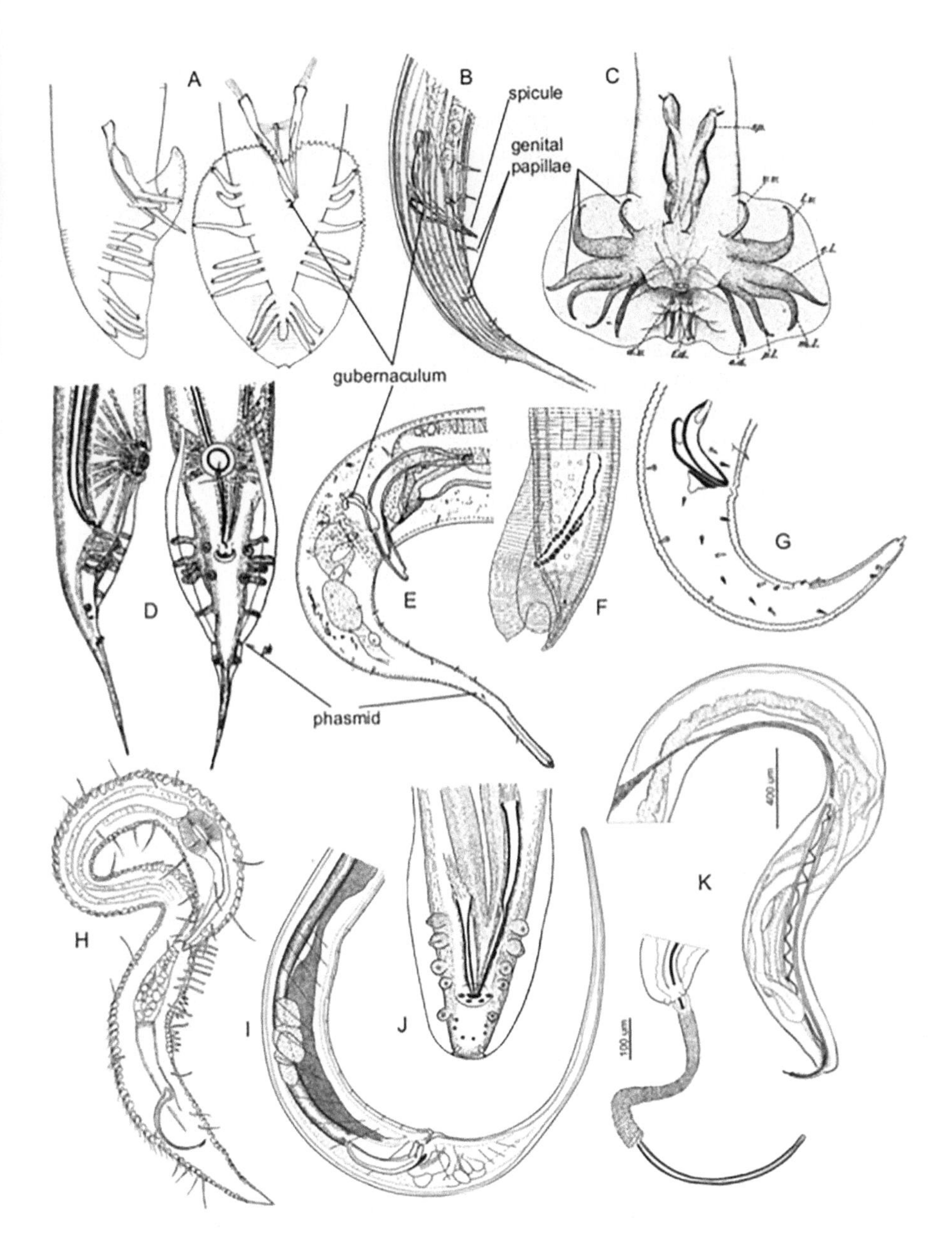

Figure 31.5 Variation in nematode male tails. (A) *Caenorhabditis remanei* (Rhabditidae) lateral and ventral from (Sudhaus and Kiontke, 2007), (B) *Mononchoides macrospiculum* (Diplogastridae) lateral from Troccoli et al., 2015), (C) *Cooperia curticei* (Strongylidae) ventral after (Chitwood and Chitwood, 1974), (D) *Heterakis gallinarum* (Ascarididae) lateral and ventral after (Chitwood and Chitwood, 1974), (E) *Semitobrilus pellucidus* (Tobrilidae) lateral from (Yunliang and Coomans, 2000), (F) *Dolichodorus cobbi* (Tylenchidae) lateral from (Golden et al., 1986), (G) *Plectus parietinus* (Plectidae) lateral from (Holovachov, 2006), (H) *Epsilonema rugatum* (Desmodoroidea) whole animal lateral from (Leduc and Zhao, 2016), (I) *Iotonchus stockdilli* (Mononchoidea) lateral (Yeates, 1987), (J) *Dirofilaria immitis* (Onchocercidae) ventral from (Chitwood and Chitwood, 1974), (K) *Trichuris bainae* (Trichuridae) lateral with spicules inside of animal and extruded with a spicular sheath from (Robles et al., 2014).

scavenger cells, since they are capable of phagocytosis (in large nematodes like *Ascaris*), or endocytosis (*C. elegans*).

The **excretory system** lacks cilia that could produce a flow of body fluids. It can consist of a single cell or a group of cells. In either case, there is one ventral excretory pore. In Chromadoria, the excretory system is relatively complex with a large excretory canal cell (H-shaped in rhabditids) two gland cells of unknown function, a duct cell and a pore cell. The duct is lined with cuticle. The main function of this system might be osmoregulation. In Enoplia and marine members of Chromadoria, the many glands in the cuticle may also serve excretory or secretory functions.

Information for this section is synthesized from several sources: Chitwood and Chitwood, (1974), Maggenti (1981), Schmidt-Rhaesa et al., (2013), and Bird and Bird (1991).

REPRODUCTION AND DEVELOPMENT

The stem species of nematodes was gonochoristic with males and females. Reproduction of all non-parthenogenetic nematodes involves copulation and internal fertilization. The ancestral **female reproductive system** consisted of two opposed ovaries, gonoducts and uteri that are connected to a common vagina and have a common vulval opening in the middle of the animal. This arrangement is modified in several lineages so that the vulva can be either far anterior or posterior, there can be only one gonad branch, or both gonads can lie in parallel. The germ cells are located at the distal end of the ovary and progressive stages of oocyte development are seen along its length. The gonoduct comprises a sperm storage area, the spermatheca, and a uterus. The **uteri** connect to the vagina. Oocytes are paused at Prophase I and are triggered to complete meiosis on fertilization. Egg-laying is facilitated by specialized vulva dilator muscles. Some species are viviparous, releasing early stage larvae.

The ancestral **male reproductive system** consists of two testes with a common vas deferens that leads to the cloaca in the posterior of the animal. Chromadoria have only one testis, which is frequently reflexed. The ancestral male nematode possessed copulatory organs: two spicules that are inserted into the female vulva and a gubernaculum in the dorsal wall of the cloaca that is thought to guide the spicules (Figure 31.5). Spicules and gubernaculum are secreted by specialized cells in the cloaca during the last larval stage. Their shape and size are species specific. Sensilla called genital papillae are usually found in the male tail (Figure 31.5). The number of genital papillae was probably nine bilateral pairs in the stem species of Rhabditina and their homologies can be assigned by shared developmental origins (Fitch and Emmons, 1995). However, the number of genital papillae varies considerably in other nematodes: some with many and some with apparently none. **Spermatozoa** are non-flagellate and move in an amoeboid fashion. For sperm motility – as well as signaling oocyte maturation and gonadal sheath contraction – nematodes use a novel protein, major sperm protein (MSP), in place of actin (Miller et al., 2001).

Except in Enoplia, there is considerable variation in reproductive mode. Self-fertile androdioecious hermaphroditism has evolved multiple times, and parthenogenesis has evolved several times in Rhabditina (Kiontke and Fitch, 2005). One clade (*Diploscapter*) is among the longest lived parthenogenetic lineages known (Fradin et al., 2017). There are also species with "three sexes" (hermaphrodites, females and males; Kanzaki et al., 2017). In some species, different modes of reproduction are present in different generations ("heterogony"; see Figure 31.9B). Sex determination can be chromosomal or environmental. In Rhabditina, an XX-XO chromosomal system is prevalent, while in Spirurina both XX-XY and haplodiploid systems are also present. In Tylenchina, many plant parasitic species have an environmentally controlled sex ratio, with males only produced under stress.

Most of our understanding of nematode **development** comes from studies in *C. elegans* (Rhabditina, Rhabditomorpha), but comparative studies have revealed considerable diversity across nematodes, particularly with respect to early embryogenesis (Schulze and Schierenberg, 2011). For each character highlighted below, its state in *C. elegans* is described, followed by a description of relevant variation and inferred nematode ancestral state, if possible (**Figures 31.6, 31.7**).

Caenorhabditis elegans

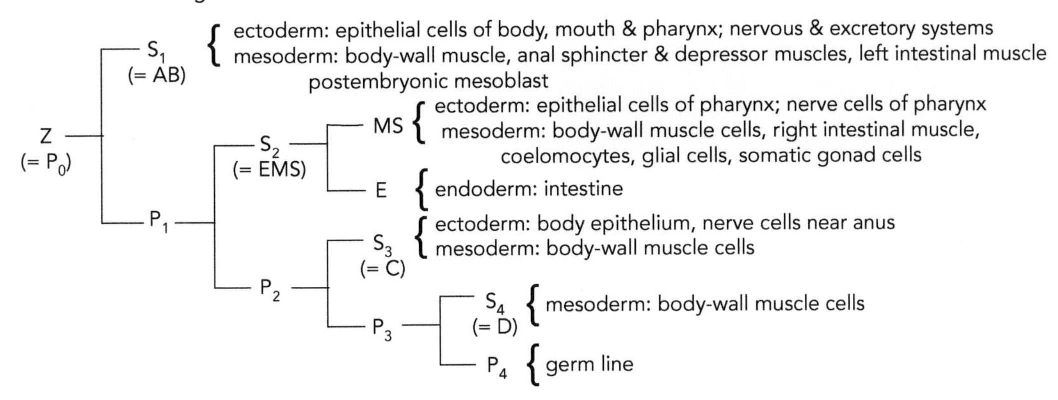

Romanomermis culicivorax

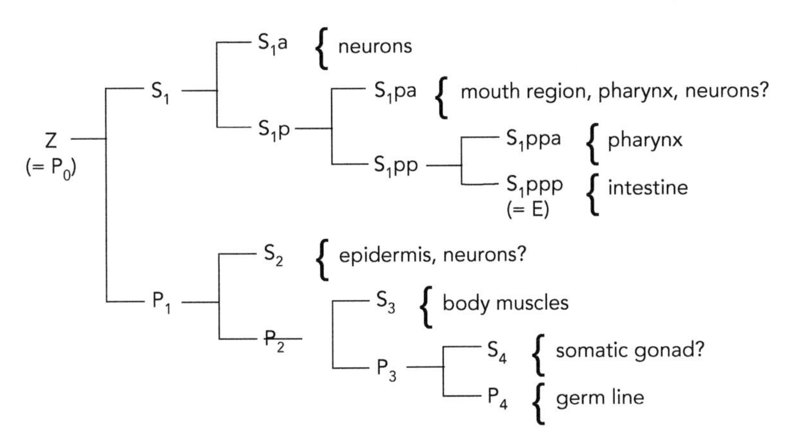

Figure 31.6 Early founder cell lineages for two nematodes, *C. elegans* (Rhabditina) and *R. culicivorax* (Mermithida). S and P designate lineages that are somatic and germ line precursors, respectively. Suffix "a" indicates an anterior daughter; suffix "p" indicates a posterior daughter. Cell designations in parentheses (AB, EMS, E, C, and D) are more often used to designate the founder cells in *C. elegans* and other rhabditids. (Adapted with permission from Nielsen, 2012.)

Zygote polarity: In *C. elegans*, sperm entry point determines the posterior pole of the zygote and thus of the entire embryo (Goldstein and Hird, 1996; Rose and Gönczy, 2014). There is considerable variety among nematodes in the way zygote polarity is established, and the ancestral state is not known (Goldstein et al., 1998; Lahl et al., 2006).

First division: In *C. elegans*, the first division produces cells of unequal size and asymmetric, determinate fate, with the anterior cell, AB, larger than the posterior cell, P_1, the latter giving rise to the germ line (Figure 31.5). In Enoplia as well as Mononchida (Dorylaimia) species, the first division produces cells of equal size and mostly indeterminate fate (Voronov, 1999; Brauchle et al., 2009; Voronov and Panchin, 1998; Figure 31.7). Equal cleavage with mostly indeterminate fate is shared with outgroup taxa including Nematomorpha and is thus plesiomorphic. Note that equal cleavage is not always correlated with indeterminate fate; e.g., the equally sized blastomeres of *Hypodontolaimus* (Chromadorida) have asymmetric, determinate fate (Malakhov, 1994).

Cleavage leading to 4-cell embryo: During the second division in *C. elegans*, the spindle of the posterior cell, P_1, rotates to become perpendicular to that of AB, leading to a T-shaped arrangement of cells; the eggshell constricts these cells into a rhomboid arrangement (Rose and Gönczy, 2014). By this stage, all three body axes are determined. Up to six different early cleavage patterns exist in other nematode groups (Schulze and Schierenberg, 2011; Malakhov, 1994; Brauchle et al., 2009), some variable even within the same species. The simplest, H-type arrangement (divisions without spindle rotations such that both

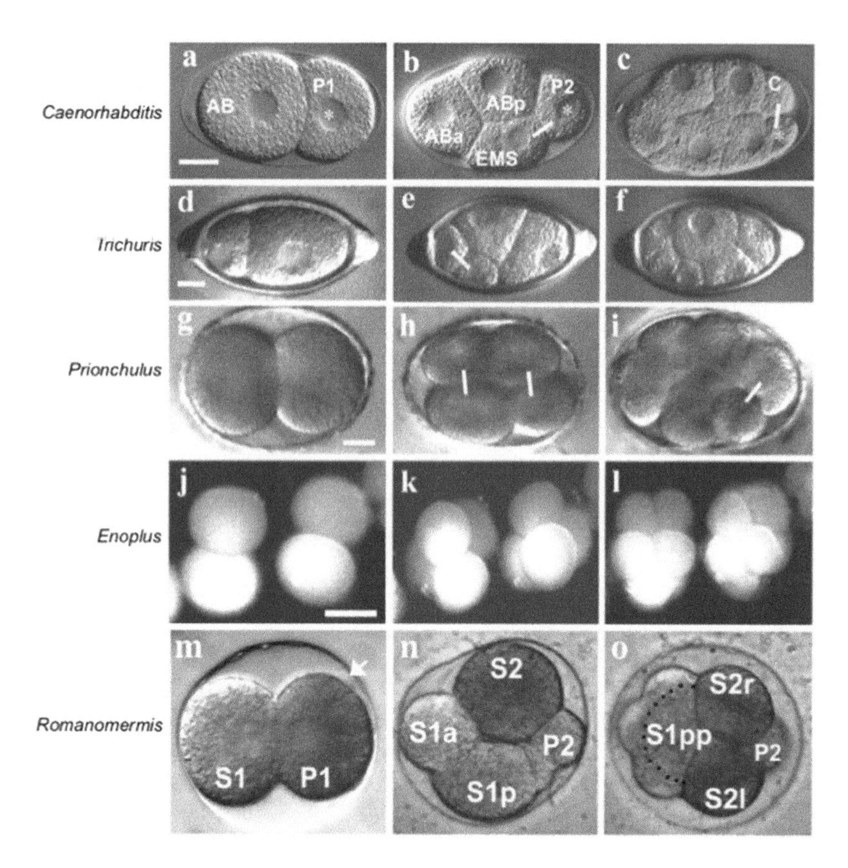

Figure 31.7 Variation in cleavage among different nematodes. In each panel, anterior is oriented to the left and dorsal is up, except in j–l, where orientation is arbitrary. (a–c) *C. elegans*. (a) The zygote, Z (= P_0), has divided unequally to form a larger somatic cell S_1 (= AB) (anterior) and a smaller germ line precursor cell P_1 (*, posterior). (b) At the four-cell stage, S_1 has divided equally with a transverse spindle orientation and P_1 has divided unequally to produce S_2 (= EMS, ventral) and P_2 via a longitudinal spindle orientation (constriction by the eggshell makes it seem as though these divisions have occurred in the same orientation). (c) At the eight-cell stage, germ line precursor cell P_2 has divided unequally to produce S_3 (= C). (d–f) *Trichuris muris*. (d) At the two-cell stage, an extremely unequal first cleavage has generated a smaller anterior and a larger posterior cell. (e) At the four-cell stage, both daughters have performed an equal division. (f) At the eight-cell stage, an approximately equal division has occurred. (g–i) *Prionchulus* sp. (g) The two-cell stage results from an equal first cleavage. (h) At the four-cell stage, both daughter cells have divided equally via a transverse spindle orientation. (i) At the eight-cell stage, one cell has performed a germ line-like asymmetric division. (j–l) In *Enoplus brevis*, equal early divisions give no indication for the presence of a germ line, which appears to be determined much later. Epi-illumination is used for *E. brevis* because these embryos are not transparent. Lines connecting cells indicate particular sister cells. Scale bars: (a–i) 10 μm, (j–l) 100 μm. (m–o) *Romanomermis culicivorax*. (m) Two-cell stage. (n) Four-cell stage. (o) Eight-cell stage. (Adapted with permission from Schierenberg, 2006.)

second divisions are perpendicular to the first; Figure 31.7j) is probably ancestral, as it is shared both by Clade II species and outgroup representatives such as the tardigrade *Hypsibius exemplaris* (Schulze and Schierenberg, 2011; Malakhov, 1994; Brauchle et al., 2009; Voronov and Panchin, 1998).

Generation of founder cells (progenitors of germ layers and differentiated cell types): Within just four rounds of cleavage, the *C. elegans* embryo establishes all six founder cell lineages, which are invariant and determinate, and occupy canonical positions in the embryo (Figure 31.6; Sulston et al., 1983). Besides segregation of determinants by lineage, signaling between cells is also important for fate determination (Rose and Gönczy, 2014). Except for P_4 (germ line), S_4 (= D, producing some of the body-wall muscle cells) and E (intestine), the other founder cells contribute to several tissues and germ layers. This invariant lineage, canonical blastomere arrangement, and early determination of all founders are highly derived within Nematoda. Other species in Chromadoria have similar cell lineage patterns, but with considerable differences in timing, sequence and degree of "regulation" versus "determination" (Sulston et al., 1983; Borgonie et al., 2000; Boveri, 1899; Dolinski et al., 2001; Houthoofd et al., 2003; Vangestel et al., 2008). Early-diverged Chromadoria even show within-species variation in blastomere arrangements and lineages, indicating early cell fates are determined positionally instead of lineally Schulze et al., 2012). *Romanomermis culcivorax* (Clade I) has six founder cells, but tissues are formed by adding circumferential rings of cells, reminiscent of a segmentation process (Schulze and Schierenberg, 2008, 2009). In Enoplia, (e.g., *Prionchulus, Pontonema, Enoplus* and *Tobrilus*), asymmetry of cell divisions is only seen in later stages, the founder cells arise in variable spatial positions, early embryogenesis appears to be more regulative than determinative, and the germ line does not appear to be segregated early (Schierenberg, 2005). In all clades, there is early determination of E, but it is derived from the anterior blastomere of the two-cell stage in Enoplia and Dorylaimia (or either blastomere in *Enoplus*; Schierenberg, 2005). In Nematomorpha, cleavage patterns are indeterminate with indefinite blastomere arrangements. Regulative development with variable spatial blastomere arrangements and mostly indeterminate cell lineages giving rise to founders, possibly with the anterior blastomere biased to give rise to endoderm, thus appear to be plesiomorphic for nematodes.

Isolation of germ line from soma: The germ line founder (P_4) is set aside early from the somatic founder cells in *C. elegans* and in most other nematodes. There is apparently no such early isolation in Enoplida species (Voronov and Panchin, 1998; Voronov, 1999) or in dioctophymatids of Dorylaimia, in which the germ line progenitor first appears as a bilateral pair of cells from different blastomeres instead of from a single medial progenitor (Malakhov, 1994). However, the germ line does appear to be set aside early as P_5 in *Tobrilus* (also Dorylaimia; Schulze and Schierenberg 2011). Thus, it is probable that early germ line isolation from soma evolved early within Nematoda.

Establishment of bilateral symmetry: In *C. elegans* and all other nematodes except Enoplia, bilateral symmetry is established by symmetric left-right divisions of cells aligned along the midline. In Dorylaimia (e.g., *Tobrilus*), only three somatic midline cells have left-right symmetrical divisions, suggesting that bilateral symmetry must be established late, possibly via cell–cell interactions. It is likely that the latter mode is plesiomorphic to nematodes.

Gastrulation: In *C. elegans* and in most nematodes of Dorylaimia and Chromadoria, gastrulation begins with a slight movement of the two daughters of E into the very small blastocoel, eventually resulting in a blastopore bounded by mesoderm (Nance et al., 2005). The P_4 germ line precursor is then passively internalized, the blastopore lengthens a bit and most of the remaining myoblasts (from C and D lineages) and then AB-derived pharyngeal precursors internalize (Nance et al., 2005; Chihara and Nance, 2012). In *Tobrilus*, a large blastocoel is formed into which the gut precursors move, followed by other cells, generating an archenteron (Schulze and Schierenberg, 2011). This latter type of gastrulation is likely to be plesiomorphic for Nematoda.

Blastopore closure: In Rhabditina, the blastopore closes very early and does not form an elongated slit; mouth and anus openings are formed secondarily. In Spirurina, a slit-like blastopore forms, is usually closed in the middle and the remaining anterior opening forms the mouth, as in many Protostomia (Malakhov, 1994). All Enoplia and Dorylaimia studied (e.g., *Tobrilus* and *Romanomermis*) have a slit-like blastopore which closes in the center, resulting in the future mouth and anus forming from the anterior and posterior blastopore openings; this "amphistomy" is thus likely to be ancestral in Nematoda.

Postproliferation morphogenesis: In *C. elegans*, as in all nematodes, most proliferation is completed after closure of the blastopore and epiboly results in epidermal cells covering the embryo, forming circumferential bands around the embryo (Chisholm and Hardin, 2005). Elongation of the embryo is accomplished by actinomyosin contraction in these bands against the internal hydrostatic pressure of the embryo (Priess and Hirsh, 1986). In the early stages of elongation in *C. elegans*, most of the epidermal ("hypodermal") cells fuse to make a large syncytial sheath covering most of the animal (Podbilewicz and White, 1994; Shemer and Podbilewicz, 2000) such large hypodermal syncytia are not formed in Clades I and II. The resulting worm shape is then maintained by secretion of cuticle.

Juvenile development: All nematodes go through direct development (Malakhov, 1994). Like other Ecdysozoa, all nematodes progress through postembryonic development via molts. These molts define stages, with the first larval stage carrying the cuticle first secreted by the embryo, and the adult corresponding to the fifth stage. Some species molt within the egg and emerge as second stage or third stage larvae. Molts allow increase in body size which may be associated with changing food source, as the larger mouth opening (and teeth, where present) permit access to different food items. Molts also allow development of stage-specific cuticular structures like teeth and alae. Some zooparasitic nematodes may increase their body length 100-fold as adults by restructuring the cuticle and underlying tissue without going through additional molts.

DISTRIBUTION AND ECOLOGY

Nematodes are hugely abundant and display a surprising biodiversity despite their largely invariant body plan. Although approximately 30,000 species of nematodes have been described, this is likely to be a significant underestimate of the actual number (Lambshead and Boucher, 2003; Lambshead, 1993; Lorenzen, 1994). As an example, since collection efforts intensified, the number of known species in the *Caenorhabditis* genus has jumped more than two-fold just in the last decade (Kiontke et al., 2011; Stevens et al., 2019). To illustrate nematode abundance and habitat diversity, American nematologist Nathan Cobb imagined that "if all the matter in the universe except the nematodes were swept away, our world would still be dimly recognizable ... we would find its mountains, hills, vales, rivers, lakes, and oceans represented by a film of nematodes ... locations of various plants and animals would still be decipherable ... in many cases even their species could be determined by an examination of their erstwhile nematode parasites" (Cobb, 1915). Indeed, the most interesting aspect of nematode biodiversity is their adaptations to extraordinarily diverse, often extreme, environments and their interactions with other organisms (Kiontke and Fitch, 2013). Nematodes are found in terrestrial and marine sediments globally, including the deep sea and high mountains (**Figure 31.8**; Van den Hoogen et al., 2019). They

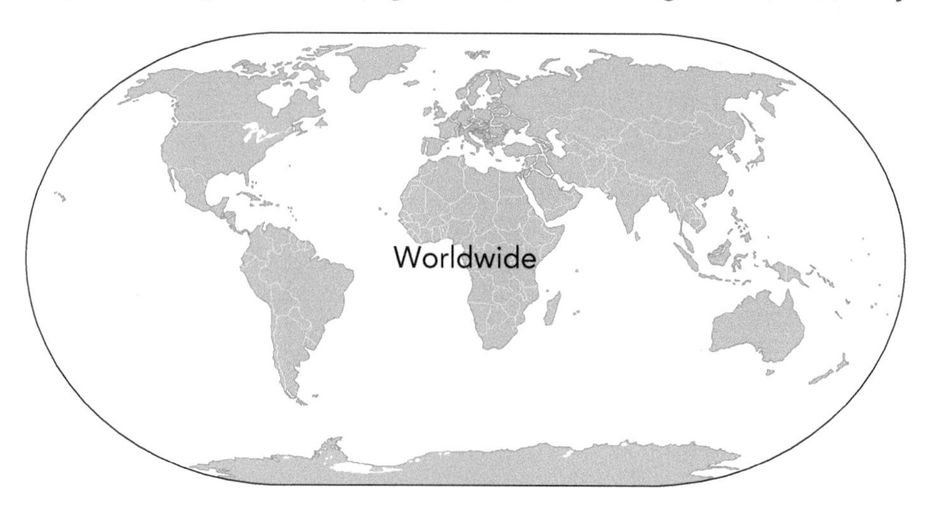

Figure 31.8 Distribution of nematodes. Nematodes are distributed globally and found in a multitude of environments and substrates.

are found in arctic and antarctic ecosystems, and thus have evolved protection against freezing. In arid systems, nematode species use anhydrobiotic toolkits to survive and are resistant to high surface temperatures. Nematode species are abundant in high energy sediments, such as littoral deposits, and have been found in percolating groundwater in deep mines. Species adapted to high salinity and low pH have been described from salt pans, vinegar production works and soda lakes. Because of their spatial as well as temporal diversity in ecosystems, and the presence of ecological succession in species groups (Rehfeld and Sudhaus, 1989), nematodes have been useful bioindicators in environmental monitoring and forensics (Van den Hoogen et al., 2019; Bongers and Ferris, 1999; Szelecz et al., 2018).

Many free-living nematodes, especially of the clade Rhabditida, form interactions with other animal species that can be highly specific, and in some cases essential (**Figure 31.9**). Such interactions can be for dispersal. A specialized non-feeding and developmentally arrested third larva with distinct morphology, the dauer larva, mounts a phoretic carrier that takes it to a new habitat, where it dismounts and resumes development if conditions are favorable (see Figure 31.9A). The dauer larva is an evolutionary novelty of the Rhabditida that allows for long-distance travel and probably canalized the evolution of parasitism in this clade (Sudhaus, 2010). Some free-living, microbivorous Rhabditina species live in close association with other animals throughout their life cycle, e.g., they live in the brood chambers of wild bees, the tunnels of wood-boring beetles, the brood balls of carrion or dung beetles, or the nests of termites, ants and even mice. Dauer larvae of these species enter a phoretic relationship with their hosts but are otherwise bacterivorous. Some parasites have multi-host lifecycles where an intermediate host species is colonized by third-stage larvae. These intermediate hosts then act as phoretic carriers of the parasite.

Nematodes are remarkable for the repeated evolution of parasitism. Vertebrate parasitism evolved at least five times, plant parasitism three times and in at least seven lineages, parasites or parasitoids of invertebrates evolved. The modes of parasitism in Nematoda are structured around molt transitions. In Rhabditina, strongylomorph parasites, including human hookworm and many devastating gut and tissue parasites of farm animals, transition into their vertebrate hosts as third stage infective larvae, homologous to dauer larvae. These dauer larvae establish the infection by simply being ingested or by actively seeking out a host and penetrating its skin. Fliarial nematodes (Spirurina) infect their definitive vertebrate hosts via an arthropod vector (commonly mosquitoes, tabanid flies, ticks, or mites). The vectors acquire infection by ingesting first stage larvae, which develop and arrest as third stage, dauer-like larvae before reintroduction to the definitive vertebrate host. In *Strongyloides* parasites (Tylenchina) the adults in the vertebrate host gut pass eggs into the environment. These either develop directly into infective third stage larvae or go through one cycle of free-living growth and reproduction before becoming infective larvae. In Enoplina and Dorylaimina animal parasites, the association between lifecycle transitions and the third stage is not as strong. In Trichocephalida (Dorylaimida; the whip worms and trichina worm of pigs), the infective stage is the first stage larva, but the third stage is involved in important within-host site selection. In Mermithida (Dorylaimida; parasites of many arthropods), the larvae grow in the host before emerging as third stage larvae and completing development to adult as non-feeding, free-living individuals. Plant parasitic nematodes can be divided into ectoparasites, which "graze" on roots and root hairs, and endoparasites, which burrow into host tissues and often induce specific host reactions to form feeding sites or galls. Ectoparasites consume host material at all stages of development (Perry and Moens, 2013). In endoparasites, especially in Tylenchina, the transition from the soil to the host is initiated in the third larval stage.

Free-living nematodes can be characterized along an *r*-strategist–*K*-strategist spectrum, often called the colonizer–persister spectrum in nematological

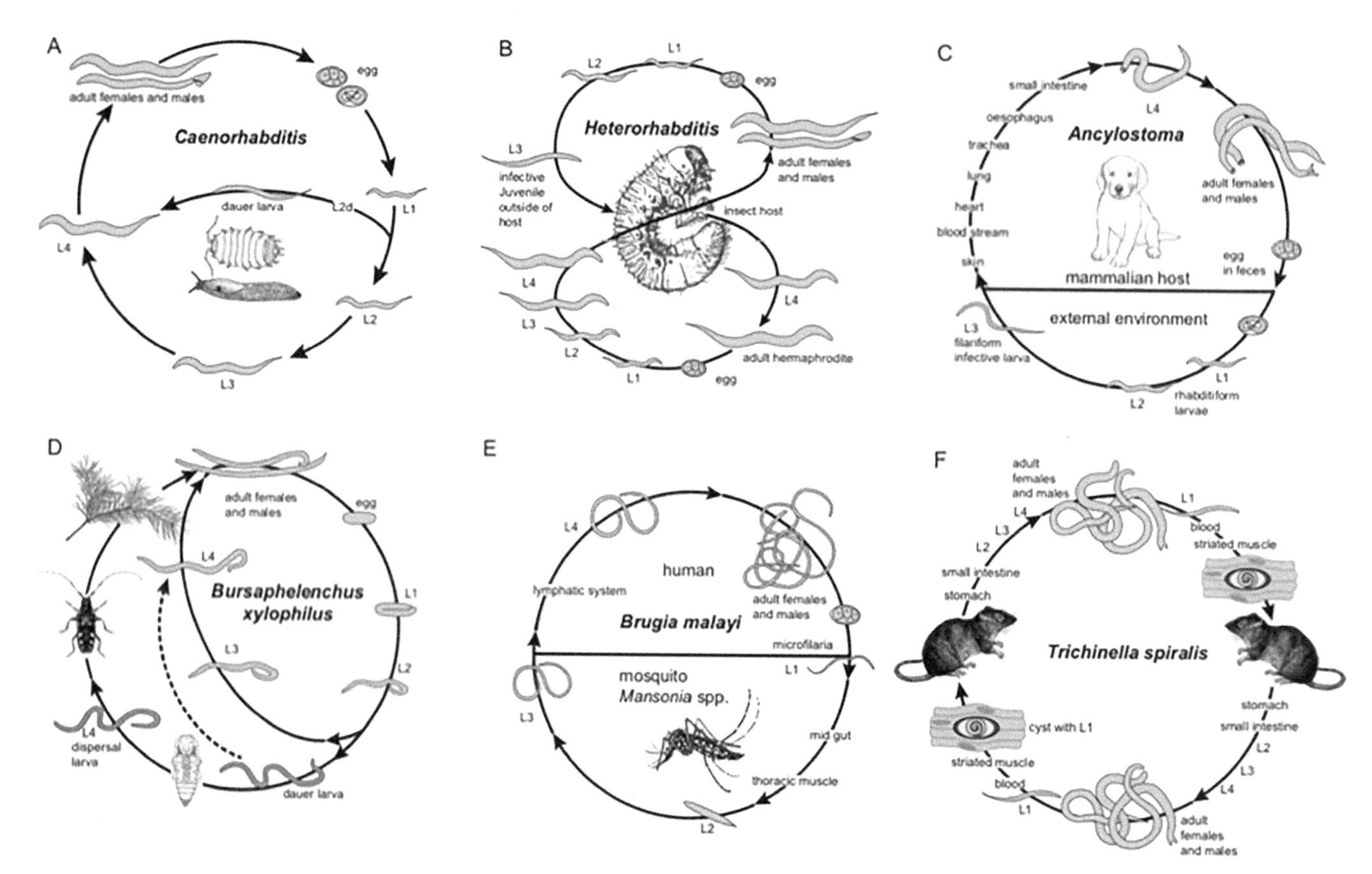

Figure 31.9 Life cycles of selected nematodes. (A) *Caenorhabditis* species (Rhabditomorpha). Normal development progresses through four larval stages and takes about three days. Under unfavorable conditions that include crowding and absence of food, the animal can enter an alternative pathway through a specialized thirds stage larva, the dauer larva that is non-feeding and non-ageing, resistant to harsh environmental conditions, and morphologically and behaviorally distinct. The dauer pathway is triggered during the L1 stage and passes through a morphologically and physiologically distinct L2d stage. The dauer larva serves as a dispersal stage. In most rhabditid species, including *Caenorhabditis* species, the dauer larvae are phoretic on other invertebrates, such as slugs for *C. elegans* and isopods for *C. remanei*. Dauer larvae mount a carrier and are thus transported to new environments. If the conditions are favorable, the dauer larva may leave its phoretic carrier and resume development. (B) *Heterorhabditis* species (Rhabditomorpha). These nematodes are insect pathogens. They have an obligate mutualistic relationship with the bacterium *Photorhabdus luminescens*. An infective juvenile (homologous to the dauer larva) enters an insect host and releases *Photorhabdus* bacteria from its gut. The bacteria grow inside the host, killing it rapidly. This species is heterogonic, i.e. the infective juvenile develops into a hermaphrodite. Her offspring become hermaphrodites, females and males. There can be 2 to 3 generations inside one host. When food becomes scarce, hermaphroditic infective juveniles develop that store the symbiotic bacteria in their gut before dispersing into the soil. (C) *Ancylostoma* (hookworm) species. *Ancylostoma* is a genus of the vertebrate parasitic Strongyloidea (Rhabditomorpha). Adults of these species reside in the small intestine of their mammalian host where they are attached to the intestinal wall and feed on blood. Eggs are passed with the host's feces into the environment, where embryonic development is completed and the so-called rhabditiform larva hatches. It quickly goes through two molts and develops into a filariform infective larva (homologous to the dauer larva) that detects a new host and penetrates its skin. From the skin, the larvae enter the blood stream and are carried to the heart and the lungs. Here, they penetrate through the pulmonary alveoli into the bronchiae and reach the mouth. When they are swallowed, they enter the digestive tract and settle in the small intestine, where they molt into L4 and finally adult males and females that can live for several years. Hookworm infections are among the most debilitating tropical diseases, causing severe anemia and subsequent stunting of physical and intellectual development. (D) *Bursaphelenchus xylophilus*, the pine wood nematode (Tylenchomorpha). This species feeds on plant cells in live pine wood, causing the death of the pine tree, and on fungi in dead wood. When food becomes scarce, dauer larvae develop. In the presence of an insect vector (typically pupae of a wood-boring beetle), these dauer larvae molt into the L4 dispersal stage, mount the emerging beetle and are transported to another tree. Under favorable conditions, dauer larvae can also skip this phoretic phase and molt into feeding L4 larvae. The pine wood nematode is native to North America but has been introduced to Europe and Asia where it causes extensive damage to forests of non-resistant trees. Other tylenchomorphs of economic importance are the root-knot nematodes (*Meloidogyne*, *Heterodera* and *Globodera* species) and root lesion nematodes (*Pratylenchus*). (E) *Brugia malayi* (Spiruromorpha). *B. malayi* is one of three species that cause lymphatic filariasis in humans. The adults lodge in the lymphatic system, leading to blockage and subsequent disfiguring and disabling swelling of limbs and genitalia. Female *B. malayi* produce microfilariae (L1 larvae) that circulate in the blood stream and are taken up by the intermediate host, mosquitos, with a blood meal. In the mosquito, the larvae make their way from the midgut to the thoracic muscles, where they molt to L2 and then develop into the infective L3 larvae. These migrate through the haemocoel to the proboscis of the mosquito. When the mosquito bites a human, these larvae are injected into the bite and are thus transmitted to the next definitive host. (F) *Trichinella* sp. (Trichinellida). The natural hosts of *Trichinella* species are rats and pigs and other associated animals. Humans and domestic animals can also be infected. Infection begins when an animal ingests meat that contains encysted *Trichinella* L1 larvae. When the meat is digested, the larvae are released and move to the small intestine, where they rapidly go through 4 molts to adult males and females. The adults burrow into the mucosa and females release L1 larvae into the bloodstream. The larvae make their way to the striated muscles and penetrate muscle fibers, where they encyst. When these cysts are ingested by a predator or a scavenger, the cycle begins anew.

literature. The *r*-strategists are boom-and-bust organisms, proliferating rapidly on ephemeral, rich habitats, and producing many long lived or dispersing propagules. *K*-strategists invest more into growth before reproduction and tend to live in more stable habitats. Just as free-living species' life cycles are structured around the maximization of Darwinian fitness, parasites have evolved strategies that maximize the likelihood of successful colonization of future hosts. These strategies can often contain non-intuitive components. For example, *A. suum* is a large (20 cm) gut parasite of pigs. It infects new hosts through direct ingestion of eggs from the environment. These eggs hatch in the gut, but rather than remaining in this, their final site, the larvae proceed to burrow through the gut wall, enter the mesenteric blood supply and transition to the lungs. Having developed to third stage larvae, they burst out of the capillaries into the tracheal system, migrate up to the throat, and are swallowed back into the gut where they grow to adults (Anderson, 2000). This seems like an arcane developmental route, but evolutionary comparisons between pairs of nematodes that do and do not have a tissue migration phase showed that the migration is associated with increased adult size, and thus fecundity (*A. suum* females produce 100,000 eggs per day), and so plays a key part in the success of these parasites (Read and Skorping, 1995).

Nematodes impact human life in many ways. Most fundamentally, they are important components of many ecosystems. Many soil and sediment biota are dominated by nematode individuals (Platnova and Gal'tsova, 1985) and they likely play key roles in sediment nutrient cycles. Their use as bioindicators of soil quality suggests that they play key roles in generating soil quality. They are even important in breaking up bacterial biofilms that can otherwise block septic systems. As parasites they moderate the diversity of ecosystems and thus promote biodiversity. More directly, plant parasites compete with humans for the photosynthetic products of our crop plants, causing at least $120 billion of losses each year (Singh et al., 2015; Jones et al., 2013). Parasites of farm animals cause huge losses also. For example, estimates of their impact on UK sheep farming alone exceed £84 million. Disease in farm animals (and in humans) is controlled by the use of a small set of drugs. Resistance to these drugs, including multidrug resistance, is growing, and especially prevalent in the common parasites of sheep and cattle.

The burden of nematode diseases in human populations is still high. Human ascariasis caused by *Ascaris lumbricoides* (Spirurina) affects over 800 million people, and human hookworm caused by *Ancylostoma duodenale* (Rhabditina) affects over 430 million. These infections occur largely in lower and middle income countries, and with other soil-transmitted nematodes they were responsible for an estimated health burden of 5.18 million disability adjusted life years (DALYs) in 2010 (Pullan et al., 2014). Filariases, caused by *Onchocerca, Brugia,* and relatives (Spirurina) caused 2.7 million DALYs in 2010 and 2.07 million in 2015 (Mitra and Mawson, 2017). The global cost of these human diseases has prompted efforts to reduce or eliminate infections both regionally and globally, and major mass drug administration campaigns focused on soil transmitted infections are underway (Jourdan et al., 2017). Global climate change affects the distribution of disease organisms; for filarial nematodes the spread of vector species, particularly mosquitoes, is bringing the risk of existing and new zoonotic filariases to temperate climates (Okulewicz, 2017).

PHYSIOLOGY AND BEHAVIOR

As nematodes possess no special gas exchange or circulatory organs, these functions are carried out by diffusion and probably the movement of pseudocoelomic fluid. Globin proteins exist in the pseudocoelom and elsewhere, suggesting that oxygen transport and storage occur despite the small size of many nematodes (Geuens et al., 2010).

Undoubtedly the most amount of information concerning nematode behavior comes from studies in *C. elegans*. Such studies take advantage of the fact that its behaviors are relatively simple and its nervous system, like all Rhabditida, contains a constant number of neurons with known connectivity (Jarrell et al., 2012; White et al., 1986). Here, we only mention some examples. For recent reviews, see WormBook (https://www.genetics.org/content/WormBook).

The longitudinal arrangement of muscles described above (Anatomy) is responsible for the typical sinusoidal locomotion of nematodes (Jarrell et al., 2012). The movement is facilitated by an undulating wave of alternating muscle contractions of the muscles on the dorsal and ventral side that, when travelling from head to tail, results in forward movement. This is similar to a snake, except that the nematode undulates along the dorsal-ventral axis. Reverse locomotion is facilitated by backward waves. The head muscles, however, are arranged differently to allow bending in three dimensions, presumably important for feeding. Additional types of movements can occur, such as deep bends bringing head and tail together, sharp bends of the tail (particularly in males) and coiling/spiraling. Different types or gaits of locomotion depend on the medium and the nematode's sensory abilities (Korta et al., 2007).

Probably the most complex behaviors are involved in copulation (Dusenberry, 1980; Barr and Garcia, 2006; Hart, 2006). Males can sense the presence and absence of females via diffusible signals. Depending on their nutritional status, males will leave a food source to search for a mate. Besides long-range cues, there are short-range recognition cues that may include chemical as well as mechanosensory stimuli. Once a male has located the approximate location of the vulva with his tail, the spicules are used to probe and find the vulva. Spicules are then inserted, and sperm are ejaculated. Females not only show attraction to males, but also display selective behaviors to resist mating by some males, such as "sprinting" and expulsion of sperm by uterine contractions.

Work in *C. elegans* revealed that it is exquisitely sensitive to its environment, displaying remarkable behavioral plasticity. It was discovered that *C. elegans* has short- and long-term memory in the form of habituation to sensory stimuli and also capable to learn and remember environmental features that predict food quality or adverse stimuli. Because of the genetic tools available for this species, the cellular and molecular basis of these behaviors have also been worked out at least partially (Ardiel and Rankin, 2010). The sensitivity of *C. elegans* to the environment is reflected in the very numerous sensory receptor genes present in the *C. elegans* genome. With over 1,200 sensory receptors, this nematode appears to have a richer sensory experience than even canids which have only ~300 receptor genes.

Nematode lifespans vary widely. Free-living, *r*-strategist, terrestrial bacteriovores have rapid lifecycles in optimal conditions, developing from egg to fertile adult in 2–3 days, and living for only 2–3 weeks. Other nematodes, such as terrestrial plant parasites, are closely tied to the life cycles of their hosts and have one generation per year. Animal parasites similarly have lifespans adapted to the host: insect parasites have rapid development while parasites of long-lived vertebrates (such as humans) can have similarly extended lifespans. *Onchocerca volvulus*, the river blindness parasite (Spirurina), can live for up to 10 years. Because many nematodes can slow or stop development (and apparently aging) under adverse conditions, or diapause in anhydrobiotic or cryobiotic states, many species may have ecological life cycles that are much longer than that measured under optimal conditions. For example, *Panagrolaimus davidi* (Tylenchina), an Antarctic species, has an egg-to-adult life cycle of 7 days in optimal laboratory conditions, but may only be unfrozen for a couple of weeks a year in nature (Wharton et al., 2003). Nematode eggs can diapause for long periods. Cysts containing arrested, embryonated eggs, derived from the sclerotized bodies of females of *Globodera* (Tylenchina) plant parasites, can persist in soils for half a century or more.

GENETICS AND GENOMICS

Caenorhabditis elegans was the first multicellular organism to have its genome sequenced (*C. elegans* Sequencing Consortium, 1998). Since then, well over 100 nematode genomes have been sequenced, most along with their whole-animal transcriptomes (see Databases and Collections below; International Helminth Genomes Consortium, 2019). The most represented taxon is *Caenorhabditis*. Average genome size is ca. 80 million base pairs, with ca. 18,000 protein-coding genes. For several species, population genomic datasets have been generated (Cook et al., 2017).

POSITION IN THE ToL

Relationships between nematodes and other animals have traditionally been as controversial as the name applied to the phylum itself (Nemata or Nematoda). Recent molecular studies place Nematoda together with its sister group Nematomorpha as the closest relatives of Panarthopoda (Arthropoda, Onychophora, Tardigrada) in Ecdysozoa (Giribet and Edgecombe, 2017). (See also the list of synapomorphies with Nematomorpha listed at the beginning of this chapter.) The presence of a nerve ring surrounding the anterior digestive system (the pharynx) has been suggested to be a feature that joins Priapulida, Kinorhyncha, Loricifera, Nematomorpha, and Nematoda (as Cycloneuralia), but this grouping has not been recovered in molecular phylogenies. One finding in several molecular analyses of Ecdysozoa that has had conflicting support is a grouping of Tardigrada with Nematoda plus Nematomorpha, though this has been attributed to artifacts in analysis of molecular data prone to homoplasy and substitution biases (Campbell et al., 2011).

DATABASES AND COLLECTIONS

- *C. elegans* and other nematodes genetics/genomics database (WormBase): https://wormbase.org
- *C. elegans* anatomy/images database (WormAtlas): https://www.wormatlas.org
- *C. elegans* and other nematodes information (WormBook): http://www.wormbook.org and https://www.genetics.org/content/wormbook
- *Caenorhabditis* Genetics Center living strains collection (https://cgc.umn.edu/)
- *C. elegans* wild strains (https://elegansvariation.org/)
- Parasitic nematode genomes and transcriptomic resources https://parasite.wormbase.org
- Parasitic nematodes genomes: http://www.nematode.net
- Q-Bank nematodes database: https://qbank.eppo.int/nematodes/
- RhabditinaDB, database for taxonomy and phylogenetics of Rhabditina (rhabditina.org): https://wormtails.bio.nyu.edu/fmi/webd/RhabditinaDB?homeurl=http://wormtails.bio.nyu.edu/Databases.html
- WoRMS, World Register of Marine Species (http://www.marinespecies.org/aphia.php?p=taxdetails&id=799)
- Several microscope slide collections of fixed nematodes also exist; e.g., at the University of Ghent, Belgium (http://www.nematodes.myspecies.info/), and the USDA Nematode Collection at Beltsville, Maryland (https://nt.ars-grin.gov/nematodes/).

CONCLUSION

Nematodes have provided important model systems since the earliest studies of development and cytogenetics because the embryos develop *ex utero* and most are transparent with canonical, mostly determinative cell lineages. Using *Parascaris equorum* (an intestinal parasite of horses), Theodor Boveri provided cytogenetic evidence supporting August Weissmann's hypotheses regarding meiotic reduction of chromosomal material into haploid germ cells and the reconstitution of diploidy by fertilization (Boveri, 1888; Brenner, 1974). Seventy years later, Sydney Brenner developed *C. elegans* for genetic investigations of development and behavior (Brenner, 1974). Since then, this biomedical model system has amassed a huge array of powerful investigative tools and depth of knowledge about the mechanisms of living systems, garnering three Nobel prizes. This knowledge has served to reveal even more about the profound depth of our ignorance regarding nematodes and living systems in general.

REFERENCES

Anderson, R. C. Nematode parasites of vertebrates: Their development and transmission. 2nd edn, 650 (CABI Publishing, 2000).

Andrássy, I. A Nematodák evolúciója és rendszerezése. A Magyar Tudományos Akadémia Biológiai Tudományok Osztályának Közleményei 17(1): 13–58. Pub. by MTA Biológiai Tudományok Osztálya, Hungary (1974).

Andrássy, I. Evolution as a Basis for the Systematization of Nematodes. Pub. by Pitman Publishing, London, England (1976).

Ardiel, E. L. & Rankin, C. H. An elegant mind: Learning and memory in *Caenorhabditis elegans*. Learning and Memory 17, 191-201, doi: 10.1101/lm.960510 (2010).

Barr, M. M. & Garcia, L. R. Male mating behavior. In WormBook, 1-11, doi: 10.1895/wormbook.1.78.1 (2006).

Bird, A. F. & Bird, J. The structure of nematodes. (Academic Press, 1991).

Blaxter, M. Nematodes: The worm and its relatives. PLoS Biology 9, e1001050, doi: 10.1371/journal.pbio.1001050 (2011).

Blaxter, M. & Koutsovoulos, G. The evolution of parasitism in Nematoda. Parasitology 142 Suppl 1, S26–39, doi: 10.1017/S0031182014000791 (2015).

Blaxter, M. L. & Robertson, W. M. The cuticle. In Physiology and biochemistry of free-living and plant-parasitic nematodes, 25–48 (CABI Publishing, 1998).

Blaxter, M. L. D. L., P.; Garey, J. R.; Liu, L. X.; Scheldeman, P.; Vierstraete, A.; Vanfleteren, J. R.; Mackey, L. Y.; Dorris, M.; Frisse, L. M.; Vida, J. T.; Thomas, W. K. A molecular evolutionary framework for the phylum Nematoda. Nature 392, 71–75, doi: 10.1038/32160 (1998).

Bongers, T. & Ferris, H. Nematode community structure as a bioindicator in environmental monitoring. Trends in Ecology and Evolution 14, 224–228 (1999).

Borgonie, G., Jacobsen, K. & Coomans, A. Embryonic lineage evolution in nematodes. Nematology 2, 65–69, doi: 10.1163/156854100508908 (2000).

Boveri, T. Die Befruchtung und Teilung des Eies von *Ascaris megalocephala*. Jenaische Zeitschrift für Naturwissenschaft 21, 423–515 (1888).

Boveri, T. Festschrift zum sebenzigsten Geburtstag von Carl von Kupffer, 383–430 (Gustav Fischer, 1899).

Brauchle, M., Kiontke, K., MacMenamin, P., Fitch, D. H. A. & Piano, F. Evolution of early embryogenesis in rhabditid nematodes. Developmental Biology 335, 253–262, doi: 10.1016/j.ydbio.2009.07.033 (2009).

Brenner, S. The genetics of *Caenorhabditis elegans*. Genetics 77, 71–94 (1974).

Burr, A. H. et al. A hemoglobin with an optical function. J Biol Chem 275, 4810–4815, doi: 10.1074/jbc.275.7.4810 (2000).

Campbell, L. I. et al. MicroRNAs and phylogenomics resolve the relationships of Tardigrada and suggest that velvet worms are the sister group of Arthropoda. Proceedings of National Academy of Sciences of United States of America 108, 15920–15924, doi: 10.1073/pnas.1105499108 (2011).

Chihara, D. & Nance, J. An E-cadherin-mediated hitchhiking mechanism for *C. elegans* germ cell internalization during gastrulation. Development 139, 2547–2556, doi: 10.1242/dev.079863 (2012).

Chisholm, A. D. & Hardin, J. Epidermal morphogenesis. WormBook, 1–22, doi: 10.1895/wormbook.1.35.1 (2005).

Chitwood, B. G. & Chitwood, M. B. Introduction to nematology. (University Park Press, 1950).

Chitwood, B. G. & Chitwood, M. B. Introduction to nematology. (University Park Press, 1974).

Cobb, N. A. Nematodes and their relationships. Yearbook of the United States Department of Agriculture 1914, 457–490 (1915).

Cook, D. E., Zdraljevic, S., Roberts, J. P. & Andersen, E. C. CeNDR, the *Caenorhabditis elegans* natural diversity resource. Nucleic Acids Research 45, D650–D657, doi: 10.1093/nar/gkw893 (2017).

De Ley, P. & Blaxter, M. In The biology of nematodes (ed. D. Lee) Ch. 1, 1–30 (CRC Press (imprint of Taylor & Francis), 2002).

De Ley, P. & Blaxter, M. L. In Nematology monographs & perspectives, vol. 2 633–653 (E. J. Brill, 2004).

Dolinski, C., Baldwin, J. G. & Thomas, W. K. Comparative survey of early embryogenesis of Secernentea (Nematoda), with phylogenetic implications. Canadian Journal of Zoology 79, 82–94, doi: 10.1139/cjz-79-1-82 (2001).

Dorris, M., De Ley, P. & Blaxter, M. L. Molecular analysis of nematode diversity and the evolution of parasitism. Parasitology Today 15, 188–193 (1999).

Dusenberry, D. B. In Nematodes as biological models. Volume 1. Behavioral and developmental models. (ed. B. M. Zuckerman) Ch. 3, 127–158 (Academic Press, 1980).

Fitch, D. H. A. & Emmons, S. W. Variable cell positions and cell contacts underlie morphological evolution of the rays in the male tails of nematodes related to *Caenorhabditis elegans*. Developmental Biology 170, 564–582, doi: 10.1006/dbio.1995.1237 (1995).

Fitch, D. H. A., Bugaj-Gaweda, B. & Emmons, S. W. 18S ribosomal RNA gene phylogeny for some Rhabditidae related to *Caenorhabditis*. Molecular Biology and Evolution 12, 346–358, doi: 10.1093/oxfordjournals.molbev.a040207 (1995).

Fradin, H. et al. Genome architecture and evolution of a unichromosomal asexual nematode. Current Biology 27, 2928–2939 e2926, doi: 10.1016/j.cub.2017.08.038 (2017).

Geuens, E. et al. Globin-like proteins in *Caenorhabditis elegans*: in vivo localization, ligand binding and structural properties. BMC Biochemistry 11, 17, doi: 10.1186/1471-2091-11-17 (2010).

Giribet, G. & Edgecombe, G. D. Current understanding of Ecdysozoa and its internal phylogenetic relationships. Integrative and Comparative Biology 57, 455–466, doi: 10.1093/icb/icx072 (2017).

Golden, A. M., Handoo, Z. A. & Weehunt, E. J. Description of *Dolichodorus cobbi* n. sp. (Nematoda: Dolichodoridae) with morphometrics and lectotype designation of *D. heterocephalus* Cobb, 1914. Journal of Nematology 18, 556–562 (1986).

Goldstein, B. & Hird, S. N. Specification of the anteroposterior axis in *Caenorhabditis elegans*. Development 122, 1467–1474 (1996).

Goldstein, B., Frisse, L. M. & Thomas, W. K. Embryonic axis specification in nematodes: evolution of the first step in development. Current Biology 8, 157–160, doi: 10.1016/S0960-9822(98)70062-4 (1998).

Haag, E. S., Fitch, D. H. A. & Delattre, M. From "the Worm" to "the Worms" and back again: the evolutionary developmental biology of nematodes. Genetics 210, 397–433, doi: 10.1534/genetics.118.300243 (2018).

Hart, A. C. Behavior. In WormBook Vol. 1.87.1 (eds V. Ambros & The *C. elegans* Research Community) (WormBook, 2006).

Holovachov, O. Morphology and systematics of the order Plectida Malakhov, 1982 (Nematoda) PhD thesis, Wageningen University, (2006).

Holterman, M. et al. Phylum-wide analysis of SSU rDNA reveals deep phylogenetic relationships among nematodes and accelerated evolution toward crown clades. Molecular Biology of Evolution 23, 1792–1800, doi: 10.1093/molbev/msl044 (2006).

Houthoofd, W. et al. Embryonic cell lineage of the marine nematode *Pellioditis marina*. Developmental Biology 258, 57–69, doi: 10.1016/S0012-1606(03)00101-5 (2003).

International Helminth Genomes Consortium. Comparative genomics of the major parasitic worms. Nature Genetics 51, 163–174, doi: 10.1038/s41588-018-0262-1 (2019).

Jarrell, T. A. et al. The connectome of a decision-making neural network. Science 337, 437–444, doi: 10.1126/science.1221762 (2012).

Jones, J. T. et al. Top 10 plant-parasitic nematodes in molecular plant pathology. Molecular Plant Pathology 14, 946–961, doi: 10.1111/mpp.12057 (2013).

Jourdan, P. M., Montresor, A. & Walson, J. L. Building on the success of soil-transmitted helminth control: The future of deworming. PLoS Neglected Tropical Diseases 11, e0005497, doi: 10.1371/journal.pntd.0005497 (2017).

Kanzaki, N. et al. Description of two three-gendered nematode species in the new genus *Auanema* (Rhabditina) that are models for reproductive mode evolution. Science Reports 7, 11135, doi: 10.1038/s41598-017-09871-1 (2017).

Kiontke, K. et al. Trends, stasis, and drift in the evolution of nematode vulva development. Current Biology 17, 1925–1937, doi: 10.1016/j.cub.2007.10.061 (2007).

Kiontke, K. & Fitch, D. H. A. The phylogenetic relationships of *Caenorhabditis* and other rhabditids. In WormBook Vol. 2005 (ed. The *C. elegans* Research Community) (WormBook, 2005).

Kiontke, K. & Fitch, D. H. A. Nematodes. Current Biology 23, R862–864, doi: 10.1016/j.cub.2013.08.009 (2013).

Kiontke, K. C. et al. A phylogeny and molecular barcodes for *Caenorhabditis*, with numerous new species from rotting fruits. BMC Evolution Biology 11, 339, doi: 10.1186/1471-2148-11-339 (2011).

Korta, J., Clark, D. A., Gabel, C. V., Mahadevan, L. & Samuel, A. D. Mechanosensation and mechanical load modulate the locomotory gait of swimming *C. elegans*. Journal of Experimental Biology 210, 2383–2389, doi: 10.1242/jeb.004572 (2007).

Lahl, V., Sadler, B. & Schierenberg, E. Egg development in parthenogenetic nematodes: Variations in meiosis and axis formation. International Journal of Developmental Biology 50, 393–397, doi: 10.1387/ijdb.052030vl (2006).

Lambshead, P. J. D. Recent developments in marine benthic biodiversity research. Recent Developments in Benthology 19, 5–24 (1993).

Lambshead, P. J. & Boucher, G. Marine nematode deep-sea biodiversity: Hyperdiverse or hype? Journal of Biogeography 30, 475–485 (2003).

Leduc, D. & Zhao, Z. Phylogenetic relationships within the superfamily Desmodoroidea (Nematoda: Desmodorida), with descriptions of two new and one known species. Zoological Journal of the Linnean Society 176, 511–536, doi: 10.1111/zoj.12324 (2016).

Lorenzen, S. The phylogenetic systematics of free-living nematodes: Translation with an introduction by H. M. Platt. (The Ray Society, 1994).

Maggenti, A. General nematology. (Springer-Verlag, 1981).

Malakhov, V. V. Nematodes: Structure, development, classification and phylogeny. (Smithsonian Institution Press, 1994).

Miller, M. A. et al. A sperm cytoskeletal protein that signals oocyte meiotic maturation and ovulation. Science (New York, NY) 291, 2144–2147, doi: 10.1126/science.1057586 (2001).

Mitra, A. K. & Mawson, A. R. Neglected tropical diseases: Epidemiology and global burden. Trop Med Infect Dis 2, doi: 10.3390/tropicalmed2030036 (2017).

Nance, J., Lee, J. Y. & Goldstein, B. Gastrulation in *C. elegans*. WormBook, 1–13, doi: 10.1895/wormbook.1.23.1 (2005).

Nielsen, C. Animal Evolution: Interrelationships of the Living Phyla. 3rd edn, (Oxford University Press, 2012).

Okulewicz, A. The impact of global climate change on the spread of parasitic nematodes. Annals of Parasitology 63, 15–20, doi: 10.17420/ap6301.79 (2017).

Perry, R. N. & Moens, M. Plant nematology, 2nd edn, (CABI Publishing, 2013).

Platnova, T. A. & Gal'tsova, V. V. Nematodes and their role in the meiobenthos [Nematody i ikh rol' v meiobentose]. (Oxonian Press Pvt. Ltd., 1985).

Podbilewicz, B. & White, J. G. Cell fusions in the developing epithelial of *C. elegans*. Developmental Biology 161, 408–424 (1994).

Poinar, G. O., Jr. The geological record of parasitic nematode evolution. Advances in Parasitology 90, 53–92, doi: 10.1016/bs.apar.2015.03.002 (2015).

Priess, J. R. & Hirsh, D. I. *Caenorhabditis elegans* morphogenesis: the role of the cytoskeleton in elongation of the embryo. Developmental Biology 117, 156–173 (1986).

Pullan, R. L., Smith, J. L., Jasrasaria, R. & Brooker, S. J. Global numbers of infection and disease burden of soil transmitted helminth infections in 2010. Parasite Vectors 7, 37, doi: 10.1186/1756-3305-7-37 (2014).

Read, A. F. & Skorping, A. The evolution of tissue migration by parasitic nematode larvae. Parasitology 111 (Pt 3), 359–371, doi: 10.1017/s0031182000081919 (1995).

Rehfeld, K. & Sudhaus, W. Die Sukzession der Nematoden im Kuhfladen: Gesetzmäßigkeiten und Wege zu einer Kausalanalyse. Verhandlungen der Gesellschaft für Ökologie (Göttingen 1987) 17, 745–755 (1989).

Robles, M. D., Cutillas, C., Panei, C. J. & Callejon, R. Morphological and molecular characterization of a new *Trichuris* species (Nematoda: Trichuridae), and phylogenetic relationships of *Trichuris* species of Cricetid rodents from Argentina. PLoS One 9, e112069, doi: 10.1371/journal.pone.0112069 (2014).

Rose, L. & Gönczy, P. Polarity establishment, asymmetric division and segregation of fate determinants in early *C. elegans* embryos. In WormBook (ed. The *C. elegans* Research Community), 1–43, doi: 10.1895/wormbook.1.30.2 (2014).

Schierenberg, E. Unusual cleavage and gastrulation in a freshwater nematode: Developmental and phylogenetic implications. Developmental Genes and Evolution 215, 103–108 (2005).

Schierenberg, E. Embryological variation during nematode development. In WormBook 1(55), 51, doi: 10.1895/wormbook.1.55.1 (2006).

Schmidt-Rhaesa, A. et al. Nematoda. Vol. 2 (Walter de Gruyter GmbH, 2013).

Schulze, J. & Schierenberg, E. Cellular pattern formation, establishment of polarity and segregation of colored cytoplasm in embryos of the nematode *Romanomermis culicivorax*. Developmental Biology 315, 426–436, doi: 10.1016/j.ydbio.2007.12.043 (2008).

Schulze, J. & Schierenberg, E. Embryogenesis of *Romanomermis culicivorax*: An alternative way to construct a nematode. Developmental Biology, doi: 10.1016/j.ydbio.2009.06.009 (2009).

Schulze, J. & Schierenberg, E. Evolution of embryonic development in nematodes. EvoDevo 2, 18, doi: 10.1186/2041-9139-2-18 (2011).

Schulze, J., Houthoofd, W., Uenk, J., Vangestel, S. & Schierenberg, E. *Plectus*: a stepping stone in embryonic cell lineage evolution of nematodes. EvoDevo 3, 13, doi: 10.1186/2041-9139-3-13 (2012).

Shemer, G. & Podbilewicz, B. Fusomorphogenesis: Cell fusion in organ formation. Developmental Dynamics 218, 30–51, doi: 10.1002/(SICI)1097-0177(200005)218:1<30:AID-DVDY4>3.0.CO;2-W (2000).

Singh, S., Singh, B. & Singh, A. P. Nematodes: A threat to sustainability of agriculture. Procedia Environmental Sciences 29, 215–216 (2015).

Stevens, L. et al. Comparative genomics of 10 new *Caenorhabditis* species. Evolution Letters 3, 217–236, doi: 10.1002/evl3.110 (2019).

Sudhaus, W. & Fitch, D. Comparative studies on the phylogeny and systematics of the Rhabditidae (Nematoda). Journal of Nematology 33, 1–70 (2001).

Sudhaus, W. & Kiontke, K. Comparison of the cryptic nematode species *Caenorhabditis brenneri* sp. n. and *C. remanei* (Nematoda: Rhabditidae) with the stem species pattern of the *Caenorhabditis elegans* group. Zootaxa, 45–62 (2007).

Sudhaus, W. Preadaptive plateau in Rhabditida (Nematoda) allowed the repeated evolution of zooparasites, with an outlook on evolution of life cycles within Spiroascarida. Palaeodiversity 3, 117–130 (2010).

Sulston, J. E., Schierenberg, E., White, J. G. & Thomson, J. N. The embryonic cell lineage of the nematode *Caenorhabditis elegans*. Development Biology 100, 64–119 (1983).

Szelecz, I. et al. Comparative analysis of bones, mites, soil chemistry, nematodes and soil micro-eukaryotes from a suspected homicide to estimate the post-mortem interval. Scientific Reports 8, 25, doi: 10.1038/s41598-017-18179-z (2018).

The *C. elegans* Sequencing Consortium. Genome sequence of the nematode *C. elegans*: a platform for investigating biology. Science 282, 2012–2018, doi: 10.1126/science.282.5396.2012 (1998).

Troccoli, A., Oreste, M., Tarasco, E., Fanelli, E. & De Luca, F. *Mononchoides macrospiculum* n. sp. (Nematoda: Neodiplogastridae) and *Teratorhabditis synpapillata* Sudhaus, 1985 (Nematoda: Rhabditidae): nematode associates of *Rhynchophorus ferrugineus* (Oliver) (Coleoptera: Curculionidae) in Italy. Nematology 17, 853–966 (2015).

van den Hoogen, J. et al. Soil nematode abundance and functional group composition at a global scale. Nature 572, 194–198, doi: 10.1038/s41586-019-1418-6 (2019).

van Megen, H. et al. A phylogenetic tree of nematodes based on about 1200 full-length small subunit ribosomal DNA sequences. Nematology 11, 927–950, doi: 10.1163/156854109X456862 (2009).

Vanfleteren, J. R. & Coomans, A. Photoreceptor evolution and phylogeny. Zeitschrift Zool Syst Evolut-Forsch 14, 157–169, doi: 10.1111/j.1439-0469.1976.tb00934.x (1976).

Vangestel, S., Houthoofd, W., Bert, W. & Borgonie, G. The early embryonic development of the satellite organism *Pristionchus pacificus*: differences and similarities with *Caenorhabditis elegans*. Nematology 10, 301–312, doi: 10.1163/156854108783900267 (2008).

Voronov, D. A. & Panchin, Y. V. Cell lineage in marine nematode *Enoplus brevis*. Development 125, 143–150 (1998).

Voronov, D. A. The embryonic development of *Pontonema vulgare* (Enoplida: Oncholaimidae) with a discussion of nematode phylogeny. Russian Journal of Nematology 7, 105–114 (1999).

Weismann, A. The germ-plasm: A theory of heredity. (Charles Scribner's Sons, 1893).

Wharton, D. A. A functional biology of nematodes. (Croom Helm Ltd., 1986).

Wharton, D. A., Goodall, G. & Marshall, C. J. Freezing survival and cryoprotective dehydration as cold tolerance mechanisms in the Antarctic nematode *Panagrolaimus davidi*. Journal of Experimental Biology 206, 215–221, doi: 10.1242/jeb.00083 (2003).

White, J. G., Southgate, E., Thomson, J. N. & Brenner, S. The structure of the nervous system of the nematode *Caenorhabditis elegans*. Philosophical Transactions of the Royal Society of London. Series B, Biological Sciences 314, 1–340, doi: 10.1098/rstb.1986.0056 (1986).

Yeates, G. W. Distribution of Mononchoidea (Nematoda, Enoplea) in pasture soils, with description of *Lotonchus stockdilli* n. sp. New Zealand Journal of Zoology 14, 351–358, doi: 10.1080/03014223.1987.10423005 (1987).

Yunliang, P. & Coomans, A. Three species of Tobrilidae (Nematoda: Enoplida) from Li River at Guiling. China Hydrobiologia 421, 77–90 (2000).

TARDIGRADA

Vladimir Gross and Georg Mayer

CHAPTER

Infrakingdom	Bilateria
Division	Protostomia
Subdivision	Ecdysozoa
Phylum	Tardigrada

CONTENTS

General Characteristics

At least 1,230 described species (~16% marine, ~84% limnoterrestrial)

Up to 2,654 estimated total species (87% completeness for limnoterrestrial species, 19% completeness for marine species)

Microscopic animals with a cosmopolitan distribution (including species endemic to Antarctica)

Found in mosses, lichens, and soil as well as marine and freshwater sediments

Can undergo cryptobiosis to tolerate extreme environmental conditions

No economic value to date

Synapomorphies

- five-segmented body consisting of a head and four trunk segments
- posterior-most legs face backward
- two pairs of connectives linking the brain with the first trunk ganglion
- buccopharyngeal apparatus
- miniaturized body
- reduced Hox gene cluster

HISTORY OF TAXONOMY AND CLASSIFICATION

Initial sightings of tardigrades (**Figure 32.1A**) surprised early researchers in the 18th century, who inevitably compared these animals to other enigmatic creatures commonly found in the same habitat, such as polyps and the "wheel animals" (i.e., rotifers; Goeze 1773; Spallanzani 1776). Although Eichhorn (1781) writes that he first saw a tardigrade in 1767, the first recorded observation of a tardigrade is generally attributed to Johann August Ephraim Goeze, who in 1773 described in detail the tiny animal he called *kleiner Wasserbär*

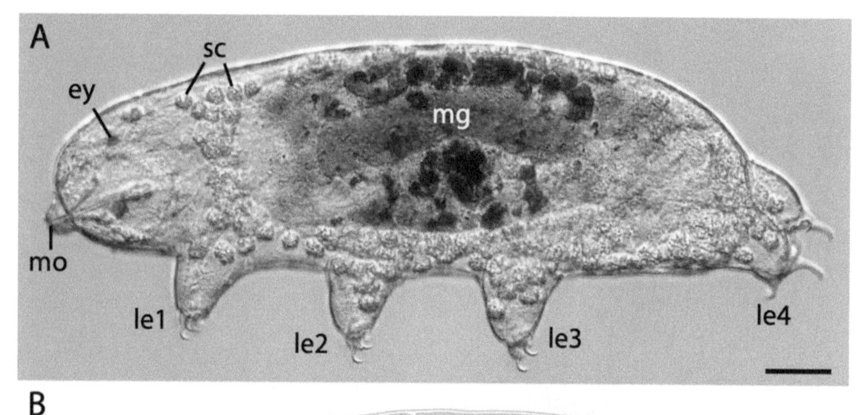

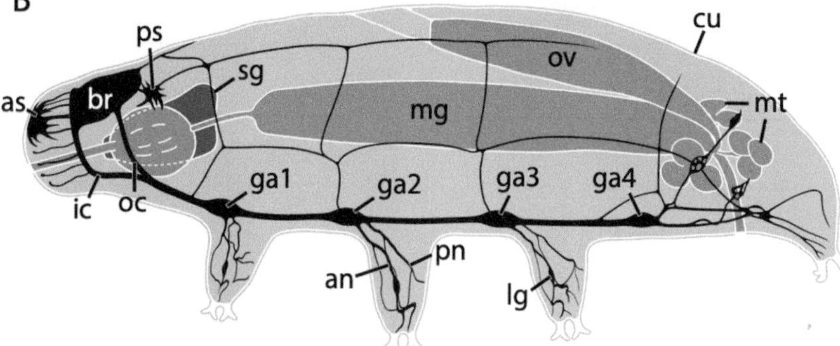

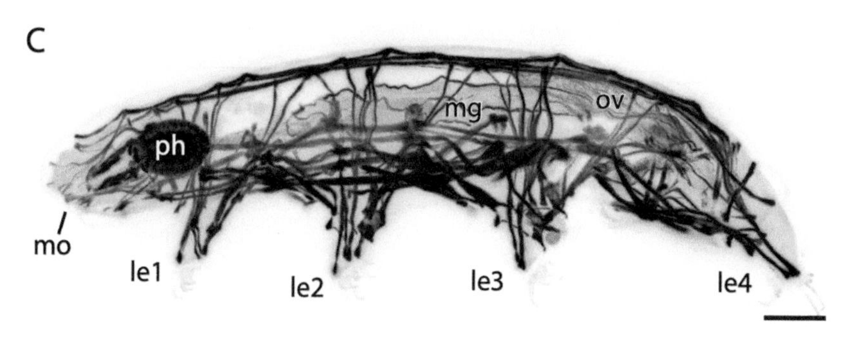

Figure 32.1 Anatomy of a tardigrade. Lateral view; anterior is left, dorsal is up in all images. (A) Light micrograph of the eutardigrade *Hypsibius exemplaris*. (Modified after Martin et al. 2017.) (B) Generalized schematic diagram showing the internal anatomy of a eutardigrade. Nervous system depicted in black; various organs in gray. (Modified after Mayer et al. 2013b.) (C) Fluorescent f-actin labeling showing the musculature of the eutardigrade *Hypsibius exemplaris*. Maximum projection of a confocal z-stack. an, anterior leg nerve; as, anterolateral sensory field; br, brain; cu, cuticle; ey, eye; ga1–ga4, trunk ganglia 1–4; ic, inner connective; le1–le4, legs 1–4; lg, leg ganglion; mg, midgut; mo, mouth; mt, Malpighian tubules; oc, outer connective; ov, ovary; ph, pharynx; pn, posterior leg nerve; ps, posterolateral sensory field; sc, storage cells; sg, salivary gland. Scale bars: 25 μm (A), 20 μm (C).

("small water bear") due to its resemblance to a miniature bear. Shortly thereafter, in 1776, Lazzaro Spallanzani compared its movement to the slow but determined crawling of a tortoise, thereby calling it *il Tardigrado* ("slow walker"), after which the taxon was officially named Tardigrada by Doyère (1840). Müller (1785) classified water bears as mites (which were considered insects at the time) based on the "shape, number and use of their feet." He therefore named his animal *Acarus ursellus* in recognition of both its mite-like and bear-like qualities. Since then, the tardigrades have been placed alternatively with the "Aschelminthes" (Crowe et al. 1970), Nematoda (Yoshida et al. 2017), and Arthropoda (Giribet et al. 1996). Nowadays they are considered to be a part of the Panarthropoda (Tardigrada + Onychophora + Arthropoda; Giribet and Edgecombe 2017).

The Tardigrada (common name: "water bears") today consists of over 1200 described species (Guidetti and Bertolani 2005; Degma and Guidetti 2007; Degma et al. 2018) divided into two main groups: the Eutardigrada (Richters, 1926) and the Heterotardigrada (Marcus, 1927; **Figure 32.2**). A third group, Mesotardigrada (Rahm, 1937), was described based on drawings of *Thermozodium esakii*, a tardigrade that Gilbert Rahm (1937) reported to have a combination of eutardigrade and heterotardigrade characters. However, no voucher specimens of this species exist and all subsequent attempts to find it at or near the *locus typicus* have failed, with explanations for the confusion ranging from misinterpretation of the original collected specimens to changes in habitat conditions (Grothman et al. 2017). Consequently, the Mesotardigrada is considered *nomen dubium* and is not generally accepted by tardigradologists today (Grothman et al. 2017).

Four fossil specimens found in Siberian deposits dated to the Cambrian and showing an "Orsten"-type preservation share similarities with extant tardigrades (Maas and Waloszek 2001). These specimens possess only three pairs of legs, with one of the four specimens additionally showing anlagen of a fourth pair (Maas and Waloszek 2001). The fossils most closely resemble the ectoparasitic arthrotardigrade *Tetrakentron synaptae* in shape, but probably represent a sister group of extant tardigrades (Maas and Waloszek 2001).

The Heterotardigrada is subdivided into the Echiniscoidea and the Arthrotardigrada (Figure 32.2). The latter is considered to be paraphyletic, as the former is nested within it (Fujimoto et al. 2016). All arthrotardigrade species are marine except for *Styraconyx hallasi* (Kristensen, 1977), which is from a warm spring in Greenland and is said to display the most plesiomorphic characters compared to other tardigrades. Since tardigrades probably originated in the ocean, the Arthrotardigrada likely represent the body plan most closely resembling the last common tardigrade ancestor. The Echiniscoidea, on the other hand, consists mostly of armored limnoterrestrial species and a few unarmored marine forms.

The Eutardigrada is subdivided into two monophyletic groups: the Apochela and the Parachela (Figure 32.2; Degma et al. 2018). Eutardigrades comprise the vast majority of described species despite a relatively uniform external

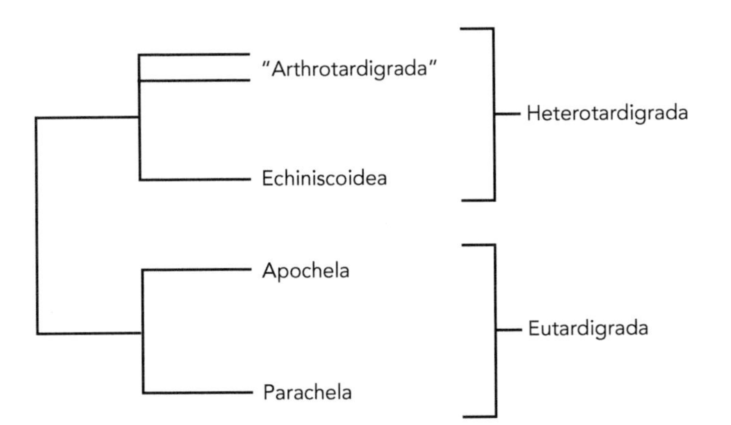

Figure 32.2 Internal relationships of the major tardigrade subgroups. The "Arthrotardigrada" is paraphyletic, as the Echiniscoidea is nested within it. (Topology according to Guil and Giribet 2011, Bertolani et al. 2014, and Fujimoto et al. 2016.)

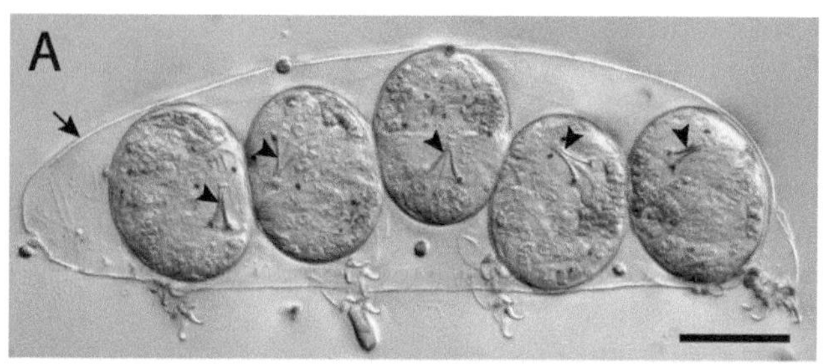

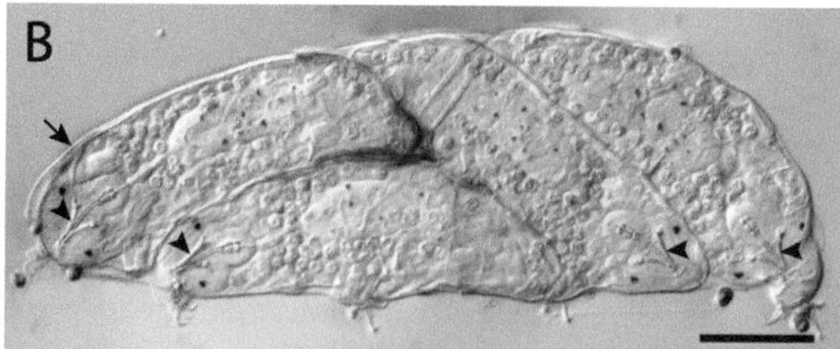

Figure 32.3 Egg deposition and hatching in the eutardigrade *Hypsibius exemplaris*. Anterior is left, dorsal is up. Many eutardigrade species lay eggs into the exuvium during molting. (A) Five developing embryos inside an exuvium. (B) Freshly hatched tardigrades must find their way out of the mother's exuvium after hatching. Arrowheads indicate the buccopharyngeal apparatuses of developing embryos (A) and hatchlings (B). Arrows indicate the mother's exuvium. (Modified from Gross et al. 2015.) Scale bars: 50 μm.

morphology compared to heterotardigrades. Some of the best-studied model tardigrade species belong to the Parachela, such as *Hypsibius exemplaris* (Gąsiorek et al. 2018; **Figures 32.1A,C; 32.3A,B; 32.4A–D**), *Ramazzottius varieornatus* (Bertolani and Kinchin 1993), and the *Macrobiotus hufelandi* species complex (Kaczmarek and Michalczyk 2017). The Apochela contains only ~40 species, the vast majority of which belong to the genus *Milnesium*.

EXTERNAL AND INTERNAL MORPHOLOGY

The adult tardigrade body can reach lengths of up to 800 μm and consists of five segments: a head and four trunk segments (Figure 32.1A–C; Nelson et al. 2015; Smith and Goldstein 2017). Each trunk segment bears one pair of unsegmented, ventrolateral walking legs called lobopods; the posterior-most pair faces backward and is used for climbing or clinging to the substrate (Greven and Schüttler 2001). Each leg is equipped distally with a number of claws (generally four individual claws in heterotardigrades or two double claws in eutardigrades), but elaborations such as adhesive disks or pads are also present in some arthrotardigrade species (Gross et al. 2014). Due to the miniscule body size, dedicated respiratory and circulatory systems are absent (Czerneková et al. 2018).

The **central nervous system** is represented by an anterior dorsal brain and a rope ladder-like ventral nerve cord, with one ganglion per trunk segment (Figures 32.1B; 32.4A–D; Mayer et al. 2013a,b; Martin et al. 2017). The brain occupies ~1% of the total body volume (Gross et al. 2019a). Each ganglion is connected longitudinally via somata-free connectives and transversally via commissures or a central fiber mass (Mayer et al. 2013b). Peripheral nervous

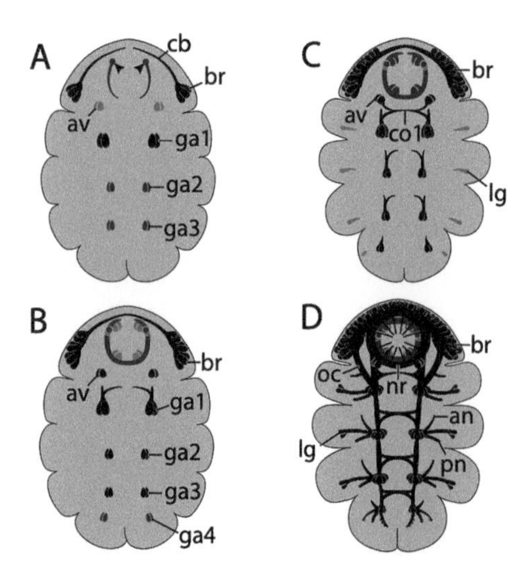

Figure 32.4 Nervous system development in the eutardigrade *Hypsibius exemplaris*. Dorsoventral view; anterior is up in all images. (A) The nervous system develops from anterior to posterior, beginning with the head and first trunk ganglion. The anlagen of the stomodeal nervous system also arises early (arrowheads). (B) Brain cells arise from lateral to dorsal in the head. The fourth trunk ganglion is the last ganglion to form. (C) Leg ganglia appear after the trunk ganglia are established. Connectives as well as horizontal and contralateral commissures begin to grow anteriorly from each ganglion. (D) Anterior and posterior leg nerves form, and the stomodeal nervous system with its nerve ring and neurites is established. (Modified from Gross and Mayer 2015.) an, anterior leg nerve; av, anteroventral cells; br, brain; cb, developing central brain neuropil; co1, commissure of the first trunk ganglion; ga1–ga4, trunk ganglia 1–4; lg, leg ganglion; nr, nerve ring; oc, outer connectives; pn, posterior leg nerve.

structures include the stomodeal/stomatogastric nervous system (Zantke et al. 2008; Mayer et al. 2013a; Gross and Mayer 2015), various cephalic sensory organs or sensory fields (Walz 1978; Dewel and Dewel 1996; Mayer et al. 2013a), and nerves innervating the legs and genital opening or cloaca (Mayer et al. 2013b). The eyes resemble inverse pigment-cup ocelli and are reported to consist of both ciliary and rhabdomeric receptor cells (Kristensen 1978; Dewel et al. 1993; Greven 2007). Each eye lies within the contours of the brain, with a direct neural input into the dorsolateral brain lobe but no connection to the epidermis (Greven 2007).

The **muscular system** is made up of single-celled muscles (Shaw 1974) that can be categorized into somatic and visceral musculature, as well as musculature of the buccopharyngeal apparatus (Figure 32.1C; Marchioro et al. 2013; Nelson et al. 2015). Somatic muscles consist of longitudinal (grouped into dorsal, lateral, and ventral groups), dorsoventral/transverse, and leg muscles (Marchioro et al. 2013). Ring and diagonal muscles are absent. The leg musculature is complex and, although the arrangement is unique to each of the four legs, the vast majority of leg muscles are serially repeated (Gross and Mayer 2019). Striation patterns include cross-striation (all muscles in arthrotardigrades; buccopharyngeal muscles in eutardigrades; "herringbone" leg muscles in marine eutardigrades) and an intermediate smooth/oblique striation (all other eutardigrade muscles) (Walz 1974; Crisp and Kristensen 1983; Schmidt-Rhaesa and Kulessa 2007; Halberg et al. 2009; Schulze and Schmidt-Rhaesa 2011; Marchioro et al. 2013). The pharynx consists of myoepithelial cells arranged radially around a Y-shaped lumen (Figure 32.1C; Eibye-Jacobsen 2001).

The **digestive system** consists of an anterior subterminal (in Eutardigrada and Echiniscoidea) or ventral (most Arthrotardigrada) mouth, the buccopharyngeal apparatus (consisting of the buccal tube, pharynx, and a pair of stylets), a thin esophagus, a voluminous midgut, and a hindgut, which ends in a cloaca (in Eutardigrada) or anus (in Heterotardigrada) (Figure 32.1B; Hyra et al. 2016; Gross et al. 2018; Gross et al. 2019a). The sharp stylets are used to pierce the cells of the potential food item (generally algae, fungi, or animal prey), which is then ingested by the sucking action of the muscular pharynx. Digestion occurs inside the midgut, which is full of microvilli (Greven 1976; Gross et al. 2018) and lined by a peritrophic membrane (Raineri 1985). The cells of the midgut are constantly replaced via two proliferation zones at the anterior and posterior margins of the midgut (Hyra et al. 2016; Gross et al. 2018). In many species, the contents of the midgut are often solely responsible for any observable pigmentation. **Excretory function** is achieved by Malpighian tubules (in most eutardigrades; Figure 32.1B), which also secondarily function as osmoregulatory organs in the marine-adapted eutardigrade *Halobiotus crispae* (Møbjerg and Dahl 1996).

The **reproductive system** consists of a large, unpaired dorsal gonad that is attached anteriorly to the dorsal and dorsolateral body wall via two ligaments and, in eutardigrades, opens into the digestive tract at the transition between the midgut and hindgut (Figure 32.1B; Poprawa et al. 2015). In dioecious species, the gonopore is positioned medially on the ventral side between the third and fourth leg pairs, anterior to the anus (Dewel and Dewel 1997). A pair of sperm ducts is present in males while females often possess large seminal receptacles (Hansen et al. 2012). Some species display sexual dimorphism, manifested in the shape of the gonopore, body size, and primary clava morphology (Renaud-Mornant and Deroux 1976).

REPRODUCTION AND DEVELOPMENT

Several different reproductive modes are observed in tardigrades. The vast majority of marine species are dioecious, with the remainder being hermaphroditic (Nelson et al. 2015). Limnoterrestrial species can be dioecious, hermaphroditic, or, most prevalently, parthenogenetic, whereby males are rare or completely absent (Nelson et al. 2015). Tardigrades are egg-laying animals, and eggs are deposited either freely into the environment or into the shed cuticle during molting (Figure 32.3A). The eggshells of freely laid eggs often exhibit ornate elaborations that also serve as important taxonomic characters, as in many species of *Macrobiotus* (Kaczmarek and Michalczyk 2017). Developmental time varies depending on the species and environmental factors like temperature, but generally lasts several days. In the model eutardigrade *H. exemplaris*, for example, hatching occurs after ~4.5 days (**Figure 32.3B**; Gabriel et al. 2007; Gross et al. 2015). In this species, segments appear to arise simultaneously (Gross et al. 2017), although at least the nervous system develops in a gradient from anterior to posterior (Figure 32.4A–D; Gross and Mayer 2015). External morphogenesis is largely completed by ~50 hours into development, which is also when muscle twitching is first detectable (Gabriel et al. 2007; Gross et al. 2017).

Postembryonic development is direct, with no true larval stage in the sense that the freshly hatched animal does not undergo a radical metamorphosis, nor does it occupy a different habitat than the adult. Tardigrades molt periodically throughout life in order to grow in size and, in some species, to lay eggs or defecate (Nelson et al. 2015). The first-instar forms of some heterotardigrade species display differences in claw configuration and/or gonopore morphology (Bertolani et al. 1984). The same is true of first instars of apochelans, which are referred to as hatchlings to differentiate them from second instar juveniles (Morek et al. 2018). The total lifespan of tardigrades in the active state has been estimated to be up to 2 years, while cryptobiotic specimens have been revived after over 30 years of being frozen (Tsujimoto et al. 2016).

DISTRIBUTION AND ECOLOGY

Tardigrades are cosmopolitan and may be found in almost any temporarily or permanently wet habitat, which are generally divided into marine, freshwater, and terrestrial/limnoterrestrial (**Figure 32.5**). Marine species are typically interstitial while a minority exhibit specialized lifestyles, such as on barnacles (e.g., *Echiniscoides rugostellus*; Perry et al. 2018), sea cucumber appendages (*Tetrakentron synaptae*; Cuénot 1892), or in deep-sea manganese nodules (*Moebjergarctus manganis*; Bussau 1992). Muddy or anaerobic sediments are often devoid of tardigrades, as these animals are sensitive to low oxygen levels.

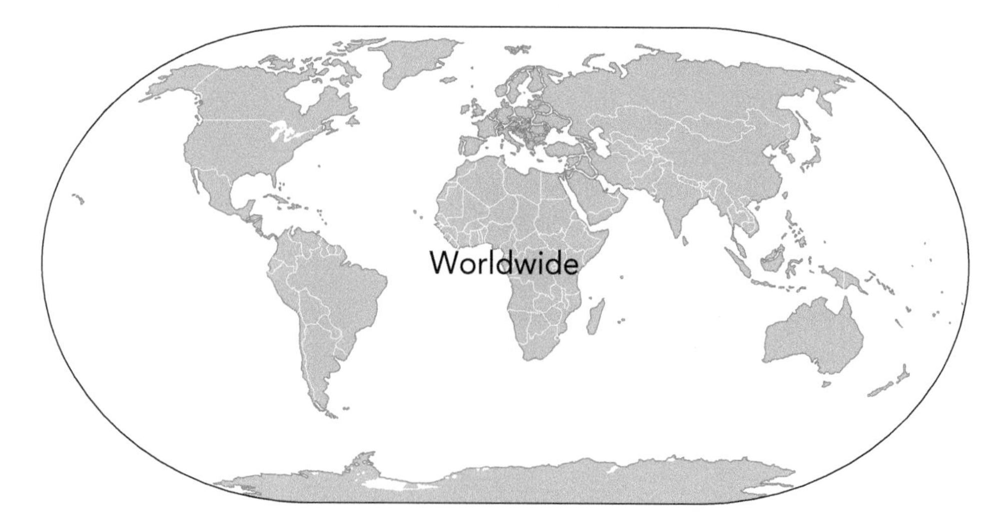

Figure 32.5 Distribution map for Tardigrada. Tardigrades are cosmopolitan and may be found in almost any temporarily or permanently wet habitat, which are generally divided into marine, freshwater, and terrestrial/limnoterrestrial.

Limnoterrestrial species are commonly found in lichens, mosses, and other cushion plants, as well as in topsoil and leaf litter. These species are able to survive extended dehydration and extreme environmental conditions by entering cryptobiosis, a form of suspended animation. True freshwater species are few in number but may be found on underwater plants and algae, or in the top layer of submerged sediment. Although freshwater species are generally weaker cryptobionts than their limnoterrestrial relatives, some are able to undergo cryptobiosis after preconditioning (Wright 1989) or, alternatively, form cysts and thereby reduce their metabolism in order to wait out unfavorable conditions (Guidetti et al. 2011; Møbjerg et al. 2011).

PHYSIOLOGY AND BEHAVIOR

Behavioral studies are scarce, partly because of the impracticality of observing tardigrades in their natural habitat. Observations of mating and reproduction represent the majority of behavioral studies in tardigrades and were first recorded over a century ago (von Erlanger 1895; Baumann 1961; Bingemer et al. 2016; Bartel and Hohberg 2019; Sugiura et al. 2019). The most detailed and complete description of copulation is for two species of Macrobiotidae, which lay eggs freely into the environment (Sugiura et al. 2019). Mating in these species has been subdivided into five stages, beginning with tracking and courtship behavior, followed by ejaculation and finally contraction of the female body in order to facilitate the capture of spermatozoa (Sugiura et al. 2019). Mating in other species seems to follow similar patterns, irrespective of whether fertilization occurs internally like in Macrobiotidae (Sugiura et al. 2019), externally but within the shed cuticle like in the isohypsibiid *Isohypsibius dastychi* (Bartel and Hohberg 2019), or externally in the environment like in the arthrotardigrade *Batillipes noerrevangi* (Kristensen 1979). In all cases, the male actively seeks out and is directly attracted to females that are ready to mate. In the eutardigrade *I. dastychi*, localization of the female by the male decreases exponentially with distance, suggesting that a pheromone gradient may be established in the environment by the female (Bartel and Hohberg 2019).

Excretory function in tardigrades is accomplished most likely by the Malpighian tubules in eutardigrades and the "segmental glands" in heterotardigrades (Greven 1982; Węglarska 1987; Dewel et al. 1992; Møbjerg and Dahl 1996; Pelzer et al. 2007; Møbjerg et al. 2018). This hypothesis is based on ultrastructural data that strongly point to the Malpighian tubules as the site of active ion transport and urine formation and modification, although direct evidence is still lacking (Greven 1982; Pelzer et al. 2007). The Malpighian tubules of the eutardigrade *Halobiotus crispae* are thought to additionally play an osmoregulatory role, as this species is secondarily adapted to the marine environment (Kristensen 1982). It should be noted that the tardigrade Malpighian tubules were named as such due to their positional correspondence to structures of the same name in mites (Plate 1889), i.e., at the transition between the midgut and hindgut, but while these structures may serve the same function in both animal groups, their homology is considered to be unlikely (Greven 1982).

Behavioral studies involving photoreception in various tardigrade species have produced mixed and often contradictory results. The heterotardigrade *Echiniscoides sigismundi*, for example, is positively phototactic, i.e., it actively moves toward light (Marcus 1928b), while the eutardigrade *Dactylobiotus dispar* actively avoids light (Marcus 1928a). Other species were observed to react differently to light depending on their age, as in the case of *Hypsibius convergens* (Baumann 1961) or *Macrobiotus hufelandi* (Beasley 2001). Any affinity or aversion to specific wavelengths of light remains unknown.

Tardigrades are best known for their ability to tolerate unfavorable environmental conditions via a latent state for up to 30 years (Nelson et al. 2015;

Tsujimoto et al. 2016). This process, called cryptobiosis, can be induced by a number of conditions including low temperatures, osmotic stress, and desiccation. The resistant state that occurs upon entering cryptobiosis, called a "tun," offers protection against these and other hazards, such as radiation and vacuum (Nelson et al. 2015; Hengherr and Schill 2018; Jönsson et al. 2018; Schill and Hengherr 2018). The molecular mechanisms behind cryptobiosis are only beginning to be understood but the pace of studies is increasing quickly, facilitated by the recent sequencing of several tardigrade genomes (Arakawa et al. 2016; Hashimoto et al. 2016; Koutsovoulos et al. 2016) together with transcriptomic (Yoshida et al. 2017; Kamilari et al. 2019) and functional studies (Kondo et al. 2015; 2019; Hashimoto et al. 2016; Hering et al. 2016). Current analyses point towards various molecular drivers of stress resistance in different species. Genes encoding various proteins that offer protection against oxidative stress (superoxide dismutase) or osmotic stress (NFAT5), for example, have been expanded in tardigrades, while genes in the peroxisome pathway have been reduced (Yoshida et al. 2017). Additional gene families specific to tardigrades have also been implicated in cryptobiosis (Yoshida et al. 2017). These "tardigrade-specific" genes, such as cytosolic abundant heat soluble (CAHS), secretory abundant heat soluble (SAHS), mitochondrial abundant heat soluble (MAHS), and damage suppressor (Dsup) gene families are highly expressed in the eutardigrade *R. varieornatus* (Yamaguchi et al. 2012; Yoshida et al. 2017). However, transcriptomic analyses indicate that the "tardigrade-specific" genes are restricted to eutardigrades and are missing from the heterotardigrade *E.* cf. *sigismundi* (Kamilari et al. 2019). This implies that different evolutionary lineages of tardigrades may employ unique mechanisms to tolerate stress (Kamilari et al. 2019).

GENETICS AND GENOMICS

Complete genomes are available from at least three species, all of which are eutardigrades (**Table 32.1**; Arakawa et al. 2016; Hashimoto et al. 2016; Koutsovoulos et al. 2016; Bemm et al. 2017), while a complete mitochondrial genome has been sequenced from the heterotardigrade *Echiniscus testudo* (Arakawa 2018). The full genomes range from ~75–105 Mb and are distributed across 5–7 chromosome pairs, each of which are ≤1 μm in length (Ammermann 1967; Rebecchi et al. 2002). Approximately 14,000–20,000 genes are present in the genome (Table 32.1), depending on the species, with the difference largely attributed to losses rather than gains of genes compared to the tardigrade ancestor (Yoshida et al. 2017). Only five of the ancestral ten Hox genes (*labial*, *Hox3*, *Deformed*, *fushi tarazu*, and *Abdominal-B*) are present in the tardigrade genome and these are either located on different scaffolds (in *H. exemplaris*) or are each separated by several non-Hox genes (in *R. varieornatus*), indicating low levels of synteny (Smith et al. 2016; Yoshida et al. 2017). Expression analyses and

TABLE 32.1 Complete genome sequences are currently available from three tardigrade species, all of which are eutardigrades. From Gross et al. (2019b)

	Hypsibius exemplaris **(Arakawa et al. 2016; Koutsovoulos et al. 2016)**	***Ramazzottius varieornatus*** **(Hashimoto et al. 2016)**	***Milnesium tardigradum*** **(Bemm et al. 2017)**
Taxonomy ID	232323	947166	400877
Assembly version	nHd.3.1.5	Rv101	—
Assembly size (bp)	104,154,999	55,828,384	75,059,434
Number of scaffolds	1421	199	6654
Gene count	20,076	13,920	19,401

alignments of Hox gene domains with those of arthropods and onychophorans indicate that a large middle body region was lost in the tardigrade lineage, suggesting that the entire tardigrade body is homologous to just the head and one abdominal segment of arthropods (Smith et al. 2016).

POSITION IN THE ToL

The position of tardigrades has long been a matter of heated debate, partly because of the implications for arthropod evolution. It is clear that tardigrades belong to the Ecdysozoa (molting animals), but where they belong within this group remains contentious (**Figure 32.6**; summarized in Smith and Goldstein 2017). While several molecular studies maintain that tardigrades are sister to the Nematoda or Nematoida (Yoshida et al. 2017; Arakawa 2018) they are generally accepted to be members of the Panarthropoda (Tardigrada + Onychophora + Arthropoda), and within this group either sister to Arthropoda (generally supported by morphological and neuroanatomical data) or to Onychophora + Arthropoda (supported by molecular studies) (Smith and Ortega-Hernández 2014; Giribet and Edgecombe 2017).

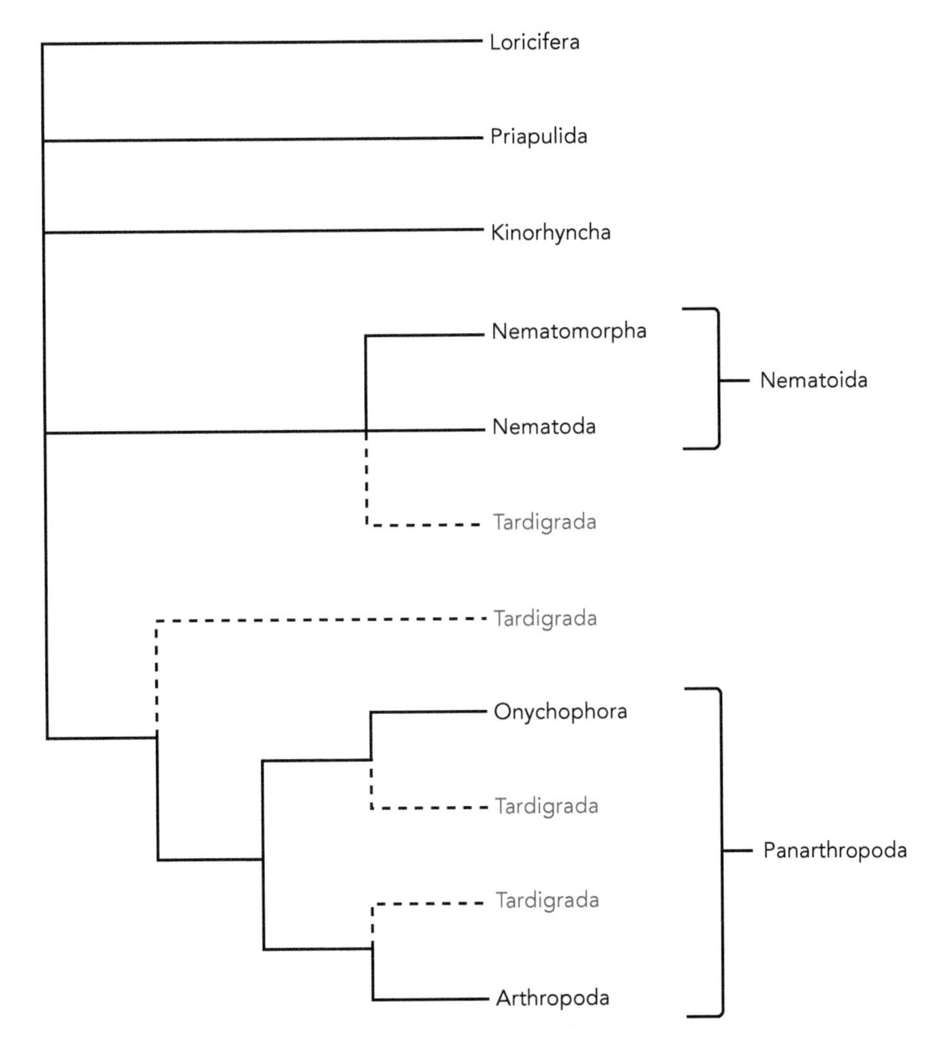

Figure 32.6 Alternative hypotheses regarding the position of Tardigrada within the Ecdysozoa. Tardigrades have alternatively been placed as the sister group of Arthropoda (Tactopoda hypothesis), Onychophora (Lobopodia hypothesis), Arthropoda + Onychophora (Antennopoda hypothesis), or Nematoda/Nematoida (paraphyletic Panarthropoda). Summarized in Giribet and Edgecombe 2017. (Topologies based on Budd 2001, Dzik and Krumbiegel 1989, Ma et al. 2009, and Borner et al. 2014.)

DATABASES AND COLLECTIONS

Active cultures of several tardigrade species are maintained privately in laboratories around the world. Live cultures are commercially available from Carolina Biological Supply Company (www.carolina.com).

A regularly updated checklist of all tardigrade taxa is maintained at www.tardigrada.modena.unimo.it (Guidetti and Bertolani 2005; Degma and Guidetti 2007; Degma et al. 2018), which also contains a list of the most recent taxonomic keys for several tardigrade taxa. A database of older, sometimes harder-to-find works is maintained at www.tardigrada.pub. An interactive map of all published records of marine tardigrades is made available in Kaczmarek et al. (2015). The Tardigrada Newsletter (www.tardigrada.net/newsletter) is a comprehensive source of tardigrade-related news, recent publications, links to other tardigrade websites, and contact information of tardigradologists all over the world. The complete genomes of *H. exemplaris* and *R. varieornatus* can be explored at ensembl.tardigrades.org.

CONCLUSION

The ability of tardigrades to tolerate extreme conditions has been recognized since the beginning of tardigrade research, but the mechanisms and effects of cryptobiosis are only beginning to be uncovered. Recent breakthroughs in the field of tardigrade extremotolerance have identified a number of important molecular players that, together with improvements in imaging techniques that allow new ways of visualizing tardigrade anatomy, have renewed interest in tardigrades, both public and scientific. Recent projects that successfully transformed human cells with tardigrade stress genes promise that further efforts to uncover the mechanisms and factors involved in cryptobiosis may sooner or later contribute toward biomedical research for use in human medicine.

REFERENCES

Ammermann, D., 1967. Die Cytologie der Parthenogenese bei dem Tardigraden *Hypsibius dujardini*. Chromosoma (Berlin), 23, 203–213.

Arakawa, K., Yoshida, Y. and Tomita, M., 2016. Genome sequencing of a single tardigrade *Hypsibius dujardini* individual. Scientific Data, 3, 160063.

Arakawa, K., 2018. The complete mitochondrial genome of *Echiniscus testudo* (Heterotardigrada: Echiniscidae). Mitochondrial DNA Part B, 3, 810–811.

Bartel, S. and Hohberg, K., 2019. Experimental investigations on the partner-finding behaviour of *Isohypsibius dastychi* (Isohypsibiidae: Tardigrada). Zoological Journal of the Linnean Society, 1–9.

Baumann, H., 1961. Der Lebenslauf von *Hypsibius* (*H.*) *convergens* Urbanowicz (Tardigrada). Zoologischer Anzeiger, 165, 123–128.

Beasley, C.W., 2001. Photokinesis of *Macrobiotus hufelandi* (Tardigrada, Eutardigrada). Zoologischer Anzeiger, 240, 233–236.

Bemm, F., Burleigh, L., Förster, F., Schmucki, R., Ebeling, M., Janzen, C.J., Dandekar, T., Schill, R.O., Certa, U. and Schultz, J., 2017. Draft genome of the eutardigrade *Milnesium tardigradum* sheds light on ecdysozoan evolution. bioRxiv.

Bertolani, R., Grimaldi de Zio, S., D'Addabbo Gallo, M. and Morone de Lucia, R.M., 1984. Postembryonic development in heterotardigrades. Monitore Zoologico Italiano, 18, 307–320.

Bertolani, R. and Kinchin, I.M., 1993. A new species of *Ramazzottius* (Tardigrada, Hypsibiidae) in a rain gutter sediment from England. Zoological Journal of the Linnean Society, 109, 327–333.

Bertolani, R., Guidetti, R., Marchioro, T., Altiero, T., Rebecchi, L. and Cesari, M., 2014. Phylogeny of Eutardigrada: New molecular data and their morphological support lead to the identification of new evolutionary lineages. Molecular Phylogenetics and Evolution, 76, 110–126.

Bingemer, J., Hohberg, K. and Schill, R.O., 2016. First detailed observations on tardigrade mating behaviour and some aspects of the life history of *Isohypsibius dastychi* Pilato, Bertolani & Binda 1982 (Tardigrada, Isohypsibiidae). Zoological Journal of the Linnean Society, 178, 856–862.

Borner, J., Rehm, P., Schill, R.O., Ebersberger, I. and Burmester, T., 2014. A transcriptome approach to ecdysozoan phylogeny. Molecular Phylogenetics and Evolution, 80, 79–87.

Budd, G.E., 2001. Tardigrades as 'stem-group arthropods': The evidence from the Cambrian fauna. Zoologischer Anzeiger, 240, 265–279.

Bussau, C., 1992. New deep-sea Tardigrada (Arthrotardigrada, Halechiniscidae) from a manganese nodule area of the eastern South Pacific. Zoologica Scripta, 21, 79–91.

Crisp, M. and Kristensen, R.M., 1983. A new marine interstitial eutardigrade from East Greenland, with comments on habitat and biology. Videnskabelige Meddelelser Fra Dansk Naturhistorisk Forening, 144, 99–114.

Crowe, J.H., Newell, I.M. and Thomson, W.W., 1970. *Echiniscus viridis* (Tardigrada): fine structure of the cuticle. Transactions of the American Microscopical Society, 89, 316–325.

Cuénot, L., 1892. Commensaux et parasites des Echinodermes. Revue Biologique du Nord de la France, 5, 1–22.

Czerneková, M., Janelt, K., Student, S., Jönsson, K.I. and Poprawa, I., 2018. A comparative ultrastructure study of storage cells in the eutardigrade *Richtersius coronifer* in the hydrated state and after desiccation and heating stress. PLoS One, 13, e0201430.

Degma, P. and Guidetti, R., 2007. Notes to the current checklist of Tardigrada. Zootaxa, 1579, 41–53.

Degma, P., Bertolani, R. and Guidetti, R., 2018. Actual checklist of Tardigrada species. http://www.evozoo.unimore.it/site/home/tardigrade-tools/documento1080026927.html. Accessed 30 June 2018.

Dewel, R.A., Dewel, W.C. and Roush, B.G., 1992. Unusual cuticle-associated organs in the heterotardigrade, *Echiniscus viridissimus*. Journal of Morphology, 212, 123–140.

Dewel, R.A., Nelson, D.R. and Dewel, W.C., 1993. Tardigrada. In: Harrison, F.W. and Rice, M.E. (Eds), Microscopic anatomy of invertebrates, vol. 12, 143–183 Wiley-Liss, New York.

Dewel, R.A. and Dewel, W.C., 1996. The brain of *Echiniscus viridissimus* Peterfi, 1956 (Heterotardigrada): a key to understanding the phylogenetic position of tardigrades and the evolution of the arthropod head. Zoological Journal of the Linnean Society, 116, 35–49.

Dewel, R.A. and Dewel, W.C., 1997. The place of tardigrades in arthropod evolution. In: Fortey, R.A. and Thomas, R.H. (Eds), Arthropod relationships, 109–123. Chapman & Hall, London.

Doyère, M., 1840. Mémoire sur les Tardigrades. Annales des Sciences Naturelles, Zoologie, Série 2, 14, 269–361.

Dzik, J. and Krumbiegel, G., 1989. The oldest 'onychophoran' *Xenusion*: a link connecting phyla? Lethaia, 22, 169–181.

Eibye-Jacobsen, J., 2001. Are the supportive structures of the tardigrade pharynx homologous throughout the entire group? Journal of Zoological Systematics and Evolutionary Research, 39, 1–11.

Eichhorn, J.C., 1781. Beyträge zur Naturgeschichte der kleinsten Wasserthiere, die mit blossem Auge nicht können gesehen werden und die sich in den Gewässern in und um Danzig befinden, Berlin and Stettin.

Fujimoto, S., Jørgensen, A. and Hansen, J.G., 2016. A molecular approach to arthrotardigrade phylogeny (Heterotardigrada, Tardigrada). Zoologica Scripta, 46, 496–505.

Gabriel, W.N., McNuff, R., Patel, S.K., Gregory, T.R., Jeck, W.R., Jones, C.D. and Goldstein, B., 2007. The tardigrade *Hypsibius dujardini*, a new model for studying the evolution of development. Developmental Biology, 312, 545–559.

Gąsiorek, P., Stec, D., Morek, W. and Michalczyk, Ł., 2018. An integrative redescription of *Hypsibius dujardini* (Doyère, 1840), the nominal taxon for Hypsibiodea (Tardigrada: Eutardigrada). Zootaxa, 4415, 45–75.

Giribet, G., Carranza, S., Baguñà, J., Riutort, M. and Ribera, C., 1996. First molecular evidence for the existence of a Tardigrada + Arthropoda clade. Molecular Biology and Evolution, 13, 76–84.

Giribet, G. and Edgecombe, G.D., 2017. Current understanding of Ecdysozoa and its internal phylogenetic relationships. Integrative and Comparative Biology, 57, 455–466.

Goeze, J.A.E., 1773. Über den kleinen Wasserbär. In Bonnet, K. (Ed), Herrn Karl Bonnets Abhandlungen aus der Insektologie, 367–375. Gebauer, Halle.

Greven, H., 1976. Some ultrastructural observations on the midgut epithelium of *Isohypsibius augusti* (Murray, 1907) (Eutardigrada). Cell & Tissue Research, 166, 339–351.

Greven, H., 1982. Homologues or analogues? A survey of some structural patterns in Tardigrada. In Nelson, D.R. (Ed), Proceedings of the third international symposium on Tardigrada, 55–76. East Tennessee State University Press, Johnson City, TN.

Greven, H. and Schüttler, L., 2001. How to crawl and dehydrate on moss. Zoologischer Anzeiger, 240, 341–344.

Greven, H., 2007. Comments on the eyes of tardigrades. Arthropod Structure and Development, 36, 401–407.

Gross, V., Miller, W.R. and Hochberg, R., 2014. A new tardigrade, *Mutaparadoxipus duodigifinis* gen. nov., sp. nov. (Heterotardigrada: Arthrotardigrada), from the Southeastern United States. Zootaxa, 3835, 263–272.

Gross, V. and Mayer, G., 2015. Neural development in the tardigrade *Hypsibius dujardini* based on anti-acetylated α-tubulin immunolabeling. EvoDevo, 6, 12.

Gross, V., Treffkorn, S. and Mayer, G., 2015. Tardigrada. In Wanninger, A. (Ed), Evolutionary developmental biology of invertebrates 3: Ecdysozoa I: Non-Tetraconata, 35–52. Springer, Vienna.

Gross, V., Minich, I. and Mayer, G., 2017. External morphogenesis of the tardigrade *Hypsibius dujardini* as revealed by scanning electron microscopy. Journal of Morphology, 278, 563–573.

Gross, V., Bährle, R. and Mayer, G., 2018. Detection of cell proliferation in adults of the water bear *Hypsibius dujardini* (Tardigrada) via incorporation of a thymidine analog. Tissue and Cell, 51, 77–83.

Gross, V. and Mayer, G., 2019. Cellular morphology of leg musculature in the water bear *Hypsibius exemplaris* (Tardigrada) unravels serial homologies. Royal Society Open Science, 6, 191159.

Gross, V., Müller, M., Hehn, L., Ferstl, S., Allner, S., Dierolf, M., Achterhold, K., Mayer, G. and Pfeiffer, F., 2019a. X-ray imaging of a water bear offers a new look at tardigrade internal anatomy. Zoological Letters, 5, 14.

Gross, V., Treffkorn, S., Reichelt, J., Epple, L., Lüter, C. and Mayer, G., 2019b. Miniaturization of tardigrades (water bears): morphological and genomic perspectives. Arthropod Structure and Development, 48, 12–19.

Grothman, G.T., Johansson, C., Chilton, G., Kagoshima, H., Tsujimoto, M. and Suzuki, A.C., 2017. Gilbert Rahm and the status of Mesotardigrada Rahm, 1937. Zoological Science, 34, 5–10.

Guidetti, R. and Bertolani, R., 2005. Tardigrade taxonomy: an updated check list of the taxa and a list of characters for their identification. Zootaxa, 845, 1–46.

Guidetti, R., Altiero, T. and Rebecchi, L., 2011. On dormancy strategies in tardigrades. Journal of Insect Physiology, 57, 567–576.

Guil, N. and Giribet, G., 2011. A comprehensive molecular phylogeny of tardigrades — adding genes and taxa to a poorly resolved phylum-level phylogeny. Cladistics, 27, 1–29.

Halberg, K.A., Persson, D., Møbjerg, N., Wanninger, A. and Kristensen, R.M., 2009. Myoanatomy of the marine tardigrade *Halobiotus crispae* (Eutardigrada: Hypsibiidae). Journal of Morphology, 270, 996–1013.

Hansen, J.G., Kristensen, R.M. and Jørgensen, A., 2012. The armoured marine tardigrades (Arthrotardigrada, Tardigrada). Scientia Danica Series B Biologica, 2, 1–91.

Hashimoto, T., Horikawa, D.D., Saito, Y., Kuwahara, H., Kozuka-Hata, H., Shin-I, T., Minakuchi, Y., Ohishi, K., Motoyama, A., Aizu, T., Enomoto, A., Kondo, K., Tanaka, S., Hara, Y., Koshikawa, S., Sagara, H., Miura, T., Yokobori, S.-I., Miyagawa, K., Suzuki, Y., Kubo, T., Oyama, M., Kohara, Y., Fujiyama, A., Arakawa, K., Katayama, T., Toyoda, A. and Kunieda, T., 2016. Extremotolerant tardigrade genome and improved radiotolerance of human cultured cells by tardigrade-unique protein. Nature Communications, 7, 12808.

Hengherr, S. and Schill, R.O., 2018. Environmental adaptations: cryobiosis. In: Schill, R.O. (Ed), Water Bears: The Biology of Tardigrades, vol. 2, 295–310. Springer Nature Switzerland, Cham, Switzerland.

Hering, L., Bouameur, J.-E., Reichelt, J., Magin, T.M. and Mayer, G., 2016. Novel origin of lamin-derived cytoplasmic intermediate filaments in tardigrades. eLife, 5, e11117.

Hyra, M., Poprawa, I., Włodarczyk, A., Student, S., Sonakowska, L., Kszuk-Jendrysik, M. and Rost-Roszkowska, M.M., 2016. Ultrastructural changes in the midgut epithelium of *Hypsibius dujardini* (Doyère, 1840) (Tardigrada, Eutardigrada, Hypsibiidae) in relation to oogenesis. Zoological Journal of the Linnean Society, 178, 897–906.

Jönsson, K.I., Levine, E.B., Wojcik, A., Haghdoost, S. and Harms-Ringdahl, M., 2018. Environmental adaptations: radiation tolerance. In: Schill, R.O. (Ed), Water Bears: The Biology of Tardigrades, vol. 2, 311–330. Springer Nature Switzerland, Cham, Switzerland.

Kaczmarek, Ł., Bartels, P.J., Roszkowska, M. and Nelson, D.R., 2015. The zoogeography of marine Tardigrada. Zootaxa, 4037, 1–189.

Kaczmarek, Ł. and Michalczyk, Ł., 2017. The *Macrobiotus hufelandi* group (Tardigrada) revisited. Zootaxa, 4363, 101–123.

Kamilari, M., Jørgensen, A., Schiøtt, M. and Møbjerg, N., 2019. Comparative transcriptomics suggest unique molecular adaptations within tardigrade lineages. BMC Genomics, 20, 607.

Kondo, K., Kubo, T. and Kunieda, T., 2015. Suggested involvement of PP1/PP2A activity and *de novo* gene expression in anhydrobiotic survival in a tardigrade, *Hypsibius dujardini*, by chemical genetic approach. PLoS One, 10, e0144803.

Kondo, K., Mori, M., Tomita, M. and Arakawa, K., 2019. AMPK activity is required for the induction of anhydrobiosis in a tardigrade *Hypsibius exemplaris*, and its potential up-regulator is PP2A. Genes to Cells, 00, 1–13.

Koutsovoulos, G., Kumar, S., Laetsch, D.R., Stevens, L., Daub, J., Conlon, C., Maroon, H., Thomas, F., Aboobaker, A.A. and Blaxter, M., 2016. No evidence for extensive horizontal gene transfer in the genome of the tardigrade *Hypsibius dujardini*. Proceedings of the National Academy of Sciences of the United States of America, 113, 5053–5058.

Kristensen, R.M., 1977. On the marine genus *Styraconyx* (Tardigrada, Heterotardigrada, Halechiniscidae) with description of a new species from a warm spring on Disco Island, West Greenland. Astarte, 10, 87–91.

Kristensen, R.M., 1978. On the fine structure of *Batillipes noerrevangi*, Kristensen 1976. 1. Tegument and moulting cycle. Zoologischer Anzeiger, 197, 129–150.

Kristensen, R.M., 1979. On the fine structure of *Batillipes noerrevangi* Kristensen, 1978 (Heterotardigrada). 3. Spermiogenesis. Zeszyty Naukowe Uniwersytetu Jagiellońskiego Prace Zoologiczne, 25, 97–105.

Kristensen, R.M., 1982. The first record of cyclomorphosis in Tardigrada based on a new genus and species from Arctic meiobenthos. Journal of Zoological Systematics and Evolutionary Research, 20, 249–270.

Ma, X., Hou, X. and Bergström, J., 2009. Morphology of *Luolishania longicruris* (Lower Cambrian, Chengjiang Lagerstätte, SW China) and the phylogenetic relationships within lobopodians. Arthropod Structure & Development, 38, 271–291.

Maas, A. and Waloszek, D., 2001. Cambrian derivatives of the early arthropod stem lineage, pentastomids, tardigrades and lobopodians – an 'Orsten' perspective. Zoologischer Anzeiger, 240, 451–459.

Marchioro, T., Rebecchi, L., Cesari, M., Hansen, J.G., Viotti, G. and Guidetti, R., 2013. Somatic musculature of Tardigrada: phylogenetic signal and metameric patterns. Zoological Journal of the Linnean Society, 169, 580–603.

Marcus, E., 1927. Zur Anatomie und Ökologie mariner Tardigraden. Zoologische Jahrbuecher Abteilung fuer Systematik Oekologie und Geographie der Tiere, 53, 487–588.

Marcus, E., 1928a. Zur vergleichenden Anatomie und Histologie der Tardigraden. Zoologische Jahrbuecher Abteilung fuer Allgemeine Zoologie und Physiologie der Tiere, 45, 99–158.

Marcus, E., 1928b. Zur Ökologie und Physiologie der Tardigraden. Zoologische Jahrbuecher Abteilung fuer Allgemeine Zoologie und Physiologie der Tiere, 44, 323–370.

Martin, C., Gross, V., Hering, L., Tepper, B., Jahn, H., Oliveira, I.S., Stevenson, P.A. and Mayer, G., 2017. The nervous and visual systems of onychophorans and tardigrades: learning about arthropod evolution from their closest relatives. Journal of Comparative Physiology A, 203, 565–590.

Mayer, G., Kauschke, S., Rüdiger, J. and Stevenson, P.A., 2013a. Neural markers reveal a one-segmented head in tardigrades (water bears). PLoS One, 8, e59090.

Mayer, G., Martin, C., Rüdiger, J., Kauschke, S., Stevenson, P.A., Poprawa, I., Hohberg, K., Schill, R.O., Pflüger, H.-J. and Schlegel, M., 2013b. Selective neuronal staining in tardigrades and onychophorans provides insights into the evolution of segmental ganglia in panarthropods. BMC Evolutionary Biology, 13, 230.

Møbjerg, N. and Dahl, C., 1996. Studies on the morphology and ultrastructure of the Malpighian tubules of *Halobiotus crispae* Kristensen, 1982 (Eutardigrada). Zoological Journal of the Linnean Society, 116, 85–99.

Møbjerg, N., Halberg, K.A., Jørgensen, A., Persson, D., Bjørn, M., Ramløv, H. and Kristensen, R.M., 2011. Survival in extreme environments – on the current knowledge of adaptations in tardigrades. Acta Physiologica, 202, 409–420.

Møbjerg, N., Jørgensen, A., Kristensen, R.M. and Neves, R.C., 2018. Morphology and functional anatomy. In: Schill, R.O. (Ed), Water Bears: The Biology of Tardigrades, vol. 2, 57–94. Springer Nature Switzerland, Cham, Switzerland.

Morek, W., Stec, D., Gąsiorek, P., Surmacz and Michalczyk, Ł., 2018. *Milnesium tardigradum* Doyère, 1840: the first integrative study of interpopulation variability in a tardigrade species. Journal of Zoological Systematics and Evolutionary Research, 1–23.

Müller, O.F., 1785. Von dem Bärthierchen. In: Füßly, J.C. (Ed), Archiv der Insectengeschichte, 25–31. Zürich.

Nelson, D.R., Guidetti, R. and Rebecchi, L., 2015. Phylum Tardigrada. In: Thorp, J.H. and Rogers, C. (Eds), Ecology and general biology: Thorp and Covich's freshwater invertebrates, 347–380. Academic Press, Cambridge.

Pelzer, B., Dastych, H. and Greven, H., 2007. The osmoregulatory/excretory organs of the glacier-dwelling eutardigrade *Hypsibius klebelsbergi* Mihelčič, 1959 (Tardigrada). Mitteilungen aus dem Hamburgischen Zoologischen Museum und Institut, 104, 61–72.

Perry, E.S., Rawson, P., Ameral, N.J., Miller, W.R. and Miller, J.D., 2018. *Echiniscoides rugostellatus* a new marine tardigrade from Washington, U.S.A. (Heterotardigrada: Echiniscoidea: Echiniscoididae: Echiniscoidinae). Proceedings of the Biological Society of Washington, 131, 182–193.

Plate, L., 1889. Beiträge zur Naturgeschichte der Tardigraden. Zoologische Jahrbücher, Abteilung für Anatomie und Ontogenie der Tiere, 3, 487–550.

Poprawa, I., Hyra, M. and Rost-Roszkowska, M.M., 2015. Germ cell cluster organization and oogenesis in the tardigrade *Dactylobiotus parthenogeneticus* Bertolani, 1982 (Eutardigrada, Murrayidae). Protoplasma, 252, 1019–1029.

Rahm, G., 1937. A new ordo of tardigrades from the hot springs of Japan (Furu-Yu section, Unzen). Annotationes Zoologicae Japonenses, 16, 345–352.

Raineri, M., 1985. Histochemical investigations of Tardigrada. 2. Alkaline phosphatase (ALP) and aminopeptidase (AMP) in the alimentary apparatus of Eutardigrada. Monitore Zoologico Italiano, 19, 47–67.

Rebecchi, L., Altiero, T. and Bertolani, R., 2002. Banding techniques on tardigrade chromosomes: the karyotype of *Macrobiotus richtersi* (Eutardigrada, Macrobiotidae). Chromosome Research, 10, 437–443.

Renaud-Mornant, J. and Deroux, G., 1976. *Halechiniscus greveni* n. sp. tardigrade marin nouveau de Roscoff (Arthrotardigrada). Cahiers de Biologie Marine, 17, 131–137.

Richters, F., 1926. Tardigrada. In: Kükenthal, W. and Krumbach, T. (Eds), Handbuch der Zoologie, vol. 3, 1–68, Walter de Gruyter & Co., Berlin and Leipzig.

Schill, R.O. and Hengherr, S., 2018. Environmental adaptations: desiccation tolerance. In: Schill, R.O. (Ed), Water Bears: The Biology of Tardigrades, vol. 2, 273–294. Springer Nature Switzerland, Cham, Switzerland.

Schmidt-Rhaesa, A. and Kulessa, J., 2007. Muscular architecture of *Milnesium tardigradum* and *Hypsibius* sp. (Eutardigrada, Tardigrada) with some data on *Ramazottius oberhaeuseri*. Zoomorphology, 126, 265–281.

Schulze, C. and Schmidt-Rhaesa, A., 2011. Organisation of the musculature of *Batillipes pennaki* (Arthrotardigrada, Tardigrada). Meiofauna Marina, 19, 195–207.

Shaw, K., 1974. The fine structure of muscle cells and their attachments in the tardigrade *Macrobiotus hufelandi*. Tissue and Cell, 6, 431–445.

Smith, F.W., Boothby, T.C., Giovannini, I., Rebecchi, L., Jockusch, E.L. and Goldstein, B., 2016. The compact body plan of tardigrades evolved by the loss of a large body region. Current Biology, 26, 224–229.

Smith, F.W. and Goldstein, B., 2017. Segmentation in Tardigrada and diversification of segmental patterns in Panarthropoda. Arthropod Structure and Development, 46, 328–340.

Smith, M.R. and Ortega-Hernández, J., 2014. *Hallucigenia*'s onychophoran-like claws and the case for Tactopoda. Nature, 514, 363–366.

Spallanzani, L., 1776. Il Tardigrado, le Anguilline delle tegole, e quelle del grano rachitico. Opuscoli di fisica animale, e vegetabile, vol. 2, 222–253. Modena.

Sugiura, K., Minato, H., Suzuki, A.C., Arakawa, K., Kunieda, T. and Matsumoto, M., 2019. Comparison of sexual reproductive behaviors in two species of Macrobiotidae (Tardigrada: Eutardigrada). Zoological Science, 36, 120–127.

Tsujimoto, M., Imura, S. and Kanda, H., 2016. Recovery and reproduction of an Antarctic tardigrade retrieved from a moss sample frozen for over 30 years. Cryobiology, 72, 78–81.

von Erlanger, R., 1895. Zur Morphologie und Embryologie eines Tardigraden (*Macrobiotus macronyx* Duj.). Biologisches Centralblatt, 15, 772–777.

Walz, B., 1974. The fine structure of somatic muscles of Tardigrada. Cell and Tissue Research, 149, 81–89.

Walz, B., 1978. Electron microscopic investigation of cephalic sense organs of the Tardigrade *Macrobiotus hufelandi* C.A.S. Schultze. Zoomorphologie, 89, 1–19.

Węglarska, B., 1987. Studies on the excretory system of *Isohypsibius granulifer* Thulin (Eutardigrada). In: Bertolani, R. (Ed), Biology of Tardigrades, selected symposia and monographs, 15–24. U.Z.I. Mucchi, Modena.

Wright, J.C., 1989. Desiccation tolerance and water-retentive mechanisms in tardigrades. Journal of Experimental Biology, 142, 267–292.

Yamaguchi, A., Tanaka, S., Yamaguchi, S., Kuwahara, H., Takamura, C., Imajoh-Ohmi, S., Horikawa, D.D., Toyoda, A., Katayama, T., Arakawa, K., Fujiyama, A., Kubo, T. and Kunieda, T., 2012. Two novel heat-soluble protein families abundantly expressed in an anhydrobiotic tardigrade. PLoS One, 7, e44209.

Yoshida, Y., Koutsovoulos, G., Laetsch, D.R., Stevens, L., Kumar, S., Horikawa, D.D., Ishino, K., Komine, S., Kunieda, T., Tomita, M., Blaxter, M. and Arakawa, K., 2017. Comparative genomics of the tardigrades *Hypsibius dujardini* and *Ramazzottius varieornatus*. PLoS Biology, 15, e2002266.

Zantke, J., Wolff, C. and Scholtz, G., 2008. Three-dimensional reconstruction of the central nervous system of *Macrobiotus hufelandi* (Eutardigrada, Parachela): implications for the phylogenetic position of Tardigrada. Zoomorphology, 127, 21–36.

CHAPTER
33

MINOR ECDYSOZOAN PHYLA NEMATOMORPHA, PRIAPULIDA, KINORHYNCHA, LORICIFERA

Andreas Schmidt-Rhaesa

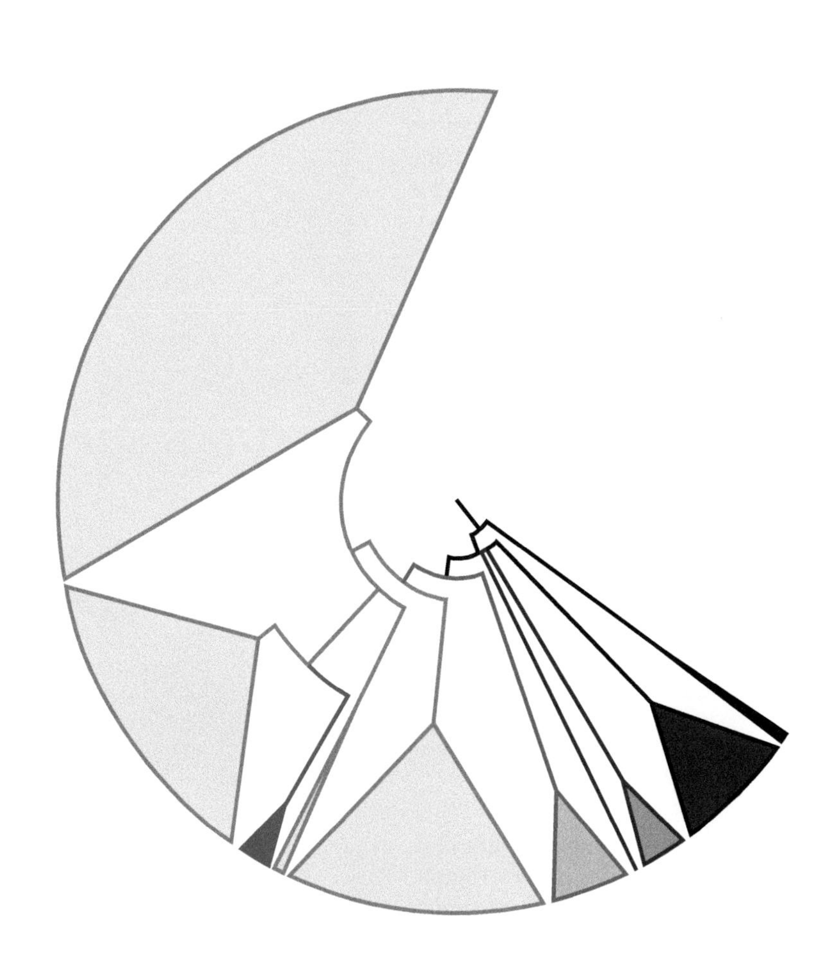

CONTENTS

Synapomorphies

Priapulida
- scalids in 25 longitudinal rows
- pharyngeal teeth

Kinorhyncha
- trunk divided into 11 segments (– zonites)
- neck region with placids which function as closing apparatus
- mouth cone with oral styly

Loricifera
- adult with lorica
- long and thin scalids
- larva (Higgins larva) with paired toes

Nematomorpha
- cuticle in adults with layers of thick fibers
- characteristic larva with rings of hooks and stylets
- parasitic life style of adult worms

General Characteristics

About 660 species (Nematomorpha: 360, Priapulida: 21, Kinorhyncha: 250, Loricifera: 30)

Microscopic to macroscopic animals with bilateral symmetry, marine (Priapulida: Kinorhyncha, Loricifera) or parasitic (Nematomorpha)

No commercial value

Ecological importance limited

HISTORY OF TAXONOMY AND CLASSIFICATION

Nematomorphs and macroscopic priapulids have been known for some time. Free-living nematomorphs were for a while combined with mermithids, which are large nematodes with a similar external appearance and life style. Parasitic nematomorphs in the past have been regarded as intestinal worms, together with other parasitic worms, predominantly nematodes. Priapulids were originally lumped together with sipunculids and echiurids, as Gephyrea. The smaller species of priapulids as well as kinorhynchs and loriciferans were discovered after microscopic techniques had developed and after the microscopic life in marine sediments (called meiofauna) had come into focus. Loriciferans, among the smallest metazoan animals, were described as late as 1983 by Reinhardt M. Kristensen.

Nematomorpha, Priapulida, Kinorhyncha, and Loricifera were included into a paraphyletic assemblage under different names (Nemathelminthes, Aschelminthes, and Pseudocoelomata) and have in the past been associated with other taxa (Nematoda, Rotifera, Acanthocephala, Gastrotricha, and others). Molecular and morphological analyses hypothesize that this assemblage falls into two clades named Cycloneuralia (Nematoda, Nematomorpha, Priapulida, Kinorhyncha, and Loricifera) and Gnathifera (Gnathostomulida, Rotifera, and Acanthocephala). Cycloneuralia, although seemingly well supported by a characteristic type of brain, appears to be not monophyletic, as one clade, Nematoida (= Nematoda + Nematomorpha), may be more closely related to arthropods. The other clade is called Scalidophora (= Priapulida + Kinorhyncha + Loricifera). Scalidophora, Nematoida, and Arthropoda are included in the Ecdysozoa.

ANATOMY, HISTOLOGY, AND MORPHOLOGY

Priapulids, kinorhynchs, and loriciferans have a body divided into a trunk and an introvert that can be withdrawn into the trunk (**Figure 33.1**). A neck region may additionally be present. The introvert is covered with cuticular scalids,

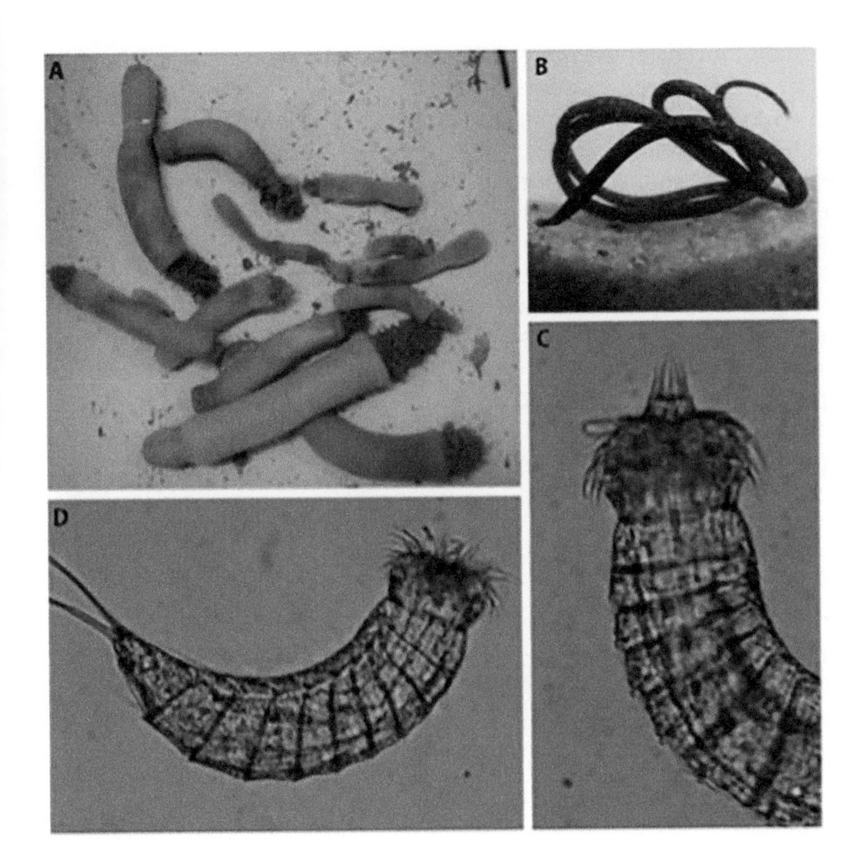

Figure 33.1 (A) Specimens of *Priapulus caudatus* (Priapulida) in different sizes (see 1 cent coin as comparison). The introvert is extended to different degrees and the caudal appendage is well visible. (B) Female specimen of *Chordodes* sp. (Nematomorpha), the anterior end is to the right, the posterior to the left. (C, D) Undetermined kinorhynch. The introvert is maximally everted in (C), showing the mouth cone with oral styly.

which range from small, cone-like structures (Priapulida) to longer, bristle-like structures (Kinorhyncha and Loricifera). Each scalid includes receptor cells and therefore has, besides locomotory function, a sensing function.

In priapulids, scalids are arranged in 25 longitudinal rows on the introvert. At the posterior end of priapulids, there may be an appendage with respiratory function or, in the genus *Tubiluchus*, a long muscular tail. Kinorhynchs have a segmented trunk, cuticle, and internal musculature with 11 separate trunk segments. Scalids are longer than those in priapulids, but also arranged in a pentaradial symmetry. Loriciferans have a rigid cuticle encircling the body much like an encasement, called the lorica. Their scalids are long, numerous, and different types can be distinguished. The introvert is usually withdrawn with the aid of retractor muscles and everted by contraction of the trunk musculature to increase the internal pressure of the body. Body cavities are absent in loriciferans, tiny or absent in kinorhynchs, and large in macroscopic priapulids. The body cavity is lined by an extracellular matrix, thus it is a primary body cavity. The brain in all three mentioned taxa is circular around the pharynx and has, with few exceptions, nuclei anterior and posterior of the neuropil. The ventral nerve cord can be paired (Kinorhyncha and Loricifera) or unpaired (Priapulida) and segmental ganglion-like swellings are present in kinorhynchs. The intestinal tract starts with a terminal mouth opening, followed by a muscular pharynx. In priapulids, the pharynx is covered by cuticular teeth. Following the pharynx is a straight midgut leading to a short hindgut and ending in a terminal or subterminal anus. Excretion is by protonephridia, these are of characteristic shape in each taxon. All taxa are dioecious. Gonads in nematomorphs extend over the entire body and join the intestinal system in a cloaca. Scalidophorans generally have sac-shaped gonads and no or few additional reproductive structures. The cellular epidermis is covered by a cuticle, which includes α-chitin and is molted several times during development.

In nematomorphs, the tiny larva somewhat resembles scalidophorans in having a trunk and an introvert, but as there are significant differences, these introverts may not be homologous. Hooks in nematomorph larvae are solid cuticular structures and do not include receptor cells. They are hexaradial in arrangement instead of pentaradial as in scalidophorans. Adult nematomorphs are long and slender, from several centimeters up to more than one meter in length with a diameter of maximally 2 mm. There is a marine genus, *Nectonema*, which can be recognized by having two longitudinal rows of cuticular bristles. Freshwater nematomorphs (= Gordiida) often have a sculptured cuticle, but no further appendages or extensions. The cuticle is molted once during development and changes from a permeable thin structure to a complex thick layer. Circular musculature is reduced and the longitudinal musculature forms an almost closed layer around the body circumference. In several species, the intestinal tract is partly reduced (in *Nectonema* the posterior part is reduced an in several gordiids the mouth opening). In gordiids, the posterior part of the intestine joins the gonoducts to form a cloaca. A pharynx is not present. Gonads are tube-like and produce large amounts of gametes.

REPRODUCTION AND DEVELOPMENT

With the exception of few loriciferans, all taxa treated here are gonochoristic and reproduce sexually, with the exception of one nematomorph species. This one species is *Paragordius obamai*, which was shown to reproduce by parthenogenesis. In general, there are still open questions concerning reproduction and early development in each taxon.

Priapulids develop with a larva that has a lorica, which is an encasement-like thickening of the cuticle. The lorica has a species-specific structure. There is one larval stage without a lorica between hatching and the first lorica-larva. The species *Meiopriapulus caudatus* reproduces directly and by vivipary.

The early development of kinorhynchs is still largely unknown. Eggs seem to be deposited individually and early postembryonic stages are unknown. What is known is that there are several juvenile stages, which develop by adding segments and some changes in morphology. The number of juvenile stages appears to be fixed at six.

Over the years, it was shown that loriciferans possess an astonishing variety of life cycles and different life stages across species. These include different larval types (Higgins larva, ghost larva, mega-larva), postlarval stages, and different reproductive strategies such as parthenogenesis, paedogenesis, and the alternation of sexual and asexual reproduction.

Freshwater nematomorphs (= Gordiida) possess a small larva which is morphologically very different from the adult. These larvae passively infect a variety of aquatic hosts, in which they encyst. Development proceeds when this intermediate host is eaten by a final host, which is usually a terrestrial insect or millipede. The details of this life cycle and especially the transmission of larvae are still partly unknown. In the final host, nematomorphs develop to vermiform juveniles which finally molt to adults. Marine nematomorphs (genus *Nectonema*) also possess a larval stage, but their life cycle is even less known than the freshwater one.

DISTRIBUTION AND ECOLOGY

Priapulids, kinorhynchs, and loriciferans are present in all oceans (**Figure 33.2**). Few species occur in the intertidal waters, most species are present in the subtidal waters down to the deep sea. Macroscopic priapulids burrow through the sediment, preferably mud, while the smaller species live in the interstitial system between the sand grains. Some loriciferans are known from anoxic parts of the ocean.

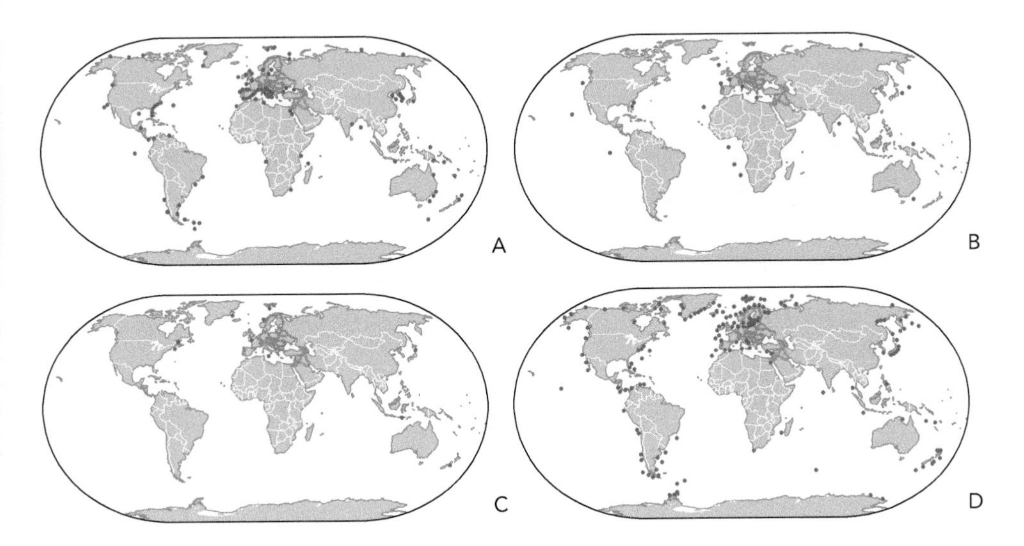

Figure 33.2 Distribution maps for Kinorhyncha (A), Lorcifera (B), marine Nematomorpha (C), and Priapulida (D).

Marine nematomorphs are known from some spots across the world oceans. Terrestrial species are found on all continents except Antarctica. They are parasites which leave their host at maturity. Terrestrial species drive their host toward water by still unknown mechanisms. Hosts may survive the infection, but have, depending on the size relationships between host and parasite, more or less atrophied gonads and fat bodies, which will influence their future fitness. Marine nematomorphs have been found most often in crabs caught around some spots in the North Atlantic and on very few occasions in other marine regions.

PHYSIOLOGY AND BEHAVIOR

Most conspicuous in the biology of (terrestrial) nematomorphs is that they are able to alter the behavior of their hosts. Although they develop in water and infect aquatic hosts first; they eventually parasitize terrestrial hosts (few exceptions are known). The transition is not fully understood yet. Parasitized hosts actively seek water and jump into it to release the mature nematomorphs. To date it was not possible to show what creates this change of behavior.

Little is known concerning the behavior or physiology of priapulids and kinorhynchs and almost nothing is known about loriciferans, which rarely are collected alive. The Baltic Sea species *Halicryptus spinulosus* (Priapulida) is known to tolerate low oxygen concentrations and some sulfide concentrations. Some loriciferans were found to live in completely anoxic regions of the Mediterranean Sea and probably the lack of mitochondria is one adaptation to this.

GENETICS AND GENOMICS

The taxa treated in this chapter have been represented in genetic studies with only few species per taxon. Genomic data are present for *Priapulus caudatus* (Priapulida) and transcriptomic data are available for the priapulids *H. spinulosus* and *Tubiluchus* cf. *corallicola*, the kinorhynch *Pycnophyes kielensis*, the loriciferan *Armorloricus elegans*, and the nematomorph *Paragordius varius*. Single gene sequences are available for a broader range of species, most of which are 18SrDNA sequences.

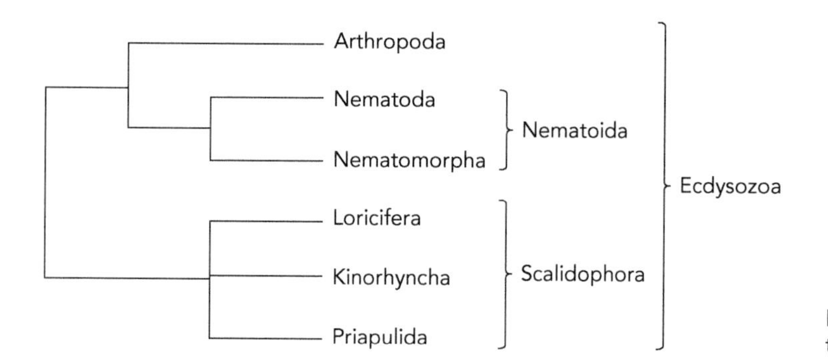

Figure 33.3 Phylogenetic scheme for the taxa discussed in this chapter.

POSITION IN THE ToL

Molecular analyses as well as the presence of a cuticle with α-chitin and a development pattern including molting of this cuticle, speak for a closer relationship of priapulids, kinorhynchs, loriciferans and nematomorphs with arthropods in a higher category called Ecdysozoa (**Figure 33.3**). Though the shape of the brain argues for a monophyletic arrangement of these groups uniting priapulids, kinorhynchs, loriciferans, and nematomorphs with nematodes (= taxon Cycloneuralia), this taxon is not supported in molecular analyses and instead nematodes and nematomorphs (= taxon Nematoida) are more closely related to arthropods. Priapulids, kinorhynchs, and loriciferans are united in a taxon Scalidophora. Important autapomorphies of Scalidophora are a division of the body into trunk and eversible introvert, and the presence of scalids on the introvert. Some authors regard nematomorphs as closely related to scalidophorans, but besides the molecular data a number of arguments favor a closer relationship between Nematoda and Nematomorpha (e.g., the long and thin body shape, presence of dorsal and ventral epidermal cords, absence of circular musculature and protonephridia, and more.

DATABASES AND COLLECTIONS

Nematomorphs and scalidophorans are spread over collections of the entire world. The most comprehensive database for the marine species is the World Register of Marine Species (WoRMS) (www.marinespecies.org), for European terrestrial nematomorphs it is Fauna Europaea (www.fauna-eu.org). The community of researchers working on priapulids, kinorhynchs, and loriciferans meets every 3 years for a "Scalidophora meeting".

CONCLUSION

All four taxa presented here certainly belong to the very understudied ones, a fate shared with several animals of little obvious economic or medical importance. Nevertheless, the study of such animals is of great importance to understand all parts of the big picture of nature. And it is fascinating, because still many open questions await answers and new species, structures and other topics await discovery.

REFERENCES

Adrianov, A.V., Malakhov, V.V., 1999. Cephalorhyncha of the World Ocean. KMK Scientific Press, Moscow.

Danovaro, R., Dell'Anno, A., Pusceddu, A., Gambi, C., Heiner, I., Kristensen, R.M., 2010. The first metazoa living in permanently anoxic conditions. BMC Biology, 8, 30.

Dunn, C.W., Hejnol, A., Matus, D.Q., Pang, K., Browne, W.E., Smith, S.A., Seaver, E., Rouse, G.W., Obst, M., Edgecombe. G.D., Sørensen, M.V., Haddock, S.H.D., Schmidt-Rhaesa, A., Okusu, A., Kristensen, R., Wheeler, W.C., Martindale, M.Q., Giribet, G., 2008. Broad phylogenomic sampling improves resolution of the animal tree of life. Nature, 452, 754–749.

Hanelt, B., Thomas, F., Schmidt-Rhaesa, A., 2005. Biology of the phylum Nematomorpha. Advances in Parasitology, 59, 243–305.

Hanelt, B., Bolek, M.G., Schmidt-Rhaesa, A., 2012. Going solo: discovery of the first parthenogenetic gordiid (Nematomorpha: Gordiida). PLoS One, 7: e34472. doi: 10.1371/journal.pone.0034472.

Hanelt, B., Schmidt-Rhaesa, A., Bolek, M.G., 2015. Cryptic spesies of hairworm parasites revealed by molecular data and crowdsourcing of specimen collections. Molecular Phylogenetics and Evolution, 82, 211–218.

Kristensen, R.M., 1983. Loricifera, a new phylum with Aschelminthes characters from the meiobenthos. Zeitschrift für Zoologische Systematik und Evolutionsforschung, 21, 163–180.

Kristensen, R.M., 1991. Loricifera. In: Harrison, F.W., Ruppert, E.E. (Eds.), Microscopic anatomy of invertebrates, Volume 4. Aschelminthes. Wiley-Liss, New York, pp. 351–375.

Kristensen, R.M., Higgins, R.P., 1991. Kinorhyncha, In: Harrison, F.W., Ruppert, E.E. (Eds.), Microscopic anatomy of invertebrates, vol. 4: Aschelminthes. Wiley-Liss, New York, pp. 377–404.

Laumer, C.E., Bekkouche, N., Kerbl, A., Goetz, F., Neves, R.C., Sørensen, M., Kristensen, R.M., Hejnol, A., Dunn, C.W., Giribet, G., Worsaae, K., 2015. Spiralian phylogeny informs the evolution of microscopic lineages. Current Biology, 25, 2000–2006.

Neves, R.C., Reichert, H., Sørensen, M.V., Kristensen, R.M., 2016. Systematics of phylum Loricifera: identification keys of families, genera and species. Zoologischer Anzeiger, 265, 141–160.

Schmidt-Rhaesa, A. (Ed.), 2013. Handbook of Zoology: Gastrotricha, Cycloneuralia and Gnathifera. Volume 1: Nematomorpha, Priapulida, Kinorhyncha and Loricifera. De Gruyter, Berlin.

Schmidt-Rhaesa, A., Panpeng, S., Yamasaki, H., 2017. Two new species of *Tubiluchus* (Priapulida) from Japan. Zoologischer Anzeiger, 267, 155–167.

Sørensen, M., Pardos, F., 2008. Kinorhynch systematics and biology – an introduction to the study of kinorhynchs, inclusive identification keys to the genera. Meiofauna Marina, 16, 21–73.

Thomas, F., Schmidt-Rhaesa, A., Martin, G., Manu, C., Durand, P., Renaud, F., 2002. Do hairworms (Nematomorpha) manipulate the water-seeking behaviour of their terrestrial hosts? Journal of Evolutionary Biology, 15, 356–361.

Van der Land, J., 1970. Systematics, zoogeography, and ecology of the Priapulida. Zoologische Verhandelingen (Leiden), 112, 1–118.

PHYLUM XENOTURBELLIDA

CHAPTER 34

Hiroaki Nakano

Infrakingdom	Bilateria
Division	Bilateria or Deuterostomia
Phylum	Xenoturbellida

CONTENTS

General Characteristics

Six valid species, more species probably present

Bilateral animals, mostly found on the deep-sea floor

Simple body plan lacking anus, centralized nervous systems, or excretory systems

Likely to be sister group to the Acoelomorpha, forming the clade Xenacoelomorpha

Xenacoelomorpha has been suggested as a sister group to Nephrozoa or as a member of the deuterostomes

No commercial value

Ecological importance probably small

Synapomorphies

- ring and side furrows with possible sensory functions
- frontal pore and ventral glandular network
- statocyst with motile flagellated cells inside
- lack of reproductive organs such as gonads, gonopore, or copulatory organs
- dense string leading into a cup-shaped structure at the proximal end of the epidermal cilium

HISTORY OF TAXONOMY AND CLASSIFICATION

The oldest existing specimen of *Xenoturbella* was collected in 1878 from the Swedish west coast by August Wilhelm Malm, a Swedish zoologist. Although he did not make scientific descriptions of the animal, the specimen is stored and exhibited at the Gothenburg Natural History Museum. The first scientific report of the animal was in 1949 by Einar Westblad (Westblad 1949), based partly on the specimens collected by Karl Alfred Sixten Bock (both Swedish zoologists). Due to its morphology, the animal was regarded to be a member of the Turbellaria within the phylum Platyhelminthes, and thus named *Xenoturbella bocki* ("strange turbellarian"; greek "xeno" = strange, peculiar, with the specific name "*bocki*" named after Bock) (**Figure 34.1**).

X. bocki for long remained the single species of Xenoturbellida, until a second species, *X. westbladi* (specific name "*westbladi*" named after Westblad), was reported in 1999 (Israelsson 1999). However, haplotype network analyses using cytochrome-c-oxidase subunit I (COI) strongly suggested that it is a junior synonym of *X. bocki* (Rouse et al. 2016).

In 2016, four new species were reported from the Eastern Pacific: *X. monstrosa*, *X. churro*, *X. profunda*, and *X. hollandorum* (Rouse et al. 2016). These new findings revealed unknown diversity within the group, such as body size in some species reaching 20 cm and an organ called the ventral glandular network.

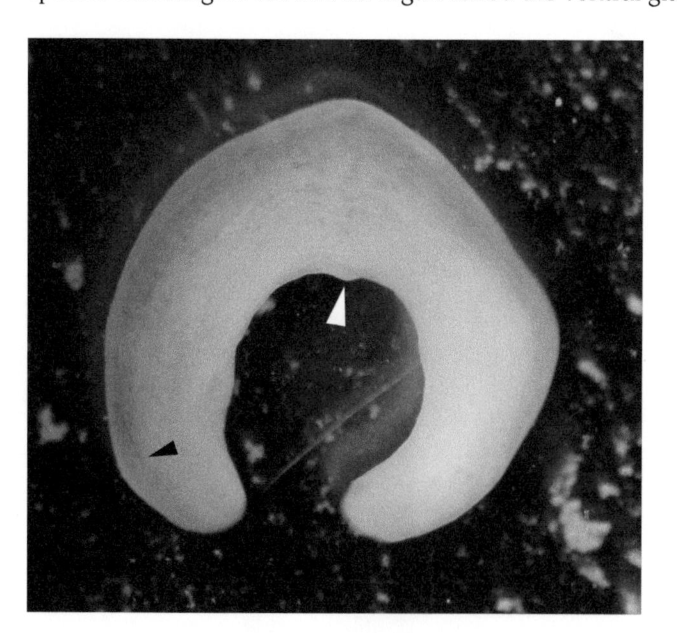

Figure 34.1 *Xenoturbella bocki*. This specimen, about 2.5 cm long, was collected in Gullmarsfjord on the west coast of Sweden. The anterior is to the lower left. White arrowhead: ring furrow, black arrowhead: side furrow.

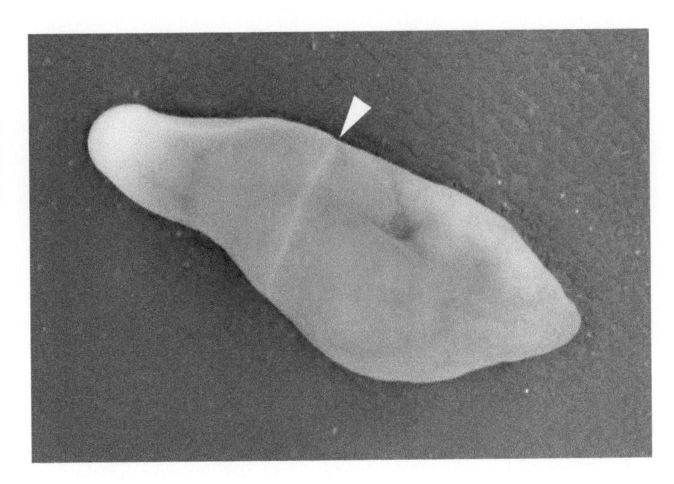

Figure 34.2 *Xenoturbella japonica*. This specimen, about 4 cm long, was collected along the coast of Izu Peninsula, Japan. The anterior end, to the upper left, is distinguishable with its white color. White arrowhead: ring furrow.

Subsequently, another species, *X. japonica*, was reported from the Japanese coast, bringing the species number to six (Nakano et al. 2017, 2018) (**Figure 34.2**). Studies on this species showed another previously unreported organ from *Xenoturbella*, the frontal pore, which was found to be also present in *X. bocki* and remained undiscovered for over 60 years.

ANATOMY, HISTOLOGY, AND MORPHOLOGY

Xenoturbella is a marine worm arranged in a bag like morphology with three layers (Westblad 1949) (**Figures 34.3**, **34.4**). The outer layer is composed of the epidermis, intra-epidermal nerve net, and the basal lamina. Ultrastructure of the epidermal cells and its ciliary structures have been observed in detail (Franzén and Afzelius 1987, Pedersen and Pedersen 1988, Lundin 1998). The nerve net is suggested to be uniformly distributed throughout the body and centralized nervous systems such as brains or nerve cords are lacking (Westblad 1949, Raikova et al. 2000, Stach et al. 2005). However, some concentrations of nerve cells are observed around the ring furrow, side furrow, and statocyst (Westblad 1949, Ehlers 1991, Israelsson 2007a, Nakano et al. 2017, Martín-Durán et al. 2018). The middle

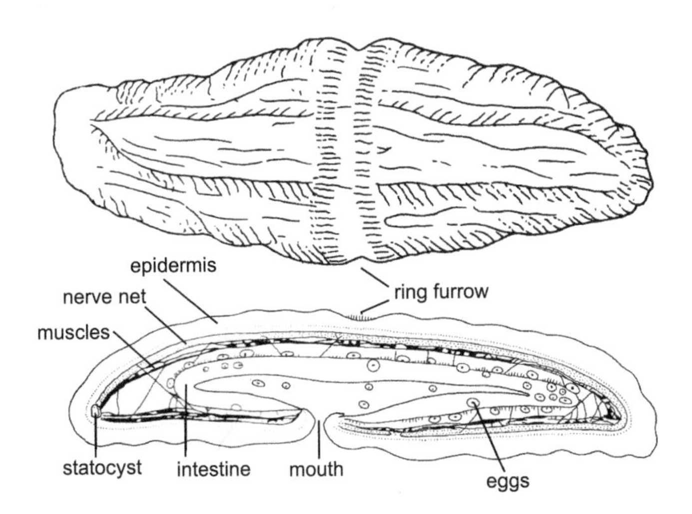

Figure 34.3 **General morphology**. The top drawing shows a dorsal view, and the bottom drawing shows a sagittal section of *X. bocki*. Anterior is to the left in both. Modified from Westblad, 1949.

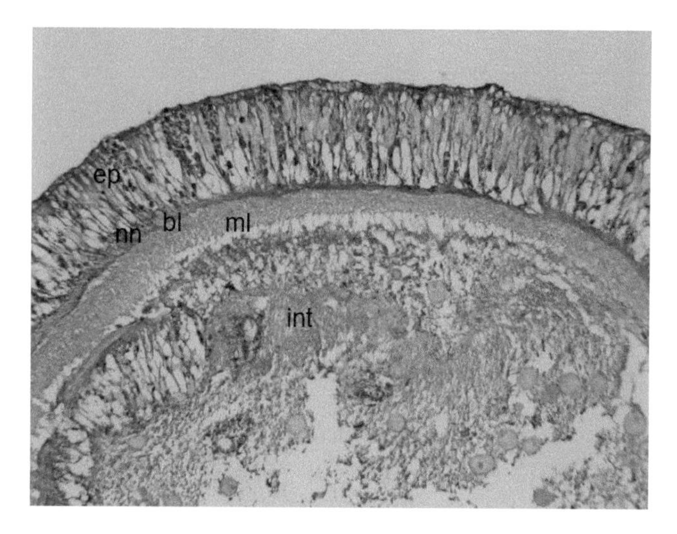

Figure 34.4 Internal morphology. This transverse section of the dorsal part of a *X. bocki* specimen shows the arrangement of the epidermis (ep), intraepidermal nerve net (nn), basal lamina (bl), muscle layer (ml), and the intestine (int).

layer is the muscle layer, with circular, longitudinal, and radial muscles. The inner layer is the intestine. The mouth opens at the midventral position, but anus is absent. Organized reproductive organs such as gonads have not been reported.

The ring furrow encircles the body at the middle like a belt, and the side furrow starts at the anterior tip and continues down both lateral sides (Figures 34.1–34.3). They have been suggested to be sensory organs based on their morphology, thickening of the nerve layer beneath these organs, and gene expression (Westblad 1949, Martín-Durán et al. 2018).

A statocyst is present near the anterior tip of the animal (Westblad 1949, Ehlers 1991, Israelsson 2007a, Nakano et al. 2017), with a concentration of the nerve net present around this organ (Figure 34.3). It is a vesicle with several motile flagellated cells inside the cavity. Its function has been suggested as a georeceptor, but endocrinal functions have also been suggested (Israelsson, 2007a).

A frontal pore opens at the anteroventral tip, ventral to the side furrow, and continues into the ventral glandular network (Nakano et al. 2017, Maeno et al. 2019) (**Figure 34.5**). This epidermal branching structure is externally visible in five species. It continues until near the posterior tip in *X. monstrosa*, *X. churro*,

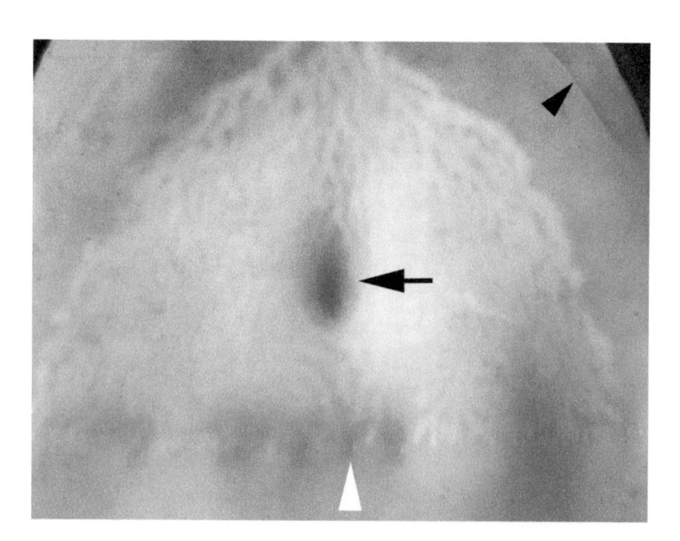

Figure 34.5 Ventral glandular network. The picture shows the anteroventral part of a *X. japonica* specimen, showing the conspicuous white branching structure. White arrowhead: ring furrow, black arrowhead: side furrow, black arrow: mouth.

and *X. profunda* (Rouse et al. 2016). In *X. hollandorum* and *X. japonica*, it stops at or anterior to the ring furrow (Rouse et al. 2016, Nakano et al. 2017). The structure is also present in *X. bocki*, but is not externally identifiable (Nakano et al. 2017).

REPRODUCTION AND DEVELOPMENT

Sexual reproduction

The breeding season of *X. bocki* is winter (Westblad 1949, Nakano et al. 2013, Nakano 2015). *X. bocki* is suggested to be hermaphoroditic (Westblad 1949), whereas *X. profunda* is reported as gonochoric (Rouse et al. 2016). *Xenoturbella* do not possess reproductive organs such as gonads or genital pore, and gametes are present at various sites within the body. The mode of sexual reproduction remains unobserved, but the sperm morphology of *X. bocki* suggests external fertilization (Obst et al. 2011) and eggs have been observed to be released from ruptures in the body wall (Nakano et al. 2013).

Sperm of *X. bocki* consists of a round head without a separate midpiece and a long 9 + 2 single flagellum (Obst et al. 2011). Eggs are about 205 μm in diameter with a transparent layer enclosing the egg (Nakano et al. 2013). At fertilization, the fertilization envelope forms between the egg and the transparent layer.

Development

Developmental stages have been reported for only *X. bocki*, and even in this species, only fragmented information is available. Embryos undergo holoblastic radial cleavage, and begin to rotate inside the fertilization envelope using cilia (Nakano et al. 2013, Nakano 2015). The oval-shaped swimming stage embryos, sometimes referred to as larvae, are uniformly ciliated (Nakano et al. 2013). The only externally observable organ is the apical tuft at the anterior end, with mouth, vestibule, anus, blastopore, or ciliary bands all absent (**Figure 34.6**). Concerning internal morphology, the embryos possess a thick epidermis consisting of a single layer of columnar cells, with the internal cavity mostly filled with spherical undifferentiated cells. Developing muscle cells and nerve cells are present between the epidermis and the undifferentiated cells. No digestive systems or coeloms are present. About five days after hatching, the embryos start to glide

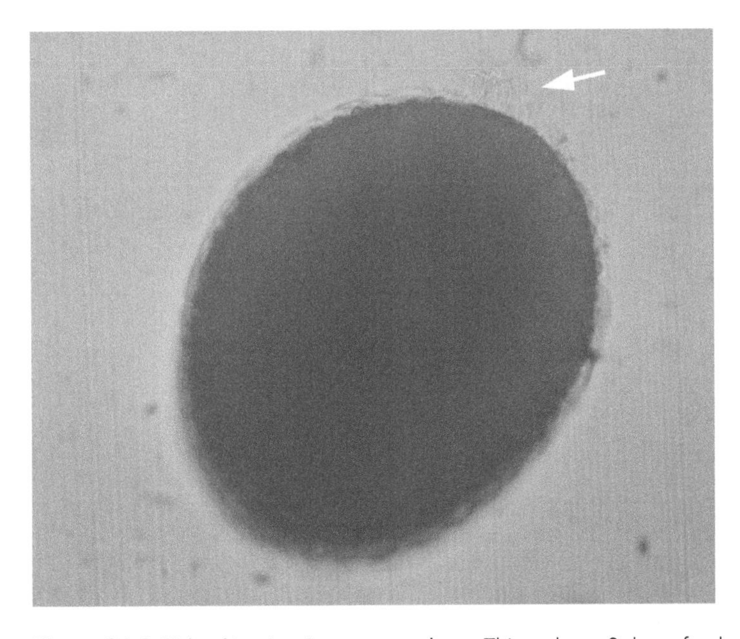

Figure 34.6 ***X. bocki*** **swimming stage embryo.** This embryo, 3 days after hatching, is uniformly ciliated and swims using the cilia. White arrow: apical tuft.

on the substrate and contract using muscles. Stages between cleavage and hatching, such as gastrulation, and the growth process following five days after hatching, such as the formation of the mouth and statocyst, remain unknown.

Vegetative reproduction

There have been no definitive reports of vegetative reproduction for *Xenoturbella*. Splitting of injured animals have been observed (Nakano 2015), but it remains to be seen if the two parts can regenerate into two whole animals.

DISTRIBUTION AND ECOLOGY

Xenoturbella bocki has been collected mainly from the Swedish west coast from the muddy sea bottom at about 50–200 m depth. There are also reports from the Norwegian and Scottish coasts (Westblad 1949, Israelsson 1999), suggesting a wide distribution in the North East Atlantic (**Figure 34.7**). *X. japonica* was collected at 380–560 m depth from the Western Pacific roughly 600 km apart (Figure 34.7), showing a wide distribution along the Japanese Pacific coast (Nakano et al. 2017, Maeno et al. 2019).

The four species from the Eastern Pacific were collected from depth of 631–3,700 m along the coasts of United States and Mexico (Rouse et al. 2016) (Figure 34.7). *X. monstrosa* was collected over 2,500 km apart, showing that some species have a very wide distribution. Despite repeated visits to the same site where numerous specimens were once collected, no additional *X. monstrosa* individuals were further found, suggesting that the habitat of *Xenoturbella* may change over time. *X. churro* was found just 30 cm from a *X. monstrosa* individual, showing that different *Xenoturbella* species can cohabitate at a same environment. Observations of these Eastern Pacific species in their natural habitat using ROV showed that they are often found near cold-water methane seeps, hydrothermal vents, or whale-falls, suggesting an ecological relationship with the dense organism communities at these sites.

All the habitats of *Xenoturbella* reported so far are either the deep sea or areas known for its biodiversity of deep-sea organisms at a relatively shallow depth (e.g., Gullmarsfjord, Sweden and the Japanese Pacific coast). Therefore, other species of *Xenoturbella* may inhabit the deep-sea floor and future faunal surveys of the deep sea may reveal undiscovered xenoturbellid species

Contamination of bivalve DNA has been reported from five of the six species (Bourlat et al. 2003, 2008, Rouse et al. 2016, Nakano et al. 2017), strongly suggesting that they feed on bivalves. Although no instances of feeding behavior have been reported, bivalve larva and sperm have been found from inside xenoturbellids (Israelsson 1999), supporting this hypothesis.

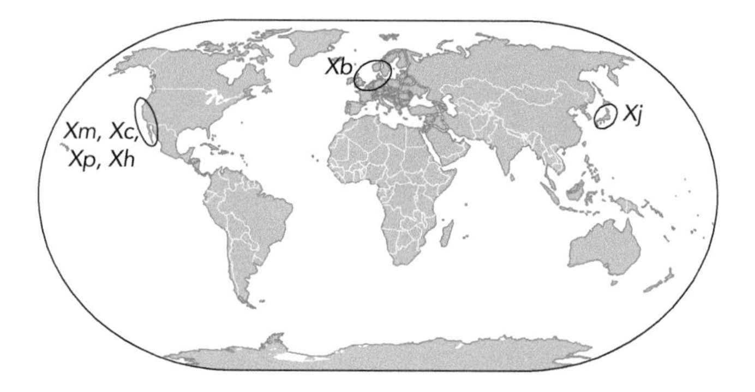

Figure 34.7 ***Xenoturbella* distribution.** Xenoturbellids have been reported from three regions around the world. None of the known species have been found from multiple regions. *Xb*: *Xenoturbella bocki*, *Xj*: *X. japonica*, *Xm*: *X. monstrosa*, *Xc*: *X. churro*, *Xp*: *X. profunda*, *Xh*: *X. hollandorum*.

BEHAVIOR AND PHYSIOLOGY

Xenoturbella glides upon a substrate using its cilia while excreting mucus from the frontal pore. Swimming has not been reported. It can roll up into a ball using internal musculature and can stay in that form for several months without feeding. *In situ* video recordings by ROV showed that *X. monstrosa* specimens are present on top of mud at the deep-sea floor (Rouse et al. 2016). However, *X. bocki* specimens in laboratory aquarium are capable of digging tunnels deeper than 15 cm, and it is possible that they live deeper down below the sea bottom, especially the larger Pacific species.

Little is known on *Xenoturbella* physiology. Two species of endosymbiotic bacteria and an infecting orthonectid species have been reported from xenoturbellids (Israelsson 2007b, Kjeldsen et al. 2010, Nakano and Miyazawa 2019), but their effects on the host animals remain unstudied.

GENETICS

Whole nuclear genome sequences are not publicly available for any xenoturbellid species, but there are ongoing genome sequencing projects. Transcriptomic analyses have been performed on some of the species, and have been used mainly for phylogenomic studies and identification of *Hox* genes and nervous system related genes (Bourlat et al. 2006, Philippe et al. 2011, Perea-Atienza et al. 2015, Gavilán et al. 2016, Cannon et al 2016, Rouse et al. 2016, Brauchle et al. 2018, Thiel et al. 2018). Mitochondrial genomes for all six species show the same gene order and direction, ranging from 15,210 to 15,404 bp (Perseke et al. 2007, Bourlat et al. 2009, Rouse et al. 2016, Nakano et al. 2017). Phylogenetic analyses using the mitochondrial genomes suggest that the six species can be divided into two clades (**Figure 34.8**). The first group consist of three species reported from the Eastern Pacific, *X. monstrosa*, *X. profunda*, and *X. churro*. These species are collected from depth of 1700–3700 m, and their body size are about 10–20 cm. The second group shows a wider distribution, consisting of *X. hollandorum* from the Eastern Pacific, *X. japonica* from the Western Pacific, and *X. bocki* from the Atlantic Ocean. These species are collected from shallower depth than the first group at about of 50–650 m, and their body size are smaller at about 2–6 cm.

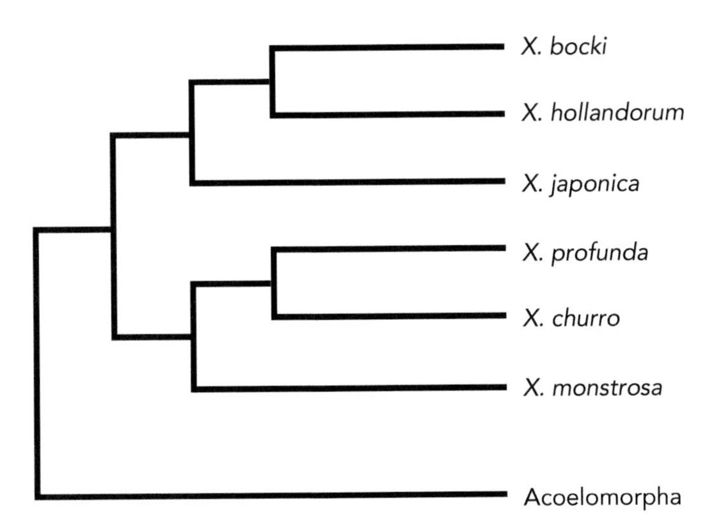

Figure 34.8 ***Xenoturbella* diversity.** The six species are divided into two clades based on phylogenetic analyses using mitochondrial genomes.

POSITION IN THE ToL

Most phylogenetic analyses suggest that *Xenoturbella* is a sister group to the Acoelomorpha, forming the clade Xenacoelomorpha (Hejnol et al. 2009, Philippe et al. 2011, Haszprunar 2015, Cannon et al. 2016, Robertson et al. 2017, Philippe et al. 2019). Analyses on their mitochondrial genome and some large scale phylogenomic studies suggest that Xenacoelomorpha is a member of the deuterostomes (Philippe et al. 2011, 2019, Robertson et al. 2017), whereas other large scale phylogenomic studies suggest that Xenacoelomorpha is a sister group to the Nephrozoa (all other extant bilaterians) (Hejnol et al. 2009, Cannon et al. 2016). The ongoing genome sequencing projects shall be useful for clarifying the positions of *Xenoturbella* and Xenacoelomorpha in the ToL.

DATABASES AND COLLECTIONS

A large collection of *X. bocki* samples is present at the Swedish Museum of Natural History, Stockholm, Sweden. Specimens of the four eastern Pacific species are mostly lodged at the Scripps Institution of Oceanography Benthic Invertebrate Collection, La Jolla, California, USA (Rouse et al. 2016), and those of *X. japonica* are deposited at the National Museum of Nature and Science, Tsukuba, Japan (Nakano et al. 2017). Information on *Xenoturbella* is present in online databases encompassing metazoans (e.g., World Register of Marine Species), but there are no public databases focusing on this animal at the moment.

CONCLUSION

Research on *Xenoturbella* has so far been focused on a single species, *X. bocki*. Further research on the other five species, as well as discovery of new species, shall provide new and exciting information concerning the evolution of this animal group. One of the major remaining questions for *Xenoturbella* is its phylogenetic position. If the currently more accepted theory with *Xenoturbella* (or Xenacoelomorpha) as the sister group to Nephrozoa is true, studies on this animal may uncover new insights on the morphology, physiology, life history, and genomes of early bilaterians and its evolution from a non-bilaterian ancestor. If *Xenoturbella* turns out to be a deuterostome, it will represent a unique situation in which a metazoan group underwent extreme secondary loss of significant organs such as coeloms and excretory organs with no extant species being parasitic, sessile, or microscopic. In this case, studies on this animal shall yield important insights into changes in morphology and genomes accompanying simplification events during its evolution.

REFERENCES

Bourlat, S.J., Nielsen, C., Lockyer, A.E., Littlewood, D.T.J., Telford, M.J., 2003. *Xenoturbella* is a deuterostome that eats molluscs. Nature, 424, 925–928.

Bourlat, S.J., Juliusdottir, T., Lowe, C.W., Freeman, R., Aronowicz, J., Kirschner, M., Lander, L.S., Thorndyke, M., Nakano, H., Kohn, A.B., Heyland, A., Moroz, L.L., Copley, R.R., Telford, M.J., 2006. Deuterostome phylogeny reveals monophyletic chordates and the new phylum Xenoturbellida. Nature, 444, 85–88.

Bourlat, S.J., Nakano, H., Åkerman, M., Telford, M.J., Thorndyke, M.C., Obst, M., 2008. Feeding ecology of *Xenoturbella bocki* (phylum Xenoturbellida) revealed by genetic barcoding. Molecular Ecology Resources, 8, 18–22.

Bourlat, S.J., Rota-Stabelli, O., Lanfear, R., Telford, M.J., 2009. The mitochondrial genome structure of *Xenoturbella bocki* (phylum Xenoturbellida) is ancestral within the deuterostomes. BMC Evolutionary Biology, 9,107.

Brauchle, M., Bilican, A., Eyer, C., Bailly, X., Martínez, P., Ladurner, P., Bruggmann, R., Sprecher, S.G., 2018. Xenacoelomorpha survey reveals that all 11 animal homeobox gene classes were present in the first bilaterians. Genome Biology and Evolution, 10, 2205–2217.

Cannon, J.T., Vellutini, B.C., Smith III, J., Ronquist, F., Jondelius, U., Hejnol A., 2016. Xenacoelomorpha is the sister group to Nephrozoa. Nature, 530, 89–93.

Ehlers, U., 1991. Comparative morphology of statocyst in Platyhelminthes and the Xenoturbellida. Hydrobiologia, 227, 263–271.

Franzén, A., Afzelius, B.A., 1987. The ciliated epidermis of *Xenoturbella bocki* (Platyhelminthes, Xenoturbellida) with some phylogenetic considerations. Zoologica Scripta, 16, 9–17.

Gavilán B., Perea-Atienza, E., Martínez, P., 2016. Xenacoelomorpha: A case of independent nervous system centralization? Philosophical Transactions of the Royal Society B, 371, 20150039.

Haszprunar, G., 2015. Review of data for a morphological look on Xenacoelomorpha (Bilateria incertae sedis). Organisms Diversity and Evolution, 16, 363–389.

Hejnol, A., Obst, M., Stamatakis, A., Ott, M., Rouse, G.W., Edgecombe, G.D., Martinez, P., Baguna, J., Bailly, X., Jondelius, U., Wiens, M., Muller, W.E.G., Seaver, E., Wheeler, W.C., Martindale, M.Q., Giribet, G., Dunn, C.W., 2009. Assessing the root of bilaterian animals with scalable phylogenomic methods. Proceedings of the Royal Society B, 276, 4261–4270.

Israelsson, O., 1999. New light on the enigmatic *Xenoturbella* (phylum uncertain): ontogeny and phylogeny. Proceedings of the Royal Society B, 266, 835–841.

Israelsson, O., 2007a. Ultrastructural aspects of the 'statocyst' of *Xenoturbella* (Deuterostomia) cast doubt on its function as a georeceptor. Tissue and Cell, 39,171–177.

Israelsson, O., 2007b. Chlamydial symbionts in the enigmatic *Xenoturbella* (Deuterostomia). Journal of Invertebrate Pathology, 96, 213–220.

Kjeldsen, K.U., Obst, M., Nakano, H., Funch, P., Schramm, A., 2010. Two types of endosymbiotic bacteria in the enigmatic marine worm *Xenoturbella bocki*. Applied and Environmental Microbiology, 76, 2657–2662.

Lundin, K., 1998. The epidermal ciliary rootlets of *Xenoturbella bocki* (Xenoturbellida) revisited: new support for a possible kinship with the Acoelomorpha (Platyhelminthes). Zoologica Scripta, 27, 263–270.

Maeno, A., Kohtsuka, H., Takatani, K., Nakano, H., 2019. Microfocus X-ray CT (microCT) imaging of *Actinia equina* (Cnidaria), *Harmothoe* sp. (Annelida), and *Xenoturbella japonica* (Xenacoelomorpha). Journal of Visualized Experiments, 150, e59161.

Martín-Durán, J.M., Pang, K., Børve, A., Lê, H.S., Furu, A., Cannon, J.T., Jondelius, U., Hejnol, A., 2018. Convergent evolution of bilaterian nerve cords. Nature, 553, 45–50.

Nakano, H., 2015. What is *Xenoturbella*? Zoological Letters, 1, 22.

Nakano, H., Lundin, K., Bourlat, S.J., Telford, M.J., Funch, P., Nyengaard, J.R., Obst, M., Thorndyke, M.C., 2013. *Xenoturbella bocki* exhibits direct development with similarities to Acoelomorpha. Nature Communications, 4, 1537.

Nakano, H., Miyazawa, H., Maeno, A., Shiroishi, T., Kakui, K., Koyanagi, R., Kanda, M., Satoh, N., Omori, A., Kohtsuka, H., 2017. A new species of *Xenoturbella* from the western Pacific Ocean and the evolution of *Xenoturbella*. BMC Evolutionary Biology, 17, 245.

Nakano, H., Miyazawa, H., Maeno, A., Shiroishi, T., Kakui, K., Koyanagi, R., Kanda, M., Satoh, N., Omori, A., Kohtsuka, H., 2018. Correction to: A new species of *Xenoturbella* from the western Pacific Ocean and the evolution of *Xenoturbella*. BMC Evolutionary Biology, 18, 83.

Nakano, H., Miyazawa, H., 2019. A new Species of Orthonectida that parasitizes *Xenoturbella bocki*: Implications for studies on *Xenoturbella*. The Biological Bulletin, 236, 66–73.

Obst, M., Nakano, H., Bourlat, S.J., Thorndyke, M.C., Telford, M.J., Nyengaard, J.R., Funch, P., 2011. Spermatozoon ultrastructure of *Xenoturbella bocki* (Westblad 1949). Acta Zoologica, 92, 109–115.

Pedersen, K.J., Pedersen, L.R., 1988. Ultrastructural observations on the epidermis of *Xenoturbella bocki* Westblad, 1949; with a discussion of epidermal cytoplasmic filament systems of invertebrates. Acta Zoologica (Stockholm), 69, 231–246.

Perea-Atienza, E., Gavilán, B., Chiodin, M., Abril, J.F., Hoff, K.J., Poustka, A.J., Martínez, P., 2015. The nervous system of Xenacoelomorpha: a genomic perspective. Journal of Experimental Biology, 218, 618–628.

Perseke, M., Hankeln, T., Weich, B., Fritzsch, G., Stadler, P., Israelsson, O., Bernhard, D., Schlegel, M., 2007. The mitochondrial DNA of *Xenoturbella bocki*: genomic architecture and phylogenetic analysis. Theory in Biosciences, 126, 35–42.

Philippe, H., Brinkmann, H., Copley, R.R., Moroz, L.L., Nakano, H., Poustka, A.J., Wallberg, A., Peterson, K.J., Telford, M.J., 2011. Acoelomorph flatworms are deuterostomes related to *Xenoturbella*. Nature, 470, 255–258.

Philippe, H., Poustka, A.J., Chiodin, M., Hoff, K.J., Dessimoz, C., Tomiczek, B., Schiffer, P.H., Muller, S., Domman, D., Horn, M., Kuhl, H., Timmermann, B., Satoh, N., Hikosaka-Katayama, T., Nakano, H., Rowe, M.L., Elphick, M.R., Thomas-Chollier, M., Hankeln, T., Mertes, F., Wallberg, A., Rast, J.P., Copley, R.R., Martinez, P., Telford, M.J., 2019. Mitigating anticipated effects of systematic errors supports sister-group relationship between Xenacoelomorpha and Ambulacraria. Current Biology, 2, 1818–1826.

Raikova, O., Reuter, M., Jondelius, U., Gustafsson, M., 2000. An immunocytochemical and ultrastructural study of the nervous and muscular systems of *Xenoturbella westbladi* (Bilateria inc. sed.). Zoomorphology, 120,107–118.

Robertson, H.E., Lapraz, F., Egger, B., Telford, M.J., Schiffer, P.H., 2017. The mitochondrial genomes of the acoelomorph worms *Paratomella rubra*, *Isodiametra pulchra* and *Archaphanostoma ylvae*. Scientific Reports, 7, 1847.

Rouse, G.W., Wilson, N.G., Carvajal, J.I., Vrijenhoek. R.C., 2016. New deep-sea species of *Xenoturbella* and the position of Xenacoelomorpha. Nature, 530, 94–97.

Stach, T., Dupont, S., Israelsson, O., Fauville, G., Nakano, H., Kånneby, T., Thorndyke, M., 2005. Nerve cells of *Xenoturbella bocki* (phylum uncertain) and *Harrimania kupfferi* (Enteropneusta) are positively immunoreactive to antibodies raised against echinoderm neuropeptides. Journal of the Marine Biological Association UK, 85, 1519–1524.

Thiel, D., Franz-Wachtel, M., Aguilera, F., Hejnol, A., 2018. Xenacoelomorph neuropeptidomes reveal a major expansion of neuropeptide systems during early bilaterian evolution. Molecular Biology and Evolution, 35, 2528–2543.

Westblad, E., 1949. *Xenoturbella bocki* n.g, n.sp.: a peculiar, primitive turbellarian type. Arkiv för Zoologi, 1, 11–29.

DEUTEROSTOMIA

HIGHER TAXON

6

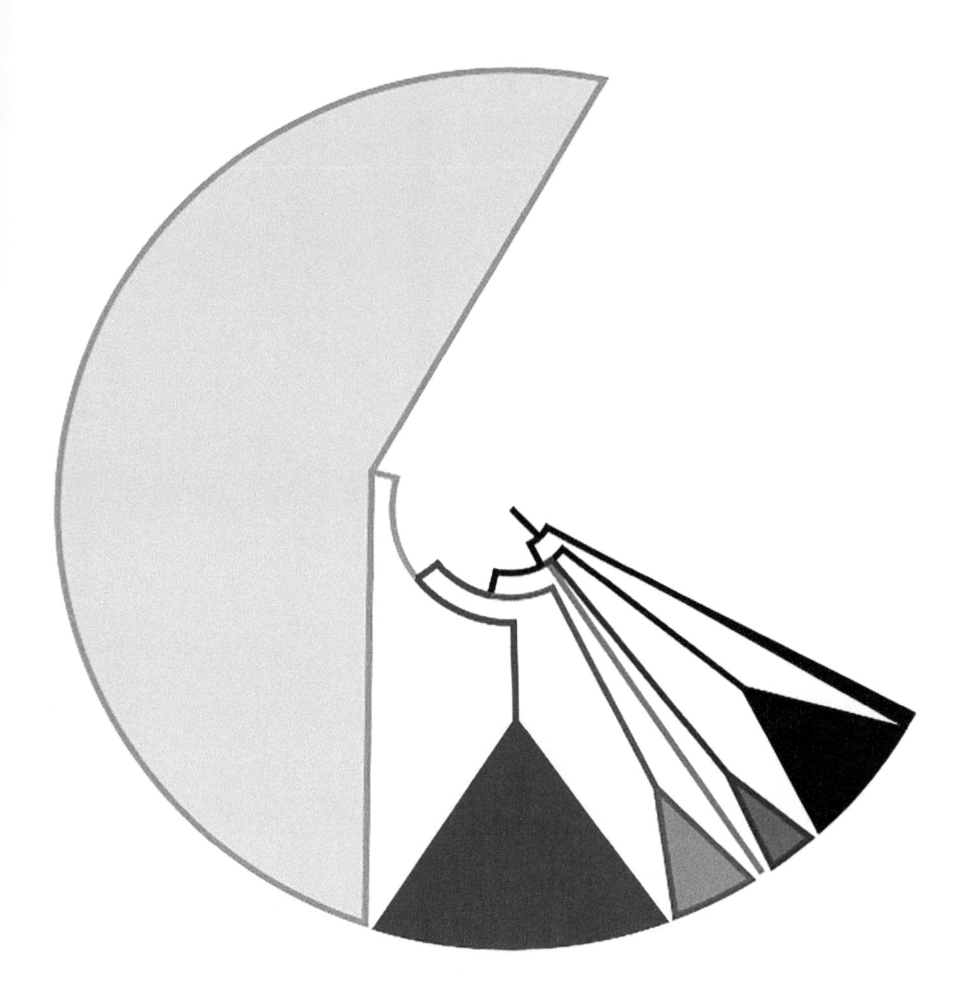

Deuterostomia means second mouth in Greek and this taxon includes a large group of animals. It is a superphylum of the higher division Bilateria and is the sister taxon of the superphylum Protostomia. In some schemes, Deuterostomia and Protostomia together make up the higher clade Nephrozoa. The figure shows the developmental progression of protostomes and deuterostomes. The developmental formation of the mouth and the anus in these two groups creates a clear boundary between Deuterostomia and Protostomia. Basically, deuterostomes develop their mouths at the opposite end of the embryo where the blastopore forms. In contrast, protostomes develop their mouths at the end of the embryo where the blastopore first develops. This basic developmental difference is a very precise indicator of membership in these two higher taxonomic entities.

There are three major clades of deuterostomes:

- Chordata (vertebrates and tunicates)
- Echinodermata (sea urchins, sea cucumbers)
- Hemichordata (acorn worms and graptolites)

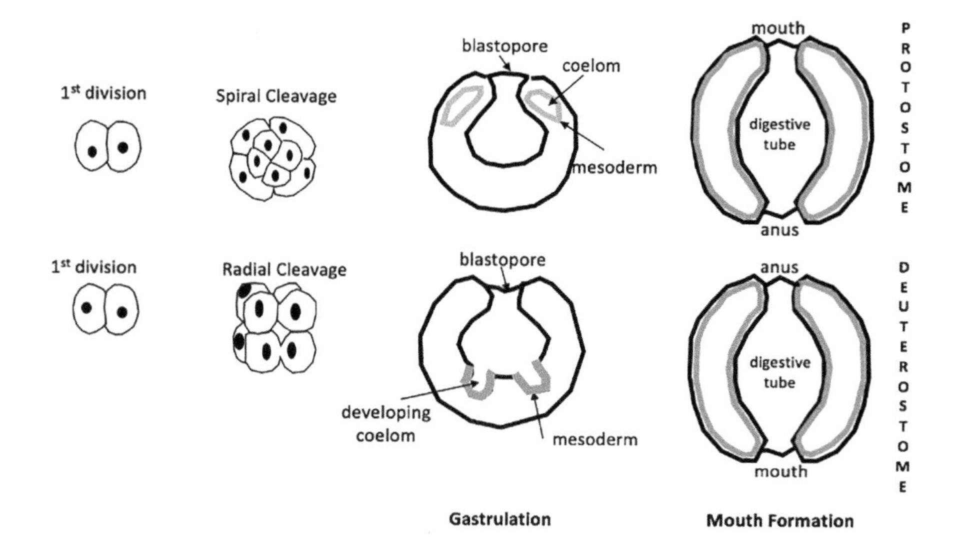

General Characteristics

The blastopore (the opening at the bottom of the forming embryonic gastrula stage) becomes the anus, whereas in protostomes the blastopore becomes the mouth

CHAPTER

35

PHYLUM ECHINODERMATA

Sabine Stöhr

Infrakingdom	Bilateria
Division	Deuterostomia
Phylum	Echinodermata

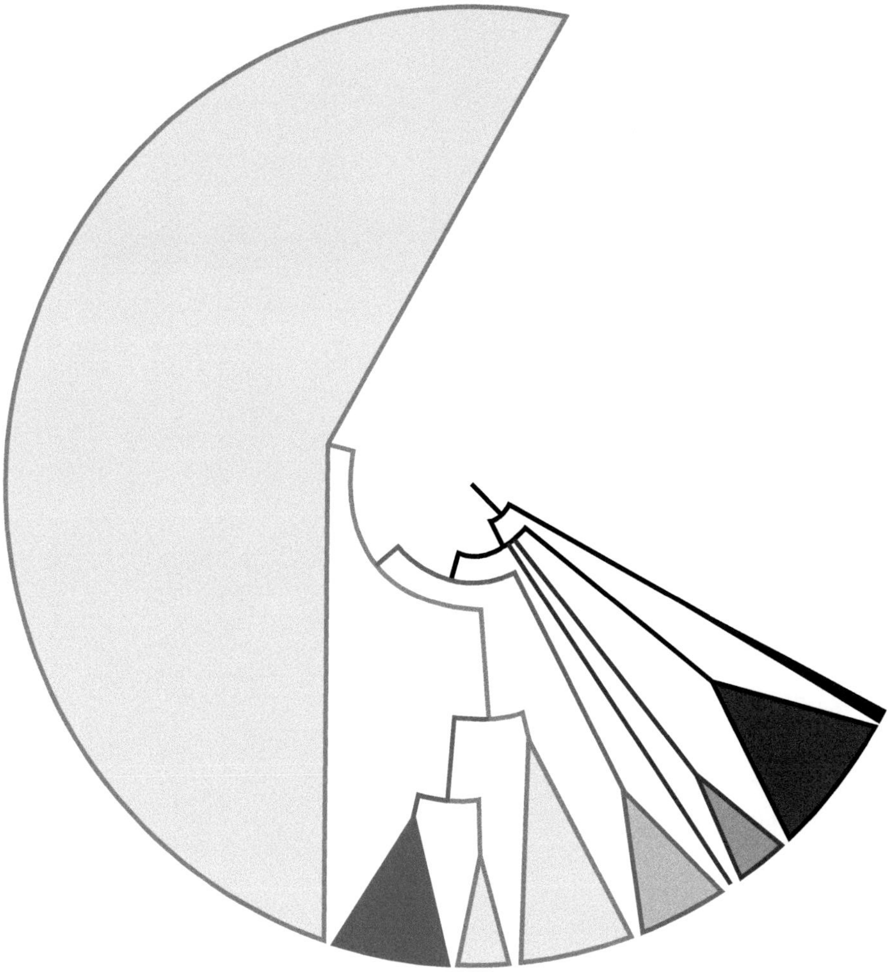

CONTENTS

Synapomorphies

- pentamerous radial symmetry in adults
- mesodermal calcitic endoskeleton
- mutable collagenous connective tissue
- water vascular system

General Characteristics

Pentamerous radial symmetry in adults

Mesodermal calcitic endoskeleton

Mutable collagenous connective tissue

Water vascular system

More than 7000 living species described, along with more than 15,000 fossil species

Numerically dominant in many marine faunas

Five living classes, approximately 16 extinct classes

Distributed worldwide in all marine habitats, at all depths, but rarely in brackish conditions

Free-living, adults mostly benthic

Commercial fishery of sea urchins and sea cucumbers for human consumption

Rich fossil record, dating back to the Lower Cambrian (>540 Ma)

INTRODUCTION

Echinoderms are a moderately species-rich phylum, but some species occur in large numbers and are conspicuous components of many faunal communities. They possess limited osmoregulatory capacity and are therefore restricted to marine environments with few exceptions that are adapted to or can tolerate brackish conditions (WoRMS Editorial Board 2019). The five living classes (Crinoidea, Asteroidea, Ophiuroidea, Echinoidea, and Holothuroidea) are a small remnant of the rich diversity of forms and lifestyles that appeared (and disappeared) during the long evolutionary history of this phylum (Wray 1999). Modern echinoderms differ morphologically from their Paleozoic ancestors as a result of the Permian-Triassic extinction event and the following diversification. They are disparate in body plan, life histories and ecology, and each of the five classes will be treated separately on the following pages, after a general introduction of shared features. Despite recent advances in the study of the evolution, genetics and morphology of echinoderms, many aspects—particularly life history, development, physiology, and ecology—remain poorly understood for most of the known species.

HISTORY OF TAXONOMY AND CLASSIFICATION

The current most widely accepted phylogenetic hypothesis (**Figure 35.1**) groups the classes Ophiuroidea (brittle and basket stars) and Asteroidea (sea stars) as sister taxa in the subphylum Asterozoa, the Holothuroidea (sea cucumbers) and Echinoidea (sea urchins) as subphylum Echinozoa, and the Crinoidea (sea lilies and feather stars) at the base of the tree, sometimes also named Crinozoa (Telford et al. 2014; O'Hara et al. 2014; Reich et al. 2015).

ANATOMY, HISTOLOGY, AND MORPHOLOGY

All living echinoderms originate from a five-rayed (pentaradiate) ancestor but species with higher order radial symmetries have evolved in asteroids and ophiuroids. The number of appendages (but not body symmetry) is increased by branching in some Ophiuroidea and many Crinoidea. Among extinct forms, other types of symmetries (e.g., bilateral, spiral, asymmetrical) were present. The typical echinoderm body has two surfaces, the oral surface (where the mouth is) and the opposite aboral surface, but the orientation and extent of these surfaces vary greatly among classes. The **body wall** of an echinoderm consists of several layers, a cuticle, the epidermis, the dermis, a muscle layer, and the coelom lining. Embedded in the dermis lies the **skeleton**, which is composed of numerous

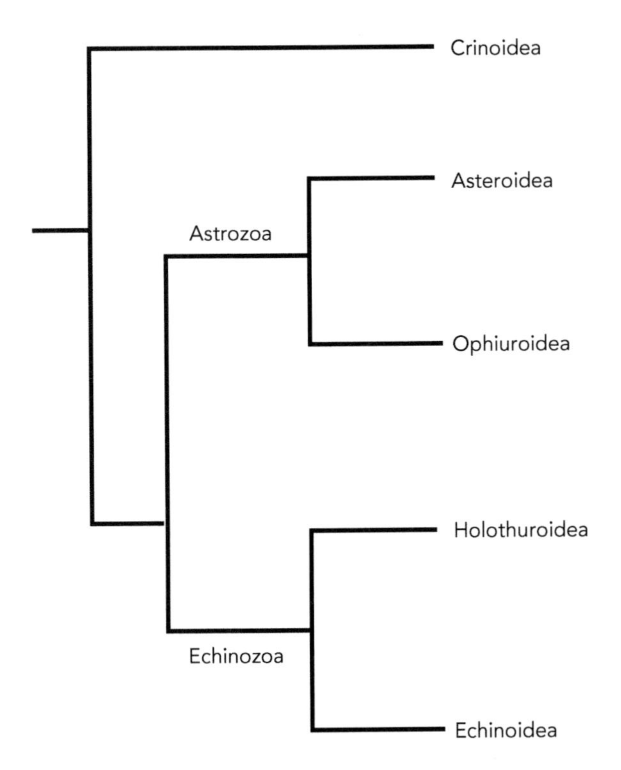

Figure 35.1 Interrelationships of the living echinoderm classes.

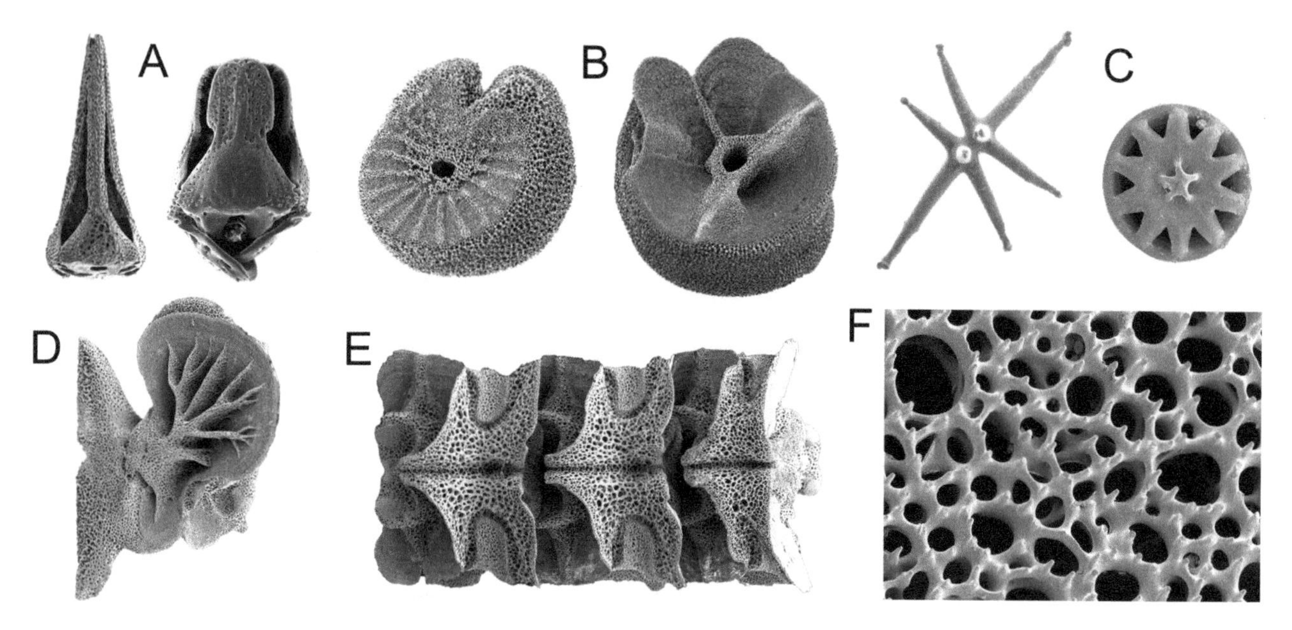

Figure 35.2 Examples of echinoderm skeleton (ossicles) and stereom structure. (A) echinoid pedicellariae, (B) crinoid arm pieces with articulation surfaces, (C) holothuroid dermal spicules, (D) ophiuroid half-jaw, (E) articulated ophiuroid arm vertebrae, and (F) ophiuroid stereom (Sources: A, A. Kroh, B, C. Messing, C, M. Reich, D–F, S. Stöhr.)

pieces, called ossicles, with many specializations and novelties (**Figure 35.2**). It consists of calcite with high magnesium content. Each ossicle consists of polycrystalline calcite with a mesh-like structure that is called stereom, infused with an organic matrix. A remarkable feature of all echinoderms is the **mutable collagenous tissue** in the dermis that connects skeletal parts with each other and muscles with skeleton. This tissue is under neural control and can alter its state within seconds or minutes, changing reversibly from soft to stiff (for example, to allow the animal to lodge itself into hiding places), or disintegrating irreversibly to autotomize body parts in response to external stress (e.g., attacking predators) or in asexual reproduction.

Echinoderms have a large coelomic body cavity that houses the gut/stomach and the gonads (**Figure 35.3**). The coelom wall is ciliated. A unique apomorphy of echinoderms is the **water vascular (or ambulacral) system**, which consists of fluid-filled coelomic tubes. From a circumoral ring canal a radial canal extends into each arm where side branches end in tube feet (podia) that have locomotory function, collect food particles and transport these to the mouth, can take up dissolved nutrients, and are used for gas exchange and excretion (Figure 35.3). The water vascular system connects to the outside water through a tube called stone canal that opens in a porous ossicle called madreporite (sieve plate) in asteroids, ophiuroids and echinoids. Crinoids possess pores but no madreporite and holothuroids have an internal madreporite without connection to the outside. On the circumoral ring canal we find **Tiedemann's bodies** which produce coelomocytes (amoeboid cells). These collect waste products and transport them to the tube feet and papulae (body wall extensions in asteroids), where they are expelled. **Polian vesicles**, elongated coelomic expansion sacs, open into the ring canal and serve to regulate water pressure in the tube system. The body fluid thus consists mostly of seawater (except in holothuroids), with excretory cells and some proteins. Another tube system, the **hemal or sinus system**, also consists of a circumoral ring canal (the hyponeural ring), which is situated below the ambulacral ring, enclosing the hyponeural hemal ring vessel. Radial and lateral canals with hemal vessels radiate into the arms. In some groups, there are also a genital hemal ring and a gastric hemal ring. The hemal rings connect through the axial

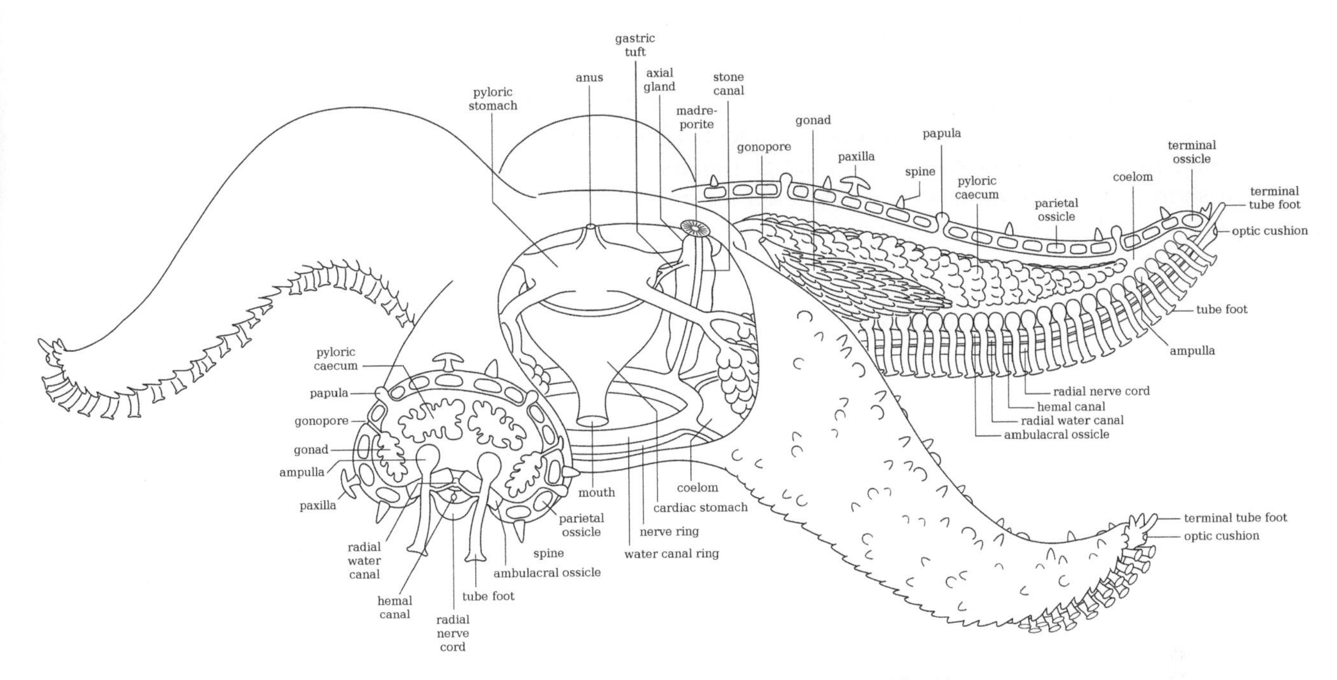

Figure 35.3 Anatomy of a sea star, showing typical echinoderm organ systems. (Source: A. Allievi.)

hemal vessel inside the axial canal, whose function is still unclear but does not seem to be excretory. The hemal system is filled with body fluid but has no role in gas exchange and is not considered a blood system, despite its name. It is believed to have a nutrient storage function but understanding of the hemal system is still poor. Respiratory pigments are generally absent but some holothuroids and ophiuroids have hemoglobin in their coelomocytes. The **nervous system** follows the same pattern with circumoral nerve ring, radial nerves and an aboral genital nerve ring, in addition to peripheral nerve nets. Sensory components in the epidermis receive input from sensory cells, and motor components in the coelomic lining innervate muscles and tube feet, but a central nerve mass is missing (except in crinoids). Ganglia (small nerve masses) are present in some echinoderms, e.g., in connection to the tube feet of each arm joint in Ophiuroidea.

All echinoderms have great **regeneration** capacities and are able to repair extensive damage. Whole arms can be regrown, a lost disc (e.g., in ophiuroids) and internal organs (e.g., ophiuroids, holothuroids) may be replaced, and some asteroids can regrow a complete animal from a severed arm without disc fragment.

Information for this section was synthesized from (Byrne & O'Hara 2017; Ruppert et al. 2004; Zueva et al. 2018) and we recommend the student consult these references too.

REPRODUCTION AND DEVELOPMENT

Reproduction in all classes is sexual, but asexual propagation (fissiparity) has evolved in asteroids, ophiuroids and holothuroids. This involves dividing of the body along a fission plane by softening and disintegration of the mutable connective tissue, followed by regeneration of the lost parts to complete two animals. In all classes except crinoids, cloning of larvae has been observed as a unique way of asexual reproduction. Echinoderms are typically gonochoric, but both sequential and simultaneous hermaphrodites are known. Most echinoderms are broadcast spawners and release their eggs and sperm into the water, where fertilization occurs, but parental care such as brooding has been found in all classes. Brooders retain the eggs inside the gonads or in special body cavities (e.g., bursae in ophiuroids), sperm enters and fertilization is internal, usually resulting in a lecithotrophic larva, but direct development without a larva has also been documented in all classes except Crinoidea. The developing progeny are released shortly after metamorphosis or nurtured for an extended period of time, depending on the species. Most broadcast spawners and some brooders reproduce during specific seasons, whereas other brooders may be sexually active during most of the year, bearing young of various age and size (e.g., the ophiuroid *Amphipholis squamata*). Asexual reproduction does not seem to follow a seasonal pattern.

Embryonic development follows the typical deuterostome pattern with radial cleavage, blastula, and gastrula, to a bilateral larva. Three paired coelomic cavities develop in the embryo through enterocoely, called protocoels (= axocoels), mesocoels (one of which becomes the hydrocoel), and metacoels (= somatocoels). Their walls form the mesoderm. Types of larvae vary by class but in general there are planktotrophic (feeding) and lecithotrophic (non-feeding) larvae (**Figure 35.4**), and complicated life cycles are common. A ciliated planktotrophic larva called dipleurula is considered ancestral for echinoderms and a feature (synapomorphy) shared with hemichordates. Several types of lecithotrophic larvae have evolved, sometimes multiple times in the same class, and one larval type may transition into another before metamorphosis to the adult form. The cilia in these larvae have locomotory function only, whereas cilia in planktotrophic larvae have both locomotory and feeding functions. Metamorphosis takes place either in the water column

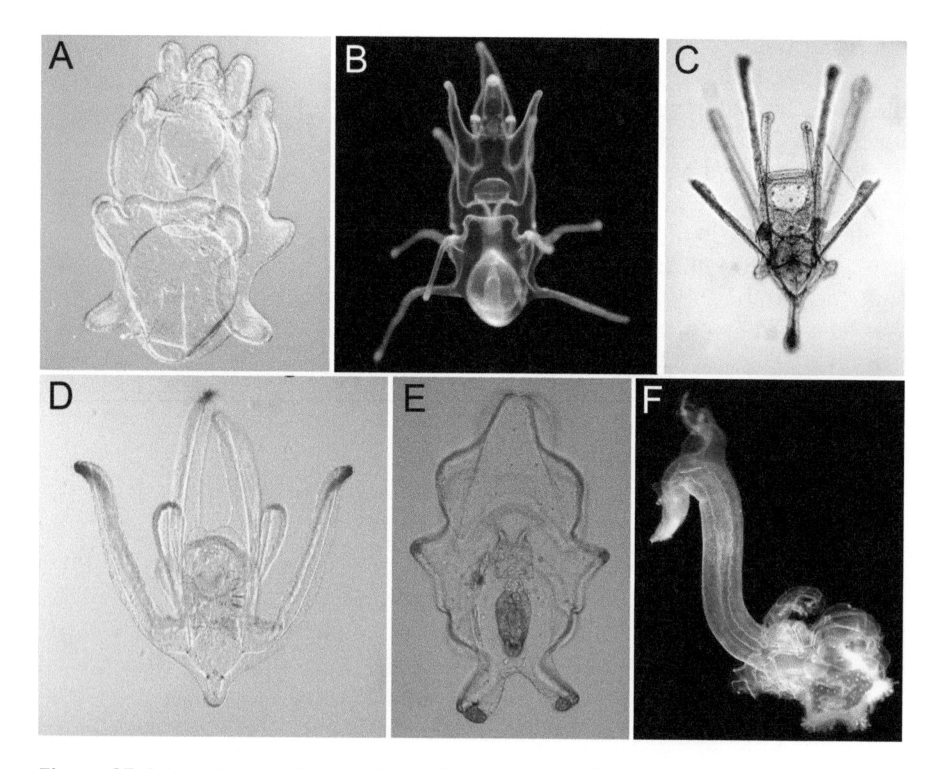

Figure 35.4 Larval types of echinoderms. (A) bipinnaria and (B) brachiolaria of the sea star *Asterias rubens*, (C) echinopluteus of the irregular sea urchin *Echinocardium chordatum*, (D) ophiopluteus of the brittle star *Ophiura albida*, (E) auricularia of the sea cucumber *Australostichopus mollis*, (F) giant bipinnaria with still attached juvenile of the sea star *Luidia sarsii*. Pluteus larvae have skeletal rods with ciliated bands as arms, whereas the other larval types lack hard skeletal parts but have ciliated bands. (Sources: A, C, D, Otto Larink, and B, Rebekka Schüller, both http://planktonnet.awi.de, license: CC BY 3.0 https://creativecommons.org/licenses/by/3.0/, E, M. Sewell, F, S. Stöhr.)

before settlement (ophiuroids, some asteroids) or on the seafloor (echinoids). During metamorphosis from a bilateral larva to a pentaradiate adult, the anterior-posterior body axis shifts (except in holothuroids) and the body twists around the new axis. The coelomic cavities undergo a complex reconfiguration, where the right axo- and hydrocoel are almost completely reduced. The left hydrocoel develops into the ambulacral system, left and right somatocoel form the body cavity and the hemal system, and the left axocoel forms axial canal and axial vessel.

Information for this section was synthesized from (Arnone et al. 2015; Byrne and O'Hara 2017; Eaves and Palmer 2003; McEdward and Miner 2001; Ruppert et al. 2004).

DISTRIBUTION AND ECOLOGY

Echinoderms are found in all oceans and seas, at depths from the intertidal to the abyss (**Figure 35.5**). They have a low capacity for osmoregulation and are therefore generally restricted to marine environments with a salinity of 30–35 ppt. Global patterns of species richness have recently been studied in Ophiuroidea, and at low latitudes (0–30°) diversity was greatest on the continental shelf and upper slope, whereas the deep sea fauna is most species-rich at higher latitudes (30–50°) in proximity to continental margins (Woolley et al. 2016). Members of all echinoderm classes have been found in chemosynthetic environments, such as hydrothermal vents, methane seeps, sunken wood, and whale cadavers.

A small number of species (e.g., the brittle star *Ophiophragmus filograneus*) are adapted to fluctuating salinities (such as can be found in estuaries) and can tolerate 10–20 ppt, at least temporarily, whereas salinities ≥ 38 ppt or below 10 ppt will eventually kill the animals (Turner & Meyer 1980). Crinoids are filter feeders, asteroids are mainly predators and scavengers, echinoids are predators, herbivores or filter feeders, ophiuroids are predators, scavengers or filter feeders, and holothuroids are filter or deposit feeders. Some species live solitary, others aggregate in large numbers (brittle star beds, sea lily gardens). Details are presented with each class.

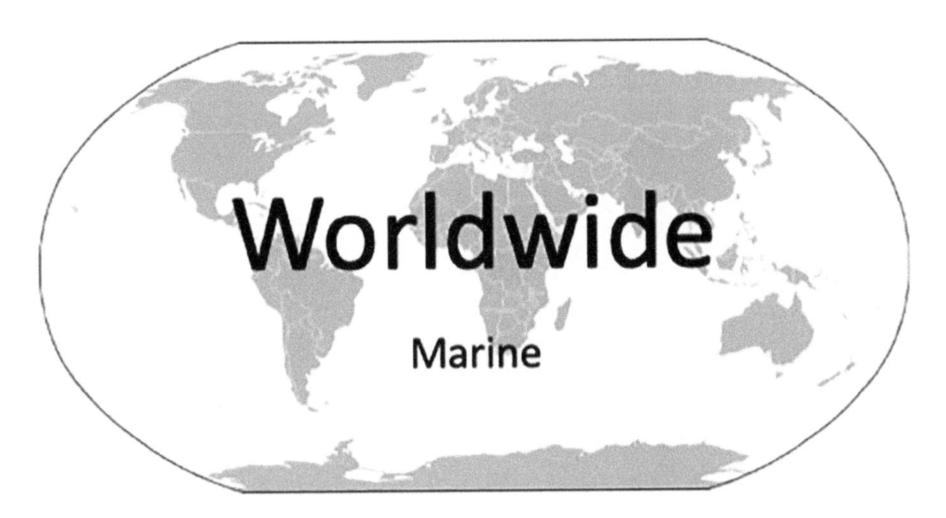

Figure 35.5 Distribution map of Echinodermata. Echinoderms are found in all oceans and seas in marine conditions, rarely in brackish waters (e.g., the Baltic Sea is devoid of echinoderms), at depths from the intertidal to the abyss.

PHYSIOLOGY AND BEHAVIOR

Basic physiological functions of echinoderm body structures are mentioned in above section on anatomy. A common stress reaction in many echinoderms is autotomy (spontaneous casting off) of body parts, e.g., arms in ophiuroids, crinoids and asteroids, dorsal discs in some ophiuroids, inner organs in some holothuroids. Transferring an echinoderm to freshwater is a method of immobilizing and killing it (by increasing the osmotic pressure in the tissues) and will often not trigger autotomy, which is desirable when collecting animals for morphological or anatomical examination.

A wide variety of behavior have been observed in echinoderms. They are generally bottom-living, either on the substrate or burrowing (e.g., irregular sea urchins, some brittle stars), but some are epizoic on or hosts of other species. To release eggs or sperm into the water, some species may take a special posture to raise their bodies from the sea floor (**Figure 35.6**). Most species do not interact for reproduction, but some brittle star species (e.g., *Ophiodaphne formata*) living on burrowing sea urchins form permanent pairs of a female and a dwarf male that attaches to the mouth of the female. Some echinoderms are active hunters of other invertebrates and fishes, while others collect particles with their appendages and feet. Further details are presented below for each class.

GENETICS AND GENOMICS

Echinoids such as the purple sea urchin *Strongylocentrotus purpuratus* have been used for over a century as model organisms in many different fields, e.g., in developmental biology, cell biology, immune response, and molecular studies. The nucleotide database at NCBI (https://www.ncbi.nlm.nih.gov) currently holds over 1.3 million records of echinoid DNA and RNA sequences, with *S. purpuratus* accounting for almost a third of all records. Asteroidea are represented with over 1.2 million records and Holothuroidea with about 500,000 records, whereas Ophiuroidea are underrepresented with regard to their high species diversity with only about 120,000 nucleotide records. Echinoderms have been of particular interest in studies of *Hox* genes, because of their long evolutionary history. *Hox* genes are important for the development and evolution of body plans and help to understand the place of echinoderms among deuterostomes and their divergence from a bilateral body organization. Advancements in molecular techniques have created renewed interest and *S. purpuratus* was the first echinoderm for which the complete genome was sequenced. At present,

Figure 35.6 Spawning positions of the sea star *Echinaster sepositus* (left) and the sea cucumber *Holothuria* cf. *forsskali* (right). (Source: H. Moosleitner.)

complete genomes of 11 species of echinoderm are available on NCBI, the echinoids *S. purpuratus*, *Lytechinus variegatus*, *Eucidaris tribuloides*, *Hemicentrotus pulcherrimus*, the asteroids *Patiria miniata*, *Acanthaster planci*, *Patiriella regularis*, the holothuroids *Apostichopus parvimensis*, *Apostichopus japonicus*, and the ophiuroids *Ophiothrix spiculata* and *Ophionereis fasciata*. The development of exon-based methods has resulted in a data matrix of 1,484 exons of 416 genes in 576 ophiuroid species, one of the largest of its kind, which was used to resolve the long controversial phylogeny of the Ophiuroidea. Transcriptome analysis is also used to understand physiological processes, such as the mechanics of mutable connective tissue in fissiparous echinoderms.

Information for this section of the chapter was synthesized from (David & Mooi 2014; Dolmatov et al. 2018; Heyland et al. 2014; O'Hara et al. 2017)

POSITION IN THE ToL

Echinodermata are deuterostome animals, sister group to Hemichordata, with which they form the Ambulacraria, sister group to the Chordata.

COMMERCIAL IMPORTANCE AND DATABASES

Echinoids and holothuroids are sought after delicacies for human consumption, particularly in Asia, and overexploitation is a growing threat to an increasing number of species from more and more parts of the world oceans. Thirteen holothuroid species have been classified as vulnerable or at risk of extinction by the IUCN. Efforts are therefore being made to breed and culture these to achieve sustainability. Harvesting of wild populations is done mainly by SCUBA diving. In sea urchins, the gonads of both males and females, both referred to as "roe," are removed and eaten raw. Sea cucumbers are either eaten raw or dried and sold as beche-de-mer or trepang. The commercially attractive species are the subjects of intense research efforts into their biology, ecology, and conservation. Echinoderms are also being investigated for pharmaceutical properties of biochemical compounds, e.g., antimicrobial or antioxidant substances. Harvesting of echinoderms that are dried and sold as souvenirs and as home decoration items, and live capture for the aquarium trade also have a negative impact on some populations and their ecosystems.

The most authoritative taxonomic database currently available is the World Register of Marine Species (WoRMS; http://www.marinespecies.org), which provides information on taxon names, classifications, authorities, synonymies, literature references and more. There are sub-registers of WoRMS with their own interfaces and additional functionality, so far for Echinoidea (http://www.marinespecies.org/echinoidea/), Ophiuroidea (http://www.marinespecies.org/ophiuroidea/) and Asteroidea (http://www.echinobase.org/entry/), to be followed by Holothuroidea and Crinoidea, as well as a digital document archive with echinoderm conference proceedings and historical documents (http://www.marinespecies.org/echinodermfiles/). Genomic information can be found at Echinobase (http://www.echinobase.org/entry/). There is also an echinoderm blog (http://echinoblog.blogspot.com). Distribution records can be found on the Ocean Biogeographic Information System (OBIS; https://obis.org), and distributions and specimens from a large number of institutions worldwide are aggregated by the Global Biodiversity Information Facility (GBIF; https://www.gbif.org). All of these resources are free to use.

Information for this part of the chapter was synthesized from (Byrne & O'Hara 2017; Conand & Byrne 1993; Eeckhaut. 2018; González-Wangüemert et al. 2018; Marmouzi et al. 2018; Reynolds & Wilen 2000; Stabili et al. 2018)

CONCLUSION

Echinoderms are considered a neglected taxon and knowledge about many aspects of their diversity, biology, anatomy, ecology, and evolution is still limited. They are charismatic members of the marine megafauna, true artforms of nature, which makes studying them a pleasure. The prevailing lack of taxonomists for most echinoderm classes opens opportunities for early career scientists.

A. CLASS CRINOIDEA

AI. PHYLOGENY AND CLASSIFICATION

Synapomorphies

- body plan: stalk and theca
- multiple stone canals
- U-shaped gut
- auricularia, doliolaria, vitellaria larvae

The living Crinoidea (sea lilies and feather stars) are the smallest class of echinoderms with fewer than 700 species, of which about 100 species are sea lilies, all others are feather stars (**Figure 35.7**). They first appeared in the Early Ordovician (~500 Ma) and were one of the dominant life-forms in the Paleozoic, and in total more than 6,000 fossil species have been described. The phylogeny of the crinoids is still being debated and no consensus between hypotheses has been reached. All living Crinoidea are placed in the subclass Articulata and a recent classification proposed four orders (Isocrinida, Hyocrinida, Cyrtocrinida, and Comatulida) and 32 families, of which the Comatulida is the most speciose group with 21 families (**Figure 35.8**). However, eight families have not been placed on a phylogeny due to conflicting or missing data. The Comatulida has recently been suggested to be paraphyletic.

Information for the Crinoidea was synthesized from (Baumiller 2008; Byrne & O'Hara 2017; McEdward & Miner 2001; Nakano et al. 2003; Ruppert et al. 2004).

AII. ANATOMY, HISTOLOGY, AND MORPHOLOGY

Crinoids are pentaradial with long arms (usually branching) extending from a cup-shaped body, called theca, which is orientated with both the mouth and the anus away from the substrate. The theca consists of a skeletal calyx, housing the gut and canal systems. Its outside is defined as aboral surface. The calyx is covered by a membrane, called tegmen, where the mouth and a conical anal papilla are found (**Figure 35.9A**), thus defining the oral surface. From the theca, a stalk extends, which attaches to the substrate with a disc-shaped holdfast or segmented cirri. The stalk-less feather stars cling to the substrate with cirri (**Figure 35.9B**). The arms bear rows of short side-branches, called pinnules (Figure 35.9A) that collect food particles from the water current. The skeleton

Figure 35.7 Crinoidea body shapes. A stalked sea lily (*Cenocrinus asterius*, left), and a stalk-less feather star (*Colobometra perspinosa*, right). (Source: C. Messing.)

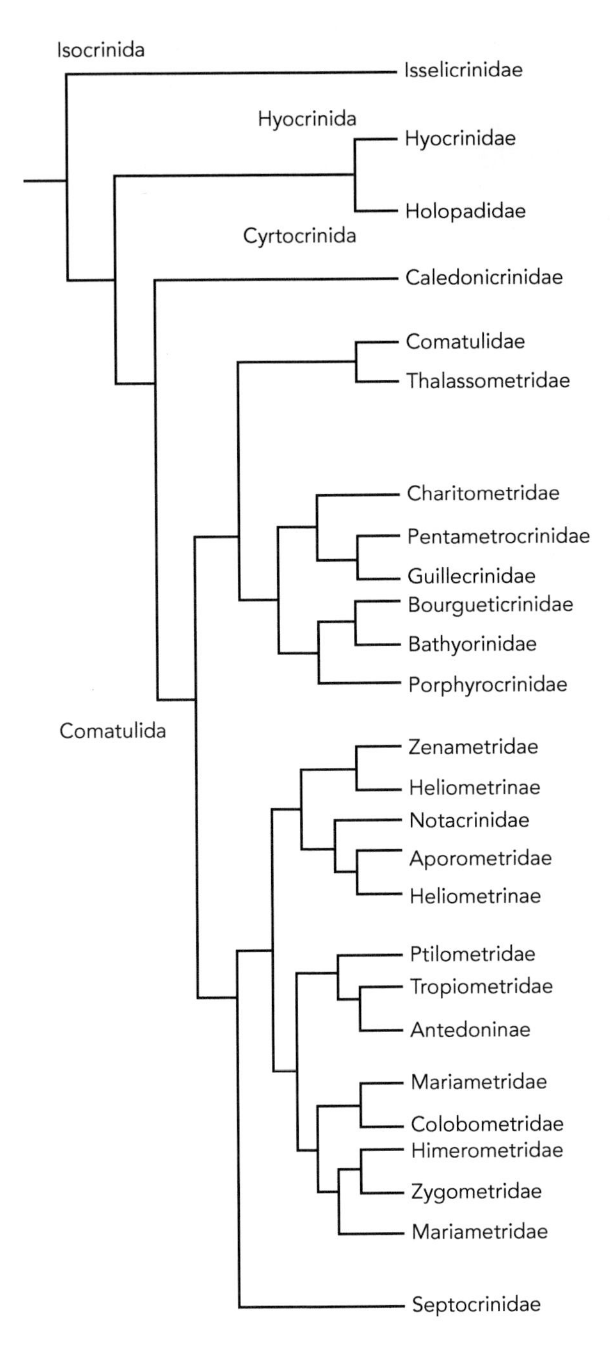

Figure 35.8 **Interrelationships of the living Crinoidea, one of several recent, preliminary hypotheses, several families are paraphyletic and eight families have been excluded due to lack of data.** (Modified from Byrne and O'Hara 2017; families from WoRMS 2019.)

consists of many specialized types of ossicles, giving a segmented appearance to stalk, arms, cirri and pinnules. Ciliated ambulacral grooves extend from the mouth across the tegmen onto the arms, and they collect and transport food. Madreporites are absent; instead, numerous hydropores in the tegmen open through short canals directly into the perivisceral coelom. The ambulacral ring has multiple stone canals that also open into the perivisceral coelom. The intestine is U-shaped, from mouth to anus, inside the calyx. Unlike other echinoderms, the nervous system of crinoids contains a large aboral ganglion. Crinoid tube feet lack adhesion discs at the ends and are arranged in triplets with different morphology and function in collecting and sorting food particles; they are also used for gas exchange.

Figure 35.9 Details of feather stars. (A) Oral view of the theca with tegmen (T), anal papilla (A), ambulacral grooves (G) running along the arms and converging on the central mouth, arms with pinnules (P). (B) Lateral view of theca with cirri (C). (Source: C. Messing.)

AIII. REPRODUCTION AND DEVELOPMENT

The majority of crinoids are gonochoric and the sexes cannot be distinguished morphologically; few hermaphrodites are known. The gonads lie in special pinnules, not inside the calyx. Most species broadcast eggs and sperm into the water column, few species brood the embryos in special pouches called marsupia. It was long believed that crinoids possessed only a doliolaria larva, but recently, a non-feeding auricularia larva was found in a sea lily. Similar to holothuroids, when present, the auricularia transforms into a non-feeding barrel-shaped doliolaria. Both larval stages have ciliated bands but no hard skeletal structures. Some brooding crinoids have a vitellaria, similar in shape to the doliolaria but without cilia. The doliolaria develops an adhesive disc and settles on the seafloor before metamorphosing to a stalked (pentacrinoid stage) juvenile. In sea lilies, the stalk is kept but in feather stars it is lost during development.

AIV. LIFESTYLE

Crinoids are semi-sessile to sessile and live in all seas with marine conditions (but not in brackish waters), from the intertidal to abyssal depths, although sea lilies are not found shallower than 100 m. Feather stars are particularly abundant and very conspicuous, brightly colored components of shallow tropical reefs. Crinoids are filter feeders that rely on water movement for transport of particles to the arms. Some feather stars can detach from the substrate and swim with undulating movements of their arms, and some sea lilies have been observed to shed the distal part of the stalk and crawl away with arm movements. In contrast, Paleozoic crinoids are believed to have been sessile and motility is thought to have evolved in response to predation. Crinoids are hosts to a variety of other animals, such as various crustaceans, snails, fish, and brittle stars, and some of these mimic the color pattern of their host to camouflage themselves. Common predators of crinoids are fishes, cidaroid sea urchins and sea stars.

SUBPHYLUM ASTEROZOA

The Asterozoa includes the living Asteroidea and Ophiuroidea (Figure 35.1), and the extinct Somasteroidea. They originated in the Early Ordovician (~510–493 Ma) and it is unclear whether all three share a common ancestor or if Somasteroidea are the shared ancestor of Asteroidea and Ophiuroidea (Mah & Blake 2012). Asterozoa have a stellate body plan with a more or less well-defined central body, called disc, from which appendages, called arms, extend in a radiating fashion.

B. CLASS ASTEROIDEA

BI. PHYLOGENY AND CLASSIFICATION

The Asteroidea (sea stars) are the second largest class of echinoderms, with almost 1,900 species and 36 families. The latest phylogenetic hypothesis suggests seven orders, among which the Velatida are paraphyletic (**Figure 35.10**). The morphologically aberrant genus *Xyloplax* has been considered as sixth class, but it appears to be an asteroid in the order Velatida.

Information for the Asteroidea was synthesized from (Byrne & O'Hara 2017; Hennebert 2010; Lawrence 2013; Linchangco et al. 2017; Mah 2019; Mah & Blake 2012; McEdward & Miner 2001; Santos et al. 2005)

Synapomorphies

- two stomachs
- gonads extend into arms
- pyloric caeca
- madreporite dorsal
- open ambulacral groove
- bipinnaria and brachiolaria larvae

BII. ANATOMY, HISTOLOGY, AND MORPHOLOGY

Asteroids are originally pentamerous, but species with up to 50 supernumerary arms are known. Body shapes (**Figure 35.11**) vary from the familiar star-like forms (e.g., *Asterias*) in which the arms converge in the center without a well-defined disc, to inflated or flat cushion-shaped or pentagonal forms (e.g., *Culcita*) with extremely short arms, and forms with a well-defined central disc and clearly offset thin arms that are superficially similar to ophiuroids (e.g., *Brisinga*). Asteroids are dorsoventrally flattened, usually orientated with the mouth towards the substrate; hence the ventral surface is defined as oral, also called actinal. The dorsal or aboral surface is called abactinal, and the border between these surfaces is demarcated by one or two rows of marginal skeletal plates. The ambulacral

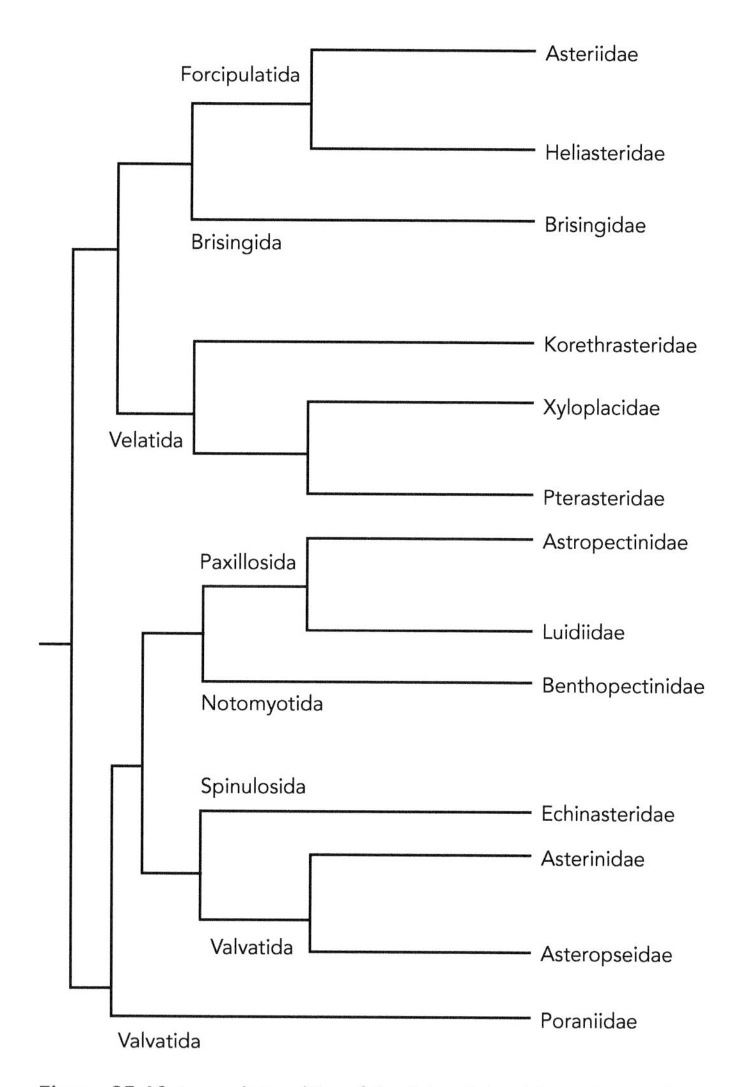

Figure 35.10 **Interrelationships of the living Asteroidea orders and representative families.** Velatida are paraphyletic and require more data. (Modified from Linchangco et al. 2017; families from WoRMS 2019.)

grooves on the oral side of the arms are open (not covered by skeletal elements) and converge on the central mouth (**Figure 35.12**). The tube feet project into the ambulacral groove and are used in locomotion, food collection, and burrowing. They operate with several groups of muscles and pressure controlling pouches, called ampullae. Tube foot morphology varies among groups, but the majority of asteroids have feet that end in a disc-like structure (absent in burrowing species), sometimes called suckers, although adhesion to the substrate is achieved with chemical secretions rather than suction. The skeleton of the asteroids is highly variable among groups, strongly calcified in some, poorly calcified or absent in others. Various types of ossicles form the abactinal disc, among them is the madreporite. The arms are composed of numerous plates and spines, which are serially repeated and give a segmented appearance. Special skeletal structures are the morphologically diverse pincer-like pedicellariae on the surface of many asteroid groups. Pedicellariae are gripping devices, which are used in feeding or defense. Umbrella-like paxillae are other skeletal surface structures found in some groups (e.g., *Astropecten*). The coelom wall has extensions from the surface in the form of soft finger-like sacs called papulae, which function as part of the excretory system and in respiration. Coelomocytes transport waste products into the papulae, from which they are

Figure 35.11 Sea stars vary in symmetry and shape. (A) There are pentaradial forms, with long arms (*Asterias rubens*), and, (B) with short arms (*Porania pulvillus*), and multi-rayed forms, (C) *Acanthaster planci*, and (D) *Luidia ciliaris*. (Sources: A, B, S. Stöhr; C, D, H. Moosleitner.)

expelled. Some asteroids possess light sensitive spots at the tip of their arms (e.g., *Asterias rubens*) that allow phototactic behavior but provide no vision.

The digestive system of asteroids consists of the ventral mouth, the muscular cardiac stomach, the pyloric stomach, a short intestine that connects to a rectum that terminates in the dorsal anus, and paired nutrient storage and enzyme excreting organs, the pyloric caeca, that extend from the pyloric stomach into

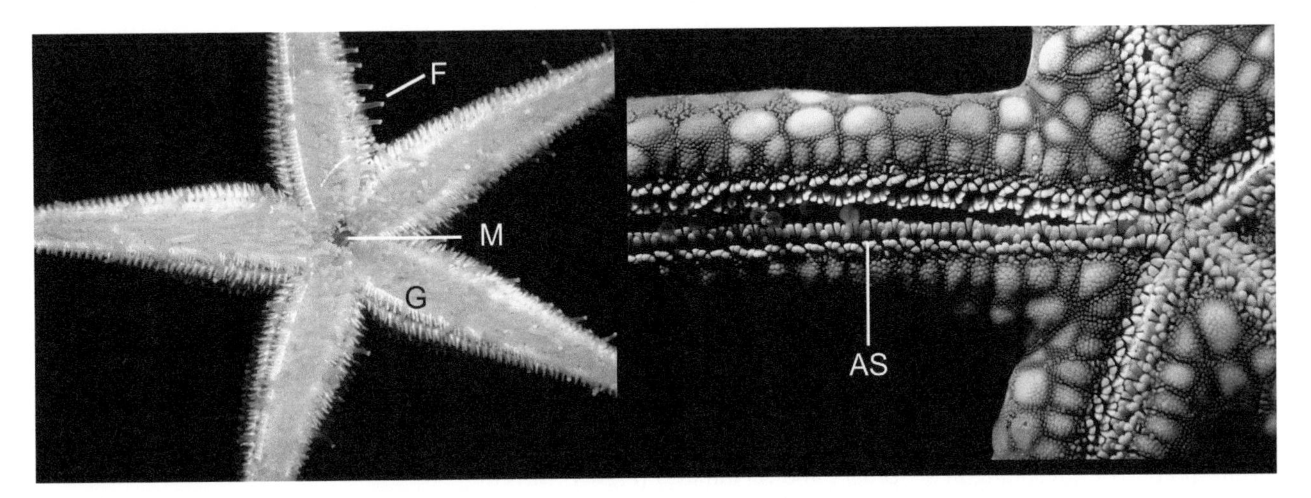

Figure 35.12 Sea stars have an open ambulacral groove (G) with numerous tube feet (F) converging on the central mouth (M), *Asterias rubens* (left) (Source: S. Stöhr.) The skeleton consists of numerous plates, typical for many asteroids are the ambulacral spines (AS) sitting on adambulacral plates along the open amublacral groove, *Nardoa tuberculata* (right.) (Source: H. Moosleitner.)

the arms. A pair of gonads extends into each arm, overlaying the pyloric caeca (Figure 35.3).

BIII. REPRODUCTION AND DEVELOPMENT

Most asteroids are gonochoric and the sexes cannot be distinguished morphologically, but some are known to be simultaneous hermaphrodites and may self-fertilize (e.g., *Parvulastra*). Asteroids reproduce mainly sexually, but asexual fission also occurs in some species (e.g., *Coscinasterias*). Most asteroids are broadcast spawners that release eggs and sperm into the water column, but intergonadal brooding has been found in some viviparous Asterinidae and these are also hermaphrodites. There are two planktotrophic larval types, a bipinnaria that develops into a brachiolaria (**Figure 35.4A, B**). Neither type has evolved independently, but they are developmental stages with slightly different morphology. They have coelom supported arms that are traced by several ciliated bands for feeding and locomotion, and they have a digestive tract with mouth, gut, and anus. The brachiolaria also has attachment structures that are used to anchor the larva to the ground for metamorphosis (e.g., *Asterias rubens*). In some species (e.g., *Crossaster papposus* and brooding *Patiriella*), the bipinnaria stage has been reduced and only a planktonic lecithotrophic brachiolaria develops. In others (e.g., *Luidia*), the brachiolaria has been lost and metamorphosis happens in the plankton (**Figure 35.4F**).

BIV. LIFESTYLE

Asteroids are free-moving, many are quite active, and occur in all seas (some tolerate brackish conditions), at all depths. They live on the substrate or burrow into sediment. Many asteroids are predators or scavengers and either swallow their food whole or extend their stomach through the mouth and digest the food in part or completely externally. They prey on a wide variety of animals, such as other asteroids, gastropods, polychaetes, ophiuroids, or holothuroids. Some species feed on algal and microbial films on the substrate, others are suspension feeders (e.g., Brisingida). The crown-of-thorns sea star (*Acanthaster planci*) has a bad reputation as destroyer of reefs but the causes of mass occurrences and their consequences are still poorly understood, despite intense research efforts. Most asteroids are harmless to humans, but may have toxins in their skin that deter predators, and *A. planci* has toxic spines that cause severe pain when the skin is pierced. Asteroids are hosts to small shrimp, epizoic gastropods, and endoparasitic barnacles and pearlfish. Predators of asteroids are other asteroids, fish, and gastropods.

C. CLASS OPHIUROIDEA

Information for the Ophiuroidea was synthesized (From Byrne & O'Hara 2017; Hendler 2018; McEdward & Miner 2001; O'Hara et al. 2014, 2017, 2018; Stöhr et al. 2018; Thuy & Stöhr 2018).

Synapomorphies

- madreporite ventral
- tube feet without ampullae
- no anus
- single stomach
- stomach and gonads restricted to disc
- bursae
- closed ambulacral groove
- ophiopluteus larva

CI. PHYLOGENY AND CLASSIFICATION

The Ophiuroidea (brittle and basket stars) are the most species-rich class among the living echinoderms with about 2,100 described species. Their phylogeny and classification have recently been revised on the basis of a large molecular study, with support from morphology, resulting in six orders and 33 families (**Figure 35.13**). The old division between basket stars and brittle stars can no longer be maintained, because the basket stars (order Euryalida) share a common ancestor with brittle stars in the order Ophiurida, based on molecular data and fossil

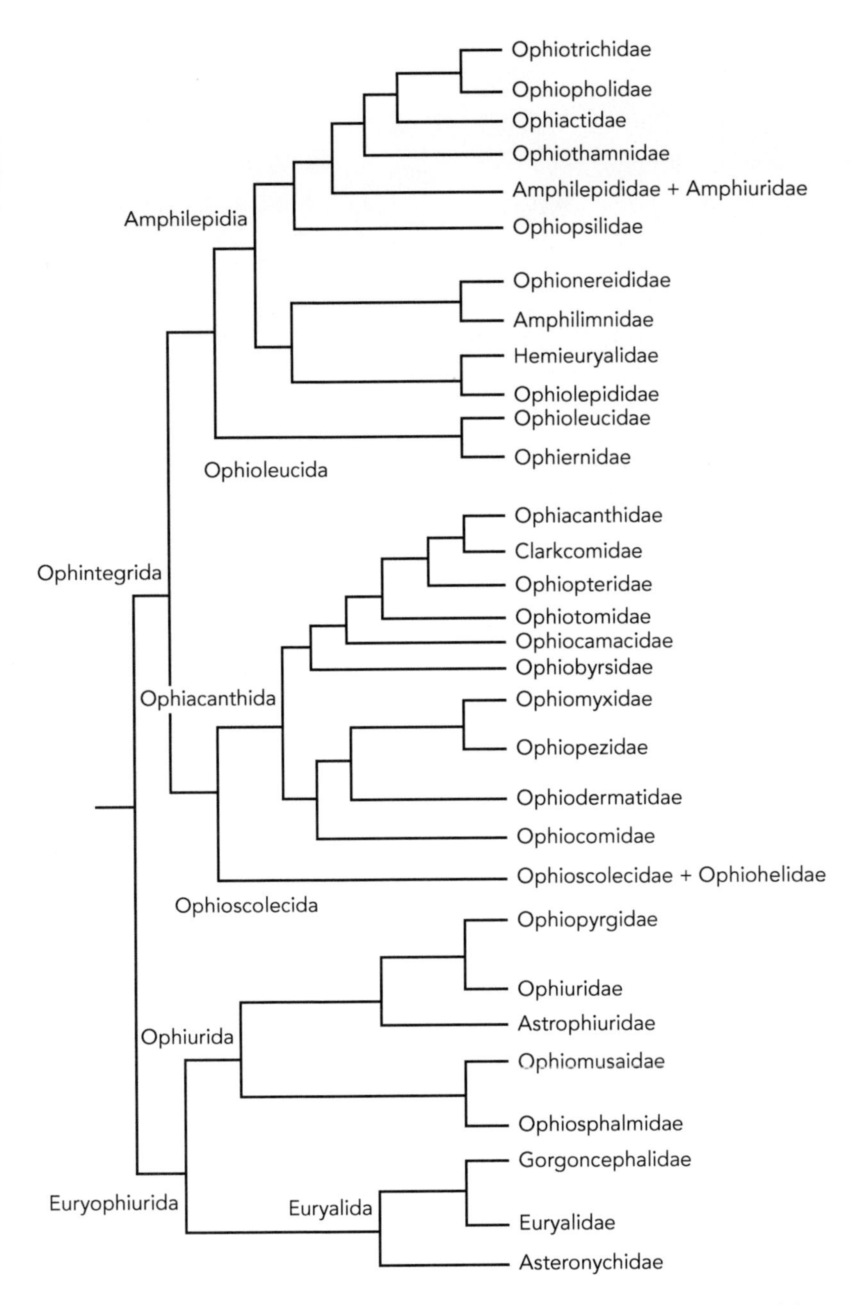

Figure 35.13 Interrelationships of the living Ophiuroidea families. Amphilepididae + Amphiuridae and Ophioscolecidae + Ophiohelidae are not fully resolved yet. (Modified from O'Hara et al. 2018.)

material. Together these form the superorder Euryophiurida, whereas all other brittle stars belong to the superorder Ophintegrida. This classification and phylogeny may need to be revised in the future, because a large number of genera have not been analyzed yet. Increasing molecular evidence suggests that the actual diversity of the ophiuroids may be considerably higher than the number of currently recognized species, because the proportion of morphologically similar species with wide geographic distribution is high, suggesting the presence of cryptic species complexes.

CII. ANATOMY, HISTOLOGY, AND MORPHOLOGY

Ophiuroids are usually pentamerous but hexamerous species are common in some genera and a few species with symmetries up to the order of ten are known.

Figure 35.14 Most brittle stars have simple arms, such as (A) *Ophiarachna incrassata* and (B) *Ophiomastix caryophyllata*, but in the order Euryalida, species with branched arms occur, (C) *Gorgonocephalus caputmedusae* and (D) *Astroboa nuda*, both in filter-feeding position, resembling feather stars. (Sources: A, D, H. Moosleitner, B, C, S. Stöhr.)

They are distinguished from most asteroids by their well-defined central disc that is clearly distinct from the thin snake-like winding arms (**Figure 35.14**). The arms are usually simple but in euryalids, species with branching arms have evolved (Figure 35.14). Ophiuroids are generally benthic and their body is usually orientated with the mouth towards the substrate, hence the oral surface is defined as ventral and the aboral surface as dorsal. The skeleton of the ophiuroids shows many specializations, such as the mouth parts, which consist of jaws (oral plates), teeth on a dental plate, various papillae, adoral and oral shields and other ossicles (Figures 35.2, 35.15). The arms consist of a large number of segments, each of which has a so-called arm vertebra in the center, which is completely enclosed by two lateral (left and right side) plates, a dorsal and a ventral plate (**Figure 35.15**). The ambulacral groove is completely closed over by the ventral arm plates (and sometimes the lateral plates) and the tube feet emerge through holes in or between these plates. Reductions of the skeleton are common in a range of species. The madreporite is ventral, in one of the oral shields in the mouth frame. It may have one, several or no hydropores and is usually larger than the other oral shields but can often not be distinguished externally at all. Euryalids and fissiparous species may have more than one madreporite.

The tube feet lack ampullae and adhesion discs and are generally not used for locomotion, but small juveniles are able to walk on their feet. Ophiuroids move by winding their arms, usually two arms lead the way and the rest of the arms and the disc follow. Contrary to asteroids, ophiuroids have a single sac-like stomach, lack an intestine, the mouth functions also as anus, the gonads are usually restricted to the disc, and pyloric caeca are absent. In many species, the stomach wall has small skeletal ossicles. Ophiuroids possess special pouches in the disc to either side of each arm, called bursae. These are primarily used for respiration but the gonads are attached to the inner wall of each bursa and

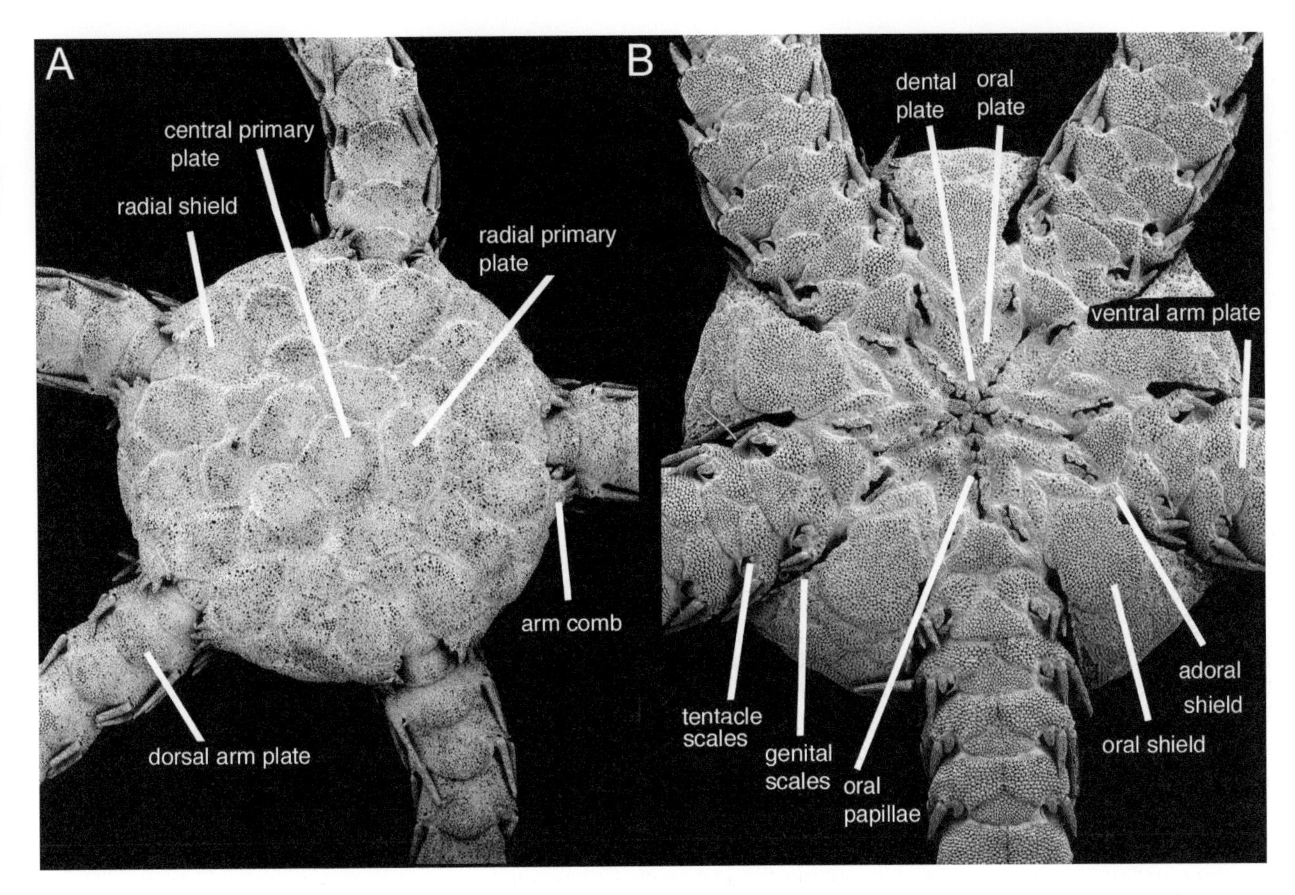

Figure 35.15 The brittle star skeleton is highly specialized (*Ophiura sarsii*): the dorsal disc is covered by scales and plates, the arm including the ambulacral groove is completely enclosed by plates, spines are lateral at the arm, the mouth skeleton consists of many pieces. (A) Dorsal/aboral, (B) ventral/oral. (Source: S. Stöhr.)

eggs and sperm are released through gonoducts that open into the bursa, and then from the bursa through slit-shaped openings into the surrounding water. Like the stomach wall, the bursa wall may contain small ossicles. In most ophiuroids, the gonads are restricted to the disc, placed at either side of each arm, except in some non-branching euryalids and very few Ophiacanthidae (e.g., *Ophiocanops*) in which they extend into the arms. Many ophiuroids have a light sensitive neural net in their epidermis that allows phototactic behavior but no vision. The previously hypothesized calcitic microlenses as an optic system have been questioned recently.

CIII. REPRODUCTION AND DEVELOPMENT

Most ophiuroids are gonochoric and only few species are sexually dimorphic (e.g., *Ophiodahpne scripta*). Some species are simultaneous hermaphrodites (e.g., *Amphipholis squamata*), which may self-fertilize, others are sequential hermaphrodites (e.g., *Ophiopeza spinosa*), and these are first male then female. The planktotrophic ophiopluteus larva (Figure 35.4D) is considered ancestral and not homologous to the echinopluteus, but in living ophiuroids, lecithotrophic development is common. Ophioplutei typically have four pairs of arms that are supported by skeletal rods and bear ciliated bands. They have a complete digestive system with mouth, gut and anus. The pluteus metamorphoses directly into a juvenile in many species, in others development goes through a lecithotrophic doliolaria stage. In some species, the pluteus does not resorb its arms and may produce a second juvenile, a process known as larval cloning. Lecithotrophic larvae may be pelagic with a reduced pluteus or a doliolaria, or benthic with a vitellaria. In brooding species, a vitellaria is found, or development is direct without a larva. Brooders retain the eggs in the bursae, fertilization is internal and the embryos develop internally. The young ophiuroids may be released shortly after metamorphosis or kept in the bursae for an extended time, depending on the species. Pluteus larvae appear to be absent in the order Euryalida. Fissiparity is found in many hexamerous species, particularly in the genus *Ophiactis*, and these species also reproduce sexually by broadcast spawning, but brooding hexamerous species appear not to divide.

CIV. LIFESTYLE

Ophiuroids are free moving. Many species have a more cryptic lifestyle than asteroids, hiding in crevices (e.g., *Ophiothrix*), burrowed in the substrate (e.g., *Amphiura filiformis*), or living on a variety of hosts such as sponges (e.g., *Ophiactis savignyi*), echinoids (e.g., *Amphipholis linopneusti*), crinoids (e.g., *Ophiomaza cacaotica*) or even on pelagic jellyfish (*Ophiocnemis marmorata*). Basket stars perch on raised structures such as boulders or large corals to access the water current for filter feeding with their multi-branched arms. Other euryalids (e.g., *Asteronyx*) cling to sea pens. Ophiuroids are hosts to polychaetes, gastropods, copepods, sponges, and myzostomatids, in both commensal and parasitic relationships. Juvenile ophiuroids may settle on adults of the same species that are not their parents (e.g., *Ophiothrix fragilis*) or crawl into the bursae of another species (in Ophiocomidae), where they stay until they reach a larger size. Feeding modes vary between species and there are predators (e.g., *Ophiura sarsii*, *Ophiarachna incrassata*), scavengers, suspension feeders (e.g., *Ophiocomina nigra*), and deposit feeders (e.g., *Amphiura* spp.). Predators of ophiuroids include fishes, asteroids, and crustaceans.

SUBPHYLUM ECHINOZOA

The Echinozoa includes the living Echinoidea and Holothuroidea (Figure 35.1), and possibly the extinct Ophiocistioidea, Helicoplacoidea, Cyclocystoidea, and Edrioasteroidea (Moore 1966). The evolutionary history of this group is still being disputed and thus its origin can either be dated to the Lower Cambrian when Helicoplacoidea first appeared or to a later time period, depending on the interrelatonships of these groups. Echinozoa have a globular or cylindrical body plan without appendages.

D. CLASS ECHINOIDEA

Information for the Echinoidea was synthesized (From Byrne & O'Hara 2017; Kroh & Smith 2010; McEdward & Miner 2001; Moore 1966; Ruppert et al. 2004; Mongiardino Koch et al. 2018; Smith & Kroh 2019).

Synapomorphies

- body plan: globular
- Aristotle's lantern
- skeleton in alternating columns of ambulacrals and interambulacrals
- echinopluteus larva

DI. PHYLOGENY AND CLASSIFICATION

Echinoids first appeared in the Middle Ordovician (~470 Ma). There are about 850 living species, classified in about 70 families, and 174 extinct families from the fossil record. Traditionally, echinoids were grouped in regular and irregular forms. Recent phylogenetic hypotheses found the irregular echinoids to be a monophyletic group, whereas the regular echinoids are not monophyletic (**Figure 35.16**).

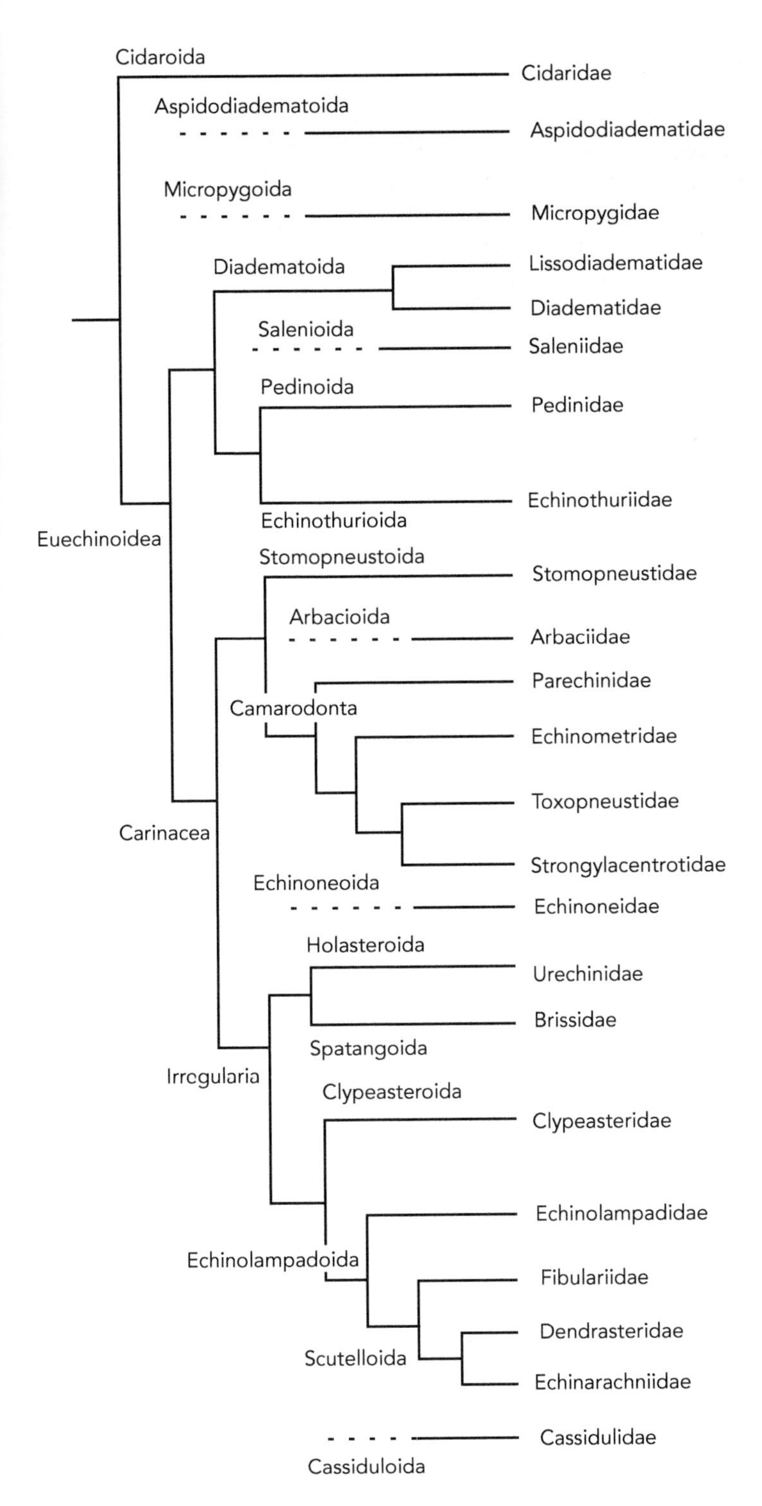

Figure 35.16 **Interrelationships of living Echinoidea orders and representative families.** Dashed lines indicate orders that were not included in the latest phylogenomic analysis and cannot be placed at this time. (Modified from Mongiardino Koch et al. 2018, incorporating Kroh and Smith 2010; families from WoRMS 2019.)

DII. ANATOMY, HISTOLOGY, AND MORPHOLOGY

The Echinoidea (sea urchins) are pentaradial with a typically globular body, either symmetrically round (pentaradial) and only slightly depressed in the regular urchins (**Figure 35.17**), or asymmetrical, secondarily bilateral and more or less flattened in the irregular urchins (heart urchins, sand dollars). Their

Figure 35.17 Sea urchin body shapes. Regular urchins: (A) *Echinus esculentus*, (B) *Diadema setosum*, (C) *Phyllacanthus imperialis*, Irregular urchin. (D) The sand dollar *Clypeaster humilis*. (Sources: A, S. Stöhr, B–D, H. Moosleitner.)

skeleton consists of many interlocking plates that form a rigid test or theca (**Figure 35.18**).

The mouth is orientated toward the substrate and most of the surface of the test belongs to the oral surface. The aboral surface is composed of the area around the anus, called the periproct, and the plates surrounding it, called the apical system, and the ambulacral and interambulacral plates downward to the periphery of the body. The apical system includes five genital plates, one of which is the madreporite, alternating with five ocular plates (Figure 35.18). The mouth opening is surrounded by a flexible membrane, the peristome, which in regular urchins is reinforced by numerous skeletal plates that support five pairs of large feet (buccal podia) and pedicellariae (Figure 35.18). In regular urchins, the mouth lies in the center of the oral surface of the test, the anus in the center of the aboral surface. In irregular urchins, the mouth is anterior, except in sand dollars, which have a central mouth, whereas the anus and periproct are posterior, outside the central apical system (Figure 35.18). The plates of the test are arranged in alternating double columns; the columns that belong to the water vascular system (clearly identifiable by the presence of external tube feet and internal ampullae) are called ambulacra and the columns between these are called interambulacra. The interambulacral plates bear tubercles to which spines and the pincer-like pedicellariae articulate, and the ambulacral plates have usually paired pores from which the tube feet extend. Pedicellariae in echinoids are morphologically highly diverse stalked appendages with valves that

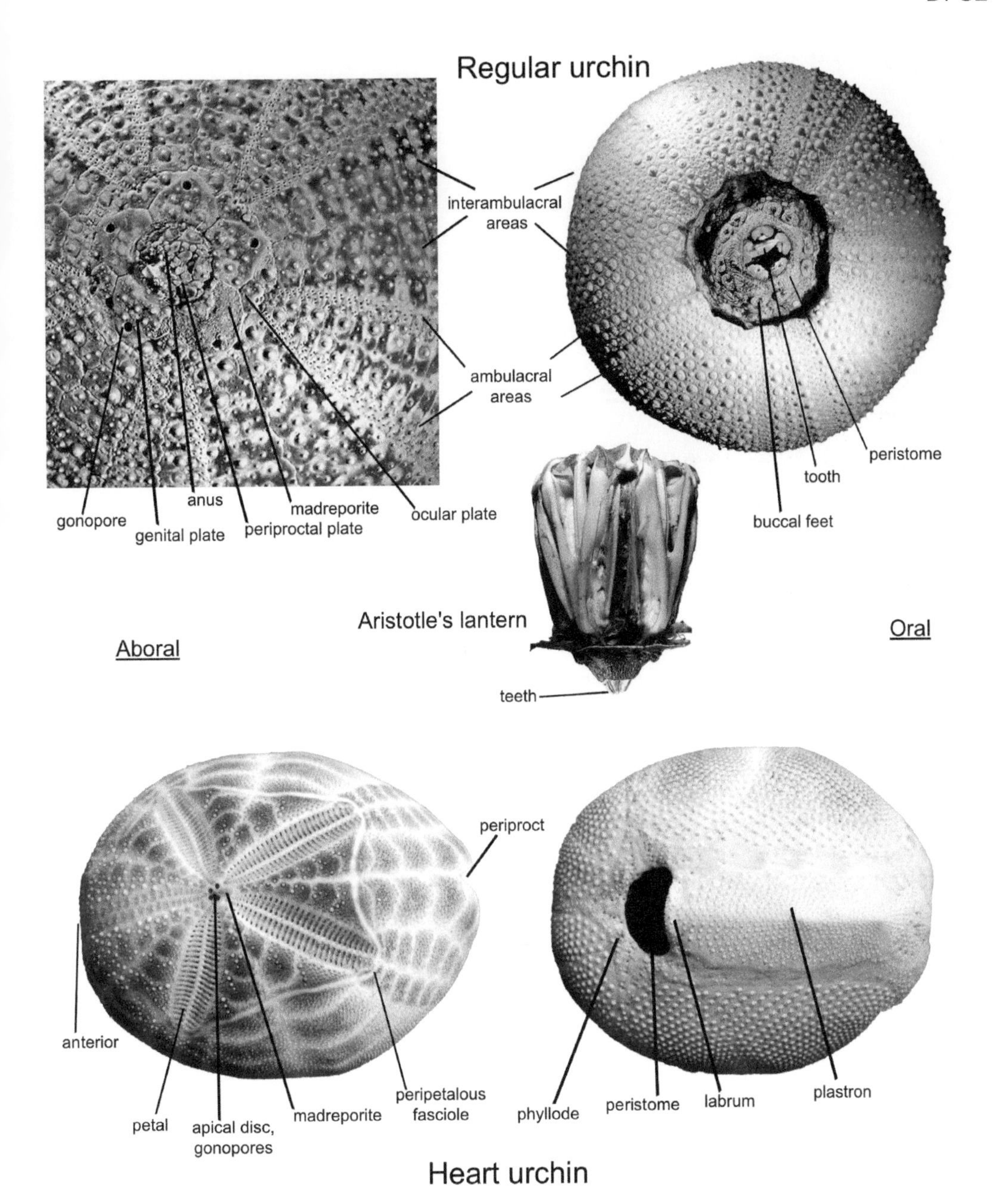

Figure 35.18 Morphology of sea urchins. The hard calcareous test consists of ambulacral areas with pores through which the tube feet extend, separated by interambulacral areas, lacking tube feet. In regular urchins, there is an apical system of specialized plates surrounding the anus and a complex apparatus of muscles and ossicles (Aristotle's lantern) for operating the teeth. Heart urchins have asymmetrical body symmetry where the anus is outside the apical system at the posterior end. (Sources: Regular urchin S. Stöhr; Heart urchin H. Moosleitner.)

can open and close (Figures 35.2, **35.19**). They are used for defense against pests and parasites and for cleaning the test surface.

Regular echinoids have a complex jaw apparatus called the Aristotle's lantern (Figure 35.18), which consists of large pyramidal jaws, teeth, and many other ossicles; ligaments and muscles attach to a skeletal support structure on the inside of the test, called the perignathic girdle. Special muscles control the up- and downward movement of the teeth. The teeth grow continuously at their inner end, as they are worn down by use. Sand dollars also have a well-developed Aristotle's lantern, but other irregular urchins lack a jaw apparatus. At the peristomial edge of the test in many regular urchins there are ten slits, the buccal notches, from which the buccal sacs extend. These function as expansion

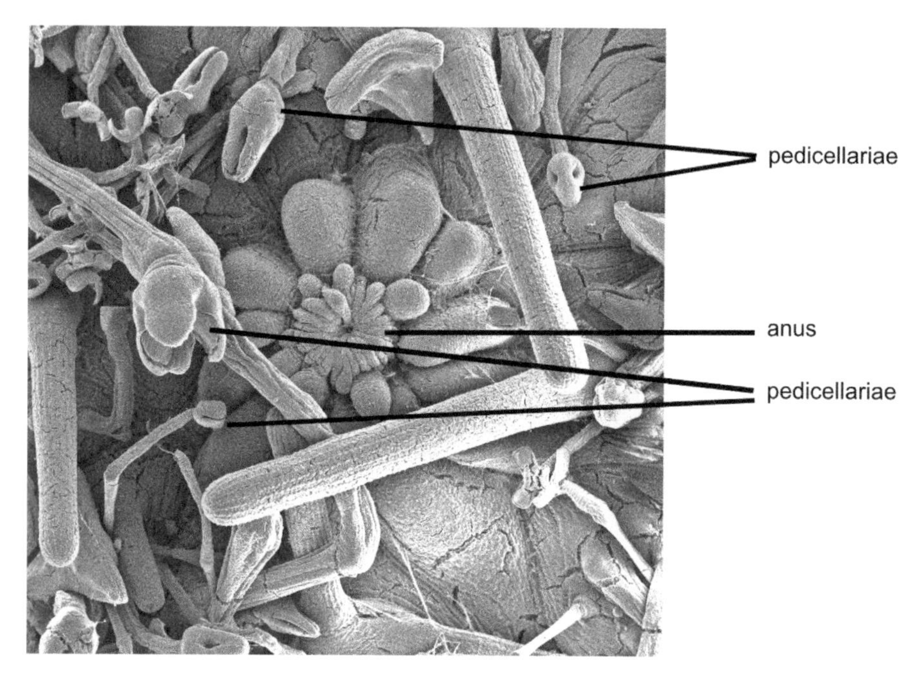

Figure 35.19 Different types of sea urchin pedicellariae, here surrounding the anus, in *Tripneustes ventricosus*. (Source: R. Turner.)

organs of the coelom in fluid control during lantern movement. Based on their appearance, the buccal sacs were long believed to have respiratory function and were called gills but this has been disproved. Irregular urchins have various specialized structures in adaptation to their burrowing habit, e.g., aboral bands of ciliary spines to create water currents inside their burrows, or food grooves that transport fine particles to the mouth.

Echinoids have a long, coiled gut, made up of a pharynx inside the lantern, leading to a stomach, followed by the esophagus and the intestine. The gonads are found on the inside of the aboral disc, connected to the genital hemal ring canal and to the genital plates with short gonoducts.

DIII. REPRODUCTION AND DEVELOPMENT

Most echinoids are gonochoric broadcast spawners. Their ancestral larva is the planktotrophic echinopluteus (Figure 35.4C), which is still found in the majority of species. It has evolved independently of the ophiopluteus. Echinoplutei typically have ciliated bands and eight arms with hard skeletal rods, but there is considerable variation among species in the number of arms. Lecithotrophic larvae have evolved in many echinoid groups and range from plutei with fewer arms, lacking a gut and ciliated bands, to ovoid larvae without arms, cilia or gut. Brooding echinoids keep the larvae externally among the spines, near the mouth, or in depressions of the ambulacral system called marsupia (e.g., *Abatus*).

Synapomorphies

- body plan: cylindrical
- feeding tentacles
- respiratory tree (water lungs)
- cuvierian tubules
- single gonad

DIV. LIFESTYLE

Irregular urchins burrow into soft sediment. Regular urchins live on the surface of various types of hard substrate, including coral reefs, and they may burrow into rock and coral with the aid of their teeth and spines. Many regular urchins eat algal matts and large kelp that they collect with spines and tube feet or graze from the substrate with their teeth. In the absence of predators, echinoids may cause severe overgrazing as has been observed off the North American west coast when sea otter populations declined and the increase of urchin populations decimated the kelp forests. Some cidaroid urchins are predators of stalked crinoids. Sea urchin spines and pedicellariae may contain toxins that can be painful to humans. Echinoids are hosts to sponges, hydroids, bryozoans, snails, ophiuroids, holothuroids, crustaceans, and fish. Predators of echinoids are fish, molluscs, and sea otters.

E. CLASS HOLOTHUROIDEA

Information for the Holothuroidea was synthesized from (Byrne & O'Hara 2017; Kerr & Kim 2001; McEdward & Miner 2001; Ruppert et al. 2004)

EI. PHYLOGENY AND CLASSIFICATION

Holothuroids (sea cucumbers) first appeared in the late Silurian (~420 Ma). The fossil record is poor, with fewer than 20 species known from body fossils, because most holothuroids have a soft body with only minute skeletal elements (Figure 35.2) that are difficult to assign to species. The latest molecular phylogeny accepts seven orders (**Figure 35.20**), but several families appear to be paraphyletic or polyphyletic and more data are needed to resolve these.

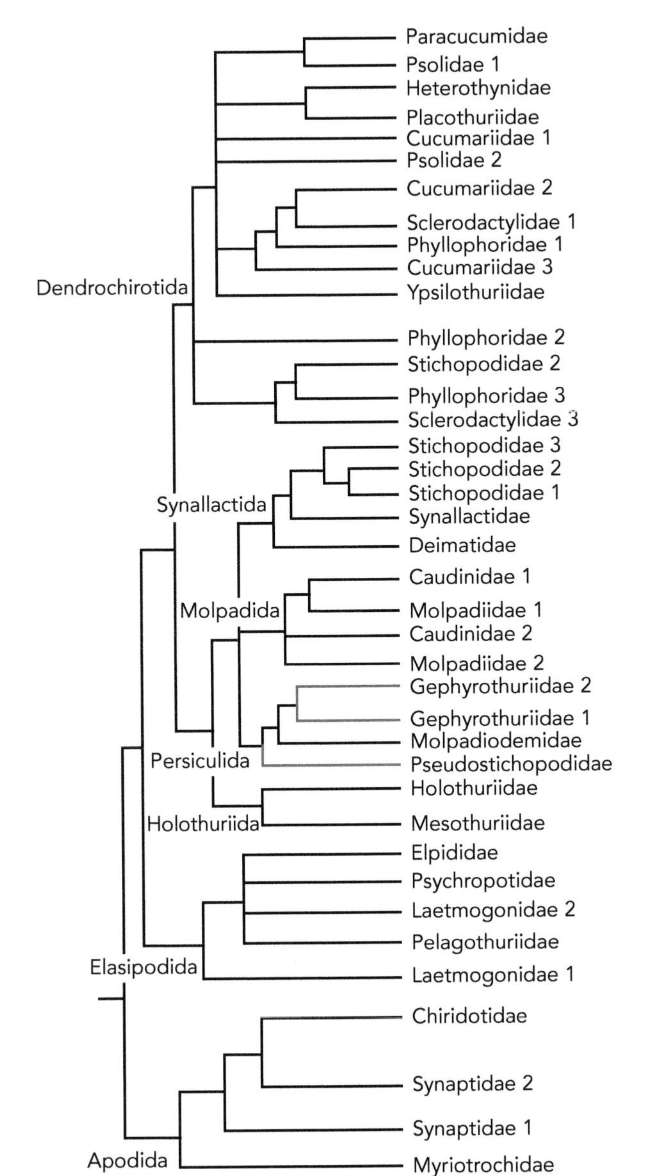

Figure 35.20 Interrelationships of Holothuroidea families and orders. Several families are paraphyletic or polyphyletic and some branches are not fully resolved due to lack of data. (Modified after Miller et al. 2017.)

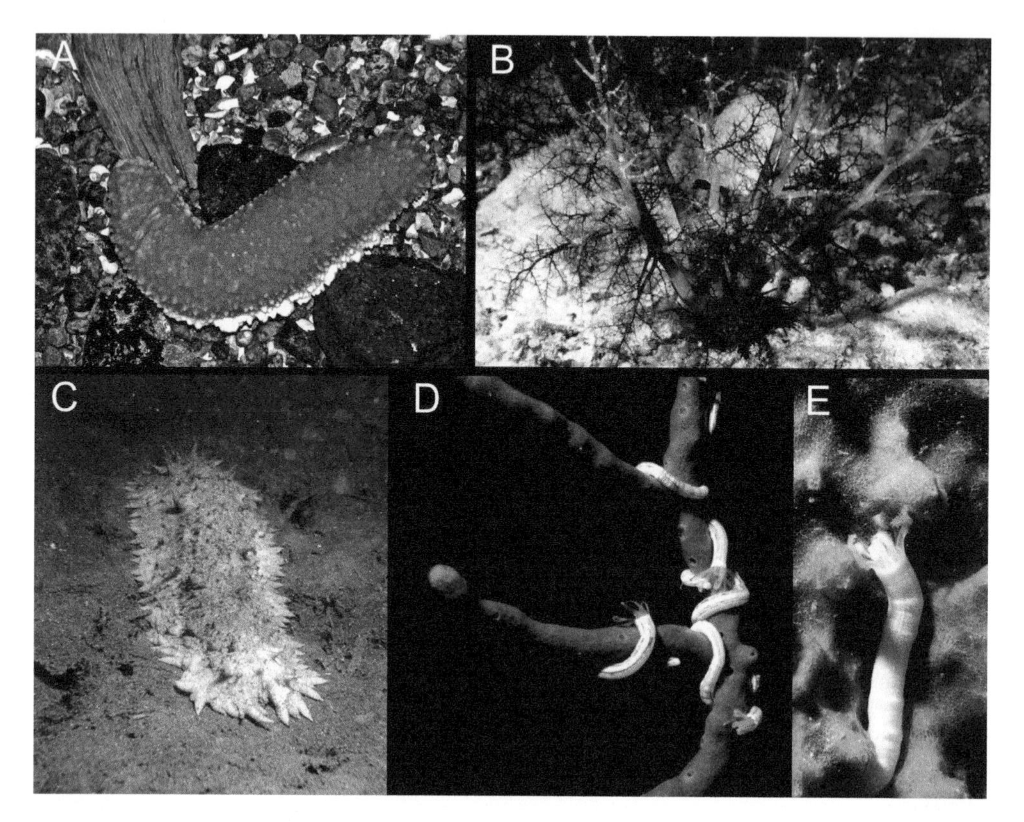

Figure 35.21 **Some sea cucumber forms.** (A) *Parastichopus tremulus*, (B) extended feeding tentacles in *Neothyonidium magnum*, (C) *Stichopus regalis*, (D, E) *Synaptula* sp. (Sources: A, S. Stöhr; B-E, H. Moosleitner.)

EII. ANATOMY, HISTOLOGY, AND MORPHOLOGY

Holothuroids have a cylindrical body, with the mouth at one end (anterior) and the anus at the opposite end (posterior). Body shapes vary and there are sausage-shaped elongated (**Figure 35.21**), U-shaped, flask-shaped, and limpet-shaped species. Holothuroids have strong longitudinal muscle bands in their integument, giving them great dexterity. The skeleton may consist of strong plates (e.g., Psolidae), but in most species, it is reduced to microscopic spicules embedded in the dermis, where they may be densely packed or sparsely scattered. The spicules display a high diversity in shapes, from simple scales with mesh-like stereom to rod, star, and anchor shapes (Figure 35.2). A ring of ten more robust skeletal elements surrounds the mouth opening and supports the contractile buccal feeding tentacles (**Figure 35.22**). These tentacles are modified buccal feet, 10–30 in number depending on the species, and more or less branched. They extend directly from the water vascular ring canal.

The holothuroid digestive system includes a pharynx inside the calcareous oral ring, a small stomach and a long S-shaped intestine (attached to the interior of the body wall by a mesentery) that leads to a cloaca (rectum) that opens through the anus (Figure 35.21). Both mouth and anus have sphincter muscles that can close the openings. From the circumoral water vascular ring canal one or several stone canals lead to one or several madreporites, which lie inside the coelom without connection to the outside. There is also at least one Polian vesicle. Five longitudinal (radial) canals extend from the oral ring posteriorly. These are embryologically interradial and not homologous with the radial canals in other echinoderms. Tube feet with ampullae are sometimes connected to these canals, but do not indicate an ambulacral region. The holothuroids have a pentaradial symmetry with secondary bilateral body plan, which may have

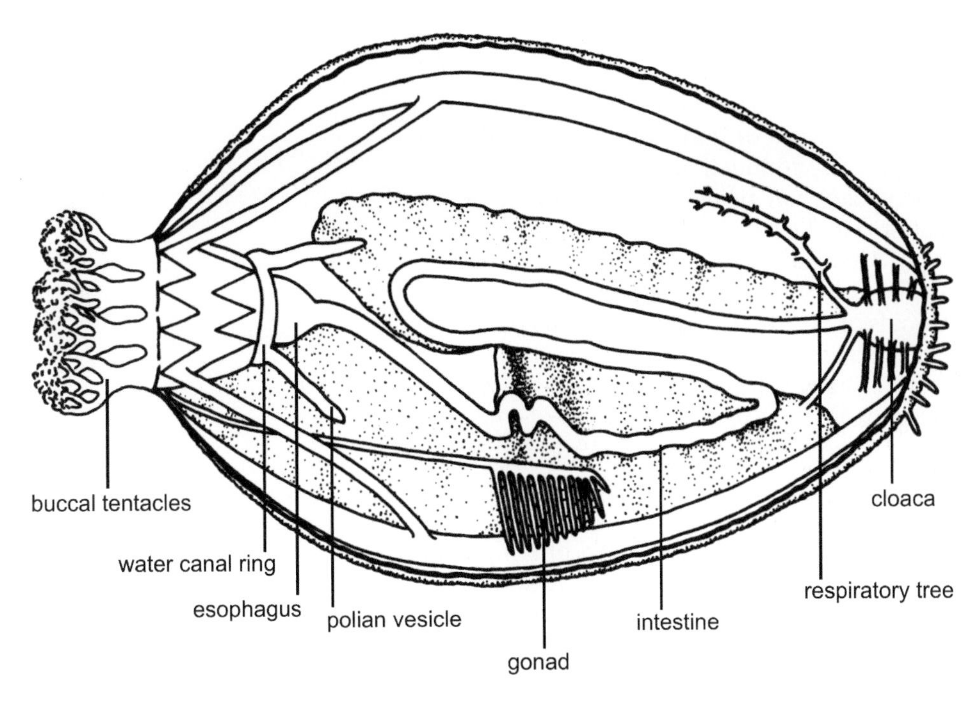

Figure 35.22 **Anatomy of a sea cucumber.** (Modified from Byrne 1985, with permission.)

well defined dorsal and ventral surfaces that do not correspond to the positions of mouth or anus. As in echinoids, most of the body surface belongs to the oral part, and only a small area around the anus is the aboral surface. Modifications of the tube feet and their arrangement over the body surface are common. Tube feet are absent in Molpadida, and Synaptulida lack both feet and longitudinal canals. Holothuroids have a single longitudinal gonad that connects with a gonoduct to a gonopore that lies dorsally near the buccal tentacles. Some species possess a branched respiratory organ and some of these also have sticky and toxic organs that produce Cuvierian tubules, both connected to the rectum. Respiratory organs and Cuvierian tubules can be expelled through the anus as a defense mechanism, called evisceration. These organs are later regenerated.

EIII. REPRODUCTION AND DEVELOPMENT

Holothuroids are generally gonochoric and the sexes are externally similar, except in brooding species that have dimorphic genital papillae. Asexual reproduction by fission occurs in some species (e.g., *Holothuria abra*), but the mechanisms that trigger it are unknown. The ancestral holothuroid larva is the auricularia (Figure 35.4E), an ovoid lobed larva with a single ciliated band tracing the body contour. It is usually planktotrophic but in some species it may be non-feeding, and in the majority of holothuroid orders it is absent, replaced by non-feeding larvae. The auricularia has a complete digestive tract with mouth, gut and anus. During development, the auricularia transforms into a doliolaria larva, which eventually metamorphoses into a juvenile sea cucumber (pentactula stage), and all these stages were originally pelagic. The majority of holothuroids have lecithotrophic development with barrel-shaped doliolaria or vitellaria larvae, which may be pelagic or benthic. Vitellariae are characterized by being overall ciliated, whereas the cilia on doliolariae are organized in transverse bands, but intermediate forms are known. Brooding has been observed, in some species with a reduced auricularia, in others direct development without a larva is assumed, but more studies are needed to determine the mode of development.

EIV. LIFESTYLE

Holothurians are free-moving, although some species are rather sedentary, and typically live on the surface of all types of bottoms or burrowing in soft sediment. They move by working their tube feet or by contractions of their body wall musculature. Some species (e.g., Elasipodidae) can swim by contracting their body wall muscles, others have a posterior veil-like structure composed of fused webbed tube feet (e.g., Pelagothuriidae). They feed by ingesting the substrate and extracting food particles and microbes or by filter-feeding suspended particles from the water column. Holothuroids are hosts to molluscs, crustaceans, sponges, and other echinoderms. An unusual association involves pearlfish that live inside the coelom of sea cucumbers, usually without harming their hosts, but some species eat the sea cucumber gonads.

ACKNOWLEDGMENTS

I am very grateful to the following people, who generously provided images: Horst Moosleitner, Charles G. Messing, Richard L. Turner, Andreas Kroh, Mary Sewell, Mike Reich, and Maria Byrne. Larvae images with CC BY license were downloaded from Planktonnet by the Alfred-Wegener-Institut (http://planktonnet.awi.de), taken by Otto Larink and Rebekka Schüller. Alessandro Allievi created the sea star illustration. I would also like to thank Rich Mooi for many stimulating discussions and proofreading of the manuscript.

REFERENCES

Arnone, M.I., Byrne, M. & Martinez, P. (2015) Echinodermata. In: Wanninger A. (Ed.), *Evolutionary Developmental Biology of Invertebrates*. Springer, Vienna, pp. 1–58.

Baumiller, T.K. (2008) Crinoid ecological morphology. *Annual Review of Earth and Planetary Sciences* 36, 221–249.

Byrne M. 1985. Evisceration behaviour and the seasonal incidence of evisceration in the holothurian *Eupentacta quinquesemita* (Selenka). *Ophelia* 24, 75–90.

Byrne, M. & O'Hara, T.D. (Eds.) (2017) *Australian Echinoderms*. CSIRO Publishing, Melbourne.

Conand, C. & Byrne, M. (1993) A review of recent developments in the world sea cucumber fisheries. *Marine Fisheries Review* 55, 1–13.

David, B. & Mooi, R. (2014) How Hox genes can shed light on the place of echinoderms among the deuterostomes. *EvoDevo* 5, 22.

Dolmatov, I.Y., Afanasyev, S.V. & Boyko, A.V. (2018) Molecular mechanisms of fission in echinoderms: Transcriptome analysis. *PLoS One* 13, e0195836.

Eaves, A.A. & Palmer, A.R. (2003) Reproduction: Widespread cloning in echinoderm larvae. *Nature* 425, 146.

Eeckhaut, I. (Ed.) (2018) Beche-de-mer Information Bulletins 38, 1–100. (available from https://coastfish.spc.int/en/publications/bulletins/beche-de-mer/483-beche-de-mer-information-bulletin-38).

González-Wangüemert, M., Domínguez-Godino, J.A. & Cánovas, F. (2018) The fast development of sea cucumber fisheries in the Mediterranean and NE Atlantic waters: From a new marine resource to its over-exploitation. *Ocean and Coastal Management* 151, 165–177.

Hendler, G. (2018) Armed to the teeth: a new paradigm for the buccal skeleton of brittle stars (Echinodermata: Ophiuroidea). *Contributions in Science* 526, 189–311.

Hennebert, E. (2010) Adhesion mechanisms developed by sea stars: A review of the ultrastructure and composition of tube feet and their secretion. In: J. von Byern, J. and Grunwald, I. (Eds.), *Biological Adhesive Systems: From Nature to Technical and Medical Application*. Springer Vienna, Vienna, pp. 99–109.

Heyland, A., Hodin, J. & Bishop, C. (2014) Manipulation of developing juvenile structures in purple sea urchins (Strongylocentrotus purpuratus) by morpholino injection into late stage larvae. *PLoS One* 9, e113866.

Kerr, A.M. & Kim, J. (2001) Phylogeny of Holothuroidea (Echinodermata) inferred from morphology. *Zoological Journal of the Linnean Society* 133, 63–81.

Kroh, A. & Smith, A.B. (2010) The phylogeny and classification of post-Palaeozoic echinoids. *Journal of Systematic Palaeontology* 8, 147–212.

Lawrence, J.M. (Ed.) (2013) *Starfish: Biology and Ecology of the Asteroidea*. The John Hopkins University Press, Baltimore.

Linchangco, G.V., Foltz, D.W., Reid, R., Williams, J., Nodzak, C., Kerr, A.M., Miller, A.K., Hunter, R., Wilson, N.G., Nielsen, W.J., Mah, C.L., Rouse, G.W., Wray, G.A. & Janies, D.A. (2017) The phylogeny of extant starfish (Asteroidea: Echinodermata) including *Xyloplax*, based on comparative transcriptomics. *Molecular Phylogenetics and Evolution* 115, 161–170.

Mah, C.L. (2019) World Asteroidea Database. Available from: http://www.marinespecies.org/asteroidea/ (accessed on March 11, 2019).

Mah, C.L. & Blake, D.B. (2012) Global diversity and phylogeny of the Asteroidea (Echinodermata). *PLoS One* 7, e35644.

Marmouzi, I., Tamsouri, N., El Hamdani, M., Attar, A., Kharbach, M., Alami, R., El Jemli, M., Cherrah, Y., Ebada, S. S. & Faouzi, M.E.A. (2018) Pharmacological and chemical properties of some marine echinoderms. *Revista Brasileira de Farmacognosia* 28, 575–581.

McEdward, L.R. & Miner, B.G. (2001) Larval and life-cycle patterns in echinoderms. *Canadian Journal of Zoology* 79, 1125–1170.

Miller, A.K., Kerr, A.M., Paulay, G., Reich, M., Wilson, N.G., Carvajal, J.I. & Rouse, G.W. (2017) Molecular phylogeny of extant Holothuroidea (Echinodermata). *Molecular Phylogenetics and Evolution* 111, 110–131.

Mongiardino Koch, N., Coppard, S.E., Lessios, H.A., Briggs, D.E.G., Mooi, R. & Rouse, G. (2018). A phylogenomic resolution of the sea urchin tree of life. *BMC Evolutionary Biology* 18, 189.

Moore, R.C. ed. (1966) *Part U: Echinodermata 3 Treatise on Invertebrate Paleontology*. University of Kansas Press, Boulder.

Nakano, H., Hibino, T., Oji, T., Hara, Y. & Amemiya, S. (2003) Larval stages of a living sea lily (stalked crinoid echinoderm). *Nature* 421, 158–160.

O'Hara, T.D., Hugall, A.F., Thuy, B. & Moussalli, A. (2014) Phylogenomic resolution of the Class Ophiuroidea unlocks a global microfossil record. *Current Biology* 24, 1874–1879.

O'Hara, T.D., Hugall, A.F., Thuy, B., Stöhr, S. & Martynov, A.V. (2017) Restructuring higher taxonomy using broad-scale phylogenomics: The living Ophiuroidea. *Molecular Phylogenetics and Evolution* 107, 415–430.

O'Hara, T.D., Stöhr, S., Hugall, A.F., Thuy, B. & Martynov, A. (2018) Morphological diagnoses of higher taxa in Ophiuroidea (Echinodermata) in support of a new classification. *European Journal of Taxonomy*, 1–35.

Reich, A., Dunn, C., Akasaka, K. & Wessel, G. (2015) Phylogenomic analyses of Echinodermata support the sister groups of Asterozoa and Echinozoa. *PLoS One* 10, e0119627.

Reynolds, J.A. & Wilen, J.E. (2000) The sea urchin fishery: Harvesting, processing and the market. *Marine Resource Economics* 15, 115–126.

Ruppert, E.E., Fox, R.S. & Barnes, R.D. (2004) *Invertebrate Zoology*. 7th edition. Brooks/Cole, Belmont, CA.

Santos, R., Haesaerts, D., Jangoux, M. & Flammang, P. (2005) Comparative histological and immunohistochemical study of sea star tube feet (Echinodermata, Asteroidea). *Journal of Morphology* 263, 259–269.

Smith, A.B. & Kroh, A. (2019) The Echinoid Directory. World Wide Web electronic publication. *The Echinoid Directory. World Wide Web electronic publication*. Available from: http://www.nhm.ac.uk/our-science/data/echinoid-directory/index.html (accessed on March 27, 2019).

Stabili, L., Acquaviva, M.I., Cavallo, R.A., Gerardi, C., Narracci, M. & Pagliara, P. (2018) Screening of Three Echinoderm Species as New Opportunity for Drug Discovery: Their Bioactivities and Antimicrobial Properties. *Evidence-Based Complementary and Alternative Medicine*. Available from: https://www.hindawi.com/journals/ecam/2018/7891748/ (accessed on March 26, 2019).

Stöhr, S., O'Hara, T.D. & Thuy, B. (2018) World Ophiuroidea Database. *World Ophiuroidea Database*. Available from: http://www.marinespecies.org/ophiuroidea (accessed on September 19, 2018).

Telford, M.J., Lowe, C.J., Cameron, C.B., Ortega-Martinez, O., Aronowicz, J., Oliveri, P. & Copley, R.R. (2014) Phylogenomic analysis of echinoderm class relationships supports Asterozoa. *Proceedings of the Royal Society B: Biological Sciences* 281, 20140479.

Turner, R. & Meyer, C.E. (1980). Salinity tolerance of the brackish-water echinoderm *Ophiophragmus filograneus* (Ophiuroidea). *Marine Ecology Progress Series* 2, 249–256.

Thuy, B. & Stöhr, S. (2018) Unravelling the origin of the basket stars and their allies (Echinodermata, Ophiuroidea, Euryalida). *Scientific Reports* 8, 8493.

Woolley, S.N.C., Tittensor, D.P., Dunstan, P.K., Guillera-Arroita, G., Lahoz-Monfort, J.J., Wintle, B.A., Worm, B. & O'Hara, T.D. (2016) Deep-sea diversity patterns are shaped by energy availability. *Nature* 533, 393–396. https://doi.org/10.1038/nature17937.

WoRMS editorial board (2019) World register of marine species. *World register of marine species*. Available from: http://www.marinespecies.org (accessed on February 22, 2019).

Wray, G.A. (1999) Echinodermata. Spiny-skinned animals: sea urchins, starfish, and their allies. *Tree of Life web project*. Available from: http://tolweb.org/Echinodermata/2497/1999.12.14.

Zueva, O., Khoury, M., Heinzeller, T., Mashanova, D. & Mashanov, V. (2018) The complex simplicity of the brittle star nervous system. *Frontiers in Zoology* 15, 1.

CHAPTER

36

PHYLUM HEMICHORDATA

Kenneth M. Halanych

Infrakingdom	Bilateria
Division	Deuterostomia
Phylum	Hemichordata

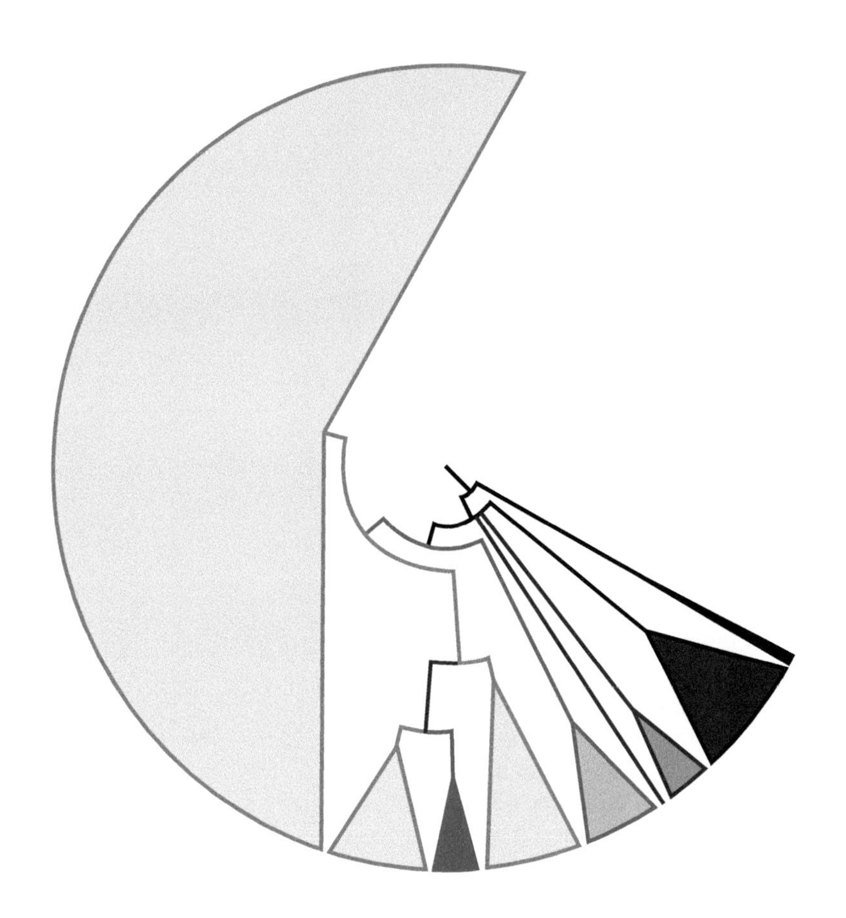

CONTENTS

General Characteristics

Hemichordates have a tripartite body and comprise two clades

Enteropneusta (Gegenbaur 1870) comprise the acorn worms with 108 valid species

Pterobranchia (Lankester 1877) are a clade of colonial filter feeders with 23 valid species

The vermiform enteropneusts are mainly infaunal or epifaunal deposit feeders

Pterobranchs form erect or encrusting colonies usually in high-flow areas

They live from the intertidal to the over 8,000 m depth

They have a coelomic body organization

No commercial value

Ecological importance in certain environments

Autapomorphies

- tripartite body plan with a distinct collar region
- buccal diverticulum projecting anteriorly from collar region
- molecular characters based on 18S rDNA, mtDNA genomes and phylogenomic data

HISTORY OF TAXONOMY AND CLASSIFICATION

Eschscholtz was the first to formally collect hemichordates when *Ptychodera flava* (Eschscholtz 1825) was discovered in the Marshall Islands. Eschscholtz thought this new animal to be a sea cucumber and his work remained relatively unknown. The discovery of *Balanoglossus clavigerus* (Delle Chiaje 1829) attracted more attention. The gill slits flanking the gut in these vermiform creatures were not described until Kowalevsky's work (1866) leading Gegenbaur (1870) to coin the name Enteropneusti (Greek – εντερο: "entero," *intestine*; πνευμον: "pneumon," *lung*). In contrast, pterobranch hemichordates were discovered slightly later. Collections of *Rhabdopleura normani* (Allman 1869) by G.O. Sars in 1866 have typically been thought to be the first collection of pterobranchs. However, *Cephalodiscus* (M'Intosh 1882) was collected during the *Erebus* and *Terror* expedition to the Antarctic over 20 years earlier, in either 1841 or 1842, and went unnoticed in museum collections (Ridewood 1912,1921). Because these new pterobranch animals were colonial and built a coenecium, or exterior casing, they were grouped with previously known colonial animals, including bryozoans and hydroids. Lankester (1877), who considered these animals as bryozoans, created the term Pterobranchia (Greek –πτερον: "ptero" *wing or feather*; Old French – branche, *gill*) in recognition that the featherlike tentacles also acted as gills. Not until Harmer's (1887) work recognizing the importance of the gill pores in *Cephalodiscus*, were pterobranchs considered as related to acorn worms. Just prior to Harmer's work, Bateson (1885) had erected Hemichordata to ally hemichordates with chordate animals. In contrast, Metschnikoff (1881) argued that the system of water handling channels in some acorn worm larvae resembled the ambulacrarial system of echinoderms. He correctly inferred hemichordates and echinoderms to be sister taxa in Ambulacraria (see Halanych 1995, Cannon et al. 2014). The current scheme used today of two major clades or classes, Enteropneusta and Pterobranchia, within Hemichordata was the work of Willey (1899a, b).

Currently hemichordates comprise Enteropneusta and Pterobranchia (Tassia et al. 2019; Halanych et al. 2019). Enteropneusta contains 108 valid species within 4 recognized families. The 23 species of pterobranchs are assigned to 2 genera *Cephalodiscus* and *Rhabdopleura*.

ANATOMY AND MORPHOLOGY

Although both enteropneusts and pterobranchs have three body regions, their general morphology is markedly different (**Figures 36.1** and **36.2**). Acorn worms range in size from a few millimeters (meiofaunal) to almost a meter in

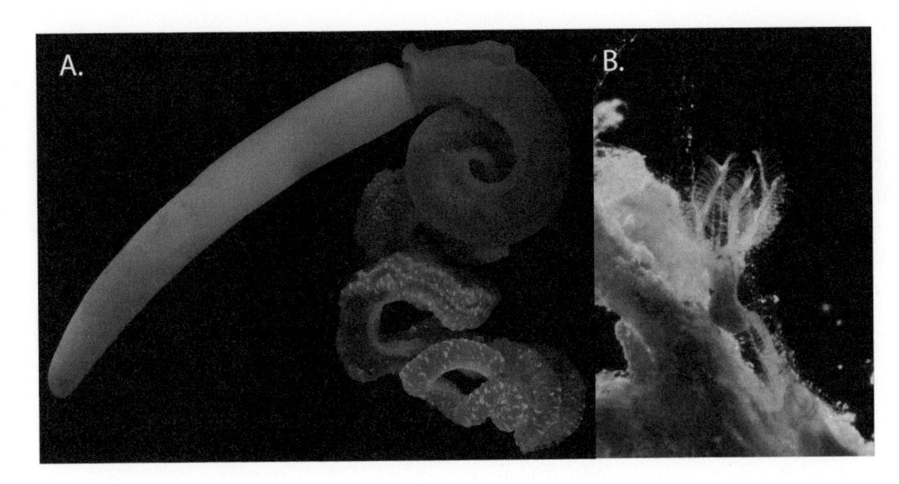

Figure 36.1 (A) *Saccoglossus kowalevskii*, an example of an enteropneust. Photo credit Michael Tassia. (B) *Cephalodiscus gracilis*, and example of a pterobranch zooid and coenecium. (Photo credit Ken Halanych.)

length in the deep sea, but most familiar forms range 5–15 cm long and 1–3 cm wide; their bodies are soft and extensible making exact measurement difficult. Enteropneusts have a cylindrical to spherical shaped proboscis followed by a well-developed collar region. In some of the deep-sea enteropneusts, the collar region can possess a pair of relatively large lateral projections. Posterior to the collar is a long trunk that has a series of ciliated gill slits on either side. In contrast, individual pterobranchs range in size from 1 to 10 mm but their colonies can be up to 25 cm in diameter. Pterobranchs possess a flattened ovoid proboscis called a cephalic shield that connects to a small collar region bearing tentaculated arms used in filter feeding. The trunk is elongate and houses a U-shaped

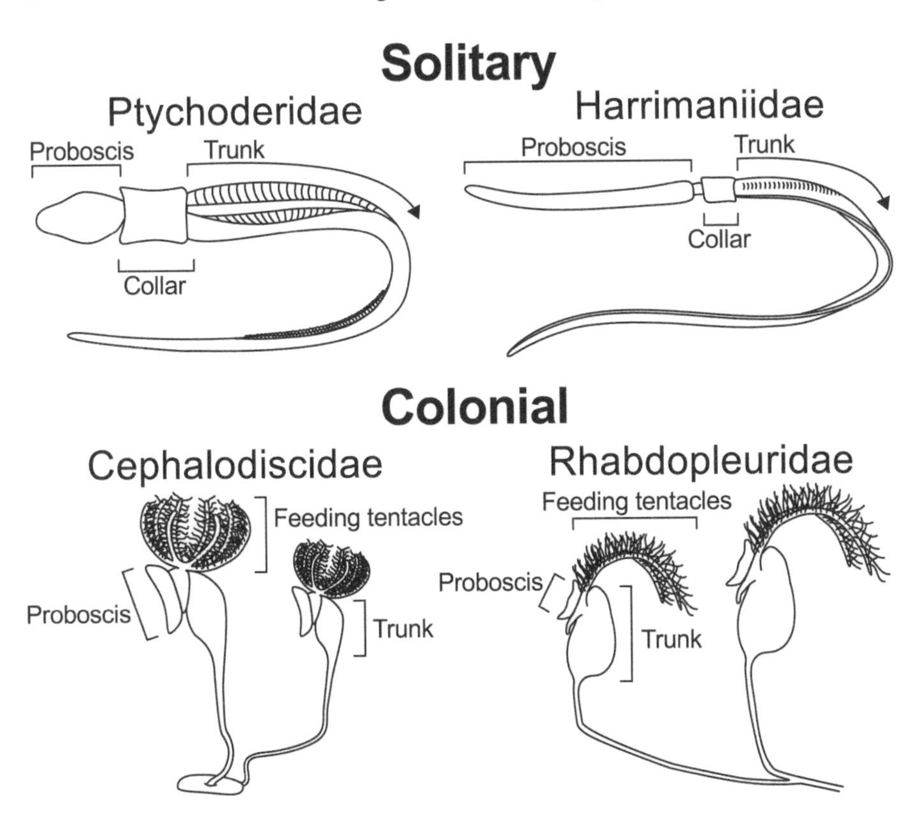

Figure 36.2 Generalized body plan of hemichordates. The top panel shows typical worm-like morphology for enteropneusts. The bottom panel shows the two types of colonial filter feeding pterobranchs. (This image is from Tassia et al. 2016 and was redrawn from Rychel and Swalla 2009.)

digestive track with the anus somewhat anteriorly placed on the dorsal side. This posterior region of the pharynx (near the anterior of the trunk) possesses a ciliated gill pore in the case of *Cephalodiscus,* but *Rhabdopleura* lacks this structure. Posteriorly, the trunk tapers into a long stolon that connected the zooids with other individuals in the colony. Pterobranchs live inside a self-made house, the coenecium, that is made up of secretions by the cephalic shield. Depending on the species of pterobranch and the amount of water flow, the coenecium can be more of a low-lying creeping form or a large erect structure.

Gill slits of hemichordates are conspicuous and were one of features used to originally place them within chordates. The homology of gill slits between hemichordates and chordates is not in question, and phylogenetic analyses show gill slits to be a deuterostome symplesiomorphy (Halanych 1995, Cannon et al. 2014, Tassia et al. 2019). Both the morphological development (Gillis et al. 2012) and the underlying molecular mechanisms (e.g., *Pax 1/9*; Ogasawara et al. 1999) involved in gill slit development are conserved in deuterostomes. The number of acorn worm gill slits can vary from one gill slit (*Meioglossus psammophilus*) to well over a hundred (*Balanoglossus aurantiacus*). Moreover, acorn worm gill slits can possess cartilaginous support structures.

Hemichordates tend to be soft bodied and break apart easily. The epidermis is generally ciliated but cilia are particularly dense on the proboscis and collar regions of acorn worms as they aid in food handling and particle sorting. The cephalic shield of pterobranchs is densely ciliated on the ventral region which is used for tube building as a locomotory surface. The feeding arms and tentacles also contain cilia organized to aid particle capture, sorting, and transport (Halanych 1996). Distribution of glandular (e.g., mucous producing) cells generally mirrors the distribution of cilia. The mucous is used for locomotion and particle capture, but in acorn worms it may also serve defensive purposes, although more empirical evidence is required. Whereas all acorn worms likely have the ability to produce mucous throughout the length of the body, some species of deep-sea Torquaratoridae can produce sufficient quantities to build mucous tubes (Halanych et al. 2013).

During development, hemichordates possess three sets of coelomic cavities (i.e., proboscis protocoel, paired collar mesocoels, and paired trunk metacoels). Adult worms retain paired coeloms in both the collar and trunk regions. In the proboscis, enteropneusts typically retain a single unpaired coelom, but in pterobranchs this coelom is reduced. Within the buccal cavity, enteropneusts and pterobranchs both possess an anterior buccal diverticulum sometimes referred to as the stomochord. Earlier workers homologized this structure to the chordate notochord. However, this has been an intensely debated hypothesis (Bateson 1886, Silén 1954, Annona et al. 2015).

Nervous systems of hemichordates were classically considered to be diffuse nerve nets, but the actual situation is a bit more complex. Throughout most of the body, nerves are intraepidermal with no clear pattern or organizational scheme. In acorn worms, intraepidermal nerves converge along the dorsal and ventral apices of the trunk into roughly organized trunk nerve bundles. By comparison, nerves are better delineated and more internalized in the collar region where they form the collar nerve chord. This increased complexity may be due to the presence of other complex structures in this region (e.g., the heart–kidney complex and buccal diverticula). The nervous system in pterobranchs has not been well-studied, but intraepidermal nerves seem to converge into ganglia at the base of tentaculated arms in the collar region. Nerve tracks (as opposed to organized chords) radiate from these ganglia particularly along the mid-ventral line of the trunk (Rehkämper et al. 1987).

Both pterobranchs and enteropneusts possess a blood vascular system. In pterobranchs, the blood vessels are best developed in the arms and cellular products can be seen moving through the vessels under light microscopy. Acorn worms' blood vessels are arranged mainly as dorsal and ventral tubes that run anteriorly to posteriorly. In addition, enteropneusts have a pericardium and heart that sit dorsally on the stomal chord. Blood moving from the heart moves anteriorly through a glomerulus where excretory products are filtered out and eventually excreted through a pore on the proboscis. The blood vascular system is also well developed in association with the gill slit facilitating oxygen exchange; many enteropneusts live in muddy environments where oxygen is likely low.

REPRODUCTION AND DEVELOPMENT

Hemichordates are dioecious, but a few cases of hermaphroditic pterobranchs have been reported. Development of acorn worm species within *Saccoglossus* and *Ptychodera* have been best studied (Bateson 1885, Morgan 1891, Burdon-Jones 1952, Lowe 2008, Röttinger et al. 2015, Tagawa et al. 1998), and very little information on early development in pterobranchs is available. Acorn worms within Harrimaniidae undergo direct development whereas members of Ptychoderidae and Spengelidae have indirect development with tornaria larvae. Although the developmental mode of Torquaratoridae has not been studied (beyond the ability of some to brood; Osborn et al. 2013), their phylogenetic placement predicts indirect development.

Early development of enteropneusts studied to date is similar through the gastrula stage and cleavage has been characterized as holoblastic and radial. For direct developing acorn worms, the blastopore closes off after gastrulation and the embryo begins to elongate. The boundary between the proboscis and the rest of the body is the first to arise which is also when coelomic sacks arise from lateral pockets of the archenteron (Kaul-Strehlow and Stach 2013). Subsequently, the demarcation between the collar and the trunk forms followed by a progression of gill slits developing from anterior to posterior along the trunk. In contrast, indirect developing enteropneusts will form a ciliated larva that hatches after the protocoel develops but before the archenteron connects with the epidermis to form the mouth. At this stage, the organism is considered a tornaria larva which, over time, develops an increasingly intricate series of grooves that aid water flow and feeding over its ciliated surface. As the larva gets ready to metamorphose, the anterior end begins to repattern and a collar emerges in the mouth region. At this point, the posterior region of the larva begins to bulge and elongate to form the trunk region of the animal. In terms of homologizing larval and adult body, most of the tornaria larva turns into proboscis (anterior to the larval mouth) or collar (around and just posterior to the larval mouth) whereas the trunk region is only the most posterior of the larva. Some enteropneust, particularly the ones with indirect development are known to have robust regenerative capabilities (Luttrell et al. 2018). Despite these abilities to regenerate, asexual reproduction is not an acorn worm characteristic.

Information on pterobranch reproduction is limited (but see John 1931, Lester 1988). Larvae are yolky and often retained inside the coenecial tubes, making observation of early embryology difficult. Larvae that are starting to develop adult features have been observed inside the tubes and the swimming ability of the larvae is not really known. A competent larva will settle to form a progenitor of a new colony. In the case of *Cephalodiscus*, whether a larva can settle within the same coenecium is unknown. Once the progenitor settles and builds a tube, it will begin to bud off new individuals. In addition to sexual development, pterobranchs build colonies through asexual reproduction. *Rhabdopleura* colonies form in a manner where each individual has an individual tube. As a result, a common stolon connects all the zooids of the colony. Although some *Cephalodiscus* asexually reproduce in this manner, several species produce unconnected clusters of individuals that live within a more open coenecium.

Each of these clusters typically has one to several mature feeding zooids that support new zooid development. At some point, the common attach organ holding these clusters together can split resulting in new clusters.

DISTRIBUTION AND ECOLOGY

Hemichordates are found globally in marine and estuarine ecosystems and range in depth from intertidal to < 8,000 m. Because of their fragile nature they are easily broken or destroyed by mechanical collection. Thus, acorn worms are best known from intertidal regions. However, there are number of undescribed meiofaunal taxa, and work in the deep sea during the past 15 years has yielded several new species (e.g., Holland et al. 2005, Deland et al. 2010, Osborn et al. 2013). Most acorn worms are deposit feeders, although *Saxipendium coronatum* can live on recent basalt flow near hydrothermal vents that contain little sediment. The worms can be meiofaunal, infaunal, or epifaunal in habitat preference. Pterobranchs colonize high-flow areas, presumably to aid their filter feeding. Although a few species north of the equator are known, pterobranchs are primarily found in the southern hemisphere and are especially abundant and diverse around Antarctica. These colonial animals are typically found in cold (deep waters or high latitude) or shallow warm (tropical reef) waters; however, pterobranchs are usually not found in temperate regions. Several species form their colonies on calcium carbonate shells or reefs. **Figure 36.3** shows the general distribution for hemichordates.

PHYSIOLOGY AND BEHAVIOR

Physiological studies of hemichordates are limited and tend to focus on morphology of the gas exchange and excretion systems. Both groups have a glomerulous structure which serves to filter blood (Halanych et al. 2019, Tassia et al. 2019). Examinations of physiological tolerances to environmental parameters, energetics, and endocrine systems are wanting.

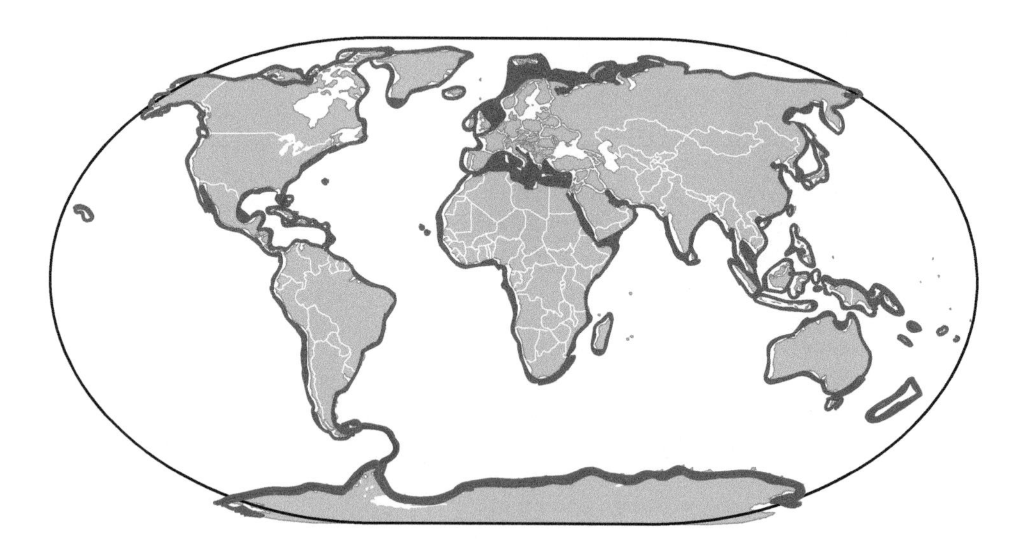

Figure 36.3 Distribution map for modern hemichordates. Red color is for Enteropneusta and blue for Pterobranchia. Enteropnuest are denizens of the deep-sea globally although species names have not been confirmed or assigned to many of the collected animals. For the sake of viewing the figure, the entire deep sea was not colored. Enteropneusts are found most coastal areas in the proper sediment, whereas pterobranchs are more restricted. (For more information on distribution see Supplementary Table 1 from Tassia et al. 2016.)

Outside of feeding biology few formal studies of hemichordate behavior have been undertaken and most of what is known is antidotal. Acorn worms tend to be either epifaunal, as with many of the deep-sea torquaratorid forms, or infaunal as exemplified by near-shore and coastal harrimaniids and ptychoderids. (Note that the recognition of meiofaunal forms is more recent, and in situ observations are needed.) Epifaunal forms will glide across the sediment surface deposit feeding with their proboscis and the ciliary fields on the collar engaged in food particle sorting. Some have the ability to secrete mucous tubes, presumably for protection, or can swim for short distances by twisting their body back and forth. Interestingly, tube building and escape behaviors may be the oldest known behaviors in the fossil record dating from the Cambrian and still observed in moderns forms of the same taxon (Halanych et al. 2013). Infaunal worms will dig tubes within the sediment that they maintain. *Saccoglossus* worms are often presented as an example in invertebrate biology textbooks for building and maintaining a U-shaped tube. However, these worms are mainly deposit feeders and will stick their anterior end out of the tube to gather sediment. In case of pterobranchs, the feeding behavior is dictated by the need to extend the ciliated arms and tentacles into a water current for filter feeding. To do this, they will position themselves at the opening in their coencial tube. Pterobranchs use muscular control of tentacles and ciliary sorting to capture and process food particles (Halanych 1993).

GENETICS AND GENOMICS

Two hemichordate genomes have been published as of early 2021 (Simakov et al. 2015). *Saccoglossus kowalevskii*'s genome is about 758 Mb compared to *Ptychodera flava*'s 1,229 Mb genome which also harbors considerably more heterozygosity. Most knowledge of specific genes in these animals comes from either phylogenetic or developmental studies that have employed the candidate gene approach (Lowe et al. 2003). Phylogenetic studies have been largely based on either the ribosomal genes (Halanych 1995, Cannon et al. 2009, 2013), mtDNA genomes (Perseke et al. 2011, Li et al. 2019) or phylogenomic studies (Cannon et al. 2014). Based on available reports, both the nuclear and mitochondrial genomes appear to be typical of those found in other metazoans in terms of size and gene content. However, the translation code for both hemichordate and pterobranch mitochondrial genomes do contain amino acids that use a non-standard translation code (Li et al. 2019). For years, pterobranchs proved difficult to sequence which may have been due to substitutional bias in nucleotide composition which was observed in the mtDNA genome (Perseke et al. 2011, Li et al. 2019). Datasets for phylogenetic analyses can be found in Cannon et al. (2009, 2013, 2014), Perseke et al. (2011), and Li et al. (2019).

POSITION IN THE ToL

Hemichordata is the sister taxon of Echinodermata. Together these groups are referred to as Ambulacraria which the sister taxon to Chordata with Deuterostomia. Morphological characters (e.g., ciliated gill slits, post-anal tail, etc.) were used to originally place hemichordates within, and then as sister to, Chordata. In contrast, several molecular studies starting with 18S ribosomal gene data (Halanych et al. 1995, Cameron et al. 2000), and later mitochondrial genome data (Perseke et al. 2011, Li et al. 2019) and transcriptomic data (Cannon et al. 2014) show a close relationship between hemichordates and echinoderms vindicating Jefferies (1986) interpretation of gill slits in fossil echinoderms and Metschnikoff (1881) arguments about ambulacral-like systems in acorn worm larvae.

The evolutionary relationships among recognized hemichordate genera are shown in **Figure 36.4**.

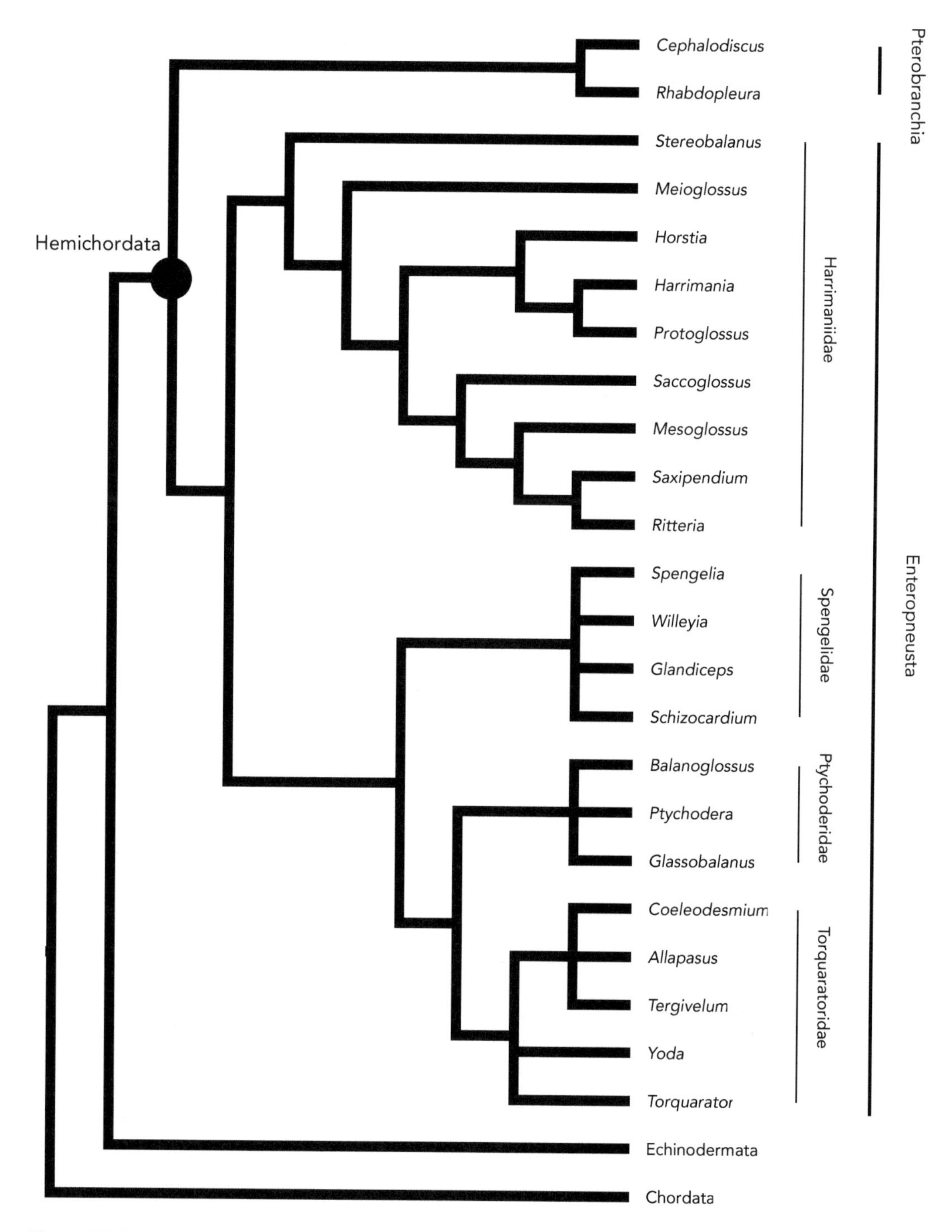

Figure 36.4 The current understanding of Hemichordata's position with Deuterostomia as well the relationships of recognized genera within Hemichordata. (Tree is a consensus from multiple sources: Cannon et al. 2009, 2013, Li et al. 2019, Cannon et al. 2014 and Tassia et al. 2019.)

DATABASES AND COLLECTIONS

Although hemichordates are a lesser-known group, some resources have been developed, in part because of their importance for comparative studies to chordates. Hemichordates are exceptionally fragile. They fragment and disintegrate easily upon collection and preservation and, thus collections in natural history museums are limited (with the exception of fossil graptolite forms). Tassia et al. (2016) highlight some of the global diversity for the group. A list of species can be found at the WORMS (World Register of Marine Species), Hemichordates World Database page (http://http://www.marinespecies.org/hemichordata). The Integrated Taxonomic Information System (ITIS) Standard Reports Page for Hemichordates is also another source (https://www.itis.gov/servlet/SingleRpt/SingleRpt?search_topic=TSN&search_value=158616#null). Lastly, resources are available based on genomic projects in hemichordates (https://groups.oist.jp/molgenu/hemichordate-genomes; https://www.hgsc.bcm.edu/other-invertebrates/acorn-worm-genome-project).

CONCLUSION

Hemichordates are undervalued as models for deuterostome biology. Whereas echinoderms, and especially sea urchins, have been subject to considerable comparison to chordate lineages, enteropneusts provide a more appropriate comparison to chordate systems for several reasons (e.g., a clear adult anterior-posterior axis, developmental modes more similar to chordates, functioning gill slits, neural expression patterns that are easier to homologize to chordates, etc.). Pterobranchs, on the other hand, offer comparison to colonial tunicates. They will also be the last major deuterostome body plan to have their nuclear genome sequenced.

Part of the reason hemichordates have been undervalued is the lack of basic biological knowledge in part because they are very soft bodied making them difficult to work with. In the case of pterobranchs, they are not easy to collect. Species that occur in the shallow tropics tend to be encrusting. Thus, one must have a search pattern in mind when looking for them and it can be hard to obtain enough material. In contrast, cold water species can form large colonies but ship time and deep water equipment is needed for collection. Please note that deep water acorn worms are fragile in nature which is why they were first described from photographs (Holland et al. 2005).

REFERENCES

Allman G.J. (1869) *Rhabdopleura normani*. Allman, *nov. gen. et sp. Report of the British Association for the Advancement of Science* 1868: 311–312.

Annona G., Holland N.D. & D'Aniello S. (2015) Evolution of the notochord. *EvoDevo* 6(1): 30.

Bateson W. (1885) The later stages in the development of *Balanoglossus kowalevskyi*, with a suggestion as to the affinities of the Enteropneusta. *Quarterly Journal of Microscopical Science* 25: 81–122.

Bateson W. (1886) The early stages in the development of *Balanoglossus*. *Studies from the Morphological Laboratory in the University of Cambridge* 2: 131–160.

Burdon-Jones C. (1952) Development and biology of the larva of *Saccoglossus horsti* (Enteropneusta). *Philosophical Transactions of the Royal Society B: Biological Sciences* 236(639): 553–590.

Cameron C.B., Garey J. & Swalla, B.J. (2000) Evolution of the chordate body plan: New insights from phylogenetic analyses of deuterostome phyla. *Proceedings of the National Academy of Sciences of the United States of America* 97(9): 4469–4474.

Cannon J.T., Kocot K.M., Waits D.S., Weese D.A., Swalla B.J., Santos S.R. & Halanych K.M. (2014) Phylogenomic resolution of the hemichordate and echinoderm clade. *Current Biology* 24(23): 2827–2832.

Cannon J.T., Rychel A.L., Eccleston H., Halanych K.M. & Swalla B.J. (2009) Molecular phylogeny of Hemichordata, with updated status of deep-sea enteropneusts. *Molecular Phylogenetics and Evolution* 52(1): 17–24.

Cannon J.T., Swalla B.J. & Halanych K.M. (2013) Hemichordate molecular phylogeny reveals a novel cold-water clade of harrimaniid acorn worms. *The Biological Bulletin* 225(3): 194–204.

Deland C., Cameron C.B., Rao K.P., Ritter W.E. & Bullock T.H. (2010) A taxonomic revision of the family Harrimaniidae (Hemichordata: Enteropneusta) with descriptions of seven species from the Eastern Pacific. *Zootaxa* 2408: 1–30.

Delle Chiaje S. (1829) Memorie sulla storia e notomia degli animali senza vertebre del Regno di Neapel. *Napoli*, 4: 1–72.

Eschscholtz F. (1825): *Berichtüber die zoologische Ausbeute während der Reise von Kronstadt bis St. Peter-und Paul*. Oken's Isis.

Gegenbaur, C. (1870). *Grundzüge der vergleigchenden Anatomie, Zweite Auflage*. Engelmann, Leipzig.

Gillis J.A., Fritzenwanker J.H. & Lowe C.J. (2012) A stem-deuterostome origin of the vertebrate pharyngeal transcriptional network. *Proceedings of the Royal Society of London B: Biological Sciences* 279(1727): 237–246.

Halanych, K.M. (1993) Suspension feeding and the lophophore-like apparatus of the pterobranch hemichordate *Rhabdopleura normani*. *Biological Bulletin* 185: 417–427.

Halanych K.M. (1995) The phylogenetic position of the pterobranch hemichordates based on 18S rDNA sequence data. *Molecular Phylogenetics and Evolution* 4(1): 72–76.

Halanych K.M. (1996) Convergence in the feeding apparatuses of lophophorates and pterobranch hemichordates revealed by 18S rDNA. *Biological Bulletin* 190: 1–5.

Halanych K.M., Cannon J.T., Mahon A.R., Swalla B.J., & Smith C.S. (2013) Modern Antarctic acorn worms form tubes. *Nature Communications* 4: 1–4.

Halanych, K.M., Cannon, J.T. & Tassia, M.G. (2019) Hemichordate: Pterobranchia. In: Schmidt-Rhaesa A. (Ed.), *Handbook of Zoology, Miscellaneous Invertebrates*. 9: 267–282. De Gruyter, Berlin.

Harmer S. (1887) Appendix to W. C. M'Intosh Report on *Cepholodiscus dodecalophus*. In *Challenger Reports*. XX (62): 39–47.

Holland N.D., Clague D.A., Gordon D.P., Gebruk A., Pawson D.L. & Vecchione M. (2005) "Lophenteropneust" hypothesis refuted by collection and photos of new deep-sea hemichordates. *Nature* 434(7031): 374–376.

Jefferies, R.P.S. (1986) *Ancestry of the Vertebrates*. British Museum (Natural History).

John C.C. (1931) *Cephalodiscus*. *Discovery Reports* 3: 223–260.

Kaul-Strehlow S. & Stach T. (2013) A detailed description of the development of the hemichordate *Saccoglossus kowalevskii* using SEM. *Frontiers in Zoology* 1053(1): 1–32.

Kowalevsky A. (1866): Anatomie des *Balanoglossus*. *Mem. Acad. Imper. Sci. St. Petersbourg* 10(7).

Lankester R. (1877) Rhabdopleuridae. In: Kukenthal W., Krumbach T. (Eds.), *Handbuch der Zoologie. Order of Branchiotrema: Pterobranchia*. Walter de Gruyter Broch, H., Berlin & Leipzig, 1–32.

Lester, S.M. (1988) Ultrastructure of adult gonads and development and structure of the larva of *Rhabdopleura normani* (Hemichordata: Pterobranchia). *Acta Zoologica* 69(2): 95–109.

Li, Y., Kocot K.M., Tassia M., Cannon J.T., Bernt M., Halanych K.M. (2019) Mitogenomics reveals a novel genetic code in Hemichordata. *Genome Biology and Evolution* 11: 29–40.

Lowe C.J. (2008) Molecular genetic insights into deuterostome evolution from the direct-developing hemichordate *Saccoglossus kowalevskii*. *Philosophical Transactions of the Royal Society B: Biological Sciences* 363(1496): 1569–1578.

Lowe C.J., Wu M., Salic A., et al. (2003): Anteroposterior patterning in hemichordates and the origins of the chordate nervous system. *Cell* 113(7): 853–865.

Luttrell, S.M., Su, Y.-H. & Swalla, B.J. (2018) Getting a head with *Ptychodera flava* larval regeneration. *Biological Bulletin* 234:152–164.

M'Intosh W. C. (1882) Preliminary notice of *Cephalodicus*, a new type allied to Prof. Allman's *Rhabdopleura*, dredged in H.M.S. 'Challenger'. *Annals and Magazine of Natural History (5th Series)* 10: 337–348.

Metschnikoff, V. (1881) Uber die systematishe Stellung von *Balanoglossus*. *Zoologischer Anzeiger* 4: 139–143.

Morgan T.H. (1891): The growth and metamorphosis of tornaria. *Journal of Morphology* 5: 407–458.

Ogasawara, M., Wada, H., Peters, H. & Satoh, N. (1999) Developmental expression of Pax 1/9 genes in urochordate and hemichordate gills: insight into function and evolution of the pharyngeal epithelium. *Development* 126: 2539–2550.

Osborn K.J., Gebruk A.V., Rogacheva A. & Holland N.D. (2013) An externally brooding acorn worm (Hemichordata, Enteropneusta, Torquaratoridae) from the Russian Arctic. *Biological Bulletin* 225(2): 113–123.

Perseke, M., Hetmank, J., Bernt, M., Stadler, P. F., Schlegel, M. & Bernhard, D. (2011) The enigmatic mitochondrial genome of *Rhabdopleura compacta* (Pterobranchia) reveals insights into selection of an efficient tRNA system and supports monophyly of Ambulacraria. *BMC Evolutionary Biology* 11(1): 134.

Rehkämper, G., Welsch, U. & Dilly, P.N. (1987). Fine structure of the ganglion of *Cephalodiscus gracilis* (Pterobranchia, Hemichordata). *Journal of Comparative Neurology* 259(2): 308–315.

Ridewood, W.G. (1912) LXV.—On specimens of *Cephalodiscus nigrescens* supposed to have been dredged in 1841 or 1842. *Annals and Magazine of Natural History* 10(59), 550–555.

Ridewood, W.C. (1921) XLII.—On specimens of *Cephalodiscus densus* dredged by the 'Challenger'in 1874 at Kerguelen Island. *Journal of Natural History* 8(46): 433–440.

Röttinger, E., DuBuc, T. Q., Amiel, A. R. & Martindale, M. Q. (2015) Nodal signaling is required for mesodermal and ventral but not for dorsal fates in the indirect developing hemichordate, *Ptychodera flava*. *Biology Open* 4: 830–842.

Rychel A.L. & Swalla B.J. (2009) Regeneration in Hemichordates and Echinoderms. In: Rinkevich B. & Matranga V., (Eds.). *Stem Cells in Marine Organisms*. Springer-Verlag, New York, 245–266.

Silén, L. (1954) Reflections concerning the "stomochord" of the Enteropneusta. *Journal of Zoology* 124(1): 63–67.

Simakov O., Kawashima T., Marlétaz F., et al. (2015) Hemichordate genomes and deuterostome origins. *Nature* 527: 1–19.

Tagawa K., Nishino A., Humphreys T., & Satoh N. (1998) The spawning and early development of the Hawaiian acorn worm (Hemichordate), *Ptychodera flava*. *Zoological Science* 15(1): 85–91.

Tassia, M.G., Cannon, J.T. & Halanych, K.M. (2019) Hemichordate:Enteropneusta. In: Schmidt-Rhaesa A. (Ed.), *Handbook of Zoology, Miscellaneous Invertebrates*. 9: 283–311. De Gruyter, Berlin.

Willey A. (1899a) Enteropneusta from the South Pacific, with notes on the West Indian species. *Willey's Zoological Results* III: 32–335.

Willey A. (1899b) Remarks on some recent work on the Protochorda, with a condensed account of some fresh observations on the Enteropneusta. *Quarterly Journal of Microscopical Sciences* 42: 233–244.

ADDITIONAL READINGS

Benito J. & Pardos F. (1997) Hemichordata. In: Harrison F.W. & Ruppert E.E. (Eds.) *Microscopic Anatomy of Invertebrates*. Wiley-Liss, New York, 15–101.

Dawydoff C. (1948) Classe des Enteropneustes. In: Grasse P.P. (Ed.) *Traite de Zoologie, Vol. XI*. Masson et Cie, Paris, 369–453.

Hadfield M.G. (2002) Phylum Hemichordata. In: Young C.M., Sewell M.A., & Rice M.E. (Eds.) *Atlas of Marine Invertebrate Larvae*. Academic Press, San Diego, 553–564.

Halanych, K.M., Cannon, J. T. & Benito, J. (2012) Pterobranchia. In McGraw Hill, (Ed.) *McGraw Hill Encyclopedia of Science & Technology, 11th Edition*. 14: 586–587. McGraw Hill Publishers.

Hyman L.H. (1959) Phylum Hemichordata. In: *The Invertebrates*. McGraw-Hill, New York, 72–207.

Maletz, J. (2014) The classification of the Pterobranchia (Cephalodiscida and Graptolithina). *Bulletin of Geosciences* 89(3): 477–540.

Tassia, M.G, Cannon, J., Konikoff, C.E., Shenkar, N., Halanych, K.M. & Swalla, B.J. (2016) The global diversity of the Hemichordata. *PLoS One* 11: e0162564.

Welsch U. (1984) Hemichordata. In: Bereiter-Hahn J., Matoltsy A.G., & Richards K.S. (Eds.), *Biology of the Integument 1: Invertebrates*. Springer-Verlag, Berlin, 790–799.

PHYLUM CHORDATA

TOPIC INVERTEBRATE CHORDATES

Linda Z. Holland

CHAPTER

37

Infrakingdom	Bilateria
Division	Deuterostome
Phylum	Chordata
Topic	Invertebrate Chordates

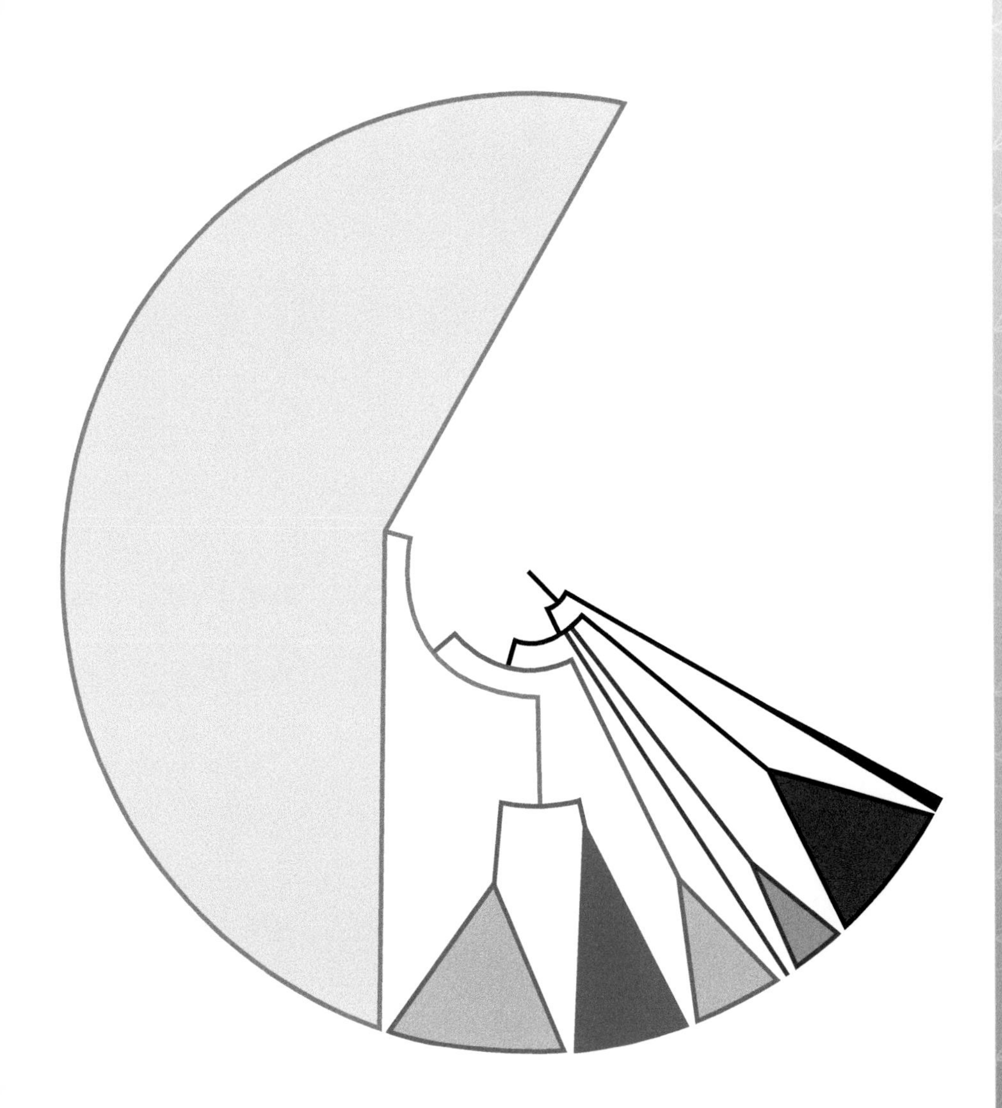

CONTENTS

Synapomorphies

- kidneys with cyrtopodocytes
- large repertoire of green fluorescent proteins
- eggs with green fluorescent protein
- asymmetric larvae with mouth on the left, gill slits skewed to the right
- club-shaped gland in the larva
- larval tail fin with rays including long ciliary rootlets

Two invertebrate subphyla, the Cephalochordata (amphioxus or lancelets) and the Tunicata (tunicates or urochordates), together with the Vertebrata comprise the phylum Chordata, which is characterized by the presence of a notochord and dorsal hollow nerve cord (**Figure 37.1**). Cephalochordata will be considered first, followed by a discussion of the Tunicata.

A. SUBPHYLUM CEPHALOCHORDATA

General Characteristics

Numbers of valid species: Currently 32 cephalochordate species are recognized: 26 *Branchiostoma* species, 3-4 *Asymmetron* species, and 2 *Epigonichthys* species

Numbers of cryptic species: There may be three or more cryptic species of *Asymmetron* and/or *Epigonichthys*

Key features: Marine animals from 1–6 cm long with chevron-shaped paraxial muscles, endostyle, notochord, dorsal, hollow nerve cord with anterior swelling (the cerebral vesicle) pharyngeal gill slits, filter feeders, unduplicated genomes ~ 500 mb, And considerable synteny with vertebrate genomes. Evolving more slowly than the slowest evolving vertebrate known – the elephant shark

Ecological importance: Abundant in shallow marine habitats with a sandy bottom. They are major consumers of phytoplankton, At least one species is an intermediate host for tape worms. Amphioxus is preyed upon by sting rays

Commercial value: Not significant at present. Once there was major amphioxus fishery off the coast of Xiamen, China. Large numbers of amphioxus were collected with shovels and sieves of bamboo. The animals were typically dried for human consumption. However, due to habitat destruction, today, the fishery is vastly reduced or nonexistent, and amphioxus is protected off Xiamen

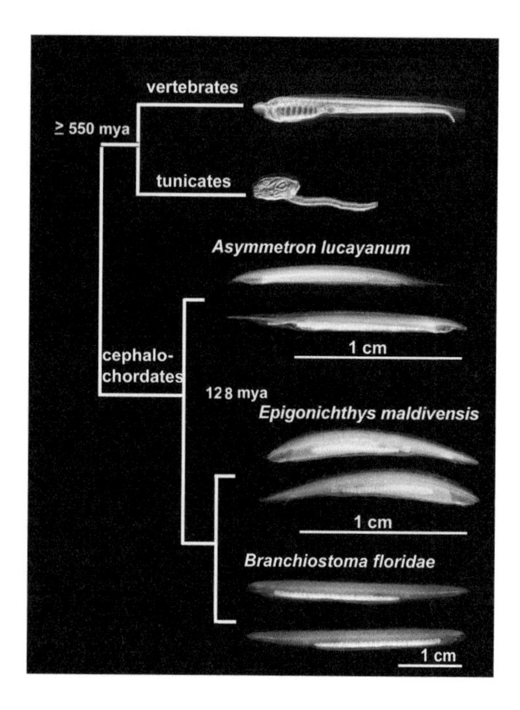

Figure 37.1 Chordate phylogeny. Chordates radiated in the Cambrian, at least 550 million years ago (mya). *Asymmetron* split from the *Epigonichthys-Branchiostoma* clade about 128 mya. Tunicates are represented by the larva of the appendicularian *Oikopleura dioica* and vertebrates by the ammocoete larva of the sea lamprey *Petromyzon marinus* (anterior of both is to the left). For the three genera of amphioxus, a view of the left side is shown above a view of the right side. (*P. marinus* photo courtesy of Jeramiah Smith. Photo of *Epigonichthys maldivensis* reprinted from Igawa et al., 2017.)

HISTORY OF TAXONOMONY AND CLASSIFICATION

Cephalochordates were first described by P.S. Pallas in 1774, who, thinking they were allied to the molluscs, called them *Limax lanceolatus*. However, in 1834, Costa recognized their vertebrate affinities and named the European species *Branchiostoma lubricum.* In 1836, Yarrell renamed it *Amphioxus lanceolatus.* Because of the rules of nomenclature, the first species became *Branchiostoma lanceolatum* (Pallas, 1774). Amphioxus then became a common name. There are three valid genera of amphioxus: *Branchiostoma* with 26 named species, *Asymmetron* with 3-4 named species and several cryptic ones (Kon et al., 2007; Igawa et al., 2017), and *Epigonichthys* with 2 species and possibly additional cryptic ones (Figure 37.1) (Igawa et al., 2017). The genera *Epigonichthys* and *Asymmetron* were initially thought to be identical and variously called one name or the other. However, they are now recognized as distinct species on the basis of mitochondrial DNA sequences (Kon et al., 2007). *Asymmetron* is sister to the *Branchiostoma/Epigonichthys* clade (Figure 37.1). Estimates of divergence time between *Asymmetron* and the *Branchiostoma/Epigonichthys* clade range from ~120 mya (Yueet al., 2014) to ~40 mya (Igawa et al., 2017) depending on the method used.

ANATOMY/MORPHOLOGY

As cephalochordates are evolving very slowly, the morphology of all three genera is similar. They resemble small, jawless fish lacking fins and eyes and with chevron-shaped paraxial muscles extending nearly to the anterior tip of the animal plus a pharynx with bilateral series of numerous gill slits (**Figure 37.2**). Their maximum adult size ranges from about 1.5–6 cm long, depending on species. They have the characteristic chordate apomorphies of a dorsal, hollow nerve cord and a notochord (Figure 37.2).

Muscles and nerves: The notochord in amphioxus, unlike that in other chordates, is composed of modified striated muscle cells in which the myofibrils extend from left to right (Flood, 1967; 1975). The segmental paraxial muscles lie on either side of the notochord and, like the notochord, extend the full length of the animal. The notochordal and paraxial muscles send processes to the nerve

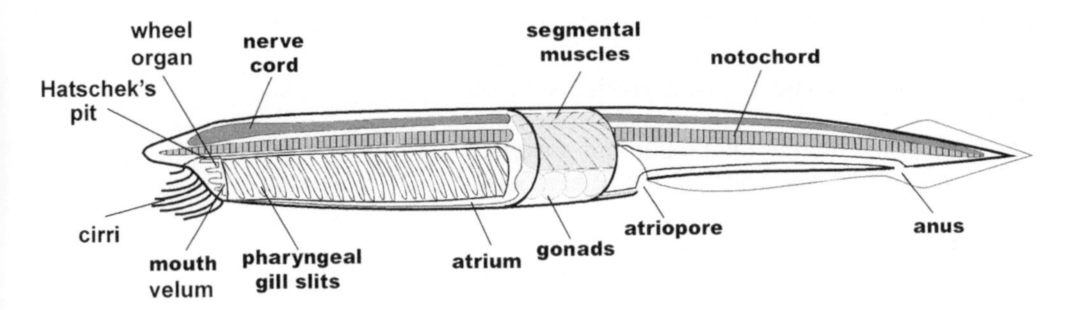

Figure 37.2 Schematic drawing of an adult of *Branchiostoma floridae* showing the major features. Cirri over the mouth are motile and function to exclude large particles from the mouth. Not shown are the endostyle in the ventral midline of the pharynx, the vascular system and the kidneys associated with the dorsal part of the branchial bars. Anterior to left.

cord to synapse directly with muscle neurons in the cord itself (Flood, 1966; 1968). In addition, the pterygial muscle in the wall of the atrium can contract to expel gametes from the atrium or large particles from the mouth. The nerve cord lacks ventral roots, but has numerous dorsal nerves. These nerves lack ganglia. Their axons pass between the myotomes. Branches of the dorsal nerves connect with the nerve plexus on the pterygial muscles. As the left and right myotomes are offset by half a segment with those on the left shifted anteriorly of those on the right, the nerves are similarly offset (Wicht and Lacalli, 2005). There are also muscles associated with the larval gill slits that can cause the gill slits to close. The nerve cord contains a slight anterior swelling, the cerebral vesicle but lacks other morphological divisions. On the basis of microanatomy and comparisons of gene expression in developing amphioxus and vertebrates, it has been suggested that the amphioxus nerve cord has homologs of the vertebrate diencephalic forebrain, midbrain, hindbrain and spinal cord, including the *zona limitans intrathalamica* and midbrain/hindbrain boundary; however a homolog of the telencephalon is likely lacking.

Photoreceptors: Even though amphioxus lacks image-forming eyes, the amphioxus nerve cord has four types of photoreceptors. Two, the organs of Hesse and the Joseph cells are microvillar. The former are scatted ventrally along the length of the nerve cord posterior to the forebrain while the latter are located dorsally in the presumed midbrain. The frontal eye at the anterior tip of the amphioxus nerve cord and the lamellar body, located dorsally in the forebrain homolog, are both ciliary photoreceptors. As the structure of the lamellar body with its whorls of expanded ciliary membrane is virtually identical to that of the pineal in the larval lamprey, it has been suggested that the amphioxus lamellar body is homologous to the vertebrate pineal. The organs of Hesse and Joseph cells contain melanopsin and are maximally sensitive to light in the blue part of the spectrum (470 nm); the spectral sensitivity of the other types of amphioxus photoreceptors is unknown (del Pilar Gomez et al., 2009). Both the organs of Hesse and the frontal eye are associated with pigment cells. Amphioxus has no other pigment cells, which is not surprising as it lacks neural crest–cells which migrate from the edges of the neural plate in vertebrate embryos and give rise to numerous tissues including much of the cartilage of the head, parts of cranial ganglia and cells of the adrenal medulla as well as most pigment cells (Gammill and Bronner-Fraser, 2003).

Digestive system: Amphioxus has a tubular gut with an anterior mouth and a posterior anus that skews to the left side. The mouth opens into the pharynx, perforated bilaterally by numerous gill slits. The gill slits open into the atrial cavity, which forms at metamorphosis by folds of the body wall extending ventrally and fusing in the ventral midline. The atrial cavity opens posterior to the gill slits via the atriopore. At the posterior end of the pharynx, there is a digestive diverticulum that extends forward alongside the pharynx. The endostyle lies in the ventral midline of the pharynx. It binds iodine and is considered homologous to the vertebrate thyroid gland. It secretes mucous that traps foods particles and is carried posteriorly by cilia to the intestine and digestive diverticulum. Anterior to the mouth are the wheel organ and Hatschek's pit, considered a homolog of the vertebrate adenohypophysis. Movable buccal cirri at the anterior end of the animal can inhibit very large particles from entering the mouth and pharynx. Just posterior to the wheel organ is a velum with a central hole, the velar mouth, which also regulates the movement of food into the pharynx.

Kidneys: Amphioxus has numerous kidneys or nephridia, containing cells termed cyrtopodocytes, each associated with the dorsal part of each gill bar. The homologies of amphioxus kidneys are uncertain as they have features in common both with vertebrate kidneys, specifically the pronephros, and with invertebrate kidneys or protonephridia (Holland, 2017). The nephridial lumen opens at the dorsal edge of each gill slit.

Circulatory system: There are dorsal and ventral blood vessels. Anteriorly the dorsal vessels are paired. Blood vessels associated with the gill bars connect the dorsal and ventral blood vessels. There is a vessel underlying the endostyle and associated with the digestive diverticulum. There is no endothelium. Blood cells are scarce (Rahr, 1981; Jefferies, 1986). At least the sub-intestinal vessel is contractile and can reverse its direction of pumping. As this vessel expresses the heart marker *Nkx 2.5* (*tinman*) during development, homology with the vertebrate heart has been proposed (Holland et al., 2003).

Ectodermal sensory organs: Adults are not uniformly ciliated but have several types of ectodermal sensory cells. Primary or type I sensory cells, which are likely mechanosensory, are widely scattered over the epidermis. They extend axons to the central nervous system (CNS) and at least some are a hold-over from the larva stage. Type II cells lack axons (Lacalli, 2004). They typically appear during metamorphosis. In addition, there are ventral pit cells with short cilia along the ventral part of the atrium (Stokes and Holland, 1995b). At the anterior end of the animal are putative pressure-sensitive cells, the corpuscles of de Quatrefages. They are primary sensory cells enclosed in a capsule and are innervated by the rostral nerves of the CNS (Baatrup, 1982).

Reproductive system: Sexes are separate in amphioxus, although hermaphrodites have occasionally been noted (Yamaguchi and Henmi, 2003). The gonads mature when adults are about half their maximum length. *Branchiostoma floridae* has 26 gonads on each side associated with the atrium. *Asymmetron* and *Epigonichthys* have gonads only on the right side. At spawning, gametes move from the gonads into the atrial cavity and are shed out of the atriopore.

REPRODUCTION, EMBRYOLOGY, AND LARVAL ANATOMY

Amphioxus species typically spawn in summer as the seawater warms up. In *B. floridae* each individual will spawn at 10–14 day intervals from late May through August. The breeding season for cold-water species is limited to about one month in June. On spawning days, the oocytes undergo the meiotic divisions in the early afternoon, arresting at second meiotic metaphase. At the same time, sperm, which grow in a syncytium, individualize. Spawning is triggered by sunset. The animals, first males, then females emerge from the sand about half an hour after sunset and shed eggs and sperm before burrowing back into the sand. It is not known what determines the day of spawning in *Branchiostoma*; however, for *Asymmetron*, spawning occurs on a few days just before each new moon. If kept in the laboratory on a 14 hour day/10 hour night cycle and an artificial moonlight regime and well fed, *Asymmetron* will reliably spawn a few days before the new moon (Holland, 2011). In contrast, spawning in *Branchiostoma* does not depend on moonlight. However, *Branchiostoma* can be induced to spawn by an increase in temperature. In *B. floridae*, ripe animals kept for three weeks at 17°C will typically spawn 24 hours after being shifted to 25°–28°C (Holland and Li, 2021). Development to metamorphosis occurs in the plankton.

B. floridae is in continuous breeding cultures in several laboratories in the U.S.A., Europe, Taiwan, and China. As it is a warm water species, it only requires about 3 weeks until metamorphosis and 3–5 months to sexual maturity. Embryos are easy to raise in petri dishes if fed twice daily on phytoplankton. Brown algae such as *Tisochrysis* are optimal. Methods for microinjection of eggs and mutagenesis have been developed (Holland and Onai, 2011; Hu et al., 2017). Cold water species of *Branchiostoma* have not been put into breeding cultures as their larval life is several months long. Attempts to raise *A. lucayanum* through metamorphosis have not yet succeeded.

Development is direct with only a very gradual metamorphosis (**Figure 37.3**). Sperm have an acrosomal granule that undergoes exocytosis and adheres the sperm to the vitelline layer of the egg. Fertilized eggs undergo a cortical reaction with exocytosis of cortical granules into the perivitelline space. Complete elevation of the fertilization envelope requires about 20 minutes. Meanwhile, the second polar body is emitted. Shortly after fertilization, the germ plasm, which is rich in endoplasmic reticulum and mRNAs for germ cell markers including *Vasa* and *Nanos*, coalesces at the vegetal pole of the zygote (Holland and Holland, 1992; Zhang et al., 2013). At the same time, the fertilizing sperm migrates to the vegetal pole. Then the sperm nucleus, accompanied by a cloud of mitochondria, migrates towards the animal pole to join with the maternal chromosomes that have migrated to one side of the animal pole. The zygote nucleus then moves to the center of the single-cell embryo.

First cleavage begins at the animal pole with the cleavage furrow expanding vegetally. Cleavage in amphioxus is equal and indeterminant or regulative with cell fates being decided only gradually during embryogenesis. The first two cells if separated each develops into a half-sized larva. Since the germplasm is segregated into a single cell at first cleavage, if the two initial blastomeres are separated, only one will inherit the germ plasm (Zhang et al., 2013). At gastrulation, the cell containing the germplasm comes to lie in the anterior mesendoderm. Then, the cell divides into a small group of germ cells that, at the early neurula stage, migrate to the tail bud at the posterior end of the larva. There, they continue to divide, and small groups of cells associate with the dorsal aspect of each muscular somite that pinches off sequentially from the tail bud. In *Asymmetron*, as noted above, only the germ cells that associate with the somites on the right persist.

Gastrulation occurs by simple invagination. At the mid-late gastrula stage, all the ectodermal cells develop a single cilium. The embryos then rotate within the

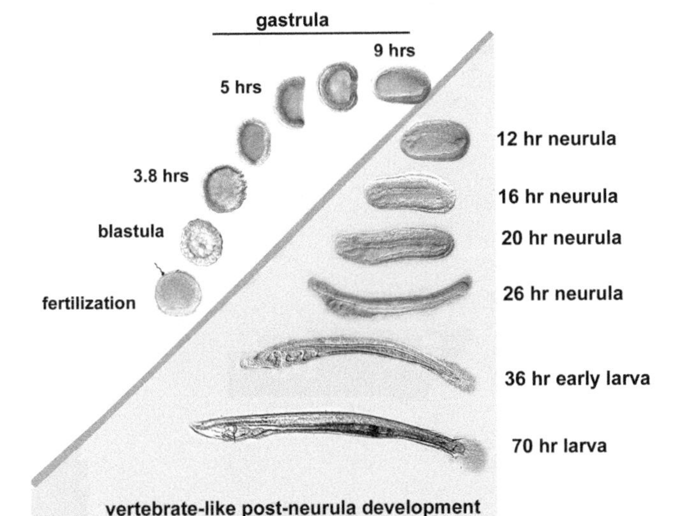

Figure 37.3 Development of *Branchiostoma floridae* at 25°C. Early development is invertebrate-like. The eggs, about 130 µm in diameter, are fertilized at second meiotic metaphase. The fertilization envelope, which increases the diameter of the embryo to about 450 µm is not shown. Gastrulation is by simple invagination. The blastopore becomes the neurenteric canal connecting the nerve cord and posterior end of the gut. From hatching, which occurs at the early neurula stage, about 9 h after fertilization, development is more like that in vertebrates. Animal pole to the left before the neurula stage. Anterior to the left from the neurula through larval stages.

fertilization envelope. Neurulation begins with a flattening of the dorsal side of the embryo. The dorsal mesendoderm develops three anterior-posterior pleats. The two dorsolateral ones will form the anterior somites, which pinch off from anterior to posterior, and the central one forms the notochord. Hatching occurs at the early neurula, which has three somites on a side and a single photoreceptor in the center of the CNS. Somites continue to pinch off from the dorsolateral mesendoderm until 8–12 are formed; as the embryo elongates, additional somites, notochord and nerve cord develop from the tail bud. The larvae retain cilia until metamorphosis.

When the larvae have about 20–25 somites, the mouth opens on the left, and gill slits begin to form on the right in *Branchiostoma* spp. and more ventrally in *Asymmetron* (Holland and Holland, 2010). The larvae have a specialized group of long cilia, the ciliary tuft or oral papilla, in front of the mouth (Andersson and Olsson, 1989; Lacalli, 2004). Gill slits are added sequentially behind the first, from anterior to posterior. At the two gill slit stage, a second photoreceptor, the frontal eye, develops at the anterior tip of the nerve cord. There us a single larval kidney (Hatsheck's nephridium) and a ciliated pit in the ectoderm on the left just anterior to the mouth, which fuses with an anterior diverticulum of the larval gut (Hatsheck's anterior, left diverticulum), and expresses genes such as *Pax6* that are characteristically expressed in the adenohypophysis. On the right side of the pharynx, there is a larval secretory organ, the club-shaped gland, that undergoes apoptosis at metamorphosis (Holland et al., 2009b) and an endostyle, which moves to the floor of the pharynx at metamorphosis. The endostyle secretes mucus and is a homolog of the vertebrate thyroid.

Metamorphosis, during which the highly asymmetrical larva becomes largely symmetrical, begins in *B. floridae* when there are about 9–11 gill slits; that is about 2–3 weeks at 30°C. Metamorphosis can be induced and synchronized in late larvae by application of T3 thyroid hormone or Triac (Paris et al., 2008). The first indication of metamorphosis is the formation of the atrium from folds of the body wall that extend ventrally and fuse in the ventral midline. Next, a second row of gill slits forms on the right above the first row. The first row migrates to the left and a tongue bar grows ventrally to divide each gill slit into two. The mouth, on the left in larvae, moves to the front of the animal and sensory cirri form. They can bend over the mouth to prevent ingestion of very large particles. The ciliated pit plus Hatsheck's anterior left diverticulum migrate anteriorly, coming to lie dorsally in front of the mouth to form the "wheel organ" and Hatschek's pit, thought to be homologous to the vertebrate adenohypophysis (Sahlin and Olsson, 1986). Additional kidneys form, each associated with the dorsal part of the gill bars. The last structure to form is a gut diverticulum that grows anteriorly from the posterior end of the pharynx. The diverticulum stores food (unicellular algae and other small particles) and may function in digestion. Although this has been called the "liver", there is scant evidence for homology with the vertebrate liver. At metamorphosis, most of the ectodermal cilia are lost. New probably sensory cilia (the ventral pit cells) develop along the ventral part of the atrium. The newly metamorphosed animals abandon a pelagic existence and burrow tail first into the sand.

Adult amphioxus can regenerate the tip of the tail and the anterior tip of the animal (Andrews, 1893; Somorjai et al., 2012). At present, it is unknown whether there are pools of pluripotent stem cells or whether cells at the cut surface dedifferentiate and then redifferentiate or whether there is some combination of the two.

ECOLOGY AND PHYSIOLOGY

Habitat/Ecology: As adults, most species of amphioxus live buried in the sand in comparatively shallow water (e.g., bays or other regions with relatively little wave action) with only their extreme anterior ends sticking out. However, one species, *Epigonichthys inferum* has been found at a depth of 229 m near a whale-fall (Kon et al., 2007). As noted above, amphioxus is only known to emerge from the sand after dark to spawn. Populations of *B. floridae* can contain up to 1200 individuals per square meter of sand (Stokes, 1996; Stokes and Holland, 1996).

Parasitism: Branchiostoma floridae can harbor plerocercoid larvae of a tapeworm, the adults of which live in the gut of a sting ray, which is the major predator of adult amphioxus in Tampa Bay, Florida (Holland et al., 2009a)

Immunity: Like other invertebrates, cephalochordates lack adaptive immunity. Not surprisingly, they have an expanded repertoire of innate immune receptors (Litman et al., 2010; Dishaw et al., 2012). Many of these genes have arisen via domain-swapping, suggesting new functions. Moreover, variable region containing chitin-binding proteins (VCBPs), consisting of immunoglobulin V (variable) and chitin binding domains, mediate recognition of surface proteins of bacteria, etc., through the V domains. The gut has been implicated as the major site of the innate immune response (Huang et al., 2011). However, as amphioxus may well eat bacteria as well as phytoplankton, it is difficult to distinguish the role of the gut epithelium in mediating innate immunity from a role in feeding. Bays such as Old Tampa Bay in Tampa, Florida, USA where amphioxus live may have very high bacterial counts, especially in summer. Moreover, amphioxus is subject to infection with a red bacterium, most likely a *Rhodobacter* that is endemic in populations worldwide and can kill individuals that are stressed. The *B. floridae* genome does contain a chitinase (Guerriero, 2012), and it is thought that the peritrophic membrane, an extracellular sheath that encloses fecal pellets, contains chitin (Nakashima et al., 2018).

Feeding: Both larval and adult amphioxus are filter feeders. Cilia around the gill slits draw water and particles in through the mouth. The current exits the gills into the atrial cavity and is expelled out the atriopore. Food is extracted from the current of water by a mucus net secreted chiefly by the endostyle. In larvae, the club-shaped gland also appears to contribute mucus. In the adult, ciliated tracts, the peripharyngeal bands, move the mucus secreted by the endostyle down the sides of the pharynx and into the midgut, where the strand is rotated by a ring of long cilia. Food also enters the digestive diverticulum in adults, which has been reported to take up phytoplankton by phagocytosis. The animals feed continuously; however, if food is absent from the seawater, young larvae will stop feeding and never recommence even if food is provided.

DISTRIBUTION AND BEHAVIOR

Geographical distribution: Cephalochordates are very widespread in inshore, marine environments, occurring in tropical, sub-tropical, and temperate waters (Poss and Boschung, 1996; Lambert, 2005). *Asymmetron* and *Epigonichthys* are circum-tropical/sub-tropical, while *Branchiostoma lanceolatum* occurs as far north as Norway (Kon et al., 2006; Kon et al., 2007). Cephalochordates have not been described from arctic or Antarctic waters. It is likely that their range will be extended as more sampling is done (**Figure 37.4**).

Behavior: As noted above, all species of amphioxus live buried in the sand with only their mouths slightly protruding. They are known to come up into the water column only when spawning. As soon as gametes are shed, they re-burrow. The instinct to burrow is acquired only at metamorphosis. The newly hatched early neurula is uniformly ciliated, and, as soon as the first photoreceptor develops in the nerve cord, the embryos swim up and toward the light. When the mouth, first gill slit and anus open, the larvae hang in the water column, due

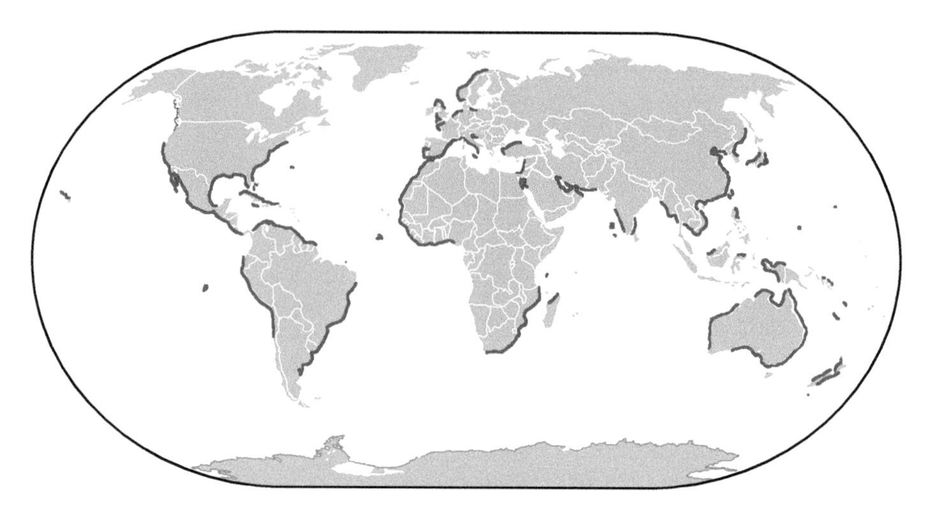

Figure 37.4 Worldwide distribution of cephalochordates (amphioxus) indicated in red. They live burrowed in the sand in shallow water wherever the habitat is suitable—typically near shore in fine to coarse sand with little organic material. Many of the gaps in distribution on the map are most likely due to insufficient sampling. Red dots in the ocean coincide with islands.

to the beating of ectodermal cilia, and orient at an angle toward the light (Stokes and Holland, 1995a). At metamorphosis, most of the cilia are lost and the young adults burrow in the substratum.

GENOMICS AND GENETICS

The genomes of three species of *Branchiostoma* (*B. floridae, B. belcheri,* and *B. lanceolatum*) and one species of *Asymmetron* (*A. lucayanum*) have been sequenced and fully or partially assembled (Putnam et al., 2008; Yue et al., 2016, Marlétaz et al., 2018; You et al., 2019). *B. floridae* has 38 chromosomes and *A. lucayanum* 34 (Holland et al., 2015). Their genomes, on the order of 500–600 Mb, have not undergone the whole genome duplications that occurred in the vertebrate lineage and have retained considerable synteny with vertebrate genomes. Comparisons indicate that amphioxus genomes are evolving more slowly than that of the elephant shark, which is the slowest-evolving vertebrate known. For example, amphioxus has a single cluster of 15 Hox genes that has synteny with the 4 clusters of vertebrate Hox genes. This high degree of synteny has allowed the reconstruction of an ancestral chordate genome with 17 linkage groups (i.e., chromosomes; Putnam et al., 2008). Surprisingly, in spite of splitting over 100 mya, *A. lucayanum* and *B. floridae* can hybridize and generate viable larvae with 36 chromosomes and characteristics intermediate between those of their parents (Holland et al., 2015). There is a relatively high level of polymorphisms—1–2% in coding regions and >5% in non-coding regions. Transposons, of which there are 30 superfamilies, make up 28% of the *B. floridae* genome. This is greater diversity than in vertebrates. None of the families of transposons has undergone a major expansion (Cañestro and Albalat, 2012). An analysis of microRNAs (miRNAs) revealed at least 245 miRNAs in *Branchiostoma* (Zhou et al., 2012). Functions have been suggested in regulation of developmental genes and players in the innate immune system. Comparisons between *Asymmetron* and *Branchiostoma* genomes have proven highly useful for identification of regulatory elements. Genomes of *Branchiostoma* species are so conserved that comparisons among them reveal too many conserved non-coding regions for regulatory regions to be distinguished. Techniques for mutagenesis including TALEN and CRISPR/Cas9 have been developed in amphioxus (Su et al. 2020; Holland and Li, 2021). Homozygous mutants of *Branchiostoma floridae* can be obtained within 2 years of initial microinjections into eggs or 2-cell embryos.

POSITION IN THE ToL

Because amphioxus is very vertebrate-like but simpler, during much of the 20th century, it was considered to be the sister group of vertebrates, and tunicates, because of the comparatively simple body plan of some and, tunicates, in particular their larvae, were generally accepted to be the sister group of the Cephalochordata plus Vertebrata. However, molecular phylogenetic analyses with large numbers of nuclear genes showed that Cephalochordata is the sister group of Tunicata plus Vertebrata (sometimes termed Olfactores) (Delsuc et al., 2006; Delsuc et al., 2008). The cephalochordate and vertebrate genomes are evolving relatively slowly. In contrast, those of tunicates are evolving particularly rapidly (Edvardsen et al., 2005; Berná and Alvarez-Valin, 2014; Yue et al., 2014). The slow evolutionary rate of cephalochordates may explain why the three modern genera are very similar to one another and not greatly different from fossil cephalochordates from the Cambrian such as *Pikaea* or fossil jawless fish such as *Myllokunmingia* or even the larvae of modern lampreys (Figure 37.1). In contrast, as will be discussed below, the fast-evolving tunicates have undergone considerable divergence in body plans and life styles.

B. SUBPHYLUM TUNICATA

General Characteristics

Numbers of valid species: There are three classes of tunicates: Appendicularia, Thaliacea, which is divided into three orders (Salpida, Doliolida, Pyrosomida), and Ascidiacea (B3; **Figures 37.5**, **37.6**). There are 70 named species of Appendidcularia, 13 named species of salps, 5 collected but undescribed and an estimated 8 additional species, 25 species of doliolids, and three genera and an uncertain number of species of pyrosomes. Ascidiacea have over 3,000 named species

Numbers of cryptic species: Unknown, but may be considerable

Ecological importance: Tunicates are important herbivores, eating microscopic plants and excreting fecal pellets that provide nourishment for deeper-living organisms. Ascidians occur in large numbers in inshore environments, while salps occur in large numbers in both nearshore and open ocean environments.

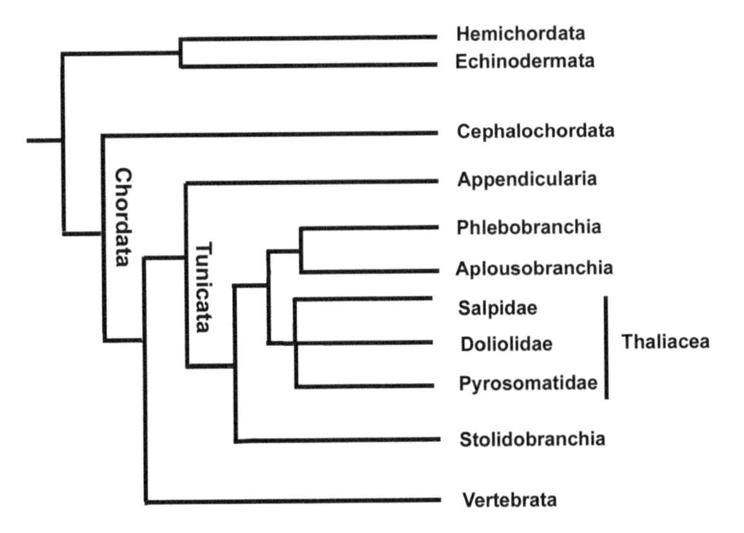

Figure 37.5 Tunicate phylogeny. Ascidians are paraphyletic. The clade including the orders Phlebobranchia plus Aplousobranchia is sister to the monophyletic Thaliaceans. The order Stolidobranchia is sister to the Thaliaceans plus the Phlebobranch/Aplousobranch clade. (After Delsuc et al., 2018; Kocot et al., 2018.)

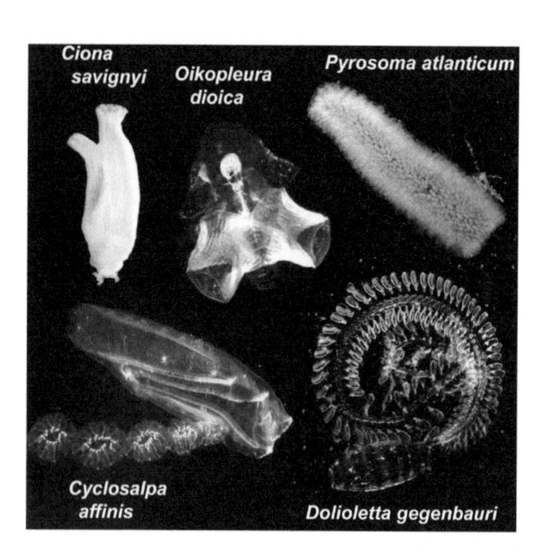

Figure 37.6 Representatives of the five main groups of tunicates. The ascidian *Ciona intestinalis* (order Phlebobranchia), the appendicularian *Oikopleura dioica*, the pyrosome (*Pyrosoma atlanticum*), the doliolid (*Dolioletta gegenbauri*) and the salp (*Cyclosalpa affinis*). The appendicularian is within its house, which functions in buoyancy and to concentrate food particles. The "knobs" on the pyrosome are each single zooids. The doliolid picture is a nurse with a stolon bearing Trophozooids (gasterozooids) on each side. The central row of phorozooids on the stolon is not clear. The salp is an asexually reproducing solitary form with a chain of aggregates each destined to reproduce sexually. (Reprinted with permission from Holland, 2016.)

Synapomorphies

- presence of an extracellular tunic. The tunic, which surrounds the zooids, can be very thick in some ascidians but is relatively thin in appendicularians and thaliaceans. The tunic contains cellulose, synthesized by cellulase, likely acquired by lateral transferase from bacteria (Nakashima et al., 2004)
- the appendicularian house also contains cellulose
- highly diverse body plans
- most species are hermaphroditic
- thaliaceans have alternation of sexual and asexual generations

Appendicularians and doliolids are smaller, but can also occur in large numbers. Pyrosomes are rarely abundant, but in 2017 a bloom of pyrosomes occurred off the northern Pacific coast of the United States and Canada. Much of "marine snow" is created by the dead carcasses of pelagic tunicates as well as discarded appendicularian houses. Ascidians are important fouling organisms as they colonize any available surface from docks, to ropes to boat bottoms. In addition, because they attach to boats, they can become widely dispersed; as a result, many species of ascidians are considered invasive

Commercial value: No commercial value is known for appendicularians or thaliaceans. One species of ascidian is aquacultured for human consumption in Japan. A natural product extracted from an ascidian is marketed as an anticancer drug

HISTORY OF TAXONOMY AND CLASSIFICATION

Tunicates were first described by Aristotle. They were initially considered to be molluscs. It was Kowalevsky in 1870 who first recognized the affinities of tunicates and vertebrates (Kowalevsky, 1870). Since tunicates, unlike cephalochordates and vertebrates, are evolving very rapidly, it is not surprising that they are particularly diverse and that their evolutionary history and, hence, their classification, have been debatable.

Phylogenetic analyses based on morphology or 18s rDNA initially placed tunicates basal to amphioxus plus vertebrates (Cameron et al., 2000; Winchell et al., 2002; Cameron, 2005; Stach, 2008). This proved problematic when comparisons of gene expression in embryos of ascidians, amphioxus and vertebrates showed that ascidians appeared to have some vestiges of structures such as migratory neural crest cells (Jeffery et al., 2004; Jeffery, 2006, Abitua et al., 2012), that vertebrates have but amphioxus clearly lacks (Holland et al., 1996). This problem disappeared with phylogenetic analyses based on concatenated nuclear genes. The first such analysis showed that tunicates and not amphioxus are the sister group of vertebrates, and that, rather than having evolved from doliolids, appendicularians are the sister group of all other tunicates (Delsuc et al., 2006) as had earlier been proposed on the basis of gamete morphology (Holland et al., 1988).

Because the five groups of tunicates are so very diverse both in morphology and ecology, each group is considered separately below.

CLASS APPENDICULARIA

Appendicularians were described in the early 1800s and misidentified as molluscs. It was not until 1851 that Huxley recognized that appendicularians are tunicates (reviewed in Lemaire and Piette, 2015). Appendicularians were once called "larvaceans" because of the mistaken idea proposed by Walter Garstang (Garstang, 1928, sometimes as 1929) that they had evolved by neoteny from an ancestral doliolid. In the widely accepted scheme based chiefly on morphology that was popularized by Alfred S. Romer, a sessile, tentaculate filter feeder gave rise to echinoderms and pterobranch hemichordates, then the tentacles or arms of such a hemichordate were lost in the ancestral ascidian, whose larvae gave rise to amphioxus and vertebrates by neoteny and whose adults evolved into ascidians and thaliaceans, with, as Garstang proposed, appendicularians then evolving from doliolids (Romer, 1967). However, the molecular phylogenetic analyses cited above have demonstrated that appendicularians are the sister group of all other tunicates (Delsuc et al., 2018). Therefore, the ancestral tunicate was probably a free swimming organism more like an appendicularian or possibly a cephalochordate than an ascidian or thaliacean.

All appendicularians are pelagic filter feeders and all have a similar body plan consisting of a trunk and a motile tail (Figures 37.1, **37.6**, **37.7A**). These are the only tunicates to retain a nerve cord and a tail in the adult. In the others, the CNS consists solely of a ganglion and associated nerves (Lacalli and Holland, 1998, Osugi et al., 2017). The only appendicularian that has been studied extensively is the single dioecious species *Oikopleura dioica* as it has been in laboratory culture since the early 1980s. In fact, O. *dioica* is the only dioecious tunicate; all others are hermaphrodites. Hermaphroditic species of appendicularians have so far defied attempts to put them into laboratory culture.

Ecological importance: Although most appendicularians are small, they can be extremely numerous. Their discarded houses, fecal pellets and dead bodies are a major part of "marine snow" cycling nutrients from surface waters to the deep sea.

Commercial value: Appendicularians have no known commercial value.

Synapomorphies

- they secrete a "house" that serves in filter-feeding and provides buoyancy
- alone among the tunicates, they retain the tail as adults
- all species but one, *Oikopleura dioica,* are sequential hermaphrodites
- alone among tunicates, their eggs have cortical granules that undergo exocytosis at fertilization

HISTORY OF TAXONOMY AND CLASSIFICATION

Appendicularians were first thought to be related to "medusae" (Chamisso, 1821). Subsequently, they were classified as "molluscs" (Mertens, 1830) or "zoophytes" (Quoy and Gaimard, 1833). Finally, Huxley (1851), recognized that appendicularians are tunicates. The monograph by (Fol, 1872), in which he noted that appendicularians have a "permanent larval form" is most likely the genesis of the term "larvaceans" for the group. The class Appendicularia has a single order, Copelata with three families–the Oikopleuridae with 11 genera and 37 species, the Fritillariidae with three genera and 30 species and the Kowalevskaiidae with one genus and two species.

ANATOMY/MORPHOLOGY

With the exception of the mesopelagic *Bathochordaeus,* appendicularians are under 1 cm in length. The length of the trunk ranges from about 1 mm in *Oikopleura* to 3 cm in *Bathochordaeus* (Figures 37.6, 37.7A). The trunk is elongate in fritillarids. The gonad(s) are located at the posterior end of the trunk; in hermaphroditic species, the testis is at the posterior terminus of the trunk. The mouth, which is surrounded by mechanosensory cilia, is at the anterior end of the trunk. Food enters the mouth and is carried through the pharynx into the looped gut that opens ventrally anterior to or at the same level as the single pair

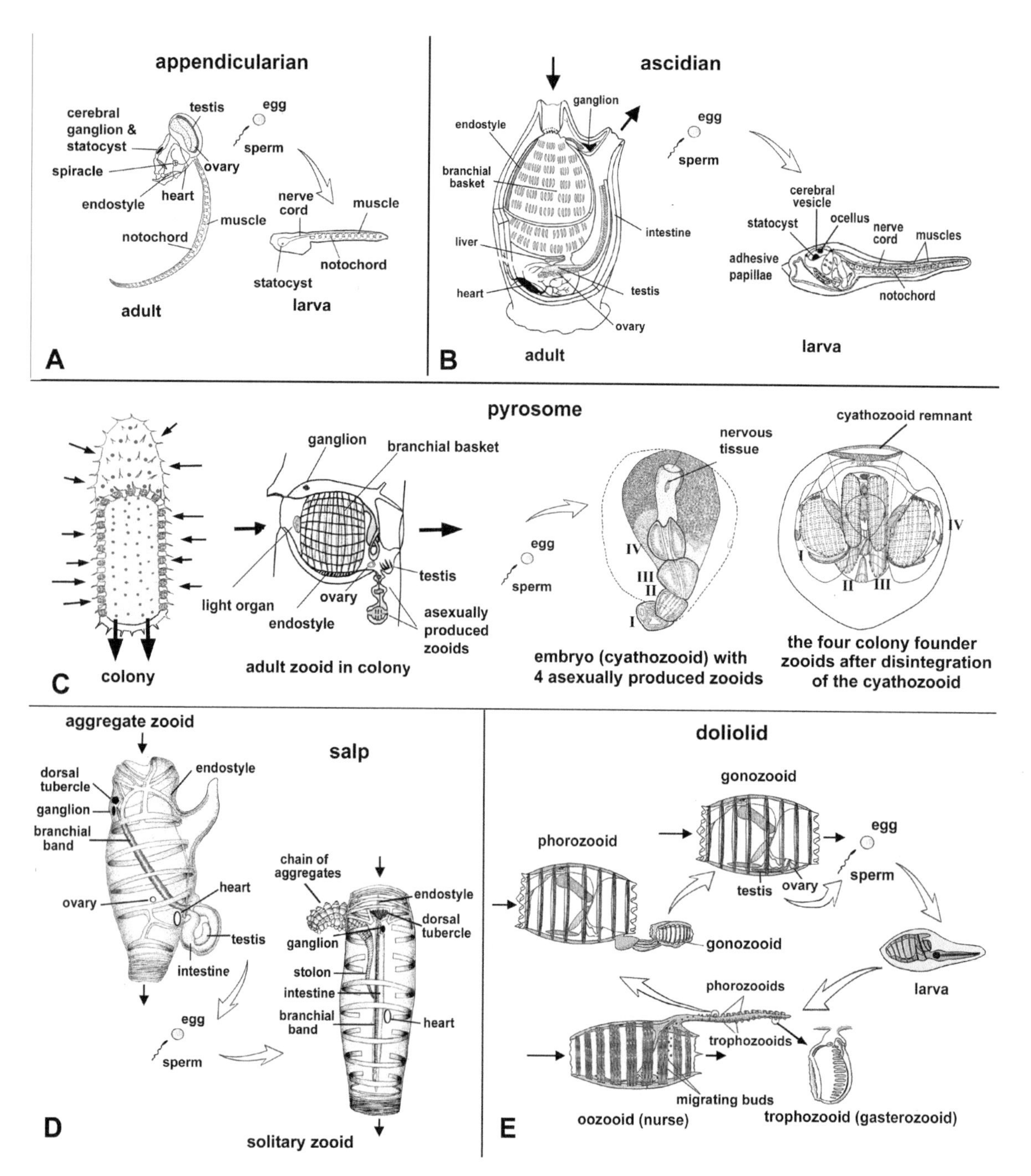

Figure 37.7 Diagrams of the five major groups of tunicates showing major anatomical features. Tunicates have hearts that periodically reverse the direction of the blood flow. (A) Appendicularians, most of which are only a few millimeters long, are exclusively pelagic. They secrete a series of houses that provide buoyancy and serve in filter feeding. None is colonial. All except *Oikopleura dioica* are hermaphroditic. Larval development is direct. (B) Solitary ascidian. Adult ascidians are sessile. Coloniality has evolved repeatedly. Arrows indicate the flow of water in the incurrent siphon and out the excurrent siphon. Ascidians are typically simultaneous hermaphrodites but are self-sterile due to incompatibility of gamete surface proteins. The larva is motile. After a short life in the plankton, it attaches to the substratum by the adhesive papillae. The tail is then resorbed, the branchial basket develops and the nervous system is remodeled. C, D, E. Thaliaceans. (C) Pyrosome colonies, that can reach several meters long, consist of thousands of very small zooids. They move slowly through the water by jet propulsion. Each zooid in the colony reproduces asexually, enlarging the colony, and sexually, producing an embryo termed a cyathozooid that is brooded within the colony. Each cyathozooid is resorbed after it buds off four zooids; this founder colony is expelled from the parent colony to start a new one. (D) Salps have a biphasic life cycle where the solitary zooid repeatedly buds off chains of aggregate zooids. Each aggregate, which typically has just one egg, is fertilized as soon as the zooid expands and begins pumping water in the mouth and out of the cloaca. The fertilized eggs develop directly into small solitary zooids and are then expelled out of the cloaca. Sperm are produced about the time the embryo matures. Doliolid sperm undergo an acrosome reaction. In other thaliaceans and ascidians the acrosome is either vastly reduced or absent. After shedding sperm, the zooids die. (E) Doliolids have an extremely complex life cycle. The larva develops into a non-feeding oozoid or nurse with a stolon bearing two types of zooids – trophozooids, that only feed and stay attached to the stolon, and phorozooids, which break off the stolon. The phorozooids then bud off gonozooids, each of which develops an ovary and a testis. They are distinguished from salps by muscle bands that are continuous around the body. (Reprinted with permission from Holland 2016.)

of gill slits. An endostyle is located mid-ventrally in the pharynx. The endostyle, which binds iodine and is considered a homolog of the vertebrate thyroid, is relatively small in the Fritillaridae and absent in Kowalevskiidae (Ericson et al., 1985). The pharynx contains a dorsal ciliated band in the Oikopleuridae and Fritillaridae and four longitudinal groups of ciliated papillae in the Kowalevskiidae (Carlo et al., 2003). True mesenchyme is absent in appendicularians. The tubular heart, which is absent in the Kowalevskiidae, has a muscular wall on the left and a non-muscular wall on the right and is adjacent the gut (Figures 37.6, 37.7A) (Onuma et al., 2017). The heart beat reverses direction periodically as in other tunicates. There is no clear kidney in appendicularians. The CNS consists of a cerebral ganglion, a statocyst and a tail nerve cord. Unlike in tadpole larvae of ascidians, nerve cell bodies exist throughout the length of the nerve cord. At hatching, the notochord, which lies ventral to the tail nerve cord, consists of a stack of about 20 cells with central nucleus and a vacuole; cells continue to be added as the adult grows (Søviknes and Glover, 2008). After hatching, the vacuoles in adjacent cells fuse to form a central continuous lumen with notochord cells dorsal and ventral to the lumen. The notochord is flanked by muscle cells. The haemocoel in the tail contains two or more possibly secretory subchordal cells, depending on the species (Fredriksson and Olsson, 1991).

REPRODUCTION, EMBRYOLOGY, AND LARVAL ANATOMY

O. dioica eggs are fertilized at first meiotic metaphase (**Figure 37.8**) (Holland et al., 1988). At fertilization, there is a cortical reaction with exocytosis of cortical granules but without obvious elevation of the fertilization envelope. The sperm has a typical acrosome and forms an acrosomal tubule that fuses with the egg plasma membrane (Holland et al., 1988). There are some cytoplasmic movements after fertilization but none resembling the segregation of the myoplasm as in ascidians (Fujii et al., 2008; Nishida, 2008; Stach et al., 2008; Nishida and Stach, 2014). Development is direct from egg to adult. Cleavage in appendicularians is determinant and development is mosaic. Cell lineages have been mapped and are similar to those in ascidians (Nishida, 2008; Nishida and Stach, 2014). However, gastrulation occurs one cell division earlier in *O. dioica*. Hatching can occur as early as 6 hrs after fertilization. The first house is made by 24 hrs post-fertilization. At the end of development, the tail undergoes a 90° shift.

In hermaphroditic appendicularians, the sperm are expelled first and later the eggs are dispersed by rupture of the gonad. The animals then die. An evolutionary adaptation to a short life cycle is that, at least in *O. dioica* and likely in other species, the gametes grow in a syncytium. The entire female germ line is in a single multinucleate cell termed the "coenocyst." The nuclei divide within this cell and then about half become polyploid while the remaining ones begin meiosis. Each meiotic nucleus is partially enclosed in a plasma membrane. Ring canals connect the future oocytes with the general cytoplasm, which moves through the ring canals into the oocyctes. Ultimately the polyploid nuclei degenerate and the oocytes become individualized (Ganot et al., 2007a; Ganot et al., 2007b). Typically each female *O. dioica* produces a few hundred eggs. The sperm also develop in a syncytium and are likely individualized shortly before spawning (Onuma et al., 2017).

ECOLOGY AND PHYSIOLOGY

Except for the genus *Bathochordaeus*, most appendicularians are quite small—the trunk being about 1–2 mm long. They secrete a house and beat the tail to draw a current of water through it. The house provides buoyancy and contains

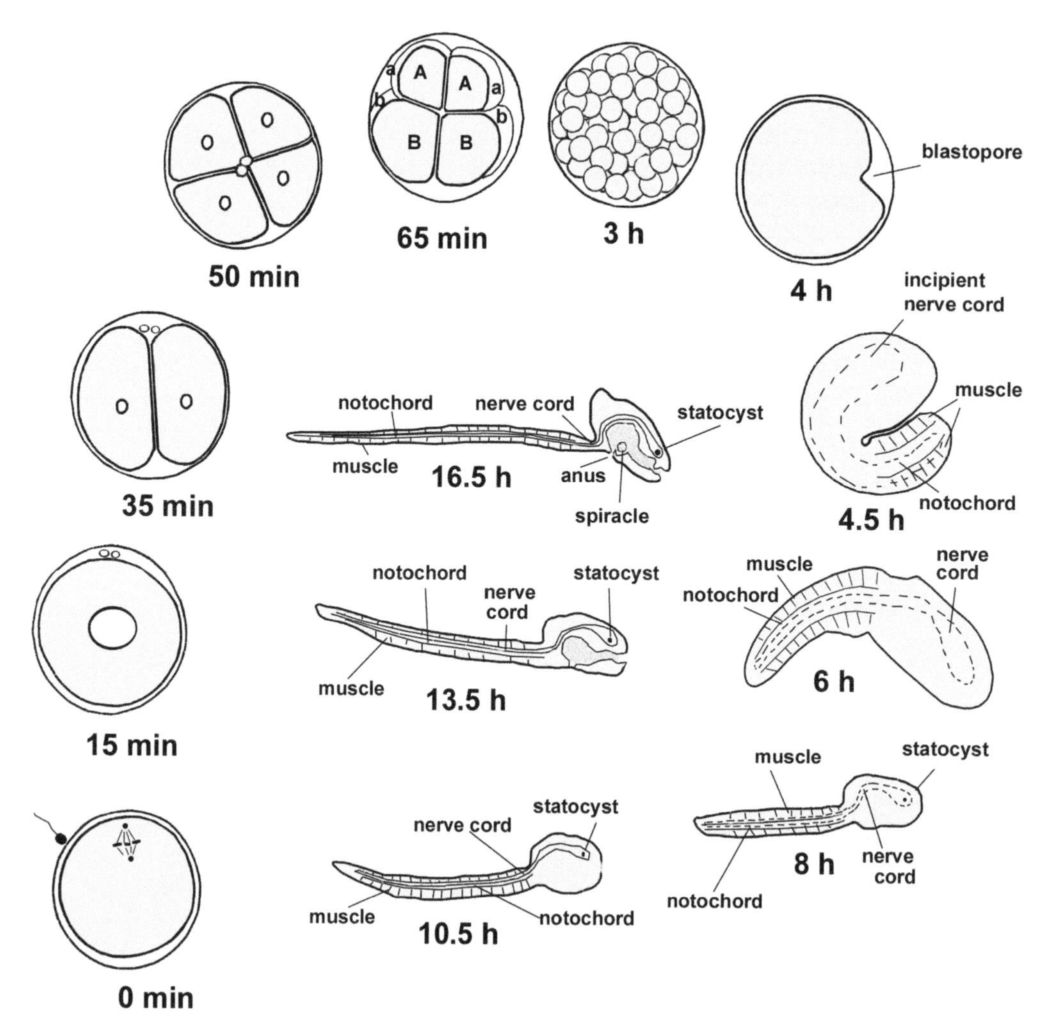

Figure 37.8 Development of the appendicularian, *Oikopleura dioica*. Eggs are fertilized at first meiotic metaphase. Cortical granules undergo exocytosis and the sperm undergoes an acrosome reaction. The meiotic divisions are completed within 15 min at 23°C resulting in two polar bodies. The first four cleavages are equal. A dorsal view of the 4-cell stage showing the two polar bodies is shown at 50 minutes. Cell lineages are fixed as shown at 65 minutes. Nuclei are omitted from 65 minute on. Cell lineages not shown from 3 hours on. By 4.5 hours at the early tadpole stage, the major tissues are delineated. The embryo rapidly elongates as tissues differentiate. In the final stage of development in *O. dioica*, the tail rotates 90° about the long axis. Early development in ascidians is similar to that in appendicularians, although gastrulation occurs one cell division later, and the larval tail does not rotate. There are nerve cells in the appendicularian tail nerve cord, but none in that of ascidian larvae. Development in thaliaceans is more derived. (After Nishida, 2008.)

an elaborate set of filters that concentrate the particulate food, ranging from viruses to phytoplankton (Lombard and Kiørboe, 2010). The house in *O. dioica* can have a diameter up to 8–10 mm, while those of *Bathochordaeus*, adults of which have a trunk 1 cm or more in length, can be up to 1 m in diameter. When the house becomes clogged, the animal discards it and creates another by inflating the house rudiment located around the trunk. The animals also leave their houses to spawn. They then swim up to the water surface where they can be aggregated in windrows. Appendicularians are very abundant. Over 3,000 individuals per liter of spawning *O. longicauda* have been documented (Alldredge, 1982). Once the animals spawn, they die. Moreover, the spawned-out individuals and discarded houses form a major component of marine snow taking nutrients from surface waters to deeper layers of the ocean (Lombard and Kiørboe, 2010; Zervoudaki et al., 2014). The life span of appendicularians is typically short. For *O. dioica* it ranges from 5–10 days depending on temperature and food. That means that when environmental conditions are favorable, large populations can build up quickly. *O. dioica* has been put into continuous breeding cultures (Fenaux and Gorsky, 1979; Martí-Solans et al., 2015).

Appendicularians can form a major part of the diet of fish and cnidarian medusae (D'Ambra et al., 2013; Bachiller et al., 2018). They are parasitized by certain dinoflagellates (Alf and Enric, 2006). Gregarine protozoa can inhabit the gut (Leander et al., 2006). Ciliates have been found in the house.

DISTRIBUTION AND BEHAVIOR

Appendicularians occur widely from tropical waters to the arctic and antarctic (Hunt and Hosie, 2005; Swadling et al., 2010; Basedow et al., 2014; Deibel et al., 2017). They are mainly in shallow water, but *Bathochordaeus* and *Mesochordaeus* live in the mesopelagic zone (200–1,000 m deep). They live inside the house, beating their tails periodically to maintain water flow through the filters. When the filters of the house become clogged with particles, they swim out of the house and inflate a new one that is preformed around the trunk. At spawning, at least the shallow water species leave the house and swim to the surface.

GENOMICS AND GENETICS

O. dioica has the smallest genome (~72 Mb) of all tunicates sequenced to date. It is estimated that 39% of the 17,000–18,000 genes are transpliced (Danks et al., 2015). That is several genes are transcribed together, a short spliced-leader (SL) RNA is then linked to the 5′-ends of these mRNAs that are then cut up into mRNAs for individual genes. Transplicing occurs in many eukaryotes and prokaryotes, although it is absent from cephalochordates and rare in vertebrates where it occurs by a non-SL method (Lei et al., 2016). Although it is debatable whether trans-splicing evolved repeatedly, the consensus is that it may have had an ancient origin and been lost independently in several taxa (Krchňáková et al., 2017).

The reason for such a low number of genes in appendicularians appears to be gene loss. For example, *O. dioica* has lost a number of Hox genes including *Hox 3* and *Hox 5–8* (Edvardsen et al., 2005). It has also gained some new genes by duplication (e.g., two paralogs of *Pax2/5/8* compared to a single *Pax2/5/8* gene in amphioxus. In addition, *O. dioica* has evolved 80 unique proteins termed oikosins, many generated by gene duplication, that form specific regions of the house. They are secreted by the trunk epithelium, termed the oikoplast that consists of large polyploid cells. The new house wraps around the trunk of the zooid until the animal discards the old house. The beat of the tail helps to inflate the new house as soon as the old one is jettisoned. Sulfation of the proteoglycans in the house rudiment is necessary for inflating the house rudiment into the new house (Thompson et al., 2001; Hosp et al., 2012).

POSITION IN THE ToL

Molecular phylogenetic analyses with concatenated nuclear gene sequences have consistently placed appendicularians as the sister group to all other tunicates, albeit with a very long branch (Delsuc et al., 2018; Kocot et al., 2018) (Figure 37.5). This is in agreement with findings from morphological studies that while eggs of other tunicate classes have lost cortical granules and their sperm do not form an acrosomal tubule, appendicularians, like ambulacrarians and cephalochordates, have retained both structures, suggesting that appendicularians indeed did not evolve by neoteny from larvae of another tunicate (Flood, 1978; Holland et al., 1988).

CLASS THALIACEA

General Characteristics

The class Thaliacea has three orders (Pyrosomidae, Doliolidae, and Salpidae)

Recent molecular phylogenetic analyses place Thaliacea as monophyletic in spite of the large differences in anatomy among the three orders

They are the sister group of aplousobranch plus phlebobranch ascidians (Figures 37.6, **37.7C–E**) (Delsuc et al., 2018; Kocot et al., 2018)

All thaliaceans are pelagic filter feeders and reproduce both asexually and sexually. Sexually reproducing forms are hermaphroditic

ORDER PYROSOMATIDA

General Characteristics

Pyrosomes are strictly pelagic

They occur in large cylindrical colonies composed of hundreds or thousands of small zooids that closely resemble those of ascidians (Figures 37.6, 37.7C). However, as in other thaliaceans, the mouth and cloaca are at opposite ends of the zooid and not adjacent as in ascidians

All the mouths face outward and the cloacas inward. A current of water is drawn in the mouth by the beating of cilia on the branchial basket and expelled through the cloaca. As the colony is closed at one end and open at the other, it moves through the water by slow jet propulsion. Colonies can reach up to 10 m or more long. A penguin has been found trapped inside one large colony

Smaller colonies have occurred in large blooms such as off the northwest coast of the USA in the winter of 2017–2018 (Marlétaz et al., 2018)

HISTORY OF TAXONOMY AND CLASSIFICATION

Pyrosomes were first discovered by Péron in 1807 (cited in Piette and Lemaire, 2015). Subsequently, in 1815, Lesueur recognized that pyrosomes were tunicates (Lesueur, 1815). Because of similarity in body plans between pyrosomes and ascidians, it has been postulated that ascidians gave rise to pyrosomes (reviewed in (Piette and Lemaire, 2015). This remains a possibility.

ANATOMY/MORPHOLOGY

A pyrosome colony develops from an initial cluster of four zooids that grows out of the rudimentary embryo (Figure 37.7C). The colonies grow by asexual budding. Each pyrosome zooid has a branchial basket terminating posteriorly in a looped gut. The gut empties into the cloaca. Pyrosomes are hermaphroditic. The ovary is located along one side of the branchial basket and the testis, branched like a bunch of bananas, is adjacent the cloaca. At the mouth end of the zooid there is a light organ containing luminous bacteria (Figure 37.7C). When a colony is disturbed, a wave of luminescence migrates over the colony. Buds form serially near the posterior end of the colony. Individual zooids are typically a few millimeters long. The nervous system consists of a ganglion, a statocyst and an associated neural gland, which opens into the pharynx, together with eight pairs of peripheral nerves that innervate the muscles of the mouth and cloacal sphincters and the branchial basket inter alia. The heart is midventral between the posterior end of the endostyle and the gut. Like hearts in other tunicates, the beat periodically reverses. Organs giving rise to blood cells have been described in the dorsal mesenchyme of some pyrosomes. The blood moves towards the sub-endostylar sinus and then flows across the branchiae to the dorsal sinus. It then moves back to the ventral sinus and the posterior opening of the heart.

REPRODUCTION, EMBRYOLOGY, AND LARVAL ANATOMY

Pyrosomes have a complex alternation of sexual and asexual generations. Sexual reproduction has not been directly observed in pyrosomes. Spermatogenesis takes place in the lobes of the testis with the spermatogonia and spermatocytes at the tips of the lobes and the spermatids where the lobes come together. The sperm are highly elongate with the mitochondrion wrapped in a spiral around the nucleus (Holland, 1990). The single oocyte is evidently fertilized internally; it is not known if the sperm from some zooids in a colony can fertilize eggs from other zooids in the same colony. The fertilized zygote develops directly into a rudimentary organism termed a "cyathozooid." The cyathozooid has a neural tube that forms and then shrinks, a heart, peribranchial cavities, a pharynx with an endostyle, a cloaca, and a gut rudiment (Figure 37.7C). However, it lacks a buccal siphon and gut. Since the eggs are very yolky, this organism is stretched out on the yolk surface. The cyathozooid asexually buds a chain of four zooids and is then shed from the colony to form the core of a new colony. Each of these four primary zooids buds additional zooids, and they, in turn bud more, resulting in a large colony.

ECOLOGY AND PHYSIOLOGY

Large blooms of pyrosomes have been documented off Taiwan and the northwest coast of the USA; the aftermath of such blooms is massive death and sinking to the sea bottom (Kuo et al., 2015; Marlétaz et al., 2018). Their fecal pellets and carcasses are a major contributor to carbon flux from the surface to the deeper layers of the ocean. Like appendicularians, pyrosomes are filter feeders. They are preyed upon by fish [e.g., tuna, slickheads (*Alocephalus*) and barrelfish for which they can be ~90% of the diet], seals and sea turtles (Parker, 2011). There are no reports of parasites of pyrosomes, although the amphipod *Phronima* is known to eat out the inside of a pyrosome colony and take up residence in the remaining gelatinous "shell" (Hirose et al., 2005).

DISTRIBUTION AND BEHAVIOR

Pyrosomes occur throughout the world ocean in temperate and tropical waters as well as in the sub-antarctic (Pakhomov et al., 2000; Décima et al., 2019). As noted above, the colony moves through the water by slow jet propulsion. Depending on the species, colonies can reach up to 10 m or more in length.

GENOMICS AND GENETICS

The genome size is unknown. Transcriptome sequences of a pyrosome are available (Kocot, et al., 2018).

Commercial value: None

POSITION IN THE ToL

Molecular phylogenetic analyses place pyrosomes as sister group to doliolids plus salps in a monophyletic clade of tunicates (Govindarajan et al., 2011; Kocot et al., 2018) (Figure 37.5).

ORDER DOLIOLIDA

General Characteristics

Doliolids are the sister group of salps (Figures 37.5, 37.6, 37.7E)

The zooids are typically small, up to 5 cm long depending on the species and the stage

They inhabit near-shore marine waters worldwide and recently *Dolioletta gegenbauri* been put into continuous laboratory culture (Walters et al., 2019a)

In doliolids, like pyrosomes and salps, the mouth and cloaca are at opposite ends of the zooid. Doliolids move slowly through the water propelled by the current generated by cilia on the branchie. The zooids are completely encircled by muscle bands

HISTORY OF TAXONOMY AND CLASSIFICATION

Doliolids are the sister group of salps (Figure 37.5). They were first recognized as tunicates in 1833 (Quoy and Gaimard, 1833).

ANATOMY/MORPHOLOGY

Doliolid zooids are typically small, up to 5 cm long depending on the species and the stage (Figures 37.6, 37.7E). The zooids are barrel-shaped, encircled by muscle bands. Cilia beating on the branchial bands draw water in through the anterior mouth and out through the posterior cloaca. The zooids have several ciliated mechanosensory cells associated with the borders of the mouth and cloaca. The oozoids have a statocyst innervated by a peripheral nerve. The structure of these sensory organs varies from species to species. There is a neural gland located below the ganglion. The anterior end has a narrow canal that leads to the funnel-shaped vibratile organ that in turn opens into the pharynx (reviewed in detail in Drach, 1948). The endostyle secretes mucous that traps food particles and is moved into the gut by the peripharyngeal bands of cilia (Bone et al., 1997).

Doliolids have a very complex alternation of sexual and asexual generations (Braconnot, 1974; Godeaux et al., 1998). The oozoid or nurse, produced sexually by union of egg and sperm does not itself feed. Rather, it produces a stolon with three rows of buds. The outer two rows, the trophozoioids or gasterozooids, feed on small particles including phytoplankton. Their anatomy is modified with a very large mouth and an expanded gut. The center row of buds, the phorozooids, may break off the stolon. They in turn bud off gonozooids, that are hermaphrodites producing one or a few eggs and sperm. Eggs are shed first, before the testis matures. In some instances, the gonozooids (termed gonophorozooids) can also bud asexually, and the phorozooids can bud more phorozooids as well as gonozooids. For the coastal *Dolioletta gegenbauri* it was estimated that the larva metamorphoses into an oozooid within four days at 20° C and that the oozooid can produce 70 phorozooids per day over 20 days with the oozooid lifespan being about 21 days (Deibel and Lowen, 2012). These times depend on food availability.

Although most doliolids live at shallow depths and are filter feeders, recently, mesopelagic and bathypelagic doliolids have been found with submersibles. The bathypelagic *Pseudusa bostigrinus* has a highly modified body plan in conjunction with the absence of phytoplankton at depths of 1166–1890 m and has evolved carnivory. The phorozooid, the only stage found, is bell shaped like a hydromedusa, with four bands of muscles and bears buds. Muscle contractions propel them through the water, but they are often stationary with a wide-open mouth pointed upwards. It lacks the ciliated peripharyngeal bands of cilia present in other doliolids and has a very small endostyle. One muscle band acts as a sphincter to close the mouth, which traps particles raining down. In addition they eat eggs of other species and crustaceans. The more typical mesopelagic *Doliolula equus* is interesting in that some individuals have orange pigment patches and are luminescent. In addition, it has commensal ciliates and is parasitized by the amphipod *Phronima*. It usually has a hydroid polyp attached to the rim of mouth or buccal opening (Robison et al. 2005).

REPRODUCTION, EMBRYOLOGY, AND LARVAL ANATOMY

Doliolids are hermaphrodites. Eggs are shed first, before the testis matures. In some instances, the gonozooids (termed gonophorozooids) can also bud asexually, and the phorozooids can bud more phorozooids as well as gonozooids. For the coastal *Dolioletta gegenbauri* it was estimated that the larva metamorphoses into an oozooid within four days at 20° C and that the oozooid can produce 70 phorozooids per day over 20 days with the oozooid lifespan being about 21 days (Deibel and Lowen, 2012). These times depend on food availability.

The testis of the gonozooid is long (Figure 37.7E) and the sperm themselves are elongate with a distinct acrosome that undergoes exocytosis but does not produce a long, acrosomal tubule (Holland, 1989). The ovary typically produces fewer than five eggs. Fertilization probably occurs either shortly before or shortly after the eggs are expelled from the cloaca. The eggs of *Dolioletta gegenbauri* develop within a large fertilization envelope. Development is direct. Embryos of several species have been described. The larva of *D. gegenbauri* has a small, immotile tail that is resorbed before hatching. The zooid hatches in about 2.5 days. The larva of *Doliolum denticulatum* develops a long immotile tail and is elongate without a clear trunk by 36 hours. By the third day, a trunk emerges and the tail is resorbed. A notochord may be present transiently. In all cases, hatching occurs after the adult is fully formed.

ECOLOGY AND PHYSIOLOGY

With the exception of the mesopelagic doliolids noted above, doliolids are pelagic filter feeders. Their diet is chiefly phytoplankton (Walters et al., 2019b). However, doliolids can eat copepod eggs. The current of water going in through the mouth and out through the cloaca not only propels the animal through the water, it also brings in food. Occasionally, the circular muscles contract and the animal jerks forward. Doliolids, like salps and pyrosomes function in the oceanic food web to take nutrients in the form of fecal pellets and dead carcasses from surface waters to deep waters. Large blooms have been documented, and it is estimated that a single gonozooid can produce over 700,000 progeny over a generation time of 50 days, resulting in a doubling of the population every 2.5 days. Predators include sapphirinid copepods and fish. There is evidence from DNA sequencing that they may be parasitized by alveolate protozoans in the Apicomplexa (Walters et al., 2019b).

DISTRIBUTION AND BEHAVIOR

Doliolids are very widely distributed in the world ocean, and are particularly abundant in temperate and sub-tropical oceans. They can occur at densities as high as 10^4/cubic meter (Walters et al., 2019b).

GENOMICS AND GENETICS

Nothing is known about nuclear genomes or transcriptomes of doliolids.

POSITION IN THE ToL

Phylogenetic analyses with 18S rDNA places doliolids as the sister group of pyrosomes + salps within a monophyletic Thaliacea (Tsagkogeorga et al., 2009; Kocot et al., 2018) (Figure 37.5).

ORDER SALPIDA

General Characteristics

Salps are pelagic marine animals with an alternation of sexual and asexual generations (Figures 37.6, 37.7D)

A sexually produced form, the oozoid or solitary form, can daily bud off chains of 100 or more gonozooids or aggregates

They can occur in blooms of extremely large numbers

Commercial value: None

HISTORY OF TAXONOMY AND CLASSIFICATION

Salps were first described in 1775 from material collected on the ill-fated expedition from Denmark to Arabia Felix (Yemen) from which only one man returned alive (Forskal and Niebuhr, 1775). Salps were recognized as being related to ascidians by Cuvier in 1804 (Cuvier, 1804). In phylogenetic analyses with large sets of nuclear genes, salps and doliolids are sister groups and most closely related to the phlebobranch + aplousobranch clade of ascidians (Figure 37.5) (Delsuc et al., 2018; Kocot et al., 2018).

ANATOMY/MORPHOLOGY

Depending on the species, adult salps range from about 1 –10 cm long (Figures 37.6, 37.7D). Salps, like doliolids, have an anterior mouth, a posterior cloaca and muscle bands, but the muscles are not continuous ventrally as they are in doliolids. The zooids move through the water by contracting the muscles—in effect, pumping water through themselves. They have a diagonal branchial band with a single pair of gill slits, one on each side and a ventral endostyle that is ciliated. The heart is located ventrally and, as in other tunicates, periodically reverses the beat. The blood vessels, like those of other invertebrate chordates, lack an endothelium. There are numerous cell types in the blood. Some salps (e.g., *Cyclosalpa*) have luminous organs; it is unknown whether or not the luminescence is caused by symbiotic bacteria. The digestive system consists of a gut that is twisted into a U or O shape with an expanded central region termed the stomach. The nervous system consists of a dorsal ganglion underlain by a neural gland. During development, the nervous system is initially tubular, but develops into a ganglion with a pigment spot forming part of a photoreceptor or eye. Peripheral nerves radiate out from the ganglion and innervate the muscles and other structures (Lacalli and Holland, 1998). The neural gland or duct has been suggested to be homologous to the neurohypophysis (e.g., possibly homologous to part of the vertebrate pituitary) or alternatively a hydrostatic organ controlling fluid flow into the blood spaces. The neural gland has two cavities that open into the pharynx. Anterior to the ganglion is the vibratile organ that is ciliated and opens into the pharynx; its function is unknown. The testis in sexually reproducing forms (aggregates) is located within the loop of the gut. The ovary in these hermaphroditic forms is located posteriorly in the lateral body wall.

REPRODUCTION, EMBRYOLOGY, AND LARVAL ANATOMY

Salps have only two types of zooids. The solitary form buds off chains of aggregates, which in turn reproduce sexually. The aggregate zooids at the tip of a chain begin pumping first, followed linearly by the remaining zooids in the chain. In most species, each aggregate has a single oocyte in the ovary. Sometimes there are more, but never more than two develop (Daponte and Esnal, 1994). The ovary is within the body wall, surrounded by single layer of follicle cells that connects to the body wall by a so-called "oviduct." When the chain is shed, the zooids at the tip expand and start pumping first, followed in turn by the more posterior zooids. As soon as pumping begins, sperm shed by an older aggregate zooid are taken into the pharyngeal cavity and home in on the spot where the 'oviduct' joins the pharyngeal epithelium. It was long a mystery how the sperm could traverse the oviduct and reach the egg as the oviduct appeared to be solid until after fertilization, when the oviduct shortens and a lumen containing sperm opens up. This mystery was solved by electron microscopy. The sperm have a long corkscrew-shaped head surrounded by the mitochondrion (Holland, 1988). The sperm homes in on the endoderm exactly where it joins the oviduct, which is indeed solid (Holland and Miller, 1994; Boldrin et al., 2009). Exactly how the sperm locates the proper spot is not known, but the possibility has been suggested that the cells near the oviduct release a sperm attractant (Boldrin et al., 2009). The sperm then burrow through the center of the cells of the oviduct to reach the oocyte. Development is direct without a larval stage. The fertilized egg cleaves into two, four and eight cells. Then, the cells separate and non-germinal cells termed calymmocytes, apparently derived from the follicle cell layer, move in between them. At the eight and sixteen-cell stages, the blastomeres appear to divide asymmetrically. The smaller cells, as well as the calymmocytes, eventually degenerate, and the blastomeres come back together and develop into a small adult solitary form (Brien, 1948; Sutton, 1960). This very baroque mode of early development has only been observed with light microscopy of paraffin-embedded specimens. It should be confirmed with techniques offering higher resolution. During embryogenesis, the embryo remains attached to the parent by a thick placenta, derived from maternal tissue. Posterior to the placenta, a structure termed the eleoblast develops from the embryo. It contains mesenchyme and blood cells. It appears to give rise to at least mesenchyme and muscles of the oozoids (aggregates) as they bud from the stolon. When the zooid is mature, it breaks through the body wall and exits the parent via the cloaca to begin reproducing asexually by budding aggregates from a stolon located in the posterior part of the parent. Once the solitary form has left its parent, the testis matures and sperm are shed. The zooid then dies.

ECOLOGY AND PHYSIOLOGY

Salps undergo a diel migration going down in the water column to 200 m or more and rising to the surface at night. The chain of aggregates is typically shed by the solitary form when the salps are at the surface. Salps play a major role in global ocean ecology as they are relatively large zooplankton and because of asexual reproduction, large populations can accrue fairly rapidly when conditions are favorable. In addition, salps eat phytoplankton and other small particles; both fecal pellets and dead zooids are colonized by bacteria and viruses. Finally, salps are preyed upon by a number of other organisms. These include the amphipod, *Phronima* that takes up residence inside the salp, eating out its insides and using the test as a house, shrimp (Fanelli and Cartes, 2008), jellyfish (Tilves et al., 2018), fish (Diaz Briz et al., 2017), corals (Mehrotra et al., 2016), and sea turtles (Henschke et al., 2016).

DISTRIBUTION AND BEHAVIOR

Salps occur widely distributed in the world ocean. One species, *Salpa thompsoni,* is a major component of the zooplankton in Antarctic waters, while blooms of *S. aspera* and *Cyclosalpa bakeri* have been described from the Gulf of Alaska (Li et al., 2016). Numerous species occur in the temperate through sub-tropical oceans (Nogueira Junior et al., 2015). As noted above, they occur in large aggregations, typically undergoing a diel migration, coming up towards the surface at night and down during the day. Chains of aggregates released at night will gradually break up into individual zooids. They are major consumers of phytoplankton, cycling carbon in the form of fecal pellets and dead carcasses into the deeper ocean (Trueblood, 2019).

GENOMICS AND GENETICS

A preliminary genome assembly and transcriptomes of aggregates of one salp species *Salpa thompsoni* from Antarctic waters have been reported (Jue et al., 2016; Batta-Lona et al., 2017). Seasonal and age changes were determined. The genome size is estimated to be about 600 Mb (Jue et al., 2016). Like other tunicates, *S. thompsoni* appears to be evolving very rapidly.

POSITION IN THE ToL

Phylogenetic analysis with 18S rDNA sequences place salps as the sister group of pyrosomes within a monophyletic Thaliacea (Figure 37.5) (Tsagkogeorga et al., 2009)

CLASS ASCIDIACEA

General Characteristics

Ascidians are the most numerous and best known of the tunicates

They are exclusively sessile (Figures 37.6, **37.7B**)

Most live in shallow water and are filter feeders

They frequently colonize docks, ropes and boat bottoms. As a result of colonizing mobile surfaces, some species have been very broadly dispersed around the world and are often labeled "invasive species."

Commercial value: Ascidians including *Halocynthia roretzi* are aquacultured for human consumption in Japan and Korea. In addition, an ascidian is used for fishing bait in Australia. Ascidians have been the focus of much natural products chemistry. Hundreds of compounds including aromatic alkaloids, cyclic peptides and depsipeptides have been isolated from ascidians. Many of these are apparently synthesized by symbiotic bacteria, including cyanobacteria (Watters, 2018). Of all the compounds isolated from ascidians, only a few are being marketed commercially as anti-cancer drugs. One is Ecteinascidin (ET-743, trabectedin) from *Ecteinascidia turbinata*. ET-743 is FDA approved and has the trade name Yondelis®. A second compound, Aplidin® (dehyrodidemnin B, plitidepsin)—from *Aplidium albicans*—has orphan drug status [22]. Both are marketed by PharmaMar (Madrid, Spain). However, isolation of these compounds can require extremely large quantities of ascidians. The bacterium responsible for ET-743 has not been cultured, and the chemical synthesis of these compounds is complex and as yet not cost-effective—reviewed in (Watters, 2018).

Ascidians have a negative commercial value as biofouling organisms. They hinder aquaculture of mussels by colonizing the shells and competing for food. They slow boats by colonizing boat bottoms. Commercial boats once had paint with tributyltin on the bottoms. However, as tributyltin is very toxic to marine organisms in general, it has been banned by the International Convention on the Control of Harmful Anti-fouling Systems on Ships of the International Maritime Organization. Other chemicals such as cybutryne are in use, but they may also kill marine organisms in general as they leach out of the paint.

HISTORY OF TAXONOMY AND CLASSIFICATION

Ascidians were first recognized as animals and not plants by Aristotle. Linnaeus grouped solitary ascidians with the molluscs and colonial ones with the zoophytes (reviewed in Brien, 1948). By 1816, it was recognized that both solitary and colonial ascidians belonged to the same group, which Lamarck in 1816 called Tunicata (Lamarck, 2013, reprint of original 1816). It was (Kowalevsky, 1870), who recognized that tunicates were chordates. There are over 3,000 named species of ascidians. The orders of Ascidians have been particularly controversial. By the late 1800s some researchers recognized two orders or subclasses – the Pleurogona with one order or suborder (Stolidobranchia) (Perrier, 1899) and the Enterogona with two orders or suborders (Aplousobranchia and Phlebobranchia) (Della Valle, 1899). However, as molecular phylogenetic analyses with sets of nuclear genes have shown the Enterogona to be paraphyletic (Figure 37.5), the names Enterogona and Pleurogona are no longer accepted usage (Shenkar et al., 2009), and the Stolidobranchia, Aplousobranchia and Phlebobranchia are considered to be orders.

Aplousobranchia plus Phlebobranchia is the sister group of Thaliacea while the order Stolidobranchia is the sister group of the Aplousobranchia and Phlebobranchia plus Thaliacea (Delsuc et al., 2018; Kocot et al., 2018), a conclusion also reached in an analysis based on 18s rDNA (Tsagkogeorga et al., 2009) (Figure 37.5). Phylogenetic analysis with concatenation of 13 mitochondrial proteins also revealed ascidians as paraphyletic, but allied the Thaliacea as the sister group of the Phlebobranchia within the Aplousobranchia/Phlebobranchia clade (Rubinstein et al., 2013). These analyses are not far from the idea based on morphology that thaliaceans evolved from ascidian ancestors and are most closely allied to the aplousobranch family Polycitoridae (Brien, 1948). The take-home message of these analyses is that Thaliacea should probably not be considered a tunicate class, but rather an order of the Class Ascidiacea on a par with the three orders of sessile ascidians.

ANATOMY/MORPHOLOGY

Ascidian zooids have an endostyle, a U-shaped gut, a branchial basket with incurrent and excurrent siphons on the upper surface of the zooid (Figures 37.6, 37.7B). The endostyle secretes mucus that entraps food in the branchial basket and is carried into the stomach. Ovaries and testes are typically located near the loop of the gut but in some species are in the post-abdomen. The heart, located adjacent the intestine or in the post-abdomen periodically reverses the beat as in other tunicates. The blood contains several cell types. The myocardium consists of striated muscle. There are also smooth muscles that extend from the incurrent siphon to the base of the abdomen and some transverse muscles that form sphincters around the incurrent and excurrent siphons. When zooids are touched, the siphons contract causing water to squirt out–hence the common name of "sea-squirts." The nervous system consists of a ganglion typically located between the siphons and peripheral nerves. Around the openings of the siphons are numerous ectodermal sensory cells.

REPRODUCTION, EMBRYOLOGY, AND LARVAL ANATOMY

Most ascidians reproduce both sexually and asexually, although some such as *Ciona* sp. reproduce exclusively sexually. Asexual reproduction is typically by budding. Some ascidians bud from a stolon, while others bud directly from the body wall of the parent. The process of budding is related to the ability of ascidians to regenerate. A pivotal experiment showed that if zooids of the ascidian *Perophera*, which reproduces by budding from a stolon, are irradiated such that their cells cannot divide and then injected with blood cells from another individual, the blood cells will reconstitute a viable organism (Freeman, 1964). As a result, ascidians have been used as a model for stem cells (Laird et al., 2006; Kassmer et al., 2016). Asexually reproducing ascidians can form large colonies that encrust on hard surfaces. Coloniality appears to have evolved several times over, as has loss of the larval tail (Racioppi et al., 2017).

Ascidian zooids are typically hermaphroditic. Some species such as *Ciona* release sperm and eggs into the seawater (Figure 37.8). Other species such as *Botryllus* and *Botrylloides* brood the larvae, and release a very large larva about 3 mm long that has already begun to form buds (Zaniolo et al., 1998). Although many ascidians release sperm and eggs simultaneously, self-sterility in at least *Ciona* is ensured by incompatibility of proteins on the egg coat and sperm surface (Yamada et al., 2009; Otsuka et al., 2013). *Ciona* species are the most commonly studied ascidians as they are solitary and readily raised in the laboratory. Although they tend to be self-sterile, some individuals show varying degrees of self-fertility, facilitating genetic experiments.

Ascidian development is mosaic or determinant. That is cell fates are determined very early in development such that if one of the first two blastomeres is killed, the other develops only into the tissues it would normally make in an intact embryo (Chabry, 1887). Development is best known for two solitary ascidians, *Ciona intestinalis* (now called *Ciona robusta*) and *Phallusia mammillata* (Figure 37.8). The eggs are shed at first meiotic metaphase. Sperm preferentially enter the animal half of the egg and two polar bodies form. The sperm entry point serves as a focus for a cortical contraction and a wave of calcium release from internal stores that occurs at the same time as cytoplasm containing determinants for most of the tail musculature (i.e., the myoplasm) coalesces near the vegetal pole (Sardet et al., 2007). As the embryo has a fixed cell lineage, inductive interactions between cells and tissues are minimal (Roure and Darras, 2016). Within 24 hours of fertilization, the embryo gastrulates, and a neural tube and tail form. As the larva of *C. robusta* does not feed, the endoderm does not develop into a pharynx with gill slits and a gut but remains an endodermal strand. The larval CNS consists of a sensory vesicle or cerebral vesicle with a statocyst and photoreceptor plus a visceral ganglion and tail nerve cord. There are only about 300 cells of which 100 are neurons in the larval *Ciona* CNS. There are no neurons in the tail nerve cord. The notochord, which has just 40 cells, is present only in the tail. During development, a lumen forms that is continuous throughout the length of the notochord. On either side of the notochord is a series of muscle cells, 36 in all (Passamaneck et al., 2007). The larvae have numerous ectodermal sensory cells. At the anterior tip of the larva are adhesive papillae that stick the larva to the substrate as metamorphosis begins. Tadpole larvae of some solitary ascidians such as *Ascidiella aspersa* develop a pharynx with gill slits as well as a heart before settlement.

Metamorphosis: In *Ciona*, the non-feeding tadpoles settle onto a suitable substratum such as a petri dish, rock, rope or pier, affixing by the anterior adhesive papillae. The tail retracts, in part due to contraction of the epithelium and/or notochord cells and in part due to apoptosis of most of the tail cells. The primordial germ cells, located in the tail, migrate anteriorly. *Ciona* then develops a pharynx with a perforate branchial basket, endostyle, circular gut, heart. The nervous system is remodeled into a ganglion. Within 24–48 hours after settlement, the zooids begin feeding. *Ciona robusta* (*intestinalis*) has been put into continuous laboratory culture and mutant lines have been generated (Veeman et al., 2011).

ECOLOGY AND PHYSIOLOGY

As noted above, all ascidians are sessile, although most have a pelagic larva. Most live in shallow water and are filter feeders. The larvae settle wherever there is a suitable substratum such as rocks, boat bottoms, mooring lines, docks or floating debris such as ocean currents carried from Japan after a major tsunami to the northwest shore of the U.S.A. (Therriault et al., 2018). Ascidians are often associated with bacteria that secrete numerous compounds that have been tested for antibacterial or anticancer activities (Ayuningrum et al., 2019).

DISTRIBUTION AND BEHAVIOR

Ascidians are present throughout the world ocean. More than 70 species have been described from Antarctic waters and nearly as many from the Arctic (Segelken-Voigt et al., 2016). Most species are filter feeders. A current generated by branchial cilia moves water through the incurrent siphon into the branchial basket where particles, chiefly phytoplankton, adhere to mucus that moves into the gut. The current exits through the excurrent siphon. Although most species occupy shallow waters, some ascidians in the families Molgulidae and Octanemidae inhabit deep oceans and are carnivorous. Thus, carnivory in which the incurrent siphon is modified into lips, has probably evolved more than once in ascidians. These ascidians have enlarged oral apertures that can engulf prey such as crustaceans and assorted detritus. Their gill slits are either small and non-penetrant or absent altogether. At least one octanemid, *Megalodicopia hians,* spawns eggs and sperm that develop into typical tadpole larvae.

GENOMICS AND GENETICS

Ascidians are evolving particularly rapidly. Ascidian genomes range from about 170 mb for *Ciona* to ~ 194 mb in *Botrylloides* (Blanchoud et al., 2018) to about 580 mb for *Botryllus schlosseri* (De Tomaso et al., 1998). The ANISEED database (www.aniseed.cnrs.fr) gives access to the annotated genomes of nine ascidians as well as gene expression patterns. It also includes epigenomics datasets, such as RNA-seq, ChIP-seq and SELEX-seq (Brozovic et al., 2018). The *Ciona robusta* (formerly *C. intestinalis*) genome has received the most attention. It has about 20% of its genes in operons. Most operons include just two genes; the maximum number of genes in an operon is six (Zeller, 2010). Ascidian genomes have a high level of polymorphism (Nydam and Harrison, 2010). Ascidians have lost a number of genes, exemplified by the *Hox* genes. Two ascidians, *C. robusta* and *Halocynthia roretzi* both have 9 *Hox* genes compared to 15 in amphioxus. The lost genes are not the same *Hox* genes as lost in *Oikopleura*. Eight of the *Hr* Hox genes are orthologous to *Ci-Hox1, 2, 3, 4, 5, 10, 12* and *13*. Interestingly, in *H. roretzi*, the 9 genes are on a single chromosome, whereas in *C. robusta*, they are split onto two chromosomes. Ascidians have between about 300 and 500 microRNAs (miRNAs), most of which are unique to ascidians (Wang et al., 2017). In addition, ascidians have a robust repertoire of transposons, some of which have been used to generate mutant lines (Permanyer et al., 2003). Both the TALEN and CRISPR/Cas9 methods for site-specific mutagenesis have proven effective in *Ciona* (Treen et al., 2014; Pickett and Zeller, 2018).

Like other invertebrates, ascidians lack adaptive immunity and have an expanded repertoire of genes involved in innate immunity. The gut has been implicated as the major site of the innate immune response. Ascidians, like amphioxus, also have chitin associated with the fecal pellets (Liberti et al., 2018). The chitin synthase gene is expressed in the gut; there is evidence that the chitin protects the gut from infection by bacteria (Nakashima, et al., 2018).

At least one species of ascidian, *Botryllus schlosseri,* has evolved a type of immunity such that if two genetically unrelated colonies touch each other, they, in effect, fight each other, whereas, if two genetically related colonies touch each other, they fuse. This immune-like response was found to be mediated by an allorecognition system controlled by allele-sharing at a single, highly polymorphic locus (Cohen et al., 2017). Tunicates are unique among animals in synthesizing cellulose, which is a component of the test. The enzyme synthesizing cellulose in *Ciona robusta,* Ci-CesA, has both a cellulose synthase domain and a cellulase domain. The ancestral tunicate likely acquired this gene by lateral transfer from bacteria (Nakashima et al., 2004).

POSITION IN THE ToL

Phylogenetic analyses with nuclear genes have shown that ascidians are paraphyletic. Aplousobranchia plus Phlebobranchia is the sister group of Thaliacea, while the order Stolidobranchia is the sister group of the Aplousobranchia and Phlebobranchia plus Thaliacea (Delsuc et al., 2018; Kocot et al., 2018) (Figure 37.5).

DATABASES

Cephalochordata: *Branchiostoma floridae* genome: https://genome.jgi.doe.gov/Brafl1/Brafl1.home.html

Branchiostoma lanceolatum genome: https://metazoa.ensembl.org/Branchiostoma_lanceolatum/Info/Index

Branchiostoma belcheri genome:

http://genome.bucm.edu.cn/lancelet/

See also https://www.genome.jp/kegg-bin/show_organism?org=bfo

And https://www.uniprot.org/proteomes/UP000001554

EST sequences are deposited in GenBank www.ncbi.nlm.nih.gov.

Branchiostoma belcheri: http://genome.bucm.edu.cn/lancelet/ This database also has *B. floridae* genome, and annotated transcripts and proteins of both *B. floridae* and *B. belcheri* as well as ESTs of *B. belcheri.*

Asymmetron lucayanum transcriptome assembly https://www.ncbi.nlm.nih.gov/bioproject/235900

MicroRNAs of *B. floridae, B. belcheri, Ciona intestinalis* (now *Ciona robusta), Ciona savignyi, Oikopleura dioica* at MirBase: http://www.mirbase.org/index.shtml

Tunicates:

https://www.aniseed.cnrs.fr/This database has genome and expression data for 13 species of ascidians.

Ghost database has genomic and cDNA expression data for *C. intestinalis* https://integbio.jp/dbcatalog/en/record/nbdc00082

Salpa thompsoni genome dataset: https://www.bco-dmo.org/dataset/675040

Oikopleura dioica genome browser http://www.genoscope.cns.fr/externe/GenomeBrowser/Oikopleura/

See also Oikbase: (http://oikoarrays.biology.uiowa.edu/Oiko/)

CONCLUSION

Although cephalochordates and tunicates are closely related, they offer stark contrasts in rates of evolution. There are only three genera of cephalochordates. They are evolving extremely slowly and are very similar in morphology, behavior and habitat in spite of having split at least 120 mya. If there ever were any other species of cephalochordates that did not burrow in the sand and only emerge after dark, they would likely have become extinct after jaws evolved in the early Silurian, about 443 mya. In contrast, tunicates are evolving extremely rapidly and have diversified to a remarkable extent. Since vertebrates are evolving almost as slowly as cephalochordates, it is likely that the common ancestor of chordates was not evolving especially fast. The unanswered question is what happened that enabled tunicates to begin evolving rapidly. Were the evolutionary constraints first lifted on the genome that then allowed the morphology to diversify or were the constraints first lifted on the morphology that in turn lifted the constraints on the genome. One possible factor is the difference between regulative development in amphioxus and vertebrates with gradual specification of cell fates vs mosaic development in tunicates with early decision of cell fates and a fixed cell lineage. With gradual specification of cell fates, a change in the developmental program until quite late in development would probably result in mis-specification of cell fates and non-viable embryos. However, with early decision of cell fates, once cell fates become fixed, a change in the later developmental program might change the embryos but not prevent viability. It is the proverbial chicken and egg problem: which came first in tunicates–lifting of the constraints on genome evolution or the early decision of cell fates that in turn allowed the constraints on genome evolution to lift? Without an undersea time machine allowing us to return to the Cambrian when cephalochordate, tunicate and vertebrate lineages split, we may not solve the conundrum.

REFERENCES

Abitua PB, Wagner E, Navarrete IA, Levine M. 2012. Identification of a rudimentary neural crest in a non-vertebrate chordate. Nature 492:104–107.

Alf S, Enric S. 2006. Seasonal occurrence and role of protistan parasites in coastal marine zooplankton. Marine Ecology Progress Series 327:37–49.

Alldredge AL. 1982. Aggregation of spawning appendicularians in surface windrows. Bulletin of Marine Science 32:250–254.

Andersson E, Olsson R. 1989. The oral papilla of the lancelet larva (*Branchiostoma lanceolatum*) (Cephalochordata). Acta Zoologica (Stockholm) 70:53–56.

Andrews EA. 1893. An undescribed acraniate: *Asymmetron lucayanum*. Studies of the Biological Laboratories of Johns Hopkins University 5:213–247 + pl. XII-XIV.

Ayuningrum D, Liu Y, Riyanti, Sibero MT, Kristiana R, Asagabaldan MA, Wuisan ZG, Trianto A, Radjasa OK, Sabdono A, Schaberle TF. 2019. Tunicate-associated bacteria show a great potential for the discovery of antimicrobial compounds. PLoS One 14:e0213797.

Baatrup E. 1982. On the structure of the Corpuscles of de Quatrefages (*Branchiostoma lanceolatum* (P)). Acta Zoologica (Stockholm) 63:39–44.

Bachiller E, Utne KR, Jansen T, Huse G. 2018. Bioenergetics modeling of the annual consumption of zooplankton by pelagic fish feeding in the Northeast Atlantic. PLoS One 13:e0190345.

Basedow SL, Zhou M, Tande KS. 2014. Secondary production at the Polar Front, Barents Sea, August 2007. Journal of Marine Systems 130:147–159.

Batta-Lona PG, Maas AE, O'Neill RJ, Wiebe PH, Buklin A. 2017. Transcriptomic profiles of spring and summer populations of the Southern Ocean salp, Salpa thompsoni, in the Western Antarctic Peninsula region. Polar Biology 40:1261–1276.

Berná L, Alvarez-Valin F. 2014. Evolutionary genomics of fast evolving tunicates. Genome Biology and Evolution 6:1724–1738.

Blanchoud S, Rutherford K, Zondag L, Gemmell NJ, Wilson MJ. 2018. De novo draft assembly of the *Botrylloides leachii* genome provides further insight into tunicate evolution. Science Reports 8(1):5518. doi: 10.1038/s41598-018-23749-w.

Boldrin F, Martinucci G, Holland LZ, Miller RL, Burighel P. 2009. Internal fertilization in the salp *Thalia democratica*. Canadian Journal of Zoology 87:928–940.

Bone Q, Braconnot JC, Carré, C, Ryan KP, 1997. On the filter-feeding of *Doliolum* (Tunicata: Thaliacea). Journal of Experimental Marine Biology and Ecology 214:179–193.

Braconnot J-C. 1974. Reality of sexual cycle of pelagic Tunicata, *Dolium nationalis* Borgert, with first description of larvae. Comptes Rendus Hebdomaires des Scéances de L'Academie des Sciences Série D 278:1759–1760.

Brien P. 1948. Embranchement des Tuniciers: Morphologie et Reproduction. in: Traité de Zoologie: Anatomie, Systématique, Biologie: Échinodermes, Stomocordés, Procordés. Grasse PP. editor, 553–930, Masson et Cie, Paris, France

Brozovic M, Dantec C, Dardaillon J, Dauga D, Faure E, Gineste M, Louis A, Naville M, Nitta KR, Piette J, Reeves W, Scornavacca C, Simion P, Vincentelli R, Bellec M, Aicha SB, Fagotto M, Guéroult-Bellone M, Haeussler M, Jacox E, Lowe EK, Mendez M, Roberge A, Stolfi A, Yokomori R, Brown CT, Cambillau C, Christiaen L, Delsuc F, Douzery E, Dumollard R, Kusakabe T, Nakai K, Nishida H, Satou Y, Swalla B, Veeman M, Volff J-N, Lemaire P. 2018. ANISEED 2017: Extending the integrated ascidian database to the exploration and evolutionary comparison of genome-scale datasets. Nucleic Acids Research 46:D718–D725.

Cameron CB. 2005. A phylogeny of the hemichordates based on morphological characters. Canadian Journal of Zoology 83:196–215.

Cameron CB, Garey JR, Swalla BJ. 2000. Evolution of the chordate body plan: new insights from phylogenetic analysis of deuterostome phyla. Proceedings of the National Academy of Sciences USA 97:4469–4474.

Cañestro C, Albalat R, 2012. Transposon diversity is higher in amphioxus than in vertebrates: functional and evolutionary inferences. Briefings in Functional Genomics 11:131–141.

Carlo B, Francesca C, Paolo B. 2003. Alimentary tract of Kowalevskiidae (Appendicularia, Tunicata) and evolutionary implications. Journal of Morphology 258:225–238.

Chabry L. 1887. Embryologie normale et tératologique des Ascidie. Paris. Felix Alcan. Paris.

Chamisso Av. 1821. De animalibus quisbusdam e classe Vermium Linneana. In: Chamisso Av, Eysenhardt CW editor. Circumnavigatione Terrae, auspicante Comite N Romanzoff, duce Ottone de Kotzbue, annis 1815–1818, peracta observatio Fasciculus secundus, reliquos vermes continens. Nova Acta Academiae Caesarae Leopoldino-Caroinael 10(2). 543–574, pl. xxxi.

Cohen CS, Saito Y, Weissman Irving L. 2017. Evolution of allorecognition in botryllid ascidians inferred from a molecular phylogeny. Evolution 52:746–756.

Cuvier G. 1804. Mémoire sur les Thalides (*Thalia* Brown) et sur les Biphores (*Salpa* Forskal). Annales du Muséum d'histoire naturelle Paris 4:360–382.

D'Ambra I, Graham WM, Carmichael RH, Malej A, Onofri V. 2013. Predation patterns and prey quality of medusae in a semi-enclosed marine lake: implications for food web energy transfer in coastal marine ecosystems. Journal of Plankton Research 35:1305–1312.

Danks GB, Raasholm M, Campsteijn C, Long AM, Manak JR, Lenhard B, Thompson EM. 2015. Trans-splicing and operons in metazoans: Translational control in maternally regulated development and recovery from growth arrest. Molecular Biology and Evolution 32:585–599.

Daponte MC, Esnal GB. 1994. Differences in embryological development in two closely related species: *Ihlea racovitzai* and *Ihlea magalhanica* (Tunicata, Thaliacea). Polar Biology 14:455–458.

De Tomaso AW, Saito Y, Ishizuka KJ, Palmeri KJ, Weissman IL. 1998. Mapping the genome of a model protochordate. I. A low resolution genetic map encompassing the fusion/histocompatibility (Fu/HC) locus of *Botryllus schlosseri*. Genetics 149:277–287.

Décima M, Stukel MR, López-López L, Landry MR. 2019. The unique ecological role of pyrosomes in the eastern tropical Pacific. Limnology and Oceanography 64:728–743.

Deibel D, Lowen B. 2012. A review of the life cycles and life-history adaptations of pelagic tunicates to environmental conditions. ICES Journal of Marine Science 69:358–369.

Deibel D, Saunders PA, Stevens CJ. 2017. Seasonal phenology of appendicularian tunicates in the North Water, northern Baffin Bay. Polar Biology 40:1289–1310.

del Pilar Gomez M, Angueyra JM, Nasi E. 2009. Light-transduction in melanopsin-expressing photoreceptors of amphioxus. Proceedings of the National Academy of Sciences USA 106: 9081–9086.

Della Valle A. 1899. Tunicata. in: Zoologischer Jahresbericht fur 1898, Herausgeben von der Zoologischen Station zu Neapel, Mayer P. editor, 47–52, Friedlander und Sohn, Berlin.

Delsuc F, Brinkmann H, Chourrout D, Philippe H. 2006. Tunicates and not cephalochordates are the closest living relatives of the vertebrates. *Nature* 439:965–968.

Delsuc F, Philippe H, Tsagkogeorga G, Simion P, Tilak M-K, Turon X, López-Legentil S, Piette J, Lemaire P, Douzery EJP. 2018. A phylogenomic framework and timescale for comparative studies of tunicates. BMC Biology 16:39.

Delsuc F, Tsagkogeorga G, Lartillot N, Philippe H. 2008. Additional molecular support for the new chordate phylogeny. Genesis 46:592–604.

Diaz Briz L, Sánchez F, Marí N, Mianzan H, Genzano G. 2017. Gelatinous zooplankton (ctenophores, salps and medusae): an important food resource of fishes in the temperate SW Atlantic Ocean. Marine Biology Research 13:630–644.

Dishaw LJ, Haire RN, Litman GW. 2012. The amphioxus genome provides unique insight into the evolution of immunity. Briefings in Functional Genomics 11:167–176.

Drach P. 1948. La notion de procorde et les embranchements de cordes. in: Traité de Zoologie: Anatomie, Systématique, Biologie: Échinodermes, Stomocordés, Procordés, Grasse PP. editor, 545–551, Masson et Cie; Paris, France.

Edvardsen RB, Seo H-C, Jensen MF, Mialon A, Mikhaleva J, Bjordal M, Cartry J, Reinhardt R, Weissenbach J, Wincker P, Chourrout D. 2005. Remodelling of the homeobox gene complement in the tunicate *Oikopleura dioica*. Current Biology 15:R12–R13.

Ericson LE, Fredriksson G, Ofverholm T. 1985. Ultrastructural localization of the iodination centre in the endostyle of the adult amphioxus (*Branchiostoma lanceolatum*). Cell Tissue Research 241:267–273.

Fanelli E, Cartes JE. 2008. Spatio-temporal changes in gut contents and stable isotopes in two deep Mediterraean pandalids: influence on the reproductive cycle. Marine Ecology Progress Series 355:219–233.

Fenaux R, Gorsky G. 1979. Culture technique for appendicularians. Annales de l'Institut Océanographique 55:195–200.

Flood PR. 1966. A peculiar mode of muscular innervation in amphioxus. Light and electron microscopic studies of the so-called ventral roots. Journal of Comparative Neurology 126:181–218.

Flood PR. 1967. The notochord of amphioxus: A paramyosin catch-muscle? Journal of Ultrastructure Research 18:263 only.

Flood PR. 1968. Structure of the segmental trunk muscles in amphioxus, with notes on the course and "endings" of the so-called ventral root fibers. Zeitschrift für Zellforschung und Mikroskopische Anatomie 84:389–416.

Flood PR. 1975. Fine structure of the notochord of amphioxus. Symposium of the Zoological Society of London 36:81–104.

Flood PR. 1978. The spermatozoon of *Oikopleura dioica* Fol (Larvacea, Tunicata). Cell andTissue Research 191:27–37.

Fol H. 1872. Études sur les Appendiculaires du détroit de Messine. Memoires de la Société de Physique et d'Histoire Naturelle de Genève 21:445–449.

Forskal P, Niebuhr C. 1775. Descriptiones animalium, avium, amphibiorum, piscium, insectorum, vermium/quae in itinere orientali observavit Petrus Forskål. Post mortem auctoris edidit Carsten Niebuhr. Adjuncta est materia medica kahirina atque tabula maris Rubri geographica. Hauniæ [Copenhagen]:ex officina Mölleri.

Fredriksson G, Olsson R. 1991. The subchordal cells of *Oikopleura dioica* and *O. albicans* (Appendicularia, Chordata). Acta Zoologia (Stockholm) 72:251–256.

Freeman G. 1964. The role of blood cells in the process of asexual reproduction in the tunicate Perophora viridis. Journal of Experimental Zoology 156:157–183.

Fujii S, Nishino T, Nishida H. 2008. Cleavage pattern, gastrulation, and neurulation in the appendicularian, *Oikopleura dioica*. Development Genes and Evolution 218:69–79.

Gammill LS, Bronner-Fraser M. 2003. Neural crest specification: migrating into genomics. Nature Reviews Neuroscience 4:795–804.

Ganot P, Bouquet J-M, Kallesøe T, Thompson EM. 2007a. The *Oikopleura* coenocyst, a unique chordate germ cell permitting rapid, extensive modulation of oocyte production. Developmental Biology 302:591–600.

Ganot P, Kallesøe T, Thompson EM. 2007b. The cytoskeleton organizes germ nuclei with divergent fates and asynchronous cycles in a common cytoplasm during oogenesis in the chordate *Oikopleura*. Developmental Biology 302:577–590.

Garstang W. 1928 (sometimes as 1929). The morphology of the Tunicata, and its bearings on the phylogeny of the Chordata. Quarterly Journal of Microscopical Science 72:51–187.

Godeaux J, Bone Q, Braconnot JC. 1998. Anatomy of Thaliacea. In: The Biology of Pelagic Tunicates, Bone Q. editor, 1–24, Oxford Univ. Press, New York.

Govindarajan AF, Bucklin A, Madin LP. 2011. A molecular phylogeny of the Thaliacea. Journal of Plankton Research 33 843–853.

Guerriero G. 2012. Putative chitin synthases from *Branchiostoma floridae* show extracellular matrix-related domains and mosaic structures. Genomics Proteomics and Bioinformatics 10:197–207.

Henschke N, Everett JD, Richardson AJ, Suthers IM. 2016. Rethinking the role of salps in the ocean. Trends in Ecology and Evolution 31:720–733.

Hirose E, Aoki MN, Nishikawa J. 2005. Still alive? Fine structure of the barrels made by *Phronima* (Crustacea: Amphipoda). Journal of the Marine Biological Association of the United Kingdom 85:1435–1439.

Holland LZ. 1988. Spermatogenesis in the salps *Thalia democratica* and *Cyclosalpa affinis* (Tunicata: Thaliacea): An electron microscopic study. Journal of Morphology 198:189–204.

Holland LZ. 1989. Fine structure of spermatids and sperm of *Dolioletta gegenbauri* and *Doliolum nationalis* (Tunicata: Thaliacea): implications for tunicate phylogeny. Marine Biology 101:83–95.

Holland LZ. 1990. Spermatogenesis in *Pyrosoma atlanticum* (Tunicata: Thaliacea: Pyrosomatida): Implications for tunicate phylogeny. Marine Biology 105:451–470.

Holland LZ. 2016. Tunicates. Current Biology 26:R146–R152.

Holland LZ, Gorsky G, Fenaux R. 1988. Fertilization in *Oikopleura dioica* (Tunicata, Appendicularia): Acrosome reaction, cortical reaction and sperm-egg fusion. Zoomorphology 108:229–243.

Holland LZ, Holland ND. 1992. Early development in the lancelet (= amphioxus) *Branchiostoma floridae* from sperm entry through pronuclear fusion: presence of vegetal pole plasm and lack of conspicuous ooplasmic segregation. Biological Bulletin 182:77–96.

Holland LZ, Li G. 2021. Laboratory culture and mutagenesis of amphioxus (*Branchiostoma floridae*), in: Carroll DJ, Stricker SA (Eds.), Developmental Biology of the Sea Urchin and Other Marine Invertebrates: Methods and Protocols. Methods in Molecular Biology 2219. Springer US, New York, NY, 1–29.

Holland LZ, Miller RL. 1994. Mechanism of internal fertilization in *Pegea socia* (Tunicata, Thaliacea), a salp with a solid oviduct. Journal of Morphology 219:257–267.

Holland LZ, Onai T. 2011. Analyses of gene function in amphioxus embryos by microinjection of mRNAs and morpholino oligonucleotides. in: Vertebrate Embryogenesis: Embryological, Cellular, and Genetic Methods, Pelegri FJ. Editor, 423–438, Humana Press; Totowa, NJ.

Holland, N.D., 2011. Spawning periodicity of the lancelet, *Asymmetron lucayanum* (Cephalochordata), in Bimini, Bahamas. Italian Journal of Zoology 78:478–486.

Holland ND. 2017. The long and winding path to understanding kidney structure in amphioxus–a review. International Journal of Developmental Biology 61:683–688.

Holland ND, Campbell TG, Garey JR, Holland LZ, Wilson NG. 2009a. The Florida amphioxus (Cephalochordata) hosts larvae of the tapeworm *Acanthobothrium brevissime*: natural history, anatomy and taxonomic identification of the parasite. Acta Zoologica (Stockholm) 90:75–86.

Holland ND, Holland LZ. 2010. Laboratory spawning and development of the Bahama lancelet, *Asymmetron lucayanum* (Cephalochordata): fertilization through feeding larvae. Biological Bulletin 219:132–141.

Holland ND, Holland LZ, Heimberg A. 2015. Hybrids between the Florida amphioxus (*Branchiostoma floridae*) and the Bahamas lancelet (*Asymmetron lucayanum*): developmental morphology and chromosome counts. Biological Bulletin 228:13–24.

Holland ND, Panganiban G, Henyey EL, Holland LZ. 1996. Sequence and developmental expression of *AmphiDll*, an amphioxus Distal-less gene transcribed in the ectoderm, epidermis and nervous system: insights into evolution of craniate forebrain and neural crest. Development 122:2911–2920.

Holland ND, Paris M, Koop D. 2009b. The club-shaped gland of amphioxus: export of secretion to the pharynx in pre-metamorphic larvae and apoptosis during metamorphosis. Acta Zoologica (Stockholm) 90:372–379.

Holland ND, Venkatesh TV, Holland LZ, Jacobs DK, Bodmer R. 2003. *AmphiNk2-tin*, an amphioxus homeobox gene expressed in myocardial progenitors: insights into evolution of the vertebrate heart. Developmental Biology 255:128–137.

Hosp J, Sagane Y, Danks G, Thompson EM. 2012. The evolving proteome of a complex extracellular matrix, the *Oikopleura* house. PLoS One 7:e40172.

Hu G, Li G, Wang H, Wang Y. 2017. *Hedgehog* participates in the establishment of left-right asymmetry during amphioxus development by controlling *Cerberus* expression. Development 144:4694–4703.

Huang S, Wang X, Yan Q, Guo L, Yuan S, Huang G, Huang H, Li J, Dong M, Chen S, Xu A. 2011. The evolution and regulation of the mucosal immune complexity in the basal chordate amphioxus. The Journal of Immunology 186:2042–2055.

Hunt BPV, Hosie GW. 2005. Zonal structure of zooplankton communities in the Southern Ocean South of Australia: results from a 2150km continuous plankton recorder transect. Deep Sea Research Part I: Oceanographic Research Papers 52:1241–1271.

Huxley TH. 1851. Remarks upon *Appendicularia* and *Doliolum*, two genera of the Tunicata. Philosophical Transactions of the Royal Society of London 2:595–605, pl. xviii.

Igawa T, Nozawa M, Suzuki DG, Reimer JD, Morov AR, Wang Y, Henmi Y, Yasui K. 2017. Evolutionary history of the extant amphioxus lineage with shallow-branching diversification. Scientific Reports 7:1157.

Jefferies RPS. 1986. The Ancestry of the Vertebrates. London: British Museum (Natural History). 376 pp.

Jeffery WR. 2006. Ascidian neural crest-like cells: phylogenetic distribution, relationship to larval complexity, and pigment cell fate. Journal of Experimental Zoology 306B:470–480.

Jeffery WR, Strickler AG, Yamamoto Y. 2004. Migratory neural crest-like cells form body pigmentation in a urochordate embryo. Nature 43:696–699.

Jue NK, Batta-Lona PG, Trusiak S, Obergfell C, Bucklin A, O'Neill MJ, O'Neill RJ. 2016. Rapid evolutionary rates and unique genomic signatures discovered in the first reference genome for the southern ocean salp, *Salpa thompsoni* (Urochordata, Thaliacea). Genome Biology and Evolution 8:3171–3186.

Kassmer SH, Rodriguez D, De Tomaso AW. 2016. Colonial ascidians as model organisms for the study of germ cells, fertility, whole body regeneration, vascular biology and aging. Current Opinion in Genetics & Development 39:101–106.

Kocot KM, Tassia MG, Halanych KM, Swalla BJ. 2018. Phylogenomics offers resolution of major tunicate relationships. Molecular Phylogenetics and Evolution 121:166–173.

Kon T, Nohara M, Nishida H, Sterrer W, Nishikawa T. 2006. Hidden ancient diversification in the circumtropical lancelet Asymmetron lucayanum complex. Marine Biology 149:875–883.

Kon T, Nohara M, Yamanoue Y, Fujiwara Y, Nishida M, Nishikawa T. 2007. Phylogenetic position of a whale-fall lancelet (Cephalochordata) inferred from whole mitochondrial genome sequences. BMC Evolutionary Biology 7:1–12.

Kowalevsky A. 1870. The kinship of ascidians and vertebrates. Quarterly Journal of Microscopy Science 10:59–69.

Krchňáková Z, Krajčovič J, Vesteg M. 2017. On the possibility of an early evolutionary origin for the spliced leader trans-splicing. Journal of Molecular Evolution 85:37–45.

Kuo C-Y, Fan T-Y, Li H-H, Lin C-W, Liu L-L, Kuo F-W. 2015. An unusual bloom of the tunicate, *Pyrosoma atlanticum*, in southern Taiwan. Bulletin of Marine Science 91:363–364.

Lacalli T, Holland L. 1998. The developing dorsal ganglion of the salp *Thalia democratica*, and the nature of the ancestral chordate brain. Philosophical Transactions of the Royal Society of London B 353: 1943–1967.

Lacalli TC. 2004. Sensory systems in amphioxus: a window on the ancestral chordate condition. Brain Behavior and Evolution 64:148–162.

Laird DJ, De Tomaso AW, Weissman IL. 2006. Stem cells are units of natural selection in a colonial ascidian. Cell 124:647–648.

Lamarck J-B. 2013 (reprint of original 1816) Histoire naturelle des animaux sans vertèbres; Chapter 4, Tuniciers. Cambridge University Press.

Lambert G. 2005. Ecology and natural history of the protochordates. Canadian Journal of Zoology 83:34–50.

Leander BS, Lloyd SAJ, Marshall W, Landers SC. 2006. Phylogeny of marine gregarines (Apicomplexa) — *Pterospora, Lithocystis and Lankesteria* — and the origin(s) of coelomic parasitism. Protist 157:45–60.

Lei Q, Li C, Zuo Z, Huang C, Cheng H, Zhou R. 2016. Evolutionary insights into RNA trans-splicing in vertebrates. Genome Biology and Evolution 8:562–577.

Lemaire P, Piette J. 2015. Tunicates: exploring the sea shores and roaming the open ocean. A tribute to Thomas Huxley. Open Biology 5:150053.

Lesueur C-A. 1815. Memoire sur l'organisation des Pyrosomes. Bulletin de la Société philomathique de Paris. 5:70–74.

Li K, Doubleday AJ, Galbraith MD, Hopcroft RR. 2016. High abundance of salps in the coastal Gulf of Alaska during 2011: A first record of bloom occurrence for the northern Gulf. Deep Sea Research Part II: Topical Studies in Oceanography 132:136–145.

Liberti A, Zucchetti I, Melillo D, Skapura DP, Shibata Y, De Santis R, Pinto MR, Litman GW, Dishaw LJ. 2018. Chitin protects the gut epithelial barrier in a protochordate model of DSS-induced colitis. Biology Open 7:bio029355.

Litman GW, Rast JP, Fugmann SD. 2010. The origins of vertebrate adaptive immunity. Nature Reviews in Immunology 10:543–553.

Lombard F, Kiørboe T. 2010. Marine snow originating from appendicularian houses: Age-dependent settling characteristics. Deep Sea Research Part I: Oceanographic Research Papers 57:1304–1313.

Marlétaz F, Firbas PN, Maeso I, Tena JJ, Bogdanovic O, Perry M, Wyatt CDR, et.al. 2018. Amphioxus functional genomics and the origins of vertebrate gene regulation. Nature 564:64–70.

Martí-Solans J, Ferrández-Roldán A, Godoy-Marín H, Badia-Ramentol J, Torres-Aguila NP, Rodríguez-Marí A, Bouquet JM, et.al. 2015. *Oikopleura dioica* culturing made easy: A Low-Cost facility for an emerging animal model in evodevo. Genesis 53:183–193.

Mehrotra R, Scott CM, Hoeksema BW. 2016. A large gape facilitates predation on salps by *Heteropsammia* corals. Marine Biodiversity 46:323–324.

Mertens CH. 1830. Beschreibung der *Oikopleura*, einer neuen Mollusken-Gattung. Mémoires de l'Académie impériale des sciences de St.-Pétersbourg. 6:205–220, 202 pl.

Nakashima K, Kimura S, Ogawa Y, Watanabe S, Soma S, Kaneko T, Yamada L, Sawada H, Tung CH, Lu T-M, Yu JK, Villar Briones A, Kikuchi S, Satoh N. 2018. Chitin-based barrier immunity and its loss predated mucus-colonization by indigenous gut microbiota. Nature Communications 9(1):3402. doi: 10.1038/s41467-018-05884-0.

Nakashima K, Yamada L, Satou Y, Azuma J-i, Satoh N. 2004. The evolutionary origin of animal cellulose synthase. Development, Genes and Evolution 214:81–88.

Nishida H. 2008. Development of the appendicularian *Oikopleura dioica*: Culture, genome, and cell lineages. Development, Growth & Differentiation 50:S239–S256.

Nishida H, Stach T. 2014. Cell lineages and fate maps in tunicates: Conservation and modification. Zoological Science 31:645–652.

Nogueira Junior M, Brandini FP, Ugaz Codina JC. 2015. Diel Vertical Dynamics of gelatinous zooplankton (Cnidaria, Ctenophora and Thaliacea) in a subtropical stratified ecosystem (South Brazilian Bight). PLoS One 10:e0144161.

Nydam ML, Harrison RG. 2010. Polymorphism and divergence within the ascidian genus *Ciona*. Molecular Phylogenetics and Evolution 56:718–726.

Onuma TA, Isobe M, Nishida H. 2017. Internal and external morphology of adults of the appendicularian, *Oikopleura dioica*: an SEM study. Cell and Tissue Research 367:213–227.

Osugi T, Sasakura Y, Satake H. 2017. The nervous system of the adult ascidian *Ciona intestinalis* Type A (*Ciona robusta*): Insights from transgenic animal models. PLoS One 12:e0180227–e0180227.

Otsuka, K., Yamada, L., Sawada, H. 2013. cDNA cloning, localization, and candidate binding partners of acid-extractable vitelline-coat protein Ci-v-Themis-like in the ascidian *Ciona intestinalis*. Molecular Reproduction and Development 80:840–848.

Pakhomov EA, Perissinotto R, McQuaid CD, Froneman PW. 2000. Zooplankton structure and grazing in the Atlantic sector of the Southern Ocean in late austral summer 1993: Part 1. Ecological zonation. Deep Sea Research Part I: Oceanographic Research Papers 47:1663–1686.

Pallas PS. 1774. Spicilegia Zoologica, Vol 1, Fascicle 10. G. A. Lange, Berlin.

Paris M, Escriva H, Schubert M, Brunet F, Brtko J, Ciesielski F, Roecklin D, et.al. 2008. Amphioxus postembryonic development reveals the homology of chordate metamorphosis. Current Biology 18:825–830.

Parker D. 2011. Oceanic diet and distribution of haplotypes for the green turtle, *Chelonia mydas*, in the central north Pacific. Pacific Science 65:419–431.

Passamaneck YJ, Hadjantonakis A-K, Di Gregorio A. 2007. Dynamic and polarized muscle cell behaviors accompany tail morphogenesis in the ascidian *Ciona intestinalis*. PLos One 2:e714.

Permanyer J, Gonzàlez-Duarte R, Albalat R. 2003. The non-LTR retrotransposons in *Ciona intestinalis*: new insights into the evolution of chordate genomes. Genome Biology 4:R73.

Perrier E. 1899. Traite de Zoologie. Fasicle V. Amphioxus-Tuniciers. Paris: Masson et Cie eds. Libraries de l'Academie de Medicine. Pp. 2138–2355.

Pickett CJ, Zeller RW. 2018. Efficient genome editing using CRISPR-Cas-mediated homology directed repair in the ascidian *Ciona robusta*. genesis 56:e23260.

Piette J, Lemaire P. 2015. Thaliaceans, The neglected pelagic relatives of ascidians: A developmental and evolutionary Enigma. The Quarterly Review of Biology 90:117–145.

Poss SG, Boschung HT. 1996. Lancelets (Cephalochordata: Branchiostomatidae): How many species are valid? Israel Journal of Zoology 42 Supplement:13–66.

Putnam NH, Butts T, Ferrier DEK, Furlong RF, Hellsten U, Kawashima T, Robinson-Rechavi M et.al. 2008. The amphioxus genome and the evolution of the chordate karyotype. Nature 453:1064–1071.

Quoy JRC, Gaimard JP. 1833. *Voyage de découvertes de l'Astrolabe exécuté par ordre du Roi, pendant les années 1826–1827–1828–1829, sur le commandement de M. J. Dumont d'Urville.* Paris: J. Tastu. 304–306.

Racioppi C, Valoroso MC, Coppola U, Lowe EK, Brown CT, Swalla BJ, Christiaen L, Stolfi A, F. R. 2017. Evolutionary loss of melanogenesis in the tunicate *Molgula occulta*. EvoDevo 8: Article number:11.

Rahr H. 1981. The ultrastructure of the blood vessels of *Branchiostoma lanceolatum* (Pallas) (Cephalochordata. I. Relations between blood vessels, epithelia, basal laminae, and "connective tissue". Zoomorphology 97:53–74.

Robison, BH, Raskoff, KA, Sherlock RE. 2005. *Ecological substrate in midwater: Doliolula equus*, a new mesopelagic tunicate. Journal of the Marine Biological Association of the United Kingdom 85 (3): 655–663.

Romer AS. 1967. Major steps in vertebrate evolution. Science 158:1629–1637.

Roure A, Darras S. 2016. *Msxb* is a core component of the genetic circuitry specifying the dorsal and ventral neurogenic midlines in the ascidian embryo. Developmental Biology 409:277–287.

Rubinstein ND, Feldstein T, Shenkar N, Botero-Castro F, Griggio F, Mastrototaro F, Delsuc F, Douzery EJP, Gissi C, Huchon D. 2013. Deep sequencing of mixed total DNA without barcodes allows efficient assembly of highly plastic ascidian mitochondrial genomes. Genome Biology and Evolution 5:1185–1199.

Sahlin K, Olsson R. 1986. The wheel organ and Hatschek's groove in the lancelet. *Branchiostoma lanceolatum* (Cephalochordata). Acta Zoologica (Stockholm) 67:201–209.

Sardet C, Paix A, Prodon F, Dru P, Chenevert J. 2007. From oocyte to 16-cell stage: Cytoplasmic and cortical reorganizations that pattern the ascidian embryo. Developmental Dynamics 236:1716–1731.

Segelken-Voigt A, Bracher A, Dorschel B, Gutt J, Huneke W, Link H, Piepenburg D. 2016. Spatial distribution patterns of ascidians (Ascidiacea: Tunicata) on the continental shelves off the northern Antarctic Peninsula. Polar Biology 39:863–879.

Shenkar N, Gittenberger A, Lambert G, Rius M, Moriera da Rocha R, Swalla BJ, Turon X. 2009. Ascidiacea World Database. Enterogona. (http://www.marinespecies.org/ascidiacea/)

Somorjai IML, Somorjai RL, Garcia-Fernàndez J, Escrivà H. 2012. Vertebrate-like regeneration in the invertebrate chordate amphioxus. Proceedings of the National Academy of Sciences USA 109:517–522.

Søviknes AM, Glover JC. 2008. Continued growth and cell proliferation into adulthood in the notochord of the appendicularian *Oikopleura dioica*. Biological Bulletin 214:17–24.

Stach S. 2008. Chordate phylogeny and evolution: a not so simple three-taxon problem. Journal of Zoology London 276:117–141.

Stach T, Winter J, Bouquet JM, Chourrout D, Schnabel R. 2008. Embryology of a planktonic tunicate reveals traces of sessility. Proceedings of the National Academy of Sciences USA 105:7229–7234.

Stokes MD. 1996. Larval settlement, post-settlement growth and secondary production of the Florida lancelet (=amphioxus), *Branchiostoma floridae*. Marine Ecology Progress Series 130:71–84.

Stokes MD, Holland ND. 1995a. Ciliary hovering in larval lancelets (= amphioxus). Biological Bulletin 188:231–233.

Stokes MD, Holland ND. 1995b. Embryos and larvae of a lancelet, *Branchiostoma floridae*, from hatching through metamorphosis–growth in the laboratory and external morphology. Acta Zoologica Stockholm 76:105–120.

Stokes MD, Holland ND. 1996. Life-history characteristics of the Florida lancelet, *Branchiostoma floridae*: some factors affecting population dynamics in Tampa Bay. Israel Journal of Zoology 42 Supplement:67–86.

Su L, Shi C, Huang X, Wang Y, Li G. 2020. Application of CRISPR/Cas9 Nuclease in Amphioxus Genome Editing. Genes 11:1311.

Sutton MF. 1960. The sexual development of *Salpa fusiformis* (Cuvier). Journal of Embryology and Experimental Morphology 8:268–290.

Swadling KM, Kawaguchi S, Hosie GW. 2010. Antarctic mesozooplankton community structure during BROKE-West (30°E–80°E), January–February 2006. Deep Sea Research Part II: Topical Studies in Oceanography 57:887–904.

Therriault TW, Nelson JC, Carlton JT, Liggan L, Otani M, Kawai H, Scriven D, Ruiz GM, Clarke Murray C. 2018. The invasion risk of species associated with Japanese tsunami marine debris in Pacific North America and Hawaii. Marine Pollution Bulletin 132:82–89.

Thompson EM, Kallesøe T, Spada F. 2001. Diverse genes expressed in distinct regions of the trunk epithelium define a monolayer cellular template for construction of the oikopleurid house. Developmental Biology 238:260–273.

Tilves U, Fuentes VL, Milisenda G, Parrish CC, Vizzini S, Sabatés A. 2018. Trophic interactions of the jellyfish *Pelagia noctiluca* in the NW Mediterranean: evidence from stable isotope signatures and fatty acid composition. Marine Ecology Progress Series 591:101–116.

Treen N, Yoshida K, Sakuma T, Sasaki H, Kawai N, Yamamoto T, Sasakura Y. 2014. Tissue-specific and ubiquitous gene knockouts by TALEN electroporation provide new approaches to investigating gene function in *Ciona*. Development 141: 481–487.

Trueblood LA. 2019. Salp metabolism: temperature and oxygen partial pressure effect on the physiology of *Salpa fusiformis* from the California Current. Journal of Plankton Research 41:281–291.

Tsagkogeorga G, Turon X, Hopcroft RR, Tilak M-K, Feldstein T, Shenkar N, Loya Y et. al. 2009. An updated 18S rRNA phylogeny of tunicates based on mixture and secondary structure models. BMC Evolutionary Biology 9:187.

Veeman MT, Chiba S, Smith WC. 2011. *Ciona* Genetics. In: Pelegri FJ editor. Vertebrate Embryogenesis: Embryological, Cellular, and Genetic Methods. Totowa, NJ: Humana Press. 401–422.

Walters TL, Gibson DM, Frischer ME. 2019a. Cultivation of the marine pelagic tunicate *Dolioletta gegenbauri* (Uljanin 1884) for experimental studies. JoVE:e59832.

Walters TL, Lamboley LM, López-Figueroa NB, Rodríguez-Santiago ÁE, Gibson DM, Frischer ME. 2019b. Diet and trophic interactions of a circumglobally significant gelatinous marine zooplankter, *Dolioletta gegenbauri* (Uljanin, 1884). Molecular Ecology 28:176–189.

Wang K, Dantec C, Lemaire P, Onuma TA, Nishida H. 2017. Genome-wide survey of miRNAs and their evolutionary history in the ascidian, *Halocynthia roretzi*. BMC Genomics 18:314.

Watters DJ. 2018. Ascidian toxins with potential for drug development. Marine Drugs 16:162.

Wicht H, Lacalli TC. 2005. The nervous system of amphioxus: structure, development, and evolutionary significance. Canadian Journal of Zoology 83:122–150.

Winchell CJ, Sullivan J, Cameron CB, Swalla BJ, Mallatt J. 2002. Evaluatng hypotheses of deuterostome phylogeny and chordate evolution with new LSU ad SSU ribosomal DNA data. Molecular Biology and Evolution 19:762–776.

Yamada L, Saito T, Taniguchi H, Sawada H, Harada Y. 2009. Comprehensive egg coat proteome of the ascidian *Ciona intestinalis* reveals gamete recognition molecules involved in self-sterility. Journal of Biological Chemistry 284:9402–9410.

Yamaguchi T, Henmi Y. 2003. Biology of the amphioxus, *Branchiostoma belcheri* in the Ariake Sea, Japan. II. Reproduction. Zoolological Science 20:907–918.

You L, Chi J, Huang S, Yu T, Huang G, Feng Y, Sang X, Gao X, Li Ta, Yue Z, Liu A, Chen S, Xu A. 2019. LanceletDB: an integrated genome database for lancelet, comparing domain types and combination in orthologues among lancelet and other species. Database 2019.

Yue J-X, Kozmikova I, Ono H, Nossa CW, Kozmik Z, Putnam NH, Yu J-K, Holland LZ. 2016. Conserved noncoding elements in the most distant genera of cephalochordates: The Goldilocks principle. Genome Biology and Evolution 8:2387–2405.

Yue J-X, Yu J-K, Putnam NH, Holland LZ. 2014. The Transcriptome of an amphioxus, *Asymmetron lucayanum*, from the Bahamas: A Window into chordate evolution. Genome Biology and Evolution 6:2681–2696.

Zaniolo G, Manni L, Brunetti R, Burighel P. 1998. Brood pouch differentiation in *Botrylloides violaceus*, a viviparous ascidian (Tunicata). Invertebrate Reproduction and Development 33:1–23.

Zeller RW. 2010. Computational analysis of *Ciona intestinalis* operons. Integrative and Comparative Biology 50:75–85.

Zervoudaki S, Frangoulis C, Svensen C, Christou ED, Tragou E, Arashkevich EG, Ratkova TN, Varkitzi I, Krasakopoulou E, Pagou K. 2014. Vertical carbon flux of biogenic matter in a coastal area of the Aegean Sea: The importance of appendicularians. Estuaries and Coasts, 37:911–924.

Zhang QJ, Luo YJ, Wu HR, Chen YT, Yu JK. 2013. Expression of germline markers in three species of amphioxus supports a preformation mechanism of germ cell development in cephalochordates. EvoDevo 4:17.

Zhou X, Jin P, Qin S, Chen L, Ma F, 2012. Systematic investigation of amphioxus (*Branchiostoma floridae*) microRNAs. Gene 508:110–116.

EPILOGUE

THE SERIOUS STUDENT OF INVERTEBRATE ZOOLOGY

Bernd Schierwater and Rob DeSalle

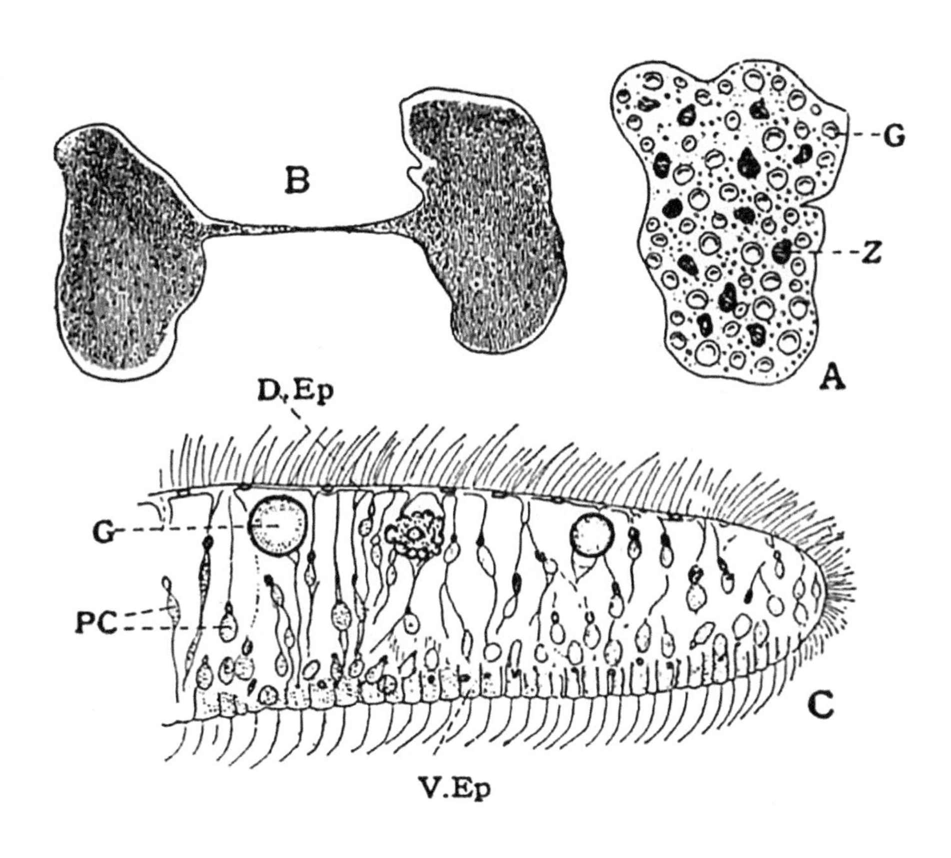

CONTENTS

Dobzhansky's famous quote from 1973 "nothing in biology makes sense except in the light of evolution" demonstrates his basic awareness of biological life on this planet. His quote is more newsworthy and relevant than ever today. No other part of the Tree of Life (ToL) is more suited to learning about evolution than the Invertebrate Animals (of course we are quite biased in making this statement). The correct grouping and relationship assignments of more than a million described species could lead us to a better or even deeper understanding not only of how animal body plans (bauplans) are built and how the underlying genetic information relates to morphological changes, but also of how certain diseases in humans develop, how global warming affects life on earth, and how humans can address the steadily increasing number of threats to nature. Thus we feel strongly that the serious student of Invertebrate Zoology (IZ) should study the whole of IZ and make this wonderful part of our biodiverse world part of her career. Invertebrates offer students with primary interests in developmental biology, genetics, ecology and/or evolution, ample variation to make substantial progress in the study of life on our planet. Many students and teachers we hope will use this book as a state-of-the-art reference not only to look up terms and to get a quick overlook over specific taxa, but also as a study and course book and enjoy an exciting demonstration of modern integrative biology. In addition, there are many ways a student can use IZ as a springboard into other areas of biology. We recommend that a student who is interested in, say, neurobiology learns and knows the invertebrate animal nervous systems inside and out, and the only way to do that is to be a serious student of IZ. And to expand this idea, some of these marvelous invertebrates have remarkable biological characteristics that make them excellent subjects for studying human-based diseases and human biology topics like stem cells. For instance, one of our favorite animals – Placozoa – are now being used to study cancer and stem cell regulation processes in space experiments.

SERIOUS STUDENTS

First and foremost, we should address why a serious student would want to focus on Invertebrate Zoology. Understanding the diversity and evolution of life on this planet should be reason enough to make anyone a serious student of IZ, but alas one has to pay rent and buy beer. Throughout this textbook, we have asked our authors to also point to economic, ecological or social items that involve invertebrates. Perhaps the most common ones mentioned are those involved in medical situations. Many of the invertebrates we describe herein are parasites and in that way have a direct impact on human health while others are model systems for biomedical research. Knowledge of these invertebrates is noble as it contributes to human well-being. Invertebrates can be used as indicator species for understanding climate change, our biggest human-caused phenomenon that we as a society need to confront. Unfortunately one of the most promising groups in this context, the insects, are incredibly hard hit by human-caused extinction; many insect species die out before they have been scientifically described. Much of the importance of IZ comes down to the continuation of basic science as an important endeavor in human societies.

Basic science has been the bulwark of modern life; without it much of what we know about human diseases and disorders (from genetic to physiological and psychiatric malfunctions) would not be possible. For instance, as we write and compile this textbook, the world is in the midst of a pandemic caused by a virus called SARS-Cov-2, the coronavirus that is the cause of COVID-19. Basic biology is at the heart of nearly every remedy, mitigation or response humans can make to this pandemic. We won't mitigate infectious diseases without understanding evolution, and IZ is one of the best ways to enter the evolutionary realm.

DUELING SCIENCE

The astute reader of this text might have noticed that discrepancy, discordance and incongruence of studies on the same organisms are highly evident in Invertebrate Zoology. This might turn the serious student of IZ off, because after all science is about truth, right? Well, this notion of science is actually incorrect. Science is about the best explanation one has for the natural phenomenon being observed. And the best explanation of anything is heavily dependent on the data at hand. We point to the sister of all other metazoans (SOM) problem we discuss at length in Chapter 4. Early on, Porifera were the undisputed SOM, based on morphology compared to the most recent non-metazoan animal. Then *Trichoplax adhaerens* was discovered crawling on the side of an aquarium in Berlin and the simplicity of this animal led to a consideration of placozoans as the SOM. While simplicity doesn't mean ancestral, Placozoa do offer a good look into a possible ancestral body-plan with which scientists can work further to best understand animals. For some time Placozoa and/or Porifera were the only games in town, until phylogenomics came along. Large molecular datasets when treated certain ways with models also propped up Ctenophora as the SOM. This strange result, which could not be reconciled easily with animal form and development, started a new debate about the SOM. Better put, this result we believe has more to do with the particular sensitivity of model-based computational tools compared to traditional comparative morphology and hence is an important contributor to clarity in modern systematics. The provisionary nature of science informs us along the way of the best approaches and methods to get to the best explanation. It also results in interesting and unexpected discoveries, which we would not have approached if everything was settled on.

For instance, the longstanding suggestion that when we use the term Crustacea we are talking about a monophyletic group, is now thought to be rejected because Hexapods (insects, etc.) are imbedded in other so-called crustaceans. This new thinking about ,rustaceans has led to the renaming of most of the groups involved and a need for reassessment of the evolution of the group. Dispute is not a bad thing in science; in fact, it is the lifeblood of science. Our advice to the student here is to be aware of the hypotheses, their historical development and current datasets that are used, in order to attempt to get at the best explanation for how animals evolved on this planet.

LEARN THE ANATOMY

Today most students are focused on the molecular biology, DNA sequencing and bioinformatics of organisms. While phylogenomics has added greatly to our understanding of the relationships of Invertebrate Animals, it has also convinced us that a new morphological perspective on these animals is important. Here is why. From 2008 to 2019, several phylogenomic datasets were generated (all of them available on the textbook website) and as we pointed out above Ctenophora became a focus as the SOM for some computational biologists. Anatomy was somehow relegated to a lost art and considered an afterthought in many of these studies. Many researchers preferred to generate their phylogenies based on the phylogenomic data and then interpret the anatomy based on the phylogenomic tree. All of this happened while there were at least six to seven morphological data matrices in the literature (also included on the textbook website) relevant to the SOM. Why not simply combine or concatenate the morphology with the molecules? We showed (Neumann et al., 2020) that, when the concatenation exercise is accomplished, the morphology can have a significant effect on the topology of a phylogenomic tree. Because there are oftentimes hundreds to thousands more characters generated by a molecular dataset than by a morphological one, we examined tree topology under different morphology weighting schemes and showed that very slight weighting of morphology resulted in changing topologies in a number of the existing datasets.

LEARN THE SYNAPOMORPHIES, ESTABLISH HOMOLOGY, UNDERSTAND EVOLUTION

Many of the animals we describe in this text have fantastic morphologies and behaviors. The beautiful intricacies of rotifers is legend; the profound bizarreness of barnacles is historic (Darwin was confused by them); and the novel anatomical structures of members of small groups like kinorhynchs, acanthocephalans and ectoprocts are stunning. But science is about explanation and the best way to apply explanation to IZ is to understand common ancestry. Specifically, for each group of animals discussed in this text as monophyletic there is also a set of characters that can be hypothesized to be characteristic of the common ancestor of that group. These shared characters are called synapomorphies and discovering the set of shared derived characters of a group is not only an exciting endeavor, it makes comparative zoology scientifically sound.

Connected to the synapomorphies as a serious student of IZ you will need to discover and characterize the overarching glue of comparative biology – so-called homology. This oftentimes poorly understood concept has a checkered history in comparative biology. The famous British biologist, John Maynard Smith, once said that homology was a term "ripe for burning" because of the confusion it caused biologists. But on the other hand, Gary Nelson, the less famous New York systematist, once said that "nothing in evolution makes sense except in the light of homology" (an obvious borrowing of our Dobzhansky's words at the outset of this Epilogue). The definition for homology we need and apply here is "a character of two or more taxa is homologous if it originated from the same (or corresponding) character of the last common ancestor." Getting this concept straight is a *conditio sine qua non* for any biology text.

We hope too that we have relayed to the student in this text the enthusiasm that all 40 authors have for the study of invertebrates. In addition to the amazing depth of information from these authors, we hope that the serious student of IZ will also recognize the excitement, enthusiasm and true inquisitiveness that can be generated around this amazing collection of organisms. We also hope that the serious IZ student will realize that there are more new questions in the discipline than answers that have been offered by invertebrate zoologists in the past two centuries. This open-endedness of IZ, we hope, is its biggest attraction to the serious student.

Bernd Schierwater, Hannover, Germany
Rob DeSalle, New York, USA

INDEX

Note: Locators in *italics* represent figures and **bold** indicate tables in the text.

D

R

S